THE GULLIVER FILE

But, I had another reason which made me less forward to enlarge his Majesty's dominions by my discoveries; to say the truth, I had conceived a few scruples with relation to the distributive justice of princes upon those occasions. For instance, a crew of pyrates are driven by a storm they know not whither; at length a boy discovers land from the top-mast; they go on shore to rob and plunder; they see an harmless people, are entertained by kindness, they give the country a new name, they take formal possession of it for the king, they set up a rotten plank or a stone for a memorial, they murder two or three dozen of the natives, bring away a couple more by force for a sample, return home and get their pardon. Here commenceth a new dominion acquired with a title by *divine right*. Ships are sent with the first opportunity; the natives driven out or destroyed, their princes tortured to discover their gold; a free licence given to all acts of inhumanity and lust; the earth reeking with the blood of its inhabitants: and this execrable crew of butchers employed in so pious an expedition, is a *modern colony* sent to convert and civilize an idolatrous and barbarous people.

From: *Gullivers Travels,* by Jonathan Swift.
Everyman's Library Edition, London, 1991, page 315.

But, I had another reason which made me less forward to enlarge his Majesty's dominions by my discoveries; to say the truth, I had conceived a few scruples with relation to the distributive justice of princes upon those occasions. For instance, a crew of pyrates are driven by a storm they know not whither; at length a boy discovers land from the top-mast; they go on shore to rob and plunder; they see an harmless people, are entertained by kindness, they give the country a new name, they take formal possession of it for the king, they set up a rotten plank or a stone for a memorial, they murder two or three dozen of the natives, bring away a couple more by force for a sample, return home and get their pardon. Here commenceth a new dominion acquired with a title by *divine right*. Ships are sent with the first opportunity; the natives driven out or destroyed, their princes tortured to discover their gold; a free licence given to all acts of inhumanity and lust; the earth reeking with the blood of its inhabitants: and this execrable crew of butchers employed in so pious an expedition, is a *modern colony* sent to convert and civilize an idolatrous and barbarous people.

From: *Gullivers Travels*, by Jonathan Swift.
Everyman's Library Edition, London, 1991, page 315.

THE GULLIVER FILE

THE GULLIVER FILE

Mines, people and land: a global battleground

Roger Moody

Minewatch, 1992

The Gulliver File
Mines, people and land: a global battleground
by Roger Moody

Published in 1992 by Minewatch, 218 Liverpool Road, London NI ILE, UK.
and WISE-Glen Aplin, PO Box 87, Glen Aplin Q 4381, Australia.

Distribution:
Sales to bookshops: Pluto Press, 345 Archway Road, London N6 5AA, UK.
Sales to the mining industry and libraries: Uitgeverij Jan van Arkel, A. Numankade 17,
3572 KP Utrecht, the Netherlands.

Moody, Roger
The Gulliver File. Mines, people and land: a global battleground /
by Roger Moody – London: Minewatch

ISBN 90 6224 999 X
ISBN 0 7453 0607 1

Production: International Books, an imprint of Uitgeverij Jan van Arkel,
A. Numankade 17, 3572 KP Utrecht, The Netherlands.
Cover illustration: Loek Koopmans
Cover design: OptimaForma, Nijmegen
Printing cover: SSN, Nijmegen
Printing: Bariet, Ruinen

Printed in the Netherlands

CONTENTS

GRATITUDE

First, thanks must go to Jan and Jill Roberts – co-workers and co-founders of CIMRA (Colonialism and Indigenous Minorities Research/Action) – without whom the seed-idea of *Gulliver* would never have taken root. They – and later members of the CIMRA collective – supported me personally through what must have seemed many fruitless (if not incomprehensible) hours of labour. Next in chronological order, my deep gratitude must go to Lin Pugh: like many young Australians, she breezed into Europe fourteen years ago on the grand tour (anti-nucléaire). Unlike most others, she stayed – helping establish the World Information Service on Energy (WISE) and, above all, *Kiitg* (the *Keep it in the Ground* newsletter). The latter was, for several years, the only global source of critical information on uranium mining and what the companies were up to. It was thanks to Lin's endeavours that WISE-Amsterdam adopted *Gulliver* as a project, securing funding at a crucial stage from a Netherlands charitable foundation, NOVIB.

But, like Topsy, the giant outgrew its original dress; it turvied and became unmanageable. For a period during the early 'eighties, all writing ceased, although material was still being filed. Then, a white knight appeared in the form of WISE-Glen Aplin, one of the organisation's national relays. This Australian group agreed to take over publishing responsibility for the book, and to try to secure the necessary funding: it launched an international appeal which, fairly quickly, gained a very encouraging response from Australia itself. At this point, too, I had the enormous good luck to become acquainted with Mick Licarpa, whose innumerable hours devoted to typing out my dishevelled manuscripts were matched by his uncanny precision in checking sources, proper names, and corporate cross-holdings. Mick was plagued with ill-health through the years I knew him, but rarely complained; he finally succumbed to a heart attack in 1989. Wherever his spirit is now floating (though he himself repudiated this entire notion) I am sure it will continue to find errors in this work, which no-one else would even guess at!

The WISE collective consists of a remarkable family: Tineke, Thijs, Olav, and Monica Muurlink. All of them devoted some of their labours to *Gulliver*. Thijs revived faith (at least my faith) in the project and set out determinedly to back it with funds. Olav did much of the original typesetting and some of the indexing. But it was Monica who took on the brunt of the gargantuan task. By immersing herself in every aspect of the work; acquiring major skills in computing; and putting thousands of dollars of her own earnings towards the author's expenses on the book, she evinced a confidence which literally saved the whole enterprise (at least in its final form). If this book is dedicated to anyone, it is to her; although, in her self-effacing fashion, she would probably rather credit went elsewhere. Monica herself asked for the following credits to be included, which I am very happy to do.

"In Australia, WISE-Glen Aplin's work on the *Gulliver File* has been supported by a network of campaigning groups, trade unions, and a large number of individuals. We need to recognise that most of this support – both practical and financial – came from the little people and struggling movement groups, in direct contrast to the corporate systems this book exposes. Moral support and encouragement from many people proved a decisive factor in keeping alive the project over so many years, when often there was little else going for it.

In particular I need to acknowledge Dave Keenan, Movement Against Uranium Mining in Sydney (especially Murray Matson), Awareness Education, Eileen and Tom Edgar, The Pacific Peacemaker Fund, Penny Coleing, Joan Shears, Greenpeace (Sydney), and the three other WISE-Glen Aplin people, whose work was crucial: Thijs, Olav and Tineke Muurlink."

Monica came to England for several months in 1990, to ensure that the book would finally go to press. At this point, Minewatch agreed to take over publishing responsiblity and raise the very considerable printing costs. It was Minewatch which invited Uitgeverij Jan van Arkel – a noted Netherlands ecological publisher – to design the work and move it into production. Printing costs were raised from the Department of Energy and Ecology in the Netherlands Ministry of Development Co-operation, from a British charitable foundation, the Gillet Trust, and the Margaret Laurence Fund (Toronto).

We would also like to acknowledge the substantial financial contributions of WISE Glen Aplin, WISE Amsterdam, the Polden-Puckham Charitable Trust, and the NOVIB, without which publication of GULLIVER would not have been possible.

All of these are to be thanked for their invaluable contributions to the final unrolling of the *Gulliver* saga.

In particular, I owe debts of great gratitude to Minewatch co-workers – to Albert Beale (for unravelling large chunks of glossary, and dedicated proof-reading and typesetting which went beyond the call of duty!); to David Arnott (for burning many night-hours putting corrections onto disk); to Carolyn Marr (for her freely-given time in making the geographical index); to Paul Thomas (for badly needed computer advice, given night or day); and, above all, to Christine Lancaster. It was Christine who took responsibility for putting *Gulliver* into its final form. The task went well beyond her worst nightmares, but she accomplished it without rancour or false heroics, and still managed a smile at the end!

In addition, my gratitude goes to Simon Chambers, for the work he did on the RTZ Group entries (in the context of the book *Plunder!*) which enormously cut the amount of final editing time. It goes to the members of Partizans (People against RTZ and its Subsidiaries) who supported the project, even though it fell well outside their immediate remit. And to all at Uitgeverij Jan van Arkel, for the care with which they have prepared the final book.

A work that has taken one-and-a-half decades to come to fruition not only pays conventional costs: some outdatedness, galloping inflation, changes in the potential audience, loss of original enthusiasms. There are also hidden prices to be paid – battered reputations, conflict over disbursement of dwindling funds, and even loss of long-standing friendships.

In answer to the inevitable question – "Was it worth it?" – my own answer has to be: "No."

Those less closely connected to *Gulliver* will evaluate it on other grounds, and, I hope, may give a more positive response.

Roger Moody,
February 1992

INTRODUCTION TO THE GULLIVER FILE

Four years before Ronald Reagan took over the US presidency, the American Mining Congress foreshadowed the rapaciousness with which his administration would view both the people and the environment. At its 1981 San Fransisco Convention, the country's mining leaders (all, obviously, male) pooled their concerns about poor demand for metals, the oversupply of minerals, and the higher working and capital costs of complying with "stringent environmental regulations."

Anticipating measures which were later to be taken by Reagan's Secretary of the Interior, James Gaius Watt, the convention called for public lands to be opened up to exploration and possible mining. It also demanded free access for multinational corporations to the international seabed. And it came out against the holding of buffer-stocks in metals – one of the key elements of the emergent strategies by South nations, to bring the North towards more equitable and moral international commodity agreements.

Above all, the convention slammed the environmentalists (they were not usually dubbed "greenies" then), the growing anti-nuclear campaign, and the incandescent Native American movement. Charles Barbour, chairperson of ASARCO, and vice-chair of the American Mining Congress, estimated that these organisations had put an extra 15 cents onto the cost of producing every pound of refined metal in the USA. It was fully accepted, said Barbour, that "the protection of human health was a primary consideration," but "standards relating to animal and plant life should be justified by cost-benefit analysis."

Barber then made the analogy which gave the title to this book: "Like Gulliver, the mining industry is a robust giant held down by a million silk strings."

In fact, the genesis of the book *Gulliver* preceded the 1981 American Mining Congress by around three years. Initially it was conceived as a brief exposé of the corporate links between uranium mining companies, and the dominance in world supply of "yellowcake" by a handful of key actors: RTZ, Cogéma, Eldorado Nuclear, Kerr McGee, Exxon, United Nuclear, Urangesellschaft, ERA, and others. Certainly this original concept has continued to dominate the text over intervening years: hopefully there is not one major uranium mining project during the period 1975-1985 which has not been covered in detail in these pages.

However, it became clear a decade ago – not least because of the aggressive stances being adopted by the global mining industry – that *Gulliver* might do more than serve those opposing uranium mining. As the nature of specific corporate strategies became clearer (at least to this author), and as cooperation increased with land-based (particularly indigenous) communities around the world, the groups sponsoring this research (see *Acknowledgments* section) saw the need for a much more comprehensive study. This study would look at all major mining operations with an impact on indigenous people, and examine the most powerful minerals corporations with an extremely critical eye. It has to be admitted that this book, despite taking well over a decade to bring to fruition, still does not measure up fully to these exacting criteria. Nonetheless, it is the only work of its kind. Because it is published by an organisation which serves a growing, viable, international network, Gulliver is also a "work in progress" which invites its readers to contribute further documentation, corrections, and critical comments. Minewatch is committed to expanding the data-base on which this book is based, and making it available as widely as possible. There may be only one Gulliver, but there can be (and are) thousands of Lilliputs!

What you now have in your hands is a tool (in Ivan Illich's sense of the term). It is not expected that anyone will read *Gulliver* from cover to cover at one sitting. But nor is it intended solely as a work of reference. It is hoped that readers will make links between various corporate endeavours, and see for themselves how certain trade practices (transfer pricing, locating in offshore tax havens, manipulating foreign currency exchanges, repatriation of tax-free profits, employing non-union or "migrant" labour, etc) bring riches to multinationals at the expense of the rest of us. You may thus begin to question the very necessity of "the mining industry". It is not that this huge sector – with such vast tangential and peripheral operations – is entirely inimical to human needs, or unhearing of human demands. Indeed, in the years since the American Mining Congress of 1981, minerals producers have increasingly been forced to recognise the brutalising impacts of the extractive process, and the reality of some of its worst practises. Rather, the truth is that – by being organised primarily along corporate lines, with decisions taken according to an industrial-ising, as opposed to a conservationist, or rural-revitalisation, agenda – mining cannot support its own best intentions, nor fulfil its most sustainable expectations.

Can there, indeed, be such a phenomenon as "sustainable mining" – the buzz-word which has reverberated around the mining world since environmental aspects of the industry first came under global attack in the late 1980s?

In 1991, the United Nations organised a conference in Berlin on mining and the environment. This was the first forum to raise, in any concerted fashion, fundamental questions about the minerals sector and the environmental impacts of mineral extraction. It also looked at the disparity between developing "northern" standards (aimed at reducing toxic discharges, global warming and "amenity" disturbance), and the ecologically unacceptable ones inherited by administrations and communities in

Africa, South America and parts of the Pacific. (Not to mention the veritable holocaust caused by unbridled coal, uranium, and other mining in eastern Europe over a period of more than forty years.) Conference participants provided some intriguing half-answers to prevailing questions (such as the need for "green groups" to participate in environmental impact assessments, or the allocation of mining taxes to those primarily affected by mineral operations). However, they failed to properly define the *responsibility* for the historical imbalances which have enabled a bare fifth of the world's human population to metaphorically gobble at Top Table, while the remainder not only forages for the crumbs, but delivers the very feast. (The disparity between western-style per capita minerals consumption, and third-world derived minerals production, is even starker than that which obtains in the agricultural sector).

The conference also ignored the inequalities *within* specific regions, such as Western Australia, or northern Queensland, the "four corners" region of the USA, the Philippines Cordillera, the altiplano of Bolivia (and many other places). Here, traditional land-holders – who may or may not support mining on their own account – provide the essential material basis for the state's extractive industry. (As custodians of the soil, and all it yields, indigenous peoples are the veritable "bankers" from whom mining companies steal the capital.) Yet, indigenous communities rarely derive any direct benefit from mining themselves, while suffering all its worst effects.

Moreover, mining industry practice lags well behind its newly-emergent theorists, notwithstanding the recent formation of the ICME (International Council on Metals and the Environment), or the heady rush by almost every western mining major to position "environmental excellence" alongside "enhancing shareholder value" in its annual report. At the time of publishing this book, the only organisations to try to put mining *per se* on the programme of the 1991 United Nations Conference on the Environment have been its supposed detractors (including Minewatch, publishers of this book).

At root, this could well be because the very concept of "sustainable mining" is a chimera. Replacing the contours of ripped-up ranges may serve an aesthetic purpose, but can never compensate for the loss – even "temporary", but is it ever? – of part of an eco-system, or indigenous peoples' sacred places. "State of the art" tailings containment systems run foul of human error (as at the Key Lake dam in Canada) or "Mother Nature" (as with the habitual releases of contaminated water from the Ranger mine in northern Australia). And "recycling" of waste dumps carries with it its own problems of toxicity.

The truth is that "sustainability" implies something quite different for those at the sharp end of the bulldozer than it does for those in the driving seat. The scraping of topsoil, or removal of forest cover, can irreplaceably interrupt agricultural cycles; test drilling can interfere with precious aquifers; road building and site construction drives away game – and all this even before mining starts.

Clearly, the further underground that mineral extraction proceeds, then the less damaging some of its environmental and social impacts may be. But the difference is often more apparent than real. Longwalling in the USA may be projected as the acme of miner safety, but it has resulted in considerable surface instability, notable cave-ins, and increased dust levels for miners. Sub-surface gold extraction in the Philippines came to an abrupt halt in 1990 when Benguet Corporation, the country's largest mining company, itself argued that the recent earthquake demonstrated the inherent instability of the workings.

It is difficult not to conclude that an industry which is market-led, geared to an extractive (rather than substitutive or recycling) philosophy, and which has customarily viewed the whole globe as its oyster, will never take on board concerns now being expressed by environmental organisations – even the most moderate ones. Evidence of this is the reaction in Tasmania to a recent report by the Combined Environment Groups which argues for the preservation of the wilderness zones in ex-

change for miners' access to "non-contentious" areas. (Chris Sharples, *Minerals Supplies and the National Estate in Tasmania: Achieving a Balance.* Combined Environmental Groups, Hobart, March 1991). Instead of settling for a third of the cake, the industry – in league with the Liberal opposition – is threatening a bigger encroachment on protected lands than ever before.

However, there are those associated with mining whose reaction to the new wave of environmentalism is neither knee-jerk hostility, nor dictated solely by self-interest. There are also those on the other side of the chain-link fence whose prime concern is not the exclusion of pristine areas from all human intervention, but the rights of land-based communities to decide for themselves what they do in their own territory, so long as they have demonstrated "stewardship" of their own resources: the options clearly can include mining. Common ground between erstwhile combatants can now be found in organisations like the Association of Geoscientists for International Development (AGID) and Small Mining International (SMI), as well as UN agencies such as UNEP, UNDP ... and even the World Bank. (In 1990, the World Bank delivered a heavy warning to the Guinean government that it would not support the Mount Nimba iron ore project which is located on a World Heritage site.)

Ironically, these new alliances may endorse the very types of mining which industry people are characterising as environmentally unacceptable – specifically *garimpo*, or artisanal, enterprises. This is not to suggest that small-scale mining is ecologically pure: on the contrary, it can be enormously damaging, with its use of mercury to amalgamate gold, or "rocket" dredges that rip out river banks. Rather, it is to propose that local people have to be entrusted with decisions about their own lives. In the final event, it is they, not the multinational corporations – nor, indeed, the "get rich quick" entrepreneurs that engage in illegal border-hopping, as in Venezuela and Guayana – who carry the can for future generations. If, having considered all the alternatives (including simply being left to their own devices) these communities decide on mineral extraction, then it is their choice. But the decision cannot be pre-empted, nor even foreseen.

For example, information in late 1990 from Bougainville suggested that, while the islanders were united in rejecting the re-entry of the expelled mining company CRA, they are far from decided about the merits of selling their "independent" resources to another operator: some want to revive the industry, others want nothing more to do with it, and are turning to renewed cocoa production instead.

Similarly, in Alaska, some native people appear well satisfied with their role as equal partners with mining companies, under the Alaska Native Claims Settlement Act (ANCSA) of 1971. Others, however, now repudiate ANCSA and are calling for complete sovereignty over their lands. As a 1991 ECOSOC report summarised the divisions within the Alaskan communities over Cominco's Red Dog mine:

"For the indigenous people who are managers and employees of NANA [the native corporation which has shares in Red Dog] the ... mine is a source of great pride, an enterprise of global significance that they helped build ... For some other indigenous people in north-western Alaska, the mine is a threat to their health and traditional way of life." (*Discrimination Against Indigenous Peoples: Transnational Investments and Operations in the Lands of Indigenous Peoples,* Report of the United Nations Center on Transnational Corporations, pursuant to the Subcommission resolution 1990/26, Geneva, July 17 1991.)

A new organisation, Minewatch was set up in 1989 to work with a wide array of land-based peoples who need to make their own decisions on mining issues. It is essentially neither anti-mining nor uncritically pro-Green. It currently has a membership of some 90 NGOs and communities – from Alaska to Zaïre – and national groups in Ireland (Minewatch Ireland) and New Zealand (Minewatch Aotearoa). While it has so far concentrated on mitigating long-standing damage (working with an intermediate technology group in Sierra Leone which is

trying to rehabilitate agricultural land ravaged by diamond mining) and on forestalling future land and forest loss to mining in established mineral zones (such as in India, Indonesia, and the Philippines), it is increasingly being called on by those in "virgin" areas, such as Burma, Venezuela, and Panama.

While views on the question posed above ("Can there be sustainable mining?") may differ within the Minewatch network, there seems to be agreement on one crucial issue now under discussion within the industry: who is to blame for the undisputed messes caused by mineral extraction to date; and will it be western-based corporate endeavours, or radical alternatives, that prevent them from happening in the future?

The "South" groups with which Minewatch collaborates are emphatic that environmental ills visited upon us all – of which mineral extraction and downstream processing are a critical component – must not be attributed to poorer parts of the world simply because they are saddled with antiquated systems and machinery, or lack of capital. Such ethnocentric thinking is already being translated into a false dichotomy between "environmentally unsound" third world mining projects and "technically superior" western ones: a version of the age-old syndrome whereby the culprits blame the victims.

From which source, we might ask, have these irresponsible mining operatives in South America, or Africa, got much of their finance, equipment and expertise in the first place? And who is buying and processing their output? Into whose automobiles, computers, and supersonic aircraft, is this bauxite, iron ore, or titanium, metamorphosed? Into which coffers do both the direct profits from extraction, and a succession of added values, usually tumble? (The uranium industry presents a "worst case" example of these propensities. Uranium has been exploited primarily on indigenous lands, such as the Four Corners of the USA, Australia's Arnhemland, the Aïr region of Niger, or the lakes and forests of northern Saskatchewan. Yet, the actual land-holders (or custodians, as many prefer to be called) have often received nothing in exchange for the loss of soil and sub-soil resources, have never benefitted from the electricity derived from the uranium, and have sometimes been at the receiving end of the worst consequences of nuclear power: tailings piles in New Mexico emitting radon "daughters" for hundreds of thousands of years to come, and nuclear test sites at Maralinga, South Australia, becoming uninhabitable for Aboriginal people.)

If "the polluter pays" is a principle to be applied both globally and retrospectively, then how much larger the bill should be for companies like Anglo-American (high-grading along the Namibian coastline), RTZ (responsible for the ravages of Richards Bay), or BHP (dumping heavy metals wholesale into the Fly River from its Ok Tedi mine), than it should be for the *garimpeiros* of Brazil, let alone the "pocket miners" of the Philippine Cordillera.

Gulliver was started as a modest project, fourteen years ago, with the naive expectation that, by better understanding who controlled mining, land-based communities would be empowered to intervene against mine plans, or insist on better ones. Clearly a lot more than this was required: hence the development of a monster "dossier" which hopefully now enables the reader to evaluate the intentions of specific companies against their records. (Interestingly, as the book was going to press, legislators in Wisconsin passed a piece of preliminary legislation, making it mandatory for mining companies to submit an account of any delinquencies, before being allowed to explore or mine in the state).

But this is still only "scraping the surface." We are beginning to understand that tropical forests can only be "sustained" – or their products "recycled" – by allowing full land and resource rights to forest-dwellers. We are now dimly perceiving the critical importance of "biodiversity" as inseparable from the diverse peoples who practise and preserve it.

So, surely, we have to concede that the primary role in deciding where, what, and how, to mine, must lie with those on whose land the minerals are to be found. Such a concession will be viewed with alarm by captains of the industry, because, to date, mining's practitioners have been cavalier (to say the least) in their attitude to local communities and dismissive of most small-scale community extractive techniques. The proposal will also not sit well with some technicians for international environmental protection who – ironically, while supporting methods to strengthen the global diversity of plants and animals – still believe that "biggest is best" when it comes to mineral extraction. (This tendency – evinced in corridors as disparate as those of the World Wide Fund for Nature(WWF), the ICME, and the United Nations Development Program(UNDP) – can lead to carving "mineral zones" out of some parts of the planet, while designating "protected eco-zones" in others).

Only those most profoundly affected by mining are likely to approve this proposal without severe qualification. Just as Amazonian Indian communities were accused of holding back the development of Brazil during the 1960s and 1970s, or Aboriginal organisations of sequestering Australia's mineral-rich land during the 1980s, so newly-empowered rural cooperatives in Africa, or land-owner associations in the Pacific, will be similarly indicted. They will be portrayed as "jeopardising" the advancement of the teeming poor in the bidonvilles and shanty-towns of the rest of the world, for their own narrow, pretended, cultural integrity.

This is an ethno-centric, Eurocentric, and indeed industrio-centric view. According to it, indigenous peoples have little inkling of the world around them, small care for the welfare of others, no inclination to share their own (fast-dwindling but genuinely self-renewable) resources, while world managers – whether from the World Bank, the United Nations or British Petroleum – comprehend every nook and cranny of our biosphere, are overwhelmingly solicitious for the world's impoverished dependants, and redistribute their profits for the benefit of all humankind.

To put it mildly, growing evidence (including much within the pages which follow) suggests that this is hardly the case.

LILLIPUTIA

How to find your way around Gulliver without being a Jahoo

Even if it has not been laid out with silken strings attached, it should be fairly easy for the reader to find their way around *Gulliver*. Essentially, the book is a compendium of information about mining companies. Fourteen years of labour have wreaked their toll, however – on style, if not content. Thus you will find very short – virtually elliptical – entries, cheek-by-jowl with essays which could fill (and in one case, that of RTZ, *have* filled) a book. You will also discover that some companies no longer exist (though they did at the time of the original research), and that allusions to certain activities or subsidiaries in one entry are contradicted in another.

As pointed out in the "notes on the headcharts" section (below) such discrepancies are the result of several different factors: in particular, the fact that companies' entries were updated at different times between 1987 and 1991. By and large, the bigger the company, the more up-to-date it is. (Unfortunately, most corporate reference books suffer from the same disability – not least, the *Financial Times Mining International Yearbook* which, unlike *Gulliver,* depends entirely on data supplied by the companies themselves.)

There may also be outdated addresses in the "Contact" section at the end of many entries but, as far as possible, these details have been checked in the year before going to press. Where there was some uncertainty, but the information was probably still accurate, the address has been left in rather than cause a potential source of information or support to be cut off.

Important notes on the headcharts

The information contained in the "headcharts" displayed at the top of each entry has been checked and double-checked from numerous sources by several different people. Most of these sources will be "industry" – in other words, dependent on information provided by the companies themselves. However, in view of the widespread practice adopted by bigger corporations of concealing certain holdings (both in themselves and in their subsidiaries), and the equally widespread use by many investors of "nominees" to disguise the origin and extent of their investment, all headchart data must be regarded with caution. In particular, percentages should not be treated as gospel, even where they are so finely tuned (*i.e.* 13.654%!) as to imply absolute precision. They may refer to the proportion of equity held in a subsidiary only at a certain date (for example, before the capital of that company was increased through a major share issue) and may or may not carry voting rights. For example, the distribution of voting and non-voting shares in Rössing Uranium Ltd is a crucial factor in determining who controls, and benefits from, its operations: but these differences are reflected in the text of the entry, not the chart. Nonetheless, where companies are linked by a straight line (without an accompanying percentage figure) – thus indicating full ownership and/or control by the company at the top of the line of the company which is at the bottom – I am confident that this information is correct. *At least*, it *would* have been correct at the time of the typing-up of that particular entry. But, since *Gulliver* is a massive work which has taken years to prepare, some entries were completed a long time before others. At best, I have tacked on *addenda* to the text of many entries (even though the headchart shows an earlier situation – see below); but others remain in their original form.

Because a headchart is a snap-shot of the situation at one moment, it can conflict with parts of the continuing narrative of the text. The following is an example of such an anachronism. During the preparation of *The Gulliver File*, one of the biggest conglomerates in mining, Consolidated Goldfields (CGF), was involved in a fight-to-the-death against an even bigger one, Anglo-American. The wheeling and dealing which occurred during 1987-89 resulted in a change of ownership of GFSA (Goldfields of South Africa) and the loss of Newmont's major South African holdings: eventually the Hanson Trust took over CGF. However the headcharts reflect the situation before this battle royal took place.

So, some data on the headcharts should be treated with circumspection: I urge that information on a specific company taken from the headchart should be used in conjunction with the entry which follows; also (as with any compendium of this type) information on a company should be checked against other entries on the same company (using the comprehensive index). Readers should also carefully check the *dates* of source material (using the bibliographical references at the end of each entry).

As with other categories of information in the *File*, I ask readers who are aware of changes in the headchart data to send us details – preferably with copies of the newspaper article or book page in which the new information is featured.

Most headcharts are simple to interpret. The full name of the company featured in the entry is printed in UPPER CASE CHARACTERS to the right of the page, together with a code for its country of incorporation (see the list of country abbreviations given in a following section). The names of other companies having a controlling interest (though not necessarily *the* controlling interest) in the featured company will be found printed above the name of the company itself. In most cases, these will be bigger companies, but sometimes families (*e.g.* the Rockefellers) or groups and institutions (especially insurance companies). Other companies wholly or partly controlled *by* the featured company are shown, generally, below the main name. In some instances however, and especially where a merger

has taken place, these companies will be shown alongside.

All companies which have their own entry in the *File* are underlined in all headcharts. The position of subsidiary or associate companies in the chart does not necessarily denote their size or their importance to the featured company. I have tried to place the more important subsidiaries higher in the chart, but restrictions of layout have sometimes meant this could not be achieved.

Where a name appears in the headchart or index in *italics* it indicates a mining project or mine.

As already mentioned, percentage figures indicate, so far as is known or can be estimated, the holding which one company has in another. Often, two figures will be found, at either end of one connecting line, *e.g.*:

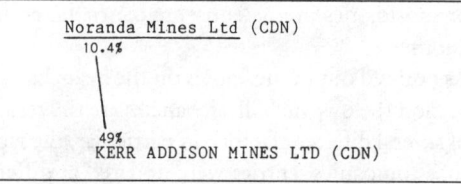

This simply shows that the two companies have cross-holdings in each other; of course it does not indicate that, in this example, Kerr Addison Mines has any day-to-day control over Noranda, and such holdings (by "children" in their "parents") usually only assume importance when attempts are made by unwelcome "predators" to take over the parent company.

Unbroken direct lines show that the company at the top owns or effectively controls 100% (or the percentage shown) of the company below. Where this line is dotted, it indicates that I believe the holder to maintain effective control but cannot, through lack of data, translate this into a percentage.

Lastly, there are sometimes question marks at one end or the other of these ownership/control lines. Here, the holding company or interest does not possess total control of the "held" company, but I have been unable to determine to any degree of accuracy just what it does own.

Using the company index

Using the index really is quite simple! All companies which have a text in the bulk of the book also have their own company number. This number is displayed both at the head of each entry, and in the left hand margin of the company index. *It does not refer to a page number.* So, if you wish to find a company, you can either look directly in the text (where companies are listed in alphabetical order), or trace it through the company index. The numbers *following* a company name in the index are references to other entries in the book where mention of that company is also found.

Should the company not have its own write-up, but nonetheless be mentioned in the text, it will be indexed without an identifying number to the left; but the numbers to the right will identify entries where it is mentioned. (There are *many* more companies dealt with in the book than those which give their name to an entry.)

For example: the reader who wants to find which companies PUK is involved with, will look up PUK in the index and discover that company numbers 14, 23, 30, 61, 76, ... etc, are listed. Instead of looking at PUK itself, the reader could simply refer to these other entries in the index, discovering a cornucopia of linkages (Alianza Petrolera Argentina SA, Amax, Amok, Bechtel, Boroquimica SA, etc).

Companies will sometimes be known by various, slightly different, names – their full official name (usually as in the headchart); their common, frequently abbreviated, name (often as in the title of the entry); perhaps other names pertaining to earlier or later versions of the company title. Where necessary, these differing versions will appear in the index, cross-referenced to one major version of the name.

All companies are also entered in a geographical index according to their country of incorporation, and according to places where they have major projects. On occasion it has not been possible to discover exactly where something is going on: the index remains silent on that point. While federal states are broken down into their components, it has not always been possible to locate a specific site: in this case, the operations are referenced under the country. (Companies active in several parts of a federal country will be listed under that country's name as well as under the relevant states.) Naturally, not every locale is indexed: there seemed little point in listing places where exploration has yielded nothing, or where every non-mining subsidiary is to be found, let alone where the company chair spends his holidays! (Having said that, I could hardly ignore the late Armand Hammer's frequent jaunts to the USSR and China.)

The glossary

As a work which commenced life determined to lift the lid off the uranium industry, it is hardly surprising that the book's glossary contains much nuclear terminology. But I have also tried to define basic terms used in all hardrock mining, and included definitions which more properly relate to "economics" or "commerce". The glossary is by no means exhaustive (and, once again, readers are invited to let Minewatch know of other terms they encounter in the text which they would like to see defined). Nor is it objective – if, by that word, we mean that the descriptions are agreed upon universally. On the contrary, I have allowed myself a deal of discussion about certain controversial terms, especially "multinational". The glossary is intended, then, to be both discursive and an integral part of the whole book. It has its own introductory notes which give further information.

Units of measurement and related terminology

Terms used for units of measurement should be treated cautiously. Drawing as it does on numerous sources of information from all over the world, *The Gulliver File* has – unavoidably – sometimes mixed up its standards. The same company's entry might contain references to the grade of uranium in ore, variously, in percentages, in pounds per ton, and in kilogrammes per tonne; production figures could be in pounds, kilogrammes, tons or tonnes –

generally according to the terminology used in the relevant source material.

More perplexing, it has not always been possible to distinguish between the metric tonne, the (in British parlance) "imperial" ton (or "long ton"), and the US ton (or "short ton"): in the case of large quantities, the difference can be significant. The fault lies mostly with the sources used, which are themselves not always clear.

More details of the definitions of, and relationships between, such units will be found in the list of units and their abbreviations given later. Note that we have tried to avoid using abbreviations for units in a way that would clash with the officially recognised abbreviations of SI (Systèm International) Units – for example we have tried to avoid using "t" for "ton" since "t" is the SI abbreviation for "tonne". However, ambiguities of this sort in source material might sometimes have been carried over into the text. The use of "mt" to mean "metric ton" (*i.e.* a tonne) – as found in some mining literature – is a related cause of confusion which we have tried to overcome.

One particular problem relates to a major US mining publication, the *Engineering and Mining Journal (E&MJ):* it has used the symbol "M" to mean a thousand (not a million!). Because of this peculiarity, there could be some misinterpretation of quantities when information has come from this source, particularly in the case of the earlier references to editions of *E&MJ*.

The titles from Mining Journal Publications *(Mining Journal, Mining Magazine,* and *Mining Annual Review)* – probably the most authoritative in the world – have adopted the metric system: most of their data is given in SI units. However, some information from these sources is also in non-SI units, especially in the case of material up to the early 1980s.

Lastly, there is the traditional problem with the word "billion". In US usage it means a thousand million. In Europe, it has historically meant a million million; but the US usage is now quite common in Britain, especially in financial statistics. So, once again, please interpret data according to the source of the information.

Bibliographical references and their abbreviations

Gulliver's bibliography does *not* purport to be comprehensive. It only gives the most common sources used, or quoted, in the entries – one list of magazines, newspapers, and other serial publications, another of books – with the abbreviations which are used for them in the numbered references after company entries. In the case of the first list, the relevant date or issue number or equivalent is given with the abbreviation in the reference. Suggestions for further reading are sometimes also to be found at the end of an entry. The most important book yet published on the world uranium industry is not listed at all (although I drew on it a little in the later stages). This is Terence Price's *Political Electricity – What Future for Nuclear Energy?* (Oxford University Press, 1990). A decidedly partisan work, which rates nuclear power as the next best thing to sliced white bread, it is nevertheless full of fascinating information.

Other abbreviations

This guide to *The Gulliver File* has already referred to several specialised lists of abbreviations which follow: those used in the headcharts for the country of incorporation of each company; those used for various units and measurements; and those used for publications referenced in the text. But there are many other more general (though still somewhat technical) abbreviations used in the book – for the component states of federal countries, different sorts of nuclear reactors, company terminology, and much more. All these are gathered in one remaining abbreviations list, which is intended to leave nothing but the most everyday term unexplained.

ABBREVIATIONS USED IN THE REFERENCES

Journals & magazines

AFR	*Australian Financial Review*, Sydney, Australia.
Age	*The Age*, Melbourne, Australia.
AkwN	*Akwesasne Notes*, Mohawk Nation at Akwesasne, New York State, USA.
BBA	*Birch Bark Alliance* (Ontario Public Interest Research Group) Peterborough, Canada.
BHPS	*Black Hills Paha Sapa Report*, Rapid City, USA.
CAFCINZ Watchdog	*CAFCINZ Energy Control Watchdog* (CAFCA) Christchurch, New Zealand (Aotearoa).
DT	*Daily Telegraph*, London, UK.
E&MJ	*Engineering and Mining Journal*, Chicago, USA.
Energy File	*Energy File*, Vancouver, Canada.
FT	*Financial Times*, London, UK.
FT Min	*Mining* (Financial Times International Yearbooks) London, UK. Previously called *Financial Times Mining International Yearbook* (*MIY*).
Gua	*Guardian*, London, UK.
INON	*International Newsbriefing on Namibia* (Namibia Solidarity Campaign) London, UK.
Kiitg	*Keep it in the Ground* (WISE) Amsterdam, Netherlands.
MAR	*Mining Annual Review* (Mining Journal Publications) London, UK.
MIY	*Financial Times Mining International Yearbook*, London, UK. Subsequently called *Mining* (*FT Min*).
MJ	*Mining Journal*, London, UK.
MM	*Mining Magazine*, London, UK.
MM Survey	*Mining Magazine Survey of Mines / Annual Review*, London, UK.
MMon	*Multinational Monitor*, Washington, USA.
NARMIC	*NARMIC Paper*, Washington DC, USA.
NFP	*Nuclear Free Press* (Ontario Public Interest Research Group) Peterborough, Canada.
Nukem Market Report	*Nukem Market Report* (Nukem GmbH) Hanau, Germany.
Obs	*Observer*, London, UK.
Press	*West Australian Press*, Perth, Australia.
R&OFSQu	*Analysis of Rand and OFS Quarterlies* (*MJ* supplement) (Mining Journal Publications) London, UK.
Reg Aus Min	*Register of Australian Mines* (Louthean Publications and Lodestone Press) Leederville, Western Australia.
RMR	*Raw Materials Report*, Stockholm, Sweden.
SunT	*Sunday Times*, London, UK.
Uranium Redbook	*Uranium: Resources, Production and Demand* (OECD/IAEA) Paris.
US Surv Min	*US Survey of Mines* (Engineering and Mining Journal) Chicago, USA.

Wall St J	*Wall Street Journal*, New York, USA.
WISE NC	*WISE News Communiqué* (World Information Service on Energy) Amsterdam, Netherlands.

Books etc.

Abrahamson & Zabinski	Dean Abrahamson & Edward Zabinski, *Uranium Mining in Minnesota* (University of Minnesota Center for Urban and Regional Affairs) Minneapolis, USA, 1980.
AMIC	*AMIC Survey* (Aboriginal Mining Information Centre) Healesville, Australia, 1982.
Atom's Eve	Mark Reader (ed), Ronald Hardbert & Gerard L Moulton, *Atom's Eve: Ending the Nuclear Age* – An Anthology (McGraw Hill) New York, USA, 1980.
Big Oil	*Big Oils Move into Mining*, USA, 1983.
BOM	List of mining projects (US Bureau of Mines) Washington DC, USA (various dates – as given in reference).
Davies *et al*	Rob Davies, Dan O'Meara & Sipho Diamini, *The Struggle for South Africa:* A Reference Guide to Movements Organisations and Institutions (Zed Press) London, 1984.
EPRI Report	*Foreign Uranium Supply: Project 883, Final Report* (Electric Power Research Institute) Maryland, USA, 1978.
Global Reach	Richard J Barnet & Ronald E Muller, *Global Reach: The Power of the Multinational Corporations* (Jonathan Cape) London, UK, 1975.
Goldstick	Miles Goldstick, *Uranium Mining in Canada* (British Columbia Survival Alliance) Cornwall, Vancouver, Canada, 1981.
Greenpeace Report '90	Ciaran O'Faircheallaigh, *Uranium Demand, Supply and Prices – 1991-2000* (Greenpeace) Brisbane, Australia, 1990.
Intern Min/Met Rev	*International Minerals/Metals Review* (McGraw Hill) New York, USA, 1982.
Lanning & Mueller	Greg Lanning & Marti Mueller, *Africa Undermined* (Penguin) Harmondsworth, UK, 1979.
Lynch & Neff	Michael C Lynch & Thomas L Neff, *Political Economy of African Uranium and its Role in International Markets: Final Report March 1982* (Massachusetts Institute of Technology) Boston, USA, 1982.
McGill & Crough	McGill & Crough, *Indigenous Resources Rights and Mining Companies in North America and Australia* (Department of Aboriginal Affairs) Canberra, Australia, 1986.
Merlin	LJ Merlin, "Canadian Uranium Industry" in *Uranium Supply and Demand: Proceedings of the Second International Symposium held by the Uranium Institute* – London 1977 (Mining Journal Books) London, UK, 1977.
Moss	Norman Moss, *The Politics of Uranium* (André Deutsch) London, UK, 1981.
Neff	Thomas Neff, *The International Uranium Market* (Ballinger) Cambridge, Massachussetts, USA, 1984.

Nuclear Barons	Peter Pringle & James Spigelman, *The Nuclear Barons: The Inside Story of How They Created Our Nuclear Nightmare* (Michael Joseph) London, UK, 1982.
Nuclear Fix	Thijs de la Court, Deborah Pick & Daniel Nordquist, *Nuclear Fix* (World Information Service on Energy) Amsterdam, Netherlands, 1982.
Pringle & Spigelman	See *Nuclear Barons*.
Reece	Ray Reece, *The Sun Betrayed* (Black Rose Books) Montreal, Canada, 1979.
Roberts	Jan Roberts, *Massacres to Mining: The Colonisation of Aboriginal Australia* (Dove) Blackburn, Australia, 1981.
Rogers & Cervenka	Barbara Rogers & Zdenek Cervenka, *The Nuclear Axis* (Julian Friedman) London, UK, 1978.
San Juan Study	Office of Trust Responsibilities, Bureau of Indian Affairs Lead Agency, *Uranium Development in the San Juan Basin Region: Final Report* (US Department of the Interior) Albuquerque, USA, 1980.
Sklar	Holly Sklar (ed), *Trilateralism: The Trilateral Commission and the Élite Planning for World Management* (Black Rose Books) Montreal, Canada, 1980.
Sullivan & Riedel	Raymond P Sullivan & Donald W Riedel, *Utility Involvement in Uranium Exploration and Development: A Growing Trend* (IAEA – paper at Buenos Aires symposium) Vienna, 1979.
Tanzer	Michael Tanzer, *Race for Resources* (Monthly Review Press) New York, USA, 1980.
UN NS-36	Brian Wood (ed), *International Seminar on Role of Transnational Corporations in Nambia* (UN – Document NS-36) New York, USA, 1982.
WDMNE	Stopford, Dunning & Haberich, *The World Directory of Multinational Enterprises* (Macmillan) New York, USA, 1982.
Weiss	David Weiss, *The Hour is Late* (University of California at Santa Cruz – Senior thesis for Boards of Environmental Studies and Sociology) Santa Cruz, USA, 1979.
Who's Who Sask	Walter Davies, *Who's Who in Saskatchewan* (One Sky) Saskatoon, Canada, 1980.
Yellowcake Road	W Davis (ed), *Corporate Uranium in Saskatchewan: The Yellowcake Road* (One Sky) Saskatoon, Canada, 1981.

COUNTRY ABBREVIATIONS

For reasons of space, the country of incorporation of companies shown in the company head-charts is usually given by means of a standard abbreviation. This is the list of those abbreviations, drawn from the (British) Automobile Association's list (1988). *NB:* No political inference is to be drawn from the choice of country name or abbreviation.

A	Austria	ETH	Ethiopia
ADN	South Yemen	F	France
AL	Albania	FJI	Fiji
AUS	Australia	FL	Liechtenstein
B	Belgium	GB	United Kingdom
BD	Bangladesh	GBA	Alderney, Channel Islands
BDS	Barbados	GBG	Guernsey, Channel Islands
BG	Bulgaria	GBJ	Jersey, Channel Islands
BH	Belize	GBM	Isle of Man
BR	Brazil	GCA	Guatemala
BRN	Bahrain	GH	Ghana
BRU	Brunei	GR	Greece
BS	Bahamas	GUY	Guyana
BUR	Burma	H	Hungary
C	Cuba	HK	Hong Kong
CDN	Canada	HKJ	Jordan
CH	Switzerland	I	Italy
CI	Ivory Coast	IL	Israel
CL	Sri Lanka	IND	India
CO	Colombia	IR	Iran
CR	Costa Rica	IRL	Irish Republic
CS	Czechoslovakia	IRQ	Iraq
CY	Cyprus	IS	Iceland
D	Germany, Federal Republic of	J	Japan
DDR	German Democratic Republic	JA	Jamaica
DK	Denmark	K	Kampuchea / Cambodia
DOM	Dominican Republic	KWT	Kuwait
DY	Benin	L	Luxembourg
DZ	Algeria	LAO	Laos
E	Spain	LAR	Libya
EAK	Kenya	LB	Liberia
EAU	Uganda	LS	Lesotho
EAZ	Tanzania	M	Malta
EC	Ecuador	MA	Morocco
ES	El Salvador	MAL	Malaysia
ET	Egypt	MEX	Mexico

MS	Mauritius	RU	Burundi
MW	Malawi	RWA	Rwanda
N	Norway	S	Sweden
NA	Netherlands Antilles	SD	Swaziland
NIC	Nicaragua	SF	Finland
NL	Netherlands	SGP	Singapore
NZ	New Zealand / Aotearoa	SME	Surinam
P	Portugal	SN	Senegal
PA	Panama	SP	Somalia
PAK	Pakistan	SU	USSR
PE	Peru	SWA	Namibia
PL	Poland	SY	Seychelles
PNG	Papua New Guinea	SYR	Syria
PY	Paraguay	T	Thailand
RA	Argentina	TG	Togo
RB	Botswana	TN	Tunisia
RC	Taiwan	TR	Turkey
RCA	Central African Republic	V	Vatican City
RCB	Congo	VN	Vietnam
RCH	Chile	WAL	Sierra Leone
RI	Indonesia	WAN	Nigeria
RIM	Mauritania	WD	Dominica
RL	Lebanon	WG	Grenada
RM	Madagascar	YU	Yugoslavia
RMM	Mali	YV	Venezuela
RN	Niger	Z	Zambia
RO	Romania	ZA	South Africa
ROK	South Korea	ZRE	Zaïre
ROU	Uruguay	ZW	Zimbabwe
RP	Philippines		

UNITS OF MEASUREMENT

As explained above in the "Lilliputia", or guide to using *The Gulliver File*, units for physical quantities are mostly quoted as found in the source material listed in the references. They are a mixture of the standard, internationally recognised, metric units (Systèm International, or SI, units) together with older units which are still common, particularly in English-speaking countries. And these latter, older, units can themselves be defined differently in different countries.

This section attempts to list all the units commonly found in the *File*, together with their abbreviations. It includes definitions of SI units for those not familiar with them, and defines non-SI units in terms of SI units for those *only* familiar with SI usage.

Note that the abbreviation for the plural of a unit is, properly, the same as the abbreviation of the singular.

SI prefixes

Readers may find some *derived* SI units (*i.e.* multiples and sub-multiples of the basic units) in the text, which are not listed explicitly in the tables below. The following list gives the standard prefixes (which are added to the basic units to provide the names of the derived units), together with their abbreviations and their meaning. These abbreviations are, similarly, combined with the abbreviations of the basic units to give the abbreviations of the derived units.

Quantity

deka	da	10 x
hecto	h	100 x
kilo	k	1000 x
mega	M	1,000,000 x
giga	G	1,000,000,000 x
tera	T	1,000,000,000,000 x
deci	d	0.1 x
centi	c	0.01 x
milli	m	0.001 x
micro		0.000,001 x
nano	n	0.000,000,001 x
pico	p	0.000,000,000,001 x

Length

metre	m	the basic SI unit of length
foot	ft	0.3048m
yard	yd	3ft *or* 0.9144m
mile	mi	1760yd *or* 1.609km

Area

are	a	100 square metres
hectare	ha	10,000 square metres
square yard	sq yd	0.8361 square metres
acre		4840 square yards *or* 0.4047ha
square mile	sq mi	640 acres *or* 2.590 square kilometres (259.0ha)

Volume

cubic foot	cu ft	0.02832 cubic metres
cubic yard	cu yd	27 cubic feet or 0.7646 cubic metres

Capacity

litre	l	0.001 cubic metres
gallon (UK)		volume occupied by 10 pounds weight of water *or* 0.005,943 cubic yards *or* 4.536 litres
gallon (US)		0.8327 UK gallons *or* 0.004,949 cubic yards *or* 3.777 litres

Weight

kilogram	kg	used, rather than the gram, as the basic SI unit of mass or weight
gram	g	0.001kg
tonne	t	1000kg (a "metric ton")
ounce	oz	28.35g
pound	lb	16oz *or* 0.4536kg
ton (UK)		2240lb *or* 1.120 US tons *or* 1.016t (a "long ton")
ton (US)		2000lb *or* 0.8929 UK tons *or* 0.9072t (a "short ton")

Miscellaneous (including electrical and nuclear)

joule	J	SI unit of work or energy, defined as 1 newton-metre (where a newton is the SI unit of force – that force which accelerates 1kg at 1 metre per second per second)
watt	W	SI unit of power, defined as 1J per second; when referring to usage of electricity, usual unit is the kilowatt (kW); when referring to the capacity of power stations generating electricity, usual unit is the megawatt (MW)
kilowatt-hour	kW-hour	common unit of electrical energy (3,600,000J)
becquerel	Bq	SI measure of the rate of spontaneous nuclear transformation of a radioactive nuclide: defined as one disintegration per second
curie	Ci	SI unit of activity, used prior to the becquerel *(q.v.)*: 37,000,000,000Bq
sievert	Sv	SI unit for dose equivalent (in joules per kilogram) of exposure to radiation, adjusted to take account of the effectiveness (in causing biological damage) of that particular sort of radiation. There was an earlier unit, the rem: 1Sv = 100 rem
pH		measure of acidity or alkalinity (basicity); 7 is neutral. Defined as $\log_{10}$ of the reciprocal of the concentration of aqueous hydrogen ions – hence lower numbers represent greater acidity and higher numbers greater alkalinity.

Alphabetical list of abbreviations for units of measurement

Explanations of some of the common abbreviations are given here – where necessary, see the preceding lists for the definitions of the terms themselves or of their component parts. Note that some abbreviations used in the text of the book for SI units of measurement are not given in this list if they are simply derived from other abbreviations by the use of standard SI prefixes. So, if you can't find what you're looking for here, check the prefix list above.

Bq	becquerel
Ci	curie
cu	cubic (in cu ft, cu m, etc)
ft	foot
g	gram
ha	hectare
J	joule
kg	kilogram
km	kilometre
kW-hour	kilowatt-hour
l	litre
lb	pound
m	metre
mi	mile
mt	"metric ton", *i.e.* tonne (used, confusingly, in some sources)
MW	megawatt
oz	ounce
sq	square (in sq yd, sq km, etc)
Sv	sievert
t	tonne (in some sources, may be used to mean one or other sort of non-metric ton)
t/d	tonnes per day
t/m	tonnes per month
t/y	tonnes per year
t/yr	tonnes per year
tpd	tons per day *or* tonnes per day
tpy	tons per year *or* tonnes per year
W	watt
yd	yard

OTHER ABBREVIATIONS

Some specialised categories of abbreviations have already been given elsewhere: the section on units of measurement included its own abbreviations list; there is a listing of the abbreviations used in the headcharts for the country of a company's incorporation; and there is a section explaining the abbreviations used for newspapers, magazines, books, etc, in the bibliographical references at the end of each entry. That leaves many other abbreviations used in the text – such as for the component states of federal countries, for various types of nuclear reactor, for company and commercial terminology in different countries, for names of groups and organisations, and so on. These are all given here in one alphabetical list. Where the fuller version might itself need some explanation, it is marked with an asterisk (*) and will be found in the Glossary.

A/S	Aksjeselskap*
AAPC	Australian Aluminium Production Commission
AB	Aktiebolag*
ABC	Australian Broadcasting Corporation
ACTU	Australian Council of Trade Unions
ADC	Aboriginal Development Commission (Australia)
AG	Aktiengesellschaft*
AGM	Annual General Meeting
AGR	Advanced Gas-Cooled [nuclear] Reactor
AIAS	Australian Institute of Aboriginal Studies
AIDESEP	Asociacion Interetnica para el Desarrollo de la Selva Peruana (Peru)
ALP	Australian Labor Party
AMIC	Australian Mining Industry Council
AMIC-Them	Australian Mining Industry Council
AMIC-Us	Aboriginal Mining Information Centre
ANC	African National Congress (of South Africa)
APT	Additional Profits Tax (in Papua New Guinea)
BBC	British Broadcasting Corporation
BC	British Columbia, Canada
BPk	Beperk*
BRA	Bougainville Revolutionary Army
BV	Besloten Vennootschap met beperkte aansprakelijkheid*
BWR	Boiling Water [nuclear] Reactor
CAFCA	Campaign Against Foreign Control of Aotearoa
CAFCINZ	Campaign Against Foreign Control of New Zealand

---

*   See the Glossary for further explanation.

Candu	Canadian Deuterium [nuclear] Reactor
CANUC	Campaign Against the Namibian Uranium Contract
CEASPA	Centro de Estudios y Accion Social Panameno (Panama)
CFC	chlorofluorocarbon
Cia	Companhia*
Cia	Compañia*
Cie	Compagnie*
CIMI	Conselho Indigenista Missionario (Brazil)
CIMRA	Colonialism and Indigenous Minorities Research and Action
Co	Company*
COMARE	Committee on Medical Aspects of Radiation in the Environment*
Corp	Corporation*
DAA	Department of Aboriginal Affairs (Australia)
EC	European Community – formerly European Economic Community (EEC)
EEC	European Economic Community (see EC)
ERMP	Environmental Research and Management Programme* (Western Australia)
FBR	Fast-Breeder [nuclear] Reactor
FIRB	Foreign Investment Review Board (Australia)
FoB	free on board*
Ges	Gesellschaft*
GmbH	Gesellschaft mit beschränkter Haftung*
HTR	High-Temperature [nuclear] Reactor
IAEA	International Atomic Energy Agency*
ICME	International Council on Metals and the Environment*
ICRP	International Commission on Radiological Protection*
IDAF	International Defence and Aid Fund for Southern Africa
Inc	Incorporated*
INFCE	International Nuclear Fuel Cycle Evaluation*
IWGIA	International Workgroup on Indigenous Affairs (Denmark)
JV	joint venture*
K	Kina (Papua New Guinea currency)
KLC	Kimberley Land Council
LME	London Metals Exchange*
Ltd	Limited*
Ltda	Limitada*
Ltée	Limitée*
LWR	Light-Water [nuclear] Reactor
M	million (of sums of money)
MELSOL	Melanesian Solidarity Front
MNC	multinational company/corporation*
MUF	material unaccounted for*
MUN	Mineworkers Union of Namibia
NGO	Non-Governmental Organisation
NL	No Liability*
NLC	Northern Land Council (Australia)
NM	New Mexico, USA
NPT	Nuclear Non-Proliferation Treaty*
NRPB	National Radiological Protection Board*

NSW	New South Wales, Australia
NT	Northern Territory, Australia
NUM	National Union of Mineworkers
NUNW	National Union of Namibian Workers
NV	Naamloze Vennootschap*
NWT	Northwest Territories, Canada
NY	New York, USA
Oy	Osakeyhtiöt*
PLC/plc	public limited company*
ppm	parts per million
PT	Perushaan Terbetas*
PTy	Proprietary*
Pvt	Private*
PWR	Pressurised Water [nuclear] Reactor
Qld	Queensland, Australia
SA	Sociedad Anónima*
SA	Société Anonyme*
SA	South Australia
SAIRR	South African Institute of Race Relations
Sdad	Sociedad*
Sdad	Sociedade*
SL	Sociedad Limitada*
Sté	Société*
SWAPO	South-West African People's Organisation of Namibia
Tas	Tasmania, Australia
TGWU	Transport and General Workers Union, Britain
TNC	transnational company/corporation*
TUC	Trades Union Congress, Britain
U_3O_8	uranium oxide*
UDI	Unilateral Declaration of Independence (Southern Rhodesia – later Zimbabwe, 1965)
UF_4	uranium tetrafluoride*
UF_6	uranium hexafluoride*
UKAEA	United Kingdom Atomic Energy Authority
UNDP	United Nations Development Programme
UNEP	United Nations Environment Programme
US	United States (of America)
USA	United States of America
USA	United Steelworkers of America, in the USA (sometimes, USWA)
USWA	See USA
Vic	Victoria, Australia
WA	Western Australia
WDLC	Western Desert Land Council, Australia

GLOSSARY

This is your glossary of mining, trade, economic, and technical terms. Please use it!

The mining industry *does* possess a few volumes which are larger – if not more comprehensive – than the one you are now holding in your lap. One of these is the *Dictionary of Mining, Mineral and Related Terms*, published by the US Bureau of Mines in 1968, and reprinted in 1990. This 1274-page book is a fabulous work, in more than one sense of the term: it is a collection of fables – though not falsehoods – and also a fascinating journey in its own right through mining history.

Much of the mining-related part of this glossary is drawn from that work, although some more recent mining terms are not to be found in it ("solution mining", for example, or "tributing" in the South African sense). Just as that *Dictionary* will give the patient reader an insight into the enormous business which mining involves (every description among the scores of thousands contained there relates to some distinct mining practice, somewhere in the world, at some time past or present), so – it is hoped – this glossary will provide a brief introduction to some of the realities of mining, which every non-miner ought to know.

It is, indeed, a sobering thought that, while the human activities encompassed by food production (the most important economic activity on this planet) have passed into the language (including that of geography syllabi), the rubric on which the second most important economic activity is based remains privy to a comparative few. I hasten to add that the purpose of this glossary is not to be definitive – let alone exhaustive: it relates directly to the material in the remainder of this book, rather than the mining industry as a whole. Nonethelesss, the reader is urged not to pass lightly over this part of the work. "Know your enemy!" is a sound rule for those who see the effects of mining and wish to counter them in a concerted, informed, fashion. "Know your industry!" is an equally good behest to be made to those who do not necessarily reject mining practices, but wish to scrutinise them intelligently.

As well as defining mining terms which are common in the *File*, and some corporate business terminology, I have also included a large amount of material on uranium and nuclear power (for which I am indebted to Miles Goldstick's *Nuclear Words and Terms* – published by the World Information Service on Energy, Amsterdam – whence many entries in this glossary have been adapted). The reasons for this will be obvious.

I only hope that the items themselves will be equally transparent.

Please note that, when a term in a glossary entry appears in **bold**, that term itself has an entry in the glossary, which should be consulted for fuller information. Names of companies which have an entry in the *File* appear in the glossary in *italics*.

acid mine drainage – drainage which occurs when iron pyrites are exposed to weathering (in mines, waste dumps, or **tailings** heaps) or when sulphide minerals break down under the combined effect of oxygen and water – releasing sulphuric acid, with consequent harm (sometimes incalculable) to rivers and underground water supplies.

Aksjeselskap (A/S) – Norwegian designation for a **company** with shareholders.

Aktiebolag (AB) – in Sweden or Finland, a **company** with shareholders.

Aktiengesellschaft (AG) – in the German Federal Republic, an entity roughly equivalent to a British **public limited company**; subject to more rigid laws of membership and accountability than a **Gesellschaft mit beschränkter Haftung**. The same term with similar meaning is used in Austria and Switzerland.

alluvial deposits – minerals which have been washed out of one place and, carried by water, are now found in a river bed, lake bed, etc, or in an area previously covered by water.

alluvial mining – the exploitation of alluvial deposits.

alpha particle – a positively-charged particle made up of two neutrons and two protons (the nucleus of a helium atom) which is emitted by certain **radioactive** material. The emission process is called alpha decay. The alpha particle can cause ionisation, and it is the largest of the atomic particles emitted by radioactive material. It cannot easily penetrate clothing or skin, but is highly dangerous if emitted by something which has been inhaled or ingested. (See also **radiation, ionising**.)

anomaly – any geophysical departure from normal, thus an indication of possible **mineralisation** (presence of a mineral).

assay – an analysis of a substance containing a mineral or metal, carried out to ascertain its purity or its proportion in a given sample.

assets – just about anything owned by a company, which features on its balance sheet, both movable (*e.g.* equipment) and immovable (*e.g.* land), legitimately acquired or not.

associate company – a company over which another company has influence and, in certain areas (*e.g.* **joint ventures**), a measure of control. The investment of the latter in the former is usually between 20% and 50%.

bacterial leaching – the use of bacteria (or algae) to extract minerals such as uranium, molybdenum, radium, selenium, or lead, from **ore** heaps (see **heap leaching**) or mine waters.

beach sands (or **mineral sands**) – sand dunes, either coastal or located inland (as a relic of former seas), from which strategic minerals such as titanium and zircon can be extracted after the processing of ilmenite, rutile, and monazite. The **radioactive** elements **thorium** and **uranium** are themselves associated with monazite. Such sands may also contain so-called "rare earths" (yttrium, *et al*), which are neither *that* rare, nor earths! These groups of metals/minerals are distinguished by their radioactive properties and their usage – in oxide or metal form – in the aerospace, "hi-tech", and nuclear power industries.

beneficial shareholder – the person or **company** behind a **nominee shareholder** or, more generally, the person or body which benefits from an interest of which it is not the owner.

Beperk (BPk) – South African designation for a **company** with shareholders.

Besloten Vennootschap met beperkte aansprakelijkheid (BV) – (or just Besloten Vennootschap) – in Belgium and the Netherlands, equivalent of a British **private limited company**.

beta particle – the fragment emitted by some **radioactive** material in a process called beta decay. It is either an electron, which is negatively charged and referred to as beta-minus decay, or a positron, which is positively charged and called beta-plus decay. It can cause ionisation. (See also **radiation, ionising.**)

boron – an element which is a powerful absorber of slow neutrons. Because of this property, it is used in steel alloys for making nuclear reactor control rods. (See *RTZ.*)

by-product – a secondary mineral (or minerals) extracted alongside the substance primarily sought. **Uranium** has been found, or mined, as a by-product of gold (*e.g.* by *ERGO),* phosphates, copper, nickel, manganese, vanadium, coal, and beryllium; and has been extracted from **tailings** or solutions left behind after mining (*e.g.* copper leach liquors).

capital – See **share capital.**

carcinogen – a cancer-causing substance.

cartel – a combine, "trust", "club", or loose combination of producers, which seeks to influence the price of raw materials or of services and/or commodities, or their availability (itself a major factor in price-fixing), by controlling resources, regulating outputs, or setting purchase patterns. The essence of a cartel is that it limits, or excludes, competition; usually by acting in a highly secretive fashion, and denying its own very existence. Ironically, some corporate bodies which have set up, or dominated, minerals cartels also espouse "free trade" (*e.g. RTZ*). For details of the uranium cartel see also *Rio Algom, Westinghouse, Gulf Oil, TVA,* and *MKU.*

Committee on Medical Aspects of Radiation in the Environment (COMARE) – a British governmental group, set up to investigate all environmental aspects of **radiation** from artificial sources.

Compagnie (Cie) – in Belgium, France, Luxembourg, designation for a **company.**

Companhia (Cia) – in Portuguese-speaking countries, designation for a **company.**

Compañia (Cia) – in Spanish-speaking countries, designation for a **company.**

company – a business, usually a legally incorporated one, which is a separate legal entity from the people owning it; abbreviated "Co" in the name of specific companies. The term is used in many English-speaking countries. In the UK it usually has limited liability (*i.e.* its shareholders may not be personally liable for the company's debts, beyond the value of the shares they hold), and may then be a **private limited company** or a **public limited company.**

concentrate – a (usually ground-up) mineral product obtained by physical or chemical separation from **ore**, having eliminated most of the "waste". Common physical methods are flotation separation and gravity separation. (For **uranium** concentrate, see **yellowcake.**)

conglomerate – a large, multi-company, business organisation which usually consists of a **holding company** and a group of subsidiaries involved in widely varying activities.

corporation – USA term for a **company**, without separate legal forms to distinguish between a private company and a public company (*i.e.* one in which the public is free to buy shares); may be abbreviated to "Corp" in the name of a specific corporation.

cut-off grade – the lowest grade at which mining is considered economically feasible.

daughter product – a nuclide into which a radioactive nucleus transforms itself by radioactive decay from a "parent product". For example, radon-222 is the "daughter" product of radium-226.

depleted uranium – uranium in which the proportion of the **isotope** uranium-235 is lower than that found in naturally occurring **uranium**. It is an inevitable by-product of the **enrichment** of uranium.

diversification – expanding a company's operations into areas not previously associated with its "core" activities. For example from oil into mining (see *Exxon, Arco, Mobil, RTZ, AAC, CSR*, amongst many others).

dosimeter – (or **dosemeter**) – a device used to measure a dose of **ionising radiation**. The measurement may be by means of an ionising chamber, blackening of photographic film, or the extent of a chemical reaction in a solution.

downstream processing – what happens to a raw material, or mineral, once it has been extracted from the ground, and before it becomes an end-product for sale on the market; or, what is still to happen to a material in further processes beyond the stage it has currently reached.

drift mining – the working of rich **alluvial deposits** by underground mining methods.

drilling – the initial investigation of an underground mineral deposit by obtaining a sample of **ore** on which a chemical or radiological **assay** can be based, so giving more accurate information than any preceding aerial or surface survey. Diamond drilling consists of the extraction of a core, or cylinder, of rock; other methods include rotary-mud and rotary-air drilling. Drilling equipment is carried on rigs, which have improved in mobility, diminished in size (a boon for access to remote or heavily-forested areas), and advanced in accuracy over the last 25 years.

enriched uranium – See **uranium**, and **enrichment**.

enrichment – a process, also called isotopic separation, by which the proportion of one **isotope**, **uranium-235**, is increased above that in naturally occurring uranium in one part of a quantity of uranium, leaving a lower-than-naturally-occurring proportion in the remaining part. The portion with the increased proportion of uranium-235 is called enriched uranium; the portion with the decreased proportion (consisting therefore almost entirely of uranium-238) is called depleted uranium. Much of the processing of uranium, from the **milling** of the **ore** onwards, is geared towards this process. The natural proportion of uranium-235 is 0.71%; for some nuclear reactors the proportion needs to be 2-3%; for some military purposes it needs to be enriched to over 90%. One enrichment method is gaseous diffusion, operating on **uranium hexafluoride** (or hex); the plants which do this are amongst the largest (and greatest electricity-consuming) industrial facilities in the world. Gas centrifuges are a more efficient method: they also use hex. A third method is laser separation.

Environmental Research and Management Programme (ERMP) – a research programme into the environmental and social aspects of the Argyle diamond mine in Western Australia, commissioned in 1982 by Australia's Aboriginal Development Commission.

equity – the value of a company's assets after all outside liabilities (other than to shareholders) have been allowed for. A company ultimately belongs to its "true owners" – those who, in theory, take the greatest risk and are entitled to the company's reserves and profits – *i.e.* its ordinary shareholders.

feasibility study – a study carried out by, or for, a company, to try to evaluate likely costs, revenues, equipment requirements, production rates, and engineering problems of a proposed mining project. The study may, or may not, include environmental and social factors (usually not).

ferrous – having a high proportion of iron in its composition (of metal or metal-bearing mineral).

fertile material – a material which can be transformed into a **fissile material** (or fissionable material). The two main fertile nuclides are uranium-238 and thorium-232, which respectively form plutonium-239 and uranium-233 (both fissionable by thermal neutrons).

film badge – a type of personal dosimeter in the form of a badge, worn for a known period of time, containing photographic films shielded by different materials, permitting discrimination between beta and gamma exposure. Since **radiation** darkens the film, it provides a record of radiation exposure.

fissile material – a material made up of heavy atoms which can be split into pieces, emitting energy. (See **fission**.) A material is said to be "fissile by" (or "fissionable by") either "fast neutrons" or "thermal neutrons" (the latter also called "slow neutrons"), depending which energy level a neutron needs in order to cause the atoms of the material to split. Thorium-232 and uranium-238 are fissile by fast neutrons; uranium-235 and -233, and plutonium-239, are fissile by slow neutrons.

fission – the splitting of the **nucleus** of an atom into lighter fragments, accompanied by the release of energy and generally one or more neutrons. Fission can occur either spontaneously or as the consequence of absorption of a neutron. If a substance undergoes fission spontaneously, releasing energy in the form of radiation, then it is called **radioactive**. The main fissionable nuclides are uranium-235, plutonium-239, and uranium-233. The intensity of the fission reaction depends on many factors, including the degree of **enrichment** of the fuel, its mass, and its physical arrangement in a given volume.

flotation – a method of mineral separation in which a froth, created in waste by a variety of chemical reagents, "floats out" some minerals remaining in the waste, allowing others to sink.

free on board (FoB) – pertains to goods (*e.g.* ore) which are put on board a ship free of all expense and shipping costs for the buyer.

fuel chain and **fuel "cycle"** – the sequence of operations involved in supplying fuel for nuclear power generation and nuclear weapons. The most common civil sequence begins with **uranium** exploration, proceeds to uranium mining and **milling**, refining, **enrichment**, fuel fabrication, **fission** in a reactor, reactor waste storage, and ("finally") reactor decommissioning.

futures – See spot sales.

gamma ray – short wavelength electromagnetic radiation of great penetrating power, released by some nuclear transformations. It can penetrate through the human body. Gamma rays are emitted by many radionuclei that undergo alpha or beta decay. As well, many nuclides become gamma emitters from a neutron-gamma reaction. (See also **radiation, ionising**.)

Gesellschaft (Ges) – in German-speaking countries, designation for a **company**.

Gesellschaft mit beschränkter Haftung (GmbH) – in the German Federal Republic, an entity roughly equivalent to a British **private limited company**, without restriction on the number of members: hence they often include some very large institutional and private investors. The same term is used with similar meaning in Austria and Switzerland. See also **Aktiengesellschaft**.

grade – classification of **ore** according to the proportion of the desired material within it. Grades are usually published as ratios or percentages (*e.g.* 10kg/tonne silver-in-ore, or uranium grading at 0.03%).

half-life – the time it takes for half of any amount of a **radioactive** substance to undergo decay. The process of radioactive decay is independent of temperature, pressure or chemical condition. Half-lives range from less than a millionth of a second to millions of years. Natural uranium has a half-life of about 4500 million years, or roughly the age of the earth.

hardrock – mineral deposits which are **igneous** or **metamorphic**, as opposed to **sedimentary**.

heap leaching – the extraction of minerals from heaps of **ore** by pouring chemical or biological reagents over them. An acid or cyanide solution is commonly used (especially on ore containing gold or copper): for this process, the ore is heaped on supposedly impermeable pads. See also **bacterial leaching**.

heavy metal – a metal which reacts readily with dithizone (diphenylthiocarbazone) – principally zinc, copper, cobalt and lead, but also bismuth, cadmium, gold, indium, iron, manganese, mercury, nickel, palladium, platinum, silver, thallium and tin. Many of these metals are particularly injurious to living organisms (including humans) when taken up by plants and fish and passed into the food chain.

high-grading – working the richer parts of an **orebody**, at the expense of the less rich deposits. The practice (which derives its name from a term used during the Gold Rush days to denote theft) has been condemned as irresponsible, since it restricts the working life of a deposit and may involve inordinately heavy cost to the environment (see *AAC*).

holding company – a **company** which controls (usually through majority shareholdings) a number of other companies, often thus benefitting from having access to a much larger amount of capital than it possesses itself.

Incorporated (Inc) – in the USA, designates a legally constituted **corporation** or **company**.

igneous – (of rocks and minerals) formed from a molten state, *i.e.* from volcanic eruptions.

in situ **leaching** – See **solution mining**.

International Atomic Energy Agency (IAEA) – an organisation set up in 1957 under the auspices of the United Nations. It has the conflicting roles of promotion and regulation of the nuclear industry.

International Commission on Radiological Protection (ICRP) – a British-based body without any national or international official status, set up originally to monitor the medical use of X-rays. However, its recommendations on upper limits for radiation exposure are now often (though not always) accepted, both by national governments and international bodies. They are also criticised as too high by many – *e.g.* Catherine Caufield in her book *Multiple Exposures* (Penguin) 1989.

International Council on Metals and the Environment (ICME) – formally launched in Canada in 1990, to offset mounting criticism of mining operations, it comprises representatives of many (but not all) leading mining corporations, including *Noranda, Boliden, Cominco, Phelps Dodge, ACFC-Union Minière,* and *RTZ* (a late joiner).

International Nuclear Fuel Cycle Evaluation (INFCE) – a multi-national study under the auspices of the **International Atomic Energy Agency** in the late 1970s, intended to control proliferation. (See also **Nuclear Anti-Proliferation Act**.)

inventory – in the case of mining, the stock of raw materials held (such as uranium) by a **utility**, government or other body. The material is often held against future needs or future sales.

isotope – a particular version of the atoms of a given element, differing from another isotope of the same element only in the number of neutrons in the nucleus. All isotopes of an element share the same chemical properties, but they differ – even if only slightly – in their physical properties: in particular, one isotope may be **radioactive** but another not. An isotope which emits radiation spontaneously is called a radioisotope. Only some elements have different isotopes; where different isotopes do exist, it is not necessarily the case that any of them is radioactive. Isotopes of the same element are differentiated by reference to their atomic weight – *e.g.*, uranium-235 and uranium-238.

joint venture (JV) – a project of two or more companies which have combined together for a specific purpose. This practice (which is increasingly common in the mining industry) can cut costs, improve "efficiency", and reduce various (financial) risks.

leaching – removal of a mineral by a solvent, or by a reagent that reacts with the mineral being sought, to form a compound containing the mineral. Water is a common solvent, sulphuric acid and hydrochloric acid are common reagents. (See also **heap leaching**, **bacterial leaching**, and **solution mining**.)

Limitada (Ltda) – in Portuguese-speaking countries, designation for a limited liability **company**.

Limited (Ltd) – in English-speaking countries, designation of a limited liability **company**.

Limitée (Ltée) – in Canada, designation of a limited liability **company**.

lode – virtually synonymous with, but more poetic than, **vein**. It is an ancient miners' term which is a corruption of "lead", meaning a mineral formation by which miners were led to their discoveries. It is used to describe a flat, or stratified, mass of ore.

London Metals Exchange (LME) – the London commodities market for trade in aluminium, nickel, copper, lead, zinc, and tin.

longwalling – underground mining of coal along a large expanse of "face" from which the material is hauled, allowing the workings then to collapse. Although an old practice (which possibly originated in Shropshire, England, in the late 17th century) it has recently been revived in the USA as cheaper, safer and more efficient than conventional methods; however it can lead (and has led) to disastrous cave-ins of the ground and buildings above.

material unaccounted for (MUF) – material which has "gone missing" from uranium processing, or **enrichment**, or fuel fabrication plants, and may well have been stolen. Usually used in reference to **trigger material** – *i.e.*, material of sufficient quality/quantity to trigger safeguards under the **Nuclear Non-Proliferation Treaty (NPT)**. (See *Numec, MKU, Kerr McGee.*)

merger – the combining of the assets (and, usually, the operations) of two or more companies into one unit. This is generally arranged to the mutual satisfaction of those controlling the respective companies, and is therefore different from a takeover. A takeover occurs when a stronger company mounts a successful bid for the shares of a more vulnerable company, which often has tried to resist the bid. (See *CGF, Texaco, RTZ.*)

metallurgy – extracting metals from ores and refining or processsing them for human use.

metamorphic – (of minerals) deriving from rocks which have been subject to great pressure, high temperature, and chemical alteration.

milling – the grinding and crushing of **ore** to remove unwanted constituents, at the point closest to the mine. "Waste" from milling is usually called tails, or **tailings**.

mineral sands – See beach sands.

mineralisation – the presence of minerals in an orebody. (The word can also apply to the *process* of the formation of minerals from organic matter.)

multinational company/corporation (MNC) – (or **transnational** company/corporation

(TNC)) – a collection of corporate entities established under the laws of different countries, which effectively work together across national borders as one coordinated unit. Though existing as a united entity to the extent of having a single headquarters (in the "host" country), a true MNC can be distinguished from simply a national company with overseas branches/subsidiaries by the fact that most of the national units making up the MNC are capable of existing as independent companies. MNCs have been the subject of millions of words, attempting to analyse, describe, condemn, and defend their existence and activities. They generally operate on the same principles in the various countries they work in, as though not bound by the laws or conventions of any one country – for example, when it comes to "posting" their tax returns, situating their processing plant, employing their workers and paying them fair rates, or observing environmental criteria which smaller companies have to follow. Generally, the leaders of MNCs (who are, almost without exception, western-educated and western-influenced, irrespective of the country the MNC is nominally based in) hold a "one world market" view of natural and human resources – in which tariff barriers come down (or, conversely, go up when more convenient); labour forces move (or are moved) freely across frontiers; raw materials are extracted at their cheapest rate and "add value" inexorably, as they pass **downstream** though the links in the corporate chain; and the end product reaches the consumer market at a highly-inflated cost (which puts a bigger margin of profit in the hands of the shareholders of the MNC). This profit is then used to expand operations, either within a given region, or in new, untapped zones; or to better "integrate" the production of a given material or consumer product; or to diversify into new products and, hence, new markets. This view of MNCs is, in broad terms, not disputed by government leaders – many of whom mistrust (and seek to limit the activities of) MNCs, seeing in them threats both to domestic control of resources and people, and to "national identity". This view of MNCs is also that of most of their

opponents, drawn from innumerable communities and NGOs across the globe. The main critique of the MNC given by these opponents is that the corporations refine the capitalist *raison d'être* ("buy cheap and sell dear") to an intolerable degree – one which puts at risk local control of resources; regional co-operation among struggling economies; workers' rights and trade union organisation; the development of appropriate technologies; and, above all, a view of the world as an immensely diverse and complex network of co-operating communities, where trade is based on satisfying mutual needs, not aggrandising a miniscule élite or further lining the pockets of an affluent few. A distinction between transnational corporations and multinational corporations was initially made in the 1960s, when researchers sought to contrast corporate enterprises effectively based in one country and owing their allegiance to bankers, politicians, investors and shareholders there (MNCs) – the best example probably being the Japanese *sogo shosha* – with corporate conglomerates which have effective bases, and allegiances to élites, in various countries, thus transcending national constraints (TNCs). However, the TNC, in this sense, has never thoroughly developed. The huge "natural" disasters caused by MNCs in the last decade – for example the Bhopal outrage (see *Union Carbide)* – demonstrate that, whether they like it or not, what they do "abroad" can be made accountable to multinationals "back home". The runaway corporation, owing allegiance to nothing and nobody but itself, is, thankfully, proving to be a disappearing species. In a real sense, the truly transnational movement is represented by MNCs' opponents – international trade unionists, anti-apartheid activists, environmental pressure groups, land rights movements, peace campaigners, and radical "third world" lobbies. Nonetheless, the adaptability and power of the MNC (not least its power to corrupt through buying into, or buying off, the opposition) has given rise to the single biggest bloc of economic manipulability in the world today. The multinational creates dependency in a far more subtle and insidious fashion than the state, disguising the existence of subsidiaries in a manner which the CIA and the KGB should envy, and controlling consumer choice with a complexity that needs full-time monitoring across this globe in order to keep up with changes in marketing. The MNC today dispenses largesse in the form of scholarships to poor students in the third world; endows universities and colleges (for example *RTZ); funds* political parties; sits at table with government leaders *(AAC);* practises its own brands of east-west diplomacy (the late Armand Hammer of *Occidental);* funds substantial military research; controls biodiversity by limiting the availability of vital seed stocks *(Shell);* and – more recently – projects itself as the epitome of all things "green". In the final event, however, MNCs are still nominally and legally responsible to their ordinary (voting) shareholders. What they do is based on consent – although there are examples of corporations which seek to impose their will by force (for example *CGF/GFSA* and *RTZ* with their private armies in southern Africa). If the contemporary MNC portrays itself as indispensable, it is largely because the rest of us have failed to free ourselves from the omniscient, omni-powerful patrimony (and there are no MNCs effectively run by women) which MNCs epitomise.

Naamloze Vennootschap (NV) – Dutch equivalent of a British **public limited company.**

National Radiological Protection Board (NRPB) – British "watchdog" agency which sets supposedly safe limits for radiation exposure and which may conduct enquiries into instances of alleged over-exposure.

natural fission reactor – a nuclear reaction caused by a natural concentration of uranium sufficient to generate its own **fission** process. To date only one such "reactor" has had its existence verified, at Oklo in Gabon.

No Liability (NL) – in Australia, designation of a limited liability **company.**

nominee shareholder – someone in whose names shares are registered but who is only a nominee or trustee for the **beneficial shareholder,** so concealing the latter's identity. This makes virtually impossible investigation of the real power-holders in certain companies; this has especially been the case with those companies doing business in South Africa. The British government, under the 1948 Companies Act, has wide powers to investigate nominees; and, under more recent legislation, a **company** can itself demand the names of those holding disguised shares in itself. But this right does not extend to ordinary shareholders or the public at large.

non-ferrous – not containing significant amounts of iron (of metal or metallic ore).

Nuclear Anti-Proliferation Act – a measure introduced by the Carter administration in the USA in 1977, which amended the country's 1954 Atomic Energy Act to try to establish strict criteria for the handling and export of US-origin nuclear material. The aim was to prevent further proliferation, following the International Nuclear Fuel Cycle Evaluation. (See also **Nuclear Non-Proliferation Treaty.**)

Nuclear Non-Proliferation Treaty (NPT) – properly the Non-Proliferation of Nuclear Weapons Treaty – signed in 1968. Coming into force two years later, it aimed both to popularise nuclear power and to prevent non-nuclear weapons states from acquiring the means to manufacture nuclear weapons. Most objective analysts would now view the NPT as, at best, a benign paradox: the fact that Iraq is a signatory, while Israel, South Africa, India, China and Pakistan have refused to sign, speaks for itself. Equally important, though less remarked upon, is the sad fact that the nuclear weapons powers were under no obligation themselves to disarm, though intentions towards that end were incorporated in the Treaty as a *quid pro quo* for the non-nuclear states forswearing the development of military nuclear capacity from the "civilian" nuclear technology which was shared with them under the terms of the Treaty. Certain substances suitable for military nuclear uses, which should trigger safeguards under the NPT, are called **trigger materials.**

open-pit mine – (or **open-cast** or **open-cut**) – surface workings, exposed to daylight, varying in size from a small quarry to the biggest human-made excavations on earth. (See **underground mine.**)

option – a contract which carries with it the right to buy or sell within a given period.

ore – material, usually obtained from mine workings, from which one or more minerals can be extracted. The term "reserves in ore" denotes estimated quantites of the mineral(s) which could be extracted from a given amount of material: the ore, in this case, has usually not yet been extracted from the earth.

orebody – a body of **ore** still in the ground.

Osakeyhtiöt (Oy) – in Finland, designation of a company.

overburden – the earth, vegetation, loose rock, forest cover ("waste", in the mining industry misnomer) which has to be removed to give access to ore-bearing strata.

Perushaan Terbetas (PT) – in Indonesia, designates a **company.**

plutonium – a highly toxic, heavy, **radioactive,** metallic element. It is an extremely dangerous substance because of its **radioactivity** and the fact that, when ingested as an oxide or other compound, it deposits in the bone and is excreted only very slowly. Metallic plutonium is not absorbed by digestive organs. Inhalation of only a few thousandths of a gram may lead to death within a few years, and much smaller quantities can cause lung cancer after a latent period of about 20 years.

Private (Pvt) – in India and Zimbabwe, legal designation of a private **company.**

private limited company – in Britain, a **company** which does not raise money by making shares available to the public (unlike a **public limited company**). So there is no way, through the structure of the company, that a private company can be held to account.

privatisation – the sale of government-owned **equity** in a nationalised or state-controlled industry to private investors. (See *CEGB, Péchiney.*)

Proprietary (Pty) – in Australia and South Africa, designates a limited liability **company.**

prospect – a place where minerals are expected but not proved. (Hence prospecting generally precedes exploration.) Prospecting is carried out to define the nature, extent and value of an orebody, using fairly sophisticated means, such as aerial surveying, magnetometry, soil testing, and seismic probes.

public limited company (plc) – a company, in British law, which offers shares to the public on the stock exchange. Under the 1985 Companies Act, a plc may have an unlimited number of shareholders and must make its accounts, together with some other information, available. (Previously, the designation "limited" (Ltd) was used for either a public or a **private limited company;** now, "Ltd" is used only for the latter.) Anyone buying shares in a plc can theoretically wield some influence over it at general meetings. However, a plc is not necessarily "accountable" and, through the use of front companies, **associate companies,** or companies registered in tax havens with laxer legal requirements (such as Bermuda, the Cayman Islands, Hong Kong, Singapore, Luxembourg, Netherlands Antilles, the Channel Islands, the Isle of Man), it can still keep its financial transactions secret (see *AAC).*

radiation – the propagation of energy through space or matter. There are two basic forms of radiation: particle (or corpuscular) radiation and electromagnetic radiation. Electromagnetic radiation consists of radiant energy travelling in waves at the speed of light; it is classified according to wavelength. The main types, ranging from short to long wavelength, are: **gamma rays**, X-rays, ultraviolet light, visible light, infrared light, microwaves, and radio waves. Gamma rays and X-rays are types of electromagnetic radiation involved in radioactivity. Particle radiation consists of energetic particles. In the case of spontaneous decay of radionuclei, particle radiation consists mainly of **alpha particles** and **beta particles**. Some types of electromagnetic radiation, and all forms of particle radiation, have the ability to induce ionisation, either directly or as a secondary effect: see **radiation, ionising**. When a substance emits radiation, the *process* is called **radioactivity**.

radiation, ionising – radiation which can deliver energy in a form capable of knocking electrons off of atoms, turning them into ions. On interaction with matter, high-intensity ionising radiation puts atoms or molecules into a more excited or interactive state and thereby promotes chemical reactions which would otherwise occur more slowly or not at all. **Radioactive** nuclides pose the greatest threat to human health when they are inhaled or ingested. However, radiation-emitting fragments can be so small that they can enter the skin via the many sweat pores and follicles all over the body. There are three main types of ionising radiation: **alpha particles**, **beta particles**, and **gamma rays**. The biological damage of alpha radiation is considered to be about 20 times that of the same absorbed dose of gamma or beta radiation. Gamma radiation can be as harmful as beta radiation but can travel greater distances. Minute dust fragments which contain alpha-, beta-, and gamma-emitters can be transported great distances by wind and water.

radiation, natural background – the total radiation from cosmic radiation and terrestrial

radioactive materials that is not human-made. These materials include radon, uranium, potassium, and other trace elements. Human activites which disturb the ground – such as uranium mining and milling – make these substances more hazardous since they become more easily available to life forms via uptake in air, water and food. Not all "background" radiation is natural; much is generated by previous human activities such as the routine operation of nuclear reactors, and fallout from nuclear weapons.

radioactive – radiation-emitting.

radioactivity – the emission of **radiation** by a radioactive substance. It occurs when the nuclei of atoms undergo a transformation, such as **fission** (splitting into two or more parts).

radium – a substance which has an especially dangerous radioactive **isotope**, which is present in uranium and thorium mill **tailings**. Radium is easily dissolved and thus spread by water flow; it is harmful to life forms even at low concentrations. (One of the products of its **radioactive** decay is the even more dangerous **radon** gas.) Radium-226 (radioactive, with a **half-life** of 1600 years) is the most plentiful isotope of radium; it is the fifth decay product in the uranium-238 decay series. Other isotopes of radium include radium-223 (half-life 11.1 days), the fifth decay product in the uranium-235 decay series; and radium-224 (half-life 3.64 days), the fifth decay product in the thorium-232 decay series.

radon – a gaseous element which has an extremely dangerous radioactive **isotope**. The radioactive radon-222 (half-life 3.82 days) is the most plentiful isotope of radon; it is a derivative of uranium-238 and the immediate **daughter product** of radium-226. Other isotopes are radon-223 in the uranium-235 decay series, and radon-220 in the thorium-232 decay series. There are three main reasons why radon is so dangerous. *Firstly* because it is a gas: it can thus be breathed into the body, and it can be carried hundreds of miles by the wind, so affecting large numbers of people. Radon is the only gas which occurs in the uranium decay series. Unnaturally large amounts of radon are continually coming out of the ground at uranium mine and mill waste areas. *Secondly* because it emits the most harmful type or radiation – alpha radiation. *Thirdly* because it has a short half-life, rapidly producing extremely hazardous daughter products. The first four decay products of radon have a total half-life of less than one hour, so producing a lot of radiation while in a position to cause damage – hence the prevalance of lung cancer among uranium miners.

recovery – the rate at which a required mineral is extracted at the **milling** or concentration stage of mining, or the amount so recovered. The former can refer to the ratio of the mineral extracted (or concentrated) to the total amount of the mineral present; or it can refer to the ratio of the mineral extracted or concentrated to the amount of **ore** processed.

red cake – the vanadium **concentrate** in a **milling** operation.

reprocessing – the chemical separation of "spent" or used nuclear fuel into its constituent parts, enabling some reusable material (including **isotopes** of **uranium** and **plutonium**) to be recovered for possible future use after further processing. Reprocessing used fuel increases the amount of radioactive waste that has to be disposed of.

reserves and **resources** – terms sometimes used interchangably in mining circles. By and large, reserves are fairly well explored and defined amounts of ore, from which a given amount of mineral extraction can be expected. "Proven ore reserves" are established after three-dimensional drilling or excavation, and are much more certain than "probable reserves". Resources are, customarily, implied reserves, which further exploration and drilling should either increase or diminish. "Probable resources" are a little more specific than "resources" but still require a great deal of investigation before they can be classed as reserves.

rights issue – an invitation to existing company shareholders to extend their holding of shares. A common way of increasing a company's capital.

sedimentary – (of rocks and minerals) formed by the accumulation of sediment from water or air, characterised by strata (or stratification). For example coal (formed from the sedimentation of animals and plants), oil and gas.

share – one of a number of equal portions of the nominal (or "issued") capital of a company, entitling the holder to a proportion of the distributed profits (usually as dividends). Shares may or may not carry voting rights at meetings to control the management of the company. (See *Rössing*.)

share capital – (or **capital**) – the value of all shares issued, or authorised to be issued, by a **company**. The amount cannot be extended without a resolution passed by shareholders at a general meeting authorising the board to do so.

Sociedad (Sdad) – in Spanish-speaking countries, designation for a **company**.

Sociedad Anónima (SA) – in Spanish-speaking countries, designation for a **company** similar to a British **public limited company**.

Sociedad Limitada (SL) – in Spanish-speaking countries, designation for a limited liability **company**.

Sociedade (Sdad) – in Portuguese-speaking countries, designation for a **company**.

Société (Sté) – in French-speaking countries, designation for a **company**.

Société Anonyme (SA) – in Belgium, France, Greece, Luxembourg and Switzerland, designation for a **company** similar to a British **public limited company**.

solution mining – (or *in situ* leaching) – a process whereby an acid or alkaline solution (often highly toxic, such as cyanide) is pumped through boreholes at high pressure, and often at high temperature, into a mineral deposit, resulting in a slurry or solution from which the mineral can then be extracted. Solution mining has been quite widely used in the uranium industry; several projects (*e.g. Beverley*) have rejected use of the process because of its potentially damaging consequences.

spot market – See spot sales.

spot sales – transactions in cash, for transfer of the commodity immediately or in the near future (*i.e.* on the spot market). (This is in contrast to dealing in "futures", which is a commitment to supply or receive a commodity months or years in the future for a price paid now.) The spot market price is often lower than that secured for long-term contracts; it is comparable to the "exchange value" for uranium set by *Nuexco.*

strip mining – similar to **open-pit** mining, but usually used to describe surface coal extraction, and often covering a considerably greater area than open-pit mining for other products. (See *Peabody.*)

strip ratio – (or **stripping ratio**) – the number of tons (or other unit) of **overburden** ("waste") which have to be removed, at an **open-pit** mine, to permit the mining of one unit of **ore**.

subsidiary (company) – a **company** controlled, in its overall business, by another company, the latter often called a "parent" company. Control is usually exercised through the acquisition of more than 50% of the **shares**, though this is not necessarily the case (see *AAC, RTZ, CRA, Mitsubishi).*

tailings – or **tails** – the solid material left over from an ore **milling** process, after the required mineral(s) have been extracted, pertaining especially to **uranium** milling where the tailings are a fine sand. (Large volumes of tailings are produced in the uranium milling process: hundreds of tons of waste for every ton of **yellowcake** produced. These tailings contain about 85% of the total radioactivity of the ore, including about 99% of the radium; the tailings also contain almost 100% of the heavy metals in the ore.) Tailings may be stored or disposed of – chemically treated or untreated – in a variety of ways: from simple dumps at the mine/mill site; to specially-constructed tailings "ponds"; to being thrown unceremoniously and untreated into the nearest sea or river (see *BHP).* Tailings, including uranium tailings, have been used in house construction in Canada and the USA, as well as for road in-fill and for other purposes. In the USA, the Mill Tailings Act of 1978 urged that all uranium tailings should be put back in the mine and sealed; this has not happened. Additionally, the milling process often produces twice as much liquid waste as tailings: liquid wastes have a great impact on the surrounding environment as they can carry contamination long distances via streams, rivers and lakes. The tailings themselves can be carried along by the liquid waste.

takeover – a forced **merger**, when one **company** buys up the shares of another.

test-drilling – sinking holes into an **orebody**, to recover samples which are then subject to physical or chemical tests (**assays**).

thorium – an element of which a naturally-occurring **isotope** (thorium-232) is a **fertile material**, producing **uranium-233**. Thorium can be extracted from monazite, which is found in beach sands in a number of countries – particularly Australia, Brazil, India, Malaysia, South Africa; but also in potentially economic quantities in Madagascar, Mauretania, Sri Lanka, the USA, Liberia, Thailand, Zaïre and elsewhere. There is a thorium-fuelled reactor in Colorado,

USA, and several companies in the USA have (or had) thorium-producing plants, used for nuclear fuel production – including *General Atomic* and *Numec*. The *RTZ* company Thorium Ltd, which stopped processing monazite in 1975, was originally founded in 1914 to supply thorium nitrate for use in gas mantles. Thorium is also to be found in uranium mill **tailings** in Canada. *Rhône-Poulenc* is the world's single biggest supplier of monazite.

transfer pricing – a process of artificially redistributing a company's costs between different stages of processing its products, and/or between the company's operations in different countries, for financial advantage. Although the process can be quite complex, and of its nature it is a highly secretive procedure, in essence it is quite simple. Materials, goods, or services, are priced by a company at an untrue, low rate in a high-tax zone, in order to escape heavy duties, and the difference in **value added** is made up by inflating production charges at another stage in a low-tax or tax-free zone. Thus, true production costs and product value are disguised in a way that suits the company best. The larger the corporation, the more massive the tax evasion can become. (See *Alusuisse, AAC,* and *CRA.*)

transnational company/corporation – See multinational company/corporation.

tribute – a system, historically, whereby a syndicate of mines delivered ore to an owner at an agreed price. In the South African context, it is a system whereby one mine allows its ore to be used by another on a contractual basis.

trigger material – material of a form suitable for nuclear **fission,** or for one-step conversion to such a form, the presence of which should prompt ("trigger") the implementation of safeguards under the **Nuclear Non-Proliferation Treaty.** In practice this has usually not happened (see *Nukem).* Uranium concentrate (yellowcake) is not a trigger material; **uranium hexafluoride** is. (See also **material unaccounted for.**)

underground mine – a mine with the extraction area not visible from the earth's surface, usually accessible only by way of a vertical or diagonal shaft (as opposed to an **open-cast mine**). The workings can be very extensive and deep (in the case of South African mines for example) and – as in Canada – can stretch under rivers and lakes.

upstream processing – stages in the procedure from obtaining the raw material, to producing the final product, which have taken place prior to the current stage.

uranium – a dark grey, metallic element, discovered in 1789. Naturally occuring uranium is a mixture of three **isotopes:** uranium-234 (0.01%), uranium-235 (0.71%), and uranium-238 (99.28%). Uranium is the densest non-human-made element. Uranium ore normally contains only a few hundredths of a percent of uranium, though some extremely high-grade ore in Saskatchewan, Canada, contains up to 60% uranium. Uranium is both chemically and radiologically toxic: it poses a health hazard as a heavy metal as well as because of its **radioactivity.** Uranium-235 is an essential isotope for fission-based nuclear energy and weapons, since it is one of three primary fissionable materials, and the only one which occurs naturally; it has a **half-life** of 703.7 million years. (The proportion of uranium-235 in natural uranium is too low for most uses for which uranium-235 is wanted: see **enrichment.**) Uranium-238 is also indispensable to the nuclear industry, since it is one of the two **fertile materials** that can be used for the production of fissionable material; it has a half-life of 4468 million years. Uranium-238 yields the primary fissionable material plutonium-239; another important characteristic of uranium-238 is that it will itself fission under fast-neutron bombardment.

uranium hexafluoride (UF$_6$) – (or **hex**) – a uranium compound which is a white solid at room temperature, vaporising at 56.5 degrees C. It is used in the nuclear fuel and nuclear weapons manufacturing processes because it is the simplest compound of uranium which can be easily vaporised: the main methods of **enrichment** of uranium require a uranium compound in gaseous form, in order that the differences in atomic weights of the different uranium **isotopes** can most easily be used to separate them out. Hex is the final product of the refining process before enrichment, and (enriched uranium) hex is the feed material for fuel fabrication – it is therefore one of the more common uranium compounds to be transported from plant to plant. But the physical and chemical properties of hex set it apart from the vast majority of radioactive materials: not only is it radiologically and chemically toxic – like other compounds of uranuim – but the chemical hazard of hex is especially great. It is corrosive, and it forms a highly toxic cloud on contact with air or water.

uranium oxide – general name for a compound of **uranium** and oxygen. The important ones are: UO_2, U_3O_8, and UO_3. The dioxide is used in nuclear fuel elements in certain types of reactor. U_3O_8 is the major constituent of uranium **concentrate**, or **yellowcake**.

uranium tetrafluoride (UF_4) – (or **green salt**) – an intermediate product in the production of **uranium hexafluoride** from uranium ore, and of uranium metal from hex after enrichment.

utility – in the context of this book (and particularly in North America), a private or publicly-owned corporation providing electricity, which owns its own generating plant.

value added – the increase in price, if not value, of a resource or raw material as it passes through various hands, or stages of production or manufacture. Each "middle person" makes their own profit: hence the finished or delivered item may be much more costly than originally conceived. Companies which are highly integrated in their **downstream processing** may, effectively, control the market in certain goods and set whatever price they judge the consumer will stand: for example, De Beers's control of diamond production from the mine to the cutting/polishing shop (see *AAC*).

vein – a zone or belt of mineralised rock, clearly defined from other rock. (See also **lode**.)

yellowcake – (or **uranium concentrate**) – the marketable product from a uranium mill. It is a fine, sand-like material which is insoluble in water. It consists of between 70% and 90% of one of uranium's oxides (U_3O_8), the rest being mostly uranium decay products and heavy metals. It is usually produced from uranium **ore** by putting the ore though a mill using the following process: crushing; grinding; leaching with sulphuric acid or sodium carbonate-bicarbonate; separation by filtration, decantation, or centrifugation; further separation by a solvent extration or ion exchange process; and finally precipitation by neutralisation with ammonia, magnesia or caustic soda. After this processing, the resulting product is a solid – usually canary yellow in colour (though it may be dark brown or black) – and ranging in consistency from granular to powder.

THE GULLIVER FILE

1 AAEC/Australian Atomic Energy Commission

It operates the Federal Australian Government's uranium stockpile at Lucas Heights, New South Wales.

Until ERA was formed in September 1980, AAEC was one of the owners of the Ranger uranium mine. At that point, the AAEC's interest – along with those of the Australian government, EZ Industries (see EZ Electrolytic Zinc), and Peko Wallsend was transferred to ERA.

In the mid 1970s, AAEC bailed out the Mary Kathleen mine, and was granted a minority ownership; Mary Kathleen was closed in late 1982.

AAEC also helped out Queensland Mines, when delay in opening the Nabarlek mine threatened to jeopardise Japanese contracts for nearly 200 tons. Queensland Mines paid the AAEC for the uranium loan – and the AAEC repaid it when Nabarlek uranium began to be shipped in 1981 (1).

The AAEC in 1976 announced uranium occurrences at its Austatom uranium prospect, 28km west of the Jabiluka and Ranger mines in the Northern Territory. Although further drilling occurred in 1977, no results were announced (2).

The AAEC also held 10.7% of the Ngalia Basin Exploration JV along with CPM, Urangesellschaft, and Agip Nucleare, which in 1981 was sold to Cocks Eldorado, SC Ex, and Offshore Oil (2).

In 1986, the AAEC was formally replaced by the Australian Nuclear Science and Technology Organisation (ANSTO), whose brief is not only to build on the AAEC's commitment to nuclear power but to expand the organisation into non-nuclear fields as well (3).

A year later, after a fire broke out at Lucas Heights destroying a laboratory cell used for processing isotopes (4), it looked as if ANSTO would be co-operating with the Chinese régime on a project by which the Asian country would take West German nuclear waste, to be buried in Synroc (synthetic rock) developed in Australia (5).

References: (1) *MIY* 1982. (2) *Reg Aus Min* 1981. (3) *Nuclear Spectrum*, Vol. 3 No. 2, Australia, 1987. (4) WISE NC 272, 10/4/87. (5) FoE Victoria, 11/88.

```
AAEC / AUSTRALIAN ATOMIC ENERGY COMMISSION (AUS)
                                    \
                                 41.64%
                                 MKU (AUS)
```

---

2 AAR Ltd

```
          CSR (AUS)
              \
           AAR LTD (AUS)
              /
         Minad (AUS)
```

This is a natural gas and oil production company with interests in Australia, Indonesia, and elsewhere. It holds 44% of the major Hail Creek coal deposit, Queensland.

It has a JV with CRA, which holds 25% of Hail Creek coal (1) through CRA's IOL subsidiary; and with MIM and Teton at Honeymoon.

As of 1977 (2) it was uranium prospecting in: Billeroo, South Australia (SA), Lake Frome, Balcoona Creek, Wertaloona, Lake Namba, and Curnamona (SA); Westmoreland, Ironhurst, Mt Talbot, Fraser Peak, Dagworth, and Newcastle Range (all in Queensland).

In 1980 AAR acquired IOL's 3.8% interest in Honeymoon (SA) (1, 3) which was planned to be South Australia's first operating mine (4). See MIM for details.

AAR is also part (12.75%) of a consortium exploring for uranium in northern Queensland at Westmoreland. Other partners in the JV are Urangesellschaft 37.5%, Queensland Mines 40%, and IOL 9.75%. The venture is being vigorously opposed by north Queensland Aborigines: see Queensland Mines (5). There has been emphatic resistance to Honeymoon: see MIM for details.

AAR held 75.5% together with Teton Australia in the Gould's Dam or Billeroo uranium prospect, 150km west-northwest of Broken Hill in South Australia. As of March 1981 reserves were 1260 tonnes of contained uranium oxide (6).

References: (1) *MIY* 1981. (2) *Reg Aus Min* 1976/77. (3) *FT* ?/9/79. (4) *FT* 28/10/81. (5) Roberts. (6) *AMIC* 1982.

3 Aberfoyle Ltd

This is a holding company, exploring throughout Australia and operating tin mines in Tasmania and New South Wales.

It acquired a uranium prospect in Tarcoola, South Australia, in 1979 (1); this was in conjunction with the French government agency Afmeco (2).

Of late Aberfoyle has suffered from strikes and the imposition of tin quotas (it is Australia's second biggest tin producer) with production at its major Que River prospect in Tasmania (90% owned) nearly half down in the first half of 1984, compared with the same period of 1983. Aberfoyle's explorations – mostly in JVs – continue "to be encouraging" though nothing has been heard recently of its uranium dabblings.

A newly incorporated subsidiary company, Australian Diamond Exploration, is now participating (25%) with Ashton Mining (who are partnered with RTZ/CRA in the infamous Lake Argyle diamond prospect) and AOG Minerals in diamond exploration throughout Australia (3).

In 1987 Aberfoyle commenced a US$112 million investment programme in a zinc/lead/silver mine in north-west Tasmania, where a "massive" deposit had been located (4).

References: (1) *AFR* quoted in *MJ* 21/12/79 p. 526. (2) *Reg Aus Min* 1981. (3) *MJ* 29/6/84. (4) *MJ* 23/10/87.

4 A.C.A. Howe Exploration Inc

```
            Pancon (AUS)
                  /
A.C.A. HOWE EXPLORATION INC. (CDN)
```

In 1984, A.C.A. Howe's head Peter Howe told a gathering in London that "exploration [was] awakening from a long dull slumber" as he opened a London office to be headed by none other than Dr Calvert Armstrong who discovered the huge Jabiluka deposit (see Pancontinental) when he was Howe's chief geologist (2).

A.C.A. Howe is situated in Canada. According to its own blurb, it "specialises in all forms of mineral exploration and mine evaluation".

Apart from being involved in discovering the Jabiluka deposit, the company evaluated the Wheal Jane mine in Cornwall (see RTZ) which led to its successful reopening by RTZ. It has carried out numerous other evaluations in the UK, France, Sweden, Spain, Portugal and West Africa (3).

In the late 1970s the company gained an exploration licence for uranium in the Central African Republic, to cover several regions (1).

In 1988 Claude Resources engaged the company to do a pre-feasibility study on its JV with Corona Corp to exploit the large Seabee gold deposit in the La Ronge area of Saskatchewan (4).

References: (1) *Uranium Redbook* 1979. (2) *FT* 1984 [date mislaid]. (3) *MM* 12/82. (4) *E&MJ* 12/88.

```
            Cominco Ltd (CDN)
                  /
Cominco Australian Pty (AUS)
                  /
                 47%
               ABERFOYLE LTD (AUS)
                  /
      Cominco Exploration (AUS)
```

5 ACM/Australian Consolidated Minerals

This works through joint ventures with other companies in nickel, copper, gold, diamond, and uranium searches (1); HQ is Perth, Western Australia (WA).

It joined in 1978 with Japan's PRNFC and Australia's Command Minerals in a uranium search at Walling Rock, Murchison, WA, identifying several anomalies with the use of ACM's "satellite identification" techniques (2); it holds there 26% of the JV. Recent exploration, however, has proved "disappointing" and is to be restricted to the Ida Valley area (3).

With 13%, one of the partners of the Ashburton JV together with Command Minerals, Nickelore, and West Coast Holdings, it also partners Getty Oil (60%) in a uranium search of the Pilbara, WA, region since 1979 (4), an area of strong Aboriginal land claims and cultural importance (see CSR). Exploration was hampered there until 1980 by wet conditions (3); results for 1981 are not known.

By March 1983, ACM had completed acquisition of the large Big Bell gold property formerly owned by Nickelore (5).

In 1985, due to the huge costs of developing Big Bell, ACM sold 50% in the project to Placer Development (7). By then its interests were very much centred on gold, especially on the Golden Crown and Westonia prospects in Western Australia which were opened officially in just one week in 1986 – a "first" in Australian mining history (8). By the middle of the following year, ACM had cleared its development debt on these projects, and had also ventured into gold exploration elsewhere (9).

Over the next three years, ACM was developing a gold project in Tasmania (10), had purchased a majority stake in the UK-based Paringa Mining and Exploration Co plc, giving it access to a minority stake in North Flinders Mines, which operates the Granites gold mine in the Tanami desert district of the Northern Territory, and also has interests in the US, southern Africa and Indonesia (11).

It has an important JV with Outokumpu to de-

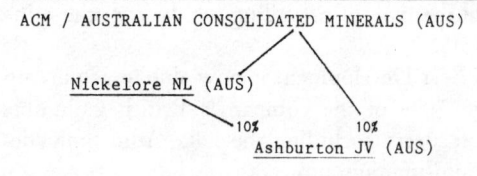

velop the Mount Keith nickel deposit in Western Australia, to provide the Finnish company with ore for its Kokkola roasting plant (12).

References: (1) *MIY* 1981; *Reg Aus Min* 1976/77. (2) *FT* 20/11/78. (3) *Reg Aus Min* 1981. (4) *FT* 12/7/79. (5) *MJ* 20/5/83. (6) *MJ* 20/12/85. (7) *MM* 7/86; *MJ* 30/5/86; see also *MJ* 18/10/85. (8) *MJ* 7/8/87; see also *MJ* 25/9/87. (9) *E&MJ* 3/88. (10) *MJ* 10/11/89. (11) *MJ* 17/11/89; *MJ* 3/11/89; *MJ* 4/5/90.

6 Adanac Mining and Exploration Ltd

It owns 87 mineral claims in the Atlin district of British Columbia, Canada, – although Placer Development has an option to acquire 70% of these (1) – mostly molybdenum leases.

In 1981 Adanac proposed opening a molybdenum mine at Ruby Creek, British Colombia, which would involve the extraction and stockpiling of uranium (as a byproduct) – thus violating the 1979 provincial government moratorium on uranium mining (see Canada) (2).

In late 1981 Adanac were pushing ahead with their molybdenum mine in British Columbia, despite the ban on uranium mining and despite the deposit containing important proportions of uranium.

The proposed mine would dump more than 500lbs of uranium and 1000lbs of thorium tailings daily into the Yukon River – the source of virtually the whole of the drinking water for the Yukon. Teslin, a mainly native town, is directly

Conwest Exploration Co. Ltd (CDN)

26%

ADANAC MINING AND EXPLORATION LTD (CDN)

in the path of prevailing winds from the mine (3).

Placer Development is now able to acquire up to 70% of the company's British Columbia prospects, including the Lake Atlin molybdenum/uranium project (4).

References: (1) *MIY* 1982. (2) *Kiitg* 1-2/1981; *MJ* 12/2/82. (3) *Transitions* Winter/1981 (quoted in *BBA* Winter 1981). (4) Annual Report 1982 quoted in *MJ* 12/2/82.

7 Afmeco [Australian-French Metals Corp] Pty Ltd NL

It is associated with CEA (France); uranium exploring in late 1970s on Aboriginal areas near Alice Springs, central Australia (1).

In 1980, Afmeco was among several foreign groups wanting to explore for uranium and presumably other minerals in the Alligator River region, the uranium province of Australia's Northern Territory (2).

Although the exploration carried out by Afmeco in the Canning Basin (Western Australia) between 1978 and 1983 only revealed one area of uranium mineralisation which might be potentially exploitable, the company's operations were quite extensive, leading, as one geologist put it, to "...significant advances in the understanding of the geology of the Devonian and Carboniferous classic sedimentation in the area..." – which also happens to be of major importance to Aboriginal people (3).

References: (1) Roberts. (2) *FT* 23/9/80. (3) Peter Botton, Uranium Exploration in the Canning Basin: A case study from "The Canning Basin", Proceedings of the Canning Basin symposium, Perth, 6/84.

```
                            CEA (F)
                          /
`·°MECO [AUSTRALIAN-FRENCH METALS CORP.]
                          PTY LTD NL (AUS)
```

8 Afrikander Lease Ltd

```
The Anglo-American Corp. of South Africa Ltd (ZA)
                              44.56%
                   AFRIKANDER LEASE LTD (ZA)
```

Around half of the company's total land holdings are leased to Vaal Reefs which has financed the entire capital requirements of the company's mine in Western Klerksdorp since April 1979 (1). A decision to exploit the lease for uranium proved unwise by 1982, when Vaal Reefs and Afrikander suspended uranium treatment operations and turned to gold instead. The uranium section of the plant was put on a care and maintenance basis in 1983 (1) and has not been worked since (2).

Under a royalty agreement between Vaal Reefs and Afrikander, concluded in 1979, Vaal Reefs would process around 15,000 tons of ore per month from the lease, for which it agreed to pay a 5% royalty on revenue from sales to Afrikander plus further royalties if profits exceeded 30% of revenue; financing was also provided for the new mine. This had substantial tax advantages for both companies as Afrikander did not pay tax rates for gold mines and Vaal Reefs could deduct capital expenditures from its taxes (3). In 1983, Afrikander got R50,000 minimum as a uranium royalty (1).

Afrikander has inevitably been a victim of the falling uranium market, and development of the uranium lode on its land was postponed after several studies and discussions for more than three years, from 1975 (4). Even when the decision was finally taken, an element of confusion remained (5).

There has also been speculation as to whether Vaal Reefs has used its own lower cost uranium production to contribute to Afrikander's sales, including one contract extending from the end of 1984 to 1990 (6).

References: (1) *MIY* 1985. (2) *MAR* 1984. (3) *MJ* 23/3/79. (4) *MM* 5/79; 2/78. (5) *Nuclear Fuel* 1/2/82. (6) *MJ* 24/7/81.

9 Aggressive Mining Ltd

Together with Esso Minerals Canada (60%). Aggressive owns a gold/silver/base metals claim in the Gaspe area of Quebec; Curtin has various prospects in Sudbury, Ontario (seat of the notorious Inco smelter).

Aggressive itself also holds uranium prospects in the Sudbury area, none of which appears to have gone beyond an exploration stage (1).

References: (1) *MIY* 1985.

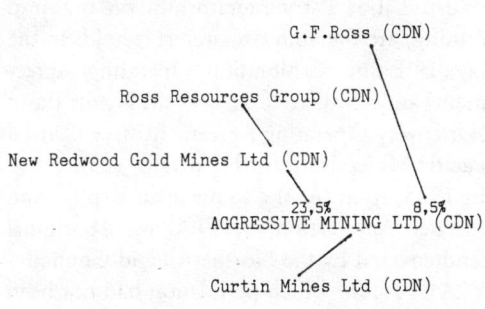

10 Agip Nucleare SpA

Agip Nucleare is responsible for nuclear fuel "cycle" activities on behalf of ENI, Italy's huge state-controlled energy/mining/manufacturing enterprise. These activities include "exploration for, mining and supply of, uranium ores. Operations downstream of the reactor. Research and development of renewable energy sources. Conservation and rational use of energy *(sic)*" (1). Italy's uranium exploration was originally in the hands of a company called Somirem, which was absorbed into Agip Nucleare after ENI's foundation. At that time (it appears) Agip SpA was responsible for uranium mining and exploration (2); later, the nuclear operations of ENI were consolidated in Agip Nucleare.

There seems to be some confusion, however, in mining circles as to the exact delineation of responsibility between Agip SpA and Agip Nucleare: for example, in 1979 *Mining Journal*

called Agip SpA "the uranium mining arm of ENI" (3). Agip has an Australian subsidiary which seems variously to be known as Agip Australia Pty Ltd and Agip Nucleare Australia Pty Ltd (4). It is also not clear whether Agip Nucleare International is technically a subsidiary of Agip SpA or Agip Nucleare similarly with Agip Uranio. In practice, however, the distinctions do not matter: everything nuclear connected with Agip is effectively done under the auspices of ENI.

In 1982, ENI contracted huge losses, the responsibility for which fell squarely on its mining subsidiaries Samim, Anic, and Agip Nucleare. The change was primarily due to "the nuclear sector that [is] suffering from the failure of the country's nuclear construction programme to go ahead as expected. Furthermore the group has made a substantial investment in stocks of uranium which had to be devalued at year's end to prevailing market prices" (5).

That year, Agip cancelled an important contract with Madawaska Mines (6) on which basis the Madawaska uranium mine near Bancroft, Ontario, had been resuscitated in 1976. The mine was originally taken out of production in 1964, and rehabilitated nine years later (7). Despite a design capacity of 320 tonnes/year U_3O_8, there were problems with supplying ore of sufficient grade to the mill, and production was still only 230 tonnes/year by 1979. Additions to the workforce led to higher production, but all to little avail: as Agip retrenched, so Madawaska went under.

Nor was this without a lot of recrimination between Agip and the Canadian industry over setting of the price. Originally Agip's contract was for 26 million lbs U_3O_8, at 5% less than the world market price (8), US$32/lb (7) (though this appears to have initially covered about 1.5 million lbs) (9).

The Atomic Energy Board of Canada (AECB) insisted that the price should be fixed at prevailing market levels. Although Agip contested the AECB's ruling and took the matter to the Federal Court of Appeal, it lost: the renegotiated price was US$42/lb (9).

Agip's other uranium ventures in the 1970s and early 1980s took it across the world:

- It joined Noranda in a "vigorous campaign" for uranium in the Great Bear Lake area of the North West Territories (NWT) of Canada (10) and also in Saskatchewan (11).
- In 1974 it signed a contract with the Bolivian nuclear energy agency, COBOEN, to explore over a 50,000 sq km area (12).
- At the same time, it held two exclusive exploration licences over Sokoto state and parts of Bauchi, Borno and Gongola states in Nigeria (12).
- It has explored for uranium in Guinea (13).
- In 1979 it concluded an agreement with the Zambian government for a uranium search, under which Agip would bear the costs and the government receive at least one tenth of the production (14). But although drilling started in the Gwembe valley later that year (15), the company pulled out at least of the north-west of the country after attacks by "terrorist gangs" in late 1984: its 200 employees were withdrawn (16).
- Another uranium exploration programme in the Congo Republic was declared "unsuccessful" in early 1983 (17).
- Agip Uranio holds 6.53% of Somair in Niger a share reduced from 8% when the country's president, Hammani Diori, enlarged the national participation in Somair (7).
- Also in Niger, Agip concluded a contract for 275 tonnes/year U_3O_8 with the option for an additional 275 tonnes/year, but the right to reduce deliveries by 50%-75% (18). In 1981 Agip did not take up its share of production and its concession (with Onarem, Cogema and Japan's Atomic Energy Commission) was that year transferred to PNC and Onarem (19).
- Similarly, in Gabon, Agip has consistently taken up amounts of uranium less than the contracted maximum thus "aggravating" the problem of Comuf's viability (7). Agip is not, however, directly involved in uranium production in this west African state.
- Agip Nucleare (Australia) attempted during the 1970s to promote several major Aus-

tralian uranium prospects. In the Alligator River region it was part of AUREC (20). At the Mount Hogan uranium prospect 322km south-west of Cairns, in 1981, Agip was the operator and driller (Central Coast Exploration the owner). Agip was also able to earn up to 50% interest (21) in the Spear Creek area and the Carlton Hills district of Queensland where, in 1981, Agip Australia had a JV agreement with Mary Kathleen Uranium on a total of eleven leases (24).

Currently its major Australian interests (though dormant, thanks to the discredited, partial Australian Labor Party moratorium on uranium mining) are the 46.84% interest it holds in the Ngalia Basin (Exploration Operating Agreement) and a similar 38.73% in the Ngalia Basin (Discovery Operating Agreement) (see Central Pacific Minerals for further details) (23).

In 1985, Agip (Aus) also made an Exploration Licence Application, or ELA, on Aboriginal land covered by the Northern Land Council – ELA 3114; for which permission had not been obtained from the Aboriginal land owners (24). Agip's search for uranium on its home ground was concentrated on the Alps and Sardinia (25). At Val Seriana, a small mine was being proposed around the Novazza deposit in the late 1970s (26), though the reserves had still only been "staked" several years later (5). Airborne spectrometer surveys have been conducted over Sardinia notably with the assistance of Hunting Geology and Geophysics Ltd (27), but no economically viable resources had been reported there by 1983 (5).

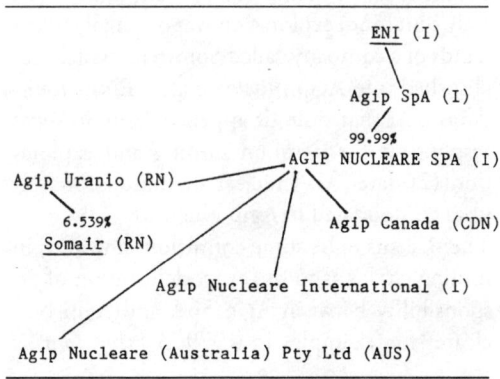

In 1978 Agip signed a contract with the USSR agency Technabsexport to cover the enrichment of uranium for the period 1980-83 (3).

In the last few years Agip's commitment to uranium seems to have tailed off considerably. However, other Agip companies have taken major stakes in a copper-nickel deposit at Radio Hill in the Australian Pilbara (28), and the very large steam coal deposits to be found in Queensland's Bowen Basin (29).

References: (1) *WDMNE*. (2) *Big Oil*. (3) *MJ* 16/2/79. (4) *Reg Aus Min* 1981. (5) *MJ* 30/12/83. (6) *FT* 24/3/82. (7) Neff. (8) *Foreign Uranium Supply*, NUS Corporation, Rockville 1978. (9) *FT* 13/9/78. (10) *News of the North* Save the North Program, 20/6/80, Regina. (11) *Yellowcake Road*. (12) *Uranium Red Book* 1979 OECD/IAEA. (13) *FT Energy Report* London 1978; and *FT* 13/3/79. (14) *FT* 5/4/79. (15) *MJ* 5/10/79. (16) *MJ* 16/11/84. (17) *MAR* 1984. (18) *Nuclear Fuel*, Washington, 3/1/83. (19) *MAR* 1982. (20) *FT* 23/9/80; and *A Slow Burn*, FOE Victoria, No. 1, Sept 1980. (21) *Reg Aus Min* 1982. (22) *MJ* 17/4/81. (23) *Reg Aus Min* 1983-84. (24) Northern Land Council documentation 1985. (25) *MJ* 20/6/81. (26) US Survey of Mines, *E&MJ* 11/79. (27) *FT* 15/5/78. (28) *MJ* 11/8/89. (29) *MJ* 6/10/89; see also *MJ* 1/7/88 and *E&MJ* 6/88.

11 AG McKee (USA)

This is an engineering company which, in 1979, won a contract (with Union Minière and another Belgian company, Traction et Electricité) to develop Algeria's Hoggar mountain uranium (1); McKee was to provide the ore processing plant (2).

However, nothing further has been heard about these companies with regard to the Hoggar mountain; instead, Belgian Mining Engineering was reported to be consulting with the Algerian government in 1981 (3).

The company was also responsible for construction and development of the Rössing uranium mine in Namibia (4).

References: (1) *MJ* 6/7/79. (2) *MIY* 1980. (3) *Le*

Monde 12/6/81. (4) *Sharing* (World Council of Churches bulletin on transnational corporations) 12/81.

12 Agnew Lake Mines Ltd

Agnew Lake Mines Ltd holds a 90% interest in the Agnew Lake uranium mine near Espanola in the Elliott Lake region west of Sudbury, Ontario; a rehabilitated property which came into production in mid-1977, using bacterial leaching (1). The mine soon encountered problems, however, due to coarseness of ore and a changeover to flood leaching in 1978, which delayed underground production (2). Recovery rates sunk to less than 40% capacity (3).

In 1979 the company decided to close the mine, maintaining "salvage" leach operations from then on, and as long as they were economic (3). During 1980, however, production appears to have improved, with the closure provision resulting in a net credit of C$10.9 million (4).

At 31 December 1979 more than 700,000lbs of U_3O_8 remained to be paid to Eldorado Nuclear under a 1976 loan agreement with Uranerz Exploration and Mining (Canadian subsidiary of Uranerzbergbau GmbH of Bonn) which took a 10% undivided interest in Agnew Lake in 1974 and was thereby entitled to 10% of the production (1).

In late 1982, Kerr Addison announced that it would finally phase out Agnew Lake because of "declining solution grades and weak uranium prices". Production would cease in early 1983, with 54 workers laid off. The major part of the "environmental" programme was expected to be completed within the following year (5).

References: (1) *MIY* 1981. (2) *MJ* 18/5/79. (3) *FT* 28/9/79. (4) *FT* 20/3/81. (5) *MJ* 19/11/83.

```
Kerr Addison Mines Ltd (CDN)
          \
     AGNEW LAKE MINES LTD (CDN)
```

13 Agrico Chemical Co (USA)

This phosphates producer was to co-operate with Freeport-McMoran on recovery of uranium from its phosphoric acid facilities at Donaldsville, Louisiana; start-up was expected in mid-1981 and presumably deferred (1).
In 1983 Agrico entered a JV (49%) with the Sri Lankan government to develop phosphates on the island. Although small-scale mining has already started, Agrico said it won't put in additional money until the third world government has assured it of "certain tax holidays and relief from import duties on equipment and export tariffs" (2).
Agrico and Eppawala Phosphate Pvt. Ltd (state owned) later formed a US$40 million JV to exploit these deposits, but by 1990 the project had foundered on high processing costs (acid in the phosphates) and low market prices (3).

References: (1) *FT* 10/10/79; *MAR* 1981 p. 358. (2) *E&MJ* 3/83. (3) *MAR* 1989.

14 Alianza Petrolera Argentina SA (RA)

Along with Sasetru SA, Minera Sierra Pintada, Inalruco SA and Pechiney Ugine Kuhlmann, it is developing the Sierra Pintada orebody in the Mendoza province of Argentina (1); see Minera Sierra Pintada.

References: (1) *MM* 1/80; *Nucleonics Week* 15/5/80.

15 Alkane Exploration NL (AUS)

Primarily a gold exploration company, its current projects embrace several in New South Wales, including a JV with Golden Plateau NL (1). Alkane has also held coal, oil and diamond claims, specifically in the Kimberley region of Western Australia (2).
In the late 1970s it was partnered with the CEA subsidiary Afmeco (80%) in a uranium exploration JV at Batchelor in the Rum Jungle region of the Northern Territory (3) about which little more has been heard since.

References: (1) *MAR* 1985. (2) *Reg Aus Min* 1981. (3) *EPRI Report.*

16 Allan Capital Corp (USA)

Its interest in the Bison Basin, Wyoming, uranium prospect was purchased in 1980 by Energy Capital Ltd (GB) (1).

References: (1) *FT* 14/7/80.

17 Allied Corp

Allied Corporation is an important US producer of amorphous metals and alloys, used in conserving heat in transformer coils etc. It was one of the three companies contracted to build the huge Barnwell nuclear fuel reprocessing plant before President Carter suspended work on it in 1977; the other major partners were Gulf Oil and Royal Dutch Shell (1).
In 1982, Allied took over Bendix (2). Following this acquisition – "one of the most complex take-over battles in US corporate history" (3) – Allied (which now ranks among the top 30 of *Fortune's* 500 most powerful companies) "went on the warpath" (3) to expand its European interests. The *Financial Times* speculated that Ed Hennessey, Allied's redoubtable chief executive, was prepared to shell out US$1 billion to acquire a chemicals company (4).

References: (1) *Bulletin of the Atomic Scientists* 3/84. (2) *MAR* 1984. (3) *Gua* 5/5/84. (4) *FT* 5/5/84.

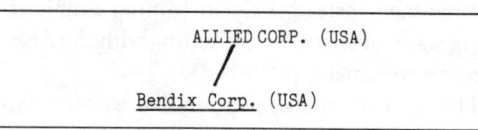

ALLIED CORP. (USA)

Bendix Corp. (USA)

18 Allied Nuclear Corp

Federal American Partners (USA)
86.9% stock
ALLIED NUCLEAR CORP. (USA)

Formerly controlled by Western Nuclear which sold its interest to Federal American Partners in 1977, Allied Nuclear acquired three quarters of a million lbs of uranium oxide, and Federal American was able to expand its uranium interests in the Gas Hills of Wyoming (1).

References: (1) *MJ* 29/7/77.

19 Allstate Explorations NL (AUS)

This is a gold, coal, base metals and diamond exploration and production company (1). It holds 50% interest with Getty Oil in the Daintree, Queensland uranium prospect. In 1980 the companies decided it was no longer feasible for uranium (2).

References: (1) *Reg Aus Min* 1981. (2) *Annual Report* 1980.

20 Alpha Nuclear Co (CDN)

Alpha, in Ontario, has developed a personal dosimeter *(sic)* for uranium miners to gauge radiation emissions from radon and radon ions in uranium mines. Attached directly to the miners' batteries it is plugged into a central computer at the end of a workshift, and computes the radiation dosage received (see also Rio Algom/Denison at Elliot Lake) (1).

References: (1) *MJ* 27/2/81.

21 Alusuisse/Schweizerische Aluminium AG

It is one of the world's largest bauxite and alumina production companies which (like all the major aluminium mining companies) is fully integrated, ie operates smelters and producer's works, to manufacture finished products (1). It has a large number of subsidiary companies throughout Europe and Scandinavia. Alusuisse subsidiaries mine bauxite from southern France to the Mokanji Hills in Sierra Leone.

Alusuisse holds 7.5% indirect participation in Interamericana de Alumina CA which plans to operate an alumina plant on the Orinoco River in Venezuela – an area previously free from industrial development and used by Indians. Bauxite has recently been discovered in the upper reach of the Orinoco.

In 1978/79 it was participating in uranium exploration in the Athabasca uranium basin, Canada (2).

But its most important uranium venture is at Bakouma in the Central African Republic (formerly the Central African Empire until the overthrow of self-styled "Emperor" Bokassa). After a delay in development of the deposits in 1974, Alusuisse came up with a feasibility study which suggested that uranium could be mined at the rate of 800 tons a year in the late 1980s – although US$500 million would be necessary just for access routes and transportation. Reserves at Bakouma have been variously estimated at 16,000 and 18,000 tons of uranium, by open-pit methods.

In Australia its wholly-owned Swiss Aluminium Australia Pty Ltd holds 70% interest in the Gove, Northern Territory, bauxite mine, along with a consortium led by CSR. This is an area of great cultural significance to Aboriginal Australians (fuller account: see CSR).

Opposition to Alusuisse has come from environmentalists in Switzerland who forced the company to commit SFr250,000,000 over 1981-1983 to update and improve its pollutive smelters in Switzerland; and from Aboriginal people at Gove.

More recently Alusuisse has been under investigation by the Icelandic government on suspicion of evading taxes by manipulating the price of imported alumina from Gove. This example of the notorious practice of "transfer pricing" (condemned in 1981 by public figures as disparate as Pope Paul and Robert Mugabe) is worth treating in some detail. The following account comes from *Earth* (Friends of the Earth newspaper):

"Alusuisse's Gove alumina is shipped to Iceland to be smeltered in a smelter owned by another of the company's subsidiaries, Icelandic Aluminium Company. Swiss Aluminium Australia – which exports the alumina from Australia – declares the export price at the time of shipment but it turns out to be much lower than the price registered when the raw material arrives in Iceland. "Markups between January 1974 and May 1980 averaged 54.1%. As a result ISAL can declare annual operating losses, or only marginal profits, thus avoiding paying tax. Total profits before tax for this eight-year period were US$300,000 on net sales of no less than US$510 million" (3).

In spite of this the chairman of Alusuisse could announce in 1977 that losses in Britain had been offset by satisfactory results from Icelandic and Norwegian operations!

Founded in 1888, the company was initially set up – as a source frankly puts it – "to exploit the availability of cheap power in Switzerland". Smelters were first built in neighbouring countries, while large US operations were added after World War 2. Then the company "integrated backward" into bauxite mining and alumina production, mainly in Australia and West Africa (4).

In 1978, it joined Musto Explorations in evaluating uranium on their claims in northern Saskatchewan's Key Lake area in Canada (5).

As the smallest of the six integrated producers which "used to dominate the world aluminium industry" (6), Alusuisse plunged deeply into the red in 1985 and 1986, with losses totalling more than one billion Swiss francs (SFr1.3 billion) (7). In a reorganisation plan which in-

volved sacking the company's chair and chief executive officer (6) and selling off some of its primary aluminium (especially smelting) interests as well as its share in Switzerland's Kaiseraugst nuclear power stations (8), the company moved into profit in 1987 (9).

At the beginning of 1990 the company renamed itself Alusuisse-Lonza, reflecting its newfound emphasis on the chemicals branch of its operations (10) and announced it would be making two new acquisitions in the near future as part of what the *Mining Journal* called its "new, aggressive acquisitions policy" (11).

Despite relinquishing some of its smelting and refining interests three years before, the company announced it would be increasing capacity at the Gove alumina plant (11).

In 1986 the Australian government announced that it would be instituting a special investigation into the pricing and export of bauxite and alumina by Alusuisse's subsidiary operating the 70% share in the Gove plant, which had been supplying bauxite to the parent company without paying Australian taxes since the project commenced (12).

Further reading: Tobias Bauer, Greg Y. Crough, Elias Davidson, Frank Garbely, Peter Indermaur, Lukas Vogel, *Silbersonne am Horizont – Alusuisse – Eine Schweizer Kolonialgeschichte*, Limmat Verlag Genossenschaft, Zurich, 1989. (French version: Alusuisse 1888 - 1988, Editions d'en bas, Lausanne 1989) This book outlines the colonial history of the company, and deals in part with its penetration of Aboriginal land, and transfer-pricing practices in Australia.

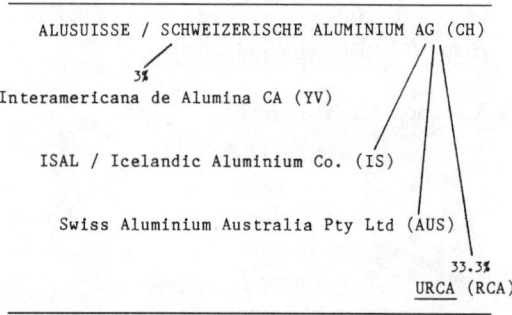

```
ALUSUISSE / SCHWEIZERISCHE ALUMINIUM AG (CH)

              3%
Interamericana de Alumina CA (YV)


    ISAL / Icelandic Aluminium Co. (IS)


   Swiss Aluminium Australia Pty Ltd (AUS)

                                33.3%
                               URCA (RCA)
```

References: (1) *MIY* 1985 (2) *MJ* 18/5/79. (3) *Earth* (Melbourne, Friends of the Earth) 1/81. (4) *WDMNE.* (5) *FT* 12/6/78. (6) *FT* 17/1/86. (7) *FT* 26/9/89. (8) *FT* 21/3/86. (9) *FT* 10/9/87; see also *FT* 7/4/88, *FT* 26/9/89. (10) *FT* 14/3/90. (11) *MJ* 16/3/90. (12) *FT* 10/4/86.

22 Amalgamated Rare Earth (CDN)

It was exploring for uranium in the Cavendish, Cardiff and Mountmout township areas of Ontario in 1979, with 15% of its finances supplied by Imperial Oil. It was expecting to get a licence to mine from the Atomic Energy Board of Canada in October 1979 (1), but no further news.

References: (1) *BBA* No. 4 1979.

23 Amax Inc

"The company [Amax] has offered us a share in the mine ... our share is death", Rod Robinson of the Nishga Tribal Council, British Columbia, 1980.

America's largest mining company, Amax is the world's biggest producer of molybdenum (1) (a crucial ingredient in the hardening of steel especially for weapons systems) and has been involved in virtually every metal on the planet. Its uranium interests were substantial in the 1950s and 1960s (2) but now appear confined to prospecting in Saskatchewan (3) and coping (or not coping) with the legacy of radioactive tailings from its Climax Uranium Co subsidiary. These were offered free to contractors in the Grand Junction, Colorado, area, in the 1960s and were used to build some 5,000 homes, schools and churches (4).

In the decade 1972 to 1982, Amax built up its assets base from under US$2 billion to well over US$5 billion. It diversified dramatically, entering JVs with many other multinational corporations, and spreading its tentacles all over the world. As the *Mining Journal* pithily put it, "This continued philosophy of growth [philos-ophy of continual growth?] was founded on the belief that economic expansion would be maintained throughout the world, and that the company's earnings growth would outstrip the cost of servicing the enormous debt load that it was taking on" (5). By 1984 it was quite clear that Amax would have to make drastic cutbacks in order to maintain its pre-eminence in certain fields. After criticising Pierre Gousseland, the chief executive who took over from Ian McGregor in the 1970s, for failure to cut costs, the new Chief Executive, Allan Born, started on a draconian programme of slashbacks (6). Born's strategy was certainly radical. In his first year, he reduced the Amax workforce by nearly 10,000 jobs (7). Some of the company's assets in US oil and gas reserves were sold off to Britoil (8). The company got rid of its stake in Zambia Consolidated Copper which had been an embarrassment for some years (9). It sold its tungsten interests to its subsidiary Canada Tungsten Mines primarily a mine in the North West Territories (NWT) of Canada. This mine was closed in 1986 and put up for sale (10). But Amax retained its stake in the Hemerdon mine in Devon, England (see below) (11).

The Florida phosphates interests were up for sale in early 1986 (12). It also surrendered its important 25% interest in the Mount Newman JV in the Pilbara region of Western Australia to BHP (13). This is an iron mine that has long adversely affected Aboriginal people and encroached on their land (14, 15). In March 1986, Amax announced it would sell its 50% interest in the Buick lead/ zinc mine and Boss Smelter (Missouri) to its partner Homestake (16, 17). At the end of 1985, Amax's 50/50 copper JV with Anaconda at Twin Buttes, the Anamax company, ceased operations (100) after a lengthy strike (18), and the company's phosphates interests were on the market (16).

With losses that year running at the almost unbelievable rate of US$2 million per working day *(sic)* a total loss of US$621 million was chalked up for 1985. Amax announced that it would henceforth concentrate on three major areas of exploitation: molybdenum, coal and aluminium (19).

Even here, the going has not been smooth. The Henderson molybdenum mines in Colorado were closed in 1982 (18), to be reopened a year later (20) only to suffer another reduction in the workforce in 1985 (21) and again in 1987 (22). The Mount Emmons mine, also in Colorado, ceased production following protests by local people (23). This was after the Mount Tolman mine in Washington State had got the cold shoulder in 1983 (5, 24).

The Kitsault molybdenum mine in British Columbia stopped production in 1983 as well (25) and from 1984 onwards the company's molybdenum mines were operating at only 50% of capacity (26, 27). However, by 1988, Amax was still the world's lowest cost producer of this metal, and well poised to take advantage of any improvement in the market (28). In early 1988 the company decided to permanently close Kitsault, and the underground operations at Climax, near Leadville, Colorado. While the open-pit operations at Climax (suspended in 1986) could be re-opened at the time of writing,

Amax's only major molybdenum producer is Henderson (28).

When Amax acquired part of the Howmet aluminium subsidiary of Pechiney in 1984, it became the third largest aluminium producer in the USA (29). The acquisition was intended to reverse the corporation's flagging fortunes (30). "... one of those rare but beautiful transactions in [sic] which everyone benefits" enthused Gousseland at the time (31): at the same time, Amax took a 25% stake in the controversial Becanour smelter in Quebec (32). Certainly aluminium has not been doing so badly: its Alumax subsidiary (formerly Amax 50%, 45% Mitsui and 5% Nippon Steel, now owned 100% by Amax) reported net earnings of US$63 million in 1985 (16). The JV had no less than 93 plants and fabricating facilities in the US, and 15 in the Netherlands, France, West Germany, Canada, Mexico and Britain in 1984 (33). Its South Carolina smelter was due to re-open in 1986 (34). By this date, Amax was

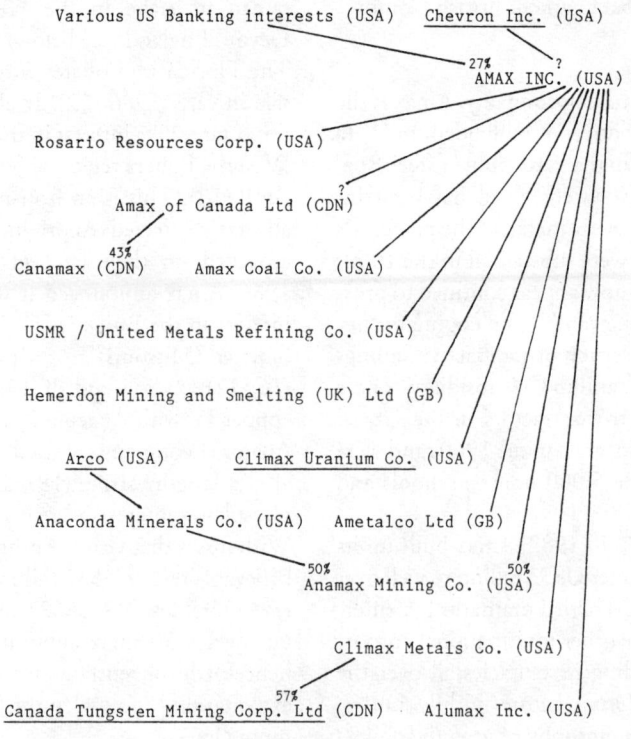

in talks with its Japanese partners, to buy out their share in Alumax (35).

Amax's US coal mines produced nearly 35 million tons of coal in 1985 (16). With 3.76 billion tons of reserves, the company is the third largest producer in the US (18). Amax has a 50/50 JV with Petrofina of Belgium called Finamax, that supplies coal to utilities in the US and Europe (36). It has also recovered its interest in nickel (after initially declaring it would get out altogether) (37) and entered a JV with BRGM (Coffremmi-Amax 49%, BRGM 41%) to exploit nickel deposits in New Caledonia (16).

In the last few years, the corporation has become one of the USA's major producers of magnesium (16), and is now the country's joint-first producer of potash (16).

At Golden Grove in Western Australia its 31.15% interest (along with Esso Exploration and Production Australia, EZ and Aztec Exploration Company which Amax took control of in 1986) (38) promises dividends from a new, rich multi-mineral deposit (16). It has a gold/silver JV at Mount Waihi in New Zealand (a project that has attracted opposition from environmentalists) (39) which evinced good assays in 1985 (16).

Amax also recently signed an agreement with Chile's state company Corfo, to produce boric acid, potassium and lithium – a crucial element in the operation of nuclear power plants – from the Atacama salt flats in the northern part of the country (40), a proposal first mooted in 1984 (41). The amount of the metal produced will be decided by the Chilean junta's nuclear energy authority CCHEN. With Amax's help Chile's contribution to the world lithium market "could be more than 50% [of lithium] available" by 1992 (42).

Amax started life as the American Metal Company in 1887. This company changed its name to American Metal Climax, after acquiring Climax Molybdenum Co 70 years later (43). Real growth did not start until after the second world war, when Amax built up its supremacy in molybdenum and its important tungsten interests. However, in its early years, as the American Metal Company, it was instrumental

in financing the important RST (Rhodesian, later Roan, Selection Trust) which was formed by the US financier Chester Beatty in 1923 (44). In 1930, Beatty sold 43% of RST to American Metal, and in 1970 the Zambian government bought out 50% of Amax's RST copper holdings in Zambia, whereupon Amax formed RST International, which later became Amax International Ltd (44). Botswana RST (Botrest) was founded in 1968 to raise capital for the extensive Selebi-Pitwe copper-nickel deposit; 18 years later, Botrest was re-structured, with Anglo-American extending emergency funding to Amax (16). As of 1986, Amax Inc held 29.8% and Anglo-American 12.05% in Botrest (25).

Socal's bid for Amax in 1981 was at the time the biggest ever attempted in corporate history – US$3.9 billion for the 80% of Amax which Socal didn't own (45). Socal's first take-over attempt was made in 1975, when the company acquired a 20% stake. There has been little doubt how important the mining company would have been to Socal/Chevron – at the time it had a good financial position, and the majority of its interests were in the USA (45) although it needed cash (46). The degree to which Amax diversified in the mid 'seventies, to prevent the Socal take-over, is still a point of debate (45). In any event, Ian McGregor, then Amax's chief executive and chairman, strongly supported Socal's 1981 bid – not surprisingly, since he stood to make a substantial killing on the increase in value of his 89,000 Amax shares (47). (The US Securities Exchange Commission later indicted a friend of McGregor's, Thomas Reed, for insider-dealing in Amax shares, after he learned of the pending Socal offer, but there has not yet been proved any direct connection between Reed's illegal action and conversations McGregor had with Reed at the time) (48).

Other shareholders also supported Socal's wooing, and were reportedly outraged when the Amax directors rejected it: a number of them filed suit against their own board as a result (2). Over the period 1957-1967, Amax diversified rapidly into oil and gas, metal fabrication, and back into southern Africa, with a purchase of 29.6% in the Tsumeb copper mine in Namibia and 17.3% in the country's O'okiep copper producer (44). A 7.7% stake in Palabora was acquired but sold later in 1968 (44). During this period Amax was also a funding member of the influential Capricorn Africa Society, which sought to build a multi-racial "independent" Africa by preserving the values and stakes of white liberalism (49). Five years later, Ian McGregor was sitting on the Council of Foreign Relations study group on US policy towards Africa (50) and helping to finance the Africa-America Institute, later revealed to be a CIA conduit to select pro-western African trade unions and politicians, such as Tom Mboya (51).

From 1969 until 1977 Amax was under the chairmanship of Ian McGregor and capital expenditure shot through the roof as did profits. McGregor – later to become the chairman of the British National Coal Board and ruthlessly close down British coal pits leading to the most famous industrial action in recent British history – operated a hands-on style of management, keeping close control of every Amax operation (52). Since he arranged for only US$400 million of depreciation and depletion provision during his period as chief executive, it is little wonder that, within a few years of his relinquishing control, Amax was in "full scale retreat" (52, 53).

Amax had clearly over-reached itself after 1970s expansions into metals and fuels, particularly coal and so-called natural resources including forest products (43). It had to write down its African copper and nickel interests in 1977 (43). Although its investments in Australia, especially in Western Australia, since 1963, had been extensive, and in 1979 it set up Amax Australia Ventures Ltd, to consolidate its projects on the continent (54), a plan to naturalise its "down under" assets was shelved in 1981 (55) because of the depressed mining outlook and a drop in share values.

There then followed the cutbacks and closures

already outlined in this article. By 1986 "the bleeding has stopped" claimed Amax's president Allan Born (56) and the company's 1987 earnings were "the best since 1981" (57) thanks largely to Alumax earnings and gold earnings from the Sleeper mine in Nevada (57, 58). Another massive restructuring, instituted that year, resulted in the disposal of the company's 47% stake in ACM (59) and take-over of full control of Aztec, its erstwhile partner in the Golden Grove JV in Western Australia (38). It sold its petroleum subsidiary, Amax Petroleum Corp (60), and disposed of the El Mochito lead/zinc mine in Honduras (61, 63) the main project operated by its Rosario Resources subsidiary, which it had acquired in 1979, though the Sandinista government of Nicaragua later nationalised some of its assets (62).

The parent corporation reduced its equity share in Amax Gold in late 1977, although Amax Gold increased its share in Canamax to 43% (61). In a mood of rare buoyancy, Canamax announced that it would open "a gold mine a year" from 1987 onwards. That year it opened the Bell Creek mine at Timmins, Ontario (64). It also increased its stake in the Ketza River gold project (65) and took over full ownership of its other main Canadian gold venture, the Kremzan prospect, north-east of Wawa, Ontario (66). In early 1988 Allan Born could report that Amax was, once again, viable, and headed for profit (67).

Amax prides itself on its environmental innovations – such as the development of viable wetlands from some of its coal slurry disposal sites (68) and its institution of the so-called CRP (Colorado Joint Review Process for Major Energy and Mineral Resources Development Projects) which was meant to soften the edges of the planning process for its Mount Emmons mine, in Gunnison County, Colorado, in the late 1970s and early 1980s (69). (The CRP was, rather, a euphemism for curtailing the public review process – a point not lost on local residents and ranchers who successfully prevented Amax opening a tailings disposal site for this molybdenum mine in 1980) (70). In 1987, the company received an award for Excellence in Mining and Reclamation for its Ayrshire mine, Indiana (63).

But this is pretty thin icing on a rotten cake. Amax has customarily only spent a fraction of its capital outlay on environmental protective measures: under 5% in 1974, for example – possibly less than was spent on a public relations exercise to evade public opposition and the law (71). In another example of Amax's disregard for the environment, research conducted by the University of Washington into one Amax proposal indicated that the cost of rehabilitating scarred land would be US$3000 per acre – some thirty times what the company was prepared to spend (72).

Apart from RTZ, Amax has probably been criticised longer and harder than any other mining company mentioned in this *File* and it is only possible to summarise here some of the major issues:

- The Hemerdon Mining and Smelting company is now wholly owned by Amax (73). It bought out half the share in the project from Billiton (Shell) in 1985 (74). The company then submitted revised plans for this tungsten mine – originally boasted as potentially the biggest tungsten mine in western Europe (75) and the only open-pit mine in England (76). These plans were approved in 1985 (77, 33). Nonetheless the deposit lies only two miles from the Dartmoor National Park, just north-east of Plymouth. Despite the promise of 350 jobs (78) in one of the most depressed areas of Britain, local opposition to the mine has been considerable. According to the Dartmoor Preservation Society, the tungsten lies under several ancient burial grounds and monuments, while the projected tailings from the original project (15,000 tonnes of waste per day, used for a tailings dam at Crownhill Downs) would create a huge blotch on the landscape (76).

- As already pointed out, the role of Amax during its expansionist years, 1967-1977, cannot be separated from the presence and power of Ian McGregor – nineteenth century captain

of industry and chairman of an organisation called Religion in American Life which popularised the slogan "The family which prays together stays together" (79). He was appointed to be head of the British National Coal Board in 1983, partly because of his reputation as a strike-breaker among the coal-mining unions of America. McGregor's major strike breaking act was in 1975, when, alone among nine major companies operating coal mines in the western USA, Amax refused to negotiate a multi-employer agreement, allowing miners to move between various mines without losing benefits. Not only did Amax defeat the strikers, it also successfully reduced the power of the main mining union, the United Mineworkers of America (UMWA), and increased the presence of non-union workforces to 60% of the total in US coal mines (79).

Amax also scored a win in 1980 when a US court ruled that the UMWA could not insist on unionisation at the new mines then being opened up by the company (79).

- The UMWA has also co-operated with local people and environmentalists in opposing Amax's plans to strip-mine in Tennessee in 1975. The Smartt *(sic)* Mountain project consisted of a proposed 10,000-acre strip mine in Van Buren, Sequatchie and Bledsoe counties. The Concerned Citizens of Piney, assisted by a group called Save Our Cumberland Mountains (SOCM) examined the company's statements regarding the ten areas of impact – economic, environmental and social. In every case the study pinpointed "major inconsistencies" between "what Amax would have the people of Tennessee believe and what the record show Amax actually doing in other states. Moreover [it] reveals a pattern of repeated non-compliance with the law and a disregard for local citizens in other states in which Amax operates" (80). Although Amax promised that more than 300 people would be employed at Piney, SOCM showed that Amax had successfully reduced employment below this level in all its other projects – including those with a much greater capacity. Although the company promised to "successfully" reclaim land, concurrently with mining, Amax's record in Kentucky in 1974 and Perry County Illinois in 1975 suggested the opposite. While Amax promised its Piney project would not pollute "any of the waters of the state of Tennessee" (81), between 1972 and 1976 in Illinois and Kentucky it had been guilty of illegal contamination – in one case despite nine previous orders of non-compliance issued against the company (82). SOCM also detailed six instances where Amax had failed to comply with the law over environmental protective measures, in Oklahoma, New Jersey, Missouri, and Washington state. SOCM successfully defeated Amax's plans (83).

- In 1978 the Greenland (Inuit) Labour Union identified Amax as a particular example of the potential domination of their country by multinationals. At the time, Amax, through its subsidiary Arctic Mining Company, held a very substantial molybdenum lease at Mestersvig, in the eastern part of the world's largest island: Amax did not proceed with plans to open up this deposit on a commercial scale (84).

- Several north American Indian nations have clashed with Amax during the seventies. In 1975, the Papago sued Amax's copper subsidiary, Amax Copper Mines, and Anamax Mining Co (partly owned by Amax) for damage caused to the tribe in the usage of excess water in the Upper Santa Cruz Basin, Arizona. Local mesquite and willow trees were destroyed as a result, and wells went dry (85).

By the late 1970s the Northern Cheyenne and the Crow nation of Montana had also entered suit against the company over coal rights: the Crow in an attempt to get the best deal (86), the Cheyenne in an attempt to recover control over their own land and void a lease which had given Amax, together with Peabody Coal, Chevron, and Continental Oil, rights to strip-mine, sub-lease, lay roads, and build chemical and power plants, over much of a 243,000-acre area of Cheyenne territory (87).

At the same time, the Colville Confederated Tribes were faced with Amax's plans to blast open their sacred Mount Tolman, in order to extract nearly a thousand million lbs of molybdenum and a substantial amount of copper over a 43-year period (88). The project would produce nearly one thousand million tons of tailings, and slightly less waste rock, in its life (death) time; 600 feet of the top of the mountain would be lopped off; 8000 gallons per minute of water would be drained from the region – reducing ground water recharge in the Last Chance basin to one third of its then current level; numerous local fauna and flora would be threatened or destroyed (89). Dust levels would rise alarmingly and molybdenum poisoning – potentially carcinogenic – would pose a constant danger (88).

Amax eventually withdrew from Mount Tolman after traditional Colvilles (native Americans) joined with non-native environmentalists and other local people, in the Preservation of Mount Tolman Alliance (88).

- One of the most prolonged and tenacious battles against Amax by an indigenous people has been the courageous struggle waged by the Nishgas of British Columbia to gain recognition of their traditional land claims, and protect their people and land from the poisons emanating from Amax's Kitsault mine.

The Nishga are among the few native Canadians who have never relinquished their traditional rights or extinguished them in exchange for reserves or cash. These 2200 West Coast Indians who have lived in the valley of the Naas River "since time immemorial" have conducted a quest for recognition for 112 years – perhaps the longest single fight for legal recognition of any Canadian people (83).

In 1979, Amax was given permission to dump around 100 million tonnes of mine tailings from its Kitsault molybdenum mine into the waters of Alice Arm – a confluence of two important salmon rivers, the Kitsault and the Illiance (90), used by the Nishga for food and commercial fishing (83). Up to

8000 times the normal permissible limit of toxic and heavy metals (including arsenic, cadmium, copper, iron, manganese, lead, zinc, mercury and radium-226) would be deposited at the rate of 12,000 tonnes a day of tailings (91, 92). This method was pushed by Amax in preference to land disposal which would cost an additional US$20 million – although neither method of disposal was considered by independent researchers to be viable or safe (90). An independent study commissioned by the Nishga showed that the dumping would have (to quote a local MP, Jim Fulton). "... a severe impact on the fisheries ... and constitute a genocidal effect on the Nishga" (83). The Nishga's call for a public inquiry was ignored by the British Columbia government, which contravened the regulations of its own Pollution Control Board (91) and the Metal Mining Liquid Effluent Regulations (90). Little wonder that the Nishga thought they were being used as guinea-pigs by the company (93). The tribe went to court to challenge the legality of the permit in early 1981 (91).

The following year Amax got permission to emit into the air most of the same pollutants it was daily pumping into the waters of Alice Arm. An appeal against the permit by the Nishga, supported by church leaders, environmentalists, and local politicians, was dismissed by the government (94). A little later, the Kitsault mine was temporarily shut down on economic grounds, while the Nishga demanded prosecution of Amax for polluting the surface waters of their river and contravening even the over-lenient provisions of the special Federal Order-in-Council, passed in 1979 (95). Early in 1983, the newsletter of the Canadian United Food and Allied Workers Union, *The Fisherman,* revealed collusion between Amax and certain officials in the environment service, facilitating permits for the Kitsault tailings disposal plan, and pre-empting moves by the Nishga (96). Two months later, the British Columbia Ombudsman recommended the government set up a public inquiry to examine the permit issued to Amax, claiming the Nishga should have been given the opportunity to challenge it (97).

With the mine still at a standstill, the Nishga appear to have won their battle (if only by default) and in 1988 Kitsault was permanently closed. The damage caused to Alice Arm and the Nishga people has yet to be determined.

• The most intense opposition faced by Amax occurred in 1980 when it attempted to drill on the Noonkanbah pastoral lease in Western Australia (WA), on the land of the Yungngora people, who had acquired it in 1976 from the Aboriginal Lands Fund Commission and within three years pulled it back from the brink of ruin (98). By 1978, because of the almost complete absence of land-rights provisions in Western Australian law, no less than thirty companies had trespassed on Noonkanbah station, staking out more than 600 mining claims (98). That year, Amax took the lead though CRA, the RTZ subsidiary, was also prominently involved in attempts to mine (99). Amax's bulldozer obliterated two burial sites and one ceremonial site, while Aboriginal lease holders were involved in a dance festival and the inauguration of the Kimberley Land Council. The community's main concern was the sacred Pea Hill goanna dreaming, supposedly sacrosanct under the Western Australian Heritage Act of 1972. Although Amax agreed to move away from one of the sites in this dreaming area, it soon started drilling on another. On 28 March 1980, the reactionary state government of Charles Court gave permission to Amax to drill in this second zone. This action precipitated the greatest outrage from Aboriginal people and their supporters that Australia has seen (probably before or since). First of all, the WA police protected the Amax operators, claiming intimidation – although the company was forced to withdraw after a few days (101). Three months later, however, Amax, together with CSR, moved a drilling rig 2200 kilometres from Eneabba in the direction of Noonkanbah, in a paramilitary operation that has few parallels in mining history. This convoy was met with protests not only along the route, but at Noonkanbah itself, when 250 Aboriginal people tried to block its entry to the station. Eighteen protesters staged a sit-in at Amax's office in Perth, and many others were arrested trying to stop the convoy. The Australian Confederation of Trade Unions (ACTU) imposed an immediate ban on work at Noonkanbah – signalling the first time the trade unions had done more than mouth platitudes about Aboriginal rights (98). In the USA, fifteen native American men and women protested outside Amax's US headquarters, and several African diplomats at the United Nations privately expressed regret at the events, hinting on a possible boycott of the 1982 Commonwealth Games. (In the event, no such boycott occurred.)

An Aboriginal delegation from the National Aboriginal Conference addressed the UN in Geneva, and virtually every major Australian newspaper condemned the actions of Amax and the Court government (98). In the context of the developing national Aboriginal land rights struggle, Noonkanbah can hardly be over-estimated: it galvanised the formation of a National Federation of Land Councils, compelled the National Aboriginal Conference into a more radical position, brought Aboriginal issues to a world-wide audience, and, perhaps most important of all, finally sunk the image of Aboriginal people as passive victims in the face of inevitable "progress". In this respect, whether or not Amax learned prudence in its dealings with indigenous people (there is some evidence it did), its actions inadvertently gave the Aboriginal land rights movement its greatest fillip of recent years.

In 1988, the relationship between Chevron and Amax changed dramatically, when Chevron offered the mining company its (then) 15.4% of common stock in Amax, and Amax accepted (102). A little later, the Amax board instituted a stock purchase plan, to protect itself against any hostile future bids (103). Around the same time, Amax formed a new subsidiary, Climax Metals Co (a return to its roots?) to consolidate

its molybdenum interests and its majority interest in Canada Tungsten Mining Corp Ltd (104).

Like many other metals and mining businesses, in the latter half of the 'eighties Amax restructured itself, selling off most of its "non core" assets and concentrating on aluminium, gold and coal. However, it still holds important natural gas interests in the USA, and has not dropped molybdenum (105, 106) now consolidated in the Climax Metals Company (107, 108). Moreover, it has been re-investing in coal production which in 1988 was already the third largest in the USA (109) although it consists of steam coal, sold largely to electrical utilities (107).

Alumax Inc, the company's aluminium subsidiary, is itself the third largest integrated aluminium business in the USA, now expanding into Europe and Quebec (through its 25% in the controversial Becancour smelter) (111). Amax Gold (Amax 87%, although shares were being offered to the public in 1988) (112) holds a number of properties in Canada through Canamax Resources (Amax 47%) (107, 113), and 28% in the Waihi JV in Aotearoa (New Zealand) about 70 miles south-east of Auckland (114) where one of its partners is ACM.

ACM is also a partner in the Golden Grove polymetallic sylphide deposit in Western Australia, with Aztec Mining (115), a company in which Amax has acquired a majority (52%) interest and is regarded as Amax's vehicle for "re-entry" into Australia (107). This was a region from which it looked like withdrawing in the mid 'eighties, when it sold its 25% stake in Mount Newman iron ore and its 47% share of ACM (105). Aztec holds 32% of the Harbour Lights gold mine and several other licences.

But it is Amax Gold's Sleeper mine in Nevada which is the jewel in its crown (116) ensuring Amax's position as the USA's third biggest gold producer. The Sleeper mine earned itself the unenviable reputation as a "bird killer" thanks to the company's failure to protect the wildlife from the tailings pond (it admits to 88 deaths in 1988 alone) (117). However, the company claims Sleeper will become a model of its kind: "an oasis in a rather desolate area" (118). Notwithstanding this, the company sees *in situ* gold leaching as the technology of the future (117). And Amax's President Allen Born in 1989, when asked what Amax's position was on environmental control, could only volunteer that "something must be done to preserve our natural resources before it is too late" (117). In 1989 Amax tried to buy out Falconbridge of Canada, but was prevented by a counterbid from Noranda and Trelleborg of Sweden (119). The following year, it also tried to buy out 55% of Peabody Coal, but withdrew in the face of the Hanson Trust (110).

Further reading: Fact Sheet: *Amax in Randolph County,* West Virginia, February 1987.

A study of American Metal Climax Inc, by Concerned Citizens of Randolph County.

The Amax Record Elsewhere, A study prepared for the Concerned Citizens of Piney, by Save Our Cumberland Mountains 15/7/76.

"How McGregor's Men broke the US Miners' Union". *Guardian* 5/11/84.

Bruce Johansen and Roberto Maestas, *Wasi'chu,* Monthly Review Press, New York and London 1979.

Massacres to Mining, by Jan Roberts, Dove Communications, Victoria, 1981.

The Nation (US), Vol. 23, No. II, 11/10/80.

Aborigines and Mining Companies in Northern Australia, by Ritchie Howitt and John Douglas, APCOL, Chippendale, 1983.

"Amax: A Century of Change" in (120), which gives a useful, if completely uncritical, potted history of Amax.

Contact: Coromandel Peninsular Watchdog, Box 51, Coromandel, Aotearoa (New Zealand)

Kimberley Land Council, Box 382, Derby 6728, Australia.

Project North, 80 Sackville St, Toronto, Ontario M5A 3E5, Canada.

References: (1) *MJ* 6/3/87. (2) *Big Oil.* (3) *Yellowcake Road.* (4) *High Country News,* Colorado 10/3/78. (5) *MJ* 8/4/83. (6) *FT* 11/11/85. (7) *FT* 8/4/86; *Metal Bulletin,* London, 10/9/85; *FT* 7/9/85. (8) *FT* 7/2/84; *Gua* 2/2/84. (9) *FT* 14/6/84. (10) *MJ* 22/8/86; *FT* 8/4/86. (11) *Aurora,* NWT, Canada, 1/84. (12) *MJ* 1/1/86, *MJ* 12/12/86. (13) *MJ* 18/4/86. (14) Roberts. (15) *MJ* 27/12/85. (16) *MAR* 1986. (17) *MJ* 14/3/86. (18) *MIY* 1985. (19) *MJ* 7/2/86. (20) *FT* 17/11/83. (21) *FT* 2/8/85. (22) *MJ* 6/3/87. (23) *Mining Monitor,* New Zealand, No. 42, 24/2/84. (24) *FT* 25/1/83. (25) *MIY* 1987. (26) *MM* 10/86. (27) *MJ* 30/1/87. (28) *FT* 10/2/88. (29) *MJ* 9/3/84; *MJ* 25/5/84; *MJ* 18/5/84. (30) *FT* 7/2/84. (31) *FT* 21/10/83. (32) *MJ* 7/8/87; *Times,* London, 7/2/84. (33) *MAR* 1985. (34) *MJ* 2/7/86; *FT* 19/7/86. (35) *MJ* 2/7/86. (36) *MJ* 23/3/84. (37) *FT* 9/10/85. (38) *E&MJ* 9/87. (39) *Mining Monitor,* NZ, 24/2/84. (40) *FT* 3/3/86. (41) *FT* 24/8/84; *Gua* 5/9/84; *MJ* 31/8/84. (42) *MM* 2/86. (43) *WDMNE.* (44) Lanning & Mueller. (45) *FT* 7/3/81. (46) *FT* 9/3/81. (47) *Obs* 3/4/83. (48) *Obs* 2/9/84. (49) Bloch and Fitzgerald, *British Intelligence and Covert Action,* Derry, 1983. (50) W Minter, *The Portugese in Africa,* Penguin Books, 1972. (51) "The CIA as Equal Opportunity Employer in Africa", *Ramparts,* San Franscisco, no date – as quoted in (44). (52) *FT* 29/3/83. (53) *FT* 10/2/83; *Gua* 21/10/83. (54) Ritchie Howitt, *CRA & Amax,* duplicated study, self-published, 1982. (55) *FT* 23/5/81; *FT* 15/9/81. (56) *MJ* 13/2/87. (57) *MJ* 6/11/87. (58) *FT* 14/2/85. (59) *MJ* 17/4/87; *FT* 9/3/87; *MJ* 5/2/88. (60) *MJ* 1, 8/1/88. (61) *MJ* 6/11/87. (62) *E&MJ* 11/83. (63) *MJ* 3/7/87. (64) *MJ* 2/10/87. (65) *MJ* 6/2/87. (66) *MJ* 19/12/86; *MJ* 31/10/86; *MJ* 27/3/87. (67) *MJ* 12/2/88. (68) *MM* 9/85. (69) *FT* 27/6/80. (70) Information from The High Country Citizens Alliance, PO Box 1066, Crested Butte, Colorado 81224, USA, 1980. (71) *Coal Age,* USA, 10/74. (72) *Western Coal Rush,* published by Corporate Information Center, Interfaith Center on Corporate Responsibility, Washington DC, USA, 1/75. (73) *MJ* 17/1/86. (74) *MJ* 19/4/85. (75) *Gua* 5/4/84; *FT* 4/4/84. (76) *Gua* 8/10/82. (77) *FT* 16/8/85. (78) *FT* 8/10/81. (79) *Gua* 5/11/84. (80) *The Amax Record Elsewhere,* A study prepared for the Concerned Citizens of Piney, by SOCM, 15/7/76. (81) Amax Petition for Hearing before Water Quality Board of Tennessee, 1976. (82) Records of Kentucky Department of Environmental Protection and Natural Resources, quoted in (80). (83) Project

North, *Newsletter,* Vol. 5, No. 8, Toronto, Oct/Nov 1980. (84) Statement by Odak Olsen, leader of the Greenland Labour Union, quoted in *IWGIA Newsletter,* Copenhagen, 12/78. (85) United States of America vs City of Tuscon et al, Civ. 75-39 before US District Court for the District of Arizona. (86) *AkwN,* Akwesasne, Late Spring 1978. (87) Bruce Johansen and Roberto Meastas, *Wasi'chu,* Monthly Review Press, New York 1979. (88) Bruce Johansen, "Indians fight Amax", *MMon,* 7/80. (89) "Mount Tolman Fact Sheet", Preservation of Mount Tolman Alliance, date unknown. (90) *Mining Organiser,* Vol. 1 No. 3, Madison, 3/81. (91) *MJ* 15/5/81. (92) *MJ* 13/2/81. (93) Project North, *Newsletter,* 11/81. (94) Project North 10/82. (95) Project North 12/82. (96) *The Fisherman,* Canada, 28/2/83. (97) Project North, May/June 1983. (98) Roger Moody, "Killing Dreamtime", *Native People's News,* London, No. 5, Spring 1981. (99) Press Release by Yungngora Aboriginal Community, Noonkanbah Station, 25/5/79. (100) *MJ* 1/11/85. (101) *Kimberley Land Council Newsletter,* 4/80. (102) *FT* 23/5/88, *MJ* 27/5/88. (103) *MJ* 10/6/88. (104) *E&MJ* 3/88. (105) *MJ* 24/2/89. (106) *FT* 15/2/89. (107) *MJ* 1/7/88. (108) E&*MJ* 3/88. (109) *FT* 19/2/88. (110) *MJ* 25/5/90. (111) *MJ* 13/4/90; see also *FT* 7/3/89. (112) *MJ* 8/7/88; see also *MJ* 5/6/87. (113) *MAR* 1989. (114) *E&MJ* 12/87. (115) *MJ* 14/4/89. (116) *MJ* 20/5/88; *MJ* Supplement on gold 19/5/89. (117) *MJ* 8/9/89. (118) *MJ* 19/5/89. (119) *FT* 4/8/89; *FT* 29/8/89; *FT* 2/9/89. (120) *MJ* 8/4/88.

24 Ambrosia Lake Uranium

It merged with Kerr-McGee in 1962 (1).

References: (1) US Congress Subcommittee on Energy 1975.

```
        Kerr-McGee Corp. (USA)
                        \
        AMBROSIA LAKE URANIUM (USA)
```

25 American Copper and Nickel Co

Homestake's copper exploration partner in the USA (1), it also held a uranium lease in the Black Hills of Lawrence County, South Dakota, in 1979/80 (2).

References: (1) *MIY* 1982. (2) *BHPS* 4/80.

```
        Homestake Mining Co. (USA)
                        \ 60%
        AMERICAN COPPER AND NICKEL CO. (USA)
```

26 American Lake

Acquired by Kerr-McGee in 1960 and merged (1).

References: (1) US Congress Subcommittee on Energy, 1975.

```
        Kerr-McGee Corp. (USA)
                 | merger
        AMERICAN LAKE (USA)
```

27 American Nuclear Corp

It owned a uranium property with Federal Resources Corp (40%) at Gas Hills, Wyoming, which was leased by the Tennessee Valley Authority (TVA). The mine was, in 1979, sufficiently delayed by state regulations for it to cancel its contract to supply uranium concentrate to Florida Power and Light Co. The company then reapplied to the State Department of the Environment to place the mine in production (1). Mining stopped due to a drop in uranium prices in 1981 (2).

It was reportedly also exploring in Texas up to 1979 (3).

With the sale of Federal American Partners' interests in the Gas Hills uranium mine to TVA, American Nuclear retained royalty on the uranium and the partners' interest in the uranium mill (4).

References: (1) *Wall St J* 23/3/79. (2) *MJ* 20/3/81. (3) *US Surv Min* 79. (4) *MJ* 7/1/83.

```
          AMERICAN NUCLEAR CORP. (USA)
                       /
                   60%/
        Federal-American Partners (USA)
```

28 American Uranium

It is a broker/dealer in uranium, issuing certificates of ownership to all buyers of uranium stocks, stored in the company's own warehouse in Missouri. Set up in October 1982, this marks the first time that a commodities trading centre has been established for uranium in the USA. AU's uranium can be sold to anyone with the cash, though a licence is required to "possess" it. Comments Nancy Dunne: "Until now, US investors have been able to participate in the uranium market only by buying shares in publicly quoted uranium companies or by purchasing mineral rights to uranium deposits." Now, apparently, you can get hold of the stuff as easily as a gun – and perhaps for a similar aim? (1). In 1983, American Uranium leased a number of its non-uranium claims to Vernon Stratton and Co of South Dakota (2).

References: (1) *FT* 21/10/82. (2) *E&MJ* 2/83.

```
        Edlow International Ltd (USA)
                         \
                          \
                AMERICAN URANIUM (USA)
```

29 Amoco Minerals Ltd

A subsidiary of Standard Oil of Indiana (Amoco), the fifth largest US company, it is involved in mining, extracting and processing many minerals worldwide. It has extensive interests in the Ok Tedi (PNG) project; in seabed mining (through Ocean Minerals Co); in copper (in the Yukon and in Arizona); in molybdenum (Idaho); and in gold (South Dakota and Nevada). It currently holds 50% of the Rio Blanco Oil Shale Co (synthetic fuels) and is involved in experimental tar sands projects in Utah and Alberta (1).

It also has extensive mineral exploration licences in New Zealand, including the Coromandel which is "pepper-potted with its activities". In the early 1980s it was prosecuted successfully for discharging filthy water from an old mine into a stream near Waitekauri (2).

Amoco's uranium explorations, until the merger, were in the hands of Cyprus Mines; exploration in Colorado indicated 20 million lbs of uranium in 1977 (3).

Cyprus Mines is a 51% partner with Westinghouse's wholly-owned Wyoming Mineral Corp (49%) at the Hansen project in Colorado, and according to the Uranium Information Network (USA) has proposed operating a mine at Fremont. The Hansen project has now been deferred (4).

It has a JV with Cyprus Pima Mining Co (mainly copper production) owned 50.01% by Amoco, 25% by Union Oil Co, and 24.99% by Utah International Inc

In 1979 Cyprus Mines controlled Cyprus Anvil (63%); this passed to Amoco on the merger with Standard Oil of Indiana. Canada's Foreign Investment Review Agency, however, later refused to allow Amoco to acquire Cyprus Anvil which in 1981 was acquired by Dome Petroleum (5).

Standard Oil was established in 1899 as part of the huge Standard Trust monopoly but hived off in 1911 as a result of an anti-trust decree when it became an oil company through various diversifications and acquisitions (notably Avisun chemicals interests, bought from Sun Oil in 1968). Standard's predominant interests have been in the USA. However, worldwide exploration was stepped up in the 1970s, and by 1983 it had acquired 50% in the important Detour goldmine in Ontario, Canada, as wel as five Kentucky coal mines. Its mineralogists are currently ranging as far afield as Australia,

Canada, Chile, Colombia, Mexico, Papua New Guinea and New Zealand, carrying the Amoco torch (6).

In the late 1970s, Amoco acquired over 100,000 acres of land in Wisconsin (Florence, Forest and Marinette Counties) for mineral – including uranium – exploration (7).

A 1984 article in the influential journal *Afrique-Asie* accuses the new Guatemalan régime of Oscar Mejia Victores of deliberately implementing a plan to "free the territory occupied by Indians, for the benefit of oil companies, to which, under the new petroleum law of 26 September 1983, the government has given extremely favourable financial privileges". In particular, the Guatemalan régime would take only 15% of the value of extracted oil, as opposed to a previous 55%. *Afrique-Asie* singled out Amoco as one of three companies entrenched in Guatemala which stood to benefit from this new regulation (8).

In 1984, *Kainai News,* a north American Indian newspaper, reported that the company had failed to pay royalties on 1.38 million barrels of oil taken from Arapaho and Shoeshone lands; 72 Indian land-owners from the Wind River reservation in Wyoming were suing Amoco for the return of the oil and US$41 million (9).

In 1985, Amoco spun off its Cyprus Minerals subsidiary, enabling it to become completely independent (10).

In 1988 Amoco took over debt-laden Dome Petroleum of Canada, to become that country's biggest single oil producer (11). That same year it was ordered to pay damages of no less than US$85 million to 90 plaintiffs, including French fisherpeople, hoteliers and others, in compensation for the infamous *Amoco Cadiz* disaster, which resulted in an oil slick of 58 million gallons discharging on to 200 miles of northern French coastline in 1978 (12).

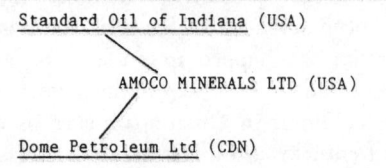

Standard Oil of Indiana (USA)

AMOCO MINERALS LTD (USA)

Dome Petroleum Ltd (CDN)

Perhaps, learning a little from this incident, the company has increased its provision against the future cost of environmental programmes in the refining and marketing business (13), at the same time as selling off much of its downstream and retail businesses in Italy, Australia, India and Britain (14).

Amoco has a 30% stake in the much-criticised Ok Tedi mine of Papua New Guinea (see BHP).

Further reading: "Amoco's Expanding Mineral Interests" in *MJ* 15/5/81.

Contact: Center for Alternative Mining Development Policy, 1121 University Ave, Madison, Wisconsin 53715, USA.

References: (1) *MJ* 15/5/81. (2) *CAFCINZ Watchdog* 1981. (3) *MJ* 2/12/77. (4) *MIY* 1981. (5) *MJ* 7/8/81. (6) *WDMNE.* (7) *Daily Cardinal,* Wisconsin, USA, 20/2/80. (8) *Afrique-Asie,* Alger, 4/84. (9) *Kainai News,* reported in Onaway magazine, Spring 1984. (10) *FT* 18/7/88. (11) *FT* 21/4/87; *FT* 12/2/88; see also *FT* 24/4/87, *FT* 25/6/87 and *FT* 15/7/88. (12) *FT* 12/1/88; see also Ken Gourlay, *Prisoners of the Sea,* Zed Press, 1988. (13) *FT* 20/10/89. (14) *FT* 16/9/89.

30 Amok Ltd

One of the most successful of all uranium mines, the Cluff Lake project entered 1986 with target output achieved (1), phase two of its construction complete ahead of schedule (2), and plans to recover gold and uranium from mill residues approved by the Federal Canadian Atomic Energy Commission, and the Saskatchewan Department of the Environment (1).

However, due to the extremely high level of uranium-in-ore, Cluff Lake is one of the world's most dangerous mines. Its conception and construction have also infringed upon and endangered indigenous land claims throughout the last ten years.

For these reasons, its original start-up date (1979) was put back two years to await the outcome of a provincial inquiry into uranium mining. After 16 months, the Bayda Inquiry gave the mine a highly-controversial bill of health (3).

The Cluff Lake deposit is situated within the Carswell Dome structure of the Athabasca Basin, 50 miles south of Lake Athabasca itself. It was discovered by Mokta as the result of an airborne survey. It was soon realised that the Carswell structure is extremely unusual – an astrobleme, possibly produced by extra-terrestrial impact (4). From the beginning, it was the "D" orebody which attracted special attention, with its extremely high uranium assays (upwards of 700lb per ton of ore in parts), and its ancillary minerals: including gold, bismuth, lead, cobalt and selenium. (The gold alone is concentrated in up to 1.6 ounces per ton) (4).

The "N" orebody, with similar reserves to "D" of around 5000 tonnes, averages 0.3% uranium (5), while the Claude deposit holds around 4800 tonnes, and the "O-P" deposit another 1500 tonnes with higher concentrations of up to 0.7% (6).

As of the beginning of 1985, total reserves were put at 38 million lbs of U_3O_8 averaging 0.5% uranium (7). Reserves at the time of the Cluff Lake Inquiry had, in fact, been rated 45% higher (8).

By October 1981, the "D" orebody had been completely mined, creating a stockpile for surface production. Milling began in October 1980 and continued until 1984 (9).

Although the second phase had originally been planned for start-up in 1981 or 1982, with mining of Claude, followed by the "N" orebodies (10), it did not in fact commence until late 1984, when the open-pit lode at Claude and "O-P" (an underground deposit) came under the pick and shovel.

A new mill had to be constructed to handle ore from the second phase, since the uranium was of a more conventional grade. (Milling capacity of 230,000 tons/year ore, producing between 850 tons and 1270 tons uranium/year in ore.) The performance of both these deposits "surpassed expectations", delivering 490 tons of uranium in the first half of 1985 (2). By the sec-

ond half of that year, output was supplemented by production from the Dominique-Peter orebody.

That year, Amok decided to try to bring into production the "N" and "N-40" orebodies by the 1990s (2). In 1987, the company also received permission to reprocess radioactive tailings held in concrete vaults at the mine, after coming up with a permanant solution to the problem of leaky and cracking vaults: the residues are to be blended with tailings from Claude and "O-P" to lower the radioactivity below the permitted maximum, and then deposited in a large pond "where they will be further diluted and stored indefinitely" (11).

Apart from its current seven mining leases, covering nearly 500,000 acres of the Carswell Lake zone, Amok has three further claims and six leases covering another 75,000 acres to the north of the current mine (7). The company has (through its Famok subsidiary) also been involved in exploration JVs with Eldorado and SMDC at Fond du Lac, and with Ontario Hydro (12).

Formed at the end of 1967, Amok had a prime aim to exploit the Carswell Lake deposit, although it started its exploration with a very low profile: three years before anyone knew about it according to the British Columbia magazine *Energy File* (13). Now owned 37% by Mokta, 38% by Cogema and 25% by Pechiney, the original ownership was slightly different. In 1980, Mokta held only 17% (14); it was increased to 25% the following year (15). At the same time, Cie Française des Minerais d'Uranium SA (CFMU) held 20%. In 1979, the SMDC took 20% of the Cluff Lake mine

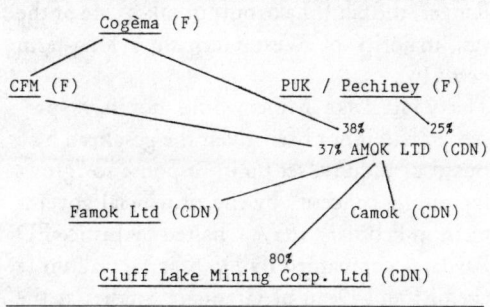

69

(3), after speculation that it might take up to another 10% more (16). The stake cost the SMDC US$66.9 million, and Amok was not displeased (16).

By the time Cluff Lake uranium started coming onto the market, a glut had developed in the world uranium supply and, in 1981, revenues· from uranium were nearly half (45%) of those projected by the Bayda Inquiry (17). Envisaging this, Saskatchewan's Premier, Allan Blakeney, in 1979 toured Europe and Australia, touting "staged development" (a moderated version of the old uranium cartel) while Canadian industry personnel were racing to beat Australian uranium onto the market (18).

In fact, the gloomiest forecasts didn't prove correct, and by the end of that year, the first four years' production from Cluff Lake had been sold to France and West Germany (19). After 1984, most of the production from the mine has been earmarked for France (20).

In 1974, the Canadian government had enunciated a safeguards policy on the supply, *inter alia,* of uranium, which – though not imposed until the beginning of 1977 – made nuclear trade with France unacceptable (4).

After negotiations over the end use of uranium, and a supposed safeguards policy was in place, the Canadian government lifted the ban in 1981, thus – very conveniently – permitting Amok to fulfil its contracts (21).

The Phase I production was divided equally between Cogema and an unnamed German utility, with the SMDC and Amok splitting the revenues according to their equity share of the project (6). It has been estimated that no less than 26% of West Germany's uranium needs are met by Amok. Together with contracts at Ranger, the Cluff Lake output takes care of the vast majority of West Germany's long-term needs (6).

The Cluff Lake Mine would not have proceeded had it not been given the go-ahead by a Board of Inquiry, set up in response to "growing public concern" by the provincial government in February 1977. Chaired by Justice ED Bayda, accompanied by Dr Ken McCallum (a chemist and Dean of Graduate Studies at the University of Saskatchewan) and Dr Agnes Groome, a professor of education at the University of Regina, the Inquiry's briefs were to determine whether local, economic, environmental, social and occupational health impacts could warrant the refusal of a licence; whether the potential misuses of Saskatchewan uranium were such as to justify specific precautionary measures, or halt the development altogether. Although the deadline for ending the hearings was set at November the same year (22), Bayda did not in fact report until June 1978. Almost on the day the report was published, the Blakeney government announced that it accepted its broad conclusions, and that uranium mining, within certain parameters, would proceed (23). Envisaging yearly royalties of around half-a-billion dollars over the next ten years, the main local press welcomed the decision (23), while the mining industry, though not jubilant, was greatly pleased, blithely predicting that the environmental and safety measures could be accepted "quite readily" (23).

Bayda stipulated that the Amok mine could proceed, so long ·as additional safety measures were incorporated into the project. (The company's plans were found to contain "a number of inadequacies".) Following Phase I development – ie after three years – a comprehensive impact assessment study was to be prepared before further development could proceed; as many native and northern residents "as possible" were to be employed at the mine; an environmental protection fund was to be established to take in revenue from uranium companies, purchasers and provincial royalties. (This fund would finance research and environmental protection measures which might be needed in the future.) A Northern Development Board was to be set up to oversee regional development; royalties were to be shared between the province and the north to ensure the north received its "fair share" of benefits; the uranium was to be sold under long-term contracts "wherever possible", with annual changes in price to ensure a fair return to the province; future uranium development would be "programmed" to avoid a glut situation or a future "boom-and-bust" cycle (23).

The Bayda board found that current maximum radiation standards were "adequate", and neatly side-stepped the issue of nuclear wastes, by arguing that Saskatchewan would not face these problems anyway, while the rest of the world "accepted" such risks (24). The report rejected arguments about the potential of alternatives to nuclear energy (25) and claimed that, since nuclear proliferation was already widespread, for Saskatchewan to withhold its small cache of yellowcake wouldn't matter one way or the other (24).

These arguments were met with scorn and derision. The Saskatchewan Council for International Co-operation was "astonished" that "... the Cluff Lake Board of Inquiry not only gave carte blanche to uranium development but also announced that the "morality" of the issue was of no concern to them because this province has no nuclear reactors of its own and because our contribution to the nuclear stockpile is insignificant in world terms. This kind of moral ethic is surely the stuff of which nuclear wars – all wars are made" (26).

The Roman Catholic Bishops and Catholic Women's League, the Human Rights Association and the United Church, all came out in opposition to implementation of the Bayda Report, and its facile assertions. But the most significant objections came from native Canadians.

Up until the end of 1977, native peoples within the province, mainly grouped under the Federation of Saskatchewan Indians, engaged in intense debate as to the value of mining on their claimed lands.

In January 1977, Allan Campbell of the Federation had asked the Prince Albert and Meadow Lake district representatives and Chiefs for their views to be made known to Bayda (27). "Will the people have a chance to share in the ownership and control of mineral resources?" asked Campbell. What will be the social impacts of a "very large and sudden development"? What damage will occur to the environment? And what will be the long term problems of "wastes

that remain dangerous for 250,000 years and can be used to poison people or produce atomic bombs"? (27).

A few months later, the Northern Municipal Council (NMC), representing the interests of a large number of Cree and Chipewyan (but not the 6000 Treaty Indians in the area) delivered a singeing attack on the manner in which decisions had been taken about uranium development, and the assumptions underlying the government's position.

Pointing out that the provincial and federal governments had "not even begun serious negotiation about our rights as local people" and that there had been no consultation over the Amok project, the Council saw the Cluff Lake mine as the thin edge of a potentially highly-destructive wedge. While the NMC presentation to the Bayda Inquiry gave no answers, its rhetorical questions provide a model for the criteria by which native people need to judge such "developments". "If development threatens to destroy the positive aspects of our culture and social life, it is always possible the 'economic' cure may be worse than the 'poverty' disease" (28).

Three months later, a consultation group commissioned by Bayda to comment on the socioeconomic aspects of Cluff Lake – "and other proposed uranium developments with emphasis on Northern Saskatchewan" – stressed that native self-determination, specifically the settlement of outstanding land claims, must precede decisions about non-traditional development. It concluded that "While there may be some benefits from wage employment resulting from uranium mining and mineral development generally, the exact proportion of which is uncertain at this point, the remaining aspects appear largely detrimental in nature to the northern region and native peoples. These impacts ... include impacts on their traditional pursuits and activities (*i.e.* hunting, trapping, fishing), social impacts (*i.e.* influx of a large group of non-northerners), cultural impacts, and impacts on community structure" (29).

Within a few months of the end of the Inquiry, representatives of virtually all indigenous groups potentially affected by Cluff Lake and other uranium projects in the Athabasca basin had come out against a green light for Amok.

On 7 October 1977, the Prince Albert and Meadow Lake District Chiefs unanimously adopted a resolution which, while pointing out that they could not come to firm decisions about the impact of further uranium mining on the Indians of Northern Saskatchewan, opposed all future uranium projects until "land selection by Bands with unfulfilled Treaty land entitlement is completed; the Treaty rights of Hunting, Fishing, Trapping and Gathering are guaranteed against violation; the Treaty rights for health, economic development, resource management, are assured; we have the time and resources to carefully examine the many serious questions related to the uranium industry". The Chiefs also announced their refusal to join the Bayda Inquiry, since its terms of reference would "... mean that Indian people and their governments (Chiefs and Councils) have already concluded that further uranium development is an acceptable and desirable course of action... This is not the case." In a pithy condemnation of the narrowness of the Bayda brief, the Chiefs stated: "It is clear to us that the Cluff Lake Inquiry is not dealing with the question of whether or not uranium mining should expand in Northern Saskatchewan, but simply with the question of how" (30).

At an August meeting of Status and Non-Status Indians, and Metis, predominantly from the communities of La Loche, Turnor Lake, Buffalo Narrows, St George's Hill, Dillon and Patuanak, similar resolutions were passed and a ten-year moratorium on all uranium development called for (31).

In December, the Indians of N-22 area – the area immediately affected by Amok's proposals, and inhabited by more than 200 native people – told the Bayda Board that they, too, didn't want to see their land turned into "... a commodity, an object to be exploited for its natural resources". Acting chair of N-22, Jimmy Deranger of Grand Council Treaty No. 8, promised the Dene were "... prepared to fight back like a cornered animal, rather than give up

the freedom we had had for centuries in the north country" (32).

When the Bayda report was finally published, and it became clear that native opposition would not prevail, the indigenous ground began to shift. David Ahenakew of the Federation of Saskatchewan Indians expressed concern that treaty claims – especially for the Fond du Lac and Stoney Rapids Indians – would be held up, but said he had no real opposition to uranium development, so long as "environmental concerns are adequately addressed and land entitlement issues [are] not allowed to go by the board" (33). Although the Treaty and Carswell Lake Dene were still mainly opposed to the project – and continued to be into the early 1980s – because their land claims had not even been recognised, let alone settled (34), other native groups were campaigning for employment at the mine.

Under new surface agreements, Amok had promised to employ at least 50% northern or native people in its operation. By the beginning of 1982, some 40% of the labour force was indigenous (35), and the Federation of Saskatchewan Indians had ceased making waves (35). Lyle Bear, the native labour co-ordinator for Amok/Cluff Mining, was vigorously defending the contribution Amok was making to employment of northerners and justifying its status as a non-Union employer. "If there was a collective bargaining agreement at Cluff Lake" said Bear, "I'm afraid it would be more likely to supersede the surface lease..." (36).

One of the major questions raised by opponents, particularly native ones, to the Amok mine, related to the influx of white workers – with their trail of booze, gambling, whoring and conspicuous spending – that has so blighted traditional communities in the Canadian north. By 1982, Amok was employing the "Seven in – Seven out" option – flying in for a week's work, followed by a week's recreation. But, though the company has made some effort to serve isolated Indian communities in this fashion, even the company recognises that this can amount to a diluted version of the South African contract labour system, with fathers separated from their families, for the sake of cash employment (37).

Even as Lyle Bear was promising that Amok would employ at least 50% native people in the workforce, there were predictions that the company would have to lay off most of its workers, in order to avoid polluting Claude Lake (38). A year later Cluff Mining's own figures showed the shrinking proportion of jobs which had been going to indigenous people. In 1979 56% of jobs at the mine went to northerners (including native employees). A year later the figure was down to 50%, and between 1981 and 1982, as the project reached full production, it dropped to 47%. Only six of the supervisory jobs were reserved for regional inhabitants that year, and another 8 for technicians and clerics: in other words, the majority were employed as the "shock troops" in the most hazardous uranium mine environment in the world (39).

June 1983 saw formal approval being given to Phase II of Cluff Lake – but without the full public hearings which many citizen groups had demanded, and a "highly-technical EIS (Environmental Impact Statement)" which, according to the Regina Group for a Non-Nuclear Society (RGNNS) "even a branch of the government environment ministry found it necessary to criticise" (40).

Most important, the EIS gave the go-ahead for full-blown development of Carswell Lake uranium with virtually no consideration of the continuing impact on indigenous people, or means by which they could control future operations (40).

By this time, non-native opposition to uranium mining (Key Lake moved to full production in 1983) had become clamorous. In a highly-significant, indeed unique, action, the leaders of five major Christian churches called for a moratorium on uranium mining in Saskatchewan as "a significant contribution to world peace" (41).

The Saskatchewan Young New Democrats called for a halt to uranium mining at their 1980 annual convention; the youth wing of the Saskatchewan Progressive Conservative Party also demanded cessation of government spending on uranium mining (42).

However, the provincial NDP – the government party which had approved Cluff Lake in such obscene haste – remained solidly in favour of mining. This is hardly surprising, considering that its reputation as a provider of jobs and company income depended through the years 1979-1982 to a large extent on ensuring that Amok's plans advanced without a hitch. In 1980 – the crucial year that the mine reached full output – various branches of the NDP administration and Amok joined in a virtually criminal collusion, to keep facts on safety violations from public attention, and neutralise an increasingly dissatisfied – and threatened – workforce.

Full details of these scandals were divulged in 1980 by the Saskatchewan independent monthly newsmagazine *Briarpatch,* after Margaret Swanson quit as Amok's radiation health officer and planner. Margaret Swanson had, initially, been prepared to credit Amok with sincerity and diligence.

She stipulated that workers should use only pressurised, air-conditioned, lead-lined cabs, ordered monitoring equipment, safety masks, and both short- and long-term dosimeters. Ms Swanson also instituted a three-month programme of radiation protection education. The company was delighted with her proposals and gave her full authority "to close down the mine if conditions were unacceptable" (43).

But that was before the Bayda Inquiry had delivered its conclusion, and at the time when the facade of overriding concern about safety had to be kept before the public. Soon after the mine was given its go-ahead, Margaret Swanson found her safety measures thwarted, challenged, or simply neglected. Her authority was "whittled away" and, after succumbing to pressure, she left the job.

Much of what the company had promised was incorporated into the 40-page lease agreement with the provincial (NDP) government. There was still hope that this would be implemented, and any violation could be challenged.

Within the next two years, these hopes were

dashed. According to *Briarpatch's* special report:

- Although Amok had agreed to abide by the strict terms of the provincial Occupational Health and Safety Act and its Radiation Health and Safety Act (tougher than federal provisions then in place), there was no strong union (as at Elliot Lake – see Rio Algom) to enforce the implementation of the Acts, and they were imposed on Amok only in the form of a contract. The province could sue for breach of contract, but the company "... could rest comfortably in the almost certain knowledge that its NDP partner was not about to close down the whole operation over the breaking of a couple of regulations" (43).

- When the mine started operating, *none* of the trucks had the agreed insulated or pressurised cabs, there was only one safety mask for every twenty workers, large areas of the ore face were left exposed, it wasn't sprayed to control dust, nor was there any monitoring of the dust. And the education programme agreed with Margaret Swanson was abandoned.

- In June 1980, workers walked off the site when their complaints about cabs went unheeded. They returned after the management gave in, but the largest machine in the pit still went without special protection. At this point, the Labour department's Occupational Health and Safety Branch gave Amok a special exemption.

- Individually-carried dosimeters were allowed to read up to twice the allowable dose, without any preventative action being taken: indeed, so-called radiation control officers openly ridiculed the devices.

- By July, the management's attitude to the workforce, and particularly the large proportion of native workers at the ore-face, "had become a bad joke", with dissidents being told bluntly, "If you don't like it, quit!" Tests were being conducted on the especially dangerous high-grade ore, even though "none of the required health and safety features of the milling operation was in place", there was no ventilation system, and no scanner device to test for radiation after washdown. A second strike occurred, when construction workers at the mill walked out.

- The following month, workers got their grievances together in a huge list which they presented to the Occupational Health and Safety Committee – a somnolent body dominated by the management. After the peremptory sacking of one worker, others formed the Cluff Lake Workers Association. Despite attempts to intimidate them at mine level, the Workers Association managed to send a delegation to Amok HQ in Saskatoon. This consisted of a number of native workers, supported by their white fellows. Unfortunately the next few days proved extremely disappointing – if not a fiasco – as the workers were kept waiting by management, a representative from the United Steel Workers of America [USA] found his overtures rejected, especially by the native member of the three-person monitoring committee, and disillusioned native workers headed north. After minimal concessions had been made, one workers' representative announced a victory, while others returned to the mine, their hands laden with empty promises.

The demand for an independent investigation of complaints was not acceded to (44), and it was left to the Saskatoon Citizens for a Non-Nuclear Society (SCNNS) to make public the five key violations of the Amok management, including:

i) making false entries regarding workers' radiation doses;

ii) refusing to give back pay for the time the workers were away, due to illegal dismissal;

iii) failing to ensure practical safety standards (ie compulsory wearing of safety masks).

The following year, four workers were splashed in the face by yellowcake slurry, while clearing a blockage at the base of a settling tank: they were removed from the job, given hospital tests, then returned, because the amounts of yellowcake they'd swallowed were allegedly "minimal" (45).

This was not the first "accident" to arouse public concern. In June, a 22-litre drum of yellowcake spilled on an Air Canada passenger flight, because it hadn't been properly sealed (46). The Atomic Energy Board of Canada (AECB) withdrew Amok's export licence, only to return it promptly a week later (44), after Air Canada stated that such traffic was a "normal" occurrence (46).

More serious have been problems associated with the tailings disposal at Cluff Lake. The Bayda report acknowledged that the "potential hazards of the tailings pile will remain for several hundreds of thousands of years" (47), but that proper measures to separate out not only the uranium, but also thorium-230 and radium-226 would be "most complicated and expensive" (48). It did not recommend that the technology should be developed to do this – only that "further research" was necessary."

By early 1983, Cluff Mining had to admit that some of the 10-tonne "flower pots" used to contain the radium-226 in tailings were beginning to leak through hair-line cracks caused by northern winters (39). Some 200 of the pots had developed cracks (40). In April, five of the barrels which had simply been stacked on top of each other in an open shed, instead of being buried as recommended by Bayda, tipped over, spilling 2.5 tonnes of radioactive sludge into the soil (49). Amok complacently relocated "a large number of containers *within* the storage area" (author's italics) and undertook to collect any further spillage in "the sump at the end of the storage pad". It promised to investigate "remedial action" during the next spring thaw (50). Although a similar spill is not known to have occurred since, the storage was still causing anxiety the following year, when the Inter-Church Uranium Committee pointed out that "the tiniest fraction of radium-226 can cause bone cancer" and that Amok had certainly not complied with Bayda's requirement for such tailings to be stored in "permanant stable containment" (51).

On 29 September 1983, more than 9000 litres of contaminated water was released after a pipeline broke. Nor was this the first time such a rupture occurred: in November 1981, 19,000 litres tipped onto the ground, when a waterline,

designed to return waste water from the settling pond to the plant for re-use, broke (52).

As this entry of the *File* went to press, it was learned that Maisie Shiell – who had taken issue with both Amok and the Saskatchewan Environment Department over the 1981 radioactive spills – planned to take both the company and the Department of the Environment to court. This was for breaking the Environmental Assessment Act by proceeding with its new gold recovery plant (1) without public assessment of the EIS, an operation Ms Shiell (and others) believe to be hazardous to workers and the environment (53).

Amok's 1987 output was up slightly on 1986, to 900 tons uranium, while the company in March that year began recovering gold and uranium from its Phase 1 leach residues. The first gold bar was poured in April and production by the year end was more than 6000 ounces (55). Permission was granted for the facility by the AECB and the Saskatchewan Department of the Environment (56).

In 1988 uranium output at Cluff Lake was somewhat higher than in the previous two years (1016 tonnes) and the life of the pit was put at about twenty years at current output (57).

Further target: Cluff Lake uranium is tranpsorted (as of 1983) by North Star Transport and Sinco Truckline to the Reimer terminal in Saskatoon (54).

Contact: Regina Group for a Non-Nuclear Society, 2138 McIntyre St Regina, Saskatchewan, S4P 2R7.

Saskatoon Citizens for a Non-Nuclear Society, Box 8161, Saskatoon, Saskatchewan.

Inter-Church Uranium Committee, Box 7724, Saskatoon, Saskatchewan, S7K 4K4.

Carswell Lake Dene Support Committee (1980), 134 Avenue F South, Saskatoon, Saskatchewan.

Saskatchewan Environmental Society, Box 1372, Saskatoon Saskatchewan S7K 3N9.

Further reading: Miles Goldstick, *Voices from Wollaston Lake; Resistance Against Uranium Mining and Genocide in Northern Saskatchewan*, Earth Embassy and WISE, Vancouver, Stockholm and Amsterdam, 1987.

References: (1) *MAR* 1987. (2) *MAR* 1986. (3) *FT* 31/7/79. (4) *EPRI Report* 4/78. (5) *Foreign Uranium Supply Update,* 1980, Rockville Md., NUS Corp, see also *MJ* 18/9/81. (6) Neff. (7) *MIY* 1987. (8) *FT* 10/9/79; SMDC Annual Report 1981; Public Accounts, Province of Saskatchewan 1981. (9) *MJ* 28/9/84. (10) B Merlin, *Canada's Uranium Production and Potential,* Uranium Institute, June 1977. (11) *MJ* 20/3/87. (12) *EPRI Report* 1981. (13) Energy File, Vol. 2, No. 4. (14) *MJ* 4/7/80. (15) *Yellowcake Road.* (16) *FT* 13/6/79. (17) *The Economics of Uranium in Saskatchewan,* Inter-Church Uranium Committee, Saskatoon, June 1982. (18) *Nuclear Newsletter,* Saskatoon, Vol. 3 No. 11, 21/11/79. (19) *Toronto Globe and Mail,* 14/8/79. (20) *FT* 10/9/79. (21) *MJ* 10/4/81. (22) *Should Uranium Stay in the Ground?* broadsheet published by the Saskatoon Environment Society, Saskatoon, 1977. (23) *Star-Phoenix,* Saskatoon, 9/6/78. (24) *Star-Phoenix* 8/6/78. (25) *Nuclear Newsletter,* Saskatoon, Vol. 2, No. 10, 21/6/78. (26) Roger Moody, "Uranium Mining to go ahead in Canada", Gemini News Service, London, GN 2342, 1978. (27) Letter from Campbell, Federation of Saskatchewan Indians, Regina, 21/1/77. (28) A presentation by the Northern Municipal Council for "Overview" – Phase 1: The Cluff Lake Board of Inquiry, Regina, Saskachewan, 28/4/77. (29) *Progress Report: Commentary on Socio-Economic Impact Analysis of The Cluff Lake and other Proposed Uranium Developments with Emphasis on Northern Saskatchewan,* prepared for Cluff Lake Board of Inquiry by Hildebrand, Mc Neal, and Culbert, 21/7/77. (30) Joint Statement by the Meadow Lake and Prince Albert District Chiefs, 7/10/77. (31) *New Breed* No. 2, Association of Metis and Non-Status Indians of Saskatchewan, AMNSIS, Regina, Sept-Oct 1977. (32) *Natotawin,* Beauval, 1/12/77. (33) *Star-Phoenix* 9/7/78. (34) *Natural People's News,* London, No. 2, Spring 1980: see also *Dene Yahtee* No. 4, Regina, 11/12/79. (35) Heather Ross, personal communication to CIMRA, London, 13/1/82. (36) *Metis Newsletter,* Yellowknife, 18/12/81. (37) *Business Review,* Regina, Spring 1982. (38) *Star-Phoenix* 16/3/82. (39) *Nuclear-Free Press,* Peterborough, Summer 1983. (41) *Christian Leaders call for a halt to Uranium mining for the sake of peace,* pamphlet, Inter-Church Uranium Committee, Saskatchewan 1983. (42) *Kiitg,* Amsterdam, May 1980. (43) *Briarpatch,* Saskatchewan, Special Supplement, "Amok and the NDP: Production for Profit, Not for People!" by Murray Dobin, 16/10/80. (44) *Newsletter,* RGNNS, 8/80. (45) *Star-Phoenix* 27/11/81. (46) *Star-Phoenix* 28/6/80. (47) Bayda Inquiry Report, 1978, p. 110. (48) Bayda Report, p. 111. (49) *Briarpatch,* July/Aug 1983. (50) Letter from BM Michel, Senior and Operations Vice-President, Amok Ltd, to Saskatchewan Land Protection Branch, Saskatoon, 2/5/83. (51) Document by Bob Regnier, Inter-Church Uranium Committee, Saskatoon, 11/1/84. (52) *Star-Phoenix* 18/11/81. (53) *Newsletter,* Saskatoon Environmental Society, Apr/May 1987; *Toronto Star* 12/4/87; see also WISE Communique, N.C. 277, Amsterdam, 24/7/87. (54) *Uranium Traffic,* Saskatoon, 1981, update July 1983. (55) *MAR* 1986. (56) *MAR* 1987. (57) *Greenpeace Report* 1990.

31 Ampol [Petroleum] Ltd

In an extraordinary game of bid and counter-bid between 1979 and late 1980, Pioneer Concrete finally won control of Ampol Petroleum, with the result that it now controls the Nabarlek uranium mine in Australia's Northern Territory. (References to this battle royal are given below) (1).

Ampol's main revenue now comes from its uranium holdings, rather than oil (2): uranium had become Ampol's most profit-making sector by mid-1982 (3).

Ampol is Australia's only home-based petrol company; it markets petrol under its own name. During 1980, Ampol Exploration's oil sales improved (4).

It is involved with TVQ-O, the Brisbane TV station formerly owned by Ansett (5).

Ampol owns three highly profitable tourist resorts in the Northern Territory of Australia. At Katherine Gorge it has (according to the Sydney-based Aboriginal Land Rights Support Group ALRSG) debased and exploited the Djuan Aborigines, who live in corrugated iron

```
     Pioneer Concrete Services LTD (AUS)
                    88.2%
         AMPOL [Petroleum] LTD (AUS)
                        14.6%
                49.17%
         Ampol Exploration Ltd (AUS)
```

shacks just outside the National Park gates, while "Ampol rakes in about US$15,000 each week from its monopolised accommodation-kiosk-boat tour business" (6).

In 1987, Pioneer drastically restructured its operations, after acquiring 41% of Giant Resources. As a result, Pioneer, which now owned 88% of Ampol, expected to own between 90% and 100%. The restructured group also has a share in Noranda Pacific.

In March 1988, Pioneer announced that it had made an agreement in principle to explore for more uranium in the Nabarlek area, now that the first of the contemporary Australian uranium mines had finished milling (7).

By 1988, Ampol (now 100% controlled by Pioneer International, as the parent company renamed itself) had an active drilling programme for oil in off-shore Australian waters and Papua New Guinea (at Iagifu-Hedinia) as well as in China (8). That year, Pioneer announced that it would be selling off all its mining interests in Australia and Canada in order to concentrate on its concrete, cement and aggregate interests, as well as Ampol's petrol refining and marketing (9).

For effects of mining at Nabarlek and the resistance by Aborigines and others, see Queensland Mines Ltd.

Contact: Aboriginal Land Rights Support Group, 262 Pitt St, Sydney, NSW 2000, Australia.

References: (1) *FT* 26/9/79, 16/10/79; *The Age*, Melbourne, 29/11/79; *DT* 8/10/80; *FT* 1/11/80, 6/11/80, 7/11/80, 26/2/81. (2) *FT* 27/8/81. (3) *FT* 25/8/82. (4) *FT* 21/10/81. (5) *FT* 8/10/80. (6) ALRSG *Newsletter* (Sydney) 8/82. (7) *FT* 25/3/88. (8) *The Australian* 27/12/87. (9) *FT* 27/9/89; *MJ* 29/9/89.

32 The Anglo-American Corp of South Africa Ltd

Stepping off a spaceship, making a whirlwind tour of global mining, then examining the index to *The Gulliver File*, a visitor from outer space.may be forgiven for assuming that – if spaceship earth is fuelled by uranium, its banking system solidly based on gold, its leaders of fashion luxuriously bedecked in diamonds and platinum, and the most crucial decisions about its mineral resources taken in London and Johannesburg – only two names need be recorded to take back to Mars or Pluto. One of these is RTZ – in terms of market capitalisation and influence, by far the most powerful mining conglomerate this side of the solar system.

The other is Anglo-American (AAC) – in terms of value (for its assets and production) a bigger swimmer in the milky way than RTZ (1), but handicapped by its identification with the apartheid state.

In a perfect world for mining, perhaps there would be only one name our alien interloper would need to retain: RA (Rio-Anglo, not to be confused with Rhoanglo) or ATZ (Anglo-Tinto-Zinc). Or perhaps just Mine Inc *(All mine, indeed!)*.

In fact the links between the AAC and RTZ go back a long way, and are deeper than either corporation would admit. The extent of AAC's holding in the British corporation has never been revealed, but it is often contended that the Oppenheimers and Rothschilds have considerable secret or indirect holdings in RTZ. Charter Consolidated certainly had an admitted 8% through the 1970s (2). They are also linked in several important projects – notably the Argyle diamond mines on Aboriginal land in the Kimberleys of Western Australia, and the Palabora copper uranium mine in South Africa. RTZ and AAC front companies share domiciles in the town of Zug, Switzerland, through which they launder sensitive wares. They co-operated in evading UN and British sanctions against the Ian Smith régime in Rhodesia, and have combined to cheat the Zimbabwean and Botswanan governments of revenue from nickel operations,

77

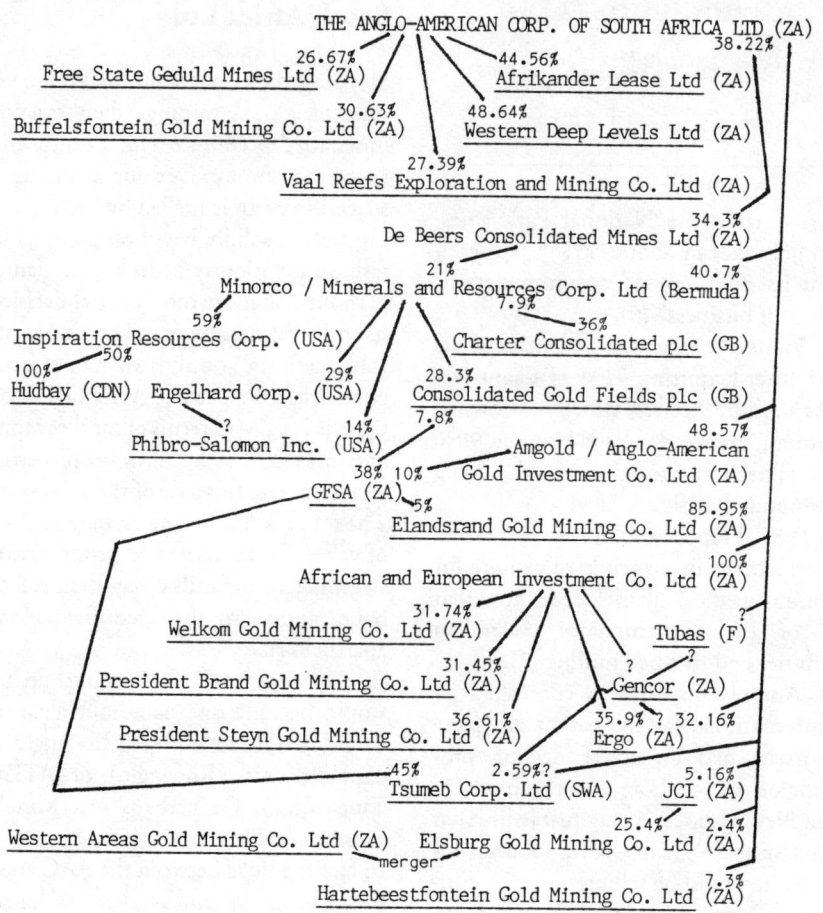

THE ANGLO-AMERICAN CORP. OF SOUTH AFRICA LTD (ZA)

26.67% — Free State Geduld Mines Ltd (ZA)
44.56% — Afrikander Lease Ltd (ZA)
38.22%
30.63% — Buffelsfontein Gold Mining Co. Ltd (ZA)
48.64% — Western Deep Levels Ltd (ZA)
27.39% — Vaal Reefs Exploration and Mining Co. Ltd (ZA)
34.3% — De Beers Consolidated Mines Ltd (ZA)
21% — Minorco / Minerals and Resources Corp. Ltd (Bermuda) — 40.7%
7.9%
59% — Inspiration Resources Corp. (USA)
36% — Charter Consolidated plc (GB)
100% — Hudbay (CDN)   50%
29% — Engelhard Corp. (USA)
28.3% — Consolidated Gold Fields plc (GB)
7.8%
? — Phibro-Salomon Inc. (USA)   14%
48.57% — Amgold / Anglo-American Gold Investment Co. Ltd (ZA)
38% 10% — GFSA (ZA)   5%
85.95% — Elandsrand Gold Mining Co. Ltd (ZA)
100% — African and European Investment Co. Ltd (ZA)
31.74% — Welkom Gold Mining Co. Ltd (ZA)   Tubas (F)   ?
31.45% — President Brand Gold Mining Co. Ltd (ZA)   ?   ? — Gencor (ZA)
36.61% — President Steyn Gold Mining Co. Ltd (ZA)   35.9% ? 32.16% — Ergo (ZA)
45% — Tsumeb Corp. Ltd (SWA)   2.59%?   5.16% — JCI (ZA)
Western Areas Gold Mining Co. Ltd (ZA)   Elsburg Gold Mining Co. Ltd (ZA)   25.4%   2.4%
merger
7.3% — Hartebeestfontein Gold Mining Co. Ltd (ZA)

through an elaborate "transfer pricing" scheme (see RTZ) (3). More recently, Rio Tinto Zimbabwe (Riozim) and Anglo-American Zimbabwe (Anzim) signed a JV agreement with Robertson's subsidiary Plateau Mining, to exploit platinum reserves in the central Hartley complex of the country's Great Dyke region (421). What distinguishes Anglo-American from RTZ is not simply their origins (the one in north American and British investment in South Africa's fabulous diamond and gold wealth, the other in older metals and uranium in Europe and Australia) but their corporate aims and style of management. Both diversified extraordinarily at a certain point in their development – AAC some years before RTZ, but along simi-

lar lines. However, while RTZ has tightened its portfolio and consolidated its hold on a large proportion of western world mineral resources, AAC's diversification is hardly at an end.

AAC is quite clearly dominated, as it always has been, by one family, the Oppenheimers, whose pretended liberalism – a highly developed, finely tuned mask on its desire to maintain white dominance of South Africa's natural resources – has been widely trumpeted around the world. There are numerous heads of state or trading blocks who have been graced with the apparition of an Oppenheimer stepping from his private jet to greet them. And from the Russians to the ANC (African National Congress) of South Africa, there is hardly one politically important dinner table at which they have not sat. While RTZ's directors keep the lowest of profiles, AAC's maintain the highest. While AAC's "hands on" management ensures that every major group within the Oppenheimer domain is aware of, and fits into, the global strategy, RTZ purports to give comparative freedom of decision and operation to its subsidiary and associate companies. In the final event, however, this difference may be a cosmetic one. Both corporations try to gain management control, even of projects in which they nominally hold a minority share (AAC in the Morro Velho gold mine, for example [see below]; RTZ in Rössing and Palabora). They are led by the same breed of men (even where, as in South Africa, they may differ marginally over such matters as limited black and coloured elections, or when to recognise the NUM). From his/her vantage point in outer space, our Martian or Venusian would be hard put to it, to distinguish between the human suffering caused by AAC's violent confrontations with the black NUM and RTZ's long-standing denial of Aboriginal land rights in Australia; or to detect a real difference between the environmental degradation caused by the Rössing mine and by Consolidated Diamond Mines' (CDM) diamond excavations along the Skeleton Coast.

The Anglo empire

"We're living in a country controlled by cartels and lobbyists and a government only too keen to satisfy the lobbyists at the expense of 25 million people." This statement was made in 1986, not by the ANC, nor a leader of COSATU, but Robin McGregor, the editor of the country's *Who Owns Whom.* "The distribution of wealth in this country is zero," declared McGregor (4). Three years before it had been revealed that AAC companies owned no less than 56% of the Johannesburg stock exchange's total share value. That year, directly administered companies within the Oppenheimer conglomerate were worth R10 billion at a conservative estimate: non-administered companies even more (R17 billion) (5). AAC produced nearly half of

South Africa's gold, 43% of its uranium and 25% of its coal (6). Within the next year or so, its investment was to increase by 30% and its trading profits improve by nearly half (46%) (7). By the end of 1984, the corporation owned 69% of the total capital invested in the country's mining; around a quarter of all the assets held by South Africa's "Top 100" non-state corporations. It had an acknowledged 831 subsidiaries and associate companies (of which nearly 200 were outside South Africa), and investments in just over 500 others (8).

AAC's diversification outside of South Africa and gold and diamonds took place in the 1960s, as the company used a hugely increased cash flow from its monopoly over mining (9) to buy into manufacturing, property and finance companies (10). By the end of the decade, the corporation's financial muscle was such that one academic, PW Hoek of the University of Pretoria, warned in a secret report that AAC was a threat to the state itself, especially in view of the low-tax position its operations enjoyed (11). Over the next decade, AAC (mainly through Minorco) acquired more than one hundred companies worldwide (12). It came to control three of the top four South African mining houses, six of the top ten finance houses, the largest investment Trust, the second largest property company, the second largest merchant bank, the largest transport company and the country's fastest growing car manufacturer. It was second only to the government as an employer of labour (13). Through its dominating interest in Johannesburg Consolidated Investment (JCI), AAC acquired a 40% equity in the Argus group of newspapers (the Johannesburg *Star* and other publications) and in 1971 control of the South Africa Association of Newspapers (now Times Media Ltd) (3, 10, 14). Through Minorco's control of IRC (see below) AAC later came to own a 71% interest in Madison Resources Inc, which owns 35% of the Arcata Corp – probably the biggest printers in the USA (used by Simon and Schuster, McGraw Hill, *Time, Newsweek, Readers's Digest* and others) (15). Through the Bowater group of Britain, AAC has gained control of Mondi

Paper, the largest pulp, paper and board producer in South Africa (3, 10). It also owns 50% of HL and H Timber Holdings Ltd – a supplier to Mondi, and the mining industry's main provider of pit props. Through Natal Tanning Extract, AAC owns or leases more than 100,000 hectares in south-east South Africa for forestry, sugar cane and tanning extract (3). AAC's steel, iron and engineering operations include Scaw Metals – a major supplier to the mining industry and parastatals like Escom and South Africa Railways (3); Boart International – which uses De Beers industrial diamonds to manufacture a wide variety of tools and equipment (3); Highveld Steel and Vanadium Ltd, established with the co-operation of Newmont and Davy (a major British engineering company) (see Davy McKee) (10). A JCI-administered company, Consolidated Metallurgical Industries (CMI) runs the world's largest ferrochrome production plant, drawing on chromite supplied by other mining companies (3, 16). In early 1989, Highveld and Samancor (Gencor) formed a JV (probably with Taiwanese capital) to manufacture ferrochrome and thus contributing further to the western world's dependence on South African ferrochrome output (17).

African Explosives and Chemical Industries (AECI) is run jointly by AAC and Britain's Imperial Chemical Industries (ICI) (10), although it has its origins in an outfit established by De Beers in 1924 to supply explosives to the mining industry. It is now the world's largest producer of commercial explosives and has played a crucial role in arming South Africa's security forces and army (3).

So, if you are shot in South Africa, the chances are the guns were primed by AAC. If you take a pint in a local bar, the chances are it was brewed by South African Breweries (SAB), a highly diversified company which runs the Holiday Inn hotel chain, Amalgamated Retail and Edgars Stores, and which is indirectly controlled by AAC through the Premier Group, a food milling, drugs and fisheries conglomerate in which AAC and JCI hold 47% (3, 18). If you're white, you may be paying a mortgage to Amaprop – one of South Africa's largest property owners. And if you have the "fortune" to live in a stone house, the bricks almost certainly came from Tongaat-Hulett (AAC: 38.4%) a diversified industrial group into textiles, food, aluminium processing (19) – and one of the world's biggest manufacturers of bricks (3). If you work down one of AAC's mines, your meat, vegetables, fruit and other products may well come from Amfarms, the only food company administered by a mining house (3). If you have the even greater fortune to own a car, it will either be a Toyota, or come from Samcor, the country's second biggest manufacturer: Samcor inherited the interests of Chrysler in South Africa, and more recently the Ford Motor Co (when US pressure forced Ford to pull out in 1987); it is owned 76% by AAC and AMIC (3, 20), with 24% in the hands of a "worker-controlled" Trust (21).

In 1983, AAC also owned 43% of Asea Electric SA (25% was owned by the Swedish parents) (22): that year part of the holding was sold, so that AAC could gain a foothold in the electronics industry via Altech of Japan (23).

Many of these companies are administered by the Anglo-American Industrial Corporation (*AMIC*), although some of the most powerful (ie SAB) are not directly controlled in this fashion. In 1979 *AMIC* alone had more than a hundred subsidiaries and associates and by the early 1980s was controlling more than the combined GDPs (Gross Domestic Products) of the Southern African Development Co-ordination Conference (SADCC) (8). *AMIC* has been crucial to the corporation's diversification and its vertical integration, especially as the threat of sanctions became a reality and AAC's more visible exports stood the risk of being targeted (3).

Control of the AAC/De Beers empire undoubtedly lies with the Oppenheimer family, even though it nominally (through E Oppenheimer and Son) holds only 8% of AAC. Together, the family stake and De Beers's holding in AAC makes their combined position unassailable. Though AAC's stake in De Beers is appreciably smaller, its world standing, access to governments, and the huge funds on which it can call, make it highly unlikely that De Beers would try to slip from under the Oppenheimer thrall. Other institutional shareholders in both companies have negligible power.

AAC is essentially a mining finance house with direct control over the funds held by holding companies or group subsidiaries (24). Though such ownership is officially well under 50%, through hidden subsidiaries and holdings, cross-linked directors, pension fund investments and other devices, AAC has a hole-shot position. Thus is created "an endless circle of ownership, participation and shared facilities" (25). Administered companies – even where AAC's holding is well under half – are implicitly dominated by AAC; "controlled associates" in theory stand at an arm's length from AAC but in practice consent to fit in with the dominant *modus vivendi;* related companies are those in which AAC is neither the dominant shareholder (Harry Oppenheimer once defined this as below 30% but, depending on other holdings, the threshold could be even lower) (24, 25), nor runs the management (3).

"The ability of Anglo to maintain control over many companies without taking a majority shareholding depends to a large extent on its position in the financial markets" (3). And here AAC is unbeatable. Ernest Oppenheimer founded his own bank, Union Acceptances, and by 1968 this was the seventh largest in the apartheid state (10). Morgan Guaranty, Citibank, the Union Bank of Switzerland, Banco Commerciale Italiano, Deutsche Bank, Banque de Paris and Standard Chartered have all invested in AAC's operations – the last named being a key agent in the sale of gold from the company's mines (13). But it is the link with Barclays National (Barnat) that has been most important in recent years. Under pressure from the anti-apartheid and disinvestment movement at home, the British bank started to hand over control of its South African assets in 1985. When the Bank finally pulled out the following year, AAC acquired 22.5%, De Beers 7.5% and

Southern Life Assurance (an AAC company) another 25%. More than twenty billion Rand in assets, and around 17 billion Rand in deposits, passed to AAC for an eightieth (527 million Rand) of that figure (3).

Diamonds: the Emperor's old clothes

"De Beers sells illusions," the *Economist* declared in 1987 (26). Once the world wakes up to the fact that little pieces of compacted carbon are no more than vanity mirrors, while their "industrial" version can be readily sustituted by other, humanmade artefacts, De Beers will go down the tubes. Or, more appropriately, down the kimberlite pipes. But that day is a long way off. This is not simply because of the protection De Beers has enjoyed from the South African Treasury (whereby the corporation has supported long-term state loans, and the régime permitted De Beers to raise funds abroad to finance overseas developments) (27). Nor is it simply due to the huge amount the company spends on perpetuating the Grand Illusion (more than a hundred million dollars a year in advertising), and the willingness of old customers (particularly in the USA) and newer ones (particularly in Japan and Taiwan) to succumb to it. It is also because, bringing down De Beers would threaten a trade on which a number of third world countries, particularly India, Zaïre and Botswana, have come to depend.

Attempts have been made to get from under the De Beers sway but, to date, they have failed. The most notable example is that of Zaïre which, after pulling out of the De Beers Central Selling Organisation (CSO) in 1981 because De Beers was taking too heavy a cut, was forced to rejoin two years later (28). The CSO flooded the Indian market with cheap gems, slashed the price of boart (industrial diamonds) (29), promoted smuggling which funnelled off perhaps half of the country's output (30, 117), and seduced CRA, another prospective competitor, into the CSO by taking the vast majority of output from the new Argyle diamond mine (see below).

The Zaïre lesson has been well learned by other countries which might be tempted to go it alone. Botswana has the world's richest single kimberlite pipe at Jwaneng, and 15.5% of the world's output of diamonds (31), making it globally the world's third largest producer overall, and the second most important gem producer (31). It is theoretically in a good position to market its production outside of the De Beers cartel. However, in 1987, after twenty years of association with the CSO and threats to break away, it almost literally climbed into bed with the Oppenheimers. It secured a 2.6% stake in De Beers itself, with two members of the board, and 50% control of Debswana, with the South African oligopoly (32).

Angola (which recently dissolved its diamond company (33) dominated by De Beers), Sierra Leone – and Argyle Diamond Sales itself (34) – have all been rumoured to be taking steps to sell their production outside the CSO. But, so far, the monopoly is holding. The Indian government continues, much to its embarrassment as an anti-apartheid state, to sell most of the output from its numerous cutting workshops to the CSO (35), while the Beijing régime in 1986 forged secret links with De Beers through a London-based company, Chichester Diamond Services (36). Israel – a very important diamond cutting and marketing centre which takes 30% of De Beers' diamonds directly (37) – welcomes the Oppenheimers with open arms (33).

It takes De Beers's covert relationship with the Russians to best illustrate how AAC/De Beers can achieve what no other cartel has ever been able to effect, for such a long period of time. In 1966 large consignments of high-quality diamonds began appearing on the international market from Siberian mines. These posed an immediate threat to De Beers, which soon despatched one of the clan (Sir Philip Oppenheimer) to offer the Russians participation in the cartel. The pact was sealed in conditions of great secrecy (38). Ever since then – and despite occasional hiccups – the Russian State Diamond Trading Organisation has sold a large portion of its output (worth US$1 billion annually) through the CSO. After the Sharpeville massacre in 1960, the Russians officially boycotted South African trade and officially De Beers declared the contract annulled (40). In fact, it was only public admission of the link which disappeared from view. Ever since then, contracts have been maintained between the world's oldest opponent of apartheid and South Africa's chief financial architect. The Oppenheimers have frequently visited Moscow, Russian mining engineers have looked at De Beers operations in South Africa, and De Beers miners have gone to Siberia (a less fractious journey, one would expect, than that of most dissidents). At the annual "Platinum Dinner" held in London's Savoy Hotel, the two parties meet to renew their diamond sales agreement, and (almost certainly) co-ordinate their platinum sales, as the world's two largest producing countries of this precious metal (3, 38).

The prospect of international sanctions, especially in the USA, which loomed large in 1987, has certainly worried De Beers more than the remote chance that the USSR would cancel its lucrative contracts with the CSO. US investigations into the cartel also promise to be far more thorough and damning than they were in Britain (41) at the time when AAC (through Minorco) attempted to take over Consolidated Goldfields (see below). However, for some years, De Beers has taken steps to ensure that its diamond sources and shipping routes are thoroughly disguised (42). De Beers diamonds can reach the USA through intermediary companies in Switzerland, Israel (43) and Britain, and innocuous-sounding staging-posts like the Isle of Man. (In 1988 De Beers set up three Manx companies registered in Zürich with no apparent connection to De Beers, called Pacini Ltd, Diamanx Products Ltd, and Manxtal Cutting Tools Ltd) (44). Through deals like Debswana, it has increased third world countries' dependence on the CSO, and deepened the reluctance of pro-sanctions campaigners to damage fragile economies (45). It has also channelled money (impossible to say how much) to governments that are critical of sanctions. For example, in 1987 it donated £ 47,500 to the Conservative Central Office (CCO): this was laundered (like its diamonds) through the Diamond Trading Company (Pty) Ltd in London, and the BUI (46).

De Beers was founded in 1880 by that epitome of imperialism, Cecil Rhodes, as an amalgamation of the undertakings of De Beers Mining Co Ltd and Kimberley Central Diamond Mining Co Ltd Within a decade, thanks to the financial assistance of the Rothschilds, Rhodes had built up the country's strongest mining conglomerate, and controlled the most important diamond discoveries in the "new world" (29, 47). Rhodes died in 1902, the year Ernest Oppenheimer arrived in South Africa. De Beers had been expanding in South Africa itself (particularly at Bultfontein and Koffiefontein) and Oppenheimer, through Consolidated Diamond Mines (CDM), was to buy up the world's biggest diamond deposits in German South West Africa (Namibia) – with the financial help of the ubiquitous JP Morgan (63). After being elected to the De Beers board, and later (1929) becoming its chairman, Oppenheimer bought a 30.4% interest in De Beers, while De Beers took a 33.1% interest in AAC (29). In the 1970s, AAC and Rand Selection Corporation (Randsel) merged their holdings (48, 49), thus increasing the De Beers share of AAC. In 1982, AAC bought 10 million shares in De Beers (50). Since then, AAC has held 34% of De Beers, and De Beers has held 38% of the Anglo group.

The CSO was set up in 1930, as the marketing arm of De Beers. Although 30% of De Beers's assets lie outside the diamond industry, most of these are in AAC or Anglo subsidiaries and associates (such as Minorco). De Beers Industrial Diamonds (Ireland) markets all the company's natural and synthetic industrial diamonds, while its Industrial Diamond Division manufactures them (in South Africa, Eire and Sweden) (3). De Beers Consolidated Mines itself has numerous mines in South Africa and Lesotho, and is prospecting throughout the continent (51). Consolidated Diamond Mines

of South Africa (Pty) Ltd runs the Namibian mines (51).

De Beers's own diamond production has waned in recent years, especially in South Africa (52), though tending to pick up in Namibia (53). This has increased the urgency with which De Beers has gone after diamonds elsewhere. The year 1980 was a crucial one in the history of the company (39). An unprecedented recession set in and sales only picked up gradually over the next five years (54). De Beers cut the number of its clients, reorganised its management at CSO, and became more flexible in its contracts (for example, not insisting that clients sell only through the CSO). Although record profits and prices were recorded in 1987 (58, 84) and the following year sales had rocketed by 88% (34, 56), an increasing proportion of the output was coming from mines other than those controlled by De Beers. A study carried out by the Honolulu-based East-West Center in 1988 concluded that the De Beers share of the world market – variously estimated at 74%-85% (57, 59) – would have dropped to 63% by the turn of the century (57, 58).

However, said the report's authors, the cartel would continue to be effective, even as the twenty-first century loomed round the corner: such was the compulsion of voluntarism – by which producers consented to be tied into the CSO system – and such was the power of De Beers's unique classification and stockpiling systems (58). By and large, De Beers pays producers incentive prices when market prices are weak, and, in order to avoid the over-production that could then result, runs a quota system among its key suppliers. It dominates the trade in rough diamonds, through buffer-stock managers in Kinshasha, Antwerp and elsewhere, who can buy up excess supplies from independent traders when the market is slack. The middle people in the diamond trade – the rough-diamond cutters – are also held in check, by an extraordinary and elaborate system of "sights" held ten times a year, usually in London, but sometimes in Johannesburg, Lucerne or Kimberley. There, a buyer must accept or reject a complete box of stones selected by De Beers: buyers cannot select at will (60). The system is not foolproof, of course – the near-collapse of 1980 demonstrated that. Nor could it be replicated in any other industry, such is its reliance on what is essentially a mirage. Indeed, if all the customers of De Beers one day decided their necklaces and tiaras weren't worth paste, and put them up for sale, the cartel would sink for ever. "Selling a diamond is the last thing De Beers wants you [the public] to do", it has aptly been said (29). In that event the young boy would indeed be vindicated, and the Emperor revealed without any clothes.

Anglo ... bestrides the world ...

"Economically [Southern Africa] is still in the hands of the successors to BSAC, ie Anglo-American/De Beers who control most of the nickel, coal and diamond production, about a third of the ferrochrome output and a bit less of copper and cobalt" (61).

... in Africa

It was inevitable that, as the Oppenheimer dynasty took root in South Africa, it should seek investments elsewhere in the British empire. Thus, AAC expanded into Rhodesia (later the Central African Federation) and what are today Botswana and Namibia. In Zimbabwe alone by the mid-1980s AAC owned "the entire sugar industry apart from two small refineries, dominates mining, and has commercial interests in four commercial banks, six finance houses, three building societies and numerous other industries ... exerting enormous political pressure too. Anglo-American with its web of interests in the economy has been particularly successful in lobbying the Government. It has persuaded the World Bank and the cabinet to finance a new power station ... " (62).

AAC had managed to gain control of all three of Botswana's major mines by the 1970s, as well as its breweries, freight services, and milk marketing (3, 24), while, on the eve of independence of Northern Rhodesia, AAC companies – and

the British South Africa Company (BSAC) (in which AAC held an important share) – virtually held the Zambian mineral economy in hock. After independence, the country's resources were drained by creating Charter Consolidated out of BSAC, Central Mining and Consolidated Mines, and the setting up of Zamanglo (later Minorco) in Bermuda (422).

"Within the corporation" wrote Greg Lanning and Marti Mueller in their classic study of mining in Africa, "the whole of central and southern Africa is regarded as an Anglo fiefdom" (24).

First investment by the Oppenheimers outside South Africa took place in the Rhodesian copper belt in 1924. Four years later, Rhodesian Anglo-American (Rhoanglo) was formed to exploit the "first and biggest" copper field in the British empire (3). Long before the setting up of the Central African Federation, out of Northern and Southern Rhodesia and Nyasaland in 1953, Ernest Oppenheimer had supported the amalgamation of the northern copper belt with the relatively impoverished south, in order to maintain white dominance of the economy in both parts of the region. Three years before this, Oppenheimer had shifted the offices of Rhoanglo, Rhokana, Nchanga and its Broken Hill operations to Salisbury. Within a few years, the AAC group had acquired the Wankie colliery, set up Rhodesian Alloys (with British steel maker John Brown), taken over the state's Iron and Steel Corporation, and diversified into sugar, citrus and cement (24, 63). The move also reduced considerably the British tax liability of the companies involved (63).

Oppenheimer made much of the facilities afforded AAC's black workers in the Federation, and contrasted them with the discrimination and penury to which apartheid had subjected them further south. His motives, to say the least, were mixed: building black mining villages and allowing the operation of black unions served as a safety valve for nationalist agitation. The one thing which Ernest and son Harry could not stomach was black majority rule and "the transformation of Rhodesia into an exclusively African country" (64).

As pressure for independence mounted within the Federation during the late 1950s, AAC, which had previously supported the recalcitrant imperialist lobby at home and abroad, began switching its support from the white United Federal Party, to the nationalists in the north (65). In Southern Rhodesia, however, the story was different: AAC, like RTZ (through its large Empress Nickel mine), and another British conglomerate, Lonrho, stood to benefit from Ian Smith's unilateral declaration of independence (UDI) and either opened new mines or expanded existing facilities, while the régime clamped down on capital and dividends remittance (24). AAC developed its Bindura nickel mine during this period: it was producing 630,000 tons of nickel by 1971 (66). At the same time, much of the mineral production found its way out of the colony, despite supposed UN sanctions. Among the many fronts set up by the mining companies to disguise this theft was Anglo's Swiss agency, SALG.

But it is AAC's role in supplying oil to Smith's band of outlaws which shows its true colours during the fifteen years of UDI. Using its Freight Services (now Rennies Freight Services or Renfreight) subsidiary, it transported Shell oil from Mozambique to Rhodesia and, when independence for the Portuguese colony threatened, set up a complex swap arrangement with SASOL in South Africa, thus freeing South African oil for Rhodesian consumption (3, 24). AAC's role was fully revealed (or as fully as might have been hoped) in the British government's Bingham Report of 1978. By this time Freight Services had spawned no less than 50 subsidiaries and acquired the respected British forwarding agency Davidson Parks and Speed (without the knowledge of its British directors) (67, 68).

When it was clear that the Smith régime was soon to fall, AAC – not surprisingly, and using the pragmatism it was later to adopt in its homeland – began courting the black nationalist leaders of the future Zimbabwe. Oppenheimer took many behind-the-scenes initiatives and ferried black leaders to meetings with Smith in his personal jet (24).

AAC's penetration of Zimbabwe's economy has increased since independence in 1980 – notwithstanding the early determination of Prime Minister Robert Mugabe to send packing those companies (RTZ, AAC and Union Carbide in particular) which had made fat profits under the Smith regime (3). In 1980, AAC had directors on the board of 82 Zimbabwean companies and exerted control over almost half of them (3). Three of the country's largest concerns – Bindura Nickel, Zimbabwe Alloys and Hippo Valley – are administered by Amzim. Moreover, AAC operates the country's only coal mine, the Hwange (Wankie) colliery, and has interests in a wide range of other products and services, including agriculture, timber, food-processing, milling, ethanol and financial services (3). Through RAL Holdings and AECI of South Africa, it controls the production and distribution of fertiliser.

In concert with RTZ and Centametall in Zug, Switzerland, AAC has also been involved in an elaborate system of "toll refining" the country's nickel production which, through transfer pricing, has resulted in the country losing perhaps 20% of its mineral income (3).

The nickel "scam" involved Botswana's Selebi-Phikwe copper/nickel mine, recently called "the biggest haemorrhage" in Anglo's body, with accumulated debts of £300 million by 1988, and its sales revenue amounting to less than the annual charges on its deferred interest (69). Owned by Botswana RST (a JV between AAC and Amax) the mine has not paid taxes to the Botswana government, and royalties were suspended in 1981. The ore was initially sent to Amax's refinery in Louisiana, to bring that plant up to capacity, much to the chagrin of the Botswana government which naturally wished the refinery built closer to the mine (3). In 1988, the ore was being shipped to Falconbridge's Norwegian refinery (70).

Botswana's only major new mining development for the next few years (37) is, not surprisingly, partially owned by AAC. The government has brought in Soda Ash Botswana (SAB) and AECI to develop the Sua Pan soda ash deposits. AECI (jointly controlled by AAC and

De Beers) holds 52% of SAB, with the Botswana government retaining 48% (37).

AAC has had (or still has) operations in much of the rest of Africa, including Angola, Ghana, Kenya, the Ivory Coast, Malawi, Nigeria, Tanzania, Zaïre and Namibia. Although AAC was prospecting for uranium in the 1970s, Anglo's main interest in the South African-occupied territory is in diamonds and gold. Consolidated Diamond Mines (CDM) owns the world's richest source of gem diamonds, at Oranjemund on the Skeleton Coast, which has been the object of one of the largest civil engineering projects in the southern hemisphere (3). The company contributes up to 16% of Namibia's gross domestic product and 23% of its income (72). The Thirion Commission, set up in 1982 to examine Namibia's mining industry, excoriated CDM for a range of malpractices, from tax evasion and transfer pricing, to a "scorched earth" policy, whereby the diamond gravels were deliberately overmined ("high-graded") to exhaust their potential before independence (73, 74). Thirion also found that – "almost unbelievably" (74) – the Diamond Board of South West Africa allowed De Beers to export and assess the value of all its Namibian gems without any check by the administration: hardly surprising, given that the Secretary of the Board was also the Company Secretary of CDM (74)! The Commission made wide-ranging recommendations, including that the Diamond Board should be entirely reconstituted and take possession of all mined diamonds; that mining companies should be prevented from monopolising and profiting from leases which they don't mine. Interestingly, Thirion also advocated small-scale mining as an alternative to the depradations of CDM, with a data bank and possibly a mine school being established (74). CDM came out fighting (75), seeking to make a distinction between the "good mining practice" which all companies followed in a sellers' market, and the overmining of which it had been accused. It claimed that the life of the Namibian deposits was being continually extended by prospecting, and that it could provide information on taxation, so long as this was

in camera: something which Justice Thirion rejected (75). The South African-appointed Turnhalle administration later exonerated CDM of the Thirion charges, though many observers were convinced by them (3, 76). In 1989, CDM announced an expansion, with new mines at Auchos and Elizabeth Bay (near Luderitz) capable of producing nearly 300,000 carats (72).

CDM is also linked with AAC in a gold prospect near Karibib, in south-west Namibia. Called the Navachab project, it first came to public attention in early 1987, when the Namibian Water Affairs Department began searching for scarce water supplies for a future mine (77). Later, Namibia's Secretary of Water Affairs predicted that the cost of water for this low-grade deposit, with a 13-year life, would alone be around US$10 million (78, 80). Initial ownership of Navachab was said to be: AAC 53.34%, CDM 33.3% and Amgold 13.36% (79). Later it was announced that AAC and Rand Mines, in a JV called Metall Mining Corporation, had acquired 20% of Navachab, which suggests that Rand bought into the project, rather than that AAC had surrendered control (80).

In the Americas

Although US Big Money has progressively – if slowly – pulled out of southern Africa in the last decade (81), South African capital has flown in the other direction quite dramatically, and resulted in South African control of existing North American institutions (82). By 1981 AAC was the largest single foreign investor in the whole of the USA (8). Five years later, it had acquired 143 separate investments in northern America – 106 in 32 US states, and 37 in Canada, including 24 manufacturing businesses and an investment banker (83). In 1988, AAC and GFSA's overseas holdings represented more than 30% of their total deployed investments, and Amgold's investments were particularly lucrative, thanks to their listing on "many international exchanges and AAC's quality as a gold 'fund'" (81).

However, Amgold's investments have not led to

any dramatic new mining developments, although Minorco and IRC did announce in 1988 a limited partnership to explore for gold in North America, called the Western Gold and Mining Co Limited Partnership (Westgold) (84) which has been producing from an open-pit mine at Austin, Nevada, and an off-shore placer deposit in Alaska (85).

In contrast, AAC's mineral exploration in South America has yielded important potential riches and given the corporation key leverage in the economies of Brazil and Chile in particular, but also Peru and Argentina. The corporation's vehicle for penetration of South America was initially Empresas Consolidas Sudamericanas, a Panamanian company representing the Hochschild interests, in which AAC bought 40% in 1981 (3). Three years later, AAC, De Beers and Minorco bought out the remaining 40% and, with that purchase, acquired most of the assets of Consolidated Mining and Industries (86). The following year, Empresas Sudamericanas contolled virtually all AAC's South American interests (87). Two years later, and in the face of Brazil's ostensible banning of military, academic and sporting links with South Africa (3), Empresas Sudamericanas was renamed the Anglo-American Corporation of South America (AMSA), with a subsidiary, Ambras, handling the Brazilian assets. Initially the take-over gave AAC access to Brazil's Codemin ferronickel mine, and interests in fertilisers, carbon black, industrial phosphate, gypsum, niobium, iron ore, cashew nuts, ranching, banking and explosives. AAC's strategy was to purchase existing and potential gold mines – a move foreshadowed in 1973, when AAC established Ambras in Rio de Janeiro and made the former head of the Cabora Bassa dam project in Mozambique its Brazilian chief (88). AAC also ensured De Beers's continuing monopoly over world diamond resources, by gaining the support of Brazilian entrepreneur Agosto Azevedo Antunes, who had previously worked with Hanna and Bethlehem Steel (88). By 1982 AAC had become the biggest holder of foreign prospecting rights in the country (88), and by

1989 was estimated to have 45,000 square kilometres under concession (89).

The Brazilian government in the late 1970s set up two commissions to look at the role of foreign mining and business interests in the country. Their findings boasted one of the longest titles ever given to a government report (90). However AAC had smoothed its path by wining and dining the editors of *Jornal do Brasil* and *O Globo,* and Harry Oppenheimer was received by President Geisel. The government decided that AAC was a good bet and would bring the country much-needed capital with which to develop its fledgling mining industry (88).

It was the Morro Velho gold mine in Minas Gerais that was to prove AAC's springboard into the country. In 1979 Ambras bought 49% in Siderugiga Hime, part of the Bozano Simonsen banking group, led by a former minister, Mario Simonsen (3). Bozano Simonsen's 1% share in Morro Velho was fleshed out to 51%. This majority ownership was too much even for the Brazilian régime which, in 1980, refused government funding to develop Morro Velho and two years later took steps to reduce AAC's holding. AAC now appears effectively to own 49% of Morro Velho (91).

Morro Velho's contribution to Brazilian gold production has been dramatic. From a failing performer in the 1970s, it was turned around to become the country's most important producer. In 1986, it delivered some 7 tonnes of glittering riches – about half of the country's total production (71), and two years later 9 tonnes (16). AAC also spent US$60 million exploring a gold prospect at Jacobina in Bahia, which was incorporated into Morro Velho (88).

In 1988 Ambras was granted an option by the Brazilian company TRV Mineração Ltda to earn 50% in a deposit in Goais (92). But it is another deposit in the same state, at Crixas, which may prove even more extensive than Morro Velho's, though initial production will be around 3 tons/year (80). Called the "largest primary gold deposit discovered in Brazil" (16) Crixas has reserves of around 70 tons (16, 92). AAC bought out Kennecott's 50% option on the project (now held by Amisa) and is partnered at Crixas by Morro Velho and Inco Gold (93, 230).

Crixas opened in late 1989, while the Jacobina mine underwent a major expansion at roughly the same time, tripling its ore output (423). Meanwhile, AAC's exploitation of older gold workings, in a huge complex around the Nova Lima area of Minas Gerais, is yielding a modest, but growing, output from deposits at Mina Grande, Faria, Queiroz, Raposos and Cuiaba (423).

In early 1990, Morro Vehlo's managing director, Juvenal Felix, announced that – with exploration costs running at up to US$5 million a year – the company was likely to be opening up a new bedrock mine in the Amazon rainforest region (423).

Two other aspects of AAC's dominance of Brazil's gold production should be noted. One is that the company has been not merely a sleeping partner, but instrumental in supporting the Brazilian government policy of removing small-scale and indigenous miners (*garimpeiros*) from gold workings, which might get in the way of the corporate operations. This was spelled out by AAC in its house magazine in 1984: "... the activities of the unemployed-turned-goldminers is seriously hindering the development of some capital-intensive mining methods. The nature of Brazil's gold deposits is such that mining might be carried out in some areas using open-cast pit methods, while in other areas only underground methods are feasible. Either approach requires that the area be clear of small-time miners" (94).

The second aspect has almost gone unnoticed, though it dramatically underscores both the apartheid régime's connections with AAC and the continued close collaboration between South Africa and the Brazilian government, typified by trade aid and nuclear exchanges in the 1970s, which a return to democracy has apparently little disturbed. In late 1988, the South African régime despatched a team of mining engineers to develop Brazilian gold mining techniques. It was a "low profile" visit, said South Africa's Ambassador to Brazil, so as not to embarrass the south Americans (95). "Low profile"

or not, the exchange produced rich dividends for AAC in Brazil. Several Brazilian technicians now train in South Africa at AAC's expense and, with the help of the parent company, a substantial technology centre has been opened up in Nova Lima, where 20 technicians are studying the development of bacteriological processing of gold (423). The total staff at the Nova Lima complex is a staggering 6200 people – making it perhaps the most significant corporate gold centre outside of South Africa.

Establishing Empresas Sudamericanas gave AAC an entrée into Chile's biggest privately-owned copper company, Empresas Minera Mantos Blancos, which is second in terms of output only to Disputada (16). Initially AAC held 72% of the mine with the Hochschilds, the World Bank holding 12%, and various Chilean investors the other 16% (96). More recently Cominco and Falconbridge moved in to share Mantos Blancos with the IMF (World Bank) and the AAC (16). The mine underwent a major expansion in 1988 with new roads and exploration (16).

AAC is also partnered with Cominco (26%) and the Chemical Bank (16) in the Marte gold prospect near Salar de Maricunja (97), which was due to begin production in 1988 (98). Other Chilean prospects include Inca do Orro quite close to Salar de Maricunga, a silver-gold deposit near Marte (91) and other deposits at Soledad, Escondida, Valy, Malia, Cacique and Caspiche (91). In 1988 AAC was also drilling for gold at Lobo, in a JV with Anaconda (91). To round off this resumé of the corporation's southern American forays, we should take note of the Arcata silver mine owned by Empresas Sudamericanas in Peru (103); the fertiliser operations of Petrosur in Argentina (103); and the fact that the two most active gold exploration companies in Argentina are St Joe (Fluor) and AAC, with interests in the Hualcamayo gold deposit in La Rioja, and elsewhere (100).

..."Down Under"

Both for political and practical reasons, AAC has maintained an "extremely low profile"

(101) in Australia and the Pacific region. The political motives derive from an obvious sensitivity to strong government-led revulsion against apartheid. (During much of the mining industry's expansionist phase of the late 1970s and the early 1980s in Australia, Prime Minister Fraser was rattling his sabre against the evils of the South African régime while doing little about discrimination against Aboriginal communities.) The practical reason is quite a simple one: its own exploration ventures have never been successful (102). For instance, it managed to miss locating the Argyle deposit in the late 1970s even though its Stockdale subsidiary was busy in the same region (103), and its Blue Spec gold project had a "brief and disastrous career" (103) before being shut down in the 1970s.

Anglo's strategy has, therefore, been to walk quietly, act stealthily and buy as much as possible into existing Australian-led ventures. Certainly Renison Goldfields, a subsidiary of Consolidated Gold Fields (UK) has been a target of AAC, not only for its considerable gold reserves, but its majority (62.24%) share in the world's biggest single mineral sands producer, Associated Minerals Consolidated (AMC) (104). The dramatic failure of AAC's protégé Minorco to gain control in 1988/89 of Renison's parent company (see section below) may mean that AAC will concentrate on buying into smaller projects in future.

Among its Australian interests have been a 40% share in the Mount Morgan gold tailings reclamation project managed by Peko-Wallsend in Queensland, which it acquired in 1983 (105) and a half share in the Walhalla (north-east Victoria) exploration venture started three years later (101). In 1984 Steetley Industries sold out to AAC, giving the company industrial mineral sources in New South Wales, Queensland and Victoria, as well as in South Australia (106). The same year, the company reported low grade alluvial deposits on a prospect in New Zealand's South Island which it didn't deign to develop due to the low price of gold (107).

Another tailings venture in Fiji seems not to have been developed; nor does a "promising" porphyry copper deposit at Namosi, on the is-

land, which AAC located in the 'seventies along with Amax and Preussag (108). Its major interests in the region now seem to be the Kaltails project, along the famed Kalgoorlie "golden mile" in Western Australia, which involves extracting gold from ten dumps in the area (102) but only at marginal profit. In 1987 AAC floated AA Pacific with the Kaltails project as its most important (60%) subsidiary: the Western Australian state government in 1987 acquired 15% of the venture (through Goldcorp) and Poseidon (a company in which AAC has a significant stake) (118) later moved in (109). By early 1988, the project had run into opposition from environmentalists and was not finding the water it required (102). The scheme finally started up in 1989 (109).

AAC's Stockdale subsidiary (owned by De Beers, with Ashton Mining holding 28%) was one of the first companies attracted to the Kimberleys, by the hope that its mineralogy (similar to its namesake in South Africa) would yield fabulous gems. (It also explored a nickel deposit near Turkey Creek in the same region in 1977) (110). Stockdale's explorations covered the Pilbara (111), the Daly River Aboriginal reserve and Keep River in the Northern Territory, and the Nullagine diamond field (Australia's oldest) which is partly covered by the Jigalong Aboriginal reserve (112). Stockdale soon acquired 70,000 square kilometres of leases in the West Kimberley region. Its explorations at this time were so extensive that the chair of Northern Mining (later the junior partner in the Ashton Diamond JV) was quoted as saying he saw "Oppenheimer under every bed!" (111).

The Aboriginal people of Oombulgurri, where Stockdale came to concentrate its search, were solidly opposed to the company's incursions (111, 113) especially after it trepassed onto the Forrest River Reserve (114). They eventually banned the company from their land (115). Stockdale is still active in the region, particularly at Bow River reserve, and is allied with AOG (28.33%) and Aberfoyle in the Australian Diamond Exploration JV, managed by Ashton Mining in the Northern Territory (NT) (116). But during the last decade its parent has been able to vicariously cash in on the world's largest diamond mine.

The ostensible vehicle for the process was Ashton Mining Ltd, which is controlled by Malaysia Mining Corp Bhd (46.3%), then owned substantially by Charter Consolidated plc (13.83%). Ashton controls, along with CRA, the Argyle Diamond Mines JV (for further details see CRA). Since both Ashton Mining and CRA are longstanding friends of De Beers, it is not surprising that, when De Beers itself mounted a concerted public relations campaign to bring the Argyle deposit under its feoff (117), it met little opposition from the two major companies involved. More surprising is that Australian politicians (of both major parties) should have succumbed – after initial posturing against apartheid – to its blandishments. In fact, it was only when two researchers, Jan Roberts and Les Russell (Boolidt Boolithba) exposed the AAC hidden strings behind both Ashton and CRA (at the time Charter also held at least 8% of RTZ, which controlled CRA) that opposition to De Beers began to register in Australia. By then it was too late. The CSO had shaken hands with CRA and was to end up with 75% of the gem, and 95% of the industrial diamond output from the most important diamond find of the twentieth century (3, 115, 120). Evidence suggests that CRA is offering the Argyle output to De Beers at prices lower than it might have secured had it gone the way of Zaïre and broken away from the CSO (121). There are also grounds for believing that De Beers was determined to secure control of marketing from Argyle, not simply because it threatened the CSO's hegemony, but because it needed an alternative supply with which to "blackmail" any black South African government, should it seize AAC's own diamond resources in South Africa (121).

Anglo's output: gold

"Gold remains the central focus of the group" commented the *Mining Annual Review* in 1989, referring to the AAC itself. Although this is a narrow view of what constitutes the Oppen-heimer dominion, there is no doubt that South Africa remains the world's largest gold producer, and AAC is far and away the apartheid state's biggest producer. AAC companies (including crucially the two gold producers in which AAC holds major shares, GFSA and JCI) contribute around 70% of the country's production (3).

Nonetheless, the price of gold has fluctuated considerably since its big "high" in 1980 and, in the past two or three years, has experienced a real decline (122). South African costs have been rising – especially labour costs as a result of black trade union activity, ore grade falling, and mines are being plumbed to deeper and more dangerous depths (123). Gold provided nearly half of the apartheid state's export earnings from 1980 to 1987, but the country's share of western world output declined from around 70% to 44% over the same period. In an attempt to offset the slide, a South African government tax commission proposed tax reforms and incentives to gold mining in 1988 – although they did not find favour with everyone (124).

The imposition of sanctions – particularly by the USA – and the withdrawal of US investment (125) has naturally worried AAC and other South African producers or investors (one of which, the Cookson group, put the bulk of its holding into a new company with AAC in 1986 in order to sidestep any bans) (126). However supplies continue to be "laundered" – through Zürich to Italy (127). Huge amounts of gold, both official and "illicit", have entered Taiwan. In 1988 the country received officially 330 tonnes, but the actual amount could have been twice as much (128). Little wonder that South Africa's public relations front, the World Gold Council, has been expressing considerable anxiety that South African production, rather than declining, could be outstripping demand, thus contributing to the decline in prices (127). Nonetheless AAC's strategy has been to press ahead with expansions which would have made any smaller company blench, or (perhaps literally) fall apart at the seams. In 1985 it opened up "the largest single mining project to be undertaken in the world" (129) when it expanded production at Western Deep Levels to South No.1 shaft, at a cost of US$27 million (130, 135). The following year, the company announced that its exploration boom would last a decade, with seven new gold mines opening in the Orange Free State and West Witwatersrand, another one likely in Bloemhoek/Welgelegen, and yet another in the Kalkoenstroom/Doornrivier area (131). Ergo also increased its gold and uranium tailings treatment output (132). In 1987, Rand Mines and AAC opened up another gold mine near the eastern Transvaal town of Barberton, inside the Kangwane "homeland" (133). A "significant break with tradition" occured in 1988 (134), when AAC brought on stream a small heap-leaching project at Carletonville, in the west Rand (134, 135). The same year Vaal Reefs and Western Deep Levels underwent further expansions (16, 34). Though secure in its pre-eminence at home, AAC has also aggressively searched for and exploited untapped deposits elsewhere in the world. To this end, as mentioned, it has managed to cash in on the biggest Brazilian gold project (Morro Velho), move in on highly prospective deposits in Chile and Argentina (16) and recently open a new gold mine in Namibia (Navachab) (16).

So long as AAC enjoys its highly lucrative strategic links with Johnson Matthey and Engelhard (sole suppliers of gold bullion to the London market) (3), so long as sanctions continue to be partial and ill-applied, and so long as "discreet" (3) links remain with the USSR and other eastern European producers, AAC's dominance will not really be under threat.

The origins of AAC as a gold producer date back to 1937, when the West Rand Investment Trust (WRIT) was established to group together the principal gold mining interests of AAC and its associate companies. A few years later, the Orange Free State Investment Trust Ltd (OFSIT) was formed for a similar purpose, and in 1972, the two companies merged to become Anglo American Gold Investment Co Ltd (Amgold) (136). Although Amgold is not AAC's sole gold finance company (others in-

clude New Central Witwatersrand Areas, DAB Investments, and New Wits Ltd) (3), it is the key vehicle by which the corporation has secured its aegis over the country's major gold mine. In effect, what looks like "investment" is either management control or, through interlocks with other part-owned holding companies, effective control. As *Raw Materials Report* ably demonstrated in a study released in 1985, if AAC's nominal investment in the country's gold mines is compared with its real ownership, the latter usually exceeds (sometimes quite dramatically) the former. (In the case of Western Deep Levels, for example, by nearly 7%) (137). Crucial to this strategy has been AAC's majority ownership of JCI, and its considerable holding in GFSA. Until GFSA sold some of its equity to the Rembrandt Group in 1987 (138) (partly in order to avoid sanctions) (3) and until the Minorco/Consgold battle royal focussed world attention on Anglo's relationship with its only rival mining finance house in South Africa (see below), AAC and Consgold between them held around 70% of GFSA. (Moreover, even after the Rembrandt take-over, there was speculation that this huge conglomerate could offer its GFSA holding to AAC – "Rembrandt's conservative labour policies go down well with conservative politicians," remarked the *Financial Times* at the time) (138, 139).

The largest single producer in the Oppenheimer domain is Freegold, centred around Welkom in the Orange Free State. Free State Geduld, President Brand, President Steyn, Welkom, and Welkom Holdings, were amalgamated, bringing five mines under one corporate structure, along with eight metallurgical plants (140). (Initially a sixth mine, Jeanette, had been included) (141). AAC holds 48.6% of Amgold which itself owns around 20% in each of the mines, as well as having direct holdings in each of the companies (ranging from 2.2% in Free State Geduld to 16% in President Steyn as at 1986) (142). The mines are run by Freegold (Free State Consolidated Gold Mines) but two other holdings companies, Orange Free State Investment (Ofsil) and Welkom Gold Holdings (Welkom) hold 51% and 5% respectively of Freegold, while Welkom itself owns 30% of Ofsil (143, 148). Initially announced as a practical arrangement – pillars between the existing mines could come down and be mined (144) – AAC's motives in making the merger were clearly more devious and self-serving.

Mining chiefs in South Africa immediately welcomed news of the merger (145). During 1984 and 1985 the five mines delivered 110.3 tons of gold, which was 10% of the western world's total production (142, 143, 148). AAC claimed Freegold could be producing 133 tonnes/year until well into the next century (144, 146) and when the merger was sealed in 1985, the London stock market toasted its success (147). The South African government initially objected to the merger, since it reduced income tax (148), but later approved it. However, US shareholders weren't happy (150), at least initially. (British and north American shareholders are not permitted to hold more than 5% in any gold stock and they feared the scheme would commit them to holding funds in mines of variable quality (151). In the event, they rallied round) (148, 150).

By 1988, Freegold was showing good results, despite the 1987 NUM-led strike (152). But the alarm bells were already being sounded: Johannesburg Stockbrokers, Mathison and Hollidge, warned that South Africa's gold mining industry was "… in an advanced state of senescence". Production costs at US$260 an ounce were more than a quarter higher than those in the USA (69).

Anglo's output: platinum

Even if AAC's predominance in gold production slides, there are still two mineral resources over which AAC will continue its global dominance. Of course, the Oppenheimer clan's interests in a vast range of other materials and services cannot be gainsaid, but it could only hold the rest of the world to ransom over the supply of diamonds and platinum. Platinum in recent years has enjoyed an unprecedented boom, as demand for a variety of purposes – ranging

from autocatalyst manufacture, to new legal tender coins, to a tremendous upsurge in Japanese platinum jewellery – has outstripped production (152, 153, 154). In 1988, the vast majority of the western world's platinum (and platinum group metals) output came from just one country – South Africa (2,560,000 ounces out of a total of 2,800,000). The only major non-western producer, the USSR, supplied a mere 400,000 ounces (16, 153).

Of the four platinum producers in the apartheid state, Rustenburg – with three mines in the Bushveld, the Atok mine (Lebowa Platinum) (155) in the Lebowa "homeland", and several other deposits under its control (153) – is far and away the world's biggest producer. However the company is more secretive than most: production and sales figures are not issued (164). It is well geared to take up the expected 30% increase in production over the next decade (157), and to continue dominating the world's refining and sales markets through Engelhard and Johnson Matthey (see below).

Rustenburg is owned 23.79% directly by AAC, 8.25% by Lydenburg Platinum Ltd and 32.6% by JCI. JCI is nominally 48.2% owned by Anglo – itself more than sufficient to give the Oppenheimers control. In fact other holdings certainly increase Anglo's dominance of JCI to well over 50% (3).

Anglo's output: coal

Despite the rising costs of South African production – freight charges in particular – AAC is the country's largest producer of coal. In 1988 and early 1989 it confounded predictions that exports would be hit by costs and sanctions when it increased sales and cut expenses (for example, awarding wages to black miners well below the inflation rate) (16).

Over the last five years, Amcoal and associated AAC collieries have fared well, except for a dramatic fall in exports during 1987 (16, 37, 158, 162). 1985 provided a particular boost to AAC's fortunes, leading the *Financial Times'* Kenneth Marston to ask, "if the South African economy is in such bad shape, how is it that

Anglo turned in its best-ever profits of R990.4 million in the year to March 31st?" (159).

South Africa's coal reserves are vast – they could last another 200 years at current output (164) – and are AAC's second most important export after gold (164). But export figures do not represent the full importance of the mineral to Anglo's fortunes, since the majority of production is absorbed by three South African parastatals: Escom (the Electricity commission) – AAC's most important single customer (3), Sasol (the coal to oil producer which has been crucial to the régime's side-stepping of oil sanctions) (161), and Iscor, the steel producer (16, 100).

AAC's involvement in coal mining stemmed originally from the need to supply cheap fuel for gold mining – it was not until the 1970s that both the export and internal boom commenced (162). The Anglo American Coal Corp (Amcoal) controls thirteen collieries in the Transvaal, Orange Free State (OFS) and Natal. (In July 1987 it opened the New Vaal colliery in the Transvaal, which is one of the largest in the country, producing 15 million tonnes a year) (163). GFSA itself runs three collieries, and a fourth with Gencor (3), while JCI through Tavistock Collieries – also administers three mines, and a fourth with Total (3). In addition, AAC itself runs directly the Vierfontein Collieries which supply the power station of the same name (3).

Since South African coal has to compete with new suppliers from Australia, Columbia and elsewhere (159), and is the only mineral to have been substantially affected so far by international sanctions (45), long-term prospects are not entirely rosy (164). However, a prevailing favourable US Dollar/Rand exchange rate has worked to the company's advantage, and, while Denmark, France, West Germany and Japan have instituted partial or complete bans, Spain, the Netherlands, Israel, Britain, and most far eastern markets, continue to import (16).

In 1987 the South African coal industry – representing Gencor, Rand Mines as well as the Amcoal/AAC collieries – set up a Bureau in London to promote sales to the British CEGB

(165). The move was supported by several Tory MPs, four of whom in 1989 visited South Africa and exonerated the industry's industrial relations. In a rather surprising (though somewhat oblique) condemnation of their report, the *Mining Journal* pointed out that the MP's had given "no indication of any black workers actually being promoted to positions of responsibility directly senior to a white miner" (166). "What is certain," continued the *Mining Journal* "is that the report's authors have hardly added to their prestige by failing to admit within the report that all four are Conservative MPs" (166).

It is also important to note that AAC's Zimbabwean mining corporation, Amzim, administers and 23% owns the country's major Hwange (Wankie) colliery and runs Botswana's Morupule Colliery which supplies Botswanan power stations (3).

Anglo's output: uranium

Along with JCI, AAC controls some 71% of South Africa's uranium production. However, sanctions were instituted from the beginning of 1987 against direct imports into the US and Canada of southern African uranium. They began taking their effect that year and in 1988. In addition, legislation implementing the Free Trade Act between the US and Canada freed an important log-jam in uranium supply to the world's biggest consuming country and led to a drop in prices (16, 167). In 1986 the major AAC uranium producers had rescheduled their uranium contracts before the ban took effect and, as the balance between supply and demand started to gain a certain equilibrium (168), Freegold (169), Vaal Reefs and Western Deep Levels all chalked up a handsome profit (168), with the Ergo plant treating just under 20 million tons of tailings and producing 160 tons of uranium (171).

Two years later, however, South African production was in the doldrums with firm future sales of only 15,000 tons uranium (against 41,000 expected three years before) (172). That year Driefontein, Randfontein, the Harmony mine, and the Chemwes project, all ceased uranium production (16). These four combined had produced around a fifth of Nufcor's output that year (173).

Although Ergo opened a new treatment plant at Daggafontein in a JV with East Daggafontein Mines in late 1987, the company had no new plans to produce yellowcake until 1997 (174). Similiarly, Sallies (South African Land and Exploration Co Ltd – AAC effective ownership around 45%) was contemplating its own extraction of uranium from slimes in 1984, but four years later there was nothing (128).

Nonetheless, AAC has benefited from the parlous state of the uranium industry in one major respect: Vaal Reefs mine (always by far the country's largest producer) picked up contracts from failed producers, had a higher recovery grade, and coughed up 1886 tons of uranium in 1988 compared with 1677 the year before (16). AAC (19.1%) and De Beers (9.5%) also have a share in the RTZ-managed Palabora mine, which is a minor uranium producer (176).

But long gone are the days when "almost in a fairy godmother fashion, uranium has come to the rescue of many gold mines at a time when the gold price has slumped" (177). In the late 1970s AAC's uranium activity was in the ascendancy: it had exploration projects in Saskatchewan and Manitoba (though no significant discoveries appear to have been made) and in Zimbabwe (Rhodesia) in the Kariba area (178). In the Karoo area, AAC had a JV with Enusa; in Namibia it was exploring for uranium with Minatome and Aquitaine around Swakopmund, and two areas around Luderitz (108).

The prospect for AAC's uranium production is not good, despite the status of Vaal Reefs, and the fact that an unknown proportion of the output has (notwithstanding sanctions) been exported to Europe, processed into uranium hexafluoride (UF$_6$) and then shipped to the USA (see Rössing for more details). Tighter US legislation, industrial action in Europe, combined with open access to Canadian supplies, and a continued surplus of uranium on the open market, may mean that South Africa's uranium export industry all but ceases to exist within the next ten years. (As long ago as 1965, the South African government had proposed a trilateral deal whereby the US, South Africa and the Congo would barter uranium and De Beers diamonds for agricultural equipment. When the US government discovered who was involved it stopped the deal on anti-trust grounds) (179).

Of course there is still an internal market, supplied by Nufcor, for the apartheid state's own "civil" and military reactors. As one of the major partners in Nufcor (29), the AAC will benefit from the South African nuclear industry, so long as there is one.

The Minorco saga

Sometimes dubbed AAC's "overseas flagship" (175), Minorco was to be the Oppenheimer vehicle for the biggest attempted take-over in mining history in 1988/89, when the company tried to increase its already considerable minority holding in CGF (Consolidated Gold Fields) and mount a full bid. The Minerals and Resources Corporation (Minorco) had its origins in the company Ernest Oppenheimer established in 1928 to finance the expansion of the Rhodesian copper belt (3). Headquarters were transferred to Northern Rhodesia (Zambia) in 1954, and a decade later the company was renamed Zamanglo (Zambian Anglo-American) when Northern Rhodesia achieved independence.

Before independence, Zamanglo had enjoyed virtually unrestrained repatriation of its profits. Even after independence, thanks to adjustments made to its accounting policies, the company managed to export about three-quarters of its profit in dividends (3). Thanks to tax concessions, freedom from exchange controls, good management contracts, and generous compensation granted when the Kaunda government took over 51% of the company – now renamed Minorco – it could shift its base to the tax haven of Bermuda. It had, declared Kaunda, "... taken out of Zambia every ngwee that was due to them" (180).

By 1980, Minorco had built up an appreciable stake in Consgold (181). The following year its

94

assets were £2 billion – larger than its parent's (US$2.5 billion). Within the next four years, Minorco's forays into North America (and to a lesser extent South America) secured for it: 22% of Phibro-Salomon, 29% of Engelhard Corporation, 60% of Inspiration Resources Corporation (IRC), 36% of Charter Consolidated, and between 11% and 25% of Anglo-American do Brasil, as well as 10% in Empresas Sudamericanas Consolidadas (99). It also held 10% of the Anglo-American Investment Trust (Anamint), 60% of Imetal, considerable minority interests in Australian Anglo-American and half of Zambian Copper Investments (156).

In 1983 Minorco had reduced its holding in Phibro-Salomon to 22.3% – arousing speculation that it might use the realised cash to buy a controlling stake in CGF: both CGF and Minorco denied the rumour (182). Two years later, Minorco sold 10 million shares in Phibro-Salomon, reducing its interest to 14% (183) and was swimming about in US$400 million cash (184). Phibro was still Minorco's biggest asset – around 40% of its net value – and despite undertaking to pull out of South Africa, Phibro refused to buy back the shares held by Minorco (185). Then, in what the *Mining Magazine* termed an "incestuous transaction", Minorco in 1986 bought 49% of Adobe Resources from IRC (Minorco then held 59% of IRC) (186). The following year Minorco (now held 39% by AAC and 21% by De Beers) reduced its equity in IRC somewhat further – to 56% (187). In 1987, it sold its share in Anamint to De Beers – thus enabling the world's premier diamond exploiter to secure a firmer hold on its diamond trading companies (188). (Two years earlier, *Raw Materials Report* had speculated that Minorco might join up with Union Minière, since the companies held diamond interests in common – notably through Sibeka, and in the Argyle diamond JV in Western Australia) (137).

1987 was an important year for Minorco. After an intricate and hugely lucrative tax "scam" was exposed by Dutch krakers (squatters) and the Aktiegroep Splijt Apartheid, who "raided" the company's small Amsterdam office disguised as bankers and left with hugely incriminating documents (189), it changed its domicile from Bermuda to Luxembourg (becoming Minorco Societé Anonyme of Boulevard de la Petrus). However, the main reason for the move was not further tax evasions: rather it was to guard against the impact of sanctions against AAC's assets in southern Africa (187, 190). Possibly too, Luxembourg was chosen because, like the insignficant town of Zug in Switzerland, it harboured other nominal corporations intent on evading existing sanctions (such as Nulux, jointly owned by Nukem and RTZ Mineral Services). The company also sold all its remaining shares in Phibro for US$808 million – a "major breakout" by the "Anglo American empire" as the *Mining Journal* called it (191) which, coming before the big stock market crash of October that year, "effectively doubled the buying power of an already impressive nest egg" (192).

With assets valued at around three billion US dollars, cash holdings alone of one billion dollars (193) and operations in six continents (160), Minorco was now solidly in the Big Time. It said it wanted only "major active stakes" in future projects, with a sharpened focus on precious metals: Luxembourg would give a "more realistic" base for running its expansion programme (192). It was therefore hardly surprising that, within the year, it was on the rampage after one of the biggest prizes in mining: Consgold.

Before turning in some detail to this particular, extraordinary saga – one which apparently put paid to AAC's attempts to dig for itself an unassailable bunker in North America, Britain and the Pacific region – it is important to look briefly at its earlier march through some vital western financial institutions, primarily in the USA. Within months of buying an initial minority stake in Consgold in 1980 (see later), AAC was using Minorco to take over and build up a substantial stake in Engelhard Minerals and Chemicals Corporation. In 1981 Engelhard was split, and its commodities trading arm, Philip Brothers, amalgamated with Salomon Brothers into a new entity, Phibro-Salomon

(194). At one fell swoop, the world's biggest publicly-owned marketer of raw materials had gone to bed with America's largest private investment bank and the world's largest bond-trading company (3). Commented the *Financial Times*, "[Observers say] [it] will be the most formidable trading house in the world, combining skills in virtually all commodities, financial and physical, with a client list that includes governments and most of the biggest corporate names around the globe" (195). The Oppenheimer initial share in Phibro-Salomon was 27%, reducing, as we have seen, over the next six years, until all the shares were sold by 1987. Apart from using Phibro as a cash vehicle, the Oppenheimers had to smart under criticisms from the growing anti-apartheid lobby in the USA (3).

Not so for Engelhard itself – which has proved too vital as a processor of precious metals (3), a world leader in speciality chemicals and metallurgical techniques (196) with even a kaolin project in Georgia (84). The original link-up with Anglo has been called an event of "decisive importance" which marked "... the restoration of foreign confidence in apartheid capitalism" (8).

AAC and Engelhard have benefitted mutually from their partnership. The company's eponymous founder was a friend of both Presidents Kennedy and Johnson (as well as being a model of Ian Fleming's Goldfinger in the James Bond novel of the same name) (3): little wonder then that, in 1986, despite the US Treasury's ban on the import of krugerrands to the USA, it was Engelhard which got a new minting contract to supply gold bullion (198).

But perhaps the most important single aspect of this relationship has been Engelhard's production of catalytic converters for motor vehicles, a market which in this new "green age" has boomed. It is Rustenburg's platinum which is the major source for the American corporation (3), as well as being the world's single biggest supplier (153).

Engelhard first acted jointly with AAC as far back as 1958 when they took over Central Mining – an important South African gold pro-

ducer with interests in Canada – to prevent GFSA doing the same thing (3). Later, Engelhard assisted AAC in taking up a 14.5% stake in Canada's Hudson Bay Mining and Smelting (Hudbay).

Hudbay was eventually to become a subsidiary of Inspiration Resources (199, 200), with exploration ventures in the Flin Flon-Snow Lake region of Manitoba (copper-lead-zinc-silver) and gold exploration elsewhere (201). Initially, however, the Canadian company was to have been a subsidiary of Plateau Holdings, a company set up to facilitate capital access to US and Canadian markets for Minorco and AAC (202).

By 1983, IRC was Minorco's main "operating arm" (203), with copper mines in Arizona (200), ownership of Inspiration Coal, Terra Chemicals and Trend International Ltd – which drills for oil and gas in North America, Paraguay, Indonesia and the North Sea (3, 199). However, IRC has not been a runaway success story. Although its interests remain in Trend and Terra, and it has acquired Danville forestry and printing businesses, Minorco had to "save" the company in 1984 by pumping in US$100 million so that IRC could purchase Madison Resources' oil interests (204). More recently, Minorco itself had to buy back the oil and gas investments (3). Both companies formed a partnership in 1988 to search for gold in North America (but which major resource company by this time hadn't done so?) (205) – though this is no indication that IRC will ever rise above the middle ranks.

In 1988 and 1989, Minorco participated in "the biggest, longest-running, and probably the most complex, take-over bid yet seen in the UK" (206), when it tried to buy out Consolidated Gold Fields (CGF/Consgold). The bid was officially launched in September 1988, and set at nearly three billion pounds (£2.9 billion) (207). News of it had reached CGF several weeks beforehand (208). Consgold's reaction was immediate and combative. Within a month, it was demanding an inquiry into insider dealing in its shares (209, 251); had asked President Reagan to block the bid – on the grounds that important rutile, monazite and zircon reserves would pass into South African hands (210); and had taken Minorco to court in New York for violation of the anti-Trust acts (211). The European Community said it would investigate whether EEC rules had been breached (212); the South African régime launched its own inquiry into the bid (213); Australian Prime Minister Bob Hawke attacked the Minorco strategem, as did the Governor of Ohio *(sic)* – both on the grounds that important strategic resources would come under South African control (211, 213, 214). The Australian associate company of Consgold, Renison Goldfields, threatened to sue both its parent and Minorco, alleging breach of the Australian Companies and Acquisitions code (215, 236). And, at the end of October that year, the British secretary for Trade and Industry referred the whole matter to the Monopolies and Mergers Commission (216), (239).

The battle was to grind to a conclusion eight months later, when Lord Hanson offered just over £3 billion for Consgold, after agreement that Minorco would sell its near-30% stake in the British-South African company to the rapacious Knight (217).

In the meantime, a battle of words and documents had been enjoined between the two erstwhile fellow-travellers over apartheid terrain, which left no expletive unturned (218). In presenting itself as a maiden about to be siezed by the 'Orrible Oppenheimers, Goldfields' own appalling history of exploitation in South Africa came under scrutiny (219). Hence, in its early days fighting off Minorco, Goldfields desperately attempted to present an anti-apartheid front, and a readiness to sell off its interests in GFSA (220). But it was Minorco and Oppenheimer/De Beers which undoubtedly came off worst in the cerebral encounter. No-one could be left in any doubt at the end of the day that Minorco was AAC's major overseas vehicle for expansion and sanctions avoidance, and more importantly that the Oppenheimer empire was firmly rooted in the mire of South African racialism.

In terms of historical fact, the connection between AAC and CGF has been more than a flirtation. That formidable, damnable imperialist, Cecil Rhodes, founded both CGF and De Beers using capital acquired from exploiting cheap black labour, and cheap black coal. The notorious Jameson Raid, intended by Rhodes to unseat Boer power in southern Africa, was financed by CGF's ill-gotten gains, and it was only when that criminal escapade ended in dismal fiasco that control of CGF ostensibly passed to London (221). From 1968 onwards, AAC and CGF parried each other's thrusts, as the Oppenheimers stalled at least four attempts by Goldfields to take over another South African mining house (222). During this period, both the companies diversified increasingly, and moved capital out of South Africa to projects overseas (221).

Probably in response to a widespread rumour that Gencor was preparing to take over CGF/GFSA itself (223), De Beers conducted a much-criticised (224) dawn raid on the London stock market in early 1980, and built up a major stake in Goldfields (181). The following year, when AAC/De Beers underwent a major restructuring, this 28.9% share was passed to Minorco (194, 225). Now Minorco became AAC's "major merchant venturer" outside South Africa (226), and clearly posed a massive threat to Goldfields. According to Congold's chair, Rudolph Agnew, Harry Oppenheimer at that point promised that AAC wouldn't increase its holding in Goldfields to more than 30% (222), and – apparently amicably – the companies swapped directors (221). An "uneasy peace" prevailed from 1981 to 1984 (221), but frictions had already emerged when AAC began stalking Newmont in 1981. By then, the American company – already the single largest US exploiter of black southern African labour (3) – was passing into the Goldfields camp. When the Texan corporate raider, T Boone Pickens, lurched onto the horizon in 1986, Newmont allowed CGF to build up its equity to 49.3% (227).

By 1986, relations between the two predators were at "rock bottom" (222). According to Agnew, AAC was threatening to sell its Goldfields stake elsewhere, so Goldfields was forced into "merger" talks with Minorco (228). Agnew demanded that Minorco's South African ownership be scaled down. "We argued very strongly," he claimed, "that serious companies did business in serious cities and this did not mean Zug" (222).

In 1987, Minorco published its own version of the 1986 discussions with Goldfields. Declaring itself "very surprised" that Goldfields had not agreed to a merger at this time, Minorco accused CGF of allowing GFSA and Driefontein themselves to acquire up to 7.8% in Goldfields as a "buffer" against the AAC overseas arm (229). (It was also clear that if Minorco and GFSA chose to act together, their stake in Goldfields would climb over the 29.9% level which, under British take-over rules, would compel them to make a full bid for the London company.)

In order to reduce its huge debt, Newmont also sold off its holdings in O'okiep, Palabora, Tsumeb and other southern African companies. While the Gamsberg zinc deposit changed hands from Anglo to GFSA, Newmont's interest in Palabora passed to AAC and De Beers (16, 93, 187, 231). Over the next two years, Newmont Gold (80%) was to develop into one of the world's most important producers – and the biggest in the USA. While the American management tried to maintain a fairly disinterested stance during the Minorco bid for its parent company, it clearly did not favour being thrown to the wolves yet again: Minorco said it would sell Consgold's 49% holding in Newmont, if it gained control of CGF (232). The British Department of Trade and Industry (DTI) had been in no doubt about the insidiousness of the 1981 take-over. "In acquiring a quarter of the company's shares, Anglo misled Congold's management, manipulated the London Stock Exchange and flouted the British Companies Act which subsequently had to be amended" (233). The secret buying, declared the DTI, had started in 1979, with AAC using secret subsidiaries and holding companies, particularly De Beers. Though Harry Oppenheimer allegedly feared a US take-over of Consgold (234), Anglo's real concern was that Old Mutual (which controls Barlow Rand) or Sanlam (Federale Mynbou) would mount a bid.

Minorco's ostensible objective in grabbing Consgold was spelled out in September 1988 by its chair, Sir Michael Edwardes – formerly at the helm of British Leyland and Chloride. Consgold's 38% (direct and indirect) investment in GFSA would be sold, as would CGF's 22% of Driefontein, its 26% in Kloof, 16% in Doornfontein, 38% in Deelkraal, and its 46% holding in the Northam platinum mine (235). If not sold outright, CGF's equity in Newmont would be "substantially reduced" (221). Minorco would be built into one of the world's "leading natural resources companies" (221). Financial commentators were not so sanguine. Some viewed the bid as simply a massive asset-stripping operation, designed to break up CGF once and for all, and which "thoroughly undervalued the company" (80). Others were convinced that – largely because of AAC's costly gold mining in South Africa and poor record at finding new deposits in North America – it wanted access to Renison Goldfields' highly-valued new deposits in Australia and Papua New Guinea. Certainly Renison was aware of the threat, and so was Prime Minister Namaliu of Papua New Guinea. "We cannot allow the apartheid régime to benefit from our rich resources," he declared in 1987 – referring specifically to Renison's thirdpart holding in the Porgera deposit (227). Renison's chair, Max Roberts, the following year expressed similar fears that the company's assets in Tasmania (the Mount Lyell copper mine), the Philippines and Indonesia (where RGF holds 75% of the Kola tin mine, boosting its share of western world tin output to 10%) (237) might go down the same nefarious path (236). In October 1988, RGF went to court in Australia in an attempt to stop Minorco (238).

But it is Renison's key role in the supply of mineral sands (it is the world's largest producer, delivering 30% of the west's titanium oxide, 40% of its monazite, and 45% of its zircon) which could have been Minorco's greatest prize. AAC has no mineral sands operations worth speaking of (merely a 5% holding in

Gencor which owns the vital Richards Bay Minerals deposits dominated by RTZ). Just after Minorco made its play for Consgold, a detailed report on the world mineral sands industry, prepared for the Dublin-based company Kenmare, showed that world demand for zircon would outpace supply until 1994, as the strategic mineral found new applications in electrical ceramics (and the world's nuclear industries). Its price was calculated to increase dramatically, while in 1988 it was reaching no less than A$2000/tonne on the spot market – ten times the price at the start of the decade (239, 240).

Initially the Monopolies and Mergers Commission (MMC) looked as if it was only going to scrutinise Minorco's play for mineral sands (197). Later, it was clear that the MMC would investigate the "insider dealing" which had brought the South African company to within grasp of one of Britain's three most important mineral producers (RTZ and Lonrho being the other two). (Consgold's British assets – especially the Amey Roadstone Construction Group (ARC) – were an admitted major plum in the pie into which Little Micky Edwardes wanted to get his thumbs) (241).

The DTI had started in 1986 its investigation of American Barrick Resources Ltd's build-up of 9.7 million shares in Consgold (228). Despite the name, Barrick is a Canadian company, with an important gold prospect at Carlin in Nevada, next to Newmont's own claim (242). Although Barrick maintained that the 1.37% interest it had acquired in September 1988 (243) was a "passive" one, it was a company which did not seem entirely unconnected with AAC. Early the next year, as Minorco made its second bid for Goldfields, it was revealed that the London stockbrokers James Capel – solidly pro-Minorco (244) – had breached the United Kingdom Take-over Code, when it bought, sold, then repurchased Consgold shares on behalf of Barrick (245). In June, James Capel and Smith New Court were charged with refusing to divulge who they bought shares for – ie American Barrick (246). Most of the speculation over secret share build-ups in Consgold

centred on Oppenheimer's Swiss connections and in particular Vadep of Barr (part-owner of Britain's popular "Tie Rack" retail outlet) (247), which by late 1988 had built up 1.8% of Minorco's equity. Vadep's sole director is to be found on the board of six De Beers companies (248). In addition, Capricorn Trustees held six million Minorco shares, and a London-based investment outfit called Primadonna (sic), aligned to AAC, held further assets in the company (248, 249). Early in 1988, Consgold was protesting that Swiss interests in CGF, plus other dubious "friends" allied to Minorco, could secure more than a 50% control in the London company (250). In October Consgold demanded an inquiry into how a Swiss Bank co-operative (including a 75% owned subsidiary of the British Midland Bank, with undoubted Oppenheimer alliances) could have secured two million Consgold shares (209, 252). At this time, CGF issued a document attacking AAC, claiming Minorco's take-over would result in the dismemberment of the company (253). It also released to selected journalists a formidable batch of documents, mostly secured in the USA under the Freedom of Information Act, alleging De Beers assistance to the Nazi régime during the Second World War, and attempts to cut the vital flow of industrial diamonds to the allied "war effort" (254). The British financial press speculated that CGF would also be looking for a so-called "white knight" to stave off the South African threat: RTZ was mentioned, but discounted once it over-stretched itself in buying BP Minerals (255) (see RTZ); so were the Hanson Trust and Newmont itself (256).

Meanwhile, CGF was promising it would pull out of South Africa altogether (257), and filed documents in the US maintaining that the Oppenheimers were bent on control of the world's gold market and posed a direct threat to domestic US resources (particularly Newmont's) (258). In late October, the New York District Court issued a temporary injunction against Minorco raising its stake in CGF to more than 30% (259, 315).

1989 began with the MMC confirming it

would expand its probes to include the relationship between Minorco and AAC (260). The Oppenheimers by now owned between 60% and 70% of Minorco (261) and possibly as much as 71% (262). In its own investigation of the company, the London-based Anti-Apartheid Movement identified Oppenheimer interests of around 46% in Consgold, not 40% as AAC would have the world believe (261). Although Minorco soon moved in on Charter Consolidated (in which it holds 36%, while Charter itself holds 4% of Minorco), purporting to take it out of the Oppenheimer camp (263) and presage a reduction in AAC's grip on Minorco, anti-apartheid researchers were convinced that the Oppenheimers could gain control of more than half of Consgold (221, 263).

True Battle Royal was the order of the day in the first six months of 1989. In early 1989, the Chemical Bank of the US said it would no longer finance companies in South Africa – but excluded its facilitation of the Minorco bid, with the facile argument that the company didn't have direct South African interests (264). As rumours circulated that the MMC would clear the bid, so long as Minorco sold off the Newmont and Renison acquisitions (265), the Luxembourg company confirmed its re-entry into the lists, and was negotiating with both Rembrandt (which had recently acquired substantial stakes in GFSA) and Gencor, to dispose of their joint 38% holding in GFSA (266).

The British Office of Fair Trading was starting to examine whether De Beers shouldn't be referred to the MMC for its dominance of the diamond industry (267), but, almost immediately afterwards, the MMC cleared Minorco for the final assault (268, 274). In a unaminous finding which British newspaper, *The Independent,* called "breathtakingly dismissive" (269), and which the British Labour Party deplored (270), the MMC refused to await the findings of the DTI into AAC/De Beers oligopolistic practices, declaring that neither the British gold nor strategic metals markets would be adversely affected by a take-over, and that any threat of an increase in boycotts against British operations once taken over by Minorco

(specifically ARC) was not sufficient to hold up the take-over (270, 271).

Two weeks later, the European Economic Community's (EEC) informal inquiry – which had concentrated on the threat to platinum markets – also cleared the bid, although it demanded that Minorco should sell Northam, the platinum field developed by GFSA (269, 272). The US Appeals Court was not so willing to be led by the nose: it demanded that the bid should be considered further (273). Meanwhile Minorco was preparing a new, record offer of £3.2 billion (274). Edwardes confirmed that Minorco would break up Consgold beyond recognition: Consgold retorted that this was "financial terrorism" (275). A few days later, in a vain attempt to sweeten the offer, Oppenheimer, joined by three other Minorco directors, left the Luxembourg board (276). Consgold then issued its Defence document (277), but refused to confirm CGF's asset value (278). Early the next month, the State of Michigan announced that it would sell all its Minorco shares – representing 5.4% of Minorco's holding in Consgold (279). By the middle of March, GFSA and Driefontein said they would refuse any conditional bid for their holdings (280), and the first closing date for the offer passed with only 0.2% of CGF shareholders accepting the Oppenheimer blandishments (281). On 22 March, the US Appeals Court confirmed its injunction against Minorco, saying it "... is hard to imagine an injury to competitors more clearly of the type the anti-trust laws were intended to prevent" (282). Nevertheless, the US Committee on Foreign Investment found that a take-over wouldn't bar access by US companies to strategic minerals (280, 283). Consgold now declared that, if its integrity were preserved, it would issue shares with a guaranteed good return, or a special cash dividend attached (170). The pledge didn't cut much ice, was later thrown into doubt (284), and then dropped (285).

Minorco's final bid for Goldfields was slated at £15.50 a share – over fourteen times CGF's expected 1989/90 earnings (286). Naturally CGF rejected it – more importantly, an estimated half of CGF's institutional shareholders were reckoned to feel the same way (287).

Confronted by an offer from Minorco to set up three separate US companies, run by court-appointed Trustees, the US Appeals Court's Judge Mukassey countered: "The Anglo group's record in circumventing legal restrictions and engaging in anti-competitive behaviour is not reassuring" (288).

For the next month, the scene of action shifted to New York (289). Although Minorco had taken 54.8% of Consgold by the end of April (290, 291), both Consgold (290) and Newmont (292) were determined that it would not prevail – this, notwithstanding a UK Take-over Panel's demand that CGF should drop its New York action (293). By May, Newmont had become the key player in the game. Just as it announced it was recapitalising Peabody Coal (49.9% owned) (294), Minorco seemed desperately to be seeking a buyer for the CGF holding in the American company: this could clear it for a final assault on Consgold itself (295).

As the US Appeal Court refused yet again to lift its injunction, Minorco apparently admitted defeat (296, 303). The stock market wiped £100 million off the value of Consgold shares, and the company was landed with a £50 million end-of-battle bill (297).

On 22 June, Lord Hanson, after a "friendly chat" with Consgold's Rudolph Agnew, stepped in to pick up the (gilded) pieces, offering £3.5 billion (298), which was accepted by CGF.

As a result, Minorco got lots and lots of money (around one billion pounds) with which it could buy up Charter Consolidated if it wished (299) – a company in which Hanson himself had acquired 7% by 1984 (300). Dubbed "one of the most fearsome corporate predators on either side of the Atlantic", Hanson had already swallowed up British Ever Ready batteries and the UDS retaium.g chain, as well as Imperial Tobacco in 1986, balancing this with sell-offs of acquisitions like Smiths Corona, John Collier Menswear, and the Courage brewery (299). Hanson looked likely to strip Consgold just as Minorco would have done: retaining ARC and

selling the GFSA share to Rembrandt (301), or to Gencor, itself owned 28% by Rembrandt (298). Gencor had recently boasted spectacular new growth, including the acquisition of Mobil Oil USA (298). Renison would be auctioned off – possibly to WMC or CRA (298), as would the shares in smaller South African companies (301).

Had it all been a soufflé? Or worse, a massive diversion of public attention from the real issues? One curious anomaly to the affair was that Consgold and its supporters argued, almost in the same breath, that Minorco posed a threat, as a Trojan horse for Oppenheimer/South African penetration of its British, North American and Australasian interests, and was an asset-stripper of the worst kind. Either Minorco was an outpost of the "Evil Empire", or (as Edwardes fervently protested) it was simply running its own ship: it would be difficult (though not impossible) to have it both ways. In the event, Hanson's intervention was far from that of the stereotypical "White Knight": the British peer was primarily interested in making money. The South African Emperor was interested in expanding control, dominating key markets and evading the consequences of future possible sanctions. However, Minorco was certainly squeezed into a corner – not by the half-hearted and defective rationalisations of the British Monopolies and Mergers Commission (paper tiger if ever there was) but by the zealously defended protectionist complexions of the US judiciary. So, in the final event, had Minorco gained control of Consgold it would almost certainly have divested itself of the key strategic interests it acquired, whether it really wanted to or not. The global mining map will probably not look much different in a few years time than if Oppenheimer had won the Consgold battle after all. And what of Consgold's main lines of defence? It postured variously as a Saint on the road to Damascus, a damsel in distress, a Great and Inviolate British Institution, and a bulwark against apartheid. None of these images fit. Despite its Pauline conversion to disinvestment from South Africa, Consgold held on to the bulk of its apartheid interests through the whole period of the Minorco gameplay. Its claim that Oppenheimer's attentions were wholly unprovoked and undesired sits uneasily alongside its willingness to negotiate with AAC after the 1981 Minorco bid, without securing any defences against a future bid for the whole company. As for its claim that CGF is compatible with BP, Lonrho or RTZ, this could never be seriously entertained for an outfit which was founded in South Africa, secures so much of its profit from the region, and whose investments (apart from ARC) have contributed virtually nothing to the British people – not even the British Exchequer. But the biggest lie was the proposal made in all seriousness by Consgold's Rudolph Agnew, that CGF was a different kind of employer of South African labour than its main rival; that the anti-apartheid lobby should regard it as an ally. "Which is the worse of two gold evils?" asked the *Guardian* in a pointed editorial in 1988, concluding that there was little to choose between them (302). Consgold operates like AAC in South Africa, the paper declared in May 1989, and has similar interests. "... It is regarded by some experts as a worse employer of black labour than Harry Oppenheimer's Anglo American empire..." (303). The anti-apartheid movement was not taken in. In a report issued by End Loans to South Africa (ELTSA) in October 1988, conditions at the Tsumeb copper mine in Namibia were depicted as "shameful" and "appalling", with workers poisoned by arsenic, lead and cadmium discharges, and conditions at their hostels among the most degrading in the country (304). At Consgold's AGM that year, campaigners mounted a spoof boxing match between "Rippling Rudolph Agnew" on the one hand and "Mad Mike Edwardes" on the other – with "PW Botha as referee" (305).

Unfortunately, amid the hullaballoo and sparring, one key point got lost: rather, it was never made. The encroachment of Oppenehimer's empire into Europe would undoubtedly have secured vital dividends for the South African apartheid state, and helped perpetuate the régime (306). But of themselves, these returns may have been only marginally more important

than the dividends already filling the coffers from AAC's existing international trading enterprises.

If we want increased access by the world's poorer communities to its resources, then concentrating the production of certain strategic minerals or precious metals under one corporate body is something devoutly to be feared and resisted. However, AAC is already in a key-hole position when it comes to platinum supply. And the world's largest producer of titanium and zircon is now RTZ, whose track-record in southern Africa and among indigenous peoples worldwide is as bad as, if not worse than, that of AAC. De Beers's dominance of global diamond production is ably assisted by RTZ, through its associate company CRA in the Australian Kimberleys. It was perhaps only in its search for new gold deposits where AAC had sustained a very poor track record, and – through its entrée into Renison and Newmont – that the western world's biggest single producer could have recouped its position as a leading supplier (307). Yet from early 1989 there was no question that Renison would leave the Anglo camp (308), while Newmont's resistance to a Minorco takeover was never in doubt. (But, as already pointed out, when the American Company took the decision to disinvest its southern African interests in 1988, its stake in the important Palabora mine passed to AAC which, once again, found itself in harness with RTZ.)

In fighting off the predations of a manifest despot, the institutional, protectionist, and, to an extent, anti-apartheid forces which lined up behind Consgold may have simply been pawns in a game – whereby the most crucial mineral assets on the planet passed into the hands of a clutch of more shadowy, but no less damaging, corporate imperialists, RTZ and Hanson being chief among them.

A Charter for Consolidation

It is through Minorco, with a holding of 36%, that AAC controls the operations of Charter Consolidated. Formed in 1965 as a merger between a "rag bag of assorted companies" (309) including the British South Africa Company (BSAC), Central Mining and Investment Corp (an Engelhard company), and Consolidated Mines Selection (associated with Anglo since 1917), it was to become AAC's main holding company until Minorco took over that role in the early 1980s, "eas[ing] the penetration of areas where Anglo American's direct South African connections could be embarrassing" and would impede financial support from bodies such as the World Bank (309). Charter therefore advanced into the USA, Britain (a joint project with Imperial Chemical Industries – ICI – in potash), Canada, France, Australia, Malaysia, Portugal and Mauritania (3, 24).

The Oppenheimer strategy was to secure at least 10% through the British-registered company in each investment Charter made, in order to benefit from double tax relief. (The trigger for establishing the company seems to have been AAC's need for it to place its initial Hudbay stake with a non-South African company) (3).

In its first three years of operation, Charter accrued a capital of US$324 million. Its investments have included a significant minority interest in RTZ (at least 8%, reduced considerably in the 1970s), the Cape Asbestos group of companies in both South Africa and Britain (24) and participation in a consortium in Zaïre's Shaba province during the 1970s which was later shelved (24). As already noted, it had until 1987 a substantial (13.8%) stake in the Malaysian Mining Corporation which controls Ashton Mining.

During 1983 and 1984 Charter embarked on an expansionist phase within Britain which gained it control of a major Scottish coal-mining equipment company, the country's most important precious metals refiner. Anderson Strathclyde was the firm for which Charter made a bid that was strongly resisted: in its defence document, the Scottish company exposed Charter's connections with AAC (six out of fifteen directors were Oppenheimer appointees, two of the others directed Anglo companies) in no uncertain fashion (310). Charter won control, thanks partly to the efforts of the Kuwait Investment Office (KIO), itself an ally of Phibro-Salomon (3), but was forced to cut back on its operations in Britain to a certain extent (3).

Charter's investments currently include 33.5% in Anglo American Corp of Zimbabwe, nearly 30% of Rowe and Pitman – stockbrokers to the British Queen – 75% (Umetco 25%) of a Portuguese tin and wolfram mine (200), joint ownership with AAC of the north Yorkshire miner, Cleveland Potash (16, 200), and the Alexander Shand coal mining company. (It also holds 3.8% in Minorco itself.)

But the jewel in its dubious crown is Johnson Matthey plc (35%), a company with which AAC has long connections dating back to 1884, when JM assayed the first Witwatersrand gold find (3). One hundred years later Johnson Matthey Bankers collapsed, after illegaly exporting money from Nigeria and making a number of bad loans. As JM's largest single shareholder, Charter offered to inject new money into the ailing giant, so long as the Bank of England and other banks took the dead wood off its hands. The ploy succeeded – a remarkable coup given that Charter should have been held responsible for JMB's bad deals, not speculating out of them.

Johnson Matthey plc owns, jointly with Rustenburg Platinum (itself substantially owned by AAC and JCI), the capital of Matthey Rustenburg Refiners Pty Ltd (MRR), which operates two platinum refineries: one at Wadeville, South Africa, and the other at Royston, Hertfordshire, in England (200): this plant is the world's largest of its kind. MRR is also building a refinery in Bophuthatswana, one of the South African "homelands" (311). Johnson Matthey is in the curious position of refining copper for the Oman Mining Co LLC and shipping back the gold and silver extracted for sale in the country's souks (200). JM is the sole international sales agent for Rustenburg Platinum: it has the industry at its fingertips (153).

The Oppenheimers: agents of apartheid or seeds of its destruction?

Whole volumes have been inscribed (313)

which tackle the question: is the Oppenheimer dynasty (which no-one denies) responsible for perpetuating the world's most entrenched system of racial inequality (which few would deny), or is it the country's prime economic agent for change?

Certainly modernisation (high technology, advanced communications, free movement of capital across borders) leads to new relationships between bosses and workers, creates new economic classes (black entrepreneurs, shareholders, managers) and leads to certain political shifts. But the process itself is not a liberating one. (It could be argued that the Nazification of Germany in the 1930s and '40s was a prime example of modernisation) (314).

The real question to ask is whether the structural changes which the AAC has definitely brought about in the South African economy have led or will lead to an empowerment of the black majority. The corporation's record of recent oppression of the NUM (examined in detail below) reveals the proverbial mailed fist underlying the velvet glove. Certainly, AAC is neither an agent of the South African government (in many respects the opposite) (315), nor an advocate of an unchanging apartheid system. From its earliest days the need to bring cheap, plentiful labour to the gold and diamond mines and, more recently, to keep it there, brought AAC into conflict with the white miners (312). But it is not an advocate of black majority rule, let alone one accompanied by nationalisation of the mines or banks. The Oppenheimer recipe for peace in southern Africa is a pragmatic bundle of strategies and alliances calculated neither to promote a truly independent non-aligned South Africa, nor alienate a nascent black entrepreneurial class; it is aimed at convincing the outside world (particularly in the USA and Europe) that sanctions will hurt black workers and risk damaging the institutions required for a viable non-apartheid, free enterprise state (316). Perhaps, above all, it is intended to project the Oppenheimers as the true face of liberal, white capitalism. What is good for AAC must be good for South Africa. Little wonder that the Afrikaner/Broederbond-domi-nated body politic has rarely found common cause with the clan, and – as the days of reckoning for the régime come thicker and faster – appears to find AAC almost as threatening as the ANC. (As in 1985, for example, when a team of businessmen led by Gavin Relly, chair of AAC, met exiled leaders of the African National Congress) (317).

It is important not to view the influence of the Oppenheimers in southern Africa as a single band, continuing uninterrupted over the past eighty or ninety years. Ernest was more closely identified with the institutions of apartheid than his son Harry, and there are differences of style and emphasis in public declarations between him and his own son Nicholas. Ernest Oppenheimer supported racially separate housing, the compound system, bantustans: indeed most of the rubric of so-called "petty apartheid". In a speech in 1957, he declared there were "very good grounds for discrimination in South Africa. There [are] different backgrounds, and the legitimate right of the white worker to have his standard of living protected" (318). While he supported blacks having "an effective share in government," this was clearly to be government of the lower echelons. "Above all," he warned, "we should not put uneducated people, still in a semi-barbarous state, in charge of a developing country like South Africa." A "self-respecting Native middle class," he added, was "the perfect guarantee against lawlessness and Communist agitation" (318).

Oppenheimer was fairly obsessed with the dangers of communism, especially one wearing a black skin and destroying the zealously guarded institutions of white South Africa. "Black nationalism is a major danger to the unity, security, and property of South Africa," he once stated (63, 319).

His son Harry hardly differed. Though he was to take a lead in opposing the Group Areas Act (because it put a brake on economic growth using black labour) and a formative role in the Progressive Party (later the Progressive Federal Party, PFP) (8) his watchword was "reform not revolution". In 1984, he claimed that universal suffrage would lead to "chaos and disorder": votes should be restricted to "an élite" (320).

Harry Oppenheimer retired from De Beers in 1984, having pulled the cartel out of its worst trough this century. His son, Nicholas, became deputy chair of De Beers and chair of the CSO. Julian Ogilvie Thompson moved to the main seat at the diamond corporation (321). Gavin Relly, an Oppenheimer man for most of his life, became chair of AAC in 1982.

Zach de Beers became leader of the Progressive Federal Party in 1988 (322).

Between them, Relly and Zach de Beers have consolidated the main planks of their predecessors: increased support (through bodies like the Urban Foundation) for black enterprises and cleaning up the townships (323); opposition to sanctions, (324, 325); campaigning against the Group Areas Act and similar apartheid legislation inhibiting the movement of black labour (168, 325) but refusing to support majority black rule (326).

Labour without love

"...the fruits of [their] labours which have, over the years, filled the vaults of Fort Knox, bought houses for the Oppenheimer family, endowed Rhodes scholarships, and provided Elizabeth Taylor with her finery..." (328).

The price of labour is the biggest single component in the working costs of gold (24). AAC's political complexion – its attitude to government and institutionalised apartheid – has been continually dictated by its need to attract supplies of black labour to the mines, keep them relatively supine, and keep them cheap.

Confronted by an implacable, racist, white miners' union in its early days, and the escalating costs of plumbing ever greater depths for gold, AAC set its teeth against the colour bar (312). The confrontations culminated in a white strike in the early 1920s, which led to the bloody repression of the so-called "Rand Rebellion" (312). From 1911 until 1969, the price of employing white labour in the mines went up 70% – objectionable in itself, but scandalous

when we realise that black workers in all that time got no pay increase (329).

AAC's response to the "white threat" was to employ foreign and migrant labour, encouraging the removal of black families from their land, and, through the iniquitous pass laws and Group Areas Act, to maintain a black labour pool near the mines (327). After the Sharpeville Massacre of 1960, however, it was clear to the Oppenheimers that relations between the miners and workforce could never be the same again. If the black labour pool were to be preserved, and the confidence of the rest of the world (the world of gold investors and buyers) not to be shattered, some black demands must be satisfied. "By the late 1960s, growth had become incompatible with the maintenance of labour policies" (312). Within the next decade, as the gold price increased in world markets, AAC realised that a new skilled black labour force would be required, and wage levels were increased for the "top" 15% of African workers (312).

For a while the recruitment of foreign labour (from Angola, Mozambique, Malawi and elsewhere) and the importation of Chinese workers, answered the need. But between 1974 and 1982, for a variety of reasons (riots, the unreliability of workers from outside, and the independence of the ex-Portuguese territories), non-South African labour diminished. AAC had set itself against the other major mining companies and the government, by withdrawing from wage regulations in 1971 and unilaterally increasing black wages (8). Three years later, as the corporation decided to cut down on immigrant labour, it declared it would open negotiations with black trade unions (330) – a move which was solidly opposed by the régime (331). That year, although Gavin Relly was still advocating the use of migrant labour – with the appalling toll in family separation and human suffering that involves (332) – AAC advocated building houses for black families near the mines (312). Four years later, as the labour shortage eased, the reforms recommended by the Wiehahn Commission – primarily the recognition of black unions and establishment of black training courses (3, 312) – were implemented. The white Mineworkers' Union (MWU) reacted angrily and went on strike. (The MWU's paper, *Die Mijnwerker*, dubbed AAC the "Advancement of Africans Corporation") (333). For the time being, the colour bar remained in place.

During the early 'eighties, AAC mounted a campaign to scrap the infamous 1911 Native Labour Regulations Act – which, *inter alia,* prevented black miners obtaining blasting certificates and qualifying as skilled miners (334) – and to abolish racial discrimination in employment (335). Nicholas Oppenheimer declared that the régime had not kept its promise to allow the corporation to build homes for senior black workers (334), and AAC set up LITET (Labour-Intensive Industries Trust) to increase job opportunities for black workers (334). In 1983, Relly reaffirmed AAC's demand that influx controls must be abolished, because "restricting the supply of unskilled people enhances the demand for skilled labour, and therefore its price" (336).

That year, Relly also spoke of a "fascinating transition from paternalism to bargaining, with all the implications it has on the psychology both of labour and management" (337).

The black miners' strike in 1984 was to transform this "fascination" into the first of AAC's massive dilemmas.

After the South African Chamber of Mines (COM) initiated negotiations with the NUM over the annual wage award, its offer of a 10% rise (as distinct from 25% demanded by the black trade union) had been followed by stoppages at two Anglo mines: Goodehoop and Kriel. In a demonstration against the COM's "package" at the Company's Vryheid Coronation colliery, a black miner was shot dead (338). Fears were expressed in the industry that the NUM – albeit only 15% of the total workforce in the mines then unionised – might embark on its first legal strike, bringing non-unionised workers with them: "Problems for owners thus created hardly bear thinking about," remarked the *Financial Times* (338).

Within a month, the "nightmare" was growing apace. During the country's first legal strike at the President Brand and Western Holding's mines in September (both managed by AAC), police used tear gas to drive strikers from dormitories, and beat them so badly that many required hospitalisation: 13 miners later sued the apartheid state's Minister for "Law and Order" for more than US$1 million (339).

The industrial action quickly spread to five other Anglo mines, and in the ensuing clashes with police, ten miners were killed and hundreds injured (340).

Four months later, in early 1985, the NUM made the apartheid system in mining – specifically job reservation for whites, and the refusal of blasting certificates for black workers – their main bargaining counter with the COM for that year. By then membership of the NUM had grown to 110,000 (339). In April, after the NUM demanded a 40% increase in black wages across the board (119), Anglo sacked four miners at Vaal Reefs. Five hundred miners protested, only to be met by police called in by the management. At least one miner was hit by birdshot or rubber bullets, and died (340). Thousands of mineworkers then struck at Vaal Reefs and Clarksdorp. AAC's response was to sack almost the entire workforce at Vaal Reefs – some 14,400 out of around 16,500 (341). Many of the miners were put on buses to be returned to the Transkei, Ciskei, Lesotho, Swaziland and Mozambique (342). Very soon after, as bombs damaged two of AAC's offices in Johannesburg (341), the company made an offer to reinstate most of the sacked workers, but not before "weeding out" what it called "dissident elements" from those re-applying for jobs (343). NUM leader Cyril Ramaphosa countered that he "... was not aware of miners who were guilty of intimidation and violence" (344). The corporation estimated its loss of gold during this strike at one ton (345), although the losses were to reach four tons by the year end (346). In an interview with the *Financial Times,* Ramaphosa accused AAC of deliberately trying to break the NUM, but failing: forcing AAC to re-employ the majority of sacked workers indicated Union success. The key issue in the dispute, said Ramaphosa, was job reservation for whites. He predicted violence would increase (347). Within a fortnight, another 19 mine workers had died in what was officially described as "inter-tribal fighting" among Xhosa mineworkers at President Brand mine (348).

In July, as the COM offered a wage increase of between 14.1% and 19.6%, the NUM membership balloted in favour of strike action rather than accept the offer (349, 350). However, individual mines appeared to be offering differing increases (349, 351), while the white Council of Mining Unions (CMU) accepted an 11% across-the-board increase instead of its original 20% demand (351). Vaal Reefs – working at 70% of normal levels since May – began actively trying to recruit replacement workers for the 14,000 sacked earlier in the year (163).

By the beginning of August – despite speculation that it was weakening (353) – the NUM had decided on mass strike action to enforce its minimum 22% wage increase claim. By then, it claimed 150,000 members and was recognised at nearly three-quarters of the 40 gold and coal mines in the country (354). Around 80% of its membership was in AAC mines (355). The NUM also threatened a boycott of all white businesses near mines, unless the state of emergency was lifted within 72 hours – and a national strike if the Botha régime attempted to repatriate the 40% of the workforce recruited from outside the Republic, in response to overseas sanctions (356).

The strike was narrowly averted when the NUM accepted new offers made by AAC which were 2.8% above the level unilaterally enforced by the COM, and therefore close to the minimum 22% the NUM had demanded (357).

However, within three months, mineworkers were taking strike action in non-AAC mines, and at the end of February 1986, 19,000 mineworkers came out once again at Vaal Reefs, protesting at the arrest of ten colleagues charged with murder of four team leaders at the mine, and in solidarity with strikers elsewhere (358). It was unclear whether the NUM officially supported this action, given that it was aimed primarily at the police authorities, rather than management. A sympathy strike also started at the Goedehoop colliery, site of the first legal industrial action in 1984 (346). The Vaal Reefs strike was over without major violence within a few days, after talks between the NUM and management (359), although demands for release of the detained men were not met (360). Then, within a fortnight, another fourteen people were killed, and many injured, in what was, once again, officially described as "inter-tribal clashes" between Basotho and Xhosa miners at Vaal Reefs (361). The following month, AAC reported that it was beginning to recover from the effects of the previous year's strikes – Western Deep Levels having managed to restore underground workings badly damaged at that time, although lower gold grades were still limiting production (362).

That summer, the government took draconian measures, under the state of emergency, against the black trade unions in general, imprisoning nearly 1000 militant unionists, including the vice-president of the NUM who was also the president of the Congress of South African Trade Union (COSATU). This followed a "final offer" by the COM of wage increases of between 15% and 20%, some 10-15% below the levels demanded by the NUM (which had already modified its demands for a 45% increase that year). The NUM met in secret to decide what action to adopt (363), as 1200 miners came out at the Finsch diamond mine of De Beers, following strike action a week earlier at four other De Beers mines (364).

In an event overshadowing even the deaths at the hands of the police, in September 1986 black mineworkers were plunged into mourning as 177 gold miners died in the second biggest mine disaster in the country's history, at the Kinross gold mine in the Transvaal, operated by Gencor. A stay-away protest at this holocaust, by 325,000 miners and 275,000 other black workers, in early October, demonstrated not only the gathering strength of the NUM (365) but also the anger with which black workers regard their exploitation in the most dangerous areas of the mines. As a report by a researcher at the University of Witwatersrand, Jean Leger, pointed out: the black death rate in mines is not only far higher than that of whites (a ratio of five to two) but has not materially altered since 1941 (1.96 per thousand that year, and 1.62 per thousand forty-three years later). Although one of the major NUM demands (as evinced in the 1985 strikes, specifically at AAC mines) had been the abolition of job reservation, in fact the industry would grind to a halt were this to be rigidly applied. What happens, reported Leger, is that black workers work in white jobs where whites cannot be found, but without sufficient training or safety measures (366).

Hopes of a settlement between the NUM and the COM rose in October as the union agreed wage increases of up to 19.5% with AAC's gold/uranium producer, Ergo (367). A little later, the NUM lowered its threshold to a 24% demand across the board, while AAC, JCI, Gencor and Rand Mines' gold division increased their offer half a point up to 23.5% (368). In November, yet more miners died in a dispute over the opening of a beer hall at Vaal Reefs (which had been opposed by the NUM) (369). Within ten days, 20 more miners had died at Vaal Reefs during factional fights; the NUM was quite clear that the oppressive, claustrophobic conditions imposed on migrant workers in the single-sex compounds made frustration and tension, leading to violence, inevitable (39).

By the beginning of 1987, more than hundred mineworkers had died from this so-called "intergroup" violence (372) – three-quarters of them at mines operated by AAC. Opinions as to why the violence broke out naturally differed. A researcher for COM, Kent McNamara, partly blamed the NUM for imposing a "you are with us or against the struggle" ultimatum, which deepened tensions already existing between South African-born workers, and foreign workers from neighbouring states (372).

As early as November 1985, AAC and the NUM had set up a joint commission of inquiry into the group fighting at Anglo mines. Entitled *Reaping the Whirlwind,* the resultant report was leaked to the *Johannesburg Weekly Mail* in early

1987. The company argued that tensions in the workplace led to violence outside – hardly a novel thesis! But the tensions were ascribed sometimes to "agitators or more radical elements", although the company agreed that linking job status to a particular ethnic group could give rise to conflict. What the company failed completely to recognise (or admit) was that the NUM, far from provoking rivalries imposed by the hostel, migrant labour, and job reservation systems, had simply enabled them to be expressed more openly. In opposing apartheid *per se,* the NUM was providing the only formula for avoiding internecine and self-destructive violence in future. Anglo was clearly blaming the messenger for the message of doom, because it was unable and unwilling to take on the government itself where it mattered.

In its contribution to the commission report, the NUM cogently argued that not only the migrant system and its attendant evils, but also exploitation by management, were at the root of the violence. "Conditions in the hostels are not only dehumanising but also impose a system of controls upon individuals that make migrants vulnerable to manipulation. Unable to take responsibility upon themselves, workers are put in a defensive situation which predisposes them to violence. The only source of security then becomes the home group and the tribe, and it is the exploitation of this base which gives the violence an ethnic or tribal character" (372).

The NUM also claimed that, while AAC had ostensibly been moving away from compounds on ethnic lines, it "deliberately stalled or even reversed" integrated accommodation, soon after the NUM rose in strength. Not only had AAC been actively trying to destabilise the NUM, it stereotyped a particular branch by the ethnic majority represented within it, thus deliberately fanning sectarian flames. An even more serious allegation made by the NUM was that AAC employed a special force, the Emergency Protection Unit, whenever there were incidents of mine violence, and that it attacked workers along ethnic lines – at Vaal Reefs they "knew which blocks to attack and which ones to leave out" (372). In addition, terror groups known as the "Russians" had been used by the management. At Western Deep Levels they had dressed as Sothos and attacked Xhosa-speaking miners: it was this which precipated the so-called "faction fighting" in 1986. The NUM also claimed that Anglo had set up a quasi-terrorist organisation called Fito, which threatened the lives of NUM workers, specifically at President Steyn mine, causing the violence there at the end of 1986.

The NUM called for an end to the migratory system and, in the meantime, for AAC to stop exploiting ethnicity within the workforce (372).

In February 1987, NUM President James Motlatsi, revealed that the union was now the biggest in the country (total signed up membership being 344,000, and paid-up membership 227,590). He pledged a total fight against the migratory labour system, and the *induna* (tribal control) system operating at the mines (352). At its annual Congress a little later, the Union re-affirmed its support for the African National Congress (ANC), for sanctions against South Africa, and nationalisation of the mines (373), although the latter would be approached with some caution (374).

The NUM went into 1987 lodging a 55% pay rise with the COM, and demanding a declaration of intent by the Chamber that the migrant labour and hostel system would be disbanded; the *indunas* would be replaced with elected worker representatives (374); and there be direct consultations on the introduction of mechanisation in mines which (as at Randfontein and Western Areas) threatened to reduce the workforce by between a quarter and a third (375).

For its part, the COM went into 1987 having failed to get the militant white unions to agree to the dropping of job reservation (375), and far from united in its attitude to the iniquitous migrant labour system (375).

By July, the recipe for further protests and bloodshed was compounded when, after granting the white mineworkers wage increases of 13.5-14.5% against their demand of 20% (163), COM offered increases of only up to 17.5% to the NUM, which had been demanding a hike of 30% (scaled down from its initial demand of 40-55%). In a patronising statement issued on 23 June, the President of the COM, Peter Gush – also a director of AAC – refused to condemn the migrant labour system outright, arguing that, while the Group Areas Act needed to be removed, "considerable progress" had been made in enabling black employees "who wish to become home owners" to do so. While regretting the "inevitable" plight of the large proportion of the workforce remaining in hostels, the President of the COM gushed lyrically about the "vital economic stimulus to the territories and countries of southern Africa" provided by black wages, "... a lifeline which is currently worth in the region of 1 billion Rand a year, and constitutes an important stabilising socio-political factor in the region" (376). Notwithstanding this "lifeline" the COM rejected the NUM's demands, fixed the wage increases at between 17% and 23% (377) and set the scene for what was to become the biggest strike action seen in South Africa since the Rand Rebellion of 1922-24 – this had been the occasion when white miners rose against the liberalisation of the colour bar, and in consequent riots suffered 250 dead and numerous imprisoned and wounded (378).

August 9 was chosen by the NUM as the strike day: around 200,000 black mineworkers would come out at 28 gold mines and 19 collieries (379, 380). Although the NUM was manifestly opposed to apartheid, demands were scrupulously formulated not to give the régime an excuse to intervene (380). Added to the wage demand were demands for thirty days annual leave, improved danger pay, death benefit, and an officially recognised Soweto Day on 16 June (the only demand which could be construed as political) (381). The NUM also counselled strikers to return to their homes during the strike – thus intensifying pressure on the companies, and distancing unionists from any violence which might be incited (381). A demand by the NUM to the COM that it would not use private security forces, call in the police, or deprive miners of food during the confrontation, was pointedly ignored by the COM (382). Peter Gush, in a combative statement, denied that the management had ever used violence to strike-break: "Violence has been used to force people to strike," he declared (383).

The "trial of strength" (384) duly started on 9 August. The NUM claimed that 42 of the 46 mines where the dispute was declared had responded to the strike call, and that 340,000 miners had joined in. Even the COM acknowledged that up to 230,000 miners had downed tools (358). On the first day, there were several clashes but no deaths. At Vaal Reefs, according to the NUM, workers were forced underground at gunpoint. Peter Gush maintained the NUM had been locking out workers unsympathetic to the strike, and agree that security officials "had to open fire with rubber bullets on strikers at two [Anglo] mines during attacks" (385). On the third day of the action, as arrests were gathering pace – 78 of them at Vaal Reefs for conspiracy to murder (386) – AAC would not even acknowledge which of its mines had been affected by the strike (32). "The Bosses get richer, the workers get poorer, now is the hour!" read the banner headline on the "strike special" of the *NUM News* (385). In an interview with the *Guardian,* Cyril Ramaphosa revealed that he had never been a miner, but from his early days was affected by the plight of mineworkers – his grandfather having worked in the "Big Hole" of De Beers in the early part of the century. "I felt that the oppression our people were subjected to was really concentrated in the mines" (387).

On 14 August, AAC called in police to end what it called an "illegal" sit-in at the Ergo plant, and used rubber bullets to "control" the situation at Western Deep Levels, where it claimed 700 armed men had attacked the mine security forces: it also claimed that strikers were beginning to return to work (388).

As fighting broke out at Vaal Reefs, and an independent monitoring group revealed AAC had already lost 90 million Rands (around £27 million) in the previous week (389), the NUM and AAC met to try to resolve the issue of "violence", in a Johannesburg hotel (390). Talks

ended abruptly, however, when news reached the NUM of a police attack on strikers outside President Steyn: Ramaphosa accused AAC of perfidy in calling in police, precisely when talks were in progress (391). "Anglo American is a treacherous, cowardly and ruthless organisation" he declared (392). The following day, 2000 miners ignored a threat by AAC – dropped on their heads by helicopter in the form of leaflets – that they would lose their jobs at No. 6 shaft at Vaal Reefs unless they returned to work (393). By then, AAC was losing US$8 million a day in profit (394).

At AAC's annual general meeting on 20 August, Gavin Relly claimed that the industry had done its best to "facilitate the emergence of independent free trade unions" (392), and that this was not incompatible with the company's intransigence on bargaining (395). At the GFSA Libanon mine, the following day, a worker was killed as security forces tried to compel strikers back to the pits (396). By then some 250 miners had been injured in clashes with police (392).

With the strike nearing its third week, both the NUM and the COM sat down to talk, the AAC somewhat relenting on its previously implacable attitude (397): it extended sacking deadlines at both Western Holdings and Vaal Reefs (398). The talks once more ended abruptly – although the NUM modified its minimum wage demand to 27%, the COM was not prepared to talk about pay, danger money, or a Soweto anniversary holiday (399). The NUM's response was to vote unanimously to continue the strike, several hundred men rejoining it after earlier threats of dismissal, thus confirming the NUM's allegation that previous return-to-works were mainly in response to management pressure (400).

It was at this point that the velvet glove fell away decisively from Anglo's mailed fist. As 3000 workers staged a sit-in at Western Deeps – the very shaft where workers had been given an ultitmatum to "dig or get off the pit" (401) – the company summarily sacked 16,000 miners, threatening more than 18,000 others. The *Financial Times* correspondent in Johannesburg,

Jim Jones, appropriately commented: "Anglo-American, which considers itself the most liberal of the mining houses, has adopted a divide and rule policy towards the strikers. Its strategy is to reduce support ... by selective dismissals, or threats of dismissal, but this has not yet produced any obvious weakness of the strikers' determination" (402). On 28 August, AAC appears to have, once again, relented somewhat in its attitude (it was becoming decidedly schizophrenic – but maybe that was policy too) and temporarily halted the dismissals. "If we fired all the unionists, who would we have left to talk to...?" one official asked (403). On 30 August the NUM reluctantly decided to return to work (404), as, once again, AAC threatened mass dismissals (405). The same day AAC claimed that 90% of the workforce had returned to the mines (405).

In an indicative, and important, statement just after the strike, AAC's head of labour relations, Bobby Godsell, reiterated his belief that recognising the NUM, and at the same time fighting it tooth and claw, was logical and appropriate. "... what South Africa needs is not a black/white love-in, but a situation where blacks and whites have shared institutions and collective bargaining ..." he declared. What both AAC, the industry, and the NUM should jointly fear, he said, is low-cost gold production from outside South Africa, which is taking away the state's share of the market (406). Here, encapsulated, was AAC's most intellectualised view of labour relations under apartheid: no question that control of mining as a vital sector of the economy should remain outside the hands of the black majority; no question that the real threat to the industry was an enemy without, rather than the state within.

No fewer than 10,000 black workers had been laid off at Vaal Reefs alone between 1986 and 1987 (407); 1987 had seen 118 work stoppages and strikes, and 110 miners had died as a result (408). Once again, however, the carrot accompanied the stick. In late 1987, AAC offered free shares to employees, in an "unprecedented" scheme designed to give the black workforce a 5% interest in AAC (409). Though, by the middle of 1988, the corporation claimed that 64% of its employees had taken up the offer (410), Cyril Ramaphosa declared: "It stinks ... what matters is not wealth in the future but wages now" (411).

Amendments were finally made to the Mines and Works Act in 1987, enabling people of all races to acquire a blasting certificate (412). The "scheduled person" definition was scrapped (413) but only to be replaced by a new concept of "competent person", necessitating new qualifications (such as language and literacy) (407) while an advisory committee was set up as a "screen" (413), against the reality that one step had been taken forward only for another to be made back. Amendments to the Group Areas Act, which would still have legalised forced, removals and the creation of whites-only areas, were opposed by black political parties the following year (414).

In the first half of 1988, AAC offered to reinstate just over half of the 18,000 workers still unemployed after the 1987 strikes, and to compensate some of them for loss of earnings. Reportedly, AAC and the NUM agreed on a code of conduct on workplace violence and union accountability (415). In early May, 18,000 workers again demonstrated at the mines – this time against the new Labour Relations Amendment Bill, which would allow companies to sue unions for illegal strikes, bar secondary picketing (solidarity strikes), and weaken the powers of the Industrial Court (416). A month later, no fewer than two million black workers stayed away in a 3-day protest against this legislation – without doubt the biggest non-violent demonstration in South African history (417). However, due to an interdict brought by AAC against NUM officials in the Orange Free State, relatively few miners stayed out (417).

In July, the NUM and the COM reached agreement for 1988 pay rises of between 13% and 16.5% – 5.5% lower than demanded by the NUM (418). The NUM was naturally "not entirely satisfied" but admitted that management attitudes had considerably stiffened over the past twelve months (419).

This, then, was the view from the ground –

from the underground, in fact – of South Africa's biggest corporate entity, as it battled to preserve its power, and control events during the traumatic years 1984-88. As this entry is being written, the saga is far from over. Indeed, it may only be beginning, as the NUM reflects on the ability of the industry, and AAC in particular, to agree to wage hikes and ameliorative measures, combined with a willingness to employ violence and throw thousands of miners on the scrap-heap, thus undercutting the NUM as a political force. A fundamental question for the NUM must continue to be: to what extent do you accept the parameters of advancement laid down by your employers, when their own interests are so deeply divided? On the one hand, the AAC is dependent on the oppressing state, and often co-terminous with it; on the other hand the AAC is dependent on black labour, and therefore forced to negotiate with it (420).

Part of the answer to this question must derive from a historical analysis of AAC as a monopoly power. Part must lie in deciding what alternative deployment of capital and exploitation of labour is open to the corporation should the final confrontation be forced by the black workforce. (This in itself begs the question of whether a militant black union could survive without the hands that both bite and feed it.) Part of the answer must come from an examination of AAC's own attitude to labour vis-à-vis the state: put simply, has the leadership of AAC ever demanded a reformation of the apartheid system, which does not simply free surplus labour for its own ends, but empowers workers and their communities to take control of their own future?

The answer to that, it is quite clear, is a resounding "No!".

Further reading: The bibliography on AAC, its associated operations, the Oppenheimers, and the mining industry in South Africa, is very considerable. In one way or another, the corporation has attracted more attention from researchers than any other mining company. The first works named in this list are indispensable for a full understanding of the strategies employed by AAC both inside and outside South Africa. The reader is also referred to the many articles quoted in the text of this entry, which are not repeated here, to save space.

David Pallister, Sarah Stewart, and Ian Lepper, *South Africa Inc*, originally published by Simon and Schuster, London, 1987, revised and undated for Corgi Books, London, 1988. (Not only the required text for an understanding of AAC, but surely one of the best books ever written on the operations of a multinational. This entry in *The Gulliver File* owes a great deal to the work.)

Duncan Innes, *Anglo American and the Rise of Modern South Africa*, Heinemann Educational Books, London, 1984. (A Marxist interpretation of the rise of AAC as supreme practitioner of monopoly capitalism. Recommended.)

Greg Lanning and Marti Mueller, *Africa Undermined: A History of the Mining Companies and the Underdevelopment of Africa*, Penguin Books, Harmondsworth, 1979. (Invaluable for its tracing of the expansion of the Oppenheimer influence throughout southern Africa. An updated version is now sorely required.)

Merle Lipton, *Capitalism and Apartheid: South Africa 1910-1984*, Gower/Maurice Temple Smith, London, 1985. (Just as it purports, an incisive and readable study of the relationships between the apartheid state, the white and black workforce and the companies.)

Also worth reading: Sandy Boyer, *Black Unions in South Africa*, Africa Fund, New York, 1982.

Brian Bunting, *The Rise of the South African Reich*, Penguin, Harmondsworth, 1964, revised 1969.

Gwendolen M Carter, *The Politics of Inequality: South Africa since 1948*, Thomas and Hudson, London, and Praeger, New York, 1958.

Simon Cunningham, *The Copper Industry in Zambia: Foreign Mining Companies in a Developing Country*, Praeger, New York, 1981.

Robert H Davies, *Capital, State and White Labour in South Africa, 1900-1960*, Harvester Press, Brighton, 1979.

Rob Davies, Dan O'Meara and Sipho Dlamini, *The Struggle for South Africa: a reference guide to Move-*

ments, *Organisations and Institutions,* (two volumes), Zed Books, London, 1984.

Fion De Vletter, *Migrant Labour in the South African Gold Mines: An Investigation into Black Worker Conditions and Attitudes,* ILO, World Employment Programme, Migration for Employment Project, 1972.

GV Doxey, *The Industrial Colour Bar in South Africa,* Oxford University Press, London, 1961.

Edward Jay Epstein, *The Diamond Invention,* Hutchinson, London, 1982.

Ruth First and Johnathan Steele with Christabel Gurney, *The South African Connection: Western Investment in Apartheid,* Penguin Books, Harmondsworth, 1973.

SH Frankel, *Capital Investment in Africa: Its Cause and Effects,* Oxford University Press, London, 1938.

Sir Theodore Gregory, *Ernest Oppenheimer and the Economic Development of Southern Africa,* Oxford University Press, Cape Town, 1962.

Richard Hall, *The High Price of Principles: Kaunda and the White South,* Penguin Books, Harmondsworth, 1969.

Joseph Hanlon, *Beggar Your Neighbour: Apartheid Power in Southern Africa,* Catholic Institute for International Relations and James Currey, London and Indiana University Press, Indiana, 1986.

Anthony Hocking, *Oppenheimer and Son,* McGraw-Hill, New York, 1983.

Edward Jessup, *Ernest Oppenheimer: A Study in Power,* Rex Collings, London, 1979.

Frederick A Johnstone, *Class, Race and Gold,* Routledge and Kegan Paul, London, 1976.

Fred Kamil, *The Diamond Underworld,* Allen Lane, London, 1979.

Ruth Kaplan, *The Anglo-American Corporation of South Africa Ltd, Investments in North America,* The Africa Fund, New York, 1982.

William Minter, *King Solomon's Mines Revisited: Western Interests and the Burdened History of Southern Africa,* Basic Books, New York, 1986.

Vella Pillay, *The Role of Gold in the Economy of Apartheid South Africa,* UN Centre Against Apartheid, New York, 1981.

Carroll Quigley, *The Anglo-American Establishment,* Books in Focus, New York, 1981.

Ann Seidman, Neva Sediman Makgetla, *Outposts of Monopoly Capitalism: Southern Africa in the Changing Global Economy,* Lawrence Hill/Zed Books, London, 1980.

Anthony Sampson, *Black and Gold: Tycoons, Revolutionaries and Apartheid,* Hodder and Stroughton, London, 1987.

Geoffrey Wheatcroft, *The Randlords: The Men who made South Africa,* Wiedenfeld and Nicholson, London, 1985.

Francis Wilson, *Labour in the South African Gold Mines, 1911-1969,* Cambridge University Press, 1972.

F Yachir, *Mining in Africa Today Strategies and Prospects,* Zed Books, 1988.

Contacts: Your local anti-apartheid group; or Anti-Apartheid Movement, 13 Mandela St, London NW1. (They can also supply a list of international, regional and local anti-apartheid groups and associated organisations.)

The Struggle for South Africa: A reference guide to movements, organisations and institutions (Rob Davies, Dan O'Meara, Sipho Dlamini, Zed Books, London, 1984), contains the most comprehensive list of organisations concerned with various aspects of AAC and capitalism in southern Africa (as at 1984) but, unfortunately, no addresses.

Minewatch, 218 Liverpool Rd, London N1 1LE, England (tel 071-6091852) regularly monitors the operations of all AAC's main companies.

References: (1) *Raw Materials Study Group Survey,* Stockholm, 1987; see *FT* 2/10/89. (2) *Janes World Mining: Who Owns Whom,* Janes Yearbooks, London, 1970. (3) David Pallister, Sarah Stewart, Ian Lepper, *South Africa Inc, The Oppenheimer Empire: The secrets behind one of the most influential organisations in the world,* revised edition, Corgi Books, London, 1988. (4) *Weekly Mail,* Johannesburg, 7-13/3/86. (5) *FT* 11/8/83. (6) *MJ* 6/7/84; see also RMR, Vol.3 No.2, 1984. (7) *MJ* 6/12/85. (8) Rob Davies, Dan O'Meara, Sipho Dlamini, *The Struggle for South Africa: A reference guide to movements, organisations and institutions,* Zed Books, London, 1984, (9) Duncan Innes, *Anglo-American and the Rise of Modern South Africa,* Heinemann Educational Books, London, 1984. (10) Anthony Sampson, *Black and Gold: Tycoons, Revolutionaries and Apartheid,* Hodder and Stroughton, London, 1987. (11) *Die Afrikaner* 20/2/70. (12) *MMon* 9/88. (13) Tanzer. (14) See also *Weekly Mail,* Johannesburg, 9-15/8/85. (15) *South* magazine, London, 7/86. (16) *MAR* 1989. (17) *MJ* 21/4/89. (18) *FT* 18/10/83. (19) *FT* 19/10/83. (20) *FT* 1/2/84; *FT* 8/3/85. (21) *FT* 25/11/87; *FT* 20/7/87. (22) *FT* 13/10/83. (23) *FT* 27/10/83. (24) Lanning & Mueller. (25) *Financial Mail,* Johannesburg, "Inside the Anglo Power House", 4/7/69. (26) *Economist,* London, 10/1/87. (27) Brian Bunting, *The Rise of the South African Reich,* Penguin Books, 1969. (28) *AFR* 10/3/83. (29) *Raw Materials Report* Vol. 2 No. 3, 1983. (30) *FT* 25/3/83. (31) *MJ* 4/8/89. (32) *FT* 13/8/87. (33) *Diamond Review,* Vol. XVIII, No. 10, 31/10/86. (34) *MJ* 10/3/89. (35) *MJ* 3/10/88. (36) *FT* 29/9/86. (37) *MAR* 1988. (38) *African Defense,* 2/86. (39) *FT* 14/1/87. (40) *De Beers Annual Report,* Johannesburg, 1963. (41) *Gua* 2/1/89; *Times* 1/2/89. (42) *Times* 4/8/84; see also *MJ* 21/11/86. (43) *MJ* 8/8/86. (44) *Gua* 8/10/88. (45) *FT* 9/6/88. (46) *Labour Research,* London, 7/88. (47) *WDMNE.* (48) *De Beers Annual Report* 1976. (49) *FT* 11/12/87.

(50) *FT* 15/12/82. (51) *MAR* 1987. (52) *FT* 13/8/87; *MJ* 6/3/87. (53) *Independent* 15/7/89; *MJ* 5/5/89. (54) *MJ* 21/11/86; *FT* 19/3/83; *E&MJ* 6/84; *FT* 16/3/85; *FT* 22/8/84. (55) *MJ* 18/3/88; see also *E&MJ* 5/87. (56) *FT* 7/7/88. (57) *FT* 16/12/89. (58) Charles Johnson, Martyn Marriott, Michael von Saldern, "The World Diamond Industry 1970-2000" in *Natural Resources Forum,* USA, 5/89. (59) *MJ* 6/3/87. (60) *Economist* 10/1/87. (61) *MM* 3/89. (62) *Gua* 24/3/83. (63) See also Sir Theodore Gregory, *Ernest Oppenheimer and the Economic Development of Southern Africa,* Oxford University Press, Cape Town, 1962. (64) Ernest Oppenheimer, Speech to the Duke of Edinburgh's Study Conference on Human Relations in Industry, Oxford, 1956. (65) Patrick Keatley, *The Politics of Partnership,* Penguin Books, 1972. (66) *MAR* 1972. (67) *Gua* 27/3/84. (68) See also TH Bingham QC and SM Gray, *Report on the Supply of Petroleum and Petroleum Products to Rhodesia,* London, 9/78. (69) *Business,* London, 8/88. (70) *FT* 24/6/88. (71) *E&MJ* 3/87. (72) *FT* 4/8/89. (73) *Windhoek Advertiser* 11/3/86; *WA* 7/3/86; *WA* 13/3/86; see also P

Thirion, *Commission of Inquiry into Alleged Irregularities and Misapplication of Property in Representative Authorities and the Central Authority of South West Africa,* Windhoek, 1986. (74) *Namibian* 14/3/86. (75) *Rand Daily Mail* 1/8/84. (76) See also *MJ* 14/2/86 and Granada Television *World in Action,* 1988. (77) *MJ* 30/1/87. (78) *E&MJ* 3/37; see also *MJ* 9/10/87. (79) *Windhoek Observer* 26/5/87; *Windhoek Advertiser* 7/10/87. (80) *E&MJ* 7/89. (81) *MJ* 24/6/88. (82) David Kaplan, *The Internalisation of South African Capital,* African Affairs, 10/83. (83) *Los Angeles Weekly,* quoted in (15). (84) *MJ* 1-8/1/88. (85) *MJ* 10/10/88. (86) *MJ* 16/11/84; *FT* 15/11/84. (87) *FT* 11/7/85. (88) *Raw Materials Report,* Vol. 3 No. 1, 1984. (89) *MJ* 6/5/89. (90) Republica Federativa do Brasil, Congresso, Camara dos Deputados, *Comissão Parlamentar de Inquérito para investigar o comportamento e as influências das impresas multinacionais e do capital estrangeiro no Brasil, Relatorio e concluses,* Diario do Congresso Nacional (Secção 1 – Suplemento) 1976; *Comissão Parlamentar de Inquierito destinada a evaluar a politica mineral brasileira, ibid* 1978. (91) *MJ* 6/5/88. (92) *MJ* 13/3/87. (93) *MJ* 15/4/88. (94) *Optima,* Vol. 32, No. 4, 12/84. (95) *FT* 19/11/88. (96) *E&MJ* 12/84; see also *MJ* 1/7/83. (97) *E&MJ* 10/87. (98) *E&MJ* 6/88. (99) *FT* 5/10/82.

(100) *MJ* 12/2/88. (101) *MJ* 28/3/88. (102) *FT* 20/10/87. (103) *FT* 17/11/84. (104) *E&MJ* 8/89. (105) *MJ* 25/3/83; *MJ* 1/7/83. (106) *FT* 24/1/84. (107) *AAC Annual Report 1984,* Johannesburg. (108) *AAC Annual Report 1976,* Johannesburg. (109) *MJ* 4/8/89. (110) *Age* 30/1/77. (111) *North Queensland Messagestick,* 10/81. (112) *Tribune* 25/10/77. (113) *Tribune* 29/3/78. (114) *Age* 18/3/78. (115) Roberts. (116) *Mining* 1987. (117) *Gua* 8/7/83. (118) *FT* 8/2/89. (119) *FT* 29/4/85. (120) *FT* 22/12/82. (121) Jan Roberts' personal communication to author, 9/89. (122) *MJ,* Gold Supplement, 13/5/88. (123) *MJ,* Gold Supplement, 19/5/89, "Gold '89 – the outlook for supply and demand"; see also *MJ* 10/6/88, *MJ* 21/4/89, *MJ* 10/3/89, *MJ* 31/3/89. (124) *FT* 7/12/88; *MJ* 29/1/88; see also (122). (125) *FT* 19/7/88. (126) *FT* 2/12/86. (127) *Gua* 9/3/89. (128) *Sallies Annual Report,* Johannesburg, 31/12/88. (129) *MJ* 25/6/86. (130) *E&MJ* 9/86. (131) *E&MJ* 11/86. (132) *MJ* 27/6/86. (133) *E&MJ* 9/87. (134) *MM* 4/88. (135)

MJ 4/4/88. (136) MJ 3/4/87. (137) RMR, Vol. 3
No. 2, 1985. (138) FT 22/9/88. (139) See also FT
1/7/88. (140) MM 1/86; MJ 22/11/86; see also MJ
27/12/83. (141) MJ 7/12/84. (142) MJ 22/11/85.
(143) MJ 29/11/85. (144) FT 16/12/85. (145) FT
21/2/84; see also FT 5/12/84, FT 8/12/84. (146) FT
9/2/85. (147) FT 26/11/85. (148) FT 23/11/85.
(149) FT 21/11/85.
(150) FT 29/1/86. (151) Gua 13/2/85. (152) E&MJ
3/88. (152) Johnson Matthey, Platinum 1988, London, 1988. (153) Johnson Matthey, Platinum 1989,
London, 1989. (154) See also FT 13/6/88, FT
31/1/86, FT 27/7/88. (155) FT 13/10/87; E&MJ
12/87. (156) MJ 2/11/84; FT 12/6/84. (157) MJ
6/3/87. (158) FT 30/11/84; FT 13/11/85. (159) FT
30/11/85. (160) FT 6/7/88. (161) Rogers & Cervenka. (162) MJ 13/6/86. (163) E&MJ 7/87. (164) FT
10/5/88. (165) Obs 6/12/87. (166) MJ 23/6/89.
(167) see also EP Gush, chair of Vaal Reefs, statement in Annual Report, 31/12/88. (168) MJ 4/4/86.
(169) MJ 25/7/86. (170) FT 5/4/89. (171) MJ
27/6/86. (172) UX Report, Vol. 2 No. 43, 24/10/88.
(173) Nuexco Annual Review, 4/89. (174) E&MJ
12/87. (175) FT 13/12/84. (176) MJ 15/4/88, MJ
15/7/88. (177) Rand Daily Mail 16/8/76. (178)
OECD/IAEA, World Uranium Potential, Paris,
1978. (179) The Star, Johannesburg, 8/5/65; Rand
Daily Mail 29/5/65. (180) African Development,
Zambia supplement, London, 10/73. (181) MJ
15/2/80. (182) MJ 8/7/83. (183) FT 25/10/85; see
also MJ 5/7/88. (184) FT 28/10/85. (185) Metals
Bulletin 6/9/85. (186) MM 1/86. (187) MJ 25/9/87.
(188) FT 14/7/89. (189) Gua 9/9/86; Bluf, Amsterdam, 8/86; for a full resumé of this scam, see (3).
(190) See also Statement by Minorco, Bermuda; FT
18/9/87. (191) MJ 2/10/87. (192) MJ 15/1/88.
(193) MJ 25/3/88. (194) FT 10/10/81. (195) FT
8/4/81. (196) Engelhard Corp Annual Report 1986.
(197) Gua 1/11/88. (198) MMon 1/86. (199) MJ
10/6/83.
(200) MIY 1987. (201) FT 23/8/84. (202) FT
7/3/83. (203) FT 12/6/84. (204) FT 20/10/84; FT
17/10/84. (205) E&MJ 2/88; MJ 25/12/87. (206)
FT 18/5/89. (207) MJ 23/9/88; FT 23/9/88. (208)
Gua 14/11/88. (209) Times 3/10/88. (210) FT
6/10/88. (211) FT 12/10/88. (212) FT 27/10/88.
(213) MJ 14/10/88. (214) See also FT 18/10/88.
(215) MJ 21/10/88. (216) FT 26/10/88; Gua

26/10/88. (217) Obs 25/6/89. (218) FT 18/11/88.
(219) Gua 22/9/88. (220) Gua 7/11/88; Gua
14/9/88; FT 14/8/88; FT 14/9/88; FT 31/10/88.
(221) FT 22/9/88. (222) Paul Johnson, Consolidated
Goldfields: A centenary portrait, 1960-1980, Weiden-
feld and Nicholson, London, 1987. (223) FT
29/1/87. (224) MJ 8/8/80. (225) MJ 5/3/81. (226)
FT 20/10/81. (227) FT 11/10/87. (228) MJ
12/12/86. (229) FT 7/10/87. (230) See also MJ
23/10/87; MJ 6/5/87. (231) MJ 8/4/88; see also MJ
5/2/88, MJ 19/2/88, FT 8/4/88, MJ 15/7/88. (232)
FT 22/11/88; FT 2/11/88. (233) Dept of Trade and
Industry, Consolidated Gold Fields Ltd, Investigation
under Section 172 of the Companies Act, 1948, London, 1980. (234) Times 14/4/80. (235) Independent,
London, 23/9/88. (236) FT 19/10/88. (237) FT
28/9/88. (238) Gua 19/10/88. (239) FT 1/11/88.
(240) See also FT 14/10/88. (241) FT 20/10/88;
Gua 20/10/88; MJ 4/11/88; FT 6/3/89; Gua
23/12/88; see also MJ 28/10/88, FT 24/10/88, Gua
6/3/89, SunT 5/3/89. (242) Obs 23/4/89; MM
10/89. (243) FT 27/9/88. (244) Obs 26/2/89. (245)
Obs 16/4/89; FT 18/4/89. (246) Gua 14/6/89; FT
14/6/89. (247) Gua 23/11/88. (248) Gua 14/10/88.
(249) Gua 24/10/88; Obs 20/1/89.
(250) Independent 24/4/88. (251) Gua 26/9/88.
(252) See also FT 29/9/88, FT 3/6/89. (253) Obs
16/10/89; Gua 17/10/88; FT 17/10/88. (254) Consgold file on the AAC/Minorco bid, London, 1988,
released for limited circulation, copy available from
The Gulliver File author for a negotiable fee. (255)
FT 4/1/89. (256) FT 30/9/88. (257) Obs 30/10/88.
(258) FT 20/10/88. (259) FT 25/10/88; Independent
25/10/88. (260) Obs 1/1/89; FT 3/1/89. (261) Gua
11/10/88. (262) FT 10/10/88. (263) Brian Bolton,
South African Investment in Britain, ISKRA Research, 56 Old Rd, Bromyard, Herts, England,
1988. (264) FT 7/1/89. (265) Gua 23/1/89. (266)
FT 25/1/89; see also FT 27/1/89. (267) FT 1/2/89.
(268) MJ 3/2/89. (269) Independent 18/2/89. (270)
Gua 3/2/89. (271) FT 3/2/89. (272) Gua 18/2/89;
FT 18/2/89. (273) FT 19/2/89. (274) MJ 17/3/89;
Obs 19/2/89; MJ 24/2/89; see also FT 21/2/89, Gua
21/2/89. (275) Gua 24/2/89. (276) FT 27/2/89.
(277) MJ 17/3/89; see also Gua 4/3/89. (278) FT
10/3/89. (279) MJ 3/3/89. (280) MJ 31/3/89. (281)
FT 18/3/89; see also Obs 19/3/89. (282) Independent
24/3/89. (283) FT 23/3/89; FT 25-26/3/89. (284)

FT 12/6/89; FT 13/6/89. (285) FT 4/7/89. (286) FT 11/1/89. (287) FT 11/4/89; Gua 11/4/89. (288) FT 19/4/89. (289) Gua 25/4/89; see also FT 24/4/89, FT 25/4/89, FT 26/4/89. (290) FT 27/4/89. (291) See also Gua 27/4/89. (292) FT 28/4/89. (293) Gua 10/5/89. (294) FT 12/5/89; see also FT 13/5/89. (295) FT 13/5/89. (296) FT 17/5/89; see also FT 15/5/89. (297) Gua 18/5/89; FT 18/5/89. (298) FT 5/7/89. (299) Gua 23/6/89. (300) Obs 22/7/84. (301) Gua 6/7/89. (302) Gua 22/9/89. (303) Gua 17/5/89. (304) ELTSA, A Kaffir is just a Kaffir, London, 10/88. (305) Gua 8/11/88. (306) Gua 22/9/88. (307) FT 21/3/89. (308) FT 22/3/89; FT 8/2/89. (309) Richard Hall, The High Price of Principles, Penguin Books, London, 1973. (310) Anderson Strathclyde brochure to shareholders, 17/3/83. (311) FT 21/1/88. (312) Merle Lipton, Capitalism and Apartheid: South Africa 1910-1984, Gower/Maurice Temple Smith, London, 1985. (313) See especially (3, 9, 312). (314) Detlev JK Peukert, Inside Nazi Germany, Conformity: Opposition and Racism in Everyday Life, Penguin Books, 1989. (315) FT 26/10/88. (316) Gua 13/7/88. (317) FT 13/9/85. (318) Star 6/8/57. (319) Anthony Hocking, Oppenheimer and Son, McGraw-Hill, New York, 1973. (320) Manchester Weekly Guardian 26/2/84. (321) Gua 28/11/84. (322) Obs 3/4/88. (323) FT 18/11/85; MJ 6/6/86. (324) Relly statement to AAC AGM, 1976. (325) FT 11/7/86. (326) FT 10/6/86. (327) Bernard Magubane, The Political Economy of Race and Class in South Africa, Monthly Review Press, New York, 1979. (328) RW Johnson, quoted in (327). (329) Francis Wilson, Labour in the South African Gold Mines 1911-69, Cambridge University Press, 1972. (330) Speech by Ernest Oppenheimer to Institution of Personnel Management, roneoed, 1974. (331) Rand Daily Mail 9/10/75. (332) Frederick A Johnstone, Class, Race and Gold, Routledge and Kegan Paul, London, 1976. (333) Geoffrey Wheatcroft, The Randlords: The men who made South Africa, Wiedenfeld and Nicholson, London, 1985. (334) Gua 22/12/83. (335) FT 21/12/83. (336) X-Ray, London, 11-12/83. (337) AAC advertisment in FT 19/10/83. (338) FT 3/8/84. (339) E&MJ 3/85. (340) Obs 28/4/85. (341) Eve Standard, London, 30/4/85. (342) Eve Standard 29/4/85. (343) FT 30/4/85; MJ 3/5/85. (344) Gua 1/5/85. (345) Statement by Vaal Reefs and Southvaal Holdings, Johannesburg, 3/5/85. (346) FT 26/2/86. (347) FT 1/5/85. (348) FT 13/5/85. (349) FT 1/7/85.

(350) MJ 19/7/85; FT 10/6/86. (351) E&MJ 7/85. (352) FT 27/2/87. (353) FT 6/8/85. (354) SunT 4/8/85. (355) FT 30/8/85. (356) FT 5/8/85; E&MJ 9/85. (357) FT 30/5/85. (358) Weekly Mail, Johannesburg, 28/2-6/3/86. (359) FT 28/2/86. (360) MJ 2/3/86. (361) Gua 17/3/86. (362) FT 22/4/86. (363) FT 2/7/86. (364) FT 9/7/86. (365) FT 2/10/86; FT 3/10/86. (366) FT 18/9/86. (367) FT 7/10/86. (368) FT 25/10/86. (369) FT 25/11/86; FT 9/12/86. (370) E&MJ 1/87. (371) FT 14/1/87. (372) Weekly Mail, Johannesburg, 23/1-29/1/87. (373) AfricAsia, London, 4/87. (374) FT 11/3/87. (375) FT 22/6/87. (376) MJ 3/7/87. (377) MJ 17/7/87. (378) HJ Simons and RE Simons, Class and Colour in South Africa 1850-1950, Penguin, 1969. (379) FT 4/8/87. (380) MJ 7/8/87. (381) FT 8/8/87. (382) Gua 8/8/87. (383) Obs 9/8/87. (384) FT 10/8/87. (385) Gua 11/8/87. (386) Gua 13/8/87. (387) Gua 14/8/87. (388) FT 15/8/87; Gua 15/8/87. (389) Gua 17/8/87. (390) FT 18/8/87; see also MJ 21/8/87. (391) FT 19/8/87. (392) FT 21/8/87. (393) FT 20/8/87; Gua 20/8/87. (394) Newsweek 24/8/87. (395) Gua 21/8/87. (396) FT 22/8/87. (397) Gua 26/8/87; FT 26/8/87. (398) Gua 25/8/87. (399) Gua 27/8/87. (400) FT 27/8/87. (401) Gua 28/8/87. (402) FT 28/8/87. (403) FT 29/8/87. (404) Gua 31/8/87. (405) FT 1/9/87. (406) FT 3/9/87. (407) Business Magazine, London, 8/88. (408) MJ 13/5/88. (409) FT 27/11/87; MJ 4/12/87. (410) MJ 17/6/88. (411) FT 27/9/87. (412) Chamber of Mines 98th Annual Report; see MJ 24/6/88 and MJ E&MJ 9/88. (413) MJ 22/7/88. (414) FT 3-4/9/88; Gua 27/9/88; Gua 30/11/88. (415) MM 5/88; see also MJ 29/1/88. (416) Gua 13/5/88. (417) MJ 10/6/88. (418) FT 5/7/88; FT 6/7/88; FT 6/8/88. (419) MJ 8/7/88. (420) See FT 23/6/86. (421) MM 11/89. (422) M Bostock, C Harvey (eds), Economic Independence and Zambian Copper, Praeger, New York, 1972. (423) MJ 19/1/90.

33 Anglo-Bomarc Mines Ltd (CDN)

This is a British Columbia chartered company, formerly Anglo Northern Mines Ltd and Bomarc Mining Co Ltd, which has been exploring for uranium in northern Saskatchewan. It holds investments in Great Hercules Resources (1).

References: (1) *Who's Who Sask.*

34 Anglo-United Development Corp

It holds a gold property in Yellowknife, North West Territories (NWT), and other prospects in Canada, but main exploration is in Eire through Munster Base Metals which holds 54 exploration permits, of which 16 are wholly-owned (1).

Its Donegal uranium exploration started in 1978, drilling the following year with fairly encouraging results (2). The next year, "very encouraging results" were reported (3) and "hopes for a large tonnage" mentioned in the company's 1980 annual report, especially at the Finn prospect.

Anglo-United and Central Mining Finance (Charter Consolidated) were reported in a JV in Connemara, west of Galway, in 1979 (4); Irish Base Metals, another Northgate affiliate, was also at Fintown and Thomastown, Co Kilkenny (4). One Irish director of Anglo-United was brother to Peter McAlear, Eire's Minister of Energy in 1979 (4). By the late 1980's, Anglo-United, together with other uranium prospecting companies, had withdrawn from Donegal due to a fall in uranium price and – undoubtedly – local opposition. Resistance throughout Eire to uranium mining has been strong to the extent that the company's half-yearly report in mid-1980 observed that its "considerable enthusiasm" had been blighted "to some extent by the strong opposition to uranium exploration from a local, but vociferous, anti-nuclear lobby."

Northgate sold its interest in the company in 1982 to a group of European investors, and the company has moved into production as opposed to simply exploration: in particular by buying the Gowen anthracite mine in Pennsylvania (USA). Much of the company's activities in Eire were through a JV in which Pennaroya and Preussag were partners (5).

In 1986, there was another change of ownership when the British Anglo United plc bought out Toronto Anglo-United Development Corp: Anglo-United's main shareholder (with 29.9%) was Hillsdown Holdings (6) until late 1988 (7). That year, Anglo enhanced its sizeable British coal interests by buying into another British company, Burnett and Hallamshire (8). This was then renamed NSM, a company which continues to hold important acreages in Britain and the USA which it exploits for coal mining or lucrative "waste management" landfill (9). In early 1990, however, Anglo said it wanted to sell its NSM stake (10). Meanwhile, the company had astonished hardbitten British businesspeople by raising City funds to buy out the Coalite group (notorious for its interests in the Falkland Islands, which it has been said to virtually own) for just under half a billion pounds (11). Throughout the next nine months Anglo asset-stripped what it saw as Coalite's "peripheral" operations (12) to repay the debt incurred over its David-eats-Goliath take-over: Coalite's quarrying activities were gobbled up by Anglo American's Charter Consolidated (13).

Contact: DUC Straboy, Glenties, Co Donegal, Eire.

References: (1) *MIY* 1981. (2) *FT* 15/6/78. (3) *MJ* 1/8/80. (4) Personal communication from Just Books, Belfast. (5) *MJ* 17/3/82. (6) *FT* 4/2/87. (7) *FT* 14/10/88. (8) *FT* 23/1/87. (9) *FT* 23/2/90. (10) *FT* 20/3/90. (11) *FT* 23/7/89; *MJ* 30/6/89; *Gua* 25/5/89. (12) *FT* 4/10/89. (13) *FT* 17/1/90.

```
ANGLO-UNITED DEVELOPMENT CORP. (CDN)


Munster Base Metals Ltd (IRL)
```

35 Anschutz Mining Corp

Its HQ is in Denver. It has been reported exploring for uranium in Texas (1).

In 1981 it discovered uranium at an undisclosed site in Paraguay (Anschutz 50%), along with Keko (ROK) (25%) and Taiwan Power Co (RC) (25%) (2).

While the Paraguay project has clearly fallen on stony ground, in 1989 a Panamanian subsidiary of Anschutz, called South American Placers, had embarked on gold exploitation in the renowned Madre de Dios region of Andean Peru, where it was installing a gold dredge, together with local partners, under the JV name of Carisa. Despite further plans to increase production in this remote, rainforest region, and talks with the World Bank for a loan in 1986, Carisa has remained a relatively small operation (3). It is, however, Peru's largest producer of alluvial gold, and – as would the Paraguayan uranium project – it works land and rivers claimed by indigenous people.

In 1984, the major Peruvian Indian organisation, AIDESEP, condemned the presence of mining companies in Madre de Dios and called for their removal (4).

References: (1) *E&MJ* 11/78. (2) *MJ* 3/7/81; see also *MAR* 1985. (3) *E&MJ* 12/89. (4) *Voz Indigena* Vol. 3 No. 11, July/Aug 1984.

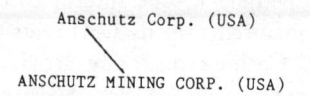

```
                Anschutz Corp. (USA)

ANSCHUTZ MINING CORP. (USA)
```

36 ANZ/Australian and New Zealand Banking Group

"...the most important link in the financing of the Australian uranium industry" (1).

Formed in 1970, ANZ was by the 1980s the 12th largest company in Australia and makes a greater profit than any other private trading bank in the country. Through its completely-owned subsidiary ANZ Nominees, it has important interests in all of Australia's current major uranium projects:

Pancontinental: At 31 August 1979, ANZ held over 2 million shares making it the largest shareholder in the company.

Western Mining Corp Ltd: No. 2 shareholder.

Peko Wallsend: No. 3 shareholder.

Oilmin NL: No. 3 shareholder.

CRA: No. 4 shareholder with 7.5 million shares at the beginning of 1980.

EZ Industries: No. 5 shareholder with more than two million shares in 1979.

In addition to many interlocking directorships, ANZ is the sole banker for EZ, Petromin Oil and Transoil; it's a banker for Western Mining, for CRA, and for Mary Kathleen Uranium as well as for Peko Wallsend.

In early 1990, ANZ agreed to merge with National Mutual Life Association, Australia's second biggest non-bank financial institution. This deal, the biggest ever in the country's financial sector, has created ANZ-NM, with combined assets of not many cents less than one hundred billion US dollars (2).

References: (1) *Uranium Investment Dossier*, MAUM, Victoria 1981. (2) *FT* 3/4/90.

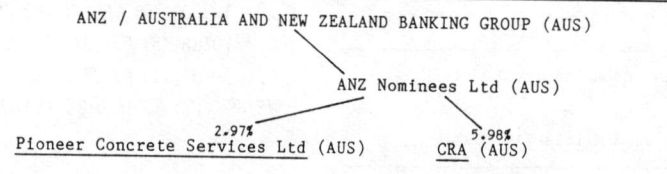

```
ANZ / AUSTRALIA AND NEW ZEALAND BANKING GROUP (AUS)

                          ANZ Nominees Ltd (AUS)
                  2.97%                    5.98%
Pioneer Concrete Services Ltd (AUS)      CRA (AUS)
```

37 AOG Minerals Ltd

The company was formed to allow its parent, the AOG Corporation, to concentrate on petroleum. It had two uranium prospects (all its projects are at the exploration stage).

Together with Union Oil Development it had 50% of the Litchfield prospect between the Daly and Finniss Rivers in the Rum Jungle area of the Northern Territory.

With Uranerz as a partner, it also held 50% of the Mount Fitch prospect in the Rum Jungle district where about 1400 tonnes of uranium oxide have been located (1).

An Aboriginal land claim has been made on the area. This claim – by the Finniss River community – was the subject of several High Court actions during 1981-83, with representations by the Northern Land Council, CRA, Peko Wallsend, local graziers and the Northern Territory government (2).

The company has now completely withdrawn from uranium exploration – it sold its interests in Rum Jungle to Uranerz and Kumagai Gumi Co Ltd (of Japan) in 1985 (3). Nonetheless it continues to hold a 28.33% share in Australian Diamond Exploration Ltd along with Aberfoyle and Ashton Mining Ltd, and has 35% in the JV in the Keep and Daly river areas of the Northern Territory. It also has a JV with Esso Exploration and Shell Australia – precious metals prospecting in northern Queensland (3).

In 1987 AOG sold its diamond interests, and changed its name to Australmin Holdings Ltd, after it acquired all the shares of Australmin Pacific Ltd.

Its current prospects include interests in French Polynesia, chromite in New Caledonia (Kanaky), with gold exploration ongoing in West Papua (Irian Jaya) (4).

With restructuring of the group, Elders CED Ltd and National Investment Group acquired

```
        AOG Corp. (AUS)
              /
    AOG MINERALS LTD (AUS)
```

substantial interests, along with AOG, in Australmin (4).

References: (1) *Reg Aus Min* 1981. (2) *Reg Aus Min* 1983/84. (3) *MIY* 1987. (4) *MIY* 1990.

38 Aquarius Resources Ltd

```
        Powerco (USA)
              /
    AQUARIUS RESOURCES LTD (CDN)
```

Originally the Driftwood Oil Corp, Aquarius is searching for uranium in northern Saskatchewan. Powerco is reported to control Aquarius (1). The company was also involved in a JV for uranium exploration with Eldorado Nuclear (2).

References: (1) *Who's Who Saskatchewan.* (2) Goldstick.

39 Aquitaine Australia Minerals (Pty) Ltd

A subsidiary of France's second most important, state-owned, oil company, it has JVs in gas (Queensland) (1) and nickel exploration (Western Australia) (2).

It had a JV with Jimberlana Minerals (AUS) and Pan D'Or Explorations (AUS) to explore for uranium in the Northern Territory in 1979; Aquitaine was to earn up to 50% interest (3).

In 1981 ELF expanded its non-petroleum exploration activity through its Australian subsidiaries. Among its JVs are one with MIM at Lady Loretta, Queensland (lead, zinc, and silver), and with St Joe Minerals (USA) in Western Australia (4).

In the late 1980s, together with Total Mining Australia and Urangesellschaft, Elf Aquitaine Triako Mines (as a JV) opened up a trial uranium mine at Manyingee near Onslow on the west coast of Western Australia. The companies used in situ leaching techniques, with an

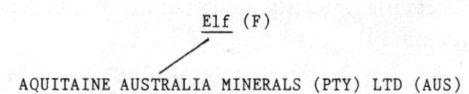

```
              Elf (F)

AQUITAINE AUSTRALIA MINERALS (PTY) LTD (AUS)
```

alkali solution consisting of hydrogen peroxide, sodium bicarbonate and hypochlorite. Although claimed to be simply a test (since no mining licence had been granted) the experiment lasted 169 days, involving total injection into the orebody of 40.5 million litres of leaching solution. Environmental and pro-Aboriginal organisations were asking by 1986 where the one and a half tonnes of uranium produced at Manyingee had been sent (probably Ranger) and where the groundwater monitoring records were to be found (if indeed they were kept at all). Enquiries revealed nothing, except a wall of hostility by the companies (Total told one enquirer, when asked if the company had commenced Stage Two of its operations "that is none of your business") and bland assurances from the government (5).

Anti-uranium activists who visited the mine at this time were able to photograph containers on site, and reported large piles of waste with virtually no protection for the public and wildlife (6).

References: (1) *FT* 28/10/80. (2) *Reg Aus Min* 1976/77. (3) *FT* 2/5/79. (4) *MJ* 2/7/82. (5) *Chain Reaction,* Melbourne, No. 47 Spring 1986. (6) Rick Humphries, *Manyingee Uranium Deposit,* Australian Conservation Foundation, undated.

40 Aquitaine Mining Corp

An affiliate of ELF, in 1979 it had a 50% interest in the Lakeview uranium prospect in south central Oregon, USA, with Polaris Resources. The prospect is located close to the White King mine which produced nearly half a million

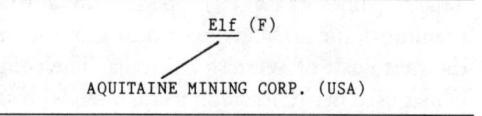

```
              Elf (F)

AQUITAINE MINING CORP. (USA)
```

pounds of uranium in the 1950s and was operated up to 1979 by Phelps Dodge (1).

References: (1) *MJ* 10/8/79.

41 Aquitaine of Canada

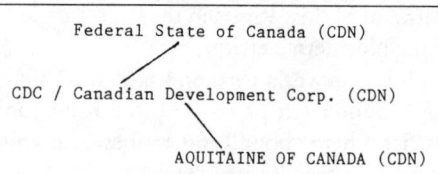

```
    Federal State of Canada (CDN)

CDC / Canadian Development Corp. (CDN)

            AQUITAINE OF CANADA (CDN)
```

Formerly a subsidiary of France's second most important, state-owned oil company, Elf-Aquitaine, it was purchased by the Canadian Development Co in 1981, in what one US oil executive called "Canadian rape of US oil and mining interests in Canada" (1).

In the early 1980s Aquitaine of Canada began exploring for uranium at Vaughan, near Windsor, Nova Scotia – despite the objections of local citizens exemplified by the Citizens' Action Group to Protect the Environment (CAPE), the Fundy Area Citizens' Committee and the Uranium Committee of the Ecology Action Centre in Halifax. The company – among others – stepped up its public relations programme to meet a feared moratorium on uranium mining, which in fact soon followed (2).

Contact: CAPE (Citizens' Action Group to Protect the Environment), c/o Centre Burling PO, Hants County, Nova Scotia, Canada.

References: (1) *FT* 30/6/81. (2) *Globe and Mail,* Toronto, 4/7/81.

42 Aquitaine SWA

It was exploring for uranium in Namibia in 1977, along with several other multinational companies including UCI Union Corp (now Gencor) and Falconbridge (1). In November of

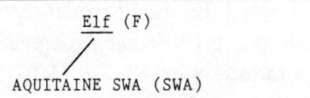

```
        Elf (F)
         /
AQUITAINE SWA (SWA)
```

the same year, Aquitaine claimed its activities in Namibia were confined to exploring, not mining: this was in reply to suggestions that it was interested in General Minings' (now Gencor) Langer Heinrich uranium find (2).

References: (1) *MM* 7/77. (2) *Marches Tropicaux* 1977.

43 Arco/Atlantic Richfield Co

Chaired by the ubiquitous, art-loving, Robert O Anderson, Arco is a "major integrated oil company" with interests in petrochemicals and – through Anaconda – in copper, bauxite and uranium mining, and fabrication of derived products (1).

It has eleven operating companies, including Arco Solar Industries, Arco Transportation Co and Arco Coal Co (1).

"Fundamentally an energy company" (2), today's huge corporation – 21st in terms of total revenue in 1981 among western multinationals (3), and 13th in terms of worldwide sales in the early 1980s (1) – has its roots in the Atlantic Refining Co, formed in 1870 and absorbed into the world's biggest monopoly, Standard Oil, before achieving "independence" in 1911. In 1966 Atlantic merged with Richfield Oil Corp, a relatively small Los Angeles refining company which was forced through financial problems into receivership in 1931, but managed to become a major oil marketeer over the next thirty years (1).

Atlantic Richfield established its modern capital base with a major oil discovery at Prudhoe Bay on the Alaskan North Slope in the late 1960s. During the 1973/74 oil "crisis" and with crude flowing free from Alaska, Arco established the policy which has kept it thriving (mostly) since: independence from imported oil; balancing domestic production and refining/marketing activities; and forward integration in the oil industry and diversification. To achieve these aims, it sold its major British and Canadian operations in the mid-1970s, consolidated its refining operations to four facilities, and acquired the Anaconda Copper Co in 1977.

The Arco/Anaconda merger was one of the first modern acquisitions by an oil company of a traditional mining outfit. It enabled Arco to become "a major actor" (4) in the mining industry and boosted its status from the 15th to the 12th largest industrial enterprise in the USA. The take-over was achieved, however, only in the face of prolonged and almost unprecedented opposition from the US Federal Trade Commission (FTC) which argued for several years that the merger would hamper competition in the copper and uranium oxide industries. The Federal Court of Appeals ruled against the FTC, but an unruffled FTC pursued its anti-trust case, with the result that in October 1979 Arco was barred from making new investments in the copper industry for a minimum of 5 years and a maximum of 10. Anaconda was given 5 years to divest the majority of its half interest in the Anamax Copper Refining Co (a divestiture still not achieved as the *File* went to press) (4).

By the time Anaconda slipped into Arco's grasp it was severely weakened by the nationalisation of its huge Chilean copper assets under the Allende government, and it was in great financial trouble despite its status as the 3rd largest copper producer in the USA after Kennecott and Phelps Dodge (4).

However, it brought to Arco huge copper, coal and uranium reserves. In 1980, for example, Arco rated 6th in the world, among oil companies, as a coal producer. As a copper producer, its output was exceeded – among oil majors – only by Sohio, Pennzoil and Standard Oil of Indiana (Amoco) (4). The same year, only one oil company exceeded its uranium output – Kerr-McGee – while it rated 11th among the top 20 uranium producers of the world with a production of 2025 tons U_3O_8 (5). A year later, it rated 4th among all uranium producers in the USA, exceeded only by Kerr-McGee, Utah International and Exxon Minerals (5). It also held the 8th largest lode of uranium in the San Juan basin area (6) and had become the 2nd most important uranium miller in the USA (2nd, once again, only to the infamous Kerr-McGee) (7).

Noted native American woman writer Winona La Duke has commented that "basically the history of Anaconda competes in 'ugliness' with Arco's history of genocide against Indian and other native peoples. The two are a perfect pair"(8).

Arco, in its early days, was certainly one of the oil companies which benefited from the theft of Indian land in Oklahoma (purportedly guaranteed to native Americans "as long as the grass shall grow") in the late 19th century (8). One hundred years later it was again being accused of robbery – this time by the FBI, IRS and various government agencies. Up to US$80 million in oil royalties was reported stolen by oil companies, including Arco, in the form of royalties not declared to Indians on the reservations (9). In contrast, Arco has for some years been charging excessive rates, along with other oil companies, for the piping of oil from the Alaskan North Slope: at least, that was the opinion of the Federal Energy Regulatory Commission, which in late 1984 reached agreement with the corporation on lower rates (10).

Always a company to pare its expenses, whatever the human cost, three years earlier it suddenly ditched Nigeria by breaking an agreement to buy crude oil from the Third World state, when oil on the spot market became cheaper. Commented an Arco spokesperson blandly: "We just couldn't justify the price" (11).

At around the same time, the corporation was trampling on sacred sites in the Northern Cheyenne reservation: an act which incited a lawsuit from traditional leader of the Northern Cheyenne, Charles White Dirt (12).

Arco's brushes with indigenous communities have not been restricted to mainland USA, for the corporation is one of the multinationals currently implicated in the Arctic Pilot Project (APP), a scheme to transport liquified natural gas by tanker which has been condemned by all the leading native peoples' organisations in the Arctic circle, led by the ICC (Inuit Circumpolar Conference) (13). When hearings conducted by the Canadian National Energy Board into the US$2 billion APP commenced in early

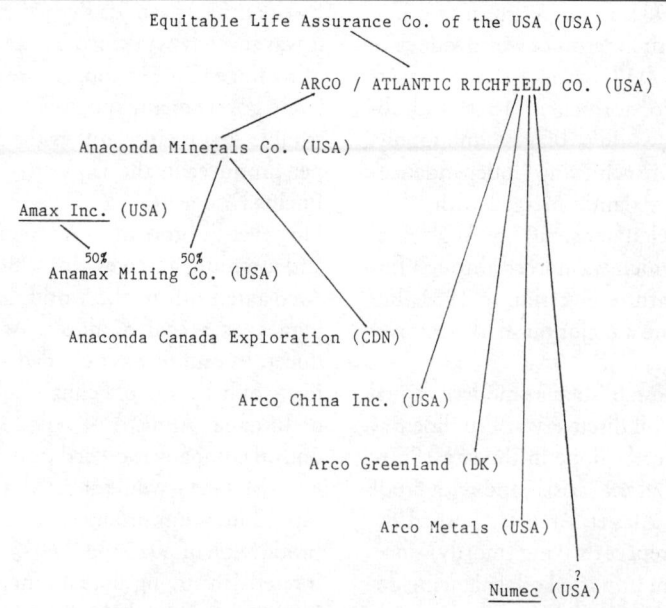

```
                    Equitable Life Assurance Co. of the USA (USA)


                              ARCO / ATLANTIC RICHFIELD CO. (USA)

          Anaconda Minerals Co. (USA)

    Amax Inc. (USA)
           50%        50%
         Anamax Mining Co. (USA)

              Anaconda Canada Exploration (CDN)

                              Arco China Inc. (USA)

                        Arco Greenland (DK)

                              Arco Metals (USA)
                                          ?
                              Numec (USA)
```

1982, both environmental and Inuit ("Eskimo") groups pointed to the threat posed by the tankers – particularly their noise – to animal life in the region (14). And at a native peoples' seminar on the APP in late 1982, Arco was specifically singled out for its deep penetration of Inuit land in Alaska (*viz* Prudhoe Bay), Greenland and northern Norway.

In northern Norway Arco has been exploring for minerals on Sami ("Lapp") land without any control by the indigenous land-users (15).

By mid-1984 an oil exploration agreement between Arco and the Danish and Greenland governments concerning the Jameson Land area in eastern Greenland was reported nearing completion, but thrown into jeopardy when the ruling Siumut party demanded "unilateral" (ie nationalist) control over the country's mineral resources (16).

Meanwhile the company withdrew from another venture of potentially disastrous impact to indigenous peoples: it pulled out of the Jari venture, masterminded by the US industrialist Daniel Ludwig (and later abandoned), to grow rice and other wholly inappropriate cash crops in the Amazon jungle (17). However, that didn't stop the Anaconda subsidiary advertising for a mineral investment manager and regional geologist to explore in "pre-Cambrian shield provinces" of Brazil – one of the main geographical locations for uranium.

Apart from a few minor forays (such as an attempt, heartily rejected by local communities, to exploit uranium on a 600-acre lease in upstate New York) (18), Arco's mineral ventures have been concentrated in Anaconda.

This company's activities have a long and ignoble history. It was accumulating capital (as were Homestake and Kennecott) on Ojibway, Lakota and Crow Indian lands in the 19th century, long before it moved its operations elsewhere (19). By 1926 it was (along with Kennecott, Asarco, Union Minière and RTZ) a participant in a copper cartel known as Copper Exporters Inc, which increased the income of its participants by 84.4% in the space of only a couple of years (20).

By then it had entrenched itself in two of the biggest copper lodes in the world: in Montana and in the Chilean Andes (21).

At Butte, Montana (Anaconda's largest open pit copper mine) (2), the company controlled labour and land through its participation in a minerals trust so powerful that, at the turn of the century, a special session of the state legislature had to be called when it laid off three-quarters of the industrial workforce (22). As the Butte mine expanded, so – using Montana's "eminent domain" law (*i.e.* the rich get richer, the poor, poorer) – Anaconda forced the relocation of farming families and trapped the local community into dependence on its operations (21). Anaconda controlled a good chunk of the local media, too: in 1959 it owned no less than half of Montana's daily newspapers (22).

By 1929, US investment in Chile had exceeded US$400 million, almost all of which was in the hands of Anaconda and Kennecott. According to Eduardo Galeano "in half a century, [these two multinationals] bled Chile of US$4 billion sent to their home offices under various headings; yet by their own inflated figures they invested no more than US$800 million, almost all in profits taken from the country" (23).

It was Anaconda which developed what is now the Chuquicamata copper mine (the country's largest producer), and mined for ten years – through its Andes Copper subsidiary – the appropriately named Cerro Indio Muerto (Dead Indian Mountain): this mine, the El Salvador, is now controlled by Codelco's copper division (24).

When an elected government nationalised the copper mines in 1970, the socialist president, Salvador Allende, claimed Anaconda had "garnered US$79 million in profits from Chile – the equivalent of 80% of its global profits, although its investment in Chile was less than one-sixth of its total investments abroad"(23).

Of all the Chilean companies, Anaconda suffered the worst: prior to the nationalisation it had already had to sell 51% of its operations to Codelco (4). Although the Anaconda workforce, like Kennecott's, was paid wages well below their equivalent in the USA, Anaconda at least seems to have deliberately created a quiescent and well-paid Chilean labour élite which could be counted on when the nationalisation took effect (25). At that point, Anaconda got compensation but sustained considerable financial "injury" (4).

How far it was involved in toppling Allende is open to question (unlike the role of ITT), but for more than 20 years the company employed, as its vice-president for Chilean operations, a former right-wing Chilean senator and ambassador to the USA (25).

But, the bigger they come, the harder they tend to fall.

Still suffering from its expulsion from Chile, Anaconda couldn't contend with the copper slump that set in over the next decade. Its sugar daddy Arco was growing during this period: revenues increased from US$12.74 billion in 1978 to US$26.99 billion in 1982 (26) thanks to North Slope oil production (27), while in 1980 the company shelled out US$3 billion for further expansion (28). Anaconda, however, laid off 200 workers at its big Berkely Pit copper mine at Butte in late 1981 (29), after an announcement that it was unlikely to reopen its Montana copper refining and smelting facilities following a strike and additional financial constrictions, heightened by demands from the Environmental Protection Authority (EPA) that it clean up both its act and its operations. The EPA accused Anaconda of "misleading" the agency by overstating the costs it was prepared to sustain meeting anti-pollution regulations (most of the costs, said the EPA, were going to Anaconda facilities not directly related to demands of the Clean Air Act).

Anaconda decided to close down and ship its copper concentrates to where the legislation was less environmentally conscious and the air even less clean – i.e. Japan (30, 31). This move aroused the ire of environmentalists and mining companies alike. US industry cited the EOA strictures as an example of how pollution control regulations can "backfire" (32); the USA (United Steelworkers of America trade union) claimed the pollution issue was largely an excuse for Anaconda to move to greener – and more lucrative – fields, thus creating a variation

of the penetration of US labour markets by Japanese business. The decision was taken after lengthy pressure by Arco/Anaconda on Montana congressmen to make concessions to the company, and cost 1500 US jobs. Arco promised new jobs for some workers and set up a derisory US$5 million "community adjustment fund" (32).

Two years later, with pre-tax losses running at US$539 million over a three-year period, production at Twin Buttes, Arizona, was cut by half (26), while at Butte itself – site of what was once called "the richest hill on earth" – operations were suspended in June the same year (33). As the dawn of 1985 broke over Atlantic's not-so rich field, Arco decided to absorb a complete write-off of its interest in Anaconda to the value of US$785 million (34).

Copper wasn't Anaconda's only casualty of the early 1980s. At the beginning of 1981 the company announced that it would phase out open pit uranium mining at its sites near Grants, New Mexico, because of a 25% decline in prices and a run-down of ore; underground mining would continue until 1983 (35). In fact, mining was suspended in 1982, as the company realised it had enough ore to fulfill contracts (4), and, by the end of the following year (like every other producer in this once-heartland – deadland? – of the world's uranium industry), Anaconda had ground to a halt in the Grants region (36). Left behind are huge radioactive tailings dumps from what, for a long time, was the world's biggest open-pit uranium mine (6) – bang in the middle of the Laguna Indian reservation. The Jackpile mine and the Bluewater mill, which were employing no less than three-quarters of the Laguna tribe's workforce (about 450 people) in the late 1970s, have created well over 14 million tons of tailings (37).

The Jackpile deposit was "discovered" by a Navajo sheep-herder named Paddy Martinez on his sheep range in 1950. After geologists from the Atomic Energy Commission (AEC) established that this was, indeed, a potentially fabulous find, Anaconda swiftly moved in, centring its operations on the small Laguna pueblo. The Bureau of Indian Affairs (BIA) soon persuaded the Laguna Tribal Council to sign a lease with Anaconda "for as long as the ore is producing in paying quantities"; there were no provisions for environmental studies or protection, and soon the Jackpile mine (named after its operator) expanded into the agricultural Paguate valley as a large strip-mining operation, working seven days a week around the clock (38). The Rio Paguate, main source of water for the native Americans in the valley, which runs under the uranium deposit, began turning a phosphorescent green (39) and, within twenty years, five miles of fertile land had been transformed into a "lunar landscape" (21). Sheep began drinking contaminated water and chewing irradiated alfalfa, and stomach cancers among the Laguna people began to rise (39). Several sacred mesas were destroyed, mine workers have gone deaf through blasting and, above all, the native community has been betrayed by both the company and the BIA-influenced Tribal Council. Although an utterly derisory 2% was promised to the Laguna as a royalty value of the uranium at the market, much of that was never paid (21). Anaconda, in any event, was able to deduct the amount from taxes payable on income to the federal government (40). Moreover, there was no compensation for loss of agricultural land; no compensation for the construction of a highway; and a promise not to drill below the village of Paguate was broken (40).

A mining industry article in 1957 tried to present the "other" side of the Jackpile Paguate picture. Comment would be superfluous:

"Early negotiations by The Anaconda Company with the Lagunas established the right of the latter to benefit from the development of the orebody that was found to exist on the land granted to them by the Government of the United States of America. This right was not restricted to royalties or monetary proceeds which would normally be received by any owner who leases or otherwise makes his property available to production. It was recognized that they should receive benefits individually from employment wherever such was possible.

"Three years ago these people were completely unfamiliar with the work and equipment associated with a mining operation. Today, at Jackpile, one will find Laguna Indians efficiently holding their own as heavy equipment operators. They have been taught to operate shovels, cranes, trucks, drills, bulldozers and other types of equipment common to open-pit mining. All equipment operation is handled by these men, under the direction of a few competent and experienced supervisors. To a marked degree the success of the Jackpile mine as a sizeable, low-cost operation, is due to the aptitude of these people upon whose land a deposit was found, which has and will contribute much to the uranium requirements of the United States" (41).

The "uranium requirements" referred to were, initially, wholly for nuclear weapons for the US military: Anaconda goes down in history as the first major provider of US nuclear fuel for the cold-war build-up of the 1950s and beyond.

The Jackpile Paguate mine has not been Anaconda's only uranium venture: by 1980 it operated three mines in Grants, one of which (P-7/10) was within the limits of the Canoncito Indian Reservation (6). It has also held a permit, apparently not implemented, on Navajo land at Crownpoint, New Mexico (22). Jackpile Paguate has, however, dwarfed all other Anaconda/Arco nuclear operations. Condemned on several occasions for contaminating groundwater by the EPA (42), noted for the high rate of birth defects among the surrounding population (43), it exemplifies the social and environmental destruction which large-scale mining – especially for uranium – can visit upon an impoverished and relatively powerless community.

The collapse of Arco/Anaconda's copper interests shouldn't suggest that the corporation is no longer interested in metals. On the contrary, it is still quite heavily invested in molybdenum, silver and soda ash, while it recently entered gold exploration – along with RTZ, BP and

Anglo-American – in the new Eldorado of Brazil (44).

However, Arco's extensive aluminium interests have passed into the hands of Alcan; this includes Arco's share in the Aughinish alumina refinery in County Limerick, Eire, and various plants in the USA (45), and follows a long drawn out battle by Alcan to gain a foothold in the USA as a primary aluminium producer (46). It also follows considerable losses sustained by Arco Metals (a subsidiary formed in 1982 out of Anaconda Aluminium Co and Anaconda Industries) and opposition by the US Justice Department, citing anti-trust laws (47, 48). It is not known whether a proposal to site an aluminium smelter in Newfoundland is still going ahead (44): in 1983 progress on this was dependent on the world aluminium market (49).

By the beginning of 1985 Arco was still in possession of its Anamax copper subsidiary (which has also produced uranium from copper ores), a JV with Amax that the US Federal Trade Commission insisted should be sold by late 1984 when Arco took over Anaconda (2).

The corporation's other major non-oil and non-metal interest is coal. The Black Thunder mine in Wyoming's Powder River Basin (also a major uranium region) is one of the largest surface coal mines in the USA. Anaconda holds a 34% interest in the new Curragh surface coal mine in Queensland (where the company is operator) and a 15.4% interest in Blair Athol, also situated in the "sunshine state" (2). Its major new coal hopes are centred on indigenous Kalimantan (formerly Dutch Borneo) where, with its JV partner Utah International (50%) it recently signed agreements with the military régime (50, 51) and is rated as one of the most likely companies-to-succeed in mining coal in the region (44). Four years earlier Arco had entered into an agreement with Interchar SA, a Colombian company, to exploit coal in the Cerrejon basin of the country (52). Together with the Australian Moonie (sic) Oil Co it has been involved in a coal liquefaction project at Gelliondale, Victoria, in recent years

(53) where a huge 2.3 billion tonnes of wet brown coal have been identified (54).

Arco's early interests in solar energy – by mid-1978 it had "bought and merged itself deeply" (55) in the market – have not been consolidated. It used solar panels in its Wyoming coal mine in the early 1980s (50), but its only other major project seems to be the setting-up of a photo-voltaic electricity generating plant in California, whose power is sold to Southern California Edison and fed through the grid (57).

In 1980 Arco joined with LM Ericsson of Sweden to form a new company selling wire and cable products as well as telecommunications equipment in the USA (58).

Wall Street speculated in early 1984 that Arco would bid for Gulf Oil after Texan longhorn T Boone Pickens Jr grabbed 20% of Gulf's shares. In the event Gulf went to Chevron, and Arco was cheated of an opportunity to become America's 3rd largest oil company after only Exxon and Mobil.

After Petro-Canada took over Arco's Canadian subsidiary, Atlantic Richfield Canada in the early '80s (59), Arco has, like other oil companies, been looking for ways to prevent a takeover of its own assets. In early 1985 it proposed merging into one of its own wholly-owned subsidiaries, a re-incorporating which would eliminate cumulative voting in the election of directors – b coulso abolish the right of shareholders to call special meetings and propose amendments to the certificate of incorporation (60).

In early 1987 Anaconda finally agreed with the Laguna Pueblo people to "reclaim" 2656 acres of land which it had "disturbed" (devasted) through thirty years of operating the Jackpile Paguate three open-pit and nine underground mines. The reclamation plan, approved by the Bureau of Land Management, includes not only the mines, but 32 tailings piles, 23 subgrade ore stockpiles, four topsoil stockpiles and 66 acres of buildings and roads. The plan was expected to take ten years to fully implement at a cost of US$43.5 million (61).

The year before, the company finally settled its long-standing dispute with the US Supreme

Court over back Alaskan taxes and agreed to pay £234 million (62).

Arco Coal moved into Queensland in no uncertain fashion in 1988 when it became the marketing manager of the Emerald coal mine, the state's first new underground mine for many years (63). The mine is a high-quality coking- and steam-coal producer, with low sulphur content, and is a JV between Arco Coal Australia (50%), Kennecott Exploration (Australia) Ltd (22.5%) and Lend Lease Resources Pty (22.5%) (64).

The same year, Arco became a minority partner in a major coal venture in north-eastern Venezuela, called Carbones del Guasare SA (63). In 1967 Arco had started oil exploration in various parts of Indonesia in a highly sucessful off-shore venture in the north-west Java Sea. In the two decades since, it has ventured into the Malacca Strait, north-east Kalimantan (from which it later withdrew) and elsewhere. In 1985 it decided to proceed with a major new development in the Java Sea, 60 miles north of Jakarta, the Bima oil field (65).

Then, in 1990, it announced it had acquired 28.125% in the Keapala Burung Selatan A production sharing contract, inside West Papua (Irian Jaya, Indonesia), along with Indonesian companies headed by Pertamina (66).

The Anaconda mines in Butte Montana ("the richest hill on earth") were sold to MRI (Montana Resources Inc) in the late 'eighties, and Anaconda finally headed for demise, as the company's assets were bought out by MRI, to be joined in 1989 by Asarco in a roughly 50/50 JV (67).

Further reading: Norman Girvan, *Copper in Chile: A Study in Conflict between Corporate and National Economy,* Institute of Social and Economic Research, University of the West Indies, Jamaica, 1972. (Although written before the complete nationalisation of Chilean copper, effected by President Allende in 1971 – a move which led to the bloody overthrow of the government and civil war – this study is an excellent introduction to the dominance over the copper resources and economy effected by Anaconda until

the late 1960s. It also gives a brief, but comprehensive, picture of Anaconda's overall development.)

References: (1) *WDMNE*. (2) *MIY* 1985. (3) *MM* on 12/12/83. (4) *BOM*. (5) *Market shares and individual company data for US energy markets: 1950-1981: Discussion paper 014R,* Washington 9/10/81 (American Petroleum Institute). (6) *San Juan Study*. (7) *Statistical data of the uranium industry,* Grand Junction, 1982 (US Dept of Energy). (8) *AkwN* Summer 1979. (9) *SunT* 26/4/81. (10) *FT* 29/12/84. (11) *Gua* 21/4/81. (12) *AkwN* Winter 1982. (13) *IWGIA Newsletter,* København 4/82 & 6-10/82. (14) *FT* 3/2/82. (15) IWGIA Newsletter (København) 3/83. (16) *FT* 7/7/84. (17) *FT* 9/7/81. (18) *Kittg* 2/82. (19) Winona LaDuke: *Military neocolonization in North America ... and the land is just as dry,* New Haven, 1981 (Harvard University Dept. of Economics mimeographed paper). (20) Lanning & Mueller. (21) Weiss. (22) Bruce Johansen & Roberto Maestas, *Wasi'chu: The continuing Indian wars,* New York, 1979 (Monthly Review Press). (23) Eduardo Galeano: *Open Veins of Latin America: Five centuries of the pillage of a continent,* New York, 1973 (Monthly Review Press). (24) *E&MJ* 11/84. (25) Richard J Barnet & Ronald E Muller, *Global Reach: The power of the multinational corporations,* Jonathan Cape, London, 1975. (26) *MJ* 8/8/83. (27) *FT* 27/8/82. (28) *FT* 8/1/80. (29) *FT* 30/12/81. (30) *MJ* 10/4/81. (31) *FT* 18/12/80. (32) *FT* 6/2/81. (33) *E&MJ* 2/83. (34) *FT* 16/1/85. (35) *MJ* 23/1/81. (36) *E&MJ* 3/85. (37) *Tailings piles in New Mexico: Report by the New Mexico Energy and Minerals Department, January 1979,* Albuquerque, 1979 (New Mexico Energy and Minerals Dept). (38) *BHPS* 7/79. (39) Winona LaDuke Westigaard: *What is Anaconda?,* unpublished research paper 1978 (mimeographed), quoted in: Weiss. (40) Winona LaDuke Westigaard: *Anaconda and the Laguna people,* unpublished 1978 (mimeographed) & *Anaconda: The industrial snake* in: *AkwN* Summer 1979. (41) *Mining Congress Journal,* USA (place unknown), 6/57. (42) *High County News,* Wyoming, USA, (place unknown), 10/3/78. (43) Statement by Nick Franklin, Secretary of the Department of Energy and Minerals for the State of New Mexico, quoted in: *New Times* (Albuquerque?) 27/11/78. (44) *MAR* 1984. (45) *MJ* 25/1/85. (46) *MJ* 13/1/84. (47) *FT*

20/6/84. (48) *FT* 6/10/84. (49) *MJ* 11/11/83. (50)
MJ 18/5/84. (51) *MJ* 15/6/84. (52) *FT* 23/7/80.
(53) *MM* 12/81. (54) *FT* 30/7/80. (55) Reece. (56)
MJ 9/1/81. (57) *FT* 31/10/83. (58) *FT* 11/3/80.
(59) *FT* 30/6/81. (60) *FT* 20/3/85. (61) *E&MJ*
2/87. (62) *FT* 14/1/86. (63) *MAR* 1989. (64) *MJ*
27/11/87. (65) *FT* 26/2/85. (66) *Jakarta Post*
19/4/90. (67) *E&MJ* 1/90.

44 Argor Explorations

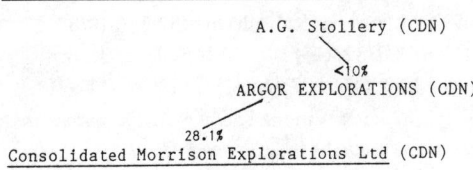

With its HQ at Toronto, it has been exploring
for uranium in northern Saskatchewan (1).

References: (1) *Who's Who Sask.*

45 Argosy Mining Corp

It has carried out geochemical and scintillometer surveys at a uranium holding in the
south-west of Eire, in Allihies (1).
Exploration in Eire is now dormant (2).

References: (1) *MM* 5/81. (2) *MAR* 1982.

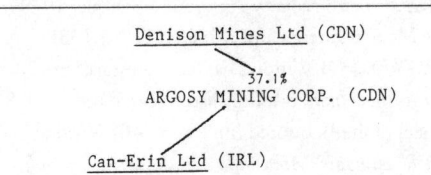

46 Arizona Public Service Co (USA)

A US utility involved under its own name, and
with El Paso Electric and Public Service Co of
New Mexico, in uranium development from
1976 onwards. No known production from
any reserves (1).

References: (1) Sullivan & Riedel.

47 Arrowhead Uranium

Arrowhead was bought out by the Rare Metals
Corp in 1954, and merged with El Paso Natural Gas in 1955 (1).

References: (1) *US Congress Sub-committee on Energy
1975.*

```
         El Paso Natural Gas Co. (USA)
                    \ merger
                     \
                  ARROWHEAD URANIUM (USA)
```

48 Asamera Ltd

Based in Calgary, it operates in Indonesia, the
Middle East, Singapore, South America, the
US, and northern Saskatchewan (1).
At its Keefe Lake-Henday JV in northern Saskatchewan it first reported "exciting" anomalies
in 1978 (2), while at Dawn Lake (also northern
Saskatchewan) similar results were reported in
1979 (3).
Its JV partners in northern Saskatchewan are
Saskatchewan Mining Development Corp
(50%), Kelvin Energy (6.5%), and Reserve Oil
and Minerals (7.5%) (4).
In late 1982 the company finished its drilling
programme at its Dawn Lake uranium property
where it reported an increase in reserves by
more than 50% (5).
When uranium exploration in Canada reached
a peak in the late 1970s, Asamera was among
the top ten corporations in terms of expenditure
(a total of US$90 million was spent in 1978)
(6).
Asamera's share in the Waterbury JV was purchased in 1982 by Idemitsu Kosan of Tokyo
(7).

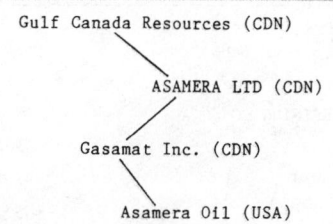

```
     Gulf Canada Resources (CDN)


              ASAMERA LTD (CDN)


        Gasamat Inc. (CDN)


            Asamera Oil (USA)
```

In 1987, Asamera was reorganised, with all its North American mineral interests being incorporated in Asamera Minerals Inc (8). Its main mineral venture is the Cannon gold mine in Washington State, in which it holds 51%, although it still holds sizeable exploration leases worldwide.

The same year, Gulf Canada Resources Inc, controlled by the Reichmann family (see Brinco, and Olympia and York Investments), gained effective control of Asamera (9) after Gulf Canada Corporation was itself split into three bodies, including Gulf Canada Resources (10).

Although not owning any of the equity, Asamera is the operator of the Dawn Lake uranium project, which now looks like not commencing until at least the 21st century – if then (11).

References: (1) *Who's Who Sask.* (2) *FT* 2/8/78, *MJ* 15/12/78. (3) *MJ* 9/11/79. (4) *FT* 5/12/78. (5) *E&MJ* 1/83. (6) Goldstick, 1981. (7) *MM* 11/84. (6) Goldstick, 1981. (7) *MN* 11/84. (8) *NJ* 10/6/87. (9) *FT* 29/3/88. (10) *FT* 18/6/87. (11) *Greenpeace Report* 1990.

49 Asarco [American Smelting and Refining Co]

Asarco (formerly the American Smelting and Refining Co) has penetrated the Australian mining giant MIM like a slithy toad: quietly but efficiently (for details, see MIM). Just as well – since 1982 was its worst year on record (US$74.1 million down as against profits of US$50 million in 1981) (1).

During 1982 the company also sold its 33% share in Revere Copper and Brass, one of its major entrées into solar power in the 1970s, and liquidated Sunworks Inc (80%), another spoiler in the solar market.

Asarco is one of the world's great silver miners, but it also deals in copper, lead and zinc as well as smelting and recycling them. While its main operations are in the USA, it also mines and refines in Peru, Mexico, Canada (asbestos in Quebec) and Bolivia where it holds a majority (58%) interest in a Bolivian company formed to operate the Quioma lead/zinc/silver mines (2, 3). It also has several interests in Australia (besides its share in MIM) – in fact, it gained control of a lead/silver lode at Mount Isa, Queensland as long ago as 1930 (4).

Its JVs include ones with Amax and Arco/Anaconda (Anamax) in the Eisenhower Mining Co (27%); with Newmont and Phelps Dodge it formed Southern Peru Copper.

Although Charles Barber, president of Asarco, is now immortalised as the man whose wisdom gave this book its title, the company is mainly involved in uranium exploitation only through its share of MIM. Between 1969 and 1972, however, along with Denison, Cominco and Uranerz, Asarco reportedly "took an interest" in uranium exploration in the north of Guyana; it is not known whether this encroached on the land of the Akawaio in the Upper Mazaruni region (5).

Along with Phelps Dodge and Newmont the company also has copper mining operations on Papago native land in southern Arizona (6). The entire southern Arizona copper belt – which produces two-thirds of America's copper – used to be within the Papago reservation, before discovery of the metal when Congress "simply passed a law declaring that the land no longer belong [to them]" (6).

Its control over burgeoning solar power in the USA is incontestable. As a major producer of cadmium, it has its foot in the door of photovoltaic cell production under development by Shell/SES, Texas Instruments and other firms (7).

In 1977, the corporation received no less than 20% (some US$1.3 million) HUD (US gov-

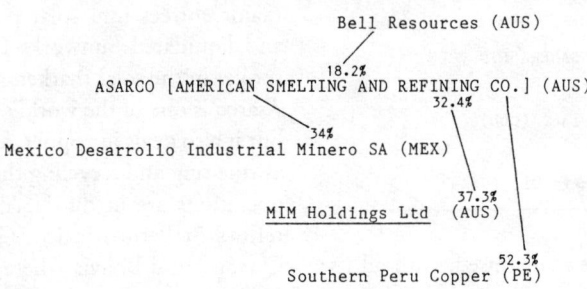

Bell Resources (AUS)
18.2%
ASARCO [AMERICAN SMELTING AND REFINING CO.] (AUS)
32.4%
34%
Mexico Desarrollo Industrial Minero SA (MEX)
37.3%
MIM Holdings Ltd (AUS)
52.3%
Southern Peru Copper (PE)

ernment) third cycle funding of solar power (7). As Ray Reece has put it:

"It would take some digging to find a more classic example of the tightening corporate grip on America's energy future, including the bantam solar industry, than that provided by Asarco ... With revenues last year (1978) in excess of a billion dollars, Asarco isn't so much a corporation, as a conglomerate of corporations, the bulk of which are engaged in one form or another of mining and metals processing. Asarco placed an ad in the September 27, 1976, issue of *Barron's Weekly* which proclaimed "Electric Power from the Sun – A reality with the help of Silver". The copy alluded to the government's massive power-tower program due to consume up to half a million square meters in heliostats whose reflective properties derives from a coating of silver – "the world's largest producer" of which is Asarco. The firm ranks fifth in the world production of copper as well, explaining its majority holdings in Revere Copper and Brass, which abides in turn among the nation's top ten manufacturers of solar collector systems with Revere's featuring copper absorber plates, naturall" (7).

Among the company's latest plans was a proposal, submitted to the US Forest Service and Montana state government in 1984, to open a silver/copper mine near Rock Creek in the Cabinet Mountains scheduled Wilderness Area of the Manisku National Forest, Montana (8). Asarco's profits were down in 1983 (9), and by the end of 1984, with several of its mines closed, it had to write off US$216 million of its assets (10). In 1984 Asarco also shut down its Tacoma smelter in Washington State, USA, because of environmental regulations as well as depressed prices and shortage of concentrates for "profitable" processing (11).

These drawbacks did not stop (indeed, probably encouraged) Western Australian "entrepreneur" Robert Holmes à Court in acquiring 9.9% in Asarco and making a bid for 50%, mainly to get a big chunk of MIM, through his company Bell Resources. Asarco filed a lawsuit against Holmes à Court, citing violations under the US Securities and Exchange Act (12).

The mid 'eighties were a parlous time for Asarco and it ran into heavy losses (13). A cost-cutting programme, combined with a strategy to turn the company into a fully-integrated producer of copper, was having considerable success by the end of the decade (14) as Asarco bought Ray Mines off Kennecott (15), expanded its Mission complex in Arizona (16), acquired 49% of Montana Resources Inc (17), and consolidated its interests in lead, silver (18) and gold (19). It sold its coal strip mines in Illinois in 1990 (20) but still bears legal responsibilty for litigation and remedial measures involving what it calls its "legacies from the past" – notably eight US sites where its mines and processing plants are still a "continuing source of environmental and occupational concern" (21). In 1988 it also purchased 34% in MEDIMSA (Mexico Desarrollo Industrial Minero SA), a major copper miner and smelter (22).

References: (1) *MJ* 18/2/83. (2) *MAR* 1984. (3) *MIY* 1985. (4) *BOM.* (5) *Uranium Redbook* 1979. (6) Tanzer. (7) Reece. (8) *MJ* 1/6/84 (9) *MJ*

15/4/83. (10) *FT* 31/1/85.(11) *MJ* 6/7/84. (12) *FT* 12/3/85. (13) *MJ* 22/11/85. (14) *E&MJ* 9/89. (15) *MJ* 14/11/86. (16) *MJ* 3/3/89; *MJ* 4/12/87. (17) *MM* 5/90. (18) *E&MJ* 9/88. (19) *MJ* 14/4/89. (20) *E&MJ* 1/90. (21) *MJ* 14/4/89. (22) *MJ* 20/4/90.

50 Ashburton JV

A JV in uranium exploration at Pilbara, a largely Aboriginal area in Western Australia (1).

References: (1) *FT* 12/7/79.

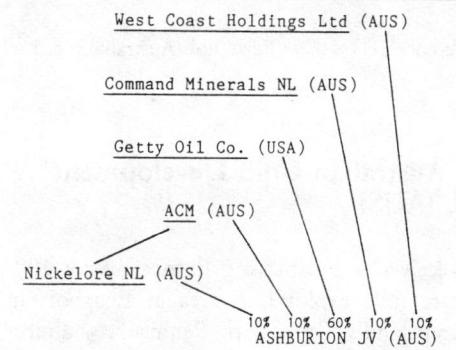

51 Ashland Oil Inc (USA)

Ashland is said to be an "independent petroleum refiner with manufacturing interests"; it has links with Kaiser Aluminium which acquired its Filtrol Corp in 1982 (1).
It was reportedly exploring for uranium in Texas during 1978 (2).

References: (1) *MJ* 20/11/81. (2) *E&MJ* 11/78.

52 Atlas Corp

During the mid-'seventies, Atlas operated a uranium/vanadium mill below minimum Environmental Projection Agency (EPA) standards. It was forced temporarily to close this mill in July 1977 for "routinely releasing radio-nucleides and other toxic material into the Colorado River" (1). It opened the plant after changing the configuration of the tailings dam to "prevent pollution of the Colorado River by chemical discharge" (2).

A high-grade deposit at Moab, Utah, (grading 16lb per ton) was reported in early 1979 (3), and in April of the same year the company gained "tentative approval" to start underground mining at its Snow Mine in Emery County, Utah (4).

It processed its own and other companies' uranium ore at its Moab, Utah, plant: in 1979 more than a million pounds of uranium concentrate (5); this was raised in 1980 to about one and a half million pounds (6).

At the end of 1982, the company laid off 110 mine and mill personnel at its operations in Moab – more than a third of the remaining workforce. Earlier in 1982 it laid off 175 workers (7).

In mid-1982 Atlas reported it would buy certain uranium properties from Exxon in Garfield County, Utah; Atlas refused to divulge the size of reserves or how much it paid for them (8). In 1983, however, the company was reported to have commissioned a feasibility study for a new uranium/vanadium mine near Ticaboo, Utah (see also Plateau Resources) (9).

By 1984, Atlas reported that the Bullfrog uranium properties, acquired from Exxon in 1982, had an average ore grade "significantly higher than the average in the USA", with important quantaties of vanadium. Preparations were in hand to develop these, build a processing plant, and pay royalties on future production to Exxon.

Further uranium deposits were also being "actively prospected" in 1984, particularly in the Moab area of Utah and Colorado (10).

In September 1987 Atlas announced that it would be closing its uranium operations in the western USA in order to focus on precious metals exploration – especially gold in Nevada (11,

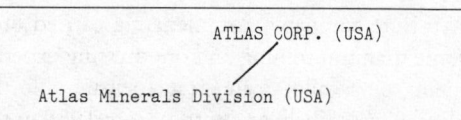

12). The following year Atlas began decommissioning and reclaiming its Moab mill – which had cost it about US$2 million to maintain on stand-by (13).

Thus ended the yellowcake saga of what, by the 1970s, had been the US's biggest single uranium producer (12).

Contact: ISSUE (Interested in Saving South Utah's Environment), POB 963, Moab, Utah 84532, USA.

References: (1) *Future Newsletter* 30/9/78. (2) *MJ* 12/8/77. (3) *FT* 12/10/79. (4) *MJ* 13/4/79. (5) *MIY* 1981. (6) *MIY* 1982. (7) *MJ* 26/11/82. (8) *MJ* 27/8/82. (9) *MAR* 1984. (10) *MIY* 1985. (11) *E&MJ* 10/87; see also *E&MJ* 6/89. (12) *MJ* 16/2/90. (13) *E&MJ* 1988.

53 Atomic Minerals Division of the Indian Geological Survey

In the 1990's this organisation reported uranium mineralisation over a 165km belt in southern parts of Karnataka and Andhra Pradesh, containing about 73,000 tons of indicated inferred reserves.

Other exploration is continuing in the Singhbhum belt of Bihar (a major tribal region) and parts of Rajasthan and Madhya Pradesh (1).

References: (1) *MJ* 13/7/84.

```
              Republic of India (IND)
                    /
       Indian Geological Survey (IND)
                    \
          ATOMIC MINERALS DIVISION (IND)
```

54 Atomienergia OY (SF)

A private company, Atomienergia carried out some uranium mining and ore dressing experiments on a pilot scale near Joensuu, North Karelia (yes, Sibelius country!) from 1958 until 1961. No further such operations were carried out but exploration continued – though mostly by the government Geological Survey (1). See also Outokumpu.

References: (1) *Uranium Redbook* 1979.

55 Aurec/Australian Uranium Resources Exploration Co Ltd (J)

Set up by C Itoh (70%), Furukawa (10%) and Sumitomo (20%) to explore for uranium in Australia's Northern Territory in the 1970s (1).

References: (1) *A Slow Burn* (FoE/Australia) No. 1, 9/74.

56 Australian Gold Development NL (AUS)

It received a prospecting licence over a 100-square mile exploration area at Ebagoola in Queensland's Cape York Peninsula – almost certainly on Aboriginal-claimed land. No further details: AGD was delisted at the stock exchange in 1972, though it reapplied in 1977 (1).

References: (1) *MJ* 12/8/81.

57 Avian Mining (AUS)

It was drilling for uranium in Niue in the Pacific in the late '70s, spending a small amount of money over a long period (US$250,000 over five years) without any apparent results. At the time, and in the context of gaining riches from uranium, Premier Robert Rex spoke of Niue becoming "another Nauru" (a Pacific island where the islanders gain reputedly the highest *per capita* income in the world from phosphate mining) (1).

Opposition to the project came from Niueans living in New Zealand; Niue, supposedly the world's largest coral reef, was a New Zealand colony from 1901 to 1974, when it became an independent territory but remained a protectorate of New Zealand (2).

Contact: CAFCA (Campaign Against Foreign Control of Aotearoa), POB 2258, Christchurch, New Zealand.

References: (1) *Povai,* Vanuatu 4-5/78. (2) *Los Angeles Times,* quoted in: *New York Times* 7/9/79.

58 Baltimore Gas and Electric Company (USA)

It participated, with the Intercontinental Energy Corp, from 1974 until at least 1979 in uranium exploration (1). No known production.

References: (1) Sullivan & Riedel.

59 Barlow Rand [Barrand] Ltd

Barlow Rand (previously known as Barrand) was formed in 1971, when the "fast-growing industrial conglomerate" Barlow's took over control of Rand Mines Ltd (in which previously the Anglo-American Corp had a 30% controlling interest). Rand Mines then became a wholly-owned subsidiary.

Rand Mines was formed by Alfred Beit and Julius Wernher in 1892 and soon grew into the greatest mining company in the world (1). It was Beit who "revolutionised" South African gold mining, by developing the group system of mining companies and attracting foreign investment to the burgeoning industry. Rand was originally not a mining company at all – but a resource for financial and engineering expertise put at the service of individual mines, which were then themselves absorbed as subsidiaries into the empire. Only at the point when the mine became viable were shares released to the public. So successful was this new form of "safe" finance-raising that European investors provided no less than £125 million between 1887 and 1913 to develop the group's deep-level mining operations (1).

However, by the 1960s the Rand properties were not among the more profitable of South Africa's gold producers (60% of the workforce, for example, was foreign labour from Portuguese-occupied Mozambique and Malawi). In 1974 Barlow's made a bid for control of the Union Corp, which was withdrawn when GFSA entered the picture. (Control of Union Corp eventually passed to General Mining – see Gencor.)

Thomas Barlow and his sons established their one-man import agency in Durban in 1902. Twenty-five years later the group acquired the franchise to distribute Caterpillar's earth, (and

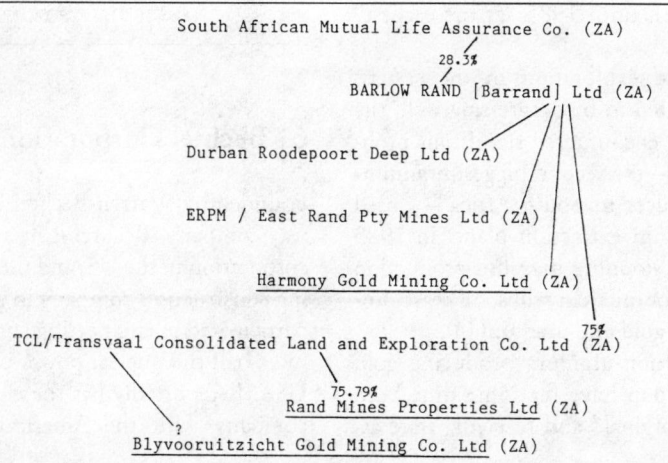

South African Mutual Life Assurance Co. (ZA)
28.3%
BARLOW RAND [Barrand] Ltd (ZA)

Durban Roodepoort Deep Ltd (ZA)

ERPM / East Rand Pty Mines Ltd (ZA)

Harmony Gold Mining Co. Ltd (ZA)

75%
TCL/Transvaal Consolidated Land and Exploration Co. Ltd (ZA)

75.79%
Rand Mines Properties Ltd (ZA)

?
Blyvooruitzicht Gold Mining Co. Ltd (ZA)

people-) moving equipment in southern Africa, "a business which was to prove the backbone of the group's history in the next 40 years" (2). After going public in 1941, Barlow's grew rapidly through conglomeration into timber, paint and steel.

By the late 1970s the Barlow Rand Group had enviable investments: not only in Harmony, one of the major gold mines in the world, but also in the largest chrome producer in South Africa and in the apartheid republic's largest lime producer (1).

In 1983 it increased its holdings in the large TCL – itself boss of the important asbestos producers, Cape Blue Mines and Cape Asbestos – and also absorbed minority share interests in the four Rand gold mines that Barlow previously didn't own (3). It is now a large, diversified group whose operations not only encompass mining but also manufacturing, distribution, agricultural and other interests in South Africa, Britain, the USA and Europe (3).

However, its fortunes have fluctuated in the past few years. While the group's coal mines continue to increase their overall production rates, this has been "solely because of increasing offtake from new tier collieries by the state-owned electricity utility, Escom" (4). Two new collieries are planned to supply coal to new power stations in South Africa in the 1990s (4). Production problems at Barlow's chrome mines were substantial in 1983, because of weak markets, while a new fluorspar mine was not expected to re-open until 1985 "at the earliest" (4).

And, while gold exploration in the eastern Transvaal appeared to be progressing well, the four gold mines encountered significant problems. Harmony – the second biggest uranium-from-gold producer in South Africa – closed one of its uranium extraction plants in 1983 and looked like stopping uranium production altogether. Blijvooruitzicht will soon cease production of both gold and uranium (4).

The other two non-uranium producing gold mines in the group have for some time been classified as "marginal" and receiving state assistance (5).

Barlow Rand draws on financial services provided by Consolidated Goldfields Services Ltd, a branch of Consolidated Gold Fields, and it has an exploration JV (through TCL) with Shell Coal in the so-called Manhattan Syndicate (3).

From 1984, carnivores in England will risk "eating South African", since J Bibby, the huge British supplier of animal foodstuffs, has been taken over by Barlow Rand, in the first major South African investment in Britain for years (6).

References: (1) Lanning & Mueller. (2) *WDMNE*. (3) *MIY* 1982. (4) *MAR* 84. (5) *FT* 19/10/82. (6) *Gua* 8/9/84.

60 Basin Oil Exploration Ltd

This is a member of the Conwest group of companies, together with Adanac, Chimo Gold Mines, International Mogul Mines, Central Patricia Gold Mines, and others (1).

It was part of Bond Corporation's attempt to get into the Beverley uranium grab (see also Beverley Uranium Prospects) through a consortium headed by Endeavour Resources (2).

References: (1) *MIY* 1981. (2) *FT* 15/9/79.

Conwest Exploration Co. Ltd (CDN)

98.5%
BASIN OIL EXPLORATION LTD (CDN)

61 Bechtel Corporation

Founded by Warren Bechtel in 1898 as a railway company, Bechtel is now the 25th largest corporation in the US, and the largest engineering construction company in the world. It "has participated in engineering and constructing almost half the nuclear power plants built in the USA. It reportedly has the closest links of any company with the American Central Intel-

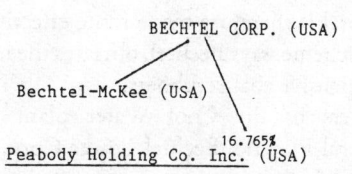

```
              BECHTEL CORP. (USA)

  Bechtel-McKee (USA)          \
                          16.765%
  Peabody Holding Co. Inc. (USA)
```

ligence Agency (CIA), excluding the CIA's own front companies" (1).

The US Agency for International Development (AID) – the "aid" wing of the CIA – has awarded Bechtel a US$4.3 million technical services contract to "help" several third world countries to "develop" their fossil fuel resources; the countries include Morocco, Bangladesh, Costa Rica, and Jamaica (2).

The appointment of George Schultz as US Secretary of State in June 1982 brought light to bear on the secrecy with which the Bechtel Corporation operates, for Schultz is a former president of Bechtel, while US Secretary of Defense Weinberger was formerly chief counsel to the corporation, and other right-wing US officials (such as Richard Helms of the CIA under Nixon, and Secretary of the Department of Energy, W Kenneth Davis) also held high office in Bechtel.

Schultz remains chairman of the company but the finance is firmly in the hands of the Bechtel family which owns the majority of the company's shares.

In order to buy the silence of former top executives it gives them huge handshakes on their retirement – something which has turned many former executives literally into overnight millionaires (3).

Philip Habib, US special "peace" negotiator, is also a consultant of Bechtel, hired by none other than George Schultz in 1981 (4). After Schultz's appointment there were allegations among US congress members that the company would "tilt" against Israel because of Bechtel's Middle East interests (16% of its 1981 revenue came from the oil states, especially from Saudi Arabia where it is turning an indigenous fishing village into an industrial city of 300,000 inhabitants at Jubail) (3).

Bechtel built Da Nang, the US air-force base in South Vietnam; the Alaskan oil pipeline; was involved in construction on RTZ's Bougainville copper project in Papua New Guinea; and assisted Cominco in its plans to open a mine on Aboriginal land in the Canadian high arctic (5). Pancontinental will also use Bechtel's dubious services in its construction at the Jabiluka uranium mine in Australia's Northern Territory (1).

Most recently Bechtel linked up with RTZ (through US Borax): it is to build the Quartz Hill molybdenum mine in Alaska from 1984 on. This proposed mine has been the subject of strenuous environmentalist objections in the USA (see RTZ).

Bechtel also owns 17% of the huge Peabody Coal Co which is currently the biggest threat to Navajo farmers in the Big Mountain area (see Peabody Coal).

It is also the corporation which put the San Onofre reactor on *backwards* (6).

Bechtel was targeted, among half a dozen corporations, for special attention by the Black Hills Survival Gathering in South Dakota, USA, during 1980.

Bechtel holds 21% of InterNorth Inc, which, with another US Company, Energy Transmission Systems, in 1983 was planning to transport coal slurry from Wyoming's Powder River basin area – already ravaged by uranium mining – to southern central US (7).

Among Bechtel's notable mishaps are the following:

- The Humboldt Bay, California, plant was one of the first operating nuclear power plants to be shut down permanently, in 1977, after it was discovered it was sitting directly on top of an earthquake fault.
- Consumers Power Co of Michigan sued Bechtel for US$300 million in 1974 when its Palisades plant broke down shortly after it started operation. Bechtel agreed to a US$14 million settlement.
- Portland General Electric Co sued Bechtel for US$32 million after severe leaks in the steam generator tubes of its Trojan nuclear plant shut it down, and the discovery that it

did not meet earthquake standards set by the NRC. Bechtel countersued, and an out-of-court settlement was reached in 1981.

- The 420-ton reactor vessel of its San Onofre Unit 2 was installed 180 degrees backwards in 1977: this was not discovered for seven months, a fact other nuclear engineers found incredible. San Onofre also sits near an earthquake fault. Unit 1 has been shut down by the NRC until it meets federal seismic standards after being almost constantly plagued with other mechanical problems preventing its operation.
- Bechtel designed the first privately-owned fuel reprocessing plant in West Valley, New York, which shut down in 1972 after six troubled years. During its operation there were repeated radioactive leaks into the air and into a creek that feeds into Lake Erie. Left behind are 600,000 gallons of high-level radioactive wastes buried underground in leaking tanks and 163 tons of irradiated fuel sitting in the spent fuel pool.
- Four Bechtel plants are among those having the most severe accidents in the last 10 years, according to a new NRC report: Rancho Seco, California; Turkey Point 3, Florida; Point Beach 1, Wisconsin; and Davis-Besse, Ohio (8).

Among its most recent contracts, Bechtel was awarded work on the new £10 million Gold Quarry gold mine in Nevada, operated by Newmont. Bechtel has also been awarded smaller contracts to build gold mines in Nevada (9). Meanwhile, Bechtel's Los Angeles Power Division (called LAPD but *not* to be confused with the LA Police Department!), together with Bechtel Petroleum, is completing the USA's first coal-gasification plant in southern California. Vaunted by the company as an operation which may launch a "new breed of power plant", the Cool Water scheme involves grinding coal, which is mixed with water to form a dirty "slurry", pumping this to a gasifier, combining it with oxygen, and converting it into gas (along with pollutant sulphur) suitable for a Combined Heat and Power (CHP) turbine. According to Bechtel, since "clean up" happens

before combustion, sulphur removal and waste disposal is therefore made more effective. This new scheme, says Bechtel, obviates the necessity of expensive coal-scrubbers.

Owners of the Cool Water plant include General Electric, Bechtel Power Corp, an unnamed Japanese consortium, and the Electric Power Research Institute – all are notable proponents of nuclear power.

Within the first two years of its operation the programme ran into financial snags – but the new US Synthetic Fuels Corp, created by US Congress in 1980, then stepped in with a US$120 million commitment in price support (10).

Bechtel is responsible for evaluation of the expansion of the Chuqui copper project, controlled by the Chilean state mining company Codelco (11).

A feasibility study contract for a coal mine and slurry pipelines in the Zhungeer coal field of Inner Mongolia (China) was awarded to Bechtel in June 1983 (12).

Bechtel in 1981 put forward proposals for the Trans-Sahara pipeline, after ENI and the Algerian government initially discussed the idea in 1972. Bechtel entered the arena just as Phillips Petroleum and BP were pulling out – "no coincidence", as the *Financial Times* has put it (13). The same year, Bechtel and Morrison-Knudsen formed a JV to construct the huge Ok Tedi mine in Papua New Guinea. Bechtel embraced this scheme – of enormous logistical difficulty – with its customary ethnocentric brashness. Everything "from beans to bulldozers" would be shipped to the site, and an entire township was constructed on indigenous lands to house 2500 people, with temporary camps for another 4000 workers, not to mention an airstrip, schools, hospital and other provisions for the Ok Tedi workforce.

Bechtel's biggest problem in construction of Ok Tedi to date has been the unfortunate tendency of Mother Nature to undo their best work: in January 1984 some 50 million cubic metres of hillside slid into the Ok Ma valley "precisely in the area where Bechtel MKI engineers had planned the dam to contain tailings from the gold recovery process" (14).

At its peak, Bechtel was employing 3270 workers, of whom 70% were from Papua New Guinea itself. Carl Perkins, Bechtel's project manager at Ok Tedi, has called the scheme "a megaproject, a one-of-its-kind challenge, something that provided much more than just a new mining complex"(15).

The Sami of northern Scandinavia, having been defeated in their heroic attempt to stop construction of the Alta-Kautokeino dam (20), may now also be the "beneficiaries" of another assault on their land, caribou routes and culture. Bechtel in 1980 took part in a preliminary survey for the construction of a gas pipeline across huge tracts of Sami land, a major aim of which is to provide fuel for military (NATO) use. As of the beginning of 1984, the project was still under (serious) discussion (16).

In early 1984, Bechtel also put in a bid for the faltering Scott Linlithgow shipyard in Scotland, against a bid by Trafalgar House (17). This came at a time when Bechtel urgently needed to diversify and consolidate operations. It also coincided with a vigorous defence of Bechtel's record on nuclear power plant construction made by Bechtel's old friend RTZ (in the person of its chief executive Alistair Frame) at the Sizewell Inquiry in England into the siting of a nuclear reactor (18). But as the shipyard's 2000-plus workers decided to occupy their workplace, to guarantee its future, Bechtel withdrew its take-over offer (30).

However, in the long run – according to the *Financial Times* – Bechtel's future may lie "across the Pacific in Indonesia, Malaysia and above all China. If and when the Japanese make their expected full-scale landing in the US, they are likely to meet a lot of Bechtel men going the other way" (19).

Just before the cataclysm of Three Mile Island cut off the US nuclear industry in its prime, Bechtel had constructed some 23% of all US reactors (46 in all) and engineered some 31% of the total (64 in all) – more than twice as much as its closest rival (21).

Bechtel, together with Alexander Gibb and Partners (GB), are carrying out feasibility studies for the Aye Koye aluminium smelter in Guinea – a project long mooted by foreign aluminium giants (especially Alusuisse) (22), but not pulled back onto the drawing board until the death of Guinea's leftist president Sekou Touré (23).

China's National Coal Development Corporation in 1984 signed a 15 year "pact" with Bechtel, setting up a US$3 million JV called China American International Engineering, which will undertake work outside and inside the Peoples' Republic, involving coalmines, pipelines, civil engineering, "energy" and communications projects (24).

Bechtel built Freeport's copper concentrate pipeline to take copper concentrate from the mine at Tembagapura (West Papua (Irian Jaya), Indonesia) to the coast some 110km away for shipment to Japan and Germany (25). In 1981, this was the largest such pipeline in the world (26). The line was cut by the OPM freedom movement in 1977.

In August 1984, the corporation was awarded the contract to manage construction of Alcoa's US$1500 million aluminium smelter at Portland, Victoria, Australia. Work started in November 1984 and the smelter was operational in 1987. Bechtel was responsible for engineering, procurement, site construction and "generally related activities" at the smelter (27).

The site is part of the traditional land of the Aboriginal Gunditj-Mara people, who strenuously fought – using court claims, direct action, occupations and sabotage of Alcoa's machinery – to stop the smelter being built (28). Under such pressure and because of the high electricity price charged by the Victorian state government, the project was mothballed in 1982. The Labour government in the state – despite token gestures towards Aboriginal land rights – revived the project soon after coming to power in 1983, and will in fact take 25% stake in it (29).

• In late 1984, Bechtel was awarded a US$40 million contract to build a phosphate rock slurry pipeline over mountains from Bernal

(Utah) to Rock Springs (Wyoming), to feed Chevron's fertiliser operations (31).

- Meanwhile another slurry pipeline – for coal – in which Bechtel had been part of a construction consortium with Internorth as the operator, was cancelled because of "legal and regulatory delays" (32).
- In 1983, Bechtel was awarded a contract by the Gabonese government to help it draw up its five-year plan, focusing on petrol, manganese, lead, talc and uranium (33).
- Study of Utah's high-volatile sub-bituminous coal deposits, wiht a view to converting them to synthetic crude oil ("syncrude") led Bechtel in 1985 to conclude that the state could be commercially producing by the early 1990s (34).
- The following year Bechtel gained a half billion pound contract to construct a thermal power station for the Turkish government (35).
- As its expansion into plant automation, waste treatment, space, defence and power co-generation continued in 1986, the company booked in around half a billion dollars worth of new business. (However, Bechtel is not required under US law to disclose its earnings) (36).
- 1986 also saw Bechtel commissioned to carry out a feasibility study into developing the Algerian port of Algeciras as a "superport" (37).
- In 1987, the International Indian Treaty Council reported that Bechtel had approached the Chickaloon Traditional Village Council in Alaska, with a view to building a waste incinerator on tribal lands. Although ash from the plant has been classified as hazardous waste by the US government, Bechtel apparently told the tribal members that it could be used as "fertilizer for their gardens" (38).
- Venezuela was the site of another major Bechtel venture in 1988 as the company was commissioned to carry out a feasibility study for the country's Guayana province aluminium smelter – one of the largest of its kind in the world (with technology supplied by Pechiney) (39).

- The same year, the huge Escondida mine in Chile (in which RTZ has a 30% share) began construction. Bechtel carried out the on-site studies, although Fluor Daniel has provided the engineering and construction management services (40).

Further reading: Jim Riccio, "Incompetence, Wheeling and Dealing: The Real Bechtel" in *MMon*, Washington, 10/89.

References: (1) *Time and Energy* No. 7, Australia. (2) *MJ* 19/2/82. (3) *Gua* 29/6/82. (4) *FT* 26/7/82. (5) Personal communication from Heather Ross (advisor, Treaty Council No. 9, Ontario); *NJ* 15/1/82. (6) *Nuclear Free Press* summer 1982. (7) *MJ* 28/1/83. (8) *Critical Mass Journal*, USA, 9/82. (9) *MM* 5/84. (10) *Bechtel Briefs* 1-2/84. (11) *E&MJ* 6/83. (12) *MJ* 1/7/83. (13) *FT* 3/5/83. (14) *MJ* 8/6/84. (15) R Jackson: *Ok Tedi, the pot of gold*, University of Papua New Guinea, 1982. (16) 12/12/83 letter from the Norwegian Embassy in Bonn (D) to the Gesellschaft für bedrohte Völker, quoted in *Pogrom* No. 107, mid 1984. (17) *FT* 22/2/84, 18/2/84. (18) *Transcript of proceedings of the Sizewell Inquiry* 9/11/83. (19) *FT* 16/11/82. (20) *Native Peoples' News*, London, Nos. 8, 9. (21) *Electrical World*, USA, 15/1/79. (22) Ronald Graham: *The aluminium industry and the Third World*, London, 1982 (Zed Press). (23) *FT* 11/5/84. (24) *FT* 12/4/84. (25) *West Papua: Obliteration of a People*, London, 1983 (Tapol). (26) *Pacific Islands Yearbook* 1981. (27) *MJ* 17/8/84; see also *MJ* 19/10/84, *MM* 12/84. (28) *Native Peoples' News*, London, Nos. 7-10. (29) *FT* 1/8/84. (30) *Gua* 11/2/84; *Gua* 3/3/84; *Gua* 3/4/84. (31) *MJ* 9/11/84. (32) *FT* 3/8/84. (33) *MJ* 22/4/83. (34) *MJ* 1/3/85. (35) *FT* 29/1/86. (36) *FT* 13/3/86. (37) *FT* 21/1/86. (38) *Treaty Council News*, Vol. 7 No. 3, San Francisco, 12/87. (39) *E&MJ* 6/88. (40) *E&MJ* 10/88.

62 Beisa Mines Ltd

To great fanfare, Gencor announced the Beisa mine in 1978. It was to be the first primary uranium producer in South Africa for some years (1). On 5 March 1982, the mine was offi-

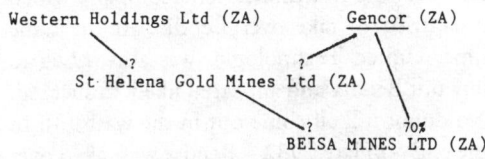

```
Western Holdings Ltd (ZA)        Gencor (ZA)
              \?              ?/          \
        St Helena Gold Mines Ltd (ZA)     \
                         \?        70%\
                         BEISA MINES LTD (ZA)
```

cially opened. "Rain seems to follow me around," wrote George Milling-Stanley, recalling that occasion. "And sure enough, it was overcast and damp ... when with the other guests I flew in to the official opening of the Beisa uranium mine in South Africa's sunny Orange Free State."

But not only the weather surrounding Beisa has been overcast. Within barely two years, Beisa was closed down after riots, flooding and market collapse. "A drastic step, something of a rarity in South Africa ... forced on Gencor by the weakness of the market for uranium," mourned Milling-Stanley in May 1984 (2).

Capital costs for Beisa were high at £122 million (easily three times that for a conventional gold mine). Situated at the Palmietkuil and Kalkoenkrans farms in the Orange Free State, it was also the first uranium/gold venture to open up since the 1950s apart from those adjoining existing mines (1).

The Beisa deposits are close to both the Beatrix and St Helena mines, owned by Gencor. In August 1981, St Helena announced that it would acquire Beisa's mining assets and the right to extract and sell all base and precious metals from the area (3). Since then, Beisa's production has been subsumed under St Helena's accounts to provide a tax write-off for St Helena (4).

The gold section of Beisa was reportedly the first to use on a commercial scale a new carbon-in-pulp extraction technology (3).

Soon after opening, the mine was seriously hampered by large quantities of water leaking into the workings (5). In the first 9 months of that year, 175 tons of uranium oxide came out of Beisa (6). A year later, both tonnage and grade of uranium had improved (7). But an explosion at the mine killed sixteen people. The workers refused to go underground, claiming

conditions were unsafe, and R100,000-worth of damage (£58,000) was caused to mine property. Five miners were arrested in protests in April, after dismissals by the mine management for refusal to go underground: the company claimed that about 500 workers "had decided to quit their jobs" because of the explosion (8).

In the first quarter of 1984 the mine delivered 153 tons of uranium oxide, at an average grading of 0.53kg/ton (9).

In their influential report on the African uranium industry for the Massachusetts Institute of Technology (MIT), Lynch and Neff gave two reasons why Beisa would succeed while the other primary uranium producer in South Africa – Afrikander Lease, worked by Vaal Reefs, an Anglo-American subsidiary – had failed. First, because it contracted sales for a substantial amount of its output at favourable terms (10); second, because its costs appeared to be low (11).

What could not be predicted by the MIT authors, however, was that the uranium market would plummet so considerably in the years 1981-1983. Nor that Beisa would experience severe underground flooding and faulting (12) which meant that it remained below full capacity.

As it is, the shareholders of St Helena and the South African taxpayers are the main casualties of the mine's closure; the former because they've never derived any income from Beisa, the latter because St Helena Mines was allowed to deduct its R220 million capital expenditure from its own tax liability (2).

Although Beisa closed in 1982, Gencor opened up a new gold mine in the same reefs, in 1988, called Oryx: there are no plans for uranium production (13).

References: (1) *FT* 20/7/78. (2) *FT* 5/5/84. (3) *MM* 11/81. (4) *Nuclear Fuel* 28/9/81. (5) Rand and OFS quarterlies in: *MJ* 4/82. (6) id in: *MJ* 10/82. (7) id in: *MJ* 10/83. (8) *Gua* 12/4/83. (9) Rand and OFS quarterlies in: *MJ* 4/84. (10) *FT* 30/8/78. (11) Lynch & Neff. (12) *MJ* 4/5/84. (13) *NM* 5/88.

63 Belgian Mining Engineering Co

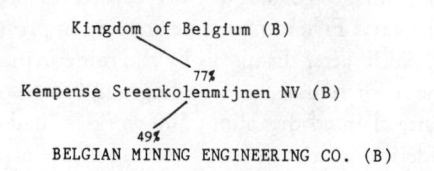

It was reported to be consulting with the Algerian government in 1981 over development of the country's Hoggar Mountain uranium (1). There is no further information, as none of the other bidders – AG McKee, Traction et Electricité and Union Minière – was reported, in 1979, to have won the contract for the Hoggar mountain development (2).

References: (1) *Le Monde* 12/6/81. (2) *MJ* 6/7/79.

64 Bendix Corp

Bendix, though best known for washing machines, hasn't in fact made one for 40 years. It concentrates on automobile parts, aerospace, industrial tools, and military equipment. It also has an unusually large amount of spare cash which could mean it will look for a take-over opportunity – as it did some years ago when it took over 20% of Asarco (which it later profitably sold) (1).

It is a private contractor working for the US Federal government, and as such took unauthorised rock samples from the Chippewa Cliff reservation (possibly for uranium sampling) using the name Derry, Michner and Booth (DMB) of Golden, Colorado. They told tribal police they were fossil collectors. They also did water sampling at Bad River (2).

In late September 1982, the company tried to

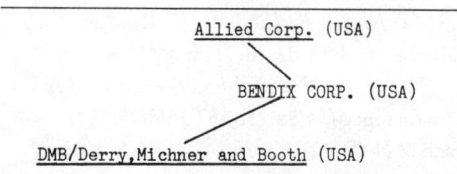

take over the Martin Marietta company which itself tried to take over Bendix. At the same time, United Technologies was also trying to buy out Bendix and appeared likely to succeed. No doubt it'll all come out in the wash (3)! In the end – in late 1982 – Bendix was taken over by Allied Corp, an important US producer of amorphous metals and alloys (4).

References: (1) *FT* 21/6/82. (2) *AkwN* Winter 1981. (3) *FT* 17/9/82. (4) *MAR* 1984.

65 Benguet Corp

The country's oldest mining company, one of the five most profitable industrial corporations in the Philippines (1983 being its most profitable year ever) (1), Benguet is also the country's largest gold producer (nearly 30% of total production in 1983) (2). It is the country's biggest producer of refractory chromite: indeed it is the world's leading producer of this metal, although production was stepped down in the late 1970s (2).

Benguet's best-known chairman was Jaime Ongpin, a "liberal" who criticised the government; "sympathised" with the anti-Marcos New Peoples Army ("Our people see these NPA people regularly," Ongpin was quoted as saying in 1984. "We've talked with them. We have to deal with them.") (3); and who fell foul of the Marcos régime over the supply of copper feed to the state's Pasar smelter. Benguet wanted to continue its contract with Mitsubishi rather than become a partner in Pasar (4). Although it held out for some months, in summer 1984 it backed down from a plan to challenge the government in court and agreed to deliver the required tonnage (5) – but in early 1985 the Philippines Supreme Court supported its original decision (6).

Nonetheless, Ongpin's brother Robert was the régime's Minister of Industry, and blood is thicker than all the waters of Benguet province: there has never been any serious challenge by the company or the régime to each other's power (7). Indeed, when the régime devalued

```
Republic of the Philippines (RP)          Allen & Co (USA)
                                      ▸ 70%    30%
                                      BENGUET CORP. (RP)
```

the peso in 1983, Benguet especially stood to benefit (8): "We have pesos coming out of our ears," remarked Jaime Ongpin in mid-1984 (3). Benguet operates four adjacent underground gold mines in Benguet province, just north of Baguio City; its open-pit copper/gold mine at San Marcelino, Zambales, contributes nearly 50% of the company's consolidated earnings (9).

It is also a diversified company whose non-mining activities are becoming increasingly important: construction, steel alloy castings and machinery find a market both at home and abroad (3). Its wholly-owned Benguet Management Corporation administers interests in timber, lime, trucking, shipping, agribusiness and real estate (3).

For "real estate" read "tribal land" – for Benguet encroaches on more Igorot territory than any other mining corporation in the Philippines. Both its Itogon gold operations and its Balatoc mill are on Ibaloi land (6).

In 1975 Marcos decreed that only mining companies could sell gold, a decision which especially hit the Benguet people who depended for much of their meagre income on their own gold panning operations.

The previous year, Bontoc peasants from the mountains of Mainit had risen to defend their lands against Benguet. They protested at the eventual pollution of their rivers, the destruction of their virgin forests, the killing of all fish and other food resources, and the loss of irrigation for their rice fields. They cited the copper mine in Baguio which had destroyed the surrounding flora and fauna as far as the adjoining provinces hundreds of miles away, despite "anti-pollution" devices. "When the mines have been exhausted of their gold, we shall have lost our land and our fields shall have been wasted. What will happen to us and our children?" Unheeded by government officials, they set up walls and barricades on the roads to block the

mining personnel. Then, after several rebuffs, they attacked the company's camp, with the women taking the initiative, driving the company off their land.

It was but a brief respite: the 1975 legislation, prompted by Benguet and seven other major companies, is estimated to have deprived 20,000 families of their main source of livelihood (10).

Benguet employs nearly 20,000 people (1), of whom between 45% and 50% at the mine-face are tribal people (6). Recently tribal employees at Benguet's Baco-Kelly mine spoke out against conditions they and their families had to endure in service of the company. Contrasting them with those of expatriate engineers, Andoy (22 years old), Jun (24) and Mang Tonio (54) spoke about families crammed into rooms no more than 8 by 10 feet; lack of health facilities; a contract quota which effectively amounts to slave labour; and "preventive suspension", a means of laying off miners suspected of theft without supplying any evidence.

The miners are affiliated to the National Federation of Labor Unions which, in January 1980, after a strike by miners, won a pay increase.

"Our aspirations are simple. We do not wish to get rich. We only want justice and equality. We want to be treated as equals ... We want to claim our right to a full life" (11).

Although the Philippine Chamber of Mines has been the most prominent explorer for uranium in the country, several companies have joined the yellowcake search (12). Benguet was joined by Getty Oil in the late '70s in a uranium exploration JV which, by 1983, had yielded few results (13) and was replaced by a gold exploration JV that still continues (14).

After the ousting of Marcos and his wife in 1985, both the ownership and nominal control of Benguet changed. Benjamin "Kokoy" Romualdez had originally purchased a 66.8% block of shares in Benguet Corp in 1975 from the Ayala Family group (their best-known company is the San Miguel brewing corporation). In 1986, after allegations of corruption, Romualdez fled the Philippines and his shares were sequestered by the Philippine Commission for Good Government, under the Aquino government (15).

At this point, the Aquino administration put several of its own key figures onto the company's board. Ongpin, who had been closely associated with Romualdez, stepped into a new pair of shoes as Aquino's finance minister, to be replaced at Benguet by Delfin Lazaro (16) (who in the early 1970s had worked for Conoco; Ongpin died in 1987).

Meanwhile, Benguet Exploration, the corporation's main exploration subsidiary, had passed into the part-control of Independent Resources Ltd, of Perth, Western Australia, with 33.5% of the shareholdings (15). (Independent Resources Ltd also holds 22.39% of Consolidated Gold Mining Areas Ltd, a company with investments in Jingellic Minerals NL, and several gold mining projects near Kalgoorlie) (17). Another major shareholder in Benguet Exploration Inc is Ricardo Silverio, a Marcos associate who appointed Jack Rodriguez as chairman in 1987. Both Rodriguez and his wife Sonya are closely aligned with anti-communist organisations. (Sonya is also a leader of the Christian fundamentalist Spiritual Action Movement). Under Rodriguez's chairmanship the company has received a sizeable, but unknown, investment from General Singlaub and probably the Nicaraguan "contras" (Singlaub is head of the World Anti-Communist League) (15).

In 1988, Benguet Exploration also commissioned BHP Engineering to conduct a feasibility study for treatment of gold tailings left behind by small-scale gold miners in Boringnot, Davao del Norte (18). Benguet Exploration operates four groups of mineral properties in the Baguio-Benguet district, including a gold mine on Panaon Island, southern Leyte, and the Kingking gold-copper project at Pantukan (where it is partnered with Energy Corp and Tetra Management Corp, a subsidiary of Independent Resources) (17, 19).

In 1986, Benguet Corp announced that it would be increasing its gold output to a minimum of 10,000 ounces a month (20). The following year its production had quadrupled (15).

After 4000 workers went on strike in late 1989, "paralysing" its operations – the workers had been threatened with dismissal during a "retrenchment" operation (21) – Benguet announced that is would be generating its own power for the gold mines and its Dizon copper-gold mine in Zambales (22). Then, the following year, the company was faced not only with losses from the major earthquake of mid-1990 (23), but resistance from Igorots (indigenous Filipinos of the Cordillera) and others (including small-scale miners) incensed at the environmental pollution caused by Benguet, its invasion of their territories, and threats of dislocation to their settlements. A petition submitted by the residents of the *barangays* (councils) of Tuding and Tocmo, in March 1989, accused Benguet of "scraping" away mountain sides crucial to the Kankanaey tribal people, grossly violating "their mountains and destroy[ing] the natural water system" as well as causing the "displacement of large numbers of families engaged in small agricultural and mining activities". The Pollution Adjudication Board (PAB) ordered Benguet to stop mining in the Itogon mountain region (specifically at Antamok Vein, Loakan Vein, and 3-Vein concessions), after it found "exceedingly high values of dissolved solids [are] dumped from the open-pits and flowing as well as polluting and gravely silting the Antamok river" (24).

Benguet has also proposed building a huge tailings dam at Sitio Liang on the Antamok river, and constructing a diversion tunnel from the river to the dam. The following year the department of Environmental and Natural Resources (DECR-CAR) ordered Benguet to stop its activities in the region and pay a fine of P50,000 for violating environmental regulations. Meanwhile the local people had mobilised behind barricades to prevent Benguet continuing its polluting and life-threatening work – specifically on the Grand Antamok Project. Commented a representative of the Itogon Small-Scale Miners and Panners Foundation "... instead of suspending either operations, they are now operating on a 24-hour schedule" (25). By this time, some 3500 local people had signed a petition demanding the cessation of all Benguet's open-pit activities (25).

Further reading: Sterling Seagrave, *The Marcos Dynasty*, Macmillan, London, 1989. (Excellent backgrounder on the growth of Benguet and its interlinks with the Marcos régime.)

Minewatch: "Comments on the Grand Antamok Project, London 1991.

"Beyond 13 Years: Facts about the Grand Antomok Project" Task Force on Open-Pit Mining, Baguio 1990

Contacts: Cordillera Resource Center for Indigenous Peoples Rights, Rm 314, Laperal Session Rd, Baguio City, 2600 Philippines; tel 442-4175.

Cordillera Environmental Concerns Committee, Rm 314, Laperal Session Rd, Baguio City, 2600 Philippines.

Cordillera Peoples' Alliance, Lockbox 596, Garcom, PO Box 7691, DAPO 1300, Domestic Rd, Pasay City, Philippines.

References: (1) *MJ* 27/7/84. (2) *MAR* 1984. (3) *FT* 14/6/84. (4) *MJ* 9/3/84. (5) *FT* 7/8/84. (6) *E&MJ* 3/85. (7) Personal communication from Philippines Support Group, London, early 1985. (8) *MJ* 6/1/84. (9) *MM* 11/84. (10) Felix Razon & Richard Hensman, *The oppression of the indigenous peoples of the Philippines*, København, 1976 (IWGIA document No. 25). (11) *Tribal Forum*, Manila ?/84. (12) *MJ* 11/1/80 & 16/3/79. (13) *E&MJ* 9/85. (14) *MIY* 1985. (15) *Birds of Prey – Gold and Copper Mining in the Philippines: A political economy of the large-scale mining industry*, Report prepared by Stuart Howard for the Center for Labor Education, Assistance and Research (CLEAR), Baguio City, and RDC-Kaduamo, Baguio City, 10/88. (16) *FT* 10/3/86. (17) *MIY* 1990. (18) *E&MJ* 10/88. (19) See also *MM* 10/86. (20) *MJ* 21/3/86. (21) *MJ* 22/9/89. (22) *E&MJ* 3/90. (23) *FT* 1/8/90. (24) *Bantayan*, Cordillera Environmental Concerns Committee, Vol. 2 No. 1, 7-
9/89. (25) *Northern Dispatch*, Northern Information Network, Baguio, Vol. 2 No. 10, 9/3/90.

66 Bercel Investments Inc (CDN)

It was involved in uranium exploration in northern Saskatchewan. No further information available (1).

References: (1) *Who's Who Sask*.

67 Beverley Uranium Project

This is situated in South Australia, and otherwise known as Lake Frome.

It is administered by the South Australian Uranium Corporation, and owned by Western Nuclear (subsidiary of Phelps Dodge) 50.00%, Oilmin NL (AUS) 16.66%, Petromin NL (AUS) 16.66%, Transoil NL (AUS) 16.66%.

These are estimated reserves of around 16,000 tonnes of U_3O_8 and further reserves of around 7000 tonnes in a separate orebody, at fairly high grading. Finance to develop the site was to be provided by Western Nuclear, with *in situ* leaching. Hydrological testing was underway in 1981 (1).

Between 1982 and 2011 the partners had expected to spend US$515 million in the building, running and rehabilitation of the mine, producing just over 11,000 tons of uranium oxide.

A second well field was to be built in 1984, and a central processing plant and full-scale produc-

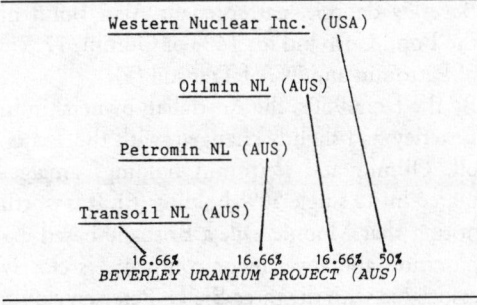

tion was to commence in 1986. Five years would be given to rehabilitation (2).

However, when the Australian Labor Party came to power in the state in 1983, the mine was suspended, and – of 1984 – despite a severely compromised ALP policy on uranium mining, it looked as if the mine would never be opened. The partners applied for a "retention lease" on the site in March 1983 (3).

In 1982, CANE, the South Australian (SA) anti-nuclear group, published a critique of the Draft Environmental Impact Statement on Beverley, prepared by the four partners.

CANE criticised the project from a number of points of view:

• The price assumed for sales of yellowcake would be twice the then-current spot price, with no guaranteed contracts: the economic viability of the project was therefore "in serious doubt".

• The partners' bland assurances that contamination of underground aquifers would not occur during mining was belied by contradictory statements and evidence that contamination in the Willawortina aquifers was likely. The extent of seepage of radioactive water from the aquifers could not be accurately determined.

• The partners' assurance that in situ leaching was a well-tried and tested method of mining was not borne out by experience in the USA. In particular, seepage from Wyoming Mineral's Lakewood, Colorado mine had caused "unexpected groundwater contamination", revealing previously unidentified sources of leakage and other problems (4).

In 1979, in order to gain an entry into the Beverley deposit, entrepreneur Alan Bond of the Bond Corp bid for 14% of Oilmin, 17.5% of Petromin and 8% of Transoil (5).

By the late 1980s, the Australian ownership in Beverley had slightly changed, with the Transoil, Oilmin and Petromin holdings amalgamated into a single 50% holding (6). It is worth noting that Moonie Oil, a Brisbane-based exploration and investment company, is closely associated with the three Bjelke-Petersen clones (as of 1983, holding 25.8% in Transoil NL,

23.2% in Oilmin NL and 9.1% of Petromin NL, (7).

References: (1) Reg Aus Min 1981. (2) MM 10/82. (3) Reg Aus Min 1983/84.(4) CANE, Adelaide, SA (5) FT 15/9/79. (6) Greenpeace Report '90. (7) Financial Times Oil and Gas Annual, 1983.

68 BHP/Broken Hill Pty Co Ltd

Australia's largest company, with major interests in energy exploration, it now holds Utah's uranium interest. In the late 1970s it was reported "leading" the development of solar power in Australia (1) – little more has been heard about this since.

Founded in 1885 to exploit the fabulous deposits at Broken Hill, its 177,000 shareholders now ostensibly control huge resources of iron ore, manganese, coal, oil, gas and parts of the manufacturing sector (2).

In a deal called "absolutely scandalous" by the Australian Labor Party's science and technology spokesperson, Barry Jones, BHP acquired control of Utah International in early 1983, and virtually complete ownership the following year (3).

Meanwhile, Robert Holmes à Court, through two of his subsidiary companies, had made a daring bid for BHP and secured a toe-hold in the giant's lair (4).

The company's exploitations of Aboriginal land include:

• Taking a 100-year lease on part of the Groote Eylandt Reserve, to mine a vast manganese deposit, after the local mission concluded a deal with BHP. In 1977 the Northern Land Council demanded a re-negotiation of the lease – which had only paid 1.25% in royalties directly to the local people – along with protection of sacred sites, and Aboriginal women, and other social and welfare provisions (5). Seven years later a report for the Australian Institute of Crimonology established Groote Eylandt Aborigines as the "most imprisoned" of all aboriginal Australians (6).

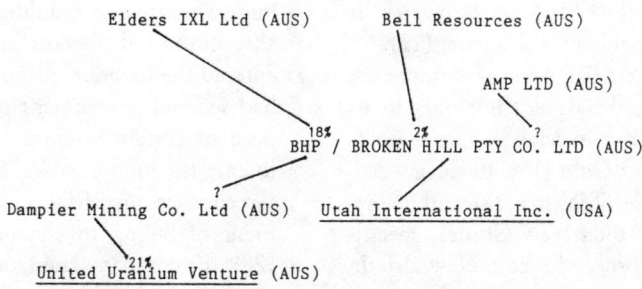

```
        Elders IXL Ltd (AUS)        Bell Resources (AUS)

                                            AMP LTD (AUS)

                    18%              2%              ?
                    BHP / BROKEN HILL PTY CO. LTD (AUS)

                  ?
  Dampier Mining Co. Ltd (AUS)    Utah International Inc. (USA)

                21%
      United Uranium Venture (AUS)
```

- Taking possession of a large exploration lease on the Ombulgurri Aboriginal Reserve in the Kimberleys (Western Australia) – despite promising the community that they wouldn't (7).
- Taking another lease to look for diamonds in the same region in 1979, in a JV with ACM, Command Minerals, and West Coast Holdings (8).

With its new-found thirst for US take-overs, BHP in late 1984 made a bid for the Energy Reserves Group of Kansas, so as to possess an oil and gas asset with which to offset foreign exploration expenditure (9).

After attempting to become the biggest shareholder in BHP, and acquiring more than 30% of the company's shares by late 1987, Robert Holmes à Court reduced his holding in November 1987 by 2.5% (10). Alan Bond bought out Bell Resources, after Holmes à Court's dramatic post-stockmarket collapse in October 1987, then the company itself bought back a majority of the holding, leaving Bond with 10% – later 5% (11). However, 18% of the shares are still held by Elders IXL in a friendly arrangement with BHP (12).

In 1987 BHP and Utah formalised their alliance with the formation of BHP-Utah Mineral International Corp, thus becoming "one of the largest mining and minerals operations in the world", with major operations in Japan, Europe, the US and South America (13). Two years later, thanks to the tie-up between Elders IXL and BHP in Berwick Pty Ltd, Mitsubishi obtained a sizeable stake in BHP (in a fairly complex arrangement, whereby Berwick passed on to Mitsubishi shares received as dividends from BHP as part of its 50/50 JV with Elders IXL) (14).

Although the company sold its mining interests in South Africa in 1986 (after a large-scale mid-decade restructuring of interests) (15) and has continued to dispose of, or put on the back burner, other relatively small projects (for example, the Macraes gold project in New Zealand which it sold to Macraes in 1990 (16), its Kalimantan coal interests, and the Samarco iron ore mine (49%) in Brazil about which little has been heard in recent years) (17), its late 1980s projects are among the world's most important. They include 57.5% in the Escondida copper mine in Chile (potentially the world's third biggest), with JV partners RTZ (30%), a Mitsubishi-led consortium (10%), and the International Finance Corporation (2.5%) (18). (This is a mine which by early 1990 was well ahead of schedule (19) with 75% of its concentrates output earmarked for Japanese, West German and other smelters.) By the turn of the decade BHP was also one of the western world's fifth largest producers of coal (20), joined with Meekatharra Minerals at Ballymoney, County Antrim, Northern Ireland (21), and owning the Navajo mine on Navajo land in New Mexico, and others.

From 1985 onwards BHP has also been expanding considerably its iron ore interests, with the purchase of CSR and Amax's assets in Mount Newman (22), the development of the Iron Duke deposit in South Australia (23) and the Telfer mine in the Pilbara (Western Australia) (BHP 30%, Newmont Australia 70%) (20).

In early 1990 BHP also bought the 70% of Mount Goldsworthy Mining Associates it didn't own, thus making it sole owner (23).

As with other Australian mining companies, BHP has been lured by gold: not only in its "home territory", but in Mali, where it holds 65% of a significant Gold JV with the government near the village of Syama (24), and on several prospects in Papua New Guinea, specifically the Wala prospect in East New Britain (25).

The company has also ventured into mineral sands: it holds a licence to explore for these minerals near Maralinga – the site of 1950s and 1960s British bomb tests – which it negotiated with the Aboriginal landowners, the Maralinga Tjarutja (26).

In the past two years, community protests against BHP's projects have developed to such an extent that it is now one of the most criticised of any company in the industry.

- In 1989 BHP acquired land and mining rights to an iron sand deposit at Maioro, on the Waikato Heads, south of Auckland, Aotearoa (New Zealand). Since this is the site of important Maori burial grounds, the Ngati Te Ato, representing local traditional land-holders, has vociferously opposed the licence and called on the government's Waitangi Tribunal (ostensibly set up to settle land claims) to compulsorily acquire the land, as the sacred sites are not adequately protected (36).

- BHP is the major Australian company (and one of the biggest of all multinationals) with oil exploration permits in Burma – some of which seem to encroach on areas claimed by indigenous nations (such as the Karen). Despite a concerted campaign by advocates of democratisation in Burma, and an appeal for the company to negotiate any leases with the democratic movement, which won the 1990 elections, the company's chair Sir Arvi Parbo has adamantly refused to suspend the agreements, or talk with the democratic movement (27).

- Greenpeace divers in early 1990 blocked off a waste outlet at BHP's Port Kembla plant,

claiming that the company was dumping huge amounts of cyanide, zinc, ammonia, chromium and phenols into a creek which entered the harbour. Although the company had secured government permission to dispose of certain amounts of dangerous effluent (including 62kg of cyanide) and Greenpeace primarily accused the government of failing to change the regulations (28), the company later admitted that it had been illegally discharging cyanide into the local waters (29).

- As the Gemco (Groote Eylandt Mining Company, BHP 100%) continues to produce record outputs of manganese – whose fortunes are directly dependent on the steel industry (30) – so Aboriginal members of the Angurugu community, and trade unionists, have stepped up their fight to reduce the dust hazards from mining (31). In 1987, health workers identified what they called the "Angurugu Syndrome" – in other words manganese poisoning – which affected considerable numbers of Aborigines and workers (32).

- In October 1989, the Jawoyn people of Coronation Hill in the Kakadu National Park (Northern Territory of Australia) won a major battle to stop BHP from mining a rich deposit of gold, platinum and palladium – at least for twelve months (33). The deposit was to be found in what the Jawoyn call "Sickness Country" – a site rich in uranium and also highly significant to Aboriginal people (it is close to several Bula and dreaming painting sites) (34). BHP had originally insisted that the majority of the Jawoyn approved of mining, in a claim clearly refuted by mid-1989 (35). As things stood at the beginning of 1990, Stage 3 of the Kakadu National Park has had its mining zone cut from 2252 square kilometres to only 37 sq km (33), while the Jawoyn are fairly confident that their traditional land has been protected (36).

- Unfortunately, the people of Papua New Guinea, living near or downstream of the Ok Tedi mine (BHP 30%, Amoco 30%, West

German consortium 20% and PNG government 20%) have not been so fortunate. The mine, which had experienced huge delays, controversy between the mining partners and PNG government, strikes, and several environmental disasters including cyanide spills (34), entered its final phase of production as a copper producer in 1989 (38). It has had a major impact on local people and the environment for many miles around: the Ok Tedi gold/copper mountain was lowered by 150 metres; by 1988, more than 4000 people had thronged into the village of Tabubil, and more than 200km of roads have been driven through the rainforest (39).

Then, in a highly controversial decision condemned by many inside and outside Papua New Guinea, the mining partners gained permission from the PNG government to dump tailings directly into the 1100km-long Fly River system on the grounds that constructing a tailings dam was "not feasible" due to dangers of land slippage (such as the major collapse of 1984) (40). This decision is one of the most fateful and dangerous ever made in the history of mining in the Pacific region. Already in 1986, a monitoring programme had revealed occasional high levels of heavy metals and cyanide up to 80km downstream of the mine (with cyanide content up to 50 micrograms per litre). Toxicity tests carried out since 1984 showed that the tailings are highly toxic, with 9-day LC50s as low as 0.4% and 0.1% for freshwater fish and shrimps. In practical terms this means that half the fish will die (or are dying) when tailings are diluted 250 times, and half the shrimps will die when tailings are diluted 1000 times – as happens 50km downstream of the mine (41). In 1985 the PNG government condemned the partners for not constructing a permanent tailings dam and gave them until January 1990 to comply. With a change in government and the pressures caused by the closure of the Bougainville mine (see CRA) the central administration finally relented on this crucial provision.

Further reading: for an excellent overview of the impact of the Ok Tedi mine on the people and the environment, see David Hyndman, "Ok Tedi: New Guinea's Disaster Mine", *The Ecologist,* Vol. 18 No. 1, 1988.

Contact: (in Aotearoa) Ngati Te Ato, c/o Nganeko Minhinnick, Box 250, Waiuku, South Auckland, Aotearoa (New Zealand).
(re Burma) Burma Peace Foundation, 218 Liverpool Rd, London N1 1LE; tel 071-700 3032.

References: (1) *MM* 2/78. (2) *FT* 7/11/83. (3) *NJ* 12/10/84. (4) *FT* 10/2/84. (5) *Age* 13/10/77. (6) *Age* 22/2/84. (7) Roberts. (8) *Age* 13/5/79. (9) *MJ* 16/11/84. (10) *FT* 18/11/87. (11) *FT* 4/8/88. (12) *FT* 26/7/88. (13) *E&MJ* 5/87. (14) *FT* 8/89 (exact date unknown). (15) *FT* 19/12/86. (16) *E&MJ* 3/90. (17) *E&MJ* 5/87. (18) *MJ* 6/4/90. (19) *FT* 23/6/89; *MJ* 6/4/90; see also *E&MJ* 9/88. (20) *MJ* 1/12/89. (21) *MJ* 7/7/89. (22) *FT* 1/10/85. (23) *E&MJ* 6/89. (24) *FT* 17/8/89; *MM* 5/90; see also *MJ* 10/3/90, *E&MJ* 5/89, *E&MJ* 6/89. (25) *MJ* 23/6/89. (26) *MJ* 11/5/90. (27) Letter from CJ Austin, BHP Director of Public Affairs, to David Arnott, Secretary, Burma Peace Foundation, Melbourne, 27/7/90; *Age* 13/3/89; Letter from Garry Woodward, University of Melbourne, to Sir Arvi Parbo, 22/3/90, and 12/4/90 response from Arvi Parbo; see also letter from Garry Woodward, 26/4/90. (28) *Australian* 23/2/90. (29) *TNT* No. 339, London, 20/2/90. (30) *FT* 3/8/89; see also *MM* 5/89, *E&MJ* 12/88. (31) Inspection Report 78: *Groote Eylandt Manganese Mine,* ACTU/Victorian Trades Hall Publication. (32) *Land Rights News,* Alice Springs, 6/87. (33) *New Scientist* 14/10/89. (34) *Land Rights News* Vol. 2 No. 12, 1/89; see also *Land Rights News* Vol. 4 No. 4, 9/87, *Land Rights News* Vol. 2 No. 5, 12/87. (35) *Sydney Morning Herald* 6/5/89. (36) *Pacific News Bulletin,* 1/90. (37) *FT* 1/9/89; *FT* 3/10/88; William S Pintz, *Ok Tedi: Evolution of a Third World Mining Project,* Mining Journal Books Ltd; R. Jackson, Ok Tedi: Pot of Gold, University of PNG. (38) *E&MJ* 9/89; see also *E&MJ* 8/88. (39) *FT* 5/7/88. (40) *FT* 6/10/89. (41) *The Siren* 31/7/86. (42) Pacific Concerns Resource Centre, *Action Alert,* 21/5/90.

69 Bicroft (CDN)

A company set up in the 1950s to mine uranium from Ontario for export for nuclear weapons construction in the USA (1); presumed defunct.

References: (1) *BBA* No. 4 1979.

70 Big Indian Uranium Corp (USA)

Not known to be involved in uranium mining, in June 1983 it was evaluating reopening the Glory Hole *(sic)* mine above Central City, Colorado along with Chain Mines Inc (1).

References: (1) *E&MJ* 6/83.

71 Blyvooruitzicht Gold Mining Co Ltd

Founded in 1937, it has been involved in gold mining, as well as extracting uranium, silver and osmiridium as by-products (1). Ownership in the early 1970s was Anglo-American (15.6%), GFSA (11.9%), Barlow Rand (8%), Sicovam (11.6%) and Barclays Bank (40%) among others (2). In the intervening years Barlow Rand has gained control of the company through a large (but unknown) holding by Transvaal Consolidated.

The company's lease, acquired from the apartheid government of South Africa, is situated in the Oberholzer district of the West Rand in Transvaal; its uranium plant is at Eastdene township (3). Uranium production markedly increased when the plant was extended in 1977 (4), but within seven years the company was announcing that it would close towards the end of 1984 "unless [there is] a substantial increase in contract prices" (5).

And, indeed, after producing nearly 300,000kg of uranium in 1983 – making it number seven out of South Africa's producers – it announced that it was to close at the end of 1984 (6). However, in October that year the company said that it would begin developing the lower-grade gold and uranium ore in the Main Reef, at a cost of US$3 million (7).

In the last quarter of 1984 it delivered just over 38,000kg of uranium (8), and nothing was produced in 1985. It therefore seemed as if the mine had come to the end of its yellowcake road: nonetheless it had delivered 15,000 tons of the deadly metal in the 35 years until then (9).

Virtually nothing is known about the company's contracts, except that it landed a contract with an unknown purchaser in 1979 to supply a "substantial quantity" of uranium – other sources say "most of the output" (10) – over the next 7 years, in return for an interest free loan of R16.8 million (1).

By 1990 not only had Blyvooruitzicht come to the end of its yellowcake road, but the mine itself was (as the *Mining Annual Review* put it) "in deep water", with no more than five years probable life left (11).

References: (1) *MIY* 1985. (2) Lanning & Mueller. (3) *MIY* 1985. (4) *MJ* 4/11/77. (5) *MJ* 24/9/82. (6) *MAR* 1984. (7) *E&MJ* 10/84. (8) Rand and OFS quarterlies March 1985 in: *MJ* 26/4/85 supplement.(9) *MJ* 5/10/84. (10) Lynch & Neff. (11) *MAR* 1989.

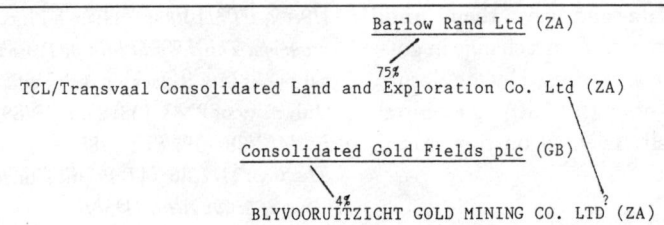

72 BNFL/British Nuclear Fuels Ltd

Set up in 1971 (on 1 April – hence the common form of address: "British Nuclear Fools"!), it owns all Britain's nuclear fuel production facilities including (and especially) those intended for military use: the Springfields "hex" plant and fuel-rod production plant, the Capenhurst enrichment plant, the Windscale reprocessing plant, and the two military reactors at Chapelcross and Calder Hall (1).

BNFL also holds, on behalf of the British Government, a roughly one-third interest in Urenco, and is responsible for the actual management of the Capenhurst enrichment plant (2).

As part of the British Government's Civil Uranium Procurement Directorate it has offered to explore for uranium in Zambia (3); it also has had a uranium exploration venture in Niger which is now presumed to be redundant (4).

BNFL's Windscale reprocessing and high level waste "storage" plant has been the subject of enormous and growing criticism, and also of a major government inquiry in 1977/78 to which many anti-nuclear groups contributed and which became the focus for a revival of the British anti-nuclear movement.

In the past eight years, several major awards have been made to the widows of workers dying as a result of radiation leakages from Windscale (5). Indeed there have been more than 300 accidents at the plant (now renamed "Sellafield" to disguise its wretched past) since 1950: the plant itself had to be shut down in 1973, after an "accident" which contaminated 35 workers (6). This was not the first time that a major disaster threatened: in 1957, spent fuel caught fire and released up to 20,000 curies of radioactivity into the atmosphere – some 400 times what was released in the 1979 incident at Three Mile Island (7). It was this fire which, in 1984, prompted the (then) Irish premier Charles Haughey to link an extraordinary incidence of Down's Syndrome births across the water, at Dundalk, to the fallout (8). Two doctors writing in the *British Medical Journal* agreed with the evidence of gross abnormality, but would not link it with radiation or the proposal that fallout materials from the fire actually crossed the sea (9).

In 1977 BNFL paid compensation to the widows and dependents of five former Windscale workers, without admitting liability: the highest award of £67,000 was settled out of court to avoid detrimental publicity. However, five years later, in an agreement generally welcomed by trade unions, automatic compensation was agreed "... in the few cases where there is sufficient possibility that death might be attributed to occupational radiation overdose" (10).

But the biggest criticisms of BNFL in recent years have undoubtedly stemmed from its lethal discharges into the Irish Sea and to the local community.

In 1983, Yorkshire TV (YTV) televised evidence that operations at Windscale were linked to the abnormally high cancer rate among children on the Cumbria cost: the leukaemia rate among children under ten, a mile from the plant, was no less than 10 times the national average, according to Dr Philip Day of Manchester University. The most likely sources for the contamination were house dust in which plutonium was found, among other nuclear metallic traces, and the beaches in the area. BNFL emphatically denied the accusations, calling the study "incomplete" (11).

The government's response was to immediately appoint a Commission under Sir Douglas Black, a former president of the Royal College of Physicians, to investigate the cancer rates in Windscale's locality. Foreshadowing the Commission's ambiguous, inconclusive and incomplete findings when it finally reported the following summer (12), the London *New Statesman* agreed that, while there may be no "definite link [between cancer rates and Windscale

```
          The Crown of Great Britain
          and Northern Ireland (GB)


BNFL / BRITISH NUCLEAR FUELS LTD (GB)
              33.3%
Urenco / Uranium Enrichment Co. (GB)
```

discharges], ... they must remain the prime suspect until some other locally-occurring carcinogen can be identified" (13).

The environmental action group Greenpeace Ltd was, however, in no doubt. Soon after the YTV revelations, it began trying to block the exit pipes into the Irish Sea through which the plant's effluent is carried, sending divers down – at obvious risk to their health – to try to identify the point of discharge. A court ban on their action soon followed (14), and a few days later the group was fined £50,000 (plus costs) for defying the injunction. While agreeing that BNFL was carrying on "a controversial business," Justice Comyn declared it was "... tragic how many people feel – either with an air of martyrdom or with an air of defiance – that they can take on the might of the law" (15).

Within less than two years, it was BNFL which fell foul of the law. While the company continued in the intervening months to claim that no harm has been or could ever be caused by contaminated effluent, it was being taken to court. Juggling his figures, Con Allday (sic) for BNFL maintained that less than a tenth of a curie of radioactivity had been discovered washed up on Cumbrian beaches over an eight-month period compared with "routine releases" from Windscale of 67,000 curies a year which "in turn is much less than the permitted levels" (16). In July 1985, BNFL was found guilty by a Carlisle Crown Court jury of not keeping the level of radioactivity in discharged nuclear waste "as low as reasonably achievable" (the so-called *alara* principle), not giving the public adequate warning of the radiation risk, and not keeping records of radioactive material discharged into the sea. The charges all related to one tiny period of Windscale's operations when the plant was shut down in November 1983 and Greenpeace was busy trying to plumb the lethal outlet. Police questioned a hundred BNFL staff during their investigations (17). BNFL was fined only £10,000 for its transgressions, with £60,000 costs, reflecting the fact that the judge could not find any danger to the public had resulted from the company's "haphazard and casual style of management" (18).

Ironically, only two days before the verdict, Greenpeace announced that half a million curies of plutonium-241 had been dumped into the Irish Sea by BNFL up to the end of 1982, breaking down into the highly dangerous americium-241 (19).

Nor is it only the Windscale plant which has leaked deadly liquor. Also at the end of 1983, *Time Out* magazine revealed that, three years previously, in May 1980, a breached waste pipe from the Springfields fuel fabrication plant caused such widespread damage to a local farm that the soil had to be creamed off to a depth of 18 inches and BNFL decontamination squads worked for 48 hours to neutralise the potential damage (20).

In late 1983 – as the Thatcher régime's mania for abolishing nationalised companies reached its height – BNFL was briefly mooted as a possible candidate for privatisation. A 49% sale of assets was apparently considered – until it was calculated that the cost of de-commissioning and sanitising the old military reprocessing site at Windscale would cost as much (about £200 million) as any sale might achieve (21).

Between 1978 and 1981 the Stop Urenco Alliance in Britain and anti-nuclear organisations in the Netherlands campaigned vigorously against the Urenco enrichment company, organising demonstrations jointly at BNFL's Capenhurst plant and at Almelo in the Netherlands. In the early '80s, Australian groups also became alarmed at the emergence of an industry-led campaign to site a uranium hexafluoride processing plant in South Australia.

BNFL held 35% of the Uranium Conversion JV, set up in June 1981, for which it would supply the expertise, with BHP (35%), Roxby Management Services (ie WMC and BP) (35%), and the South Australian government (5%), supplying the rest. With the arrival of the Labour government in South Australia in early 1983 the project was presumed dormant (22).

The latter half of the 'eighties proved to be a period of unprecedented boom for BNFL, and one in which it met more condemnation for its dangerous activities than the British nuclear industry had been confronted with in its previous forty years.

Boasting in 1985 that it had never missed one delivery date in forty years, and was exporting its "skills" "to most countries in the world" (23), BNFL announced that the Springfields fuel fabrication plant had secured orders "well into the next century". Two years later, the company joined up with its partners in the THORP (Thermal oxide reprocessing plant – destined to become one of the biggest, if not the biggest, nuclear fuel reprocessing and waste-emitting plants in the world) to form British Nuclear Technology. This is a consortium set up to "exploit an international market for the technology of spent fuel reprocessing and radioactive waste management" (24). A few months later, BNFL announced it would put £10 million into the funding of a development agency in West Cumbria (25). In 1989, another £10 million was promised towards a mixed oxide (uranium/plutonium) fuel development programme with an eye to the West German market (26). BNFL has long-term reprocessing contracts with West Germany well into the next century (27). Since the abandonment after vigorous public protest of the Wackersdorf plant in Bavaria, the West German industry had turned to BNFL as a means of reprocessing around half the spent nuclear fuel that would have gone to Wackersdorf. One argument that played a part in ensuring the collapse of the Wackersdorf scheme was that BNFL and Cogema of France could offer lower costs in reprocessing – a point which, as the *Financial Times* reported, led "some environmentalists [to] fear that [there are] lower safety standards in Britain and France" (28).

Meanwhile, BNFL's safety record was under relentless attack. Radiation leaks from its Windscale/Sellafield reprocessing plant were ackowledged on several occasions during 1986 (29, 30), leading the Irish government to call for a halt in the discharges (30, 31). Nuclear waste pumped into Rivacre Brook, from BNFL's Capenhurst enrichment plant prompted Greenpeace, Friends of the Earth and local residents to demand a proper investigation (32). In

January the following year, the British government allowed BNFL to make aerial discharges of radiation which were, on average, seven times the highest levels previously discharged from the plant – discharges which, according to Cumbrians Opposed to a Radioactive Environment (CORE) "can be measured at the European monitoring stations in Poland, Spain and West Germany" (33).

In June that year, a government report concluded that the Windscale/Sellafield and Dounreay reprocessing plants were probably responsible for the high level of leukaemia among small children living close to the plants (34), and the families of eighteen children with leukaemia proceeded to sue BNFL for damages (34, 35). By 1990, thirty-five families had gone to court.

In 1988, a report from Sir Richard Doll, head of the Imperial Cancer Research Fund's Cancer Epidemiology and Clinical Trials Unit at Oxford University, confirmed evidence collected for many years by scientists like Alice Stewart, Rosalie Bertel and Karl Z Morgan, that the radiation limits for low-level exposure had been set consistently at far too low a level: Doll concluded that it should be lowered by a factor of three. Despite this, the National Radiological Protection Board (NRPB) could reveal that about 500 workers at Sellafield were still being exposed to radiation doses above the old limits (37). Fears were now being raised that childhood leukaemia clusters, identified near nuclear installations in Britain, might be related to the hazardous nature of the work carried out by their fathers (38).

Early the next year, a report for the Medical Research Council (MRC) Epidemiology Unit, published in the *British Medical Journal*, work carried out under the direction of Professor Martin Gardner, concluded that the risk of the child of a BNFL/Sellafield worker getting leukaemia – through damage to the parent's sperm cells – was nearly seven times the norm (39). In response to this news, Dr Roger Berry, BNFL's health and safety director, suggested that worried workers might opt "not to have children" (40).

151

Meanwhile, BNFL's internecine connections with the apartheid régime and the nuclear-military complex had been under fire. In 1987, the company was attacked – by the governor of Alaska and indigenous people in the region – for proposing to fly large quantities of reprocessed plutonium to Japan, over the Arctic circle. Fears were expressed that terrorists might hijack the cargo, which could be modified for weapons grade fuel production (4). After public protests, the plan was dropped. In March 1988, the EEC announced that South African (and Namibian) uranium continued to enter Europe, mainly through the West German port of Bremerhaven, of which a small portion (about 20 tonnes) had gone to BNFL (42). Eight months later, the British government finally admitted that 1100 tons of uranium oxide, comprising a second, secret contract between Rössing Uranium and BNFL, had been delivered to Britain, classified as "N" (meaning it could be used for military purposes) (43). And, in 1989, it was revealed that BNFL had supplied a Japanese-origin consignment of reprocessed fuel to the USA, where it was probably used at Savannah River for a military reactor producing tritium (44). BNFL's role as a member of the Urenco consortium – which had taken a back seat to its domestic ascendancy as Britain's most dangerous parastatal polluter – came under scrutiny in 1989, when Urenco joined an American consortium comprising Duke Power, Northern States Power, Louisiana Power and Light, and Fluor Daniel, to establish a privately-owned uranium enrichment plant in Louisiana, probably the first of its kind anywhere. The scheme – first announced in 1988 when David Fishlock, the pro-nuclear *Financial Times* correspondent, called it "...a red flag for those who claim that nuclear power in the US is as good as dead" (45) – has been attacked by a wide coalition of local environmental and black neighbourhood groups (46).

Further reading: CORE, *A Report on the Sellafield Aerial Discharges* (from address below).

Contact: CORE, 98 Church St, Barrow-in-Furness, Cumbria, Britain; tel 0229-33851.

CANT, c/o Wietze Weiland, Rt.1, Box 190B, Homer, Louisiana 71040, USA.

Greenpeace Ltd, 30 Islington Green, London N1, Britain; tel 071-354 5100.

Greenpeace (London), 5 Caledonian Road, London N1, Britain; tel 071-837 7557.

Friends of the Earth Ltd, 26 Underwood St, London N1, Britain; tel 071-490 1555.

The Dublin Clean Seas Campaign, Box C 137, 15 Upper Stephen St, Dublin 8, Eire.

Stop Urenco Campaign Netherlands, c/o Milieudefensie, Damrak 26, 1012 LJ Amsterdam, Netherlands; tel 020-221366.

References: (1) Durie & Edwards, *Fuelling the nuclear arms race,* London, Pluto Press, 1982. (2) *Background Papers 1-4,* London 1978, Stop Urenco Alliance. (3) *FT* ?/5/79. (4) *EPRI Report.* (5) WISE News Communiques 78/81, *passim.* (6) *Obs* colour supplement ?/?/84. (7) *MM* 7/84. (8) *Gua* 22/10/82. (9) *British Medical Journal,* London, 11/8/84. (10) *Gua* 22/10/82. (11) *SunT* 30/10/83; *Gua* 31/10/83; *Windscale, The Nuclear Laundry,* prod James Cutler, Yorkshire TV, screened 1/11/83. (12) New Statesman, London, 18/11/83. (13) *New Statesman,* London, 18/1/85. (14) *FT* 23/11/83. (15) *FT* 2/12/83. (16) *Gua* 23/8/84. (17) *FT* 10/8/84. (18) *Gua* 24/7/85; *FT* 24/7/85. (19) *Gua* 23/8/84. (20) *Time Out,* London, 1/12/83. (21) *Obs* 2/10/83. (22) *Anti-Nuclear Times,* Adelaide: CANE, late 1982. (23) *Lancashire Evening Post* 21/8/85. (24) *FT* 23/7/87. (25) *FT* 13/11/87. (26) *FT* 24/10/89. (27) *FT* 3/8/89. (28) *FT* 26/7/89. (29) *FT* 17/2/86; *FT* 19/2/86; *FT* 3/3/86. (30) *FT* 18/2/86. (31) See also WISE N.C. 285.2888, 274.2333. (32) *Peace News* 9/10/87. (33) WISE N.C. 1/4/88. (34) *Independent* 18/3/89. (35) See also *Gua* 2/1/90. (36) *Independent* 16/2/90; *FT* 3/10/89. (37) *Gua* 29/6/89. (38) *Gua* 22/7/89. (39) *British Medical Journal,* London, No. 300, 2/90. (40) *Independent* 22/2/90. (41) *Gua* 21/11/87. (42) *Gua* 14/3/88. (43) *Gua* 9/11/88; see Rössing entry. (44) *Gua* 1/5/89. (45) *FT* 15/4/88. (46) WISE N.C. 336, 20/7/90; see also *Groundswell,* US, Spring, 1990.

73 Bokum Resources (USA)

In 1978, the company submitted a plan to the New Mexico Environmental Improvement Division (NMEID) for a mine and mill at Marquez, near Albuquerque, New Mexico (NM), involving placing uranium tailings in a large pond in the middle of the Rio Grande river-bed. The plan was rejected by NMEID. A revised plan to place tailings in pits excavated in non-porous shale, but in the same river-bed, later found favour with NMEID but was opposed by local people (1).

In the original agreement, Long Island Lighting Company (Lilco) was to provide US$50 million to Bokum while the company provided 3 million lbs of uranium concentrate to Lilco between 1986 and 1989 – this in addition to 7 million lbs already contracted (2).

In 1978, a citizens' proposal, sponsored by the Sandoval Environmental Action Community (SEAC), to gain a moratorium in Sandoval County was rejected by the commissioners on grounds that no case had been proved of an "imminent danger" under which their police powers could be invoked (3).

In another development, Bokum was reported to have sunk a shaft at Ambrosia Lake, NM (4). By summer 1980, the Marquez mine and mill were 80% completed. SEAC led the fight against the company's plans: in May that year it entered a writ of mandamus against NMEID, challenging the legality of a ground water discharge permit granted to Bokum (5). A year later the suit was dismissed.

At the same time a New York grand jury was investigating allegations that Lilco officials had been bribed with Bokum stock to sign the original, ill-fated Bokum contract (6).

Opposition has come from the Sandoval Environmental Action Community and the Santo Domingo native Americans in the Rio Grande who in 1979 called on all state inhabitants to resist uranium mining in the valley (7).

In 1984, the US Circuit Court of Appeals upheld a decision taken by the New Mexico District Court to place Bokum into involuntary bankruptcy after a suit by the Long Island Lighting Company and other major creditors. This was because the Marquez uranium mine had never been brought into production (Lilco had contributed quite a lot of money towards the project in exchange for potential uranium supplies) (8).

References: (1) *New York Times* 25/11/79. (2) *FT* 3/8/78. (3) *Albuquerque Journal* 14/12/78. (4) *US Survey of Mines* 1979. (5) *SEAC News* 8/80. (6) *SEAC News* Spring 1981. (7) Statement from Santo Domingo Pueblo Government Office, NM 87052, USA. (8) *E&MJ* 2/84.

74 Boliden AB

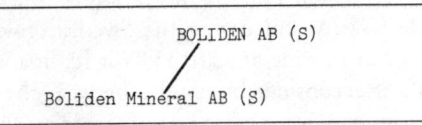

This is a large Swedish mining, smelting, and chemicals company with fifteen mines and extensive production of copper, zinc concentrates, and other metals (1).

With LKAB it was, for some years, investigating extraction of uranium from alum shales at Ranstad; work halted because of price depression in 1981, though LKAB was reported continuing. In May 1976, Boliden and Ytong AB got a contract to survey Marke province (2).

Boliden holds 50% of Mincon Corp (a US smelter); 55% in Atlantic Nickel Mines (Canada); and is associated in a lead smelter in West Germany with Preussag AG (3).

Extensive and vocal opposition to mining at Ranstad has been expressed by local people and the national anti-nuclear movement for years.

In 1987 the Swedish government ordered Boliden to cut sulphur dioxide emissions from its Ronnskar smelters (the only copper smelters in the country) by 50%, with the result that the company scaled down its unrefined and blister copper production, while re-investing in its smelter and a lead plant. Although the company initially vigorously opposed the state measures, it later concluded they had had a positive effect – in particular by reducing costs (4).

By then Trelleborg AB had become the majority shareholder in Boliden with nearly 41% of voting rights and 51.2% of shares. Boliden had also acquired Greenex A/S, operator of the Black Angel zinc/lead mine in Greenland.

This mine, situated at Maarmorilik, in the Bay of Uummannaq on the country's west coast, was formerly owned by Cominco, which intended to close it in 1986. Boliden envisaged – to use the graphic metaphor of the *Engineering and Mining Journal*– mining Greenex's reserves "till the pips squeak" (5). (One notable fact is that the indigenous/Greenland workforce at the mine has been dropping considerably) (5).

By early 1988 Boliden was pleading considerable poverty and urging the Swedish government to provide at least 35% of its financing (6). After considerable conflict between the two parties, the government finally agreed on an aid package comprising around one-sixth of the company's investment budget (7).

From that point onwards, Boliden has expanded beyond exploration in Sweden – indeed abandoned grass-roots exploration altogether – in favour of exploiting known deposits. It now has interests in Spain (in Andaluza de Pirites or APIRSA), Portugal, Saudi Arabia (purchased from Gränges in an area also explored by RTZ) (8), and in a high-grade zinc deposit in Burkina Fasso at Perkoa. It also acquired in 1989 the Bald Mountain gold and base metal deposit in northern Maine (purchased from Chevron) (9) and formed a joint venture with Allis Chalmers (10). In Ghana it is involved in supplying equipment to the Teberebie Goldfields project operated by Pioneer Group Inc (11).

References: (1) *MJ* 10/7/81. (2) *MJ* 3/6/81. (3) *MIY* 1981. (4) *FT* 26/6/87; see also *MJ* 3/7/87. (5) *E&MJ* 2/88. (6) *MJ* 6/11/87; *MJ* 25/12/87. (7) *MJ* 25/3/88. (8) *E&MJ* 7/89. (9) *MJ* 29/9/89; see also *MM* 8/89. (10) *E&MJ* 3/88. (11) *MJ* 21/7/89.

75 Bond Corp

Generally regarded as Australia's highest flying corporate kite, Alan Bond has aggressively been taking over stakes in several major Australian mineral and energy projects since 1979.

However, in June 1982, he began selling off his stakes in the Cooper Basin "synfuels" project – perhaps the country's most "exciting" new energy scheme – to cover losses incurred in his other pies like Simplicity patterns and his purchase of Swan Brewery. Swan lager is now Bond's major interest, it seems – and the chairman has his eyes on South-East Asia for major expansion. Swan's under-utilised assets would, he said in June 1982, be used to "help pump a tremendous volume of beer" into this part of the world (1).

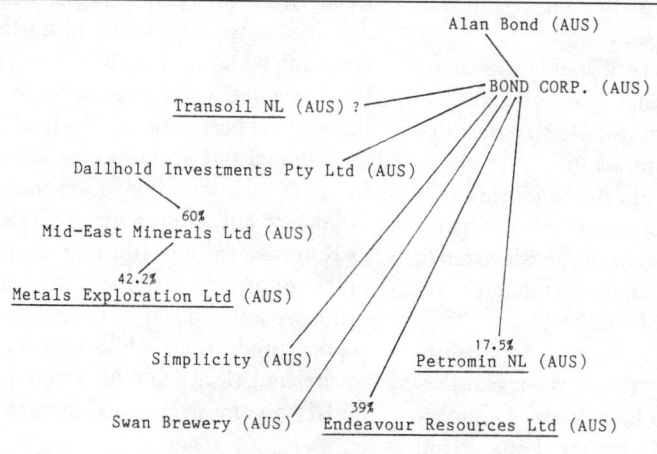

In May 1982, the company was all set to get a loan from the EEC to expand its mines in northern Sweden (2).

In April 1984 Alan Bond, through Swan Brewery, bid for Winthrop Investments through which he could control North Kalgurli, since Winthrop controls 38% of Mid-East Minerals which holds 34% of Metals Exploration which itself controls 29% of North Kalgurli (3). The bid was successful.

He also acquired virtually all of Pacific Copper, and the leading Canadian energy company, Sulpetro (4).

Bond also purchased all the share capital of Northern Mining NL from Endeavour, in 1982 and sold a 39% interest in the company to the Western Australian government the following year, thus giving the WA government access to the Argyle diamond project (5).

After a series of financial disasters and mismanagement in the late 1980s, the Bond Corporation was facing liquidation at the beginning of the following decade (6).

References: (1) *FT* 3/6/82. (2) *FT* 19/5/82. (3) *FT* 17/4/84. (4) *FT* 29/10/83. (5) *MIY* 1985. (6) *FT* 31/7/90.

76 Boroquimica SA (RA)

It was part of a consortium, along with Companhia Naviera Perez Companc, Noranda Mines and Rio Algom, bidding to develop the subsidiary ore deposits at Sierra Pintada in Argentina (1) in late 1979 where there is an estimated 14,500 tonnes of uranium (2). However, by mid-1980 this consortium had lost out to a French-Argentinian group headed by Minera Sierra Pintada and PUK (3).

References: (1) *MM* 1/80. (2) *IAEA/OECD Red Book* 1982. (3) *Nucleonics Week* 15/5/80.

77 Bow Valley Industries Ltd

It has a 20% or 25% share* with Esso Resources Canada (50%) and Numac Oil and Gas (25%) in the Midwest Lake uranium prospect in northern Saskatchewan which was planned to produce more than 4 million pounds of uranium a year from 1986 (1); reserves in 1979 were estimated at 97 million pounds of uranium (2).

Labrador Mining and Exploration (LMX) and Hollinger Mines (which part controls LMX) held talks on transfer of Bow Valley shares to them; but in 1979 these talks were reported to have broken down (3).

In 1987, in order to re-establish its "upstream" interest in exploration (after privatisation in Britain, which effectively denied it new domestic oil and gas ventures) (4), British Gas plc bought 51% of the equity of Bow Valley, but with only one-third of the voting control, in order to conform to Canadian government

* According to *FT* 13/9/78 this was a 25% stake; according to *FT* 16/2/79 it was a 20% share.

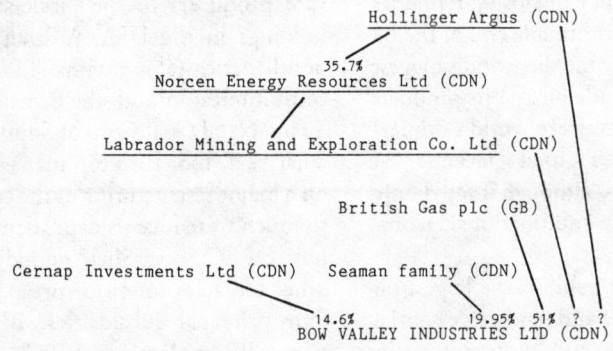

regulations on foreign investment. In the event of a planned take-over of Bow Valley, however, British Gas would effectively hold the cards (4). After taking over Ocelot Industries' substantial gas reserves in south-west Saskatchewan in 1988 (5) the company entered a JV with the WX syndicate (Ennex International 50%, Westfield Minerals 25%, and Dominion Mining 25%) to explore 75 precious metal properties, mostly in Nevada, in late 1989 (6).

In mid-1988 Denison announced plans to sink an exploratory shaft at Midwest Lake, as the first step in a three-year C$18 million programme to test ground conditions and mining methods. That year the project received provisional environmental approvals. The partners at Midwest were considering producing "by the mid-1990s" from drill-indicated resources in excess of 20,000 tons U grading 1% (7).

References: (1) *Fifteenth Mining Activity Survey* 1/82 p. 79. (2) *FT* 16/2/79. (3) *FT* 20/2/79. (4) *FT* 22/12/87. (5) *FT* 4/6/88. (6) *MJ* 22/9/89. (7) *MAR* 1989.

78 BP/The British Petroleum Co plc

By far Britain's largest company, with around 1500 subsidiaries (1) and a market capitalisation of little less than £10,000,000,000 (£9,482 million to be exact) (2), BP also has the world's greatest secure oil reserves (3) and is generally acknowledged to be "the best company in the world at finding oil" (4).

Thanks to its recent policy of closing refineries and pursuing the more profitable end of the oil market, BP has now become the world's biggest "spot" trader in both crude oil and its products (5). In 1981 it was the western world's 8th largest company with a total revenue of US$52,229 million (6) – although it rated only 41st among the "most multinational" companies (7).

Its American subsidiary Sohio is the largest oil producer in the USA (8) and has been essential to the parent's growth and profitability (in 1982, for example, it provided no less than 78% of BP group profits, and 60% the following year) (9).

Incorporated in 1909 as the Anglo-Persian Oil Co (name changed to Anglo-Iranian Oil in 1935 and to British Petroleum in 1954), BP was half bought-out by the British government at the start of World War 1 to ensure oil supplies for the Royal Navy. BP remained a "crude oil" company until after World War 2, when it began looking for new sources of oil (notably in Nigeria) and establishing refineries in Europe. In 1965 it became the first company to strike oil in the North Sea. Four years later it discovered oil on the Alaskan North Slope. Possession of crude supplies in the USA, and its purchase of Arco's marketing and refining assets in 16 States, gave the company "leverage to make a major entry to the USA" (10). Its US assets were merged with those of Sohio giving it an immediate 25% stake, and – in 1978 – control (of 53%). The following year BP won no less than nine of twenty bids (worth more than US$100 million) for oil exploration and production contracts in Alaska's Beaufort Sea. Although these were won with the co-operation of four Alaskan native corporations (11), the company's stake in the Prudhoe Bay field and its participation in the 800-mile Trans-Alaska Pipeline (12) have given rise to alarm and opposition among many native and environmental groups in North America (12) (see Arco for further information).

Diversification into other resources started in the mid-'70s, primarily into coal. By 1980 it had become one of the world's ten biggest private producers of the black stuff, thanks to holdings in the USA, Australia, Canada and South Africa (along with CFP/Total it held a third interest in both the Ermelo mine and the Trans-Natal Coal Corp of South Africa) (10).

That year, too, BP's top management decided on a major restructuring of the corporation, not so much to reduce its dependence on oil in the post-OPEC, post-Shah period as to give its other tentacles room to spread. It then set up four principal subsidiaries: BP Oil International, BP Exploration, BP Chemicals Interna-

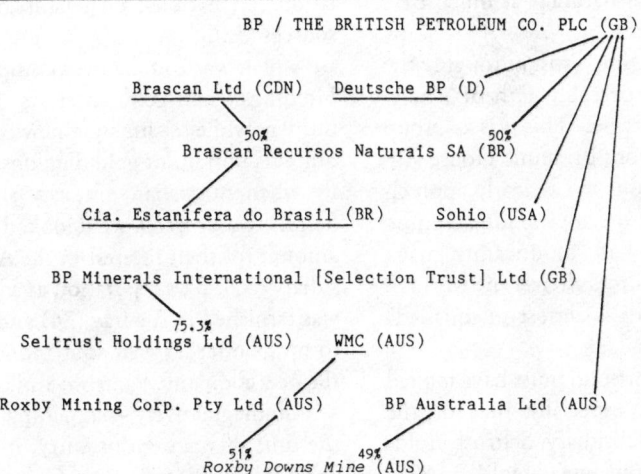

BP / THE BRITISH PETROLEUM CO. PLC (GB)

Brascan Ltd (CDN)   Deutsche BP (D)

50%   Brascan Recursos Naturais SA (BR)   50%

Cia. Estanífera do Brasil (BR)   Sohio (USA)

BP Minerals International [Selection Trust] Ltd (GB)

75.3%
Seltrust Holdings Ltd (AUS)   WMC (AUS)

Roxby Mining Corp. Pty Ltd (AUS)   BP Australia Ltd (AUS)

51%   49%
Roxby Downs Mine (AUS)

tional and BP Minerals International (13). (The last named effectively took over the interests of Selection Trust in 1982.) In 1981, to finance this new rationalisation, BP launched the biggest rights issue ever recorded to that date in Britain: £600 million (14).

By the time the company stepped into the 'eighties under its new chairman, Peter Walters (15), its diversification strategy was paying some dividends: it could boast of making dog food in the Netherlands, broiling chickens in Belgium (Pingo Oven-Ready), fish farming in Norway and developing computer software in the USA (16). However, its backbone remains oil and, to a lesser extent, the mineral interests it acquired at the beginning of this decade. On the eve of further reprivatisation of the company in 1983 – when half a billion pounds was raised through the sale of 7% of the government's holding in the company ("... BP is the quick cash that no Chancellor can resist") (17) – BP remained a corporation "whose fortunes rest on oil and gas. Diversification into coal and minerals ... should dilute this dependence, but are not likely to prove major profit-earners until the 1990s" (17).

A year later – after the company failed to strike oil in the South China Sea (18) and its Mukluk oil field in Alaska (the most expensive oil well ever drilled) (19) proved as dry as dead bones (20) the company still relied on North America

and the North Sea for its biggest turnovers (21). But Alaska had ceased to be "the jewel in BP's American crown" (22). The 80% of its profits still accruing from "upstream" activities were now to be made in "a range of discoveries and not just two [oil] fields" (21).

BP's move into minerals in the late 1970s was spectacular by any yardstick. Through its acquisition of a majority interest in Sohio by 1978 and control of Selection Trust two years later, BP's mineral interests made the corporation "one of the bigger mining enterprises in the world" (10).

BP's involvement with Sohio began in 1970 when the British company started drilling for oil on Alaska's North Slope and secured a stockholding in the American major. By the early 1980s Sohio held more than half the entire reserves of the main reservoir of Prudhoe Bay, comprising more than 25% of the USA's proved stocks of crude oil (10). Although Sohio's mining interests were relatively small, by the turn of the decade it had "so much money it doesn't know how to spend it" (23). So when Kennecott copper (still reeling from the expropriation of its Chilean assets in 1971 and from low copper prices, but holding 20% of all US copper reserves) looked like providing an avenue into large scale mining, Sohio snapped it up: since the acquisition took place long after BP had secured its majority stake in

Sohio, the decision was naturally as much BP's as its American partner's.

The take-over of Selection Trust at roughly the same time fitted in impeccably with BP's new-found modes and strategies. This was a "group whose style the men from Britannic House [BP headquarters in London] must clearly appreciate" (24). Not one of the top-ranking mining houses, it nonetheless had "an uncanny knack of finding natural resources all over the world ... the number of its finds is almost unequalled" (24).

Indeed, the Seltrust portfolio must have looked extremely impressive, especially that of the 75%-held Australian subsidiary, Seltrust Holdings Ltd (technically now owned by BP's 100% subsidiary Auselex Investment Holding (Pty) Ltd). There was a 47% beneficial interest in the Agnew nickel mine in Australia (later 60%) (73) and the Teutonic Bore copper, zinc and silver deposit; a 50% interest in a Nevada, USA, gold mine (Occidental Minerals holding the other 50 %) and a majority (64%) stake in Les Mines Selbaie, a copper, zinc, silver and gold project in Quebec, Canada. What with diamond mines in Namibia, a 34% holding in the Unisel gold mine of South Africa and various other stakes in Africa and North America, Seltrust gave BP a mineralogical foothold in three continents. It also helped defray some of BP Minerals' fixed costs (10).

Yet Seltrust was an "overpriced" acquisition, and three years later "benefits [had] yet to be seen at the bottom line" (5) despite its 5% in the very promising Mt Newman iron ore project in Western Australia (25) and the extremely promising Temore gold prospect in New South Wales (26, 27) held by Seltrust as to 56.25% (28).

By the beginning of 1983, Seltrust Holdings was proving something of a plugged nickel (29): net loss by the end of the year was A$10.6 million, exploration was reduced, and no profit foreseen in 1984 (30). In April 1984, new ways were mooted to correct the situation (31). The move followed the announced sale of Selco, Seltrust's Canadian subsidiary, to BP Resources Canada, thus helping to satisfy Canadian regulations on national ownership of national resources (32).

BP's plan was to take over completely Seltrust Holdings' non-gold interests and its debt burden, while setting up a new company to acquire its important gold diggings. Fur began to fly when Australian minority shareholders in Seltrust were offered what looked like a derisory amount for their interest in the Australian subsidiary (33). BP's reputation as a "good citizen" was tarnished in the fray (34) and it was forced to up its offer, as well as abandoning plans for the new company, Paragon, and more modestly – if unimaginatively – launching Newco. As of the time of writing this entry, it looked as if a Perth businessman, Laurie Connell, would end up with the lion's share of Newco (35) and as if BP's down-under reputation would remain distinctly under-down for some time to come.

Meanwhile, BP's other mineral interests in the former colony continue to be held by BP Australia – or, rather, BP Minerals Australia, which was set up in 1983 (36). The interests include the Stuart Shelf prospect – a copper, zinc and silver project at Benambra in Victoria, which was reporting high copper values in the late '70s (37) – and, above all, the Olympic Dam/Roxby Downs uranium mine. All three projects are conducted with WMC.

It was the Olympic Dam deal with WMC (51%) which finally launched BP into both Australian minerals and uranium mining on a major scale. (A uranium lode at Coppermine in Canada's Northwest Territories has remained in abeyance (38), although BP's explorations for uranium in the high arctic continue to be of concern, especially to the Inuit and MicMaq nations) (39). BP, along with other oil majors, was ready to move into uranium areas on Aboriginal land in Australia's Northern Territory in the late '80s, but was stopped by the moratorium imposed on exploration in 1980 (40).)

BP's major contribution to the development of Roxby Downs has been the A$50 million it shelled out in 1979 for the feasibility study that has caused so much controversy, particularly as it affects the Kokatha people on whose land this unprecedented mineral deposit is situated (41).

Just what BP would get out of the arrangement with WMC was not that clear – and it has become increasingly likely that BP will severely limit any future expenditure as the mine proceeds. In what the *Financial Times* ingenuously called "an elegant arrangement from Western Mining's point of view", BP could be liable for a final expenditure of A$110 million or more (41). The company can hardly be said (unlike its Australian partner) to view uranium mining as a moral obligation, or the Olympic Dam deposit as absolutely crucial to British nuclear power. In a briefing paper on western European energy supplies published in late 1984, BP anticipated a "considerable" expansion of western nuclear power, but, while claiming the EEC region possessed only 6% of assured supplies, estimated that outside sources offered "considerable security" (42).

In view of the opposition which its share in Australian uranium has evoked from the anti-nuclear movement – including two major blockades, international Aboriginal protest, and boycotts of its service stations on both sides of the globe – it may well be reconsidering its future obligations to WMC.

In 1983 BP became the first non-Swedish company to get exploration and mining rights in that country when the Swedish government approved an application by BP to join LKAB in exploration of the south central part of the country between Falun and Derebro: BP Minerals would get 25% of any of the recovered ore (so far uranium has not been mentioned as a possible reward) (43).

BP bought out troubled Dome of Canada's oil interests in 1982 (44) and two years later was on the point of bidding for Johnson Matthey, the precious metals and chemicals company partly owned by Anglo-American, a company with the sole refining rights to the world's largest platinum mine – Rustenberg in South Africa (45).

The company is also one of the major deep sea consortia. Although its participation in the Kennecott consortium is only 12%, Sohio's is 40% (46).

The company's ventures in "new" types of energy production have been fairly limited compared with, for example, Exxon or Arco. In 1980 it joined CRA and BHP in a JV to develop the Rundle oil shale deposits in Queensland, Australia (47), but later withdrew, as did other big companies (48). In 1984, Sohio Shale also pulled out of the huge (US$2.5 billion) Paraho-Ute oil shale project in NE Utah, USA (49).

BP Solar Systems announced in 1983 that it would build Britain's largest solar power plant at Marchwood, Hants, putting up half the money itself and getting the other 50% from the EEC; BP has also agreed to assess the solar power potential for Britain, on behalf of the EEC's energy and research development programme (50).

Its approach to other new technologies has also been fairly cautionary. Although shelling out more than £2 million in the early '80s to back "revolutionary" new scientific research (51), it abandoned its early ventures into single cell protein proliferation ("biomass") in the 1970s (52), and years later still found no buyer for the biotechnological process it had developed (53). Apart from the mounting and sustained opposition to BP's participation in Roxby Downs, the company has aroused most ire for its operations in support of apartheid, for the potentially disastrous ecological implications of its oil drilling and transportation in the Beaufort Sea, and for its trespass on beauty spots back home. BP has had more than 20 South African subsidiaries, including three mining companies (1). One of these, Unisel (now owned and managed by Gencor) is a moderate producer of gold (25). Another mine, Middelburg, in which BP had a share, started coal production in 1985; in the early '80s a telex leaked from BP's London HQ showed that the company had been attempting to get round various European part-embargos on South African coal by disguising its origins (54).

When BP took over Seltrust in 1980, the Anti-Apartheid Movement announced a "major international campaign" against the deal on the grounds that as well as deepening British involvement in South Africa (specifically with purchase of Charter Consolidated's 26% interest in the company), it gave Anglo-American an entrée into North Sea oil, thus violating the oil sanctions imposed on South Africa since 1973 (55).

BP's Seltrust holding also gave it a major slice of the Tsumeb copper mine in Namibia, as well as – through the South West African Selection Trust (Pty) Ltd – a stake in other base metals in the occupied territory (56).

It's also important to note here that BP was among several oil majors which supplied oil to the Smith régime in Rhodesia through its South African subsidiaries, in gross violation of UN sanctions, during the 1970s (57, 74).

Much closer to home soil, BP's plans to expand its Wytch Farm oil field in Dorset by plumbing the depths of Poole Harbour (58) met with tremendous opposition from environmental and nature conservation groups because of the potential threat to wildlife and holiday-making. After Friends of the Earth, the National Trust ("we will fight to the last ditch" – or dytch?), the Nature Conservancy Council *et al* waded in against BP (59), the plans were dropped (60).

Surprisingly, BP's other controversial operations haven't roused half as much opposition: perhaps because there are no known endangered reptiles in Rondonia, Brazil, where BP had extensive plans to mine gold (61). Or because there are no well-tramped beauty spots in East Kalimantan (formerly Borneo) where the company is now mining coal in one of the last extensive areas of primary tropical forest in the world (62). Or because the company took ecological advice before plundering the Atabian pine plantations of Fiji, despite the opposition of a large part of the population to multinational appropriation of a key resource (6).

Potentially the most devastating of the company's recent involvements stems from its participation in the Great Carajas project, which has been criticised around the world for its impact on indigenous peoples, fauna, flora and the Amazon rainforest (63).

As part of its "continuing strategy of diversification" (64), BP formed BP Mineração in 1980 to explore for gold, copper, zinc and lead in Brazil. A year later BP bought half of Brascan's 99% holding in Brascan Recursos Naturais, the Brazilian tin mining and smelting group, Brazil's second largest tin producer. This gave the company entrée to Rondonia (65). A few months later BP announced that it would spend US$40 million in Brazil on oil and mineral exploration – about a tenth of its world-wide total – and would be prepared to participate in the Grand Carajas project, "the world's largest untapped [mineral] deposits" (66).

In 1983, BP Mineração laid out its Brazilian blueprint for the following three years: it would spend US$100 million developing gold deposits south-east of Porto Velho in Rondonia; put US$25 million into cassiterite mining in Rondonia and Amazonas; and tap other minerals in several other states, including Mato Grosso (a major tribal region), Minas Gerais and Goias. No less than 60 geologists have been employed by BP in Brazil, and more than 600 mining permits at one time were held across this huge land (43).

If we pass briefly over some of the other dubious activities of BP in the past few years – for example, its attempt ("scuppered" by the West German government) to dump 7500 tons of toxic waste in the sea (67) and its participation in an alleged chemicals price fixing cartel along with Shell and ICI (68, 69) – perhaps its most objectionable home-based activity is the work of wholly owned Scicon. Bought by BP in 1981, this high technology subsidiary was the subject of a spy scandal in 1983 when, according to the FBI, its work on the US intercontinental ballistic missile programme was photocopied by an employee and promptly despatched to Moscow (70).

Scicon supplies expertise to both banking and industry, but a third of its turnover in Britain comes from military contracts. The highly classified US subsidiary Systems Control Inc, based in Silicon Valley, California, depends for no less than three-quarters of its work on offence contracts; it also provides on-site support for the US Air Force, information on military transportation and "battle management" (71).

No doubt Scicon, as one of BP's major new enterprises, conforms to chairman Walters's view of the multinational being "... an effective force for good in other societies because it [can] help remove social practices which obstruct progress". The claim was made by BP's chair in an address given in 1983, arranged by the Christian Association of Business Executives to hear Britain's most powerful executive justifying "the morality of wealth creation". Free enterprise is the "most effective instrument available to man in his efforts to reduce poverty", noted Mr Walters; one consequence of the "moral snobbishness of some who denigrate those who make money is that millions may now be needlessly hungry and suffer from political oppression ... because of a refusal to take advantage of the most effective means of promoting development and creating wealth" (72).

In 1987, BP took over the remainder of Sohio which it did not own, and could at last bring the US subsidiary "under orders", after some drastically bad investment decisions during the 1970s. The merger puts BP in third place among the world order of oil companies (only Exxon and Royal Dutch/Shell now being bigger), and made BP "a truly international company" (75). It presaged aggressive exploration to extend oil holdings on the Alaskan North Slope and pressure on the US Congress to open up the Arctic National Wildlife Reserve for exploration (75).

Starting in 1987, the Kuwait Investment Office (KIO) began building up a stake in BP (76), which raised alarums in the English establishment. The Monopolies and Mergers Commission investigated, but could find nothing untoward, and within several months, the KIO had built up a nearly £1 billion stake – amounting to 22% of the company's share capital (77). Although the KIO did not ask for a seat on BP's board, it nonetheless in theory holds sufficient shares to attempt a block on any major motion put to BP's shareholders. Ironically, much of the money used by the KIO to invest in BP had itself come from BP in the days before its huge assets in Kuwait were nationalised (78).

In early 1988, BP had not only built up its BP Gold subsidiary into the second biggest holder of gold reserves in the US (second only to Newmont) (79) but "snatched" the majority in Britain's largest independent oil producer, Britoil, by taking over the stake held by Arco (80).

Finally, in 1989, after a couple of years of secret negotiations, BP announced that it was selling off the majority of its minerals assets to RTZ (see RTZ). As it entered the new decade, BP retained control of its Canadian gold assets and (perhaps surprisingly) its share in the Kaltim Prima coal mine in Indonesian Kalimantan (see CRA). Despite a willingness to shed its 49% share of Roxby Downs to RTZ, legal action taken by Western Mining Corporation (WMC) prevented this happening, so BP continues to cause alarm to Aboriginal people, and environmentalists at large (see WMC).

Contact: Partizans, 218 Liverpool Rd, London N1 1LE, Britain.

References: (1) *Who Owns Whom 1984,* Dun & Bradstreet, London. (2) *Obs* 4/11/84. (3) *FT* 30/11/82. (4) *Gua* 8/12/83. (5) *FT* 20/8/84. (6) *MMon* 12/82. (7) *WDMNE.* (8) *Gua* 15/10/83. (9) *FT* 7/12/83. (10) *BOM.* (11) *Gua* 13/12/79. (12) *FT* 19/10/79. (13) *FT* 17/12/80. (14) *Gua* 19/6/81. (15) *FT* 28/11/81. (16) *SunT* 21/6/81. (17) *SunT* 28/8/83. (18) *FT* 21/2/84. (19) *FT* 30/10/83. (20) *FT* 6/12/83. (21) *Obs* 19/9/84. (22) *Gua* 31/10/83. (23) *Gua* 22/8/80. (24) *SunT* 22/6/80. (25) *MIY* 1985. (26) *MJ* 23/11/84. (27) *FT* 5/5/84. (28) *FT* 31/7/84. (29) *FT* 9/9/83. (30) *MJ* 18/5/84. (31) *FT* 7/4/84. (32) *FT* 26/8/83. (33) *FT* 18/1/85. (34) *FT* 22/1/85. (35) *FT* 1/3/85. (36) *FT* 8/11/83. (37) *FT* 8/6/78. (38) *Kiitg* 4/80. (39) *Native Peoples' News,* London, No. 9 Summer 1983. (40) *FT* 23/9/80. (41) *FT* 28/7/79. (42) *MJ* 28/9/84. (43) *E&MJ* 4/83. (44) *FT* 12/8/82. (45) *Gua* 20/10/84. (46) *RMR* Vol. 1 No. 4 1983. (47) *FT* 12/2/80. (48) *MAR* 1984. (49) *FT* 27/2/84. (50) *FT* 14/1/83. (51) *New Scientist,* London, 21/4/83. (52) *SunT* 15/8/82. (53) *FT* 20/4/83. (54) *Links,* Manchester, NW TU Anti-Apartheid Committee, Spring 1982. (55) *Anti-Apartheid News,* London, 9/80. (56) *Transnational corporations and other foreign interests operating in Namibia,* Paris 1984, UN. (57) *Obs* 25/6/78. (58) *Gua* 13/4/83. (59) *Gua* 6/8/84. (60) *FT* 14/11/84.

(61) *FT* 19/1/84. (62) *Tapol* No. 5, London, TAPOL 1983. (63) *ARC Newsletter,* Boston, Anthropology Resource Center, 3/83. (64) *FT* 2/10/80. (65) *FT* 5/1/81. (66) *FT* 8/2/82. (67) *Gua* 11/1/82. (68) *FT* 18/10/83. (69) *Gua* 18/10/83. (70) *SunT* 23/10/83. (71) Scicon propaganda sheet (produced by BP), supplement to *BP Shield,* London, BP, Winter 1983. (72) *BP Shield,* London, BP, Autumn 1983. (73) 23/11/84. (74) Martin Bailey & Bernard Rivers, *Oilgate: The sanctions scandal,* Hodder & Stoughton, London, 1979. (75) *FT* 22/7/87. (76) *FT* 20/11/87. (77) *Gua* 11/7/80, *FT* 21/6/88. (78) *Gua* 20/11/87; *FT* 21/6/87. (79) *FT* 11/7/88. (80) *FT* 23/1/88; *FT* 12/12/87; *FT* 24/2/88; *FT* 9/12/87; *Gua* 10/12/87.

79 Brae Co

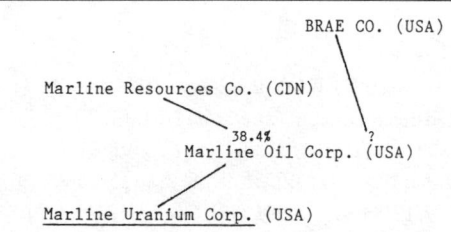

With its HQ in Delaware, USA, it has an undisclosed minority interest in the Marline Oil Corp which is exploring in the Athabasca region of Saskatchewan in Canada (1) and in Virginia, USA.

References: (1) *Yellowcake Road.*

80 Brascan Ltd

"Known and feared for its take-over expertise and financial strength" (1), Brascan had been trying to raise its stake in the important "natural resource" company Noranda, in the teeth of strong resistance from Noranda's board, since 1979. In August 1981 it succeeded in launming Noranda's largest shareholder.

A major investment in Brascan was later acquired (2) by Peter and Edward Bronfman (cousins to Samuel Bronfman, who owns Seagrams, the world's largest distillers – which bid for Conoco in 1981) after Brascan's management failed to gain control of the Woolworth supermarket chain (3).

The Bronfmans are firmly allied to the Patino family, notorious for building up their mining empire on the slavery of the tin mines (4). They also control Trizec Corp, Canada's second largest property company; took over the Ernest W Hahn group (USA) in 1980 (5); and own 96% of Patino's Brazilian subsidiary, Companhia Estanifera do Brasil (Ceasbra) (6).

34% of Patino's stake in Edper was acquired by buying out Northgate's share in early 1982, but Northgate still holds 6% indirectly in Brascan through a holding in Patino (7).

Mitchell Sharp, Canada's Foreign Secretary in 1968/74, has been a director of Brascan – and was one of the founding members of the Trilateral Commission (8).

Together with Noranda, Brascan did exploratory work in the Duddridge Lake area of Saskatchewan in the mid-'70s – a uranium province previously unknown but close to the Wollaston Lake area (9).

Brascan, formerly in a JV with BP, controls one of Brazil's three biggest tin mines based on three deposits (Santa Barbara, Jacunda and Alto Candeias). In Rondonia in 1984, Brascan Recursos Naturais brought on stream the country's first hard rock tin mine (10).

In 1985 Brascan restructured itself, in particular to cover the large debts occasioned by its 46% interest in Noranda (11): it set aside C$140 million to cover costs of restructuring (11).

Brascan enlarged its energy interests the following year by buying up Conrad Black's 41% interest in Norcen Resources Ltd (12).

Two years later, the company said it would raise its interest in Noranda (possibly to 55%) and aimed at securing 50% or more in all affiliate companies "in order to consolidate results on the group's balance sheet" (13). As of 1990 its main affiliates were the John Labbat Ltd food and beverage group (41% "fully diluted" as the *Mining Journal* put it – presumably without in-

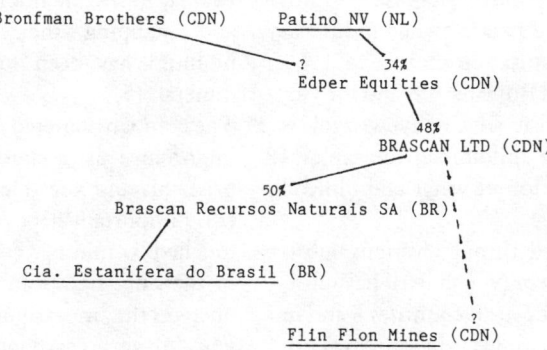

```
Bronfman Brothers (CDN)        Patino NV (NL)

                            ?          34%
                        Edper Equities (CDN)

                                   48%
                        BRASCAN LTD (CDN)

                   50%
        Brascan Recursos Naturais SA (BR)

    Cia. Estanifera do Brasil (BR)


                                        ?
                        Flin Flon Mines (CDN)
```

tending the pun) and the financial services activities of Trilon Financial Corp (13).

As of 1988 it also held 10% in Scott Paper (the producer of "Scotties" and similiar products) which came in for worldwide criticism in 1989 for its involvement in a huge pulp and paper plantation project in West Papua (Irian Jaya): the company later withdrew from the project (on its admission, because of informed criticism by Friends of the Earth, Survival International and the Indonesian Human Rights Group, Tapol) (14).

Brascan's interest in Westmin at the beginning of 1990 was quoted as 62% (13), but probably its major new fields of interest have centred on Brazil. In 1983, it formed a JV with Caraiba Metais SA – Ind e Com, called Pedra Verde, and started producing copper from a mine and smelter in Bahia (15). It also joined with BP in setting up CESBRA (Cia Estanifera do Brasil) – formerly Brascan Recursos Naturais – the country's second largest tin producer, whose sole production is currently from the Santa Barbara cassiterite mine in Rondonia, with smelting at Volta Redonda. The Santa Barbara mine is operated by Mineração Jacunda Ltda and its operations – in particular its alleged destruction of a large part of the Jamari National Forest – were the subject of major condemnation in the London *Sunday Times* in mid-1989 (16). The *Sunday Times* research was vigorously attacked by BP and Cesbra (17), whose president Claudio Galeazzi claimed that the *garimperiros* in the Bom Futuro area had caused more damage in 18 months than Cesbra had caused over thirty

years (17). (Galeazzi did not point out that a large part of the feed for the company's Volta Redonda smelter actually comes from the Bom Futuro miners) (17).

The main object of the *Sunday Times* attack was BP, a British company. In fact, even before the article appeared, BP had agreed to sell-on its stake in Cesbra to RTZ. One possible outcome of the Murdoch press's new-found (and probably expedient) concern for the rainforests of Brazil was that RTZ dropped the Jacunda mine like a hot potato. Brascan now owns 99% of Cesbra (17).

References: (1) *FT* 8/1/80. (2) *MJ* 21/8/81. (3) *FT* 7/1/80. (4) *Patino the Tin King.* (5) *FT* 22/5/80 (6) *FT* 5/11/80. (7) *FT* 13/1/82. (8) Sklar, 1980. (9) *Canadian Mining Journal* 4/76. (10) *MAR* 1984. (11) *FT* 3/11/85. (12) *FT* 24/2/86. (13) *MJ* 10/11/89. (14) *Down to Earth,* Survival International newsletter and *Tapol,* 1989-90 *passim.* (15) *MIY* 1990. (16) *SunT* 18/6/89; see also *Weekend Australian* 24-25/6/89, *MJ* 23/6/89. (17) *MJ* 5/1/90.

81 BRGM/Bureau de Recherches Geologiques et Minières

The French are past-masters at neo-colonialist

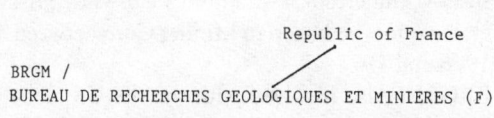

```
                              Republic of France

BRGM /
BUREAU DE RECHERCHES GEOLOGIQUES ET MINIERES (F)
```

penetration which nobody recognises – until it's too late. BRGM is a state-owned industrial and commercial organisation created in 1959 by merging the Mines Bureaux (sic) and the Geological Research Bureau (sic). Its avowed role is to "promote at home and abroad the exploration for and exploitation of water and mineral resources ..." (1).

Its work is conducted through various subsidiaries, including Serem (which itself has subsidiaries in various third world countries – and in Australia, where it holds tin interests in Queensland) (2), Coframines, Comilog, Sofremines and others with similar innocuous-sounding acronyms. It also has two subsidiaries which operate in New Caledonia, called Chromical and Cofremmi (1).

Its only known current uranium involvements are in exploration of extensive areas of the Ivory Coast, in tandem with Sofremines, and in Rwanda (3).

However, its other ventures cover almost every non-fuel mineral, including copper prospecting in Argentina and Peru, tin and tungsten in Malawi, iron ore in Senegal, a JV with Krupp *et al* in Cameroon (iron ore), exploration of the upper Affole region in Mauritania and the central Red Hills area of the civil-strife torn Sudan, mapping of Oman, dressing ornamental stones in Saudi Arabia, producing sulphides in Cyprus, and – oh yes – it does have antimony/tantalum mines in France itself! (3)

In 1980 Amax and BRGM were sued by Patino when BRGM formed a company called Societé de Promotion des Mines, into which it placed the assets in the New Caledonian nickel/cobalt project which were previously owned by Cofremmi, and in which Patino had an £8.5 million investment. Because of lack of progress (and profit) from the venture, Patino wanted to pull out, invoking an original agreement with BRGM which the French Bureau claimed didn't allow Patino to withdraw (4).

BRGM undertook a uranium field-work programme for the Nigerian Mining Corp between 1976 and 1978.

BRGM is also involved with the the US Geological survey and the government of Saudi Arabia, not to mention Riofinex, in exploring and mapping the Arabian shield, where uranium has been found along with other minerals (5).

The company entered the 1990s confirming its importance as a large many-tentacled enterprise, playing key roles in the exploitation of fairly important deposits in areas relatively untouched by mining. For example, it is involved in the only significant gold mine in Senegal, operates the "most significant mine" in Burkina Faso (6), and is carrying out a mining inventory of Honduras. In 1987 it negotiated an exclusive three-year prospecting licence for rare earths in the Machinga district of Malawi (7); in Cameroon it is also investigating a large rutile deposit (6) as well as carrying out a mining inventory of the north-west region of the country (8).

The JV with the Sudanese government and Total in the Red Sea Hills area of the Sudan continues to progress (6, 9), and indeed gold has been the major target for the company in recent years (10) with prospecting for the precious metal under-way in the Central African Republic, Guinea, Mali, Niger, Malawi, Gabon, Uganda (6) and Canada (11).

It has entered a phosphates JV with Cominco (24.5%) and Australmin (formerly AOG) in French Polynesia (11) and is partnered with another French monolith, Pechiney, in developing tin and tungsten deposits in Orissa, northern India (6) – while in the south and central part of the country it is exploring for diamonds (mainly in Andhra and Madhya Pradesh) (12).

While continuing to explore, especially for gold, on its home ground (the Massif Central of France has been "heavily prospected by BRGM and Cogema" commented the *Mining Magazine* in February 1990), BRGM is also clearly poised to forage further and further afield as the decade wears on.

At the end of 1989, it announced that it would be carrying out "another gold and uranium project" in the Kerio Valley of western Kenya, after an initial survey yielded "very encouraging results" (14).

References: (1) *MIY* 1982. (2) *Reg Aus Min* 1982.

(3) *MAR* 1984. (4) *FT* 30/4/80. (5) *MIY* 1985. (6) *MAR* 1989. (7) *MJ* 23/10/87. (8) *MJ* 18/4/86. (9) *MAR* 1989. (10) *MJ* 31/10/86. (11) *MIY* 1990. (12) *MM* 5/89. (13) *MM* 2/90. (14) *MM* 12/89.

82 Brinco Ltd

Originally the British Newfoundland Corporation (name changed in 1971), Brinco was given huge exploration and mineral rights over Labrador (50,000 square miles) and Newfoundland (10,000 square miles). It also part-controlled the Churchill Falls (Labrador) Corp when it negotiated sale of power from Churchill Falls to Hydro-Québec in a deal which "scalped" the people of Newfoundland. Brinco got an exemption from sales and gasoline tax during the construction of the project, at a cost of US$50 million lost revenue to the people of the province. Brinco's power during the construction was said to be "almost absolute": visitors couldn't get off a plane at the Falls without permission of the company – local trappers needed a pass (1). In June 1974, Brinco made another killing by

selling its interest in Churchill Falls (Labrador) Corp back to the government (2).

Currently Brinco is involved in exploration and development of oil, gas, coal, uranium, asbestos, limestone, and zinc (2).

Brinco's intentions to open up a uranium mine on land owned largely by Inuit ("Eskimo") and settlers caused an outcry which has reverberated around the world and resulted in a temporary withdrawal of Brinco from the scene.

In 1977, Brinex (ie Brinco, then controlled by RTZ with a minority holding by Bethlehem Steel) found uranium reserves at Kitts-Michelin, near the east Labrador seaboard, together with Urangesellschaft. The companies rated their find "insufficient" to justify mining (3). A year later they discovered several boulders with a high uranium content (4), and within a year had announced their intention to build a full-scale open-cut mine at Michelin, and an underground mine at Kitts; 1.3 million pounds of uranium oxide would, the companies reckoned, be mined over 15 years (5).

In autumn 1979, Brinex exercised its option to transfer Urangesellschaft's 40% interest in Kitts-Michelin to a third party, in the shape of

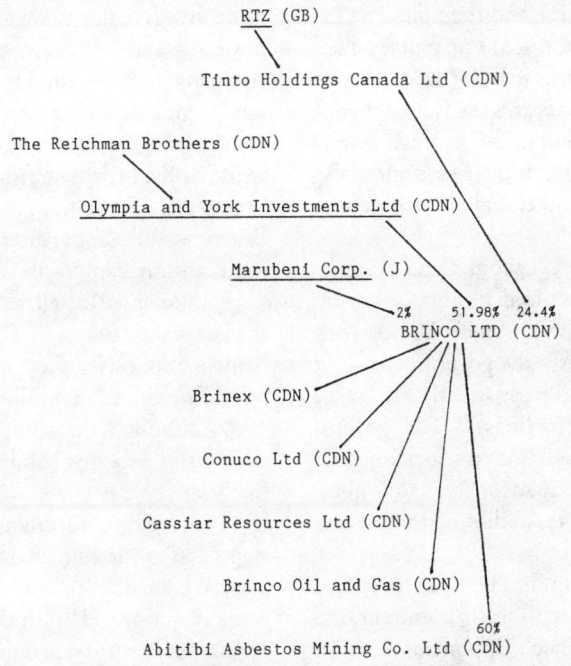

the American utility Commonwealth Edison, in return for which Commonwealth Edison would buy 18 million pounds of the uranium and an all-weather road would be constructed to Goose Bay, 85 miles away (6).

Early in 1980, however, the market for nuclear power slumped further and, as the *Financial Times* put it, "the project became the victim of the uncertainty in the nuclear power industry" (7). US utilities were also "thought" to be amply supplied with uranium – a fact confirmed a year later when the market virtually crashed as US utilities sold off their surplus.

More important, resistance to Kitts-Michelin had built up since 1978 (8) and culminated in a demand for public hearings on Brinco's environmental impact statement (a revised one) in 1979: this was instead of a government inquiry, since Newfoundland – the governing body for semi-colonised Labrador – makes no provision for environmental hearings as such.

Brinco was trounced by the local people. The Naskapi Montagnais (or Inuit) were particularly concerned about the proposed road and its adverse impact on hunting. The Inuit and settlers from Postville and Makkovik, the two settlements closest to the proposed mines, united to oppose both the mine and the mill. The Labrador Inuit Association sent one of its spokesmen to see RTZ's mine at Elliot Lake: despite company blandishments and the fact he'd gone away in favour of mining, he returned solidly opposed. The Labrador Inuit Association also enlisted the help of other Labrador organisations and experts (9).

On 29 May 1980, the Newfoundland Premier, Brian Peckford, announced a moratorium on uranium mining in the province. Brinco (Brinex) had failed to satisfy the government "... it can and will safely and permanently dispose of the waste materials (tailings)". The government's report described Brinco's social impact studies as "woefully inadequate" and questioned the economic feasibility of the scheme (10).

While opposition to the mine – for its potentially devastating impact – was paramount, the major concerns of Brinco's opponents were re-linquishing control over their own lives and development, and allowing the thin edge of an ugly industrial wedge to enter their land-based economy (11).

Despite this setback, Brinco carried on as if nothing had happened. In 1981, it announced results of a new feasibility study into the mine: estimated reserves were 10,886 tons of U_3O_8 and production was scheduled to start in 1984 at an annual rate of nearly 600 tons (12). Actual development did, however, await "market improvement" (13).

In the meantime, Brinco sought to double its total assets by the mid-'80s and ventured into oil exploration in Texas (14); acquired 50% of a gold mine in Manitoba, along with New Forty-four Mines; entered a JV with Suncor exploring for coal in Nova Scotia; and, in 1981, started a 50/50 JV with Weldwood of Canada to develop a team-coal project on Vancouver Island (15).

While many groups were involved in the struggle against Brinco at Kitts-Michelin, major co-ordination was through the Labrador Inuit Association.

It is worth noting that, at the time Brinco submitted its environmental impact studies for the mine, it was majority-owned by RTZ, the most powerful and "experienced" uranium mining company in the world. Despite this, a few thousand people – and only a few hundred in the communities of Postville and Makkovik – managed (with a little help from the declining market) to turn the giant ... .

Brinex was also apparently participating in a JV for uranium exploration with the SMDC at Cree Lake and Russell Lake, Saskatchewan in the late '70s (16).

Brinco also owns the Cassiar asbestos mine which produced a million tonnes of ore in 1982, although the company's investment in the Abitibi Asbestos Mining JV was written off that year (17).

In 1983 the EEC threatened to ban the importing of asbestos fibre, a move which stood to imperil Canadian producers – of which Brinco was a major one (19). In the US, new curbs on asbestos exposure were opposed by the Asbestos

Information Association, a group of US and Canadian miners and asbestos product manufacturers (18).

By 1984 Brinco's major operation, its Cassiar asbestos mine accounted for no less than 93% of the group revenue (oil and gas providing the remainder) (20).

Contact: Labrador Inuit Association, PO BOX 70, Nain, Labrador A0P 1L0, Canada.

References: (1) *As if People Mattered*, Labrador Resources Advisory Council, 10/77. (2) *MIY* 1981. (3) *MJ* 9/9/77.(4) *FT* 16/8/78. (5) *MJ* 25/5/79. (6) *MM* 9/79 p. 281. (7) *FT* 24/1/80. (8) Various press communiqués of the Labrador Inuit Association; personal communications with Rev Leslie Robinson, Makkovik. (9) *Project North Newsletter* 1979 passim. (10) Statement of the Newfoundland Government, issued 29/5/80 (see *FT* 4/6/80). (11) LIA submission to public inquiry, 1979 (available from LIA or CIMRA). (12) *MJ* 24/4/81. (13) *MM Survey* 1982 p. 79. (14) *FT* 16/4/81. (15) *MJ* 1/1/82. (16) *Yellowcake Road*. (17) *MJ* 3/6/83. (18) *MJ* 25/5/84. (19) *E&MJ* 3/83. (20) *MJ* 22/6/84.

83 Brominco (CDN)

It was formed, with its HQ in Montréal, in 1976 by the amalgamation of Naganta Mining and Development Co Ltd, Nemrod Mining Company Ltd, Timrod Mining Company Ltd, Valdex Mines Inc, NSUC, First Orenada Mines Ltd, and Jolin Bourlamaque (1).

It has explored for uranium in northern Saskatchewan; no further details.

References: (1) *Who's Who Sask.*

84 BRPM/Bureau de Recherches et Participations Minières

This is a general prospecting body which has been involved in oil explorations in Morocco (1). In 1970 it joined with the United Nations in a reconnaissance of the country for uranium.

Four years later, BRPM had made "considerable efforts in the matter of exploration" (2); integrated geological training teams were set up and two project zones – Upper Mouloua and the High Western Atlas (an area predominantly used by Berbers) – were selected for further investigation. A uranium-bearing sandstone level was discovered at Imin Tanout in Marrakesh province in the second zone and drilling commenced in the early 1980s (2). Thorium has also been located in this region (3).

BRPM's extensive exploration of the High Atlas, the surrounding region, and the eastern Rif, continued into the 1990s. Its most important exploration project currently is probably the Douar Lahjar, south of Marrakesh, which is described as a "massive sulphide" lode, and which BRPM is investigating along with the Moroccan company Monium Nord Africain (4).

Its exploration for uranium, along with precious and base metals, continues apace (5), and it recently agreed to carry out a broad-based geological and mining programme in Iraq (4).

It is also important to point out that BRPM has numerous majority-owned (or significantly owned) mining subsidiaries throughout the country, producing iron ore, copper, gold, silver, anthracite, lead, zinc and cobalt, with plans to exploit galena, barium, sulphate, fluorine (when the price is right), and possibly potash and magnesite (5).

References: (1) *MAR* 1984. (2) *Uranium Redbook* 1979. (3) *Middle East Economic Digest* 12/2/84. (4) *MAR* 1989. (5) *MIY* 1990.

85 Brünhilde/Gewerkschaft Brünhilde (D)

A mining syndicate which was set up to mine

uranium at Menzenschwand in the Black Forest, it was refused permission to mine after 15 years exploration (1).

In 1988, around 1000 local people demonstrated against the effects of the company's processing plant at Menzenschwand, where ten local children have developed leukaemia. They claim that radioactive contamination of the local river, which provides drinking water for the entire neighbourhood, had reached 1400 times the levels set by the West German authorities.

Another facilitiy, at Ellweiler (2), is also under investigation as a spin-off from the so-called Nukem scandal: one of the waste barrels which "disappeared" after processing at the Mol plant in Belgium ended up at Ellweiler (3).

Menzenschwand was producing around 30 tonnes of uranium a year, in "precommercial" exploitation, and is an underground mine, thanks to state denial of surface mining permits – and further expansion was planned (4).

The 1988 demonstration spotlighted the fact that contamination of the area surrounding Menzenschwand and Ellweiler had probably given rise to ten cases of leukaemia, including two children who died (5).

A year later, the company announced that the mine would be closed and uranium milling halted, although "uranium-containing residues" would continue to be treated (6); the West German Rhenish Palatinate government cited high radon emmissions as the main reason for the closure (7). Later that year, a symposium on low-level radiation and health, held near Ellweiler, revealed that waste from the mill had been used as building material (offered free to local people – including the NATO war headquarters for Central Europe!) (8). Research carried out by physician Schmitz-Fenerhake on children in the area had revealed a threefold greater risk of their contracting leukaemia than normal (8, 9). And yet, by April 1990, the operating licence had still not been withdrawn for the Ellweiler mill (10).

Contact: Initiative gegen Atomlagen, St Wendel, Ureweiler Str 29, D-6690 St Wendel, West Ger-

many; and BI Gegen Uranabbau in Menzenschwand, c/o Peter Diehl, Schulstr 13, 7881 Herrish-Ried, West Germany (tel 0776-41034).

Kreisverwaltung Birkenfeld, z.H. Herrn Stefan Schupp, Schlossallee 11, D-6588 Birkenfeld, West Germany; tel 06782-15318.

References: (1) *Nuclear Europe* 9/83. (2) WISE NC 285.2893, 286.2897. (3) *TAZ* 22/2/88. (4) *Nuclear Fuel* 9/3/87, 9/2/87. (5) WISE NC 288 4/3/88. (6) *Badische Zeitung* 3/5/89. (7) WISE NC 313 2/6/89. (8) *TAZ* 4/11/89. (9) See also WISE NC 321 17/11/89. (10) TAZ 17/4/90; *Der Spiegel* 19/2/90; see also WISE NC 331 27/4/90.

86 Brush Wellman Inc (USA)

In 1978, it announced "environmental studies" into the recovery of uranium from beryllium at its mill in Delta, Utah. It also staked claims over more than 12,000 acres for uranium and other mineral exploration. It entered into an agreement with an un-named concern to explore and possibly develop some 20,000 acres in Juab County, Utah, possibly containing uranium and other minerals (1).

By the beginning of 1989, the company had worked around 2000 of the 12,000-13,000 acres under licence. It had also bought up Williams Gold Refining Co, and Metals Engineering Co (2).

References: (1) *MJ* 10/3/78. (2) *MAR* 1989; *MIY* 1990.

87 Buffelsfontein Gold Mining Co Ltd

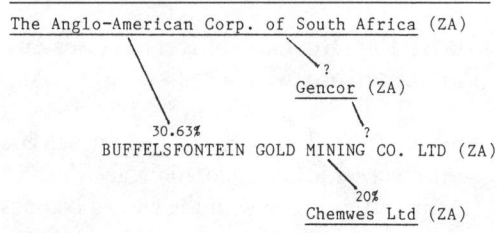

It was the 3rd biggest producer of uranium in South Africa: its 1983 output was 611,000kg U_3O_8 at an average grade of 0.197kg per tonne (1) and its production for the nine months ending March 1985 just over 500 tons (2), bringing a net income of around R17 million (2).
A "tribute" arrangement with Vaal Reefs to mine on its property commenced in 1977, but ended in 1983 (3). However, in the same year Buffelsfontein acquired the Beatrix mine in the Orange Free State as a "means of gaining valuable tax savings on Beatrix's establishment costs" (1, 4).
Buffelsfontein operates three extraction plants: one for uranium and another two for the extraction of pyrite concentrates and the manufacture of sulphuric acid as a byproduct of its uranium operations (5). In 1982 a 200t/day acid plant extension – to supply a major part of acid for uranium leaching – began operations (5).
Buffelsfontein also holds a minority in the Chemwes scheme (Stilfontein 80%), to which it has contributed about 5 million tonnes of uranium-bearing slimes on the mine property (5) until it stopped in 1983 (6).
Buffelsfontein is one of the more shadowy of the apartheid gold and uranium producers. Its uranium contracts are not known (7). Owned 25.2% by Anglo-American, 4% by Anglovaal, 17.5% by Gencor and 14.4% by Sicovam in 1972 (8), it is controlled and managed by Gencor while Anglo-American's holding has increased; technically this is only 8.7% (5) but actually more like a third (9).
In the last quarter of 1984 its gold operations secured a record profit (more than R80 million at pre-tax level) (10), and its uranium operations are among the more viable in South Africa.
Buffelsfontein uranium production fell gradually between 1986 and 1989 (from 656 tonnes to just over 500 tonnes), but it probably had around another seven years life, with potential production of around 3000 tonnes of contained uranium (11).

References: (1) *MAR* 1984. (2) *R&OFSQu: MJ* 26/4/85. (3) *MAR* 1985. (4) *MJ* 19/8/83. (5) *MIY*

1985. (6) *MJ* 8/4/83. (7) Lynch & Neff. (8) Lanning & Mueller. (9) *RMR* Vol. 3 No. 2 1985. (10) *FT* 17/1/85. (11) *Greenpeace Report* 1990.

88 Buka Minerals Ltd

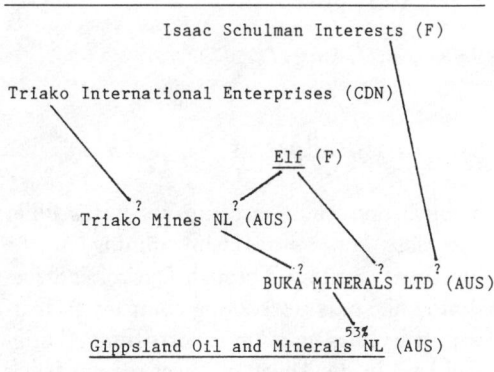

Buka has many JVs with Triako and several important projects of its own, including base metal ventures at Broken Hill and others on the islands of Bougainville and Buka in Papua New Guinea (1). Its major shareholding interests combine those of Isaac Schulman's group of companies and the Elf-Aquitaine mini-empire (1).
Together with Western Nuclear Australia and Pancontinental Mining, it lay claim to several potential uranium leases in the Northern Territory in the 1970s, including Buka leases 145, 552, 671 and 672 (2).
Through its interest in Gippsland (53%), Buka has a share in the Georgetown uranium/base metal project in Queensland (1).
Elf Aquitaine Triako Mines changed its name in the 1980s to Triako Resources, and together with Total Mining Australia (82%) and UG of West Germany (8%) it holds 10% of the Manyingee deposit (3).

References: (1) *Reg Aus Min* 1981. (2) *A Slow Burn*, FoE/Australia-Victoria, No. 1, 9/74. (3) Rick Humphries, *Manyingee Uranium Deposit*, Australian Conservation Foundation, undated.

169

89 Burlington Gold Mines Ltd (CDN)

It held a surface exploration permit for uranium in the Nelson mining division of British Columbia in February 1979; presumed as surrendered (1).

References: (1) *Energy File* 3/79.

90 Burns Philp Ltd

Though not a mining company, Burns Philp has close connections with mining in the Northern Territory. Through Territories Stevedoring, the only stevedoring company in Darwin, it handles uranium exports from Aboriginal land in the Alligators River region. It has also been supplying hydrated lime for use in the tailings dams at Ranger and Nabarlek (1).
"The Burns Philp empire has carried the Australian flag throughout the South Pacific," say Bishop and Wigglesworth. It specifically controlled copra exports from Vanuatu before it became independent in 1980, when the Vanuatu Co-operative Federation challenged its control (2).

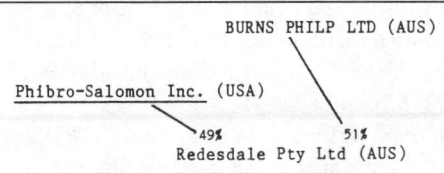

```
                         BURNS PHILP LTD (AUS)


Phibro-Salomon Inc. (USA)

                   49%           51%
                   Redesdale Pty Ltd (AUS)
```

Further reading: Bishop and Wigglesworth, *A Touch of Australian Enterprise: The Vanuatu Experience*, IDA, 73 Little George St, Fitzroy, Victoria 3065, Australia.
James E Winkler, *Losing Control: TNCs in the Pacific*, PCC/Lotta Pacifica, 1983.

References: (1) Personal communication with IDA, 5/82. (2) Bishop and Wigglesworth, *op cit*.

91 Burwest Inc (USA)

It has explored for uranium in Texas (1).

References: (1) *E&MJ* 11/78.

92 CAG/Construtora Andrade Gutierrez (BR)

This development company was behind the US$300 million uranium mine and 500 tonne/year processing plant to be located in the north of the Somali Republic, which was announced in late 1984 by the financing company Soarmico. Construtora AG would use a separation process for yellowcake developed by Nuclebras, though by 1986 the project looked increasingly doubtful (1).
In mid-1989 the company announced that it "need[ed] to diversify by investing in mining in Brazil and abroad" (2). It would be looking for diamonds, gold, uranium, phosphates, iron and titanium. Shortly afterwards, the 51% state-owned Uranio do Brasil signed a contract with the Andrade Gutierrez conglomerate to exploit the Lagoa Real uranium deposits, with Andrade spending US$10 million on a feasibility study and another US$100 million on a processing complex (3). Lagoa Real has an estimated 193,000 tonnes of uranium, second largest in the country (4).

References: (1) *MAR* 1986. (2) *E&MJ* 5/89. (3) *MJ* 1/9/89; see also *MJ* 13/1/89. (4) *MJ* 6/10/89. (5) See also *E&MJ* 10/89.

93 Callahan Mining Corp

The major revenue for the company is from two silver mines in the Coeur d'Alene mining district of Idaho. Callahan owns the Galeana mine

```
         CALLAHAN MINING CORP. (USA)

              80,3%
         Pinnacle Exploration Inc. (USA)
```

run by Asarco (which gets 50% of the takings, splitting off 12.5% to Hecla) (1) and the majority of the Caladay project, in which Hecla again has an interest (2).

The company – classifying itself as a "natural resource" enterprise – also has oil and gas interests in both the USA and the Canadian Arctic islands (3).

Its subsidiary Pinnacle Exploration also had a 15% carried interest in the Homestake Pitch uranium project (3).

References: (1) *MIY* 1985. (2) *MAR* 1984. (3) *MIY* 1982; see also *MAR* 1989, *MIY* 1990.

94 Calvert Gas and Oils Ltd (CDN)

It was formerly Premier Border Gold Mining Co.

It has explored in northern Saskatchewan (1).

References: (1) *Who's Who Sask.*

95 Camflo Mines Ltd

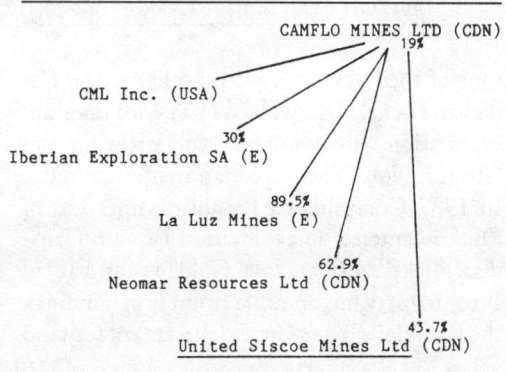

It explores and develops minerals, oil, and gas in Canada and the USA. It has also explored for precious minerals in the Sierra Morena in southern Spain, and has interests in several other Canadian companies (1).

It is exploring for uranium in Saskatchewan, and holds 33.3% interest with United Siscoe and Getty Minerals in a large uranium prospect

in the Kasmere Lake area of north-western Manitoba (2).

Three directors of United Siscoe are also directors of Camflo, including the chairman, RE Fasken (2).

In 1984 more than two-thirds of the company's revenue came from gold mining (3).

References: (1) *Who's Who Sask.* (2) *MIY* 1981. (3) *MIY* 1985.

96 Camindex (CDN)

It was exploring for uranium in the Anstruther and Burleigh townships of Ontario in the late '70s; no mining (1). No further information.

References: (1) *BBA* No. 4, 1979.

97 Campbell Resources Inc (CDN)

Formerly Campbell Chibougamau, this company is active in exploring and mining for copper, gold, oil, gas base, and precious metals in Canada, Alaska, and Mexico.

It was part of Norcen's JV at Blizzard, British Columbia, with E & B Explorations, and Ontario Hydro (see Norcen Energy Resources Ltd). It also had a 15% net interest in a JV with Norcen, to explore for uranium across Canada. It was part of a JV exploring for tin and tungsten in the Yukon and in British Columbia with Canadian Nickel (Inco) and Billiton Exploration Canada (1).

See Norcen.

References: (1) *MIY* 1981.

98 Canada Tungsten Mining Corp Ltd

This is Canada's only producer of tungsten (1) and is also known to have held a property near Schefferville, northern Quebec, on which a possible uranium oxide mineralisation with

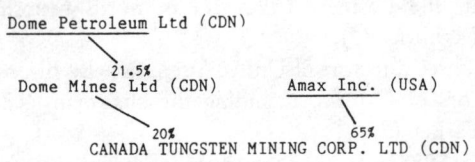

```
Dome Petroleum Ltd (CDN)
              \
            21.5%
    Dome Mines Ltd (CDN)        Amax Inc. (USA)
                  \                  /
                20%                65%
        CANADA TUNGSTEN MINING CORP. LTD (CDN)
```

very high grades of uranium (one find approaching 40%) was located by provincial government geologists (2). No further development appears to have taken place.

It has also participated with Dome Petroleum in exploring for oil and gas in the Beaufort Sea with potentially disastrous consequences for the native peoples of the Canadian north and for marine life in the Arctic (3).

In early 1983 the company shut down its mining operations – throwing 200 people out of work. It said it wouldn't reopen "unless and until there was a significant improvement in market conditions" (4).

References: (1) *MIY* 1981. (2) *FT* 28/7/78. (3) *MIY* 1982. (4) *FT* 23/12/82.

99 Canada Wide Mines Ltd

As a 100%-owned subsidiary, it not only technically owns, (since 1979), but also operates, Esso Canada's 50% stake in the Midwest Lake JV uranium project in Saskatchewan, along with Numac and Bow Valley (25% interest each). Completion date was expected to be 1986; mine and mill studies have been under way (1). Estimated reserves are 2 million tons of rock averaging 1.25% uranium oxide; ore grade was revised downwards from 3.4% in the course of 1979 (2), nonetheless, with an estimated 26 million kilograms of uranium oxide

```
Esso Resources Canada Ltd (CDN)
                     \
          CANADA WIDE MINES LTD (CDN)
                    /
                  40%
          Midwest Lake JV (CDN)
```

the deposit is moderately important. In December 1981, however, Canada Wide Mines announced that development of Midwest would be held up because of poor uranium markets: commencement had been scheduled for 1982 with production in 1985 (3).

References: (1) *MM* 1/82. (2) *FT* 11/1/79; *MIY* 1982. (3) *Star-Phoenix* (CDN) 4/12/81.

100 Canadian Dyno (CDN)

The company was set up in the 1950s to mine uranium in Ontario for export to the USA for nuclear weapons construction (1); presumed defunct.

References: (1) BBA No. 4, 1979.

101 Canadian Nickel Co

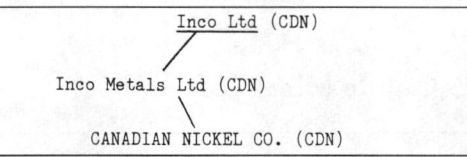

```
         Inco Ltd (CDN)
              /
    Inco Metals Ltd (CDN)
              \
        CANADIAN NICKEL CO. (CDN)
```

One of Inco's exploration subsidiaries (1), Canadian Nickel was involved in a JV for uranium exploration with Eldorado Nuclear in the late '70s (2). No known exploitation of uranium.

In 1987 Consolidated Rambler Mines Ltd of Ontario entered an exploration JV with Canadian Nickel's affiliate Inco Gold (a subsidiary of Inco) to carry out an exploration programme in the Rambler's Bett's Cove claim area of Ontario (3).

References: (1) *MIY* 1990. (2) Eldorado Nuclear Annual Report, 1979. (3) *MIY* 1990.

102 Canadian Occidental Petroleum Inc

It owns joint property with Inco Ltd at

```
Occidental Petroleum Corp. (USA)

CANADIAN OCCIDENTAL PETROLEUM INC. (CDN)
```

McClean Lake, northern Saskatchewan, on which significant uranium mineralisation (more than 0.1% in places) has been discovered (1): there is an estimated 6350t of U_3O_8 (2).

It also held an exploration licence for the Vernon Mining Division of British Columbia up until 1979 (3) and was involved in a JV for uranium exploration with Eldorado Nuclear in the late 1970s (4).

During the 1980s the company's exploration programme came to depend more on oil and less on gas, as it moved into western Alberta, Peru and Bolivia (5).

At the end of the decade, the company was still in a JV with Inco Ltd and Minatco (Total Cie Minière SA's subsidiary) holding mineral properties (including uranium prospects) in Saskatchewan (6).

References: (1) *MJ* 5/10/78, 9/5/80. (2) *MAR* 1981. (3) *Energy File* 3/79. (4) Eldorado Nuclear Annual Report 1979. (5) *FT* Oil and Gas 1983. (6) *MIY* 1990.

103 Canadian Superior Exploration Ltd

It has been exploring for uranium in Saskatchewan (1).

References: (1) *Who's Who Sask.*

104 Canadian Superior Mining (AUS) Pty

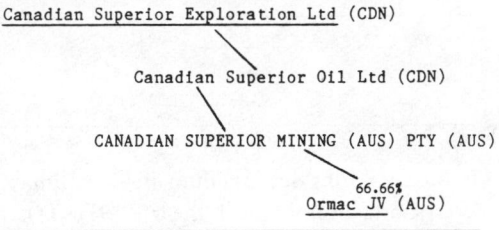

It has conducted uranium and base metals exploration in New South Wales, Western Australia, and in the Northern Territory, both in its own right and as part of JVs (1), specifically in Ormac JV where it holds two-third against Ocean Resources' one-third (for further details see Ocean Resources NL).

No work has been carried out in the area since

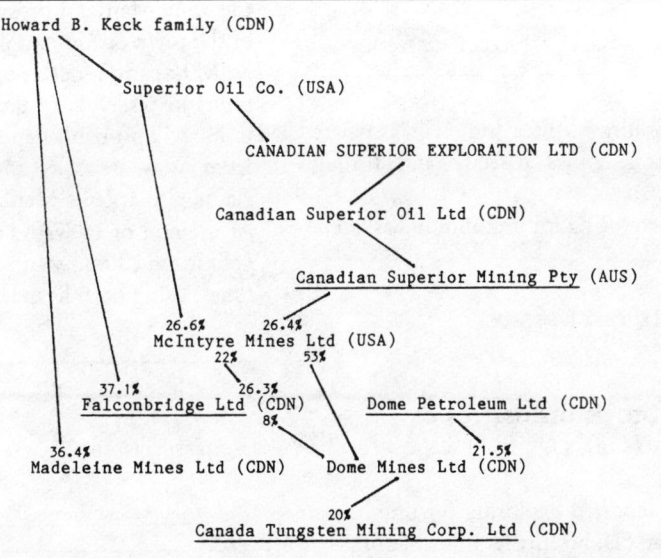

1977 due to tenure problems with Aborigines from the east Alligator River area. The Northern Land Council has not reached agreement with the JV, though the federal government indicated in its 1977 policy statement on uranium that the area would be available for mining (2).

References: (1) *Reg Aus Min* 1981. (2) *AMIC* 1981.

105 Can-Erin Ltd

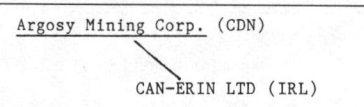

It was exploring for uranium in the Allihies, West Cork, area of Eire in the late 1970s (1).

References: (1) Personal communication from Cork anti-nuclear group, 4/3/80.

106 Canray Resources Ltd

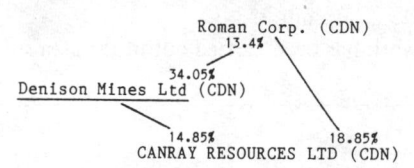

Formerly Goldray Mines Ltd, it is associated with Denison Mines directly and through Roman Corp
It has been exploring for uranium in Saskatchewan (1).

References: (1) *Who's Who Sask.*

107 Can-Stock Industries & Resources (CDN)

It has been reported exploring for uranium in Saskatchewan (1); no further information.

References: (1) *Who's Who Sask.*

108 Caraday Uranium Mines Ltd (CDN)

Exploring for uranium in the Cardiff and Faraday areas of Ontario in the late 1970s (1); no further information.

References: (1) *BBA* No. 4, 1979.

109 Carbrew Exploration (CDN)

It was exploring for uranium in the Cardiff area of Ontario in the late '70s; no mining (1). No further information.

References: (1) *BBA* No. 4, 1979.

110 Carpentaria Exploration Co Pty Ltd

It holds 6.6% of the Yarramba uranium prospect near Honeymoon, South Australia (1).
A dispute between the partners in the prospect slowed exploration during 1982 and 1983 (2).
It also holds 3500 tonnes of U_3O_8 in reserve at the Lake Maitland prospect, 90km north-west of the town of Agnew, Western Australia. (3).
MIM has announced possible economically exploitable reserves at Lake Maitland, but exploitation will probably depend on how (if ever) the Lake Way deposits and the Cultus Pacific claim, also at Lake Maitland, proceed (2).
At the end of 1987 the company opened up a gold mine 100km south of Townsville, Queensland, using both heap leaching and carbon-in-

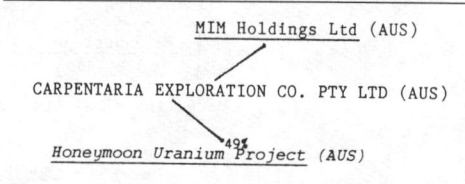

174

pulp extraction, while the following year it commenced gold mining at Tom's Gully mine in the Northern Territory (3).
See Sedimentary Uranium NL (qv).

References: (1) *Reg Aus Min* 1981. (2) *Reg Aus Min* 1983/84; *AMIC* 1982. (3) *MIY* 1990.

111 Carr Boyd Minerals Ltd

This Western Australian mineral (gold, copper) exploration company (1), has co-operated with Alcoa, Amax Exploration, CSR (on the Carr Boyd nickel prospect), Geopeko, and CRA (2). In 1979, it entered a JV for uranium prospecting with Esso Australia Ltd and Otter Exploration NL at Pandanus Creek, Northern Territory; Esso was enabled to earn a 51% interest in the prospect by executing and financing all exploration work by April 1986, and had (in 1979) an option to earn a further 4% by completing environmental impact studies (3).
It is also involved in the Prairie Downs uranium prospect at Ballards Well, Western Australia (WA).
JV partners in other projects have included WMC, Esso, CRA, Geopeko, and Marathon Oil (4).
In early 1988, Carr Boyd proposed that Ashton Mining Ltd (28% holder of the world's biggest diamond mine, managed by CRA at Lake Argyle, WA) should increase its stake in Carr Boyd from 21.9% to 30% – thus enabling the company to use more capital to expand its gold and other activities (5). Not long afterwards, Ashton raised its share of Carr Boyd to 47% (6, 7).
However Carr Boyd wanted Ashton's share limited to 30% (8), and when Ashton took over Hill Minerals in 1988, thus boosting its share to

47%, Carr Boyd took Ashton to court. (Eventually the legal wrangle was settled, with Ashton guaranteeing that it would be "supportive" of Carr Boyd, while allowing it "independence") (6, 9).
Meanwhile Carr Boyd increased its stake in the important Harbour Lights gold mine to 92% by buying out Aztec's 32% share (10) and progressed with investigating diamond deposits both in the USA (at Kelsey Lake, Colorado) and in a JV with De Beers subsidiary Stockdale Prospecting Ltd at the Ellendale prospect of Western Australia – a region of strong Aboriginal claims (see AAC) (5, 11).
By mid-1990, however, it was apparent that Ashton was far from happy to remain in a back seat – presumably because Carr Boyd's diamond interests fitted well with its own – and it made a share offer for all the remaining shares in Carr Boyd that it didn't own (12).
Carr Boyd's most important venture for the 1990s is likely to be the Mount Weld rare earths deposit, near Laverton, Western Australia. The company's prospect in this region has been financed by the University of WA which has apparently devised a new method of mechanically processing high purity metal oxides, as opposed to the traditional, more expensive chemical treatment (7).This process purportedly gives a titanium product of 99.992% purity (7). The Mount Weld deposit contains more than 1.6 million tons of rare earths with a 26.1% total grade value (13, 14) making it allegedly the world's richest deposit of its kind (13). It is also said that the deposit is exceptionally low in thorium (15).

References: (1) *Reg Aus Min* 1981. (2) *Reg Aus Min* 1976/77. (3) *MJ* 18/5/79. (4) Annual Report, 1981. (5) *MJ* 1-8/1/88. (6) *MJ* 11/8/89. (7) *MJ* 17/11/89. (8) *MJ* 5/5/89. (9) See also *MJ* 5/5/89. (10) *MJ* 23/6/89. (11) *MJ* 30/10/87. (12) *Metal Bulletin* 14/6/90; see also *MJ* 13/4/90. (13) *MAR* 1990. (14) see also *E&MJ* 1/90, *MJ* 1/6/90. (15) *E&MJ* 5/89.

```
            CARR BOYD MINERALS LTD (AUS)
                          /        \
Australian Ores and Minerals Ltd (AUS)
                                    43%
                          Hill Minerals Ltd (AUS)
```

112 CCHEN/Comisión Chilena de Energía Nucleara

The Chilean junta – one of the most repressive régimes operating in the world in recent years – was seriously involved in promoting nuclear power for a decade. Chile possesses two research reactors, one of which was supplied by the British General Electric Company and Fairey Engineering (part of the Trafalgar House group of companies, also involved in South Africa) (1), the other was provided by Junta de Energia Nuclear (JEN) of Spain. CCHEN was involved in constructing both reactors, the first of which has received its enriched uranium from the USA, with BNFL the fabricator (2).

With help both from the IAEA and JEN, CCHEN began exploring for uranium in the 1970s: in 1975 a contract between the Chilean régime and the United Nations Development Programme (with the IAEA) initiated the survey of 105,000 square kilometres of the country, and an investigation of the feasibility of extracting uranium from copper (2). Four major areas were delineated: the Sierra Gorda, Chuquicamata/Exotica (in Famagusta province), the Tento-Marde in central Chile, and the Carrocal area in Copiaco (3).

By 1980, the Pudahuel mining company announced plans to begin extracting uranium from copper at the Cascada mine in northern Chile (4), and feasibility studies were announced for the Chuquicamata mine (5).

In 1986 the US mining giant Amax signed a contract with the Chilean state production company CORFO, to produce lithium from the Atacama salt flats in the north of the country (6). The amount of lithium produced and sold will be controlled by CCHEN With Amax's assistance, the country's contribution to the world lithium market "could be more than 50% ... by 1992" (7).

```
                        Republic of Chile
 CCHEN /
 COMISIÓN CHILENA DE ENERGÍA NUCLEARA (RCH)
```

References: (1) *Leveller* magazine No. 12, London, Feb 1978. (2) *Nuclear Fix.* (3) *MJ* 1/7/77. (4) *MM* 12/80, *FT* 7/9/80. (5) *MM* 2/80. (6) *FT* 3/3/86. (7) *MM* 2/86.

113 CDC Oil and Gas Ltd

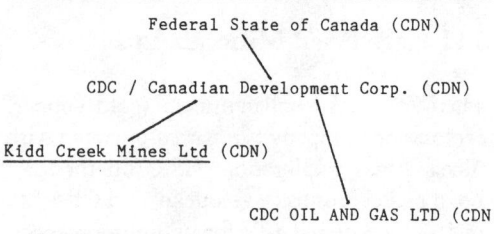

```
            Federal State of Canada (CDN)


     CDC / Canadian Development Corp. (CDN)

Kidd Creek Mines Ltd (CDN)

                      CDC OIL AND GAS LTD (CDN)
```

It is a wholly owned subsidiary of Canadian Development Corporation, exploring for uranium in northern Saskatchewan (1).

In 1981 CDC gained a 37% control over certain Canadian assets owned by Texasgulf in a complex merger between Elf and the US/Canadian company (for fuller details see Texasgulf) (2).

CDC is part of the Canterra Energy Ltd group which in the early 1980s set up an *in situ* pilot plant to process oil sands in the Athabasca region of Alberta – a project vigorously opposed by the Cree Treaty Council No 6.

Contact: Cree Treaty Council No 6, Box 960, Lac La Biche, Alberta T0A 2CO, Canada.

References: (1) *Who's Who Sask.* (2) *BOM.*

114 CDR Resources Inc (CDN)

Formerly Highland Lode Mining Ltd, it has sub-opted some properties to Denison Mines. It has been reported exploring for uranium in northern Saskatchewan (1).

References: (1) *Who's Who Sask.*

115 CEA/Commissariat à l'énergie atomique

"I do not recognise any moral arguments. I knew how to make a bomb and I made it", commented Yves Rochard, chief scientist in CEA's secret post-war military department (1). Completely controlled by the French state, the CEA was established in 1945 by General Charles de Gaulle. Since that time, the country's "civil" and military nuclear programmes have advanced inextricably together (2). Indeed it was not until 1976 that Cogéma was established as a subsidiary to deal primarily with "civil" uranium supplies and associated nuclear fuel facilities. As can be judged by the 1976 chart below, Cogéma is a single, if major, "cog" in the huge wheel of the CEA: an organisation which certainly "has no equivalent outside France" (2).

Formally, the CEA has five major responsibilities:

- to advise the government on domestic and overseas nuclear programmes in the context of exports and non-proliferation obligations;
- to maintain responsibility for most research and development programmes on nuclear reactors and their fuel processes;
- as a shareholder in Framatome-Novatome, to take direct involvement in the construction of French nuclear plants;
- through its numerous affiliates and subsidiaries, notably Cogéma, Eurodif, and SGN, to develop all aspects of nuclear fuel production and services, including prospecting and working uranium deposits, enrichment, fuel fabrication, reprocessing, and the packaging, storage and transport of waste;
- through its Institute of Nuclear Protection and Safety (IPSN) to provide technical "support" for the safety responsibilities of the Central Service for Safety of Nuclear Installations (SCSIN) (2).

This bland list of responsibilities conceals more than it reveals. Of the CEA's annual budget of around FFr16.2 billion (for 1985), just under half is devoted to researching, developing and testing of bombs and warheads for the French *force de frappe*. The rest goes not only on the development of enrichment services (in which CEA/Cogéma is a world leader) and the Fast Breeder reactor (Superphénix) – the most public and controversial of its "civil" developments – but on a diversification programme which puts it ahead of all its western equivalents.

It has been developing semi-conductor technology since 1972 under the auspices of the Efcis company (now absorbed into the Thomson electronics group); its Leti laboratory has joint electronics ventures with Rhone Poulenc; its research and development links with industry now involve CEA in robotics, fluid separation in food, perfume and the oil industry; scanning equipment for diagnosing diseases; bio-technology and microchips (3); as well as solar technology. In 1979, CEA (55%) joined with Total/CFP and the Alsthom-Atlantique and CGE groups to expand the Thémis solar-powered station (4).

Although the 1985 head of the CEA, Gerard Renon, rejects the long-pointed accusation that the monolith is a "state within a state" (3), the fact remains that CEA is not responsible to any regulatory agency, and – for at least 1951-1970, the most critical years in the emergence of France as a "renegade" nuclear power – it virtually spent what it wanted. According to Francis Perrin, High Commissioner of the CEA in charge of scientific activities during that period: "We were given a statute of autonomy analagous to that of Renault. The difference was that Renault was there to make money for the state, and the CEA was there to spend it" (3).

```
                    Republic of France

CEA / COMMISSARIAT À L'ÉNERGIE ATOMIQUE (F)

    Afmeco (AUS)

        Cogéma (F)

          Novatome (F)

            Sofratome (F)
```

Power in the nuclear industry in France is confined to an extremely small group of around 20 white males (not a woman or immigré in sight) (2), most of whom have trained at the École Polytechnique and slip into each others' shoes with ease. For example, Renon took over as CEA's Administrator General in 1983, when Michel Pecqueur shifted to the helm of Elf-Aquitaine (3), while Georges Besse, who set up the Pierrelatte military uranium enrichment plant in the late 1950s and later headed Cogéma, is currently the chairman of Renault, and was the chairman of Pechiney (1).

The CEA was the scientific brainchild of the legendary Frédéric Joliot-Curie. Initially he had enormous obstacles to overcome in creating a state agency with only one tenth the resources of the British Atomic Energy Authority (AEA) and one hundredth the size of the US Atomic Energy Commission (AEC) (5). In the immediate post-war years, the British and Americans controlled most supplies of uranium (including the French deposits in the Belgian Congo); French supplies of heavy water had been kept in Canada; even French scientists had trouble in returning home from their war-time labours on the Manhattan project in the USA (5).

Joliot started by building a research reactor, and sending "young geology students and old Resistance fighters armed with Geiger counters to roam the French empire, especially Africa, in search of uranium" (5). Although he publicly maintained that France wasn't interested in nuclear weaponry, the reactors inherited from the allied programme included weapons-grade plutonium reactors which, theoretically, could give the country enough plutonium in five years to make its own bomb.

By 1948, the US was openly attacking Joliot (an avowed communist) and he was losing friends within his own country. When the USSR exploded its first bomb in 1949, his fate was sealed. Driven to declaring that "The French people ... never shall make war against the Soviet Union" (6), he was soon dismissed from government service. By then, his CEA team had extracted its first plutonium.

Not only Joliot, but a number of his left-wing confrères, were also sacked, thus leaving the way open for the take-over of the CEA by the Corps des Mines (7).

In 1951, the Corps des Mines – which comprises the ten best graduates of the École Polytechnique – hijacked the CEA, and Pierre Guillaumat, a staunch conservative, took the helm. Guillaumat not only forged the CEA as it essentially stands today, but also founded Elf Aquitaine (5). It was Guillaumat who entrenched the uranium industry in both France and the colonies, as well as creating a new industrial department in the CEA. It was also the ubiquitous Guillaumat who started work in earnest on the French nuclear bomb – despite having no formal go-ahead from the government (8). Exploiting the ambivalence of Prime Minister Mendès France, who fondly believed that civil and military aspects of nuclear technology could be firmly separated, Guillaumat proceeded with a bomb programme (9). In February 1955, the Mendès-France government fell. Exploiting the uncertainties of the transition period, Guillaumat signed a secret agreement with the new Ministry of Defence and Atomic Energy to build a third, plutonium-producing reactor. Its purpose could by then hardly be in doubt – although the very day permission for the reactor was granted, the new Fauré government told the French public that it had not even decided to study how to make a bomb, let alone build one: the doubling of the CEA's budget would be for increased electricity production (5). Fauré's successor, Guy Mollet, discovered the secret bomb project in 1956 and wanted to halt it on the spot. But the Gaullists threatened to withdraw their support from his shaky coalition, and after the Suez crisis erupted – demonstrating France's vulnerability to "blackmail" by a third-world non-nuclear power – Guillaumat was granted a new secret protocol which led directly to the testing of the country's first atomic bomb in the Sahara on February 13th 1960 (5).

During these same years, Guillaumat was pushing France's gas-graphite reactor, as an alternative to US light-water technology which was soon to dominate the world market, and to gain a foothold in Europe (5). By the late 1960s, it was only a hard core of CEA scientists which seemed opposed to propagation of the light-water reactor. André Girarud was appointed head of the CEA in 1970, by which time the CEA had apparently lost its battle over reactors. To add to the CEA's humiliation, the new Prime Minister, Georges Pompidou, decided to downgrade the CEA as an institution, making it henceforth responsible to the Ministry of Industry, not the Prime Minister.

Giraud adapted to the newly imposed structure of the CEA, and was more responsible for propelling the *commisariat* into total involvement in the nuclear fuel process than any other single person. By the time he left the CEA in 1978, it had become "the only fully-integrated nuclear corporation in the world" (5): a nuclear "major" deliberately fashioned by Giraud along the lines of Exxon, Gulf, Westinghouse – companies with involvement not only in uranium mining, but also many other aspects of the control of fuel supplies. Little wonder Giraud had declared, on taking over the CEA, "We are in the position of Shell in about 1910. It is up to us to build a Shell of the Atom" (10).

Initially, Giraud chose to build up the CEA's reprocessing technology and forged an alliance with the British (later BNFL) and the Germans (Nukem) to exchange technical know-how (5). He then launched an attack on the US monopoly of uranium mining and marketing, as well as enrichment. During the 1960s the aggressive French exploitation of its African colonies had secured excessive supplies of uranium for its failed gas-graphite reactor programme (which required natural uranium as a feedstock). American dominance of the enrichment market meant that these stocks would have to be sold at less than the price of extraction. It is not surprising that the first meeting of the uranium cartel, which came to transform the uranium industry in the early 1970s, was held under Giraud's auspices within the CEA (11).

André Petit, an employee of the CEA, was appointed by the cartel to be their co-ordinator of the false worldwide "bids" which would carve up the uranium world for the French, Austra-

lian, British, Canadian and South African members (11). The CEA's objective of keeping Westinghouse reactors out of Europe was also served by the cartel's operations – while bringing the CEA into an alliance with Gulf, which had similar aims.

The CEA began exploring for yellowcake in France very soon after the war's end, focussing on known uranium indications and a few small occurences found during radium exploration (12). In 1948, the large deposits at La Crouzille were discovered; within the next seven years, deposits in Limousin, Forez, Vendée and Morvan were being mapped. Over the next thirty years (until just after the formation of Cogéma), some 270,000km^2 had been prospected in France at a cost of US$185 million (12).

As already pointed out, the CEA was going full-out for deposits in its African colonies during this period. In Gabon, the CEA started prospecting as early as 1947, although it was not until 1955 that the Mounana deposit was discovered in the Franceville sedimentary basin: by the end of 1957, nearly 6000 tons of uranium oxide had been identified (13).

A year later, Comuf was set up, and commercial mining commenced in 1961 (14). Within a decade, the original Mounana deposits had been exhausted; by then the Oklo deposit had come on stream (13).

The CEA carried out reconnaissance work in Niger between 1954 and 1956; two years later, it discovered the Azelik and Madaouela deposits. The Arlit deposit was located in 1966 and brought into production five years later (13).

Exploration commenced in 1947 in the Central African Empire (now Republic), with little success. Using improved techniques, and prospecting the entire eastern half of the country, the CEA later located the Bakouma deposit (15). Joining with CFMU (Compagnie Française des Minerais d'Uranium – a company in which Imetal, other Rothschild interests, Mines de Kail St Therese SA and the CEA have joint interests), and the CAR government, the URBA consortium was set up in 1969, to exploit Bakouma. It ceased operations in 1971 (15), but in 1974, URCA was founded, with Alusuisse joining the CAR and CEA/CFMU. Within two years, Cogéma and Imetal replaced CEA/CFMU (16).

The CEA's other explorations in the heady post-war years encompassed virtually all its African possessions: Morocco, the Western Sahara and Mauritania (17); Senegal since 1957 (18); since 1959 in Cameroon (15); up to 1963 in Madagascar (15); the Ivory Coast from 1947 to 1951 (19).

During the 1960s Uruguay also called on CEA help (along with that of the IAEA and CNEA of Argentina) in trying to locate uranium reserves (18).

Although it was the British and US Americans who provided the expertise and finance for South Africa's uranium-from-gold programmes during the post-war years, the CEA has also been an important customer. In the late 1970s the CEA was reported to have secured long-term supplies of uranium, in return for help in financing a major gold and uranium programme developed by Randfontein Estates. Costing more than US$100 million in capital investment, the project was well advanced by 1977. Designed to treat 250,000 tonnes a month of gold and uranium ore, the CEA had reportedly agreed a very favourable price of US$27/lb (with increases built in, as costs rose) for the entire mine output (20).

By 1985, probably "inspired" reports in the western press maintained that the CEA/Cogéma was no longer dependant on these supplies – estimated to be around 900-1000 tons a year – although a Cogéma spokesman was at pains to maintain that "the contract was a strictly commercial affair and France will honour it" (21).

In 1986, a former High Commissioner of the CEA admitted that France had built Israel's Dimona reactor and nuclear fuel reprocessing plant, knowing it would be used for plutonium extraction and in defiance of an agreement with the post-war US government (22).

Further reading: M Beulayque, "Les Mines d'Uranium d'Arlit Republique du Niger" in *Annales des Mines,* Paris, March 1972, pp 29-42.

References: (1) *FT* 6/3/85. (2) IAEA *Bulletin,* Vienna, Autumn 1986. (3) *FT* 20/2/85. (4) *FT* 27/6/79. (5) Pringle & Spigelman. (6) Lawrence Scheinman, *Atomic Energy Policy in France under the Fourth Republic,* Princeton Universaty Press, 1965. (7) John Ardagh, *The New France,* 1970; Dominque Dejeux, "Le Corps des Mines ou une Nouvelle Mode d'Intervention de l'État", Paris (mimeo), 1970; Erhard Freidberg and Dominique Dejeux, "Fonctions de l'État at Rôle des Grands Corps: Le Cas du Corps des Mines," in *Annuaire International de la Fonction Publique,* Paris, 1972-73; Jacques Kosciusko-Morizet, *La "Mafia" Polytechnicienne;* FF Ridley, "French Technocracy and Comparative Government", *Political Studies,* 1966; Philippe Simonot, *Nucléocrates,* 1978; Ezra N Suleiman, "The Myth of Technical Expertise". *Comparative Politics,* Vol. 10, Oct 1977; Jean-Claude Thoenig, *L'Ere des Technocrates,* Paris, 1973. (8) Bernard Goldschmidt, Les Rivalites Atomiques, Fayard, Paris 1967. (9) Philippe Simonet, *Les Nucléocrates,* Grenobles, Presses Universitaires, 1978. (10) *L'Express,* Paris, 22/3/71. (11) *International Uranium Supply and Demand,* Hearings before the Subcommittee on Oversight and Investigations of the Committee on Interstate and Foreign Commerce, Vol. 1, House of Representatives, US GPO, 1977. (12) IAEA/OECD *Red Book* 1979. (13) *EPRI* 1978. (14) *FT* 12/5/82. (15) Cogéma Annual Report 1980. (17) Dave van Ooyen, "Uraniummijnbouw in Frankrijk," Doctoral thesis, University of Amsterdam, 1982. (18) IAEA/OECD *Red Book* 1982. (19) *Nuclear Fix.* (20) *MJ* 22/7/77. (21) *DT* 30/7/85; see also *FT* 11/9/85. (22) *Nucleonics Week* Vol. 27, 16/10/86.

116 CEGB/Central Electricity Generating Board

CEGB is part of the British Government's Civil Uranium Procurement Organisation.

In both 1981 and 1982, the CEGB was strongly criticised: first by the government's Monopolies and Merger Commission, then by Dr Gor-

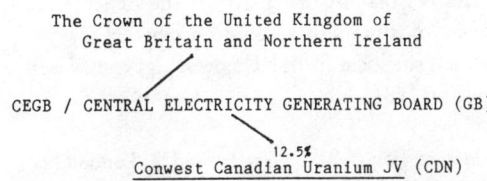

The Crown of the United Kingdom of
Great Britain and Northern Ireland

CEGB / CENTRAL ELECTRICITY GENERATING BOARD (GB)

12.5%

Conwest Canadian Uranium JV (CDN)

don MacKerron, for wildly underestimating the cost of nuclear (as opposed to coal-fired) power stations in England and Wales. MacKerron said the CEGB was "misleading and systematically optimistic as to the economies of nuclear power" (1).

Early in 1982, the CEGB assisted the South African régime in its efforts to make a bomb, by sending a senior British official to help develop the apartheid state's nuclear power programme. Tony Gopsill, former deputy manager of the Hinkley Point A nuclear power plant, was seconded to Escom (the South African state electricity generating service) to contribute his special experience of Magnox reactors. Twenty other experts went with him from British industry to assist in South Africa's development of coal-fired electricity.

CEGB is part of the Conwest Canadian Uranium Exploration JV (see Conwest), and also of a consortium exploring for uranium at Tcheli, Niger (2). It was also an unsuccessful bidder for the government holding in the Ranger mine which was later sold to Peko Wallsend and EZ Industries.

In Britain, campaigning has centred on the CEGB as a consumer of Namibian uranium: there have been two nation-wide Days of Action involving demonstrations at CEGB offices throughout Britain (1980 and 1981). Some anti-nuclear groups have also urged CEGB consumers to withhold that proportion (approximately 10%) of their electricity bills which goes towards the costs of generating nuclear power.

Together with Westmin (until 1981, Western Mines Ltd /CDN/) the CEGB is financing half of an exploration programme for uranium (in return for a 50% project equity) covering prospects in Saskatchewan, Ontario and the North-

west Territories, in Canada (see Westmin) (3). Its interest fell to 30% in 1981 (4).

In late 1976, Noranda signed an agreement with the CEGB to explore for uranium between Rabbit Lake and Key Lake, northern Saskatchewan (CDN), with the SMDC (5).

Recent research shows that the CEGB is responsible for no less than three-fifths of Britain's entire output of "acid rain" pollution (basically sulphur dioxide [SO_2] and nitrogen oxides). Britain is also the biggest acid rain producer in western Europe. The CEGB supplies about 90% of all airborne SO_2 from British power stations, a total of 2.4 million tonnes a year – more than that produced by any other western European organisation (6).

"In fact its airborne SO_2 output is larger than that of several European countries, including Austria, Belgium, Bulgaria, Denmark, Finland, Greece, Hungary, Ireland, Netherlands, Norway, Portugal, Romania, Spain, Sweden and Switzerland. If the CEGB were to declare UDI and take on national status, it would be the fifth largest SO_2 polluter in western Europe" (7).

To date the CEGB has rejected any significant desulphurisation measures (despite a plethora of reports trying to show it's not an easily quantifiable problem) as it has also rejected the estimate that it would cost no more to reduce the SO_2 pollutive emissions by 30% (the UNECE [United Nations Economic Commission for Europe] agreed target) than it would to build the Sizewell nuclear power station (7) .

In April 1988, it was also announced that the CEGB would continue its uranium exploration in the Kayelekera region of northern Malawi, which had commenced in 1983 (8).

In 1986, BB Drilling, on behalf of the CEGB, completed a 5500metre drilling programme; the area of exploration has been reduced to 560 square kilometres (9).

Between 1988 and 1990 the CEGB went through the most important upheaval in its history, when the Conservative government of Margaret Thatcher decided to privatise the British electricity industry and, effectively, destroy the Central Electricity Generating Board (10). Initially the nuclear power generation side of the CEGB was also to be sold to private companies. But, due to "revelations" (not new to anyone in the anti-nuclear movement) of the huge and disportionate costs of continuing the British nuclear power construction programme, the government later decided to withdraw nuclear power generation from privatisation (11). A few months later, the Energy Secretary, Cecil Parkinson, indicated that, after commissioning the four pressurised water reactors already in various stages of planning (Sizewell B, Hinckley B, Sizewell C and Wylfa B), nuclear power construction would probably come to a halt (12). This decision throws into severe question the CEGB's recent uranium contracts. In 1987 it reached a "tentative" agreement with WMC at Roxby Downs for the supply of 700,000 lbs of uranium a year (13), a contract which was confirmed the following year. Then the CEGB, in answer to a critic, argued that the company "... shares your concern for the interests of the Aboriginal people, and so has a policy of purchasing its raw materials only from companies which it considers to be of high integrity and which are expected to have adequately addressed all the sensitive issues involved, before offering their product for sales. The two companies [ie WMC and BP] concerned with the Roxby project fall into that category" (and cows jump over the moon) (14).

In the meantime, however, the CEGB had signed another deal with two US companies, Everest Minerals Corp and Energy Fuels (Nuclear), to supply 3.5 million pounds of uranium over a 10-year period starting in 1987, at a base price believed to be just under US$25 per pound (some US$9 dearer than the spot price at the time) (15).

In mid-1989, the CEGB purchased Everest Minerals, and with it the Hobson and Highland uranium prospects (CEGB 75%, Cogéma/Interuran 25%), giving it access to low-cost and easily expandable supplies (16). Shortly afterwards, it took a 20% stake in the controversial Kiggavik uranium deposit, operated by UG (qv) with a stake by Kepco, a deal which Simon Roberts of Friends of the Earth in London called "... just another example of the CEGB's blatant disregard for the delicate balance of the natural environment" (17) and which the local Inuit have been vigorously opposing (18).

Contact: FoE (GB), 26 Underwood St, London N1 7JQ, England;
Northern Anti-Uranium Coalition [NAUC–Inuit word for "No"], c/o Keewatin Regional Council, Box 185, Rankin Inlet, NWT, X0C 0G0, Canada.

References: (1) *Monopolies and Merger Commission Report on the CEGB*, UK Government Stationery Office, London, 1981; Gordon MacKerron, *Nuclear Power and the Economic Interests of Consumers*, Electricity Consumers' Council, 1982 (summarised in: *FT* 28/6/82). (2) *FT* 9/2/81. (3) *MJ* 22/7/77; *FT* 21/2/80. (4) *MIY* 1982. (5) Merlin, Uranium Institute symposium, 1977 (offprint). (6) Royal Commission on Environmental Pollution: 10th Report, London 1984 (HMSO). (7) Steve Elsworth, *Acid rain*, London, 1984 (Pluto Press). (8) *MM* 4/88. (9) *MJ* 3/7/87. (10) *FT* 4/1/88; *FT* 4/2/88; *FT* 5/2/88; *Gua* 28/7/88; *FT* 26/2/88; *FT* 21/4/88. (11) *FT* 25/7/89; *Gua* 1/12/89. (12) *Gua* 21/6/89; see also *FT* 10/11/89, *Independent* 31/10/89. (13) *FT* 9/1/87. (14) BT Seabourne, Administration Branch, CEGB, to Miss BM Baxter, London, 20/6/88. (15) *FT* 5/11/85. (16) *Nuclear Fuel* 1/5/89. (17) *Gua* 28/9/89. (18) WISE N.C. 319 20/10/89.

117 Cenex Ltd (CDN)

Formerly New Joburke, Cenex owned a property near Uranium City, Saskatchewan, in 1978 which Eldorado Nuclear had previously developed, containing a few hundred tons of uranium oxide. Most of this was contracted in 1979 to be delivered to Swedish Nuclear Fuel Supply at C$52 per pound for ten months (1), with an option on a further 500,000 pounds between 1980 and 1982. Small reserves were also proven, at the same time, around the former Lake Ginch property (2).

According to a Saskatoon, Saskatchewan, newspaper (3), the company opened a mine in early

1979 or, rather, reopened one of the properties it had worked in the fifties. However, the 500,000 pounds sales contract had fallen through at the time.

In 1979 the company was placed in receivership and put up for sale. By that time it had delivered 110,000 pounds of uranium concentrate and resold 300,000 pounds (4).

References: (1) *FT* 18/1/79. (2) *MJ* 4/8/78. (3) *Star-Phoenix* 20/1/79. (4) *Yellowcake Road*.

118 Central and Southwest Fuels Inc

Between 1974 and 1979 it participated with Nuclear Dynamics in uranium exploration (1). No known further exploits.

References: (1) Sullivan & Riedel.

```
Central Power and Light Co. (USA)
                       \
CENTRAL AND SOUTHWEST FUELS INC. (USA)
```

---

119 Central Coast Exploration Ltd (AUS)

This is a gold, uranium, and oil shale exploration company with current interests in three uranium prospects, all of them in Queensland.

In the Georgetown area (about 300km southwest of Cairns) CCE has been proceeding with a JV at Maureen with Getty Oil, who in 1980 were pessimistic that the project would go to mining due to its small extent and the difficulty of mining cheaply.

Nearby, at Mount Hogan, CCE hold prospects being drilled by AGIP.

At Woolgar, 320km west of Townsville, CCE has a 50% interest in a JV with Esso Exploration and Production Australia which has so far not revealed much uranium of apparent importance (1).

Now controlled by Barrack Mines Ltd, one of Australia's most important gold producers (with an interest variously quoted as 82.5% and 95%) (2), Central Coast entered a JV with Pancontinental Mining in 1987 at the Croydon gold prospect in Northern Queensland (2).

References: (1) *Reg Aus Min* 81. (2) *MIY* 1990.

120 Central Pacific Minerals NL (AUS)

This is an Australian minerals exploration company whose main projects are at Rundle, Queensland. (1), where it part-owns the huge Rundle oil-shale deposit with Southern Pacific Petroleum; and in the Ngalia Basin, near Alice Springs, an Aboriginal uranium area. The latter is a JV with AGIP Nucleare (Australia) (41.22%) and Urangesellschaft Australia (41.22%) – CPM held, in mid-1980, 17.56% of Ngalia (2). In mid-1981 the Australian Atomic Energy Commission (AAEC) sold its 10% stake in Ngalia JV to a consortium comprising Offshore Oil, Southern Cross Exploration, and Cocks Eldorado (3); it is not known when the AAEC had acquired this stake. There had been about 1000 tons U_3O_8 identified in the Ngalia Basin, at Bigryli and Walbiri (4). Drilling commenced in September 1981 (5).

The company has a JV with Exxon (USA) in a pilot project at Rundle (2).

By 1983, the Ngalia partners' share in the JV had been reorganised, so that Central Pacific now held 27.62%, Urangesellschaft Australia 25.54% and Agip Nucleare (Australia) 46.84%, under what is technically known as an "Exploration Operating Agreement" (6).

Meanwhile, Central Pacific was reporting on its other Ngalia JV (with Agip (Australia) 38.73%, Urangesellschaft 33.52%, Offshore Oil 7.88%, Southern Cross Exploration 2.25%, Gulf Resources Ltd 1.12%) that reserves had been upgraded to nearly 3 million kgs of U_3O_8, but that activity was being reduced because of the slumped market.

This second Ngalia JV – called the Discovery Operating Agreement – in which Central Pacific holds 16.5%, is at a more advanced stage than the other JV in the same area (6).

References: (1) *FT* 9/8/80. (2) *MJ* 18/7/80. (3) *MJ* 10/7/81. (4) *MJ* 3/7/81. (5) *MJ* 15/1/82. (6) *Reg Aus Min* 1983/84.

121 Central Patricia Gold Mines Ltd

This merged with Conwest in 1982 (2). Central Patricia was exploring for uranium in Saskatchewan in the late 1970s (1).

References: (1) *Who's Who Sask.* (2) *MIY* 1985.

```
Conwest Exploration Company Ltd (CDN)


CENTRAL PATRICIA GOLD MINES LTD (CDN)
```

122 Centromin/Empresa Minera del Centro del Peru

A Peruvian state mining company which operates and explores throughout the country, it reported uranium in the Morococha region, Central Sierra, (1) in 1979.

The company's main operations are in copper mining: the Cabriza project was part-constructed by Davy McKee (2).

Since 1979 Peru has opened its doors to foreign companies for mineral searches; the role of Centromin and other state agencies seems reduced as a consequence. In early 1982 the government "welcomed" foreign uranium prospectors, especially in Puno province, near the Bolivian border (3). A mining law introduced in 1981 facilitated JVs with overseas companies (4).

By 1987, Centromin was optimistic that a deposit at Macusani, in Puno, would be capable of

```
                Republic of Peru (PE)

CENTROMIN /
    EMPRESA MINERA DEL CENTRO DEL PERU (PE)
```

producing 500 tonnes a year of uranium by the turn of the decade (5).

By the late 1980s Centromin operated seven mines, eight concentrators, two smelters, four refineries, a coal washer and two railway lines in Peru. It had become the most comprehensive of Peru's mining companies, working mostly polymetallic deposits, with exploration extending to the Morococha, Yauricocha and Cabriza regions (6).

In support of the government's policy of encourgaging small- and medium-scale mining (7) it had moved into the Madre de Dios region, intending to install a 100 cubic metres/hour suction dredge in the Madre de Dios river. This region has been the subject of pleas and speeches by the indigenous organisation AIDESEP against exploitation by mining companies, delivered by Everisto Nugkuag (now President of COICA, the Co-ordinating Committee for indigenous peoples in the Amazon basin) (8).

The election of centre-left prime minister Alberto Fujimori in 1990, paved the way for limited mining reforms during the next decade supposed to boost the mining industry as a whole, but force inefficient producers like Centromin, to conserve resources and increase production (9). The new national policy was intended to support small and medium-scale producers; but also involved reviving underutilised capacity at the archaic Centromin mines (10) and encouraging foreign investment. Early in 1991 the government announced a modest privatisation of twenty-odd companies including three mining ventures (11).

One precondition for stimulating foreign interest was to curb militancy among Peruvian mineworkers, especially those employed by Centromin. While the mining companies maintain that the workforce had been infiltrated by Sendero Luminosa (Shining Path) mao-ist guerillas – who, during much of 1989, were indeed bombing mine installations (12) – there is no doubt that pay and working conditions have generally been appalling. A series of strikes occurred through 1989, and again in

1990 after Fujimoro's announcement of "austerity" measures (13).

References: (1) *FT* 30/3/79. (2) *MAR* 1982. (3) *MJ* 12/2/82. (4) *MJ* 24/7/81, 28/8/81. (5) *E&MJ* 9/87. (6) *MIY* 1990. (7) *MAR* 90. (8) Speech by Evaristo Nugkuag to the United Nations, quoted in *IWGIA Newsletter* Nos. 43 & 44, 9-10/12/85; and speech by Evaristo Nugkuag on receiving the Right Livelihood Award 1986 in Sweden, quoted in *IWGIA Newsletter* No. 49, 4/82. (9) *FT* 26/7/90, *FT* 27/7/90 (10) *FT* 21/9/90 (11) *MJ* 15/3/91 (12) *MJ* 21/7/89, *MJ* 19/1/90 (13) *FT* 18/8/89, *FT* 16/1/90, *FT* 16/5/90, *FT* 16/10/90, *MJ* 14/8/90.

123 Century Gold Mining Ltd

This holds 80% of the uranium prospect at Hutterfield Township, Queensland with Copconda (subsidiary of Australia's York Resources); no further information (1).

References: (1) Annual Report of York Resources, 1981.

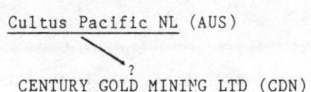

```
Cultus Pacific NL (AUS)
              |
              | ?
CENTURY GOLD MINING LTD (CDN)
```

124 CF Industries (USA)

Constructing two plants with IMC in Florida with potential to produce uranium from phosphates, due to start up in 1980/81 but presumed delayed (1).

References: (1) *MAR* 79.

125 CFM/Cie Française de Mokta

In 1986, Imetal surrendered Mokta to its own majority shareholder, Cogéma (1), with the result that Cogéma is now by far the world's biggest producer and supplier of uranium after Cameco. Until 1986, Cogéma had vied with RTZ for this highly-dubious honour. But let's take the story back a few years.... In 1980, Imetal merged with its subsidiary, Mokta, thus more than doubling its earnings over the previous year (2). Two years later, Imetal's interests were radically restructured, with the huge parent company gaining complete ownership of Mokta, after the transfer of Sté Minière et Metallurgie de Penarroya SA's Mokta holding to Imetal (3). Three years earlier, Imetal had offered nine Imetal shares for every two Mokta shares it could purchase, up to a deadline at the end of 1985 (4).

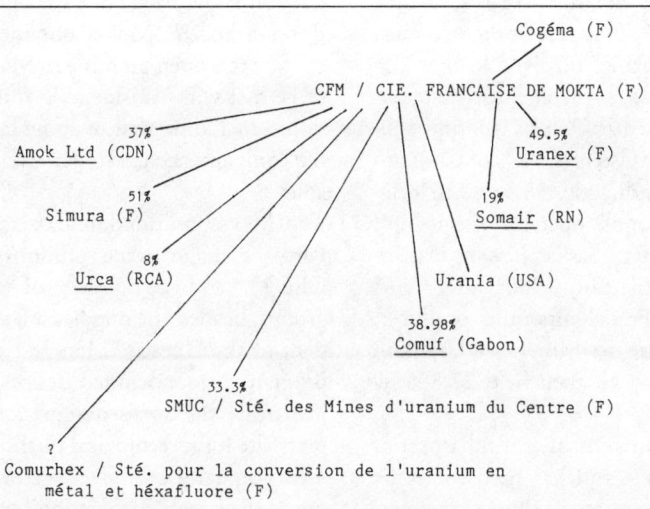

Several years earlier (1976), the commercial (ie non-mining) interests of Mokta had already been transferred to Imetal's Minemet. Pennaroya is Imetal's most important subsidiary (5), but Mokta has been the group's key uranium producer. These two companies have recently been the only ones involved in mining in the Imetal empire. Although Pennaroya's share of the highly important Neves Corvo copper deposit was sold to RTZ in 1985 (1), CFM continues to hold significant shares (14.72%) in the major Gabonese manganese mine, run by COMILOG (1), in Cia Andaluza de Minas, a Spanish iron ore project (61%), the Peruvian lead-zinc producer, Huaron (6), as well as interests in Morocco and French gravel and construction materials (7).

In 1980 the company decided to proceed with mining of two iron ore deposits in Brazil, through Mineração dos Gualaxos, itself a subsidiary of the 99.9% owned Mokta subsidiary Bramok – if market conditions allowed (6). Nothing is known about later developments at these deposits.

Mokta also announced in mid-1979 that it was to spend US$1 million exploiting talc deposits in the Brazilian state of Minas Gerais, following completion of the favourable feasibility studies (8). The company's interests in this region are now grouped under Imetal's wholly-owned Mirra Serra subsidiary (9).

Along with Minatome, the EPRI identified Mokta in the late 1970s as one of the two most active French companies involved in uranium exploitation (11). Mokta then held 7.6% of Somair – later increased to 9.4% (10), and subsequently to 19%. It also held 28% of Comuf – now, directly and indirectly, 38.98%. Its original 25% stake in Amok, operator of the Cluff Lake mine in northern Saskatchewan, was increased to 37% in the early 1980s, after CFM absorbed Imetal's French uranium producer, Campagnie Française des Minerais d'Uranium (CFMU) (20, 12), itself then held 27.8% by Cogéma (13).

Uranium is by far the most successful aspect of Mokta's operations – and has been for some years. In 1978, exploitation of this unique met-al accounted for 78.6% of its total earnings (10), and in 1979, 73.9% (6). Although its foreign uranium sales are not consolidated (3), its interests in Canada, Gabon and Niger are certainly more important than those in French uranium mines. In 1979 alone, 70% of its earnings derived from Gabon and Niger: since then, Mokta's share of Amok's production notably increased this proportion (6). Nothing, either, is known about its subsidiary Urania in the USA (14) – presumed to be a marketing company – while its erstwhile 8% interest in URCA seems dormant.

But this is not to minimise Mokta's importance as a domestic uranium provider. In 1983, uranium production in France as a whole was 3271 tons. Cogéma took 76% of the output, with the remainder split between Total/CFP and Minatome, and Mokta (15). In 1980, Mokta produced 243 tons from its French mines (14); in 1984, it mined 3,167,000 tonnes of uranium-in-ore (down by around 1,500,000 tonnes from 1982 (16). Its annual uranium capacity is 300 tonnes.

Until 1985, Mokta operated three uranium mines in France: at Hyverneresse (central France), Le Cellier and La Croze in the Lozère/Aveyron region, the ore being despatched to a treatment plant at Langogne (Lozère) (14, 17). However, in 1987 the two Lozère mines were due to close down, their production being balanced (18) by a new mine at Bondon-Lozère. Mokta obtained a licence to operate this open pit mine in March 1986, only to be met with considerable and vocal opposition – including that of some landowners that the company needed to win over for additional land.

The Mayor of Bondons-Lozère, the Socialist Party, a major trade union federation (the CFDT), and a number of conservationist groups, headed the opposition, while the President of the Unesco "Man and the Biosphere" programme condemned action which would jeopardise the area's designation as an exemplary site for its ecological qualities (namely the Tarn gorge and Tarn river headwaters). According to the Lozère Association for the Study and

Protection of the Environment (ALEPE), only four people were found to be in favour of the mine in a local opinion poll – and 3000 against (19).

Apart from its interests in the Cogéma uranium company SMUC (Société des Mines d'Uranium du Centre), and its majority holding in SIMURA, Mokta also holds an unknown portion of the shares of Comurhex (Société pour la conversion de l'Uranium en Metal et Hexafluore) controlled by Pechiney with a minority holding by Cogéma.

Contact: ALEPE, c/o Michelle Sabatier, La Salle Pruneut, 48400 Florak, France.

References: (1) *MAR* 1987. (2) *FT* 25/2/81. (3) *MJ* 27/7/84. (4) *MJ* 25/7/80. (5) *La Tribune*, 18/1/86. (6) *MJ* 4/7/80. (7) *MJ* 11/4/80. (8) *FT* 17/7/79. (9) *FT Mining* 1987. (10) *MJ* 29/6/79. (11) EPRI, 1978. (12) *MJ* 1/4/80. (13) van Ooyen. (14) *MJ* 5/2/82. (15) *MJ* 13/7/84. (16) *Mining* 1987. (17) *Uranium Redbook* 1979. (18) *MAR* 1986. (19) WISE *NC 251, 2/5/86*. (20) *MM* 11/81.

126 CFP/Total/Cie Française des pétroles SA

Founded in 1924 at the request of the French government by a group of French industrialists and bankers led by Ernest Mercier, CFP's origins are to be found in the Turkish Petroleum company (later the Iraq Petroleum Co). The French government acquired a 25% stake in CFP in 1930, later expanding it to 35% (with 40% of the voting rights). After the second world war, a large marketing network was established, first in the French colonies of Africa, then France and Europe. In 1954, the Total brand name was adopted for all the group's products and there was expansion into other countries. Over the next twelve years CFP/Total acquired petrochemical and chemical interests and, in the early 1970s, diversified into uranium, coal mining and solar energy. In 1974 it acquired Hutchinson-Mapa, France's largest producer of industrial rubber, then in 1977 Société Rousselot, the largest domestic producer of glues and adhesives, and one of the world's largest manufacturers of gelatine (1).

As France's largest oil group, the CFP/Total mini-empire has colonies in numerous countries; not only virtually all French possessions and ex-possessions, but the USA, Canada (where, over a period of ten years, it transformed itself from a relatively speculative frontier-orientated company into a much larger and integrated concern) (2), throughout Europe, Japan, Indonesia, the North Sea, South Africa, and (of course) the Middle East (1).

It can become confusing – and not the least for this author – trying to determine exactly which company in the CFP group controls which particular project. By and large the term Total attaches to ventures in which CFP/Total has an interest exceeding 50%, but it can also be used in the case of companies eligible for consolidation, in which CFP has a lower interest (1).

In the early 1980s TCM (Total Compagnie

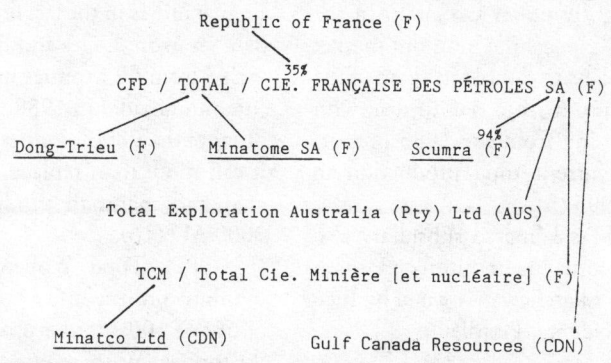

Minière) took over all of of the company's uranium interests (3), though Minatome, which CFP acquired fully in 1982, continues to be the company name most closely associated with certain projects, notably in Australia.

A re-commitment to uranium, made by CFP/Total in 1984, was all the more remarkable for following a huge decline in the group's revenues (4) and the restructuring of its oil business (5). Not only did CFP secure full control over Minatome, it also bought up Dong-Trieu and a little later expanded its French explorations, with the award of a prospect permit in the Vienne and Haute-Vienne regions of western France (6).

By early 1985, CFP had announced plans to bring into production two new French uranium mines (at Loges and Bernadan, in the Massif Central) to replace two other mines in the same area nearing exhaustion (3), and to look for a mine in the USA. (In the late 1970s CFP had acquired a uranium lease on the Spokane native American reservation in Washington State, which does not appear to have proceeded any further) (7).

Currently, TCM operates several uranium mines in central and southern France producing roughly 500 tonnes/year in concentrates (about 15% of French production). It has permits to explore about 4000km^2 of French territory, primarily in the Massif Central, but also Brittany, Veron, and the southern Vosges. Its prime overseas interest is the 10% holding in Rössing uranium which allowed it to purchase between 500 and 1000 tonnes/year until 1986. Most TCM sales are on a long-term contract basis, at prices slightly under US$30lb. In the long term, TCM "expects the uranium market to improve" and plans to double its share of the western world's output from 3% to 6%, with export sales increasing from the 1985 proportion of 10% of current total production to around 50% by 1995 (3).

In early 1985, CFP's Minatco subsidiary also bought into very promising uranium deposits at Wollaston Lake, Saskatchewan, owned by Inco and Canadian Occidental Petroleum (8).

In 1979, CFP/Total bought 20% in a solar energy venture set up with the CEA (9).

It also has an interest in Inspiration Coal, a US producer jointly owned by Minorco (AAC) and Hudbay, and is involved with another South African company, JCI, with which it entered an agreement in 1981, to export coal to France from the apartheid state (10).

In 1988, Total Resources (Canada) Ltd – wholly owned by CFP/Total and CFP's 55% subsidiary Total Erickson – made an offer for the entire common stock of Getty Resources, a Canadian gold producer whose principal asset is a 49% stake in the Tundra gold mine, Nevada (51% owned by Noranda) (11). Getty Resources is not to be confused with Getty Oil or Getty Mining (qv).

Total Minerals Corp has acquired the West Cole uranium solution mining project, formerly operated by Tenneco in southern Texas (12). Total Energold Corp not only has extensive gold interests in deposits and mines in several Canadian provinces – British Columbia, North west Territories (with Noranda), the Yukon (with Agip Resources), and Ontario – but through its wholly-owned subsidiary Sovereign Explorations Inc has begun exploration in the USA, in Nevada, Montana and Idaho (13), near Gold Star (14).

The company's 1988 uranium production was 495 tonnes uranium metal comprising 15% of total French production (13). Its total mining permits amounted to 1667 square kilometres of exploration, 18 square kilometres of mining and 127 square kilometres of mining concessions in mainland France (13). Production at its French mines in the last few years (from Bernardan, Gouzon, Loges and Bertholene mines) has varied from 623 tonnes in 1986 to 820 tonnes (uranium oxide) in 1988. With proven reserves of more than 8000 tonnes, the company claims it can more than replace what it has extracted (15) and maintain production at least until 2000 AD (16).

In 1987, Total reopened the West Cole uranium project in Texas, with a nominal capacity of 100 tonnes: production in 1988 was 90 tonnes. With reserves of 550 tonnes contained uranium, life for this small mine is set at approximately five years (16).

Another US uranium mine, the Alta Mesa, is due for a start-up in 1992, as a solution mine with a capacity of 100 tonnes. Containing reserves of 4500 tonnes, the life of this mine has been put at around 45 years (16).

In recent years the company's Manyingee project (Total Mining Australia 82%, UG 8%, Triako Resources 10%), in the West Pilbara region of Western Australia, has been the subject of heated criticism within the country. The company carried out trial mining after its EPA assessment in 1983 had to be re-submitted to the licensing authority (17). Less than 500kg of uranium oxide was extracted during this trial period and stored on-site (17, 18).

References: (1) WDMNE. (2) FT 8/10/80. (3) E&MJ 2/85. (4) FT 6/1/84. (5) FT 27/12/84. (6) MM 11/84. (7) Big Oil. (8) MJ 15/2/85. (9) FT 27/6/79. (10) FT 27/2/81. (11) MJ 1-8/1/88. (12) E&MJ 4/88. (13) MIY 1990. (14) E&MJ 7/89. (15) CFP Annual Reports, 1985, 1986, 1987. (16) Greenpeace Report 1990. (17) Shyama Peebles, Uranium Mining in Western Australia, Goldfields Against Serious Pollution (GASP), Kalgoorlie, 28/3/89. (18) WA Senate Hansard, Questions without notice, 5/5/86

127 Chaco Energy

It is the 50/50-owner of the Hope Mine with Ranchers Exploration; the mine produced only 20 tons of uranium up to June 1980 (1) and was destined for closure the same year (2). It also operated a uranium exploration programme with Ranchers (3).

References: (1) MAR 1981, p 363. (2) MJ 1/2/80. (3) MIY 1981.

```
Texas Utilities Fuel Company (USA)

                CHACO ENERGY (USA)
```

128 Charter Consolidated plc

"A wondrous hybrid of industrial holding company and investment trust, Charter Consolidated seems to enjoy the drawbacks of both and the advantages of neither" (9).

This was the main vehicle for Anglo-American's international operations in the '60s. Created in 1965 by Harry Oppenheimer "out of a rag-bag of assorted companies" (1), including the British South Africa Co, its aim was to "... ease the penetration of areas where Anglo-American's direct South African connections could be embarrassing and ... improve the possibility of support from bodies such as the World Bank" (2).

Charter Consolidated has invested heavily in mining, financial, industrial and commercial companies throughout the world and also undertakes worldwide exploration.

In 1979 a re-structuring of the company took place, with Charter selling back to Anglo-American the majority of its southern African investments, except for Zimbabwe, and Charter acquiring Anglo-American and De Beers interests in several other companies; Charter also offloaded a portion of its shareholding in Minorco, Anglo-American's main overseas arm.

Charter's activities in Australia were in 1981 joined with other Anglo-American interests in the country into Australian Anglo-American (in which Charter held, in 1982, 15%) (3).

The 1979 reorganising, according to the Mining Annual Review, ensured a much-needed cash injection into Charter, but also meant that "its role as a force in the mining industry has been much diminished" (4).

The company's key investments are now in industry (Anderson Strathclyde, which provides mining equipment, is 100% owned; it has 27.9% of Johnson Matthey, a major metal refining company), civil engineering and construction (where it has 100% of Alexander Shand, 50% of Cleveland Potash), – and in mining, with its 33.5% investment in Anglo-American Corp Zimbabwe and its 7.9% in Minorco. It also created the Malaysian Mining Corp which earns Charter handsome profits for

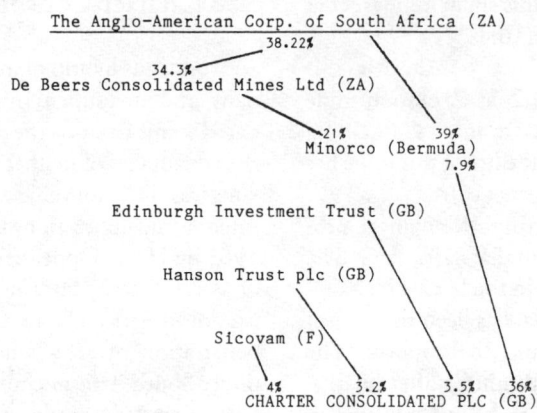

twiddling its thumbs (£5.5million on sales of £35million in 1984) (5).

Charter had a JV with RTZ developing the South Crofty tin mine in Cornwall and engaging in other explorations in the Royal Duchy. It also had a JV in Costa Rica exploring for copper along with Alusuisse and Mitsubishi (6).

There is only one known recent instance of Charter Consolidated exploring for uranium, and that was with Minatome in the late '70s and early '80s around the St Austell and Bodmin granites of Cornwall, using car-borne scintillometer surveys and soil geochemistry. No significant deposits have been reported (7).

However, in 1977 Charter Consolidated set up a JV with John Brown Engineers & Constructors Ltd called Charter-CJB Mineral Services Ltd to do a feasibility study on the Hoggar mountain uranium deposits in southern Algeria (8).

Charter's 14% share in the Malaysian Mining Corp also enables it to profit (through MMC's 46.3% control of Ashton Mining which itself has 38.2% in the Argyle Diamond Mines JV) from the plunder of Aboriginal diamond-rich land in the Kimberley region of Western Australia (see RTZ) (10).

As Minorco moved to consolidate its control over associated companies under the chairmanship of Michael Edwardes in 1988, so Charter moved to "tighten its grip" (11) over its major investment, Johnson Matthey, the precious metal refiners and the world's biggest marketer of platinum. This involved putting two of Charter's senior management in posts at Johnson Matthey.

In 1989, Charter also spent £76 million on acquisitions, mostly in the quarrying business (12), although it sold its half-share in Cleveland Potash Ltd (operating Britain's only potash mine) by disposing of its Shand Construction Group subsidiary. This leaves Cleveland Potash completely controlled by AAC of South Africa (13).

References: (1) Lanning & Mueller. (2) Richard Hall, *The High Price of Principles,* Harmondsworth, (Penguin), 1973. (3) *MIY* 1982. (4) *MAR* 1980. (5) *Gua* 14/8/84. (6) *MAR* 1984. (7) *MM* 11/82. (8) *MM* 8/77. (9) *FT* 28/6/84. (10) *FT* 26/6/84. (11) *FT* 8/12/89. (12) *Investors Chronicle* 22/6/90. (13) *MAR* 1990.

129 Chemwes Ltd

This is one of the three companies set up to produce uranium from gold-bearing slimes in South Africa, and the only one not owned by

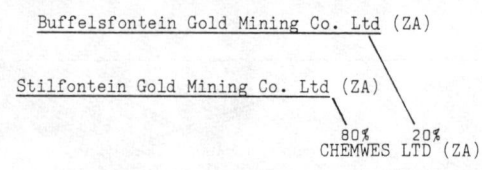

Anglo-American. Situated on Stilfontein's property, with a nominal yearly through-put of 3,240,000 tons, it treated tailings which were provided by Stilfontein and Buffelsfontein. However, in 1983 Buffelsfontein stopped supplying (1) and apparently began producing its own uranium.

Total slimes treated in 1983 reached nearly 4 million tonnes at 0.15kg/ton; 1982 uranium production was 670 tons U_3O_8, putting Chemwes well behind Anglo-American's JMS but well ahead of Ergo (2). For the first quarter of 1985 Chemwes produced only 67 tons, following a cutback announced by Stilfontein (3).

Unlike Ergo, Chemwes was established only as a uranium treatment plant, with the output split between Buffelsfontein and Stilfontein according to their equity in the scheme. Comment Lynch and Neff:

> "This lopsided share is a remnant of the quota system used in the 1960s, when the CDA [Combined Development Agency – a joint British/US uranium purchasing agency] found itself with an excess of contracted uranium, and sought to reduce it by stretching the deliveries over time. However, the quotas assigned to some producers were too small to allow economic production, so they sold their "quotas" to other mines, which could then produce at a comfortable level. Stilfontein sold its quota to Buffelsfontein and stockpiled its own uranium-bearing tailings. Subsequently, in 1978, it decided to develop a plant to process these slimes, along with some from Buffelsfontein. The particular administrative set-up results partly from a desire to be treated for tax purposes as a chemical operation, rather than a gold mine" (4).

Chemwes's contracts appear to be primarily with Nufcor to supply the South African Koeberg-1 reactor (4).

In early 1985, R346,000 had been approved for capital expenditure and not yet spent; but only R95,000 was for "commitment in respect of contracts placed" (3).

In 1989 the South African state electricity utility, Escom, terminated its uranium contract with Chemwes: this effectively ensured closure of the plant (5). A year before, Chemwes had reportedly stopped supplying uranium to an unnamed, major foreign customer which announced that it no longer wanted yellowcake, in the face of increasing sanctions (or calls for them). Indeed, when asked why the customer had pulled out, JC Janse van Rensburg, managing director of Gencor's Western Transvaal East Rand and Baverton mines, said simply "Sanctions", while maintaining that Chemwes considered the contract still operative and would continue producing (6).

Further reading: MA Ford *et al*, "The first six years of the Chemwes uranium plant" in *Journal of South African Institute of Mining and Metallurgy*, Vol. 87, No. 4, 4/87, pp 113-124.

References: (1) *MIY* 1985. (2) *RMR* Vol. 3, No. 2 1985. (3) *R&OFSQu: MJ* 26/4/85. (4) Lynch & Neff. (5) *MAR* 1990. (6) *Star*, Johannesburg 20/10/88.

130 Cherry Hill Fuels Corp

This is a subsidiary of the GPU Service Corp which itself represents Jersey Central Power and Light and Metropolitan Edison in uranium exploration in the late '70s (1). No further information.

References: (1) Sullivan & Riedel.

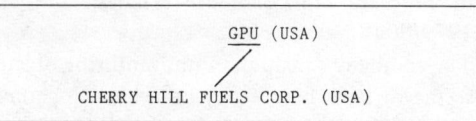

```
                GPU (USA)
                  /
     CHERRY HILL FUELS CORP. (USA)
```

131 Chevron Inc

When, in May 1911, the huge Standard Oil monopoly of the Rockefeller empire was split – by the US judiciary – into 32 separate concerns, the Rockefellers shifted their interests into five of these: Socal/Standard Oil of California,

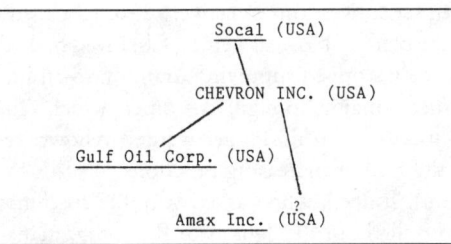

```
            Socal (USA)
                |
        CHEVRON INC. (USA)
        /               \
Gulf Oil Corp. (USA)      \
                           \
                          ? |
              Amax Inc. (USA)
```

Amoco/Standard Oil of Indiana, Mobil/Standard Oil of New York, Exxon/Standard Oil of New Jersey and Arco/Atlantic Richfield. Post-Watergate investigations showed that, even in the late '70s, Nelson Rockefeller, grandson of founder of Standard Oil of Ohio John Davison Rockefeller – and Vice-President of the USA at the time – still held no less than 2.01% of Chevron: some US$85million worth of stock (13).

Chevron was reported in 1979 to be planning uranium mining at Panna Maria, Texas in 1980 (1); by mid-1982 it was the only uranium project operating in Texas (2).

It was also exploring in Saskatchewan (3) and in Newfoundland (4) for uranium.

It has been exploring for copper and uranium in the South Kardofan region of the Sudan but with inconclusive results; other uranium explorers there are Minatome and Minex (5).

In late 1982 it announced it would join with Peko Wallsend to develop the large copper-gold deposit at Parkes, New South Wales (Australia); Chevron will take 39% (6).

It also held uranium permits in Perkins County in the Black Hills of South Dakota, USA, in 1979/80 (8).

The company's main uranium venturing of late seems to have been in Spain where, together with other multinationals (notably Esso), it has been exploring various provinces including Segovia, the Vuar valley (Sevilla), Molina de Aragon, Cuenca del Tajo, Soria and Extremadura (7).

It was in partnership with the Spanish state company JEN (40%), Westinghouse and the Bilbao Bank (combined 20%) that Chevron began exploring in the Comarca d'Osona, Catalunya, in March 1979.

Three Spanish technicians from JEN were discovered working in the area in June, and after the removal of a local farmer, some 3000 people attended a public meeting to protest against the dangers of uranium mining. Meanwhile the Osona Anti-Uranium Committee had been formed, combining the opposition of 20 villages in the region.

Police dispersed a demonstration of 5000 people that same June, and the following month 6000 came to a "Long Live the Land" festival, organised by the village of L'Esquirol in the heart of the uranium district. Local town-halls also joined the movement.

Chevron finally stopped its operations, under public pressure, in August that year (9).

In late 1983, Chevron succeeded in getting an Environmental Impact statement for its Rock Springs phosphate project approved by the states of Wyoming and Utah – despite acknowledged adverse impacts on water, air, wildlife, "visual resources" and land use (10).

The following year, Chevron gained permission to drill one well in the native Canadian area of Liard, Yukon Territory, with 64 out of 104 jobs reserved for northerners including "Indians" (11).

The same month, Chevron was forced to abandon its oil drilling operation in the southern Sudan, after three workers (a Briton, a Kenyan and a Filipino) were killed by the liberation movement for the southern Sudan, the Anyanya, which has been fighting for many years for true autonomy from the Arab, and Moslem, dominated north and against the vacillations, the redivision of the country and an increasingly draconic Islamic rule by General Numeiri (12).

In 1987, Chevron announced that it would be tapping oil in the Southern Highlands of Papua New Guinea, shipping it by pipeline 350 miles to Port Moresby through tropical rainforest. The Rainforest Action Network mounted a campaign aimed at ensuring that Chevron does not destroy virgin habitat if the project goes ahead (14).

In 1988, Chevron Resources, a subsidiary of the oil company, announced that it would soon open up a new mine at Rhode Ranch, McMullen County, Texas, south of Panna Maria: the property is jointly owned by Chevron and Minatome. With an average grade of 0.25%, the deposit is estimated at 2700 tons uranium (15). Chevron estimated that production at Mount Taylor could continue for up to 18 years, with total reserves of 15,400 tons of uranium (15, 16).

In late 1989, Chevron woke up to the fact that 8.8% of its stock had been purchased by Hugh Liedtke's Houston-based oil company, Pennzoil. "Rough, opportunistic, suspicious of the big integrated majors that lord it over the industry" as the *Financial Times* put it (18), Liedtke sent shivers down Chevron's spine. "... the San Francisco company ... behav[ed] as if barbarian hordes [were] camping on the Bay Bridge" (18). Considering the fact that Pennzoil had sued Texaco into bankruptcy only two years before (19), Chevron's reaction was perhaps only a little overwrought. Meanwhile, Amax bought back its own holding in Chevron (20). In 1987 Chevron signed a deal with the Department of Indian and Northern Affairs in Canada, giving it exploration and drilling rights in the MacKenzie Valley – the first such concession in fifteen years (21). Chevron had already acknowledged that it must have the support of the local, mainly Dene, community (22) and the JV with the Fort Good Hope community appears to be one of the most equitable ever signed between an indigenous community and a resources extraction company: allowing the community 20% in any production revenues and the appointment of half the membership of four JV decision-making committees (23). "We can stop an operation any time we feel like it, when something goes wrong," commented Chief Charlie Barnaby, the following year (24). Chevron's uranium production was consolidated in 1988 when it started up the Rhode Ranch mine in Texas with Total (45%). This is an open-pit mine containing just over 3000 tonnes of uranium oxide (25) which the two companies had acquired from Anaconda and its partners in May 1987 (26). In 1989 Chevron's Panna Maria mill in the same state produced around 1.5 million pounds of uranium oxide which had been shipped from the mine (30) a distance of one and a half thousand kilometres (26).

The Mount Taylor uranium mine was put on stand-by in early 1990, while Chevron sought a buyer: the Japanese customers which played a role in keeping the mine open for some time, with a contract for 3 million pounds of U_3O_8 in October 1988, had taken delivery of the order and there was not much prospect that they would bale the foundering operation out (27).

Homestake, which toll milled Mount Taylor's ore at is Grants mill, together with a small quantity of its own dwindling supplies, said it too would suspend operations along with Chevron (28): the combined production of Chevron and Homestake was more then 2 million pounds U_3O_8 in 1989 (27).

The closure of Mount Taylor caused Chevron to write off some US$23 million of its investment (28). It also seemed to indicate the final collapse of primary uranium mining in New Mexico – probably the most consistent supplier of nuclear fuel for war and so-called "civil" purposes in the history of the Nuclear State. Only Rio Algom and Homestake remain, treating minewater and recovering uranium from the leachings around Ambrosia Lake (27). (Rio Algom has purchased Kerr McGee's uranium leases in the area (qv). Chevron also has another JV with Homestake on a gold project at the Golden Bear mine in north-west British Columbia, which opened in early 1990 (28).

Chevron's major mineral projects for the 1990s were its 50/50 JV with Manville Corp, producing platinum at the Stillwater mine in Montana, and its Pittsburgh coal operations (28). In mid-1990 it also agreed to a JV with Ivernia West, whereby the Irish company would earn 47.5% on Chevron's propects comprising some 38 licence areas in the central Midlands region of Eire, some of which are near Conroy Petroleum's recent lead/zinc discoveries at Galmoy (17).

In Papua New Guinea, Chevron Niugini advanced work on its Lake Kutubu oil prospect in the Southern Highlands region, which it owns along with BP (25%), Picon Petroleum (14.983%), BHP Petroleum (12.5%), Merlin Petroleum (6.25%), Oil Search (10.017%) and Merlin Pacific (6.25%). Chevron's application for an oil licence was made in May 1990 – the first application to produce oil commercially from Papua New Guinea (28).

Contact: Comité Antiuranium de Vic, carrer de Saint Juste 1, 3ʳ, E-VIC (Barcelona), Spain.

References: (1) *US Surv Min* 79. (2) *Nukem Market Report* 8/82. (3) *Who's Who Sask.* (4) *Energy File* 4/80. (5) *MJ* 11/6/82. (6) *FT* 15/9/82. (7) *MAR* 1984. (8) *BHPS* 4/80. (9) Comité Antiuranium de Vic, quoted in *Kiitg* 8/81. (10) *E&MJ* 9/83. (11) *Yukon Indian News* 24/2/84. (12) *Gua* 6/2/84,14/2/84; *FT* 6/2/84; also *Pogrom* 1983/84 *passim.* (13) "The Rockefeller Empire" in, *Lords of the Realm* No. 1, London, 1983. (14) Rainforest Action Network, *News Alert,* No. 17, 8/87, San Francisco. (15) *MJ* 10/7/87. (16) see also *E&MJ* 7/87, *MJ* 31/7/87. (17) *MM* 5/90. (18) *FT* 11/12/89. (19) *FT* 8/12/89. (20) *MJ* 11/8/89. (21) *Native Press,* Yellowknife, 13/11/87. (22) *Native Press,* Vol. 16, No. 19, 10/10/86. (23) *Native Press* 10/7/87. (24) *Native Press* 29/4/88. (25) *Greenpeace Report* 1990; see also *MM* 5/89. (26) *MJ* 10/7/87. (27) *E&MJ* 3/90. (28) *MAR* 1990.

132 Chimo Gold Mines Ltd

As part of the Conwest group, it is involved (with 27%) in an extensive exploration programme throughout Canada, together with Conwest (30%), CCF/Consolidated Canadian Faraday (25%), and International Mogul Mines (18%) (1).

References: (1) *MIY* 1981.

Conwest Exploration Co. Ltd (CDN)
       \
       CHIMO GOLD MINES LTD (CDN)

---

133 China Creek Uranium Consortium (CDN)

As of 1979 it was exploring for uranium in the Castelgar area of British Columbia; now presumably defunct (1).

References: (1) *Energy File* 3/79.

134 Chinook Construction and Engineering Ltd (CDN)

As of 1979, it held an exploration licence for uranium in the Grand Forks area of British Columbia. Presumed defunct (1).

References: (1) *Energy File* 3/79.

135 Cigar Lake JV

The Cigar Lake joint venturers are the same as those involved in the Waterbury JV: Cigar Lake is part of that prospect area in northern Saskatchewan. They are: Cogéma (of Canada) 37.375%, SMDC (Canada) 50.75%, and Idemitsu Uranium Explorations Canada Ltd 11.875%. Cogéma has been the project operator since 1980.

As of the end of 1984, Cigar Lake's reserves were estimated at 88,000 tons of uranium stretching over nearly 2000metres and grading averaging 8.49%; significant mineralisation has also been discovered at Tibia Lake, 800metres north of Cogar.

The JV partners estimate that between C$250million and C$350million is needed to bring the project into production – intended to start by 1993 and to rival that of Key Lake (which at the end of 1984 was the largest producing uranium deposit).

Difficulties are acknowledged, however, including radiation dangers because of the high grade of ore, and, although the second stage of the Cluff project is being monitored by the JV, there has never before been an underground uranium mine so high in potential radioactivity

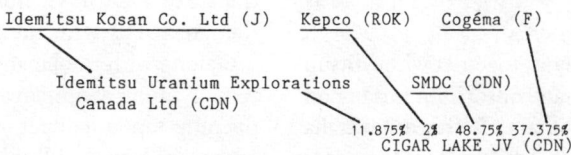

Idemitsu Kosan Co. Ltd (J)   Kepco (ROK)   Cogéma (F)

Idemitsu Uranium Explorations   SMDC (CDN)
Canada Ltd (CDN)

11.875% 2% 48.75% 37.375%
CIGAR LAKE JV (CDN)

(grades are more than 10 times higher than at Cluff Lake) (1).

In 1987, the Cigar Lake Mining Corp received approval from the Saskatchewan Ministry of the Environment to proceed with a US$40 million underground test mine, construction starting immediately: a 500metre deep shaft being sunk in 1988, and the orebody itself located the following year (2).

(The two-year test period will include site development and a feasibility study: up to 5500 tonnes of ore will be mined during the study period. Nuexco predicts that commercial production will start from the deposit in the mid-l990s) (4).

In December 1986, the Canadian federal government announced new sets of rules on foreign ownership in uranium mines, which insist that new projects must have at least a 51% Canadian ownership, though this may be reduced if the project is Canadian-controlled (3). The first transaction concluded under the new rules was the sale of a 2% non-voting interest in Cigar Lake to Kepco, by SMDC in August 1987. As part of this sale, Kepco will secure a US$150 million contract, for supply of 170 tons a year from 1993 to 2002 (and possibly until 2012). The money brought to Cigar Lake as a result of this sale represents "a significant commitment towards future production" (3).

In 1989 Eldorado Nuclear and SMDC united their uranium interests to form Cameco which is now almost certainly the world's single largest uranium producer, and certainly the holder of the world's richest grade reserves (5).

Ownership of Cigar Lake changed with the result that in 1990: Cameco owns 48.75% of CLMC (Cigar Lake Mining Corp), Cogéma Canada owns 32.67%, Idemitsu Uranium Exploration owns 12.87%, Corona Grande (wholly owned by Cogéma) owns 3.75%.

Kepco has secured 2% in return for forward contracts of around C$150 million (6, 7).

With an estimated 175,000 tonnes of uranium oxide, grading at an average of 11.5% uranium, Cigar Lake has confirmed its status as the wonder of the yellowcake world (7, 8). However the richness of the ore continues to pose safety and environmental problems, and the partners were in 1989 still trying to establish a "safe mining method" (sic) (9). Development towards a test mine continued through 1990, with the drilling of what is to become a 1600-foot shaft (8, 10). Production is not expected before 1993, despite some more optimistic predictions (11) and feasibility studies may not be completed before 1991 (7).

Contact: Survival Office Saskatchewan (SOS) Box 9395, Saskatoon, Saskatchewan, Canada S7K 7E9

References: (1) Paper by Stephen Salaff, PhD (95 Dewson St, Toronto, M6H lH2, Canada), reprinted in *MM* 11/84. (2) *MJ* 13/11/87. (3) *MAR* 1988. (4) *E&MJ* 4/88. (5) *FT* 24/2/89; *MJ* 7/4/89. (6) *MM* 5/89. (7) *MAR* 1990. (8) *E&MJ* 3/90. (9) *MJ* 7/4/89. (10) See also *MJ* 13/11/87. (11) *MM* 5/89.

136 CIM/Cie Industrielle et Minière

It is involved in uranium and barytes mining in France. It also has investments in other mining, such as a 29% stake in phosphates in Senegal (1).

References: (1) *MJ* 5/2/82.

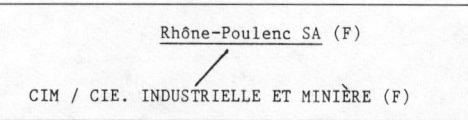

Rhône-Poulenc SA (F)

CIM / CIE. INDUSTRIELLE ET MINIÈRE (F)

137 C. Itoh

This iron ore company is, together with Mitsui, part of the Mt Newman consortium mining on Aboriginal land at Pilbara, Western Australia (other partners include Amax and CSR).

It was, until recently, involved in exploiting labour in the Philippines through the marketing of "cheap" bananas grown for multinational profit at the expense of peasant farmers; it has since withdrawn (1).

According to Kim Williamson (researcher for the Australian Labour Party's Tom Uren), in the early '70s C. Itoh entered into an arrangement with Peko Wallsend and EZ Industries to sell Ranger uranium in Japan. At the time (1973) both C. Itoh and Agip were engaged in exploration for uranium in central Arnhemland (2).

Later on it also contracted with ERA for part of the Ranger uranium for the period between 1982 and 1996, along with Kansai, Kyushu and Shikoku utilities.

In late 1983 the Chinese government was also considering bringing C. Itoh into a scheme to construct a coal slurry pipeline from Inner Mongolia to the coast (3). The same year the company announced its intention to site a subsidiary – C. Itoh Electronics – in Wimbledon, London (4).

In 1984, C. Itoh entered Indonesian west Sumatra to equip the Tanah Hitam open-cast coal mine (5).

In 1988, C. Itoh – now Japan's most successful trading house (6) – increased its holding in Mount Newman Mining Company, the iron ore mine in the Pilbara region of Western Australia, by buying out BP's 5% stake (7).

The ownership of Mount Newman is now: BHP 55%, Pilbara Iron Ltd 30%, Mitsui-C. Itoh Iron Pty Ltd 10%, and CI Minerals Aus-

tralia Pty 5% (8). C. Itoh also formed a JV in the late 1980s with the El Roble Exploracion SA, along with the Japanese mining house Nittetsu Mining Company Ltd, with a view to opening up a copper deposit in Colombia owned by Minas El Roble Ltd (9).

Together with Marubeni (16%), Sumitomo Corp (9.6%) and International Finance Corp (5%) it holds 6.4% of the Pasar copper smelting and refining company in the Philippines (see Marubeni).

And in 1989 it took 20% in Race, a leading Welsh electronics company, marking one of the "... first times a Japanese company has taken a strategic minority equity holding in a British concern" (10).

References: (1) *AMPO* (Japan) Vol. 13, No. 3, 1981. (2) *Pacific Imperialism Notebook* 5/73; *Age* 26/4/72. (3) *FT* 13/10/83. (4) *FT* 12/1/83. (5) *MJ* 18/5/84. (6) *FT* 27/5/88. (7) *MJ* 25/3/88. (8) *MIY* 1990. (9) *MAR* 1990. (10) *FT* 16/5/89.

138 Cleveland-Cliffs Iron Co

```
CLEVELAND-CLIFFS IRON CO. (USA)
                  |
                  | 30%
Cliffs Robe River Iron Associates (AUS)
```

This is primarily an iron ore company which, through subsidiaries, has part of the Robe River iron ore project at Pilbara, Western Australia, an Aboriginal area (1).

It was part of a JV with Getty Oil, Skelly Oil, and Nuclear Resources, exploring for uranium at Powder River, Wyoming (USA) (2). In April 1980 it began a uranium solution mining plant at Pumpkin Buttes, Wyoming, (3) together with Conoco.

Its uranium operations sustained a total loss in 1983 (4).

That year, exploration expenditure on uranium JVs was cut back – though uranium reserves in Wyoming were claimed to be in excess of 15,000,000 pounds (5).

Renamed Cleveland-Cliffs Inc in 1985 – one

```
                          C. ITOH (J)

Mitsui (J)      BHP (AUS)
          7%   85%        8%
          Mt Newman Mining Co. (AUS)
```

hundred and thirty-five years after its foundation – the company is the largest iron ore producer in the US to own the bulk of its reserves in fee (ie under licence) (6).

It is also the world's largest supplier of iron ore pellets (7). In 1988 it began restructuring its interests, selling its forest products assets, spinning off its oil and gas subsidiaries and hydroelectric facilities, and concentrating on its iron ore operations (7).

In 1987 it acquired Pickands Mather and Co, a subsidiary of Moore McCormack Resources Inc, with four iron ore mines in Canada, USA and Australia (8).

Cleveland-Cliffs holds around 2000 million tons of crude iron ore reserves, notably at its Tilden mine and Empire mine in Michigan, several mines in Minnesota, a mine in Ontario and one in Labrador, and the Savage River mine and pellet plant in Wynyard, Tasmania (9).

References: (1) *MIY* 1981. (2) *US Surv Min* 1979. (3) *MAR* 1981, p 357. (4) *MAR* 1984. (5) *MIY* 1985. (6) *MAR* 1990. (7) *MJ* 22/4/88. (8) *E&MJ* 1/87. (9) *MAR* 1990.

139 Cliffminex NL (AUS)

It is a gold and uranium exploration company (1) with some 22 uranium claims at Minindi Creek, near Karratha, Western Australia, where some mineralisation was reported in 1979/80 (2) and an *in situ* plant was being considered. The JV there had Keywest Explorations and Hawk Investments as partners (3).

It has also been engaged with Magnet Metals in exploration for uranium in the Eucla Basin (2) and at Chalba Creek, near Carnarvon, Western Australia (3).

The company also has interests in diamond prospecting on or close to Aboriginal land in the Kimberleys and at Lake Argyle (Smoke Creek) (4).

In 1980 it had a "stormy year" (1) when it was suspended from the Stock Exchange for failure to supply an annual report. It also came under

threat of delisting from the Stock Exchange when it held an extraordinary meeting after the Stock Exchange had demanded it be called off. The Minindi-Mooloo Downs project was put on a care and maintenance basis in 1982/83 after 432 tonnes of U_3O_8 were discovered (5).

References: (1) *Reg Aus Min* 81. (2) Annual Report of the company, 1980. (3) *West Australian* 14/11/78. (4) *Cliffminex Annual Report 1981.* (5) *Reg Aus Min* 83/84.

140 Climax Uranium Co

Now part of the Amax corporation, it is defunct as a uranium exploitation company. It was one of the formative members of the Trilateral Commission (1).

It was responsible for radioactive mill tailings left from its uranium mining at Grand Junction, Colorado over a 20-year period. The tailings were used as house building material in homes, schools and other constructions. More than six thousand buildings were contaminated, but – after a deadline for remedial state action passed in June 1980 – it seems only 740 home owners would qualify to have their properties "cleaned up". In addition, 2 million tons of tailings were on the site, to be shovelled in 1983 by the US Department of Energy "into someone else's backyard" as the TRCA Uranium/Nuclear Committee graphically put it (2).

Contact: TRCA Uranium/Nuclear Committee, POB 2932, Grand Junction, CO 81502, USA.

References: (1) Michael Garrity, *Trilateralism,* 1980 (Sklar). (2) Press release from TRCA Uranium/Nuclear Committee, Grand Junction.

```
    Amax Inc. (USA)
            \
  CLIMAX URANIUM CO. (USA)
```

141 Cluff Lake Mining Corp Ltd

Operator of the Cluff Lake mine, it is one of the world's most important (and richest) uranium producers. The Cluff Lake deposits in northern Saskatchewan were originally discovered by Amok Ltd, which set up Cluff Mining Ltd; joined in 1979 by the SMDC (1).

Although, formally, responsibility for operating the mine, and creating the consequent extremely dangerous wastes and radiation, lies with the Corporation, it is the Amok consortium which is universally identified with the project, as well as with further uranium exploration in the area. For this reason, a full account of the Cluff Lake developments are contained under the Amok entry in the *File*.

(*NB* Cluff Lake Mining Corporation is not to be confused with either Cluff Mineral Exploration Ltd or Cluff Oil.)

References: (1) *FT* 31/7/79.

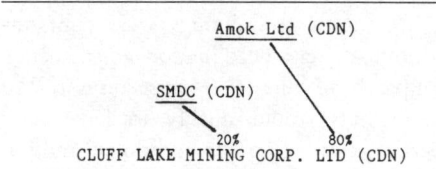

142 CMC/Continental Materials Corp

Primarily a construction industry company, it was part of a JV with Union Minière (45%), developing copper in the Santa Catalina Mountains near Tucson, Arizona (1), and had 45% interest in a JV in Lisbon Valley, Utah, (partners not known) which by 1979 had indicated 1,800,000 lbs of U_3O_8 (1, 2).

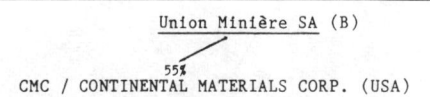

References: (1) *MIY* 1981. (2) *MJ* 16/3/79.

143 CNEA/Comision Nacional de Energia Atomica

Republic of Argentina (RA)

CNEA / COMISION NACIONAL DE ENERGIA ATOMICA (RA)

The CNEA oversees all nuclear development in Argentina. This entry does not go into any detail about the extremely contentious (and often scandalous) developments of the military dictatorship's reactor and reprocessing capacities (1). It is important to realise, however, that these programmes have to a considerable extent depended on CNEA's uranium provisions: the régime's first two reactors are fuelled by natural uranium (2).

The CNEA was formed in 1950 and put under control of the navy (2). In 1956, it took over all uranium exploration previously (since 1938, no less) performed by various other bodies (3). Within five years, its activities in the field had reached "significant levels" while its aerial scintillometry programme yielded "very satisfactory results". By the late '70s nearly 200,000 square miles had also been "logged" using geo-electrical methods (3).

Argentina's first large deposit of uranium, Don Otto in Salta province, was mined from 1962 to 1980 (2). During the '70s another two deposits were brought under production: Los Adobes in Pichinan-Chubut and Malargüe in Mendoza province (3).

Another plant, in the Sierra Grande, 70km west of Cordoba city, came on line in 1982. The Los Gigantes unit, Argentina's first mill built with private capital, is to be operated until 1997 by the firm of Eduardo Sanchez Granel y Asociados (9). Eduardo Sanchez continues prospecting for uranium around the 100 square km site; production is contracted to CNEA, initially at a cost of US$72/kg U_3O_8 (4).

By 1983 CNEA was admitting that Los Gigantes was not the dream-lode it had originally expected with uranium grades reaching 0.8% (9). In fact, its grade is closer to 0.03% (11). Production in 1983 was around 46 tons U_3O_8 from the Sanchez heap leaching ion-exchange plant.

Promising reserves were also under active exploration in the Cosquin area during the late '70s (4).

By 1982, CNEA had located more than 20 districts containing some 31 deposits of varying potential: 31,000 tonnes of reasonably assured resources were discovered, with another 15,000 tonnes in the "estimated" category (5). A 1980 reform of the 1887 Mining Code opened the door to foreign companies developing the country's mineral deposits (6); it stipulated that 75% of the uranium was to remain with CNEA, with the remainder going to the JV partner(s) (7).

Although a contract was supposedly signed in May 1980 with a French/Argentinian consortium to begin exploiting the main Sierra Pintada deposits (8) – at 24,000 tonnes the country's largest – it apparently took more than two years before details were worked out (9). By the beginning of 1985, the US$150million project was still looking for the go-ahead (10). In 1983 around 138 tonnes of uranium were produced by CNEA at its other small operations in the Salta and Mendoza areas (11).

CNEA's export contracts appear to have been confined to a deal with Brazil (a loan of 240 tonnes to Nuclebras between 1981 and 1982, to be repaid in the following two years) (12) and secret deals with Israel in the '60s (and perhaps beyond) about which almost nothing is known (13).

During 1986, Los Gigantes produced only 180 tons of U_3O_8, mainly due to problems created by the acidic effluent in San Roque Lake – the alarms accentuated by the fact that this is a major tourist area. Plans were also put in hand to develop the La Estela mine at San Luis, and a processing plant at Villa Larca (14). By the following year, the problems at San Roque had apparently been solved – at least to the satisfaction of the Cordoba authorities; the licence for the Sanchez Granel extraction and concentra-

tion plant was renewed (15, 16). New leases were also signed in 1987, with the province of Mendoza, for production from Sierra Pintada (16).

In 1988, the CNEA began to explore the possibility of using underground granite in which to store nuclear wastes: the first proposal of this kind in South America. The Swiss, West German and Canadian governments are most likely to dump their embarrassing waste in the third world country. The CNEA maintains plutonium-239 would be burned up in existing reactors (17).

Contacts between the CNEA and Brazil's CNEN in 1987 yielded a number of co-operative agreements between the two countries, including joint treatment of nuclear waste, the possible sale of Brazilian slightly-enriched uranium to its south American neighbour, and the sale of low- and medium-level enriched uranium in the opposite direction (18). Meanwhile Argentina's production of uranium in 1988 was 208 tonnes, of which around 75% came from the Los Gigantes, La Estela and San Rafael mines (19, 20). That year the feasibility study for Sierra Pintada was finally completed (20) with reserves of around 20,000 tonnes of contained uranium defined (19).

References: (1) See *Nuclear Fix,* Pringle & Spigelman, and – for a clear expose of the Nazi and West German government connection with Argentina's nuclear weapons strategy – *Newsnight Special,* BBC-TV, producer Robin Denselow, screened 19/4/82; also *FT* 22/9/83. (2) *Nuclear Fix.* (3) *Uranium Redbook* 1979. (4) *MM* 1/80. (5) *Uranium Redbook* 1982. (6) *MJ* 8/8/80. (7) *Nuclear Fuel* Washington, 28/4/80. (8) Nucleonics Week (Washington) 15/5/80. (9) *E&MJ* 12/82. (10) *E&MJ* 1/85. (11) *MAR* 1984. (12) *MJ* 29/8/80. (13) J.R.Redick: *Military potential of Latin American nuclear energy programmes.* USA, place unknown (Sage Publications) 1972. (14) *MAR* 1987. (15) *MJ* 19/12/87. (16) *MAR* 1988. (17) *FT* 8/6/88. (18) WISE N.C 279, 18/9/87. (19) *Greenpeace Report* 1990. (20) *MAR* 1989.

144 Coastal of Saskatchewan (CDN)

It was reported uranium exploring in Saskatchewan (1). No further data, although it is possible that this company is a subsidiary of Coastal Mining Co, which is a subsidiary of MA Hanna, the diversified resources company based in Ohio, which was taken over by Norcen in 1982.

References: (1) *Who's Who Sask.* (2) *MIY* 1990.

145 Cobb Nuclear Corp (USA)

Cobb ran three uranium mines in the San Juan Basin area – the New Mexico part of the Four Corners of Navajo/Hopi/Jicarilla/Pueblo/Ute native American reservation land – in the USA during the 1970s. These were Section 12, Section 14 and the Spencer Shaft. The company was also planning to sink an underground shaft at Section 10 in the Ambrosia Lake area (1).

In 1981 Cobb was a rising uranium company, reporting earnings of over one million dollars. The impact of the Three Mile Island blow-out killed its nuclear prospects (2).

Cobb bought into oil and gas in 1980, and later – with Houston Natural Gas – decided to turn to gold and re-open an old mine, the London, in Colorado.

References: (1) *San Juan Study.* (2) *AFR* 30/3/84, quoting report from the *New York Times.*

146 Coboen/Comisión Boliviana de la Energía Nucleara

Coboen discovered the promising Cotaje deposit on the western edge of the Cordillera de los Frailes, Potosi department, in 1970. Other deposits – notably Amistad and Los Diques – were being explored by the early '80s (1).

Mining was announced as beginning "soon" at Cotaje in 1981 (2) but little has been heard of the plan since. A little afterwards, the government announced that Coboen would start a five-year exploration plan, with possible JVs with outside companies (3). This appears to have been extended later (4). Taiwan was also planning to be involved (3).

In 1974 Coboen signed a contract with Agip-Uranium to prospect for, and possibly mine, uranium over an area of some $50,000 \text{km}^2$. Uranerz also got in on the act at around the same time (5).

References: (1) *Uranium Redbook* 1979. (2) *MM* 4/81. (3) *MJ* 31/7/81. (4) *Uranium Redbook* 1982. (5) *Taz,* Berlin, 10/12/81.

147 Cocks Eldorado NL (AUS)

It is part of a consortium with Offshore Oil and Southern Cross Exploration, joining AGIP (Australia), Urangesellschaft (Australia), and Central Pacific Minerals, at Lake Ngalia, Northern Territory (I).

However, Cocks had withdrawn by 1983 (2).

References: (1) *Reg Aus Min* 1981. (2) *Reg Aus Min* 1983/84.

148 Codelco/Corporacion nacional del cobre de Chile

Codelco "dominates the Chilean economy" (1). As more than a hundred opponents of the Chilean junta lay dying in the streets in 1983, the country's copper corporation was dishing up nearly US$700million in taxes to the régime out of a total revenue of nearly two billion US

Republic of Bolivia (Bolivia)

COBOEN / COMISIÓN BOLIVIANA DE LA ENERGÍA NUCLEARA (Bolivia)

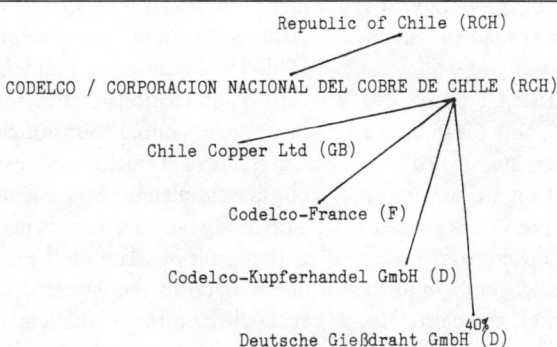

Republic of Chile (RCH)

CODELCO / CORPORACION NACIONAL DEL COBRE DE CHILE (RCH)

Chile Copper Ltd (GB)

Codelco-France (F)

Codelco-Kupferhandel GmbH (D)

40%
Deutsche Gießdraht GmbH (D)

dollars (2). By far the western world's biggest copper producer (Codelco's production in 1983 outstripped all US copper producers put together) (2), the company concentrates production on four mines, the biggest of which, Chuquicamata, was originally developed by Anaconda (a company now absorbed into Arco/Anaconda) (3). The El Teniente mine – also located in the indigenous Andes – produces 30% of Codelco's output and is the world's largest underground copper mine (3).

In 1981, a leading spokesperson for CISA (the recently formed Consejo Indio de Sud America) accused the Chilean junta of introducing draconic new landed property laws to destroy the Mapuche as a people, forcing them to uproot "... because there's a bunch of uranium underneath Mapuche land. At Temuco there was a new discovery only a few months ago" (4). Uranium exploration began in Chile in the early '70s under the auspices of the Chilean Nuclear Energy Commission (CCHEN) with help from UNDP and IAEA (5). Up in the north, in the Chuquicamata region, a pilot plant for the extraction of uranium from copper started up in 1977, and a uranium deposit was scheduled to be mined at Cascada in the early '80s (5, 6). Nothing has been heard of this project since (7). In 1985, pollution levels from atmospheric arsenic in the housing areas of Chuqui near Codelco's Chuquicamata mine, were so high that they "very nearly forced the evacuation of the women and children to the town of Calama, 18km distant" according to the *Mercurio* newspaper of Santiago.

Physicians at the mine's Roy H Glover hospital were quoted as saying that, even when a lower arsine assay prevailed, smelter fumes would blow out over the snow-covered mountains resulting in arsenic contamination of drinking water. Chuqui air measurements show 3.84 milligrams of arsenic per cubic metre, as compared with the 1.35mg/m^3 recommended by the American Conference of Government Industrial Hygienists. Dr Sergio Stoppel, Codelco's director of the hospital, while claiming that there had been no chronic arsenic problems in patients at the hospital, had himself signed chemical assays which showed that drinking water samples in more than 90% of the cases contained a greater quantity of arsenic than that allowed by international standards. (This standard is 0.05 mg/m^3 – although the Chilean arsenic standard had itself been set by the government at more than double this: 0.12mg/m^3.) Dr Stoppel also conceded that cancer rates in the non-working population of Chuqui were higher than normal and had resulted in special cancer check-ups (no mention of cancer rates among workers, or whether their families had been specifically checked.

Codelco claimed that it was spending US$155 million at Chuqui to ameliorate the situation, including the installation of a converter, and possibly installation of a flotation (or similar) process to reduce contaminants in the concentrate prior to smelting. Observers had noted that, though the Unions had no legal right of strike, the health conditions "might be used politically to justify labour unrest which would

201

be fuelled by the perception of the workers that their real salaries have been eroded to half the traditional level by their inability to strike" (8). As predictions were made that Codelco's production would peak in 1991, only to fall inexorably thereafter (9), and miners at Codelco's Salvador copper division went on strike – representing the first copper workers' industrial action in six years (10) – Chile went to the polls. At the end of the year, it said goodbye to the murderous régime of Pinochet, and elected Patricio Aylwin as President. Since then, the place of Codelco in the Chilean economy has been the subject of enormous debate both inside and outside the country. The *Mining Journal* (11) predicted that *plus ça change plus c'est la meme chose* ("Chilean mining never had it so good as under the 16 years of military rule ... some feared that Codelco's dominant role in the industry would be undermined by the entry of mining multinationals and there was talk of renegotiating some of the copper projects. But since Mr Aylwin's victory not a word has been heard on the subject") (11).

However, with Alejandro Noemi taking charge of the world's largest copper concern (with more than 90% of Chile's output under its belt) (12) considerable changes are promised for Codelco. Noemi has gone on the record as opposing the mandatory handing-over of 10% of Codelco's income to the military (13), urging the cutting-down of expansion, the laying-off of administrative workers and older miners, and improving efficiency. (Said the *Financial Times*: productivity per worker at Codelco's mines is only half what it is in the USA, while absenteeism rates are three times the Chilean average) (13, 14).

With rock bursts at El Teniente (15), a fire at Chuqui (15), declining ore grades, rising costs, and "thorny labour relations" (13), a legacy of the draconian labour laws under the Junta (11), the task of reforming Codelco must be reckoned one of the most gargantuan in the world mining industry. Noemi has publicly acknowledged that Codelco's previous strategy "... ignored costs, exploitation methods, the environment and Codelco's long-term develop-

ment", but so far little seems to have been done: 20% of the total investment over the years 1990-1994 has been set aside to bring Codelco "up to international standards" (13) – which means overhauling (but not closing down) the Chuquicamata smelter and installing a new sulphuric acid plant at El Teniente. A new tailings dam has already been constructed at El Salvador as the result of a law suit brought by fisherfolk of Chanaral in the Atacama region, after they accused the mine – which threw its waste into the sea – of creating the build-up of an artificial beach 5km long and 1.5km wide, which destroyed most of the Chanaral Bay's marine life (13).

References: (1) *MAR* 1982. (2) *MAR* 1984. (3) *E&MJ* 11/84. (4) *Natural Peoples' News* No. 6, Winter 1981. (5) *Uranium Redbook* 1982. (6) *Nuclear Fix*. (7) See eg *E&MJ* 1/85. (8) *MJ* 13/12/85. (9) *FT* 8/8/89; *MJ* 11/8/89. (10) *MJ* 22/9/89. (11) *MJ* 9/3/90. (12) *MAR* 1990. (13) *FT* 10/8/90. (14) See also *MJ* 20/7/90. (15) *Metal Bulletin* 12/7/90.

149 Codesa/Corporacion Costarricense de desarollo

Along with Costa Rica's Atomic Energy Commission (CEA), it has shown interest in uranium prospects among various igneous intrusions and volcanic assemblies on the territory of Costa Rica, but among which no significant uranium has been discovered – "so far is known" (1).

References: (1) *MAR* 1984.

```
                        Republic of Costa Rica (CR)
                                   /
CODESA /                          /
    CORPORACION COSTARRICENSE DE DESAROLLO (CR)
```

150 Cogéma/Cie générale des matières nucléaires

Created in 1976, from the Production Division of the French Atomic Energy Agency (CEA), Cogéma is now not only the world's biggest single supplier of uranium, but the only company on planet earth which offers every single stage of the nuclear process, from mining to spent fuel reprocessing (1, 2). It is the world's leader in handling and reprocessing spent nuclear fuels (3). While it does not actually build nuclear power stations, it has a strong nuclear engineering wing. While it is not known to construct nuclear missiles, its expertise is essential to the French nuclear bomb programme. For Cogéma presides over both the military and "civil" aspects of France's nuclear state: the company's reprocessing facilities at Marcoule are used for both military and non-military work (4).

Cogéma's importance as the world's leader in supplying electricity services across the whole nuclear fuel cycle can be gauged from the fact that, in mid-1986, it was servicing around 130 of the western world's operating reactors, with an installed capacity of more than 90 gigawatts: this is almost *half* the *global* total of approximately 200 GW (2). In the year 1985-86 alone, it added another 15 reactor clients to its books (5).

In addition, Cogéma has access to around 7500 tonnes a year of uranium in concentrates, the majority of which it supplies to Elecricité de France (EDF) and the French weapons programme, and a further quarter of which is shipped to foreign customers (2, 6). Although, as the uranium price began to fall in the early 1980s, the company cut back on its production slightly (2500 tonnes domestic production in 1984 reduced to 2450 tonnes the following year; 5600 tonnes from overseas mines in 1984, falling to 5250 tonnes in 1985)

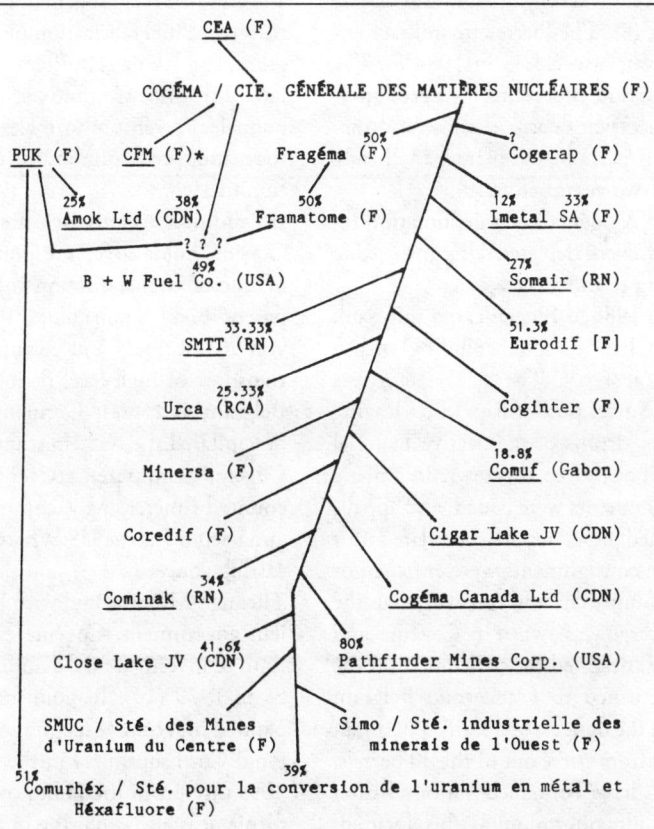

(7), Cogéma, with an assured home/government market, has been less affected by market prices than most other major suppliers. With its acquisition of CFM/Mokta in 1985, and expectations that its very high-grade deposits in northern Saskatchewan will proceed to production in the 1990s, the company's future – barring another Tchernobyl on its own doorstep – is likely to be rosy for some time to come.

Cogéma employs some 15,000 people in France and abroad (2). It converts uranium into UF_6 at the Malvesi and Pierrelatte plants operated by Comurhex. Pierrelatte is also the site of an enrichment plant using gaseous diffusion, completed in 1967. At Tricastin, Cogéma is the main partner in the Eurodif consortium – along with Belgian, Spanish, Italian and Iranian co-partners – in what is the largest enrichment plant in the world. This plant started up in 1979, reaching design capacity three years later, and in theory could supply some one hundred 900MW light water reactors (2).

However, by 1984 it was only operating at half its projected output (8). The Pierrelatte military enrichment plant was also closed in 1984 for five months – the first time in seventeen years of operation – to cut electricity costs: the low and medium enrichment parts of the plant had already been shut down two years before (9).

At Tricastin, too, Cogéma has a defluoration facility, used to convert depleted UF_6 into oxide products, for long-term storage.

On August 25th 1984, a French cargo ship sank in the English Channel, after colliding with a West German car ferry. The *Mont-Louis* was travelling to the Soviet port of Riga, after leaving Le Havre. Thirty drums of radioactive material plunged to the bottom of the sea. After many contradictory statements were issued, and appallingly misinformed news reports circulated (for example that the consignments were enriched or reprocessed uranium), it was revealed that the majority of this cargo was owned by Cogéma, and consisted of UF_6, destined for enrichment in the USSR, to be returned to France and Belgium (10). Apart from the dangers of ocean contamination – expecially from three out of the 30 barrels, which appear to have contained various fission products, including plutonium – this incident

served to reveal further aspects of the uncontrolled nuclear trade, of which Europe is the dead-centre. For example, Greenpeace Ltd revealed at the time that uranium mined in Canada, no doubt partly by Cogéma, is enriched in the USSR, made into fuel rods in the USA, sent to West German reactors, the used fuel reprocessed at Cogéma's Cap La Hague plant, and plutonium extracted after reprocessing, sent to the USA, no doubt for use in nuclear weapons (11). The London *Guardian* also revealed that shipments of UF_6 which go on a regular basis to the Eurodif plant and the Urenco plant at Almelo (Netherlands) return afterwards via Sealink, the British ferry-line, for fuel assemblage (12).

Cogéma is involved in fuel fabrication through a number of companies, notably SCN which provides assemblies for gas-cooled reactors and also produces artificial industrial diamonds (2), and Fragéma, which designs and builds light-water reactor fuel assemblies (2). Along with Pechiney and Framatome, Cogéma operates FBFC, which runs three fuel fabrication plants – at Dessel, Belgium, and Romans in France, as well as at Pierrelatte-Tricastin. Commox is a joint Cogéma-Belgonucléaire venture to make and sell MOX fuel rods, and recycle plutonium, due to enter production in 1990 (2).

In addition, Transnucléaire (Paris), in which Cogéma holds 24%, is responsible for a large part of nuclear transportation in Europe: it was the parent body Transnuklear (TN), that was involved in bribery and corruption and the illegal transport of high-level fission products between Belgium and West Germany, whose revelation precipitated the Nukem scandal of 1988 (13).

Cogéma's computer services company CISI is accounted "increasingly commercially aggressive" and itself owns the US Wharton Economics forecasting concern.

Through Minersa, and with BRGM and the Malian government, Cogéma established a consortium to investigate the diamond deposit at Keniéba in 1979 (14). Its gold interests have also expanded in recent years: the company's Pathfinder Gold Corp subsidiary in the USA has acquired a 30% interest in the Jamestown gold mine, California: it is also exploring in Spain (between Madrid and the Portuguese border), Ecuador (15) and Australia (7). The company's gold explorations in central France received a boost in 1988, when it took over Penarroya's gold interests in an area which Cogéma has exploited only too well – Limousin (16) (see below).

Cogéma was also one of several multinationals prospecting in the region of the proposed Mazaruni dam, on Akawaio (Indian) land in Guyana, during the early 1980s (88).

In Zaïre, in 1981, Cogéma headed the SMTF consortium, to investigate the Tenke-Fungurame copper deposits in Shaba province along with Anglo-Charter Consolidated, Mitsui and others (17, 31).

Cogéma has also dabbled in coal – as a partner in the Bridge Oil JV in Australia (18); although a link-up with Denison in Canada, discussed in 1982 (19), seems not to have been realised.

Another Cogéma "diversification" venture fell on stony ground in 1988, when the company lost US$44.2 million in investments in the futures market, thanks to fraud (so Cogéma claimed) by a subsidiary of its stockbrockers, Buisson (20).

Since 1974 and the oil price rise, the French state has enormously increased its reliance on nuclear-generated electricity – from only 8% in 1973, to 23% in 1983, and 66% just two years later. The intention, confirmed when Mitterand came to power in 1981 (21), was to increase the country's dependence on nuclear power to 75% by 1990 (2).

Exploitation of domestic uranium has been a crucial part of this galloping nuclearisation, and it is Cogéma which – to quote the *Financial Times* of London – provides "... the centrepiece of the ambitious ... programme" (22).

In 1986, Cogéma's three domestic mining divisions, at Vendée, Herault, and – most importantly – at La Crouzille, had a combined capacity of 3050 tonnes a year of uranium in concentrates (2).

Cogéma held by then 85% of the country's uranium reserves and operated no fewer than 53 exploration projects, covering an area of 4600 square kilometres (2); in 1980, this was the third largest such programme after the USA and Canada (23).

It was forty years earlier that uranium exploration commenced in France, and the La Crouzille deposits were located (24). By 1955, the Limousin, Forez, Vendée and Morvan deposits had been discovered (24). Within ten years, domestic production reached 1600 tons of U_3O_8 a year (25). In 1975 it had increased to 2080 tons U_3O_8, with a capacity of 2300 tons (25). The following year, the French government stepped up financial inducements to exploration (US$4.3-US$6.5 million), and by 1978 production was nearly 2000 tonnes (25), with five processing plants in operation, of which Cogéma (and Cogéma through its subsidiary Simo), operated four (at Bessines, Escarpière, St Priest, and Langogne) (25). The St Priest plant closed in 1981, to be used as a waste dump (26), and Cogéma/Simo opened up a new one at Lodève (Herault).

Uranium concentrate production increased to nearly 3000 tons in 1982 (2859 tons), and passed the three thousand mark the following year (3271 tons). By this time, Cogéma was accounting for 2495 tons of the output, and exploring in a sizeable programme (FF250 million) in the Armoricaine and Massif Central regions (27).

Of the 8560 workers employed by Cogéma in France at that time, 3000 worked in the mines, and the remaining five thousand in processing, enrichment and fuel fabrication (28).

Uranium in France derives from a fairly large number of small mines in several key areas. Without a detailed map to hand, or a more than *touristique* knowledge of central France, it is easy to confuse the mines (as this author has often done!). The following accounts summarise what is generally known about Cogéma's operations in its three divisions.

La Crouzille – Haute Vienne – Limoges

By 1980, some 10% of the Limousin region had been put under uranium exploitation: with thirty mines opened since 1949, covering some 300 square kilometres, the area of the Haute-Vienne in particular has taken on (as researcher Dave van Ooyen so aptly put it) "the form of a Gruyère cheese" (21).

Total underground workings stretch one hundred kilometres. In 1982, out of a total of eleven working mines, there were two working open-pits, at Vanacht and La Fraisse; a third, at Magnac, had been the subject of a bomb attack in 1973. Cogéma's own exploration permits at the time totalled 360 square kilometres.

Between 1949 and the opening of the Henriette mine, around 20,000 tons of uranium came out of this patchwork of mines. "Rehabilitation" of the workings, once exhausted, is minimal. One of the few mines, observed by van Ooyen, to be officially rehabilitated is the Chanteloube, near the Paris-Limoges main road (N20), which closed in 1979 after producing 270 tons of yellowcake. Out of 2,700,000 tons of ore, the vast majority (2,400,000) has simply been left by the roadside – on which a recreation park has been built (21), including six windmills as an energy exhibit (29)! By 1986, the La Crouzille mining division held 23 mining and exploration permits (a slight increase on 1982) (2). The exploration and exploitation area had also increased to more than 800 square km (813 square km) within a radius of about 20km from the Bessines concentrator operated by Simo 35km north of Limousin (2, 22). Four of the mines in the division – at Fanay Belezane, Margnac, Le Fraisse-les Gorces – account for two-thirds of production.

With a workforce of 1150, Cogéma's operations dominate the district and have profoundly affected the local economy. Although one worker was quoted (quite understandably) in 1983 as saying "The mines keep the local economy going – sub-contractors, and shops and small businesses" (22), Dave van Ooyen's own research and observations in the area at around this time confirmed that the region was being drained of people, though it was difficult to conclude that uranium mining was primarily to blame (21).

In La Crouzille, tourists can find a Uranium Service Station, as well as an Hotel Restaurant Uranium (21). In Limoges, they can sup water taken from the three lakes where Cogéma dumps its tailings – though local people are forbidden to piss there (29)! If they are "sporty" types, they can fish in the Gartempe, which has been "badly polluted by sulphuric acid" from the Bessines concentrator (29). The Haute-Vienne is a centre for

tourism, with no fewer than 17 camping sites (as of 1982), within the mining concession area, served by water from there (30).

At Razes, too, Cogéma has sited CIPRA – a training centre for foreign prospectors and technicians (21).

Vendée – Loire-Atlantique – Deux-Sèvres

The Vendée division of Cogéma – with a concentrator at L'Escarpière (25) – has been active for more than 30 years (31), with reserves at more than 25,000 tons uranium, and production (by 1985) of some 10,000 tons (31). Most of the uranium lies south-east of Nantes, though some occurrences are found near St Nazaire. The uranium zone extends over 200km along the Massif Armoricaine, and covers the *départements* of Loire-Atlantique, Maine-et-Loire, Vendée and Deux-Sèvres, with principal underground mines at L'Escarpière, le Chardon, la Commanderie (32) and Piriac. Surface mines are at Roussay. By the mid-'80s, this division was providing around one-fifth of France's production and 10% of its domestic needs (31).

More recently there has been investigation of underground deposits in granite below 300 metres (2).

Hérault – Lodève

This division employs some 400 people, and in 1985 produced 945 tonnes of uranium in concentrates (2). Initial reserves in the Hérault area were estimated at 16,000 tonnes with an assay reaching 4kg/tonne uranium, mostly underground. Surface reserves are nearing depletion, although in 1986 Cogéma expected to define sufficient underground reserves to continue production for at least another ten years (2). Uranium was first discovered by CEA geologists in 1957, but a decision to develop was not made until OPEC began organising the oil cartel in the early 1970s (2).

The Hérault ore is processed at Simo's Lodève concentrator (33).

Although there is a general impression that French uranium production proceeds with little domestic opposition, there have been notable acts of resistance over the past few years.

In 1981, more than half those who voted for the ecologist Presidential candidate, Brice LaLonde, came from Haute-Vienne where several local organisations – notably Amis de la Terre, Fédération Limousin d'Étude et de Protection de la Nature (FLEPRA), the Association Protection d'Monts Ambazac (APMA), the Association du Mouvement de Grandmont, and CLAN – have been active since the 1970s. The Grandmont group in 1981 blockaded uranium transports coming through their commune, until the barriers were removed by police (21). At Limousin in 1980, CLAN campaigned against Cogéma's test drilling outside Petits Vaux (Auriat) – drilling which might either have been for a dump for waste from La Hague, or for uranium (34).

Villagers in Hérault started resisting uranium mining in 1980, especially around the St Jean de la Blaquière mine. In May 1984, blockades were established to prevent Cogéma (illegally) gaining access to a strip of land the company had purchased from a farmer. The company broke through the barriers, and managed to commence drilling. A week later, the blockade resumed – involving some sixty people with tractors. The police dispersed the demonstrators using teargas (35), and some 300 workers belonging to the CGT (communist pro-uranium mining) union, invaded the land. As a French non-violence paper declared at the time "CGT – Cogéma – CRS [the French "terrorist" police squad] – "même combat" (36).

Other examples of local resistance have come from the Ardeche where Friends of the Earth (Amis de la Terre) campaigned in the early 1980s against research permits granted at Creyseilles, Montselques, and Gravières (34), and at Glomel, Brittany, where 700 people, mostly local farmers, started a petition against mining, addressed to the local government, and signed by more than 1800 people (37).

But the most remarkable of Cogéma's uranium interests are those centred on Canada, where the company has managed to obtain key stakes in two of the most promising uranium mines ever proposed: Cigar Lake, and Close Lake. Its important

38% stake in Cluff Lake (see Amok) meant that it received an equal share – with an un-named German utility – of uranium from Phase One of the mine (38).

The Cigar Lake Mining Corporation was formed in May 1985 by SMDC, Cogéma and other companies (Cogéma Canada, 32.625%, and a Cogéma subsidiary Corona Grande, 3.75%), by which time its fabulous reserves were estimated at 110,000 tons, with an average grade of 12-14% uranium, and 40,000 tons with an average grade of 4% (14). A revised estimate a year later earmarked some 385 million pounds of U_3O_8 at an average grade of 9% (15). Sections of uranium mineralisation on the prospect have reached the highest consistent grade ever recorded, – up to 37.7% U_3O_8, though "pockets" as high as 60% have also been recorded (40). In 1985, the chair of Cogéma, François de Wissocq, reasonably dubbed Cigar Lake Cogéma's "mine for the nineties", predicting that it would come on stream at a time when prices would be rising (41). At the time of writing, Close Lake is a prospect which is still yielding drilling results – though they are very encouraging: with grades of up to 12% U_3O_8 (42). Cogéma Canada Ltd discovered and operates this prospect, holding 41.6% of a JV in which other partners are Imperial Metals Corp, Uranerz Exploration and Mining, Geomex Minerals Inc, and the SMDC (43).

In late 1986, Cogéma submitted a proposal to the Saskatchewan DoE (Department of the Environment) for a US$50 million underground exploration programme at Close Lake, conjecturing that the deep shaft could be ready by 1988 and testing underway the following year (15). By 1989, environmental approval for this phase had not yet been granted.

Through Seru Nucléaire (now Cogéma Canada), the company has also prospected in the past with Eldorado and SDJB (25).

It is important to note that, although Canada has had a long standing theoretical opposition to supplying uranium which might end up in nuclear weapons, thanks to the involvement of Cogéma and CFM/Mokta (now, as mentioned, controlled by Cogéma) in the Canadian uranium industry, uranium has been regularly exported to France, where it could end up in the Super-Phénix Fast Breeder Reactor and thence nuclear weapons (44). Cogéma has an office in Montreal which appears to handle this trade: the Montreal Uranium Committee has mounted actions against this export (45).

In Gabon, Cogéma's prime interest is in Comuf, a company producing an average of 1500 tonnes uranium a year, about 75% of which is from underground mines: 1985 production was just under 1000 tonnes (940t) (2). Cogéma, through its new holding of Mokta, also has a 17% interest in Comilog, the country's major manganese producer which is one of the world's biggest providers of this mineral (46). Comuf works four deposits in Gabon – at Oklo, Boyindzi, Mikoulougou, and Okelo Bonga. The vast majority of their output (some 68%) goes to Cogéma, through its subsidiary Coginter (the remainder being marketed in Belgium and Japan by Uranerz) (15).

From the early 1980s Cogéma, joined in a consortium with the Gabonese government and PNC of Japan, has been involved in extensive exploration of the Bakoue region (46). Another consortium was founded in 1980 between the Gabonese government (Nord Leyou Gabon), Cogéma and Kepco of South Korea, to explore the Lastourville region, north of the original Mounana deposit (46, 47). Yet a third exploration agreement was signed in 1985, between the government and a consortium grouping Cogéma, Comuf and UG of West Germany, to explore the Haut-Ogoue region with a budget of up to FrCFA20,000 million. At the same time a JV between Mokta and Comuf began drilling in the Franceville, Akieni and Koula-Moutou regions (48).

According to Thomas Neff, because of the very large role Cogéma plays in Gabon, "... it is appropriate to regard Gabon's uranium activities as essentially part of the French system" (38). This was particularly true before 1978; when the yellowcake plant commenced production at Mounana – previously all concentrates had been shipped to France. In the past few years, Cogéma has reduced its demands on Comuf and, according to Neff, "sought to hold production down," despite

a government policy of promoting production in order to maximise revenue.

Cogéma's other main source of African uranium is Niger, a country whose economy continues to be dominated by uranium, and whose uranium industry continues to be dominated by French interests, primarily those of Cogéma (notwithstanding reductions made in Cogéma's share of Somair, under pressure from the Kountche régime in 1975 (38); Cogéma also sold part of its initial share in the other major prospect at Akouta to Enusa).

Cogéma has consistently taken more of the share of uranium prodution from the two mines than its shareholding would indicate, although Cogéma during the 1980s – as in Gabon – has been trying to reduce its dependence on African resources (38). Total sales of uranium from Niger for delivery between 1983 and 1990 amount to some 33,000 tons, of which Cogéma and other French recipients were due to receive around 21,000 tons (15). In addition, both in 1981 and 1982, France purchased additional tonnages from the Onarem stockpile, viewing it as "aid" rather than a commercial transaction (38).

Cogéma is involved in other uranium exploration in Niger (49), including Afasto-Ouest and Akborun-Azelik – the former explored by Cogéma in equal (37.5%) partnership with Onarem (15).

However, only two ventures have any realistic chance of development: Arni and Imouraren. The Arni deposit was delineated by thorough drilling in the late 1970s by SMTT (Ste Minière de Tassa N'Taghalgue), established by Cogéma and Onarem on a 50/50 basis (15, 38). Initial plans were to start production from this deposit in 1985, with full capacity in 1986 (50). Production was deferred and it looks unlikely that this deposit will proceed to commercial exploration in the near future. In 1981, after Cogéma began searching for a third partner at Arni (51), the Kuwait Foreign Trade and Contracting Co joined SMTT as an equal partner (15).

The Imouraren deposit is much larger than Arni, but original plans to develop it have been shelved indefinitely (15). Cogéma and Conoco originally held 35% each in this project, with Onarem holding 30% (38). After seeking a buyer for its share (52), Cogéma disposed of its holding to the other partners soon afterwards.

Cogéma's other exploration in Africa can be outlined as follows: in Senegambia, where exploration started in the 1950s in Senegal's oriental region with no tangible results (53), several areas continue to be examined, the most promising being at Saraya (15).

In Mali, the company is prospecting at Kenieba, Taoudeni and Hoboro-Douentza (15), and was formerly exploring with PNC of Japan (qv).

Exploration in Mauretania, where Cogéma held 10% of two prospects with Minatome, was suspended in 1982 when the war in the Western Sahara forced it to a halt (54).

Cogéma and Alusuisse formed URCA in the Central African Republic "to consider developing some 17,000 tons of [uranium] reserves" (14). Due to "infrastructural problems" and depressed prices (15), the project has lain dormant for some time (14).

Although most of Zambian uranium exploration – fairly extensive as it is – has been in the hands of PNC of Japan and Agip of Italy, Cogéma joined with Saarberg Interplan and Agip in 1980 to prospect the Gwembe area in the south of the country (55) – a project which was announced as extending to the western province, in 1981 (56). Exploration was continuing in the mid-'80s (31).

In 1981, the French and Moroccan governments announced an agreement under which Cogéma would provide the north African despot, Hassan II, with a plant for the extraction of uranium from the country's huge phosphate resources (increased considerably after the country's invasion of the Western Sahara) (57).

However, in 1984 New Scientist reported that Westinghouse was "first in line" to build the Moroccan plant, and meanwhile, although Soframtome and Framatome had won main contracts to provide a nuclear centre and power reactor to the régime – the British government was also co-operating (58).

In 1978, Cogéma signed an exploration agreement with the government of Guinée, along with its familiar partners in Africa, PNC and Agip (59), and it has also (inevitably) explored in Algeria (24).

In 1977, the French government, through the CEA, secured an agreement with South Africa, under which Randfontein Estates (JCI) would supply a major part of France's uranium needs in return for help in financing major gold and uranium development programmes. A later report – no doubt inspired by Cogéma, at a time when South African imports were coming under close scrutiny – suggests that Cogéma's Randfontein contract, for between 900 and 1000 tons a year, was actually signed in the early, rather than late, 1970s (60). In any event, it does seem that uranium imports were falling off by the early 1980s (41), although Cogéma's claims that it was "in no way dependant" on South African supplies, alongside its protestation that "the contract was a strictly commercial affair and we will honour it" (60), seem tinged with hypocrisy.

In early 1988, a Green MP from Luxembourg, Jup Weber, was leaked documents from Nulux, a subsidiary of Nukem of West Germany, showing that EDF – the huge French electricity utility which receives the vast majority of its uranium from Cogéma (9) – had drafted a contract for the delivery of 3745.5 tonnes of uranium hexafluoride to an un-named party which, it was later revealed (61), was South Africa itself. Nulux – 30% owned by RTZ Mineral Services – initially signed the deal, and there are some grounds for believing that the uranium concentrate might (at least partly) have been contracted from Rössing or Palabora mines. However, the conversion work for the uranium under this contract, covering 1980 to 1993, was to be undertaken by Cogéma's Comurhex subsidiary (62) and, since Nulux withdrew before it was implemented, it is likely that Cogéma would have provided the uranium instead of Nulux (63).

Comurhex has also long been involved in handling uranium from Namibia, although Cogéma has no known direct interests in the formerly South African-occupied country. As long ago as 1980, a Comurhex official agreed that Namibian uranium was being mixed with other source material at the Malvesi conversion plant – this in response to a request from Sweden that nuclear fuel received by the Scandinavians should be clearly "identified" (64). Again, as one of the spin-offs of the Nukem scandal which broke in 1988, anti-nuclear activists in Europe and Britain revealed that Comurhex was involved in complex "swapping" arrangements whereby the origin of uranium consignments have been concealed or illegally altered. In particular, Comurhex was at the centre of a "swap" of South African- or Namibian-origin uranium due to enter Liverpool docks, England, in 1988, when it apparently expressed willingness to provide a document showing that the yellowcake came from Canada, rather than the apartheid state (65). Shortly afterwards, Jup Weber, the Luxembourg Green MP, discovered documents implicating Comurhex in a similar title "swap" (66).

Elsewhere in the world, it is worth noting that Cogéma has, since 1969, been exploring for uranium in Indonesia, under the auspices of its subsidiary GAM. "Favourable prospects" in Kalimantan and Sumatra were claimed by 1978 (67). Those in West Kalimantan (formerly Borneo) were thought to be especially rich in 1980 (68), although a later report refers only to "traces" of uranium in the region (69).

France's main uranium prospecting partner in Indonesia during the late 1970s was West Germany. At the same time, the two governments were holding talks with the government of Guyana, with a view to prospecting the South American state (70). Cogéma secured prospecting rights in 1979 (71, 72), and apparently signed an agreement – along with Grundstofftechnik – covering the disposal of wastes from mining three years later (73). The most promising finds were, apparently, in the Essequito region, but the search was abandoned at the end of 1985, due to falling uranium prices (14).

In 1984, Cogéma, along with Pechiney, held talks over the Itataia uranium deposit in Brazil, with a view to Cogéma mining, and its French partner building, a pilot processing plant (74) at Santa Quiteria (75). Agreement was reached the following year (14).

Through Afmeco/CEA, Cogéma has long been exploring in Australia (25), and in 1988 it took up 1.25% in ERA, operators of the Ranger uranium mine (76). ERA in early 1988 signed a long term contract with EDF (77).

In the USA, Cogéma's major uranium interests are held through Pathfinder. The company's operations were cut back in the 'eighties but plans were afoot to exploit rich deposits in the Grand Canyon during the late 1980s and beyond (89). In the 1970s, Cogéma also explored the USA through its subsidiary Framco (25). The company produced some 300 tonnes of uranium concentrates at a profit of US$1.2 million in 1985 (2).

Cogéma's penetration of US nuclear fuel services gained a fillip in 1987 when, together with Framatome and Uranium Pechiney, the company took up 49% in B and W Fuel Co – a subsidiary of Babcock and Wilcox, a company with between 10% and 15% of the market share for pressurised water reactors in the USA (78). B and W Fuel took over Babcock and Wilcox's 400 tonnes a year fuel fabrication plant at Lynchley, Virginia, and was expecting to diversify downstream into the handling and reprocessing of nuclear fuels (78).

As already mentioned, most of Cogéma's uranium goes to EDF, the French state electric power company: in 1983, some 6000 tonnes with another 1000 tonnes destined for military enrichment (22). But Cogéma is also a fairly important foreign supplier: its direct customers including Japan, Sweden, West Germany, Belgium (22) and South Korea (38). In 1982, a contract was concluded with Taipower of Taiwan, to provide 2000 tons of uranium (79).

Thomas Neff, in his major study of the international uranium market, has looked at all available figures for French imports and exports from the early 1970s until 1985 (and projected into the beginning of the twenty-first century). His conclusion is that, while between 1976 and 1983 "net annual procurements were very well matched to domestic/annual requirements", supply would exceed demand thereafter. But, when allowances were made for military requirments, and unless there were major failures to meet domestic growth targets, the major problem might be not excessive supply (and by implication lack of foreign customers) but a failure to diversify out of Africa. Neff predicted increased efforts by the French state to "replace some African supply with other sources" – a strategy clearly being carried out by Cogéma in Canada. In any event, faced with surplus stockpiles, Cogéma is in the envious position of being able to reduce domestic production, or increase exports (38).

It is in a less happy position with regard to excess enrichment capacity – though it has made considerable efforts to take this up in the last few years.

In 1982, the company sent enriched uranium to India after USA President Carter suspended supplies under the 1978 Non Proliferation Act (22). The following year, Cogéma "wrested commercial superiority away from the US" in the field of enrichment, by landing contracts with five new US customers (80) – the supplies to come from the Eurodif plant.

Cogéma carries out reprocessing for Switzerland and Spain, in addition to Sweden, West Germany and Belgium (22). Its major reprocessing customer, Japan, signed a contract in 1977 (81), without waiting for the outcome of the Windscale Inquiry in Britain (Cogéma at La Hague, and BNFL at Windscale/Sellafield, were intending to jointly cope with Japan's reprocessing needs). An expansion programme was instituted at La Hague in 1985, to increase its capacity from around 1000 tonnes a year to 1600-1700 tonnes (82).

In late 1984, the company also announced plans to establish international leadership in using plutonium – or MOX (mixed oxide) fuel assemblies – in light water reactors, starting with French nuclear plants (82, 83), partly in order to burn up plutonium which might otherwise accrue (84).

Belgonucléaire and Cogéma, backed by Fragema and the manufacturing plants of Belgonucléaire and Cogéma's Melox factory at Marcoule, have also established marketing facilities for MOX fuel worldwide. (The Melox factory has a design target of 100 tonnes/year by 1995.) Both Argentina and Japan are pioneering the use of plutonium recycling in-thermal reactors (90) and MOX fuel assemblies form part of the plutonium consignments contracted to Japan. In 1985, Cogéma announced that many of these would be transported by air, as a safer proposition than sea transport (84) – a curious statement in view of its claim (in

the International Atomic Energy Agency's *Bulletin,* the same year) that it had done its utmost to guard against any mishaps affecting uranium or plutonium carried by ocean-going vessels, and that "accidents that have occured during the transport of irradiated fuel in France, as elsewhere, have never resulted in rupture or violation of a package's containment functions" (85). But then this was the year after the Mont-Louis disaster ...

The air transport plan was, apparently, not implemented at that time, but revived in · 1988, when BNFL and Cogéma jointly stated their intention to fly reprocessed plutonium (up to 45 tonnes by the turn of the century) from Scotland to Japan, as part of a 1987 contract between the two European companies and Japanese utilities. This plan attracted considerable criticism, especially from the Nuclear Control Institute in Washington (86). The plutonium flights would pass through Anchorage, Alaska, where they would re-fuel after leaving Manchester's Prestwick airport. The Governor of Alaska objected strongly to the plan (91), as did Canadian environmentalists and a Yukon MP, Audrey McLaughlin, voicing fears of many indigenous peoples in the north of the country (92).

By the mid-1980s, Cogéma was trying hard to sell surplus capacity at its Cap La Hague plant, by encouraging utilities to sign short term contracts. Already by 1985, a CEA official, Georges Vendryes, had expressed severe doubts that large-scale reprocessing to separate plutonium from spent fuel would get underway "for another twenty years" (87), as more and more uranium was coming onto the market at cheaper prices, as uranium exploration showed little sign of slowing down, and as the industry experienced considerable delays in bringing fast breeder reactors into commercial service (87).

In 1985, Cogéma could boast that – with twelve major prospecting projects in Africa, North and South America, Spain and Australia (2) – it was the only uranium company to have roughly maintained exploration efforts in the last few years at a time when its main competitors had been slashing spending because of the falling uranium price (41). Its successes since then in

Canada have further enhanced its position as world leader. So long as the French state remains the most highly nuclearised on the planet, so long as the French military régime maintains its nuclear *force de frappe*, and the various versions of French neo-colonialism, in Africa and the Pacific, remain intact, there is little doubt that Cogéma, and its parent CEA, will head the frontrunners in the uranium industry indefinitely.

The latter half of the 'eighties marked a methodical entrée by Cogéma into the world of gold, as it sent its prospectors throughout France (93) and into Spain, Canada (94), the USA and Australia (95). Its underground gold mines south of Limoges, at Bourneix and Laurieras, which the company took over in 1987, are now moderately important producers (96, 99).

In 1990 these were acquired by the Australian company Arimco NL, in return for Cogéma getting a stake of 61.3% in the Australian outfit (97). Cogéma has also been looking for gold in Ecuador (98) and Zambia where its explorations include part of the Kafue National Park (96). Here it had been involved with Agip in a JV looking for uranium, a search which yielded little of value, so it has withdrawn from uranium exploration in the country (96, 98). In fact, the late 1980s marked a number of withdrawals by Cogéma from areas once thought worthy of uranium prospecting (Senegambia for example) (98) and from mines in which it previously held an interest. In December 1988, Interuran GmbH (formerly Saarberg-Interplan Uran GmbH) announced the end of uranium mining at its Grossschloppen deposit in Bavaria (100). (Interuran is owned jointly by Cogéma Uran Services (Deutschland) GbmH, Badenwerk AG, and Energieversorgung Schwaben AG) (95).

In March 1989, Cogéma also said that it would be reducing its French workforce by 320 (to 2180) and would be stopping low-grade uranium production for economic reasons (100, 101). A year later, it announced that it would be closing its Vendée mining division for the same reason: La Escarpière mine had already closed down earlier than planned because flooding had rendered lower levels of the mine inoperable; extraction was also halted at the Piriac mine near Guérande

(102). It is also planned that the Chardon-en-Loire and La Commanderie deposits will be abandoned, while Simo, which runs the concentrator at Escarpière, is also expected to close by 1992.

Further reading: Dave van Ooyen, *Uranium-mijnbouw in Frankrijk* (duplicated) Amsterdam, 1982.

A Mouton, P Lafforgue, G Lyaudet (Cogéma) in *Industrie Minerale – les techniques,* France 10/84.

"Cogéma: Nuclear integration from ore to spent fuel reprocessing and technical services" in *E&MJ* 8/86, pp 26-30; also "Uranium Mining is Alive and Well in France" *E&MJ,* 8/86, pp 26-30.

Contact: *In France:* CLAN, c/o François et Lucette Pratbernon, 29 rue du Puy las Rodas, 87000 Limoges.

Comité de défence et de sauvergarde de la Montagne Bourbonnaise, le Puothier, 03300 La Chapelle.

Association contre les nuisances de l'exploitation des mines d'uranium de St Jean de la Blaquière, 3400 Lodève.

WISE-Paris, 5 rue Bout, I-75013 Paris.

Reseau Uranium 7, rue de l'Auvergne, F-12000 Rodez.

In UK: Partizans (People Against RTZ and its Subsidiaries), 218 Liverpool Road., London N1 ILE (tel 071-609 1852).

In Canada: Montreal Uranium Committee, Uranium Cafe Commun/Commune, 201 Milton St, Montréal, Québec.

References: (1) *FT* 26/4/85. (2) *E&MJ* 8/86. (3) *FT* 29/8/87. (4) *FT* 3/7/84. (5) *FT* 11/4/85. (6) See also *MJ* 20/7/79. (7) *E&MJ* 11/86. (8) *FT* 13/7/84. (9) *FT* 5/11/84. (10) *Gua* 20/8/84. (11) *Peace News* 7/9/84. (12) *Gua* 26/9/84. (13) *Gua* 23/2/88. (14) *MAR* 1986. (15) *MAR* 1987. (16) *MJ* 8/4/88. (17) *MJ* 13/11/81, *MJ* 9/4/82. (18) *MJ* 25/7/80. (19) *MJ* 29/2/80. (20) *FT* 1/2/88. (21) Dave van Ooyen, *Uraniummijnbouw in Frankrijk,* Amsterdam 1982. (22) *FT* 16/2/83. (23) IAEA/OECD *Red Book* 1982. (24) IAEA/OECD *Red Book* 1979. (25) *EPRI* 1978. (26) WISE Amsterdam, *Kiitg No. 7, 1980.* (27) *MJ* 13/7/84. (28) *L'Evolution générale du Nucléaire,* Cogéma, 1/82. (29) WISE Amsterdam, *Kiitg* 6/82. (30) *Tableaux Economiques du Limousin,* INSEE, Limoges, 1981. (31) *MM* 5/85. (32) See (31) pp 366-369 for a detailed description of La Commanderie. (33) see (2) for a description of the operations of this mill. (34) WISE Amsterdam, *Kiitg* Aug/Sept 1980. *See also Gua* 14/3/81. (35) *Peace News* 13/7/84. (36) *Non-violence politique*, Montargis, 6/84. (37) *TAZ*, quoted in *Kiitg* 1/83. (38) Neff. (39) *MJ* 8/11/85. (40) Miles Goldstick, *People Resisting Genocide*, Black Rose Books, Montreal, 1987 (41) *FT* 11/9/85. (42) *MJ* 3/5/85, *E&MJ* 6/85. (43) Mining 1987. (44) WISE *News Communiqué* 14/8/87. (45) Personal communication from Daniel Berman, Montreal Uranium Committee, London 1987. (46) *FT* 12/5/82. (47) *MJ* 11/4/80. (48) *MJ* 28/6/85. (49) *FT* 9/2/81. (50) *Quarterly Economic Review of Ivory Coast, Togo, Benin, Niger and Upper Volta*, Fourth Quarter 1981. (51) *Nuclear Fuel* 29/9/80. (52) *Nuclear Fuel* 18/1/82. (53) *Nuclear Fix*. (54) *MAR* 1985. (55) *MJ* 5/10/79, *MJ* 13/4/79. (56) *MJ* 15/5/81. (57) *FT* 27/1/81. (58) *New Scientist*, London, 24/5/84. (59) *FT Energy Report*, London, 1978; *MJ* 16/3/79. (60) *Daily Telegraph* 30/7/85. (61) *TAZ*, Berlin, 23/1/88. (62) Copy of draft contract between Electricité de France and Nukem Luxembourg, 1/1/79, ref. D.531.1 CPN 16.79 GDD/TIC, copy in hands of author. (63) See also *Parting Company*, Partizans, London, Nos 1 & 2, Spring 1988. (64) WISE Amsterdam, *Kiitg* 5/80. (65) Information from Greg Dropkin, representative of CANUC, at Partizans Press Conference, London, 2/6/88: documents available from Partizans. (66) information from Jup Weber, Partizans Press Conference, London 2/6/88; also conversations with Roger Moody, London July 1988. Copies of documents from Partizans. (67) *Financial Review*, Australia, 24/10/78. (68) *Nucleonics Week* 27/3/80. (69) *International Metals and Mining Annual*, McGraw Hill, New York, 1982. (70) *MJ* 10/3/78. (71) *FT* 22/2/79; *FT* 26/2/79. (72) *MAR* 1980. (73) *MJ* 5/3/82. (74) *FT* 26/4/84. (75) *MJ* 27/1/84. (76) WISE N.C. 19/2/88. (77) *MJ* 15/1/88. (78) *FT* 29/8/87. (79) *MJ* 24/12/82. (80) *FT* 11/7/84. (81) *Gua* 6/9/77. (82) *FT* 23/1/85. (83) WISE *Bulletin*, Vol. 5, No. 8, 1984. (84) *FT* 20/2/85. (85) IAEA *Bulletin*, Vienna, Spring 1985. (86) WISE N.C. 278-2387; N.C. 278-2388; N.C. 19/2/87. (87) *FT* 17/10/85. (88) Information from Audrey Coulson, to author, 1985. (89) *MAR* 1988. (90) IAEA *Bulletin*, Vienna, 4/87. (91) *Gua* 21/11/87. (92) *Native Press* 29/4/88. (93) *MM* 2/90. (94) *MJ* 12/5/89. (95) *MIY* 1990. (96) *MJ* 14/7/89. (97) *MJ* 1/6/90. (98) *MAR* 1990. (99) *E&MJ* 9/89. (100) WISE N.C. 312 12/5/89. (101) *Info Uranium*, Nos. 35, 37, 38. (102) *MM* 5/90; *MJ* 16/3/90; *MJ* 9/3/90. (103) *MM* 5/90.

151 Coluranio/Sdad Colombiana de uranio

This is a state corporation, set up in 1977 out of six state agencies, to explore for and develop radioactive mineral resources, with an initial capital of only US$750,000 (1).

In 1979, Coluranio reached agreement with the UNDP to develop exploration programmes for uranium which would continue until 1981 (when the government estimated its resources at 40,000 tons U). Others involved in the search were Enusa, Minatome, and PRNFDC (J). There were indications of uranium in Santander, Cundinamarca, and Meta provinces (2).

In the ten years since, little has been heard of Coluranio's exploits: most exploitation is now being carried out by multinationals, especially Minatome and Enusa (3).

However it did locate possibly significant uranium deposits in the Andean provinces of Cundinamarca, Boyaca, Caldas, Tolima, Hiila and Antioquia as well as in the flatland provinces of Meta, Guiania and Vaupes (3). Indigenous Colombians – particularly the Tubaroan Indians of Vaupes – occupy large parts of the surveyed area, and considerable alarm was expressed at the risks they faced (4).

References: (1) *MJ* 28/10/77. (2) *MJ* 8/1/82. (3) *Melbourne Age* 6/11/79, *El Tiempo*, Bogotá, 29/8/80. (4) *Natural Peoples' News* (London) No. 2, and *ARC Bulletin*, No.9, 1980.

Republic of Colombia (CO)

COLURANIO / SDAD. COLOMBIANA DE URANIO (CO)

152 Comaplex Resources International Ltd

```
E & B Explorations Ltd (CDN)
                    \  ?
COMAPLEX RESOURCES INTERNATIONAL (CDN)
```

The company held 42 uranium properties covering more than 3 million acres in Saskatchewan in the late 1970s, of which 26 were in the Athabasca sandstone region. Some of its biggest holdings were in the Russell Lake, Johnson Island, Key Lake, Tower Lake and White Lake areas.

The company also holds considerable exploration interests across northern America and has been exploring for uranium in British Columbia and the Northwest Territories of Canada too.

Two subsidiary companies, Wollex Exploration Ltd and Radiometric Surveys Ltd, hold uranium leases, although they are dormant companies (1). In the late '70s, Comaplex was one of Eldorado Nuclear's uranium exploration joint venturers (2).

Comaplex also has oil wells in the US in North and South Dakota (it is not known whether these are on native American land), Kansas, Texas and Wyoming.

E and B has substantial holdings in Comaplex though the largest shareholding is controlled by GF Fink, the company's president (1).

References: (1) *Yellowcake Road*. (2) Eldorado Nuclear Annual Report 1979.

153 Combined Metal Mines Ltd (CDN)

It was exploring for uranium in Saskatchewan (1).

References: (1) *Who's Who Sask*.

154 Cominak/Cie minières d'Akouta

This consortium operates the Akouta mine in Niger, a deposit in the Air Mountains discovered by the French CEA in 1966.

Original design output was 2000 tonnes of uranium concentrate per year (1) at a capital cost of 70 million dollars. Actual output (of uranium contained in magnesium uranate with a 70% uranium metal content) was 398 tons in 1978 (the year the mill began operations), 1700 in 1979, and 2200 in 1980 (2).

In 1983 and 1984 Cominak's output of uranium was approximately 2000t/yr; according to the government a large amount was being stockpiled. An agreement to supply Egypt had not been fully implemented due to Egypt's "financial problems", while new markets were being actively sought in mid-1984, especially in North Africa and the Middle East. A deal was expected "shortly" with Iran (3).

With production from Cominak at 1900 tonnes in 1989, Cominak's output was more or less what it had been for the previous three years (4) – although the *Mining Annual Review* claims that Niger's uranium production is probably underreported (5), and that France has been slowing down its contracting from both Niger and Gabon, to absorb inventories, and cut back on the high cost central African production (6).

About 60,000 tonnes of uranium reserves remain in the Akouta deposit and the associated lode at Akola (6).

References: (1) *MM* 6/77, "Japan and West Africa" in: *West Africa* 4/7/7. (2) *MAR* 1981 p 505. (3) *MJ* 1/6/84. (4) Nuexco 1989 Annual Review. (5) *MAR* 1988. (6) *MAR* 1990.

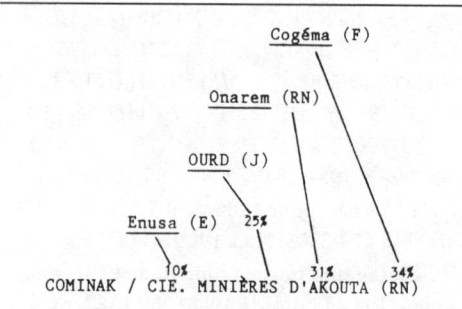

```
                              Cogéma (F)
                    Onarem (RN)
             OURD (J)
                        25%
    Enusa (E)
         10%                    31%        34%
   COMINAK / CIE. MINIÈRES D'AKOUTA (RN)
```

155 Cominco Ltd

It deals in zinc, lead, potash, mercury, copper, phosphates, coal, silver, gold, and chemical production; and is involved in smelting, refining and by-product production to international marketing "with related services" (1). It has subsidiary or associated companies in Canada, USA, Greenland, Europe, Australia, India, Japan, the Middle East, and elsewhere (eg a small chromite mine in New Caledonia, 55.5% owned, which was due to open late in 1982) (2).

Cominco discovered uranium, in the form of pitchblende, at Great Bear Lake, Northwest Territories, Canada, as early as 1932, though only radium was mined at this period (hence the name Port Radium).

Although its net profits tumbled during 1981 (3) Cominco was busy expanding from 1978 to 1980, especially in the Canadian Arctic, British Columbia (BC), and Australia (the Sullivan mine at Kimberley and explorations in Tasmania) (4). Operations with potential adverse effect on native peoples are currently the Trail refinery, BC; the Black Angel Mine (with Greenex) on the west coast of Greenland; and the Polaris mine in the Canadian High Arctic.

Cominco's diverse uranium explorations include Saskatchewan and Quebec (where a find was reported in 1978 in the Otish mountains, in a JV along with Pancontinental and the James Bay Development Corporation).

In the late '70s it was also exploring for uranium in the Northwest Territories (NWT) where its mining activities in general were the subject of an important analysis in 1977. Assessing the degree to which rents (paid on natural resource properties) actually went to the provincial government

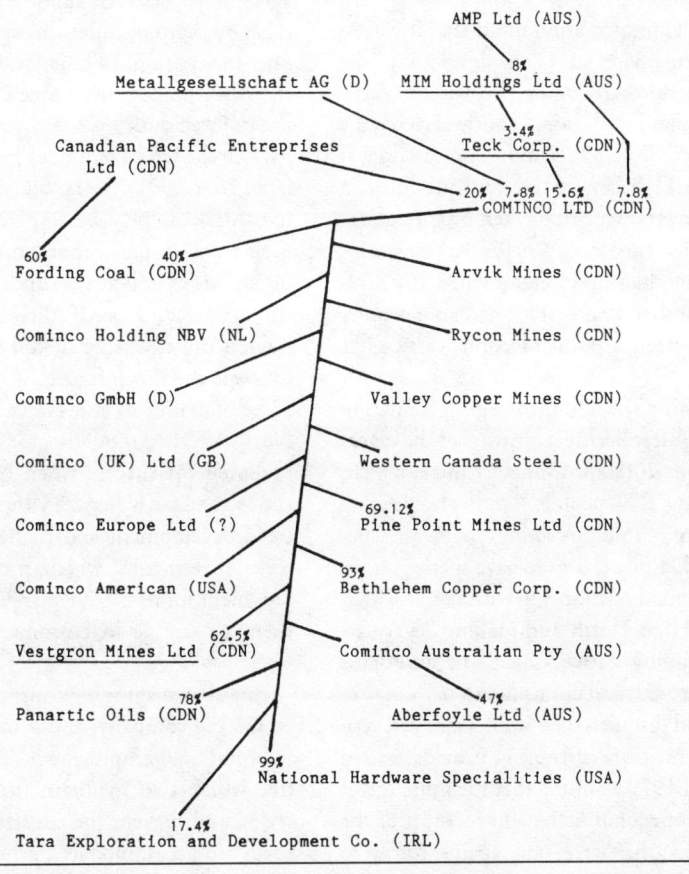

(let alone indigenous peoples), Arvin D. Jelliss estimated that between 1970 and 1974, Cominco retained more than two-thirds (73.5%) of all rent payable in the North West Territories in the form of excess profits. Concluded Jelliss:

"At the same time as the economic rents associated with Cominco's current mining projects in the NWT contribute to the viability and profitability of its long-term operations, they fail to contribute in any meaningful way to the long-term development objectives of the native people" (6).

Put bluntly: Cominco's exceptional expansion in the late '70s was based on exploitation of native peoples in its "home" country.

By 1976, Cominco – together with Esso Minerals Canada (Imperial Oil) and Urangesellschaft – had started its exploration programme on Inuit land near Baker Lake (7). These companies teamed up with the federal government to defeat the Inuit land claim in 1979 (see Urangesellschaft) (8).

In late 1979, Cominco announced that it was to make an immediate start on developing the world's most northerly mine, Polaris on Little Cornwallis Island, 90 miles south-east of the magnetic North Pole. Once in full production, it would be the 11th largest zinc-lead producer in the world, the product being taken by special ice-breaking bulk carriers. Whether these ships would be Canadian or foreign-owned (thus resulting in another drain of Canadian revenues overseas) has been a point of controversy since 1979 (9).

The Polaris mine was about to begin production in early 1982 (after Bechtel constructed the plant) producing over 200,000 tonnes of metal a year, and employing 250 people – and also leaving more than one million tonnes of waste per annum (10). Despite claims to have spent years in negotiating a good relationship with native Canadians in the High North and making "every attempt" to employ native Canadians at Polaris (11), the mine was, and continues to be, a matter of concern and disquiet. The Inuit Tapirisat, representing the majority of Inuit in Canada, issued a statement in 1979 claiming that the mine "gives no assurance of economic benefit to Inuit of the region" and that they were, once again, "forced to react to a development proposal that has already been given the go-ahead" (12).

In early 1982, the company reached agreement with the Nana Regional Corporation (owned by Alaskan Inuit) by which the Native Canadians retain an interest in the Red Dog zinc and silver deposit, should feasibility studies prove successful (13).

Among other native peoples' actions against Cominco has been a blockade by Stuart-Trembleur Native Indian Band of the British Columbia railway line near Fort James, in connection with a mercury-polluted lake, caused by Cominco (14).

Under an agreement with the Nana Regional Corporation (a native corporation set up following the Alaskan Native Claims Settlement Act in 1971), Cominco will operate the Red Dog mine and provide the money. Nana will receive a minimum of C$1M/yr, 4.5% of net smelter royalties and 25% of net proceeds. Compared with other agreements between mining companies and indigenous communities this agreement is a liberal one. In addition, as Julie Hodson (formerly with the Anthropology Resource Center in the USA) has pointed out:

"When the US government's trust responsibilities lapse after 1991, the landholdings of future Native Alaskans may be imperilled if Natives sell their stock in the corporations. The profit-making activities of Native corporations and the subsistence traditions of their indigenous stockholders must be reconciled if Native Alaskans are to shape their own future.

"The activities of the Nana corporation of the North-West Arctic Region demonstrate the potential for success when Native corporations act as mediators between the forces of disinterested development and traditional values. Nana seeks investments which provide seasonal employment for its Native shareholders, thus freeing them to engage in customary subsistence activities...

"Nana signed a mining contract with the Cominco mining company, and is training stockholders for rotational employment. This will allow Native workers to maintain their subsistence lifestyles, and prevent the construction of a "boomtown" to accommodate outside workers. Village councils meet regularly with mining officials, and Nana has reserved the right to halt operations at any time.

"Nana has also developed a regional strategy aimed at carrying Native proposals into action. It has sought village participation in developing priorities for land use, and supplies the villages with information about regional trends in mining, tourism, and other social and economic factors.

"This use of corporation resources to enhance and protect the traditional lifestyles of shareholders appears to be a viable economic strategy: Nana made a US$1.5 million profit in the last fiscal year; its earnings per share increased more than six-fold"(15).

The Red Dog deposit was due to come on stream in 1988, Cominco announced in early 1984, with an initial one million tons of ore throughput (16). In early 1983, workers from the United Steelworkers (USA) at the Trail lead-zinc mine rejected a request from Cominco for reduction in effective wages to save the company's operations. The union said it feared complying with the cutback and it would lead to lay-offs (17).

That year the company was involved in several disputes with unions resulting in strikes and strike threats: 4200 workers at the Trail smelter and Sullivan lead-zinc mine went back to work in June that year after a new contracts agreement was signed; workers in the company's gold mine in the NWT threatened to strike, because of lack of improved vacation and cost-of-living allowances; while at its potash mine in Saskatoon workers and management failed to negotiate a new contract (18).

Cominco closed 1983 with an improvement on its position in 1982 (when it sustained its first loss for 50 years). Production went over capacity at the Polaris zinc-lead mine, and the company confirmed its position as western Canada's second largest fertilizer producer (19).

Greenex started development of its Nunngarut lead-zinc mine in Greenland at the beginning of 1984 (20).

In late 1986, Cominco was parcelled out among an Australian-West German-Canadian consortium, comprising MIM, Metallgesellschaft AG, and Teck. Initially the three companies owned 31% of Cominco, and AMP another small percentage, with Canadian Pacific retaining 21% (21). However, Metallgesellschaft (after putting its main foreign mining interests in a new Canadian-based holding company called Metall Mining Corp) (22), increased its holding in MIM to around 4%, while in 1988 Teck purchased a further batch of Cominco shares, which it then intended to sell to Metall and to MIM (a quarter each) (23, 25).

In 1986 Cominco and Rio Algom "rationalised" their copper interests in the Highland Valley of British Columbia (24).

Nunachiaq Inc was set up as a holding company representing Teck Corp (50%), MIM Holdings Ltd (25%) and Metallgesellschaft AG (25%), to buy Canadian Pacific's 52% stake in Cominco (26). Further purchases over the next three years took the group's interests in Cominco to around 45% (27).

Cominco Resources International Ltd (59% owned by Cominco) (25) has been involved in recent years in quite aggressive exploration, notably in the USA (28), Turkey – where gold has been discovered near the Black Sea (29), France, Italy and other parts of Europe (30). In Eire, Cominco Ireland Ltd has an option on 30 licences, together with its subsidiary Pine Points Mines Ltd (31). But the most important of Cominco's projects for the 1990s remain its high arctic mines, and in particular the Red Dog deposit which it has been trying to start up for some time, contending with difficulties posed by the exceptionally high grade ore (32). Its Trail smelter, which produces 5.5% of the western world's refined zinc and a smaller proportion of its lead, was due to be expanded in 1989, but the plans were cancelled (33). In Chile, Cominco has a JV (42.5%) with the national company Enami, at the Quebrada Blanca copper deposit (34), and 25.7% interest in Minera Tres Cruces which operates the Marte gold mine (35). The company's Highmont mine, operated by the Highmont Mining Co, was incorporated into the Highland Valley Copper partnership in 1988 (Lornex – a Rio Algom subsidiary – with 45%, Cominco with 55%) (35, 36). Meanwhile, a company in which Cominco holds important in-

217

terests – Geddes Resources (27% owned by Northgate Exploration Ltd) (25) – has encountered strong opposition from environmental groups and the governing authorities in Alaska for its proposal to exploit the huge Windy Craggy copper deposit in north-west British Columbia. The Alaskan state government claims that the mine could cause "irreparable harm to the environment" (37) while the Tatshenshini Wild group claims that a 177km road would have to be built across the Chilkat Eagle preserve, which is a prime grizzly (bear) habitat, and the only Dall sheep winter range in British Columbia (38).

Contact: For opposition concerning Greenex: Jens Dahl, Dept of Eskimology, Copenhagen University, Norregade 20, DK-1165 København, Denmark.

References: (1) *MIY* 1981. (2) *MJ* 12/3/82. (3) *MJ* 27/11/81. (4) *MJ* 15/2/80. (5) *FT* 26/6/78. (6) *Dene Rights Newsletter* vi:l.(7) Roger Rolf *Mining Development in the Northwest Territories,* Oxfam Canada, 1976. (8) *Energy File,* BC, 4/80. (9) Statement of Liberal MP Keith Penner, 21/11/79, reported in *Native Press* 21/12/79. (10) *FT* 19/11/81, *MJ* 15/1/82. (11) *Native Press* 22/2/80. (12) *Native Press* 21/12/79. (13) *MJ* 12/1/82. (14) *Energy File* 8/79. (15) Native resource control and the multinational corporate challenge, Washington 1982 (ARC, Cultural Survival, Indian Law Resource Center & Multinational Monitor). (16) *MJ* 23/3/84. (17) *MJ* 25/3/83. (18) *MJ* 24/6/83. (19) *MJ* 11/5/84. (20) *E&MJ* 5/84. (21) *FT* 14/11/86. (22) *E&MJ* 6/87. (23) *E&MJ* 6/88. (24) *FT* 1/10/86. (25) *MIY* 1990. (26) *FT* 1/10/86; see also *FT* 14/11/86. (27) *E&MJ* 6/88; *MJ* 4/8/89; see also *MJ* 3/10/86, *MJ* 29/4/88. (28) *E&MJ* 5/88. (29) *MJ* 25/5/90; see also *FT* 10/5/90. (30) *E&MJ* 8/87; *E&MJ* 5/89. (31) *MJ* 6/10/89. (32) *FT* 26/4/90; *FT* 19/3/86; *FT* 27/7/89; *FT* 24/12/85; *FT* 22/2/90. (33) *MJ* 14/7/89; see also *FT* 24/7/90, *E&MJ* 10/86. (34) *E&MJ* 2/90; see also *MJ* 1/12/89, *E&MJ* 7/89, *FT* 20/7/90. (35) *MAR* 1990. (36) *MJ* 5/2/88; *FT* 15/1/86; see also *MJ* 4/3/88. (37) *MJ* 1/6/90. (38) *World Rivers Review,* Vol. 5 No. 2, 3-4/90.

156 Command Minerals

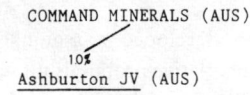

A small exploration company, partnered with PRNFDC, ACM Ltd, Nickelore, and West Coast Holdings, in uranium searches in Western Australia.
In 1976, a director of the company was convicted of ripping off Tuckenarra Pty (1).
In 1980, the company reported several diamond and uranium JV exploration programmes (2).

References: (1) *Reg Aus Min* 1976/77. (2) *Annual Report* 1980.

157 Commonwealth Edison Co

This is an Illinois-based electricity utility whose subsidiary, Edison Development Canada, in autumn 1979 bought out Urangesellschaft's 40% interest in the Brinco uranium project at Kitts-Michelin, near the east Labrador seaboard in Canada. In return for this, Commonwealth Edison agreed to buy 18 million pounds of the uranium and to construct an all-weather road to Goose Bay, 85 miles away (1) (for details of Kitts-Michelin, see Brinco).
In 1979/80, the company also contracted with Queensland Mines for a possible share in Nabarlek production (2).

References: (1) *MM* 9/79 p.281; *MIY* 1982. (2) *Nuclear Fuel* 3/3/80.

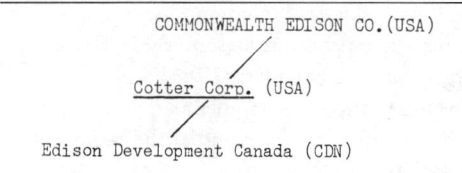

158 Companhia Estanifera do Brasil

The Tin Company of Brazil, as its name would be in English, was incorporated in 1951 to do everything imaginable to tin-based products as well as produce chemicals for soldering (1).

It has two major subsidiaries, Cia de Mineração Jacunda and Cia de Mineração Santana, and its operations centre on the Santa Barbara, Jacunda and Alto Candeias alluvial mines in Rondonia (the state named after Brazil's first official Indian protector, Marshal Rondon) (2). Cesbra's operations in the state, along with that of other companies, have encroached on various Indian lands, notably in the Aripuana Park, traditional territory of the Surui and Cinta-larga (3).

It also operates a smelter at Volta Rodonda, Rio de Janeiro, and in 1984 was bringing on stream the first hard-rock tin mine in Rondonia. This is an area where uranium has been located, although there appear to be no plans for development. (Cesbra is named in the *Financial Times* 1982 *Mining International Yearbook* as "interested" in uranium, but the reference had disappeared by 1985.)

In fact BP, through its Brazilian subsidiary BP Mineração, was supposedly the partner in Brascan Recursos Naturais, concerned with exploration *outside* any currently exploited mineral deposits (2).

The company came in for worldwide criticism in 1989 when the Sunday Times and associated Murdoch papers condemned Brascan and BP for ravaging the rainforest at Jacunda in the Jamari national forest (see Brascan). BP later sold its share in the joint venture to Brascan.

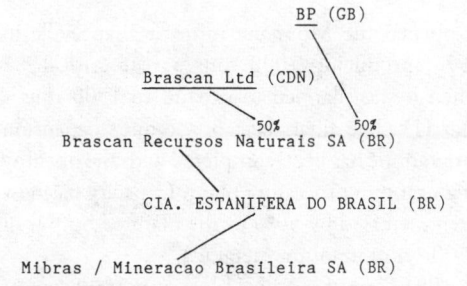

References: (1) *MIY* 1985. (2) *MAR* 1984. (3) *Survival International Review* (London) Spring 1979.

159 Companhia Portuguêsa de Radium Lda

```
United Kingdom Commercial Corp. Ltd (GB)

            CIA. PORTUGUÊSA DE RADIUM LDA. (P)
```

"It reminds me of Pope Alexander Borgia dividing the whole of the New World between Spain and Portugal!" That was reputedly the remark made by Portugal's war-time dictator Salazar when Britain and the USA started seizing known uranium deposits in the late war years (1).

One of their key prizes was the Urgeirica uranium mining district in the north of Portugal: an area which yielded radium at the end of the 19th century and was producing both radium and uranium (stockpiled) for French and Portuguese purposes until the late '20s.

In 1929, however, two Britons privately bought Urgeirica and founded the Companhia Portuguesa de Radium Lda. Until 1938, the company laboured under what J Cameron calls "extreme difficulties, both financial and technical" (2). Then, in 1942, the majority shareholding was bought out by the UK Commercial Corporation, the British war-time purchasing agency in Portugal.

For two years the 14 mining concessions in the area were worked on a care and maintenance basis, but in late 1944 "a more inspired interest was taken in CPR by the British Government ..." (2). The UKCC finally bought out minority shareholders, acquired other concessions, and by August 1945 the British government had total control, with a total of 63 concessions under its thumbs.

The company was managed in typical patrimonial fashion: with a Portuguese manager, British assistant general manager, and other senior posts equally divided between Britons and Portuguese (2).

The mines finally closed in July 1962, with the company's assets returned to the Junta da Energia Nuclear. In mid-1977 the Portuguese mining industry was transferred from the Junta to Enusa. Until the discovery of the Alto Slentejo deposits further south, the Urgeirica area was Portugal's main uranium provider, and the mill was still concentrating feed in the late '70s (3).

Total uranium in the CPR concession until its surrender was estimated at around 2000 tons from seventeen mines and other sources (eg heap leaching and slimes, see below). Of these mines Urgeirica itself was undoubtedly the biggest producer (766 tons between 1951 and 1962), followed by Bica (206 tons). Total labour force on all the mines varied between 1000 and 1200 with up to 600 men employed underground. Importantly, the total number of different names on the company's books during the period was 5800 signifying a 500% labour turnover, caused by "the many itinerant underground workers who only worked seasonally on the bigger mines ... and by the fact that on the small mines local labour was employed which did not move on to other mines when they were closed down" (2).

Although a final indemnity given the remaining workers was "considerably above the minimum legal requirement" (2) and assets, equipment and machinery were returned cleaned and serviced, neither the company nor the Portuguese government to our knowledge has issued any figures on the radiation control for these workers, let alone done any crucial health follow-up studies.

The company claims to have sealed abandoned shafts with concrete and covered tailings dams with lime after turfing over the walls. Open pits were allegedly filled in or "protected" (2).

One of the most interesting and important aspects of this little-known (but, to Britain's atomic power programme in the '50s, crucial) venture is that it appears to have been the first to use *in situ* leaching of uranium anywhere in the world (2). The company also developed heap leaching of low grade ores, using water to stimulate sulphuric acid. The process has been described by scientists in the National Chemical Laboratory at Teddington, Middlesex (4).

What is *not* known is the degree to which the solvents entered ground or river and drinking water.

References: (1) Moss. (2) *History of the evaluation*

and exploitation of a group of small uranium mines in Portugal, Buenos Aires, 1979 (IAEA SM-239/28). (3) *Uranium Redbook* 1979. (4) Miller, Napier & Wells, *Natural leaching of uranium ores and preliminary tests on Portuguese ores* in *Transactions of Institution of Mining and Metallurgy* (London) Vol. 72.

160 Compania Minera Pudahuel (RCH)

In 1980 this company announced it would produce uranium from copper mined at its Cascada copper mine (1) with 20,000kg/yr of uranium oxide coming out in October 1981. The company was also investing US$8 million in the facility and planned to sell overseas if the Chilean government didn't buy the output.

References: (1) *MM* 12/80.

161 Compania Naviera Perez Companc (RA)

This was part of a consortium (along with Boroquimica SA, Rio Algom and Noranda) bidding to develop the subsidiary orebodies of Sierra Pintada, Argentina's most important uranium deposit (1). It lost out to another consortium headed by Minera Sierra Pintada and PUK in mid-1980 (2).

References: (1) *MM* 1/80. (2) *Nucleonics Week* 15/5/80.

162 Comuf/Cie des mines d'uranium de Franceville

It opened the Mounana mine at Franceville in 1977, producing 1000 tons a year until 1982 when it was planned to expand to 1500 tons a year (1). The final stage of Comuf's expansion programme has been completed with the opening of a uranium processing plant. Capacity has now been increased by 50% to the planned output of 1500t/yr of uranium metal (2).

In 1981 Comuf produced just over 1000 tons of

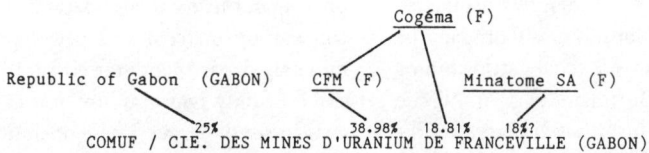

```
                              Cogéma (F)
                             /        \
  Republic of Gabon (GABON)  CFM (F)   Minatome SA (F)
                  \     25%  \ 38.98% 18.81%  18%?
         COMUF / CIE. DES MINES D'URANIUM DE FRANCEVILLE (GABON)
```

uranium (a slight reduction on 1980). Prime sales are to the French CEA as well as Japan, Belgium and Italy. A fifteen-year US$17 million loan from the European Investment Bank is being used to modernise and expand the plant (5).

In 1982, a British development education worker visited the Mounana mine together with an English friend and Gabonese nationals. She reported meeting some of the 1500 workers (all Gabonese) who mostly come from the Haute Ogoue region of the country.

Yearly health checks were held at the mine, the workers told her, but these seemed peremptory and arbitrary, and could hardly be designed in any way to monitor radiation levels on a daily basis. The workers, she observed, had no masks and were covered in uranium dust as they worked.

Despite a wage of CFA75,000 a month and free medical care – also housing, water and technical education to A level – many workers quit after only two years at the mine.

After three years some of them were observed acting "strangely": "they didn't have their minds together," as a Gabonese put it (4).

The same year, the European Investment Bank loaned nearly US$17 million to Comuf, to modernise its mine and processing plant (6).

1989 production from Comuf's mill was only 870 tonnes of uranium, indicating a fair drop (of around 7%) on the previous year. Two-thirds of output continue to go to France, with Belgium and Japan apparently still receiving some deliveries (7).

In 1990 a coup attempt was made on President Bongo as he banned the country's only political party. The London newspaper, *The Independent*, reported that the drums beat out the message: "France is Africa's friend. Africa is France's friend. Oil! Uranium! Cocoa! You want it, we got it; and its cheap!" (8).

References: (1) *MM Mines Review* 1/82. (2) *MM* 9/82; *MJ* 13/3/81, *MJ* 13/8/82. (3) *MJ* 11/9/81. (4) Personal communication with Roger Moody, 12/82. (5) *MAR* 1982. (6) *MJ* 24/7/8I. (7) *MAR* 1990. (8) *Independent* 3/3/90.

163 Conoco [Connecticut Oil Co] Inc

Subject of the "biggest take-over brawl Wall Street has ever witnessed" (1), Conoco (the ninth largest US oil company) passed into the hands of Du Pont Chemicals (the biggest US chemicals company) in summer 1981. In the previous three months Dome Petroleum, Mobil Oil, and Seagrams (the world's biggest distillers) had vied for control in a dog-bite-dog fight which made the intrigues of the Borgia's look like a quiet game of croquet.

In the process, Conoco's controlling interest in Hudson's Bay Oil and Gas of Canada (not to be confused with Hudbay/Hudson Bay Mining and Smelting) passed to Dome Petroleum (2), and Seagrams appeared to end up with 20% of Du Pont-Conoco, thus making it the largest shareholder (3).

In December 1981, however, there were "hints" that Du Pont might start selling off some of its newly acquired assets – particularly Conoco's coal (second only to Peabody Coal) to pay for its take-over spree (4).

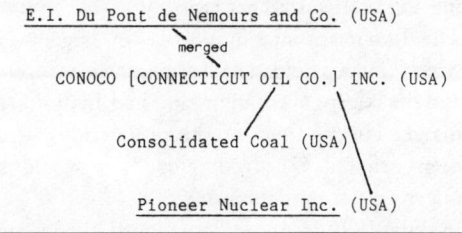

```
  E.I. Du Pont de Nemours and Co. (USA)
                 \ merged
  CONOCO [CONNECTICUT OIL CO.] INC. (USA)
                          \
           Consolidated Coal (USA)  \
                                      \
              Pioneer Nuclear Inc. (USA)
```

In April 1980, Conoco's subsidiary Consolidated Coal and the El Paso Natural Gas Company (ie Conpaso) illegally moved coal strip-mining equipment onto the Burnham part of Navajo land in New Mexico. This was two years after the Navajos had filed suit against Conpaso for failing to negotiate with them. (In 1976, the Interior Secretary rejected Conpaso's plans because they wouldn't have given sufficient royalties to the Navajos – they were later approved.) In 1980, several Navajo people occupied the site, forcing the company to negotiate (5).

Conoco holds a majority interest in the Conquista uranium mine/mill in Karnes County, Texas. In 1977, its Crown Point, New Mexico property (co-owned with Wyoming Mineral Corp) was looking for uranium buyers (designed production was up to 1 million pounds per year uranium by 1980-81) (6). Development was later deferred (7).

In early 1979, 91 native Americans – mainly Navajo – brought a law suit against various federal agencies, together with Friends of the Earth (USA), charging that they had violated the Federal Environmental Policy Act by approving development in the San Juan, New Mexico, and the Powder River, Wyoming, basins without securing environmental impact statements for the developed sites. Conoco – along with Mobil, United Nuclear, Kerr-McGee, Phillips Uranium Corporation, and Gulf Mineral Resources – entered the suit which was eventually rejected (8).

A JV exploration programme at Powder River (where Conoco had been exploring for uranium since 1961) between Conoco and the Japanese PRNFDC was announced in early 1980 (9).

Pilot solution mining operations were under way at the company's Pumpkin Butte mine in Wyoming, and a solution uranium mine was due to open at Duval City, Texas in early 1982, producing 450,000 pounds per year U_3O_8 (7).

The Imouraren uranium mine in Niger – in which Conoco holds a 35% interest with Cogéma and the Niger government (and had an option to market Niger's share of the production) – has been deferred (7). If developed, it would be Niger's major uranium mine.

In July 1982, Conoco announced that it was phasing out its minerals operations completely to concentrate on coal and petrol production. Its minerals department started in 1967, and consisted of only two uranium *in situ* leaching projects in south Texas. These operations (and milling) were halted early in 1982. Three hundred people worked in the minerals section with 50 at the section HQ in Denver (10).

This represents the first time that a major US oil company has moved out of nuclear power.

In April 1983 Conoco sold a 50% stake in the Crown Point JV with Wyoming Mineral Corp – an underground uranium project – as part of its programme to phase out mineral operations altogether (11).

Ralph E Bailey, Conoco's chairman and vice-chair of EI Du Pont de Nemours, was in 1983 elected chair of the American Mining Congress, the country's most prestigious professional mining body (12).

In the late '70s, after the formation of CERT (Council of Energy Resource Tribes), the Navajo renegotiated their lease with Consolidated Coal, producing more favourable terms for the tribe (13).

Consolidated Coal is also in a JV with Agip Carbone (Italy) to mine coal in Kalimantan (Indonesian Borneo) near the Malaysian border (14).

The corporation was represented on the Trilateral Commission by John P Austen in the late '70s (15).

Conoco has worked overtime (compared with some other "natural resource" companies) to convince the growing environmental movement that its operations are above criticism. A public declaration first conceived in 1968, and which has been regularly updated since, says that the company will "monitor public attitudes to environmental matters so that the Company's environmental policies and public statements can be responsive to those attitudes" (16). However, its Ecuadorean subsidiary has come under considerable attack for its plans to exploit oil in the Rio Yasuni area of the country, on land owned and cultivated by indigenous Huaorani communities (17). This is not the only tropical forest region which Conoco is exploring where it will inevitably have a deleterious impact on indigenous people: it is drilling for oil in Kalimati village in Bintuni, West Papua (the Indonesian-occupied territory of Irian Jaya) (18) and also has exploration concessions in several parts of the Central Highland range of the country (19). Conoco's earliest exploits in West Papua had a disastrous consequence, with the destruction of 10,000 sago trees and other vegetation around Puragi village in the 1970s (19).

Contact: AIEC, POB 7082, Albuquerque, NM 87104, USA.

References: (1) *FT* 6/8/81. (2) *ibid.* (3) *ibid.* (4) *FT* 10/12/81. (5) *Tribal Peoples Survival,* American Indian Environmental Council 5/80. (6) *MJ* 30/12/77. (7) *MM Annual Mining Review* 1/82. (8) *Navajo Times* 15/3/79. (9) *MJ* 29/2/80. (10) *MJ* 23/7/82. (11) *MJ* 15/?/83. (12) *E&MJ* 3/83. (13) *Wall Street Journal* 20/9/79. (14) *MJ* 18/5/84. (15) Sklar. (16) Statement from Conoco, originally produced 29/2/68, updated 19/10/87. (17) *Survival International Urgent Action Bulletin,* 3/90; see also Rainforest Action Network Newsletter No.16, 7/87, International Labour Reports No.39, 5-6/90. (18) Down to Earth, Tapol and Survival International, No.8, 5/90. (19) Memo from Tapol, London 6/90.

164 Consolidated Canadian Faraday Ltd

CCF's main interlock is with the minor US uranium exploration and production company Federal Resources.
In 1964, CCF closed down its Bancroft, Ontario, uranium mine. Federal Resources agreed to reactivate the mine, and incorporated Madawaska Mines (Federal Resources holding 51%) for the purpose. Production started in August 1976 after a contract with Agip Canada; uranium was, in 1980, being stored awaiting delivery. In 1979,

```
International Mogul Mines Ltd (CDN)
                  |
               31.1%
    CONSOLIDATED CANADIAN FARADAY LTD (CDN)
                     |
                   24%
          Hydra Exploration (CDN)
```

production was 695,533 lbs – to be delivered to Agip at US$50.50/lb, almost a record price (1). CCF is a participant in the Greenarctic Consortium which holds 30 million acres in northern areas of Greenland; no details are available on what it has yet found or where.

References: (1) *MJ* 14/12/79.

165 Consolidated Durham Mines and Resources Ltd (CDN)

This is an Ontario-based company involved in gold and antimony mining, and oil, gas and uranium exploration (1).
It is part of a JV with Eldorado Nuclear to search primarily for uranium near Lake George, New Brunswick (NB), including Durhams antimony mine workings at Fredericton, NB (1, 2).
No news of further developments.

References: (1) *MIY* 1981. (2) *FT* 10/11/78.

166 Consolidated Gold Fields plc

"Oldest of the major international British mining companies" (1), Consolidated Gold Fields was the creation of arch-colonialist Cecil Rhodes (founder-devastator of Rhodesia). Originally registered as Gold Fields of South Africa in 1887, Rhodes brought the company to British investors and re-incorporated it in London as Consolidated Gold Fields in 1892.
Within three years it was making fabulous profits: £2,161,778 on a capital of £1,875,000 in 1895 alone (1).
In the 1950s the company set out on a worldwide expansion programme, using its ill-gotten gains from the apartheid economy, and within the next twenty years the percentage of its profits accruing from South Africa fell from around 80% to 46% in 1974.
By 1977, it had dropped even further (to 26%), while gold, mineral sands, coal and copper mining in Australia accounted for 18% of profits, in-

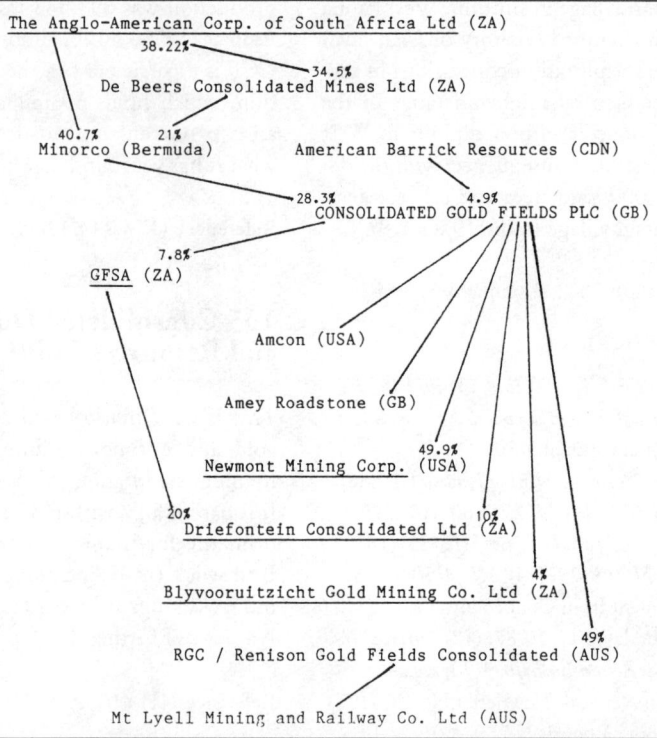

The Anglo-American Corp. of South Africa Ltd (ZA)
38.22% → 34.5% De Beers Consolidated Mines Ltd (ZA)
40.7% 21% Minorco (Bermuda)   American Barrick Resources (CDN)
28.3% 4.9% CONSOLIDATED GOLD FIELDS PLC (GB)
7.8% GFSA (ZA)
Amcon (USA)
Amey Roadstone (GB)
Newmont Mining Corp. 49.9% (USA)
20% Driefontein Consolidated Ltd 10% (ZA)
Blyvooruitzicht Gold Mining Co. Ltd 4% (ZA)
RGC / Renison Gold Fields Consolidated 49% (AUS)
Mt Lyell Mining and Railway Co. Ltd (AUS)

vestments in the United Kingdom for 37% and in North America for 19% (1).

Consolidated Gold Fields' attempts to win fabulous profits in North America had failed catastrophically by 1983. During the '70s Amcon -"a bland name with no hint of a South African connection" (2) – was set up as a holding company for the US assets, and published its own reports and accounts "even though it only had one stockholder" (2).

In 1980 the company looked for a major acquisition in the USA – there was even talk of a bid for Phelps Dodge (2). In the event, Consolidated Gold Fields secured a small stake in Newmont (3, 4).

Within two years, however, the bottom dropped out of Consolidated Gold Fields' US operations – especially of its control of a large drilling rig manufacturer, Skytop Brewster. The company folded some of its tent, leaving behind tax losses and allowances of more than US$100 million – and a hefty handshake to its rejected US chairman

David Lloyd-Jacob of US$594,000. (In this business, it also pays to lose!) (5).

Consolidated Gold Fields' then chairman Rudolph Agnew was in no doubt by 1983 that, due to its failed fortunes in America, the company should "move back to its traditional role as a mining finance house," putting an even greater emphasis on exploration than before (6).

The company's fortunes in Australia are fair – most of its operations there recently having been reorganised under the control of RGC/Renison Gold Fields. In Tasmania, RGC runs the world's largest underground tin mine; Associated Minerals Consolidated (not to be confused with ACM/Australian Consolidated Minerals) is the world's largest mineral sands producer – some of it encroaching on Aboriginal claimed land (7). Goldsworthy Mining Ltd, a large iron ore producer in Western Australia, is jointly owned by Utah Development Co, MIM and Consolidated Gold Fields (8). And in Papua New Guinea a JV between RGC, MIM and Placer Development, with a minority holding by the PNG government, has developed the large Porgera gold property.

Elsewhere in 1984, Consolidated Gold Fields was actively exploring in Chile, Canada (9) and in the Philippines, where it was evaluating a gold-rich copper deposit (6).

In Britain in the late '70s, Consolidated Gold Fields had a licence to explore copper-molybdenum deposits in Argyll, Scotland, and for nickel and copper elsewhere in north-east Scotland (10). It also has a small (2.1%) share in the Kennecott Consortium, the deep sea mining venture headed by BP (11) and now by RTZ.

The importance of South Africa to Consolidated Gold Fields – especially its controlling interest in GFSA – cannot be overestimated, though it is difficult to determine.

Consolidated Gold Fields has also been involved in uranium exploration in Namibia (12). Until 1976, it had a direct interest of 2% in Swaco/South West Africa Co which was exploiting the Berg Aukas and Brandberg West vanadium, lead, zinc and wolfram mines (1).

The company employed some 1000 Africans in conditions which the *Guardian* at the time "sharply criticised" (1). In 1976, Kiln Products Ltd took over the running of Berg Aukas – although the mine closed in the early '80s. Kiln Products is an unlisted subsidiary (13).

In 1977, Consolidated Gold Fields – quoted as "the largest British employer of African labour in South Africa" – was attacked by the bishop trustees of the Roman Catholic diocese of Westminster, who sold all but one of their 11,211 shares in it. This "unprecedented" move (14) came after the trustees had been unable to win any change in the company's policies. As the trustees' statement put it at the time: "The Consolidated Gold Fields group itself recognises that a stable work force would be more economic and efficient, but it does not share the view of the South African Christian churches on the evils inherent in the migrant-labour system and sees itself as unable to take steps to change it.

"The Gold Fields group is also concerned to maintain good labour relations, but the directors consider that the existing arrangements provide for adequate consultation between management and workers in the South African operations of the group.

"They are unwilling to encourage the formation of unions of African mine workers. Wage levels have been belatedly increased from a very low level, but, as far as can be ascertained, they are still below the accepted poverty datum line in some instances" (14).

It is debatable whether Consolidated Gold Fields' view of its investments in apartheid changed materially in the following years. By 1983, Rudolph Agnew claimed "we (have) ceased to be ambivalent about South Africa" This appeared to mean that, while not substantially increasing its investments in GFSA or directly in GFSA-managed mines, Consolidated Gold Fields was not going to substantially draw capital away from apartheid and re-invest elsewhere, especially after the American "debacle". Commented the *Financial Times* pithily: "The South African assets (Consolidated Gold Fields has decided) are just too attractive to be run down"

In 1987, a major new American gold producer, American Barrick, set up in 1983, acquired nearly 5% in CGF, which it then sold. However it bought another £31 million stake in 1988 (15).

In 1986, CGF was considerably embarrassed by revelations that GFSA (and most of the other mining companies in the apartheid state) employed a private police force, trained and equipped with weapons from a company itself set up by CGF – and in 1969, at a time when CGF held the majority of GFSA.

A Granada TV programme screened in late 1986, which spotlighted this company, also revealed that GFSA had the most patronising and insidious attitudes towards its black workforce of all South African companies. A training manual for white supervisors, for example, stated that "A Bantu will always give an answer which he thinks will please the other person". No wonder Cyril Ramaphoso, secretary of the NUM (National Union of Mineworkers) declared that "... Gold Fields is actually the worst company that we have to deal with" (16).

Perhaps partly because of these criticisms, in 1987 CGF began to cut its interest in GFSA – from 48% to 38%, selling 10% to a new holding company controlled by the South African conglomerate, Rembrandt (17).

In September 1988 Minorco launched a bid for complete control of CGF – one which was vigorously opposed by the board, and which resulted in the longest-run and possibly the most vituperative battle between major mining companies in the post-war world. Nearly a year later, the Hanson Trust, controlled by Lord Hanson, offered £3.5 billion to secure CGF, and in August that year the bid succeeded. (For a full description of this battle royal, see AAC.)

By 1990 American Barrick Resources had sold its stake in CGF (18), and Hanson had asset-stripped Consgold in a manner predicted by many of his detractors. The shares in British-Borneo Petroleum were sold even before the ink was wet on the take-over document (19). Then the 38% share of GFSA went to the Rembrandt Group, and Asteroid Pty Ltd (jointly owned by GFSA and Driefontein Consolidated) (20). All four of Amey Roadstone (ARCs)'s wholly-owned American subsidiaries were sold to CSR – though Hanson declared that the British Amey Roadstone would not go the same way as the American outfits (21, 22).

As he entered the new decade, Hanson was left holding around half of Newmont, all of Gold Fields Mineral Corp of the USA (GFMC), and about half of Renison Goldfields of Australia (23). Chalking up £1 billion pre-tax profits for 1989 (24) and a fifteen percent profit rise in the last quarter of that year (25), Hanson set about purchasing the 45.03% of Peabody Holdings Ltd owned by Bechtel and Boeing, which it didn't already control (through Newmont). It seemed that the main reason for this apparent turnaround was not any particular interest in Peabody's huge coal reserves, but a decision to "beef up" Newmont and re-value Hanson's investment in the US mining house (26). Despite a last minute bid by Amax for Peabody, Hanson had secured the coal company by mid-1990 (27).

References: (1) Lanning & Mueller. (2) *FT* 29/4/83. (3) *FT* 3/11/83. (4) *FT* 20/10/81. (5) *Gua* 5/10/83. (6) *MAR* 1984. (7) Personal communication from *AMIC*, 1981. (8) *MIY* 1982. (9) *MJ* 9/3/84. (10)

MM 11/82. (11) *Raw Materials Report* Vol. 1, No. 4, 1983. (12) Rogers & Cervenka. (13) UN NS-36. (14) *Gua* 14/7/77. (15) *FT* 10/6/88. (16) *Gua* 25/11/86. (17) *FT* 10/7/87; *MJ* 10/7/87. (18) *FT* 20/2/90. (19) *MJ* 28/7/89; *FT* 21/7/89. (20) *E&MJ* 10/89. (21) *FT* 15/4/88. (22) *MJ* 24/11/89; see also *MJ* 14/7/89. (23) *Gua* 23/11/89. (24) *FT* 30/11/89. (25) *FT* 15/2/90; *Gua* 15/2/90. (26) *FT* 30/3/90; *MJ* 23/2/90. (27) *FT* 12/5/90.

167 Consolidated Morrison Explorations Ltd (CDN)

As of the early 1980s, board interlocks existed between Consolidated Morrison and Noranda, Northgate and other companies. Three members of the Stollery family were reported to own at least 10% of Argor. The company was also involved in a project in Newfoundland which had RTZ as a manager – presumably the now-aborted Post-Makkovik uranium mine (see Brinco) (1).

It was reported exploring for uranium in Saskatchewan and holds there uranium leases in the Flin Flon area, in a JV with Saskatchewan Mining Development Corp (2).

References: (1) *Yellowcake Road.* (2) *Who's Who Sask.*

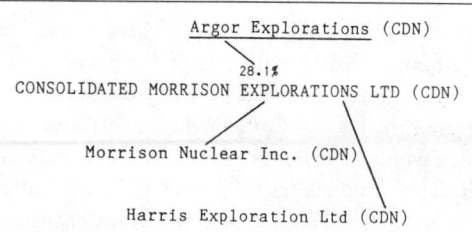

168 Consolidated Reactor Uranium Mines Ltd (CDN)

In 1979, it was reported exploring for uranium at Birch Island, British Columbia (1).

In 1980, it joined Norbaska Mines (25%) and Saskatchewan Mining Development Corporation (50%) in exploring 14,000 acres of Wollaston Lake, northern Saskatchewan (2).

References: (1) *US Surv Min* 1979. (2) *MM* 5/80.

169 Consolidated Rexspar Minerals and Chemicals Ltd

Denison Mines Ltd (CDN)

46·91%

CONSOLIDATED REXSPAR MINERALS
AND CHEMICALS LTD (CDN)

In 1977, the company carried out a feasibility study on a uranium deposit owned at Birch Island, Granite Lake, British Columbia, estimating 1.5-2 million tons of ore, averaging 1.5pounds/t: sufficient to produce 7,000 pounds/week of uranium oxide for more than 4 years at estimated costs of (then) C$27 million.

Possible start-up was projected for 1979 but opposition by environmentalists was also reported (1). The mine went into abeyance after the British Columbia moratorium on uranium mining. The property was still owned by the company the following year (2).

It also held an exploration licence in the Kamloops area of British Columbia in 1979 (3).

References: (1) *MJ* 30/12/77. (2) MIY 1981. (3) *Energy File* 5/79.

170 Consumers Gas Co (CDN)

Amalgamated now with Hiram Walker, Gooderham and Worts, Home Oil, and Cygnus Corporation, in Hiram Walker-Consumers Home Ltd, this was the fifth largest Canadian company (based on net income).

It was exploring for uranium in Saskatchewan in the 1970s (1).

References: (1) *Who's Who Sask.*

171 Conwest Canadian Uranium [Exploration] JV

Conwest organised this JV in 1975 to explore for and develop uranium throughout Canada.

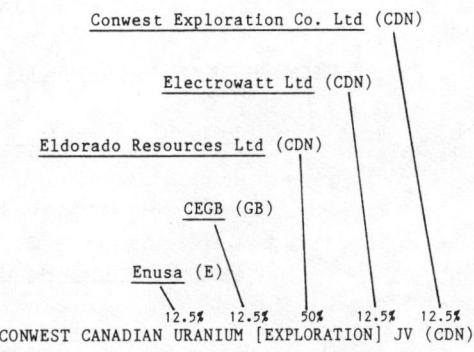

Its partners are listed in the table above; however, Conwest's own equity in the project is participated in by CCF (25% interest – the same as Conwest's), International Mogul (35%), and Chimo (15%) (1).

The JV was operating at Key Lake, Saskatchewan(2); manager is Eldorado Nuclear (1).

References: (1) *MIY* 1982. (2) *MIY* 1981.

172 Conwest Exploration Co Ltd

This is an important Canadian resource, exploration, development, production, and investment company with direct or indirect interests in a number of companies in Canada.

It is involved in an extensive exploration programme throughout Canada together with Consolidated Canadian Faraday (25%), Chimo Gold Mines (27%), and International Mogul Mines (18%), holding a direct share of 30% in it.

In 1975, it organised the Conwest Canadian Uranium Exploration JV to explore for and develop uranium in Canada (1).

It found a uranium deposit to the north-east of Kelly Lake at Geikie River East in 1977; Eldorado Nuclear was to become the operator (2).

Conwest's major mining interest is the Nansivik zinc-lead-silver mine in the Canadian high artic (owned by a subsidiary of Mineral Resources International Ltd in which Conwest holds 51%). It also owns a nickel-copper property in the Sudbury region of Ontario, in which Fal-

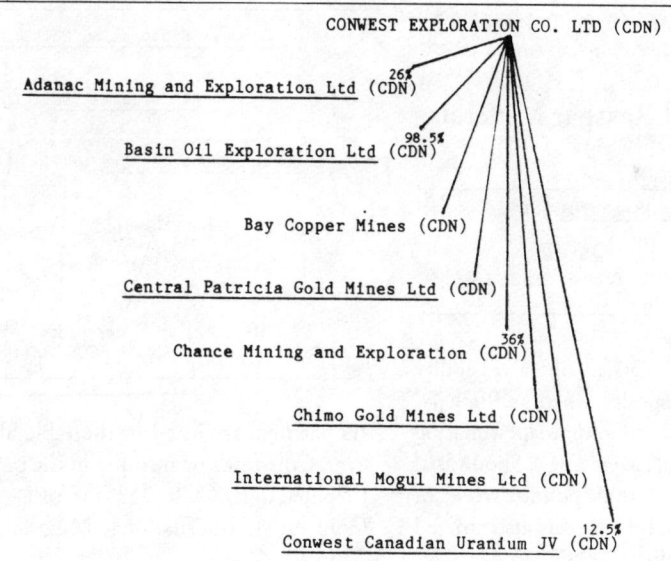

CONWEST EXPLORATION CO. LTD (CDN)

Adanac Mining and Exploration Ltd (CDN) 26%

Basin Oil Exploration Ltd (CDN) 98.5%

Bay Copper Mines (CDN)

Central Patricia Gold Mines Ltd (CDN)

Chance Mining and Exploration (CDN) 36%

Chimo Gold Mines Ltd (CDN)

International Mogul Mines Ltd (CDN)

Conwest Canadian Uranium JV (CDN) 12.5%

conbridge has an option (3), and several other gold, molybdenum and other mineral claims in Canada (4).

References: (1) *MIY* 1981. (2) *FT* 18/12/78. (3) *E&MJ* 5/89. (4) *MIY* 1990.

173 Copconda Ltd

It has been exploring for oil in the USA and Canada (1).

In 1980, it was "re-assessing" a uranium property at Hutterfield Township, Quebec, where it had started a C$130,000 drilling programme (2); it had a 20% stake in it with Century Gold Mining (3).

In 1980, York Resources was bidding for Copconda, and seemed likely to succeed (1).

References: (1) *FT* 24/7/80. (2) *MJ* 11/4/80. (3) Annual Report 1981.

174 Copper Lake Exploration Ltd (CDN)

It was exploring for uranium in the Anstruther area of Ontario in the late '70s (1); no further information.

References: (1) *BBA* no. 4, 1979.

175 Corfo/Corp de fomento de la producción

It operates Chile's oldest lead-zinc mine and has important JVs with Foote Mineral to produce lithium and with Amax to produce lithium sulphate. (It also has a JV with Amax Chemicals to produce potassium salts and boric acid) (1).

The lithium JV with Amax has to receive con-

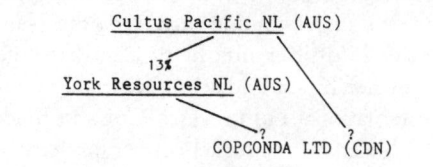

Cultus Pacific NL (AUS)

York Resources NL (AUS) 13%

COPCONDA LTD (CDN) ?

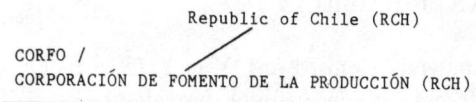

Republic of Chile (RCH)

CORFO / CORPORACIÓN DE FOMENTO DE LA PRODUCCIÓN (RCH)

sent from Chile's national atomic energy commission, the CCHEN (the country's state uranium exploration company which doesn't appear to have turned up anything very significant since the '70s) (2), due to its importance in the construction of atomic fusion fuel (1).

In 1985 it invited foreign companies to bid for two phosphate mining concessions on its property in Mejillones – a plant which is intended to produce uranium concentrate and fluorine in the future (2).

By the end of the 1980s, however, the Salar de Atacama potassium-lithium and boron project had advanced to a development stage, with the mining Joint Venturers (Corfo, Amax, and Molibdenos y Minerales) – called Minsal – having located three times more potassium, at least seven times more lithium and at least six times more boron, than there is in the Dead Sea (3). The current proposal is to use solar evaporation ponds, in sequence, to separate the various minerals (3).

References: (1) *MAR* 1985. (2) *MJ* 21/6/85. (3) *MAR* 1990.

176 Cotecna Engineering (CH)

This engineering company won a contract – in partnership with AG McKee (Britain), Union Minière (Belgium), and Traction et Electricité (France) – to develop Algeria's Hoggar Mountain deposits of uranium (1).

References: (1) *MJ* 6/7/79.

177 Cotter Corp

In the late 1970's, it was actively mining uranium in Colorado (1) where it also operated a uranium-vanadium mill with 1200t/day capacity (2).

Opposition to the mill was widespread. In summer 1978, it was revealed that the company had persistently ignored breaches of radiation control and poisoned groundwater "about halfway

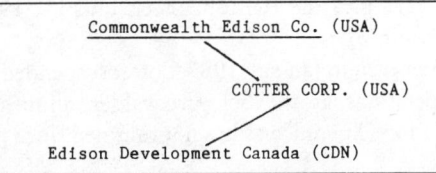

to the Arkansas River" (3). About 450 people later packed a meeting, mostly opposed to the company's plans to expand the current mill and licence a second (4).

In 1979, the company registered a prospect in Arizona (5).

In 1980, the Oil, Chemical, and Atomic Workers' Union (OCAW) – representing more than 11,000 atomic workers – initiated a campaign against Cotter's health violations at its Canon City, Colorado, mill which had affected 121 workers.

The OCAW demanded: 1) an epidemiological study of the health status of the workers by the National Institute of Occupational Safety and Health (NIOSH); and 2) inspection by the Mine Safety and Health Administration (MSHA) both of the mill and radiation from it.

In a report, released on 20th Sept. 1980, the Colorado Bureau of Investigation revealed that the Colorado Health Department had routinely notified the Cotter mill when it would inspect; allowed violation to go uncorrected for years; performed only 7 out of 12 annual inspections between 1968 and 1979; and failed to identify data falsification in the mill's exposure reports.

As of 1982, the company was also planning to open a new mine with a design production of nearly 1000t/day uranium ore at Naturita, Colorado. (6).

In January 1983 Cotter was given permission by the Colorado Health Department to sort, crush and wash uranium ore at its Schwartzwalder uranium-molybdenum mine (7) before transporting it to the company mill in Canon City. Cotter had to establish "some of the same monitoring conditions required at uranium processing mills," while water and air quality would be monitored. The mine then discharged some 600,000 gallons per day of "treated waste water" into the Ralston Creek under a 1982 permit (8).

However, in January 1987 Cotter suspended all operations at the Schwartzwalder mine and Canon City mill and has not resumed since (9).

Contact: Kay Stricklan, PO Box 1228, Canon City, CO 81212, USA.

References: (1) *Colorado Uranium Information Network* 1980. (2) *MIY* 1982. (3) *Rocky Mountain News* 26/6/78. (4) *Ibid* 2/5/79. (5) *US Surv Min* 1979. (6) *MM* 3/8. (7) *MM* 1/84. (8) *E&MJ* 1/83. (9) *E&MJ* 4/88.

178 CRA

It is hardly feasible to condense the history of CRA into less than a major book. For this is the largest company within the RTZ group – itself the largest mining conglomerate in the world – and, in 1989, it provided nearly a quarter of RTZ's profits (1). Measured by market capitalisation, CRA comes sixth among the world's top ten mining corporations (2).

RTZ may be the world's most criticised miner, but CRA is more responsible than any other of its units for the desecration of indigenous land and culture. For more than a decade, Rössing Uranium may have flagrantly violated international (UN) law, but Comalco (CRA 67%) has, for far longer, successfully cajoled, deceived and browbeaten legislators over a whole continent. Rio Algom may have been a bigger and more efficient producer of uranium, but CRA could be producing yellowcake long after Rio Algom's mines have closed. RTZ's smelting plants may have aroused more ire than CRA's – yet the Australian operations are bigger and more important. The parent company may be the most diversified mining company in the world (3), but CRA runs a close second.

Apart from the operations detailed in this entry, CRA in the last few years has held (or still holds):

• Dampier Salt – Australia's biggest producer of salt, and the world's second largest solar

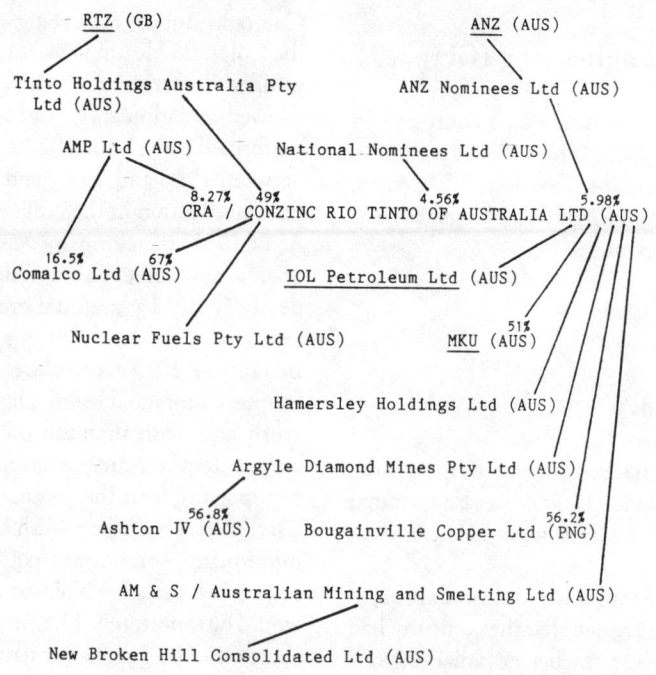

salt producer (4), 65% owned by CRA (5) (along with Marubeni, an old friend of both CRA and RTZ).

- 18% of Metal Manufacturers, Australia's leading cable group which is controlled by BICC, the British company rumoured in 1988 to be the subject of a possible take-over bid by RTZ (6).
- Australia's leading experimenter in living cell transformation, so-called "biotechnology". Biotechnology Australia Pty Ltd was formed in 1977; its first commercial product, Neogard, a pig vaccine, was released for sale in 1985 (4). CRA sold 50% of Biotechnology in 1989 to Hoechst AG of West Germany (5).
- 90% of Capital Casting, a foundry-based business in Arizona USA, purchased in 1977 (7).
- A 19.9% interest in the offshore North Herald and South Pepper petroleum prospect in Western Australia (8).
- A joint venture with Mitsubishi, formed in 1985, to explore the Oaklands coal deposit in the Riverina district of New South Wales (NSW) (9).
- A 1987 lease on large lead-zinc deposits in the north-west of Western Australia (WA), 150 km south of the town of Broome at a depth of more than 1000 metres (10).
- The Peak gold prospect, near Cobar, NSW, containing around 4.5 million tonnes of gold and other metals at fairly high grade (11).
- A sophisticated research centre in Cockle Creek, NSW, and the intention to construct another one in Perth, WA (8).
- A joint venture with Battle Mountain Gold Co (USA) (which is the main partner with RTZ in the Lihir Island gold prospect in Papua New Guinea) (12) in a precious metals project in Kalimantan (Borneo) Indonesia. CRA is the major partner and operator (60%) (13).
- Conzinc Asia Holdings, which carried out dredging for tin in Malaysia and manufactures aluminium products (5, 14). This was sold in 1989.

CRA is Australia's most important mining company: depending on how its operations and investments are evaluated, it can also be accounted Australia's second or third largest privately-owned corporation. Its major shareholder, RTZ, holds 49%: Australian public investors hold only 40% of the share capital between them (15).

Until the mid-'eighties, CRA was returning mixed, and often poor, results from its worldwide operations (16). But a decision taken to dispose of assets "not essential to the core of its business" (17) was beginning to yield results by the second half of the decade. By 1987, it had sold its timber and Melbourne property interests, and put its Forrest Gold mines and prospects in Western Australia on the market (18). In early 1988, the new company chair, John Uhrig, was telling the annual general meeting of CRA in Melbourne that, thanks to high productivity at its major mines, cash gained through disposals and the favourable exchange rate (19), CRA was now "cashed up" and ready either for a major one billion dollar acquisition (20) or a number of smaller ones (21).

1988 was a record year for the company (22) (profits were up nearly 100% on 1987's). Borrowings went down; so did corporate taxation in New Zealand and Australia, and there were "unusual" gains through a "restatement" of net future tax liabilities at lower rates (23).

It now cast around in earnest for new ventures: not in "downstream" activities, or further diversification, however. Like its parent RTZ, which had just acquired most of BP's mineral interests, it was interested in mining. Discussions about merging with two Australian coal and aluminium companies did not prove fruitful (24), but CRA was very interested in the possibility of acquiring Consolidated Goldfields' 48% stake in Renison Goldfields – should it be marketed. The company also undoubtedly held talks with RTZ about taking-over the London parent's 49% stake in the Roxby Downs uranium-copper mine (25).

In September 1989, BP withdrew its offer of the Roxby Downs share to RTZ, after Western Mining took legal action to halt the sale. However, BP Australia apparently did not prevent a later sale (409).

CRA's origins go back to the beginning of this century. The connections between the Anglo-Australian commercial elite behind the Consolidated Zinc operations, and the mining and banking interests within the Rio Tinto company, also stretch back considerably further than the Conzinc-Rio Tinto merger effected in 1962 (27), out of which came today's Rio Tinto-Zinc Corporation (RTZ). For example, in 1954 Rio Tinto acquired control of Mary Kathleen Uranium (MKU) (28) and, within the next six years, was zealously evaluating a range of Australian deposits (29).

RTZ's master plan, the brainchild of Rio Tinto's Val Duncan, honed by his colleagues Mark Turner and Roy Wright, was to "create a Rio Tinto company in each of the principal mining countries, which in turn would control a series of operations within its own territory" (30). From the start, operations and management would "as far as possible" be local, and "local participation in the equity of the overseas companies was ... essential" (30). But each regional operation would be fully backed by, and identified with, the interests of the RTZ Group in London: "... there would always be ... agreement on the aims of the individual companies and the Group as a whole" (30).

Notwithstanding the reluctant reduction of the parent company's equity in CRA to less than 50% nearly fifteen years later, the strategy laid down in 1960-62 has been relentlessly pursued ever since. CRA has always served the interests of RTZ. Or rather, as the most important and extensive mining company within the Group, CRA's investment decisions and areas of operation have usually meshed-in perfectly with those of the parent. Examples of this are numerous: the opening-up of the Weipa bauxite deposits in Cape York, Queensland, and establishment of Comalco (see below); CRA and MKU's participation in the uranium cartel; the expansion of the Hamersley and Broken Hill heavy metal fields; the opening of the Argyle diamond mine and the subsequent decision to market

most of its output through the Anglo-American/De Beers Central Selling Organisation (CSO) of South Africa; carving out the Bougainville mine as an Asian-Pacific copper producer, ensuring the Group's continued dominance in world markets.

Certainly, it has not always been an easy relationship: it is sometimes possible to detect a conflict between CRA management or executives on their home ground, and the partners in London. When CRA's chief architect, Sir Roderick Carnegie, finally left the company in 1986, handing over to John Uhrig as non-executive chair, and John Ralph as managing director (31), it was speculated that he had argued with RTZ. Carnegie had wanted CRA to bid for Australia's biggest company, BHP, and Big Daddy of St. James's Square had opposed it. While Carnegie's resignation "stunned the local community", it came as no surprise to colleagues, reported the *Financial Times* – which also asserted that Carnegie had been the main force behind moves to reduce RTZ's equity in CRA (32).

Nor have CRA's "downstream" activities, and its expansion overseas, necessarily served RTZ's best interests. Would the parent company have flirted so closely with Krupp and Kloeckner Werke in the 1980s for example? Nonetheless, CRA (more recently its wholly-owned subsidiary CRA Exploration [CRAE]) has more tenements under exploration than any other Australian mining company; its presence in key regional countries with relatively unexploited resources (such as Indonesia, Papua New Guinea, Malaysia, China) must be the envy of many Australians. Such unrivalled spread of investment and expertise has served RTZ well in its own strategy of bringing markets and mines closer together (above all in the case of Japan) (33), reducing labour, production and environmental costs, and exploiting new deposits.

The doyens of 6 St James's Square (RTZ's suitably anonymous London headquarters) have consistently denied this symbiosis between the Group and its most important subsidiary. But this is not a consistent position. With a certain smugness (or studied carelessness?) RTZ sometimes lets it be publicly known that it still holds the reins. After RTZ's holdings in CRA had fallen below 49%, the London *Financial Weekly* commented revealingly: "When CRA had troubles with industrial action at Broken Hill [in 1986] there is no question that the men from London were very much instrumental in settling the dispute. If problems with the overseas investments do arise, they are very often matched by RTZ's cultivated diplomatic skills overseas ... Sir Alistair disagrees that the London office may at times be too far removed from operations to exercise tight control" (35).

In the late 'eighties many people were disturbed at the ruthless zeal being employed by CRA, in order to gain access to valuable uranium deposits in the Rudall River National Park (WA) (*see below*). It is scarcely conceivable that the decision to try and bring this project into production was taken by CRA alone. All the company's uranium production to date has served the marketing needs of RTZ. When it was convenient to RTZ's grand plan to close MKU uranium in 1963 (despite loud protests from Australia), then it was closed.

"Australianisation" is something which CRA has ostensibly been carrying out for some years. In 1979, the Prime Minister issued a call for majority ownership of the country's natural resources to be in domestic hands (36), although, a year earlier, the government made it clear that this should not deter foreign investment (37). It took another seven years before the company complied. Various share placements and rights issues had effectively brought RTZ's equity down to just over half by the mid-'eighties. For example, CRA's increase of its stake in Hamersley Iron Ore Co in 1981, and a share-placement to Showa Denko of Japan and another company, reduced its interest from 52.9% to 52.3% in 1985 (38). It was not until late 1986, by selling 16.38 million CRA shares to Australian Mutual Provident (AMP) at A$7.50 a share, that RTZ's holding then fell to 49% (36). Even so, within a few weeks, the London company was taking up its full entitlement on a 1:8 CRA share rights issue, designed to raise

capital for expansions at Tarong, Blair Athol, and the Argyle diamond mines (39). Nor was naturalisation a major inconvenience to RTZ: a large part of CRA's debt was, at a stroke, removed from RTZ's balance sheet (36, 40), and the London Group now paid less for depreciation of its reserves (36) (because of differences between British and Australian accounting systems). Even before the pressure to "naturalise" became too much to bear, CRA and RTZ learned that this could be a strategy not to diminish, but actually increase control over domestic resources, for both their benefits. This would be by taking-over existing Australian-run enterprises (33) and offering CRA shares in exchange for shares in the Australian companies. "Naturalisation", as Ritchie Howitt commented in his incisive 1981 analysis of CRA (33), is "... one of a series of integrated corporate strategies aimed at diversifying CRA's resource base; increasing its political, economic and industrial power; and improving both its immediate and long-term profit prospects".

In 1977, CRA bid for control of two Australian companies with ownership of highly promising coal fields – AAR (Australian Associated Resources, which discovered the Hail Creek coalfield in Queensland) and Coal and Allied Industries (CAIL) (41). When the Australian Foreign Investment Review Board (FIRB) stepped in to halt the AAR take-over bid, CRA embarked on a media and letter-writing campaign, using threats and blackmail "aimed at undermining the Government's politically stable image within a week" (33). As political columnist Brian Toohey commented at the time: "... CRA has usually gone to great pains to present a public image of the good, well-mannered corporate citizen ... its recent take-over attempts, however, present a picture much more of the iron fist than the velvet glove" (408).

CRA didn't proceed with its vilification campaign, and it didn't secure AAR. But it did win, with Howard Smith, control of CAIL. It had to apologise to the Melbourne Stock Exchange for failing to inform them of the bid, and it came in for damning criticism from the press. In the *Australian Financial Review*, one journalist accused the company of "[breaking] the spirit of the law of the land, and on the face of it, the actual words of that law ... in terms of corporate behaviour it could probably dismissed by saying that all CRA has done is to commit a traffic offence. It has treated the law in a technical way, knowing that if it is prosecuted and found guilty, it can afford to pay the fine" (43). After a temporary freeze, the bid was allowed to proceed – but not before it was discovered that the chair of the FIRB, Sir William Pettingell, was himself a director of both Howard Smith and CRA! (44).

The only other occasion on which the government has rebuffed CRA's attempts at a takeover came when CRA (*i.e.* RTZ) bid for control of the Nabarlek uranium project in the Northern Territory (NT) in the early 1980s.

CRA and Aborigines

"Don't CRAp on our land!" was the demand made by the Aboriginal Mining Information Centre (AMIC) on car bumper stickers produced in 1981, and distributed throughout Australia. The slogan neatly encapsulates the reality: CRA has been more responsible for encroachment on the land of Australia's original owners and the implicit denial of their land rights, than any other mining company. In the context of RTZ's worldwide operations, its Australian subsidiary stands uniquely indicted. No less than ten Aboriginal delegates, representing several major Aboriginal organisations (Aboriginal Mining Information Centre, North Queensland Land Council, Kimberley Land Council, South Eastern Land Council, Western Desert Land Council and the National Federation of Land Councils), have attended RTZ annual general meetings over the past decade, to focus international attention on CRA's activities and plans (45).

In 1978, the first-ever Aboriginal delegation to leave Australia and speak out publicly against the effects of mining, came to Britain with a brief to address RTZ about its subsidiaries' operations *(see below)*. At the time, the US mining conglomerate, Amax, was being attacked by Aboriginal people, trade unionists, Australian Labor Party (ALP) members, and others, for its desecration of a sacred site on the Noonkanbah pastoral station. However, CRA had itself smashed very sacred ceremonial boards at Noonkanbah, and it posed a greater long-term threat than Amax, in its search for diamonds and uranium (46). The Yungngora community at Noonkanbah failed to prevent CRA from prospecting – though, out of 98 claims pegged at this time (47), only three were approved, and those were made conditional on a site survey (48). (The warden responsible for approving claims commented that Aboriginal fears were justified: "We continue to do things to the Aborigines, rather than with them") (49). Once it had proved-up the Argyle diamond deposit, CRA appeared not to be so interested in Noonkanbah.

CRA was also the first mining company in Australia against which the major land councils called a boycott, for its violation of Aboriginal land – specifically at Lake Argyle, but also at Noonkanbah (46, 50).

Yet, it was Comalco (67% owned by CRA) which promoted the "Pitjantjatjara" (sometimes known as the "Comalco") model, for negotiations between mining companies and Aboriginal people, whereby a land council or community, having received detailed prospecting plans and company intentions, identifies areas in which the company may work: not only for mining, but drilling, laying seismic lines, constructing roads, water bores and camps (51). There has been justified criticism of the "model" – mainly because it concedes the right of companies to prospect and mine in situations where land rights and a mining veto have not been granted: hence, it can increase the sense of powerlessness among Aboriginal people (52, 85).

Nonetheless, it was a marked improvement on the situation prevailing through most of Australia in the 1970s and early 1980s – let alone that which characterised Comalco's own operations in Cape York province, some forty years ago.

Is it possible then, that the leopard can change its spots? Or is it more a case of the chameleon adjusting its colours to suit differing terrain? All the evidence suggests the latter. Where CRA (or Comalco) has been confronted with strong, united, Aboriginal opposition to its intentions, it will at least make a pretence at negotiations, or its protestations of innocence will be loud and clear. (No sacred sites or sacred objects will be damagaged, declared Sir Roderick Carnegie at RTZ's 1981 AGM) (53). And, where the community lays down reasonable but unpalatable conditions, CRA – rather than practising deceit or *force majeure* – may withdraw. It did so from Oombulgurri in 1980, after the community insisted on: recognition of the Kimberley Land Council; at least three months notice of intended activities; negotiated compensation, proper training and employment; protection of sacred sites and improvement to community facilities (46, 54). But then, only the previous year, CRA had confirmed discovery of potentially the world's largest kimberlite pipe outside of South Africa, at Lake Argyle: for the time being, Oombulgurri's diamonds were not a priority (55). What CRA did to the people and land around Lake Argyle, it would undoubtedly have been prepared to do to the Forrest River people at Oombulgurri, had commercial considerations warranted. Indeed, just after CRA was beating a graceful retreat from Oombulgurri, it entered a traditional Aboriginal gathering ground at Christmas Creek and, without waiting for a sacred site survey to be completed, plunged a drill directly into a major sacred place (tjilla) called Kurungal (56).

The true face of CRA is not, then, to be judged by isolated acts of generosity (drilling for water in a parched desert community, or subsidising a Toyota), or minimal compliance with Aboriginal demands (repairing delapidated housing at Weipa, on Cape York, in the 1970s, or supplying survey maps which are calculated to confuse as much as they enlighten). CRA is to be judged by the overall impact of the company's operations on Aboriginal communities, and its secretly-expressed intentions towards Aboriginal land and land rights.

CRA has long been a key member of the Australian Mining Industry Council (AMIC – not to be confused with the Aboriginal Mining Information Centre, sometimes known as AMIC-*US!* by the Aboriginal people, to distinguish it from AMIC-*THEM!*). Although leading CRA personnel have not played such a high-profile role within the AMIC, as for example Hugh Morgan of Western Mining Corporation, it has distributed many of its self-justificatory pamphlets through AMIC (46); sponsored free booklets published by AMIC for children and, along with other AMIC members, made large contributions towards the eventual defeat of the pro-land rights Whitlam administration in 1975. There is no doubt that CRA's commercial priorities and those of AMIC – especially in the mid-1980s when land rights raised their threatening head – have been virtually synonymous.

On two occasions in the last fourteen years, the veil over CRA's "hidden agenda" for Aboriginal people has been pulled away. On both occasions the revelations were made by employees who were incensed at the gulf between the company's public declarations and its secret intent. In 1978, a Memorandum fell into the hands of CIMRA (Colonialism and Indigenous Minorities Research/Action), summarising mineral deposits discovered by CRAE on 39 Aboriginal reserves – some of which had clearly been under scrutiny for a considerable time. Among the areas listed as "top priority" for future programmes were Arnhemland, Groote Elandt and Forrest River. Daly River, in the NT, was listed as having "potential for diamonds, copper, lead, zinc and uranium". Among the reserves listed in Group 2 – "only marginally below those in Group 1" – were the West Kimberley Reserves (potentially diamondiferous) and Yuendumu. "Low mineral potential" was recorded for Jigalong Reserve in Western Australia (WA), Cundeelee (WA) among others, and "little or no mineral" potential was assessed for eleven other reserves, including Beagle Bay (WA), Aurukun-Mitchell River in Queensland (QLD) (only bauxite was considered prospective in this region), the Yalata Reserve (WA) and Mornington Island (QLD) (57). The memorandum had been compiled by Harry Evans in 1976 (the CRA geologist who "discovered" the Weipa bauxite deposits) and was, in fact, leaked to the *Melbourne Age* the following year (58). But it was not until it was publicised at the RTZ annual general meeting in 1978, that its full import was appreciated.

CIMRA, together with the major British overseas aid charity War on Want, and supported by Survival International and Greenpeace (London), issued a press release accusing CRA of deliberately focussing on remaining Aboriginal lands as soft targets (59). The furore caused by the revelation (60) forced RTZ's chair, Mark Turner, to send his vice-chair, Lord Shackleton, to Australia, ostensibly to check the allegations. In fact, Shackleton did not meet any major Aboriginal leaders during his brief visit, spent *less than an hour* at the Weipa reserve, and did not talk to any Australian press (61). RTZ also responded angrily to an Open Letter addressed to Turner by War on Want and CIMRA, accusing Comalco of forcing Aborigines off reserves and settlements, burning their homes, destroying their culture, and showing "brutal unconcern" in the face of consequent suffering (59). It sent a solicitor's letter threatening a libel action if the accusations were repeated (62): War on Want did not do so, but CIMRA continued to assert the truth of its case.

Apologists for the mining companies would argue that a Memorandum of this kind is nothing to write home about (certainly nothing to tell the press about): it is merely routine. Such cynical acceptance of the "right" of mining companies to walk onto Aboriginal land, explore and drill without consent from its owners, then blithely presume they will be able to proceed to mining when it suits their interests, should appall us just as much as a public rela-

tions exercise designed at weakening land rights campaigns of specific communities.

It was such an exercise which comprised part of a second covert action indulged in by CRA. By 1980, Greg Walker, the public affairs manager for Ashton JV/Kimberley Diamond Mines (forerunner of the Argyle Diamond Mines JV and then, as now, controlled by CRA) had commissioned two public relations companies to suggest how CRA could lobby against the setting-up of a diamond cutting industry in Australia (which would undercut arrangements it had made with Anglo-American/De Beers), and "sustain the Argyle Agreement with the Glen Hill Aboriginal community and isolat[e] this agreement from the general debate on Aboriginal land rights while encouraging community acceptance of the company's policy towards its Aboriginal neighbours". What this meant, quite simply, was that CRA wanted a public relations campaign which would convince the public that it was supporting Aboriginal self-determination while actually doing the opposite. The Glen Hill "agreement" had been signed with a selected group of Aborigines, in order not to have to deal with the demands of the whole community *(see below)*.

The company chosen for this purpose was Eric White Associates, the Australian subsidiary of the world's biggest public relations firm, Hill and Knowlton – "like Ashton Joint Venture, it was a hawk" (63). CRA's own public relations company, International Public Relations Pty (IPR), had its submission rejected (46), because it argued for securing a larger share of revenue for Aboriginal people (63). Eric White, on the other hand, was broadly in tune with Ashton JV objectives. Greg Walker urged acceptance of the White proposal – and so the world's biggest public relations company was duly brought in to White-wash Australia's biggest threat to the Aboriginal people of the Kimberleys (63). The saving in royalties to Ashton JV was promised to be "substantially greater than the proposed expenditure on the company's public relations programme next year" (3, 46, 64).

Seven years later, it was CRA's activities in the western desert which gave rise to most anxiety among Aboriginal communities *(see below)*. The Martu people have good reason to be disquieted. Several of their community visited the Argyle diamond mine, in 1988, to see what they might be letting themselves in for, and they returned absolutely determined to stop mining (that of CRA in particular) on their traditional lands.

Even before then, like almost every other Aboriginal community throughout Australia, they had heard about Weipa and Mapoon in Cape York province, Queensland: site of the world's largest open-cast bauxite mine.

These two places have become synonymous with the devastation and social breakdown caused by mining. In a lengthy report, in 1976, *Melbourne Age* reporter David Broadbent delivered one of the most powerful attacks on the human effects of mining ever made, after a visit to Cape York. "If men ever establish a base on the barren surface of Mars," he wrote, "it will look like Weipa". By then Comalco had been mining at Weipa for a decade – the result, declared Broadbent, was "acres of dead craters unrelieved by a single growing thing". The conditions at South Weipa – site of the main Aboriginal relocation – were such that "traditions have disappeared and alcohol has wreaked havoc ... the Weipa operations have caused alarm in every Aboriginal community throughout the north of Australia" (65).

A few years later, the sociologist Paul Wilson, following up an apparently routine murder investigation at Weipa in 1979, analysed more closely the reasons for the devastation. He found that the rate of imprisonment among Aboriginal people at Weipa was the highest in Australia; the rate of violent crime and self-inflicted violence (a convincing indication of grave psychological malaise) was profoundly disturbing (66). The case of Alwyn Peters, which first elicited Wilson's concern, was later to be dramatised and filmed – by the well-known Australian film-maker David Bradbury (67). What is especially alarming is that this recent film (1989) *State of shock*, graphically describes a situation which has clearly not improved over a period of ten years. Aboriginal writer Marcia Langton – confronting criticisms that *State of Shock* "reinforces stereotypes of Aboriginal drunks" – wrote in a letter to David Bradbury:

"When Mrs Jean Jimmy and Mrs Rachel Peters tell us about their country and their forced removal by police in 1963, that is a powerful explanation, if not justification, of the motives of the younger people in their destructive behaviour ... it is clear that the old ways must have provided a sense of identity ... of personhood which was satisfying, and a way of life which sustained the development and growth of the person and quality of life through to old age ... The Peters family and others were rounded up and moved from Mapoon, at the whim of a mining company. As Rachel Peters explains: 'All this suffering for nothing; there has not been any mining there since they were taken away, their homes burnt'" (68, 69).

Mapoon and Weipa

Mapoon and Weipa were two of the oldest reserves in Australia (46). "Missionised" by servants of God, many of whom were little more than racist despots, the area they comprised was cursed as the site of some 3000 million tonnes of bauxite worth an estimated A$60 billion (46). In 1957, Comalco (then a wholly-owned subsidiary of CRA's predecessor, Conzinc) opened secret negotiations with Queensland's Labor government, designed to keep all other interested parties off the newly-discovered bauxite lode. From the start, CRA/Conzinc refused to attend any meetings to discuss compensation with the Aboriginal people living on the land (44).

The same year, as a new right-wing government took power in the state, Comalco and its friends in power pushed through Parliament the Commonwealth Aluminium Corporation (Comalco) Bill. This extraordinary act of theft gave the company 2270 square miles of Aboriginal reserve land on the west coast of Cape York, and nearly 2000 square miles (1982) on the east coast. This was traditional Aboriginal land, though not classified as a reserve. The company was required to select, and retain within a decade, 1000 square miles of the western, and 500 square miles of the eastern, leases for a minimum of 105 years. The other 1270 square miles had to be relinquished by Comalco over a period of 20 years (44, 46). The royalty was set at a mere 5 cents a ton – about the lowest rate in the world. (It was doubled in 1965 (46), although it went up ten-fold to a more equitable rate in 1974) (70). Rent was set at A£2 a square mile (it rose later) – only 1/160th (one hundred and sixtieth) of the normal mining rental at the time. All timber, cattle-grazing, water and farming rights in the lease area were granted to the company; some A£8.5 million was given in government aid for infrastructure; and Comalco was exempted from the Clean Water Act.

Not one of the demands made by the Presbyterian Church of Queensland – the mission authority at the time – was accepted by the government: there was to be no compensation (because all the "natives" would be employed), no Native Welfare Fund, no exclusive pastoral rights for the mission, and no other safeguarded Aboriginal rights (71). Comalco's attitude to Aboriginal self-determination was quite clear: Sir Maurice Mawby, a director of Conzinc, told the mission in August 1957: "We aim to provide a gradual and satisfactory means of assimilating suitable natives" (72). Comalco, said Sir Maurice Mawby, was prepared to help with native education and employment opportunities – the salve to bad conscience that the missionaries most required (46). At the same time, Comalco would pay for the eviction of Aborigines from Weipa, and assist in the removal of families from Mapoon if necessary – although the company did not initially need their land.

From the very start then – and whatever the protestations of innocence by Comalco in the years to come – mining and eviction were two sides to the same dirty coin.

Comalco's initial proposal was to move Mapoon and Weipa residents to Aurukun – a reserve area some 80km south. They refused. In 1960, Comalco, on its own accord, built a settlement at Hey Point, on the mouth of the river opposite its mining headquarters. But, once again, the Mapoon and Weipa people refused to relocate.

Comalco maintained its pressure on the mission to get the Aboriginal communities out from under its feet. The Weipa people were forced into a tiny area set aside for them at Weipa South, some nine miles from the white mining town at Weipa North. This mere 308-acre plot was far removed from their traditional hunting grounds (much of which has been devastated beyond redemption, the game having fled). Their old homes were bulldozed off the beach as an "eyesore," and it was many years before anything like adequate housing, sewerage, and any pretence of training and education, were provided (46, 73). While priding itself on its "gifts" to the community, a strain of patronising puritanism has run through much of Comalco's attitude to remuneration. There is abundant evidence that cash alone has been socially destructive in many Aboriginal communities (74). Comalco has also consistently refused to consider anything approaching royalties or compensation.

However, there are two much-vaunted benefits which, the company declares, support its case for mining. The first is employment of Aborigines – intended under Queensland government policy to be the solvent for assimilation. In reality, employment has been meagre (only 10% of the workforce was Aboriginal in 1977) (75). During much of Weipa South's existence, and despite government policies of supposed equal wages, Aborigines have worked on the lowest economic rung. In 1963 they were still being classified as "handicapped, being unintegrated to the industrial demands of our society" (76). Weipa South remains the impoverished, black township, and Weipa North the modern mining town. The parallels with South Africa are all too clear (46).

A second beneficence, supposedly visited on the community by the company, is its rehabilitation of mined-out lands. It is only in the last decade that Comalco has made any real effort to compensate for some of the devastation it has caused. By 1975, only 300 hectares had been revegetated. In late 1976, Senator O'Keefe, deputy leader of the Labor opposition, reported that the exotic tree species used by the company (79) condemned the landscape to "total dead[ness]" – failing to provide food for indigenous birds or animals (77). The following year the company was compounding the error, experimenting with Burma teak, mahogany and cypress trees, aiming at planting some 400 hectares of these every year (75). In 1981 an investigator, looking at the general environmental impact of the mining, praised the company's apparently "good attitude to regeneration of mined-out areas" (it was spending about A$5000 per hectare in 1980), but found that Comalco persisted in trying to grow commercial crops, with "little success"... . "Huge fertiliser requirements make planting costs astronomical. One species of tree, an exotic called African Mahogany, is the only one showing signs of usefulness". (This investigation did find that regeneration of pre-mined vegetation was relatively successful) (78).

Comalco has made much of the climatic problems on the Cape York peninsula and the prevalence of termites. It cites drought, fire and flood (during a heavy wet season the water table can rise to the surface) as factors which "seriously limit the range of tree species which can be grown" (79). What it does not say, of course, is that mining is the prime factor in constricting the choices available in rehabilitation: though it does admit that the "ironstone mine floor which lies immediately below the relaid overburden" is so "compacted by mine machinery during the mining operation" that it has to be "ripped to allow plant penetration and development" (79).

The Australian Financial Review in 1977 predicted that, thanks to its timber operations on Aboriginal land in Cape York, Comalco by 2000 AD "will be a major timber group" (75). This was too close to suggesting that, having ripped-off and ripped-out the heart of Aboriginal traditional country, CRA was now going to

misappropriate whatever meagre fruits the barren land could support. Its public statements in the late 1970s stressed that, by starting forestry, agricultural or pastoral activities, Comalco could "obviously benefit the Weipa community and the State of Queensland," and that Weipa South council members "have expressed approval about the rate of growth in the regenerated areas, the high proportion of native species and the abundant evidence of birds and animals returning to the replanted forest areas" (81).

In fact, the majority of Aboriginal people at Weipa have expressed grave disapproval of the company's ill-fated attempts to rehabilitate the area. Joyce Hall, one of the most prominent members of the community, travelled to Europe in 1981 to testify at an international tribunal on the degree to which her land had been rendered a veritable dustbowl (82). Comalco eventually won the defamation case which it had brought against the Australian Broadcasting Commission (ABC) for similar allegations made by Joyce Hall in the Granada TV film *Strangers in their own land* (64). However, during that trial, Weipa people gave eloquent testimony on the degree to which Comalco had failed to keep its promises. Gertude Molton testified that mining had destroyed the hunting, while their homes flooded in the wet season. According to Andrew Miller, Comalco was not re-planting local varieties of plants, and animals would not return to their original habitat. Stanley Budbury said that he and his people had to walk twenty miles to catch pigs (64). Perhaps the central testimony came again from Joyce Hall. For the first time in more than twenty years, she chose to tell publicly what had happened when bulldozers first moved onto her ancestral land at Weipa. Tongue-tied by emotion, she had to leave the court at one point. When she returned, she declared:

"Since the mining come, a real sacred place was destroyed ... that is where the bodies, the deads, were put on the trees. In our own traditional ways, the bodies are not buried. They are put up into the trees until the grease is down ... left to dry before we could go back. This was destroyed when they burnt and knocked the trees down. They did not know it was there. Of course whiteman should know better. Ask first." (83)

The Aboriginal community at Mapoon was a hindrance and an embarassment: to the church, because it was becoming militant and resistant to assimilation; to the Queensland government, because it showed too many signs of being able to live independently; to Comalco because, while the company had no immediate plans to mine on mission land, it had every intention of doing so later. Indeed, the Mapoon people themselves remember being told by both the government and the Mission that "Comalco would come and turn our homes over with bulldozers, and they would dig holes all over our hunting grounds" (69).
Various mailed-fist-in-velvet-glove tactics – and more belligerent ones – were used to persuade the Mapoon people to leave "voluntarily". If a woman went off to the hospital to have her baby, she was not allowed to return home; if a man went off the reserve to find work, he was banned from coming back.

In 1963, the Missionaries packed up and went. Despite closing down all the community facilities, some seventy Aboriginal people remained. It was then that they really asserted their independence: living off the bush, ferrying-in from Weipa necessary supplies and looking after their own cattle (84). But then, under cover of darkness on November 15, 1963, armed police slunk into Mapoon, arrested the entire community, burned their homes, the church, school, stores and shops. The Elders were transported to another reserve. Some other people remained but they too were forced off within the year (69). Two years later, Alcan of Canada was to be granted a lease over much of the lands not contained within the Comalco lease (46).
Notwithstanding this act of terrorism, some Mapoon people returned to land not yet affected by mining in 1974. In order to enable the State government to close down their camp, Comalco later turned-over this part of the lease

to the state authorities. Despite closure of the community's bank account, and denial of social benefits, many of the people remained. In its one act of charity during this whole repugnant saga, by 1979 Comalco had returned the resettled area and the site of the old Mapoon mission to the people (still, of course, under the repressive tutelage of the Queensland government). Comalco continued to mine on the remainder of their land in the old Mapoon Aboriginal reserve (46).

At around this time, Comalco was laying plans to open up further bauxite mines on the Aurukun reserve, south of Weipa, along a coastal strip eight miles wide, fifty miles long and more than 750 square kilometres in area (46, 84). The plan was linked to proposed mining by a consortium of foreign companies, called Aurukun Associates (Tipperary, Billiton, Pechiney). To date, mining has not taken place at Aurukun. Nonetheless, the Associates' and Comalco's plans had severe consequences for Aboriginal people, not only at Aurukun but throughout north Queensland. Initially the Aurukun people were able to halt the operations. In retaliation, the Bjelke-Peterson regime went to the Privy Council and won permission to award mining contracts. Shortly afterwards, the régime arbitrarily cancelled the reserve status of Aurukun and Mornington Island, making them shires under control of an administrator: a return to the worst days of the early twentieth century, when Queensland had itself shown the architects of South African apartheid the way to solve their "native problem" (85). It was not until several years later, that the Queensland government modified its legislation affecting Queensland's Aboriginal people living on reserves (86, 87).

At Argyle

CRA now controls the world's single largest diamond production company, operating the world's most lucrative diamond mine. The history of the Argyle diamond mines is one of deception, derring-do, devastation and divide-

and-rule of which perhaps only RTZ's favourite son is capable. The Argyle Diamond Mines Joint Venture was originally one of many outfits, sparked by the Kimberley diamond rush of the 1970s. The Kalumburu JV was formed in 1972 by five companies, and CRA began "farming" into this group a few years later. By the end of 1976, it had secured 35% in what was then known as the Ashton Joint Venture(AJV): this gave CRA management rights (88). That year the AJV held prospecting leases covering 450,000 square kilometres (88). In 1977 and 1978, CRA secured the minority shares of both the Belgium company Sibeka (89) and Jennings Holdings. By the end of 1978, it controlled nearly 60% of the Joint Venture (90).

Meanwhile, a new Western Australia Mining Act was being pushed through, squeezing out smaller prospectors, and giving favoured status to the mining corporations (91). Throughout 1978, the company received increasingly sparkling reports of finds in the kimberlite diamond pipes located around Lake Argyle (92). No less than 26 pipes covering 600 hectares had been confirmed by the end of that year (93), and bulk sampling had commenced (94).

The value of CRA's market capitalisation took a huge leap at this point: up by half a billion dollars. As the ever-vigilant *Australian Financial Review* put it: "Expressed another way, the value of Rio Tinto-Zinc Corporation's investment in CRA has jumped A$308 million". This meant that the cost of reducing RTZ's equity in CRA to the level stipulated by the FIRB (see above) had itself increased to around A$100 million – thus making it even less likely than before (95).

It was important to CRA's interests to conceal both the value of its finds and their exact location. In a deft display of double bluff, it bought out other company leases in the Argyle area, by convincing competitors that its interests really lay there (96, 106). For as long as it could, it also refused to release details of its discoveries (97). When the Melbourne Stock Exchange took it to task, CRA agreed to publish certain details of specific finds – but only at a relatively late stage (98). At this time, too – though scarcely noticed – the AJV began talks with the Anglo American/De Beers' Central Selling Organisation (CSO) over eventual marketing (99).

Early 1979 saw the discovery of many small diamonds in the area (100). AJV was by now firmly dominated by CRA (56.8%) with Ashton Mining (46.3% controlled by the Malaysian Mining Corp Bhd) holding 24.2% and AO (Australia), itself controlled by the Malaysian government, holding 4.9% (101). Much to the chagrin of those who have long campaigned for Australian control of Australian resources, this meant that potentially the most important national mining project in the 1980s was now in foreign hands. Tanust (a subsidiary of Union Minière's Tanganyika Consolidated, or Tanks) held – and still holds – just over 9%. The only domestic equity was 5% in the hands of Northern Mining, managed by the Bond Corporation.

The following year, CRA formed Argyle Diamond Mines Pty Ltd, and two years later the AJV was replaced by Argyle Diamond Mines JV (ADMJV) and Ashton Exploration, to be managed by CRAE (88).

Towards the end of 1979, the JV announced its most exciting discovery: an "extremely rich" central kimberlite pipe on the Glen Hill-Lissadell border area (a pastoral station), comparable with any find in South Africa (102).

The Western Australian Minister of Mines approved mining – even though the AJV itself continued referring only to "sampling" (and, in order to conceal the true extent of its discovery, barred journalists from entering its prospect) (103). The area was by then massively disturbed, criss-crossed by roads and camps, the product of "booty" or "raider-type" sampling techniques (103, 104). The decision to mine was finally announced in July 1980.

Between 1982 and 1985, world diamond production (both gem and industrial) increased by more than 40% (41.4%) (105). This was almost entirely due to production from Argyle (105).

Mining of the proximal alluvials started in early 1983 and, of the Argyle pipe itself – "actually a crater, 1600m long and from 150 to 600m wide at the surface, approximately 45ha in size" (106) – in late 1985. Twenty two million tonnes of overburden had to be stripped away (107). Once mining at this deposit was on the cards, CRA began dropping a large number of its east Kimberley claims (108).

CRA was now riding a whirlwind. Its share of Argyle profits quadrupled in the first six months of 1986 (109). By 1988, the JV was admitting that its original estimate of 70 million tonnes of reserves had been "conservative" (110) and, in early 1989, the partners shelled-out nearly A\$40 million to develop deposits around the kimberlite pipe, AK-l (111).

In the late 1970s, CRA had dealt with noises off-stage from two smaller Australian companies with interests in the region. The Afro-West Company, based in Perth, claimed it had staked five areas in the middle of the kimberlite pipe, after CRA's claims had been allowed to lapse and before they were renewed. Afro-West's court challenge was voided, and in November 1981 the West Australian legislature confirmed the rights of the AJV to the whole pipe (106).

Northern Mining, the very junior partner in the whole venture, had challenged very low estimates of the value of its early discoveries. (The CSO, on behalf of CRAE and AJV gave a value of A\$11 a carat, while Northern Mining's assessor had valued them at nearly double). While this argument was never really settled, Northern Mining itself sold out to the Western Australian Diamond Trust (WADT) in 1983 (106). This was a sop to public demands that some share of the fabulous profits predicted from Argyle would go towards Australians (albeit the wealthier ones) rather than outsiders.

But, five years later, the Western Australian government sold its own taxpayers short, by putting WADT out to tender. CRA and Ashton Mining picked up the booty: CRA thereby increasing its equity in the JV to 57.8% (112). The participation of WADT in the Argyle Diamonds Mines ensured that at least a small proportion of the output was marketed outside the CSO. In 1983, CRA and Ashton Mining formed Argyle Diamond Sales (ADS) to market their lion's share of production. All the gem diamonds, except for a small proportion retained for cutting and polishing in the state, go to the CSO. Around three-quarters of cheap gem and industrial stones are also marketed through the De Beers organisation (4). The agreement with the CSO lasted until 1991 (113). When the deal with De Beers was being made, there was considerable opposition from many sectors of the Australian public: the spectre of Australia's most important gem-find being controlled by South African interests was too much for many people to bear. At this time, the Aboriginal Mining Information Centre (AMIC), and researcher Jan Roberts, exposed the links between the Oppenheimers, CRA (through Charter Consolidated's 10% holding in RTZ) and Ashton Mining (Charter Consolidated holding 28% of the Malaysian Mining Corporation Corporation and a small indirect interest in Tanks). In response, bullets were fired through the AMIC offices and its workers were chased around Melbourne (114). (To this day it is not known who was responsible for these acts of terrorism, though one AMIC worker publicly identified CRA as being responsible for hounding him personally) (115).

CRA's much more careful approach to Aboriginal communities at Oombulgurri (mentioned earlier) or its comparatively low profile at Ellendale – another potentially lucrative diamond region on Aboriginal land (85, 105) – derived primarily from its 1978 decision to concentrate on Lake Argyle. In fact, CRA has not dropped its Ellendale claim, and regards it as "highly prospective". And, despite its 1985 decision to drop a large number of claims in the east Kimberleys, the Kimberley Land Council reported then that CRA was "... still proving to be a difficult company to deal with, intent on blaming Aboriginal people for any interruptions to [its] exploration programmes, even when these interruptions are clearly caused by external market conditions" (108).

In 1982, ADM commissioned an Environmental Research and Management Programme (ERMP) (116), which admitted that five areas of spiritual significance, and no less than seventeen archeological sites, had been destroyed as a direct result of the company's activities. The ERMP also recognised the considerable fears of Aboriginal people that they would lose access to their land (about 5500ha was initially required for the mine development), and have to compete with whites for the use of their fishing holes, swimming holes, camp sites and hunting as well as social areas.

Despite this, in May the following year the Western Australian government gave permission for the mine to proceed – subject to eight conditions being met. Nine months later, the government also proposed an ex-gratia A\$1,000,000 payment to affected Aboriginal communities for a minimum of five years. But it did not address the question of who would decide the criteria for "affected" communities, nor how money could be distributed so as to maximise Aboriginal self-management.

Two of the conditions laid down in ADM's own report have still not been fulfilled: those relating to Aboriginal site protection and management, and the employment of Aboriginal people at the mine (117). Of the eight conditions laid down by the government three are worth quoting here:

4. [That] the company closely monitors social impacts, especially during the construction phase. It should cooperate with private and government agencies to overcome adverse impacts.

6. [That] an Impact Assessment group be established ... including Aboriginal groups ... with a view to further development of Government and company social programmes ... which will pave the way to avoid conflict and confrontation.

8. [That] the company consults with government and local Aboriginal groups, with a

view to changing the management of funds contributed under the "Good Neighbour Programme".

In July 1984, representatives of four major bodies (Centre for Resource and Environmental Studies of the Australian National University; Australian Institute of Aboriginal Studies; Anthropology Department of the University of Western Australia and the Academy of Social Sciences in Australia) visited the Lake Argyle region. In a report to the Kimberley Land Council (KLC) and the National Aboriginal Conference, they delivered a profound indictment of the way in which the company had failed to implement the conditions laid down for the ERPM.

Some of the ERPM's conclusions were as follows:

• The Good Neighbour Programme (GNP) has actively denied Aboriginal aspirations to land, refusing funds which could enable them to develop outstations (an essential antidote to social disintegration). The overall effect of the GNP has been to increase the company's social and political control of Aboriginal affairs, and dampen potential Aboriginal opposition. All Aboriginal people expressed a desire to be free of handouts and dependence on a European-type economic system. However, GNP funds can only be spent on capital works and the company determines what these are. Aborigines have no control over the determination of priorities for expenditure. As the number of capital items increases, so does the need for recurrent funding to maintain them, making an already poor community poorer. Exclusion of Aboriginal opinion/management reduces the people to passive victims and this, in turn, preempts effective Aboriginal leadership.

• There is now competition for fishing and recreation sports from mine workers. Important Aboriginal food sources are being exploited and, in some places, destroyed.

• Kununarra (the township nearby), is such that land suitable for residential purposes is limited. Company proposals to house staff and dependants will place housing out of reach of many Aboriginal families. One result of this has been the enforced eviction of Lily Creek people, for whom derisory compensation arrived in the form of one Toyota!

- The clear signal from the company has been that favourable decisions and fund releases depend on maintenance of good interpersonal relationships and deference to company agents. These personnel now exercise an influence over the direction of Aboriginal affairs in the region that rivals the Department of Aboriginal Affairs (DAA), or Aboriginal Development Commission (ADC), approach to block funding, and entirely eclipses the State's community welfare and Aboriginal planning authority. Since the company's Aboriginal affairs policy tends to operate against government policy, this should be a matter of concern to both State and Federal governments.

- At no time has the company entered into discussions with the communities in respect of Aboriginal employment. Aboriginal people represent 56% of the local population, but of the ADM workforce of approximately 1000, only 8 are Aboriginal, 3 of them locals, with 4 Aboriginal apprentices (government-subsidised). Training in semi-skilled occupations is nonexistent. Six people have been offered menial work for a few weeks. Wages are not equivalent to those of white workers. As a consequence, the main mine settlement is kept predominantly white.

- As pressure increases on both formal and informal systems of control, so does the sense of alienation felt by some Aboriginal youth. Nothing has been done to prevent this. The Warringarri "drop-in centre" was sold from under the young peoples' feet, and no such facilities now exist.

- ADM senior personnel demand more academic subjects in the District High School with the result that, resources being limited, other areas of the school programme will be lost. No government or company action is taken to prevent this. The number of Aboriginal children leaving the education system has increased.

- Increased tourism puts pressure on land in direct conflict with the need of Aboriginal communities. Tourist interests do not want Land Rights granted.

- Family-based community decision-making structures are breaking down. In a 34-day period, the Warmun community was required to be involved in 42 meetings. Although ADM does have a working relationship with John Toby and other men of the Mandangala community, the company does not enjoy the confidence of the majority of Aboriginal communities in the region. It is without the Aboriginal contacts needed to obtain information on which to base an impact evaluation. It has no competence in the field of Aboriginal Affairs.

- From the outset of the project it has been clear that Aboriginal groups would be profoundly affected, both by the direct effect of mining and exploration, and by the policies devised by the company to deal with the local communities, who have little recourse against actions and policy decisions of the Argyle developers. The repercussions of this major development leave the Aboriginal groups isolated, disadvantaged and powerless, without means to articulate their grievances, and with the company not subject to close government or public scrutiny. Since there has been no independent assessment of ADM, it has become increasingly urgent that Aboriginal interests are not rendered invisible (118).

As of 1986, still no effective structures to monitor the effects of the mine had been established (64). Within two years, certain changes had occurred for the better, in housing, employment and social amenities. But the "Good Neighbour" programme continues to be administered by the company. Security controls still deprive Aboriginal women of access to the Barramundi dreaming (what's left of it) (119). And the company's insensitivity to Aboriginal claims is such that, when a delegation from the Western Desert Puntukurnuparna visited Argyle in 1988 – to see for themselves what awaited them if they gave consent for mining to take place in the Rudall River region (Karlamilyi) – they were not allowed to climb the hill and view the destruction the company had wrought.

Rudall River

As for the destruction – real and potential – wrought on the Western Desert people themselves by CRA, that is a continuing story. CRA holds numerous leases (sometimes joint leases with other companies) over at least a third of the Rudall River National Park (WA): it has extended its exploration to the south of this eco-region "looking for diamonds", and had entered into a provisional "site avoidance agreement" with the Ngaanyatjarra Council, while ranging over a huge area of traditional land hugging the NT border (120).

Described as a "three act nightmare – uranium, in a National Park, on Aboriginal land" (121), the Rudall River saga evokes uncanny comparisons with many of CRA's previous exploits in Aboriginal country. There has been a strategy – as at Lake Argyle – to hive off one group of people from another, and conclude deals with them, which are then produced as "evidence" that Aborigines have agreed to exploration and mining (122). Certainly, the Strelley Mob (comprising many of the Nomads – see section on the Pilbara below) did conclude a site-avoidance agreement with Canning Resources, CRAE's front company in the Park. But in 1989, when many of them discovered that their white "leader" Don McCleod was, both figuratively and literally, in the CRA camp, and that the other Western Desert people firmly opposed all mining on their lands, they withdrew their nominal support for the company's plans (123).

There has been the customary batch of protestations from CRA that their work conforms to guidelines laid down as early as 1978, when mining conditions for Rudall River were first gazetted (124). Yet the company has broken many of the elementary rules: situating base camps close to water and rock holes, interferring with vegetation close to water sources, disturbing plants, dumping rubbish, and constructing access tracks without government permission (125). CRA has accused Aboriginal people of harassing its operations to the extent that, in 1984, it had withdrawn from part of the Park, without being able to conclude an exploration agreement with the Western Desert Land Council (126). The truth is that it had already finished its programme for that year and had neither the need, nor the volition, to reach agreement with "awkward" Aboriginal people (127).

In the last two years, CRA has also cultivated accusations that the white employees of the Western Desert Land Council (Western Desert Puntukurnuparna) are "manipulating" the Aboriginal people – a clear echo from the late 1970s when it was deliberately undermining the roles played by North Queensland and Kimberley Land Councils in supporting traditional communities (128).

The Rudall River region is the likely site of a huge minerals deposit on a par with Roxby Downs (Olympic Dam) or Arnhemland. Under conditions of considerable secrecy, CRA had been exploring the Park for no less than eighteen years (129). In 1979, it upgraded its exploration programme after discovering significant deposits of uranium, along with some copper, gold, lead, zinc, platinum and bismuth (130). By the end of 1987, CRA had spent A\$20 million exploring some 8000 square kilometres of this part of the western desert, including 6000 square kilometres within the Rudall River National Park (131). It held around 35 exploration licences, and 6 mining leases, inside or on the boundaries of the Park (132).

One of these leases, at Kintyre (Karlamilyi) (133), has been confirmed as a world-class uranium deposit, with 15,000 tonnes of resources and 20,000 tonnes of possible reserves. However, estimates of the extent and grade of the lode have been increasing over the past two years. A new drilling programme, costing A\$10 million (134), in mid-1988 revealed further deposits which could make both an open-cut, and an underground operation, viable (135). In early 1989, reserves were upped by another 1000 tonnes (136), and it has been stressed continually by CRA that Kintyre is only one part of a potentially vast uranium province (137).

Apart from being an extremely important – and unique – eco-region, rich in fauna and flora (much of it still to be recorded) (138), Rudall River has for around a decade been home for 300 or more Martu (Aborigines) from the Manjiljarra, Martujarra and Warnman people, who began returning to their traditional country after thirty and more years of enforced removal and missionisation (mainly at the Jigalong reserve). The outstation of Punmu was established in 1981 and, three years later, another at Parngurr (Pangurr) (139).

More recently it is the people at Parngurr who have had good cause to be gravely alarmed by CRA's intentions. In 1987, the company announced plans to start drilling at three sites within close distance (up to 10km) of a sacred women's dreaming, Mount Cotten (140). This, and CRA's previous activities, convinced the communities that they must stop the prospecting, until all the implications could be assessed and properly discussed. Under national and international pressure (*e.g.* from Survival International) in late 1987, the Western Australian government created an "exclusion zone" around Parngurr, from which mining would be temporarily (and possibly permanently) banned. Exploration, reconnaissance, and mining outside this zone was subject to environmental and social impact studies and site-avoidance agreements, "involving full consultation with and input from Aboriginal interests" (141).

At an historic meeting in the desert between the Martu, CRA, Uranerz, other mining representatives, tourist interests, and assorted government officials, in May 1989, the Aboriginal community made it abundantly clear that they did not want mining – especially uranium mining – under any circumstances on their lands (123).

Meanwhile, as the Australian Labor Party conducted a re-assessment in 1988-89 of its "three mines" uranium policy, CRA and the Western Mining Corporation mounted a large-scale public relations offensive, to convince the Australian public that new mines had to open, to take advantage of a "window of opportunity" scheduled for the early 1990s (137, 142).

Lead and zinc

The legendary Broken Hill ore field in New South Wales (NSW) was the site of CRA's first operations. In 1905, Melbourne businessmen set up the Zinc Corporation Ltd, to process lead and zinc residues from "the Fabulous Hill" (14, 29). Later, the Corporation became the largest producer of such concentrates in the whole of Australia (14). Its future was assured during the first World War (1917), when the British government negotiated a long-term contract for zinc concentrates (14). And, in 1949, the Zinc Corporation merged with the National Smelting Corporation – later the Imperial Smelting Corporation Ltd – to become the Consolidated Zinc Corporation Ltd (14, 29). With its head office in London, the new Conzinc had close links with the "Collins House" families of the Melbourne business establishment (143).

The 1962 merger with Rio Tinto naturally transformed some of the operations and objectives of Conzinc; but, essentially, the merger was a vehicle for confirming the penetration of British capital in Australia and facilitating the repatriation of profits. It was logical that, in 1971, seeking further control over downstream processing, the Zinc Corp and the British New Broken Hill Consolidated (NBHC) formed in 1936 (and then owned one third by CRA), should merge to become Australian Mining and Smelting (144). This gave CRA control over the Avonmouth lead and zinc smelter in Britain, and a half share in the Budel zinc smelter in the Netherlands (145).

CRA's operations at Broken Hill, held through AM&S Mining, and consisting of the NBHC mine and Cobar mine, were amalgamated in 1987. They are now known as the CZ mine: recent reserves were set at nearly 30 million tonnes of ore, grading 8% lead, 11% zinc and 70g/tonne of silver (8, 146).

While their potential is still considerable (in 1986 they were producing 6% of the western world's lead and 4% of its zinc) (147), their grades have been getting lower, and they have been beset by discontent. Morale was "at the lowest for thirty years", declared John Butcher, head of the Workers' Industrial Union (WIU), in early 1987, as the workforce at the Port Pirie smelter went on strike (36).

In 1984 the Broken Hill workforce had gone on strike for seven and a half weeks (148). Two years later they were out even longer (149), as CRA sought to double the number of shifts, impose night work, and drastically cut the workforce. This attrition was too great and, in an agreement imposed by the courts (150), CRA and North Broken Hill (NBH) ended up with more or less what they had wanted, and a reduction of 40% in the workforce (151).

The zinc-copper producer, Cobar, has also experienced problems mainly due to the fall in ore grades (8). Acquired in 1980 from Broken Hill South, it has a long history of industrial dissatisfaction which culminated in CRA threatening to close it completely, unless more productivity was offered by the workforce (152). After four years of losses, the mine was shut in 1985. An expansion programme was then carried out and, since re-opening, Cobar has proved a competent producer. Productivity per miner in 1987 was reportedly twice that of CRA's operations at Broken Hill (8).

Cobar supplies copper to the Port Kembla smelter which was due to double its capacity in March 1989 (153). Operated by AM&S Associated Smelters, a subsidiary of Electrolytic Refining and Smelting Co of Australia Ltd (ER&S), it sold out 40% of its equity to a Japanese group, led by Furukawa, in 1988 (1).

Fortunes have been worse at the Woodlawn mine near Canberra. Opened in 1979 as another lead-zinc-copper producer, ore recovery soon proved difficult (154). In 1985 CRA bought the mine from Phelps Dodge and St Joe, a subsidiary of the US company, Fluor (155). Within barely a year, the open-pit mine was scheduled to close, and underground mining to commence (156). But Woodlawn had been characterised by "poor operating performance for most of its life" (4) and, not surprisingly, CRA sold the mine in 1987 to Denehurst Ltd (8).

AM&S Associated Smelters also operates a smelter at Cockle Creek, NSW, which handles lead and zinc bullion and turns out copper sulphide, silver, gold, and cadmium, as well as sulphuric acid (14), while its Pacific Smelting Company subsidiary in Los Angeles and Memphis, USA, also recycles scrap (14).

But by far the biggest smelting operation within the AM&S network is the Port Pirie smelter operated by Broken Hill Associated Smelters (BHAS) – perhaps the world's largest lead smelter and refinery (157).

In 1974, CRA had increased its interest in BHAS from 50% to 70%, by doing a share swap with the Australian company BH South. The Labour government of the day – highly sensitive to attempts by foreign companies to control even more of Australia's resources (this was the period of the Fitzgerald Enquiry, referred to below) – put the deal under investigation. Unfortunately, the investigation committee on foreign take-overs concluded that, since CRA was already foreign-controlled and there was nothing in principle to prevent foreign equity being increased, the take-over had to be allowed (44). At this time CRA was a part of a zinc cartel, one of whose effects was to ensure that Australian manufacturing industry actually paid more for its zinc than the London market (158). Despite the passage of the Australian Trade Practices Act, designed to prevent such cartels, CRA had been operating its price-fixing ring with impunity (158).

In early 1988, CRA announced that it would amalgamate its Australian and international base metal interests into AM&S Metals Pty Ltd, which would market BHAS and Sulphide Corporation output (159). This preceded a merger between the lead-zinc mining and smelting operations of North Broken Hill Peko (NBHP) and CRA, which was completed in July 1988 (160). (North Broken Hill Holdings and Peko Wallsend had already amalgamated in 1987) (161).

The new company, Pasminco, was effectively a response to the earlier merger between the operations of MIM, Teck, and Metallgesellschaft. The only valid reason for it, commented RTZ's chief economist Phillip Crowson, was "the very real opportunity for the companies to cut their costs" (162).

As part of the deal, NBHP contributed to the new monolith (which makes Broken Hill "a one company town" said the *Mining Journal* (163)) its west coast lead, zinc, gold and silver operations, its Risdon electro-zinc smelter in Tasmania, its Elura mine at Cobar, a zinc smelter in South Australia and, of course, its Broken Hill mine (North) (164).

For its part, as well as contributing its Broken Hill operations, CRA has merged into Pasminco its Impalloy, ISC Alloys, and Pacific Zinc, operations (owned by Commonwealth Smelting Ltd) and both NBHP and CRA have dished-in A$25 million each, to up-grade the Avonmouth smelter (165) and the Budelco refinery in the Netherlands – 50% owned by Billiton (Royal Dutch Shell) (164). Broken Hill Associated Smelters is now owned 70% by CRA and 30% by NBHP (166).

Following a public share issue in early 1989, CRA's interest in Pasminco itself fell to 40% (1).

Both the Port Pirie and Avonmouth smelters have been the subject of nagging criticism over many years by environmentalists and health authorities.

The Avonmouth smelter is situated on an estuary near the major city of Bristol, in England's west country. A long-established plant producing zinc (it's the only primary zinc producer in Britain), lead bullion, cadmium and sulphuric acid, it fell on hard times in the 1970s. By 1983, with over-capacity in the world zinc industry and slack demand from customers (168), two alternatives were presented by the company (Commonwealth Smelters) to the workforce: closure or a drastic "survival plan". A third option – to join the EEC-coordinated capacity reduction programme – was never pursued (168) and, as zinc prices rose towards the end of 1983, the EEC plan was itself abandoned (169).

In the event, the survival plan was pushed through – although it meant redundancies for nearly one third of the workforce (1000 down to 700 employees). Profits soon rose (4), although further setbacks were experienced in 1985 (116, 170). By the late 1980s, Avonmouth was a comparative economic success story.

Not so, however, for much of the workforce and the communities surrounding the plant. As early as 1969, many workers were diagnosed as suffering from lead poisoning, and a government enquiry was held into the plant's operations (171). Three years later, Duncan Dewdney, RTZ's chief executive in the UK, "openly confessed that the plant's initial construction was skimped, that corners were cut and that it should have cost at least two million pounds more" (172).

In 1985, soon after the survival plan was implemented, Dr George Kazantzis of the Trade Union Congress (TUC) Institute of Occupational Health at the University of London, carried out an investigation which appeared to show increased lung cancers at the plant. Two years before, Kazantzis and a colleague had also investigated mortality from lung cancers among cadmium workers – many of whom came from the Avonmouth plant (173). The second of the two studies did not satisfy CRA, as it appeared to show that the risks were greater than they had ever admitted. There is evidence that the parent company in Britain (RTZ) took steps to suppress publication of the document (64). In any event, it was not until 1987 that the study was accepted for publication by the *British Journal of Industrial Medicine,* and it was well over a year before the report was published. This concluded that workers in the smelter – particularly those who had been employed for more than 20 years – exhibited an excess of lung cancer, although it was not possible to pin down which of the contaminants – arsenic, lead or other materials – might be responsible (174).

Meanwhile, a TV research team had been looking at health and safety at the plant. Interviewed for the programme, a co-author of the Kazantzis report, Tony Ades, confirmed that there was a definite relationship between lung cancer, exposure to cadmium, arsenic and lead workers' deaths, and the length of time they worked at the plant (175). Residents close to the Avonmouth smelter, interviewed in the programme, also expressed their opinion that the community had been put at risk from Avonmouth's continued operations (176).

The following year, across the world at Port Pirie, a health survey of 537 pre-school children drew a direct relationship between the slower mental development of four-year-olds, their blood/lead levels, and the presence of the Port Pirie smelter (177).

The Port Kembla smelter was itself under scrutiny in 1987, as the New South Wales State Pollution Control Commission declared that the plant did not conform to pollution standards (178). After an economic study of the project, CRA decided to update the smelter and expand its output (178).

Pilbara

The Pilbara region accounts for virtually all of Australia's iron ore production – itself around 10% of the world's (179). In the Pilbara, CRA's wholly-owned Hamersley Holdings Ltd is the biggest producer (8). Hamersley owns and operates the large Mount Tom Price deposit and the Paraburdoo mine, linked to Tom Price by rail (5). In 1986, it also acquired half of CSR's Yandicoogina deposit, some 140km east of Mt Tom Price at the other end of a lease held

by Broken Hill Proprietary (BHP). This prompted the *Financial Times* to remark: "... Australia's two largest companies now sit at either end of the Yandicooga deposit" (180).

In 1981, RTZ purchased Texasgulf's 50% stake in the Wittenoom and Rhode Ridge iron ore lodes (181). But, with 490 million tonnes of reserves, the Marandoo deposit (Lang Hancock interests: 50%) continues to wait for markets in Europe, Asia and the Middle East (182). In 1982 the *Financial Times* announced that Marandoo had "basically been dropped because CRA believes it could harm the viability of the Hamersley operations" (183).

CRA announced in 1986 that it would develop a new deposit at Channar, near Paraburdoo, together with the China Metallurgical Import and Export Corp (CMIEC) (5). The deal has made Hamersley the largest supplier of iron ore to the murderous Beijing regime (8), specifically the Baoshan steel complex near Shanghai (4). Production of 10 million tonnes a year, at a cost of a quarter of a billion dollars, started in 1990 (8, 185, 410).

The Australian government banned the export of iron ore from 1945 until 1960: since the Japanese would be the main customers, the embargo was mainly for political reasons (28). The notorious Australian mining "baron", Lang Hancock, persuaded RTZ's Val Duncan to invest in the Pilbara when the embargo was lifted. Using US funding, and confident of securing the market to Japan, RTZ stepped in to negotiate for Mount Tom Price (28). One of the sweeteners for the deal was an offer by the RTZ supremo to supply Western Australia with a steel mill, in return for control over the iron ore. He got the deposits, but Western Australia has never seen the steel mill (4, 44).

The US finance secured by Duncan came mainly from First Boston Bank (186), via Kaiser Steel (part of the same outfit that partnered CRA in Comalco until quite recently). Between 1979 and 1981, CRA bought almost all the shares in Hamersley that it did not then own, by issuing new shares, and thus reducing the proportion of shareholders' existing equity (33, 187).

Through a bid made by In-Situ Processes Australia, in late 1984, CRA bought out the 6.25% of Hamersley still left to devour (5, 188).

In 1988, Hamersley recorded record shipments of more than forty million tonnes of iron ore. Spurring this production was a new agreement reached with Hamersley's key customers, the Japanese steel mills (Hamersley is Japan's largest supplier, providing 18% of Japanese needs) (189). Under this agreement, firm contracts would last for up to 23 years, with fixed annual tonnages set 5 or 6 years ahead. This was, as the *Mining Journal* put it: "a watershed for the iron ore industry world-wide" (189). And, of course, particularly for CRA (190).

Through much of the 'eighties, Hamersley had been looking for new joint venture partners – to share its costs – and new markets (191), but its fortunes had not been uniformly good. In 1986, after allegations of participation in an iron ore producers cartel, Hamersley had been the last company to settle price terms with its Japanese customers (192).

On the other hand, its ascendancy in the Pilbara – indeed in the world of iron ore as a whole – should not be underestimated. The Fitzgerald Report of 1974 examined the methods by which Australian and foreign-dominated companies, operating in Australia, avoided taxation while gaining subsidies from the Federal government. It accused Hamersley of paying only A$572,000 income tax, while declaring profits of A$264 million, between 1967 and 1973 (193).

However, the Tax Commissioner later failed to ensure that Hamersley paid all its dues, when Justice Gobbo ruled that certain production processes (notably those involved in blending Mt Tom Price and Paraburdoo ores) could be exempted from sales tax (33, 194).

In addition, Hamersley has gained from the decline in the Australian dollar which occurred in the mid-'eighties: its profit rise of 60% achieved in 1985, for example, was attributed principally to foreign exchange manipulation (4).

Hamersley's holeshot position in Australia's iron ore industry, and its prestige success in securing Japanese and Chinese markets, has overshadowed its deplorable industrial relations record, its bad management practices (which, according to Richie Howitt, might have been its downfall had it worked on a less expansive and rich deposit) (33), and its disregard for Aboriginal people and their rights.

The number of strikes at Hamersley over the past two decades are almost too numerous to record: in 1976 alone, there were 157 different sets of industrial action (195). In 1988 an indefinite dispute with CRA halted all shipping of iron and salt in and out of Dampier, Cape Lambert and Port Hedland (196). It was answered by the inevitable layoffs and cutbacks to "eliminate inefficiency" (197) .

The deprivation and marginalisation of Pilbara Aboriginal people must rank alongside that of any other Australian Aboriginal community in the last fifty years. By the early years of the century, the Pilbara Aboriginal people (the Martu) – where they had not been killed – were virtual slaves of the cattle stations (46). Their situation marginally improved over the next forty years, but conditions on the pastoral leases compelled many of them to join the famous Pilbara Strike in 1946, led by Dooley and Clancy, two Aboriginal men, assisted by Don McCleod (33, 46). The strikers were determined to run their own cattle stations, and gain money through their own mining enterprises. To this end they set up Pinden (or Pindan) as a co-operative to build self-reliant settlements. Within three years, more than 700 members of the movement were living in outstations, using miners' rights to set up camp (46, 198). They formed the Northern Development and Mining Company, which soon became the largest prospecting force in the region. From the proceeds the Aboriginal people bought one station – Yandeyarra. However, the government took over this station, and turned it into what is now the only reserve in the Pilbara. Some of those who were left at Yandeyarra then set up Nomads Ltd. The Nomads still exist – centred on Strelley Station.

When mining companies – first and foremost CRA – moved into the Pilbara, Aboriginal people began to be squeezed out of their former essential roles in the economic life of the region and, as one social science student put it, the "Giants" came to "dominate" (199). Particular impact was felt by Aboriginal people on the Roebourne reserve. A white, largely male, unattached, labour force moved into Dampier in the mid-sixties, to construct the new terminal and other Hamersley facilities. As they did so, Aboriginal people – already impoverished and marginalised – shifted to the Roebourne area, putting enormous pressure on housing, other Aboriginal families, and living resources (33, 199).

Over the next few years Roebourne became a social disaster area: Aboriginal women were the victims of sexual "relief" sought by Hamersley workers, and alcohol was the only respite from tedium and degradation among Aboriginal men. "Roebourne shire in particular, and the Pilbara in general, rapidly developed an enormous contrast of wealth – extremes of "haves" and "have nots". In fact it is likely that nowhere else in Australia are these extremes exhibited in such close juxtaposition, or have developed so quickly" (33).

Hamersley refused to initiate special programmes of training and employment for Aboriginal people. Since they had to take their chance like everyone else, such false "equality" inevitably meant discrimination and increasing unemployment among Aboriginal people (33). Certainly the iron ore companies cannot be held solely to blame for much of this dissolution of culture and Aboriginal economy. Ironically, some of the legislative measures intended to equalise Aboriginal and non-Aboriginal status – such as lifting the embargo on selling drink to black Australians, and awarding equal wages to pastoral workers – in the short term worked completely against Aboriginal people (33). Nor must any analysis of the situation in the Pilbara in the past decade end on a note of hopelessness: self-reliance and Aboriginal organisation, better housing, increased employment in the urban areas, have undoubtedly helped mitigate the extremely bleak outlook of the late 1960s

and 1970s. However, the invasion by the companies, and especially Hamersley, dealt a final blow to Aboriginal prospects for self-determination. "Always the highest priority has been placed [by the mining companies] on contributing maximum profits to the Group ... . So, while State and Federal Governments must bear direct responsibility for many of the social decisions which have affected Roebourne Aborigines, it is the mining companies which made many of the most important decisions and who have benefitted economically from the oppression of local Aborigines" (33).

Comalco

The history of Comalco is essentially the story of how Australia's vast bauxite deposits were hijacked by British capitalists and their allies in the world of high finance. Nor can the astonishing rise of this ruthless band of entrepreneurs be separated from the fortunes of CRA, itself the creature of RTZ.

During the last months of the second World War, W S Robinson, a leading light in both the Collins House group of companies and the British Consolidated Zinc Corporation, together with two Australian Labor Ministers, set up an Australian Aluminium Production Commission (AAPC). In 1948, with public funds from the Federal and Tasmanian governments, the AAPC opened the Bell Bay smelter, using bauxite mainly supplied from Canada by Alcan. Over the next five years, a desultory and incompetent search for bauxite deposits was made within Australia, but little of value was recognised (200).

By 1953, Conzinc had its own men in key positions in the AAPC: these included a former Collins House mines manager, A J Keast, as general manager of Bell Bay, and Conzinc's exploration director, Maurice Mawby, as a Director. Shortly after the AAPC discontinued its own search for bauxite, Mawby instructed Conzinc to get back in the field. Within the next two years, Keast had deserted the AAPC to become Australian managing director of Rio Tinto Finance and Exploration (Riofinex). The

Treasury representative on the AAPC, Donald Hibberd, threw in the towel to become finance director of the new company formed to exploit the Weipa bauxite deposits, after they were discovered in 1955 (200, 201). Later, Mawby was to become the chair of CRA, remaining there until 1974.

Aluminium was a new and exciting prospect for the Conzinc group, but held enormous financial risks, primarily because huge amounts of electricity are required to refine the bauxite and smelt the alumina. Once the company had secured the Weipa bauxite deposits, and taken measures to throw the Aboriginal occupants off the richest major bauxite field in the world, it required cheap power. The one condition imposed by the new right-wing Queensland state government on Comalco was that it establish an aluminium refinery, and try to set up an aluminium smelter, too (200). Over the next four years, radical re-alignments took place within the world aluminium industry (202). British Aluminium – Conzinc's original partner in Comalco – sold out to Reynolds of the USA. In 1959, Reynolds itself pulled out. Conzinc looked around for a new partner, and finance, to exploit Weipa. It soon secured the interest of the Kaiser group of companies in the USA, whose 38%-owned Kaiser Aluminium and Chemical Corp was hungry for bauxite and processing plant (200). Meanwhile, the fabulously rich Oppenheimer family – owners of the Anglo-American Corporation and its numerous satellites – had used their British vehicle, Charter Consolidated, to obtain a good chunk (10 million shares) of Conzinc itself. This apartheid money enabled Conzinc to hold on to its half share in Comalco. (At the time, Charter Consolidated also held around 10% of RTZ, which it did not surrender until the 1980s).

By the end of 1960, Comalco had bought out the Bell Bay refinery and smelter for less than A$22 million (repaid over 16 years), which was far lower than its original cost to the Australian taxpayer (200).

At almost the same time, Kaiser announced a partnership with Comalco, to build an alumina refinery at Weipa, and a smelter at Bluff harbour, New Zealand. The Bluff (Tiwai Point) smelter was to be constructed amid raging controversy over its financing, costs, and environmental impact: a debate which, far from diminishing over the years, has intensified (*see below*). (Comalco had previously investigated using both the Purari River in Papua New Guinea as cheap hydro-power for a smelter (328), and Blair Athol Coal (*see below*). However, in terms of costs and availability, the Tiwai Point smelter won hands down).

In the event, the refinery was not constructed at Weipa, but at Gladstone in central Queensland. This plant has itself constituted a massive diversion of public resources into private hands. (Ownership in 1974 was: Kaiser 32.3%, Alcan 21.4%, Pechiney 20%, Comalco 13.8% and CRA 12.5%. Comalco later increased its ownership to 30.3%) (204). Not only did the plant not pay a cent in taxes in its early years but, as owners of the Weipa deposit, CRA and Kaiser paid substantially less than other customers for the bauxite delivered to the refinery (200), and the electricity consumed (205).

The plant also received ten million dollars' worth of free infrastructure at taxpayers' expense (200).

Over the next fifteen years, Comalco was to refine the strategies that have made it one of the world's leading profiteers from bauxite, and beneficiaries of the "added value" gained from refining and smelting it into aluminium. These ploys include: minimising Australian participation in the industry and supplying cheap raw materials to its paymasters (especially RTZ), and deliberately under-reporting its profits in Australia. One example of such transfer-pricing, exposed in the mid-70s, was Comalco selling bauxite at rock bottom prices to a Hong Kong subsidiary, which then sold it on to Showa Denko and Sumitomo Chemical: the bauxite, of course, never came within sight of Hong Kong, but went straight from Australia to Japan (200).

A case was brought against Comalco by the Tax

Commissioner in this instance, but it failed. Comalco's argument, that its business was not controlled by its foreign shareholders, won the day (33, 206).

Other Comalco strategies have included: declaring its "poverty" to justify further government subsidies; blackmailing the Queensland government (as it did in 1974/75) to reduce its royalty payments; lobbying against compensation to Aboriginal people for the theft of their land; making political donations; playing a leading role in the Australian Mining Industry Council (AMIC), especially in its 1985 campaign to erase Aboriginal Land Rights (64); cutting costs on refining by using plants in third world countries with cheap labour; operating few, if any, pollution controls. Comalco also bought a 20% share – increased by 17.5% in 1985 (4) – in the Euralumina refinery in Sardinia, one of the poorest parts of Italy (145). Until 1987, it had aluminium fabricating facilities, not only in New Zealand and Japan, but also in Hong Kong, Madagascar, the Phillipines and Indonesia (207).

Comalco launched itself publicly in 1970, with a share issue which left CRA and Kaiser holding 45% each of the company's equity, and the Australian "public" the remaining 10%. Just how the "public" got this foot in the aluminium door – and how "public" the share issue really was – is a tale-and-a-half (208).

The enormous profiteering which characterised the share issue was the result of a carefully-managed media "hype", and the placing of "unsecured notes" to several Australian institutions, which later gained them shares at a priviledged price. This brought in Australian Mutual Provident (AMP), Anglo-American nominees, other insurance companies, and the ANZ Bank (200). Excitement generated around the wealth locked into Weipa ensured that, on the day of issue, the share price of A$2.75 (fully paid-up by the early buyers), shot up to A$5.80 and closed at A$5.60 – thus representing a capital gain in a few hours of more than 100%. Even bigger profits were made by CRA and Kaiser (around 1000%) (200, 209).

The ordinary Australian shareholders made considerably less, however – they had forked out nearly four times as much as the companies for the same shares. Those Australians who really made a killing that day were the select band that had been offered shares before the day of issue. These "trade customers" were essentially buddies of Comalco in high office in Queensland (and a few elsewhere). "... [W]e expanded our usual list [of share recipients] to include people who've said over the years they'd like to have a chance to buy any new shares we float", Comalco blandly explained the day before the issue (210).

Such people included the Queensland State Treasurer and Acting Premier, Gordon Chalk, and his family; the Ministers of Aboriginal Affairs, Industrial Development, Works, Local Government and Electricity, Health; and the Premier of Western Australia.

While it was said of Acting-Premier Chalk's family that the only member who had not accepted shares was "the dog" (200), none of the Aboriginal community at Weipa got a look-in (let alone the dogs!). (The Queensland director of Aboriginal and Island Affairs took up 40,000 shares, "on behalf of " the 50,000 people under his protection, and 38 Aboriginal people working for the company got a small entitlement.) *The Australian Financial Review* forbade its staff to have any part in the deal; the Federal Prime Minister rejected the offer; and the Premier of South Australia, Don Dunstan, called the scheme "highly improper". However, the Presbyterian church in Melbourne, and the Anglican church in Melbourne and Sydney, accepted shares. Joh Bjelke-Peterson, a great friend of the company, and shortly to become Premier of Queensland, took up only 5 shares, but his wife later acquired 500. (A few years afterwards, the Australian Senate Select Committee on Foreign Ownership and Control of Australian Resources would find that Queensland was "the only state for which both foreign ownership and control exceeded 50 per cent" – and it was the highest in both categories!) (211).

By the late 1970s, Comalco was Australia's

leading exporter of bauxite and, with a world recovery in aluminium, it embarked on an A$80 million expansion programme which necessitated a 12% increase in mining at Andoom and Weipa (212). Production of calcined bauxite – which it first produced in 1970 – had already given Comalco half the world market (212).

In 1982, Kaiser had had its fill of Comalco, and wanted out (213). Not, however, before some of its Weipa bauxite found its way into B-52 bombers and others used in massive raids on civilians in Vietnam: many of these planes were shot down by the North Vietnamese army and the National Liberation Front and thereby recycled (200)! Kaiser sold its share of the company to CRA and AMP. CRA then came to hold 70.3%, and AMP 16.5% (214). However, CRA sold-on some 16 million of its holding to Australian investors, thus marginally reducing its control (215).

Within the next few years Comalco, like its parent CRA, was to concentrate on consolidating, expanding and controlling downstream activities and attracting new markets, as its major corporate strategy for the eighties (4). To this end, it bought 50% in Showa Aluminium Industries KK (SAL) in late 1982. Showa was a company with which Comalco had a long association, having equity in a small smelter at Kitakata, a refinery and various aluminium products. SAL had also acquired a 20.65% interest in the Tiwai Point smelter. (Showa Denko was one of the two Japanese companies culpably responsible for the mercury poisoning, known universally as Minamata disease, which killed and maimed thousands of people from the 1950s onwards (216). But, of course, Comalco was little concerned with such moral niceties: it had, after all, discussed co-financing Weipa with the Nazi war criminal, Baron von Krupp, thirty years earlier) (200, 216).

The flirtation with Showa Denko gave Comalco an unprecedented entrée into Japan (217). But SAL was not as efficiently ruthless (ruthlessly efficient?) as its weightier partner and, after a dispute between Comalco and SAL over methods of restructuring in 1985 (218), the

"Australian" company pulled out of the partnership (4). This was to cost Comalco much of its expected profits that year (219) though, in return, Comalco got back SAL's share of Tiwai, thus increasing its control over the controversial smelter, to 79.4% (220).

The following year Comalco also sold a number of businesses world-wide, including Chiap Hua Comalco, Comalco (Asia), Federal Aluminium, Hooven Comalco, and its operations in Indonesia – PT IndoExtrusions (221). Meanwhile, the company had been chalking up a five-fold rise in profits for 1983, expanding Tiwai Point again, and starting up the Boyne Island smelter (222).

In early 1985, CRA effected something of a coup, by buying into the aluminium operations of Martin Marietta in the USA. (Martin Marietta is one of the world's biggest defence contractors – Comalco certainly knows how to select its partners!). The following year a new kaolin plant was opened at Weipa, thus sacrificing even more Aboriginal subsoil (the kaolin lies under the bauxite), in a quest to "capture a major share of markets in the Asia-Pacific region" (223). In less than a year the operation was proving successful (224).

1988 saw record profits for the company, again because of reduced income tax, high prices for aluminium, and a small rise in output (225). The "mere" fact that Comalco's New Zealand's accounts are maintained in US dollars, and not NZ dollars, meant a profit increase of US$22.3 million (226). Further expansion and upgrading – to the tune of A$1 billion – was also announced (227).

A few months later, in a spring-clean of top management structure announced by CRA (228, 229), Mark Rayner moved from Comalco to become CRA's head of financial and strategic planning – only a step from the shoes of current chief executive John Ralph (230). (Interestingly too, as part of the shake-up, CRA's head of exploration moved northwards to become RTZ's head of exploration: so much, once again, for the pretence that CRA is managerially quite separate from its parent) (229).

Comalco has entered the 1990s with a lot of self-confidence (some would call it conceit), although Mark Rayner himself has predicted that the aluminium industry will enter a "trough by 1992". Despite some disastrous overseas exploits, the company has recently considered opening an alumina refinery and smelter in the USSR. It also put its support behind a proposal with BHP, the discredited Bond Corp, and the Martin Marietta Corp, to build a space launch centre at Cape York (230), which it saw as a means of "improving the availability of its Weipa bauxite operations" – although it may be guessed what such a disastrous innovation would have on the land, culture and lifestyles of the Mapoon, Aurukun and Weipa Aboriginal people. (In 1990, Comalco pulled out of this extravaganza).

For more than thirty years Comalco has been able to convince a large number of Australians that it serves the interests of "the nation" (whatever, on closer scrutiny, these can be said to be). Forgotten are the stratagems by which it seized Aboriginal land and broke up one of the longest-surviving black communities in the country; destroyed Australia's only public initiative at forming a rational and controlled bauxite industry; resisted, tooth-and-claw, attempts to ensure "added value" to its mining, by processing in Australia, rather than overseas (until the pressures became too great to resist, and the structure of the world market itself dictated the location of plant closer to both mine and consumers) (232). Forgotten are the ways in which it ensured that the interests of its corporate backers (specifically CRA/RTZ) would be served at the expense of Australian taxpayers and more commonplace investors; the manner by which it polluted and devastated natural environments, in a chain that stretches around the world.

Cape York's are not the only bauxite deposits on Aboriginal traditional land that have been the subject of intense interest by CRA or Comalco. In 1965, Amax lay claim to massive bauxite deposits on Mitchell Plateau, in the far north of the Kimberleys of Western Australia (minimum of 200 million tonnes) (46), and nearby at Cape Bougainville (around a thousand million tonnes). Three years later the company built a beneficiation plant and tested bulk samples. In 1969, the Western Australian Minister of Mines (then Charles Court, who was later to be state Premier at the time of Noonkanbah – see Amax) gave Amax a 1500 square mile lease. The company made plans to build a town, port and refinery. Twenty seven square miles were also excised from the Kalumburu Aboriginal reserve at Cape Bougainville, without any consultation with the Aboriginal owners, or consideration of compensation (233).

However, Amax failed to raise the finance needed for such a large enterprise. In 1980, CRA bought out 52.5% of the prospect, and Alcoa bought minority shares in both Mitchell River and Cape Bougainville (46). Billiton (Royal Dutch Shell) also took 10% in Mitchell Plateau, and both Sumitomo and Marubeni secured 5% (234). By 1983, Mitchell Plateau Bauxite Ltd, managed by CRA (5), had acquired another bauxite project in Western Australia, at Muchea in the Darling Ranges, northeast of Perth. None of these prospects has moved to the construction phase – "it is difficult to get enthusiastic over what will inevitably be an expensive venture", commented Rowe and Pitman in 1982 (14). Nonetheless, exploration at Mitchell Plateau in the early 'eighties had already damaged sacred sites (235).

Tiwai: a point for alarm

The controversial Tiwai Point smelter (sometimes known as Bluff, because it is situated at Bluff Harbour) is owned by New Zealand Aluminium Smelters Ltd (NZAS), which consists of Comalco NZ (owned 67% by CRA) and Sumitomo of Japan (20.6%). NZAS also has extensive downstream interests in extrusion (a 50/50 JV with Carter Holt Harvey) and fabrication. With a staff of 1750 in New Zealand – mostly at the smelter – the majority of its production goes to Asia (especially Japan and Taiwan): 27 per cent of New Zealand's exports to

Japan in 1988 were of Comalco aluminium (236).

The plant's existence has been marked by controversy for nearly thirty years: indeed, it could be claimed that it is the longest-running economic/environmental issue on the country's agenda.

Comalco came to New Zealand for cheap power to smelt alumina which derives mainly from the Weipa bauxite fields, and is refined at the Gladstone refinery. The lease it negotiated with the government in 1960 is a model of resource theft-by-stealth. Under the Manapouri Te-Anau Development Act of that year, Comalco undertook to build a power station to power a smelter. In return, the New Zealand government would grant the company exclusive rights, for 99 years, to use the waters of two of the country's most precious natural assets: Lakes Manapouri and Te Anau, situated in the Fiordland National Park. These lakes could be raised to a level which would just avoid flooding the Te Anau township – no less than eighty-four feet above its natural level in the case of Manapouri. Since Comalco would own the hydro scheme, the cost of power would be its own concern.

Within two years Comalco was complaining that it couldn't pay for the work necessary at Manapouri. In a 1963 agreement, the New Zealand government itself then undertook to construct Manapouri and sell the power to Comalco. The power station was designed by Bechtel, constructed by Utah Mining and Construction (both US companies) and eventually cost twice the original price (237).

What's more, Comalco would come to pay for its electricity at a far lower rate than that charged to everyone else in the country. Although still not officially revealed, this price was originally fixed at less than one fifth of one cent per unit: thirteen times less than the rate charged to New Zealand householders, and only one twentieth of that charged to the country's other industries and farmers.

By 1967, there were an unprecedented 264,906 signatures to a petition to stop the raising of the Lake. In 1972, the Labour party campaigned on a similar platform. Meanwhile, Comalco and its Japanese partners had built their smelter at Tiwai Point. In 1974, because of droughts, there were severe power cuts throughout the country: Lake Manapouri's level dropped – yet Comalco was able to take electricity from the national grid to compensate!

In 1980, the Campaign Against Foreign Control in New Zealand (CAFCINZ) – now Cafca (Campaign Against Foreign Control in Aotearoa) – published documents emanating from Comalco, which demonstrate the enormous lengths to which that corporation will go to ensure its own profitability at the expense of just about everyone else. These documents cover a range of issues surrounding the operations of the Tiwai Point smelter. Key among them is a telex dated 15/11/77 (IFB/MBB) from I F Borrie, Comalco's General Manager, Special Services, to M B Bennett, Comalco's Corporate Manager in New Zealand. This outlines crucial discussions between Prime Minister Muldoon of New Zealand and Mr Schlesinger, the United States Secretary of State for Energy. The US Secretary had been briefed on the issue of pricing by Comalco (namely by Cornell Maier, President of Kaiser Aluminium and Chemical Corporation). He then sent a transcript of the meeting to the company, which used it in their battle in New Zealand to keep the power price down. Prior to this, Comalco had walked out of a governmental meeting, set up to renegotiate the power price, and Muldoon had met with Schlesinger, partly to gain his support on the Comalco "problem". Comalco's campaign against the New Zealand government involved various other ruses, including a telex sent to Lord Shackleton, deputy chair of RTZ, asking him to initiate British government action against New Zealand (238).

The upshot of these various pressures (with others which have still not been revealed), and a threat in October 1977 to close down the smelter, was that Comalco finally agreed to pay 4.5 times its previous rate (only around a quarter more than it was bargaining for) and the government retreated from the 650% increase it was insisting upon (239).

In 1979, the pricing issue still loomed large. Power from Manapouri continued to be "one of the cheapest ... in the world" (240). When Muldoon came to power in the late 1970s, he promised to better the Labour opposition in squeezing concessions from Comalco (240). Once again, however, the company came out on top: while other consumers had to suffer a 45-50% increase, the smelter sustained only a 25% price hike (241).

In 1983, despite chalking up a NZ$18.1 million loss, Comalco reported a good year (242). The following year, with Tiwai Point still paying less than 2 cents a unit, and consuming one fifth of the nation's electricity supply, the government announced a 22% price rise to NZAS (243). Comalco's Don Hibberd was appalled: it would "demoralise" the management at Tiwai Point, he claimed. New Zealand farmers were incensed: the dairy industry would continue subsidising Comalco, declared the chair of the Dairy Board, although it was earning "vastly more" for the nation than the smelter. The Coalition for Open Government also rounded on Comalco (244).

Three years later the price issue remained high on the political agenda. The government set up an Energy and Minerals Advisory Committee, soliciting public opinion on various issues, including the electricity cost to NZAS (245). The previous year, the Lange government had yet again tried to squeeze more out of its recalcitrant corporate bed-partner. Comalco reacted with its heaviest sticks. It began conducting "surveys" of top government officials (246). Japanese business men at a Japan-New Zealand business gathering in Kobe threatened a boycott on future investment in New Zealand, unless the government fell in with Comalco's desires (although this was later denied) (247).

In September, Comalco and Sumitomo filed proceedings against the government in the High Court, "to protect their rights" should the government legislate an equitable pricing deal (247).

By November, Comalco was claiming it ran at a loss, citing the closure of the Goldendale smelter in the USA as evidence of the parlous state of the industry. The following month, NZAS lost its court case: the government did indeed have the right to legislate a new power price – a 100% increase was the figure mentioned (248). As CAFCA pithily put it: "New Zealand Aluminium Smelters Ltd was not yet the New Zealand House of Lords ..." (247).

In 1987, the Prime Minister David Lange ridiculed Comalco's claim that it was losing money at Tiwai Point: the company said it had chalked up a NZ$34 million loss in the first fifteen years of the smelter's operations (249). Lange declared sarcastically how "deeply grateful" he was that Comalco had kept going, despite its deprivations, "... to service New Zealand" (249).

That year, as Comalco doubled its profits (up 192%) (250) the government promised to double its charges to Comalco (251) and made another proposal, whereby Comalco would buy out the Manapouri power station, and float some shares to the New Zealand Electricity Corporation and institutional investors (406). The Guardians of Lakes Manapouri and Te Anau were considerably alarmed: its chairman was "suspicious about the sincerity of Comalco's respect for the environmental sensitivity of Lakes Manapouri and Te Anau", and worried that Manapouri's level would be raised to supply all the power used at Tiwai Point (Manapouri supplies a majority, but not the whole amount) (247). On March 25, Comalco sought a court order preventing the government from transferring Manapouri to the new Electricity Corporation (due to be inaugurated on April 1 1987). It failed.

1988 was a record year for Comalco as a whole. (According to Australian brokers, A C Goode Ltd, its debt-to-shareholders'-funds-ratio was as low as 1:13) (251). The Energy Minister Bob Tizard complained, in early 1989, that the government was not even covering its costs in supplying power to Comalco: the company could have saved as much as NZ$200 million in the previous year (251).

Through 1988 and 1989 the possible sale of the

Manapouri scheme to Comalco moved from being a nightmare impossibility to a nightmare probablity. To soften the New Zealand public for the take-over, Comalco announced that it would bring all its New Zealand operations together, in a new outfit which would hold its 79.4% share of Tiwai Point. Mark Rayner of Comalco claimed that the company was "keen ... to underline its commitment to New Zealand" (252). It also set about refurbishing its tarnished public image "from a multinational sucking money out of New Zealand to that of a local company pure as spring water" (253). TV advertisements ended with the legend: "The Power of Good" – a comment which led the Secretary of CAFCA to retort (in a letter to *The Listener*): "Comalco's history in this country is one of deceit and arm-twisting at the highest level, threats to our environment on an unprecedented scale, dubious share deals and electricity at give-away prices" (254).

The Public Services Association accused the government of back-tracking on an undertaking not to sell off public assets (255). The *Christchurch Star*, the following year, exposed Comalco's attempt at "greening" its image, and at softening public opposition to the take-over proposal: "Comalco has dressed itself up in Kiwi clothes," declared the newspaper, "by establishing a New Zealand holding company with local directors ... [but] the number of New Zealand shareholders remains extremely low. It has also set up an aluminium recycling plant capitalising on an advertising campaign promoting itself as a good corporate citizen. But the company's prime concern is to ensure that the access it enjoys to cheap New Zealand power is preserved once the electricity industry is privatised" (256). According to the *Christchurch Star*, Comalco would go along with a proposal to take a minority share in any new company, but would require an option to increase its equity to fifty per cent later.

Meanwhile, other electricity Boards in New Zealand demanded that Manapouri be tendered to them as well, and not remain the subject of a secret deal (257). The *Christchurch Star*

succinctly commented: "The time is long past when Comalco's power concession, retained at the expense of every other consumer, should have been removed ... any scheme that would give an overseas-based multinational concern the inside running on the purchase of such a strategically valuable asset as Manapouri has to be viewed with extreme wariness" (258).

Alumina from the Gladstone refinery in Queensland may also be smelted at the Bell Bay plant in Tasmania. This, too, has been the subject of controversy since Comalco bought the smelter from the Tasmanian and Federal Australian governments in 1961.

It was then powered by hydro-electricity gained from diverting the waters of the Great Lake, on the central plateau, into the South Esk river (259). The lake's level was raised, leaving its shores beachless and scarred with dead vegetation (44). Then, in 1969, the Tasmanian parliament – in the face of considerable opposition – approved the damming of Lake Pedder, and the flooding of its surrounds, including the Gordon River gorge (260).

By the early 1980s, Bell Bay was Australia's largest aluminium smelter, turning out 112,000 tonnes a year according to Comalco, and contributing about 6% of Tasmania's gross domestic product (261). But, although the smelter is a very minor contributor to state employment, it has accounted for around 40% of all electricity usage (260), dropping to around 25% in the late 1980s (256). Thirsty for more cheap electricity, Comalco enthusiastically supported proposals to dam the Franklin river in the early 1980s – the last major river wilderness area in south-eastern Australia. Thanks to an overwhelming national and international campaign, this proposal was eventually blocked.

As in New Zealand, the company has never revealed details of its secret agreement on power pricing. In 1981, it was suggested by the state Premier, that Comalco paid as little as 0.7 cents per kW/hour unit (260). As part of the price of gaining "Green" support in the 1989 Tasmanian elections, the Labor party agreed to reveal details of the deal, late in 1989 (256). It still had not done so by summer the following year.

In a comprehensive study of CRA's interests, completed in 1986, the investment advisory group McIntosh, Hamson, Hoare, Govett Ltd estimated the proportion of the company's costs that went into power generation for aluminium smelting. In 1984 this was around 9% of total operating costs (4). The average electric energy cost per unit (kW/hour) "compared favourably" with industry as a whole: at 1.12 Australian cents/unit it was less than one-third of the industry average (3.4 Aus cents). The Tiwai Point and Bell Bay costs were almost identical (Tiwai Point costs were slightly higher). Those at Boyne Island were the highest. (Comalco's interest in the Boyne Island smelter near the refinery at Gladstone, was 30% in 1989, but due to rise as the third and fourth potline came on-stream and the company took a higher proportion of the output) (262).

In 1988, Comalco and the Tasmanian government agreed to build an aluminium wheel-casting plant next to the Bell Bay Smelter (263), to sell automotive wheels to USA, Japan and Europe. This was vaunted by Comalco's Mark Rayner as a move which was good for Australia (264).

Smelters: a dirty footnote

Pollution from Comalco's refineries and smelters has long given rise to concern. Caustic soda is dumped from Gladstone's operations in the form of an ugly red sludge. In 1974, Professor Fitzgerald summarised "some recent reports [as appearing] to suggest that the environmental impact of the refinery ... could limit the region's prospects of becoming the site for continuing industrial expansion on a scale that might justify the costs of the initial take off" (193). Also of concern are the impact of boiler ash and alumina dust. The major air pollutant is sulphur dioxide – a cause of "acid rain" – whose effects vary according to the height of the stack from which they are belched out, and prevailing winds. At maximum levels of concentration, Gladstone's emmissions have been estimated as

equal to the average concentrations – from all sources – in the city of Perth (200), and this, at only two-and-a-half miles radius from the plant!

Fluorides emitted from aluminium smelters are supposed to be gathered by a variety of filters, centrifuges and water spray collectors. However, more than a decade after its operations commenced, the Bell Bay smelter reportedly had no such controls whatsoever, at around half the furnaces in the plant. Deadly fumes crept up from floor level directly into the atmosphere (44, 200).

In 1989, Australian medical authorities finally caught up with US and European studies on the links between aluminium smelting and cancer. The executive director of the Asbestos and Industrial Cancer Society, Kerry Davis, declared that "at least 39" cancer deaths had occurred among workers in the industry directly as a result of the industrial processes. This was confirmed by a specialist at the Peter McCallum Cancer Hospital, Dr Cyril Minty, who also said that as many as 10% of aluminium smelter workers could be affected by "pot room asthma" (265).

Maoris at Tiwai Point have recently complained of "severe air pollution" from that smelter, and declared that "no living plant life [is to be found] in the area". In his submission on the Ngai Tahu land claim, to the government's Waitangi Tribunal, Robert Agrippa Whaitiri complained of ships from Tiwai polluting Bluff harbour (266).

Maoris take on CRA

CRA has kept a lower profile in New Zealand exploration compared with its rampages over Australia: possibly because of rebuffs by planning authorities. In early 1984, it became the first overseas mining company to be refused a prospecting licence by the country's Planning Tribunal. It had wanted to "investigate" part of the Victoria range in Inangahua county, by blasting, excavating, and felling trees on the slopes. Passing judgment on the company's intentions, Tribunal Chair, Judge Sheppard, said CRA had "shown little concern for the environmental effects of the proposed prospecting" and "no willingness" to minimise them. A little later, CRA's prospecting in the Reefton Area on Globe Hill, Westland, was criticised, when the company bulldozed an area of historical importance (though not covered by the Historic Places Act) for a helicopter landing pad (267). The Nelson Planning Tribunal rejected the company's plans.

A more significant defeat for CRA – which partly influenced the incoming Labour government of David Lange in 1984 to introduce some restraints on large-scale mining – had occurred two years earlier. Spurred by the high gold prices of 1980, CRA tried to secure an exploration licence on traditional Maori fishing grounds, at Manaia Harbour, the Hauraki gulf and the Manaia river, all on the renowned Coromandel peninsula. Concerned that these fishing grounds would be poisoned, shellfish beds destroyed, lifestyles and traditional recreational pursuits set at risk, and economic self-sufficiency jeopardised, Maori activists began organising against CRA. Not only did they lodge formal objections to the application and secure the help of four Maori MPs, they also enlisted the assistance of the New Zealand Maori Council, the Maori section of the National Council of Churches, Maori land rights campaigners, and tribes in the Hauraki gulf. The issue was widely publicised by local media, including television. Ministers were invited by Maori traditional owners to inspect the area and determine the deleterious effects which mining would have upon it. Only CRA was excluded from invitations to visit the Maori marae, lest they use it "as a public relations exercise". Links were established with non-Maori organisations and with the Aboriginal Mining Information Centre in Melbourne. AMIC (US!) attended a CRA annual general meeting in Melbourne and, to the question: "Would CRA guarantee NZ$3,000,000 to return the land to the Maori people in its natural state after mining operations ceased?" got the timeworn, derogatory response that it was "a matter for the New Zealand government" (268). Although the campaign did not secure recognition of Maori land rights and exclusive use of their traditional territory under the Mining Act, the campaign was partly successful: CRA withdrew its application (269).

Of course CRA has had other intentions to explore, and if possible mine, in New Zealand. (In 1982 it held the largest amount of leases under application of any company, at Golden Bay, a heavy mineralised dairy and forestry, fishing, mining and tourist area in North West Nelson. It had also built a heli-pad and carried out drilling and line-cutting) (269). A later clash between the company and local farmers and conservationists occurred in 1988 in Cobb Valley and the Mount Arthur Tableland. CRA claimed that any mining would be "insignificant" – its usual threadbare "defence". The local organisation NAPSAC (Nelson Area Parks Action Committee) launched a national petition against the project – the Wharepapa Declaration – calling for the whole of North West Nelson Conservation Park to be closed to mining.

In 1991 CRA announced that it was pulling out of New Zealand exploration.

Black and gold

Australia is the world's leading exporter of coal. Not surprisingly, CRA has also been involved in producing both coking and steaming coal, though its contribution to the national effort has not been as substantial as it might have liked. (In 1986 CRA produced around 6 million tonnes, while Australia as a whole exported 92 million tonnes) (8).

Like other producers, CRA has had to suffer the effects of gross over-supply in the world market (270), and price cuts demanded by Japanese customers: in 1986, for example, it had to drop its price by around 10% (8). Also, like other producers, it has fallen foul of the unions, as it cut real wages and tried to boost productivity (8, 271).

However, early in 1989, CRA scored a coup by arranging new three-year sales with Japanese consumers, worth nearly one hundred million dollars (US$96 million) (272). This arrangement followed the federal Australian government giving approval for the controversial export of coking coal to Japan: controversial, because the currency movement left the Japanese better off than the Australians (273).

CRA's wholly-owned subsidiary Kembla Coal and Coke (KCC) runs underground mines in the Illawarra region south of Sydney (NSW), producing coking coal and some coke (4). KCC's Coal Cliff mine was at the point of closure in 1986 (4), but Coal Cliff Collieries Pty Ltd itself has an 80% interest in the Vickery JV at Gunnedah (NSW) (4). CRA has a 50.2% interest in the sulphur-steaming coal project at Blair Athol, Central Queensland, run by Pacific Coal (5). This is CRA's biggest coal project, in Australia to date – producing around 5 million tonnes, but capable of putting out 8 million tonnes a year (4).

The Tarong Coal project (in ex-Premier Joh Bjelke-Peterson's constituency in south-east Queensland) is an open-pit mine, with considerable reserves of steaming coal, and an assured contract to supply fuel for a state-owned power station (4, 14).

Controversy over coal

This bland recital of CRA's coal interests should not suggest that (for once) we have entered a non-controversial sphere of the company's operations. On the contrary, CRA's underhand attempts to take over AAR and CAIL (Coal and Allied Industries) in the 1970s, were among the more outrageous of RTZ/CRA's ploys to seize a lion's share of the nation's resources (see above).

In 1980, CRA – along with the Victorian State Electricity Commission and other companies – incensed farmers, and others in the state, when it investigated the La Trobe valley for brown coal and other deposits. A leaked Task Force report declared that this region could become "the most intensive energy producing area in the world thanks to its extensive deposits". The environmentalist magazine, *Chain Reaction*, predicted that the project could "destroy the entire La Trobe valley" (274). Undeterred by the criticisms, in 1981 CRAE stepped up its investigations of the deposit and also into the Murray Basin. Roderick Carnegie, the company's chair, waxed lyrical about the region's potential for coal-fire electric power, managed by CRA and possibly fuelling a chemicals industry – so long as the coal's high (50-60% weight) water content could be removed (275). (In 1988, CRA and Melbourne University started work on a project to turn lignite into smokeless fuel, by removing its 60% moisture content) (276).

In the last couple of years, echoes of the La Trobe controversy have been heard in the far west of Australia, where (along with Barrack Energy) CRA has proposed to deep-cut for coal in a proposed National Park, and either sell it to industrial users, or use it to fuel a A$1 million power station – possibly for a pulp mill (277). Known as the Hill River project, CRA's plans to exploit this deposit were quite advanced by 1989 (278). So, however, were strident objections to the whole scheme. One of Australia's leading conservation organisations, the Australian Conservation Foundation (ACF) (279), pointed out that the area, Mount Lesueur, contains over 800 plant species alone: perhaps 10% of the total native species in the state.

CRA blandly asserted that it is aware of the botanical importance of Mount Lesueur and is intent on ensuring "environmentally responsible development" (280). But, once again, CRA put its money on a potentially disastrous chain of consequences: destruction of part of a unique eco-region, the logging of thousands of trees, and the construction of a coal-fired power station, possibly for a highly dubious woodchip plant: thus, at a stroke, contributing both to the greenhouse effect and other atmospheric pollution.

But Australia is no longer necessarily regarded as the best bet for international coal supplies

261

(281). In any case, where cheap sources of coal are concerned, CRA has already "picked a plum" in a part of the world where environmental consciousness is weak (though rising), labour is cheap, and national legislation strongly favours the multinational at the expense of smaller, indigenous developments: Indonesia.

In Indonesia

CRA has long been deeply involved in Indonesia: as already pointed out, Comalco located some of its plant there in the early years of its bauxite exploitation. In 1975, the Indonesian régime proposed leasing one million hectares of West Papua (Irian Jaya) to the company for a large-scale integrated scheme, combining logging, timber, plywood, veneer, woodchip and pulp production (282). This project never materialised.

On the island of Kalimantan (formerly Borneo), CRA holds a half share in Indonesia's largest new coal mine, the Kaltim Prima: indeed it is likely to become the largest mine in the country, apart from the state-owned Bukit Asam coal mine, owned by PT Batubara (8). Output from Blair Athol, and the Kalimantan mine could soon make Pacific Coal, Australia's largest producer of steaming coal (283, 355).

Situated in the Samarinda district of East Kalimantan, only 20km from the coast, the Sangatta (Sengatta) coal mine is owned 50/50 by CRA and BP. Although, in 1989, BP announced the sale of its world-wide coal interest (and CRA was mooted as a possible buyer for the non-South African mines) (284), BP has held on to its share of this particular mine "for local reasons" (285). Kaltim Prima has coal reserves of at least 360 million tonnes – just under a quarter of Kalimantan's total reserves (286) – of which around a quarter is recoverable by open-cut methods (287). It is a "premium fuel with high calorific value and low sulphur" (287). Test production at the mine started in late 1988, commercial production in 1991, and is expected to increase to 4 million tonnes by 1992 (288) and 7 million tonnes three years later (288). Its major markets are expected to be in

Italy, West Germany, the Netherlands (GKE), Japan (Chugoku), Hong Kong (China Light and Power) and Taiwan (283, 289).

The Kaltim Prima partners have benefitted enormously from an "open door" policy by the Indonesian authorities, which allows "one step" approval for new projects (thus cutting out a lot of environmental and social impact studies which might be required elsewhere); guarantees unbreakable contracts; has taxation laws which are favourable to foreign investors; and allows up to 95% foreign equity ownership for a project's first ten years (290). At Kaltim Prima, the government will take a 13.5% royalty, and pay tax at 35% for the first ten years (291).

Equally important, the Kaltim Prima coal mine is central to implementation of the Indonesian régime's "transmigration" programme in Kalimantan – a programme which has been lambasted throughout the world by human rights and environmental protection organisations (including Survival International, Friends of the Earth, the *Ecologist* magazine, Tapol and others) (292).

Kaltim Prima is intended to provide jobs and encourage infrastructure for the relocation of families from the Indonesian mainland (293). Yet, full implementation of the project has required that huge areas of tropical forest be felled for roads, a power station, and a conveyor to the coast. Drastic congestion has been occurring in the local river Sengatta, with probable pollution of the water supply. Worse, there have been highly negative social impacts:

"Drunkenness, prostitution and conflicts between local people and newcomers have already reached a delicate stage", warned a local social scientist in 1989. "Large numbers of contractors have been associated with the mine and an oil lease across the river. Workers get taken on for limited periods, however. They get high wages, prices go up. When they quit, they rarely return to agriculture and, in any case, much of the good farmland has been lost to plantations and now the mine. We're seeing the birth of a floating, aimless, unskilled, population with few prospects." And a substantial proportion of these local people are indigenous to Kalimantan

They are Dayaks – whose economy, culture and livelihood, centre on the river and its banks (294).

... and gold

CRA's penetration of Indonesia doesn't stop with coal. More than ten years ago, it was investigating an alluvial gold deposit, and its parent, RTZ, reached an agreement to explore for copper and other base metals in north-west Sulawesi (295).

Both companies have also been active in Kalimantan where gold fever has dazzled both the small-time prospector and the multinational corporation, on a scale only slightly less than that currently blighting Brazil (296). An advisor to the Northern Land Council in Australia, who worked with Dayaks in 1988, came across a Rio Tinto Indonesia prospector who announced the discovery of gold in the cap of a mountain on the Kalimantan/Sarawak border: "If we want it we'll just cut the top off the mountain," he declared (297).

CRA also entered into a JV with Battle Mountain Gold in 1987, to investigate precious metals in Kalimantan (364).

Among the many multinationals invading the island (more than a hundred contracts were signed between 1985 and 1988, although many of them have lapsed) (298, 299) have been BP, Pelsart International and Pancontinental (300). In 1985, PT Rio Tinto Indonesia began exploring a prospect on the Kelian river in central Kalimantan (301): later CRA took over. With more than 30 million tonnes of ore, grading 2.2 grammes/tonne, the mine is clearly "highly prospective". (Recent figures were: 55 million tonnes of ore grading 1.9 gramme/tonne, producing 100 tonnes of gold over 11 or 30 years, depending on the scale of operations) (302).

Kelian Equatorial Mining (KEM) will benefit from favourable government legislation, which one head of a mining company has called "the best in the world" (298). Unfortunately, the conditions under which the mine will operate are not "the best" for the thousands of Dayak river-dwellers in the neighbourhood. Many of these have lost their traditional livelihood – namely, the panning of the river Kelian (a tributary of the Mahakam), and selling the nuggets for cash (303). And, according to a Dayak investigator, the project has attracted gambling, prostitution and given rise to theft – even murder – on a scale previously unknown: "Already waste from the plant has been found in the water two kilometres from the site and pollution in this part of the Kelian has meant the people have to look elsewhere for their water supplies" (297).

A feasibility study for the Kelian prospect was completed by the end of 1988 (304) but a thorough social and environmental impact study has not been carried out.

Rum Jungle: still poisoning thirty years on

Rum Jungle, in Australia's Northern Territory, was one of Australia's earliest uranium finds (though Radium Hill in New South Wales was probably the first) (305).

The first Rum Jungle find was recorded in 1912, some thirty years after it had become part of one of the far north's most colourful "frontier" mining and agricultural settlements (306). At that time, uranium was just another mineral, and it was not until after the Second World War – with a government incentive of A£25,000 for bounty hunters – that uranium was rediscovered along the Finnis river. In 1952, uranium profits were exempted from income tax, and mining was declared a "special defence undertaking": the government took over land on which the strategic mineral had been found (306). It was at this time that the Mary Kathleen deposit and Rum Jungle were discovered. (The latter by a farmer, J White, while killing kangaroos). Plans were soon being laid for underground mining at the White shaft, and seven other deposits were drilled, including the Mount Fitch orebody which was never actually brought into uranium production (306). 1952 also saw British and American experts arriving at Rum Jungle and plans for a town-site being drawn up (305, 306).

A year later, a joint British-American government group, called the Combined Development Agency (CDA), provided ten years' worth of capital on which to open the mine. Consolidated Zinc Proprietory was authorised to construct the project on behalf of the Commonwealth of Australia and, for this purpose, a wholly-owned subsidiary of Conzinc, called Territory Enterprises Ltd, was formed. The same year, the Australian Atomic Energy Authority (AAEA, later the AAEC) was set up. The AAEA took control of Rum Jungle, while Territory Entreprises was responsible for its operations (305).

Rum Jungle was mined for just over a decade, supplying yellowcake to the British and American atomic weapons programmes. When that programme was cut back in the USA, output was reduced. In 1963, the mine was closed altogether as a uranium producer, although copper mining continued until 1965 (306). Uranium seems to have been delivered some time after closure – Rum Jungle was the last uranium mine to fold in the 1960s, and it was dismantled in 1971 (306).

Three separate open-cut mines were constructed on the Rum Jungle site: besides White's cut, there was Dyson's cut, completed in 1958, and Rum Jungle South, which was not opened until 1963 (305). Construction for the White mine was carried out by the British firm of George Wimpey and Co Pty Ltd and, at one point in its operation, was even visited by that well-known wildlife enthusiast, the Duke of Edinburgh! (306).

However, the royal seal of approval did little to compensate for farmers' land loss (claims were being settled as late as 1962) (306), deplorable working conditions in its early days ("Hell Hole" was one Melbourne newspaper's appellation, after a protest strike was called in 1956) (306), and Aboriginal claims in the region. The Finnis River Claim, made by the Northern Land Council (NLC) in 1980, has been called "the most complicated yet heard in Australia" (307).

Worst of all has been the huge amount of environmental damage inflicted on a region note-

worthy for its fauna and flora. As early as 1960, an officer of the Northern Territory (NT) Administration was reporting that: "... trees along the back of one stream are dying and water holes [are] devoid of fish". Two years later, a Senior Engineer in the territory claimed that severe pollution stretched 16 kilometres from the mill, along the East Finnis river. "Heavy concentrations" of pollution were reported in January 1963, "as large numbers of freshwater shrimp ... and small fish resembling herring have been floating or lying on the banks" (305).

The Australian Senate Select Committee on Water Pollution declared two years later: "One of the major pollution problems in the NT is that caused by copper and uranium mining at Rum Jungle. The strongly acidic effluent from the treatment plant flows via the East Finnis river into the Finnis river, making the water unsuitable for either stock or human consumption for a distance of 20 river miles. Vegetation on the river banks has been destroyed, and it will be many years before this area can sustain growth" (308).

In April 1965, a water resources technical officer told his superiors that the worst period for pollution came just after floods broke down the wall holding back the effluent at the treatment plant. In reality, the early period of mining had not even been graced by a tailings dam and, when dams were built, they often got washed away by floodwaters. Not until 1961 was the situation ameliorated, when tailings were discharged into disused (but presumably quite porous) open-cuts (305).

Six years later, a Northern Territory Administration team reported that: "no significant rehabilitation has been carried out" at Rum Jungle (305). By then, the degree of poisoning caused by the mine was creating great concern in the Administration and the parliamentary opposition. However, the AAEC was unperturbed. One of its officials, Dr Warner, called the pollution "a minor, local ... problem" (305), while the Commission itself refused to take steps to revegetate the waste dumps, and would not make public a (presumably) damning report on water pollution, which the AAEC itself had carried out (305).

There was also complicity between government and company, as evidenced by a submission made in June 1971 by Mr R F Feldgenner (a senior NT official), to his Minister. Feldgenner recalled that, in 1962, the Minister for Territories had been well aware of the degree of illegal water pollution caused by the mine, but "... he was reluctant to proceed against the companies for reasons of their association with the Commonwealth in the venture". Any attempt to overcome the hazard would "involve quite unreasonable operating costs," according to another Minister (309).

By 1975, it was known that 2300 tonnes of manganese, 1308 tonnes of copper, 200 tonnes of zinc and 450 curies of radium had been released into the Finnis River, with around one quarter of the radium having found its way "probably to the sea" (310).

"About 100 square kilometres of the Finnis River flood plain have been affected by contaminants (heavy metals, uranium, radium and sulphur). In the ten kilometres of the Finnis River downstream from the mine, fish and other aquatic fauna have been almost eliminated, with the effect reducing over the next 15 kilometres downstream". Pandanus palms, water lilies and other aquatic plants had been "eliminated" (311).

CRA and RTZ refused to contribute anything towards the clean-up of Rum Jungle (312). It was left to the federal government to provide A$7.6 million to that end, with another A$16.2 million to be spent over the following six years (to around 1990), in removing heavy metals and neutralising the tailings (313, 333).

Heavy minerals

In its home state of Victoria, CRA has notched up a fairly remarkable reputation for treading on the feet of farmers, Aboriginal people and environmentalists. Its project with the most profound implications for the future is probably the proposed mineral sands mine 20km south-east of Horsham, otherwise known as WIM-150, after the area of Wimmera in which it is located. Announced by the company as the biggest mineral sands find ever made in Australia – possibly the world (314) (notwithstanding the huge deposits already found in Western Australia) – early indications were, that the deposit consisted of one billion (1,000,000,000) tonnes of contained ore in a larger deposit of nearly 5 billion tonnes (315), under some 4 metres of clay overburden (316). In 1986, CRA applied for permission to conduct exploration on a 1000 hectare site, where initial drilling had indicated monazite, xenotime, zircon, anatase and leucoxine, as well as rutile and ilmenite. (Ilmenite and rutile are used to produce titanium dioxide – mainly for the pigment industry – while rutile is utilised for the output of titanium metal which has applications in the aircraft and aerospace industries. Zirconium, processed from zircon, can be used as fuel cladding in nuclear reactors).

The Victorian state government has clearly been in favour of the project (317), although both Queensland and New South Wales state governments have banned some types of mineral sands mining (132), due to the grave risks of radioactive contamination (thorium and uranium) from tailings disposal – not to mention the huge upheaval to farmland. (Four metres depth of clay have to be removed and then 10.15 metres of sand must be dredged.) The Horsham council, backed by environmentalists, delayed the opening of the 120 tonnes/day pilot plant, with an appeal to the Planning Appeals Tribunal – but this was dismissed. In any case, the local shire council at Wimmera is in favour of the mine (298). At the beginning of 1989, CRA had permission to process 20,000 tonnes of mineral sands a year, but this was to be increased to no less than 1,000,000 tonnes a year (319).

While it looks as if WIM-150 will proceed, CRA got a drubbing in 1983, after farmers, Aboriginal people and environmentalists united to persuade two local shires (Maldon and Lexton) to exercise their powers to stop a gold-leaching project, which would have resulted in potassium cyanide and other chemicals being pumped 100 metres underground into aquifers (318, 319). These "deep leads"at Eastville not only cover nearly two thirds of underground Victoria, but are crucial in providing drinking water for livestock (320). Farmers around Mount Mitchell formed the Groundwater Protection Society (Mt Mitchell), after other farmers had initiated the Groundwater Protection Society at Eastville. During what has become known as the "Evansfor Incident", it was discovered that the company had started drilling without shire approval: the shire swiftly told CRA to stop. Whether or not it was this which was effective, or direct action – during which persons unknown smashed the company's drilling rig and painted a rude behest on the road – CRA responded and did indeed "piss off" (321).

After CRA withdrew its original plans, the company came up with another proposal to tap into the underground gold, using toxic chemicals which included hydrochloric acid (322), and that would release arsenic and copper into the groundwater (321). But this scheme, too, was eventually abandoned. In 1989, at the RTZ annual general meeting in London, the chair, Alistair Frame, gave an undertaking that RTZ (and presumably CRA) would never be involved in cyanide solution mining of the type proposed at Eastville (323).

This was not the first time CRA had tangled with Aboriginal protestors and other landholders in Victoria. In the late 1970s, CRA was among a number of companies exploring for various minerals, including uranium, in the famous Victorian Alps – an area of great, relatively untouched, natural beauty which includes the Snowy Mountains range. (CRA's lease was next door to one held by Urangesellschaft) (324). Within a few years, according to research by AMIC (US!), the company had leased an area for exploration covering one-third of the state. Not one of these leases had been secured with permission from Aboriginal people – of whom there are up to 30,000 (325). Much of

the lease area contained potential uranium deposits – a highly contentious fact raised by Aboriginal poet, Boolidt Boolitha, and others, at the RTZ annual general meeting in 1981. It was pointed out that CRA had discovered significant deposits of uranium in both north-west and central eastern Victoria, and that some of the exploration work had been conducted over Aboriginal burial sites (326).

Bougainville

It is Bougainville Copper in Papua New Guinea (PNG) which, of all the company's many enterprises, best illustrates the degree to which CRA is prepared to exploit indigenous people, and virtually wreck a major ecosystem. (The destruction of Weipa and Mapoon runs a close second).

Until 1989, Bougainville Copper Ltd (BCL) has also probably been CRA's most consistently successful subsidiary – although Comalco's returns from bauxite mining on Cape York peninsula have paid major dividends in recent years.
In 1973 BCL was not only CRA's – and RTZ's – most profitable single venture, with profits running at A$158 million, it was also the most successful company in Australian corporate history up to that time (327). The following year, 1974, it turned in another huge profit of A$140 million, due partly to the dubious practice of "high grading" (mining and processing higher grade ore while dumping lower grade material, possibly for future use) (328). By the early 1980s, Bougainville was contributing a hefty 23% of RTZ's pre-tax profits, despite representing only 9.4% of the corporation's total assets and 8% of its sales (29).
The project was set up (with some Japanese funding) (329) as a joint venture between CRA and North Broken Hill Holdings (NBHH) in 1966. By 1972 CRA held 65.6%, and NBHH 28% respectively in Bougainville Copper Ltd (330). RTZ's beneficial interest by that time was 80.7% of CRA and more than half (53%) of NBHH. (Fifteen years later CRA's interest in

266

BCL was 53.6%, with RTZ holding 49% of CRA).

Bougainville's development was underwritten by 27 British, European and Canadian banks, headed by the Bank of America. They provided nearly US$250 million by the end of 1970, with the Bank of America taking up 1% of the equity and a second syndicate of the same banks having the option to take up 2% of the equity (331).

A few months later, Mitsubishi Shoji Kaisha Ltd, with Mitsui and Co, loaned another US$50 million and signed an agreement to provide equipment worth a further US$30 million, with repayments for this stretching over ten years (332).

It is therefore little wonder that RTZ's first chairperson, Sir Val Duncan, called the mine: "the Jewel in our Crown" (44).

Discovered in 1964 by Ken Phillips, a geologist with CRAE (Conzinc Rio Tinto of Australia Exploration Ltd), the company was actually benefitting from gold discoveries that had been made near Panguna thirty years before. It was an Australian geologist working for the colonial Administration who located low-grade deposits in 1960 (334). In the early stages of testing and defining the orebody, CRA operated under a 1928 mining law which enabled it to prospect on land used by the Nasioi people for copra and cocoa cultivation, without the permission of the landowners (44). In 1966, despite opposition from some members of the PNG House of Assembly, this mining law was modified, allowing CRA virtually unhindered extension of its operations, to a ceiling of ten thousand square miles [sic] around the deposit (334, 335). The following year, CRA had confirmed the Panguna project as commercially viable. It had already spent A$4 million proving the deposit, and was to shell out another A$12 million until 1969 (334). However, it held only a two-year prospecting licence, with no guarantee that it would be allowed to mine. The company told the Administration that it couldn't foot any more bills until it got mining contract permits, and enough land on which to operate (334).

Under the 1967 Agreement, which was then negotiated with the Australian authorities in PNG, Bougainville Copper acquired virtual fiefdom over the island and its resources. The Administration was offered 20% of the equity in the new company. In return it would acquire, and offer to the company, almost any land it required (334), as well as a township, roads and port – not to mention support facilities and a tailings disposal area which could extend over 50,000 acres (334).

There is some discrepancy between historical sources as to whether BCL was offered a ten-year (44, 335) holiday from corporate taxation under the 1967 Agreement – or a more limited three-year or a five-year one (334, 336). In any event, the project was able to operate without taxation restraints during its formative years, paying no more than 1.5% royalty on the FOB (Free on Board) value of the copper produced (335).

Production from Panguna started in April 1972, after certain "teething problems" – which included landslides caused by a cyclone (328). BCL's investment in the project had, by then, cost around A$400 million. No matter: for, in the first six months of output, BCL was raking-in no less than A$1 million a day in tax-free profits (336). Its contracts at this stage included a four-year deal with Phillip Brothers of the USA (a subsidiary of Engelhard Minerals, itself an associate company of Anglo-American), and substantial arrangements to supply both European and Japanese smelters (330). The European customers included Norddeutsche Affinerie (NA) of West Germany and Rio Tinto Minera SA (RTM) of Spain (337). In 1988 further agreements were sealed with Japanese customers for long-term supply.

In 1974, CRA was forced to re-negotiate its 1967 Agreement, after Papua New Guinea gained independence. Riots broke out between Papuans and New Guineans, while outrage was expressed at the concessions CRA had obtained, and the damage it had wreaked. These events precipitated calls for Bougainvillean secession, and widespread condemnation of the company's practices.

According to mining authority, William S Pintz, the renegotiation of the agreement was a "direct result of local reaction to Panguna" and a "political backlash associated with river sediment problems" on Bougainville. It was also an opportunity for the in-coming government of Michael Somare to show its abilities and muscle (338).

The June 1972 PNG General Assembly had set out development strategies for the country (Prime Minister Somare's eight-point programme), which included avoiding the country being "in a position where we are so dependent on any other country, or on any single business or industry that we must shape our policies with the interests of that country or industry in mind, instead of our national interest" (328). Other laudable sentiments, promulgated in the Assembly, put a premium on self-reliance, locally-raised revenues, small-scale enterprises, and an equal role for women in development. (Unfortunately, none of these aims was ever thoroughly put into practice).

With the threat of renegotiation heavy in the air, Sir Val Duncan flew to Port Moresby, only to be met by "nothing!" (327): "... no red carpet, no police band, not even a minister at the airport, simply BCL's own cars" (327). RTZ's illustrious chairman spent hours trying to get hold of Somare (who was actually on the golf links at the time!) When the two parties finally met, Somare's side had the advantage of special overseas consultants on its side, including Michael Faber of the British Commonwealth Secretariat, and Stephen Zorn, perhaps the best-experienced negotiator with mining companies, anywhere in the world. For its part RTZ/CRA had little, except the original agreement (334). Albert Maori Kiki made it clear that the negotiations were not between equal partners: "... [We] are acting as a sovereign power," said Kiki, "which comprehends our role as owner of our resources, custodians of our national heritage and as a taxing authority on behalf of the people" (334). Somare was blunter: if the company didn't reach agreement, by September that year, "the agreement will be in a position to put our basic principles into practice" (334).

Understandably, RTZ soon capitulated – but not before trying to get the Australian government to loan the PNG government enough money to buy-out 50% of the mine (339). Triumphantly, Somare announced that the fledgling government had brought to heel "... one of the largest mining companies in the world – a company with big projects in Canada, South Africa, Australia, Britain, Spain and other countries and with total sales worth more than US$1500 million. But we proved our ability to stand up to the foreign interests" (334).

Over the next decade and a half, the PNG government did little to stand up to CRA, although the re-negotiated Bougainville agreement was to become something of a model for other Third World (and indigenous) peoples. In the 1970s, CRA wanted to build a A$300 million alumina refinery in the North Solomons, near large bauxite reserves: its partners were to be Mitsui (50%) and the PNG government (25%) (44). But – like the Purari hydro-electric scheme – the project petered out. CRA also considered buying into the other huge copper-gold project in PNG – the Ok Tedi mine. John Ralph, (to become CRA's chairperson on the demise of Roderick Carnegie), was especially keen on getting a slice of this particular action. The PNG government, however, didn't want "a greater role given to CRA" (338).

Throughout most of the 1970s and the 'eighties, BCL was to justify its progenitors' early investment of confidence. By 1983 the processing plant at Panguna was the world's largest copper concentrator (340). Four years later, BCL chalked up its largest profits since the *anno mirabilis* of 1974: though copper and gold production was actually lower than it had been the previous year (341).

One matter of particular concern to CRA has been the moratorium imposed by the Australian administration on mineral exploration in the North Solomons which, under the 1974 Agreement, was to be reviewed every seven years (342). A little while before this prohibition came up for its first review, BCL was expressing "considerable concern" at being prevented from "exploration ... thought to be vital in order to determine whether any other viable ore bodies exist in the Bougainville region" (343).

What has the Papuan New Guinean government got out of Bougainville? RTZ/CRA's Panguna mine was not the country's first gold prospect by any means. As early as 1528, the Spanish navigator, de Alvaro Soavedaro, had dubbed the country Isla del Oro (Island of gold), and Australian prospectors were busy in Papua with pick and shovel from the 1880s until 1914 (344). Nor, with the development of Ok Tedi, and mines at Porgera and Misima, has BCL retained its position as the biggest PNG gold producer. Should Lihir Island (RTZ: 80%) enter production, the country will be host to what has been called the richest gold deposit outside South Africa (345). Within the next five years, according to stockbrokers James Capel, PNG could become the world's fastest-growing producer of gold (346).

The mining industry has consistently regarded PNG as a state replete with unnecessary and punitive restrictions: "a welter of government controls over the operations of the private sector ... [the] legacy of the zeal of government advisors and civil servants in the early 1970s to save PNG from the perceived greed of foreign investors" (336).

Certainly it does seem, at first, as if PNG has benefitted considerably from BCL's operations. However, it is important to distinguish the pre-1974 period from the period which followed the signing of the Bougainville Copper Agreement, and to separate the benefits accruing to the central government from those denied the provincial government – let alone the traditional landowners.

The infrastructure provided by the PNG administration has been furnished at the expense of resources which could, and should, have been directed elsewhere to the benefit of the Papua New Guinean people (335). According

to BCL, the mine has been responsible for providing 45% of PNG's export earnings, as opposed to only 15% from coffee, the country's second highest earner (342). But such high reliance on one mine as a provider of foreign exchange, is hardly healthy in a rapidly developing economy (where 80% of the people live in rural areas).

There is also some debate as to the importance of BCL as a provider of internally-generated revenues. The mining industry and BCL put the proportion at 20% (344). M R Chambers puts it at more than a fifth – stating that US$660 million has been contributed to the State since 1975 (347). Other sources are more conservative, setting the figure at 16% or 17% (348).

In any event, partly to limit the economic risks, the PNG government from its early days decided to take only around 20% of BCL's equity. In 1972, a million ordinary shares were issued to the Investment Corporation of PNG – a statutory body established to promote indigenous membership of commercial enterprises in the country – while another four million shares were issued to BCL (334). The government has thus limited its access to dividends, but benefitted greatly from royalties and taxation. Crucial to PNG's taxation policy has been the concept of Additional Profits Tax (APT) (applied to the Ok Tedi project in a revised form known as Resource Revenue Taxation or RRT) (338). Under this system, while royalty taxation at 1.25% might be regarded as a concession to multinationals, "super profits" attract a much greater claw-back. APT is assessed when the company has received an annually compounded rate of return of more than 20% on the total investment (as BCL apparently did in 1987) (349). It is applied at the rate of 35% on post-Corporate Income Tax profits, raising taxation on these to nearly 60% (58%). However, it is also important to remember that the concept of APT has never been fully, or rigorously, applied to BCL (327).

Compared with the poor deals secured by other Third World governments, in negotiation (or lack of it) with mining companies, PNG's situation must certainly be counted among the more fortunate.

But North Solomons politicians, and specifically those from the Panguna region, have not been persuaded by this argument. Led by John Momis, John Kaputin and others, they have long considered the province to be sorely neglected by the central administration (327). The issue of separatism – especially by Bougainville – was to figure prominently in discussions between the Pangu Pati coalition led by Michael Somare, which came to power in 1972, and both Momis and Kaputin, in the months leading up to final independence. Agreement on a wide measure of self-government for the provinces appeared to have been reached, when BCL announced its grotesquely fat profits for 1973.

From that point on, relationships between the provincial representatives and the central government deteriorated, despite practical steps being taken towards setting up the provincial government of the North Solomons in July 1974.

Fired by separatist sentiments, especially those expressed by the Napidakoe Navitu (an organisation set up to defend land rights against the depredations of BCL), the Bougainvilleans demanded that all royalties from the mine should be returned to the province (334). The central government wasn't happy: financial independence for the provinces could lead to assertions of political independence. But, when the Bougainvilleans threatened to divert a river essential to BCL, Somare's coalition gave in. Henceforth, the whole of the 1.25% royalty was to go to the province (334, 335). Allocations for capital works, granted the next year, proved a much knottier issue. When, in mid-1975, the central government offered only a quarter of what the provincial assembly had demanded, the Bougainvilleans voted to secede. A provincial flag was raised on September 1, government offices were attacked, the airstrip was

torn up, and the mine-site itself was invaded (44, 327).

Before considering what impact the mine has had on Bougainvilleans in general, we should ask whether it has actually provided the boon in employment and opportunities widely boasted by BCL and its defenders. In the first three years of construction, the mine attracted a workforce drawn from all over the country, and further afield. Between 1966 and 1971, some 6300 newcomers entered the island (327). By 1970, out of a working population of 10,500, some 9000 were construction workers. A high (but unknown) percentage of these was recruited from outside the island (350). This created not only a differential wage system between locals and outsiders (350), but tensions which led to inter-community conflict. In 1970, a petition by 700 local villagers and company employees demanded that the company "repatriate" the outsiders. An Australian engineer working on the project during these years has attested to the apartheid-like conditions of that time (351). And, when the Australian Labor Party shadow Minister of Labour visited the mine-site in 1969, he was appalled at what he found. Accusing CRA of paying "slave wages" to black workers, he conjectured that the company's "excellent training programme" was mainly a device to secure a cheap labour pool: blacks driving trucks were getting less than a quarter of their white counterparts (334).

BCL's Employment Relations Manager at the mine in 1970, a Colonel McKenzie, was in no doubt that he preferred outsiders to potentially restive Bougainvilleans: "At present it's about 50% [ratio]," the Colonel told Richard West: "[At] the early stages we were recruiting wherever we could. But we'd prefer a proportion of 33 per cent" (28). When the construction phase came to an end, unemployed workers spilled out around the island, creating an aimless, impoverished, and self-destructive group of single men, about whom Raphael Bele, Member of Parliament for Central Bougainville, commented in mid 1973: "... [We] have so many vagrants in Bougainville. Often they paid their own expenses to Bougainville ... what a pity when a person arrives to find there are no vacancies. From there, the person's vagrancy begins; he is now included on the list of those who steal, murder and so forth. Bougainville villagers are terrified by these serious incidents; they dare not walk alone on the roads" (328).

More serious, in the long term, was the insidious substitution of imported goods for local produce (28), and consequent dangerous weakening of the indigenous economy. As Richard West and R J Jackson have both pointed out: before the establishment of BCL, Bougainville was far from being an indigent backwater, suddenly blessed with new-found wealth. Copra and cocoa cultivation was carried on in no less than 81 plantations (foreign-owned, and already employing several thousand migrant workers) (327). Not only did the mine absorb invaluable agricultural land, which Bougainvilleans might have used to grow foodstuffs, but the company itself had a policy of importing many of its provisions. Though BCL claimed to have purchased more than a million dollars' worth of fruit and vegetables from villagers (A\$1,300,000), between 1970 and the end of 1974 (355), many of these purchases were from expatriate traders (28). Local people were often not able to compete with the imports, because they could not reach the towns (lack of roads) (350). John Momis summed up the situation in 1973: "While the company makes somes attempt to buy local produce, there is no systematic attempt to replace imports by food produced by the local people themselves ... Throughout Bougainville there is a lack of incentive for people to develop their resources ... when the copper company is asked to supply [technical advice] it says that it is not prepared to favour one area at the expense of another" (328). A few months later Momis was even more damning:

> "Companies have no interest whatsoever in the welfare of the people. They tell lies to the people repeatedly, saying they are concerned about their welfare. Companies have one motive only, and that is to make profit. Many of the so-called professional people do not act in such a manner as to cause physical violence but they poison the minds of the people. They are against us, ruining us, colonising our minds, so that our people have no self-respect today. They have become tools: some of them are being trained to become very effective tools of the colonial system ... The company ... does not have any interest in the people. It would sit there in its ivory towers watching us murdering one another; it has no concern for us whatsoever. It will give you eighty dollars to be a good white-haired boy of your colonial masters" (328).

A decade and a half later, the worst fears of Bougainville landowners were confirmed. In 1988 Perpetua Serero, leader of the island's matrilineal landowners, told a visiting reporter:

> "We don't grow healthy crops any more, our traditional customs and values have been disrupted and we have become mere spectators as our earth is being dug up, taken away and sold for millions. Our land was taken away from us by force: we were blind then, but we have finally grown to understand what's going on" (348).

Four months later, echoing Serero, another traditional owner – defining her traditional land as somewhere in the bottom of the gaping mine-pit – said: "We are in the darkness at that time ... Now we see with our own eyes" (352). What the people of Bougainville see is one of the worst human-made environmental catastrophes of modern times. "Rio Tinto Zinc has more to answer for in this tiny corner of the globe than any other. The day was certainly cursed when [it] discovered copper deposits on Bougainville," said Diane Hooper in 1977 (44). "An economic godsend – and an environmental disaster" was how Philip Hughes (Head of Environmental Science at the University of PNG) dubbed BCL's operations ten years later (353). Ken Lamb, Professor of Biology at the same University, has also called the Bougainville experience "disastrous" (327). An Australian engineer, working on the mine in the early phase, was more direct. Commenting on the impact on the local people, he declared: "It's fucked them" (28).

"Take a trip down the Jaba river", invited the respected *Pacific Islands Monthly* in early l989, "where millions of tonnes of waste is dumped to become floating filth" (348). "All aquatic life in the Jaba Valley has been killed," concluded another scientist, M M M R Chambers, in 1986. "Entire villages have been moved and rebuilt on tailings down-river from the mine-site," commented the *Australian Financial Review* in late 1988, "where crops grow only after heavy application of artificial fertiliser to the highly acidic soil" (354). "When the mine first came," another Bougainvillean has declared, "everything was so new, we didn't know what to expect. The thing what we were becoming so ignorantly proud of was that it was the biggest open-cut mine in the world. At the time, our thinking was that money can be a supplement to our way of life. But now it's not only that, it is chewing and going right into the people's life and that has disturbed a traditional balance that has existed" (356).

In a central government mission to investigate the environmental effects of the mine, in 1988, the Environment Minister, Perry Zeipi, found the pollution "dreadful and unbelievable" (357). The Jaba river was "full of all kinds of chemicals and wastes" and the people had been forced to abandon traditional fishing (357).

Panguna is situated in the mountainous interior of Bougainville, virtually at the heart of the island, in an area of high rainfall (around 4800mm a year). It registers high seismicity and volcanic risk: earthquakes of 7.1 on the Richter scale are recorded roughly every ten years (347). Prior to construction of the mine, no environmental impact studies were carried out – there were no laws in the territory governing such evaluations (347). While three later pieces of legislation did incorporate some environmental regulations, they were ridiculously inadequate. The 1967 Agreement (already mentioned) obliged BCL to revegetate tailings piles and overburden; the 1971 Overburden and Tailings

Agreement was intended to compel BCL to minimize copper pollution and expedite land restoration (but it lapsed in 1980) (347); and the 1974 Agreement empowered the government to collect environmental data from the company – but didn't say what type of data.

It might as well have been that no legal attempts were made to limit environmental damage. Waste rock was simply dumped near the mine (347). From 1973 until 1983, 768 million tonnes of ore and waste was processed, rich in copper and iron pyrites. Of the 373 million tonnes of processed ore, just under 7 million tonnes was exported as concentrate, and the remainder deposited in the Jaba Valley. By 1976 alone, out of 118 million tonnes of waste rock, and 108 million tonnes of tailings which spewed out of the Panguna hole, around a third was "smeared over the valley floor and neighbouring areas", causing "serious environmental change" (327).

The tailings have continued to be released into the Jaba River Valley, and thence have leached into the Kawerong-Jaba river system, completely untreated (347). The only redress has been a short, virtually useless, pipeline to the upper reaches of the Valley (340), and engineering studies which commenced in 1987, to construct the world's largest gravity-flow slurry pipeline from the mine to the Empress Augusta Bay (358). But this innovation risked simply transferring the pollution further into the sea. In 1973, tailing discharges were running at a massive 70,000 tonnes a day – totalling 34,376,000 tonnes between January 1972 and June 1973 alone. At that time, it was estimated that about 60% of the tailings would be carried all the way along the 35km river and into the sea, while the remaining 40% would be deposited on land (359). By 1974, these tailings were eroding at some 539,000 tonnes a month (359).

It seems that around half the tailings has remained in the valley, while finer portions have been carried into the Bay (347). Contaminated with heavy metals – such as copper, zinc, cadmium, mercury and molybdenum (359) – these washes are also high in sulphur, arsenic and mercury. This sedimentation alone would have robbed the Jaba river of its aquatic life, quite apart from the chemical pollution (361).

H R Meier has declared that it will be impossible to restock the river system until long after the mine is closed (362).

By the mid-1980s, some 8000 hectares of the Empress Augusta Bay were covered with tailings, to a copper concentration greater than 500ppm (parts per million) – destroying benthic flora, but not apparently affecting fish to any great extent in the estuary (361). Nonetheless, no-one appears to doubt the profound affect that the operations have had on the major part of the Kawerong/Jaba river (363) – where the "entire length of the valley is covered by sediment up to 60m deep [sic] and 1km wide in basins". At the outflow, the river bed was more than 30 metres deep several years ago, and rising by 3.5m every year (347). Although M R Chambers claims that revegetation of tailings dumps has been relatively successful (347), this, of course, has little or no impact on the rates of leaching into the rivers. It is to be doubted that the company itself has progressed in its thinking, much beyond a statement made in the early '70s by Brian Barry, BCL's public relations manager:

> "I get pretty snakey," stomped Barry, "when I hear the conservationists complain about what we're doing to Bougainville, because I live at Toorak, which is a very nice suburb of Melbourne, and I can't go for a walk without stepping in the doggy dirt which has been left by the dear little doggies of the rich people. So why can't they worry about conservation in their own suburbs first?" (28).

Whatever the profound environmental consequences of this runaway project, the most serious physical and psychological impacts of the Bougainville mine have undoubtedly derived from the land seizure and land clearances which occurred in the late 1960s. (Of course, we should not minimise the continued loss of agricultural land and water, as the tailings area has expanded) (359). According to M J F Brown, in 1974 some 800 people lost their land rights in the tailings area alone, while 1400 people's fishing rights in the Kawerong and Jaba were jeopardised by mining operations (359). Some 220 hectares of forest was felled before mining could begin: trees were poisoned with herbicides, then chopped down and burned, while the overburden – rich in fertile volcanic ash – was hosed directly into the river (359).

BCL and its Melbourne and London masters have consistently denied any responsibility for the wave of protests and numerous acts of resistance which have dogged the mine since before its establishment. These have been portrayed as the results of secessionist tendencies in Bougainville, and conflict over the disembursement of royalties and compensation (407). Even when guerilla actions were started in earnest against BCL and CRA in late 1988, the chairperson of RTZ, Sir Alistair Frame, tried to pass over the threats as: "nothing to do with the company" (356). And, this was despite the fact that the guerilla leader, Francis Ona, was a former employee of BCL, and his followers were then wearing T-shirts sporting the emblem "Valley of Tears" – a pointed reference to Richard West's 1976 exposé of RTZ and BCL, *River of Tears*. (Ona, in a letter to the Member of Parliament for South Bougainville, David Sisito, also alleged that a white "mafia-type" network was operating within CRA, with links to the central government) (366).

Certainly John Momis – the long-standing critic of BCL, and one of PNG's ablest political leaders – is in no doubt of the responsibility that the company bears for the misfortunes of the people. In his role as Provincial Affairs Minister in the central government, Momis rounded on BCL in late 1988. He claimed that it had "blatantly disregarded the longstanding grievances of the Panguna landowners" (367), and he called for a tripartite agreement to be concluded between the landowners, the company and the government. Momis also stated that foreign traders and entrepreneurs, who have taken over many local businesses on Bougainville, were sponsored by the company through its Bougainville Copper Foundation (368). *The Australian Financial Review* itself commented, in December 1988, on the "growing and vocal opinion that [the landowners] would be better off if the company just packed up and left" (368).

Panguna lies at the heart of the land of the Nasioi people: their reaction to CRA's first explorations was to set the unequivocal tone for years to come. "Bewildered, frightened and hostile," wrote Don Woolford of this early period, "they did what they could to stop the project. Mothers put babies on survey pegs to stop the pegs being hammered in ... A village leader publicly threatened to cut his throat in protest. Workers were occasionally assaulted ... The protests were supported by the American-born Bishop of Bougainville ..." (334).

When CRA secured permission to prospect on 10,000 square miles of the island, the local fight was taken up at central level, where Bougainville representative, Paul Lapun, demanded that rights over subsoil of PNG should remain with the people, not the state. After an initial setback – and an incident when five men were jailed for expelling BCL prospectors – Lapun's motion passed by 31 votes to 21 (334).

It was a short-run victory, however: the 1967 mining lease virtually negated the landrights supposedly guaranteed the previous year (334). When construction for the port and township (at Arawa) started, it was coastal people around Rorovana who felt the brunt.

In early 1969, CRA told the Administration that it wanted to take-over 240 acres out of some 4000 acres owned by the Rorovana people around Keita. A derisory A$105 per acre, plus A$2 per felled coconut palm, was offered in compensation. The Rorovanans soundly rejected the offer.

Weak attempts at negotiating a takeover failed miserably and, on August 1 1969, backed by seventy-five riot police flown in from Port Moresby, CRA prepared to take possession of the land. In the early morning, led by District Commissioner Ashton, a hundred police carrying rifles, shields, batons and gas-masks, marched through the coconut groves onto Rorovana land. It was the women, through whom the land is held by matrilineal inheritance, who led the resistance (369):

"Three helicopters hovered above. A few hundred yards from the land a delegation of about twenty-five villagers met the party and asked for fresh talks. Ashton replied that there had been ample opportunity for talking and the survey must begin. The police moved onto the narrow stretch of beach and formed a protective barrier around the surveyors. Simultaneously about six hundred villagers materialized from the jungle undergrowth. As soon as the first marker was in, about ten women darted from among the silent spectators and tried to pull it out. A rugby-style scrimmage developed, with police pulling the women away, only to have them immediately return to the fray. As the watching villagers moved closer, the main body of police put on their masks in case tear gas was needed. At the same time more women joined the scuffle. Finally the marker was rolled free. The villagers cheered and clapped and moved off. No blows were struck and no weapons were displayed by the villagers. A few of the village men momentarily looked as if they would help the women, but they were restrained by their own people and Middlemiss, the only white face in the Rorovanan ranks. Both sides seemed happy with their day's work. From the Administration point of view, surveying had begun with only what it regarded as a token protest and no violence worth speaking of. From the Rorovanan viewpoint, a sample of village power had been offered and a symbolic victory achieved. Surveying continued over the weekend without hindrance" (334).

However, this was only the "lull before the storm".

"On Monday 4 August, CRA, its surveying finished, moved in bulldozers to clear the jungle and knock down the coconut trees. The Rorovanans stayed away. But the following morning about sixty-five villagers lined up in front of the bulldozers. They ignored several calls to move and stubbornly held their ground when police, holding their batons horizontally, tried to shepherd them off. Several rounds of tear gas were fired, but the fumes wafted away. Finally a controlled baton charge – in which the police aimed only at the legs – was ordered. In the melée one villager suffered a gashed leg. The drawing of blood apparently had a similar symbolic significance to the removal of the marker the previous Friday. As soon as it occurred the villagers moved away, leaving the bulldozers free to continue their work. Next day the bulldozing continued and Ashton told reporters he thought the crisis was over. He had spoken prematurely: politically, the crisis was beginning. Early on the Thursday hundreds of villagers, many from the other communities, walked onto the land. They moved in orderly fashion and in small groups. A few carried weapons – multi-pronged fishing spears, tomahawks, or bows and arrows. One man wore a bright yellow CRA hard hat. The build-up was watched by helicopter, and Ashton asked the company to suspend work for the day. The same thing happened on Friday. Spokesmen for the villagers said they had the support of most communities in the sub-district. Some men had walked all day to join. The Arawa villagers, well aware that their land was next on the list, were the most prominent allies" (334, 370).

From then on, opposition gathered speed. A Bougainville farmer was jailed for throwing a CRA geologist into a stream (371). The Territory's public solicitor, a strong supporter of land rights, flew to Rorovana to help the community; a new organisation, Napidakoe Navitu, was set up to bulwark the protest, and immediately gained the support of hundreds of people. In Australia itself, various white supporters of the Nasioi and Rorovanans began speaking out against the Australian administration's heavy-handedness. Incensed by photos of the tribespeople lined up against riot forces and CRA bulldozers, the Seamen's Union in four ports carried a resolution condemning the

"planned seizure of the indigenous peoples' land" (334).

At this point, the Australian government called in CRA for the first time in the crisis. With a federal election only four months off, the government and company hammered out a new offer to the dispossessed indigenous community. Total acreage of land requisitioned was reduced by 35 acres, reclassified as forty-year leasehold at A$50 an acre a year, with A$30,000 compensation and 7,000 BCL shares at par (334).

While most Bougainvilleans did not take kindly to the new offer, PNG independence was drawing near, and the fight for proper compensation was mainly conducted at a political level.

This was true, at any rate, until two decades later, in 1988. In January that year, landowners marched on the mine demanding higher compensation (357). Three months afterwards, the same landowners lodged a claim for 10 billion kina (A$14.5 billion) – a demand which few people took seriously at the time (348).

By the end of the year, however, the farmers were once again adopting direct action. In mid-November, one hundred of them set up a blockade on the roads, using heavy machinery (372). Later the same month, a small group of dissidents, led by Francis Ona, using explosives and walkie-talkies stolen from the company stores, burned BCL buildings and a helicopter, destroying vital communications and electricity installations (348, 373).

Two hundred police were drafted in to deal with this new threat (372). The BCL chair, Don Carruthers, meanwhile told the new government of Prime Minister Namaliu that it would "seriously reconsider" its future investment in PNG (including its projects at Mt Kare and Hidden Valley in Morobe province, on which it was planning to spend K200 million in 1989). Prime Minister Namaliu angrily retorted that he recalled "... numerous Australian mining companies having difficulties with local tribal landowners, environmental groups and even state and national government", without wielding a big stick (367).

The following month, production was halted as transmission lines were bombed. CRA and BCL shares fell dramatically (347), and trading closed until further notice (368). PNG Opposition leader Paias Wingti attacked Namaliu for not crushing the landowners. Squeezed on both sides, Namaliu was far from taciturn. Condemning "Rambo-style terrorism", he threatened to send in the troops with a "shoot to kill" policy. Further damage followed, but the troops still remained in base (375).

By the middle of the month, Francis Ona had met with the government, some of the saboteurs handed in their ammunition, and a curfew was declared (376). The mine returned to production and, despite the assaults, BCL recorded fat profits for 1988 (377). However, within a few weeks, as troops were drafted into Bougainville (378), attacks on mineworkers and a police patrol (379) resulted in the first deaths. These were to rise to a dozen by April. By then, the forces were operating a "shoot to kill" policy and 400 troops were based on the island (380). Although the local press speculated that Francis Ona was losing support, Father Robert Wiley, a Catholic priest from Panguna, knew otherwise: "When I talk with the people I don't get that idea at all. When I go the villages, they are praying for him" (381).

By mid-May, Bougainville copper mine was at a standstill, though the company put a brave face on its demise and began to make some repairs (382). The cost in lost production from the mine had now reached around K52 million, with the central government in Port Moresby forfeiting some K1.5 million every day in lost revenues (383).

Australia then supplied four Iroquois helicopters, ostensibly for "non-military" purposes (383). This perfidious action was to cost several innocent lives, as the aircraft – kitted-out with machine guns – strafed villages and terrorised women and children in sorties which were compared to those of US forces in Vietnam a decade earlier (384).

In September, BCL tried to re-open the mine, only to shut it down almost immediately, as a work bus was attacked by members of Ona's Bougainville Revolutionary Army (BRA) (385). At the same time, the Melanesian Solidarity Front for a Nuclear Free and Independent Pacific (MELSOL) urged a cessation of the fighting, withdrawal of central government troops, and a negotiated settlement (386).

By now, there were 1000 central government troops and police on Bougainville island, in hot pursuit of perhaps as few as a hundred Ona-led followers (387). Rumours began circulating that BCL had closed for the last time, although the company claimed that the mine was: "in a state of readiness rather than ... being ... mothballed" (388).

Reports of atrocities on the island had been appearing in many articles around the world by this time. Most (though not all of them) cited arbitrary, brutal behaviour by "Rambo"-style PNG forces, aimed at villagers, who became increasingly sympathetic to the BRA (389).

Bougainville Copper continued to "retrench" 2000 of its workers – a euphemism for sacking them (390). And, as the New Year dawned, the mine finally dropped all pretence at carrying on (391). Ona promptly set up the Republic of Mekanui and declared its independence from Papua New Guinea (392).

The central government ignored the call and petitioned the World Bank for US$130 million, to make up a deficit allegedly caused by events on Bougainville (393). Meanwhile, landowners on the mainland blockaded another Australian managed mine, Ok Tedi, demanding compensation for use of a road and increased royalties (394). The term "to bougainville" was beginning to gain currency in the Pacific region!

In February 1990, CRA evacuated all its Australian personnel (395) as reports circulated of a new "clash" between BRA and PNG forces, in which 20 "rebels" were allegedly killed (396). CRA estimated that renewed start-up of the mine would cost at least K100 million.

The withdrawal of expatriate workers effectively resulted in the abandonment of Bougainville to the BRA. Strikingly – within a fortnight of evacuation – the BRA sat down to talks with the central government (397). Over the next few months all other BCL employees pulled out, leaving the entire mine site to those sympathetic to independence (398). In May 1990, Ona declared formal secession – a move backed by a wide cross-section of Bougainvillians (399), including the provincial Premier. In response, the central government blockaded the island (400). The World Bank then agreed to increase its aid to PNG by US$280 million (401) and – after international and national pressure – Prime Minister Namaliu offered a peace deal to Bougainville which would award around K220 million in development aid to the landowners and other islanders (402).

Peace talks between the islanders and the Port Moresby government began in June (403); CRA acknowledged that the mine would not re-open for some time (404). In August 1990, an Accord was finally signed between the provincial government for the Solomons and the Namaliu government. The blockade was lifted (405). But the burning question of compensation is one that is not likely to go away, however the situation in Bougainville is ultimately resolved. It is complicated by the fact that Melanesians – like Aboriginal Australians and many other indigenous people – can never permit their land to be sold once-and-for-all. Land is a "second skin" (410), and each new generation claims compensation for any plot which was previously leased or "sold". Even if the compensation claims put forward by the older Bougainville Islanders could be settled, younger landowners would be demanding their own settlements.

To date, compensation has been paid into the Panguna landowners trust, but many islanders are dissatisifed with its composition and membership and have been seeking to have it changed (354).

The 1974 Bougainville agreement covered five areas of compensation: occupation fees; payment for restriction of access; payment for cleared land; payment for physically used land; and payments into the Trust Fund. Five year renegotiations were possible (338). Government earnings from the mine since 1972 have come to around K1 billion – including the various kinds of compensation (348, 368). Out of this, only around K1.5 million is paid annually into the landowners' Trust Fund. (The Mining Journal claims that K75 million has been paid since 1972, which averages nearly K5 million a year – a figure that presumably derives from BCL) (380).

As early as 1981, North Solomons' political leaders were expressing considerable dissatisfaction at declining royalty receipts and increasing social commitments, while seeking major changes in the allocation of revenues between central and provincial governments (338). But, it was not until 1988, that the central government began seriously to address this highly contentious issue. In June 1988 it set up a holding company called Mineral Resources Development Co. (MRDC) to purchase state equity in new mining projects (203). Early the following year, it established a local mining task force to advise the Provincial Minister for Mines and the Cabinet on important political, social and economic issues surrounding mineral exploitation in New Ireland: this task force includes the landowners of Lihir Island.

Then, in February 1989, the government announced that it would increase the proportionate share of payments to landowners from 95:5 to 80:20. A special mining lease would be agreed with each mining company, covering currency repatriation, customs duties and similar charges. While the royalty on production would still remain at 1.25%, marginal corporate income tax would rise from 35% to 58%, for project investments with an annual compounded rate of return more than 20%. The provincial governments would be compensated by a 1% charge on total export value of minerals.

However, this arrangement did not satisify the provincial governor of Bougainville (and now head of state) Joseph Kabui, who in April 1989 demanded that BCL should give A$14 million

– or even A$42 million – to the community as a "goodwill" payment (265, 411).

At the time of writing, such proposals seem highly speculative. The savage irony is, of course, that no meaningful compensation will be provided to the Bougainville people unless the mine resumes profitable production. Yet, were CRA/BCL to adequately fund the mammoth task of cleaning-up the Jaba valley and rehabilitating the devastated land, its profits would be set at nought.

Gold on Mount Kare

The Mount Kare gold prospect, discovered by CRA in the Highlands of PNG in 1986, has since become the scene of a "frontier gold rush". Worried by the reaction of local people, CRA initially proposed a small-scale mine – possibly increasing to one million tonnes ore (grading at 3-5gm/tonne gold in three million tonnes of ore) (288, 412). Its Hidden Valley gold prospect has, for the last two years, been "very close to production" (272) with the expectation that it will produce up to 200,000 ounces of gold and two million ounces of silver each year.

By mid-1989, several thousand local people at Mount Kare, joined by indigenous prospectors from outside the area, were panning almost unbelievably rich gold deposits, left by a landslide. Nuggets weighing up to ten ounces (sic) were being recovered on a regular basis, and individual miners were panning around 30/40 ounces of gold a week (413).

The PNG government was placed in a dilemma – supporting the rights of traditional landowners to the pickings, while upholding CRA's exploration rights. Eventually a new JV corporation between CRA and one group of landowners was set up.

Forays abroad: a note on Europe and the USA

In the mid-'eighties, it was becoming clear to CRA, that its reliance on over-supplied commodity products in a saturated market did not augur a stable future. It needed to move downstream, diversify geographically and conclude large, secure, long-term contracts for raw materials, to avoid the swings and roundabouts of spot trading (414).

Its first move along these lines was to prove quite spectacular and, to date, successful: its US$400 million purchase of most of the aluminium interests of the US weapons and aerospace company, Martin Marietta (MMC). Its second move ultimately fell on stony ground – or to be more precise, united opposition – between an alliance of West German Social Democrats, Grünen (Greens) and trade unionists.

CRA had flirted with two of West Germany's major steel producers, Krupp Stahl and Kloeckner Werke, in the early 1980s, when the Australian company offered to invest up to A$200 million in the supply coal for a new coal gasification plant in the Republic (412). The partners re-united three years later in a merger-vehicle called SKK (Stahl-werke Krupp Kloeckner) (there's nothing like switching around syllables to pass off an old car with new paint!). CRA would secure 35% of the new partnership in return for supplying iron ore worth around DM525 million. The German companies would get secure supplies, while CRA would undercut the market in higher-quality Brazilian ore (with its lower alumina content), gain access to direct smelting technology and later could enter directly into the steel business in Australia or Asia (414, 416).

By July 1985, however, the marriage was off. Cries of "scandal" and "irresponsible" greeted news of the deal, when Social Democrats and Greens in the Bundestag realised that the merger would restrict steel-making capacity and rolling, with the loss of around 3000 jobs. Even the government was lukewarm, and the EEC far from happy: Kloeckner having refused to submit its highly modern carbon steel-making and hot strip mill to Brussels' production quota (414). West German trade unions were also vocal in opposition to the plans (417), and it was finally declared dead in the water, when Christian Democrats in Lower Saxony and the workforce at Kloeckner's Georgsmarienhutte works near Osnabruck, dug in their heels (416, 418).

Meanwhile, Comalco had gained more than a toe-hold in the USA. After early negotiations with one of the country's biggest aluminium producers (416), the company (now 67%-owned by CRA) acquired a Kentucky rolling mill, aluminium scrap facilities, a smelter at Goldendale, and Martin Marietta Corporation's terminal at Portland, Oregon (used in 1988 to send nuclear fuel to Britain. It also secured access to MMC's new technology in smelting and aluminium processing (420), and a small, but significant, stake in the Boke bauxite mine in Guinea (421). The deal was signed and sealed in April 1985 (38, 422). In just over a year, however, Commonwealth Aluminium, the Comalco subsidiary which technically took over the Goldendale smelter (423), was locking out its workforce, as it pushed through a new work contract in the face of opposition from the USWA (United Steelworkers of America) (424). (USWA is the union which took on RTZ's Rio Algom subsidiary in its Elliot Lake Mines and won a unique health and safety contract at around this time – see Rio Algom). It was clear that Comalco wanted to shut down a plant it saw as uneconomical: shutdown costs were estimated at US$10 million (425) and, in early 1987, the company put Goldendale on the market for nearly twice that amount (US$18.7 million) (426).

Actions against CRA

From its earliest days, CRA has been the target of vociferous and prolonged opposition by numerous communities: ranging from the resistance of the Mapoon people (and their return to traditional land on the Comalco mining lease), to major manifestations in New Zealand, from 1975 onwards, against the Tiwai Point pricing deal; from the destruction of prospecting equipment (as in Victoria, when the company intended carrying out cyanide solution-mining) to what is probably the best-orches-

trated attempt to close down a mine made in recent times: the revolt by traditional Bougainville land-owners against operations at the Panguna copper mine.

Perhaps the most significant long-term campaigning has been in the hands of Aboriginal Land Council representatives, who have pioneered transnational solidarity for more than a decade. In 1978, three Aboriginal representatives from North Queensland undertook a historic visit to Britain and Europe, where they addressed press conferences, appeared on television and visited the offices of RTZ (427). In 1982, when Aboriginal representatives Shorty O'Neill (National Federation of Land Councils) and Jimmy Biendurri (Kimberley Land Council) rose at an RTZ annual general meeting to condemn the desecration of Lake Arygle sacred sites, the chair, Anthony Tuke, brought the meeting to an abrupt halt. The ensuing chaos – during which police were called to remove shareholding protestors – was reported around the world (428). Six years later, two Aboriginal representatives of the Western Desert Land Council also attended the RTZ AGM to signal the vehement opposition of the Martu to CRA's plans at Rudall River. This visit was also reported widely in the media (429).
Aboriginal people and their supporters have often attended CRA AGM's in Melbourne – starting in 1976 with questions asked by Jesuits (430). In 1987, the Northern Land Council rejected applications by CRA for certain exploration rights, on learning of the company's appalling record elsewhere (80).

With little doubt, CRA is the most feared and least loved of all Australian (and perhaps global) mining companies.

CRA in the Philippines

"The Philippines is far more densely mineralised than Australia, the tonnages are bigger and the terrain is largely unexplored. The place is wide open". That was the opinion of a *Financial Times* correspondent in 1989 (431). It was around the same time that CRA began expressing an interest in the country. True to form, within a year the Australian company had bought into a gold and copper prospect in the Mankayan region of Benguet province, high in the Cordillera. Content with nothing less than a controlling share, CRA soon took over the majority of Galactic Resources equity in the Far Southeast project (432) – giving it 30% of the joint venture and later the whole 40% share. This may still seem a mere mouse's bite of the cheddar but, under Philippines law, a foreign company may now hold more than 40% of any mine.

CRA's 60% domestic partner is Lepanto Consolidated Mining; Lepanto is itself substantially owned by foreign investors (433).

By early 1991, the Far South East project was being advertised as potentially the largest gold mine in the Philippines (434). It will be an underground mine – an unusual development for CRA, but not for its host country. While this will reduce some of the adverse social and environmental impacts of the project, it will not avoid removal of around 150 Kanka-aney people from the hamlet of Tabbak, nor the destruction of their coffee, mango and banana plantations (435) and burial sites.

The families being removed from the project site have been offered some compensation, though it is derisory in light of the fact that they are losing their traditional land. They predict that the mine will cause subsidence of the soil, landslides, and pollution of underground water supplies. Such fears have a solid foundation. Since 1936, CRA's partner, Lepanto, has built up a reputation as one of the most unscrupulous, dangerous and negligent of all Philippines mining companies. As a virtual feudal lord of the Mankayan mining district, Lepanto has ruthlessly suppressed free trade unions (436), and sacked "disloyal" workers at will (437).

Current operations at Lepanto have been heavily criticised – in particular the operations of a copper concentrate dryer which local farmers claim has ruined their crops and affected the health of their livestock and families: allegations

substantiated by a number of recent fact finding missions (438). Lepanto is also a major logging company – and its timber operations have led to the siltation of local rivers, the erosion of roads, the loss of freshwater fish and the destruction of rice paddies (439).

Perhaps worst of all, in the long term, is the legacy left by four tailings dams, two of which are in an appalling state of disrepair and one of which collapsed completely in 1986, leading to a deluge of acidic waste and heavy metals cascading into the Mankayan river. Far South East will be using a "new" tailings dam, but it is one which will probably leach into the same valley, and is being constructed in an area of heavy seismic activity and often torrential rains. It will wipe out a traditional gold-panning site.

Lepanto is almost a byword for mining disaster in the Philippines. It has done virtually nothing to clear up the mess it has caused over more than half a century, nor to adequately compensate for those affected by its noxious fumes, the poisoned waters and the ravaged agriculture. And it is this rogue stallion to which CRA has hitched its cart.

Contact: PARTiZANS, 218 Liverpool Rd, London N1 1LE, UK.

References: (1) *RTZ Annual Report* 1988. (2) Information from Morgan Stanley Capital Int., quoted in *Economist*, London, 23/7/88. (3) *RTZ Uncovered*, PARTiZANS, London, 1985. (4) *Facts about CRA, Investment Review*, McIntosh Hamson Hoare Govett Ltd, London, 3/86. (5) *MIY* 1988. (6) *FT* 11/4/86. (7) *FT* 11/10/88. (8) *MAR* 1988. (9) *MJ* 14/6/85. (10) *E&MJ* 12/87. (11) *MM* 9/87; *MJ* 17/7/87. (12) *Aus* 29/3/89. (13) *E&MJ* 8/87. (14) Rowe and Pitman, *RTZ Analysis*, London, 8/82. (15) *E&MJ* 11/85. (16) *FT* 7/11/86. (17) *Metals Bulletin*, 6/9/85. (18) *MJ* 20/3/87. (19) *FT* 3/5/88. (20) *Financial Australian*, 4/5/88. (21) *AFR* 4/5/88. (22) *FT* 7/9/88; *Independent* (London), 17/9/88. (23) *MJ* 17/3/89. (24) *Aus* 3/5/89. (25) *West Aus* 3/5/89; statement by Alistair Frame, RTZ chair at extraordinary general meeting to confirm takeover of BP

Minerals' assets, London, 1/6/89. (26) *FT* 16/12/88. (27) *FT* 16/12/68. (28) Richard West, *River of Tears*, Earth Island, London, 1972. (29) Robert Carty, *RTZ Comes to Panama: The Rio Tinto-Zinc Corporation and its plans for Developing Panama's Cerro Colorado Copper Mine*, (cyclostyled), CEASPA, Panama City, 1982. (30) Roy Wright on Sir Val Duncan, in *MM* 3/76. (31) *FT* 5/9/86. (32) *FT* 30/7/86. (33) R. Howitt, *The impact of transnational capital on the political economy of North Australia, Project Report No.1: The corporate strategies of CRA Ltd and Amax Inc., with particular reference to their impact on three W.A. Aboriginal communities*, Corporate Study Group, Sydney, 1981. (34) Information from PARTiZANS, London, 5/89. (35) *Financial Weekly*, London, 15/1/87. (36) *MJ* 17/10/86. (37) *FT* 9/6/77. (38) *MJ* 3/5/85. (39) *MJ* 7/11/86. (40) *FT* 19/8/85. (41) *Independence Voice*, 11/77. (42) *MJ* 21/10/77. (43) *AFR* 16/8/77. (44) Diane Hooper, *The Rio Tinto-Zinc Corporation: A case study of a multinational corporation*, PRIO, Oslo, 1977. (45) Information from PARTiZANS, London. (46) Roberts. (47) *West Aus* 7/11/78. (48) *Aus* 30/11/78; *AFR* 1/12/78. (49) *West Aus* 30/11/78. (50) *Age* 11/10/80; *Chain Reaction* No.21, 10-11/80. (51) Daniel Vachon and Phillip Toyne, *"Mining: the challenge of land rights"*, in (103); see also Phillip Toyne and Daniel Vachon, *Growing Up the Country: The Pitjantjatjara Struggle for their Land*, McPhee Gribble/Penguin, Fitzroy, 1984. (52) *Testimony* by Les Russell, *AMIC*, to International Tribunal on RTZ, London, 1981, summarised in *RTZ Benefits the Community*, PARTiZANS, London, 1981. (53) *N.Q. Messagestick*, late 1981; see also *CRA Annual Report*, 1979. (54) *West Aus* 20/1/77. (55) *AFR* 1/6/78. (56) *Kimberley Land Council Newsletter*, Derby, mid-1981. (57) Memorandum to J. Collier, from H.J. Evans, CRAE Pty Ltd, Melbourne, 15/12/76. (58) *Age* 7/11/77. (59) War on Want and Cimra, Press Release, London, 22/5/78; see also Open Letter, War on Want to Sir Mark Turner, chair RTZ, 23/5/78. (60) *Times* 25/5/78; *MJ* 2/6/78; *Australasian Express* 26/5/78; *Gua* 25/5/78. (61) Campaign Against Racial Exploitation (CARE), National Press Release, 25/5/78; *Cairns Post* 27/5/78. (62) Goodman Derrick and Company, London, 14/6/78. (63) *"Diamonds may be forever – but not Aboriginal Land Rights"*, David

Simpson, *Gua* 23/6/81. (64) *Dirty Bizness*, PARTiZANS, London, 1986. (65) Age 5/10/76. (66) Paul Wilson, *Black Death, White Hands,* George Allen and Unwin, Sydney, 1982. (67) *"State of Shock"*, dir. David Bradbury, Ronin Films, Canberra, 1989. (68) *The Filmmaker*, 1/89, reproduced in *Land Rights News*, NLC/CLC, 3/89. (69) J. Roberts, D. Mclean, *The Mapoon Books*, Vol.1, International Development Action (IDA), Collingwood, 1976. (70) Information from Comalco, 1975. (71) Pat Mullins SJ, *The Comalco Issue*, first draft, Parkville, 1976. (72) Letter to Church Mission, from Sir Maurice Mawby, Consolidated Zinc Corp., 21/8/57. (73) see also Comalco's own documents 1976-81. (74) Comalco Ltd, *"Aborigines and Islanders at Weipa: Notes on background and current position"*, Comalco, Melbourne, 10/76. (75) *AFR* 25/3/77. (76) Comalco Arbitration Submission, Federal Government Arbitration Court, 26/6/63. (77) Age 19/10/76. (78) A. Jeffreys' Comments on visit to Weipa, 1981. (79) *Regenerated Mined Land at Weipa*, Comalco, May 1980. (80) Personal communication to author from representative of NLC, 5/89. (81) *The Aboriginal and Islander Community of Weipa South*, Comalco Ltd, 11/80. (82) Joyce Hall testimony, in *RTZ Benefits the Community*, PARTiZANS, London, 1981. (83) Australian Supreme Court record, Comalco v. ABC, 1985, p.2338. (84) J. Roberts, D. Mclean, *The Mapoon Books* Vol.2, IDA, Collingwood, 1976. (85) Roger Moody (ed) *The Indigenous Voice, Visions and Realities*, Vol.2, Zed Books, London, 1988. (86) *Aboriginal Law Bulletin*, Vol.2 No.38, Kensington, 6/89. (87) for information on Aurukun, see (46), (65), (69), (84); *Chain Reaction* Vol.4 No.l, 1978; *Chain Reaction* Vol.4 Nos 2-3/78; *Courier Mail* 15/3/78; *Tribune* 5/4/78; *N.Q. Messagestick* Vol.3 No.2, Cairns, 10/78; *Gua* 2/6/79. (88) *MM* 11/85. (89) *AFR* 4/11/77; *West Aus* 19/12/78; *Age* 19/12/78. (90) *AFR* 19/12/78. (91) *Aus* 3/11/78; *AFR* 17/11/78; see also *Weekend Aus* 11-12/3/78. (92) *AFR* 28/3/78; *Age* 7/9/78; *AFR* 27/9/78; *Sydney Morning Herald* 7/11/78. (93) *FT* 7/9/78; see also *FT* 21/8/78. (94) *FT* 14/7/78. (95) *AFR* 14/9/78. (96) *AFR* 1/6/78. (97) *AFR* 7/9/78. (98) *AFR* 16/10/78; *AFR* 17/10/78; *AFR* 19/10/78; *AFR* 30/10/78; *AFR* 31/10/78; *AFR* 1/11/78; *Age* 1/11/78. (99) *AFR* 11/7/78. (100) *FT* 11/1/79. (101) *FT* 11/4/79. (102) *West Aus* 22-23/10/79; *WA*

6/11/79; *WA* 12/11/79; *WA* 12/12/79; *WA* 10/1/80; *Sydney Morning Herald* 5/11/79; SMH 12/11/79. (103) (ed) Nicholas Peterson and Marcia Langton, *Aborigines, Land and Land Rights,* Australia Institute of Aboriginal Studies, Canberra, 1983. (104) see also *The Bulletin* 15/1/80; *The Miner* 11/78. (105) *Diamond World Review* No.40, USA, 1987. (106) *E&MJ* 7/88. (107) *MJ* 6/12/85. (108) *KLC Newsletter,* Derby W.A. 6/85. (109) *FT* 6/9/86. (110) *FT* 5/5/88. (111) *MJ* 10/1/89. (112) *MJ* 17/3/89; see also *Aus* 4/4/89. (113) *MJ* 9/5/86; *FT* 14/10/87. (114) *Parting Company* No.9, London, 12/81; see also *FT* 20/2/82, *MJ* 26/2/82. (115) Personal communication with author, London, 1981. (116) *CRA Annual Report* 1985. (117) W. Christensen, *Aborigines and the Argyle Diamond Project* – Submission to the Aboriginal Land Inquiry, East Kimberley Working Paper No.3, Canberra, 1983. (118) R.A. Dixon, C. Elderton, J. Irvine, I. Kirkby, *A Preliminary indication of some effects of the Argyle Diamond Mine on Aboriginal Communities in the Region*: A report to the Kimberley Land Council and the National Aboriginal Conference, East Kimberley Working Paper No.8, 7/84. (119) Statement by Sir Roderick Carnegie, at RTZ AGM, London, 1985. (120) Information from Daniel Vachon, anthropologist, to author, 7/89. (121) *Times on Sunday* (AUS) 19/4/87. (122) *West Aus* 4/5/88. (123) Author's observations, Parngurr, W.A., 5/89. (124) *West Aus* 8/11/78. (125) Australian Conservation Foundation news release, Hawthorn, 15/5/88 & 16/5/88; WDLC release 23/4/89; statement by Tom Helm MLC, in *Times on Sunday* 19/4/89; see also WDLC Press Release, 16/7/87, *West Aus* 6/8/87, ACF Press Release 17/9/86 and *Sunday Territorian* 22/3/87. (126) Letter from CRAE to Minister of Mines and Energy, Perth, 11/10/84. (127) Letter from Aboriginal Legal Service of *WA* (Inc.) to Minister for Minerals and Energy, 23/10/84. (128) Statement by Alistair Frame, RTZ AGM, London, 5/89. (129) *Age* 14/8/87. (130) *Daily News* 2/9/86. (131) *West Australian Hansard,* House of Assembly, 20/10/87. (132) *Reg Aus Min* 1985-86. (133) *AFR* 4/12/85. (134) *MJ* 29/1/88. (135) *West Aus.* 21/7/88. (136) *West Aus* 21/4/89. (137) *Aus* 20/2/89. (138) Rick Humphries, *Rudall River Alert,* Perth, 16/9/86. (139) WDLC *Rudall River Fact Sheet,* 16/7/87, Statement by Nyabaru Gibbs,

Punmu, 6/87. (140) *WA* 24/7/87; *WA* 15/8/87. (141) Letter from Ernie Bridge, *WA* Minister for Aboriginal Affairs, to chairman of WDLC, 11/10/87. (142) *Herald* 3/5/88; Canning Resources Pty Ltd and CRA Exploration Pty Ltd, *Kintyre Uranium Project*, booklet produced in 1988 & 1989. (143) E.W. Campbell, *The 60 Families Who Own Australia,* Current Books, Sydney, 1963; P. Cochrane, *Industrialisation and Dependence: Australia's road to economic development*, Uni. of Queensland Press, St Lucia, 1980. (144) *MJ* 16/8/85. (145) *E&MJ* 11/86. (146) *MJ* 17/6/88. (147) *FT* 24/6/89; *FT* 3/9/86. (148) *MJ* 8/2/85. (149) *FT* 17/2/187. (150) *FT* 8/1/87. (151) Information from a Broken Hill informant to PARTiZANS, 7/87; *MJ* 20/2/87. (152) *MJ* 25/10/85. (153) *FT* 9/11/88. (154) *MJ* 3/12/82. (155) *MJ* 8/11/85. (156) *MJ* 2/11/87. (157) *FT* 8/1/87; *FT* 5/11/87. (158) *National Times* 13/6/77. (159) *MJ* 12/8/88. (160) *MJ* 10/3/89. (161) *MJ* 2/12/88; *FT* 21/9/88; see also *MJ* 25/12/87. (162) *FT* 24/6/88. (163) *MJ* 17/6/88. (164) *MJ* 24/6/88. (165) *E&MJ* 8/88. (166) *MJ* 6/11/88. (167) *Weekend Aus* 28-29/4/89. (168) Gua 9/7/83. (169) *E&MJ* 7/84. (170) *Metals Bulletin* 6/9/85. (171) *RTZ: Anti-report*, Counter Information Services; see also *Death at Work*, Work Hazards Group, Workers Educational Assoc. Studies for Trade Unionists Vol.13 No.5, 6/87. (172) *Obs* 30/1/72. (173) Armstrong and Kazantzis, *"Mortality of Cadmium Workers",* The Lancet, pp. 1424-1427, London, 1983. (174) A.E. Ades and G. Kazantzis, *"Lung Cancer in a non-ferrous smelter: the role of cadmium"*, British Journal of Industrial Medicine No.45, pp. 435-442, London, 1988; see also correspondence between the Dean of the London School of Hygiene and Tropical Medicine, C.E. Gordon-Smith and Dr Alan Dalton, Research in Health and Safety, Labour Research Dept. London, dated 17/6/88. (175) *Bristol Evening Post* 7/11/87; *BEP* 9/11/87. (176) *"What on earth is going on?",* Channel Four, screened nationwide 8/11/87. (177) *London Australia Weekly* (LAW) 12/9/88. (178) *MJ* 22/1/88. (179) *FT* 15/12/88. (180) *FT* 12/7/86. (181) *FT* 5/2/81. (182) *FT* 16/4/85; *FT* 8/5/85. (183) *FT* 8/11/82. (184) LAW 4/7/87. (185) CRA Annual Report 1988. (186) *FT* 16/12/68. (187) Statement by Sir A. Tuke, chair RTZ AGM, London, 1982. (188) *FT* 23/11/84. (189) *MJ* 19/5/89. (190) *West Aus*

20/4/89. (191) *FT* 8/5/85; *FT* 17/4/85. (192) *FT* 15/12/86. (193) T.M. Fitzgerald, *The Contribution of the Minerals Industry to Australian Welfare*, AGPS, Canberra, 1974. (194) see also *The Miner* 8/12/80. (195) *New York Times* 29/5/77. (196) *MJ* 15/4/88. (197) *FT* 3/5/88. (198) John Wilson, "*The Pilbara Aboriginal social movement: an outline of its background and significance*" in (ed) R.M. Berndt, *Aborigines in the West*, UWA Press, Perth, 1979. (199) Stockbridge et al, "*Dominance of Giants: A Shire of Roebourne Study*", Dept. of Social Work, Uni. of Western Australia, Perth, 1976. (200) J. Roberts, D. Mclean, *The Mapoon Books* Vol.3, International Development Action (IDA), Collingwood, 1976. (201) *Star* (NZ) 12/6/89. (202) R. Graham, *The Aluminium Industry and the Third World*, Zed Books, London, 1984. (203) *MJ* 1/7/88. (204) *MJ* 3/3/89. (205) *AFR* 13/7/77. (206) Greg Crough, *Transfer pricing, the High Court and the Aluminium Industry*, draft paper, Australia, 1980. (207) *E&MJ* 2/85; see also International Monetary Fund, Social and Economics Research Dept., *Subsidiaries of Multinational Corporations in the Metal Industry of South East Asia*, (date unknown). (208) *Christian Science Monitor* 23/6/70. (209) Richard West op cit (28), reproducing article by Dr A. Hall in *AFR*, "*Comalco's Capital Coup*". (210) *Telegraph* (AUS) 10/6/70. (211) Quoted in *Under Investigation*, Brisbane, 1977. (212) *AFR* 25/3/77. (213) *FT* 15/10/82. (214) *MJ* 24/12/82. (215) *MJ* 4/3/88. (216) *Minamata: words and photographs*, by W. Eugene Smith and Aileen M. Smith, Alskog-Sensorium, New York, 1975. (217) *Parting Company*, PARTiZANS, London, No.13, 5/82. (218) *Comalco Annual Report* 1985. (219) *FT* 1/3/86. (220) *FT* 20/2/86. (221) *MJ* 24/4/87. (222) *MJ* 3/3/84; *FT* 26/1/84. (223) *E&MJ* 12/86. (224) *MJ* 28/8/87. (225) *FT* 27/8/88. (226) *MJ* 3/3/89. (227) *FT* 28/2/89. (228) *MJ* 30/6/89. (229) *MJ* 7/7/89. (230) *Aus* 18/5/89. (231) *The Press*, Wellington NZ, 12/4/88. (232) *FT* 27/10/88. (233) Letter from Stan Davey, Age 20/5/69. (234) *AMIC* information, Healesville, 1983. (235) *West Australian Museums Report*, cited by *AMIC*, Healesville, 1982. (236) *The Dominion* 19/4/89. (237) *The Amazing Adventures of N.Z's No.1 Power Junky: The true story of Comalco in N.Z.*, CAFCA, Christchurch, 1977. (238) Note from D.R. Bunney, Comalco, New Zealand, 7/11/77.

(239) Comalco documents, in CAFCA Watchdog, CAFCA, No.27, 1980; see also *Christchurch Star* 29/10/77, *Press* 1/11/77, *Press* 8/11/77. (240) *AFR* 3/8/79. (241) *Press* (NZ) 24/11/84. (242) *Star*, Christchurch, 12/5/84. (243) *Star* 16/11/84. (244) *Star* 21/11/84. (245) Letter to PARTiZANS, London, from Murray Horton, Secretary CAFCA, 6/3/87. (246) CAFCA *Watchdog* No.54, 11/86. (247) CAFCA *Watchdog* No.56, 6/87. (248) *Star* 11/3/87. (249) *Press* 4/8/87. (250) *Press* 9/4/88. (251) *Evening Post* (NZ) 15/5/89. (252) *Press* 17/3/88. (253) *The Dominion* (NZ) 19/4/89. (254) Murray Horton, Secretary CAFCA, letter to *The Listener*, Wellington, 9/4/89. (255) *Press* 4/5/88; see also *Press* 29/4/88. (256) *Star* 10/6/89. (257) *Star* 12/6/89. (258) *Christchurch Star* 12/6/89. (259) *Queensland Hansard* 1957, Debate on the Commonwealth Corp. Pty Ltd Agreement Bill. (260) *MM* 3/81. (261) Comalco report, quoted in (260). (262) *MIY* 1987; see also *MJ* 2/9/88 and (226). (263) *Press* 12/3/88. (264) *AFR* 8/12/88. (265) *West Aus* 29/4/89. (266) *Press* 21/4/88. (267) *Westport News* 21/5/84; Press 21/4/84. (268) Betty Whaitiri Williams, "*How New Zealand Maoris organised against CRA and beat back the bludgers*" in (64). (269) CAFCA *Watchdog* No.40, 3/83. (270) *FT* 3/3/84. (271) *E&MJ* 5/88. (272) *E&MJ* 1/89. (273) *FT* 22/4/88. (274) *Chain Reaction*, FoE, No.21, Victoria, 10-11/80. (275) *Age* 6/5/81. (276) *E&MJ* 10/88. (277) *National Park to Mine to Pulp Mill: The Mt Lesueur story unfolds*, leaflet by Friends of Lesueur, Western Australia, 1989. (278) *MJ* 17/3/89. (279) personal communication, ACF, Perth, 4/89. (280) Aus 10/4/89. (281) *E&MJ* 2/89. (282) *Christchurch Star* 4/4/75. (283) *AFR* 9/12/88. (284) Gua 16/6/89. (285) *FT* 22/6/89. (286) *MJ* 5/5/89. (287) *MM* 2/89. (288) *E&MJ* 3/89. (289) *FT* 9/12/88. (290) *MJ* 6/11/87. (291) *FT* 9/12/88. (292) Issues of *Tapol*, Norwood, from 1984-89; see also *Down to Earth*, Survival International, London, No.3, 6/89. (293) Information from Socioeconomic and agricultural study of Sengatta area, carried out by Indonesian study group for the Indonesian government, material in hands of author, exact references not provided here to protect identity of informants, 1989. (294) "*Amazonian Battle Front Reaches Indonesia*", Roger Moody, Gemini News Service, London, 5/89. (295) *Asian Wall Street Journal* 28/2/77. (296) *Far*

Eastern Economic Review 11/5/89. (297) personal communication with author, 5/89. (298) *E&MJ* 4/89. (299) see also *MJ* 18/12/87 and *MJ* 15/1/88. (300) *MJ* 6/11/77.

(301) *MJ* 19/7/85. (302) *Kompas*, Jakarta, 12/7/89. (303) *Down to Earth*, Survival International, London, No.2, 4/89; see also *Newsweek* 15/8/88. (304) *E&MJ* 12/88. (305) *Ground for Concern, Australia's Uranium and Human Survival*, ed. Mary Elliott, FoE Australia, Ringwood, 1977. (306) Douglas R. Barrie, *The Heart of Rum Jungle*, publisher S.D. Barrie, Batchelor, NT, 1982. (307) *Northern Territory News* 27/6/81. (308) *Report of the Senate Select Committee on Water Pollution*, AGPS, 6/70. (309) Submission by R.F. Feldgenner, First Assistant Secretary (Economic Affairs NT) to Minister for Interior, Canberra, 11/6/71. (310) AAEC data, quoted in (305). (311) *Ranger Uranium Environmental Inquiry, Second Report*, AGPS, Canberra, 1977, p.75. (312) *SCRAM Journal*, Edinburgh, No.52, 2-3/86. (313) *Chain Reaction* No.39, FoE, Victoria, 10-11/84. (314) *FT* 13/1/88. (315) *E&MJ* 3/88; *MJ* 15/1/88. (316) *MM* 3/88. (317) *MJ* 14/3/86. (318) *Age* 16/5/83; *MJ* 20/8/82; *MJ* 3/8/82. (319) *Parting Company* No.2, PARTiZANS, London, 1983. (320) *AMIC* quoted in (319). (321) *FoE-Australia Bulletin* 8-9/83. (322) *Farrago*, Vol.61, 21/10/83. (323) Alistair Frame, in answer to question by Peter Lennard, shareholder, RTZ AGM, London, 1/6/89. (324) *Chain Reaction* Vol.4 No.l, FoE, Carlton, 1978. (325) Testimony by Les Russell (Boolidt Boolidtha), *AMIC*, at International Tribunal on RTZ, London, summarised in *RTZ Benefits the Community*, PARTiZANS, 1981. (326) Transcript of RTZ AGM, 21/5/81, PARTiZANS, London, 1981. (327) R. Jackson, *Ok Tedi: Pot of Gold*, Uni. of Papua New Guinea, 1982. (328) Rob Pardy, Mike Parsons, Don Siemon, Ann Wigglesworth, *Purari, Overpowering PNG?*, International Development Action, for Purari Action Group, Fitzroy, 1978. (329) *FT* 16/12/68. (330) Press Release from RTZ Corp. and NBHH Ltd, London, 31/7/72. (331) *New York Times* 23/7/69. (332) *Wall Street Journal* 19/11/69. (333) see also W. Salomons and V. Forster (eds), *Environmental Management of Solid Waste, Part II*, Springer GmbH & Co., Berlin, 1988. (334) Don Woolford, *Papua New Guinea, Initiation and Independence*, Uni. of Queensland Press, 1976. (335)

ARC Bulletin, Cambridge, USA, 3/79. (336) *MJ* 13/5/88. (337) *Raw Materials Report*, Vol.23 Stockholm, 1983. (338) William S. Pintz, *Ok Tedi: Evolution of a Third World Mining Project*, Mining Journal Books, London, 1984. (339) *Economist* 12/10/74; see also *FT* 24/8/74. (340) *E&MJ* 10/83. (341) *MJ* 19/12/88; see also *MJ* 19/9/86. (342) *MJ* 14/4/89. (343) *BCL Annual Report* 1979. (344) *MJ* 13/5/88. (345) *MM* 3/89. (346) *FT* 1/9/88. (347) M.R. Chambers, *"Bougainville Report"*, quoted in *Parting Company*, PARTiZANS, London, 6/87. (348) *Pacific Islands Monthly*, 1/89. (349) *FT* 17/8/88. (350) A.F. Mamak and R.D. Bedford, *Bougainville Nationalism*, Bougainville Special Publications, 1974. (351) F.S., personal communication to author, 1976. (352) *Age* 27/4/89. (353) Telephone interview with FoE PNG, quoted in *Parting Company*, PARTiZANS, London, 6/87. (354) *AFR* 6/12/88. (355) Age 9/12/88. (356) *Pacific News Bulletin*, Sydney, 5/89. (357) *Post-Courier* (PNG) 25/3/88. (358) *BCL Annual Report* 1986; *E&MJ* 11/87. (359) M.J.F. Brown, *"A Development Consequence – Disposal of Mining Wastes on Bougainville, Papua New Guinea"*, *Geoforum* 4 No.18. 1974. (360) B.R. Hewitt, Aleni Flores, Yann Pokei, *"Ore Mining? Or What?"* in *The Siren*, No.29, 1985. (361) J.H. Powell, *"Mining impacts on the aquatic environment with special reference to the Bougainville Copper Project"* in *Report of the Mine Rehabilitation Workshop, Bougainville*, BCL, North Solomons, 10-13/5/82. (362) H.R. Meier, *"Natural leaching of mining waste and effect on the aqueous environment"*, in *Report of the Mine Rehabilitation Workshop, Bougainville*, BCL, North Solomons, 10-13/5/82. (363) Pintz; Chambers; Powell; Jackson; Brown; et al, for full references see above. (364) *E&MJ* 8/87. (365) Alistair Frame, in response to questioner at RTZ annual general meeting, London, June 1989. (366) *The Times* (PNG), 1/3/89. (367) *Post Courier* (PNG) 30/11/88. (368) *AFR* 6/12/88. (369) *Report of UN Visiting Mission to PNG*, New York, 1971. (370) see also *SunT*, Aus 10/8/69. (371) *Aus* 7/8/69. (372) *Post Courier* (PNG) 28/11/89. (373) *FT* 29/11/89. (374) Age 3/12/88. (375) Age 7/12/88; Age 8/12/88; *FT* 8/12/88. (376) *MJ* 17/3/89. (377) *MJ* 24/2/89; *MJ* 3/3/84. (378) *FT* 22/3/89. (379) *MJ* 17/3/89. (380) *MJ* 14/4/89. (381) *Age* 27/4/89. (382) *MJ* 18/8/89. (383) *Pacific Islands Monthly*.

(384) *"the Battle for Bougainville"* by David Robie in *The Press*, Wellington 10/10/89. (385) *FT* 6/9/89, *FT* 7/7/89, *MJ* 8/9/89. (386) *Pacific News Bulletin*, Sydney 9/89. (387) *Independent* 28/10/89. (388) *FT* 1/12/89. (389) *The Press*, Wellington op cit, see also *Times* of PNG, Port Moresby 8/12/89 and *Times*, London 27/1/90. (390) *Australian* 29/12/89, *Post Courier*, Port Moresby 29/12/89. (391) *FT* 3/1/90, *MJ* 5/1/90. (392) *Times* of PNG 4/1/90. (393) *FT* 16/1/90. (394) *FT* 19/1/90. (395) *FT* 8/2/90. (396) *FT* 10/2/90. (397) *FT* 23/2/90. (398) *E&MJ* 3/90, *MJ* 13/4/90. (399) *Aus* 18/5/90. (400) *Daily Telegraph* 10/5/90.
(401) *MJ* 1/6/90. (402) *Pacific News Bulletin* 6/90. (403) Gua 28/6/90. (404) *Metal Bulletin* 23/7/90. (405) *FT* 7/8/90. (406) *Press* 11/3/87. (407) Statement by Anthony Tuke, Chair, RTZ AGM, London 1985. (408) *Newcastle Morning Herald* 21/10/77. (409) *FT* 12/9/89. (410) *Island Business*, PNG 3/4/89, see also Jean-Marie Tjibaou, *"I am always dual"*, in Roger Moody (ed.) *The Indigenous Voice, Visions and Realities*, Vol.1, Zed Books, London, 1988. (411) *MJ* 24/2/89. (412) *FT* 18/12/88. (413) *E&MJ* 6/89. (414) *FT* 26/10/84. (415) Age 4/12/81. (416) *FT* 10/7/85. (417) *FT* 18/4/85; *MJ* 3/5/85. (418) *MJ* 19/7/85. (419) *MJ* 19/10/84; *FT* 17/10/84. (420) *E&MJ* 12/84. (421) *E&MJ* 2/85. (422) *MJ* 11/1/85. (423) *MJ* 19/12/86. (424) *MJ* 8/8/86. (425) *MJ* 19/12/86. (426) *MJ* 16/1/87; *MJ* 20/2/87. (427) CIMRA Press release, London, 11/78; Press release 20/11/78; statement to RTZ from the Aborigines of North-East Australia, London, 20/11/78; Gua 3/11/78; *Australasian Express*, London, 2/11/78; *New Statesman*, London, 16/9/78. (428) *Parting Company*, No. 14, London, 1982; see also Gua 4/6/82, *FT* 4/6/82, *Times* 4/6/82. (429) *Sydney Morning Herald* 30/5/88; *West Aus* 4/6/88; Melbourne Herald 3/6/88; *New Aus Express* 8/6/88; *City Limits*, London, 5-16/6/88. (430) Age 6/5/86; *Advocate* 22/4/76. (431) *FT* 2/5/89. (432) *Metal Bulletin* 19/7/90. (433) *E&MJ* 3/90. (434) *Baguio Midland Courier* 3/2/81. (435) Information from the residents of Tabak given to the author 11/90. (436) *Birds of Prey: Copper Mining in the Philippines, a political economy of the large scale mining industry*, produced by CLEAR and RDC, Kaduamo, Baguio City, 1988. (437) *Northern Dispatch*, Baguio 20/4/90. (438) Department of Agriculture, Cordillera Administrative Region, *Report*, by Anthony Bantog et al, Baguio City, 11/89, Cordillera Environmental Concerns Committee, *Report of Fact-Finding Mission*, Baguio 1988, and similar, Baguio 1990. (439) Behis, Baltazar and Sison, *Investigation Report on the Petition-Appeal-Resolution of the residents of Barangay, Cabiten, Mankayan, Benguet against the logging operations of Lepanto Consolidated Mining Co, 6/89*.

179 CSR

While many mining corporations have diversified into other raw materials, CSR is an example of a mining company which has grown the other way round: in this case from sugar. It provides nearly all Australia's refined sugar (1). It also provides building and construction materials, chemicals – and it mines, especially coal (it holds huge steam coal deposits and other deposits in Queensland suited to synfuel conversion).

Through its own exploration group and its subsidiary AAR, CSR has searched for and exploited a wide variety of minerals, including uranium.

According to the *Mining Journal*, CSR "has long had a policy of, wherever possible, taking overall control of mining projects in which it is involved" (2).

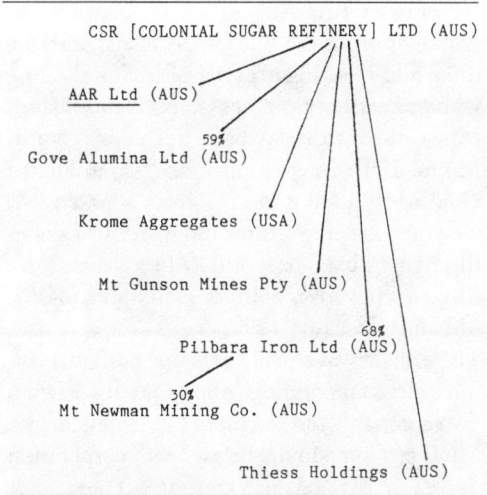

CSR [COLONIAL SUGAR REFINERY] LTD (AUS)

AAR Ltd (AUS)

Gove Alumina Ltd (AUS) 59%

Krome Aggregates (USA)

Mt Gunson Mines Pty (AUS)

Pilbara Iron Ltd (AUS) 68%

Mt Newman Mining Co. (AUS) 30%

Thiess Holdings (AUS)

Thus CSR moved into coal by acquiring Thiess Holdings in 1979; it got into copper when it acquired Mount Gunson Mines; and it got into bauxite when it took the 51% controlling interest in Gove Alumina, which in turn has a 30% right to alumina production at Gove in Australia's Northern Territory (NT) (Alusuisse has the remaining 70%) and, more important, is the sole exporter of Gove bauxite (3).

CSR benefits from the theft and depredation of Aboriginal land both at Gove, NT, and in Pilbara, Western Australia.

At Gove, Nabalco (in which CSR holds 30% of Gove Alumina), the partner with Alusuisse, was granted rights over 20,000ha of the Aboriginal Arnhemland reserve in 1969. The Aboriginal occupants weren't even consulted about the take-over. A nationally publicised court case followed in which – inevitably – Aboriginal land rights, the rights of first occupancy, lost out to commercial interests, the "rights" of first possessiveness.

CSR's Gove Alumina subsidiary was also interested in expanding the Tiwai Point aluminium smelter in New Zealand, though it faced stiff competition from a consortium headed by Comalco (in which CRA holds 67%). The existing Tiwai Point smelter has been the subject of international condemnation for the manner in which its operators have gained cheap electricity at the expense of the New Zealand consumer (see CRA) (4).

Exploitation of the Pilbara has made Australia the world's leading exporter of iron ore. Along with Hamersley Holdings (an RTZ subsidiary), Mt Goldsworthy (which includes Cyprus Mines, Utah International, and Consolidated Goldfields), CSR is in the Pilbara as part of Mt Newman Mining Company, with 30% shares through Pilbara Iron Ltd (Amax have 25%, Broken Hill 30%, Seltrust Holdings Ltd 5%, Mitsui and C. Itoh 10%) (5).

The Pilbara was a major area of Aboriginal culture, including mining (Aborigines set up their own company, the Northern Development and Mining Company), before the corporation moved in. Mt Newman consortium now holds

777 square miles belonging to Aboriginal people at Jigalong which, according to Aboriginal elders met by authoress Jan Roberts, was turned from a hill 270m high into a pit 300m deep; it was "a very sacred site" (5).

There is also evidence that, after an initial policy of employing Aboriginal workers, the Mt Newman Mining Company was "by 1970 ... making quiet efforts to get rid of both Aboriginal employees and the small group of local Aborigines who were living on the edge of the town. Local police were telling Aborigines that they were not welcome in Newman"(6).

CSR holds an interest in the Honeymoon uranium project which, though originally intended to be South Australia's first operating uranium mine, is indefinitely on hold as a result of South Australian (Labor) and Federal government policy. However, in early 1982 the Campaign Against Nuclear Energy (CANE) staged a sit-in at the mine-site which received national coverage. Aborigines also participated in that demonstration since Honeymoon lies on Aboriginal land. Two years earlier Aborigines protested vigorously against the activities of CSR as the drilling company responsible for stepping in at Noonkanbah after Amax had quit (9).

In 1981 CSR acquired virtually all the shares of Delhi International Oil (7), another company exploring for uranium, this time in Western Australia.

It was also – with Broken Hill, Peko-Wallsend, and Western Mining – part of UEGA (Uranium Enrichment Group of Australia) which was set up in January 1980 to study and, "if feasible", develop an enrichment industry in Australia (8).

The Westmoreland uranium prospect in which CSR holds 12.75% (other partners are Queensland Mines, IOL and Urangesellschaft) looks like a bum steer: only 11,400 tons of uranium have been identified and the companies say 15,000 would be required to make the project economically viable (see Queensland Mines) (10).

In early 1983 Whim Creek Consolidated, a 30% subsidiary of Northgate, bought out

CSR's minority interest in three gold JVs in Western Australia (11). By the middle of the same year, CSR's chief financial officer Gener Herbert stated in London that the company was "well placed" to make an acquisition of new interests in "industries we already know" (12). At the same time the company was producing gold from its Paringa mine treatment plant near Kalgoorlie on the "golden mile" some two months ahead of schedule (13).

By the end of the year, the company was planning to start gold and silver exploration in the Mangani region of western Sumatra, along with the Indonesian company Pagadis; a year later, however, no firm decision had been taken on opening up the mine (14).

CSR's acquisition of Delhi supposedly entitled it to "be counted among the world's multinationals" (15).

CSR's chequered past includes a period when it ruthlessly exploited slave labour, from Fiji and Polynesia, to build up the sugar industry of Fiji. The aim was to control what – by the 1870s – had become a major threat to CSR's own Australian sugar production.

The colonial government of the day prohibited the importation of "kanakas" and the use of indigenous workers – and soon CSR was misusing (East) Indians: up to 63,000 of them, only a third of whom eventually returned to India.

From the '60s onwards, CSR controlled the whole of Fiji's sugar industry by owning all its sugar mills and threatening to withdraw from the country when the workers – under threat of starvation – went on strike.

Lord Denning was commissioned to prepare a report on the industry in 1969, which finally resulted in CSR leaving Fiji – selling up to the government, at a profit to itself, and still controlling the servicing and export side of Fiji's sugar crop (16).

CSR sold Delhi to Esso Australia in 1987 (17), and most of its coal operations to Royal Dutch/Shell and Allied Industries (18), in a major disposal of its mining operations; at the same time it acquired several manufacturing companies in Australia, the US and Europe (20), including Brickworks (19). Placer Pacific also acquired several Australian properties, while CSR's Indonesian tin mining interests were hived off to Renison Gold Fields Consolidated (49% controlled by CGF) (21). However, the company announced that it still intended to hold on to its bauxite and alumina operations on Aboriginal land at Gove, in the Northern Territory, as well as its aluminium smelter at Tomago, in New South Wales (21). Billiton also (see Royal Dutch/Shell) took over CSR's gold mining and exploration interests in Indonesia.

During the 1980s CSR was pursued by victims of asbestosis caused by the operation of its Midalco subsidiary in Western Australia. By 1988, 258 damage-related suits had been taken against CSR, though only a handful of cases had been heard in the courts. In May that year a Victorian court made an award of A$680,000 against Midalco to a former worker, after he contracted the fatal lung cancer, mesothelioma; and in August, lesser amounts were awarded to other workers, against CSR itself (22). Finally in 1989 and the following year, something approaching a fair settlement was made.

Meanwhile CSR and the State Government Insurance Commission of Western Australia were engaged in a dispute as to which of them was responsible for payment of the damages to the tortured workforce and their dependents, with CSR arguing that it was not responsible for liabilities sustained by its Midalco subsidiary. The two bodies finally reached agreement in early 1989 to share the costs of compensation (23). $A15 million would be paid out by each of them to cover damages for the seven-year period of operation of the Wittenoom mine, with an additional total of $A20 million payable to claimants who worked there before 1959 (23). By the end of that year, researchers at the Queen Elizabeth II Medical centre, and the Sir Charles Gairdner Hospital, in Perth had estimated that a further 692 workers would fall victim to mesothelioma with another 183 cases of lung cancer to be expected. They predicted another 432 successful claims would be made, in addition to the 356 already accepted (24).

At the beginning of 1990, 322 workers had

been compensated (25). A few months later CSR made a partial re-entry into mining, when it bought up 48 quarries from the US ARC subsidiary of the British Hanson corporation. It also pulled out of a plasterboard JV with Redland Plasterboard (established in 1987) while retaining its Australian and New Zealand interests, through Monier PGH – a company with around half of the Australian roofing tile market (26).

Contact: CAFCA (Campaign against Foreign Control in Aotearoa), PO Box 2258, Christchurch, New Zealand,
Kimberley Land Council, PO Box 332, Derby, WA 6728, Australia.

References: (1) *MJ* 6/7/79. (2) *MJ* 27/11/81. (3) *MIY* 1981. (4) *The Comalco Documents,* CAFCINZ, 1979. (5) Roberts. (6) R. Tonkinson, *The Jigalong Mob: Aboriginal Victors of the Desert Crusade,* Cummings, California, 1974. (7) *MJ* 11/12/81. (8) Federal government statement, Canberra, 2/1/80. (9) *FT* 23/9/80. (10) *MJ* 17/12/82. (11) *FT* 14/1/83. (12) *FT* 11/5/83. (13) *FT* 6/5/83. (14) *FT* 6/4/84. (15) *WDMNE.* (16) John Wollin & John Roberts: "Fiji's sugar industry", in: *Fiji,* Melbourne, 1973 (IDA). (17) *FT* 1/4/87. (18) *FT* 29/7/88. (19) *FT* 9/12/87. (20) *FT* 9/6/88. (21) *FT* 23/5/88. (22) *FT* 9/8/88. (23) *MJ* 21/4/89 (24) *The Australian* 11/12/89. (25) *E&MJ* 1/90. (26) *FT* 1/8/90.

180 Cultus Pacific NL

This is an oil and petrol exploration company which, in 1978 – along with E&B, and Uranex – was exploring for uranium in the Billiluna-Tanami area of Western Australia (WA) near Halls Creek, E&B acting on behalf of Geomex (1): Halls Creek is a centre of Aboriginal activity and community in the Kimberley region. The project was apparently relinquished within two years (2).
The company also held four uranium prospects on the Yilgarn Shield:

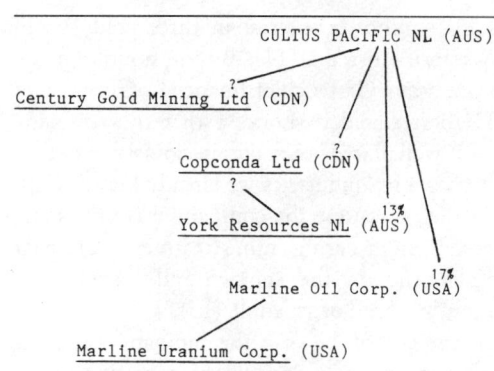

- Lake Maitland where an estimated 500 tonnes of U_3O_8 in average grading have been located;
- Lake Mason deposit, 245km north-west of Leonora, Western Australia, and less than 100km from Yeelirrie;
- Lake Raeside, 50km west of Leonora, where 1700 tonnes of uranium oxide have been discovered;
- and finally Thatcher Soak, 160km north-east of Laverton, Western Australia, where about 6000 tonnes reserves of uranium ore have been located (3).

It also held the majority interest in Robinson Range, 130km north-north-west of Meekatharra, WA, which was administered by Southern Ventures.

All five projects were participated in 70% by Cultus and 30% by Southern Ventures (4).

In Queensland, Southern Ventures and Cultus had a 50/50 uranium JV at the Lyndhurst prospect, in which a major (un-named) European utility was involved, providing finance for the 1983 exploration season. The same utility put up money for exploration of the Tanami Desert prospect, located on the Billaloona Station, in which Cultus has a 25% interest with Southern Ventures and Otter Exploration taking up 50%.

Energy Resources Group of Canada was said to be involved in this particular prospect (4).

In 1985 Cultus Pacific changed its name to Cultus Resources NL. Two years later it acquired a 69.5% interest in Marlin Oil NL and changed its name to Cultus Gold NL after selling off Marlin's petroleum interests (5).

Cultus has extensive mineral interests in Western Australia at Yarn and Edjudina, but its alluvial tin operation in north Queensland, through wholly-owned Stannum Pty Ltd, was put on hold after World Heritage listing had been made of the area (5).

Cultus Gold NL also has 20 exploration projects in the Camarines Norte area of Bicol Province, and on the Visayan Islands, both in the Philippines (5).

References: (1) *West Australian* 11/2/78. (2) *Reg. Aus. Min.* 1981. (3) *AMIC* 1982. (4) *Reg. Aus. Min.* 1983/84. (5) *MIY* 1990.

181 Cygnus Corp (CDN)

Amalgamated now with Consumers Gas, Gooderham and Worts, Hiram Walker, and Home Oil (in Hiram Walker – Consumers Home), this is the 5th largest Canadian company based on net income.

182 Cyprus Mines Corp (USA)

In 1985, Cyprus, then merged with Amoco (1), was spun off to become independent: in May 1988 it purchased the copper resources and smelting capacity of Inspiration Copper (see Anglo American) thus making it second only to Phelps Dodge as a US copper producer (2). While part of Amoco, it was involved in uranium exploration in Colorado, and with Westinghouse at the Hansen project (also in Colorado) during the 1970s. (For further details, see Amoco.)

From 1985 onwards, Cyprus moved with a speed and aggressiveness that must have shaken both its erstwhile owners and its competitors. It bought into several major prospects in the south-west USA (3) including Noranda's Lakeshore copper facilities in southern Arizona (4). It bought out Foote Mineral company from Newmont (renamed Cyprus Foote Mineral Co) which is now the western world's largest producer of lithium (5), with mines and resources in Pennsylvania, Tennessee and Nevada, and in Chile at Salar de Atacama (6). Through Windsor Minerals Inc it operates high quality talc mines in Vermont and California, and through Distribuidora Malaguena de Talcos SA (DIMTASA) it mines talc near Malaga, Spain, and operates a mill in Ghent, Belgium (7).

But it is copper, gold and molybdenum which continue to be the mainstay of its operations. Cyprus remains the USA's second largest producer of copper (4), with properties at Bagdad, Casa Grande, Miami and Kingman and Green Valley in Arizona, the Silver City mine in New Mexico, the Copperstone mine at Quartzsite, Arizona, and the 50%-held Selwyn mine in Queensland, Australia (8).

Cyprus produced more than 12 million tons of coal in 1989 from its mines in Pennsylvania, West Virginia, Kentucky, Colorado, Utah, and Wyoming (8). That year it also acquired the mineral operations of the defunct Reserve Mining Co in Minnesota, which it renamed Cyprus Northshore Mining Corp. Reserve's assets were an iron ore mine near Babbitt and a pelletising plant. However, its drawbacks include the sprawling Milepost 7 tailings basin which is one of the most pollutive features of the state: Cyprus is expected to spend at least US$30 million refurbishing the mining complex and pellet plant, while Milepost 7 has been leased to the company by the bankruptcy trustee for a minimum of 5 years (8, 9, 10).

In 1990 Cyprus also announced a JV with Mitsui Mining and Smelting Co and Mitsui and Co, whereby the Japanese partners would invest 50% in a new company to explore and develop copper, lead and zinc deposits within Cyprus' current concessions in Arizona, Colorado and New Mexico – the first such partnership between major US and Japanese mining companies (11).

The company's gold exploits have brought it into deep conflict with environmentalists, especially in New Zealand where it holds 80% (Todd Corp 20%) in the Golden Cross prospect near Waihi in the North Island. In summer 1989, Cyprus was required to post a NZ$10 million bond as part of an agreement with the local catchment board for obtaining water rights for the project. The mine is in steep hill country and is planned to have a catchment dam to trap the waste water: Cyprus will be responsible for coping with any "accidents" that occur after the 15 years proposed lifetime of the project (12). Then, a few months later, thanks to the activities of the Peninsula Watchdog group and Maori organisations, the project was halted by the High Court: the Maoris maintained that the mining tribunal which gave the go-ahead to the Golden Cross project did not take into consideration the Waitangi Treaty and its land rights provisions (13).

In June 1990, the company was given the go-ahead to open Golden Cross (9, 14) but the project had already been delayed by six months (8).

Contact: Coromandel Peninsula Watchdog, Box 51, Coromandel, New Zealand (Aotearoa).

References: (1) *FT* 26/5/88. (2) *FT* 18/7/88. (3) *E&MJ* 1/90. (4) *E&MJ* 9/87. (5) *MJ* 25/12/87. (6) Industrial Minerals 5/90. (7) *E&MJ* 12/88. (8) *MAR* 1990. (9) *MJ* 3/8/90. (10) *E&MJ* 3/90. (11) *MJ* 20/4/90. (12) *E&MJ* 7/89; see also *MJ* 28/4/89. (13) *MJ* 20/10/89. (14) *MJ* 15/6/90.

183 Dae Woo International Corp

This is an aggressive, expanding industrial corporation which, in the 1984 words of its chair Kim Woo-chong, was diversifying so as to change "from biggest to best" (1).

In the process, it has also been found guilty of criminal fraud and conspiracy in the USA for concealing the true price of various steel imports between 1980 and 1982. The Korean parent was fined US$100,000 (the maximum per-

```
DAE WOO INTERNATIONAL CORP. (ROK)
                                  /
Daewoo Industrial Co. Ltd (USA)
```

missible) and the US offshoot the same amount with an additional US$5000 sting for the US company president (2). This doesn't seem to have had much effect on US-Daewoo trade, however, for the same year the corporation entered a JV with General Motors to produce components for Opel cars in South Korea (3). And, around the same time, it landed a "stunning order" (4) with the US company Sonat to provide six drilling rigs, worth US$425M.

In 1983, Daewoo moved into uranium by acquiring a small (un-named) interest in the UG Baker Lake uranium prospect, on Inuit land in Canada's Northwest Territories (5). Mineralisation at the Lone Gull prospect is estimated to contain 14,300 tonnes uranium (6).

In the late 1980s Daewoo continued its expansionary path by forming a JV with the French state-owned Thomson CSF, the large defence and electronics group (7). It was one of the few companies to consider developing a copper mine in Myanmar (Burma) – at Monywa, some 100km west of Mandalay (8) – despite the opposition of the elected democratic parties to investment under the aegis of the vicious military dictatorship.

Contact: Burma Action Group UK, 1a, Bonny St, London NW1, England.

References: (1) *FT* 31/10/84. (2) *FT* 4/1/84. (3) *FT* 30/10/84. (4) *FT* 5/12/84. (5) *FT* 18/5/83. (6) *E&MJ* ?/84. (7) *FT* 2/8/88. (8) *MAR* 1990.

184 Davy McKee Stockton Ltd

"One of the world's largest design, engineering and construction companies serving the nonferrous, coal and ferrous mining, minerals processing and metals extraction industries, providing a total project capability. The company also has major R and D facilities where fundamental research, laboratory and pilot plant development contribute to improving the profitability of projects" (1).

Apart from refining gold in Dominica, coal cleaning in Britain, recovering precious metals

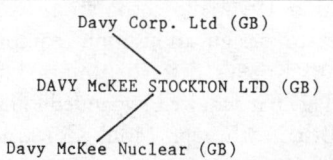

Davy Corp. Ltd (GB)

DAVY McKEE STOCKTON LTD (GB)

Davy McKee Nuclear (GB)

in India and Britain, setting up steel and iron works in New Zealand and Brazil, Davy McKee has established operations in Australia, Argentina, Brazil, Canada, India, Mexico, South Africa and the USA.

It is involved with Saarberg Interplan and the Nigerian and Moroccan governments in uranium exploration in Guinea (2).

It has also built a uranium recovery plant (uranium from gold slimes) for Johannesburg Consolidated Investment (JCI or "Johnnies") in South Africa, providing what the company claims is a "... new technique for in-solvent extraction resulting in lower capital cost and easier operation" (3).

In 1982 the company was carrying out technical and financial management of a JV to explore and evaluate uranium in Guinea in which Nigerian, Moroccan, Yugoslav and British concerns were also involved (4).

The same year, it was awarded a contract to build one of the largest flash smelters in the world for Codelco-Chile (the Junta's copper corporation) (5) and was employed to study leaching techniques for its Chuqui oxides in mid-1983 (6). Chile's Mina Norte mine was constructed by a consortium including Davy McKee; the project partially funded by the Inter-American Development Bank (7).

Homestake appointed Davy McKee as project design and engineering contractor for the McLaughlin gold project in north California (8) and it is also manager of Pancon's Paddington gold project (7).

Among recent contracts awarded to Davy McKee has been one to supply a hot stripmill and cold mill for an Alumax aluminium complex in the USA (9).

In late 1982, along with the Guinean government (50%), the Ministry of Mines and Power of Nigeria (25%), a Yugoslav company (25%)

and the Moroccan government (1.5%), Davy McKee won a small part of a contract to study Guinea's uranium deposits; more important, Davy McKee was responsible for the technical and financial management of the project. The main area covered was in Guinea's south-eastern region, where an aerial geophysical survey had already indicated "an extensive uranium field" (10).

One of the regions in which the company has gone most aggressively after new contracts has been the Philippines, where its subsidiaries have landed three major contracts: in 1980, a US$350M contract to build a fertiliser complex and a contract to engineer the major part of two nitric acid and ammonium nitrate plants (won by West Germany's Davy McKee AG, built in conjunction with Kloeckner & Co); two years later Davy Powergas Inc Pty Ltd (40.33% owned by Davy Corp and registered in India) secured a contract to build a phosphoric acid plant for the Philippines company Polyphosphates Inc (11).

In 1983 the company won a US$1.2M contract to produce a study on the second stage of the Ok Tedi mine (12) – much of this work to be carried out at its offices at Stockton, Britain (13).

In 1984, Davy McKee won a £350M contract to build a plate steel plant in Mexico – Britain's largest contract ever with South America (14).

Davy McKee Nuclear was set up in 1983 to consolidate and expand the company's "presence in the nuclear business" (15).

Davy entered the nineteen-nineties with good prospects for increasing its trade in eastern Europe (where it became the first company to instal a flue gas desulphurisation plant in a Comecon country) (16, 17) and selling itself as a green enterprise (with its waste and sewage treatment plants, and its recycling of plastic and soft drink bottles) (17).

However, its 1990 results were – as one analyst put it – "appalling" (18). Its chief executive quit ahead of time, and its new chair, Alistair Frame of RTZ, entered the boardroom confronted with a corporation in considerable disarray. Davy was one of the companies which built

RTZ's Rossing mine. (The connection between the two corporations goes back quite some distance) (19).

One of the first results of the new broom swept by the doyen of the world's largest mining company, was to negotiate the sale of Davy's substantial German interests to Metallgesellschaft. (which was also rumoured, at the time, to be interested in acquiring RTZ/CRA's stake in the Bougainville copper mine) (20). This move would give Davy the ability to concentrate on its metals engineering side, while dispensing with some of its water and waste treatment facilities – in other words moving several steps back from its ecological posturing, and more in line with those businesses that Frame himself personally would be happy to control.

- In 1987 Davy reduced its holdings in the Tata group of India (the largest family-controlled business in the country), sufficient to retain veto powers under Indian law, but which left management control to Tata (21).
- In 1989 Davy McKee was selected by Minera Northern Resources Ltd of Chile to carry out general engineering services at its important Yolanda iodine project in northern Chile (22). Later Atacama Resources and Kap Resources invited the company to expand its services to include a new prospect 30km away (23). The feasibility study was completed in 1990 and concluded that there was no practical reason against production (24).
- The company was commissioned in 1990 to carry out a final feasibility study for the Kensington gold project north of Juneau, Alaska, which is managed as a JV by Echo Bay Mines Ltd and Coeur Alaska Inc (25).
- Also in early 1990, Davy acquired part of Spie-Batignolles, the French construction concern, with which it already shared certain common interests , including chemical engineering and aluminium (26). Announcing the merger, the company said it represented "not only...a new strategic alliance aimed at 1991 [unification of EEC economic interests], it will also broaden the metal division base..." (17).

- During 1990, the corporation won a contract to design and supply equipment for South Korea's first aluminium hot rolling mill (27); it acquired Expanded Piling Group in Britain (through Monk Construction, its main construction and property subsidiary) (28); and it won a contract to design plant for Bethlehem Steel of the USA – for what was described as "one of the most significant single investments by a US steel maker in recent years" (29).

References: (1) *I MM* 11/82. (2) *MAR* 1982. (3) Advert published in *MM* 11/82. (4) *MJ* 3/12/82. (5) *MJ* 29/10/82. (6) *E&MJ* 6/83. (7) *E&MS* 6/84. (8) *MJ* 18/6/82; 19/8/83. (9) *MJ* 6/4/84. (10) *FT* 25/11/82, (11) *British Companies Operating in the Philippines,* London, 1984 (CIIR/Catholic Institute for Internal Relations). (12) *MJ* 9/9/83. (13) *FT* 10/9/83. (14) *FT* 27/4/84. (15) *Gua* 18/8/83. (16) *FT* 28/9/89. (17) *Financial Weekly* 12-19/4/90. (18) *FT* 15/11/90. (19) Transnational Corporations and their foreign interests operating in Namibia, United Nations, Paris, 4/84. (20) *FT* 7/2/91. (21) *FT* 1/7/87. (22) *MJ* 2/6/89, *MJ* 22/9/89. (23) *MJ* 11/5/90. (24) Kap Resources Ltd Yolanda Project Chile, David Williamson Associates Ltd, London 10/90. (25) *MJ* 20/1/89, *MJ* 9/3/90. (26) *FT* 28/3/90. (27) *FT* 21/6/90. (28) *FT* 11/10/90. (29) *FT* 12/10/90.

185 Dawn Mining Co

It operated a small uranium mine and mill at Ford, Washington, where in 1980 production was about 200t uranium oxide (1).

From the Midnite mine, nearly half a million pounds of uranium oxide was sold in 1980 (2). However in 1982, the Spokane Indians, on whose land the mine and mill are situated, initiated a court suit over royalties which slowed down the operation (3). The following year the

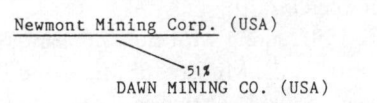

mine was inactive and the mill put on a care and maintenance basis (4).

References: (1) *MAR* 1981. (2) *MIY* 1982. (3) *Nukem Market Report* quoted in *MJ* 3/9/82. (4) *MAR* 1984.

186 Deep Bay Graphite (CDN)

It was reported exploring for uranium in Saskatchewan (1). No further information.

References: (1) *Who's Who Sask.*

187 Degussa

This is a diversified company with large interests in precious metals and chemicals (1), not to mention uranium enrichment, and exotic exploration in faraway lands. It also has major chemical interests in Brazil, Belgium, Austria, Japan, the Netherlands, the USA and elsewhere.

Its major corporate links are with Metallgesellschaft (with whom it owns Norddeutsche Affinerie, Germany's largest copper smelter, set up in 1916) and MIM, CRA and OMRD in the Frieda River consortium in the Sepik area of Papua New Guinea (2), an area exclusively used by indigenous peoples (3). It is among the first companies to be receiving gold from Ok Tedi (4).

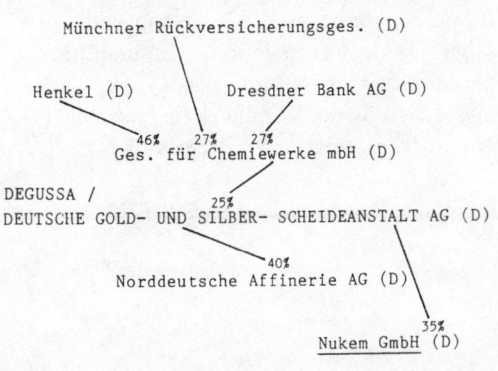

Together with Metallgesellschaft it carried out a feasibility study for the extraction of gold and silver from mine tailings at Vatukoula in Fiji in the mid 1980's (5).

In the late eighties and early 1990's, Degussa embarked on an aggressive buying programme, which resulted in its acquiring a majority share of United Silica Industrial of Taiwan (which supplies white silica to the footwear and tyre industry – Degussa is the world's largest producer of precipitated silicates) (6); the carbon black (also used in rubber goods) interests of Phillips Petroleum (7); the Manox pigments unit – which had itself been taken over by Rhone-Poulenc from RTZ Chemicals in 1990 (8). In 1990, it was also planning a merger between its pharmaceuticals subsidiary, Asta Pharma, and the West German company E. Merck (itself apparently no relation to the world's largest drugs company, Merck of the USA) (9).

Probably its major 1990 purchase was the worldwide gold and silver activities of Engelhard, a company indirectly controlled by Anglo American through its Minorco associate (10).

The following year, Degussa increased its holdings in Pancontinental from 9.5% to 14%. (Its original stake had not been revealed until the Australian government changed its foreign investment rules, making it mandatory for an investor to disclose shareholdings of 5% in a domestic company, as distinct from 10% previously.) (11)

Degussa has been embroiled in controversy both at home and abroad since the Nukem "scandal" erupted in the late 1980's – with its revelations of bribery, the "swapping" of titles to uranium shipments, and illegal export of nuclear material to countries not covered by the NNPT (Nuclear Nonproliferation Treaty) such as Pakistan and Libya (see Nukem).

As part of the strategy to rehabilitate Nukem in the wake of this scandal, Degussa took over management control of the beleaguered company in 1988 (12). However, two years later, Degussa was considering selling its 35% stake in Nukem, and also said it could dispose of its 7.5% in Ok Tedi Mining Ltd in Papua New Guinea (13).

1989 saw Degussa accused of another violation of the provisions of the NNPT, when it was cited by the US Commerce Department for selling beryllium metal (used to increase the explosive power of nuclear warheads) to India's Bhabba Atomic Research Centre without the necessary permit (14).

Meanwhile Degussa was benefitting from its share in the platinum group syndicate set up by International Platinum and Jenkim holdings of Canada, with prospecting at an advanced stage on the Big Trout deposit in Ontario, and exploration projects elsewhere in Canada (14, 15).

References: (1) *WDMNE.* (2) *MIY* 1982. (3) R. Jackson, *Ok Tedi: The pot of gold,* 1982 (University of Papua New Guinea). (4) *MJ* 6/4/84. (5) *MJ* 16/3/84. (6) *FT* 8/3/84. (7) *FT* 13/3/86. (8) *FT* 18/9/90. (9) *FT* 21/2/80. (10) *FT* 21/2/80. (11) 4/1/81. (12) *MJ* 22/1/88. (13) *MJ* 4/5/90. (14) *MAR* '90. (15) *FT Min* '90.

188 Dejour Mines Ltd (CDN)

It was involved in uranium exploration both in the High Arctic region of the Northwest Territories (NWT) (1) and in northern Saskatchewan (2) during the late '70s.

Dejour's main activity is exploring gold deposits in the NWT, Quebec and Ontario in Canada (3).

Although it holds several prospective gold properties in Quebec and Ontario, Dejour's main exploration programme is directed currently at gold-bearing zones at the Turquetil discovery, a 100,000 acre lease near Eskimo Point in the indigenous Keewatin District of the Canadian North West Territories(4).

In 1990 it also managed to arouse the ire of many across the Atlantic, when it set up a subsidiary, Omagh Minerals, to take over RTZ (Riofinex's) controversial Lack (Cavanacaw) gold prospect in Omagh, Northern Ireland. This prospect met stiff resistance from local people, who formed a highly effective opposition group to halt it. No doubt, partly because

of this opposition (which included lengthy TV and radio coverage, several major articles in British newspapers and magazines, and an Early Day Motion in the House of Commons)(5), RTZ pulled out of Omagh, claiming it had not discovered gold of sufficient value to justify its further involvement (6).

Half a year later, however, Omagh Minerals – whose managing director is an ex-employee of RTZ – took over the licence and resumed work on the open-trench exploration (7). The Omagh Gold Action Group, supported by representatives of every political party in the local council (from Sinn Fein to the 'loyalists'), deplored the development, claiming that mining would contaminate the local streams with heavy metals; dust storms would bring bad weather; prime bogland might be threatened, and irreparable damage would be done to the main local amenity – tourism (8).

Contact: Omagh Gold Mining Action Group, c/o 57 High St., Omagh, Ireland BT78 1EE.

References: (1) *News of the North* 20/6/80, quoted in: *Kiitg* 8-9/80. (2) *Yellowcake Road.* (3) *MIY* 1990. (4) *FT* Min '90. (5) *Parting Company,* PARTi-ZANS, issues 1989-1990, London, see also Tyrone Constitution 21/6/90. (6) Tyrone Constitution 16/8/90, *Ulster Herald* 16/8/90. (7) *FT* 4/12/90, Tyrone Constitution 29/11/90. (8) *Sunday News,* Belfast 27/1/81.

189 Delaware Nuclear Corp (SWA)

It was registered in 1977 – along with three other American corporations – for uranium exploration, giving its address as Volkskas Buildings, Kaiser Street, Windhoek (in Namibia) (1). No other information whatsoever.

References: (1) *Johannesburg Star* 30/7/77.

190 Delhi International Oil Corp

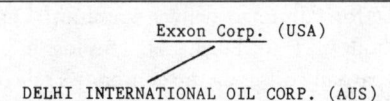

```
Exxon Corp. (USA)
       |
       |
DELHI INTERNATIONAL OIL CORP. (AUS)
```

Since November 1981, it has been virtually a wholly-owned subsidiary of the large Australian diversified natural resources company CSR (1). Early in 1982, the Australian government gave permission for Delhi (with 53.5%) to develop a uranium mine at the Lake Way property near Wiluna, Western Australia, together with VAM (46.5%). An estimated 3700t of uranium oxide is present, modestly grading at 1.5lb uranium per tonne (2).

The project is now indefinitely suspended.

Lake Way comprises about 4000 tonnes of U_3O_8, grading at 0.9kgs/tonne; the deposit is of the Yeelirrie type and extraction would be by using the carbonate leach-ion exchange method (3). It has been given federal government approval, despite its suspension on market grounds. The mine would be close to Wiluna townsite, where a large proportion of residents (out of 1500 people) are Aboriginal. The mine is also only 5kms from a community village called Ngangganawili. The Environmental Impact Statement for the project has admitted that vegetables grown in the area may become radioactively contaminated, that citrus farms may be drained of water, and kangaroo and sheep meat become unfit for consumption (4). As the nineties opened, Delhi still held, with VAM, its licence over the Lake Way project, but there was virtually no prospect (even with the relaxation of the Australian Labor Party's "three mines" uranium policy) that it would be opened (5).

References: (1) *MJ* 11/12/81. (2) *FT* 20/1/82. (3) *Reg Aus Min* 1983-84. (4) *WCIP newsletter*, Lethbridge, No. 2, 4/83. (5) Greenpeace report 1990.

191 Denison Mines Ltd

Modestly boasting itself a "diversified world energy and natural resource company ... a major international producer of Uranium, Oil, Gas, Coal and Cement Products" (1), the modern Denison was incorporated in 1973 after amalgamation with Stanrock Uranium Mines Ltd (2).

Thirteen years before, the Can-Met uranium property at Elliot Lake, Ontario was merged with Consolidated Denison to form Denison Mines Ltd (3). At the time, Denison and Rio Algom were the only operating uranium producers in Canada – thanks to a US ban on imports to boost its own producers. By 1964, employment at the two mines had dropped by one-tenth from its peak in 1958: the Canadian government created its own stockpile which helped keep Denison and its stablemate afloat (4). But it wasn't until the uranium cartel began to bite, and Denison secured spectacular contracts in the mid '70s, that its rise was assured to its cur-

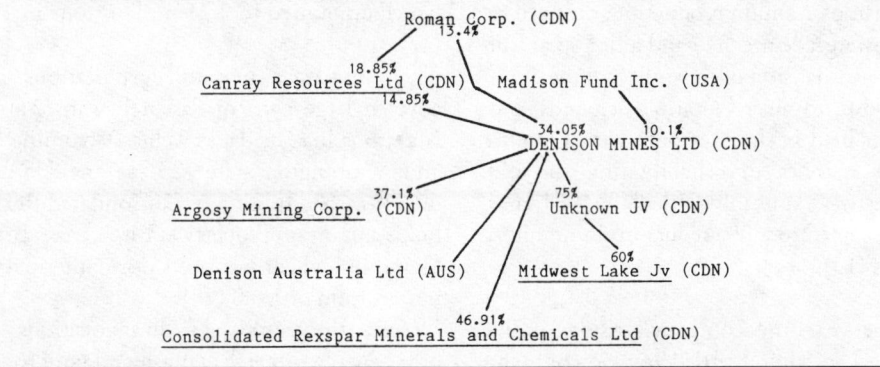

rent position as one of the West's leading purveyors of yellowcake.

In March 1974, Denison signed its first major contract of the '70s, with Tokyo Electric Power Co, providing for the sale of 40 million pounds of U_3O_8 at the rate of 4 million pounds per year from 1984 to 1993 (5) (the contract seems to have been put into operation on time) (6). Tokyo Electric's prepayment of US$10M in early 1974 covered mine development in advance of production and enabled processing facilities to be expanded to 7100 tons/day of ore (5).

A bare five months later another fat contract was signed, with Enusa of Spain, for 4 million pounds of uranium from 1975 to 1977, again with a substantial prepayment (5).

As of December 31, 1975 export sales totalled more than 80 million pounds of uranium oxide; additional supplies of uranium to Spain were to be made from the Denison/federal government joint stockpile programme (5). Then, at Christmas 1977, the company got the biggest present delivered to a uranium producer, with the world's largest uranium sales contract ever concluded (7).

126 million pounds of U_3O_8 were to be delivered to Ontario Hydro (the only operator of commercial nuclear plants in Canada) at the rate of 4.2 million pounds per year from 1980 until 2011 (10) – later extended to AD 2012 (6). The amount of advances paid by Ontario Hydro to Denison by the end of 1980 was more than US$200M (2). Little wonder that the company's profits tripled in 1978 (9). However, they declined in 1979 (10) and – due to higher costs of uranium production, the need to plumb lowergrade ore (10) and a fire up at 'tut mill (11) – expansion of the mills and the Stanrock deposit, to meet Ontario's demands, was somewhat held back. A dispute with Ontario over the manner of rehabilitating the old properties was "amicably settled" in 1981, and Denison agreed to a 9% reduction in uranium deliveries (13).

The same year, the company's rising profits were "checked" after it settled litigation brought

against it – as part of the notorious cartel case – by Westinghouse and the TVA in 1976 and 1977 for failure to deliver uranium. Denison paid out C$17.6M (13, 14). This loss in profits was partially offset by the company's sale of certain investments, notably its share in the Coalspur thermal coal property in Alberta, and higher income from oil and gas (15).

In the next two years oil and gas proved to be Denison's biggest profit-spinners, with production on stream in the Gulf of Mexico, in the Casablanca fields off-shore of Spain, and in Greece, as well as in Canada (16).

It has also taken shares in oil companies operating as far afield as Egypt and Nova Scotia, the Philippines and the South China Sea (17).

Denison diversified into coal with 28.25% of the mammoth Quintette coal mine in British Columbia, where – despite various teething problems – production was expected to reach full spate at the end of 1984 (18). Denison's partners at Quintette are Imperial Oil, Tokyo Boeki, Mitsui Mining, and Charbonnage de France (2).

By the beginning of 1982 the company's expansions at Elliot Lake were 75% complete, and a new hydro-metallurgical plant was commissioned during 1981, reaching its design capacity by December (16).

Two shafts from the old Can-Met/Stanrock mine were being rehabilitated at the start of 1983 (3, 19). A year later, the expansion was proceeding less quickly than expected (20) but underground heap leaching of low-grade "dumped" ore was producing 5 tons/month (6), a tiny contribution to normal production which amounted to 2.71 million tons in 1983 (17).

The company's uranium explorations have covered large parts of Saskatchewan (21) and Manitoba in Canada, as well as Wyoming (22) and Washington state (23) in the USA. Its 46.9% owned subsidiary Consolidated Rexspar holds a uranium property at Birch Lake, British Columbia which has been dormant since the moratorium of 1980 (24).

Despite the company's diversifications into other energy minerals, potash and gold (6) and despite, too, its purchase of Noranda's Koongarra mine in the Australian Northern Territory in 1980 (25, 26), Denison and its fortunes are still primarily identified with Elliot Lake. Numerous reports have been published (27) on the effects of mine radiation and tailing contamination from the Elliot Lake mines. The following is a summary of these. (See also Rio Algom for further information.)

- By 1974 at least 81 lung cancer deaths and hundreds of cases of silicosis were directly attributable to conditions at Elliot Lake. By January 1980 the number of acknowledged lung deaths had risen to well over 100 (28).

- Up to March 1975, 446 present and former Elliot Lake workers were identified as having lung disablilities due to dust exposure, and in January 1980 there were "dust problems of a serious nature" at both Rio Algom and Denison operations (28).

- The Ontario Workmen's Compensation Board estimated that 16 out of every 20 deaths among Elliot Lake miners were due to lung cancer (29). Stated Stephen Lewis, Ontario NDP leader: "In no individual year, from 1958 to 1974 inclusive, in no survey, quarterly, three times a year, or semi-annually, in not a single instance, did the average underground dust counts for the uranium mines of Elliot Lake fall below the recommended limits. Not one. It's really quite horrendous" (29).

- In October and November 1978, 86% of samples from the mines at Elliot Lake, taken by the Ontario Ministry of Labour, revealed free silica contamination above the standard set (28).

- Radioactive mine-dust readings taken by the provincial government around the same time showed 32% above the danger level (30). Results of the surveys were not made available to the miners themselves (31).

- Dr David Bates, chairperson of the British Columbia inquiry into uranium mining, visited Elliot Lake in the late '70s and described the Elliot Lake Health Protection Surveillance facility thus: "The general impression of this facility is that it commands so little in the way of resources that presumably the company at all levels regards the whole of the surveillance program as a very low priority activity" (32).

On the same tour Dr Bates had this to say about the quality of the environment inside one of Denison Mines' uranium mills: "The mill was very poorly maintained as far as the housekeeping is concerned. There was obviously inadequate regulation and obviously not enough care was taken either by the workers or their supervisors" (32).

- In a further submission to the Bates Inquiry, Kenneth Valentine, the National Director of Occupational Health and Safety for the United Steelworkers of America (USA), made this statement on the deplorable record of Denison and Rio Algom: "My personal observation is, you know, while I was there, that not too many of them should have been allowed to continue keeping their licence, their licence should have been pulled ... yes, there have been situations where almost everyone should have had their licence revoked ... In the final analysis, I would suggest that at the very least the AECB is guilty of criminal negligence ... And as a result of their failure, their dereliction of duties, untold numbers of people have died" (33).

- When details of the appalling lack of protection for miners became public in 1974, it triggered a three-week stoppage at the mines and forced the federal government to appoint the Ham Commission. But though new regulations were adopted in Ontario for most other mines, uranium mines continued to be governed by looser federal regulations (34).

As Stephen Lewis concluded in his submission to the Ham Commission: "The Elliot Lake story is a moral and human outrage. What is more, it's apparently not yet over" (29).

Tailings from Denison's mines have littered and petrified the landscape (rather, lakescape) at Elliot Lake for many years, as testified by photographs compiled by the Ontario Birch Bark Alliance (35).

- Elliot Lake tailings have been used as construction fill material for buildings as far away as Toronto and Montreal (36).

- A Federal/Provincial Government Task Force set up in 1976 identified fourteen houses as needing "immediate remedial action" while 311 (16% of all) needed to be "cleaned up" (36). A new building code published in 1978 limited building radiation to 0.2 WL (working levels). But, as Dr Gordon Edwards observed: "Using only data supplied by the Ontario Ministry of Housing to the Elliot Lake Environmental Assessment Board, it is shown that continuous exposure to 0.02 WL for 12 hours per day could lead to a whopping 31% increase in the incidence of lung cancer for males" (37).

- Mine waste at Elliot Lake flows into the Serpent River system and directly affects the fishing waters of Native Canadian bands downstream (38).

The international Joint Commission on the Great Lakes (IJC) identified the outflow of the Serpent River into Lake Huron as the greatest single source of radium-226 as well as large amounts of thorium isotopes. For a distance of one kilometre into Lake Huron, concentrations of radium, thorium and heavy metals are above the standards set by the IJC (39).

During the low summer flow period, liquid wastes dumped into the Serpent River headwaters range from between two-thirds and a half of the total flow (39).

- Acidic waste water also moves into ground water, raising contamination levels as the acid moves in the ground water zone (40).

Acidity levels taken at Elliot Lake in the late 70s revealed dissolved radium-226 levels well above the potable standard (41). Levels of iron, copper and uranium also exceeded those suitable for human drinking (42).

- The *Status report on water pollution in the Serpent River Basin,* delivered by the Ontario Ministry of the Environment in 1976, revealed that water downstream of Elliot Lake was so contaminated with radium that it was unfit both for human use and fish – all of which had been killed off (43).

By 1980, fish were absent from 1000 square miles of the Serpent River watershed (half the total area) and 55 miles downstream from the mines (39).

- Copper levels downstream from Elliot Lake greatly exceed the 50ppb lethal level (44).

- At Elliot Lake, over the years of mine operation, recorded types of waste releases include: unexpected seepages, unintended movement of solid wastes, dam and dyke structural failures, re-solution or erosion of barium-radium precipitate, waste pipeline breaks, pump failures, a decant tower collapse, the accidental release of a large volume of sulphuric acid, and release of fine sands in the decant effluents (39). One example is Denison's Stanrock waste area, both sides of which are bounded by dams constructed of solid mill wastes. The Stanrock dams have been a source of contamination since their construction. One failure was on November 24, 1979, when between one and two million gallons of wastes spilled into the Serpent River system (28).

On a visit to the Stanrock waste site in May 1979, chairman of the BC Royal Commission on Uranium Mining, Dr Bates noted that: "The tailings pond was clearly eroded, it was not fenced, it had been covered by lime, and a warning sign was placed beside the road to this effect, but no attempt was made to keep people off the tailings pond. The area downstream showed considerable damage where the tailings had drained into the forest and killed this. The outflow and the sludge collected in Moose Lake. It is notable that none of the tailings pond itself showed any sign of revegetation after having been abandoned for something like 15 to 20 years" (46).

- At one point in the late '70s, no less than 22 charges were laid against Denison by the Ontario Ministry of the Environment for failing to safeguard against effluents released (39). The company was acquitted on all charges because they had done what was technically feasible to avoid them.

- In 1979 the Ontario Environmental Assessment Board issued a final report on the expansion of uranium mines in the Elliot Lake area – expansions needed to fulfil both Denison and Rio Algom contracts. Among its findings, the Board slated the companies on several grounds:

- "... the Ministries of Natural Resources and Labour have no definitive criteria for evaluating the construction, materials used, and stability of tailings dams.

- ... the potential for redissolution of radium from precipitation ponds after abandonment is of great concern.

- ... the companies have been remiss in not studying backfilling as a waste management option.

- ... the process known as solidification has not undergone sufficient field trial to determine its applicability to the Elliott (sic) Lake tailings, its ability to withstand the climatic extremes, and its ability to be combined with other methods of contouring and stabilization as part of a long-term close-out process.

- [there is] little evidence to give it confidence in the use of synthetic membranes, asphalt, cement or chemical means to cover tailings areas to inhibit water infiltration in the long term.

- ... the mill tailings have the greatest potential impact on the natural environment of all the activities related to the mines expansion" (47).

In late 1980 Noranda sold its Koongarra uranium lease in the Alligator Rivers region of Australia's Northern Territory to Denison. The company considered it needed a new source of uranium to keep up with its future sales (16).

The price Denison paid for Koongarra has not been disclosed (25), but with the deal Denison acquired two orebodies: Number one containing 13,000 tonnes of U_3O_8, and the other 15,000 tonnes (inferred) of uranium oxide grading at 3.35% (48).

The Ranger Uranium Environmental Enquiry (Fox Commission) in its second report (49) recommended against mining the Koongarra deposit, describing the creek and wetland system downstream as "... so valuable ecologically [that] we would oppose in principle any mining development upstream of it". Fox also said: "Our view is that if [Noranda] operations were carried out in accordance with present plans they would themselves constitute a serious danger to the environment ... The operation of the ... mine concurrently with the Ranger and Nabarlek mines would involve bringing into the region an excessive number of Europeans. [Noranda] should therefore not be allowed to proceed" (49).

Koongarra is part of the Kakadu National Park. However, the Koongarra Project Area Act, rushed through in June 1981, just before the Australian Liberal/Country party coalition lost control of the Senate (50), enabled Denison to tap into new reserves in the National Park zone. Federal opposition (ALP) spokesperson S West at the time declared that this act made "... a farce of the definition of a National Park, particularly when Kakadu has been nominated by this government for World Heritage listing" (51).

The Alligator Rivers study, commissioned by the Australian government and the mining industry in 1977, expressed fears that mining at Koongarra could increase heavy metal and radioactive deposits flowing onto the plains, which would build up during the dry season "... to the disadvantage of the native aquatic species in the region, both plants and animals ... A further consequence could be a serious change in the quantity of water for either drinking or irrigation purposes" (52).

This study also reported that mining, increased road traffic, explosions and dust could degrade Aboriginal art sites in the region, including prehistoric sites like Anbang-bang, "perhaps older than 40,000 years" (53), which constitute "one of the world's most important storehouses of information about prehistory and the art of hunting and gathering man" (54).

In 1984 the Hawke government confirmed the de facto moratorium on uranium mines in the Northern Territory which effectively included Koongarra and Jabiluka, though not Ranger and Nabarlek (see Queensland Mines) (55).

In 1983 the Australian federal government declared its intention to incorporate Koongarra and Jabiluka into stage two of the Kakadu National Park, thus effectively preventing all mining. A US$36M "tourist package" deal was being offered to Aborigines in the region with the promise of 1300 permanent jobs and a massive increase in tourism in the area. The Northern Land Council, which functions under the Aboriginal Land Rights Northern Territory Act of 1977 (56) vigorously opposed this deal, and – it seems – a large number of traditional (Aboriginal) land owners have supported the development of both the mines in question (57).

Towards the end of 1983, Denison concluded an agreement with many of the traditional owners, which gave them a 25% stake in the project (58).

This was, however, the tail end of a lengthy process of attrition, pre-emption, seduction and (many would say) deceit. It started in 1981 with the arbitrary (and probably illegal) excision of the Denison lease from the Kakadu National Park – an action viewed with helplessness, anger and alarm by local Aborigines (59).

The *1981-82 Annual Report of the Standing Committee on the Social Impact of Uranium Mining on the Aborigines of the Northern Territory (sic!)* criticised the ability (or willingness) of government and territory institutions to convey meaningful information to Aborigines about Koongarra and other uranium projects in the region (60).

What is certain is that some Aborigines in the area have voiced their opposition to the Denison mine from the early days. On May 1 1981, a meeting of owners of the land asked: "How can it be that the minister can give our land away to a mining company without asking what we think? This is not self-determination: this is the same as always, white fellahs making their own decisions about what is good for Aborigines." They then decided to break off all negotiations with Denison (61).

One family – the Aldersons – remained adamant in their opposition to any negotiations through the whole period (62). Two of the des-ignated 29 traditional owners in early 1981 – Violet McGregor and Jessie Alderson – were opposed to deals with Denison. On 23 January 1981 they cabled the Northern Land Council (NLC) thus: "Please be advised that we will not attend further meetings. Extremely dissatisfied with NLC's conduct of Koongarra meeting and subsequent press release implying Murrunburra totally support Koongarra mine ... Request apology through media Australia wide for misrepresentation of our views and ridiculous claims of two years prior consultation on Koongarra site" (63).

By the middle of the following year, the traditional owners were still holding out, forcing the NLC to defer negotiations with Denison (64).

That December, negotiations were formally blocked when the Aboriginal Land Commissioner, Justice Kearney, advised the Aboriginal Affairs Minister that votes by three of the Aldersons against the project meant it couldn't proceed (65). On that occasion Denison offered 10% royalties to the Aboriginal community (51).

Finally, after continuing negotiations through 1982 (23), Denison agreed to restrict its mining proposal to nine leases on the original site outside the area covered by Aboriginal claims. This meant that mining could proceed without formal Aboriginal permission (66).

Denison is, generally speaking, a company which has been comparatively free from criticism, as distinct (for example) from Union Carbide, Rio Algom, Kerr-McGee, United Nuclear and other primary uranium producers.

However, in 1977 six hundred people turned out to oppose its plans to open up a US$27M open-pit uranium mine near Clearwater, British Columbia (BC) in Canada. Among those who attacked the proposal were the BC Medical Association, the United Church environmental groups, and – most remarkable of all, given their reliance on uranium for a "living" – members of the USA/United Steelworkers of America. The Steelworkers came straight to the point: "We feel we are being deceived by this company, which comes to BC to establish a uranium mine and tries to assure us that there will be no problems ... We say that the Elliot Lake experience established beyond any doubt that there will be problems."

The Federal Environment Minister also agreed that local people didn't want the mine and said it had a "disturbing" potential for environmental damage (67).

Then, in 1981, Denison – along with Eldorado Nuclear, Gulf Minerals Canada (see Eldor Resources), Uranerz Canada and Rio Algom – were charged by the Federal Canadian Government with fixing domestic uranium prices during the era of the uranium cartel (68). This was a charge which Denison vigorously rejected, maintaining that the price-fixing was at the "direction and instigation" of the Canadian Government (69) ... which indeed it was! The Canadian Government later bowed out of legal action (70).

In 1979 Denison announced a merger with the US energy concern Reserve Oil and Gas at a price of US$525M (71).

The Dome group sold its 10.1% holding in Denison in early 1983 (72).

In early summer 1984, the company announced that experiments in filling-in working stopes at its Elliot Lake mine with uranium tailings did not noticeably increase radon and radon decay product ("daughters") radiation in the mine atmosphere (73).

Denison acquired Seagull Petroleum in 1984, and reported at the end of that year that its 50% owned Quintette coal mine was expected to reach contract levels within three months (74).

In 1987, Denison, together with an un-named partner, acquired 60% of, and therefore control over, the Midwest Lake deposit: at the time of writing, it is not known whether the partners of the original JV (Canada Wide Mines, Bow Valley and Numac) reduced or sold out any of their holdings (75). Latest drilling indicates more than 20,000 tons of uranium, grading 1.1%, and Denison anticipates production in the 1990s (75, 76). Also in 1987, Denison acquired the claims of Canuc Resources, south-west of its main Elliot Lake mine, which it plans to de-

velop (75). Canuc's uranium interest had remained dormant until then.

Between 1989 and 1990, Denison underwent a massive restructuring, and cost-cutting in the face of severe losses. In late 1990, a new president, Bill James, was brought into the company to save the enterprise (77). It sold virtually all its oil and gas properties worldwide (78), and offloaded its potash interest in the 60% owned Denison-Potacan mine in New Brunswick, in early 1991 to the West German firm Kali and Salz and the French Enterprise Minière et Chimique (79). Its Quintette coal business, which had promised so well during the mid-eighties, became bankrupted, due to the squeeze on the world coal and steel markets (80) and management control of the company was taken over by Teck Corporation in March 1991 (81). Denison also sold various gold properties, despite forays into new regions such as Venezuela (82).

Bill James announced in December 1990, that Denison would be up for sale (83) and although at the time of writing, no buyer has been forthcoming, it has been rumoured that Rio Algom, its partner at Elliot Lake, and Inco, would be likely purchasers (84). James has been quoted as saying that Denison would "like to stay in uranium" (84), although the Elliot Lake operations have faced not only production cuts (down by 11.2% in 1989) (85) but higher than average costs (its 1987 shipments were, at C$44.50 a pound, some 70% above the value of Saskatchewan shipments) (86) and pressure from Ontario Hydro to reduce its prices after 1993 (87).

In 1990, Denison announced the laying-off of 28% of its workforce at Elliot Lake, and reduction in output from 3.3 million pounds a year to 2.7 million pounds by 1991 (88): this followed an announcement by Rio Algom that it too would be closing three of its mines in the region (89).

Then, in early 1991, the company said it would be laying off more than one third of the mine's total workforce of 1060, and be cutting back production even further (to 1.4 million pounds). By this time only Ontario Hydro was buying from Denison at Elliot Lake (90).

Ironically, Denison's other uranium interests seem far from foundering. Its Koongarra project has not been abandoned; while the Mid-West deposit could come into production in the near future depending on markets (Toronto-based Redstone Resources purchased a 2% gross royalty in the project from Esso Resources Canada in late 1990) (91).

And in early 1991, the company announced that it was forming a JV with Freeport Uranium Recovery Co to produce uranium oxide from two processing plants in Louisiana, with Freeport as operator and Denison as marketer (92). While Denison could still be sold and Rio Algom is clearly a putative buyer, it is interesting as a footnote to record that the former sued the latter in 1990 for allegedly causing damage to its mining activity, through adjacent operations at Elliot Lake. Denison claimed that "violent rock bursts" between 1982 and 1985 resulted from the collapse of rock pillars, due to Rio Algom's underground mining which was separated from Denison's by only a 12 metre rockwall. Denison was seeking C$35 million as compensation for ore supposedly tied up, as stabilizing work was carried out and its mining plans were changed (93).

Pressure from Ontario Hydro to reduce its prices after 1993 (94) was of no avail and in May 1991 the utility cancelled its contract, thus ensuring the closure of Denison's Elliot Lake operations by the end of 1992 (95).

Further reading: For a lucid description of the Koongarra project, its implications and impact, see P Coleing's paper for *Nuclear Alert*, Sydney, 5/1983 (Total Environment Centre).

References: (1) *MAR* 1984, front cover advert. (2) *MIY* 1982. (3) *E&MJ* 2/83. (4) Goldstick pp11-12. (5) *MIY* 1977. (6) *MAR* 1984. (7) Merlin p131. (8) *MJ* 30/12/77. (9) *FT* 13/2/79. (10) *FT* 26/1/80. (11) *FT* 24/10/80. (12) – (13) *MJ* 13/11/81. (14) *FT* 26/8/81. (15) *FT* 29/10/81. (16) *MJ* 5/2/82. (17) *MJ* 3/2/84. (18) *MJ* 10/8/84. (19) *MM* 3/83. (20) *Canadian Mining Journal* 11/83. (21) *MJ* 1/8/80. (22) *MJ* 27/7/79. (23) *MJ* 20/8/82. (24) *MJ* 30/4/82. (25) *FT* 24/6/80. (26) *FT* 14/2/81. (27) Goldstick. (28) *United Steelworkers of America and the BC Federation of Labour, January 1980, Submission to the BC RCUM, public and worker health,* Burnaby, 1980 (BC Federation of Labour). (29) Stephen Lewis, *Text of submission by ..., Ontario NDP leader, to the Royal Commission on the Health and Safety of Workers in Mines (HAM Commission), February 18, 1975,* quoted in Goldstick p26. (30) Lloyd Tataryn *Dying for a living: The politics of industrial health,* 1979 (Greenberg), quoted in Goldstick p29. (31) Goldstick p29. (32) David Bates, *Elliott [sic!] Lake, Ontario: A report to the BC RCUM, May 8-9, 1979,* quoted in Goldstick p64. (33) Statement by Kenneth Valentine to the BC RCUM, quoted in Goldstick. (34) Goldstick pp35-36. (35) *Birch Bark Alliance,* Peterborough Winter/81. (36) J Terral, *Uranium exploration: Four natural radiation hazards,* Nelson, 18/9/1979 (Kootenay Nuclear Study Group). (37) Gordon Edwards, *Estimating lung cancers or, It's perfectly safe, but don't breathe too deeply: A summary of testimony presented by ... to the Elliot Lake Assessment Board dealing with the problem of radon gas in building,* Mackay, Quebec, 1978 (Canadian Coalition for Nuclear Responsibility). (38) Information from Paul Kay to Partizans, London, 4/81. (39) Statement by Nelson Conroy, Minister of the Environment, to the BC RCUM, February 2, 1980, quoted in Goldstick. (40) JA Cherry, *Migration in groundwater of contaminants derived from surface-deposit uranium mill tailings: Submission to BC RCUM, December 1979,* quoted in Goldstick p93. (41) Goldstick pp65-66. (42) Goldstick pp71-72. (43) Status Report, Ontario Ministry of the Environment, 1976, quoted in Goldstick. (44) JB Sprague *Danger to freshwater fisheries from mine wastes and general comments on avoiding problems: Evidence to the BC RCUM, February 12, 1980,* Guelph, 1980 (University of Guelph). (45) – (46) DV Bates, *Quirke Lake Mine: Report to the BC RCUM, May 9, 1979,* quoted in Goldstick. (47) *Expansion of the uranium mines in the Elliot Lake area: A report of the OEAB,* Toronto, 1979 (Ministry of the Environment). (48) *Reg Aus Min* 1984. (49) *Fox Commission Report,* Vol. 2, Canberra, 1977 (Federal Australian Government). (50) Roberts. (51) Quoted in *CARE Bulletin,* Adelaide, 1-2/83. (52) Christian & Aldrick, *Alligator Rivers*

Study: A review report, Canberra, 1977 (AGPS), p135. (53) *Darwin Star* 16/9/81.(54) Kakadu Archaeological Working Party, *Archaeological Research in the Kakadu Park,* Canberra, 1981 (Australian National University). (55) Summary of policy in *FT* 19/11/84. (56) Ed Nicolas Peterson, Aboriginal Land Rights Handbook, Canberra, 1981 (Australian Institute of Aboriginal Studies), pp28-52. (57) *Aboriginal Land Rights Support Group Newsletter,* Sydney, 12/83. (58) *AFR* 7/7/83. (59) Colin Tatz, *Aborigines and Uranium,* Melbourne, 1982 (Heineman), p178. (60) *1981-82 Annual Report of the Standing Committee on the Social Impact of Uranium Mining on the Aborigines of the Northern Territory of the Australian Institute of Aboriginal Studies,* Canberra, 1982 (AGPS). (61) Roberts p135. (62) Michelle Sheather, *Uranium mining in Australia: Report for the NFIP Vanuatu, July 1983,* Honolulu, 1983 (Pacific Concerns Resource Center). (63) *Bunji,* Darwin No. 1, 1981. (64) *Northern Territory News,* Darwin, 16/7/82. (65) *Age* 21/12/82. (66) *AFR* 14/1/83. (67) *Nuclear Newsletter,* Saskatoon, 25/1/78 (Saskatoon Environment Society). (68) *Montreal Gazette* 8/6/81. (69) *MJ* 7/8/81. (70) *MJ* 30/12/84. (71) *FT* 18/8/79. (72) *FT* 16/2/83. (73) *MM* 5/84. (74) *FT* 24/10/84. (75) *MAR* 1988. (76) *E&MJ* 4/88. (77) *FT* 17/1/90 (78) *FT* 24/2/89, *FT* 29/7/89, *FT* 22/8/90. (79) *MJ* 1/3/91, *FT* 23-24/2/91. (80) *FT* 14/13/90, see also *MJ* 22/6/90, *E&MJ* 1/91, *FT* 15/6/90. (81) *MJ* 21/11/90, *MJ* 8/3/91, *MJ* 22/3/91, *FT* 18/3/91. (82) *MM* 5/89. (83) *FT* 21/12/90. (84) *FT* 22/1/91. (85) *MAR* 1990, *MJ* 12/5/89. (86) Greenpeace Report '90. (87) *FT* 21/3/91. (88) *FT* 9/3/90. (89) *FT* 9/3/90 see also *MJ* 16/3/90. (90) *MJ* 15/3/91 see also *MJ* 1/6/90. (91) *MJ* 23/11/90, *MJ* 12/11/90 see also Greenpeace Report '90. (92) *E&MJ* 1/91. (93) *MJ* 5/1/90. (94) *FT* 21/3/91. (95) *MJ* 10/5/91.

192 Derry, Michner and Booth

This was a front for Bendix (in their unauthorised rock sampling at the Chippewa Red Cliff reservation in Colorado) with HQ at Golden, Colorado (1).

References: (1) *AkwN*, Mohawk Nation, Winter/81.

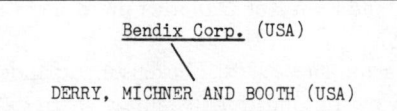

193 Deutsche BP

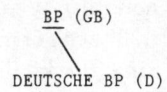

In 1981, this was BP's largest foreign subsidiary (1). For several years it had sustained considerable losses because of its unprofitable refining and marketing sectors (2). In 1979 the company acquired from Veba 25% of the huge Ruhrgas company which imports and distributes natural gas (3). In 1982 Deutsche BP's operations were cut considerably, despite a promising agreement with CFP/Total to explore for oil and gas in Algeria (4).

In 1977, the West German government awarded Deutsche BP a three-year permit to explore two uranium concessions in the Bavarian Forest, covering a total of 2920 square kilometres in Oberpfalz and Lower Bavaria (5). Little has been reported on this since.

References: (1) *FT* 15/1/81. (2) *FT* 24/1/83. (3) *FT* 6/3/79. (4) *FT* 5/8/80. (5) *MM* 11/77.

194 Dickenson Mines Ltd

It is a gold, silver, lead, zinc, gas and oil company, holding uranium interests through New Cinch, Teck (in which it holds a minority

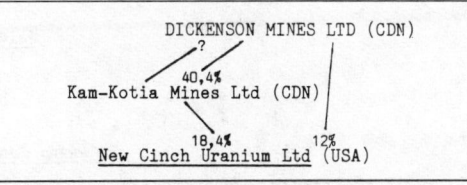

share), and Kam-Kotia, another piece of the action (1).

During the late 1980s, Dickenson expanded, mainly by consolidating its mineral interests in the Red Lake division of Ontario (through Goldquest Exploration Inc, in which it holds 59.7%), through Wharf Resources Ltd (35%) and by gaining all the assets of the Havelock Lime group of companies.

It is important to note that Wharf Resources Ltd owns an area of 4000 acres of mining claims in the sacred Paha Sapa (Black Hills) region of South Dakota, traditional land of the Lakota people. Its mining of the Foley ridge deposits in this region, by heap leaching, have yielded fairly significant amounts of gold (2).

References: (1) *MIY* 1982. (2) *MIY* 1990.

195 Dolores Bench General Partner Inc (CDN)

It was involved in uranium exploring in Saskatchewan. No further information available.

References: (1) *Who's Who Sask.*

196 Dolphin Exploration Ltd

Gippsland Oil and Minerals NL (AUS)
│ 50%
DOLPHIN EXPLORATION LTD (AUS)

This was the operating company for the Georgetown uranium/base metals project in Queensland, Australia (1).

Dolphin Explorations Ltd is 44% held by the Corona Corp of Canada, a major gold producer and explorer with projects throughout north America, Mexico and Greenland (2).

References: (1) *Reg Aus Min* 1984. (2) *MIY* 1990.

197 Dome Petroleum Ltd

In early 1982, Dome Petroleum took possession of Hudson's Bay Oil and Gas by buying it from Du Pont Chemicals in exchange for a large chunk of Conoco, acquired earlier in 1981 (1).

Though not involved directly in uranium mining, this large company has been the prime mover (with Gulf Oil and Esso) in a massive project to open up the Beaufort Sea to oil and gas production, with consequent enormous en-

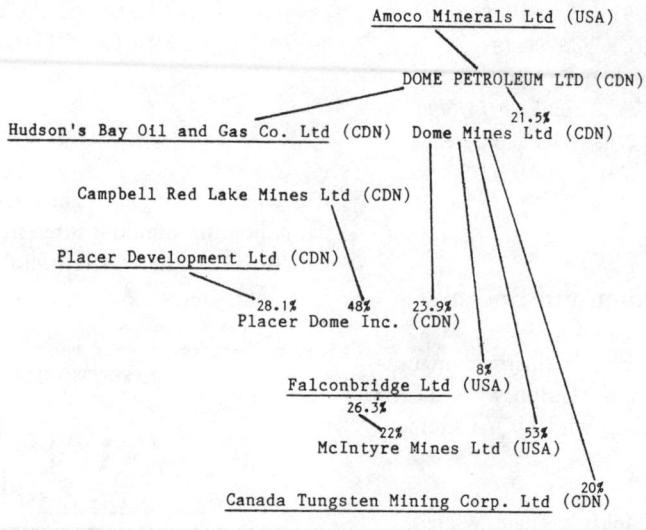

vironmental, economic and social consequences to Inuit and Dene communities in the whole of north-western Canada (2).

In mid-1988, Dome's shareholders approved its acquisition by Amoco (3), in what would be the largest merger in Canadian history (4).

References: (1) *FT* 6/8/81; 14/1/82. (2) *Project North Newsletter* 10/82, 11-12/82, 3/4/83. (3) *FT* 14/6/88, *FT* 9/6/88. (4) *FT* 15/7/88; see also *FT* 25/6/87.

198 Dong-Trieu

Although primarily a uranium mining company, centred on the mining of five deposits in the Haute-Vienne region of the Massif Central in France, Dong-Trieu also has a minority interest in a JV at Tiebaghi in New Caledonia (Kanaky), mining chrome with Inco (55%) and the Banque de Paris et des Pays-Bas (22.5%) (1).

From rampant colonialism to scarcely disguised neocolonialism, Dong Trieu is one of France's oldest mining companies. It started digging coal in what became North Vietnam in 1916, and was "producing 800,000 tonnes of good coal in 1939" according to former company head François Leger who started with Dong-Trieu in 1947 (2). It also had interests in Cambodia and Laos (3) until forced to quit Indo-China in the '50s. With compensation for its loss of North Vietnam, the Dong-Trieu partners decided to turn to uranium mining: "We couldn't afford to go into oil, and the coal industry in France was nationalised," said Leger (2).

In 1964, recoverable reserves amounting to 7000 tonnes of uranium were located in the Haute-Vienne region, near the La Crouzille deposits of Cogéma. Nine years later the company was taken over by Empain-Schneider.

```
                        CFP / Total (F)
                              /
                            /
DONG-TRIEU / CMDT / CIE. MINIÈRE DONG-TRIEU (F)
```

By 1982, Dong-Trieu was operating an open pit mine at Bernardan (Cherbois) and an underground mine at Piégut. Its uranium oxide plant at Mailhac sur Benaize, which started operations in 1979, has a design capacity of 550 tonnes/year (4): 1982 production was 400 tonnes (1), moving to 420 tonnes/year (2).

Empain-Schneider, one of the cornerstones of France's nuclear industry – it used to own Creusot-Loire, 51% holder of Framatome (see Westinghouse) – got into trouble in the early '80s and was forced to sell both Creusot-Loire and Dong-Trieu (5). The uranium deposits were picked up by Total (CFP/Total), thus promoting it to France's second biggest domestic uranium producer (Cogéma being the first) (2). All of Dong-Trieu's uranium production is sold to France's EDF electric utility (2).

Dong-Trieu remains the country's second largest uranium mining project but its operations are wholly integrated into those of its newfound benefactor (2).

References: (1) *MIY* 1985. (2) *FT* 29/12/84. (3) Dave van Ooyen, *Uraniummijnbouw in Frankrijk*, draft doctoral thesis, University of Amsterdam 9/82. (4) *Uranium Redbook* 1982. (5) *FT* 16/2/83.

199 Dow Chemical Co (USA)

It is not a uranium producer, but it has developed technology for extracting minerals (primarily magnesium) from sea water, with obvious application for the production of uranium from sea water (1).

Dow Chemical is notorious as the producer of napalm which has killed and maimed countless children, women, and men in South-East Asia, and more recently Iraq and Kuwait.

In 1984, Merrell Dow Pharmaceuticals agreed to pay US$120M into a special fund to settle claims made by 70 US plaintiffs after Dow's "morning sickness" drug Bendectin (known as Debendox in Britain) caused congenital defects in babies. Dow admitted no liability in making the payment.

Two months before, Dow became a main contributor to another fund – for US$180M – set up to meet claims by victims of Agent Orange, the devastating defoliant used in Vietnam and elsewhere (2).

Among Dow's latest "energy" involvements has been a syngas (coal gasification) project in Louisiana, USA, from 1987 onwards (3).

By 1990, Dow Chemicals was the second biggest US Chemical group (4) and the world's largest producer of magnesium (5).

References: (1) *MJ* 1/8/80. (2) *FT* 17/7/84. (3) *MJ* 20/4/84. (4) *FT* 25/7/90. (5) *MAR* 1990.

200 Driefontein Consolidated Ltd

The only uranium producer in the huge GFSA group, Driefontein operates two mines, East and West Driefontein (West Driefontein having been incorporated in Driefontein Consolidated in 1981).

Although in 1977 West Driefontein was the world's largest and most profitable gold mine (1), its uranium production was small compared with that of some other South African uranium producers, and it has fluctuated in the years since. For example, in the quarter ending June 1978 nearly 90,000kg of U_3O_8 was produced at a recovery rate of 0.12% (2). 18 months later, 71,299kg of uranium oxide was recovered from ore grading at 0.236% (3). This dropped to around 50,000kg in the quarter ending December 1982, and the grade dropped to 0.160%. More recently, in the first quarter of 1984, only 44,617kg was recovered with a grading of 0.136% (4).

In 1980 the mine had about 34,000 tonnes of recoverable uranium reserves (5). Its contracts are not known (6).

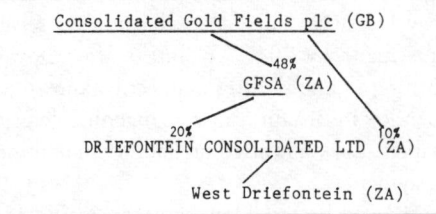

In late 1983, ten miners lost their lives at two rockbursts in mines in the GFSA group, including five who died at Driefontein, where also 57 were injured (7).

In 1990 uranium production at the mine ceased altogether (8). However, gold production was intended to last another fifty years, with the consolidation of East, West and North Drie mines into one massive underground pit (9).

In 1987 the company was owned 31% by GFSA, 11.3% by Anglo American Gold Investments and 6.8% by the Ste Interprofessionelle pour la Compensation des Valeurs (9).

References: (1) Lanning & Mueller. (2) *MJ* analysis of Rand and OFS quarterlies, 6/78. (3) Idem 3/80. (4) Idem 4/84. (5) Foreign Uranium Supply Update 1980 (NUS Corp). (6) Lynch & Neff. (7) *FT* 31/12/83, *FT* 10/2/84. (8) *MAR* 1990. (9) *MIY* 1990.

201 Dungannon Exploration Ltd (IRL)

It was part of the Sabina/E & B JV exploring for yellowcake in the Fintona area of Ireland in 1978 (1) – a venture later halted due to deterioration in uranium market (2).

References: (1) *MJ* 29/6/79. (2) *MAR* 1982.

202 E.I. Du Pont de Nemours and Co

Du Pont (the biggest US chemicals company) was established on explosives and still produces them, as well as the famed Remington rifles. Until 1947 Du Pont and the British ICI, along with the infamous IG Farben of Germany, operated a huge chemicals cartel, finally broken up by US anti-trust proceedings (1). The company's diversification in the '20s and beyond has become a model for other multinationals wishing to "control" their multifarious activities (1).

Du Pont's chairman in the early '70s, Irving

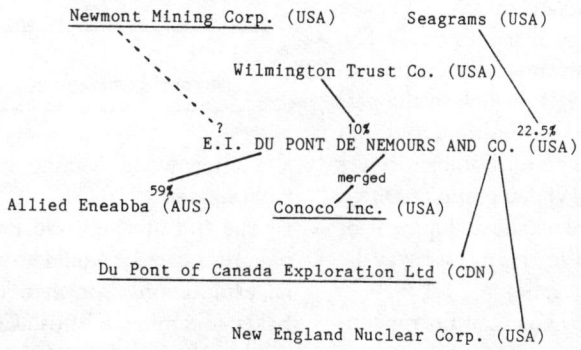

Newmont Mining Corp. (USA)   Seagrams (USA)

Wilmington Trust Co. (USA)

?   10%   22.5%
E.I. DU PONT DE NEMOURS AND CO. (USA)

merged

        59%
Allied Eneabba (AUS)   Conoco Inc. (USA)

Du Pont of Canada Exploration Ltd (CDN)

New England Nuclear Corp. (USA)

Shapiro, formed the Business Roundtable. According to Harry C Boyte in an article for *In These Times* (2), the Roundtable worked both overtly and covertly to sabotage major legislation on behalf of consumers, environmental organisations, workplace safety and full employment. It has used "... a variety of forums to propound the thesis of a "capital shortage" facing private industry". When, in 1977, it called for less government regulation of energy prices (specifically oil and natural gas), US Vice-President Mondale appeared before it to promise that the Carter administration would spur corporate investment.

When ERDA set up in 1975, Du Pont managed to get representation on no fewer than 5 of its committees (2).

In 1979, the company was the USA's number 92 defence contractor (3).

Its push into Europe started in the 1950s, and among its major corporate links are joint ownerships with Mitsui in Japan (1).

During 1981-82 it took control over Conoco in the "biggest take-over brawl Wall Street has ever witnessed" (4). In the process, Conoco's controlling interest in Hudson's Bay Oil and Gas passed to Du Pont which then exchanged it for a large chunk of Conoco from Dome Petroleum (4).

Seagrams, which now partly controls E.I. Du Pont, is the world's largest producer and marketer of distilled spirits and wines. Among its brand names are Crown Royal whiskey, Chivas Regal, Passport and Regal whiskies, 100 Pipers and John Jameson Irish whisky (1).

Du Pont's major contribution to the warmongering of the USA derives from its operation, under contract with the US Department of Energy, of the Savannah River nuclear fuel plant in South Carolina. The following report was made by the US radical group Natural Guard in summer 1982:

"The major facilities at SRP were constructed in the early 1950s. There are five nuclear reactors (three operating, one being renovated, one inoperative), two reprocessing plants for recovering plutonium and tritium, a fuel and target fabrication plant, a facility for processing and packaging tritium, a heavy water production plant, and waste storage facilities. Millions of gallons of radioactive waste are stored at SRP in 39 huge tanks, eight of which have cracked and leaked radioactive materials into the highly porous soil surrounding the plant. Cancer rates in nearby Barnwell are more than nine times the average for the rest of South Carolina, and high levels of radiation have been found in plants and animals living near the facility.

"Approximately 1000 rain-drenched people turned out on 30 May, 1982, to protest against the plant. The rally was organised by a coalition of southern anti-nuclear groups known as the Natural Guard, and it is an indication of the growing anti-nuclear movement in the southern US, where the plant is located. Speakers at the rally reflected a broad spectrum of social activists. A local farmer, McMillan, said "We thought it (the SRP) would be a boon, but all the money went north and all the radiation went south". Referring to South Carolina's nuclear power plants, McMillan said "They're going to give you energy if it kills you". The demonstration is the second recent action to focus attention on the SRP, which in the past has escaped widespread opposition, but has gone on churning out nuclear bombs for over 30 years. On March 4, 1982, a group of Dutch activists attempted to blockade a shipment of spent fuel rods destined for reprocessing at the SRP" (5).

Two years later, the Department of Energy suppressed a 1976 study carried out at SRP, which reported 11 cases of leukaemia between 1956 and 1974, five times the expected incidence among the workers studied. Du Pont had repeatedly denied that such a study even existed (6).

In late 1983, Du Pont announced that it was converting its power station at Maydown, near Derry, Northern Ireland, from oil to coal which would be imported from Scotland. The station powers a synthetic rubber plant which turns out Neoprene and Lycra products (7).

Du Pont also holds 50% of Allied Eneabba, the Western Australia producer of mineral sands (8).

Resources: *America: From Hitler to MX,* 1982, is a 90-minute film, distributed by Parallel Films (USA), which accuses Du Pont and other US companies of complicity in Nazi war production just before or during the Nazi offensive on Europe.

Contact: The Natural Guard, 18 Bluff Road, Comubia, SC 29201, USA.

References: (1) *WDMNE.* (2) [quoted in] Reece. (3) *NARMIC* 79. (4) *FT* 6/8/81. (5) *Natural Guard* (USA) quoted in *Disarmament Campaigns,* 6-7/82. (6) *Washington Post* 11/5/84. (7) *FT* 22/12/83; *Gua* 22/12/83. (8) *MJ* 4/5/84.

203 Du Pont of Canada Exploration Ltd

It held an exploration licence for the Vernon

```
E.I.Du Pont de Nemours and Co. (USA)
                    \
     DU PONT OF CANADA EXPLORATION LTD (CDN)
```

and Greenwood Mining Division of British Columbia in 1979 (1).

By the end of 1983, Du Pont of Canada Exploration stated it would leave the field of mineral exploration altogether. It would close the Baker gold mine in British Columbia which opened in 1981, due to exhaustion of reserves (2). The company had originally been set up when Du Pont itself was based in Montreal and acquired a substantial interest in Lacana which it sold in 1980 (3).

References: (1) *Energy File* 3/79. (2) *MJ* 2/12/83. (3) *E&MJ* 2/84.

204 Dynamic Mining Exploration Ltd (CDN)

This is part of an exploration programme in the Baker Lake area, Northwest Territories (NWT), together with Pan Ocean Oil (60%), Petrobec (17.1%), and Lochiel Exploration (8.6%), with a share of 14.1% (1).

The exploration programme was halted for some period by the Inuit of Baker Lake who maintained that their land claims must be settled before uranium mining commenced. Although supported by environmental groups, they failed in their court case in November 1979 (2). (For further details see Urangesellschaft). Drilling by early 1980 had indicated more than 3 million pounds of uranium oxide in the South Bissett Creek Zone (3).

References: (1) *MJ* 8/2/80. (2) *MJ* 30/11/79. (3) *MAR* 1981.

205 E&B Explorations Ltd

This is an exploration and development company, active as both operator and venturer in

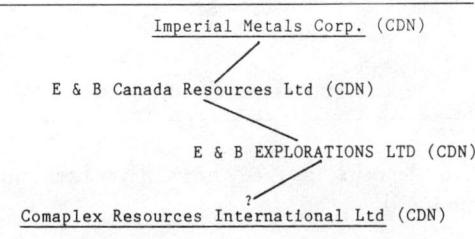

Imperial Metals Corp. (CDN)

E & B Canada Resources Ltd (CDN)

E & B EXPLORATIONS LTD (CDN)

?

Comaplex Resources International Ltd (CDN)

Canada and, to a lesser extent, in the USA and in Australia (1).

It was originally part of the Great Bear Lake, Northwest Territories, uranium consortium with Kelvin Energy in 1978-79 (2).

In 1978, it was reported to be exploring 300 square miles of Co Tyrone (Northern Ireland) as leading partner (50%) in a JV with, among others, Sabina (3).

It was part of the Norcen JV at Blizzard, Kelowna, British Columbia (BC) (4).

It was also reported, in 1979, as holding exploration licences in the Vernon, Nechako and Bennett Lake areas of BC (5).

In 1980, it was exploring in New Brunswick in a JV with Noranda and Lacana (6), and in Saskatchewan (7).

In 1980 E&B sold yellowcake to Kepco after entering into a long-term supply agreement with the company to provide some 2.2 million pounds of U_3O_8.

E&B's JV partners have been (says *Yellowcake Road*) "too numerous to list", but included Agip, Asamera, Canadian Occidental Petroleum, Comaplex, Cominco, Denison, Eldorado ... Noranda, Norcen, SMDC, Union Oil, Uranex, and Western Mines – all but the Kitchen's Inc! (8).

When uranium exploration in Canada reached a peak in the late '70s, E&B was among the top ten corporations in terms of expenditure (a total of US$90M was spent in 1978) (9).

E&B in 1980 was among the 10 most active companies exploring for uranium in Canada.

In 1981 it was exploring east of the Midwest Lake/McClean Lake belt in Saskatchewan, along the Collins Bay/Eagle Point mineralised zone (10).

In 1984 E&B Canada Resources Ltd was taken over by Imperial Metals Corp of Canada.

E&B Canada Resources operates through E&B Mines Ltd on its own account and through E&B Explorations Ltd as general manager for the six West German Geomex (originally Sedmic GmbH) drilling funds, which were taken over by Imperial Metals Corp in 1986: E&B holds an average of 36% in each of the Geomex drilling funds (11).

References: (1) *MIY* 1981. (2) *MJ* 29/6/79. (3) *FT* 7/6/78. (4) *FT* 15/5/79. (5) *Energy File* 3/79. (6) *Energy File* 4/80. (7) *Who's Who Sask.* (8) *Yellowcake Road.* (9) *Goldstick.* (10) *MAR* 1982. (11) *MIY* 1990.

206 Eastern Copper Mines NL (AUS)

This is primarily a copper miner (1) but is also into gold, silver and diamond exploration, cement production (2) and yellowcake.

Copper comes mainly from the colourfully named Mt Oxide and Wee MacGregor mines, but the company also had a copper project in Papua New Guinea (3).

Together with Aberfoyle (50%), Eastern Copper held a uranium prospect in the Aboriginal Arnhemland area of the Northern Territory. Classified as a "grass roots" prospect, and so far undelineated, it is hardly likely to move to production for a long time (4).

The company intends to bring on stream its copper mine at Laloki in Papua New Guinea in 1991 (5).

References: (1) *MIY* 1985. (2) *Reg Aus Min* 1981. (3) *MAR* 1984. (4) *Reg Aus Min* 1983/84. (5) *MAR* 1990.

207 Echo Bay Mines Ltd

This company is Canada's second largest gold producer, thanks to its rich Lupin deposit in the

```
    IU International Corp. (CDN)
              86.6%.
       ECHO BAY MINES LTD (CDN)
```

Northwest Territories (NWT) – the world's most northerly gold mine.

It has no known uranium production, but EBM achieves at least a footnote in nuclear history since, about 20 years ago, it took over Eldorado Nuclear's Port Radium (radium then uranium) mine in the NWT, developing it as a copper-silver operation. The mine closed in the early '80s, after 19 years production (1).

Echo Bay acquired the Copper Range Company in Nevada in 1985, bringing with it the world's largest heap leach gold operation (with 25% partners Homestake, and Case, Pomeroy and Co Inc) (2). It also acquired Tenneco gold mines in Nevada (3). However, its operations at the McCoy-Cove deposit in Nevada left it with egg on its face in early 1990, as it made claims of discovering large deposits of gold which, it later had to admit, simply weren't there (4).

More important, in late 1989 Echo Bay was found guilty of violating the NWT's Mine Safety Act (but fined only C$4000) after miner Ken Hunter – a native Canadian – was killed when he fell down a shaft at the Lupin mine (5). Echo Bay is now the fourth largest gold producer in north America (6).

References: (1) E&MJ 10/84; see also FT 30/8/84. (2) E&MJ 1/90. (3) MIY 1990. (4) E&MJ 1/90; FT 9/2/90; MJ 12/5/89; FT 1/3/90. (5) Native Press, Vol. 19, No. 14, 14/7/89; Native Press, Vol. 19 No. 25, 15/12/89. (6) FT 1/3/90.

208 Ecominas

This is Colombia's state mining agency which – apart from exploring for a variety of minerals (notably gold, copper, emeralds and bauxite) – had also been after uranium in Boyaca, Meta and Cundinamarca departments, areas with indigenous communities (1).

```
       Republic of Colombia (CO)

ECOMINAS / EMPRESA COLOMBIANA DE MINAS (CO)
```

No deposits in these areas have been announced.

References: (1) MAR 1984.

209 Edlow International Ltd

```
EDLOW INTERNATIONAL LTD (USA)

    American Uranium (USA)
```

It is one of the biggest – if not the biggest – transporters of nuclear material in the world (1). It has been involved with Swuco in enabling South Africa to acquire enriched uranium (see Swuco). Based in Washington DC, it is closely associated with Nuexco.

In 1987, Edlow started prospecting for titanium off the Mozambique coast. Surprisingly for a strategic mineral (especially in the hands of a company that has been complicit in evading anti-apartheid sanctions) the Mozambique government has given Edlow full rights to any minerals produced (2).

References: (1) Age 11/12/79. (2) E&MJ 7/88.

210 EG&G Geometrics (USA)

In 1983 this company supplied US$1.3M-worth of airborne radiometric equipment to enable the People's Republic of China to search for uranium to use in its power plants – and, no doubt, its weapons.

References: (1) E&MJ 1/83.

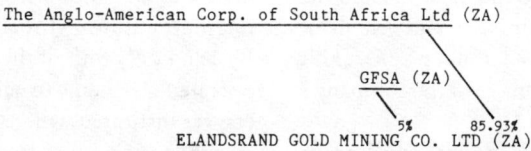

211 Elandsrand Gold Mining Co Ltd

This is a relatively recent creature (founded in 1974) of the Anglo-American empire, with expansion work – especially in the Ventersdorp Contact Reef (where uranium has been found but not exploited) – currently underway (1).

As such, Elandsrand does not produce uranium; it tributes land from Western Deep Levels, and in return allows Western Deep to mine in the rich Carbon Leader Reef.

As of February 1989, Anglo-American Corp officially held 18.66% of Elandsrand's share capital, while its subsidiary Anglo-American Gold Investment Co Ltd held 19.65%. However, according to the annual reports of both these companies, the share holding was considerably higher, with AAC and its subsidiaries holding 48% and AAGIC possessing another 23.3% (2).

References: (1) *MIY* 1985. (2) *FT Min* '90.

212 Eldorado Resources Ltd

In 1929, Gilbert Labine, a successful gold prospector, took his Eldorado Gold Mining company looking for pitchblende in the Great Bear Lake area of Canada's Northwest Territories (NWT) (1). The Port Radium pitchblende mine which Labine established began producing radium in 1932, along with small amounts of uranium for pottery and glazing (2) and copper and cobalt (3).

Between 1934 and 1939 the Port Hope refinery turned out nearly US$8M-worth of radium and silver ore concentrates (1). In 1940 the Great Bear Lake mine was shut down (3) due to lack of manpower and equipment, not to mention

flooding (2). Two years later, however, the US military badly needed an alternative to its sole uranium supply in the Belgian Congo (2) and, as the 1968 Eldorado Nuclear annual report put it:

"An urgent need for uranium in quantity arose with the inception in 1942 of the Manhattan Project, the joint British-United States-Canadian undertaking which eventually brought forth the atomic bomb. Canada's role was to supply the uranium raw material and the Government requested the re-opening of the Port Radium mine on an emergency basis, but gave no hint of the reason" (4).

In fact Eldorado Nuclear got the necessary contract at the explicit direction of General Groves, head of the US atomic bomb project (700 tons of refined uranium oxide) (2): it can have little excuse for not realising the purpose to which it would be put. An estimated one-sixth of the uranium used to blast the peoples of Nagasaki and Hiroshima to smithereens came from Eldorado Nuclear in those years (2).

While the 1968 annual report for the company also stated that "the amount of uranium provided by Eldorado for military purposes during World War II and up to 1954 is still classified information," it admitted that "the company's

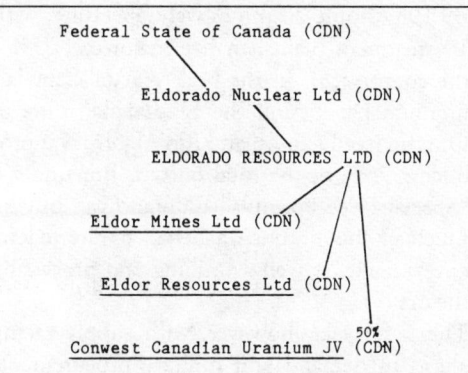

revenue from 1944 to the end of 1954 from the sale of uranium and from sales and rentals of radium was about US$82 million" (4). This represents nine times the price of the company's nationalisation in 1944 (1).

With the British government embarked on its own nuclear weapons programme and after Eldorado's yellowcake itself – a laboratory was set up in Montreal in 1943 to take the ore (23) – the Canadian government gave Eldorado the equipment and investment it needed and, in 1944, nationalised it, making it a Crown corporation (2). Called initially the Eldorado Mining and Refining Company, it soon became Eldorado Nuclear (5).

Together with the Geological Survey of Canada, Eldorado launched a uranium exploration programme throughout Canada the same year, while all private uranium prospecting in the Yukon, the NWT, and several other provinces was banned (3).

A year after establishing the Atomic Energy Control Board in 1946, the Canadian government lifted the private prospecting ban, offering a guaranteed minimum U_3O_8 price and development and milling allowances (up until 1962). The uranium rush which began in 1948 resulted in more than 10,000 discoveries of radioactive occurrences by 1956 (6).

The most important of these was located in the Lake Athabasca region of northern Saskatchewan by Eldorado in 1951. Production started at Beaverlodge (Uranium City) in 1953; it was eventually expanded to 2000 tons/day of ore (3).

By 1959, some 23 mines were operating, with 19 treatment plants, in five major districts of the country (5). In the peak year of Canada's uranium production the Elliot Lake mine in Ontario (see Denison and Rio Algom) was producing 75% of the total output. But under a "Special Price Formula" Eldorado was buying much of this output on behalf of the federal government, as well as milling and processing the ore (3).

The same year, however, with supply racing ahead of demand and military procurements dramatically diminished (6), the Canadian uranium industry virtually came to a standstill. By 1964 only four of the original 28 mines still operated. A scheme to bail out low-cost producers was instituted, whereby Eldorado and Ucan stockpiled excess ore (3). Two years later, however, the US imposed an embargo on the enrichment of imported uranium for its own uses – a move which particularly hit Canadian exports (5).

The central role of Eldorado in the early years of the Canadian uranium industry is difficult to overestimate. According to one observer:

"Eldorado became the pivot of the whole system in its dual capacity as a producer and sole sales agent for the negotiation of contracts and for the procurement and resale to foreign countries of all radioactive ores, concentrates and precipitates ... produced in Canada" (7).

It was not until 1973, with an increased demand for uranium, a partial lifting of the US embargo, support by the Canadian government for domestic producers through its Uranium Resources Appraisal Group (3) and, perhaps above all, the operation of the uranium cartel in which Eldorado and other uranium companies were intimately involved (8), that the Canadian uranium industry was "reborn" (3).

From 1974 until 1979 Eldorado Nuclear spent more than US$20M on uranium exploration, mainly through JVs with at least 27 other corporations whose expenditures came to about US$40M (8). As of the end of 1979, these JVs were with:

"Amok Limited, Aquarius Mines Ltd, Canadian Nickel Company Ltd, Canadian Occidental Petroleum, Central Electricity Generating Board, Comaplex Resources International Ltd, Consolidated Durham Mine and Resources Ltd, Conwest Exploration Company Ltd, Denison Mines Ltd, E & B Explorations Ltd, Electrowatt Ltd, Empresa Nacional del Uranio, SA, Famok Limited, Goldak Exploration Technology Ltd, Highwest Development Ltd, Imperial Oil Ltd, Noranda Mines Ltd, Norex Resources Ltd, Power Reactor & Nuclear Fuel Development, Saskatchewan Mining Development Corp, Séru Nucléaire Limitée, Shell

Canada Resources Ltd, Société de développement de la Baie James, SOQUEM, Suncan Inc, Texasgulf Canada Ltd, Uranerz Exploration and Mining Ltd" (9).

Geographically, Eldorado's major JVs in the late '70s were in the James Bay area of Quebec with Séru Nucléaire and the Sté de Développement de la Baie James in Churchill, Man with Warren Explorations, Imperial Oil and Manitoba Mineral Resources, as well as at Beaverlodge, Fond du Lac and the Lake Athabasca region of Saskatchewan (3).

The Port Radium mine closed down in 1954. Even so, its tailings ponds were continuing to pose problems as late as 1981, when gale force winds created tides that flooded four dams containing Thorium-232 and its breakdown products (10). The USA (United Steelworkers of America) have claimed that only one miner from the Port Radium mine is still living, out of those who worked five or more years at the mine (11).

Beaverlodge

The Beaverlodge mine began an expansion in 1975 designed to double production to about 2 million pounds per year uranium oxide after the '60s cutback (3). In fact, in 1976, while it increased production (to just under 2 million pounds) its earnings fell. Its ore reserves increased, however, by 383,000 tons (standing at just over 3 million tons at the beginning of 1977). The company at the time held contracts for the delivery of 13.12 million pounds of U_3O_8 up to 1986. Germany, Sweden and Japan were potential customers (12).

In 1978 the company chalked up a record US$170M profit thanks to sales of uranium concentrate from Beaverlodge; about 1.25 million pounds of U_3O_8 were produced and, despite a declining ore grade, new underground reserves were added (13, 14). However, by the first quarter of 1979 profits had slumped (15) and by the year's end the company as a whole sustained a major loss of US$1.3M (16, 17). By that time, the president of the company, Nicholas Ediger, was welcoming a decision by the Canadian government to return Eldorado to the private sector (18).

In 1980 Beaverlodge's production was marginally higher than in 1979, but the overall grade was slipping lower and lower. With depressed markets, low prices and rising costs, the company announced it would close Beaverlodge in June the following year (19).

The mine was duly shut down officially on schedule, after producing 333kg of uranium concentrates in 1982 (20, 21).

The announcement of closure came as a complete surprise to the workers in the mine, local businesses and above all the residents of Uranium City, the town built around the mine and virtually dependent upon it. While Eldorado had provided relatively generously for its residents, it was secretive and unhelpful in suggesting how it would bulwark the community against its withdrawal. A Uranium City using power from Eldorado's US$30M expansion to its Charlotte River dam could prevent the area becoming a complete ghost town, but it would inevitably be poorer and smaller (22).

The second most worrying aspects, of the affair were the medical and environmental health problems Eldorado has left. A 1980 study of both miners and non-miners in Uranium City (based on sputum samples) suggested that many residents would eventually develop cancer (23).

Liquid run-offs from the tailings dumps have contaminated Lake Athabasca, making the water unfit to drink (24).

A study carried out in 1980 on the Lorado mill, which operated at Beaverlodge between 1957 and 1960, revealed a discharge of 80 tons uranium, 220 curies of Ra-226, and many tonnes of sulphuric acid, chloride and other discharges (26). The Lorado discharge site later became a golf course – which became so highly radioactive that the Atomic Energy Board of Canada (AECB) forced it to move (22).

Tailings have also been used in the construction of homes and schools with "children and citizens [being] exposed to excess levels of radon gas" (27).

Eldorado's plan to deal with the hazards – which include millions of cubic feet of waste rock dumped beside the mine's headframe – is not encouraging. An employee in 1980 was quoted as saying there was "no way it can be cleaned as such" (22).

The loss of Beaverlodge has been more than made up in Eldorado's other ventures. In 1978 it exercised its option to increase its equity in Conwest Exploration to 50%, and put up half the funds for the Conwest Canadian Uranium exploration JV. At the same time it announced it would concentrate its exploration efforts in northern Saskatchewan, at Geikie East and Geikie West (28). (A month earlier, it was sampling for uranium underground at Consolidated Durham's New Brunswick antimony mine, with assays of the dumps showing up to 5.5lbs of uranium per tonne) (29).

Also in November 1978, its wholly owned subsidiary Eldor Resources acquired a one-sixth interest in the Key Lake Mining Corp, set up to develop the Key Lake deposits which now rank as the world's biggest single uranium producer (19). Eldor acquired the capital to buy the JV interests by borrowing and selling 2 million pounds of uranium concentrate from the government stockpile then managed by Ucan (13, 30).

In May 1982, Eldorado announced that it was purchasing Gulf Minerals' shares in return for delivery of uranium concentrates (31). This followed a loss of revenue in 1981, due to poor sales, and the purchase of Ucan's assets the same year. These assets consisted primarily of about 13 million pounds of mine concentrates plus 1.5 million pounds of concentrates outstanding on loan to Eldor Resources at a "fair market value" of US$300M (30). Partly using these assets the company bought out another Canadian company, Uranerz Exploration and Mining Ltd, thus giving it total ownership of the Rabbit Lake mine and mill in northern Saskatchewan (32), whose management Eldorado consigned to Eldor Resources. The gain of Ucan's stock-

pile also helped Eldorado finance the Blind River refinery (see below).

By mid-1984 Eldorado's fortunes were looking fairly bright. Over a ten-year period it boosted its assets eightfold to over C$915M, and during 1983 it produced record amounts of uranium concentrates totalling 1244 tons (more than double its 1982 production and reflecting the first full year of operation at Rabbit Lake). It also received 72 tons as its share from Key Lake. An "aggressive marketing programme" enabled the company to sell nearly all its production, and revenue was up 73% to C$154M (33). There were promises to develop the "privatisation" strategy announced in 1979 and the company's president, NM Ediger, described the uranium market in early 1984 as "maturing, growing and highly commercial" (34). Just before he made this statement, Eldorado launched the "most popular rights issue of the day" – a US$100M float carrying the "full faith and credit" of the Canadian government (35).

Collins Bay

Its most important new project has been the Collins Bay "B" mine, now being constructed in the vicinity of Wollaston Lake, northern Saskatchewan (36, 37, 38). Discussions with the Saskatchewan government continued in 1984 over royalties liable on this project, which will include modification to the Rabbit Lake mill at a cost of US$100M to take the additional ore. Collins Bay "B" is required to maintain a throughput of 1500 tons/day ore, though ore stockpiled at Rabbit Lake will provide sufficient mill feed until 1985 (39).

The Collins Bay "B" deposit contains nearly 4 million pounds of uranium ore on land which is occupied and claimed by Native Canadians. The Federation of Saskatchewan Indians has made it clear that the development must not proceed "... until a full-scale inquiry has been undertaken", and the Wollaston Lake Local Advisory Council (LAC) is very much opposed to it. The Northern Municipal Council, representing nine northern communities, has supported a resolution stating that the scheme – as it was broached in Gulf Canada's original proposal – was "inadequate and unacceptable" (40). The major concern is that hunting and fishing grounds will be lost or contaminated and Native Canadians discriminated against in employment policies. Since the 1981 hearings, concern has been expressed that the tailings dam will leak in the same manner as those at Key Lake (which spilled 12 times between September 1983 and March 1984) (41).

In July 1984 the Lac La Hache (Native Canadian) Band Administration also released a report on the Collins Bay "B" project which related serious problems they had experienced from the untreated wastes released by the Rabbit Lake mine into the Hidden Bay of Wollaston Lake. They say that the Collins Bay development is "even more dangerous" since the ore is covered by water, and the whole lake could become radioactively contaminated. A petition was presented which seeks recognition of the ignored aboriginal rights of the five thousand Chipewyan and Cree people of Wollaston Post and nearby communities, the immediate revoking of the licence for Collins Bay "B", and the clean up of radioactive wastes from Rabbit Lake (see Eldor Resources).

During 1983, Eldorado essentially completed its two major construction projects in Ontario: the new uranium hexafluoride conversion plant at Port Hope, and a new refinery at Blind River (33). These projects were finished in the face of massive resistance by many communities and organisations.

The company is the only uranium refiner in Canada: it converts U_3O_8 into uranium metal, uranium dioxide (for use in Candu reactors) and uranium hexafluoride (42). Its pre-eminent position has been underlined by a policy introduced in 1977 which, as well as protecting domestic uranium for home use, stipulates that, where possible, exported uranium must be upgraded to uranium hexafluoride (UF_6) just before leaving Canada (5).

In 1976, Eldorado produced 3300 tonnes of UF_6 (42). By 1980 it was producing 5500 ton-

nes of UF_6 and 1000 tonnes of UO_2(43). The following year it trundled out 6300 tonnes of uranium from Port Hope, of which 4700 tonnes went into UF_6 and 1800 tonnes into fuel for Candu reactors (44).

In 1979 the company laid out its case for another refinery besides the new one at Port Hope: with new mines coming on stream in Saskatchewan and about 80% of the new uranium production surplus to domestic requirements, another refinery would be required in late 1984, with a capacity of 9000 tonnes of uranium feed (45). After looking at two possible sites in Saskatchewan (46), one at Corman Park, 5km from the Warman township and 23km away from Saskatchewan's capital Saskatoon, was finally chosen. The company claimed the project would offer many benefits to local people; would not be hazardous; would not affect local water systems; and that "even in improbably worst-case situations, the radiation exposure to the general public is below one per cent of permissible levels" (46).

The people of Warman, including the Mennonites, rose up against the proposal (47). They initially accused the company of deception – when the land was bought for the project no mention of a refinery was made; and after claiming that it wouldn't go to any area where it wasn't wanted, in 1979 an Eldorado official announced that "even if a majority of the residents of the town opposed the refinery ... that does not mean the project would not go ahead" (48). From January 8th to 23rd, 1980, citizens voiced their opposition to Warman (49), and later that year the site was rejected by an environment assessment panel (50). On September 29th, the company announced it was dropping its options at Warman and would look elsewhere (51).

Meanwhile, Eldorado was in hot water with the new Liberal state government in Ontario as it prepared to build another refinery near Port Hope (52). The government ordered the project closed down and re-building at Blind River. Eldorado responded on June 4th, 1980, with a revised plan for all its refining operations (yellowcake to uranium trioxide) to be concentrated at Blind River, while its UF$_6$ operations would expand at Port Hope (53). The Blind River refinery promised to be "huge" (54, 55).

Eighteen bands of Native Canadians, part of the Robinson-Huron and Manitoulin Island Treaty Areas, strongly opposed the Blind River refinery (56). In a strong statement after a meeting in March, they registered their protest:

"*Now, therefore be it resolved,* that the *anishinabek* People of the Robinson-Huron Area direct the President of the Union of Ontario Indians to immediately seek the support of the National Indian Brotherhood and that they jointly use all necessary means to immediately put a stop to any further approval of the Refinery *and, be it further resolved,* that the text of this Resolution be forwarded immediately upon its adoption to the Minister of Indian Affairs, the Minister of Energy, the Minister of Mines, the Leader of the Opposition, the Leader of the NDP, and Members of the Board of AECB.

"The Chiefs of the Bands of the Robinson-Huron and Manitoulin Area note that since the proposed Refinery site is at the mouth of the Mississauga River, they are concerned for the sensitive nature of the River Delta and surrounding Marsh Land. This River represents on[e] of the last Sturgeon spawning Rivers on the Lake Huron Area.

"This Refinery will be recycling the solid radioactive wastes back to the mine tailing piles at Elliot Lake Mines and this pollution will eventually get into the Serpent River and add to the already over-polluted River system flowing by the Serpent River Indian Band.

"The Uranium, the raw material for the Refinery, will be brought into the Blind River Area from not only Elliot Lake, but Quebec and Saskatchewan as well. This yellowcake ore will be transported by rail and road transport which could result in significant accidental spills into the local environment. Trucking of recycled wastes back to Elliot Lake could also result in accidental spills.

"Responsibility for control of these wastes on the mine tailing piles will be thrown back and forth between Eldorado and Elliot Lake Mines.

"The Serpent Lake Indian Band has had experience with an Acid Plant which was abandoned on their Reserve and which represents a terrible pollution situation on their land. This situation has not been resolved, because two companies, Provincial and Federal Governments will not accept responsibility and point out that jurisdiction for this [problem] is not well defined.

"We see the situation of jurisdiction disputes being a problem in the future of Eldorado Limited at Blind River.

"The Canadian Government has failed to protect the Rights of not only the Indian People but all Canadians living in this area.

"The ANISHINABEK are committed to carry on the fight to protect the Land, The People and the future of this Area" (57).

But their protest was in vain, as was that of Algoma Manitoulin Nuclear Awareness (AMNA) who'd also mobilised against Eldorado in the area. In May, the Blind River refinery was given the go-ahead by the AECB without an environmental impact study (58). The first yellowcake went into Blind River in August 1983, from Elliot Lake. The refinery's licence includes a permit to "lose" up to 200kg of uranium dust per year (22g per hour) to the environment. And it is expected that the plant will operate at less than half its capacity of 18,000 metric tons a year (59).

In 1982, Eldorado's attempts to open the new refinery at Port Hope foundered when the AECB stipulated that the company must find a long-term effective method of reducing its heavy fluoride emissions before a further licence could be granted: its emission in summer 1981 had exceeded the maximum allowable level some 25 times. Another report, by a federal government task force, confirmed that maximum concentrates dumped in Port Hope harbour were 30 times greater than that of ore mined at Elliot Lake and that levels of radiation in invertebrates tested from the water were alarming enough to warrant an extensive dredging campaign. The levels corroborated an Ontario Ministry of Environment study in 1978 which found invertebrates within the harbour with "concentrations ranging from 128 times to 2800 times the norm for gross alpha radioactivity, 79 to 1300 times the norm for gross beta radioactivity and 86 to 460 times for Radium 226" (60). In response, Eldorado claimed that it was not subject to provincial environmental laws, being a federally owned corporation. An official for Environment Canada said: "I think it's inexcusable. I don't know what goes on in their minds. They just seem to go from problem to problem" (43).

Indeed, the Port Hope saga has been almost one problem after another.

- In May 1978, the Ontario Federal Environment Assessment Panel on the proposed Eldorado Refinery at Port Granby (one of Eldorado's earlier options) condemned the company's plans to dispose of uranium tailings (based on the Brinco subsidiary Brinex's doomed proposals for Kitts-Michelin in Labrador), concluding that the covering for the site – bentonite clay – was untried and would probably break down (Eldorado itself admitted the probability) and should be rejected because of its unknown reliability, costs, engineering difficulties and the need to retrieve the stored material (61). The Panel also noted that cattle had died and streams been polluted from other waste dumps in the area, and that fill from buildings in Port Hope had resulted in excessive radon gas build-up.

- Although Eldorado was ordered to shut down the Port Granby dump – with its 450,000 tonnes of low-level radioactive wastes situated only 25 miles from the shores of Lake Ontario – the order was lifted when Eldorado promised to improve the site and establish a proper treatment facility. The efficacy of that facility was undermined in 1980 when the treatment system at Port Granby failed and water with radium-226 levels 46 times the drinking water standard was released into Lake Ontario (61).

Eldorado seemed unbeatable, however, especially when the 1979 Nuclear Monitoring Committee (NMC) set up by the Federal Environmental Assessment Panel as an "independent citizens" watchdog to keep tabs on Eldorado" (62) was folded and a new body, the Environmental Advisory Committee (EAC) set up in its stead. The local burghers tried to get this body itself closed down in 1983 (63).

- 119 litres of acid leaked into the harbour in January 1982 (64).

- Fire broke out in December 1983 at the UF_6 plant in Port Hope. Although the US$25M blaze was extinguished in an hour, some citizens didn't wait for official warnings and left town of their own accord. A report published earlier that year on the storage and transfer of ammonia and hydrogen fluoride at Port Hope identified a previous ammonia spill of 9000kg and "poor housekeeping" practices at a plant which is located in the heart of an urban area without any protective buffer zone (65).

- Another radioactive spill occurred in Port Hope harbour on November 4th, 1984, when 680 litres of processed solution containing uranium and fluorides got mixed with cooling water and was discharged into what must now, surely, be the most radioactive harbour in the world (66).

Although the Key Lake JVs landed a US$10M uranium contract with the US Virginia Electric, Eldorado was excluded from the "hexing" work at the insistence of the purchaser in 1984. After 100 lay-offs in recent times, the exemption will cost more jobs at Port Hope. Commented President Ediger, "The worry to us is the salami effect. If the government keeps granting exemptions, you may eventually slice up the whole sausage" (66).

Uranium cartel

Eldorado played an important role in the formation of the uranium cartel. Although the first meeting of this band of thieves took place in Paris under the auspices of the CEA, and at the stimulus of RTZ, it was the Canadian government which set the ball rolling in 1981 by telling its uranium companies (above all, Eldorado) to make viable conditions for their future survival. In the event, Canada, which had just set up Ucan, got a third of the market carved out by the cartel partners (2).

In a strange twist of politics, Eldorado and Ucan were taken to court by the federal Canadian government in 1981 on price-fixing charges connected with the cartel. In 1983, though two Supreme Court judges ruled that acquitting the companies could give crown corporations "a *carte blanche* to engage in illegal activities and ... encourage other [corporate] citizens to do likewise," (65) Eldorado was ruled to be immune from the charges (67).

The company's corporate interlocks have included: the Hudson's Bay Oil and Gas Co, Brascan, Canadian Pacific, Cominco, Reynolds Metals and Phillips (1).

In July 1981 an Eldorado spokesman blew the cover on the secret Namibia connection with Japan, when – after public outcry – he admitted that the company had been refining uranium from the UN-protected territory (68).

In 1980, the company also "hexed" a controversial consignment of uranium oxide from Rio Algom, designated for Kepco (69). Later information confirmed a long-held view that the South Korean régime had been obtaining information from Canada with which to construct atomic weapons (66).

In 1956, an Eldorado employee, William Young, died of cancer after four years working at Port Hope. In an article published in 1982, his widow, Joanne Young, outlined her conviction that her husband had died from radiation exposure. She pointed out that when William began work in 1950, readings taken at some parts of the plant showed radiation levels as high as 1600 picocuries/litre in the air – more than 30 times the acceptable level in residential accommodation. Her husband had been involved in dismantling an old plant, without any protection. He died when he was only 34 years old. Concludes Joanne Young:

"I quot[e] the nineteenth century Nova Scotia journalist and reformer, Joseph Howe, as he defended himself in court against a libel suit in 1835: 'Surely a more negligent, imbecile and reprehensible body has never mismanaged public affairs'. " (62).

Opposition mounted to the development of the Rabbit Lake mine, and the Collins "B" mine, as well as plans for Collins "A" and "D", during 1986-1988, culminating in an international conference held in northern Saskatchewan in 1988, attended by delegates from scores of groups within Canada and across the world (70). In June 1986, Eldor declared the new Collins Bay mine and mill circuits at Rabbit Lake to be in commercial operation, with the Rabbit Lake mill returning to 2000 tons/year capacity (71). Production objectives for 1987 were achieved from Collins Bay "B"-zone, and Eldorado/Eldor submitted an Environmental Impact Statement for approval to develop Collins A and Collins D zones (72). At the same time, the Eagle Point deposits were being evaluated with a view to using the Rabbit Lake mill, and conventional mining methods (73). Delineation drilling was carried out on Eagle South (73) and environmental approval granted to Eldorado in early 1988 (72).

The Eagle North deposits (which Eldorado holds in JV with Noranda and SDMC), together with Eagle South, contain more than 130 million pounds U_3O_8, at an average grade of 1-2% (73). Along with Collins "A" and Collins "D", they are not only larger than Collins "B", but also deeper under Wollaston Lake (74).

Indeed, in a 1986 drilling "campaign", resources at Eagle South were delineated as containing more than 50,000 tons uranium – four times those of Collins Bay "B" zone (71).

After making a loss of C$64 million in 1986, Eldorado returned to modest profit (C$12 million) in 1987. A couple of months later, the uranium world was electrified by the news that, as the latest in a series of Federal government moves to dispose of public sector assets (75), Eldorado would be sold off, after amalgamating with SDMC (76). Initially, the new corporation would be owned 61.5% by the Saskatchewan provincial government, and 38.5% by Ottawa's Canada Development Investment Corporation (75). Privatisation would proceed in steps, last around seven years, with the whole of the new company in private hands by 1995 (75). However, various limitations on ownership would be imposed, to prevent the company ending up under overseas control: Canadian investors would have a maximum of 25% of voting shares, with non-Canadians 5%; non-Canadians would also be restricted to 20% of votes cast at shareholder's meetings (75). Due to the debt burden of both companies, the new company would initially undertake a debt issue, amounting to around 40% of net assets (76).

Further Reading: Miles Goldstick, *Wollaston: People Resisting Genocide*, (Black Rose Books) 1987.

References: (1) *Yellowcake Road.* (2) Moss. (3) *EPRI Report* 1980. (4) Annual Report 1980. (5) *Raw Materials Report*, Stockholm, Vol. 2, No. 4, 1984. (6) *Uranium Redbook* 1979. (7) WDG Hunter, *"The development of the Canadian uranium industry: an experiment in public enterprise"*, in *The Canadian Journal of Economics and Political Science*, Vol. 28, No. 3 p335, 8/62. (8) Goldstick. (9) Annual Report 1979. (10) *Kiitg*, No. 15, 8/81. (11) *Energy File* 4/79. (12) *MJ* 1/7/77. (13) *MJ* 18/5/79. (14) *FT* 19/4/79. (15) *FT* 13/5/79. (16) *Wall StJ* 1/11/79. (17) *FT* 16/11/79. (18) *FT* 2/10/79. (19) *MJ* 18/12/81. (20) *MJ* 28/5/82. (21) *MJ* 8/7/83. (22) Mark Stobbe, "Eldorado abandons Uranium City", in *NeWest Review*, Vol. 7, No. 5, 1/82. (23) Diana Ralph, "Uranium City: Omen for the North", in *One Sky Report*, Saskatoon, 5/82 (One Sky). (24) "Uranium boom in Saskatchewan", in *AkwN*, Summer/81. (25) *Kiitg* 4/81. (26) Goldstick, adapting RG Ruggles, Statement of evidence to BC RCUM. (27) *Toronto Globe And Mail* 5/76, quoted in *Yellowcake Road*. (28) *FT* 18/12/78. (29) *MJ* 17/11/78. (30) *MIY* 1982. (31) *MJ* 7/5/82. (32) *MJ* 10/12/82. (33) *MJ* 13/7/84. (34) *FT* 11/5/84. (35) *FT* 9/2/84. (36) *MJ* 16/9/83. (37) *MJ* 6/1/84. (38) *MJ* 22/6/84. (39) *MAR* 1984. (40) *Transcript of Hearings on the Gulf Minerals Collins Bay B-Zone development*, La Ronge, Saskatchewan, July 29th 1981, Vol. 4-5, quoted in Goldstick.

(41) Diana Leis, Submission to People And The Land conference, May 16th 1984 (organised by Interchurch Uranium Committee, Saskatoon, Saskatchewan) Saskatoon, 1984 (Lac La Hache Band Administration, roneoed). (42) Merlin. (43) *Nuclear Free Press, Energy Booklet*, Peterborough, Ontario [?1980/81] (OPIRG). (44) *MJ* 9/7/83. (45) *Environmental review of a uranium refinery in Corman Park R.M. Saskatchewan*, Saskatoon, 7/76 (Eldorado Nuclear). (46) *Saskatchewan site evaluation study for Eldorado Nuclear Limited, Uranium Refinery: July 1976*, Saskatoon, 1976 (Eldorado Nuclear). (47) *Energy File* 9/79. (48) *The Warman Uranium Refinery: What does it mean?*, Saskatoon, 1980 (Saskatoon Citizens For A Non Nuclear Society [SCNNS]). (49) SCNNS press release, 2/80. (50) *Nuclear Newsletter*, Saskatoon, (Saskatoon Environment Society) 18/8/80. (51) Ibid 13/10/80. (52) *MJ* 10/8/79. (53) *Nuclear Newsletter (cf* note 50) 15/12/80. (54) Ibid 23/1/81. (55) *MJ* 29/8/80. (56) *BBA* Winter/81. (57) *Yukon Indian News* 2/4/81. (58) *Nuclear News*, Saskatoon (Saskatoon Environment Society) 21/5/80. (59) *Nuclear Free Press*, Peterborogh (OPIRG) Fall/83. (60) *BBA*, Spring/82. (61) *Nuclear Free Press*, Peterborough (OPIRG) Fall/84. (62) *Ibid* Summer/82. (63) *Ibid* Summer/83. (64) *WISE Bulletin*, Amsterdam (WISE) Vol. 4, No. 5, 6/82. (65) *Nuclear Free Press*, Peterborough (OPIRG) Winter/83. (66) *Ibid* Winter/84. (67) *MJ* 30/12/83. (68) *BBA* Winter/82. (69) *FT* 28/5/80. (70) *WISE NC*, No. 295, Amsterdam, July 1988. (71) *MAR* 1987. (72) *MAR* 1987. (73) *E&MJ* 4/88. (74) Goldstick, *Wollaston: People resisting Genocide* (Black Rose Books) 1987. (75) *FT* 24/2/88. (76) *MJ* 26/2/88.

213 Eldor Mines Ltd

In 1982, Eldorado Nuclear bought out all the properties of Gulf Minerals Canada, the Gulf Oil subsidiary, together with those of Uranerz

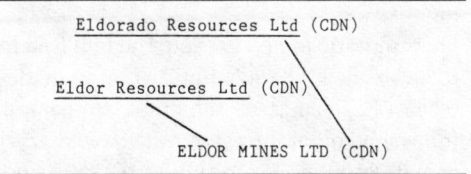

Canada. It then renamed Gulf Minerals Eldorado Resources (Eldorado Nuclear technically remaining the name of the parent company) (1). Gulf and Uranerz's Rabbit Lake uranium mines were put under the control of Eldor Mines, while control over Eldorado's Key Lake investments was assumed by Eldor Resources (1).

Until Gulf Oil, the defunct Gulf Minerals Canada's erstwhile parent, was itself taken over by Socal/Chevron in 1984, the Rabbit Lake operations (51% management control by Gulf Minerals) provided them with the majority of their yellowcake supplies, making Gulf Oil the western world's 6th largest producer of uranium oxide (2548 tons) by 1982 (2). In 1983 and 1984 production levelled at about 3.5 million pounds U_3O_8 (3). The previous year, the mine passed its "half life point" and Gulf Minerals announced it would offset its "pending exhaustion" with ore from the new Collins Bay "B" deposit (4).

Rabbit Lake

The Rabbit Lake orebody along the Wollaston Lake Fold Belt was originally discovered by the company in 1968. An early, unconfirmed, report put the grading as high as 8% uranium metal with some portions of the orebody running to no less than 65% uranium metal (5, 6). Similar reports reckoned that the new mine could boost Gulf's total reserves to more than 100,000 tons U_3O_8 (6). But despite some later discoveries of uranium in the area – Gulf in the late '70s held around 240,000 acres under claim (7) – the initial optimism was soon tempered. Reserves at Rabbit Lake by 1978 were put at a fifth of this (42million pounds U_3O_8) (7) and, soon after the mine started production in 1975 at 60% of its design capacity of 4.5million pounds U_3O_8 per year, problems began appearing.

Due to erratic mineralisation, the mill feed had to be blended (8) and a limit set on its grade of 0.6% (7). Thanks to sub-arctic temperatures (down to minus fifty for some weeks in the winter) Gulf developed (initiated?) the "labour fly-in" system, keeping the operation going around the clock by drawing the workforce from communities (including indigenous ones) within a 640km radius (8) and keeping half the employees on site at any one time (130 out of a total of 265 in 1978) (9). This arrangement clearly took a physical and emotional toll on the employees, and there were "difficulties faced by some of the wives in running their homes during their husbands' absences," as the *Mining Journal* put it at the time (9). Moreover, the company violated rules of the Canadian Labor Congress in gaining a 77-hour working week for its non-union workforce. But Gulf reported a "remarkable success in stabilising the workforce – turnover is down to an average 12%" (9), and the high returns clearly satisfied both labour and management.

The Rabbit Lake orebody was mined out in August 1984, several years earlier than expected. Gulf Minerals then started preparing the open pit as a "disposal facility" for tailings from the new Collins "B" deposit, 9km northeast. This would take both feed from stockpiled Rabbit Lake ore until 1986 (the intended startup date of Collins "B") and use a modified Rabbit Lake mill (10).

The Rabbit Lake tailings had already been dumped in Wollaston Lake since the start-up of operations in 1975 (11), and the local community, indigenous and non-native, had protested about the fact for ten years (12) .

In 1976, residents in the Fond du Lac and Black Lake communities, downstream from Rabbit Lake – which include many native people – cited unacceptably high radiation levels in the area. A Federal Radiation Protection Bureau study discovered radon and radon "daughters" ranging from 0.001 to 0.016 "working levels", emissions found "unacceptably high" by nuclear expert Dr Fred Knelman of Montreal's Concordia University (13). When no further checks were made on dangers to the communities between July 1976 and March 1980, local teachers went on strike (13). In mid-1985, a blockade organised by the Collins Bay Action Group (CBAG) and the Lac La

Hache native band council took place at the entrance to the Rabbit Lake and Collins Bay mines as part of a broadly supported Northern Survival Gathering (see Eldor Resources).

It is not clear how far Gulf Minerals' contracts have been fulfilled from Rabbit Lake, and the extent to which they will be met from Collins "B": by 1978 the company was committed to supplying nearly 20 million pounds U_3O_8 to European utilities (notably in Spain, Belgium, Switzerland, Finland and West Germany), the USA and Canada itself (7).

In any event, the Rabbit Lake production became crucial to its parent company soon after the mine was opened, when Gulf Oil began fulfilling contracts to three utilities after the United Nuclear Corporation (UNC) halted deliveries, thanks to Gulf's being cited as a member of the notorious uranium cartel (14).

Other information about Gulf Oil's participation in the cartel is to be found under General Atomic in this *File*. What is important here, is to briefly outline the relationship between Gulf Canada and its parent when, very shortly after the cartel was set up in 1972, senior executives of Gulf Minerals Canada learned of the incipient conspiracy to control the western world's uranium market. Although the French CEA and the South African, French and Canadian governments (as well as RTZ) had initially agreed to exclude both Australian companies and Gulf Oil from the uranium club, the latter forced their way in (15).

A Gulf internal memorandum later described the original cartel members as "... playing a game in which they hope to a) keep Gulf as the "new boy on the block" away from the poker table and b) block the Australians from starting production" (16). It is difficult to assess the degree to which Gulf Minerals Canada itself wished to participate in the street-corner gang. When a US executive of Gulf Minerals was asked by the congressional sub-committee, "How does all this square with your personal moral code?" he agreed that it posed a "problem of conscience" (17).

In any event, the Canadian subsidiary company felt it had to cover its involvement. Canadian federal government officials, faced with Gulf Minerals' hesitation, invited Gulf's own lawyers to draft a letter on Canadian government notepaper ordering the company to join the cartel (18).

Three years later, after Westinghouse and TVA sued Gulf Oil in the US District Court of Illinois for non-delivery of uranium, the parent company was able to argue that its Canadian subsidiaries acted "under compulsion": if they participated in the cartel, it was "in compliance with Canadian demands" (19). However, in 1976 the Canadian government passed a "gag law" preventing the removal from Canada of documents relating to the cartel (and Canadian government complicity in it). Gulf Oil was therefore in a situation of agreeing with US courts that it had participated in a cartel, claiming its Canadian subsidiaries had been "ordered" to participate, demanding the release of a fake memorandum drafted by the company's own lawyers on official Canadian government notepaper, and being strenuously opposed by "the very people who had arranged for Gulf to draft the fake documents in the first place – the Canadian Liberal Cabinet". It was indeed a "moment of supreme irony" (18).

The Canadian Supreme Court, not surprisingly, backed the federal Canadian government; the US courts found Gulf Oil guilty of a misdemeanour and fined the company US$40,000 (17). In 1982 WPPS (Washington Public Power System) also sued Gulf Minerals Canada for failing to deliver uranium (20). By 1984, the company had paid out US$70M in settlement of suits brought by US utilities (21). Then, in late 1983, the Canadian government itself dropped charges against Gulf Canada (along with Denison, Rio Algom and Uranerz) for its participation in the cartel, brought under the Combines Investigation Act (10). The federal government decided it could not proceed against privately owned corporations when state owned companies similarly charged (notably Eldorado) were not being prosecuted since they were "above the law" (22).

While embroiled with the courts, Gulf Canada

was facing opposition from farmers in the New-boyne area of eastern Ontario after a 1978 aerial survey revealed uranium mineralisation there. Without any reference to the local people, Gulf moved in, signed contracts with a number of farmers and took drilling rights on one of twelve prospects (23). Soon after, other members of the community, led by Kathryn and Marcellus White, mobilised against further exploitation and forced Gulf to withdraw. The company claimed they were abandoning the project because "there was no ore to mine". The explanation fooled no-one. "The action taken by the citizens of Leeds County serves as an example to people in other parts of the province in showing that a united front against a multinational giant like Gulf can yield results quickly" (24).

In early 1988, Eldorado was given permission to develop an underground mine at Rabbit Lake (with a minimum of 70,300 tonnes uranium) (25). Later the same year, Eldor mines received full environmental approval from the Saskatchewan government to develop both the Collins "A" and "D" deposits, and additional large reserves at Eagle Point. These were then the only environmental approvals to be granted for uranium production into the twenty first century (26). Earlier the same year, however, the Federal and Saskatchewan governments announced a merger between Eldorado Nuclear and SMDC, with their subsequent complete privatisation. Initially, the Canada Development Investment Corporation would own 38.5% and the Saskatchewan Government 61.5% of shares in the new corporation (reflecting the relative viability of the two companies). Foreign ownership in Canadian uranium companies was also eased, to allow for overseas holdings of up to 49% (27).

It was envisaged that 30% of the new company would be in private hands by 1990, 60 per cent two years later and the full amount by 1996 (28)

This means that the new company, Cameco, now has responsibility for Rabbit Lake, the Collins Bay deposits, Eagle Point, Cigar Lake, and various gold exploration projects (29) as well as base metal and industrial metal exploration (30). Although the world's biggest producer of uranium, Cameco's privatisation was delayed in 1989/90 due to weak market conditions (31).

The Rabbit Lake mill was also closed for six months during the latter part of 1989 because of the slide in the uranium price (32). Planned expansion of capacity in the area to 12 million pounds/year was held up (33), although upgrading of the mill and initial development of the Eagle Star South deposit were initiated in that period (31). The following year, the company agreed to sell a one-third stake in Rabbit Lake – with its mill, the three Collins Bay orebodies, Eagle South and surrounding claims – to Uranerz, which already owned 33.33% interest in Eagle Point North deposit (31, 34).

Cameco's estimate for future output from the Rabbit Lake mill is 5450 tonnes of U_3O_8 per year (35).

In November 1989, around two million litres of radioactive and heavy metal (radium, arsenic and nickel) -bearing fluids burst into Collins creek, which itself flows into Wollaston Lake. The seepage occurred from a faulty valve on a 10km long pipeline carrying runoff and seepage from the Collins Bay mine: loss of pressure in the pipe was recorded by monitoring equipment but not noticed until fourteen hours (sic) after the rupture (36). Although the company was said to have acted quickly to bulldoze soil around the spillway (37), public outrage was intense.

In particular, concern was expressed that drinking water and the whitefish spawning grounds in the creek and Wollaston Lake would be gravely affected by the contamination (36).

Wollaston Mayor, Emil Hansen, and Chief Edward Benonie, representing the 800 Indian community members of Wollaston Lake, called for an independent inquiry into the mining industry in northern Saskatchewan and the spill in particular. "The whole economy of the region depends on that lake" he declared. "We know the environment has changed since that mine came. Time will tell how much." Benonie also asked who would buy whitefish now that the calamity was common knowledge. Calls for a federal inquiry were echoed by Saskatchewan MP for the Prince Albert-Churchill River region, Ray Funk, who called the incident a "total breakdown of the nuclear regulatory system."

However, as anti-uranium activists, such as the Survival Office Saskatchewan, pointed out, the accident only exemplifies a situation of creeping pollution which alredy had enormous impact on the ecology. The mill releases over seven million litres of waste water per day which, though passing through settling ponds, still carry heavy metals and radioactivity to Wollaston Lake (36).

Cameco appeared in court in December 1989, to face charges of negligence arising from the spill – with a theoretical maximum penalty of $C1 million and jail sentences of three years maximum for officials (38).

In a report on the incident, the Saskatchewan Environment Minister agreed that Cameco was negligent in not maintaining a flow-alarm system, and not ensuring monitoring of the pipeline by visual and other means: however, he also said that water samples showed that the spill "clearly...had no impact on the water quality" in adjacent water courses (39).

In April, the Hatchet Lake Band of Wollaston Lake – represented by Chief Ed Benonie – was given Intervenor Status in proceedings taken against Cameco for the November 1989 disaster. Meanwhile, some 33 anti-uranium organisations in the north organised a protest meeting to support demands by the Band that a full public inquiry should be held into uranium mining in the region (40). Cameco was fined C$10,000 under the Atomic Energy Control Act of 1946 – the maximum penalty applicable to this proceeding – after pleading guilty to two charges of negligence. The Atomic Energy Control Board also told Ed Benonie that it could not propose a national inquiry into uranium mining, as it went beyond its mandate. However, Saskatchewan and Environment and Public Safety has decided to bring Cameco to court under the provincial Environmental Management and Protection Act, charging it with the release of contaminants leading to "the reasonable possibility" of water pollution; on these grounds Cameco could be fined up to one million Canadian dollars (41).

In October 1990, the Atomic Energy Control Board met with Indian representatives in the Wollaston Lake area, but a few days later renewed the Rabbit Lake operating licence for another two years.

Ed Benonie renewed his calls for a Royal Commission of Inquiry "with real power to examine the uranium industry and its impact on the people, economy, and the environment of the north." The appeal has been backed by a large number of Indian organisations and others, including the Canadian Labor Congress, the Canadian Public Health Association and Greenpeace (41).

Contact: AIM Survival Group, PO Box 8536, Saskatoon, S7K 6K6, Canada.

Collins Bay Action Group, Lac La Hache Band, Wollaston Lake, SOJ 3CO, Canada.

Green Party, Saskatoon Chapter, PO Box 9053, Saskatoon, S7K 7E7.

Chief Ed Benonie, Wollaston Lake, Saskatchewan, Canada SOJ 3CO.

Survival Office, Saskatchewan (SOS), Box 9395 Saskatoon, Saskatchewan, Canada S7K 7E9.

References: (1) MIY 1985. (2) BOM. (3) E&MJ 3/85. (4) MJ 30/4/82. (5) Canadian Mining Journal, Toronto 11/75. (6) Canadian Mining Journal, Toronto 4/76. (7) EPRI Report. (8) B Merlin, Canada's uranium production and potential, London, 1977 (Uranium Institute). (9) MJ 10/11/78. (10) MAR 1985. (11) Overthrow, New York, Summer/85. (12) The Northerner, Saskatoon, 24/4/85. (13) Briarpatch, (precise origin unknown) Canada, 4/80. (14) FT 18/8/78. (15) Pringle & Spigelman. (16) International uranium supply and demand: Hearings before the Sub-committee on Oversight and Investigations of the Committee on Interstate and Foreign Commerce, House of Representatives, Vol. 1, Washington, 1977 (US GPO). (17) Moss. (18) BBA Spring/81. (19) Nuclear Fuel, Washington, 3/3/80. (20) MJ 15/1/84. (21) FT 20/3/84. (22) MJ 30/12/83 & 13/1/84. (23) MJ 8/2/80. (24) BBA Spring/80. (25) MJ 22/1/88. (26) E&MJ 9/88. (27) MJ 26/2/88 (28) FT 24/2/88. (29) MJ 7/4/89. (30) Canadian

Mining Journal 8/90. (31) *MJ* 13/4/90 (32) *E&MJ* 3/90. (33) *MM* 5/89. (34) *Leader-Post*, Regina 13/4/90. (35) Nukem Market Report 4/89. (36) *WISE N.C.* 322 1/12/89. (37) *MJ* 8/1/90. (38) *MJ* 5/1/90. (39) *WISE N.C.* 326/7 9/2/90. (40) *WISE N.C.* 334 22/6/90, Briarpatch 5/90. (41) *MJ* 26/10/90.

214 Eldor Resources Ltd

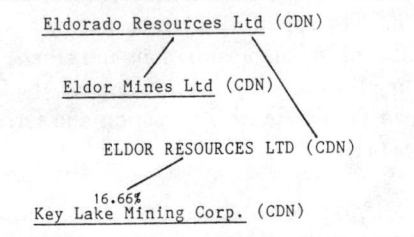

Formerly Gulf Minerals Canada, it was purchased and renamed by Eldorado Nuclear in 1982. It holds one-sixth of the Key Lake Mining Corp which operates the Key Lake mine in northern Saskatchewan (1).

In mid-1983 Eldor received approval from the Saskatchewan provincial government to develop the Collins Bay "B" deposit which it had inherited from Gulf Minerals. This was to maintain Eldor's milling of uranium in the area beyond 1985. The company was intending to triple milling capacity by the end of 1983, to which end C$114.2M was spent in 1982 (2).

Concerted opposition to the opening of Collins Bay "B" and the continued operation of Rabbit Lake has come from many groups, not least the Lac La Hache Band Administration (Native Canadian) based at Wollaston Lake.

In an Open Letter to band councils, community councils, anti-nuclear groups and other concerned groups and individuals, dated July 13th, 1984, the Band made several demands. Because of the importance of these mines to the nuclear industry and the myth that Saskatchewan native people are in favour of uranium mining, most of the Open Letter is reproduced here:

"We are concerned about the uranium mine development that is taking place on the west side of Wollaston Lake, only 20 miles across the lake from our community of 700 people ... . From 1975 to 1977 untreated wastes [from Rabbit Lake] were released into Hidden Bay of Wollaston Lake. These wastes are now "treated", however the settling ponds they pass through have no leakage-proof liners, so radioactive contaminants are still leaking into Wollaston Lake through the ground water channels. A study done by Environment Canada in 1978 determined that "the water quality of Horseshoe Lake has deteriorated considerably since 1972" and "elevated radionucleide levels were detected in the Rabbit Creek system especially in the Sedimentation Lake sediments". Tests by the Environmental Protection Services on several occasions found ammonia levels high enough to kill trout in the tailings effluent which goes into Effluent Creek and on to Hidden Bay.

Eldorado Resources is now developing the Collins Bay "B" zone orebody, 6 miles north of the Rabbit Lake Mine. This mine operation is even more dangerous because ore is under the lake. They have already built a dyke between two islands, drained the water from this area, and have begun to dig the open pit. They will soon begin to dig out the ore. The dyke is made with thin sheet steel pipes, rocks and earth and is only 4 feet above the water level. If water from the lake comes into contact with the ore in the pit, the radiation will spread to the whole lake. This dyke will be removed when the mine closes in 6 years. The radioactive tailings from this mine will be placed in the old Rabbit Lake pit. No leakage-proof liners will be installed. Radium-226 and other radioactive "daughters" will leach into the ground water channels which flow into Wollaston Lake for possibly thousands of years.

... The uranium mines that were operated by Eldorado at Uranium City, Saskatchewan, have permanently contaminated the land and water. Three lakes are now dead: nothing will grow in them. Huge piles of waste rocks and mine buildings have been left to release radiation into the air, land and water. Most of the people moved out when the mines closed in 1982. We do not want the same thing to happen here.

We have always used this land for hunting, trapping, fishing, and berry-picking. We still obtain most food from hunting and fishing. About 50 people still trap and about 50 commercial fish. The exploration and mining is destroying these livelihoods. At least 10 men have their traplines in the area where mining and exploration is being done. As Councillor Martin Josie has stated: "Now because of the white people like the Department of Northern Saskatchewan, prospectors, and mining companies, the animals are becoming scarce. Even the moose are hard to kill now and because of the mines the animals are not fit to eat." Only two men received any compensation for the losses of their traplines. Contamination of the lake may destroy commercial fishing.

Elder Helen Besskaytare has also expressed concern about the mines. "Nowadays if you kill some ducks they're skinny, not like they used to be. I hear now you can get cancer from eating the waterfowl. People used to live to be in their '60s or '70s. People lived off fish and berries. But now people are dying from cancer and we know it's from the mines. Now we can't even eat bears because we're afraid what they may have eaten from the mines. What I'm really concerned about is the kids in the future. If the water is contaminated and not fit to drink and the fish are not fit to eat, what are the children going to live on?"

Only 5 people have permanent jobs at Rabbit Lake and 5 are now employed at Collins Bay. People from Wollaston Lake are often refused jobs because they lack training, but the mining companies won't provide any training. Employment for a few people is not adequate compensation for the destruction of the environment. As then Chief Joseph Besskaystare stated: "At a meeting here in 1977 with the government and mining company officials, I told them NO to the mine because

of what it might do to the lake. In about 35 years you people will be finished mining. All the workers will go but we will still be here. After the water is contaminated, what are going to live on?"

Many workers from uranium mines have died from cancer. Over 30 miners from Uranium City and over 400 miners from the Elliot Lake, Ontario, mines have died from cancer. We do not want this to happen to Wollaston Lake people.

The people of Wollaston Lake opposed the opening of the Rabbit Lake Mine at meetings held here with the government and mining companies in 1972 and 1977. We unanimously opposed the opening of Collins Bay "B" zone at Hearings held here and La Ronge in 1981. The Government of Saskatchewan refuses to listen to us and allows uranium exploration and mining to continue.

At a community meeting here on June 13, 1984, people again expressed opposition. Band Councillor Martin Josie stated: "We, the Chief and Band Councillors, will not agree to have the Collins Bay Mine opened. This mine concerns everybody because Wollaston Lake flows everywhere, North, South, East and West. We will be very happy for any people that protest against Collins Bay "B" zone development. If people protest in other places, not only Wollaston Lake, it will be good."

The Open Letter announced a petition – to stop Collins "B" and all other uranium developments in the area – which would be presented to Neil Hardy, Minister of the Environment in Saskatchewan.

The Open Letter was signed by Hector Kkailther, Chief, and several Band Councillors. The petition made reference to "the five thousand Chipewyan and Cree people of Wollaston Post and the seven nearby communities" affected by uranium mining, whose aboriginal land rights were being violated.

In 1988, Eldorado Nuclear and the Saskatchewan Mining Development Corporation (SMDC) merged to form Cameco, now the world's biggest single holder of uranium reserves and producer of uranium. Further information about development of the Rabbit Lake and Collins Bay deposits can be found under Eldor Mines. Technically, neither Eldor Resources nor Eldor Mines exist any longer.

Contact: Lac La Hache Band Administration, Wollaston Lake, Saskatchewan S0J 300, Canada.

References: (1) *MAR* 1984. (2) *MJ* 3/6/83.

215 Electrowatt Ltd

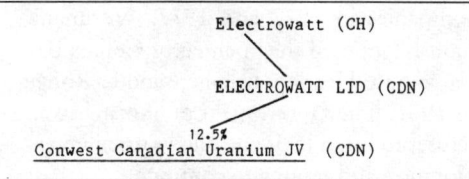

```
          Electrowatt (CH)

          ELECTROWATT LTD (CDN)

        12.5%
Conwest Canadian Uranium JV  (CDN)
```

This is part of the Conwest Canadian Uranium

Exploration JV, organised in 1975 to explore for and develop uranium throughout Canada, and operating at Key Lake, Saskatchewan, together with CEGB, Eldorado, and Enusa (1).

References: (1) *MIY* 1981.

216 Elf

Elf is France's second largest oil exploitation company (after Total/CFP) with exploration and production in Guatemala, Cuba, China, Mali, Cameroon, Ireland, Congo, Gabon, the North Sea, and – of course – France (1).

"Here is a state within a state, with its diplomatic corps, its prefects, spies, and occult finances!" – thus Daniel Schneider in a recent exposé of the activities of the company. According to him, Elf-Aquitaine "mocks the government of France and dictates its own policies over the heads of France's ministers" (2).

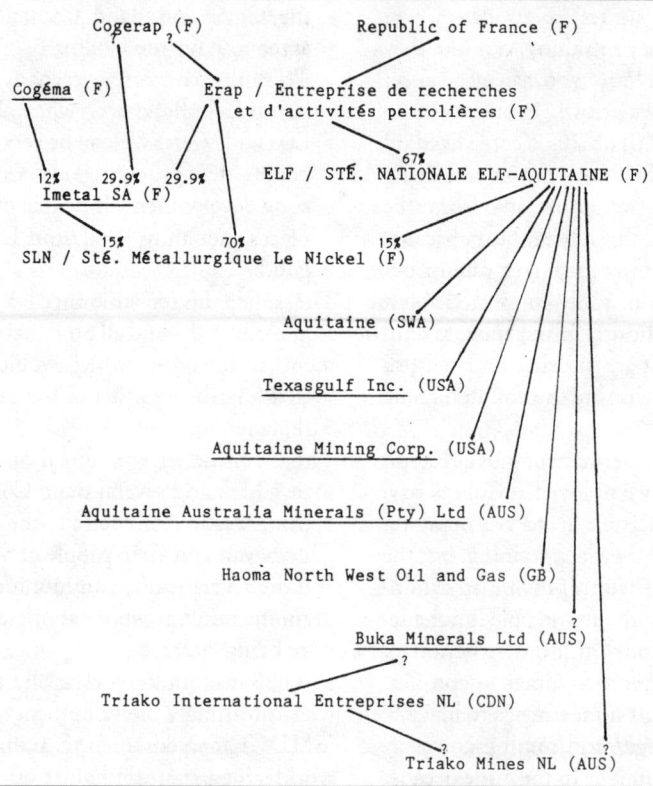

```
           Cogerap (F)                    Republic of France (F)
                   ?   ?

     Cogéma (F)        Erap / Entreprise de recherches
                         et d'activités petrolières (F)

                                      67%
      12%   29.9%  29.9%    ELF / STÉ. NATIONALE ELF-AQUITAINE (F)
      Imetal SA (F)

          15%        70%        15%
   SLN / Sté. Métallurgique Le Nickel (F)


              Aquitaine (SWA)


         Texasgulf Inc. (USA)


      Aquitaine Mining Corp. (USA)


  Aquitaine Australia Minerals (Pty) Ltd (AUS)


     Haoma North West Oil and Gas (GB)

                                         ?
              Buka Minerals Ltd (AUS)
                                    ?
   Triako International Entreprises NL (CDN)

                              ?          ?
                 Triako Mines NL (AUS)
```

It owns 15% of SLN/Société Le Nickel (along with Imetal) which benefits from French colonial occupation of New Caledonia (Kanaka). 1981 was the year of Elf's merger with Texasgulf, an internationally active Canadian company.

In early 1981, Elf announced that, together with Total, it would be spending about one and a half billion pounds on North Sea oil development (3).

In August 1981 Elf and Petrobras (the Brazilian national oil company) started an invasion of a government reserve occupied by the Satéré-Mawé tribe in Amazonia. Blame for the invasion – buck-passing – went in turn from FUNAI (the national Indian "protection" agency) to the government, then to Elf and back again. FUNAI ended declaring it could do nothing to prevent the encroachment and that, if necessary, the Satéré-Mawé would have to be moved. In response, the Satéré-Mawé have decided to ban all oil companies from their region, to prohibit anyone from working with Elf-Aquitaine, and – if Elf continues to hand out beer on their territory – to destroy the alcohol. "The Satéré-Mawé are at war", stated their representative in Manaus. According to Diffusion Inti, several support groups for the tribe have started to function in Manaus, Brasilia, São Paulo, and Rio de Janeiro.

In late 1981, Elf acquired a permit to explore for lead, zinc, copper, silver, antimony, and gold in Bretagne (Brittany) (4).

In March 1982, the company took control of Haoma North West Oil and Gas (in Britain) which is involved in energy exploration in Yorkshire and South Humberside, along with RTZ among other British companies (5).

Elf has been active in exploring for uranium through its affiliate Aquitaine Mining in the USA (6), and its subsidiaries Aquitaine (SWA) in then South African-occupied Namibia (7), and Aquitaine Australia Minerals in Australia.

In 1983, Elf took over many of the chemical interests of PUK and assumed full control of the chemical ventures jointly owned by Elf and Total with the French government, assuming responsibility for PUK's 3 billion french franc debt (8).

"The environmental impact of mining in New Caledonia is massive – 13 million tonnes of waste ore produced per annum and 22 hectares of land has to be stripped for every million tonnes of ore. The result is scarred hillsides, slag-choked rivers, and dust-covered bush. As nickel constitutes 95% of New Caledonia's exports, and the country contains one-third of all known western ore reserves, there is considerable political and strategic significance attached to smelting policies ... Many of the mining company's directors are members of the right-wing RPCR opposition party which opposes land reform and complete independence ... They fear that an independent Melanesian government with Australian backing would take over their nickel mining interests as well as possibly buying into SLN" (9).

In 1983, following huge losses by SLN, the company was restructured, with ERAP taking 70% and Imetal and Aquitaine reducing their holdings to 15% each.

As the "centrepiece" chosen by the Mitterrand government for a complete restructuring of the French heavy chemical industry, Elf has – like Pechiney – been negotiating for cheaper electricity, at the expense of the ordinary French consumer (10).

Elf is currently producing oil in northern Guatemala (11) where a dispute with its JV partner Basic Resources International SA over the quality, extent, and political feasibility of plundering the country's resources erupted in the British financial press in early 1984. Elf suggested that there were great physical and "political" difficulties associated with working in Guatemala – not least various attempts to blow up their pipeline, undertaken by liberation fighters (12). The president of Basic Resources (based in the Bahamas and possibly a company owned by the British capitalist James Goldsmith) denied this (13).

Meanwhile, opposition to Elf's continued encroachment on the Satéré Mawé reserve had been increasing internationally: Survival International France wrote to Mitterrand pointing out that a compensation payment of 5 million cruzeiros (about US$2500 at the time – enough to buy a boat which the tribe requires) was derogatory, and that actual damage to their land had been estimated by an official Brazilian government enquiry at ten times that amount (14).

Finally, in August 1984, the tribe – and the Munduruku who had also been affected by Elf's invasions – got US$150,000 between them. Ridiculously small compensation, it would seem, for blasted holes in their land and the death of four people who suffocated from fumes after handling sticks of dynamite left behind by the company in their fruitless search for oil (15).

Of late, Elf has moved into "genetic engineering" by taking shares in the large Strasbourg based company Transgene, founded in 1980. It also has its own "biotechnology" research programme – no doubt to spawn lots of little elves in our back gardens... (16).

References: (1) *MIY* 1981. (2) *Le Monde Dimanche* 20/6-5/9/82. (3) *FT* 4/2/81. (4) *MJ* 18/12/81. (5) *FT* 24/3/82. (6) *MJ* 10/8/79. (7) *MM* 7/77. (8) *FT* 15/3/83. (9) *CAFCINZ Watchdog* 3/83. (10) *FT* 15/11/83. (11) *MAR* 1984. (12) *FT* 10/11/83. (13) *FT* 19/3/84. (14) *SI News,* London, No. 1/83. (15) Gua 17/8/84. (16) *FT* 25/5/84.

217 El Paso Electric (USA)

It was involved with Arizona Public Service Co, and Public Service Electric and Gas Co, in uranium exploration in the late '70s (1). No further information.

References: (1) Sullivan & Riedel.

218 Embarcadero Corp (CDN)

It was reported exploring for uranium in Saskatchewan in the 1970's; no further information (1).

References: (1) *Who's Who Sask.*

219 Endeavour Resources

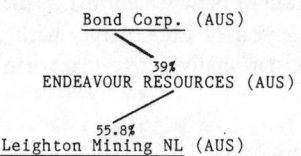

Controlled by the aggressive, possessive Alan Bond, through his Bond Corporation (1), Endeavour Resources is an oil, minerals, and tin mining and quarrying company with interests in Egypt, Papua New Guinea, the Philippines, and Indonesia, as well as the USA and many parts of Australia (2).

As long ago as 1977 it had applied for a licence to explore over 528 square kilometres of the Murray Basin, Victoria, for "Colorado type" uranium (3).

In 1979, through Endeavour, the Bond Corporation headed a consortium wanting to buy into the Beverley uranium prospects in South Australia; other partners were Leighton Mining, Basin Oil, and Reef Oil (4).

More recently, Bond tried to gain control of Northern Mining which held a 5% interest in the Ashburton JV (5), led by CRA, RTZ's Australian subsidiary, and violating Aboriginal interests at Lake Argyle, Western Australia (WA), and Glen Hill (see CRA).

Endeavour now owns Northern Mining (6), but the company's 5% stake in the Argyle diamond venture – notorious encroacher on Aboriginal sacred sites (see CRA) – was sold in 1983 to the Bond Corp, itself the owner of 39% of Endeavour. Endeavour was thus able to pay off its debts and get on with the job of plundering the earth (7).

In 1984 the company proceeded with a tin-gold alluvial mining project in the East Pilbara region, WA (8). Together with Utah it has set up a JV to explore for copper in Indonesian-controlled northern Sulawesi; called TEI (Tropic Endeavour Indonesia) it has been granted an initial 5-year contract (9).

The company's major operation is the Bluebird gold mine, 16km south of Meekatharra in WA, but due to the reduction of ore reserves, Endeavour has been exploring widely for additional deposits in the Murchison region (10).

As of 1990, Bond Corporation – despite its travails – continued to hold 38.36% of Endeavour, through its wholly-owned subsidiary Alec Fong Lim Trading Company Pty Ltd, with Sagona Pty holding 25.62% (11).

References: (1) *FT* 5/11/81. (2) *Reg Aus Min* 1976/77. (3) *MJ* 2/12/77.(4) *FT* 15/9/79. (5) *FT* 21/10/81. (6) *MJ* 12/11/82. (7) *MJ* 1/7/83; 12/8/83. (8) *MJ* 13/4/84. (9) *E&MJ* 6/84. (10) *FT Min* 1990. (11) *Reg Aus Min* 1990/91.

220 Energy Capital Ltd

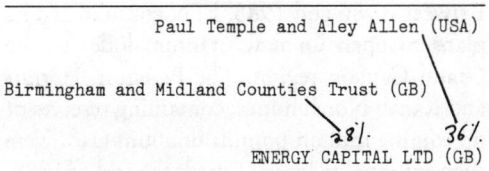

Paul Temple and Aley Allen (USA)

Birmingham and Midland Counties Trust (GB)

28/. 36/.

ENERGY CAPITAL LTD (GB)

Formerly Hamilbourne (a British brick-making company), it was partially taken over and "revived" by two US businessmen in 1980.

It acquired Allan Capital's 7.6% royalty interest in the Bison Basin, Wyoming, USA uranium prospect, and still apparently sits there – with Western Fuel and Ogle – awaiting improved markets. Bison Basin may contain 10 million pounds of uranium according to "independent geologists" (1).

References: (1) *FT* 14/7/80 (article and special supplement).

221 Energy Fuels [Ltd] (Nuclear) Inc (USA)

Formed in 1976, it contracted to supply two Swiss utilities with uranium from 1977 onwards, initially purchasing uranium ore at 15,000t/month (intending to double it by late 1978) and then providing the yellowcake from its own mines to its Blanding mill in Utah (1); the two utilities were Kernkraft Gösgen and Nordöstschweizerische Kraftwerke.

It was given permission in 1979, by the Utah state government, to start mining uranium on five adjoining properties: Hillside, Gizmo, Maybe, Vallejo and Bears Ears in San Juan County (2).

In 1981, the newspaper of the National Indian Youth Council reported that the first of these mines was complete, ore stockpiled, and Energy Fuels was now seeking a right of way across several miles of Native American reservation land. "Several vocal tribal members stopped ... an agreement being signed. Now the energy companies offer free lunches to the Indians ... [they have] offered the Tribe US$10,000 per year for five years if they approve the through traffic. While the tribal membership was reported to be "largely apathetic" to the company's encroachment, regional Native American and environmental organisations were working with Lawyers' Guild attorneys "to fight this situation" (3).

In mid-1983, despite objections from environmentalists, the US Interior Department authorised the company to drill a deep uranium exploration shaft on the Arizona strip north of the Grand Canyon.

According to *Mine Talk*, the journal of the Southwest Research and Information Center, EFN acquired this site from Western Nuclear Inc and, since it did not intend to process ore on the spot, was required to submit only a "mine plan" to the Bureau of Land Management. Transportation of ore to the Blanding mill in Utah nonetheless involved crossing a corner of the Kaibab-Paiute Reservation. Although agreement for access was obtained in 1980 from the tribe, the Thunder River league, a grassroots native American organisation based in Fredonia, called for a halt in operations until a thorough environmental impact statement could be prepared (4).

In 1985 EFN and Everest Minerals Corp managed to secure long-term contracts with the British Civilian Uranium Procurement Organisation (BCUPO) to replace the RTZ contracts

329

with Rössing Uranium which – under intense public pressure – had not been renewed at the end of 1984. EFN was to supply 1,587 tonnes of uranium to Britain for a ten year period commencing in 1987 (5).

By 1985, it was unclear whether the Blanding concentrator (due to be expanded from 2.0 million to 3.5 million short tons ore) was under construction (6) or under suspension (7).

In 1986, Energy Fuels remained the largest producer of uranium, with production of 13.8 million pounds from its Arizona strip breccia-pipe mines (8). While it continued development of several mines in northern Arizona, it has encountered strident opposition to its proposed Canyon mine on the south side of the Grand Canyon, near Flagstaff (8). The company completed its final Environmental Impact Statement for this mine in September 1986, and immediately afterwards began excavating a 17-acre site within 13 miles of the south rim of the Grand Canyon. Although great opposition was expressed to this project, the Forest Service – on whose land the mine lies – decided in favour (9). This, despite the fact that more than 200 known large archaeological sites have been disturbed as a result of the mines which were opened in this area from the 1950s onwards. According to a local group, not one of these old mines has been cleared up – tourists are exposed to 2 millirems an hour in the Powell Mountains, near the largest of these abandoned sites, the Orphan Mine. Another site just north of Tuba City, on the Navajo reservation, has radiation 20% above background levels, with rates even higher near the old mill site and the highway: nine Navajo families have moved into this area in the past five years (9).

Energy Fuels plans to sink an eight-foot shaft up to 1400 feet deep on land sacred to the Havasupai Indians, at Red Butte. The mine is along a trade route to the Hopi Mesas, and the area has religious significance not only to the Havasupai, but also Hualapai, Hopi, Paute, and Navajo peoples. The deposit is at the headwaters of the Cataract Canyon, which feeds into Havasu Creek, on which the majority of Havasupai depend. (Many were removed from the area by the Forest Service to make way for the Grand Canyon National Park in the 1970s, although an outcamp is still situated on the southern rim.) The Havasupai tribe sued the Kaibab National Forest Service in a vain effort to stop the mine while Energy Fuels has also made plans to open up two more mines on state land, west of Havasupai territory, and in the Hualapai Reservation itself.

Energy Fuels uranium would be sent to Union Carbide's Blanding Mill for processing.

It should be noted that Pathfinder (Cogéma's US subsidiary) and Rocky Mountain Energy (remamed Union Pacific Resources) are also currently evaluating breccia-pipes in the same area (8), while more than 50,000 mining claims on both rims of the Grand Canyon have been lodged in the past decade (9).

Between 1986 and 1988, EFN consolidated its plans to open up new uranium lodes in the Grand Canyon region. The Pinemut, Hermit and Kanab North mines, containing reserves of up to nine million pounds uranium (10), were then expected to open towards the end of 1989. Hermit promised to be the highest grade operation in the breccia strip. Site preparation was also started at the Canyon project, south of the Grand Canyon, where reserves were put at more than 3 million pounds of uranium grading at a fairly high 0.85% (10, 11). In fact, Kanab North and Hermit opened in 1988 (12). However, by early 1991, only the Kanab mine was still operating, with the other pits on standby or still under development. Despite owning some of the highest-grade deposits of uranium in mainland USA, and having the dubious merit of being the largest US producer in 1990 (at 2.5 million pounds U308), (13) even EFN had to succumb to the uncertainties in the market and the collapsing spot price. (The previous year, EFN's uranium- from-copper extraction plant at Bingham Canyon – the huge copper mine managed by RTZ – had also been placed on standby) (14).

Meanwhile, opposition to EFN's expansionist plans was mounting. In 1987 Earth First! demonstrators took direct action against EFN's Pigeon Mine, successfully –· if temporarily –

bringing operations to a halt through some dramatic and colourful fence-climbing and infiltration (15).

The Havasupai tribe had taken EFN to court in 1988, claiming that the company had fenced-in its sacred places, and would only promise to preserve the sanctity of the Mat Faar Tigundra site if bribed a cool US$50 million. "The company offered us money for scholarships for our children if we would stop fighting them. We said no. The land where Canyon mine will go, belongs to all the American people not to one company, one family" (16). The following year, the State District Court in Phoenix, Arizona, held hearings on the tribe's appeal of the groundwater permit for the Canyon mine (17). But, in April 1990, the company secured the permit (18).

EFN has played a curious role throughout the past decade. It has gone what its President, Gerald Grandey, calls "the extra mile" (19) to negotiate with conservationists. In 1984, it halted prospecting on some 397,000 acres of the Grand Canyon, which it then added to the Arizona wilderness Act; however the government promptly released twice that acreage of roadless land for industrial development (20).

A flash flood breached EFN's Hack Canyon site in August the same year, washing ten tons of high-grade radioactive ore down the canyon. The Environmental Protection Agency (EPA) cited the company for violation of the Clean Water Act – and EFN fairly quickly cleaned up the site to the agency's satisfaction, by removing 1500 tons of contaminated soil (20).

But there is no doubt that EFN sees itself as the vanguard, not only for the precarious uranium industry in the USA, but hardrock mining itself. Gerald Grandey has been prominent in recent years, opposing the US/Canada Free Trade Act (which has allowed cheaper supplies of Canadian yellowcake to go to US utilities) and upholding the archaic 1872 Mining Act which virtually gives away public lands to miners for a peppercorn rent (21). As chief representative of the Uranium Producers of America, Grandey has deplored the swamping of domestic produ-

cers by suppliers from eastern Europe (a sort of yellow tide?) (22).

While Grandey may be the moving force behind EFN, his acendancy is quite recent. So far as can be told from the scanty information available, EFN is a family firm, dominated by J R Adams its founder and chairman – himself the son of "uranium pioneer" Robert R Adams (23). J R A Enterprises Ltd identifies itself as a Real Estate Operator in Denver with the primary function of building apartments (24), while EFN Inc (the bracket now seems to have been dropped) was originally registered in 1978 as a subsidiary of EF Uranium Group Inc (25): a title which has not cropped up since.

Adams' fellow directors, apart from Grandey, have interests in a company called Centurion Nuclear Inc (with no known uranium exploitation), Gold Star Mining, and waste disposal/land reclamation (25).

In December 1990, the Hualapai Tribal Council finally voted to permit EFN to explore more than 190,000 acres on the reservation and to mine if uranium is found. The organisation "Hualapai for a Better Tomorrow" was determined to halt the company and reverse the decision(26). The Havasupai tribe still stands firm (27).

Contacts: Canyon Under Siege, PO Box 84, Flagstaff, AZ 86003, USA.

Rex Tilousi, Chair, Havasupai Tribe, PO Box 10, Supai, AZ 86435, USA. ph: (602) 448-2731

Hualapais for a Better Tomorrow ph: (602) 769-2423

Southwest Research and Information Center, PO Box 4524, Albuquerque, New Mexico, USA.

The Grand Canyon Coalition, PO Box 3964, Flagstaff, AZ 86003, USA.

Greenpeace USA, 1436 U Street NW, Washington DC 20009, USA.

References: (1) *MJ* 18/11/77. (2) *MJ* 13/4/79. (3) *Americans Before Columbus*, Vol. 8, No. 4, 1981. (4) *Mine Talk*, 5-6/81. (5) *MJ* 8/11/85, *Gua* 6/11/85. (6) *E&MJ* 1/85. (7) *MM* 1/85. (8) *E&MJ* 4/88. (9) Canyon Under Siege, Flagstaff, 1987. (10) *MJ* 27/11/87. (11) see also *MM* 5/89. (12) Greenpeace report 1990. (13) *E&MJ* 3/91. (14) *E&MJ* 3/90 (15) *WISE N.C.* 281 16/10/87. (16) Open letter from Delmer Uqualla, Chief of the Havasupai tribe quoted in *WISE N.C.* 314 16/6/89. (17) World Rivers Review 11-12/89 (18) *Arizona Daily Sun* 23/4/90. (19) *New York Times* 29/5/86 (20) *Sierra* 7-8/85. (21) *Denver Post* 11/8/88, *Denver Post* 16/6/88. (22) *MJ* 3/11/89, *MJ* 15/12/89. (23) *Arizona Business Gazette,* Phoenix, 1/9/86. (24) Information from Duns,1983. (25) Standard and Poor's database in S&P Register-Corporate, 6/90. (26) *Arizona Republic,* Phoenix, 7/5/90. (27) *Greenpeace News,* Washington, 30/11/90.

222 Energy Reserves Group Inc (USA)

It held uranium exploration permits in the Black Hills, Harding County, in 1979/80 (1).

References: (1) *BHPS* 4/80.

223 Energy Resources Corp (USA)

It has its HQ in Denver, and leased, in 1979, 80,000 acres near Blanding, Utah, for a proposed uranium leaching operation (1).
It was also reported, at the time, to be exploring with Hunt Oil (2).

References: (1) *MJ* 17/8/79. (2) *US Surv. Min.* 1979.

224 Enertex Developments Ltd (CDN)

It was exploring for uranium in the Cardiff area of Ontario in the late '70s (1); no further information.

References: (1) *BBA,* No. 4, 1979.

225 Enex Resources (CDN)

In 1980, it sold half its 67% interest in a uranium property in northern Saskatchewan to Union Carbide; the remaining 23% was held by Saskatchewan Mining (1). Its HQ is in British Columbia.

References: (1) *MJ* 15/8/80.

226 ENI

"At its root lies a necessary and conditioning philosophy of world presence" (1) – whatever that may mean. In practice this slogan seems to mean that this unique nationalised enterprise has burned its financial fingers, got embroiled in several major scandals, and nearly collapsed through mismanagement; all in the cause of saving domestic lame duck industry, and guaranteeing Italy energy independence.
One of the largest multinationals in the world –

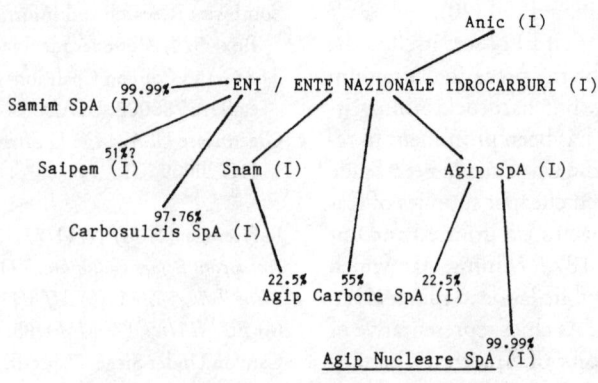

The Times Top 1000 rated it 23rd in size in 1984 (2); the *WDMNE* 24th a couple of years earlier (3) and third largest in Europe, after Royal Dutch Shell and BP (2) – its revenues proved even more impressive until the early 1980s. (*WDMNE* put it tenth in terms of world-wide sales in 1981, although *Forbes* and *Fortune* magazines rated it 17th) (4).

ENI has no fewer than 13 "sector-head" companies, under which are subsumed 285 other companies, calling on an Italian government investment of some 34,907,000,000,000 lira in 1982 (5, 6).

The more important of these are listed at the head of this entry: ENI's major concerns are oil, gas and coal (through Agip, Snam, Carbosulcis, Agip Carbone), mining and metallurgy (primarily through Samim SpA), engineering (Saipem and Snamprogetti), mechanical manufacturing (Nuovo Pignone and Savio), chemicals (Enichem), and nuclear energy (Agip Nucleare).

ENI was formed in 1953, to combine a group of five state-controlled oil, gas and petrochemical companies, the largest of which was Agip. Within a few years, under its dynamic head, Enrico Mattei, the conglomerate had built up domestic oil production and a large marketing network in Europe and Africa. In 1963 Exxon concluded a JV with ENI, and began supplying the Italian company with cheap crude oil (3).

Mattei's policy was to grant producing countries (Iran and North Africa) much higher proceeds than the oil majors were prepared to offer. This was a liberal policy pursued after Mattei's mysterious death in a plane crash in 1962. For example, in the 1970s ENI contracted with the Vietnamese government for oil exploration in a deal which earns the Vietnamese between 95% and 97% of the profits (ENI having the right to buy 50% of the oil discovered) (7). And in 1986, the current head of the company, Franco Reviglio, called for production curbs on oil, in order to support OPEC's strategy of price increases (8).

Although ENI largely prospered under Mattei, the company had only relative success in locating domestic oil and gas, and, by the time of his death, it was in heavy debt. Despite (or perhaps because of) this – and under the direction of a new chairman, influenced by the Christian Democrats (Eugenio Cefis) – ENI began diversifying, specifically into chemicals. Ten years later, its losses were mounting: it was becoming more and more dependent on state funding, and it was forced to take over the minerals and textile interests of a state-held company which was being broken up (9) – the EGAM group (3).

As the *Financial Times* aptly put it in 1982: "The power that the running of this vast concern confers on its holder means that it can never operate in a political vacuum and its whole history has been intimately linked with that of the Italian political parties" (9).

To be more precise, ENI has "tend[ed] to be a sphere of influence for the Italian Socialist Party" (6). From 1979 onwards, however, ENI began to look like a mirror image of battles within Italian state politics, as no fewer than five chairmen rose to the top post, only to be booted out (9, 10), in a period of less than four years (11).

Scandals

This period also saw some major scandals, centred on bribes paid to secure an oil deal with Saudi Arabia at advantageous prices (9), and the lending of large amounts of money to the fraudulently-run Banco Ambrosiano (12).

In 1983, the current chairman, Reviglio, was appointed to set ENI's house in order. Under his direction, the company has not avoided further scandal. A highly complex instance of speculation occurred in 1985, when the government was about to devalue the lira, and ENI sought to buy a huge cache of dollars in advance (13): the company was later cleared by the Italian Court of Accounts (13). The same year, a dam collapsed at Trento, killing 200 tourists and villagers; although not directly an ENI responsibility, ENI had earlier taken over the company which built the earthern structure (Montedison) (14).

Nevertheless Reviglio, through a tough régime

of cost-cutting, labour lay-offs (15), and refusal to support "lame ducks" in the industrial private sector (16), managed to cut ENI's deficit dramatically by 1985 (17). According to Reviglio, ENI's new role was to "... be a group of enterprises that operate with the standard of a private company but which must attain certain objectives of a national kind" (18).

In the process, ENI proposed to abandon a JV (chemicals and coal) with Occidental Oil (19); it sold off part of its stake in Saipem, its oil and gas subsidiary, in two lots (17, 20). It also "reshaped" its "rambling" network of overseas financial and operating companies to provide "total transparence" in its overseas dealings (2) (whatever that may mean – time and again, ENI's spokespeople are guilty of the worst kind of newspeak). In 1986, it floated to the private sector 18% of its Nuovo Pignone unit (22) – a heavy machinery and engineering subsidiary which had also been involved in technology for the nuclear industry (3).

But it has been the mineral sector of ENI which has provided the greatest losses to the company – "probably the worst corners of the ENI empire" (18). These losses have centred on Samim: a company which incidentally also holds 25% in equal partnership with US Steel (Essex Minerals), Union Minière, and Sun Co Inc in a seabed mining cosortium called Ocean Mining Associates (23).

But the uranium sector of ENI has also been scaled down – specifically the uranium stockpile built up for a domestic nuclear programme which has largely not materialised (12).

ENI began exploring for uranium in Italy in 1975 through its subsidiary Agip Mineraria. Within a few years it had become active in the Bergamasc Alps (specifically Val Seriana around the Novazza deposit), but also in the western and eastern Alps. It was drilling in central Italy and conducting air-borne spectrometer surveys over Sardinia (24). By 1980, through its Agip subsidiaries, it had uranium projects in Australia, Canada, Bolivia, the USA, Zambia, Guinea and Niger (25) – that year it produced some 1.8 million pounds U_3O_8 (26).

But in fact a precursor of ENI, called Somiren, had been investigating domestic uranium resources since 1955. Somiren was later absorbed into Agip Nucleare (26).

Agip Carbone and Consolidation Coal (a subsidiary of E.I. Du Pont de Nemours) have a JV to mine coal in indigenous Kalimantan, near the border with Sabah (Malaysia) (27).

As in several European countries during the late eighties, ENI was at the centre of a debate in its home country, over the merits of privatisation. In 1990, head of ENI, Gabriele Cagliari, made it clear that he was in favour of floating minority stakes in some of the group's companies, but would not seriously challenge the view of most Italian politicians, that the state should retain at least a 51% majority. By the end of the decade, seven ENI companies were already quoted on the stock exchange with speculation that Agip could join them soon (28). Cagiliari has said that he is less interested in diversifying ENI than his predecessor, but willing to adopt new activities deriving from existing concerns, including so-called "green" petrol (lead-free, presumably) recycling, water supply and refuse disposal (28).

During 1989 and 1990, the company's major preoccupation was in fighting Raul Gardini, the flambouyant head of Montedison, and a vocal supporter of privatisation, for control of Enimont, the world's ninth largest chemical company (formed in 1988) (29). After an embittered conflict over what Enimont should produce in the chemicals field, Gardini finally gave up in late 1990, resigning his presidency of the swashbuckling conglomerate, Feruzzi Finanziaria (30).

References: (1) FT 1982 passim. (2) The Times 1000 (Times Books) London, 1984. (3) WDMNE. (4) Forbes Magazine, USA, 5/7/82; and Fortune, USA, 3/5/82. (5) MIY 1985. (6) FT 7/4/86. (7) Michael Tanzer, The Race for Resources, London, 1980, Heinemann. (8) FT 20/2/86. (9) FT 26/4/82. (10) FT 5/11/82. (11) FT 27/1/83. (12) FT 11/3/85. (13) FT 26/11/85. (14) Gua 25/7/85. (15) FT 11/9/86. (16) FT 22/11/85. (17) FT 25/4/85. (18) FT 24/7/85. (19) FT 18/12/82. (20) FT 19/10/85. (21) FT 11/10/83. (22) FT 9/9/86. (23) Raw Materi-

als Report, Vol. 1, No. 4, 1983; and *MJ* 30/12/83.
(24) *World Uranium Potential* (OECD/IAEA) Paris,
1978; (25) *Uranium Red Book*, (OECD/IAEA)
Paris, 1979. (26) *Big Oil.* (27) *MJ* 18/5/84 and *MJ*
15/7/84. (28) *FT* 17/4/90. (29) *FT* 25/2/88. (30) *Independent on Sunday* 1/4/90, *FT* 23/11/90.

227 ENU/Empresa Nacional de Uranio

(Not to be confused with Enusa/Empresa Nacional del Uranio SA!)
Set up in 1977 to take over mining of uranium
from the Junta de Energia Nuclear (1), it was,
until recently, the exclusive concessionaire for
uranium exploration in the country (2).
It produced, in 1979, 120t of uranium (3), and
was intending to open up five new deposits (4),
thus expanding production to between 400 and
500t/yr (5).
In early 1981, the government announced that
it was liberalising the laws on foreign participation in uranium mining, and French, German,
and British companies were mentioned as being
interested in participating (6).
A feasibility study into a mine and mill at Nisa,
Alto Alentejo, was intended to be completed by
the end of 1982 (2). However, in mid-1981 the
Energy Minister announced a nuclear energy
"freeze" and concentration on coal as a substitute for oil (7).
In July 1980, workers struck for higher wages at
all ENU mines (8).
In 1986 ENU sent some 300,000 pounds of
U_3O_8 to Electricité de France (9).
1988 production from the Urgeirica deposit
was some 180 tons. ENU was very reticent
about supplying figures of reserves at this location, though data provided by the Portuguese
government to the OECD suggests it could
continue into the next century (10). Meanwhile
the start-up date for the Nisa (Altao Alentejo)

open-pit mine has been fixed at 1991 (10). At
the time of writing, this new mine has not been
opened.

Contact: Grupo Ecologico, Apto 567, P-4009 Porto
Codex, Portugal.
Amigos da Terra, Rua Pinheiro, Chagas 28-2-DTO,
1000 Lisboa, Portugal.

References: (1) *FT* 2/1/79. (2) *MAR* 1982. (3)
Uranium Redbook. (4) *FT* 2/1/79. (5) *MM* 3/81. (6)
New Scientist 23/4/81. (7) Reuter, 2/6/81. (8) *Kiitg,*
No. 14, 1981. (9) *Nuclear Fuel* 17/11/86. (10)
Greenpeace Report '90

228 Enusa

This Spanish national uranium company is the
main producer of uranium concentrates in the
country.
While 1980 output was an estimated 200ton-
nes of U_3O_8 (1), production capacities at
ENUSA's three mines and mills are as follows:
Cuidad Rodrigo, Salamanca, 120t/yr; El Lobo,
Caceres, 30t/yr; Saelices El Chico, Salamanca,
130t/yr.
The mine/mill at Saelices El Chico is being expanded to reach 830t/yr. A nuclear fuel factory
(UO_2) was authorised in December 1980 which
will also be constructed by ENUSA (2).
Moreover, uranium production as a by-product
of phosphoric acid was expected by 1984 at a
plant in Huelva (1).
ENUSA holds a 10% interest in Cominak, the
consortium mining uranium at Akouta, Niger.
In Colombia, South America, it associated with
the French company Minatome, the Japanese
PRNFDC, and the Colombian government's

Republic of Portugal (P)

ENU / EMPRESA NACIONAL DE URANIO (P)

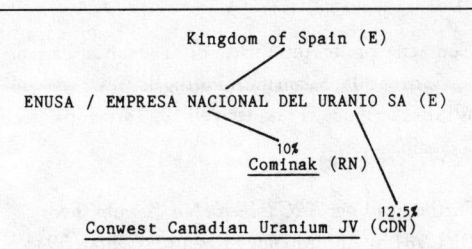

335

Coluranio, in a country-wide exploration programme planned from 1979 until 1981 (3). It has located possibly significant uranium deposits in the Andean provinces of Cundinamarca, Boyaca, Caldas, Tolima, Hiila, and Antioquia, as well as in the flatland provinces of Meta, Guiania, and Vaupes (4). Most exploitation is now being carried out by Minatome and ENUSA.

Indigenous Colombians – particularly the Tubaroan Indians of Vaupes – occupy large parts of the surveyed area, and considerable alarm has been expressed at the risks they now run (5).

ENUSA is also part (12.5%) of the Conwest Canadian Uranium Exploration JV, together with CEGB, Conwest, Eldorado Nuclear, and Electrowatt, which was organised in 1975 and operating at Key Lake, Saskatchewan.

Management of the National Uranium Exploration Plan for Spain was transferred to ENUSA during 1981, prime targets being Salamanca, Caceres and Badajoz (7).

By 1984 the company was considering a new Phosroc uranium-from-phosphoric acid unit (along with Fosforica Español) to produce some 75 tons of uranium (1), and a new mine at Quercus (at Ciudad Rodrigo, Salamanca) together with a new treatment plant (2). This was planned to open in 1985 producing 800 tons of uranium oxide a year at a cost variously estimated as US$46M (2) and US$65M (3).

Currently, the Salamanca and Badajoz mines are producing between 200 and 300 tonnes a year. Plans announced in 1987 to increase production capability to 400 tonnes in 1990 and 1000 tonnes in 1995, by expanding Salamanca further, have not been put into effect. Reserves at both deposits are around 8300 tonnes proven and around another 10,000 tonnes in the "probable" category (11).

Contact: Comité Antinuclear de Salámanca, Partado Correos 805, Salamanca, Portugal.
WISE-Tarragona, Trinquet Vell 10, Tarragona, Spain.

Further reading: ARC Bulletin No. 9, published 10/12/81 by Anthropology Resource Center, 59

Temple Place, suite 444, Boston, Mass 02111 USA.
Serrano & al, La mineria de las pizarras uraniferas de Saelices El Chico, IAEA paper produced by Enusa and presented at the International Symposium on Technical Evaluation of Uranium, Buenos Aires, 1979.

References: (1) MAR 1981. (2) WISE Barcelona, quoted in Kiitg 6/81. (3) MJ 8/1/82. (4) Age 6/11/79; El Tiempo, Bogota, 29/8/80. (5) Natural Peoples News, No. 2; ARC Bulletin, No. 9. (6) MIY 1981. (7) MAR 1982. (8) E&MJ 1/83. (9) MM 1/84. (10) E&MJ 1/84. (11) Greenpeace Report 1990.

229 Epoch Minerals Exploration NL (AUS)

It prospects, obtains and farms out mineral tenements (1).
In late 1977, it took an option on three areas near Mt Isa, Queensland, to investigate for uranium content (2).
No further information.

References: (1) Reg Aus Min 1976. (2) MJ 6/7/78.

230 ERA

(NB: Readers wanting information on the Ranger uranium mine's impact on Aboriginal people, the workforce and the environment, and its central role in the lengthy Australian debate on uranium mining, should refer to the File entry on Ranger. The ERA entry concentrates on the political context in which the mine was allowed to proceed, and its output, ownership and foreign sales.)

Political and financial aspects of ranger mine

Few people would doubt that, as Barry Lloyd put it in 1978, "It was Ranger [primarily] which elevated Australia to its current position as a future major uranium supplier" (1).

Ranger's importance lies not only in its being

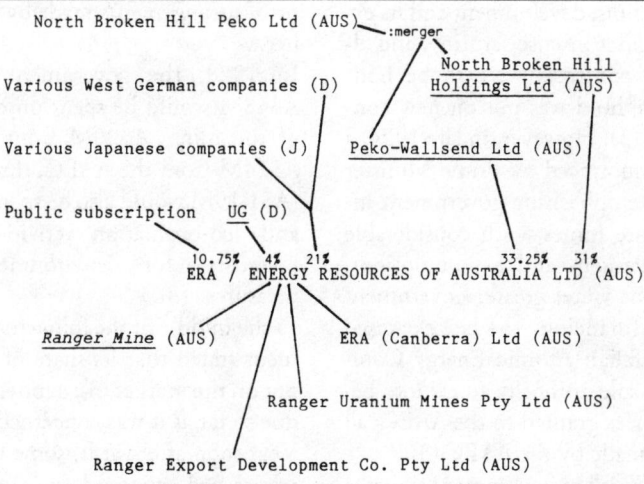

```
North Broken Hill Peko Ltd (AUS)        :merger

                                            North Broken Hill
    Various West German companies (D)       Holdings Ltd (AUS)

    Various Japanese companies (J)      Peko-Wallsend Ltd (AUS)

    Public subscription    UG (D)

              10.75%   4%  21%                    33.25%   31%
              ERA / ENERGY RESOURCES OF AUSTRALIA LTD (AUS)

    Ranger Mine (AUS)              ERA (Canberra) Ltd (AUS)

                    Ranger Uranium Mines Pty Ltd (AUS)

          Ranger Export Development Co. Pty Ltd (AUS)
```

the first mine given the go-ahead after the longest public debate on uranium mining ever held, nor simply in its negative impact on the struggle for Aboriginal land rights. This project was also of overriding importance for the Australian government in signalling to the world at large its intention to become a major participant in the development of world nuclear power and in (supposedly) the "control" of weapons proliferation. Positions adopted vis-à-vis Ranger (to a lesser extent Roxby Downs, Nabarlek, Jabiluka, Koongarra, Honeymoon, Beverley, Lake Way, Yeelirrie – the other potential uranium mines in Australia in the late 1970s) became a litmus test on a whole range of social, economic and environmental issues. Support for the Ranger project within the Australian Labor Party (ALP) precipitated a split from which (some would claim) the ALP will never completely recover: certainly it polarised "right"- and "left"-wing views on matters such as foreign investment (and control over Australia's domestic resources), the power of the mining industry (particularly in the Northern Territory) and the betrayal of the intent behind the 1976 Aboriginal Land Rights Act.

The Ranger Inquiry (Fox Committee) delivered its first Report on October 28, 1976 (the *Ranger Report,* or the *Fox Report* as it was known more colloquially). Nine months later, the Liberal (conservative) Prime Minister Malcolm Fraser announced in the Federal Parliament that uranium mining would be given permission to proceed "under strictly controlled conditions" (2). According to Fraser, the decision was motivated by a "high sense of moral responsibility to all Australians", and flowed from "four fundamental considerations": the need to reduce nuclear proliferation; to supply "essential" sources of energy to an "energy-deficient world"; to protect the environment; and to ensure that "proper provision" was made for the welfare of Aboriginal people in the Alligator Rivers Region (2).

Some ten years previously, as expectations for the global growth of nuclear power were riding high, the federal Australian government initiated what, in effect, has become the second wave of uranium development in the country's history. (For details of the first wave, see MKU.) A commercial export policy was formulated in 1967 by the Minister for National Development (3), and within less than two years, two Australian companies, EZ and Peko-Wallsend, discovered important uranium anomalies south of the Mudginberri Station, in the Alligator Rivers region of the Northern Territory (NT) (4). The project of nearly 80 square kilometres was granted special authorisation under section 41 of the Australian Atomic Energy Act (4).

With the election of a Labor government in De-

337

cember 1972, uranium development and its export was placed under intense scrutiny, and although pre-1972 contracts were to be honoured, an effective hold was put on new contracts and mining (3). However, in late 1974, a new policy was announced by Prime Minister Gough Whitlam, emphasising government involvement in future mines, with considerable control over investment and export options. The key conditions were: greater government ownership and financing to be exercised through the Australian Atomic Energy Commission (AAEC); sole authority to explore beyond current licences granted to the AAEC; all future sales to be made by the AAEC (3).

A year later, the Whitlam government signed a Memorandum of Understanding with the two Ranger partners, in an agreement described by Stephen Zorn (the international lawyer who advised the Northern Land Council on the Ranger deal) as "on a par with schemes noted only in Liberia": the Federal government would contribute 72.5% of capital funding, and the companies only 13.75% each, whereas sales would be split on a 50/50 basis (5). The AAEC would itself take a half share in Ranger (3), and receive 50% of the output (6). By then (in early 1976) the Liberal-Country Party government had already announced that the AAEC would withdraw from uranium exploration and return that activity to private enterprise. Two months later, "continuing and extending the previous government's pro-Australian policies" (7), the federal administration stated that new uranium producers must be 75% Australian-owned. However, exceptions were later to be made, if shares offered at a "reasonable" price did not attract sufficient Australian investors (7).

The 1975 Memorandum of Understanding was an obvious bugbear to the Liberal opposition, a point made by Fraser when he gave the green-light to Ranger in mid-1977. He emphasised that this special arrangement should not be thought to "... give Ranger an advantage over other mining companies", nor "accord specific marketing advantage" to the project (2). However, the Federal government would not "disturb" arrangements made by its Labor forerunner.

In 1978, the government announced that A\$48M would be spent on development costs of the mine, A\$20M from the Budget, and A\$24M from the AAEC, through borrowings. (A\$1.19M would also be spent on field research and "co-ordination activities of supervising scientists for environmental protection measures" (8).

In the middle of the following year, the government stated that its share of Ranger would be put on the market (3): a move clearly long overdue so far as it was concerned.

Very soon afterwards, some forty or fifty companies had expressed an interest in buying (9, 10). Fourteen of these tenders were under active consideration by November (11) after the deadline for bids had been extended (10, 12).

Understandably, Peko and EZ felt that they should not only be given first option (which they were) but allowed to acquire the whole project for the price of the government's participation so far (by then around A\$60M) (9); the actual market value of the mine, given that it had not even started production, being academic (13). Failing this, Peko and its junior partner EZ (there was never any doubt that, despite their equal interests in Ranger, Peko has always led on Ranger, while EZ has followed) would float a public company, thus keeping majority ownership in Australian hands. A third option was for Peko to put together a consortium of "heavyweight" Australians, notably CSR (which had for some time been keen on gaining a stake in Ranger), BHP, and others (13).

On the other hand, the Australian government and public were being left in no doubt as to where Japanese utilities stood on the issue: they wanted to buy in as quickly, and as much as possible, with the Kansai Electric Power Company leading the way (13, 14). The Northern Territory government also confirmed an interest in buying out the government's half-share (11). Among non-Pacific contenders, Esso was believed to have bid A\$165M and Denison more than A\$170M for the stake (15).

In the event, Peko won the bidding, at A$125M, with a promise of re-imbursement to the government for past and future spending on Ranger (16). Peko would combine its 25% interest with that of the AAEC in a new Australian public company, to be called Energy Resources of Australia (ERA) Ltd ERA would take over the government's commitment to fund its 72.5% of the project. Shares in ERA would be offered to public investors and major financial institutions, comprising the largest public float ever undertaken in the country (15). Overseas utilities would also be allowed to take up the remaining 25% of the equity, provided they concluded sufficient contracts to make the project immediately viable (16).

The federal government seemed pleased with the arrangement – and why not? Announcing the flotation of ERA, deputy Prime Minister Doug Anthony lauded the mix of Australian and overseas interests, promised that the assignment process would not jeopardise obligations towards Aboriginal people, nor the need for proper environmental measures, and that it would conform to the guidelines laid down the previous year for the export of uranium. Almost as an afterthought, Anthony mentioned that "the action will relieve the Government of A$200M in capital payments in relation to the project over the next two years" (16).

And what was to be EZ's role in ERA? The other Australian company had made it quite clear that it did not believe the AAEC holding should be sold at a profit, but returned to the original partners. Finally it capitulated and agreed to join with Peko in forming ERA, as well as opening the door to foreign participation (17).

Ranger ownership

ERA was incorporated on February 8, 1980, in Canberra. The interest in Ranger was transferred to the new company later that year, and a public float of A$500M was authorised in October, of which nearly A$230M (A$229M was to be repaid in respect of the federal government's and the AAEC's expenses (18). EZ and Peko would hold a 30% share in ERA, with 15% of the equity offered to the public; Japanese and West German utilities would take up the other 25% (19).

ERA was said to be seeking loan finance amounting to some US$390M, and to have obtained US$250M of this from a bank consortium headed by Continental Illinois of Chicago and J Henry Schroder Wagg of London (20). The Japanese partners were also later reported to have arranged for direct financing of US$18.8M, with a bank loan from the country's Export-Import Bank of another US$140M (21).

Some of the anti-nuclear movement predicted that ERA would be a failure, if not a disaster. FoE (Friends of the Earth) in late 1979 said it would "be, at best, a non-event, and may be one of the biggest commercial flops in history" (22). By the end of 1980 however, the book value of the company had shot up to A$1.6 billion, making it one of the top Australian commercial enterprises (3). A large chunk of initial production was under contract, and (as we shall see a little later) it was not long before the company was seeking to increase, and even double, production.

Although, as Thomas Neff has pointed out, unit costs of Ranger production are not known with accuracy, the size of the deposit, the ease of open-pit mining, and the low royalities paid to Aborigines, suggest they are below average (3): one source in the early 1980s put the costs of doubling the mine's output at A$60M, "far below (perhaps one-tenth) the cost of new production capacity at new deposits in Australia (and perhaps anywhere)" (23). On the other hand, the remoteness of the mine, high labour costs, and higher future royalties to the NT government, have probably ensured that Ranger will prove to be nothing like an economic miracle.

A crucial factor is, of course, the contract prices. While it has been generally assumed that most contracts were concluded (at least in the first couple of years) at around US$31 per pound base, some reports suggest higher charges (24), and possibly more than US$38 per pound to Japanese utilities (25).

The initial ownership structure of ERA was as follows (26): Peko-Wallsend 30.5%; EZ Industries 30.5%; Japan Australia Uranium Resources Development Co Ltd (JAURD) 10%; West German group of utilities 14%; Oskarshamnsverkets Kraftgrupp Aktiebolog (OKG) [Sweden] 1%; Australian public 14%.

JAURD itself comprised (27): Kansai Electric Power 0.5%; Kyushu Electric Power Co 2.5%; Shikoku Electric Power Co 1.5%; C Itoh and Co 1%.

The West German consortium was made up of: RWE 6.25%; UG Australia (a wholly-owned subsidiary of UG or Urangesellschaft) 4%; Inter-Uranium Australia (wholly owned by Saarberg-Interplan) 3.75%.

The public shares were initially held by 37,117 investors. Many of these, however, were also shareholders in Peko and EZ (21). Within the next four years, while the public contribution had been reduced, both Peko and EZ had slightly increased their share of ERA (up to 30.85% each in 1984) (28), although one report put Peko's at 30.96% (29). Peko's holding was said to be 33.25% by 1987: at this point, Cogéma purchased a 1.25% share (see later) (30). To date, ERA is the only major uranium company in Australia to have kept strictly to the guidelines on foreign investment laid down in 1976.

Ranger production

That grand aficionado of Aussie yellowcake, Doug Anthony, went overboard predicting that Ranger production could reach 7000 tonnes by 1985 (sic). Ranger's first year in operation was greeted with adulation by the industry (31). Output from June 1981 to June 1982 was a more modest 2677 tonnes of U_3O_8 which, though lower than capacity, included production above target for the last three months of the period. The results gave an enormous boost to EZ and Peko, both of whom had problems with their other operations. "ERA's performance must rank as one of the best corporate efforts

this year," commented the *Financial Times* (32). "Just when things seemed at their stickiest for the Lone Ranger, his faithful Indian companion Tonto [the ALP] turned up in time to tip the balance" (32).

Sales that year were put at A$146M, with a return of A$110.65M before interest, depreciation and tax. A dividend of 4 cents per share was announced. Interest on the federal government loan amounted to a hefty A$48.05M but tax was more or less completely offset by sales made from ERA to the government stockpile in return for its original loans. The price for the sales was around US$40/lb, the price fixed in the mid-seventies when uranium was riding high, and not the depressed price of US$25 to which it had sunk by the early 1980s (33).

The following year's profits (1982-83) reached nearly A$60M and ERA declared a dividend of 10 cents for the year. However the results were not as spectacular as might appear at first sight, since by then ERA had used up all its exploration expenditure – offset against tax – and had to foot a hefty A$55.97M tax charge (34). Production for the year was 3000 tonnes (35, 36). That year, the mine management predicted output could be doubled, at a "comparatively small cost", if only the ALP government, which had recently come to power, would allow the company to negotiate new contracts (34). At its 1983 conference, the ALP rejected a call to ban uranium mining, and agreed to allow Ranger (and Narbarlek) to continue producing and exporting (35). Permission was given by Prime Minister Hawke for two new contracts with US utilities to be signed (37).

This undoubtedly gave a boost to Ranger's fortunes – it reported bigger profits on somewhat smaller sales in the first six months of the following financial year (to December 1983) (37). Within the next six months the mine had coughed up more than 3000 tonnes (3098) despite an eight-week industrial dispute starting in June 1984 (38).

Production slipped somewhat the following year (39) to just over 3000 tonnes (40), and so marginally did profits: almost the same amount was repaid in interest as in the previous year

(A$47.5M) (40). Each year, 2% of ERA's revenues also go towards the Federal government's Ranger Rehabilitation Fund – set against the day when the mine closes (41). Reserves in 1985 were put at 11,000,000 tonnes of proven ore, and 224,000 tonnes of probable ore in the No. 1 orebody, grading 0.32%, with probable ore of 35.2 million tonnes in the No. 3 orebody, possible ore of 7.4 million tonnes, grading 0.21% (42, 43).

In its 1985 statement, ERA expounded on its *modus operandi*. An average of 4.5 million tons of material is mined each year, of which just over one million tons is ore ... about a fifth of the latter is either processed of stockpiled, while nearly two million tonnes comprises low mineralised ore or rockfill suitable for tailings dam construction; waste material comes to nearly two million tonnes. Ore reclaimed from the stockpiles in 1984-85 amounted to just over half a million tonnes, and a total of 1.02 million tonnes was milled.

That year, the pit in the No. 1 orebody was deepened to around 50 metres, and metallurgical recovery rates improved to 92.45%, as expansion got underway (44, 70).

By mid-1986 ERA was back on form, with the signing of new contracts (see below). "Water management problems" (a euphemism for radioactive leaks) – followed by a major leak of sulphur dioxide affecting more than 60 workers in March 1986 (45) – were "reduced" (46), and output was increased – to 3496.6 tonnes, most of which was exported to West German, Japanese, Swedish, US, Belgian and Finnish customers, at a value of A$296M (42). Ranger was claimed to be "one of the lowest cost producers in the world" (42), and Australia's reasonably recoverable resources (at less than US$80/kg) were put at 462,000 tonnes, some 29% of the western world's resources (42).

Ranger's capacity, said ERA, would be lifted to 3800 tonnes per year, and could reach 4500 tonnes by 1989 (42, 47).

Mining of the No. 1 orebody had plumbed depths of 75 metres by the end of 1987, and an exploration programme was underway in the northern part of the Ranger site (30). Production was 324 tonnes U_3O_8 (30).

Ranger contracts

ERA's contracts can be arranged into two periods: those signed before December 1972 and early 1973, when a hold was put on all new uranium deals, and those signed after 1977 – when Ranger was given the green light. Most of these were confirmed in late 1979, when the embargo on contracts was finally lifted, though there were some alterations in delivery period, amounts contracted, and pricing (due to escalation clauses) (20).

Before 1972, ERA partners Peko-Wallsend and EZ Co of Australia apparently signed only two deals: for the supply of 1300 tons to Chubu Electric Power Co from 1977 to 1982, and 2000 tons to Kyushu Electric Power Co Inc over the period 1977-86 (48).

In 1972, despite the embargo, the federal government promised that existing contracts would be honoured, using output from MKU, the AAEC stockpile, and Ranger "when opened" (3). In order to free the logjam on pre-1972 contracts, the embargo was partly lifted in 1976, to cover the pre-1972 deals (49). Just how much of the two Japanese contracts was delivered between 1976 and 1980 is not known: the majority appears not to have been. However, 181 tonnes was apparently delivered to Chubu in early 1977 (50).

In August 1977, Ranger was at last given permission to proceed. Sales would be arranged by the AAEC, together with the mine's partners, Peko and EZ (51), although the government had not then formulated its final export policy. Within a year, Deputy Prime Minister and Minister of National Resources, Doug Anthony, had issued a "long awaited statement" in this regard (1). Henceforth Australian supplies were to be governed by "comprehensive" bilateral safeguards and a "complete oversight" of uranium exports, following the completion of "environmental procedures and compliance with the government's foreign investment policy". A Uranium Export Authority would also be established (1).

However, the Authority would only function when mines other than Ranger were opened (1), and, indeed, by mid-1978, not only had the government allowed companies to resume contract negotiations before the Authority was in place (49), but also before safeguards treaties had been signed (6).

The year 1977 was one of frenetic activity, as the government paved the way for Ranger to become financially viable. It was a case of the horse following the cart; the drumming up of customers, so that by the year end Ranger could be justified on the grounds that new contracts had to be honoured (52).

Early that year, Prime Minister Fraser went to Washington. Allegedly President Carter asked his Australian counterpart to release Ranger's uranium, in order to discourage Japanese and European nuclear barons from building fast breeder reactors (53). This was the era of Carter's brave, but ultimately futile, attempt to control proliferation through the International *Nuclear Fuel* Cycle Evaluation (INFCE) – into which Australia bought a ticket via Ranger uranium. Ironically, Fox, the author of the very reports which had advanced arguments for the opening-up of Ranger and other Northern Territory deposits, was at the time himself reported criticising the lack of safeguards in both Canada and the USA (54).

In any event, Fraser had as much, if not more, reason to court Carter as the other was around. In a letter despatched to the American peanut king in February 1977, Fraser explained Australia's potential as a uranium suppier (although by then it was already widely vaunted as the yellowcake eldorado of the 1980s): "There is probably no other area of the world that is comparable to Australia's Northern Territory in its potential for significant resource expansion" said the influential Electric Power Research Institute at around the same time (7). The Australian Prime Minister also alluded to Australia's ability as a uranium supplier, to "exert influence on international development" – a sentiment close to Carter's heart (5). According to

one observer, tacit agreement on supplies from Ranger to the USA had already been reached between the two governments, before the release of the second Ranger (Fox) Committee Report in May 1977 (5).

The following month, deputy Prime Minister Anthony was publicly complaining that Ranger could not deliver its Chubu and Kyushu contracts because of delays in commissioning the mine; immediately afterwards the Japanese Ambassador held a press conference on the same issue (5). And if this was not enough, just after the Ranger final report had hit the streets in May the government was claiming that the Nuclear Non-proliferation Treaty could be "breached" if Australia did not exercise its option to supply the deadly mineral (55): a sentiment reiterated in the major statement on uranium delivered on August 25, 1977 (see above). Australia had an "obligation" to provide the resources that would enable the "rest of the world" to overcome the "energy crises", Fraser declared. To this end, Australia would not only "supply energy that will provide jobs ... heat homes ... protect standards of living and enable them to be improved ..." but also "slow the movement towards the use of plutonium as a nuclear fuel and lessen the attendant risks of nuclear weapons proliferation" (2).

By the middle of 1978, Bernard Fisk, Peko-Wallsend's general manager for uranium development, was telling the International Uranium Symposium (organised by the Uranium Institute) that the first tranche of Ranger production (3000 tonnes per year) would be "spoken for quite quickly, thereby facilitating the financing arrangements for the project, and allowing production to commence by the end of 1981" (56). At a conference in Honolulu, Mr Newman, the Minister for National Development, predicted that the "bulk of the western world's uranium orders would soon be fulfilled by Australia" (57).

After a seven-year hiatus, the federal government gave the go-ahead for Ranger sales in late 1979. The initial contract, for 2500 tons valued at between A\$160M and A\$180M (at A\$36 to A\$38 per pound), was the first signed since the early 1973 ban, and clearly announced as a "forerunner" for further contracts. Kepco was the recipient, and deliveries were intended to cover a ten-year period from 1983 (58). The contract was a coup for Peko-Wallsend, since Kepco was itself part of the CSR/Minatome/Kepco consortium bidding for the Government's share in Ranger (59). Labor's Tom Uren, a left-winger in the ALP strongly opposed to uranium mining, condemned the contract, predicting that "The Australian people will surely see the foolishness of selling uranium to such an unstable government as South Korea," and that the ALP, when in power, would cancel the sales (60).

Within six months, a batch of contracts had been announced to ERA equity partners, involving around 34,000 tons, worth more than A$2000M. Over 20,000 tons would be delivered to West German customers, RWE, UG, Saarberg-Interplan Uran (61). The Japanese customers were announced as C Itoh, Kansai Electric Power Co, Kyushu Electric Power Co, and Shikoku Electric Power Co, grouped together as the Japan Australia Uranium Resources Development Company (JAURD) (20). However, the contracted output (of around 11,000 tonnes) was "much larger than the equity [in ERA] perhaps would indicate", according to Thomas Neff. This was partly because the JAURD partners also acted as agents for other customers, and according to Neff it was "not entirely certain that all uranium involved will ultimately go to Japan or West Germany" (3).

Doug Anthony, the Federal Trade and Resources Minister, claimed that these contracts, along with those arranged with other Japanese, US, and South Korean buyers, would cover about 84% of Ranger's anticipated output: the contracts provided for price escalation adjustments during the period of delivery up to 1977 (20, 62).

The state of contracts, soon after ERA was formed, looked officially like this: 2500 tons to Kepco; 2250 tons to Indiana and Michigan Electric Co of the USA during 1982-1990; 5820 tons to UG during 1982-1996 at around

360 tons per year (63); 9094 tons over a similar period to RWE (Rheinisch Westfalisches Elektrizitatswerk AG of West Germany); 5456 tons also during 1982-1996, to Saarberg-Interplan Uran GmbH, also of West Germany; 13,413 tons to JAURD, again during 1982-1996.

A little later, a contract was announced between ERA and OKG for 3150 tons from 1982 to 1994 (48), or until 1996 (3). Synatom was also to receive 1575 tons between 1982 and 1994 – and 1011 tons was apparently earmarked to be returned to the AAEC stockpile from 1982 to 1984 (64).

The assumed price for the majority of these contracts was A$31 per pound (48).

A minimum of 1100 tons of uranium was contracted to Wisconsin Electric Power Company in late 1983, to start around 1985 and run "for several years". This was the first deal following an ALP cabinet meeting at which not only the Roxby Downs uranium project was approved, but ERA was given permission to pursue two new contracts (65).

The second of these new contracts is with Virginia Power Company, for 1400 tons. Further contracts would have to wait on the ALP conference in 1984, following which, in August 1984, ERA signed another contract with Belgium's Synatom for the supply of 500 tons of uranium (66).

The Indiana and Michigan Electric Company (wholly owned by American Electric Power Co) contract (2250 tons from 1982 until 1990) was cancelled in September that year, after the shipment of only 225 tons (66) and a suspension of the shipments at the request of the customer, in early 1983 (67).

However, the American Electric Power Service Corp signed a contract with ERA in 1985, for 2500 tons at an estimated US$500M in 1984 terms, provided all the options are exercised (68, 69).

At the same time, 675-750 tons was contracted to Pennsylvania Power and Light valued at US$50M (68).

Although these contracts appeared to be a financial coup for ERA, the Australian Financial

Review estimated that their value (based on the 1985 spot price for uranium) could be as little as US$104M: even taking into account the federal government's probable base price of A$31 per pound, they wouldn't be worth more than A$200M. The discrepancy between the two sets of figures is due to the optional provisions contained in the contracts for future contracting (66). No firm date was then fixed for implementing the contracts, although a probable start of 1987 was mentioned (66).

ERA earned A$56.5M in the financial year 1984-85 (70), and Alex Morokoff, ERA's chairman, told the company's annual general meeting that, as well as possibly doubling the mine's output over the next five years (up to 6000 tonnes per year), five new contracts had been signed up to June 1985, and a further contract with a US utility had been concluded since (71). Kepco also added a new contract to its original one, for deliveries starting in the early 1990s (69).

Another set of contracts was given the go-ahead by the federal government in 1986 (41), and at the turn of 1986-87 ERA announced four further long-term contracts with US utilities, in addition to the five already concluded (72). The following year yet another two US contracts were signed – making a total of eleven (30).

The most controversial of ERA's contracts is that with France, a deal that flatly contravenes the strict code on uranium export laid down in 1977. The 1982 ALP conference had proscribed all uranium trade with France until the Euro-nuclear state ceased atomic testing in the Pacific (73). The platform was restated two years later (when the conference also voted not to provide any subsidies, tax incentives, or compensation of any kind to the uranium industry) (74), and again in 1986. By then it was quite clear that Australian uranium could enter the French Super-Phénix (Fast Breeder Reactor) either as enriched tails, or as processed plutonium or uranium from uranium of Australian origin: a possibility confirmed in 1988, when Australian uranium had its title "swapped" and was about to go to the USA and from there probably to the Super-Phénix programme, until

the illicit deal was discovered and stopped (see Nukem for further details).

In 1986, the ALP government announced that sales to France would be resumed (75). This followed what federal Trade Minister, John Dawkins, the previous year had called a "commercial ploy" on the part of UG, to get out of its long-term contracts with ERA, by arranging a deal which the government was bound to stop. UG's deal was with the French brokerage company Enership; Dawkins maintained the uranium would not remain in France (because the French state utility Electricité de France, was "known to possess large stocks" of uranium) (63). Hence, in order not to give UG an excuse to withdraw from ERA without penalty, Dawkins gave the green light to the contract. UG strenuously denied the allegation, pointing out that it would not have paid ERA prices of more than A$40 per pound – more than double the spot price in 1985 – in order to forfeit the benefit of long-term supply commitments from ERA. It maintained that the Enership deal was covered by the Australia Euratom Safeguards Agreement (63).

Finally, in early 1988, the first direct sales to France were given the go-ahead, despite considerable opposition from within the Labor party. The federal government maintained it would otherwise have to pay a A$60M compensation charge (76). The amount of the sale was not announced (as of mid-1988) but delivery was not expected to begin immediately (77). At the same time, Cogéma announced that it had acquired a 1.25% stake in ERA (77).

Britain's role (or, rather, absence of it) in ERA has been distinctly shadowy, if not conspiratorial. Once the announcement of the formation of ERA was made, it was widely believed the British Civil Uranium Procurement Organisation (formerly Directorate), or BCUPO, comprising BNFL, the CEGB and the South of Scotland Electricity Board (SSEB) – would take part of the foreign equity, along with the Japanese and West German companies (15). The CEGB was alone expected to take about 25% of the equity.

In 1979, an Australian Parliamentary report revealed that Britain had been top of the list of potential customers for Ranger (78). In February that year, the government had asked the British Minister of State for Energy if Britain was interested in purchasing 750 tonnes of uranium per year from Ranger; five months later, Australian officials went to London and "identified an interest on the part of the UK CEGB" for 500 tonnes a year (78). British officials during the same trip had also "confirmed" a "general" interest in buying as much as 1300 tonnes from Australian sources. In fact, a draft supply agreement had been signed two years earlier, quoting this figure, although a somewhat lower amount was also mentioned (1000 tonnes) (79).

These announcements seem to have been motivated by self-serving interests on both sides. The federal government clearly wanted to land a major contract with its longest-standing trading partner as a "cachet" for further deals, (if it had been confirmed, the British uranium contract would have been the largest ever signed in Australia), while the British government needed an assured long-term supply to replace its Rössing (and probably Canadian) sources, if and when these dried up. In the light of statements made by Australian "official sources" in early 1980, it also seems that Britain wanted to bend the Australian export rules, established in 1978, and have a greater control over re-export of Ranger uranium than the Australian government could contemplate. In late 1979, a telex from the BCUPO's managing director, J Waddams, had gone out over the signature of Foreign Secretary Lord Carrington (until recently a director of RTZ) advocating to the potential West German and Japanese partners in ERA that they should form, with Britain, a new club, to market their own share of Ranger production. At that point, BCUPO was actively considering taking a 5% share in ERA (80). It is not known what the reaction of the foreign partners was to the British proposal, but Peko-Wallsend apparently made clear to its putative Japanese partners that going along with the British would jeopardise their own participation in ERA. In any event, the British initative

foundered, the Japanese joined ERA, accepting Australia's export guidelines (although it was not until July 1982, some two years after the first Japanese contracts with ERA, that the Japanese Diet ratified a bilateral safeguards treaty with Australia, as 250 tonnes of uranium were awaiting shipment) (81); BCUPO did not join ERA, nor did any of its three partners sign a supply contract from Ranger.

While it is tempting to see in the British machinations a strategy to secure uranium which might be diverted to a British nuclear weapons (i.e. Trident) programme, this is less likely than that BCUPO was trying to arrange its own mini-cartel and dictate prices to ERA.

In late 1987, North Broken Hill announced a merger with Peko-Wallsend (84) thus bringing EZ (effectively controlled by NBH) into the Peko fold, and uniting the two companies' interests in ERA.

North Broken Hill Peko Ltd now holds 86.83% of the 'A' class shares, with Rheinbraun Australia Pty Ltd holding 41.6%, UG Australia Developments Pty Ltd 26.7%, and Interuranium Australia Pty Ltd 16.7% of the 'B' class shares. Japan Australia Uranium Resources Development Co Ltd holds all of the 'C' class shares (85).

In terms of equity distribution, North Broken Hill holds 65% of ERA, with Japan Australian Uranium Development Co Ltd holding 10%, RWE 6.25%, UG 4% and Interuran 3.5% (but with Interuran now largely held by Cogema, after Saarberg relinquished 1.25% of its 3.5% holding to Cogema, and then was bought out 75% by Cogema itself) (86). Production from Ranger was 3497 tonnes in 1986, 3103 tonnes in 1987, and 3227 tonnes in 1988 (87). 1989 saw 3595 tonnes produced (88) which tipped ERA's profits down to A\$37.8 million, from its previous record of A\$63 million (89). That year, the No. 1 ore body had reached a depth of 90m below ground level and was expanding northwards. By the end of 1990, ERA admitted that reserves were being fast depleted at Ranger (with 27,464 tonnes of uranium concentrate produced to that time) and that its exploration efforts around the deposit had not yielded any major new reserves: while the No. 1 orebody was expected to be mined-out during 1991 or 1992 (91). The company was now said to be looking for new low cost deposits (92) though its major efforts have recently been put into a purchase deal with Pancontinental over its Jabiluka stake (by early 1991, NBH held 13% of Pancontinental) (91).

As of 1990, ERA had eighteen supply contracts with US, Belgian, West German, French, South Korean, Japanese and Swedish utilities (85).

Although the company maintains that the majority of its contracts are long-term, concluded at prices which fall within the Australian Federal government's "floor price" mechanism (designed to protect the industry and the economy) (91), ERA has been accused of negotiating contracts, especially with US customers, at the extremely low spot prices prevailing over the last few years, simply because the utilities refused to do otherwise. (And why should they fork out A$31 per pound when they could pick it up elsewhere for as little as A$10?). While, technically, ERA may not have been selling its own uranium at these bargain-basement prices, it is quite likely that it was entering the market, to buy at the spot price and then resell at a higher rate to some of its customers (e.g. Kepco) at the long-term price. In the light of this, various groups proposed in 1989 that uranium production should be shut down altogether in Australia, while leaving companies like ERA free to act as uranium brokers in the world at large. Commented researcher John Hallam of Friends of the Earth:

"The final irony of the matter is that a shutdown of Australian uranium production now could be achieved without adversely affecting ERA's profits, providing it's allowed to buy cheaply on spot and use the material to fill its highprice contracts – *i.e.* by leaving it in the ground." (86)

Protest

Apart from the opposition to the Ranger mine mounted by Aboriginal organisations, the Movement Against Uranium Mining, FoE, anti-nuclear groups and "dissident" Australian Labor Party members, it is worth noting that in 1981, the Uniting Church – long associated with establishing missions on Aboriginal land – decided to sell its shares in ERA, after making a A$11,500 profit: "We are involved in a genuine attempt to make amends" said the Church's Moderator (82).

In 1986, the German "Grünen" (Greens) demanded that the provincial government in Saarbrücken should ensure that Saarbergwerke AG, parent company of Saarberg-Interplan, divest itself of Saarberg's 3.75% holding in ERA (83) because of the damaging social and environmental consequences of the mining, particularly on the Aboriginal community.

Further reading: On the mine development: "Ranger Uranium Mine" *MM* 11/85 pp392-403.

Contact: MAUM, 247 Flinders Lane, Melbourne, Victoria 3000, Australia.

FoE, 222 Brunswick Street, Fitzroy, Victoria 3065, Australia.

WISE, PO Box 87, Glen Aplin, Queensland 4381, Australia.

References: (1) Barry Lloyd, "An Australian View of the Uranium Market", in *Uranium Supply and Demand*, Proceedings International Symposium on Uranium Supply and Demand (Uranium Institute) London, 7/78. (2) *Northern Territory Newsletter* 8/77. (3) Neff. (4) Peter Carroll, "Uranium and Aboriginal Land Interests in the Alligator River's Region", in *Aborigines Land and Land Rights*, ed Nicholas Petersen and Marcia Langton (Australian Institute of Aboriginal Studies) Canberra, 1983. (5) Tas Ockenden *The Uranium Hawkers*, paper included in the MAUM information pack *Energy and U* 1977 or 1978. (6) *FT* 10/1/78. (7) *EPRI* 78. (8) *Northern Territory Newsletter,* Darwin, 1978. (9) *Fin Rev* 31/8/79. (10) *FT* 27/9/79. (11) *Age* 6/11/79. (12) *Australasian Express* 26/10/79, London. (13) *Fin Rev* 31/8/79. (14) *Age* 3/9/79. (15) *FT* 19/12/79. (16) *Australasian Express* 21/12/79. (17) *FT* 24/1/80. (18) *Reg Aus Min* 1981. (19) *Courier Mail* 11/8/80, *Fin Rev* 19/8/80. (20) *MJ* 22/8/80.

(21) *Nuclear Fuel* 24/11/80. (22) *WA* 20/12/79.
(23) *Nuclear Fuel* 18/11/82. (24) *Nuclear Fuel*
30/8/82. (25) *Nuclear Fuel* 25/4/83. (26) JG
Munro, *Overview of Australian Uranium Developments,* paper presented at International Conference
on Uranium, Atomic Industrial Forum/Canadian
Nuclear Association, Quebec City, 15-18/9/81. (27)
Nuexco Monthly Report to the Nuclear Industry Nuexco 11/80, Menlo Park. (28) *Reg Aus Min* 83/84 (29)
FT 3/3/84. (30) *MAR* 1988. (31) *WA* 10/11/81.
(32) *FT* 21/8/82. (33) *FT* 21/9/82. (34) *FT*
20/8/83. (35) *MAR* 1984. (36) *FT* 19/8/83. (37) *FT*
17/2/84. (38) *MAR* 1985. (39) *MJ* 26/7/85; *MJ*
6/12/85. (40) *MJ* 13/12/85. (41) *E&MJ* 12/86. (42)
MAR 1987. (43) *MJ* 13/12/88. (44) *MJ* 6/12/85.
(45) *Tribune* 19/3/86, 26/3/86; *MJ* 21/3/86. (46)
MAR 1986. (47) See also *ERA Annual Report* 1987.
(48) Information from Australian Government,
quoted in *Kiitg,* Amsterdam, 1982. (49) *FT* 2/6/78.
(50) FoE, *Uranium Deadline,* Vol. 1, No. 4,
Aug/Sept 1977, Melbourne. (51) *FT* 25/8/77. (52)
MJ 28/10/77; see also *Gua* 7/6/77 and *FT* 8/3/78.
(53) Moss. (54) *Age* 3/11/77. (55) *WA* 27/5/77. (56)
Uranium Supply and Demand, International Symposium on Uranium Supply and Demand, Proceedings
(Uranium Institute) London, 7/88; see also *FT*
12/7/79. (57) *Australasian Express* 11/8/78. (58) *FT*
10/11/79. (59) *Fin Review* 12/11/79. (60) *Age*
10/11/79. (61) *FT* 13/9/82. (62) *FT* 19/8/80. (63)
MJ 1/3/85. (64) *Honeymoon Handbook* (Honeymoon Collective) Melbourne, 1982. (65) *FT*
15/11/83; *MJ* 9/3/84. (66) *MJ* 18/1/85. (67) *FT*
3/1/85. (68) *MJ* 11/1/85. (69) *MJ* 13/12/85. (70)
MJ 25/10/85. (71) *FT* 18/10/85; *MM* 12/85. (72)
MJ 9/1/87. (73) ALP Platform, Minerals and Energy, Clause 66, Canberra, 1982. (74) ALP Platform, Minerals and Energy, Clause 60, Canberra,
1984. (75) *Chain Reaction,* No. 47, Melbourne,
Spring 86. (76) *MJ* 19/2/82. (77) *MJ* 15/1/88. (78)
Evidence by the Australian Department of Trade
and Resources to Senate Estimates Committee 1979,
reported in *Fin Rev* 17/10/79. (79) *Age* 2/11/77.
(80) *Age* 2/2/80. (81) *MJ* 9/7/82. (82) *Age* 23/9/81.
(83) *Saarbrücker Zeitung* 9/10/86. (84) *FT* 16/8/88.
(85) *Reg Aus Min* 1990/91. (86) *Chain Reaction,*
Spring 1989. (87) ERA Annual Report 1987 and
1988. (88) *MAR* 1990. (89) *Australian* 18/8/89.
(90) *MJ* 4/1/91. (91) *Australian* 22/2/91. (92) *MJ*
4/1/91.

231 Ergo

Hailed at its inception, in 1976, as a "brilliantly
conceived project for extracting at a good profit
the gold, uranium and acid content of South
Africa's old mine waste dumps" (1), Ergo was
the pioneer of uranium/gold-from-slimes
schemes in South Africa (see also Chemwes and
JMS). Shares in Ergo were snapped up by Anglo
American and De Beers, Union Corp (now part
of Gencor), and other financial institutions, at
its official launch in 1977, leaving less than a
quarter of the share issue for the great South African public; in fact, the real value of the public
interest proved less than 10% and was "at the
highest price" (2). Initial monthly output was
to be nearly 19,000 ounces of gold, 44,000 tonnes of sulphuric acid, and 16.5 tonnes of
uranium, recovered from some 380 million
tonnes of accumulated slimes (3). Although the
1977 prospectus was "so general as to be little
use to analysts trying to evaluate the project"
(3), it was widely vaunted as escaping the risks
normally associated with gold mining operations – to the extent that one US broker said it
should be rated as a chemical venture.
Anglo-American made much of its new labour

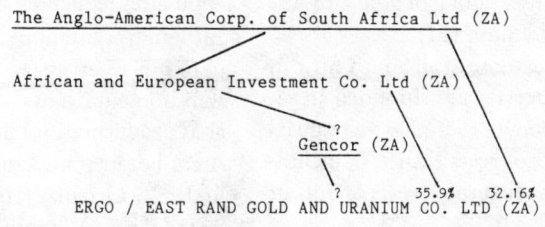

policy: Ergo would operate with a small number of trusted black and white employees – not migrant labour – and a unified wage scale and non-discriminatory attitude "to the extent permitted by existing legal constraints" (1), as chairman Harry Oppenheimer put it.

Within two years, however, the initial teething troubles became rumbles of alarm. During most of 1979 uranium recovery rates were only 72% of those planned, while recoveries from higher grade slimes were "far lower than expected" (3). In June 1981, however, Anglo American and Ergo acquired 120 million tonnes of high-grade material from Simmer and Jack mines (later incorporated into Ergo as Simmergo) (4), and two years later 12 dams owned by East Daggafontein Mines were taken over by its subsidiary Dumpco and effectively leased to Ergo (5) which has built a new treatment plant for them (4). More dams were acquired from another mine later that year (4); Ergo's profits were therefore "holding" (mainly due to gold production) by mid-1980 (6) and 292 tons of uranium – close to the original 300 tonne target – came trundling out of the flotation/concentrator over the next year (7).

Initial target treatment levels were exceeded in the financial year ending March 1982 (a total of 19.2 million tonnes) when – despite a drooping uranium market – uranium production reached nearly 300,000kg. At the time, Harry Oppenheimer reflected that "despite the vital role of nuclear energy, there seems to be little prospect for any material improvement in market conditions in the short to medium term" (8).

Uranium production was also "higher than forecast" (9) in the following year, and planned expansion looked like proceeding, without loss of dividend (10). Indeed, by March 1985 Ergo broke the 20 million ton barrier for the first time to produce nearly 7000kg of gold and just over 150,000kg of uranium (11).

Ergo's contracts – like those of all South African uranium-gold producers – are shrouded in secrecy, though it is known that sales were guaranteed "for a number of years" in 1978, including one to Iran for the first three years of output (12).

In 1987, Ergo opened its Daggafontein treatment plant, bringing to three the number of tailings plants under the company's control (13) and its gold production capacity to more than 10,000 kilograms per year (14).

However, within the next two years, as profits dropped, Anglo American adopted a "strategic planning exercise" to extend the company's life through the 1990's, by cutting back both on the gold output and working costs. It was decided to close down the flotation plant, the larger of the two acid plants, and the uranium facility by 1991, with the loss of 600 jobs. Operations at Slimmergo would be curtailed, while the carbon-in-leach facility and the small acid plant would continue at 1990 levels (15).

Between the last half of 1989 and the first half of 1990, uranium production had dropped by some 40%, and was bottoming-out at less than 100 tons per year (15).

References: (1) *FT* 29/6/78. (2) *MJ* 15/7/77. (3) *FT* 7/2/79. (4) *MIY* 1985. (5) *FT* 23/12/83. (6) *FT* 19/6/80. (7) *MJ* 3/7/81. (8) Ergo review by Chairman Harry Oppenheimer in Annual Report 6/82. (9) *MJ* 29/6/84. (10) *FT* 3/7/84. (11) *R&OFSQu MJ* 26/4/85. (12) Lynch & Neff. (13) *MJ* 20/11/87. (14) *FT Min* '90. (15) *MJ* 21-28/12/90.

232 ESI [Earth Sciences] Resources Co

With headquarters in Colorado, the company owns a uranium-from-phosphates plant at Calgary, Alberta, in Canada, which utilises phosphoric acid supplied by the Western Co-operative plant nearby. Design capacity was 40t/yr uranium. Though the plant opened in 1980, by early 1982 it had not reached full capacity (1); soon after reopening, it was indeed temporarily

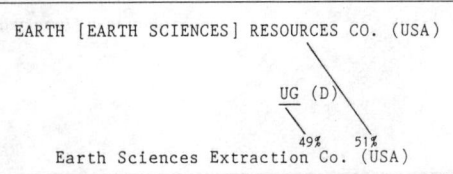

EARTH [EARTH SCIENCES] RESOURCES CO. (USA)

UG (D)

49%   51%
Earth Sciences Extraction Co. (USA)

suspended because of the state of the world uranium market, though capacity had increased by 1981 to 60t/yr (2).

However, after major equipment modifications the plant climbed to 91% operating capacity by the end of 1983, with Urangesellschaft of Canada (a subsidiary of Urangesellschaft in Germany) now having a share in the plant after helping bale out the operation. At maximum capacity the plant is intended to deliver some 150,000t/yr of phosphoric acid, from which uranium oxide is extracted.

Earth Sciences also holds the right to recover uranium from phosphoric acid produced at Western Cooperative's plant at Medicine Hat, which was closed as of late 1983.

Contracts exist between Earth Sciences and Vermont Yankee Nuclear Power and Yankee Atomic Electric of Massachusetts (3).

In 1980 Earth Sciences produced a report which concluded that 25% of America's uranium needs could be met from phosphate production. The report conjectured that – in the near future – a hundred phosphoric acid plants could produce uranium as a by-product with a total output of more than 11million pounds of U_3O_8 (4).

The Calgary plant suspended operations in August 1987 and, so far as is known, has not resumed production (5).

References: (1) *MAR* 1982. (2) *MJ* 24/7/81. (3) *E&MJ* 10/83. (4) *Kiitg* 5/80. (5) *E&MJ* 4/88.

233 Essex Minerals Co

In 1978, together with Hunting Geology and Geophysics, Essex prepared a uranium potential study of north, central, and west Africa and the Middle East. Essex provided the geologists, Hunting the uranium experts, in evaluating data from no fewer than 44 countries. Six fa-

USX (USA)
/
ESSEX MINERALS CO. (USA)

vourable uranium zones were located, one of which was being explored by Essex (1); which country?

In April 1981, Essex applied for a licence to explore for uranium in the Fingal Valley, Tasmania, Australia, in an area of 661 square kilometres previously searched by Getty Oil (2).

Together with Union Seas Inc (a subsidiary of Union Minière), and Samim (a subsidiary of ENI), Essex is part of Ocean Mining Associates which is a consortium formed to explore for polymetallic nodules on the seabed (3).

References: (1) *MJ* 15/12/78. (2) *MJ* 6/3/81. (3) *MIY* 1982.

234 Esso Exploration and Production Aus Ltd

Exxon Corp. (USA)

ESSO EXPLORATION AND PRODUCTION AUS LTD (AUS)

As part of the world's biggest corporate entity, Exxon, Esso Australia was, during the '70s, involved in uranium exploitation, apart from its activities in oil, gas, and base minerals exploration and production.

It is exploring for oil and gas in the Bass Strait region. It holds 25% in the Hall Creek coal JV; is with Amax and EZ Industries at Golden Grove, Tasmania; teamed up, in 1977, with Newmex exploration to search for minerals including yellowcake in the Ashburton gold fields region (1).

It was also said to be interested in participating in the Olympic Dam project (Roxby Downs) in 1978, an interest since then lost (2).

It also became a JV partner with two Australian oil companies in the mammoth Rundle oilshale project in Queensland (3). However, in 1981 the company announced its doubts on the feasibility of the project; the Queensland government said it would not renew the lease unless the partners spent an additional US$39 million on studies; and the project is currently on hold (4).

In late 1981, Esso announced it would take 35% in the Gloucester steam coal project in New South Wales.

In 1978, it joined with Urangesellschaft, as foreign equity partners, in the prospective uranium mine at Yeelirrie, with Esso holding 15%, Urangesellschaft 10%, and Western Mining the remaining 75% (5). In May 1982, however, the company pulled out of Yeelirrie, despite agreeing to buy another 35% of the production and sinking nearly US$20M into its planning stages (80% of the Stage 1 costs). This was, said Esso, because the project was "not economically viable under the terms of the joint venture agreement and Esso's current assessment of the world's uranium market outlook" (6). (For further information see Yeelirrie and Exxon.)

Together with Carr Boyd, and Otter, Esso Australia entered a uranium exploration JV at Pandanus Creek, Northern Territory, in 1979 (7). By the end of the eighties, Esso Australia's main commitments were top oil exploration and production in the Bass Strait oilfields (with BHP), in the Canning Basin (Queensland) and its 50% holding in the Rundle oilshale project. It also held a 35% interest in the Golden Grove base metals project in WA, as well as coal interests in Queensland and NSW. In 1989 it bought more coal assets in NSW – the Ulan and Leamington mines – while the Golden Grove prospect moved towards production (8).

In 1988, it also moved to acquire 50% of the Leamington coal mine owned by CSR, in the Hunter Valley (NSW) (9), and just over a year later, returned to coal exploration in Queensland, with a 55% interest (Mitsubishi 45%) in a major steam coal deposit near Clermont, which could produce around 6 million tonnes a year for export (10).

References: (1) *MJ* 25/11/77. (2) *FT* 29/8/78. (3) *FT* 17/6/80. (4) *MJ* 18/13/81. (5) *MJ* 18/8/82. (6) *AFR* 15/5/82. (7) *MJ* 18/5/79. (8) *Reg Aus Min* 1990/91. (9) *E&MJ* 10/88. (10) *E&MJ* 1/90.

235 Esso Minerals Canada Ltd

As part of the world's biggest corporate entity, Exxon, Esso in Canada is involved in a number of energy-exploitation projects with enormous implications for the environment and for indigenous peoples in particular. These include its syncrude project (oil sands) at Cold Lake, Alberta (a US$15 billion project) (1); exploration in the Beaufort Sea (where oil/gas exploitation in general has already been attacked by the Inuit of the Canadian and Greenland arctic for its potential damage to marine life and ecology) (2); and its explorations in the MacKenzie delta, part of a potential contribution to a future MacKenzie valley pipeline which would trespass on the lands of the Dene (3).

Imperial Oil is Canada's largest oil company.

Esso Minerals ranks among the top five minerals explorers in Canada, with interests in copper, coal, molybdenum – and uranium (4).

In 1978, it started exploring for uranium in the Bancroft area of Ontario (5).

In 1979, Mid-North Uranium concluded an agreement with Esso Minerals in respect of more than 30,000 acres of claims in the Hasbala Lake area of north-western Manitoba 6). No further details have been released on this prospect (in which Esso could earn 60%), and it is presumed "dead".

Esso Minerals was also among other companies showing continued interest in possible uranium in the Otish Mountains area of central Quebec (7), and is participating there in a JV along with Uranerzbergbau, Séru Nucléaire, Pancontinental, James Bay Development, and Soquem (8).

Esso's participation in the Midwest Lake JV uranium project in Saskatchewan is maintained through Esso Resources, subsidiary, Canada Wide Mines (9).

The company has a 49% holding in some 310 mining claims in the Chibougamau Lake area

```
Esso Resources Canada Ltd (CDN)
                 \
        ESSO MINERALS CANADA LTD (CDN)
```

of Quebec, (on which the operator is McFinley Red Lake Mines Ltd) (10).

However, in 1989 it sold out its 23% interest in the Musselwhite gold prospect in Ontario to its three former partners, Placer Dome, Inco and Corona, and relinquished its 35% equity in Les Mines Selbaie, near Matagami (NW Quebec) to Billiton International Metals, where the operator is BP Canada (11).

References: (1) *FT* 30/9/81. (2) *Project North Newsletters* 1981/82 *passim*. (3) *FT* 25/3/82; AkwN late Autumn/72. (4) *Yellowcake Road*. (5) Merlin. (6) *MJ* 13/7/79. (7) *MAR* 1982. (8) *MJ* 16/10/81. (9) *MAR* 1981. (10) *FT Min* 1990. (11) *MAR* 1990.

236 Esso Resources Canada Ltd

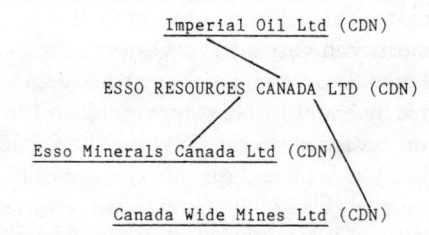

```
            Imperial Oil Ltd (CDN)


        ESSO RESOURCES CANADA LTD (CDN)


  Esso Minerals Canada Ltd (CDN)


            Canada Wide Mines Ltd (CDN)
```

As part of the world's biggest corporate entity, Exxon, Esso in Canada is involved in a number of energy-exploitation projects (see Esso Minerals Canada), but also in uranium exploring.

It has a 50% interest (technically owned, since 1979, by Canada Wide Mines, a wholly owned subsidiary) (1), in and operates, the Midwest Lake JV uranium project in Saskatchewan, along with Numac and Bow Valley (25% interest each). Completion date was expected to be 1986; mine and mill studies have been under way (2). Estimated reserves are 2 million tonnes of rock averaging 1.25% uranium oxide: ore grade was revised downwards from 3.4% in the course of 1979 (3), nonetheless, with an estimated 26,000t of uranium oxide the deposit is moderately important. In December 1981, however, Canada Wide Mines announced that development of Midwest would be held up because of poor uranium markets: commence-

ment had been scheduled for 1982 with production in 1985 (4).

The company sold to Seabright Resources Inc of Canada, some 540 claims and a 1350 tonnes per day (tpd) mill at Gays River, Nova Scotia in 1985, thus relinquishing gold claims and access to possible lead and zinc deposits (5). Fieldwork on the Obed natural sour gas field near Edson, Alberta – from which sulphur is extractable – continued during 1989-90 (6).

Test drilling for uranium at Midwest also continued during this period (6). Redstone Resources (40% owned by Franco-Nevada Mining) purchased a 2% gross royalty in the project from Esso Resources for C$3 million (7).

References: (1) *MAR* 1981. (2) *MM* 1/82. (3) *FT* 11/1/79; *MIY* 1982. (4) *Star Phoenix*, Canada, 4/12/81. (5) *FT Min* 1990 (6) *MAR* 1990. (7) *MJ* 23/11/90.

237 Everest Minerals Corp (USA)

In late 1985, Everest Minerals Corp announced a huge contract to deliver 3.5 million pounds of yellowcake to the CEGB for the British nuclear programme, over a period of 10 years commencing in 1987. A similar contract was awarded to Energy Fuels (Nuclear) Ltd of Denver (1). These contracts effectively replaced those negotiated between the CEGB and Rössing Uranium Ltd in Namibia, which had been the subject of condemnation throughout the world.

In 1984, Everest began test leaching at Exxon's uranium properties north of Douglas, Wyoming, together with Wold Nuclear Corp (Everest 70%; Wold 30%). Reserves on the 7000 acres of Exxon's property were estimated at around 25 million pounds of U_3O_8, and acquired initially by Wold (2). Although the mine was due to come on stream in "1985 or 1986" (2) it was delayed until 1988 and, at the time of writing, the project – known as Highland and West Highland – was expected to come into production that year (3). In the meantime, Everest was also increasing production at its Hobson solution mine, as it brought a Tex-l and Gruy-7b Satellite *(sic)* into operation (3), although the company permanantly closed its Mt Lucas solution mine in South Texas in 1987 (4).

In mid-1989, the British CEGB (Central Electricity Generating Board) bought up Everest Minerals, and with it the Hobson and Highland projects, with their promise of low-cost and easily expandable output (5). That year, the Highland operation produced 700,000 pounds of U308 and 150,000 pounds was turned out at Hobson (6).

References: (1) *FT* 6/11/85. (2) *E&MJ* 8/83. (3) *E&MJ* 4/88. (4) *MAR* 1988. (5) *Nuclear Fuel* 1/5/89. (6) *E&MJ* 3/90.

238 Evergreen Minerals Inc (USA)

The company held property on the Spokane Indian reservation in the late 70s (1) and began a uranium drilling programme there in 1977 along with Midnite Mines (2).

References: (1) *MJ* 21/10/77. (2) *MJ* 13/5/77.

239 Exxon Corp

Exxon's catchphrase, "Put a tiger in your tank", was the 1960s brainchild of the corporation's executive Ernest Dichter who wanted to create a "Global Shopping Center" run by Exxon (1). The sheer size of this largest of all conglomerates can be gauged from the fact that, in 1980, Exxon's sales were larger than the gross national product (GNP) of all but some twenty countries in the world (2). That year, it produced 11.4 million tons of coal, 39,000 tons of copper, and 3.6 million pounds of uranium oxide (2). Esso UK is Britain's fourth most important corporation in terms of profits, shooting to a lead position from almost nowhere in a bare twenty years (3). In 1984, it turned in a record £1 billion to its US parent (4). Esso AG in the early 1980s ranked as the twenty-ninth biggest West German industrial concern, while Esso

SA Francaise ranked as 39th in France (2). In Australia, Esso Exploration and Production is the tenth largest domestic company, while Esso Australia Ltd is number 42 (5).

Worldwide, Exxon oil and gas operations touch virtually every major resource region, ranging from JVs with AOG in Australia (6), with Royal Dutch/Shell in Mozambique in the north Rovuma basin (7), to exploitation of certain oil and gas interests acquired from WR Grace in 1985 at a cost of US$126.5M (8). In 1987, it also bid for control of Canada's Dome Petroleum (9).

Exxon has been rightly called "probably one of the two most strategic companies in the US economy" (1) (the other being General Motors). Although only number 38 in terms of defence contracts (10), by the early 1970s Exxon had acquired the largest reserves of uranium of any oil major, except Gulf (11). Its ascendancy in virtually every aspect of nuclear fuel production and enrichment has been exceeded by few other US corporations. In 1985, Exxon possessed nearly US$100,000,000,000 of world-wide oil and gas reserves, was collecting nearly US$7,000 million in chemical revenues and almost US$400 million from coal (7). The previous year it shot ahead of General Motors as America's biggest profit earner by no less than US$800 million (12). In 1986 it confirmed its position as "the biggest profit earner in corporate America" (13).

Such a scale of income accretion – unexampled in the field of human endeavour (or rapaciousness) – has not been won without "sacrifice". Indeed, the losses chalked up by the company in the field of mineral exploitation (including uranium) over the past decade might have ruined any lesser mortal entity. And the combination of ruthlessness and adaptability, which enabled the original Standard Oil Trust to achieve unique dominance over oil supplies and trading in the early years of the twentieth century, have been urgently required on many occasions in recent years.

Starting an unstoppable process, in 1951 Premier Mussadeq of Iran nationalised Exxon (and other western oil assets) only to be removed by the CIA somewhat later (14). In 1962, Ceylon nationalised Esso's oil interests, causing US aid to be cut off (1).

The company's Philippine oil interests were nationalised in 1973 (12) after the company's Philippines subsidiary had refused to sell oil to the US Navy at Subic Bay because "its overriding interest was to help enforce the world-wide Arab boycott of the United States" (1).

Venezuela also nationalised the company in 1975, while other governments, such as Nigeria's, increased their equity in Exxon's operations (12). Nor has the process come to a stop, even as the oil price has taken a dive. Esso-Papas was taken over by the Greek government in 1984, very much a "new socialist" move (15). A year later, Norway's Statoil bought up the Swedish oil and petrochemical operations of the company for US$270 million (16) and the Finnish Kemira took over Esso Chemie of the Netherlands around the same time (17).

Not surprisingly, Exxon's income remained static during most of the 1970s (12) but it picked up dramatically at the turn of the decade, as the company began diversifying at home into "synfuels", biotechnology, electric motors, and office systems; and as it consoli-

dated operations in poorer third world countries, opening their door to foreign investment. (In Brazil in 1976, for example, Exxon's ratio of surplus generated to capital employed was no less than 62.33%) (18). It also began trimming its coat elsewhere to suit its cloth – specifically its reduced income from oil (12).

The trimming continued well into the 1980s – with the selling of its office systems subsidiary in 1984 for US$100 million (19) and a reduction in capital spending by US$2.8 billion in 1986 (20). That year, Exxon drastically restructured itself, closing down Esso Europe and other affiliates based in the US, absorbing Exxon International into a "new international company" and forming a new company to coordinate coal and minerals activities (21).

By then, the company was still turning in "fabulous" profits from its upstream activities, but losing out "just about everywhere else" (22). It started buying up its own shares, as the result of a surplus of cash and diminishing opportunities to invest it profitably. As the *Financial Times* reported in 1985, Exxon had "entirely lost confidence in its ability to diversify" (22).

As Exxon passed its 120th birthday (January 10th 1990), it is remarkable that the successor

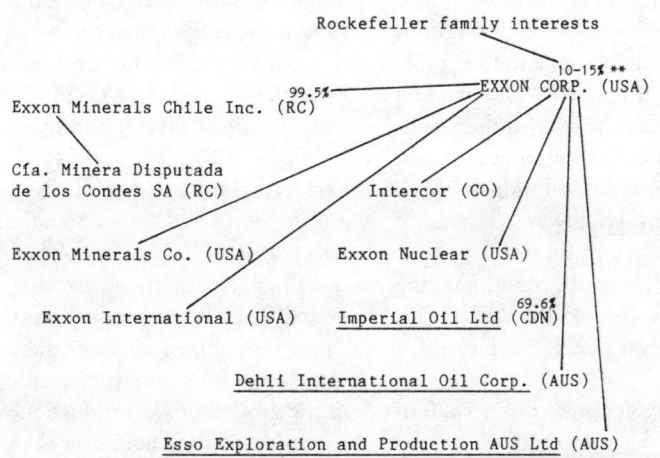

Rockefeller family interests

to Standard Oil should still be based so firmly in the exploitation of oil, and that it should still operate within the mind-set adopted by the company's founder, John D Rockefeller: preserving a strong, almost cabalistic centre; sending in "an army to tackle a problem, where competitiors would make do with a platoon" (22); deploying unprecedented aggregates of windfall profit to march on areas where few have trodden, and quitting just as dramatically when the going got tough; above all, regarding the whole globe as its oyster, and government as the tool (as in Iran and Chile) to establish itself wherever it chooses to go.

It is only within living memory that Exxon – at a cost of no less than US$100 million (1) – changed its name from Standard Oil. The question of who controls this megalith is one that has exercised a number of keen minds – not the least US Senator Lee Metcalf who, in 1972, failed to gain from the Securities and Exchange Commission a list of the company's shareholders and met with the rebuff that the information was "privileged and confidential" from the corporation itself (1).

Beyond doubt, in the latter half of the 19th century, Standard Oil was synonmyous with one man: "the history of oil [at this time] was also the personal history of John D Rockefeller, who tamed an anarchic industry and brought it under the direct control of Standard Oil" (23). In 1863, Rockefeller purchased a Cleveland oil refinery. Seven years later he formed the Standard Oil Trust, manifestly to monopolise not only the US, but the world's oil industry (24). By the turn of the century, however, strenuous competition had emerged in the shape of Gulf Oil, Texaco and Shell Transport and Trading on the home front, Nobel and the Rothschilds in Russia, and the Royal Dutch Petroleum Co in the Far East (24).

Rockefeller was hardly deterred. From 1870 to 1877, with unparalleled ruthlessness, he virtually obliterated all competition in the US. "When he began his campaign, the Standard was flanked by 15 refineries in New York, 12 in Philadelphia, 22 in Pittsburgh and 27 in the oil regions. When he finished, there was only the

Standard" (25). By 1880, Rockefeller was refining no less than 95% of all oil in the USA (25). Two years later, Rockefeller consolidated his stranglehold by forming the Standard Oil Trust, and concentrating ownership in the hands of only nine trustees. The Trust embraced forty corporations of which fourteen were wholly owned. Under this arrangement "... it was never clear who owned what or who was responsible for which actions" (25). The Trust then moved overseas, setting up subsidiaries such as the Anglo-American Oil Co, making secret purchases of stock, and engaging in a price-cutting war. Although the Trust's success was far from complete, by the 1890s "...[as] in a Conrad novel, Standard agents were rushing into the hearts of darkness everywhere, carrying their products by sampans, camels, oxen and on the backs of native bearers ... A transoceanic empire lay before them ... the Era of American Economic Supremacy had begun" (25).

By 1911, advocates of free trade and competition gained the upper hand, and the US government felt strong enough to break up the Trust, thus spawning today's Standard Oil of New Jersey (Exxon), Standard Oil of Ohio (Sohio) and Standard Oil of Indiana (Amoco). Smaller split-off operations later grew to considerable size as mainly domestic companies – companies like Atlantic (now Arco), Tide Water (to be absorbed into Getty), Continental (Conoco), Phillips and Cities Service (24).

The formal price and territorial war reached an armistice in 1928, when Standard Oil of New Jersey, BP and Royal Dutch/Shell concluded an "as is" agreement, fixing prices at the high level of the Texas/Gulf production zone. The Big Three were joined by four others over the next two decades. Despite (or because of) a decline in the growth of US production, the big seven consolidated their hold over the world market, especially the Middle East, throughout the 1950s, 1960s and 1970s. Exxon never moved from the top position in this period. Until as late as 1939, it remained the only truly international corporation among the so-called "Sisters" and on the eve of the OPEC "revolution" was still "the kingpin" (24).

Oil denoted not only control of energy resources and therefore the lifeblood of industry, but the most powerful generator of cash and thus control of US banking. John D Rockefeller and JP Morgan (see US Steel) formed an alliance more than eighty years ago which, by the 1960s, gave the Rockefeller-Morgan nexus dominance not only of America's largest banks, but key sectors of mining, airlines (24), insurance (26), philanthropy, academic institutions, and, of course, politics. Exxon's political connections and chicanery could fill a large volume in their own right. Nelson Rockefeller's influence over US policy in South America, and specifically support for right-wing régimes (27), was mirrored by the co-operation between Exxon and IG Farben – the powerhouse of the Nazi state – even when the administration had decided on war with Germany (28).

One of IG Farben's key officers was Prince Bernhard of the Netherlands who built up a considerable private stake in Exxon, and chaired the first meeting of the Bilderberg Commission in Oosterbeek, 1954. Little wonder that David Rockefeller was also a founding member of Bilderberg – this prime means of "western collective management of the world order" (29); Emilio Collado, executive of Exxon and a director of JP Morgan, was another co-conspirator (29).

Then in 1973, Rockefeller, flanked by Zbigniew Brzezinski and other prominent US Americans, including Exxon director James K Jamieson, founded the Trilateral Commission – a most important vehicle for the propagation of western values, and maintenance of western industrial power, outside of government (29).

Coal

From 1965, Exxon entered the coal industry through "extensive grass-roots" activities (2) – a euphemism for a game-plan conceived in the early part of the decade, as the threat to Exxon's position as a world oil producer from "nationalist" forces in the third world became obvious.

This was a period when Exxon moved concertedly into "safe" areas, such as Australia, the North Sea, the MacKenzie River Basin of Canada and Alaska's North Slope (30). Crucial to the Grand Plan for US energy self-sufficiency was the transfer of coal reserves – their price driven to rock-bottom because of cheap oil – to corporate control. Within a short period, particularly 1964 to 1967, under the direction of Secretary of the Interior Udall, Exxon, Conoco, Amax, Burlington Northern and Union Pacific bought up a huge share (31%) of Indian and public lands in the Four Corners region and the Great Plains, and Exxon was the biggest acquisitor of coal reserves among them (30). Indeed within a decade, Exxon was the largest holder of coal reserves among oil companies (31).

Within a few years, the company was sustaining heavy losses from its coal operations, but by the late 1970s had its act together. In 1980, it was the USA's fifth largest producer of coal (2), and by 1985 was reaping nearly four and a half hundred million dollars in annual coal income (32), with four operating mines in the USA, and – from 1981 – a JV with Denison at Quintette (16.75%) in Canada (33). Esso Australia holds 25% of the Hail Creek coking coal mine in Queensland (AAR Ltd 44%, IOL Coal Pty Ltd 2%, Marubeni Coal Pty Ltd 4%, Sumisho Coal Dev Pty Ltd 25%) which, though developed ahead of schedule (34) was still looking for customers (32). One potential customer is Hong Kong, with whom Exxon is co-operating in the construction of a coal-fired power station (35). Talks were being conducted with the Hong Kong utility China Light and Power in 1987, regarding the building of another power station on Tsingyi Island, but it is not known whether this too would be coal-powered (36).

The company's largest coal involvement by far is its 50% share in the El Cerrejon mine (owned and operated by Intercor, in which the Colombian government agency Carbocol holds the other half). Exxon developed El Cerrejon using windfall profits from its Dutch natural gas operations, and with the permission of the Dutch government (37). This will be South

America's vastest coal mine (38) with an initial investment of US$3.2 billion. "All the numbers associated with El Cerrejon are staggering" commented the President of Carbocol in 1984 (39). Seen as "setting the pattern for the steam coal industry" (40) by the time it began exporting in early 1985, the project was ahead of. schedule, raising expectations that it could provide up to 10% of the world market for coal for power (41).

But the project has been beset by controversy, especially over its phenomenal costs. In the early years, several Carbocol officials resigned, claiming that the project "endangered Colombia's long-term economic welfare", comprising "a monstrosity of enormous injury to our national interests" (42). Exxon was exempted from certain taxes, benefitting from "high profit remission rates not granted other foreign firms" (42) while, critics claimed, there was no firm guarantee of a transfer of technology, and job prospects for Colombians had been grossly exaggerated (42).

By 1986, the costs remained "crushing" – with Carbocol's foreign debt reaching no less than US$1.4 billion – and in 1987, with steam coal over-supply threatening El Cerrejon's profitability, the Colombian government was looking for a buyer for its own 49% interest (37) – Shell being one possibility (7).

Total 1987 production (9 million tonnes) was set to make Intercor the second largest supplier in the world (after Poland's Weglokoks) with predictions that it would reach the number 1 spot by 1990 (43). Carbocol concluded its first contract with the Danish Elsam in 1980 (44), and is likely to be supplying Sweden, Spain, Brazil, Korea, Japan, the Netherlands and Britain, as well as the USA (45). This, despite opposition from the US United Mineworkers Union, who campaigned against sales to US Gulf utilities in 1984 (46).

Venezuelan and Colombian indigenous organisations – notably CRIC (Colombia), MIIN (Indigenous Movement for National Identity), and ACIPY (The Indigenous Civil Association of the Yukpa People) (the latter two Venezuelan) – have expressed particular alarm at the El Cerrejon developments, not only in the indigenous regions of the Guajira and the Rio Guasare basins, but on Wayuu territory, which straddles the Venezuelan-Colombian border. In particular, an airstrip contracted by Intercor in the early 1980s "has divided our territory, destroying our territorial integrity" (47).

In 1987, the Colombian daily *El Heraldo* claimed that the ELN (National Liberation Army) had dynamited a new 600-mile pipeline around 30 times in the previous 18 months, threatening a further sabotage if Intercor and Carbocol didn't pay several million dollars in protection insurance. The company retorted that the mine "never had this problem, perhaps because historically the Guajira Desert has never been an area accessible to guerilla activity" (48).

Metals

"As the world's largest private company, Exxon commands the wherewithal to pursue mining from scratch perhaps better than any other firm" commented one observer in the 1980s (2). But despite this promise the company's mineral interests "have proven the least profitable" sector of its activities, representing "almost a 'worst case' scenario" (2). By the early 1980s, Exxon's income from its copper, lead, zinc, molybdenum and uranium interests were "clearly dreadful" (2). While investment in coal had tripled, the company cut back drastically on metals and nuclear energy development.

In 1985, the company lost US$21 million in minerals revenues (7). No wonder that, in 1987, Exxon terminated all mineral exploration in the USA, after cutting 60% off its exploration budget; in turn its exploration budget for Canada was cut by 20% and Exxon Minerals announced that it would, in future, concentrate only on the acquisition of polymetallic and precious metals properties (49).

One of its most recent Canadian ventures is a gold JV at Opapimiskan Lake, Ontario, with Dome Exploration and the Canadian Nickel Company (6). Significantly, Exxon has also announced the selling off of all its exploration projects in Australia and Papua New Guinea for US$44 million (32). This decision means that Esso will no longer hold a 50% share in the Harbour Lights gold project in Western Australia, where its partners were Carr Boyd and Aztec Exploration, and its prospects of becoming the fourth largest gold producer in Western Australia were good (50). Nor will it proceed with its Bimurra (Queensland) JV with Samantha Exploration, Buka Minerals and Samson Exploration, announced in 1980 (51).

However, Esso Resources of Australia will retain its 31.16% interest in the Golden Grove multi-metal deposit at Gossan Hill and Scuddles in Western Australia (where copper, zinc, gold and silver reserves total 36 million tons) (48) and where Esso is partnered by EZ (31.16%), Amax (31.16%), and Aztec (6.5%). City Resources of Australia has bid for all Esso's gold interests in Australia and Papua New Guinea (52).

In 1984, Esso Resources of Canada entered a JV with Lasmin Resorces and St Andrew's Goldfields to explore for three years certain claims in the Lasmin Esso Resources JV area of Cochrane, Ontario (6). While it gave up its share in a Canadian copper-molybdenum project to its partner Granduc Mines in 1985 (6), it was awaiting final decision on another molybdenum project in 1987. This is the huge (30,000 tons/day ore) US$600 million prospect at Mount Hope, Nevada – a venture which will require 10,000 acres of public land (48), disturb some 3,000 acres, and dump tailings over another 2,500 acres and some 35 miles of public right of way (53). Little wonder that this "biggest mining venture in Nevada history" (54) has still been awaiting environmental approval (48). In 1982, Exxon ceded all rights in its Pinos Altos (New Mexico) copper-zinc-silver prospect to Boliden AB (55).

Exxon's exploration programmes continue in West Germany and Spain, with reconnaissance in several other countries (6, 55), but with severe reductions in north America and Australia.

At Crandon, in Wisconsin – perhaps its most controversial project of modern times (see below) – development of a US$900 million copper-zinc-gold and lead project has been suspended.

This means that the company's most important mineral project currently in operation is its Disputada complex in Chile (El Soldado and Los Bronces) which it acquired in early 1978, when its subsidiary Exxon Minerals Chile Inc (EMCI) purchased the Disputada de las Condes SA, from the Chilean junta and private investors for US$112 million (56). Exxon's Disputada investment is the largest foreign investment in the country (57).

Los Bronces is an open-pit mine 70 miles north-east of Santiago, while El Soldado is an underground mine, 130km north-west of the capital. Both a concentrator (El Cobre) and smelter (Chagres) are close to the mine (6). Initially the project proved highly unprofitable (2). But in 1985, plans were announced to double production at El Soldado (58) with full output of 11,500 tons/day copper sulphide ore (at 1.5% copper content) (48). Meanwhile, plans to expand Los Bronces have been "downgraded" (48).

Exxon's decision to invest so heavily in Chile, following the 1973 reactionary military coup of Pinochet, was highly significant for the dictatorship: until then, its efforts to secure big foreign capital had been largely unsuccessful. Said *Business Latin America* at the time "[the investment] constitutes a public relations breakthrough as much as an economic milestone for this country. Not only is it the largest single investment by a US firm in many years, but it has been made by a large, image-conscious corporation indicating that international business is giving its blessing to the Chilean military junta" (59).

The importance of the 1978 Exxon deal for the Chilean military is underlined by the fact that, prior to Exxon's acquisition, Disputada had been losing money for nine years (54).

Synfuels

In 1979, Exxon took a turn which led directly to the diversifications and divestments of the

357

1980s. It began to invest enormously in so-called synthetic fuels: also it bid for Reliance Electric.

This new strategy was predicated on Exxon's determination to maintain control over future energy developments, without sacrificing its lead as the world's biggest supplier of oil and petroleum products. As the company's "burly" chairman, Clifton Garvin, mapped it out at the time: "... we'll be moving ahead in three broad directions ... Oil, gas and petrochemical activities will continue to be major segments of our business. At the same time we'll be investing increasingly in alternative energy areas" (60). Earlier that year, under the pressure of nationalisation from third world producers, and the revolution in Iran, Exxon had decided to become a net buyer of oil, and cut back its production by 15%. By "alternative energy" Garvin clearly meant "the transformation of coal into more acceptable liquid or gas forms and oil shale", although in 1979 he made an obligatory nod in the direction of solar power (60).

In fact Exxon had long dug into the solar market. When ERDA (the Energy Research and Development Administration) was set up by the US government in 1975, the company secured representation on no less than 15 separate advisory committees (61). Two years later, Exxon's Daystar subsidiary was awarded a total of US$1.1 million in contracts for heating and hot water systems disbursed through a government agency (HUD). One of Daystar's solar products was a solar collector "developed by aerospace engineers in a government-funded corporate laboratory [with Exxon] using its connections in Washington to subsidise the growth of its acquisition to a point where it already shares domination of the solar industry with ten or twelve other major manufacturers, and has written an "expansion" programme to suit the needs of its other corporate interests" (61). In particular, the other interests "fed" by Exxon's forays into solar power included laser-fusion and specifically a project conducted since 1972 with General Electric to "rejuvenate" spent nuclear fuel rods (61).

That Exxon was no more serious about devel-oping solar power in its own right, as a truly alternative source of energy for human needs, was exemplified in a shabby advertising campaign run in late 1976 in major American dailies, such as the *New York Times*. "Exxon answers questions about one of the newest sources of energy under the sun – The Sun!" ran the headline. Underneath, Exxon claimed that home space heating could cost up to US$20,000 using solar collectors, and around US$2,000 to provide domestic hot water "about three times more than conventional energy systems". "When will solar power become a major source of energy?" asked Exxon, answering itself: "Possibly in the next century."

These advertisments drew complaints about flagrant distortion from Senator Gary Hart (later to achieve notoriety in the Presidential nominations race of 1987) and Representative Richard Ottinger, who pointed out that Daystar was marketing hot water systems for half the price claimed by its own parent company. So – "Why is Exxon involved in solar energy?" asked Ray Reece in his exposé of the corporate seizure of US solar energy development. "[It] might simply be that Exxon intends for solar energy to be kept under wraps until fossil fuel markets are exhausted" (61).

In fact, by 1983, the company's Solar Power Corporation had failed at a cost of US$30 million because – according to one source – the photovoltaic market "would not penetrate the Third World's rural areas fast enough or profitably enough for the oil giant" (62). Exxon's vice-president John Wurmser reflected that "... no-one asks what would happen if three billion people (in the Third World) who don't have electricity today can't afford photovoltaics".

So, although Garvin forsees the future as "fundamentally electricity based" (63), Exxon has no concern for cheap or democratically controlled ways of reaching that goal, let alone "softer" energy developments which depend on truly renewable or benign sources of power and are appropriate for heating and cooking as well as lighting, industrial processes, and transport.

This was exemplified by Exxon's most significant and controversial take-over in its recent history, that of Reliance Electric. This marked an important strategic departure, since it was the company's first major acquisition, as opposed to "ground floor" development – as in uranium, coal, copper and chemicals (64).

During the 1970s Exxon was building up considerable strength in the field of advanced electronics. By the turn of the decade it was already marketing word processors, electronic typewriters, and facsimile transmission devices (64), after acquiring half a dozen small high-technology companies (65) under the aegis of its Exxon Enterprises subsidiary.

But it was not until 1979 that the writing was clearly on the (office) wall, when it bid for Reliance Electric, in order to consolidate commercial exploitation of its alternative current synthesiser (ACS).

ACS was trumpeted as a device that would control the speed of electric motors by altering the frequency and voltage of ordinary mains electricity, primarily for motors between 1000 and 2000 horse-power, but also giving "energy savings in heating, ventilation and air conditioning" (64). Reliance Electric was not only the biggest producer of electrically operated vehicles, but also a large manufacturer of electrical motors, power transmission equipment and telecommunications equipment (64).

Exxon paid US$1.2 billion for Reliance, only to be met with political opposition (66) and a charge of "reducing competition" by the Federal Trade Commission (FTC). Although cleared by the FTC three years later (67), by then the ACS scheme had run into mounting difficulties, with costs and "reliability problems" under heavy scrutiny (68). In 1986 Exxon sold off Reliance after a foray which "looked increasingly ill-advised" (69). Thus ended "perhaps the most costly and extraordinary diversification plan proposed by any of the oil majors ... to exploit a technical breakthrough in electrical motor design" (70).

Unfortunately, Exxon has not so easily let go its plans to exploit synthetic fuels, in particular oil shale – a key part of President Carter's energy independence programme in the late 1970s which, theoretically, promises self-sufficiency in oil for 75 years, from deposits in Colorado, Utah and Wyoming alone (71). It also promises open-pit mines creating "vast gashes on the countryside three miles long by one-and-a-half miles wide" (71).

The company laid out its synfuel plans in a report entitled *The role of synthetic fuels in the US energy future* published in 1980. That report acknowledged that profitability could be affected by "costly environmental standards" but projected oil shale as "by far the largest part of the programme", followed by the liquefaction and gasification of coal (72). (At the same time, Exxon's research unit funded an MIT study into the burning of coal, coal liquefaction, shale oil and heavy crude petroleum – the results were not expected until 1990) (73).

After that, Exxon invested heavily in the Colony, Parachute, Colorado oil shale project, 60% of which it bought from Arco in 1980 at a cost of US$400 million (74) and was "bankrolling" to the tune of US$3.56 million the following year (69). However, it backed out of Colony the following year (an example of "decolonisation"?) (75). But it beat CRA, BHP, and BP (among others) in the bid for a lion's share of the Rundle oil shale deposits in Queensland, Australia, (76) a JV which it shares with Southern Pacific Petroleum and Central Pacific Minerals. In 1980 a large new deposit was discovered at Rundle, promising 2.02 billion barrels of oil, at 40 litres per tonne cut-off. Early predictions that Rundle could be in petroleum production by 1984 and provide as much as 25% of Australia's domestic petroleum requirements (77) were over-optimistic. In 1984, the Rundle agreement had to be renegotiated (78) but reportedly "satisfactory" development continued through 1986 (79).

Exxon's Allsands project at Cold Lake, Alberta, Canada, currently produces around 20,000 barrels of oil per day. Exxon (Imperial Oil) dithered considerably, after Shell Canada pulled out in 1982 (80) and costs escalated (81). By 1985, a C$4 million expansion programme had been proposed, raising the company's capital expenditure at Cold Lake to C$1000 million and its output to 90 thousand barrels a day by

1988 (82). However, approval for this was still required from the Energy Resources Conservation Board (83).

Through Imperial Oil, Exxon also holds a stake in the Syncrude project in the Athabasca region of north-east Alberta – one of Canada's two commercial oil-sands projects (32).

Uranium

Exxon began mining uranium in the 1960s and its involvement has continued to the present day (12), even though its uranium and nuclear fuel fabrication interests have proved consistent financial losers (2).

Exxon Nuclear Company was founded in 1969 to manage and market nuclear products and services and consolidate the inroads the company had already made into nuclear fuel provision. As with its entrée into coal, Exxon made a clean sweep – hiring staff, purchasing leases and equipment – from the ground up, and using the best personnel available from the nuclear industry, national laboratories and the universities (2).

Within a decade Exxon Nuclear had spent some US$200 million, was employing 800 "highly-qualified personnel" and had mapped out plans for potential investment of some US$2.1 billion by the turn of the century, most of it on the development of an enrichment plant and reprocessing plant (84). Based in Richland, Washington (site of much war-time development of nuclear weapons), by the late 1970s the Exxon enrichment project had started assembling and testing its first super-speed enrichment centrifuges, with a view to developing a 3000 tonnes of SWU (separative work unit) capacity (84). Plans for a reprocessing plant – the only ones finally submitted to ERDA for approval – envisaged operations at Exxon's site near the ERDA Reservation (*sic*) in Oak Ridge, Tennessee, as starting in 1986, providing political hurdles were overcome first (84). In 1981, Exxon planned to spend US$13.5 billion on new capital projects with "nuclear" as a priority (85).

In fact, over the following decade, Exxon has abandoned its plans to build a separate ultra-centrifuge enrichment plant, and plans for a re-processing plant are on the back-burner. Nonetheless, Exxon Nuclear has pioneered development of laser-enrichment, a technology which – as described by its inventor, Dr Karl Cohen of Exxon – promises 100% "no waste" enrichment of uranium (12, 86); and together with Arco, it is a manufacturer of nuclear fuel assemblies (12). In 1986, Siemens of West Germany was holding talks with Exxon Nuclear, to gain access to its fuel elements, their technology and transfer, and possibly a stake in the US operations (87).

While the company's grand design for uranium production has also taken a nose-dive – leaving only Esso Resources Canada's 50% stake in the extremely promising Midwest Lake JV in Saskatchewan as a viable project for the future – Exxon continues to play a crucial role in the provision of uranium for US utilities. While it is not known exactly which utilities have contracted with Exxon over the past twenty years, four of them – Duquesne Light Co, Cleveland Electric Illuminating Co, the Ohio Edison Co and the Toledo Edison Co, (the Central Area Power Co-ordination Group) – in 1979 filed suit against Exxon Nuclear for alleged failure to deliver uranium concentrates (88).

Between 1983 and 1989 Exxon contracted for 2,000,000 pounds of uranium equivalent per year from Gencor (89), probably to be enriched by BNFL (90) and possibly destined for the Boston Edison Company – though Boston Edison itself has contracts for uranium supply from Palabora, the RTZ-managed mine in South Africa.

Exxon first struck yellow oil with the discovery of one of the largest uranium deposits in the US in 1968. By 1972 the Highland mine and mill in Wyoming were in operation (2) with 2000 tons per day throughput of ore (91). An underground mine was later developed, and by 1980 this mine was accounting for nearly one-tenth (9%) of total US uranium production (2). By then Exxon was the third biggest uranium producer in the country (2) and had 10% of US milling capacity (92).

This was not all. The company was already mining uranium in Texas (93) and, in short order, had the following projects underway:

- In Western Australia, a deal agreed with Western Mining Corporation (WMC) and UG for a 15% interest in the Yeelirrie mine (94, 95). Esso Exploration and Production Australia Ltd was its operating arm (96). A year later Esso Australia also took a 51% stake in the Pandanus Creek JV with Otter and Carr Boyd. It was also expressing interest in acquiring a stake in Roxby Downs (see WMC) (95) and exploring in the Northern Territory (97).
- A uranium search in the Seaward Peninsula of Alaska (98).
- Exploring for uranium in Arizona (99).
- In South Africa (along with US Steel) an exploration programme started in the late 1970s (100).
- A search started in 1978 by Esso AG in Bavaria, east of Nuremberg, at the Fichtelgebirge mountain range, bordering on Czechoslovakia, with plans to sink a shaft to test commercial viability (101). Costs were set at £8.9 million, but five times this was estimated as necessary for production (102). By the mid-1980s test drilling had revealed some 80 potential uranium lodes (103).
- Esso Minerals Niger Inc took out sizeable claims in the Tazole area near Techili (104). It also had a JV with Onarem (105).
- Exxon contracted a deal with Kerr-McGee to process 500 tons/day of its South Powder River uranium deposit in Wyoming (106).
- It drilled near West Milford, in Passaic County, New Jersey (107).
- By early May 1979 (along with Arco/Anaconda and Phillips) Exxon secured potential uranium leases (two or three dozen in all) in Norman County, and four leases in Mahnomen county, Minnesota (108).
- Drilling also started in Michigan around the same time (169).
- It obtained three leases in the Grand Canyon, thanks to the US National Park Service, and despite the Park Service itself rejecting an Exxon application (the access roads

alone would "irreparably scar the land") (109).
- It made sizeable uranium claims on land just east of Cerillo and south of Sante Fe, New Mexico (also near claims made by Union Carbide and Lone Star Mining) (110). This is presumed to be the same lease of 60,000 acres of Laguna Pueblo lands in which Exxon had gained an 87.5% interest in 1976 and located an estimated 6 million pounds of U_3O_8 (111).
- The company negotiated in 1974, and amended in 1976, an exploration permit and mining lease for uranium on a huge tract of Navajo land in New Mexico, approval for which was granted by the Secretary of the Interior in early 1977 (112). Exxon got the right to seek out uranium over 400,000 acres, of which just over 50,000 could be set aside for mining (113).

This vast acreage was not all: a year later, Exxon leased the entire Canoncito Band Navajo reservation of about 92,000 acres (114). With the acquisition of these leases – in the same area near Shiprock where the majority of Navajo and white miners contracted cancer from Kerr-McGee's operations in the 1950s and 1960s (115) – Exxon in the late 1970s became by far the largest holder of uranium leases in the San Juan belt (116).

- At another 2.5-acre site in the San Juan region, on the L-Bar ranch, east of Mount Taylor where Exxon had a 60,000 acre lease, the company planned to adopt a uranium leaching process used in Texas and Wyoming. The company's application mentioned simply a "small-scale *in situ* project ... needed to allow the feasibility of a large-scale *in situ* mining operation to be evaluated" (117).
- By 1978, high incidences of uranium were reported in the copper-zinc lode on Exxon's Crandon prospect in Wisconsin. Although the company refused to indicate its interest in uranium by-product extraction, the Center for Alternative Mining Development Policy (CAMDP) raised the issue with the state government (and was pointedly ignored) (118), and calculated that Exxon might be able to

leach no less than 17 million pounds of uranium oxide from the deposit – without a separate EIS (Environmental Impact Statement) being required (119). In 1980 Exxon also leased uranium rights to a large area (which includes the west end of Crandon), from Chicago Northwestern Railroad (120) at Forest River.

Ironically, Exxon's largest uranium losses in this period were recorded in 1978 – at a time when prices were highest (2). The spate of exploration projects announced in the late 1970s – perhaps designed at locating some magically rich, low-cost deposit, couldn't save Exxon's uranium programme. In 1979, a secret Exxon study concluded that there was no economic advantage in nuclear as against coal power (121), and in 1980, when the price of uranium slumped to its lowest for many years (122), Exxon announced plans to shut its operating mines by 1983 (123).

By then, Exxon's stay in the town of Douglas, Wyoming – near its two operating Wyoming mines – looked like being distinctly over-extended since it had not co-operated in funding local projects (124); its Highland tailings dam was leaking and repair efforts reportedly failing (125); extremely adverse radiation effects on fauna and flora near the mine were being reported by Dr Garth Kennington of the University of Wyoming – only to be ignored by the company (126).

In July 1980, the Wisconsin Legislative Council set up a sub-committee on uranium exploration safety, at which local groups demanded a moratorium. Exxon and Kerr-McGee refused to attend the preliminary hearings (127).

A year later, officials in Midway Township, Minnesota, denied Exxon permission to drill in the area (128).

Also in 1981, New Jersey banned uranium mining for seven years – a move which primarily affected Sohio, but also Exxon's claims in Passaic country (129).

A trans-European Stop Uranium Mining conference held in Fichtelgebirge, West Germany, and attended by at least forty local people, spe-cifically condemned Esso AG's uranium testing on their lands (130).

In 1982, Atlas Corp bought up Exxon's uranium properties at Bulldog, in Garfield County, Utah – a very high-grade deposit with significant incidences of uranium, where preparations were already in hand to construct a processing plant (55).

The same year, Exxon sold its 15% interest in Yeelirrie, Western Australia (131). The company stated that mining would not be economically viable under the terms of the JV agreement or its assessment of the market outlook. The announcement – greeted by WMC chair Arvi Parbo as a "major blow" (132) – followed Esso's announcement that it would take an additional 35% interest in Yeelirrie, a measure agreed to by the Federal Investment Review Board, despite its violating its own guidelines (133).

Finally, in 1983, as Exxon phased out its Wyoming operations "because of depleted reserves" (134), Everest Minerals of Texas agreed to buy most of Exxon's assets on the project, as well as its *in situ* pilot plant, while Wold Nuclear Corp of Casper, Wyoming, took over the 7000 acres of Exxon's unmined properties in the area (134).

With the close of 1983, Exxon's brash, two-decade foray into yellowcake was clearly coming to an end, at least on mainland USA.

Indigenous peoples

Exxon's operations in Ecuador (along with those of many other corporations, including Elf Aquitaine, Conoco, Occidental and BP) have caused considerable concern to human rights and indigenous support groups. Five main Indian peoples are affected by these encroachments on their lands in the Amazonian region of the country – the Quichua, Waorani, Cofan, Siona and Secoya (135). The encroachments started in the 1960s and intensified in the 1970s, driving out indigenous game, polluting the river with oil spills and destroying the fish (136). In July 1987, two missionaries, attempting to make contact with Waorani by landing in an oil company helicopter, were killed by the

Indians. Commenting on this tragic error, Survival International stated: "Oil exploration and the subsequent colonisation is without a doubt the major problem faced by Amazonian Indians in Ecuador". It asked the President of Ecuador to respect Indian lands and ensure that companies like Exxon withdrew immediately from Waorani land (137).

Esso's operations in Alaska and (through Esso Resources Canada) in northern Canada have long had a major impact on indigenous peoples in the region. In 1981 the Reagan administration voted to allow Exxon, together with Arco and Sohio, to go ahead with their plans for a 4800-mile natural gas pipeline from Prudhoe Bay to the northern mainland USA (138). (For details, see Arco entry in the *File*). But it is Exxon's operations in the Beaufort sea, and the so-called North Slope, which have most alarmed the Inuit and Dene. Esso's partners in these ventures have once again been Arco and Sohio, with Gulf (see the Gulf entry in the *File*). It was only in the mid-1980s that further large-scale exploration of the region began to fall off, with the plunge in the price of oil (139).

In some contrast to the off-hand approach Esso/Exxon has adopted with regard to the Inuit of the Beaufort Sea and the Indians of Ecuador, Esso Resources has forged what is widely credited as being a "unique" alliance with some indigenous peoples in the MacKenzie river delta. Through the seventies, the expanding Norman Wells pipeline and oilfields threatened the "livelihood and future" of the Dene in the North-West Territories of Canada (140). Then, in 1981, Esso announced that in order to draw oil from the remaining wells at Norman Wells – right under the bed of the MacKenzie river – it would build six artificial islands. Both the Dene Nation and the Department of the Environment criticised the proposals, fearing that an oil spill would obliterate all marine life in the region. "Don't play god with our lives or our livelihood," demanded Chief T'Sellie of Fort Good Hope (140). "It is beyond belief that this project has passed environmental assessment," he added, envisaging that one major spill could destroy much of the moose,

musk-rat, beaver, as well as millions of fish, on which the people depend. Within the next two years, however, the Dene began discussing a joint venture with Esso, thus "changing a position the Dene have held the last fifteen years or so" (141). Meanwhile Esso itself was "cleaning up its-act", with the adoption of oil spill contamination control measures which it claimed were virtually foolproof. These were developed with the assistance of Fort Good Dene and a consultant helping the band (142). By then Esso had reduced the number of islands to four – constructed in 1982 and early 1983 – and was allowing representatives of the indigenous inhabitants to monitor their performance (143). The same year, Shehtah Drilling was established as a partnership between the Dene and Metis and Esso, with a 2-year agreement during which the native people's loans would be repaid. Shehtah Drilling operated on the Dehcho island rig, and half its workforce was drawn from the Dene and Metis peoples (144). Even so, by 1986 the Dene were criticising the lack of training given workers on the Norman Wells project, and calling for "impact funding" of US$10.5M to be given directly to the people (145).

Exxon's companies have "a long history of involvement in exploration on Aboriginal land in Australia, dating back to the early 1970s when negotiations to undertake uranium exploration near Oenpelli in Arnhem Land were conducted" – and an agreement signed with the Oenpelli people (5). However, it has been praised for its sensitivity towards Aboriginal land rights demands (146) and its formal policy, as expressed in Esso Australia's submission to the WA Aboriginal Land Enquiry, goes further than almost any other mining company.

Esso has pledged to ensure that the "important and unique culture and traditions of the Aboriginal people" be "respected by Esso employees and reflected in the mode of company operations". It promises to brief its employees on Aboriginal history and traditions, demanding they respect Aboriginal traditions, culture and customs. It has also promised to respect sacred sites and sites of significance and to keep Aboriginal groups informed of activities potentially or actually affecting them (5).

In the Northern Territory (NT) in particular, Esso has had "numerous contacts with Aboriginal communities and Land Councils", including negotiating protection of sacred sites in the Nicholson River Land Claim area in 1979 and (until exploration stopped in 1981) consultations with the Aboriginal Sacred Sites Protection Authority at Daly River Stage Two Land Claim area, and a similar contact in the Banka Banka-Renner Springs area in 1982 (5).

During exploration of the Arafura Sea it negotiated navigation facilities and a helicopter landing pad in the Murgenella Reserve in 1983. Thereafter it stopped exploration of the NT.

In Western Australia (WA) its oil exploration leases affecting Aboriginal land are EP110 (on the Peedamulla pastoral lease near Onslow), EP104 over Dampier Land Peninsula, which includes the Beagle Bay reserve and Mowanjum mission (Esso 58.33% share), and the Lake Gregory pastoral leases and Balwina Reserve in central WA, where Esso has a 30% share while Pioneer Concrete has 14%.

At Harbour Lights, Esso signed an agreement with the Leonora Aboriginal community in 1984, while its explorations west of the Darling River, started in 1981, have involved extensive consultations with Aboriginal groups and occupants (5).

During this period, unfortunately, the parent company embarked on a project which has become a text-book demonstration of how corporate entities should *not* behave with indigenous people.

Exxon's copper-zinc deposit, discovered at Crandon, Wisconsin, in 1976, followed in the wake of a corporate "metals rush" in which around forty companies were involved (147). Exxon's prize was undoubtedly the most handsome: the location of what could be the world's largest such lode, yielding up to 125 million tons of ore (148): costs could reach US$1400 million (93) while a 600-acre site would be required for waste pond and treatment plant alone (148).

By 1977, the company had already drilled 2000 feet down at Crandon (149).

The deposit lies just two miles west of the Chippewa Sokoagon Reservation at Mole Lake, a community whose original 10,000-acre treaty lands were forfeited when an 1854 treaty was lost in a shipwreck. Although assigned a pitiful tract of land in 1937, the Sokoagon's reservation surrounds one of the most important wild rice stands in the region. The Mole Lake reservation was designed to give the people control forever of this wild rice, the water resources of Swamp Creek, and wildlife (150).

In October 1976, the Sokoagon turned down Exxon's derisory offer of US$20,000 for a mineral exploration lease, and laid claim to an area including the Crandon deposit (150).

Within the next four years, they linked up with other indigenous peoples, the Potawatomi, Oneida, Winnebago and, in particular, the Menominee, whose own reservation fifty miles downstream of Swamp Creek would be threatened by mercury, cadmium and lead pollution from the mining (151). At a conference in summer 1980, they were supported by Indians from throughout the US and Canada (152). After pressure on Exxon forced the company to finally meet with indigenous landholders (153), using funding from the state of Wisconsin the Sokoagon employed a research group, COACT, to evaluate the social, economic and environmental impact of the mine (150). COACT's report questioned the wisdom of underground mining *per se*, criticised Exxon for failing to make any "concrete commitments ... concerning job and training strategies" and asserted the right of the people to form their own businesses as "the only sure route to economic self-sufficiency" (154).

The COACT Report forced Exxon to put back its plans for about a year (148). Meanwhile, the company cancelled the prospecting stage of the mine, saving US$30 million, and attempting to advance construction by three years. Exxon claimed that a state tax of up to 20% on net proceeds over US$30 million "severely reduced" the mine's economic viability; if the laws were changed, mining could "possibly begin in 1984" (155). In fact, there is good reason to believe that, if uranium were mined as a by-product at Crandon, the tax would not be insuperable (156), and in late 1981 state moves were made to reduce the tax (157).

In 1982, Exxon filed for a mining permit with the Wisconsin Department of Natural Resources (DNR); that year the Sokoagon Chippewa, Menominee and Potawatomi tribes formed a Tri-Tribal Mining Impact committee, followed by the Wisconsin Resources Protection Council (WRPC) whose aim was to mobilise individual townships to oppose a 1981 measure permitting ground water contamination from mining (158).

At the end of that year, the Sinsinawa Dominican Sisters of Wisconsin entered a stockholder resolution at Exxon's AGM, demanding postponement of Crandon until Chippewa treaty claims had been settled. Supported by Indian delegates, the resolution got 2.5% of the vote, representing more than 16 million shares (158). Three years later, Exxon downgraded Crandon, envisaging a somewhat smaller operation in the light of adverse economic conditions (159).

Finally (we hope!), in early 1987, Exxon decided to discontinue searching for a permit to mine at Mole Lake – a nice way of saying it has put the project on the shelf. Although it gave economic reasons as the main justification, there is no doubt that the hundreds of meetings organised around the issue, prompted by an almost uniquely successful Indian/white, indigenous/conservationist lobbying process, were a key factor (160).

Exxon's dealings with the Navajo nation, although conducted during the same period as attempted to get the Crandon mine underway, were markedly more favourable to the traditional owners and land users. The Navajo-Exxon lease – almost certainly the biggest single uranium lease in contemporary USA – covered 625 square miles and stretches from Shiprock in the north almost to Toadlena in the south. Negotiations between the company and the tribal council, led by Peter MacDonald (the architect of the much-criticised CERT, or Council of Energy Resource Tribes), lasted about three years, during which the Navajos considered three options: a straight royalty payment based on production; a JV with the Navajos contributing up to 49% of the costs and capital, receiving 49% of future profits; an up-to-49% non-working interest, with Exxon paying the capital costs, and the Navajos repaying 200% of the costs out of the future net profits (161). The US Department of the Interior estimated that option two was likely to be the most profitable – bringing in between US$49M and US$71M over a ten-year span (162). The Navajo nation would be paid US$6M for exploration rights (91), and would not be held liable for the project's failure (163).

Exxon's offer was widely hailed as unique, while one mining spokesman (Homestake) found the idea of paying (a mere tribe??) several millions just "for the right to find something" distinctly "shocking" (91).

Seventeen Navajos went to court to try to prevent the agreement going ahead, when official approval for exploration was granted in early 1977 (113), and other native American groups vehemently opposed the leasing of a huge chunk of the largest remaining indigenous reservation in the USA, for the most dangerous of all forms of mining. Various Navajos pointed out the physical damage that the mine – albeit on "only" 5120 acres out of the 400,000 acreage – would cause, especially to sacred sites such as the Pena Blanca Canyon (109). A social worker for the Shiprock Bureau of Indian Affairs (BIA) Agency claimed Exxon would pull out of the project if it had to bear the real costs of social and environmental disruption (109). The Bureau of Indian Affairs, in its own EIS, pointed out that there was to be no job preference for Navajo workers, nor special housing or transportation measures. It summarised the likely environmental impacts with admirable succinctness:

"Impacts resulting from exploration will include disturbance of soils and vegetation and air quality degradation resulting from the vehicular movement and the operation of drilling equipment. If mining and milling takes place significant environmental impacts include: sub-surface water depletion, soil and vegetation disturbance, air quality degradation, interruption of the wildlife habitat, population increases, increased demands on community services and facilities, and disruption of established lifestyles and social patterns. Low levels of radioactive emissions will be found at mine and mill sites."

Moreover, in granting permission to explore, the Department of the Interior waived no less than thirteen regulations (164,112).

The best that can be said for this agreement is that it established a precedent which could be followed by native Americans should they be in favour of a mineral project. At its worst it became a divisive factor within the Navajo nation and, indeed, probably contributed to the removal of Peter Macdonald as chairman of the Navajo Nation in the early 1980s. (He was later re-elected.)

By the late 1980s, Exxon's rank as the world's biggest industrial concern had slipped. (In 1987 it was ranked second in the Fortune 500, behind General Motors. However, General Motors has nothing like the global spread of Exxon) (170). It also lagged behind Royal Dutch/Shell in terms of worldwide oil and gas revenues and resources (170) and by the following year, was trailing BP (171).

The Valdez disaster of March 1989, although not the largest oil spill in history, suddenly thrust the corporation into the public spotlight around the world as it was arraigned on five criminal counts by a Federal Grand Jury in the USA (172), bombarded by environmental organisations (173), and castigated by the citizens of Alaska who were the victims of the million gallon "spill" (174). The USA's largest pension fund tried to oust the entire Exxon board because of its failure to take preventative action against future similar disasters (175).

Even while Exxon was facing-out the massive criticism of its failures in Alaska, and defying a demand by the Alaska State House of Representatives for speedier payments of compensation, clean-up costs and criminal penalties (176), it was being accused of illegally shipping hazardous wastes to a treatment plant in Alaska

(177). The year before, an underwater Exxon pipeline at Baytown, New Jersey, ruptured, releasing more than half a million gallons of heating oil into the Arthur Kill waterway (178). Allegedly Exxon for years had "file[d] reports... that falsely indicated that the company's leak-detection system in the Arthur Kill was in good working order" (179).

The same year, the company settled US$11 million on around 200 residents of Highlands, Texas who claimed widespread illness (including high rates of cancer) due to the nearby Liberty Waste dumpsite, where they believed illegal toxics were being deposited (180).

Meanwhile, as the Valdez monstrosity continued to grab the headlines (justifiably so, considering that more than a quarter of a million birds died as a result (181) and possibly many more) (182), Exxon's mining interests continued to give cause for considerable alarm.

In 1989, Exxon sustained the worst mine safety record among the 20 largest underground producers in the USA (183).

Worldwide, by the beginning of 1990, Exxon was ranked first among multinationals as a coal producer/reserves holder (184).

Its major projects remained the Disputada mine in Chile (now usually known as Los Bronces) which underwent an expansion in late 1990 (185) despite a US$51 million net loss in 1986 (186), and the El Cerrejon project in Colombia.

In late 1988, the Los Bronces mine shut down for one month when a diversionary tunnel, intended to divert melted water from the Mapocho river (which flows through Santiago), got plugged. Since the "episode" occurred only two weeks after a major mudslide to the south of Santiago swept away a constuction camp (not an Exxon responsibility) with the loss of 30 lives, it caused "considerable public consternation" – to quote the *Engineering and Mining Journal* (185).

The Cerrejon mine – the largest in Colombia, with intended production of 60 million tonnes a year by the turn of the century (187) – has also been the subject of considerable consternation. Although not one of the notorious Colombian exploiters of child labour, no less than 32 workers have died during work at the mine in the period since 1986, from causes which are still not understood (188).

In May 1990, 3800 workers at the company's Guajira mine (managed by its subsidiary Intercor) went on strike, after Exxon refused to negotiate improved housing, a shorter working week, and better working conditions. The government responded by declaring the strike illegal (189). Walter Castillo, president of the Intercor workers' union, declared that the company's behaviour was "...completely arrogant...What we are saying to Exxon is: We are human beings. You must stop trying to block the elemental desire of every human being to satisfy their basic needs, improve working conditions and move forward" (189).

Finally it is worth recording that, prompted by the Alaskan disaster, several major environmental organisations and socially responsible shareholder groups have been demanding that US corporations adopt what are now known as the "Valdez Principles." To date, no corporation (including Exxon) has done so.

Extracts from the Valdez Principles

We will minimise and strive to eliminate the release of any pollutant that may cause environmental damage to the air, water or earth or its inhabitants. We will safeguard habitats in rivers, lakes, wetlands, coastal zones and oceans and will minimise contributing to the greenhouse effect, depletion of the ozone layer, acid rain or smog.

We will minimise the creation of waste, and wherever possible recycle materials. We will dispose of all wastes through safe and responsible methods.

We will minimise the environmental, health and safety risks to our employees and the communities in which we operate, by employing safe technologies and operating procedures and by being constantly prepared for emergencies.

We will sell products or services that minimise adverse environmental impacts and that are safe as consumers commonly use them. We will in-

form consumers of the environmental impacts of our products or services.

We will take responsibility for any harm we cause the environment by making every effort to fully restore the environment and to compensate those persons who are adversely affected.

We will disclose to our employees and to the public, incidents relating to our operations that cause environmental harm or pose health or safety hazards.

We will commit management resources to implement the Valdez Principles, to monitor and report upon our implementation efforts, and to sustain a process to ensure that the board of directors and chief executive officer are kept informed of, and are fully responsible for, all environmental matters.

We will conduct and make public an annual self-evaluation of our progress in implementing these principles and in complying with laws and regulations throughout our worldwide operations. We will work towards the timely creation of independent audit procedures, which we will complete annually and make available to the public (190).

Footnote

Apart from the examples of curbs on Exxon's power already mentioned in this entry, the following are worth noting:

• In 1986, Exxon was fined US$2 billion by the US Treasury for overcharging for oil sales in the US – the largest civil judgement awarded in the country's history (165). This followed an order by a US federal judge in 1985 that the company should repay US$895 million it had overcharged at a Texan oil field (166).

• In Malaysia in 1980, Esso Production Malaysia was accused of taking (stealing) 23,000 barrels of oil in excess of limits imposed by the government, as well as shoddy work paid by the country's oil revenues (167).

• US domestic copper producers petitioned the US government in 1984 to limit the import of Exxon copper from Chile, in order to preserve their own interests. Exxon countered that the high US costs resulted from "high wages, low-quality ore, environmental controls and the strength of the US dollar" (168).

To read: A basic introduction to the origins, growth, conceits and deceits of Standard Oil/Exxon – together with interesting vignettes of all members of the diverse Rockefeller clan – is Peter Collier and David Horowitz's *The Rockefellers, An American Dynasty* (Signet New York, 1976). See also "The Rockefeller Empire" in *Lords of the Realm*, No. 1, (Pommedor Publications, London, undated).

Further reading: On South America: "The Rockefeller Empire: Latin America", *NACLA Newsletter*, April, May, June 1979.
On Cerrejon: "The Great Colombian Coal Conundrum" *FT* 4/1/84; "Cerrejon Coal for Colombia" *E&MJ* 4/85 (pp 32J-32N); "Yo Hablo a Caracas" Seithel and Stühler, in *Pogrom* GfbV, Göttingen, No. 129, 3/87.
On Los Bronces: "Exxon trims Chile copper plans to fit cost escalation" *E&MJ*, 2/85 (pp 26-32).
On Crandon and uranium in Wisconsin: Al Gedicks, "Exxon, copper and the Sokoagon; Tightening the corporate grip in Chippewa country", *The Progressive*, US, Feb 1980 (pp 43-46); "Exxon and the Recolonisation of Wisconsin and Chile" *CALA Newsletter*, Vol 7, No.3, Madison, Feb 1979.

Contact: Arbeitskreis Uranabbau, c/o Gertrud Winkler, Bahnhofstr 37, 8664 Stammbach, Germany.
Center for Alternative Mining Development Policy, 210 Avon Street 9, La Crosse, Wisconsin 54603, USA.
Multinationals and Development Clearinghouse, PO Box 19405, Washington DC,20036, USA.

References: (1) *Global Reach.* (2) *Big Oil.* (3) *Times 1000*, Times Books London, 1984. (4) *Gua* 19/4/85. (5) McGill & Crough. (6) *MIY* 1987. (7) *MAR* 1986. (8) *FT* 4/1/85. (9) *FT* 21/8/87. (10) *NARMIC*, Philadelphia, 1978. (11) *Nuclear Barons*. (12) *WDMNE*. (13) *FT* 27/1/87. (14) Richard Barnet, *Intervention and Revolution*, New York (Mentor) 1972. (15) *FT* 7/3/84; *FT* 28/1/84. (16) *FT* 27/3/85. (17) *FT* 7/1/85. (18) *Latin America Economic Report*, Vol. IV no 2, USA. 9/1/76. (19) *FT* 30/11/84. (20) *FT* 14/3/86. (21) *FT* 19/3/86. (22) *FT* 3/7/85. (23) Harvey O'Connor, *The Empire of Oil*, Monthly Review Press New York, 1962. (24) Tanzer. (25) Peter Collier and David Horowitz, *The Rockefellers, An American Dynasty*, (Signet) New York, 1976. (26) Sklar. (27) *Nacla Newsletter*, Nos. 4 & 5 6/79. (28) Gabriel Kolko, "American Business and Germany 1930-1941", *Western Political Quarterly* 12/62. (29) Peter Thompson, "Bilderberg and the West", in Sklar. (30) Peter Wiley and Robert Gottlieb, *Empires in the Sun: The Rise of the New American West*, (University of Arizona Press), Tuscon, 1985. (31) *Atom's Eve.* (32) *MAR* 1987. (33) *FT* 12/2/81. (34) *FT* 10/11/82. (35) *FT* 11/5/83, *WDMNE*. (36) *FT* 18/6/87. (37) *FT* 18/3/87. (38) *E&MJ* 4/85. (39) *FT* 13/2/84. (40) *FT* 8/3/85. (41) *FT* 26/2/85. (42) Jorge Child in *El Spectador*, quoted in *MMon* 12/80. (43) *E&MJ* 9/87. (44) *MJ* 13/6/80. (45) *E&MJ* 9/87, *E&MJ* 7/87. (46) *FT* 4/1/84. (47) MIIN and ACIPY, Submission to the fourth Assembly of the World Council of Indigenous Peoples, Panama, 20-30 Sept 1984, quoted in *Bulletin Amerique Indienne*, No. 60, Paris, 2/86. (48) *E&MJ* 1/87. (49) *MJ* 18/7/86. (50) *FT* 20/6/84. (51) *FT* 12/6/80. (52) *E&MJ* 2/87. (53) *E&MJ* 4/83; *MJ* 25/2/83. (54) *E&MJ* 2/85. (55) *MIY* 1985. (56) *MJ* 29/1/82. (57) *MJ* 30/5/83. (58) *MJ* 7/5/85. (59) Quoted in "Exxon and the recolonisation of Wisconsin and Chile" (Community Action on Latin America (CALA)) Madison Vol. 7, No. 3, 2/79. (60) *FT* 18/12/79. (61) Reece. (62) *Renewable Energy News*, USA, 11/83, "Exxon unloading PV Business Disaster". (63) *Energy File*, British Columbia, 1979 (exact date unknown). (64) *FT* 22/5/79. (65) *FT* 8/2/82. (66) *FT* 19/5/79. (67) *FT* 5/8/82. (68) *FT* 23/3/81. (69) *SunT* 4/10/81. (70) *FT* 27/6/79. (71) *FT* 18/12/79. (72) *FT* 19/6/80. (73) *MJ* 9/5/80. (74) *FT* 15/5/80. (75) *MJ* 7/5/82. (76) *FT* 18/3/80; *FT* 29/2/80. (77) *FT* 17/6/80. (78) *E&MJ* 12/84. (79) *MJ* 5/9/86. (80) *FT* 1/5/82. (81) *FT* 30/8/81. (82) *E&MJ* 1/86. (83) *MJ* 8/11/85. (84) *Gua* 20/10/76. (85) *MJ* 8/1/82. (86) Moss. (87) *FT* 11/9/86. (88) *MJ* 21/9/79. (89) Information from Nuexco secret document dated 1983, in hands of author. (90) Greg Dropkin, *Briefing Paper on Namibian Uranium* (Canuc), London, 8/87. (91) *Business Week* 2/3/74. (92) Federal Trade Commission Bureau of Economics, *Nuclear Fuel Industry Data*, (USGPO) 4/74. (93) *MJ* 11/9/81. (94) *MJ* 9/3 /79. (95) *FT* 29/8/78. (96) *Reg Aus Min* 1981. (97) *FT* 23/9/80. (98) *E&MJ* 11/78. (99) *FUTURE* newsletter, Denver, 26/1/79.

(100) *EPRI* 4/78. (101) *FT* 25/1/80. (102) *FT* 21/11/80. (103) *Kiitg* (WISE), Amsterdam, 6/80. (104) *Niger, le pays et son marché*, No. 16, 4/79. (105) *Nuclear Fix, MAR* 1979. (106) *MJ* 9/11/79. (107) *Kiitg* 8/9, 1980. (108) Abrahamson & Zabinski. (109) *AkwN* Late Autumn/1976. (110) *Sandoval Environment Action Community Newsletter* Summer/1981, Bernalillo. (111) *The Uranium Industry in the US* (Americans for Indian Opportunity (AIO)) Albuquerque, 1976. (112) Brint Dillingham, "Exxon, Uranium and the Navajo Nation", in *American Indian Journal*, Albuquerque, 1977, (exact date unknown). (113) *Albuquerque Journal* 16/1/77. (114) Weiss. (115) Press Release from National Indian Youth Council Inc, Albuquerque, 8/78. (116) *San Juan Study*. (117) *San Juan Study*, pp 111-117. (118) *Mining Organizer*, Vol.1, No. 1, (CAMDP) Madison, May/80. (119) *Mining Organizer*, Vol. 1 No. 3, 3/81. (120) *Antigo Daily Journal* 1/5/80; *Daily Cardinal*, Madison 4/2/80. (121) *Washington Post* 6/6/87; *Esquire*, New York 19/6/87. (122) *MAR* 1981, 82. (123) *MAR* 1982. (124) Rob Kennedy, *Mining in Wisconsin and Boom Town Prosperity, The Answers to some hard questions*, (CAMDP), Madison, undated. (125) *Up North, on the Wolf,* leaflet (Save the Wolf), Elton, Wisconsin, undated. (126) *High Country News* Wyoming 10/3/78. (127) *Rich Lake Chronotype*, Wisconsin, 7/5/80; *E&MJ* 4/80, also *E&MJ* 7/80. (128) *Northern Sun News*, Minneapolis, 4/81. (129) WISE, Washington DC, 7/5/81; *MJ* 30/9/83. (130) WISE, Amsterdam, 18/6/81. (131) *Western Australian* 15/5/82; *MJ* 21/5/82. (132) *Age* Melbourne, 14/5/82. (133) *MMon* 3/81. (134) *E&MJ* 8/83. (135) *SInews*, No. 11 (Survival International), London, 1985. (136) SI press release, London, 1/8/87. (137) *UAB/ECU* 2/8/87 (Survival International), London. (138) *FT* 10/12/81, 7/10/81. (139) *Wall Street Journal*, New York, 17/3/87. (140) *Native Press*, Yellowknife, 13/2/81. (141) *Dene Nation Newsletter*, Yellowknife, 4/83.

(142) *Native Press* 26/8/83. (143) *Protect North Newsletter*, Toronto, 5/6, 1983. (144) *Native Press* 10/83. (145) *Native Press* 20/6/86. (146) Roberts. (147) *NACLA Report* 3/4, 1979. (148) *Not Man Apart*, (FoE), New York 5/83. (149) *In These Times*, US, 30/11-6/12/77. (150) Robert Gough, "Protecting a Heritage from Mining Development", *ARC Bulletin*, Cambridge, Mass 3/80. (151) Johanna Clausen, *Wisconsin Mining Impact Study*, (Dept of Revenue), Madison, 1977. (152) *Greenbay Press Gazette* 8/6/80. (153) *Rhinelander Daily News* 22/5/80; *Shawano Evening Leader* 24/4/80. (154) *Free for All* 25/6 – 9/7/80. (155) *City Lights*, US, 11/80. (156) *Daily Cardinal*, Madison, 20/2/80. (157) *MJ* 13/11/81. (158) Al Gedicks, "New Challenges to Exxon", *ARC Bulletin*, Boston, Summer/83. (159) *MJ* 31/5/85. (160) *Cultural Survival Quarterly*, Vol. 11 No. 1, Cambridge, 1987. (161) Lorraine Turner, *Navajo Mineral Development*, typescript final draft of paper, Dept of Economics, Adelphi University, 6/87. (162) *Financial Analysis of Proposed Navajo-EXXON Uranium Mining and Prospecting Proposal*, Office of Minerals Policy Development, US Dept of Interior, Washington, 28/2/75. (163) *Hard Choices: development of non-energy, non-replenishable resources*, (Americans for Indian Opportunity paper No. 10), Albuquerque, undated. (164) *CALA Newsletter* Vol. 7, No. 3. (165) *FT* 28/1/86. (166) *FT* 26/3/85. (167) *MMon* 11/80. (168) *MJ* 18/5/84. (169) *The Progressive*, USA, 2/80. (170) *Multinational Monitor* 1-2/89. (171) Council on Economic Priorities, Corporate Environmental Data Clearinghouse, *Environmental Profile of Exxon Corp*, Washington 1991. (172) *New Scientist* 10/3/90. (173) *Not Man Apart*, FOE USA, Washington 6-9/89. (174) *FT* 3-4/3/90. (175) *Gua* 5/4/90. (176) *FT* 4-5/5/91. (177) *Corporate Crime Reporter* 25/2/91. (178) *New York Newsday* 21/1/90. (179) *New York Times* 8/2/90. (180) *Toxic Times*, Summer 1990 see also *Amarillo Globe* 4/6/89 and *San Francisco Examiner* 4/6/89. (181) *Los Angeles Times* 9/12/89. (182) *Sierra* 5-6/90. (183) *MMon* 10/89. (184) *MJ* 1/12/89. (185) *E&MJ* 5/88, *MM* 9/90. (186) *MM* 11/87. (187) *Colliery Guardian* 12/89. (188) *MMon* 12/90. (189) *MMon* 5/90. (190) *FT* 27/3/91.

240 EZ

EZ is a wholly-owned subsidiary of EZ Industries, which until fairly recently was dominated by British capital as only British shareholders could hold shares in the parent company (1).

It was acquired by NBH Holdings in 1984, after protracted negotiations (2). The following year, there were rumours that NBH could make a play for ERA, the operating company of the Ranger mine, which EZ and Peko Wallsend discovered. But, in the event, NBH sold its ERA stake itself to Pioneer Concrete (3). By 1987, Peko had itself gained at least a 5% holding in NBH (4), and NBH was bidding for control of Peko (4) with which it later merged (5). Thus, both Peko and EZ – cited more than a decade ago as "the most able of the public uranium companies to finance their own development" (6) – were intimately tied in with the fortunes of NBH, which itself holds significant shares in some of Australia's most important mining corporations (including Alcoa and BHP).

EZ owns several mines in Tasmania, including the Rosebery, Hercules and Mt. Farrell mines (7, 8). Its Risdon smelter (Tasmania) – the electrolytic zinc plant which gave the company its name – commenced operations in 1971, though it has suffered from protracted industrial action in recent years (9), notably a "rolling strike" which reportedly had little effect on deliveries (10), but did affect production (11).

The Elura zinc-lead-silver mine in New South Wales was the company's "big hope for the future" in 1981 (12). Opened in 1983 (8), and beset by difficulties (9), this mine was EZ's most important new venture in the early 1980s. EZ sold its interest in the Golden Grove (Scuddles and Gossan Hill) zinc-copper-silver prospect in Western Australia in 1987 (13). (Amax also withdrew from this prospect in the 1980s: Esso now holds 35% equity) (13).

EZ has held a uranium JV with Aberfoyle (14), but its main joint interest with this fellow Australian company is now in ore-purchase from Aberfoyle's Que River mine (7).

Recently, EZ also agreed to offer an option on its 10.76% stake in the Thalanga (Queensland) base metals deposit (associated with BHP), a high-grade multi-mineral deposit, to Outokumpu Oy and Pancontinental (15).

At Coronation Hill, site of an old uranium mine in the Alligator River region of the Northern Territory (the Pine Creek geosyncline) EZ is exploring a promising gold-platinum-palladium deposit which lies in Stage III of the Kakadu National Park, where mining is intended to be permitted (16) but which is the subject of a land claim by Aborigines (13).

EZ's partnership with Peko Wallsend in ERA and its exploitation of the huge Ranger uranium deposit has been extremely important to both companies in guaranteeing their financial solvency. Final approval for the mine was given in 1978, four years after the two com-

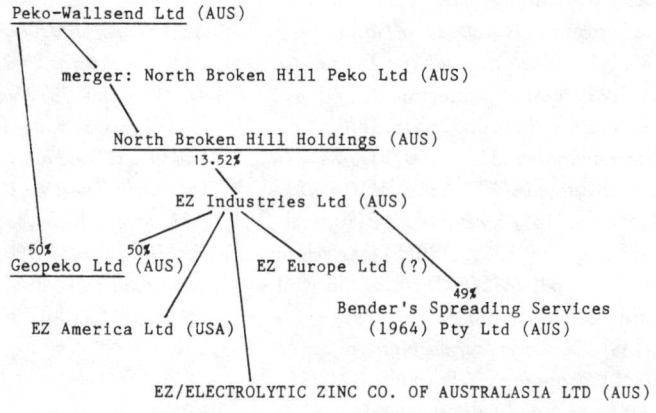

panies signed a memorandum to develop the mine with the AAEC, which then held 50% of the equity (17). This was a few months after agreement had been reached with a small number of traditional Oenpelli landowners under the aegis of the Northern Land Council.

EZ and Peko were to contribute 13.75% of the finance for 25% of the mine's output each, while making their own marketing agreements (17). Both called the agreement "a major milestone". A year later the partners signed their first contract with South Korea.

When the Federal Australian government put up the AAEC's share in Ranger for public sale – with considerable interest expressed by numerous companies – EZ opposed the arrangement, on the grounds that this lucrative share should not be sold at a profit, but re-assigned to the original owners (*ie* Peko and itself). Although Peko won the bid for the AAEC's relinquished half in Ranger, the two companies later came to an agreement to join together in forming Energy Resources of Australia Ltd (ERA) (18). And, in late 1980, EZ was given a 30.5% share in ERA (19) later increased to 30.96%. These "A" shares yielded a dividend in 1984 (20) when EZ's share of ERA brought in A$9.1M (21).

It is important to realise that, together with Peko, EZ has been exploring other areas of the Ranger uranium region, finding significant mineralisation, after the Ranger agreement was signed (22), at Barote Springs, north of the mine (23).

Two of EZ's directors, AW Hamer and GA Mackay, have been members of the ERA board, while Mackay in the late 1970s was also chairman of the Uranium Producers Forum (24).

Apart from the widespread and international opposition to Ranger, EZ was the focus of a unique dissident shareholder's intervention at its annual general meeting in 1978. Shareholders with Social Responsibility – about forty of them – tried to ask questions of the chairman, Sir Edward Cohen. Failure to answer resulted in barracking, the throwing of confetti, whistle blowing, and, eventually, a take-over of the platform, at which point police moved in to escort shareholders from the hall. One major purpose of the meeting was to pass a resolution limiting attendance at AGMs to shareholders with 100 or more shares, a manifest ploy against socially responsible shareholders. The resolution duly passed (25).

In a statement issued at an earlier demonstration against EZ, FCAATSI (Federation for the Advancement of Aborigines and Torres Strait Islanders), the Aboriginal Advancement League, FoE, MAUM (Movement Against Uranium Mining), and others, condemned the government for failing to implement some of the Ranger Inquiry (Fox Report) recommendations – including resumption of the Mudginberri and Munnarlary pastoral leases (see Peko) (26). The demonstrators demanded that all further mining developments should be suspended until the ecology of the region was adequately protected and Aborigines were assured control over the Kakadu National Park (27).

References: (1) *The Australian* 23/5/77. (2) *MJ* 15/6/84, *MJ* 29/6/84. (3) *FT* 23/4/86. (4) *FT* 16/12/87. (5) *MJ* 25/12/87. (6) Kim Williamson, *Background paper on uranium companies*, Office of Tom Uren MP, mimeographed, 1977. (7) *FT Mining* 1987. (8) *MJ* 11/11/83. (9) *FT* 1/3/84. (10) *MJ* 5/12/86. (11) *MJ* 12/12/86. (12) *FT* 26/8/81. (13) *E&MJ* 10/87; see also (8), & *MJ* 18/10/85. (14) *MIY* 1982. (15) *MJ* 23/10/87. (16) *MJ* 26/9/86, *Financial Review* 10/11/86. (17) *FT* 10/1/78. (18) *FT* 24/1/80. (19) *FT* 25/2/81. (20) *FT* 17/8/84. (21) *MJ* 28/12/84. (22) *Financial Review* 3/2/78. (23) AMIC paper, mimeographed, Victoria, 1982. (24) *Financial Review* 23/3/77. (25) *Age* 10/7/78. (26) *Ranger Inquiry, 2nd Report*, p215 (AGPO) Canberra, 1977. (27) Statement by various organisations, typewritten, Melbourne 3/2/78.

241 Falconbridge [Nickel Mines] Ltd

This is a major miller and refiner of heavy metals, especially nickel, though the company is into virtually every major metal, with operations throughout Canada, numerous subsi-

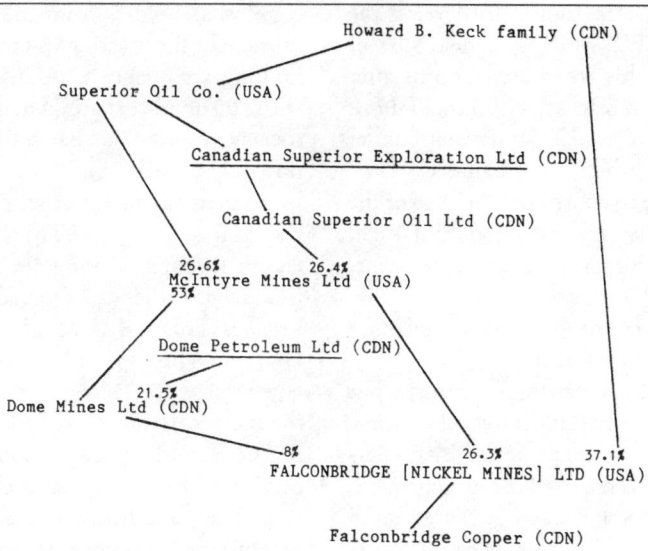

```
                              Howard B. Keck family (CDN)

       Superior Oil Co. (USA)

                     Canadian Superior Exploration Ltd (CDN)

                        Canadian Superior Oil Ltd (CDN)

            26.6%              26.4%
          McIntyre Mines Ltd (USA)
            53%

            Dome Petroleum Ltd (CDN)

               21.5%
     Dome Mines Ltd (CDN)

                   -8%                26.3%         37.1%
                   FALCONBRIDGE [NICKEL MINES] LTD (USA)

                    Falconbridge Copper (CDN)
```

diaries and interests in Chile, Zimbabwe, the Dominican Republic and elsewhere (1). It sold its controlling interest in the Oamites Mining Co of Namibia in 1982, but it still separates metals from "sludge" provided by South African companies at its refinery in Kristiansand, Norway (2).

Like Inco's operations in the Sudbury mining zone, Falconbridge's have suffered in recent years, and were only recovering in 1983 (2).

Its only known uranium involvement in recent years has been explorations in Namibia (3) which are presumed now to be defunct.

In 1984 the company closed down its oldest mine (called Falconbridge) in Sudbury, Ontario, due to its appalling safety record – four workers were killed in a rockburst in June 1984. The mine had been contributing about 10% of the company's nickel: it will not be sorely missed (4).

That year Mobil purchased Superior Oil and its Canadian subsidiary, which – between them – controlled McIntyre Mines. However, in 1985 Mobil sold its stake in McIntyre, and a direct 8% in Falconbridge (through Canadian Superior Oil) to Dome Mines Ltd, a gold producer (5).

In September 1989, Falconbridge became a wholly-owned subsidiary of a holding company equally owned by Noranda and Trelleborg AB (6). The $2.2 billion deal went through on fairly amicable terms, although Noranda and Trelleborg had to contend with a takeover attempt by Amax (7). The following year Falconbridge began selling off assets of its 51% owned subsidiary, Falconbridge Gold (8). But it has also embarked on major expansions of its nickel-copper projects in the Sudbury region of Ontario, new developments at Kidd Creek, and in the Dominican Republic, where it is the major exploration enterprise (9).

The same year that Falconbridge was taken over by Trelleborg and Noranda, Survival International (the major international organisation working for indigenous peoples rights) accused the company, together with De Beers Botswana and BP, of endangering the last bushpeople (Tsan) in the Central Kalahari game reserve of Southern Botswana. The Tsan were then being moved out of their traditional lands to make way for mining and tourist interests (10).

Contact: UN Commission on Namibia, UN Building, New York, NY 10017, USA.

Survival International, 310 Edgware Rd, London W2 1DY.

References: (1) *MIY* 1985. (2) *MAR* 1984. (3) *MM*

7/77. (4) *MJ* 29/6/84; 13/7/84. (5) *MJ* 31/5/85. (6) *FT Min* 1991. (7) *E&MJ* 10/89. (8) *MJ* 7/12/90. (9) *MJ* 14/7/89, *MJ* 11/8/89. (10) *SInews*, Survival International, No. 25, London 1989.

242 Famok Ltd

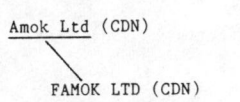

It was jointly exploring for uranium with Eldorado Nuclear in the late 70s (1); no other information.

References: (1) *Eldorado Nuclear Annual Report* 1979.

243 Faraday Uranium Mines Ltd (CDN)

Later called Madawaska Mines, this was a company set up in Faraday, Ontario, in 1949 after a local prospector discovered uranium in the Bancroft area of Ontario. The mine was opened in 1956, when Eldorado Nuclear contracted to buy its uranium output (1).
(Faraday Resources Inc is a different company owned 27% by Conwest Exploration Co Ltd) (2).

References: (1) *BBA*, No. 4, 1979. (2) *MIY* 1990.

244 Farris Mines (USA)

It was active exploring for uranium in New Mexico in 1978 (1); no further information.

References: (1) *E&MJ* 11/78.

245 Federal American Partners

It owns uranium property in the Gas Hills, 45

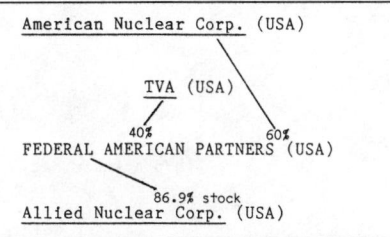

miles east of Riverton, Wyoming, which was leased by the TVA/Tennessee Valley Authority to provide uranium for its nuclear power plants. In 1979, the mine was sufficiently delayed by state regulations for it to cancel its contract to supply uranium concentrate to Florida Power and Light Co. The company then re-applied to the State Department of the Environment to place the mine in production (1).
Operations were stopped in 1981, due to a slip in uranium price, and were reported "not likely to resume before 1985" (2).
In early 1982, Federal American Partners announced an agreement in principle to sell the property to the TVA for US$4M plus a yellowcake royalty of 1% on the first 15 million pounds of concentrate from future productions (3).
In September 1982, the Department of Environmental Quality in Wyoming gave permission to the Partners to reclaim the second tailings pond at their Gas Hills mine, *in situ*: a practice largely disapproved of by the NRC (Nuclear Regulatory Commission). TVA would pay the costs (4).

References: (1) *Wall Street Journal* 23/3/79. (2) *MJ* 20/3/81. (3) *MJ* 29/1/82. (4) *MJ* 17/9/82.

246 Federal Mynbou BPK

"Afrikanderdom's first interest in mining" (1), Federale Mynbou was founded in 1953, as a small producer of coal, by Federale Volksbeleggings and Bonuskor, two Afrikaner investment companies (2). Helped by the Broederbond (the secret power behind apartheid in South Africa), Federale Mynbou soon grew in size, as-

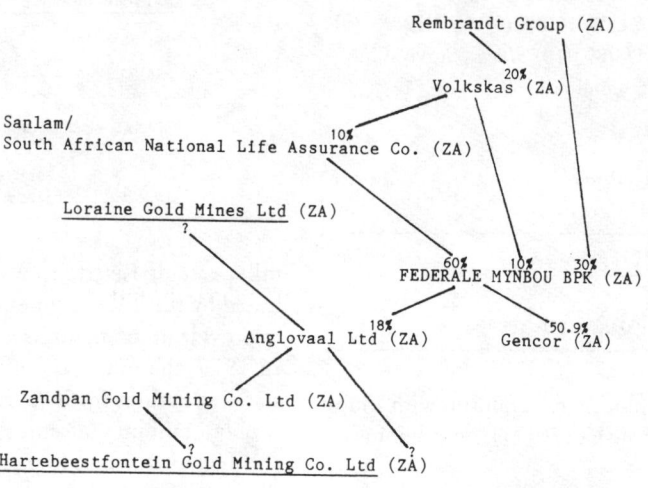

Rembrandt Group (ZA)

Volkskas (ZA) 20%

Sanlam/
South African National Life Assurance Co. (ZA) 10%

Loraine Gold Mines Ltd (ZA)
?

60% 10% 30%
FEDERALE MYNBOU BPK (ZA)

18%
Anglovaal Ltd (ZA)   50.9%
Gencor (ZA)

Zandpan Gold Mining Co. Ltd (ZA)

?   ?
Hartebeestfontein Gold Mining Co. Ltd (ZA)

sisted by its location near two Escom power plants, though its main interests remained in coal until 1964 (3).

In the meantime, Anglo-American and Federale Mynbou had jointly founded Mainstraat-Beleggings to "develop the common interests" of the two companies (2), and in 1965 they gobbled up General Mining – now the hardcore company in Gencor.

According to a recent commentary, Anglo-American "practically gave Gencor (Genmin) to Federale Mynbou Beperk, the mining subsidiary of Sanlam, [in a move] intending to bring Afrikaner capital into the gold-mining industry in the hope that this might have a "moderating" effect on some of the extremist policies of the National Party" (4).

Whatever the hopes at the start, Gencor quickly became an instrument of Afrikaner capital, especially after the murders of unarmed black demonstrators at Sharpeville caused foreign capital to leave the apartheid state like rats from a stinking pit. Sanlam, however, during the Verwoerd era, was more closely associated with the "verligte" (less rabid) branch of the National Party. Volkskas was not. And in 1982, Rembrandt and Volkskas had a humdinger of a battle with Sanlam – in the best traditions of rogues fighting over spoils – to gain control of Federale Mynbou (1).

Further reading: *RIL Rothmans Industries Ltd, Shadow Report 1984: The South African connection*, Wellington, 1984 (HART).

References: (1) *FT* 22/6/82. (2) Lanning & Mueller. (3) *MIY* 1985. (4) Rob Davies, Dan O'Meara & Sipho Diamini, *The struggle for South Africa: A reference guide to movements, organisations and institutions*, Vol 1, London, 1984 (Zed Press).

247 Federal Resources Corp Ltd

It co-owned and operated the Madawaska Mines, a uranium mining/milling complex near Bancroft, Ontario, Canada, producing uranium largely for Agip and for Canadian utilities (1) which has now closed.

The mine, originally owned by CCF/Consolidated Canadian Faraday, was closed down in 1964. It was then re-activated by incorporating Madawaska Mines. Production started in August 1976 after a long-term contract with Agip Canada for the purchase of uranium oxide. In 1977, delivery was stopped, as the contractors

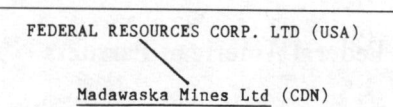

FEDERAL RESOURCES CORP. LTD (USA)

Madawaska Mines Ltd (CDN)

couldn't agree a price; uranium was stored awaiting delivery. Eventually, Agip agreed to pay US$42 a pound with an escalation clause (2). From October to April 1979, the company reports it stripped "12 million cubic yards of overburden" from five open-pit properties, to obtain just 231,000lbs of uranium concentrates (that's around 105 tonnes) (3). In 1979, production was 695,533lbs, to be delivered to Agip at US$50.50c/1b, almost a record price. In 1982, the contract was terminated (4).

Through Federal American Partners, it also co-owned a uranium property in the Gas Hills, near Riverton, Wyoming, USA, which was reactivated in 1979 (5). Operations were, again, stopped in 1981 after a period of lease to the TVA/Tennessee Valley Authority due to a slip in uranium price, and were reported "not likely to resume before 1985" (6). In early 1982, an agreement was announced to sell the property to the TVA (7). And in September 1982, TVA bought Federal's interest in Federal American Partners at a benefit of US$2.4M to Federal Resources.

In early 1983, Duke of Energy (*sic*) was bidding for control of Federal Resources: a move which would result in Federal Resources effectively controlling Duke of Energy (through a 30% holding) and the Duke having 82% control of Federal Resources (8).

References: (1) *MIY* 1981. (2) *FT* ?/9/78. (3) *MIY* 1981. (4) *FT* 24/3/82. (5) *Wall St J* 23/3/79. (6) *MJ* 20/3/81. (7) *MJ* 29/1/82. (8) *MJ* 14/1/83.

248 Felmont Oil Co (USA)

Along with Gulf States Utilities and its subsidiary Varibus, the company was involved in uranium exploration in the USA from 1976 until 1979 (1).

Felmont is a subsidiary of Homestake Mining Co (2) and has been into gold mining and exploration as well as oil production (3). Most of its oil and gas businesses were sold in the late 1980s leaving Felmont with a 25% interest in the Mountain gold mine in Nevada, and a

16.7% interest in a sulphur deposit in the Gulf of Mexico (4).

References: (1) Sullivan & Riedel. (2) *MIY* 1990. (3) *Oil and Gas.* (4) *MAR* 1990.

249 Fernandez JV

This was a JV to explore for uranium in the Ambrosia Lake area of New Mexico. In 1980, Kerr Addison was one of the partners – the others are not known (1).

References: (1) *MIY* 1982.

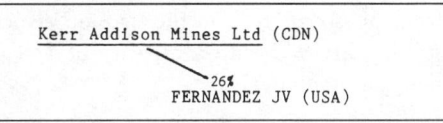

Kerr Addison Mines Ltd (CDN)

26%

FERNANDEZ JV (USA)

250 Ferret Exploration Ltd (USA)

It was reported as locating potentially "the largest uranium discovery in 10 years" at Crow Butte, Nebraska, while acting on behalf of Geomex. Two years were estimated necessary to define the prospect and determine the "best development method". Wyoming Fuel was Ferret's partner at Crow Butte.

Ferret is reported as being "one of several natural resource and exploration companies acting on behalf of the Geomex group" (1).

Ferret announced in 1988 that the results of pilot drilling at Crow Butte were so successful, it planned to move to commercial operations in 1989, along with Imperial Metals Corp (25%) (2) which itself owns 18.7% of Ferret Exploration of Nebraska (3). Crow Butte is owned 65% by Ferret and Geomex, while Uranerz owns 25% and Kepco 10% (with the right to another 10% and an option for another 10%) (4).

In mid-1990 Crow Butte was given the go-ahead. Uranium is to be mined from an area of more than 1000 hectares, involving injection with a solution of water, oxygen and sodium bicarbonate to bring the uranium to the surface through recovery wells and ion exchange extraction. The capacity of the plant is 453.6 tons of uranium and is expected to be reached some time after 1991 (5).

References: (1) *MJ* 6/2/81. (2) *E&MJ* 4/88. (3) *FT Min* 1990. (4) *Greenpeace Report* 1990. (5) *MJ* 8/6/90.

251 Fertimex/Fertilisantes Mexicanos (MEX)

In 1981 it concluded an agreement with Uramex to establish a plant to extract uranium from phosphate rock in the state of Veracruz, starting in 1983 with a capacity of 150 tonnes/year uranium (1).

References: (1) *Uranium Redbook* 1982.

252 Flin Flon Mines Ltd

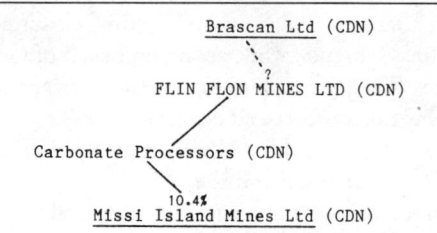

It has been in partnership with various companies – including Imperial Oil, and Golden Eagle Oil and Gas – on uranium prospects in the Flin Flon area (Smoothstone, Jackfish Creek, Lake Athabasca, Harrison River Islands) in Saskatchewan. It was also exploring in central Newfoundland (1).
It had the same board as Missi Island Mines in which it holds a minority of the stock (2).
According to *Yellowcake Road*, Brascan "held" a 7.9% interest in Flin Flon. It is possible this was lost to Westmin (see Brascan) when Westmin – which is effectively controlled by Brascan – sold a number of its uranium prospects in Canada,

or during the extraordinary Brascan shuffle of 1980.

References: (1) *Yellowcake Road.* (2) *Who's Who Sask.*

253 Florida Power and Light Co (USA)

FLORIDA POWER AND LIGHT CO. (USA)
|
Fuel Supply Services Inc. (USA)

It was involved, along with its subsidiary Fuel Supply Services, Getty Oil and TXO Minerals, in uranium exploration from 1976 to 1979 (and possibly longer) (1).

References: (1) Sullivan & Riedel.

254 Fluor Corp

Fluor is one of the world's largest construction engineering firms, building large, complex, process plants for the petrol, petrochemical, chemical, and mining – including uranium mining – industries (1).
It achieved world-wide notoriety in the 1970s when it was awarded a contract by the apartheid régime to build South Africa's oil-from-coal plant, SASOL-2. Later it was awarded similar contracts to expand SASOL-2 at a cost of £2 billion (2). In building the SASOL-2 plant, Fluor directed no fewer than 14,000 contract labourers, and gained some US$300 million of new investments over the period 1980-1984 (3). The contract labour system is one of the keystones of the whole apartheid system. SASOL enabled South Africa – despite a free-

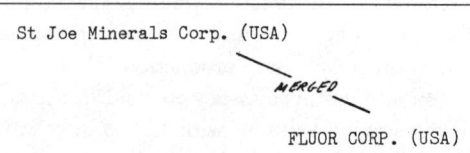

dom movement attack in 1979 – to meet, and beat, any attempts to impose an oil embargo. Later, Fluor acquired the rights to market SASOL technology in the USA (4).

Among Fluor's many projects throughout the world, the following are some of the more recent and significant:

• a contract to build part of the Alaska gas pipeline from Canada's North Slope to the Yukon (see EXXON), a project which has raised widespread concern among native Canadians in the north (2);

• a US$5 billion gas project in Saudi Arabia (4);

• the construction of the Cold Lake, Alberta, Canada, oil sands project for Esso Resources (EXXON) (4);

• a feasibility study for the development of the Ok Tedi gold-copper project in Papua New Guinea, a JV with, among others, BHP and Amoco (5);

• construction of a coal liquefaction processing and shipping facility at Shell Oil's Buckskin mine at Gillette, Wyoming, USA (6);

• development of a fertiliser industry from natural gas supplies in Mozambique (7);

• construction of an oil shale mining and refining project in Rio Blanco county, Colorado, USA, for Occidental Petroleum and Tenneco (8);

• management of a coal project at Newlands, Qeensland, Australia (9);

• project management of a zinc-lead-silver mine at Elura, western New South Wales, Australia, for EZ (10);

• a contract at the El Pauchan copper mine in Argentina (11);

• development of Chile's first lithium plant, together with Newmont's Foote Mineral Co (12);

• a feasibility study, awarded 1981, into a coal gasification plant construction on the Crow Indian Reservation at Billings, Montana, USA, together with Pacific Coal Gas of Los Angeles, and with the notorious CERT as Indian "agent" (acting for the Crow tribe); at full pelt the plant could be processing nearly 30,000 tons/day of coal (13);

• construction, along with Cementation International (a subsidiary of Trafalgar House), of RTZ's Cerro Colorado copper mine on Guaymi Indian land at Chiriqui, Panama (see RTZ), although in early 1982 considerable doubt grew about whether the project would proceed, and Fluor's contribution was reported to be "reduced accordingly" (14) (see RTZ);

• in early 1982, St Joe Minerals announced it was buying into the Sorby Hill lead-zinc-silver prospect (co-owned by MIM and ELF-Aquitaine) in Australia, not far from the important Ashton diamond venture (see CRA) (15);

• a contract, signed in May 1982, with a Chinese state corporation to modernise and expand an open-pit coal mine in Liaoning province (16).

The corporation's two major uranium projects were both started in the '70s.

Fluor was contracted by the Yugoslav government to build its first uranium mine, near Ljubljana, in the mid-'70s (17). By the end of 1980, the Yugoslav government announced its hopes to use the uranium to power its first nuclear power station at Krsko, by the middle of 1982.

In 1979, Fluor was awarded an engineering-financial study contract for the Imouraren uranium project in Niger, claimed by Fluor to become "one of the largest production installations of its kind in the world" (18). Production there, however, has since been deferred (19).

In 1981, Fluor merged with the St Joe Minerals Corp, a company with extensive interests in South America and elsewhere. The merger has made Fluor-St Joe a monolith in the mine engineering construction field, rivalled only by Bechtel.

As if this wasn't enough to unsettle the globe, Fluor in early 1981 had purchased an undisclosed number of shares in Genentech – a genetic engineering company (20). Cloned Fluors? Heaven forbid!

Fluor holds 45% of St Joe Minerals (21) and, through it, is also involved in a JV with Phelps Dodge and CRA (through New Broken Hill

Consolidated) in the Woodlawn copper-lead-zinc mine in New South Wales, Australia (22). In 1982 Fluor gained a contract to build a US$30M sponge iron plant for the Union Steel Corp (an Anglo-American Corp affiliate) (23). In mid-1983 the Chinese government awarded Fluor two engineering design contracts for coal mines in Inner Mongolia (24). A second contract was awarded by the China National Coal Co to do basic engineering on an open-pit coal mine there late in the same year (25).

It is also among seven companies which in August 1983 bid for the expansion of the Ombilin coal mine in western Sumatra, Indonesia (26).

In the late '70s, Fluor bid to build a coal gas plant on the Navajo reservation in New Mexico at an estimated cost of US$3-4 billion; up to 150 million cubic feet/day of coal would be dug there (27).

In 1984 Fluor and Saarberg Interplan were among various western companies competing for another large coal mine, this time in southern Sumatra, funded by the World Bank (28).

In another apartheid deal, Fluor was awarded the maintenance contract for South Africa's first nuclear power station, by Escom, in late 1983 (29).

Huge losses sustained by the company in 1987 necessitated a major restructuring, which included the sale of its remaining 90% holding in St Joe Gold to Dallhold Investments (30), a partitioning of the assets of Massey Coal, between Fluor and its partner Shell Oil; selling its Mocambo alluvial tin mine on the river Xingu in Brazil; and selling several other north American interests (31). Fluor Daniel, its world-wide engineering and construction subsidiary, ended the year with a backlog of nearly US$5 billion of projects. In 1987 Fluor finished constructing the Olympic Dam/Roxby Downs mine for BP/Western Mining (31).

Through St Joe Minerals, in 1986 the company formed a partnership with Homestake Lead Co (a subsidiary of Homestake Mining Co) under which each company merged its domestic lead business: amounting to six mines, four mills and two smelter complexes. The new Doe Run Co (Fluor/St Joe 57.5%, and Homestake

42.5%) is now the biggest primary lead producer in north America, with 1989 output of 22,800 tonnes of refined lead product and 4900 tonnes of zinc concentrates (32).

From 1988 onwards, Fluor's restructuring was proving successful with a "fast growing order book" and improved profitability (33). By the beginning of 1991, the company had "surged" back into a lead position, with increased orders for Fluor Daniel, strong advances in the A. T. Massey Coal operations, and gains from the sale of its Pea Ridge Iron Ore Co (34).

In 1989 Fluor was awarded a $7 million contract to manage construction of a sodium cyanide plant for Cyanco, in Nevada (35).

The following year it completed studies at Dayton Development Corp's Andacollo gold project in central Chile, identifying reserves of 1.5 million ounces of gold (36). Later that year it gained a $50 million contract to engineer, construct and manage a copper-molybdenum ore processing facility for the Los Pelambres mine in Chile. (This mine was previously owned 80% by the Midland bank of Britain, which sold half of that interest to Lucky-Goldstar International Corp of south Korea. The Cia Minera los Pelambres Ltd company is now owned 40% by Midland Montagu, 40% by Lucky-Goldstar and 20% by the British company Antofagasta Holdings) (37).

"Fluor and water – a sticky mess!" It was under this headline that the New Zealand research group, CAFCINZ (now Campaign against Foreign control of Aotearoa, or CAFCA) in 1985 revealed both the rightwing affiliations of Fluor in the USA (where it has been a major donor to the Heritage Foundation, the Media Institute, the Transnational Communication Center and the Mountain States Legal Foundation) (38) and its surrealistic (but seriously mooted) plan to buy up water in the Doubtful Sound of southern New Zealand. The purpose was to can it for the USA, in the event of a nuclear cataclysm which might contaminate the country's drinking water supplies (39). The plan was previously the brainchild of Triune Resources Corp, taken over by Fluor in 1984, and whose New Zealand frontman, Van Giels, was also behind a scheme to export H_2O from large undersea springs at Point Elizabeth on the country's west coast (40).

Little more has been heard about the proposal in the years since. Discussion about the uses of the waters in Fiordland has centred on the continuing attempts by CRA/Comalco to gain control of Lake Manapouri, as a captive source of hydropower for the expansion of its aluminium smelting capacity and its own profits (see CRA).

Contact: Native Communications Society, POB 1919, Yellowknife NWT, Canada.

Project North, 80 Sackville Street, Toronto, Ontario, Canada.

Campaign Against Foreign Control of Aotearoa (CAFCA), PO Box 2258, Christchurch, New Zealand.

References: (1) *MIY* 1981. (2) *DT* 26/2/79. (3) *Business Week*, US, 20/10/80. (4) *FT* 11/3/80. (5) *FT* 16/7/80. (6) *MJ* 5/6/81. (7) *FT* 23/11/81. (8) *MJ* 10/4/81. (9) *MJ* 24/7/81. (10) *MM* 11/80. (11) *MJ* 11/12/81. (12) *MJ* 22/1/82. (13) *MJ* 20/11/81. (14) *FT* 12/2/82. (15) *FT* 25/3/82. (16) *FT* 6/5/82. (17) *FT* 9/2/79. (18) *MJ* 1/6/79. (19) *MM* Annual Mining Review 1/82. (20) *FT* 17/1/81. (21) *MJ* 15/5/81. (22) *MJ* 3/2/82. (23) *MJ* 15/10/82. (24) *MJ* 26/8/83. (25) *MJ* 2/9/83. (26) *MJ* 19/8/83. (27) Report of CERT (Council of Energy Resource Tribes), 1978. (28) *MJ* 18/5/84. (29) *FT* 12/12/83. (30) *FT* 23/12/87. (31) *MAR* 1988. (32) *MAR* 1990. (33) *FT* 13/9/88. (34) *FT* 11/12/90, see also *MJ* 23/2/90. (35) *MJ* 4/8/89, *MJ* 18/8/89. (36) *MM* 5/90. (37) *MJ* 7/9/90, see also *MJ* 6/7/90. (38) Mother Jones 7/84. (39) CAFCINZ Watchdog, No.48 Christchurch, 3/85. (40) *Press*, Wellington 14/12/84.

255 Foote Mineral Co

```
Newmont Mining Corp. (USA)
        |
        | 87.9%
FOOTE MINERAL CO. (USA)
```

Controlled by Newmont, Foote has been an important US uranium producer. It has several alloy production plants in the USA, and an option to develop – together with Fluor – Chile's first lithium plant (1, 2).

In what has been called a "landmark decision", Foote became liable for both "punitive damages" and "compensatory damages" due to a court ruling in early 1984. Widows and children of some 31 deceased miners – killed as a result of radiation, it is claimed, in the '50s and later – sought US$1.5M for each of the 28 widows, and another US$750,000 for each of the 44 bereaved children.

The original mine was operated by the now defunct Vanadium Corporation of America (see Union Carbide) which merged with Foote Mineral in 1967. Although Foote claimed it was not liable, since the mine was not operated by them after the merger, the Utah Supreme Court ruled that the state's "wrongful death" statute could be used in this instance.

Documents recovered by the plaintiffs showed that measurements of radon in the Marysvale mine exceeded the maximum allowable by a factor of several thousand. The radiation sample, taken at the face of the longest drift on the 150-foot level revealed radon at 26,900 microcuries per litre; another sample contained 14,000 microcuries. Samples taken after the use of fans for 2 hours still showed a reading of 500 mC, and advice to install additional ventilation was not heeded (3).

Between 1963 and 1968, Foote, succeeding the Vanadium Corporation of America, operated the Shiprock uranium mill, after Kerr-McGee worked it between 1954 and 1963.

The US federal government announced plans in 1983/84 to "neutralize" the Shiprock tailings (since 1973, control of the radiated land has been in the hands of the Navajo nation). These included: relocating tailings 300 feet away from the edge of a 70-foot high escarpment overlooking the San Juan river; moving contaminated material around the pile into the pile itself; decontaminating nearby buildings; and constructing surface run-off ditches. A seven-foot thick rock cover has been proposed for the heap itself.

The tailings piles contain nearly 2 million cubic yards of uranium waste and contaminated materials (4).

Foote also owns and operates other vanadium-uranium properties on the Colorado plateau (5).

Foote joined with Union Carbide in the 1960s and 1970s in a finally successful attempt to exempt Rhodesian chrome from UN sanctions (6).

The company's lithium operations in Chile were by 1983 making an important contribution to parent Newmont's earnings (7).

In 1988, Newmont disposed of its remaining holding in Foote: it is not known who the new owners are (8).

Under the flagship of Cyprus Foote, the company is now the prime producer of lithium chemicals in the USA (9) and the world's second largest producer of lithium salt, derived from exploitation of the Atacama deposits in Chile, where Foote holds 80% of Sociedad Chilena de Litio (SCL): in 1988 SCL exported more than 7000 tonnes of lithium carbonate, representing a quarter of the global market (10).

In 1989 another Chilean outfit, in which Amax holds 63.75%, set up a huge solar evaporation project at Atacama, to produce potassium, boron and lithium salts from the brines (10).

Contact: Southwest Research and Information Center, Box 4524, Albuquerque, New Mexico 87I06, USA.

References: (1) *MJ* 21/1/82. (2) *MIY* 1979, 1981. (3) *E&MJ* 4/84. (4) *AkwN* Spring/84, quoting UPI Service, Washington. (5) *MIY* 1982. (6) Lanning & Mueller. (7) *MJ* 30/11/84. (6) Lanning & Mueller. (7) *MJ* 30/11/84. (8) *MJ* 5/2/88. (9) *MAR* 1990. (10) *MJ* 16/6/89.

256 Fosforica Española (E)

In 1983, together with Enusa, the company was considering setting up a new Phosroc uranium-from-phosphoric-acid unit to produce some 75 tons of uranium (1).

References: (1) *E&MJ* 1/83.

257 Freeport-McMoran Inc

This is a major producer of sulphur and phosphates with important interests in copper, oil, coal, gas, uranium, and chemicals (1).

Freeport-McMoran is Indonesia's sole copper producer, providing about 200,000 tonnes/year during the early 'eighties from its mine at Ertsberg in Irian Jaya, otherwise known as West Papua. West Papua was colonised by the Indonesian military in a brutal operation. Freeport distinguishes itself for being perhaps the only major foreign company to operate in this tortured land and to profit from the suppression of its people. For that alone, it deserves to be remembered.

In July 1982, the company announced it was selling virtually all of its interests in land and mineral rights at Freeport, Texas, to the Brazos River Harbour Navigation District (2).

Freeport Minerals Co pioneered the recovery of uranium from phosphoric acid in the late '70s (1). In late 1978, Freeport made its first delivery of 10,000 pounds of U_3O_8 produced at its Uncle Sam plant in Louisiana (3). This was soon followed by delivery of another 90,000 pounds (4), supposedly under long-term contracts with both a foreign and a domestic utility, both un-named. In mid-1981, Freeport reported a "setback" when a German customer cancelled its contract for delivery from its second Uncle Sam plant (9). This may well have been the same customer cited in connection with the output of the first plant.

The same year, a second plant was announced, in conjunction with a primary extraction plant to be operated by Agrico at Donaldsville, Louisiana; start-up was expected in mid-1981 but deferred (5).

By 1980, the Uncle Sam plant was producing "in excess of targets" and Freeport had another uranium-from-phosphates recovery plant up its sleeve at Faustiana, Louisiana (6). That year, the company's chairman, Paul W Douglas, esti-

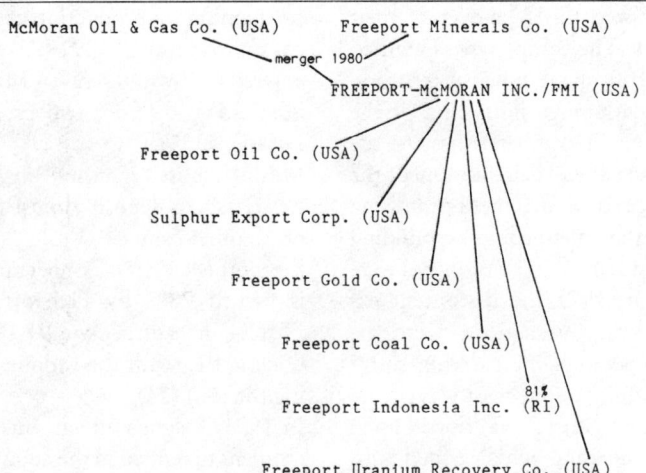

```
McMoran Oil & Gas Co. (USA)        Freeport Minerals Co. (USA)

              merger 1980

                    FREEPORT-McMORAN INC./FMI (USA)

        Freeport Oil Co. (USA)

        Sulphur Export Corp. (USA)

              Freeport Gold Co. (USA)

                    Freeport Coal Co. (USA)

                                           81%
                          Freeport Indonesia Inc. (RI)

                    Freeport Uranium Recovery Co. (USA)
```

mated that 27c out of every $1 share came from the company's uranium operations (7).

In 1981, production of uranium from Uncle Sam and Sunshine Bridge, Florida, plants fell to 60% of capacity and amounted to only 75 tons. The company sold a licence for uranium recovery technology,although it is not known to whom (8).

In late 1982, the company announced that uranium earnings had improved threefold over a year, reflecting the increase in uranium oxide production from the Uncle Sam and Sunshine Bridge phosphates-uranium plants.

More recently, Freeport concentrated exploration efforts in the USA, Australia and Chile – the prime target being gold (16).

In 1983 Freeport's Australian subsidiaries sold their entire interest in the Mt Keith and Kingston nickel properties for a sum of US$4.8M (10). Freeport's profits from uranium nearly doubled thanks to higher sales from the Uncle Sam and Sunshine Bridge phosphoric acid plants in Louisiana – a total of US$10M in profit (11). Emphasis on mining at the Ertsberg and Ertsberg East mines changed from open pit to underground production, enabling annual production to increase to an estimated 180-200 million pounds copper in concentrates (12).

In the first quarter of 1984, uranium earnings brought in US$2.5M, an increase of US$0.7M over the same period in 1983. The company's mines in West Papua made over US$1M profit for the company – at the expense of one of the cheapest labour forces in the world (13).

Metals Exploration and Freeport operated an open-cut mine near Townsville, Queensland, which was Australia's largest producer of the strategic metal cobalt, and second largest producer of nickel in 1983. The JV is known as Queensland Nickel Pty Ltd (14).

In July 1983, Freeport re-established long-term uranium oxide sales to an unspecified German utility, following a ruling by an international arbitration panel after disgreement between the two parties. Up to 800,000 pounds U_3O_8 would be supplied over a 6.5 to 13-year period from Freeport's two Lousiana recovery plants: Freeport had an option to supply 267,000 further pounds after the utility exercised its right to extend the delivery period and up the amount received by a further 250,000 pounds (1).

Freeport Indonesia (80% owned by Freeport) deserves special mention for its penetration of Indonesian-occupied West Papua and the havoc it has caused among tribal communities. Freeport was the first foreign company to invest in Indonesia after the open-door investment law introduced in 1967. Carmel Budiardjo of Tapol has described the early days:

"As the Freeport construction began, local

tribespeople resisted this take-over of their traditional lands. The company was granted full control of 100 square miles of territory, much of it the Akimuga homeland of the Amungme, Tsinga and other tribes. As the tribespeople intensified their resistance, the company concluded an agreement promising to compensate the tribespeople by building other homes and schools. The promises were not kept, and in 1977 the discontent exploded. Company installations were attacked, the copper pipeline carrying slurry more than 100 miles to the coast was cut repeatedly, and the airstrip ... was ripped up.

"This Akimuga uprising, which ended with the bombing and strafing of villages by Indonesian Air Force OV-10 Broncos, was graphically described in a small-circulation Indonesian church monthly, *Berita Oikoumene*, which also reported that at least a quarter of the Amungme people participated directly in the attacks on Freeport. This contradicts Indonesian claims that this uprising and others in Irian Jaya in the past 15 years are the work of 'riving terrorist gangs' lacking the support of the Papuan people" (17).

After the uprising, the Indonesians uprooted entire communities and sent them to barracks on the coast. By June 1980, epidemics had swept through these communities, killing more than 20% of the infant population. The people now had to depend on scarce supplies of sago and poor fishing, being deprived of traditional garden cultivation.

In 1982, Freeport announced an expanded programme on the Akimuga homelands and additional deep drilling beneath the Ertsberg East orebody (18).

Freeport restructured itself considerably in 1990, offloading $1.5 billion of assets (19), and selling its Freeport McMoran Gold subsidiary (61% owned by Freeport McMoran) to Minorco in 1991 (20). At the same time, it renamed itself Freeport McMoran Copper and Gold Inc – to reflect its greater interests in gold, especially in the Jerrit Canyon mine in Nevada (21).

Its uranium by-production at Uncle Sam plant continued through 1991 (22) and the company entered a JV with Denison Mines to reproduce uranium oxide from two processing plants in Louisiana (23).

Meanwhile its expansion on West Papua into the Grasberg deposit almost dwarfed progress on all other fronts.

Freeport McMoran Copper and Gold (FMCG) is owned 78% by Freeport McMoran Inc. FMCG, in its turn, owns 91.1% of Freeport Indonesia Inc (with the Indonesian régime owning the rest) (24).

In 1991 it signed an agreement with the government to remain in the country for at least the next thirty years (25) so long as it reduced its ownership to 40% between 2001 and 2010 (26). It announced in 1990 that it would spend more than 500 million dollars exploring a vast area (25 million hectares) of indigenous land in West Papua, in order to advance its production of copper to 52,000 tonnes a day minimum, from 1992 onwards (27).

This news was greeted with alarm from pro-indigenous groups like Survival International and Tapol, environmental organisations such as Rainforest Action Network and the Environmental Defense Fund and peoples' movements within Indonesia itself, especially the rainforest protection organisation Skephi.

The West Papuan journal *Tifa Irian* has summarised the impact of the existing Freeport operations, raising fears for the future. Its report, quoting community leaders, "... reveals the widespread resentment felt towards Freeport over the lack of compensation paid for their traditional lands. Facilities that have been provided by the company are unsatisfactory, such as the huts in Tembagapura, into which people from the Waa Valley were told to move. In contrast to the luxury homes of Freeport staff, this area is described as a 'mud town', an unhealthy slum. According to the report, a recent flood and mud-slide killed 17 people displaced by the expansion of the luxury accomodation in the town."

"The company also provided a health centre

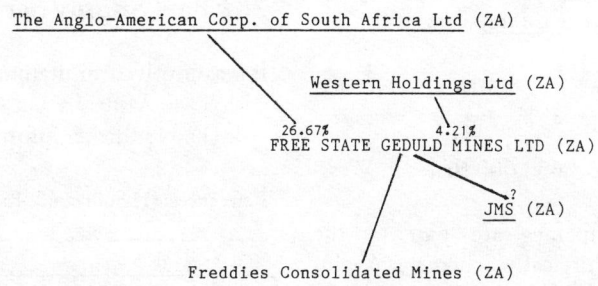

The Anglo-American Corp. of South Africa Ltd (ZA)

Western Holdings Ltd (ZA)

26.67%   4.21%
FREE STATE GEDULD MINES LTD (ZA)

?
JMS (ZA)

Freddies Consolidated Mines (ZA)

with one low-ranking civil servant who had no medical training. It built a road for local people, a school building (with no teacher or staff quarters) and provided a mobile shop. In contrast to the riches Freeport has extracted from the mountains, these offerings do not amount to much." According to a source quoted by *Tifa Irian*: "In the middle of the luxury town, the indigenous inhabitants live like beggars and are treated roughly by the company people" (27).

Further reading: *West Papua: The obliteration of a people*, London, 1983 (Tapol) – on the resistance of the Amungme and their conditions following forcible removal to make way for the Freeport mine.

Setiakawan no 4-5, 1-6/90, published by Skephi, Jakarta.

Contact: Tapol, 111, Northwood Road, Thornton Heath, Surrey, England.

References: (1) *MIY* 1981, *I FT* Min 1991. (2) *MJ* 23/7/82. (3) *FT* 23/11/78. (4) *MJ* 23/1/79. (5) *FT* 10/10/79; *MAR* 1981 p 358. (6) *MJ* 8/2/80. (7) *MJ* 22/8/80, 24/4/81. (8) *MJ* 13/8/82. (9) *MJ* 5/6/81. (10) *MJ* 19/8/83. (11) *MJ* 17/2/84. (12) *MJ* 22/7/83. (13) *MJ* 25/5/84. (14) *E&MJ* 11/83. (15) *MJ* 15/7/84. (16) *MJ* 19/11/82. (17) *Gua* 4/2/83. (18) Company report, quoted in *MJ* 19/11/82. (19) MJ 31/8/90. (20) MJ 1/2/91. (21) *E&MJ* 2/91. (22) *E&MJ* 3/91. (23) *E&MJ* 1/91. (24) MJ 18/1/91. (25) MJ 26/4/91. (26) *Down to Earth*, no 14, 8/91. (27) *Tifa Irian*, West Papua, weeks 2,3 & 4, June 1991.

258 Free State Geduld Mines Ltd

The company operates mining leases in the Welkom and Odendaalsrus districts of the Orange Free State. Although controlled and managed by the Anglo-American Corp, 40% of the company's shareholders (in the late '70s at least) were American (1).

Primarily a gold mining company, Free State also supplies uranium-bearing slimes to the JMS/Joint Metallurgical Scheme and owns one of the pyrite flotation plants used in uranium recovery by the scheme (2). In the first quarter of 1984, Free State provided 502,000 tons of slimes to JMS with a uranium grade of 0.09kg/ton – this was more than half the total delivered to the JMS for that period (3).

In 1983 the company was engaged in expansion of its mine holdings (4).

Then, in 1986, Free State Geduld joined with four other Anglo American companies in forming Freegold, the world's biggest single gold mining company with Free State Geduld taking 174 Freegold and 24 Ofsil (Orange Free State investment) shares in the new conglomerate (5).

References: (1) Lanning & Mueller. (2) *MIY* 1982. (3) *MJ* analysis of Rand and OFS quarterlies, 4/84. (4) *MAR* 1984. (5) *FT Min 90*.

259 Free State Saaiplaas Gold Mining Co Ltd

In 1981, Free State Saaiplaas merged its holdings with Western Holdings, which also acquired Welkom Gold Mining. In return, Free

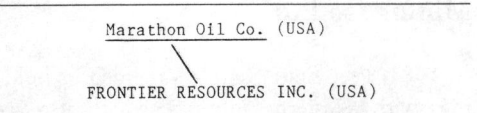

```
              Western Holdings Ltd (ZA)
                                  /
  FREE STATE SAAIPLAAS GOLD MINING CO. LTD (ZA)
```

State acquired some 3,653,000 shares in Western Holdings.

The merged companies are part of the JMS/Joint Metallurgical Scheme, Anglo-American's venture for the extraction of uranium from gold slimes.

Free State's own lease covers nearly 2000ha over farms in the Ventersburg district of the Orange Free State (1).

References: (1) *MIY* 1982.

260 Fremont Energy Corp (USA)

It was exploring for uranium in Texas in 1978 (1); it reported a "major find" of uranium to which WPPS/Washington Public Power Supply Systems (who had contracted Fremont to locate uranium to be used in its nuclear power plants) was entitled by paying Fremont a discovery bonus (2).

References: (1) *E&MJ* 11/78. (2) *MJ* 25/8/78, 9/3/79.

261 Frontier Resources Inc

It entered a JV agreement with Nuclear Dynamics and Plateau Resources to explore for uranium in eastern San Juan County, Utah, over 29,000 acres jointly owned by Nuclear Dynamics and Frontier (1).

References: (1) *MJ* 24/11/78, 12/1/79.

```
          Marathon Oil Co. (USA)
                          /
     FRONTIER RESOURCES INC. (USA)
```

262 Fuel Supply Services Inc

It was involved in uranium exploration, along with TXO Minerals and Getty Oil, in the late '70s (1). No further information.

References: (1) Sullivan & Riedel.

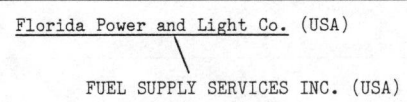

```
  Florida Power and Light Co. (USA)
                          \
     FUEL SUPPLY SERVICES INC. (USA)
```

263 Furukawa Co Ltd (J)

This non-ferrous metals smelting, chemicals and electric power company (1) has been involved with Fluor, Mitsui and C Itoh in building a copper smelter in Mexico (1); burning tyres as fuel for its Ashio smelter in Tochigi prefecture, Japan (2); and developing (with Mitsui) a new flash smelting process for PASAR in the Philippines (2).

In the 1970s it was part of AUREC/Australian Uranium Resources Exploration which explored for uranium in the Northern Territory; other partners in AUREC were C Itoh (70%), Sumitomo (20%), and the other JV partner was ENI (3).

The company has also explored in Papua New Guinea, Niger and Peru (4) and operates a copper smelter at Hibi, Okayama, Japan with Mitsui and the Nittetsu Mining Co of Japan (5).

References: (1) *MIY* 1982. (2) *MIY* 1984. (3) *A Slow Burn* (FoE/Australia-Victoria) No.1:9/84. (4) *MIY* 1985, (5) *FT Min* 1990.

264 Gardinier (USA)

It was contracting, in 1978, to set up a uranium-from-phosphates recovery plant with a design output of about 230 tonnes/year U_3O_8 at East Tampa, Florida, together with Uranium PUK (a subsidiary of PUK/Pechiney-Ugine-

Kuhlmann); another associate in this attempt was US Steel (1).

As of mid-1982, it closed its operations in Florida (2).

Gardinier was among three companies which tendered to the Moroccan government in 1979 to extract uranium from phosphates. The others were IMC and Westinghouse.

According to US intelligence reports (*ie* CIA information), Gardinier is also part of a major French consortium to construct Morocco's phosphoric acid programme and provide a nuclear extraction plant by 1995. The French company Sofratome had the contract for a feasibility study (3).

References: (1) *MAR* 1979 p.193. (2) *Nukem Market Report* 8/82. (3) *Intern Min/Met Rev.*

265 Gates Engineering Co

```
    Enserch Corp. (USA)
         /
 Ebasco Services Inc. (USA)
       /
    GATES ENGINEERING CO. (USA)
```

This mining, engineering and construction company specialises in uranium, oil shale, urban planning and other "development" projects (1).

Its brother corporation, Envirosphere, offers environmental services to the mining industry: their father Ebasco (engineering arm of granddaddy Enserch) is primarily concerned with the engineering and/or construction of energy facilities, including nuclear.

References: (1) *MM* 3/83.

266 GED Exlporations Ltd (CDN)

Once exploring, it is said, for uranium in Saskatchewan; but no further information (1).

References: (1) *Who's Who Sask.*

267 Gencor

The second largest mining house in South Africa, Gencor was formed in 1980 with the amalgamation of Unicorp/Union Corporation Ltd and General Mining and Finance Corporation Ltd. General Mining was itself taken over by Federale Mynbou Beperk in 1964 – Federale Mynbou being a creature of two Afrikaner investment companies, Federale Volksbeleggings and Bonuskor, who, in 1953, decided to form the first major Afrikaner mining company in South Africa. Federale Mynbou prospered in the '50s, to a large extent because its coal fields were located near two of the large government-owned Escom power stations (1).

Anglo-American had already taken major stakes in General Mining in the '50s: the decision to help Afrikaner capital acquire large chunks of new real estate was largely that of Harry Oppenheimer, Anglo's chairman appointed in 1957. The *Johannesburg Sunday Times* called the deal "a personal triumph for Mr Harry Oppenheimer" and "an important step forward in his proposals for closer business co-operation across the South African language barrier" (2). The brother of Hilgard Muller, South African foreign minister of the day, became General Mining's general manager.

Both Union Corporation and General Mining were founded by Germans: Adolf Goerz modified the former and the brothers Georg and Leopold Albu developed the latter. For many years both companies depended to a large extent on German financial expertise, and to a lesser extent, capital. Both the Deutsche and the Dresdner Banks invested in the two companies, though perhaps not more than 8% of their capital needs (1).

In fact, until 1974 Charter Consolidated of Britain was the largest single shareholder in Union Corporation, when it dropped out of the bidding for control of the company and General Mining secured dominance with 50.1%(1).

The streamlined Afrikaner mining-finance

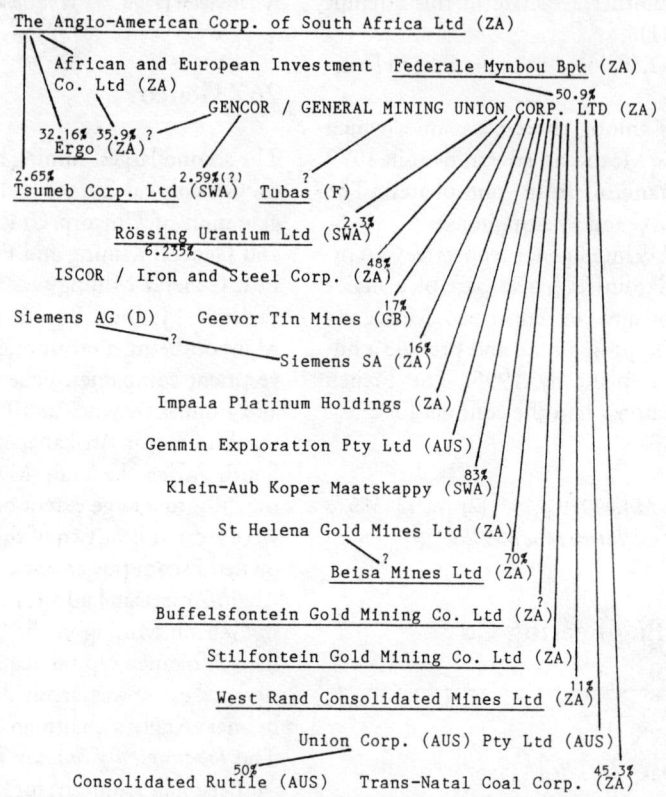

The Anglo-American Corp. of South Africa Ltd (ZA)

African and European Investment Co. Ltd (ZA)   Federale Mynbou Bpk (ZA)

GENCOR / GENERAL MINING UNION CORP. LTD (ZA)   50.9% (ZA)

32.16% 35.9% ?
Ergo (ZA)

2.65%   2.59%(?)   ?
Tsumeb Corp. Ltd (SWA)   Tubas (F)

2.3%
Rössing Uranium Ltd (SWA)
6.25%

48%
ISCOR / Iron and Steel Corp. (ZA)

17%
Siemens AG (D)   Geevor Tin Mines (GB)
?
Siemens SA (ZA)   16%

Impala Platinum Holdings (ZA)

Genmin Exploration Pty Ltd (AUS)

83%
Klein-Aub Koper Maatskappy (SWA)

St Helena Gold Mines Ltd (ZA)
?   70%
Beisa Mines Ltd (ZA)
?
Buffelsfontein Gold Mining Co. Ltd (ZA)

Stilfontein Gold Mining Co. Ltd (ZA)
11%
West Rand Consolidated Mines Ltd (ZA)

Union Corp. (AUS) Pty Ltd (AUS)
50%   45.3%
Consolidated Rutile (AUS)   Trans-Natal Coal Corp. (ZA)

house of the '80s now owns twelve mines, all primarily gold producers but with four delivering uranium as a by-product. The fortunes of Beisa, the primary uranium producer (with gold as a by-product) have been chequered: it was closed down in May 1984, a bare two years after its official opening (3).

However, the new Beatrix mine came on stream; and other uranium-from-gold producers in the group, Stilfontein, Buffelsfontein and West Rand, continued to make modest progress (4).

In 1983, the company experienced what it called "temporary difficulties" and determined to proceed with its fairly large exploration programme outside of South Africa, notably in Australia, Brazil and North America (4).

While disastrous falls in the market for blue asbestos – of which Gencor was the world's chief provider – have meant the closure of its Gefco/Griqualand Exploration and Finance Co asbestos subsidiary (4), Gencor continued to hold a roughly one-third interest in the Griqualand group of asbestos producing companies (5).

In Brazil, its gold mining prospects at Belo Horizonte continued to be developed. In summer 1984, Gencor announced that it would bring the high-grade São Bento gold mine in Minas Gerais into production by early 1987. The mine is estimated to have a life of 20 years (6).

The group's Australian subsidiary, Union Corporation (Aus), has secured 50% of Consolidated Rutile, an important producer of mineral and beach sands – an acquisition which was met with alarm from environmentalists (4) and opposition by Aborigines (7). Consolidated Rutile specialises in the production of rutile and zirconium (used in nuclear reactors).

Genmin Exploration, another Australian subsidiary of Gencor, has also entered a farm-in agreement (entitlement to up to 51%) in the Yellowdine platinum prospect 34km south-east of South Cross, Western Australia (8).

In Europe, the group owns a portion of Geevor Tin Mines in Cornwall, England, and a share in the Rubiaies mine in Spain (5). Geevor proved an unexpected boon to Gencor (and RTZ: between them the two companies held a controlling interest). As a consuming, not a producing, member of the ITC/International Tin Council, the mines were not restricted by the ceiling price imposed by Third World producers in the association. Consequently they were able to reap "rich benefits in terms of profitability" (9). Although the corporation has tried to diversify and "grow by acquisition" (it acquired an electronics subsidiary called Tedelex in 1983) (4), its base is still solidly within the apartheid system.

Apart from its gold and uranium interests in South Africa, Gencor (51%) has a JV with Union Carbide (49%) in the Tubatse ferrochrome project in the eastern Transvaal. It also operates four chrome mines in the same region (5). Its subsidiary Impala Platinum (second in importance only to Rustenburg) has maintained a price of US$465/ounce, although it has been suffering from the "recess" (10).

The company recently commenced a large-scale drilling programme in the Orange Free State, along with Anglo-American (4).

In 1983 it acquired Utah International's interest in the anthracite project in the "homeland" of Kwazulu (11), with intention to proceed to a large mine, much to the approval of Chief Gatsha Buthelezi, the Bantustan's Chief Minister, and to the chagrin of anti-apartheid groups. The mine was originally envisaged as a joint project between Gencor (through its Trans-Natal Coal subsidiary) and General Electric's Southern Sphere Holdings. It would have been the biggest investment in recent years by a US company in the apartheid state. Then the Connecticut legislature banned its state pension funds from investing in companies with interests in South Africa, and Southern Sphere sold its holding to Gencor. The mine was expected to provide about 600 jobs and last 22 years (12).

Namibia has been an area of key interest, too. Apart from investigation of the large Langer Heinrich uranium deposit located in 1975 (13), Gencor also has approximately 7% of the huge Rössing Uranium mine which it ostensibly took over from Urangesellschaft in the '70s (14).

More recently, the company has been evaluating the Tinkes uranium orebody (which is adjacent to a uranium lease held by Anglo-American) (14) from which production is planned of 1250 tons/year uranium oxide (15).

Also in Namibia, Iscor (in which Gencor has 48% interest) is the territory's most important zinc producer (13). And Gencor controls (with 83%) the Klein-Aub copper mine, also in the illegally occupied state (4).

Coal is a major interest of the company: in 1982 it held no fewer than 13 collieries, of which nine were in the Transvaal, three in Natal and one in the Orange Free State.

One of these areas, at Springbok in the northern Transvaal, has been the subject of a major investigation by the company into the possibility of extracting uranium from coal (16). Towards the end of 1979 Gencor reached agreement with Sentrachem, the major South African chemicals company, to participate in the project in return for a 10% interest in Sentrachem (which is now held 8.375% by Gencor, 1.25% by Federale Mynbou and 0.375% by Sentrust). Investigation by Gencor staff delineated four main zones, in two of which lay significant supplies of uranium. However, although the uranium could be leached, burning of the coal would produce refractory ash. Intensive laboratory tests were undertaken to attempt to directly liquify the coal, without gasification to exploit the carboniferous content of the washed coal, and to process the uraniferous ash. In 1980 it was estimated that 18 months would be required before any decision was made on this new (and apparently unique) plant, and that it would take five years before any production (17).

In 1989, Gencor established Genmin (General

Mining Metals and Minerals) as the holding company for its huge South African mineral interests (second only to those of Anglo American). At that time the gold division (Gengold) operated 14 mines, including Beatrix, Buffelsfontein, Stilfontein and Unisel and held significant interests in the São Bente gold mine in Brazil (18). Although sanctions and the dip in the gold price have, more recently, taken their toll, with the closure of some of Gencor's operating gold plant (19) and the layoffs of workers (20), a new gold project was opened in mid-1990 in the Orange Free State, at Weltvreden, with a view to comparatively cheap production starting in 1992 (18,21). Together with Oryx gold mines (owned 63% by Gencor) the company remains fairly well placed to benefit from any relaxation in sanctions against gold (22).

As the second biggest platinum producer (through Impala Platinum – 41% owned) Gencor has also benefitted from the fillip given to this so-called "green metal" by its increased use in catalytic converters for motorised vehicles. In 1990 Gencor effected a significant merger of its platinum mine at Karee in the Transvaal with the platinum interests of Lonrho, the British mining conglomerate – leading to suggestions that the two companies might be about to merger. This would have put Gencor/Lonrho among the biggest half-dozen mining houses anywhere in the world (23).

The same year Gencor bought out all Mobil's surrendered oil and gas interests in the apartheid state, which the company – under considerable pressure from anti-racists back home – had disinvested in 1989 (24). In 1989 Gencor's Trans-Natal Coal Corp operated 12 collieries in South Africa (18), while Samancor (Gencor 43.6%) continued to be the most important producer of manganese, exporting 80% of its production (25).

It also continued its mining of mineral sands, through 49% owned Consolidated Rutile (the mining company responsible for exploitation of Aboriginal land at North Stradbroke Island near Brisbane) (18) and its 25% ownership in Richards Bay Minerals, the company taken over by RTZ (through QIT) in 1989, which is one of the world's biggest producers and processors of titanium-bearing mineral sands (see RTZ).

Another operation at Richards Bay, in which Gencor holds a substantial interest, is the Alusaf smelter (the parastatal Development Corporation of South Africa has a 30.7% interest and Alusuisse holds 23% of Alusaf).

Apart from Genmin, the company has four other divisions. These are concerned with forestry products (Sappi), energy (Engen) industrial interests (Malbak) and investments (Genbel). Among Genbel's holdings are a part-share in the recently-privatised state steel corporation of South Africa, Iscor (18,26).

Although Gencor is very firmly based in South Africa, in 1990 it announced a bid for the whole of a 240,000 hectare concession in the tropical plains of eastern Bolivia, at Rincon del Tigre, where it was hoping to find platinum (27).

Gencor's reputation as a conservative, hardline, Afrikaner mining house was confirmed during the period of upheaval in the mid-eighties, when the South African National Union of Mineworkers strove to establish itself and defend workers' rights. Many of the deaths and injuries sustained by protestors at this time occurred at Gencor's mines; added to which, the company has the unenviable record of being responsible for the worst gold mining accident in South African history. 177 miners were killed at the Kinross mine in September 1986, following an officially-recognised drop in safety standards (28).

Less dramatic, but more insidious, deaths are still being caused by Gencor's legacy of asbestos mining and the operations of two subsidiaries, Griqualand Exploration and Finance Co Ltd (Gefco: 48.7% owned by Gencor and 15.4% by South African Mutual Life Assurance) and Msauli Asbes Bpk (Gencor 37.5% and South African Mutual Life Assurance 15.2%).

Figures leaked to the South African press in August 1984 revealed a "startling incidence" of asbestosis among African workers at the Penge mine, where 110 had been laid off due to asbestosis in the previous ten months. Between 1973 and 1983 some 780 of the mine's employees contracted the killer disease and fibre levels were 65 times higher than the local recommended level in the mill (which was itself about four times lower than the level set in Britain).

In July 1984, 1300 migrant workers at Penge struck for better wages, after a drastic cutback of production by Gefco (29).

Two years later, as the Black Allied Mining and Construction Workers Union (BAMCWU) demanded a ban on all asbestos mining ("Enough blood has already been coughed up on the altar of greed and in the name of profit" declared Barry Calstelman at a BAMCWU conference in June 1985), (30) it was alleged that children as young as 11 and 12 had contracted the asbestos-related disease mesothelioma, from playing on abandoned Gencor tailings dumps (30).

References: (1) Lanning & Mueller. (2) *Johannesburg Sunday Times* 30/8/64. (3) *FT* 5/5/84. (4) *MAR* 1984. (5) *MIY* 1982. (6) *FT* 22/8/84. (7) *AMIC* private communication, 1981. (8) *E&MJ* 11/83. (9) *FT* 15/6/84, 16/6/84. (10) *MJ* 8/4/83. (11) *MJ* 16/4/82; *MAR* 1984. (12) *FT* 21/9/84. (13) UN NS-36. (14) Rogers & Cervenka. (15) *E&MJ* 1/84. (16) *MJ* 14/12/79. (17) *MJ* 22/2/84. (18) MAR 1990. (19) *FT* 3/4/91. (20) *MJ* 10/7/90. (21) *E&MJ* 1/90. (22) *MJ* 21-28/12/90. (23) *FT* 17/12/90, *MJ* 19/1/90, see also *MJ* 17/8/90. (24) *FT* 26/2/90. (25) *FT* 7/11/90. (26) see also *FT* 2/11/90. (27) *FT* 13/9/90. (28) *E&MJ* 10/86. (29) *MJ* 14/9/84. (30) *Sechaba*, South Africa, 6/87.

268 General Atomic Corp

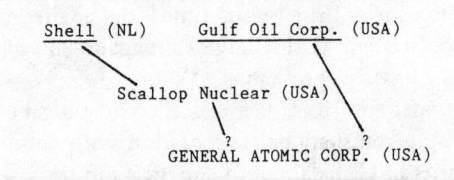

The "first company anywhere dedicated to building atomic reactors" (1), General Atomics (*sic*) was founded by Frederic de Hoffman, a former member of the Los Alamos team which constructed the Hiroshima and Nagasaki bombs (2). Later acquired by General Dynamics Corp, manufacturers of high-temperature gas-cooled (HTGR) reactors, nuclear fuels and nuclear steam systems, General Atomics was bought out by Gulf Oil in 1967 and renamed Gulf General Atomic Ltd (3). Somewhat later, half of Gulf's interests in the company was sold to Shell, but partially repurchased in 1979 (4). From 1980 until 1984, in the post-Three Mile Island era, General Atomic appears to have ended on the shelf, as it contended with complex litigation over Gulf's membership of the uranium cartel. During the previous decade it landed several contracts for reactors, for example with Delmarva Power and Light and Philadelphia Electric in the early '70s, and the Fort St Vrain HTGR station near Denver, Col, completed in 1972 (3) – though eight years later this station had still not reached full design power (5). The company also sold Indonesia its first reactor, a 250KW Triga Mark II which was operating at Bandung University in 1964. The reactor was upgraded to 1MW in 1971 (6). Ironically, finance for the project derived from sale of Sumatran coal: one in the eye (or the bunker) for those who maintain nuclear power is always clean, while coal is always 'dirty' (7)!

In 1980, General Atomic offered to build a 25MW Triga in Indonesia – so far as is known, the offer was not taken up. If it were, this would have been the largest Triga built by the company anywhere (8).

During this period, General Atomic was also testing its Doublet III fusion reactor, with the support both of the US DoE and the electric utility industry in the States (5).

Gulf was the key US protagonist in the uranium cartel which effectively cornered the uranium market and quintupled uranium prices between 1972 and 1975. Since General Atomic was both a supplier of, and a contractor for, uranium during the '70s, it ended up (like Westinghouse) being sued by utilities and itself suing uranium mining companies which didn't deliver. (Unlike Westinghouse, however, there was never any doubt of Gulf's leading role in promoting the cartel.)

By late 1978, General Atomic was in the position of having to supply, to three utilities, uranium which it couldn't secure because its own supplier, the UNC, ceased deliveries while the cartel was being investigated by a Senate committee. The shortage – some 3,500,000 lbs of uranium oxide – was (fortunately for General Atomic) partly made up by supplies from the new Rabbit Lake mine, operated by Gulf's subsidiary Gulf Minerals Canada (9).

In mid-1984, General Atomic finally settled with UNC: it paid US$130M in cash to the other company and assumed UNC's obligations to repay some 2.3 million pounds of uranium owed by UNC to a utility. At the same time, Socal/Chevron agreed to invest US$100M in UNC through the purchase of unused stock (10).

As details of the cartel became public, General Atomic entered another dispute with a supplier – this time the Johnny "M" mine, jointly owned by Ranchers and HNG Oil. General Atomic originally agreed to purchase the output from the mine and filed suit against the other two companies when, in early 1976, they tried to revise or cancel their contracts because they couldn't deliver the full amount of ore. Ranchers and HNG countered with a claim that the contract had been unenforceable and should be therefore voided.

The litigation was partly settled the same year when Gulf States Utilities (the customer) accepted delivery of some uranium. Later the three companies agreed a profit-sharing arrangement which, it was estimated, would involve Rancher and HNG paying up to US$60M to General Atomic from around 1981 onwards (11).

By early 1984, Gulf had settled its final suit with the utilities, setting up a US$30M venture capital fund as part of a US$70M settlement with the TVA (12). Three years earlier it also agreed to pay the Pennsylvania Power and Light Co US$43.9M over seven years and supply it with 2.5M pounds U_3O_8 until 1987 (13).

General Atomic's contribution to weapons programmes reared its ugly head in late 1989 in a (perhaps) surprising development, when the US Department of Energy (DoE) announced that it would be shipping 16.48 kg of 93% enriched (*ie* weapons grade) uranium to Romania, for use in a 14MW "research" reactor at Potesti – one which had been supplied by General Atomic in the 1970s: the uranium had been manufactured by the company about ten years before (14).

Before his downfall, President Ceausescu of Romania had boasted that his country had the capacity to manufacture nuclear weapons (14). The 1989 deal was arranged by Edlow Inc, whose Vice-President Rod Fisk, when asked whether he was concerned how the material would be used, retorted: "That's a question you have to address to the US government" (14).

References: (1) Moss. (2) Ed. Ruth Adams: *All in our time*, date & place unknown, quoted in Moss. (3) *Yellowcake Road.* (4) *WDMNE.* (5) *Gulf Fact Sheet 1979-80*, Pittsburgh, 1980. (6) *IAEA World List of Research Reactors*, Vienna, 1981 (IAEA). (7) Nuclear Regulatory Commission, Washington press release 14/8/81. (8) *Nucleonics Week*, Washington, 27/3/80. (9) *FT* 18/8/78. (10) *FT* 1/6/84. (11) *Wall St J* 12/3/79. (12) *FT* 30/3/84. (13) *FT* 25/10/84. (14) *Gua* 11/12/89.

269 General Electric Co (USA)

General Electric is America's largest industrial group, and until recently the second largest supplier of nuclear power equipment in the world – but in early 1982 it announced it was to stop supplying nuclear plants. Instead it would "concentrate on servicing nuclear plants and selling fuel. "In ten years time," the chairman said, "it might start selling plants again – meanwhile it was unprofitable" (1).

Founded in 1892, General Electric early became a transnational corporation with subsidiaries in Canada, South America and the Caribbean, as well as extensive interests in Europe. In a remarkable speech on the morality of multinationals, delivered in July 1980 in Virginia, Chairman Jones announced that it was "... the creation of wealth – the creation of a civilizing surplus – that is the key to a better life for all people. Poverty," Jones went on, "is the plight of all primitive or traditional peoples; not until the rise of industrial capitalism did any society have the capability to eliminate stark poverty. It is ironic that industrial capitalism, which owes so much to the Protestant work ethic and Judeo-Christian respect for personal stewardships, has now become the favorite whipping-boy for the liberation theologians" (2).

General Electric, among the biggest multinationals in the world, has links with the military which are so close that, by 1969, it had about 90 high-ranking US military officers in its employ, and its board of directors included former secretaries of Defense, Navy, and Army (3). It is implicated in the production of nuclear weapons at almost every stage, from uranium mining and milling to nuclear warheads. In 1980, it received military contracts worth more than two billion dollars. It also makes fuel rods and reactor cores, stores nuclear wastes, produces the Mark 12-A nuclear warheads – and has been the focus of an international boycott (4).

For its world-wide mineral explorations, its interests in copper, gold, molybdenum, coal, and, to a certain extent, the provision of uranium, General Electric has relied on Utah International since 1976 when the two companies merged (5).

Utah's uranium operations were conducted primarily in the USA through Pathfinder Mines, a wholly owned subsidiary of Utah, to which, in February 1982, General Electric sold 80% of its common stock (5). By the end of 1982, however, Cogéma had acquired this 80% for itself (6, 8).

A decade ago, the company was America's second largest nuclear reactor vendor (second to Westinghouse); the country's fourth uranium miller; fifth defense contractor; and seventh holder of uranium reserves (7). By early 1979 – *ie* just before Three Mile Island severed the US nuclear industry's jugular vein – General Electric had sold 30% of the total number of US nuclear reactors (9).

One of the more important actions taken

against General Electric occurred in 1980 when eight activists from the Atlantic Life Community in the USA joined workers in entering the company's "re-entry defense establishment" in Pennsylvania where they poured blood over plans and equipment and hammered the nose cones of two re-entry components from Minuteman III ballistic missiles. All were refused bail except for Dan Berrigan, a very well-known anti-war activist from Vietnam days. In a statement issued to the public and press the group stated: "We commit civil disobedience at GE because this genocidal entity is the fifth leading producer of weaponry in the US. To maintain this position GE drains 3 million dollars a day from the public treasury, an enormous larceny against the poor. We wish to challenge the lethal lie spun by GE through its motto "We bring good things to light" (10).

General Electric's military products include: "... jet engines, used in military and commercial aircraft (CF6 engine used principally in McDonnell Douglas DC10 and European Airbus A-300) ... Aerospace products include missile launch, guidance and re-entry systems, earth orbiting satellites, radar and sonar systems, armament systems, aerospace instruments and aircraft instrumentation and controls; these products are sold primarily to government agencies ... At year end 1978, General Electric and Honeywell Inc combined the world-wide operations of GE's Information Services Division and Honeywell's time sharing marketing operations that distributed the GE services in the UK, Europe and Australia. The new company is now 100% owned by General Electric" (11).

GE received no less than US$5.8M from the US Department of Environment towards a "monumental two-megawatt wind system with propellor diameter of 208 feet to be constructed in North Carolina" in 1978 (12).

By the late '70s, GE was one of the three companies most heavily involved in fast breeder research and development along with Westinghouse and Rockwell International (12).

In 1984 BHP acquired GE's Utah International subsidiary, except for an important 15%

stake which GE continued to hold in Utah's Queensland (Australia) coal interests (13,14).

Contact: Brandywine Peace Community, 51 Barren Road, Media, Pennsylvania 19603, USA.

References: (1) *FT* 9/12/81. (2) Text from Executive Reprints, Corporate Publications Warehouse, Building 705, General Electric Co, Scotia, NY 12302, USA. (3) Newsletter of WCC Programme on Transnational Corporations, 1/81. (4) *Veda News*, Oregon, USA 8/82. (5) *MAR* 1982. (6) *FT* 11/11/81. (7) *NARMIC'79*. (8) *E&MJ* 4/84. (9) *Electrical World* 15/1/79. (10) *Peace News*, London, 17/10/80. (11) *WDMNE* (12) Reece.(13) *FT* 4/4/84. (14) *FT Min* 1990.

est in the Crow Butte uranium deposit in Nebraska (3).

References: (1) *MJ* 6/2/81. (2) *FT Min* 1990 (3) *FT Min* 1991.

271 Geopeko Ltd

This is Peko Wallsend's wholly owned exploration subsidiary.

It has been exploring for uranium with EZ at Barote, north of Ranger (1), and held, in 1977, a joint uranium lease with Uranium Consolidated at Argentine, Queensland (2).

Virtually no details of the deposit have been

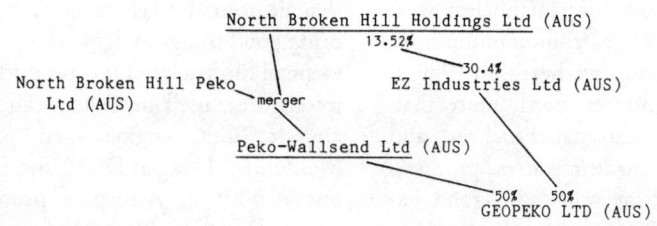

270 Geomex Minerals Inc

Imperial Metals Corp. (USA)

GEOMEX MINERALS INC. (USA)

It has been exploring for uranium in many parts of the USA (where several natural resources and exploration companies are acting on its behalf, Ferret Exploration being one of them) (1) and in Canada.

Both Geomex Development Inc and Geomex Minerals Inc were taken over by Imperial Metals Corp in 1986, increasing the number of West German-financed drilling partnerships controlled by Imperial Metals, to 14 (2).

Geomex operates a gold mine in Nevada (41.61% owned) and holds a heap leach gold property at Hay Ranch, in the same state (2). Until 1989 it also held 40% net working inter-

published in the last few years: it falls within Stage Two of the Kakadu National Park which is subject to decisions on use by the Northern Land Council (3).

References: (1) *MJ* 3/7/81. (2) *MJ* 16/9/77. (3) *Reg Aus Min* 1983/84

272 Getty Mining Co Ltd

Getty owns 49% of the Maureen uranium deposit with CCE/Central Coast Exploration (1). No field work was carried out on the deposit in 1981 or 1982 nor, it seems, since. Getty then said that if uranium markets improved, consideration would be given to an acid heap leach scheme.

References: (1) *Reg Aus Min* 1983/84.

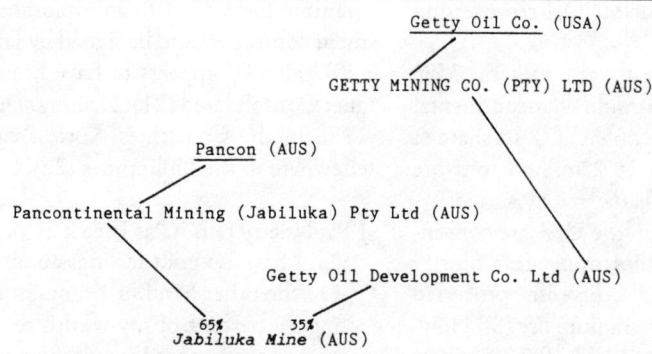

Getty Oil Co. (USA)

GETTY MINING CO. (PTY) LTD (AUS)

Pancon (AUS)

Pancontinental Mining (Jabiluka) Pty Ltd (AUS)

Getty Oil Development Co. Ltd (AUS)

65%   35%
Jabiluka Mine (AUS)

273 Getty Oil Co

Primarily an oil company, named after the billionaire art collector J Paul Getty who nominally controlled just 11.44% of the company but effectively ran it (1), it has interests in Algeria, Australia, Canada, Ireland, Liberia, the Netherlands, the Philippines, Great Britain and elsewhere.

It started expanding into minerals – especially coal and uranium – from 1973 (2). Despite minerals making no significant contribution to Getty's earnings, in mid-1981 it was still spending US$70M on the acquisition of a coal mine in Colorado, USA, and continued spending on world-wide mineral exploration projects (3).

Its partners in different projects have included Utah (BHP) at the La Escondida copper prospect in north-east Chile (1); Hanna Mining in a copper JV at Casa Grande, Arizona (4); Mitsubishi with which it has taken half shares in Mitsubishi Oil Co (2); Irish Base Metals (a Northgate subsidiary) with whom it undertook a US$6M base metals search in Eire (4) – it also paid US$38M to acquire 91 Northgate licences there (5); Comalco in JVs in the Officer Basin exploration (6); and it even entered a JV with Columbia Pictures, MCA, Paramount and Twentieth Century Fox to run a new national subscription pay-TV network in the USA.

Getty's prime involvement in the nuclear industry is through uranium mining, but in 1968

* In 1984, Texaco acquired Getty Oil: see Texaco for details.

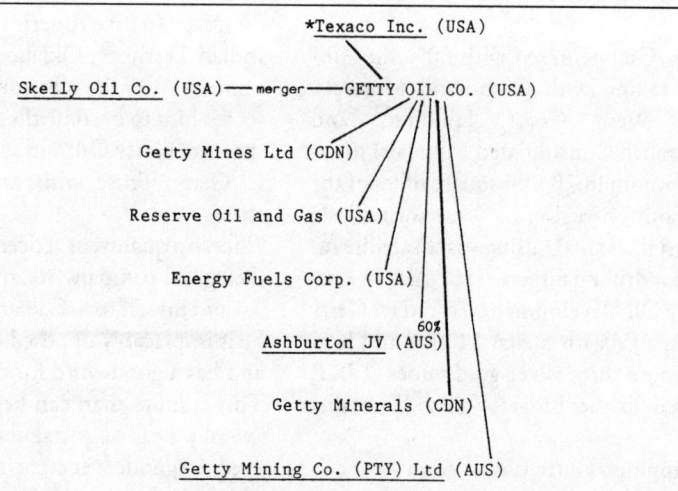

*Texaco Inc. (USA)

Skelly Oil Co. (USA)——merger——GETTY OIL CO. (USA)

Getty Mines Ltd (CDN)

Reserve Oil and Gas (USA)

Energy Fuels Corp. (USA)

60%
Ashburton JV (AUS)

Getty Minerals (CDN)

Getty Mining Co. (PTY) Ltd (AUS)

it did take a stake in nuclear fuel reprocessing too (2).

Its prime uranium interest is currently the 35% it holds in the Jabiluka mine (Pancontinental Mining holding the other 65%) (7). Its share of reserves there comes to 19.2 million tons ore grading at 0.367% U_3O_8 (6).

Its uranium operations in the USA are concentrated on the Petrotomics mine near Shirley Basin, Wyoming. In 1980, this mine processed some 1700 tons a day of uranium ore (8). However, by early 1981 it had laid off 130 miners (about one third of the workforce) (9). By 1982, the mine was running at only half capacity (10).

In 1978, together with Mitsubishi, Getty (51%) announced a JV (North American Resources Co) to explore for uranium in the San Juan and McKinley areas of New Mexico: production could start "in eight or ten years' time, after completion of US$100M-worth of mining and refining facilities" (11).

In 1979, it entered a JV with Cleveland-Cliffs, Skelly Oil (merged with Getty in 1977) and Nuclear Resources to explore for uranium in the Powder River area of Wyoming (12).

In Canada, the company was exploring in the late '70s for uranium in the Annapolis Valley of Nova Scotia (13); it was also exploring in Saskatchewan (14). In the Kasmere Lake prospect in north-west Manitoba, Getty Minerals holds 33.5% along with Camflo and United Siscoe (15).

In Australia, Getty entered with 60% the Ashburton JV (along with Command Minerals, Nickelore, West Coast Holdings and ACM/Australian Consolidated Minerals) to explore for uranium in 70,000 square miles of the Pilbara Ashburton region of West Australia (a largely Aboriginal area) "using space satellite information for drilling targets" (16). Also in Australia, Getty Oil Development Co Ltd (ie Getty Mining) has a JV with Stellar Mining and Kratos Uranium on three silver-gold mines (17). It also searched in the Fingal Valley, Tasmania (18).

In the Philippines, Getty Oil received a concession to explore the Camarines Norte area for uranium in 1977 (19): an exploration development contract was to be signed by January 1978 (20). This JV appears to have been with Benguet Consolidated (21). Drilling started in June 1978 in the Camarines Norte area as well as elsewhere in the Philippines (22).

J Paul Getty tells it "as I see it":

"... I have no guilt feelings about being rich. On the other hand, if I thought that by giving 99 percent of my wealth to the poor or some government I would help abolish poverty and human ills, I would not hesitate to do so. But suppose I did, and every man, woman and child in the world received even as much as a dollar, what good would it do? The answer is, none – and I would have nothing left to invest in productive enterprises that filled certain human needs and requirements even while creating jobs and paying salaries and taxes" (23).

Thus J Paul Getty in his autobiography: a curious, bitty, sometimes informative, often unconsciously amusing work, published in his adopted country, Britain, in 1976.

Here we learn how the late J Paul brushed shoulders with Onassis, Beaverbrook, the Duke of Windsor, Churchill, Chaplin, Nixon and many others.

He dates the formation of his empire to his father's purchase of "lot 50", an oil and gas lease on Osage (native American) land in Bartlesville Indian Territory, Oklahoma, at the turn of the century. After the oil boom, his path was firmly set for him to become the owner of four million shares of Getty Oil, and sole trustee of the Sarah C Getty Trust with another eight million shares.

"I feel no qualms or reticence about likening the Getty Oil company to an 'Empire'," he writes, "– and myself to a 'Caesar'". On the other hand he is implacably opposed to "Big Government" and has a good word for OPEC.

This is more than can be said for his views on "woolly-headed social theoreticians and dew-eyed do-gooders encroach(ing) on the preserves of law enforcement agencies and judicial systems". Or his view on capital punishment: "I am old-fashioned – or, if you prefer, arch-reactionary – enough to believe that the death penalty is the one even partially effective deterrent to murder ..." As for women, he is of course not a "women's libber", but believes they should be respected in the grand old patriarchal fashion: "Generally speaking most women seem to prefer that the male be (at least superficially) the dominant partner in a relationship."

Reeking with homespun conservative homilies, Getty finally settles for Abraham Lincoln's Gettysburg address as an earnest example of his own philosophy: "You cannot bring about prosperity by discouraging thrift. You cannot help the wage-earner by pulling down the wage-payer ... You cannot help the poor by destroying the rich". (From the Gettysburg Address – of course!) (23).

As one of the 29 companies sued by Westinghouse in the cartel case in the late 70's, Getty Oil was charged with anti-trust practices. However in early 1981, Westinghouse settled out of court with Getty – the oil company giving the nuclear giant US$13M in cash (24).

In 1988 Total Resources (Canada) Ltd, a 73% subsidiary of Total-CFP (*qv*), acquired Getty Resources Ltd, which it merged with Total Erickson Resources Ltd and renamed Total Energold Corp (25).

Total Energold has been involved in a drilling programme for gold in the Tundra region north of Yellowknife, Northwest Territories, Canada, (26) and with Davidson Tisdale Mines at the Tisdale gold property in Ontario (27).

In 1988 it was continuing to explore for uranium in northern Saskatchewan at the Russell Lake property, near Key Lake: its uranium exploration at the contiguous Taylor Lake property seems to have been abandoned that year (27).

References: (1) *FT* 28/8/81; *MAR* 1982. (2) *WDMNE.* (3) *MJ* 11/6/82. (4) *MIY* 1982. (5) *MAR* 1982. (6) *MJ* 11/6/82. (7) *Kiitg* no.22 p.16. (8) Annual Report 1981. (9) *MJ* 17/4/81. (10) *Nukem Market Report* 7/82. (11) *FT* 8/11/78. (12) *US Surv Min* 1979. (13) *Energy File* 4/82 (14) *Yellowcake Road.* (15) *MIY* 1981; *MIY* 1982; *FT* 10/5/79. (16) *FT* 12/7/79. (17) *Reg Aus Min* 1981. (18) *MJ* 6/3/81. (19) *MJ* 22/7/77. (20) *MJ* 2/12/77. (21) *MJ* /12/78. (22) *MM* (23) J Paul Getty: *As I See it*, London, 1976 (W H Allen). (24) *FT* 13/1/81. (25) *MIY* 1990. (26) *E&MJ* 10/89. (27) *E&MJ* 5/88.

274 GFSA

"Generally considered the most conservative of South African mining houses" (1), more than 80% of the company's assets are in gold, "the largest percentage exposure to the yellow metal of all the South African mining groups", despite concerted efforts to diversify in recent years (2). GFSA manages eight mines owned by seven subsidiary mining companies incorporated in South Africa. The companies are: Driefontein (operating the West and East Driefontein mines), Deelkraal Gold Mining Co Ltd, Kloof Gold Mining Co Ltd, Venterpost Gold Mining Co Ltd, Libanon Gold Mining Co Ltd, Doornfontein Gold Mining Co Ltd and Vlakfontein Gold Mining Co Ltd.

In 1977, GFSA produced nearly a quarter (21%) of South Africa's gold – West Driefontein at the time being the world's largest and most profitable gold mine (1).

The group has had some success in acquiring non-gold interests in recent years, notably its 43% in Tsumeb (operating against UN decree in Namibia) and its 25% interest in O'okiep, a major South African copper producer. However, both these mines have experienced recent cutbacks, while GFSA's JV with Phelps Dodge (49%) in the Black Mountain lead-silver-copper-zinc facility in the northern Cape (3) struggled to the end of 1983 without profits sufficient to cover the mine's debt capital (2).

In 1976, the company discovered a major alaskite uranium deposit at Trekkopje (4). This prospect – like others in Namibia apart from RTZ's Rössing mine – is currently in abeyance (2).

But the group's fortunes are firmly tied to gold, and only to a tiny extent to the uranium it recovers from West Driefontein. In 1983, there

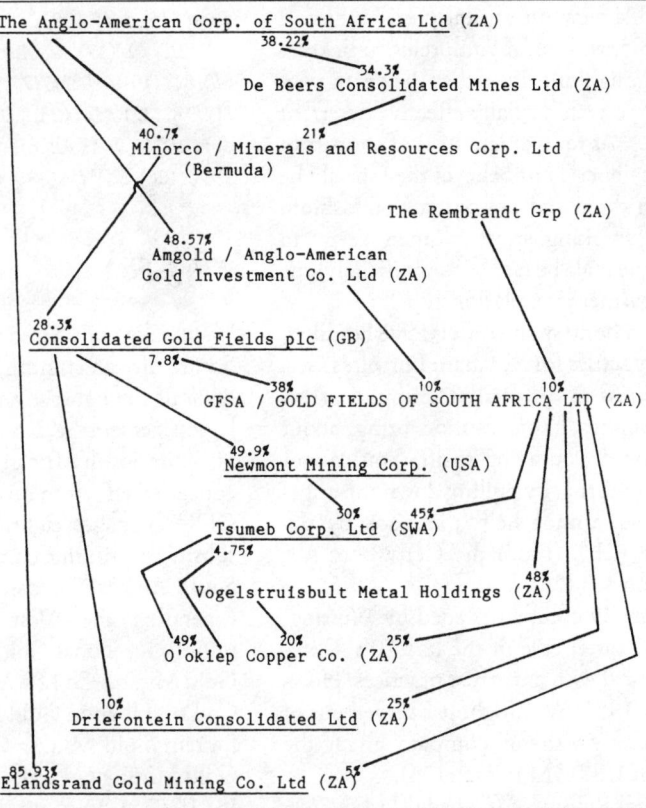

The Anglo-American Corp. of South Africa Ltd (ZA)
38.22%

34.3%
De Beers Consolidated Mines Ltd (ZA)

40.7%   21%
Minorco / Minerals and Resources Corp. Ltd
(Bermuda)

The Rembrandt Grp (ZA)

48.57%
Amgold / Anglo-American
Gold Investment Co. Ltd (ZA)

28.3%
Consolidated Gold Fields plc (GB)
7.8%

38%   10%   10%
GFSA / GOLD FIELDS OF SOUTH AFRICA LTD (ZA)

49.9%
Newmont Mining Corp. (USA)

30%   45%
Tsumeb Corp. Ltd (SWA)
4.75%

48%
Vogelstruisbult Metal Holdings (ZA)

49%   20%   25%
O'okiep Copper Co. (ZA)

10%   25%
Driefontein Consolidated Ltd (ZA)

85.93%   5%
Elandsrand Gold Mining Co. Ltd (ZA)

was speculation that GFSA would soon develop a new big gold deposit, located south of the Kloof and Driefontein properties (2).

In May 1984, the Anglo-American Corp increased its direct and indirect interest in GFSA when it acquired 4.6 million shares in GFSA from the Old Mutual Insurance group in exchange for 8.5 million Barlow Rand shares (5).

GFSA has been in the forefront of resistance to any attempts to radically raise wages for black workers or open up mines to black trade unions. As explained by Lanning and Mueller (1) this has been because GFSA has a larger proportion of poorly producing gold mines than any other South African mining company. It has been highly dependent on foreign labour, drafted in from Malawi and Mozambique (69% of its workforce in 1977). And while Anglo-American, for example, has been able to increase wages to black workers in order to re-

duce dependence on foreign miners and attract a more domestic (and domesticated) workforce, GFSA has been unable to do so.

Labour relations

The reluctance of GFSA to do more than operate strictly within the minimal legal framework was illustrated by the 1984 chairman's annual review, which is worth quoting from at length: "During the year under review there have been a number of significant developments in the manpower field in the mining industry. In particular, progress has been made in the important area of eliminating discrimination. An agreement has been reached between the Chamber of Mines and the Underground Officials Association which provides for the elimination of discrimination in occupations falling within the orbit of that Association. In the past the group has worked closely with the Association and welcomes the new developments which, *inter alia*, provide for all employees performing work falling within the scope of the Association to become members of the Association. This agreement led to the withdrawal by the Minister of Manpower of Job Reservation Determination No. 27 which had previously reserved occupation in the sampling, survey and ventilation departments for white employees only. The last remaining legal obstacle to the elimination of discrimination in the mining industry is the contentious 'scheduled person' definition of the Mines and Works Act.

"Group mines have recently apprenticed a number of black employees and it is to be hoped that they will be as successful as their coloured counterparts who were apprenticed at an earlier stage. The group's apprentice training has a high reputation for success based upon its insistence on a high standard of work being performed by its apprentices. The training programme is now non-discriminatory and entrance to it is based strictly on merit.

"Two non-white trade unions gained official recognition within the mining industry, *viz.* The Federated Mining Union and the National Union of Mineworkers. Although these unions represent a very small proportion of the non-white work force, they were entitled to participate for the first time in the wage setting process for non-white employees. It is to be hoped that this new development will evolve in a responsible manner for the benefit of all concerned. Towards the end of the year the National Union of Mineworkers was granted recognition on the Kloof mine in respect of certain classes of employees and accordingly the group was intimately involved in this new development" (6).

In 1986, CGF was considerably embarrassed by revelations that GFSA – and most of the other mining companies in the apartheid state – employed a private police force, trained and equipped with weapons from a company itself set up by CGF in 1969, at a time when CGF held the majority of GFSA.

A Granada TV programme screened in late 1986, which spotlighted this company, also re-vealed that GFSA had the most patronising and insidious attitudes towards its black workforce of all South Africa's mining companies. A training manual for white supervisors, for example, stated that a Bantu will always give an answer which he thinks will please the other person. No wonder Cyril Ramaphosa, secretary of NUM (National Union of Mineworkers) declared that "... Gold Fields is actually the worst company that we have to deal with" (7).

Perhaps partly because of these criticisms, in 1987 CGF began to cut its interest in GFSA – from 48% to 38%, selling 10% to a new holding company controlled by the South African conglomerate, Rembrandt (8).

Two years later, after Hanson plc bought out Consolidated Goldfields, ownership of GFSA changed drastically, as the British corporate pirate offloaded its South African subsidiary completely, retaining only the company's interests in the USA (through Gold Fields Mining Corp). The following year, Gold Fields American Corp itself sold four ARC (aggregates and road construction) units in the USA to CSR (9). GFSA then gained "joint control of itself," together with Rembrandt and the Asteroid Corp, which between them now own 40% of GFSA (10).

Meanwhile, Newmont (then controlled by Hanson) disposed of its own South African and Namibian interests, with the result that GFSA achieved majority ownership of O'okiep Copper (81%) and Tsumeb Copper Ltd in Namibia (9), making it the "predominant presence" in base and precious metals exploitation of the country, through the new subsidiary Gold Fields of Namibia (GFN) (11).

In October 1990, three GFSA investment companies also merged – New Wits, Selected Mining Holdings and Witwatersrand Deep (12). GFSA has fared better than perhaps any other apartheid mining company in recent years. In 1988, it posted record prices (despite falling gold production) (13) and in 1989 launched a R1000 million ($240 million) capital float for further expansion (14). The gold record was sustained in 1990, despite retrenchment (15) and weakening prices for base metals, especially at Rooiberg Tin and GFN (16). Northam Platinum Ltd (GFSA:60.4%) was listed on the London Stock Exchange that year (17).

References: (1) Lanning & Mueller. (2) *MAR* 1984. (3) *MIY* 1982. (4) UN NS-36. (5) *FT* 24/5/84. (6) R.A. Plumbridge [chair of GFSA] in: *Annual Report* 1984. (7) *Gua* 25/11/86 see also the letter from Colin Adkins, Coordinator of the Campaign for Sanctions Against Apartheid Coal, in the Gua 2/6/90. (8) *FT* 10/7/87; *MJ* 10/7/87. (9) *FT Min* 1991. (10) *FT* 15/8/89. (11) *MAR* 1990 (12) *MJ* 5/10/90.(13) *MJ* 15/7/88. (14) *MJ* 8/9/89. (15) *Metal Bulletin* 12/7/90, *MJ* 5/10/90, *MJ* 13/7/90. (16) *FT* 16/1/91. (17) *MJ* 2/2/90.

275 Gippsland Oil and Minerals NL

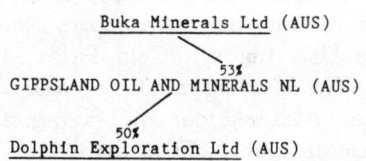

This is a medium-sized mining company with several projects, including the Georgetown uranium/base metals JV which it operates through its 50% subsidiary Dolphin.

By 1983 limited uranium resources had been outlined at the deposit, through early prospecting by Esso Exploration and Production Aus which has a 2% net profits interest. Probable reserves of over one million pounds of U_3O_8 were located (1).

The company's most recent activities have concentrated on tin prospecting (where it has a 40% JV interests with Aberfoyle in the Zeehan deposit).

References: (1) *Reg Aus Min* 1983, 1984. (2) *Reg Aus Min* 90/91.

276 Glencar Exploration (IRL)

Glencar was a 15% partner in the Fintona uranium exploration venture Ireland, managed by E & B from 1978 (1), in which Dungannon Exploration held 20% and Sabina Industries another 15% (2).

However, as opposition mounted to uranium mining in the Republic and the spot market price dropped during the 1980s, Glencar turned to gold.

In 1987 Sabina sold part of its interest in Glencar (by then renamed Glencar Explorations plc) – and itself invested in Celtic Gold plc, another Eireann mining company (3).

The following year Glencar announced the discovery of gold deposits in Eire worth an estimated £300 million (4). Together with UK-based Andaman Resources the company carried out detailed drilling at its Cregganbaun gold prospect in county Mayo, in 1989 (5). A few months later it resumed drilling on its wholly-owned Tara deposit, adjoining Tara's Navan mine, some 45km north of Dublin where large deposits of lead and zinc have been located (6). At the beginning of 1990, Glencar announced that it was finalising the purchase of a gold property in Ireland – which it did not specify (7).

By then the company had sold its minority stake in the Teberebie gold mine in Ghana, where it had been partnered with Pioneer group of Boston, Massachusetts: however it announced that it was looking for other gold properties in the West African state (8).

More important, it was soon in the business of exploiting the newly-opened states of eastern Europe, in particular the Pecs uranium mine operated by MEV, the Hungarian state-owned company, which was then producing around 600 tonnes of uranium a year (most of it destined for the Soviet Union). Glencar was called in to raise productivity at the mine and cut back on the workforce (from 7000 to 2000) (9).

Commented Brendan Hynes, Glencar's chairman at the time: "We believe Hungary offers the best immediate prospect for success in the exploration and mining industry" (10).

Glencar soon took over control of the Pecs deposit, situated in the Mecsek mountains near the Yugoslav border (11). However, the environmental hazards embraced by the MEV operation are very considerable. According to an expert on uranium mining who visited the mine in late 1990: "The main problems are the two tailings ponds which cover an area of one square kilometre each and contain a total of 15 million tons of tailings solids and 9 million tons of liquids. The ponds were erected in a plain using ring dike structures. The underground of the ponds consists of layers of clay and sand supposed to be impervious until contamination resulting from the ponds was found in observation wells near the ponds early in 1990.

At the surface of the ponds, the bare tailings material is exposed and thus subject to wind erosion and responsible for enormous radon emanations. But nevertheless there still exist no concepts for reclamation of these tailings ponds. Because of the seepage problems it might become necessary to move the tailings to better places – a very expensive task.

Another matter of concern in the long term is the now 15 leaching piles with a total of 2-3 million tons of low grade ore. They are erected on a double plastic layer of 0.8 mm thickness each to prevent seepage of the leaching liquid, but plastic layers are known to be impervious only for a rather short time. The [company] forsees a reclamation in place by covering the piles with a 30-50 cm layer of clay and revegetation. This seems to be quite insufficient.

In the short term (during continuation of mining), other environmental problems pose additional concerns: radon exhaust from the shafts, wind erosion from the leaching piles to adjacent residential areas, ore dust lost by the trucks in the road (elevated levels of radiation were found there by environmentalists) and others.

With all these unresolved problems, the legacy of uranium mining poses a huge challenge to the young democracy of Hungary" (12).

Contact: Peter Diehl, Schulst. 13, W-7881 Herrischried, FRG.

References: (1) *MJ* 29/6/79. (2) *MM* 5/81. (3) *FT Min* 1990. (4) Robert Allen and Tara Jones, *Guests of the Nation: People of Ireland versus the Multination-*

als, Earthscan publications, London 1990. (5) *MJ* 16/6/89 see also *MAR* 1990, *MM* 7/90, *MM* 8/90. (6) *MJ* 22/9/90. (7) *E&MJ* 1/90. (8) *E&MJ* 6/90. (9) *MJ* 7/9/90. (10) *FT* 20/8/90. (11) *Nuclear Fuel* 1/4/91. (12) Information from Peter Diehl, 10th October 1990.

277 Goldak Exploration Technology (CDN)

It was supposed to be exploring for the deadly metal in Saskatchewan (1).
Also, it was one of Eldorado Nuclear's JV partners in uranium exploration in the late '70s (2).

References: (1) *Who's Who Sask.* (2) Eldorado Nuclear Annual Report 1979.

278 GPU

```
GPU /
GENERAL PUBLIC UTILITIES SERVICE CORP. (USA)


            Cherry Hill Fuel Corp. (USA)
```

A consortium – including its own subsidiary Cherry Hills – representing Jersey Central Power and Light Co and Metropolitan Edison Co which was involved in uranium exploration (with no known production) from 1976 until at least 1979 (1).

References: (1) Sullivan & Riedel.

279 W R Grace and Co

Grace has been involved in uranium-from-phosphates production in Florida. Together with IMC/International Minerals and Chemi-

```
        W. R. GRACE AND CO. (USA)


  Grace Ore and Mining Co. (USA)
```

cals Corp it is operating a mine at Four Corners, Mulberry, Florida (not to be confused with the Four Corners area of Utah, New Mexico, Arizona and Colorado) with design production of 5M tons phosphate rock per year. Also, in 1979, it announced agreement with IMC to open up another mine at Manatee, north Florida, with similar output. This mine was intended for completion in 1989, though it is not known whether either of these mines was intended to be a uranium producer. Co-operation with IMC also includes phosphate plants in Manatee and Hillsborough Counties, Florida. (1).

A uranium-from-phosphates plant was operated by Uranium Recovery Corp at W R Grace's property in Bartow, Florida, which, in 1978, produced about 150 tonnes of uranium oxide (2). In 1979, Grace contracted to supply phosphoric acid to United Nuclear Corp (owner of URC/Uranium Recovery) from its Bartow complex, for uranium extraction. However, in 1980 United Nuclear filed suit against Grace alleging that the phosphoric acid did not match contract requirements, and that Grace had arbitrarily imposed conditions and limitations, making continuing operations "impracticable". The operations appear not to have started up again. This caused the suspension of Uranium Recovery's activities in January 1980 (3).

Through its wholly-owned subsidiary Grace Ore and Mining, Grace invaded the Indian territory of Rondonia in Brazil in the early '70s, along with many other Brazilian and multinational prospecting companies. This followed the discovery of cassiterite (and possibly uranium) in the late '60s (4). It also followed an episode of genocide which was horrendous even by Brazilian standards, as Brazilian land-bandits tried to destroy the Cintas Largas and force them – along with the Surui, Arara and other small tribes – into extinction (5). As a result of pressure from cattle and mining companies the original Aripuana national park (ie Indian reserve) was slashed in half, leaving most of the tribes outside the protective boundary (6). The

Cintas Largas experienced the inevitable effects of "civilisation": venereal and other diseases, dislocation, malnutrition, and being used as a pool of slave labour for the building of the new roads through Amazonia (7).

W R Grace was part of the Trombetas River Consortium (along with RTZ, PUK, Hanna and others) which originally planned to open up bauxite reserves in Para state as a part of the mammoth Carajas venture (8).

However, during the late '70s, Grace – whose roots were in Peru (1854 dates W R Grace's trading operations in the country) – began shedding its South American interests. The Peruvian government nationalised its paper and chemical assets in 1974. And, with drooping chemicals and fertiliser sales, Grace began to diversify – into fashion wear, drilling, buckskins, jewelry and restaurants, pushing fast junk foods called El Torito and La Fiesta (9).

Within the last few years, the offloading has proceeded, with Grace selling its JV (with Hanna) in Paramount Coal in 1986, and in early 1989, its 50% share in the 5 million tonne/year phosphate mine it had been operating since 1985 (though it closed from February 1986 to reopen three years later) at Four Corners, USA.

References: (1) *MIY* 1981. (2) *MAR* 1979. (3) *MJ* 11/7/80. (4) Shelton Davis, *Victims of the Miracle*, Cambridge, 1977 (Cambridge University Press). (5) Norman Lewis: "Genocide", *Sunday Times* 23/2/69. (6) *FT* 7/12/78 (special supplement on Brazil). (7) various issues of *Survival International Review, Natural Peoples' News* No. 2/3. (8) *Survival International Review* Spring/79, p22. (9) *WDMNE*. (10) *FT Min* 1991.

280 Grandad Resources Ltd (CDN)

"A 'revived' company in the Camflo-United Siscoe camp" – that's how it's described by (the Canadian) *Northern Miner*. It was exploring in Saskatchewan (1) where it had a 67% interest in a uranium project at Wilken Lake (Wollaston Lake area) (2).

References: (1) *Who's Who Sask.* (2) *Yellowcake Road.*

281 Gränges AB

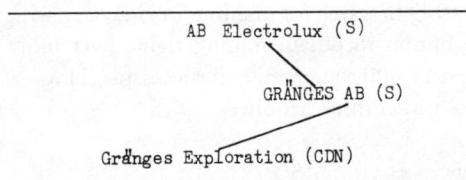

```
            AB Electrolux (S)

              GRÄNGES AB (S)

Gränges Exploration (CDN)
```

Sweden's only primary aluminium producer, it was acquired by the fridge manufacturers Electrolux in early 1981 (1).

It was exploring for uranium in Saskatchewan (2); and also involved in a copper-zinc project with Hudbay/Hudson Bay Mining and Smelting in the Flin Flon area (1).

According to the *Mining Journal*, Gränges – under the influence of Electrolux – has moved from a primary position in mining and heavy industry, railways and shipping in the late '70s to become a sub-contractor and manufacturer of finished products. In April 1983, the group agreed to sell GIM (Gränges International Mining) to Boliden, though retaining its 20% stake in the Flin Flon copper mine in Canada. Gränges's mining interests were then concentrated in its 37.5% stake in Lamco (which was 25% owned by Bethlehem Steel Corp) (4).

Along with the Canadian SMDC, Granges owns the Bigstone Lake copper-zinc deposit in Manitoba (5).

Gränges started off as the Grängesberg Co in 1896, when the iron ore mines at Grängesberg were merged with three railways linking them to the Baltic Sea. Seven years later, it acquired a half share in LKAB in Samiland (Swedish "Lapland") which was then developed into the world's largest iron ore exporter. This share was purchased by the Swedish government in 1975, and with the funds Gränges diversified into steel-making, engineering, a new mine in Liberia, and later copper and aluminium. In 1978 the group altered its structure considerably when many of its subsidiaries were transformed into the SSAB/Svenskt Stal AB in which the Swedish government holds 50% and Gränges 25% (3).

Gränges International Mining continues to operate as the manager of potentially one of the world's largest iron ore producers, the Liberian American-Swedish Minerals Co (Lamco) which has been developing high grade deposits at Mifergui in the Nimba mountains of Guinea (6).

In 1989, MIM of Australia acquired 33% of Gränges International Mining (7) and thereby control of half the company's board of directors (8).

In its turn Gränges Exploration, which was formed in 1985 after amalgamation with Pesco Resources Ltd, holds 50% of Hycroft Resources and Development Corp. Hycroft has gold prospects in British Columbia and Nevada, and is in a 29% JV with HBMS (Hudbay Mining and Smelting), Manitoba Mineral Resources Ltd and Outokumpu, at Trout Lake: the copper-zinc project which it opened up in the early eighties (7). However, its Tartan gold mine suspended operations in late 1989 (9).

Contact: Folkkampanjen Mot Karnkraft, Göran Eklöf, Box 16, 307, S-10326 Stockholm, Sweden.

References: (1) *MAR* 1981. (2) *Who's Who Sask.* (3) *WDMNE.* (4) *MJ* 22/7/83. (5) *MJ* 17/2/84 . (6) *FT Min* 1990. (7) *E&MJ* 7/89. (8) *MJ* 3/3/89. (9) *MJ* 8/3/91.

282 Great Eastern Mines Ltd (AUS)

The company used to operate the Aga Khan emerald mine at Poona (believe it or not) Western Australia; its HQ is in Melbourne (1). It was reported, in 1980, to be also raising money to examine its uranium claims at Winning Pool, WA (2).

The company has been involved in gold, diamond, emerald, oil and tungsten exploration and exploitation, mostly in Australia, though it also has oil and gas interests in the San Juan basin of the USA – a traditional Navajo (Dineh) area (3).

In more recent years, Great Eastern has concentrated on gold exploration in Western Australia and gold and platinum exploration in Ecuador, where its exploits have been channelled through its 50% subsidiary Rodecu Investments Ltd, at Leon, Balao, Aguas Calientes, Piguio Pinglio and Pinglio Uno. However, in early 1990, Great Eastern faced increasing financial difficulties and negotiated with the shady Isle of Man-registered Odin Mining & Investment Co Ltd to sell the company's Ecuadorian interests within two years. Odin holds 43.8% of the Fiji Emperor Gold Mines Ltd, a notorious exploiter of land and labour, previously managed by WMC of Australia (4).

Odin also has, through Osborne Chappel Goldfields Ltd (an associate of the Malaysian-based Osborne & Chappel International Sdn Bhd), alluvial gold holdings in Ecuador, a dredging operation in Brazil, a JV with Sibeka of Belgium exploring for diamonds along the Mabere river of the Central African Republic, as well as (in 1989) 18.7% of Great Eastern and 26% of the Fijian exploration company Nullabor Holdings Ltd (5).

In 1989, Odin lost its finance for the Los Lilenes gold project in Ecuador when the International Finance Corporation canceled its loan. Both Odin and Emperor were "under scrutiny" as auditors picked up wide variations between the carrying and market values of their long-term investments (5).

References: (1) *MAR* 1981. (2) *FT* ?/?/80. (3) *Reg Aus Min* 1981. (4) *FT Min* 1991. (5) *Reg Aus Min* 90/91.

283 Greyhawk (CDN)

This is a company set up in the '50s to mine uranium in Ontario for export to the USA for nuclear weapons construction (1); presumed defunct.

References: (1) *BBA* No. 4, 1979.

284 Grundstofftechnik (D)

It is based in Essen, capital of the Ruhr area of Western Germany.

It obtained a one-year "non-exclusive" permit in 1979 to search for uranium in Guayana, with an option to obtain mining rights over more than 11,000 square miles if successful (1).

No further information.

References: (1) *FT* 18/5/79.

285 Gulch Resources Ltd (CDN)

It has held mining properties in Black Bay, Lake Athabasca region of Saskatchewan, and in Geraldton, Ontario – uranium exploration included (1).

References: (1) *Yellowcake Road.*

286 Gulf Minerals Canada Ltd

```
        Gulf Oil Corp. (USA)
                    /
    GULF MINERALS CANADA LTD (CDN)
```

This was Gulf Oil's major minerals and uranium subsidiary which, in October 1982, was bought out by Eldorado Nuclear Ltd and renamed Eldorado Resources. The Rabbit Lake properties of the company were reorganised under the name of Eldor Mines, and its share in Key Lake and surrounding exploration lands was consolidated under a second subsidiary, Eldor Resources (1).

References: (1) *MIY* 1985.

287 Gulf Oil Corp

Not to be confused with either Gulf and Western (notorious for its exploitation of people in the Dominican Republic) or Gulf Resources

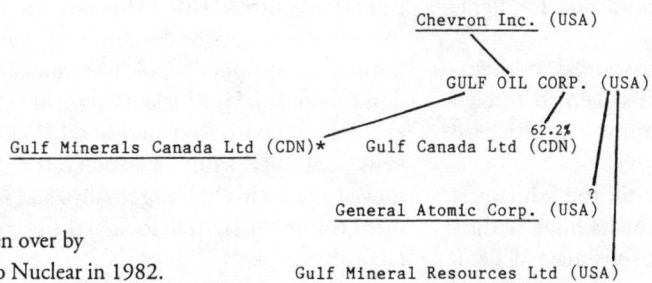

Chevron Inc. (USA)

GULF OIL CORP. (USA)

62.2%

Gulf Minerals Canada Ltd (CDN)*   Gulf Canada Ltd (CDN)

General Atomic Corp. (USA)

* Taken over by
Eldorado Nuclear in 1982.   Gulf Mineral Resources Ltd (USA)

and Chemical Corp, Gulf Oil has become an object lesson for conservative oil companies caught between the need to enlarge their traditional resource exploitation base and to diversify into unfamiliar and often hostile environments.

In 1979 Gulf was America's second most important holder of uranium reserves (1) with two million acres of mining claims and exploration permits in Canada and New Mexico (2). Market and development problems at home, however, have ensured that virtually no domestic uranium has reached the market, while its operations in Saskatchewan through its wholly owned Gulf Minerals Canada subsidiary made it the worlds 6th largest producer of uranium in 1980 (3).

Unable to replace its high-cost crude oil supply with alternative sources – despite a huge trading fleet, the 5th biggest gasoline retail outlet in the USA and potentially very profitable oil and gas reserves in the Canadian Arctic – Gulf became a prey to take-over bids from the newer brand of corporate bandit. To avoid final disintegration, it ran desperately into the arms of another of the "seven sisters". Now Gulf Oil is controlled by one of its former rivals, while its downstream operations are either in the hands of oil sheiks or up for sale; and its Canadian uranium deposits have passed under the sway of the Canadian government.

Incorporated in 1907, Gulf Oil expanded – financed chiefly by the Mellon family – by using its plentiful supplies of crude oil. It entered the Middle East in 1928 through shares in the Iraq Petroleum Co and later a JV with BP in Kuwait.

Expansion into Africa and Venezuela continued through the '60s, as did diversification into chemicals and the "middle to front end" of the nuclear chain. It purchased General Atomic in 1967, later sold half to Shell, then repurchased some of the shares (4).

Through the '70s Gulf fell on hard times. Many of its sources of crude oil were nationalised: in Kuwait in 1975 (its Italian operations were sold to the Kuwait Petroleum Corp in 1984) (5); in Venezuela the same year; in Ecuador a year later; then in Nigeria and Angola. (Ironically, Gulf managed to get back into Angola after independence and exploit oil in the Cabinda enclave; it has even lobbied in Washington against the "short-sighted policies of the Reagan administration towards the Angolan régime") (6). These take-overs wrested more than half the company's property from its control (4).

Moreover, as the corporation itself admitted, its "nuclear energy applications" in this period were "clouded by various economic, political and technological circumstances" (7). The nuclear reactor business was closed (4) – though, a few years later, General Atomic was developing a new, advanced high-temperature gas-cooled reactor at Fort St Vrain, near Denver, Colorado, also a fusion reactor with assistance from the US Department of Energy (7). However, work on the huge Barnwell reprocessing plant, which Gulf had contracted to build with the Allied Corp and Shell in 1970, to be completed by 1974 and reprocess 5.5 tons of fuel per day (8), was temporarily suspended in 1977 (9). Little has also been heard of the Gulf United Nuclear Fuels Corp, formed in July 1971 with UNC to design, make and sell nuclear fuel for light-water reactors (8).

The '70s reverberated with scandal as allegations flew about backhanders paid to foreign governments and a slush fund to elect Richard Nixon as President (4).

By the early '80s, Gulf was still the 5th largest seller of gasoline in the USA; its huge trading and transportation network subsidiary GT&T had 65 ocean-going vessels; it had profitable coal interests (it took over its first coal company as long ago as 1963, and acquired the important Kemmerer company in 1981 – apart from Kerr-McGee it was the only oil company to make profit from coal) (3); and it had vast interests in chemicals. It also owned the Europoort refinery in the Netherlands and one at Milford Haven in Wales (7). However, its profits declined by 12% in 1981, and the prices it paid for crude oil (especially from its major suppliers in West Africa) were among the highest in the world (4). Within two years it had slipped from being the world's 14th biggest multinational (4) to becoming the 19th (10). Its return on capital between 1977 and 1982 averaged only 10.2%, the lowest of all the 14 biggest US oil companies. In 1982 it began selling off loss-making refining and marketing operations in Europe (11).

Through the early '80s the biggest threat to Gulf's continued survival – in the eyes of its shareholders – appeared in the shape of fittingly named Texas oilman T Boone Pickens who bought up more shares in the company than the Mellons had, accused Gulf of mismanagement, and wanted to create a royalty trust to spin off the company's highly profitable oil and gas reserves for the benefit of shareholders. This was a move strongly resisted by Gulf's chair James Lee (who lead prayers before his board meetings) (12): he wanted to retain the cash-flow to boost the company's current position (11). Lee accused Pickens of trying to "cannibalise" his baby (13).

By early 1984 Gulf was fighting Pickens's take-over attempts in court (14) – a move it lost (15) – while the Texas booner gathered a dissident share group around him (based on his own Mesa Petroleum company) to bid for strong minority control (16). Desperate for the customary white knight to fend off the Texan hunter ("If it doesn't look like you can eat it, don't shoot at it" is a Pickens maxim) (12), Gulf entered talks with Arco and Socal (17). As Pickens – already with 13% of Gulf's shares – moved to take his holding to 20% and management control (18), Arco looked the prime marital prospect.

Socal/Chevron takeover

But, in the event, it was Socal/Chevron which took over Gulf at a price of US$13 billion: then the largest take-over deal in corporate history, dwarfing even Texaco's absorption of Getty Oil the same year. The combination of Socal and Gulf created an oil company on a par with Mobil and second only to Exxon, with annual potential revenues of nearly US$60 billion (19). Chevron overnight also became the biggest gasoline retailer in the USA (20). A month after the bid, Socal possessed some 74% of Gulf's shares (18). In the meantime, Gulf had reached agreement with the Kuwait Petroleum Corp for the sale of its Italian assets (21) and put other downstream operations, including the Milford Haven refinery and 350 British petrol stations, up for sale (22).

Within a year of the Chevron take-over the company was putting a large part of its US operations on the market and planning to cut the Gulf/Chevron workforce by 15%. "It is clear", commented the *Financial Times*, "that Chevron is wasting little time dismantling the former empire of the Pittsburgh-based Gulf" (23). However, it did not appear that Chevron, under the control of George Keller, would substantially alter Gulf's upstream operations, and – despite a US$10 billion debt thanks to the Gulf acquisition – Keller was predicting that his new-found sibling would be bringing in nearly a third of the company's earnings by 1989 (20). The most immediate (and probably lucrative) beneficiary of this unprecedented deal was certainly T Boone Pickens. Thanks to his stake in Gulf, he pickensed up a cool US$506M from the increased share price, created by his tilting at Lee's most vulnerable quarters (12). However, Pickens's ostensible concern for shareholder profits (ie his own) had the American Petroleum Institute severely worried lest the cost-cutting, cash-aggrandizing trend should cut off major investment in new oil reserves (24). A perceptive commentary in the *Financial Times*, a year after the Gulf/Chevron merger, spotlighted other dangers. As the article pointed out, raids such as those by the Bass brothers (Texaco on Getty) and Mr Pickens regularly prompt "greenmail ... a form of corporate blackmail where companies buy back their shares from predators on terms not available to the rest of the shareholders. In these cases, the corporate raiders not only make a killing for themselves, but also weaken the finances of the company." In addition, huge sums of money are spent by the old-style corporations (much of it on lawyers' fees) simply keeping the wolf at bay. One of the tactics used by corporations in the struggle has been the so-called "poison pill" – threatened companies declaring a dividend in some kind of convertible preferred stock which carries rights making hostile take-overs "prohibitively expensive" for a predator. In the view of Harold Williams, a former chair of the US Securities and Exchange Commission, this presaged "a major disaster for shareholder democracy" as it makes it impossible for shareholders (and workers) to depose managements. And, the *Financial Times* commented in 1985, "the Reagan administration does not seem at all concerned by the current upheavals ..." (25).

In early 1985, Chevron offered its 60.2% share in Gulf Canada Ltd to Canadian shareholders in return for Canadian government approval of its take-over of Gulf Canada thanks to the Socal/Gulf merger. The Canadian government did not oppose this return of highly profitable assets to Canadian hands – Gulf Canada made a profit of C$308M on sales of C$5.3 billion in 1984 (26) and it has 15% of the country's refining capacity (27). However, Olympia and York (which had earlier taken over Brinco) played a perplexing game of touch and go: initially announcing a take-over (27), then withdrawing without comment (26) and, in August 1985,

taking up nearly all of Chevron's holdings (27). Petro-Canada, the state-owned oil company, and Norcen, which had both been interested in bidding for Gulf Canada (28), said they planned to acquire some of the company's assets at a later stage (29).

Gulf's major North American oil operations centred on exploitation of the Beaufort Sea, the Arctic islands and the east coast of Canada. In 1978 Gulf Canada invested more than US$100M searching for oil, coal and other minerals in this region. At the end of that year Gulf Canada Resources was formed to handle the company's Canadian interests (7).

At Stokes Point

Five years later, this company was involved in a major controversy which reverberated far beyond the Canadian Arctic. Peter Burnet, executive director of Canadian Arctic Resources Committee (CARC) takes up the story:

"In March 1983, Gulf Canada Resources Inc applied to the Department of Indian and Northern Affairs for permission to build a marine support base at Stokes Point on the northern slope of the Yukon.

"Gulf wanted the base to supply its Beaufort Sea drilling operations, especially its huge new conical drilling unit which arrived in the Beaufort this summer from Japan. The application touched off a bitter public controversy that echoed the famous MacKenzie Valley Pipeline Inquiry of Mr Justice Thomas Berger.

"The North Slope of the Yukon is one of the most environmentally sensitive areas of northern Canada. It is the nesting area for millions of seabirds. The beluga and the bowhead whales of the Beaufort pass near the coast on their eastward migration each year.

"Perhaps most importantly, the North Slope is the calving ground for the porcupine caribou, one of North America's last great wild herds. Three different aboriginal groups – the Inuvialuit, the Indians of Old Crow, and the Dene of the MacKenzie delta – have traditionally hunted the porcupine caribou and depend upon it for food.

"For much of the last decade, aboriginal groups and conservationists fought to protect the North Slope from industrial intrusion, particularly a proposal by Canadian Arctic Gas to build a pipeline across it to Prudhoe Bay in Alaska. Two public inquiries concluded that the pipeline should not be built. Justice Berger recommended that the whole area be made a wilderness park, while the National Energy Board refused the application of Canadian Arctic Gas, saying no mitigative measures were sufficient to remove the threat to the wildlife of the area. These two decisions led the federal cabinet to withdraw the North Slope from development as the first step toward the creation of a national park.

"The Gulf application immediately raised fears that all this work would be reversed, especially when it became apparent that the Department of Indian Affairs and Northern Development (DIAND) was considering the application sympathetically. The Committee for Original People's Entitlement (COPE), representing the Inuvialut, was particularly angry, since their 1978 land claims agreement-in-principle had stipulated that Stokes Point was either to be part of a wilderness park or transferred directly to them. COPE was in the last stages of a frustrating five-year effort to negotiate a final settlement. They saw the issue as yet another attempt to undermine their agreement.

"Along with the Indians of Old Crow and the Dene, they now had to fight a battle they thought they had won years ago.

"Internal DIAND documents showed that the government was preparing to give approval to Gulf in May, without returning to cabinet to change the order withdrawing the lands from development. DIAND refused to submit Gulf's application to the Beaufort Environmental Assessment and Review Panel (EARP), claiming the base was only a temporary exploration base, and therefore outside of the panel's mandate.

"Gulf planned to spend US$60 million building the base at Stokes Point. It was to have a runway capable of handling 767s and be large enough to accommodate the whole Gulf Beaufort fleet. Few believed Gulf would cheerfully dismantle it after a few years to build another where EARP recommended.

"In late March, the members of the Canadian Arctic Resources Committee (CARC) held a press conference to announce they would begin a court action against the minister of DIAND, the Hon John Munro, if he granted the approval without a cabinet order. DIAND responded by announcing the first of several delays of any decision. The issue caught the attention of the national press, and Mr Munro's office was flooded with letters protesting any consideration of opening up the North Slope.

"As opposition to the Stokes Point application mounted, other federal departments began to take an interest. By now the controversy had transcended a simple debate between environmentalists and industrialists on the sensitivity of the North Slope. Increasingly, DIAND was seen as preparing to reverse a long-standing government policiy based upon the most exhaustive public inquiry in Canadian history."

Finally, in July that year, Munro set up a project review group comprising native representatives and the pro-"development" Yukon government. The group narrowly gave Gulf's proposal a thumbs down. Continues Peter Burnet:

"While Mr Munro's decision was celebrated by the northern aboriginal groups, CARC, and national and northern conservation groups, the episode raises serious questions about the Canadian government's ability to plan and manage Beaufort development.

"The applications were considered in isolation from several of DIAND's own policies to plan Beaufort development in a comprehensive manner that would balance development interests with those of the aboriginal peoples. The North Slope had been withdrawn from development. At least two aboriginal groups had claims in the area. The Beaufort EARP was beginning its hearings into the environmental and socio-economic impacts of Beaufort development. Yet, for almost a year, the government considered bypassing all this and allowing a major industrial development on the North Slope for the sole apparent reason that Gulf thought it would be more economical for them than to use existing facilities outside the North Slope.

"The Canadian government is both a heavy investor in Beaufort development and the regulator of that development. The Stokes Point story shows how this conflict of interest can undercut the type of sensible long-term planning that will be necessary to protect the aboriginal peoples. It also raises questions about the usefulness of public inquiries such as the Berger Inquiry and the National Energy Board hearing into projects such as the Arctic Pilot Project. Northern aboriginal peoples too often spend large amounts of time and money making serious, well documented arguments to such inquiries, only to have the same question surface a few years later" (30).

At roughly the same time, Gulf also saw its Beaufort Sea Environmental Impact Statement (EIS) – conducted jointly with Dome and Esso – criticised by the federal government for serious deficiencies. Among them was the refusal of Gulf and its partners to consider the impact of its operations on the native land claims process. Commented the Dene Nation in their own response to the EIS:

"This scenario ... is clearly based on the assumption that "development" is ultimately a good thing. However, there is no recognition that the people who live in the North might not agree. There is no consideration for the notion that Northerners have their own ideas of what they want to do with their lives and resources. Nor is there any consideration of other forms of development (ie outside the context of petroleum) which might be optional or preferable" (31).

Since then, Gulf's explorations have increased apace in the region. By early 1984, Gulf had concluded the agreements to drill wells in the Reindeer Station and Inuvik areas of the MacKenzie delta over a period of five years. (Chevron also has three wells in the same area) (32). By the end of that year, Gulf Canada Resources had announced its "most promising oil exploration results to date" in the Beaufort Sea: its Amauligak J-44 well, near Tuktoyaktuk in the Northwest Territories had a production capacity of 13,600 barrels a day, sufficient (in the view of the company's president, Harry Carlyle) to justify a pipeline through the Arctic (33). (A few months later Arden Haynes, the chair of Exxon's Canadian subsidiary Imperial Oil, declared that a pipeline recently built from Alberta to Imperial's oilfield at Norman Wells could "easily be extended to the Beaufort Sea" (34), thus raising the spectre of an arctic pipeline which the indigenous people hoped had long been put to rest.)

Gulf's treatment of other indigenous nations has been no less cavalier. Its Venezuelan subsidiary, the Mene Grande Oil Co broke an agreement giving land rights to the Kari'na Indians in its search for oil during the late 70s(35).

But its worst trespass has probably been at Mt Taylor, New Mexico, described by the Long Walk for Survival group of Native Americans as "perhaps the classic example of desecration of sacred lands" (36).

At Mount Taylor

Gulf began uranium exploration in 1967 via its newly established Gulf Mineral Resources subsidiary. A year later, it discovered significant uranium mineralisation in Saskatchewan, a find which was later to develop into the major Rabbit Lake mine (for further details, see Eldor Mines Ltd). However, it was its Mt Taylor find in 1974 which promised high dividends, it being one of the largest uranium orebodies in the USA. (By 1985, Gulf geologists reported an orebody grading an average 5lbs/ton uranium in a lode containing more than 128 million pounds U_3O_8, no less than 22% of the low-cost reserves in the whole USA) (37).

Original plans were for a mine reaching eventual production capacity of 1.42 million tons of ore per year with an average grade of 1.3% U_3O_8. Target production was put at more than 4000 tons U_3O_8 per year (38). Gulf also applied for permission to open up a mill at San Mateo with 4200 tons ore per day capacity – on a site recognised at the time as less than ideal, being "below grade, not completely lined, some fractures in area" (38).

The entire project was "less than ideal" for the Navajos. Mt Taylor is one of their four sacred mountains, and a traditional Pueblo community is located less than ten miles away at Acoma. The Chicano community at San Mateo, only a few hundred yards from the mine, has also suffered contamination of two community wells, though Gulf denied responsibility, attributing the radiation to other companies working in the area (36). The deepest uranium mine in the world with 3300 feet, Mt Taylor drew 10,000 gallons of water per minute from more than seven underground water tables. In December 1980, the Water Control Commission imposed extensive monitoring of ground water radiation levels on Gulf – measures opposed by the company on the technical grounds that they were not properly authorised by the Commissions regulations or by law (39).

As the uranium slump deepened in 1982, development of the mine was halted and the complex put on stand-by to await market improvements. The vast majority of the 360 employees were sacked (the workforce had already been halved during 1982). Up until then the mine was producing between 400 and 500 tons/day of "development ore" (total production for 1981 was 470,000lbs uranium-in-ore) (45) though none had been processed or marketed (41). Development of new orebodies was also discontinued (40).

After the Chevron take-over, however, steps were being taken to reopen Mt Taylor on a "limited basis". Chevron planned a 12-month test with a "skeleton crew" (sic) of 80 miners "to determine if production from the mine can compete in the current depressed uranium market" (37). In the absence of contracts, Chev-

ron/Gulf planned production of only a few hundred tons of ore per day, to be stockpiled.

In 1982, Gulf also leased 430 acres in Sullivan County, New York, under the impression that the zone might contain one of the biggest uranium deposits in the country (42). The Council on the Environment, an advisory group to the New York City Council, opposed the company's intentions (43) and a bill was introduced to the state legislature calling for a 10-year moratorium on yellowcake mining, sponsored by Assemblyman Maurice Hinchey and Senator John Dunne. Though the bill appears not to have passed, Gulf withdrew its plans.

There have also been reports from Norwegian anti-nuclear activists that a joint Gulf-A/S Sydvaranger company called Norwegian Gulf Exploration Co A/S has been exploring for uranium as well as gold and copper. A five-year exploration project started in 1980.

It covers the Biedjovagge area of Finmark, where uranium was located in earlier years. Significantly, Biedjovagge is in the Kautokeino municipality, source of the Alta river and site of the largest nonviolent direct action in Scandinavia of modern times, the blockading of the Alta-Kautokeino dam on Sami land (44).

References: (1) NARMIC, No. 5, 1979. (2) Yellowcake Road. (3) BOM. (4) WDMNE. (5) FT 22/2/85. (6) Gua 7/9/84. (7) Gulf Fact Sheet 1979-80, Pittsburgh, 1980 (Gulf Corp). (8) Gulf Oil: Portuguese ally in Angola, Toronto, date unknown (Corporate Information Centre), quoted in: Yellowcake Road. (9) Bulletin of the Atomic Scientists, 3/84. (10) Mmon 12/82. (11) FT 25/11/83. (12) Obs 11/3/84. (13) FT 3/11/83. (14) FT 14/2/84. (15) FT 16/2/84. (16) FT 23/2/84. (17) Obs 4/3/84. (18) FT 5/3/84. (19) FT 6/3/84. (20) FT 19/6/85. (21) FT 12/1/84. (22) FT 14/1/84. (23) FT 29/11/84. (24) FT 9/3/84. (25) FT 15/3/85. (26) FT 8/2/85. (27) FT 24/5/85. (28) FT 19/7/85. (29) FT 3/8/85. (30) Arctic Policy Review (ICC/Inuit Circumpolar Conference) Nuuk 1/84; Yukon Indian News 29/7/83. (31) Project North Newsletter, Toronto 4/83. (32) Yukon Indian News 24/2/84. (33) FT 26/9/84. (34) FT 26/4/85. (35) Survival International Review, London, Spring 1982. (36) Long Walk for Survival News-

letter, Washington 7-8/80. (37) *E&MJ* 4/85. (38) BL Perkins, *An overview of the New Mexico uranium industry*, Albuquerque, 1979 (New Mexico Energy and Minerals Department). (39) *Mount Taylor Alliance Newsletter*, Albuquerque, Vol. 2 No.1, 1981. (40) *MIY* 1985. (41) *E&MJ* 12/82. (42) *AkwN* Summer/83. (43) *Kiitg* 2/82. (44) "Uranutvinning og helseskader: Er Finnmark i faresonen?", in *Samtiden* No. 5, Oslo 1980; *Alta bilder: The 12-year struggle for the Alta-Kautokeino watercourse*, Oslo, 1981 (Pax). (45) *MJ* 12/11/82.

288 Gulf Resources Ltd

Gulf Resources and Chemical Corporation – not connected with Gulf Oil – is a major "natural resource and energy company", highly diversified into chemicals and downstream oil and natural gas activities (1).

Its only known uranium interest (through a wholly owned subsidiary) is in the Ngalia Basin in Australia, a Discovery Operating Agreement which comprises several uranium prospects on Aboriginal land in the Northern Territory, specifically Yuendumu, Bigryliy and Walbiri. It is, however, very much a minor partner with a 1.12% share whereas Agip Nucleare Australia Pty Ltd holds 38.73%, UG 33.52%, Central Pacific Minerals 16.5%, Offshore Oil 7.88%, and Southern Cross Exploration 2.25%.

Exploration by the JV was reduced in early 1983 to await market improvement (2) but appears to have been revived since (3).

Now known as GRE, Gulf Resources & Chemical Corporation is an important producer of steam coal from surface mines in Pennsylvania, USA, and until 1989 produced speciality fertilisers and salts from the brines of the Great Salt Lake, using solar evaporation. This operation was bought out by GSL Acquisition Corp of New York (4).

```
Gulf Resources and Chemical Corp. (USA)
                                     \
                    GULF RESOURCES LTD (USA)
```

Through its newly-merged Industrial Ventures Inc it produces specialty clays on around 45,000 acres of land on the Nevada/California border (5).

In mid-1985, the former company chair, A Clore, sold his 20.3% interest in the outfit to Messrs D and F Barclay, two British investors, for around $30 million (5).

References: (1) *MIY* 1985. (2) *Reg Aus Min* 1983/84. (3) Personal communication from Pat Dodson, Central Land Council, Alice Springs, 8/85. (4) *MAR* 1990 (5) *FT Min* 1991.

289 Gulf State Utitilies

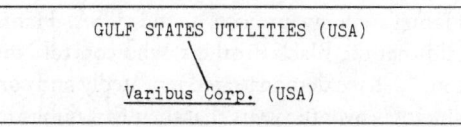

```
GULF STATES UTILITIES (USA)
                       \
          Varibus Corp. (USA)
```

Along with its subsidiary Varibus, and the Felmont Oil Co, it was involved in uranium exploration in the USA from 1976 until 1979 (1). No further information.

References: (1) Sullivan & Riedel.

290 Gunnar Gold Mines Ltd (CDN)

This is of dubious note as Canada's first major private developer of uranium which, in 1953, located a large deposit in the Beaverlodge area of Saskatchewan. It became the first beneficiary of the Canadian government's "Special Price Formula" to encourage mining of lower-grade ores, and was milling up to 2,000 tons/day ore in the late '50s (1).

When the Canadian mining industry went downhill in the late '50s and early '60s, so did Gunnar – and the mine/mill closed after 9 years (2), in 1964, leaving a broken-down site "asset-stripped for Eldorado's Uranium City" (3).

Together with Petromet Resources Ltd and Pan East Resources Inc, Gunnar holds 37.5% with

409

Mill city Gold Inc in the promising Fifteen Mile Stream gold project in Nova Scotia (4).

References: (1) *EPRI report.* (2) Goldstick. (3) *One Sky Report: Uranium City,* Saskatoon, 5/82. (4) *FT Min* 1991.

291 Hanna Mining Co

Corporate interlocks of Hanna include Bechtel (1) and, through its subsidiaries, RTZ (2).

Ten years ago Norcen was to increase its 8.8% shareholding in Hanna to 20% for US$90M cash. In April 1982, however, it failed to gain control of Hanna – a move which, according to the *Financial Times*, was a bid to get in on Hanna's energy projects in the USA. Hanna said that the Black Brothers, who control Norcen, "... have demonstrated repeatedly and convincingly over the years that their first public interest is in serving themselves and not the remaining public stockholders" (3). In the end, Hanna sold its stake (20%) in Labrador Mining and Exploration Co Ltd to Norcen (4).

Primarily an iron and nickel company, Hanna fell into the hands of George Humphrey, US Secretary of the Treasury, in the 1930s: the Humphrey-Hanna group then went on to ac-

quire Chrysler cars, thus assuring itself a market for steel (2).

Together with Bethlehem Steel the company also gained mineral rights on Inuit land in Labrador in the late '40s (5). It has JVs with Getty Oil at Casa Grande, Arizona (copper) (6), and WR Grace in south-west Virginia and eastern Kentucky (coal) (7). It was also apparently part of the Trombetas River bauxite consortium (along with RTZ, PUK, WR Grace and several other multinationals) which encroached on Indian territory in the late '70s (8). In mid-1982 a US subsidiary of Hanna in Colombia, Conicol, took a 20% stake in the country's first major mining venture, the Cerro Matoso ferronickel plant at Montelibano in north-west Colombia; Royal Dutch Shell's Billiton subsidiary will also have a 35% stake.

Hanna has been uranium exploring in Saskatchewan (1).

Although not strictly relevant to a study of uranium mining or mining's impact on indigenous peoples, Hanna's exploits in Brazil during the late '50s and early '60s (not to mention its current operations at Pocos de Caldas in Minas Gerais) are worth noting.

In 1956, Hanna bought into the British-owned gold mining company of St John D'El Rey and began exploring for iron ore in Minas Gerais.

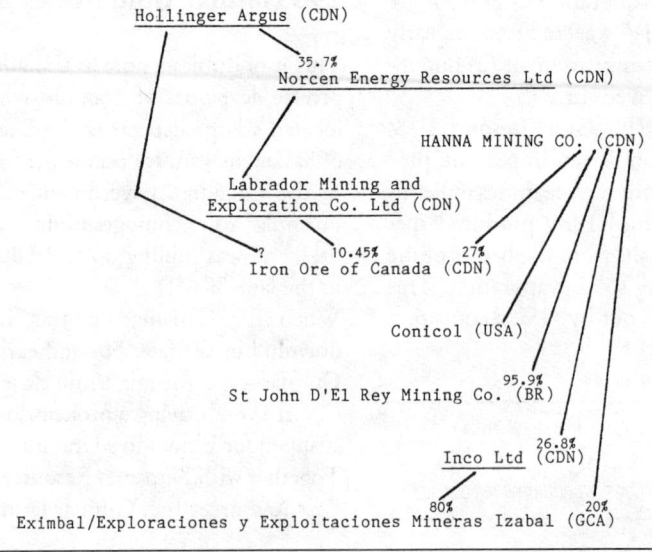

During the early '60s the Brazilian government (then critical of multinational penetration of the country's economy) investigated and expropriated Hanna's interest. After the military coup of 1964, however, the trends were reversed and Hanna expanded. "Whether Hanna Mining Company brought influence to bear on the events of April 1964 is still unknown. What is certain is that Hanna, along with several other multinational companies, directly benefitted from new mining policies ..." (9).

In June 1966, Hanna got rights to the Minas Gerais iron deposits and – by joining up with a Brazilian company, CAEMI; by influencing Brazilian bankers and politicians; and through a JV with Alcoa – became one of the most important and lucrative foreign companies exploiting Brazil (9).

In 1985 the company adopted the title M A Hanna Co as Norcen increased its holding to 28%. However, four years later, when Norcen acquired substantially all the oil and gas assets of Westmin Resources Ltd, it sold its interest in Hanna to a related company (10).

1985 also saw a major re-structuring in Hanna, as the company sought to reduce its dependence on mining and metals. (10). It reduced its interest in the Iron Ore Co of Canada (to 26.8%) and sold its 96.37% in St. John D'El Rey Mining Co plc. It effected an exchange of its 50% holding in Rapoca Energy Co for W R Grace's 50% interest in Terry Eagle Co, and the two companies sold their interest in Paramount Coal. Among its other disposals were nickel investments in Colombia (through Conicol) and Guatemala, its share of Alcoa Aluminio SA of Brazil, and its interest in the Exmibal nickel project in Guatemala which was closed indefinitely in 1981.

However, it acquired several companies prominent in polymers and plastics (including Day International Corp and Burton Rubber processing Inc) while Hanna maintained some interests in oil and gas production (10).

References: (1) *Yellowcake Road*. (2) *Last Post*, Canada, 5/73. (3) *FT* 8/4/82, (4) *FT* 6/8/82. (5) *FT* 22/6/82. (6) *MIY* 1982. (7) *MIY* 1981. (8) *Survival International Review* Spring/79, p22. (9) Several sources summarised in Shelton Davis, *Victims of the Miracle*, Cambridge, 1977 (Cambridge University Press). (10) *FT Min* 1991

292 Harmony Gold Mining Co Ltd

```
        Barlow Rand (ZA)
                    \
                     \
HARMONY GOLD MINING CO. LTD (ZA)
```

This is the biggest mining company in the Barlow Rand group, one of the largest gold mines in the world (1), and the second biggest uranium producer in South Africa until 1984 (1983 production figures were 623,600kg of U_3O_8, grading at 0.083 kg/tonne) (1). In the late '70s, significant shareholdings in the company were owned by Anglo-American (19%, since reduced to 3.62%) (2, 3), GFSA (5%), and Sivocam (16%).

Harmony is not only a gold and uranium producer but also by-produces silver and osmiridium, pyrite and sulphuric acid; some of its landholdings are used for farming (4). It holds nearly 10,000ha of mineral rights in the district of Virginia, and until early 1984 operated three uranium processing plants: Harmony, Virginia and Merriespruit.

Production at Merriespruit ceased in March 1984; this plant accounted for 28% of the total uranium produced at the mine (5). Plans to convert Merriespruit to gold production were shelved in late 1984, and the uranium plant remained in moth-balls (5, 6).

Although 1983's uranium production figures were fairly high, the grade of uranium recovered from slimes has been decreasing. In the year ending June 1975, for example, just over 3 million tons of slimes yielded just over 400,000kg of uranium grading at 0.193kg/tonne (7). Two years later, despite the increasing importance of uranium for the company, and success in selling all uncommitted uranium production, the grade had fallen to 0.12kg/tonne U_3O_8 (8).

411

In the mid 'eighties, although Harmony had a long-term sales contract with Belgium (9) and was reportedly the only mine in the Barrand group which has a chance to expand (1), its old contracts were apparently exhausted (9) and the fall in profits was forcing the parent company to reconsider the future of the operation altogether (1). In December 1988, Harmony closed down the last of its uranium plants, as it ceased production of yellowcake (10).

References: (1) *MAR* 1984. (2) Lanning & Mueller. (3) *Raw Materials Report Vol. 3, No. 2, 1985. (4) MIY* 1985. (5) *MJ* 15/3/84; 20/4/84. (6) *E&MJ* 12/84. (7) Annual Report 1975. (8) *MJ* 30/9/77. (9) Lynch & Neff. (10) *FT Min* 1991.

293 Hartebeestfontein Gold Mining Co Ltd

Like Loraine, Hartebeestfontein is managed by Anglovaal which (mainly through Zandpan Gold Mining Co Ltd) holds just over a fifth of the issued share capital (1).
A really hardy beast, the company "with its good financial position and very large proved ore reserves" has been "one of the soundest South African gold investments" (2).
In the financial year ending June 1984 it produced just under half a million kilogrammes of uranium from treated pulp (3), and at the end of 1980 the treatment plant for uranium extraction at its Klerksdorp mine had a capacity of nearly 300,000 tonnes a month (1).

In the early '80s its recoverable uranium reserves were set at 20,000 tons (4) – this from freshly-milled ore, not slimes (5) which the company appears to have ceased production of.
The company negotiated a three-year contract in 1977 and another "long-term" sales contract in 1978 (4).
Stilfontein "tributes" – in other words, pays tribute for a portion of the company's lease area, while Hartebeest itself tributes part of Vaal Reef's ill-gotten claims (1).
Anglovaal itself, while the smallest of the apartheid mining conglomerates (6), is nonetheless the seventh most important (non-state) industrial corporation in the South African economy with some 72 subsidiaries and associates (7) controlled by chairman Basil Hersov and his mate Clive Menell. Its profile is lower than that of its confederates, and it adopts a "modifying" attitude to apartheid similar to that advocated by the leaders of Anglo-American. Notwithstanding its lesser status, Anglovaal was singled out by the South African *Financial Mail* in 1981 for its "aggressive and wide-reaching acquisitions policy" (7).
Hartebeestfontein remained a uranium oxide producer until early 1991, when this entry was being printed (8).

References: (1) *MAR* 1985. (2) *MJ* 21/9/84. (3) analysis of Rand and OFS quarterlies 9/84 in *MJ* 26/10/84. (4) Lynch & Neff. (5) *Nuclear Fuel* 5/1/81. (6) Lanning & Mueller. (7) Davies *et al.* (8) *MJ* Supplement 25/1/91.

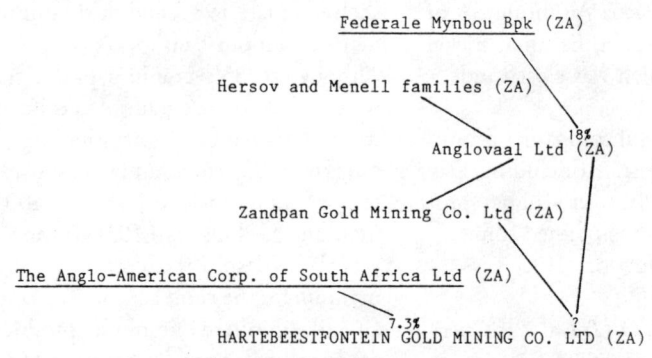

294 Hawk Investments Ltd (AUS)

Originally a property development company, it turned to exploration in 1979/80.

It was reported, as early as 1978, to be a partner in a JV for uranium exploration with Cliffminex and Keywest at Chalba Creek near Carnarvon, and at Minindi Creek near Karratha, both in Western Australia (WA). At Minindi Creek some mineralisation was reported 1979/80 (1), and an *in situ* plant was being considered (2).

In 1981, it also held a 50% interest (other partners not known) at Mombo Creek, east of Gascoyne, WA (3).

By 1983 Hawk had apparently withdrawn from Minindi Creek (4) and soon after, disappears from public view (there is no reference to it in the 1990/91 register of Australian mining).

References: (1) Cliffminex Annual Report 1980. (2) *West Australian* 14/11/78. (3) *Reg Aus Min* 1981. (4) *Reg Aus Min* 1983/84.

295 Hecla Mining Co Ltd

It is mainly concerned with lead, zinc and silver mining in Idaho, USA. It also owned 34.5% of Granduc copper mines, British Columbia, Canada, now owned by Esso Minerals Canada (1). It also has 50% of a JV with Union Carbide to mine uranium and vanadium on Hecla's Lisbon Valley property near Moab, Utah (1). In 1981 ore was being stockpiled, and commercial production was expected in 1982 (2) – up to 250 tons/day U/V (3).

Ranchers merged with Hecla (formerly owned by Amax) in 1984 (4). The mine, though, had gone on stand-by already in 1983 (5).

The Moab uranium mine remained on standby throughout 1985-86 (6). However, the company embarked on several major mining projects in the latter half of the 1980s: notably a

gold-silver-zinc-lead JV called Greens Creek JV, in Alaska, where its partners include BP Minerals America and Mitsubishi (7); also a JV with Highwood Resources Ltd of Calgary, to develop a large strategic minerals project at Thor Lake in the Northwest Territories (7).

Hecla's interest in Granduc now stands at 37.95% (8). In 1988 it agreed to acquire Cyprus Minerals clay division (9,10) and a year later it entered a JV with Agnico-Eagle to explore and develop precious metal projects, mainly in Quebec and Idaho (11).

In 1989, as the company reported mixed fortunes (10), it also announced plans to build a gold heap-leach facility at its Yellow Pine mine in Idaho (12).

References: (1) *MIY* 1982. (2) *MAR* 1981. (3) *MM* 1/82. (4) *E&MJ* 4/84. (5) *MAR* 1984. (5) *MAR* 1984. (6) *FT Mining* 1987, *MAR* 1986, 1987, 1988. (7) *MAR* 1988. (8) *FT Min* 1991. (9) *E&MJ* 9/88. (10) *MJ* 17/3/89. (11) *MJ* 19/5/89. (12) *E&MJ* 6/89.

296 Hewitt-Robins International Ltd (GB)

A manufacturer of vibrating screens used to handle uranium ore, some of which, tradenamed as Vibrex, were sold to southern France in 1983 (1).

References: (1) *MJ* 23/12/83.

297 Highwest Development Ltd (CDN)

It was involved in a JV for uranium exploration with Eldorado Nuclear in the late '70s (1); no further information.

References: (1) Eldorado Nuclear Annual Report 1979.

Ranchers Exploration and Development Corp. (USA)

MERGER

HECLA MINING CO. LTD (USA)

298 Hilliard Ltd (IRL)

It was identified by the Cork anti-nuclear group, CANG, as prospecting for uranium in Kerry, Eire, during the late '70s. Otherwise no information (1).

References: (1) CANG personal communication 4/3/80.

299 Hiram Walker – Consumers Home Ltd

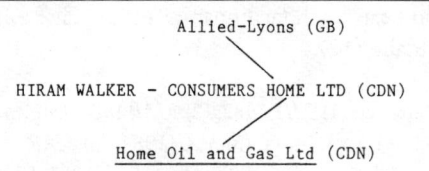

```
                Allied-Lyons (GB)
                    /
HIRAM WALKER - CONSUMERS HOME LTD (CDN)
                    \
           Home Oil and Gas Ltd (CDN)
```

An amalgam of Hiram Walker, Gooderham and Worts, Cygnus Corp, Home Oil, and Consumers Gas, Hiram Walker – Consumers Home has now grown to be the fifth largest Canadian company (based on net income). Its board interlocks with other Canadian uranium mining companies including Scurry Rainbow, Noranda, and Rio Algom. Then there's also McDonald's fast food chain and Quaker Oats, would you believe ...? (Yes, we would!)

The company has been exploring for uranium in Saskatchewan (1).

The conglomerate produces such aids to civilisation as Canadian Club and Ballantine's whisky, Courvoisier cognac, Drambuie and Tia Maria liqueurs.

In 1981/82 Hiram Walker bought out the Davis Oil company – the 8th largest acquisition of a US company in that year – at a cost of US$630M (2).

In 1987, the British food conglomerate Allied-Lyons bought a majority of Hiram Walker, and later acquired control (3) from the Reichmann brothers (Olympia and York) (4).

References: (1) *Who's Who Sask.* (2) *WDMNE.* (3)

FT 23/12/87, FT 6/1/88.(4) FT 27/11/87, FT 25/11/87.

300 HNG Oil Co

```
Houston Natural Gas Corp. (USA)
                \
HNG [HOUSTON NATURAL GAS] OIL CO. (USA)
```

Based in Texas, HNG jointly owns the Johnny "M" uranium mine at Ambrosia Lake, New Mexico, with Ranchers Exploration. Production there had slowed considerably in 1978/79 due to an accident and delay in bringing another orebody into production. Reserves are estimated at 3.1 million pounds; the ore is processed by Kerr-McGee (1).

Johnny "M" was closed in March 1982: more than 600,000lbs U_3O_8 was delivered to customers from the mine that year and in 1983 (2).

References: (1) *MIY* 1981. (2) *MIY* 1985.

301 Hollinger Argus

Hollinger Argus derives its income mainly from LMX's revenue from the Iron Ore of Canada company – and from a loan of nearly C$170M it made to Brascan (1).

In 1964, the company gained a 90-year lease to explore for minerals on Inuit and Innut land in Ntesinan (Labrador/Quebec) followed by another 60-year lease over an additional 300 square miles (2).

Via the Argus Corp (fore-runner of Hollinger Argus and considered the biggest Canadian financial empire until recently), Hollinger Argus also controls 39% of Dominion Stores (3).

As a holding company, the Hollinger group is best known for its ownership of rightwing newspapers (in particular the *Daily Telegraph* and *Spectator*) and the fairly high profile projected by the company's chair, Conrad Black.

Perhaps the most insidious and censorious of the group's recent undertakings has been to

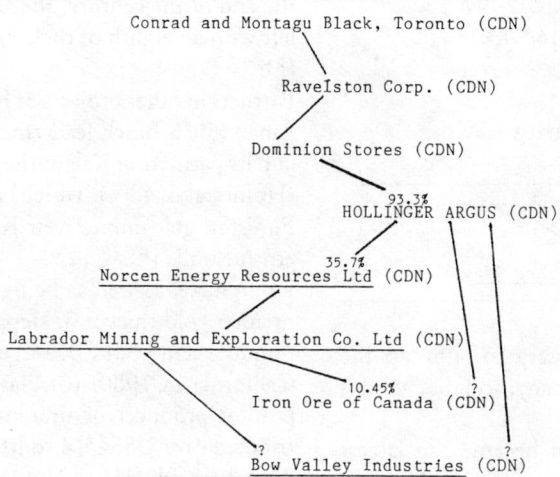

```
            Conrad and Montagu Black, Toronto (CDN)

                    Ravelston Corp. (CDN)

                  Dominion Stores (CDN)

                        93.3%
                   HOLLINGER ARGUS (CDN)

              35.7%
         Norcen Energy Resources Ltd (CDN)

        Labrador Mining and Exploration Co. Ltd (CDN)

                        10.45%       ?
                   Iron Ore of Canada (CDN)

              ?                          ?
              Bow Valley Industries (CDN)
```

buy-out the only surviving truly independent tabloid newspaper in Israel, the *Jerusalem Post* in 1989, and effectively squeeze out its liberal editorial staff, turning it into another rubber-stamp for the Israeli regime in its war on the Palestinian people (4).

References: (1) *MJ* 28/5/82. (2) RL Barsh in *Kiitg* 6/82. (3) Peter C Newman, *The Canadian Establishment* 1979, quoted in *FT* 4/6/81. (4) *Independent* 4/1/91.

302 Holly Minerals

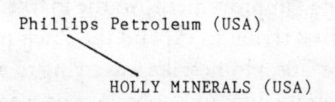

```
         Phillips Petroleum (USA)

              HOLLY MINERALS (USA)
```

It was acquired by Phillips Petroleum in 1966 and effectively merged. No current activity (1).

References: (1) *US Congress Subcommittee on Energy* 1975.

303 Home Oil and Gas Ltd

It is amalgamated now with Hiram Walker, Gooderham and Worts, Cygnus Corp, and Consumers Gas, in Hiram Walker-Consumers Home, fifth largest Canadian company (based on net income); it also had numerous interlocks with Canadian business/financial élites (1).

Home Oil had interests in Tunisia, Australia, New Zealand, the USA, Oman and Britain, with uranium exploration in Canada in the Athabasca region of Saskatchewan (2) and at Baker Lake, Northwest Territory. It was also exploring for oil in the Mesozoic Takutu Basin in the centre of Guyana (3).

Its subsidiary Scurry Rainbow (holding uranium prospects of its own in Saskatchewan) was to be incorporated into the new company Hiram Walker-Consumers Home in 1980, but this was delayed due to challenge by a minority stockholder (1).

Dome Petroleum in 1983 "farmed out" 40% of its exploratory lands agreement in Alberta, and 25% of that in the Beaufort Sea area, to Home: the Beaufort Sea has been strenuously defended against multinationals by the indigenous communities in the region (4).

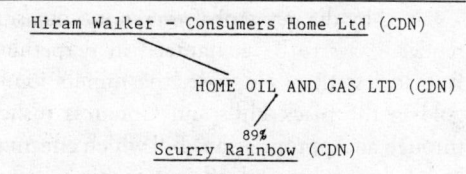

```
   Hiram Walker - Consumers Home Ltd (CDN)

              HOME OIL AND GAS LTD (CDN)

                   89%
              Scurry Rainbow (CDN)
```

References: (1) *Yellowcake Road.* (2) *Who's Who Sask.* (3) *MIY* 1982. (4) *Gua* 16/7/83.

304 Homestake Mining Co

```
          HOMESTAKE MINING CO. (USA)
                    60%
American Copper and Nickel Co. (USA)
```

Only one word is necessary to sum up the Homestake Mining Company, and that word is IGNOMINY.

Formed in 1877, it soon became the largest gold mining company in the world. Within a hundred years it was still the largest gold producer in the USA, and held the country's third largest uranium reserves too (1). By 1980 it had profited to the tune of more than one billion dollars. The Hearst newspaper empire was founded on its gains. But the whole edifice rested – quite simply – on theft: a theft about which the Supreme Court of the USA in 1980 declared:

"A more ripe and rank case of dishonourable dealings will never, in all probability, be found in our history" (2).

Homestake mines on the land of the Oglala Sioux. Moses Manuel and AC Harney, itinerant prospectors, discovered gold near Homestake's current mine at Lead, South Dakota, in 1876. A year later they sold their claim to George Hearst (his son was newspaper magnate William Randolph Hearst).

Hearst undoubtedly knew he was trespassing on the land of the Oglala, and the sacred Paha Sapa at that. And today Homestake Mining is contemptuous of that fact. In its contempt it mocks one of the most important treaties ever signed between sovereign nations: the Treaty of Fort Laramie, 1868. Under that Treaty, Lakota land – gained by the *defeat* of *Custer,* not Sioux surrender – was to be guaranteed in perpetuity. But, in less than a decade, the miners found gold in the Black Hills and Congress rushed through an Appropriations Bill which effectively ceded most Indian land to the whites. Before the end of the century, the Lakota people were left with an eighth of their original entitlement (3).

Partners in other projects of Homestake include Amax (50% Buick lead-zinc mine, Montana), and its partners in Kalgoorlie Mining Associates (Homestake 48%) which now operates the Fimiston gold mines near Kalgoorlie in Western Australia (4).

Homestake, as well as being among America's premier gold thieves (it stepped up its exploration in recent years (5), and made a gold find in California in 1980) (6), has also been an important producer of uranium. Uranium contributed over US$45M to its sales in 1980 (7).

Together with United Nuclear Corp (otherwise known as United Nuclear Partners) it operated the United Nuclear-Homestake mines in the Ambrosia Lake area at Grants, New Mexico (United Nuclear 70%, Homestake 30%).

Four mines were operating in the area until 1981. Late that year, however, Homestake announced a cut-back in mining and milling "because of depressed market conditions", affecting jobs of about 150 workers (8).

The United Nuclear Partners mill also treated ore gained from Homestake's Pitch uranium mine near Gunnison, Colorado, in which Homestake has an 85% interest. Contracts for about 4,000 tonnes of the Pitch uranium were said to have been signed by 1981 (7). In 1982, construction of a new mill at Pitch was deferred awaiting "improvement" in the market (9).

As well as trying to expand the Pitch project in the late '70s, Homestake was trying to mine and concentrate uranium ore in Los Padres National Forest near Ojai Valley, California – an activity which, according to local residents, would endanger the Lake Casitas water supply for 30,000 people.

In 1981, Homestake announced a six-million dollar project at La Sal near Moab, Utah, planned to be producing by August 1981 but probably not on schedule (10).

Homestake is also exploring for uranium in New Mexico, Colorado, Washington, Wyoming and Utah.

In its first quarter 1982 report, however,

Homestake reported a slump in profits, though 45% of them still came from uranium. About 150,000kg of uranium was produced during the first quarter of the year. The company's uranium position was "strengthened" due to its acquiring full ownership of United Nuclear's share in the three uranium mines and mills at Grants, New Mexico (11). However, by mid-1982 only the Milan mine was operating, at one-third capacity, and Grants section 23 at reduced output (12).

The most important opposition to Homestake comes from the Lakota people, specifically the Oglala Sioux who, early in 1982, sued the company over ownership of its Black Hills mine – the Lead Mine (13) – and for the gold, silver and other minerals removed from their land over the past century; and also for US$6 billion compensation for trespass and conversion of the proceeds. The suit was signed by the Oglala Lakota Sioux of Pine Ridge reservation (14).

In 1980, Miners for Safe Energy – a group representing about 700 miners at the Lead gold mine – issued a call for a state-wide moratorium on uranium mining in South Dakota, which clearly has a bearing on Homestake's activities in New Mexico and Utah too (16).

On August 13th 1982 (it was a Friday too!) the Federal District Court of South Dakota dismissed the Oglala Sioux case against Homestake. According to one of the Lakota lawyers, Mario Gonzalez, the judge had property in the Black Hills himself and had therefore been "biased and politically motivated" (15).

In Colorado, Homestake's Pitch project was the focus of considerable opposition from FUTURE (Folk United To Thwart Unsafe Radiation Emissions) in the late '70s. Homestake had been mining in the Gunnison National Forest for some time before local residents banded together and campaigned against the granting of a licence for a mill in 1977/78. Homestake planned a 1200 ton/day mill and open-pit mine a mile long and no less than 700 feet deep. FUTURE mobilised nearly 3000 signatures to a local petition; 7 environmental organisations "pitched" in to criticise the project's EIS (environmental impact statement). Granting of a

preliminary licence for the mill was delayed by citizen's action (17). By early 1982, design production for the mine was a quarter of a million tons of uranium ore per year – and final licence for the mill had not come through (18).

Westinghouse filed suit against Homestake for non-delivery of uranium as a spin-off of the cartel extravaganza – though the US District Court of San Francisco upheld Homestake's counterclaim (19).

In late 1981, the South Dakota state government introduced a 6% state levy on gross income, clearly aimed at Homestake as it affected only gold and silver producers, and Homestake is the state's sole gold-silver producer! Homestake considered it was being victimised, and took legal action against the South Dakota government for unconstitutionality. The case was dismissed (20).

In the first nine months of 1982, uranium sales totalled US$46.26 million – about a third more than in the previous year. At a meeting in London in early November 1982, when the company's shares for the first time became available in Britain and West Germany, chair Harry Conger said the company was in the "fortunate position of having uranium sales contracts (which still have a number of years to run) with about 14 utility companies. Under the terms of these contracts, the company [is] able to sell uranium at prices substantially above the current prevailing market range ..." (21).

The company was also prospecting in the Whanga-aua state forest in New Zealand (22).

After a suit for damages and compensation worth US$6 billion was rejected by the US District Court in August 1982, the Oglala Sioux tribal council filed in a federal appeals court in May 1983 for reconsideration. "Our people have lived in grinding poverty for the last 100 years," said tribal attorney Mario Gonzalez, "while other people have lived off the fruit of our land which is ours by constitution and through treaty" (23).

Homestake's uranium operations at the Pitch lease were suspended in April 1983 "due to significantly reduced demand for uranium yellowcake and the continued prospects of an over-supply" (24). Ninety workers were laid off, though the property was to be kept on a care-and-maintenance basis. Peak activity at the mine, which opened in 1978, was in 1981 when 2000 tons of ore a month were being mined (25). The company's 1983 figures included a US$5.8M write-off on its investment in the Pitch uranium project as a result of land-slip which prevented the company from mining further reserves (26).

Due to improved production and precious metal prices, Homestake in 1983 secured sufficient capital to buy itself an oil company: it merged with Felmont Oil of New York at a cost of US$400M.

At the start of 1984, though, Homestake was getting less out of its gold operations in Western Australia, less uranium from its mine at Grants – but much more out of its gold.

Its major new programme was to develop the McLaughlin project in northern California, one of the largest open-pit gold mines (27).

In late 1983, the company reported a slight "stepping up" of operations at its Grants New Mexico uranium mill "to meet some production requirements". The mill was then running on a five day week – as were Homestake's uranium mining operations in the Grants area (28).

In 1988, the corporate maverick, T Boone Pickens, made a US$1.95 billion bid for control of Homestake, causing the value of its stock (in which Pickens's company Mesa holds 3.8%) to jump (29). A few months later Homestake Mining (BC) Ltd acquired control of North American Metals Corp, of British Columbia (30).

Homestake suspended all uranium mining and milling at Grants in 1989, when its chief "tolling" customer Chevron stopped production from Mount Taylor. The company has now effectively bowed out of uranium (32).

However, it continues to be one of the major US gold producers – thanks both to production from Australia and in the Americas. In 1988 it acquired 73% of North American Metal Corp, with interests in precious metals in British Columbia, and entered an exploration JV with

Chevron in Manitoba, near Churchill. Its El Hueso mine in Chile proceeded to full production in 1989, and two new mines started production in the same year in Nevada, at Wood Gulch and Mineral Hill in Montana (32).

Through the acquisition of Felmont oil in 1984, Homestake holds 25% in a JV with Echo Bay Mines Ltd at Round Mountain Nevada: this is the world's biggest heap-leach gold operation (32) with production of more than a quarter of a million ounces of gold in 1988, and plans to expand further (33).

In 1986, Homestake bought out the 50% of the Buick mine mill and smelter then owned by Amax, and later combined these operations with those of St Joe, the Fluor subsidiary, to form the largest lead/zinc facilities in the USA – Doe Run Co (Homestake 42.5%) (33).

In 1976, Homestake decided to buy into the potentially highly lucrative "Golden Mile" – that strip of gold diggings around Fimiston and Mount Charlotte in Kalgoorlie, Western Australia which has so disfigured Aboriginal communities, compelling many families over a long period of time into becoming internal refugees or "fringe dwellers"(36).

Initially, Homestake bought a 48% stake in Kalgoorlie Mining Associates (KMA), with JV partner Lake View Pty Ltd (31). In mid-1987 North Kalgurli Mines Ltd – the vehicle for Alan Bond (the larger-than-life entrepeneur who was later to come a cropper) and now controlled by Robert de Crespigny through Poseidon Mining Management Pty Ltd – took over 51.5% of GMK in order to merge KMA's leases with those of North Kalgurli's into a "super pit" (33,34).

Two years later, GMK bid for the whole of North Kalgurli and formed a new company, Kalgoorlie Consolidated Gold Mines Pty Ltd, of which Homestake and GMK both hold 50% each (34).

Homestake's stake in Homestake Gold of Australia is actually 80% (35). But it also holds 75% of Croesus Mining NL's leases, 50% of the Lone Star JV at Tennant Creek, 80% of various gold leases in Victoria, 50% of the Mount Hope gold/base metals prospect and

418

75% of the Copper Canyon gold province in Queensland. Homestake Australia is now providing no less than a quarter of all US-accredited gold production (35).

Further reading on the Pitch project: extensive article by Ed Skinner in the *Chaffee County Times* 7/9/78.

Contact: CAFCA, PO Box 2258, Christchurch, New Zealand (Aotearoa).
Lakota Treaty Council, Oglala Lakota Nation, POB 862, Pine Ridge, South Dakota 57770, USA.
Miners For Safe Energy, POB 247, Lead, South Dakota, 57754, USA.

References: (1) *Gua* 16/8/80. (2) US Supreme Court 1980, quoted in press release by Oglala Sioux Legal Department, Pine Ridge 18/1/82. (3) *Black Hills Alliance Background Paper*, 1980. (4) *MAR* 1981. (5) *FT* 28/9/79. (6) *FT* 29/8/80. (7) *MIY* 1981. (8) *MJ* 11/9/81. (9) *MJ* 11/6/82. (10) *MJ* 20/3/81. (11) *MJ* 11/6/82. (12) *Nukem Market Report* 8/82. (13) Lawyers' press release 25/1/82. (14) Oglala Lakota Legal Rights Fund press release, quoted in *AkwN* Early Spring 1982. (15) Lakota Treaty Council press release 15/8/82. (16) *Kiitg* 6/80. (17) *Future Newsletters* 16/6/78-25/9/79. (18) *MM Survey* 1/82. (19) *FT* 4/1/80. (20) *MJ* 16/7/82. (21) *MJ* 5/11/82. (22) *CAFCINZ Watchdog* 9/82. (23) *News-Tribune*, Minnesota, 17/5/83. (24) Company record, quoted in *E&MJ* 4/83. (25) *E&MJ* 4/83. (26) *FT* 9/2/84. (27) *MJ* 4/5/84. (28) *E&MJ* 9/83. (29) *FT* 1/3/88. (30) *E&MJ* 6/88. (31) *MAR* 1988. (32) *MAR* 1991. (33) *FT Min* 1991. (34) *Reg Aus Min* 90/91. (35) *E&MJ Int. Directory of Mining* 1991. (36) For a graphic description of this process

and the result, see the film *Munda Nyuringu* directed by Bob Bropho and Jan Roberts, Melbourne, Australia 1984.

305 Honeymoon Uranium Project

This is a small but important uranium site in South Australia with an estimated 3400 tonnes of uranium oxide reserves which was due to be mined at about 450 tonnes/year from 1985.

Its ownership is: Carpentaria Exploration Co Pty Ltd, 49.0%; Mines Administration Pty Ltd, 25.5%; Teton Australia Pty Ltd, 25.5%; (3.8% of Honeymoon was originally held by CRA but sold to AAR during the reorganisation of IOL/CSR holdings in March 1980).

The companies expected to start major construction in 1982 – though in the teeth of strenuous opposition. For example, in 1981 the Broken Hill City Council (nearest city to the mine) announced that it would oppose transportation of uranium through the council's area (1).

The Australian federal government gave the go-ahead in October 1981. However, in early 1983, the South Australian Labor government decided to halt development at Honeymoon (and Beverley) taking into account – as the Minister for Mines and Energy put it – the economic, social, environmental and safety aspects of the mine. The partners were still permitted to keep a lease on the site and preserve their investment (2).

The comments – sour-grapeish as they are – of the editors of the *Register of Australian Mines* on

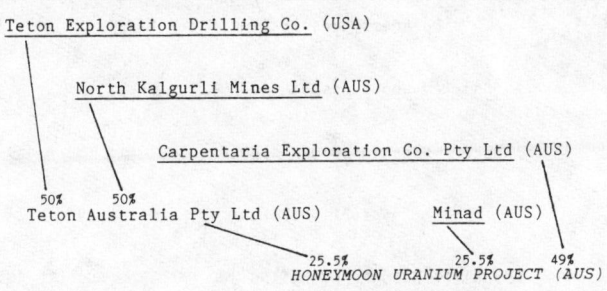

the apparent collapse of the Honeymoon project are worth quoting in full:

"Production: Honeymoon was advanced to the pilot stage following leaching trials in 1982, and became a target for the Campaign Against Nuclear Energy whose disciples gained the eye of the general press by going to the site and also claiming that leach testing had contaminated the groundwater. Though the then new State Labor Government found no evidence of this, it took the political move of deciding not to grant the partners a production licence, but made a token compromise of granting the partners a Retention Licence over the area.

"Comment: The decision was a victory for the vocal minority in the conservation movement, and while Honeymoon was being singled out because it was planning to apply a leaching technique, many have read this as a body blow to getting some of Australia's major uranium projects into operation. AAR's David Brunt commented that the project had proceeded to strict environmental conditions laid down by the SA and Canberra governments. In making the announcement, the SA Mines Department and Senator Peter Walsh both skirted the real politics behind the issue and made some amazing statements on the uranium marketplace and their responsibility to it ... this included the implication that Honeymoon was being halted to protect the future of Roxby Downs (which will have the same problems with the anti-league, even though it is the only project that can rescue SA from its prolonged economic mire) and Senator Walsh's statement that "economics alone dictate that new mining ventures at the present time would be unwise". As Australia's procrastination has already given other nations the edge on uranium marketing – aided by the effective, scaremongering campaign of the anti-uranium lobby – these sort of comments just add to the great Australian uranium marketing tragedy. The power of the public services to give government control over uranium projects has them enmeshed in a red tape that would be constrictive even in a more realistic political environment. The Honeymoon partners announced they would be seeking compensation from the SA Government" (3).

References: (1) *CANP Newsletter* 1-2/81. (2) *Tribune*, Sydney, 30/3/83. (3) *Reg Aus Min* 1983/84.

306 Hudbay

Not to be confused with Dome Petroleum's Hudson's Bay Oil and Gas, Hudbay – also uranium exploring in Saskatchewan – has been effectively Anglo-American's Canadian "arm" (1), a conduit for apartheid capital into North American natural resources. For example, along with Minorco (another Anglo-American and De Beers front, formerly registered in Bermuda) Hudbay acquired Inspiration Copper, a US

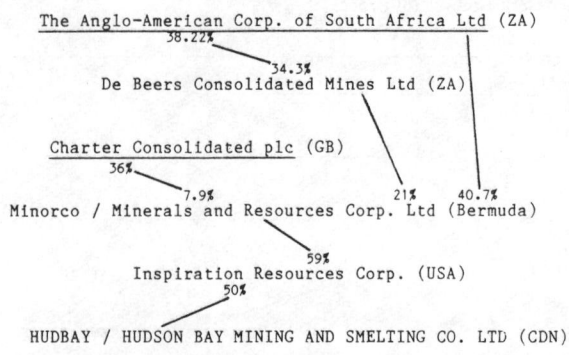

The Anglo-American Corp. of South Africa Ltd (ZA)
38.22%
34.3%
De Beers Consolidated Mines Ltd (ZA)

Charter Consolidated plc (GB)
36%
7.9%   21%   40.7%
Minorco / Minerals and Resources Corp. Ltd (Bermuda)
59%
Inspiration Resources Corp. (USA)
50%
HUDBAY / HUDSON BAY MINING AND SMELTING CO. LTD (CDN)

company, in 1978 (2). It also has investments in Du Pont, Imperial Oil, Inco, Noranda, Union Carbide (3) and wholly owns the Churchill River Power Co Ltd (operating on Inuit land in Labrador) (4).

Hudbay, as well as exploring for and producing a variety of minerals, has interests in petroleum, coal, fertilisers and agricultural products.

In November 1980 it concluded a JV arrangement (44%) with Granges and Outokumpu, among others, to open up a large copper-gold deposit at Trout Lake, near Flin Flon, Manitoba (5) which began producing in 1983 (6).

The company's properties near Whitehorse, in the "native" part of the Yukon, were closed down for "clean-up" in the same year (6).

In mid-1983 the company underwent a major restructuring which still preserved Anglo-American's control of Hudbay but allowed Minorco to reduce its voting rights, should it desire. More importantly perhaps, the reorganisation enabled Hudbay to consolidate its access to North American capital markets (7). Hudbay now holds a 50% interest in Inspiration Resources which has been reorganised into a publicly traded US company (7), but Inspiration Resources holds all the voting rights in Hudbay. Now ranked as eleventh among Canadian mining companies (8), Hudbay has developed several important new deposits in the Flin Flon region of Manitoba over the past few years: notably at the Callinan copper-zinc lode, where production began in 1989 (9), Trout Lake (where it is in a 44% JV with Granges, Outokumpu and Manitoba Mineral Resources Ltd) and on a lead reserve at Chisel Lake. The promising Namew Lake nickel deposit, which it is again developing with Outokumpu (in a 60/40 JV) (10) was plagued with excess water in 1989 (9), while the Ruttan mine was scheduled for permanent closure in 1991 (9).

Hudbay has also been forced to plan a complete overhaul of its Flin Flon copper-zinc smelter, to meet standards soon to be imposed under the 1994 Canadian Acid Rain Abatement Act. It was trying to raise the more than $130 million required for this from federal and provincial government loans (10).

References: (1) *FT* 2/5/80. (2) *FT* 5/4/79. (3) Free South Africa Committee Toronto, quoted in *Yellowcake Road.* (4) *MIY* 1981. (5) *MIY* 1982. (6) *MAR* 1984. (7) *MJ* 10/6/83. (8) *Canadian Mining Journal* 8/90. (9) *MAR* 1990. (10) *MJ* 21/7/89.

307 Hudson's Bay Oil and Gas Co Ltd

```
        Dome Petroleum Ltd (CDN)
                        /
HUDSON'S BAY OIL AND GAS CO. LTD (CDN)
```

Not to be confused with the South African-controlled Hudbay, this company passed into the hands of Dome Petroleum in early 1982 in a slick manoeuvre which resulted in Du Pont acquiring Conoco after Dome bought a large chunk of it – and exchanged it for the Hudson's Bay holding (1).

It has been exploring for uranium in Saskatchewan (2); otherwise its activities seem confined to oil and gas exploration, largely in Canada (3).

References: (1) *FT* 14/1/82. (2) *Yellowcake Road.* (3) *MAR* 1982.

308 Humble Oil (USA)

It was exploring – probably for uranium as well as oil – in Navajo (Dene) reservation land in the '70s (1).

No further information.

References: (1) Sklar.

309 Hunting Geology & Geophysics Ltd (GB)

This is one of the world's biggest radiometric "search" companies which has been heavily involved in uranium prospecting world-wide on behalf of client nations.

By mid-1977 it had flown over five million instrument line kilometres of airborne geophysics for exploration since starting in 1953. At that time more than 22 contracts had been signed in over 50 different countries (1).

Hunting "discovered" the Nabarlek uranium deposit in Australia. In 1981 it was contracted by Agip to explore for uranium in Sardinia and the Italian Alps (2).

More recently it was given a contract by the government of Mozambique to perform a minerals search (3).

It was Hunting who "sniffed out" the Puno uranium deposit in Peru (see Coluranio).

In 1978 Hunting joined Essex Minerals (a subsidiary of US Steel) in preparing a uranium potential study of more than forty countries (see Essex).

References: (1) *MJ* 27/6/77. (2) *MJ* 20/6/81.(3) *MJ* 4/6/82.

310 Hunt Oil Co (USA)

This is an oil company with interests in Australasia and South America where it has been prospecting for oil in the Peruvian southern jungle region (1).

It was reported to be uranium exploring with Energy Resources Corp (Denver, USA) in 1978/79 (2).

No further information.

References: (1) *MAR* 1982. (2) *US Surv Min* 1979.

311 Hydro Nuclear (USA)

It was exploring for uranium on Dene (Navajo) reservation land in the '70s (1).
Believed defunct.

References: (1) Sklar.

312 IAN

Primarily an exploration body, when it comes

to uranium. In March 1978 it signed a US$500,000 contract with ENUSA to explore and develop reserves in the Vaupes and Guiania areas of the country (1).

However, in late 1982 the Institute itself started a "vigorous" evaluation programme in Berlin *(sic!)* in the Department of Caldas, central Colombia, where values of around 1000 ppm uranium (*i.e.* economically recoverable under certain circumstances) have been located (2).

First metallurgical samples were sent to LKAB's Ranstad laboratory because of the similarity between the Berlin and Tasjo (north Sweden) uranium occurrences.

According to the *Mining Journal*, it may well be feasible to mine the Berlin deposit, given its proximity to the major transport centre of La Dorada and its association with phosphates (Colombia then wanted to develop a phosphates-based fertiliser industry).

Previous exploitation of the Berlin area was by Minatome – whose experience at the hands of the local people was not a happy one (3).

Ironically. as the market for uranium has worsened in the past decade, IAN's explorations have yielded little of public note, except for an important new coal field which was discovered when uranium prospecting (4).

The Paramo del Almorzadero anthracite district is situated in the Department of Santander, 320km north of the capital, Bogota, in the eastern cordillera of the indigenous Andes and the Magdalena valley.

Carboriente, the company exploiting the field, with hopes for production for export by 1993, is comprised of IAN and other state-owned ventures. In 1989, Carboriente was hoping to use World Bank loans to finance the Almorzadero project (4).

References: (1) *ARC Bulletin* 12/78. (2) *MJ* 15/10/82. (3) *El Tiempo*, Bogota, 29/8/80. (4) *MAR* 1990.

```
                    IDEMITSU KOSAN CO. LTD (J)

      Idemitsu Uranium Explorations Canada Ltd (CDN)

        11.875%                    11.875%
    Cigar Lake JV (CDN)        Waterbury JV (CDN)
```

313 Idemitsu Kosan Co Ltd

Japan's largest petroleum company, Idemitsu acquired shares in the Waterbury JV in Saskatchewan, Canada, from the former project operator, Asamera, in October 1982 (1). With 11.875% it is now part of a consortium drilling there for uranium in which Séru Nucléaire, Canada, has a major interest (33.6%) and Reserve Oil (USA) owns 3.88% while the SMDC has the controlling 50.8% stake (2).

The company is also exploring for uranium in Australia's Northern Territory (1).

In 1983, it acquired loans from Japan to start coal exploration – Japan is the world's largest coal importing country – in the USA and Canada (3).

In 1988 Idemitsu joined with the Australian company Bligh Coal Ltd (25% each) and with Rheinische Braunkohlenwerke AG and a South Korean company, Lucky Goldstar, to open up a large surface mine in Queensland near Ensham, at a cost of more than US$180 million (4).

By the turn of the last decade, Idemitsu had consolidated its interests in uranium in Canada, by securing its 11.875% stake in the Cigar Lake JV (managed by Cameco, with Cogéma and its subsidiary Corona Grande, holding the second biggest stake) (5,6).

But above all, it established itself as a major gobbler-up of coal for insatiable Japanese industrial appetites: negotiating contracts with Intercor (see Exxon) for Colombian coal (5) and with the controversial Venezuelan state company, Carbozulia, whose penetration of the land of the Yukpa and Bari Indians in the state of Zulia have been condemned by Survival International and others (7).

Currently Idemitsu's brightest black hopes centre on Australia where it holds 50% of the Ebenezer coal project in Queensland, and is partnered with BHP in the Boggabri mine project (5) in addition to the Ensham project.

References: (1) E&MJ 6/84. (2) MJ 21/10/83. (3) MAR 1984. (4) E&MJ 6/88 see also E&MJ 6/89, MAR 1990. (5) MAR 1987. (6) FT Min 1991. (7) MAR 1986, Survival International SInews, No 25 1989.

314 IMC/International Minerals and Chemical Corp (USA)

It is the world's largest private enterprise producer of phosphate rock and potash (1). IMC is also a major producer of fertiliser from phosphates, with operations in 22 states of the USA, 3 Canadian provinces and 11 overseas countries (2).

In 1979, IMC concluded eight long-term contracts with uranium utilities (two of whom were Florida Power and Light and the TVA) (3) to supply U_3O_8 over a 12-year period from its current recovery unit at New Wales, Mulberry, Florida, and two others coming on stream (1). These two plants – at Bartow and Plant City – were being constructed with CF Industries (another phosphates producer) with a projected capacity of nearly 1.3 million pounds (4). Due to start up in 1980/81, by 1982 the company made no mention of their production figures, and it is assumed they were not on stream (1).

IMC operates with WR Grace a phosphates plant in Manatee and Hillsborough counties, Florida (2).

It is also part of a uranium-from-phosphates consortium in Belgium along with Prayon, Metallurgie Hoboken-Overpelt, and Mechim (Umipray) (5).

By late 1982 its uranium operations at New Wales were (apart from Freeport McMoran)

423

Florida's only uranium production facilities. The plant reached break-even point during that year – the previous year it operated at a US$15.5 million loss (1).

The company has also been exploring in Saskatchewan (6).

The company tendered in 1981 for the construction of a phosphates-into-uranium plant to be sited at Safi, Morocco. Other companies interested were Gardinier and Westinghouse (7).

IMC also has an "interest" in genetic engineering and a JV with Genentech (8).

WR Grace sold its 50% interest in the Four Corners phosphates mine to IMC in 1988 (9). IMC also acquired an 11.1% interest in the largest phosphates producer in the Senegambia (along with Cofremines, Rhone Poulenc and others) (10).

By 1991, it was one of only seven US uranium producers left in the field – in IMC's case, at the by-product phosphoric acid facilies in New Wales and Plant City, Florida (11) where its production over the last few years has been constant at around 1000 tonnes uranium (12).

References: (1) *MJ* 8/10/82 (quoting annual report of company to June 82). (2) *MIY* 1981. (3) *MJ* 20/6/80. (4) *MAR* 1979. (5) *MAR* 1981. (6) *Yellowcake Road*. (7) *Intern Min/Met Rev.* (8) *WDMNE*. (9) *FT* 5/12/88. (10) *MAR* 1990. (11) *E&MJ* 3/91. (12) *Greenpeace Report* 1990.

315 Imétal SA

Until the early '80s this was the largest French mining company (1) – and who would be surprised at that, considering who controlled it? Perhaps Congressman McDonald summarised best the influence of the Rothschilds, not only in French commerce and overseas exploitation but the '70s uranium industry, when he addressed the House of Representatives in Washington on March 4th 1975:

"Control of the uranium cartel may be far more concentrated than would appear at first glance. As suggested by the fist clenching five arrows in the family crest, the Rothschilds of France and England have an interest in nearly every major uranium mine in the world. "Rio Tinto-Zinc, a mining company in which the French Rothschilds have traditionally been major owners, holds a controlling interest in Rio Algom mines ... The largest South African gold producer, Anglo-American Corp, is also the country's largest uranium producer, since South African uranium is produced as a by-product of gold. One of Anglo-American's associated companies, Charter Consolidated, has a Rothschild on its board, and owns nearly 10% of RTZ.

"The centerpiece of the French Rothschilds' nonferrous metal group is Imétal (formerly Le Nickel) which has a controlling share of the Mokta and Pennaroya companies, two of the largest uranium producers in France and in the former French colonies of Gabon and Niger. They also participate in joint ventures with the other large French producer, Pechiney Ugine Kuhlmann [PUK/Pechiney].

"The Rothschild presence is everything. Baron Guy de Rothschild heads Imétal. He also sits on the board of RTZ. In turn, Harry Oppenheimer, the chief executive of Anglo-American, and Sir Val Duncan, the chief executive of RTZ, sit on the board of Imétal. Only in Australia do the Rothschild companies have a relatively small share of the total uranium reserves ... In fact ... the function of the Uranium Producers Forum could be performed at a board meeting of Imétal" (2).

That statement was made fifteen years ago; pretty accurate it was then. In the intervening one-and-a-half decades, many changes – besides nationalisation – have taken place (for example, Charter Consolidated no longer has a major share of RTZ, Baron Guy no longer ostensibly directs RTZ's affairs, and Val Duncan has long gone to his eternal gardens). However, it is interesting to note that one of RTZ's most recent major projects was conducted with interests purchased from the Rothschilds: the Neves Corvo copper mine in Portugal – the country's largest – in which the British multinational pur-

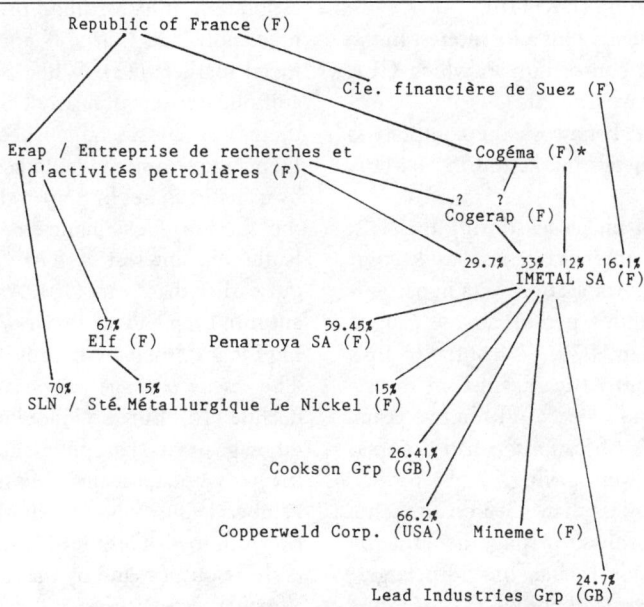

```
                Republic of France (F)

                                    Cie. financière de Suez (F)

         Erap / Entreprise de recherches et          Cogéma  (F)*
            d'activités petrolières (F)
                                          ?   ?
                                          Cogerap (F)

                                   29.7%  33%  12%  16.1%
                                         IMETAL SA (F)

                    67% (F)          59.45%
                    Elf (F)      Penarroya SA (F)

           70%       15%                     15%
      SLN / Sté. Métallurgique Le Nickel (F)

                                    26.41%
                              Cookson Grp (GB)

                           66.2%
         Copperweld Corp. (USA)   Minemet (F)

                                        24.7%
                              Lead Industries Grp (GB)
```

chased the 49% share held by Somincor, a JV of Imétal's subsidiaries Penarroya and Coframines (3).

Both RTZ's important, but shady, Minserve subsidiary and Baron Guy, have headquarter haunts in the little Swiss town of Zug (where they no doubt also meet to discuss hard times over the occasional glass of Mouton Rothschild).

Imétal was originally incorporated as Le Nickel in 1880. After reorganisation and name change in 1974, it embraced three large mining concerns: Penarroya (58.5% owned by pre-nationalisation Imétal); Le Nickel (now called Société Le Nickel or SLN, then owned as to 50% by Imétal); and Mokta (93.8% owned by Imétal in 1974) (4).

During the late '30s, the left-wing Popular Front government in France "robbed" the Rothschilds of their extensive railway holdings, and in 1940 the Vichy government of Marshal Pétain confiscated the family's fortune and stripped them of their citizenship while – as victims of Nazi persecution – they, understandably, fled the country. In post-war years, the family rebuilt its prosperity in the banking world and on further expansion into mining and metallurgy (5).

Le Nickel was formed to develop the deposits in New Caledonia (Kanaky) which have since become one of the prime targets of the FLNKS (the major liberation organisation) who managed to reduce production by 50% in early 1985 (6). Le Nickel became one of the world's biggest producers of that metal – as Penarroya was developing, through mines in France and Spain, into the leading French producer of lead and zinc. The Rothschild group held important interests in both companies; in 1971 it acquired Mokta, thus bringing considerable uranium deposits under its control. Nine years later Imétal merged with Mokta. In 1977 Imétal acquired the steel product manufacturer Copperweld, and thus consolidated, not only its downstream interests, but also its presence in the USA (7). Three years later it also acquired the Revere Smelting and Refining Co of Dallas, Texas (8). In addition, Imétal has important British interests through the Cookson group (fabricator of non-ferrous metals and chemicals in several countries) (9) and subsidiaries of Minemet, the company which handles Imétal's trading interests (9). It is also the major shareholder in the Lead Industries Group (UK) (10).

Through SLN, Imétal holds a JV interest in the Afternod deep sea consortium in which CEA and BRGM also have interests (11).

Through SLN and Penarroya the company is also involved in metals marketing in Namibia (23).

When the price of nickel fell during the '70s, Imétal's uranium interests served to keep it afloat (4). The importance of Mokta in particular to Imétal over this period can be gauged from the fact that, in 1975, 17% of the group's profits derived from Africa against an investment of only 5% of its assets. Within two years, though the group's African assets had dropped to only 1%, they were paying 32.5% profits, and a year later no less than 43% on the same investment (4). Hardly surprising, then, that by the following decade Imétal, the 38th largest multinational in terms of annual sales (6), was making a bigger proportion of profit from its foreign assets than almost any other major multinational company. In 1981 nearly all its profit derived from extra-European activity (98%), although little more than half its assets (60%) were invested there (6).

On the eve of nationalisation, Rothschild's pet dog was barking happily down a number of avenues. A new treatment plant for SLN in New Caledonia was producing, while newly discovered mineral deposits in Portugal and Australia (at Thalanga) were being investigated; a rich silver vein in Peru was also being exploited by the subsidiary Huaron mining company (9). Although 1981 saw losses sustained by SLN, its Anglo-American subsidiaries made good profits (12).

Then the Mitterrand government took over the Rothschild banking interests. Guy de Rothschild greeted the programme in high dudgeon, calling it "an extremely Marxist approach". The Baron predicted that removal of competition would induce the granting of bad credit and the operation of unprofitable industries, cushioned by what he dubbed the "Socialist prosecutors" (5).

As if to prove the Baron correct, within the fol-

lowing two years Imétal suffered a sharp deterioration in its fortunes, primarily due to depreciation charges at SLN and the downturn in metal markets (13). When SLN announced a consolidated loss of nearly US$100M for 1982, there was considerable speculation that the French government would acquire control of it by using its shares in Erap and its holding bank, the Compagnie Financière de Suez (whose Rothschild interest of 4.73% the state bought out earlier that year) (14). And indeed, by the autumn Erap had taken over 70% of SLN, leaving Elf and Imétal with only 15% each (15).

The rescue package temporarily halted SLN's decline (16), but the following year, thanks to sabotage of vital equipment at Kouaua (17) and further sabotage at the Thio mines (18) in November 1984, New Caledonia nickel production slumped. Despite this, SLN promised to increase output, and by the middle of 1985 yet a further restructuring of the company was proposed. Under this proposal, there would be an SLN New Caledonia and a newly carved metropolitan parent company in which Erap would "inject" its other non-ferrous interests, while Imétal would "bolster the sales network of the parent company" (19). During the first months of 1985, nickel production from New Caledonia appropriately see-sawed as the political situation did the same, and white violence against the Kanak liberation fighters mounted (20).

After what the directors of Imétal called a "very bad year" in 1982, the French government began investigating ways of re-organising Imétal itself. Its three major shareholders were now Erap (29.9%), Compagnie Financière de Suez (16.1%), and Cogéma (12.1%). In mid-1983 the company announced that Erap and Cogéma would form a new holding company to own another 33.3% of Imétal and to be controlled by Cogéma. Imétal would maintain control of Penarroya, and the parent company would increase its holding in Mokta and Minemet by purchasing back Penarroya shares (21).

Restructure

By mid-1984, Imétal was therefore thoroughly

re-structured, and only a shadow of its former self. Mokta was merged with Imétal, and this subsidiary, together with Penarroya, remained the company's only mining concerns (apart from various investments) (22). Its strongest subsidiary continued to be Copperweld in the USA.

By 1990, Imétal's re-structuring was virtually complete, and its withdrawal from primary mining almost absolute (24). In 1989 it disposed of its mining intersts in Spain and Africa, and sold its Minemet research centre (25). It was left with CSC Industries, Eramet-SLN (which owns a French nickel refinery) and Metaleurop, the company formed in 1988 with Preseuag holding 45% of the capital and Imétal 20% (26).

Metaleurop has become the world's largest lead and the second largest zinc producer, with around one eighth of the world's germanium production capacity, and significant portions of the world market in indium and cadmium (27). Imétal's industrial acquisitions have proceeded at a modest pace, with the 1990 purchase of Asea Brown Boveri's US subsidiary C-E Minerals for $150 million (C-E produces alumina silica calcines and other products for the semiconductor industry: it was previously controlled by the US constuction and nuclear engineering company Combustion Engineering) (28).

The world's biggest ferro-nickel producer, SLN (29) continues to be owned 15% by Imétal, with Erap holding the lion's share of 70%, and Elf-Aquitaine the remaining 15% (24). SLN, which operates four open-pit mines at Kouaoua, Poro, Thio and Nepoui, in Kanaky (New Caledonia) embarked on a major five year expansion programme in 1989 (29).

After Imétal's re-organisation in 1984, Cogema effectively came to control Imétal through a holding company (33.3%) set up with Erap, with Cogema having an additional 12% and Erap an additional 29.9% interest. (Until 1982, Amax had held 10.13%, Anglo-American subsidiary Minorco, 7.96% with smaller interests taken by Met. Hoboken and Le Populaire-Vie.) As of 1989, Erap's official share of the capital of Imétal had fallen to 5.45% with Parfinance

holding 39.48%, Euris (Cie Europeenne d'Investissement) 4.03%, Francarep 3.68% and St. Honoré Matignon 2.10% (24).

In 1985, Guy de Rothschild, chief architect of the Imétal empire recounted the formation of Imétal – "... a difficult but thrilling venture" – in his autobiography *The Whims of Fortune*. In this book he recounts, how – building on his experience as a director of the Rio Tinto company (before it evolved into RTZ) at a time when the Rothschilds held half the capital of Rio Tinto – he used SLN to recapitalise the family firm, Pennaroya. When half of SLN was sold to Aquitaine (later itself to be absorbed into Elf Aquitaine), Rothschild formed Imétal from the resulting new capital.

He also describes how, in its quest to become a truly multinational enterprise, Imétal stalked Copperweld in the USA (for its manufacturing and marketing prowess), bought up Mokta ("practically the entire capital of the largest private producer of uranium which we had been instrumental in creating years before") and took over 25% of the Lead Industries Group in Britain.

"What was the final balance of those hectic twenty years?" asked Rothschild, just before Imétal descended the slippery slope to becoming a shadow of its former self.

"Today I find it doubly sad to see Imétal, now government-controlled, staggering under the weight of the slump. All that remains for me is the remembrance of our efforts, our difficulties, our dreams..." (30).

References: (1) *MJ* 5/2/82. (2) *Congressional Record – extensions of remarks*, Washington, 4/3/75. (3) *MJ* 12/10/84; 22/3/85. (4) Lanning & Mueller. (5) *Gua* 17/11/81. (6) *E&MJ* 3/81. (7) *WDMNE*. (8) *FT* 21/5/80. (9) *MIY* 1985. (10) *MM* 11/81. (11) *RMR*, Vol. 1, No. 4, 1983. (12) *MJ* 6/11/81. (13) *FT* 4/3/82. (14) *FT* 17/2/83. (15) *FT* 7/5/83. (16) *FT* 7/10/83. (17) *FT* 26/1/85. (18) *FT* 8/2/85. (19) *MJ* 10/5/85. (20) *MAR* 1985. (21) *MJ* 15/7/83; *E&MJ* 8/83. (22) *MJ* 27/7/84. (23) *Transnational corporations and other foreign interests operating in Namibia*, Paris, 4/84 (UN). (24) *FT Min* 1991. (25) *E&MJ International Directory of Mining*, 1991. (26) *MJ*

22/4/86, *FT* 21/4/88. (27) US Dept of the Interior, Bureau of Mines, *Minerals Yearbook*, vol 3, 1988, Washington 1990. (28) *FT* 23/5/90. (29) *MAR* 1990. (30) Guy de Rothschild, *The Whims of Fortune*, Granada, 1985.

316 Imperial Metals Corp

It is a "natural resource" exploration and development company with interests in base and precious metals, oil and gas, uranium and coal. In October 1984 it planned to revive an old copper mine in Anglesey, Wales (1). A 250,000 tons/year mining and milling complex was planned (2).

In March 1984 it acquired E & B Canada Resources, more than doubling its own assets and allying itself with the Geomex partnership based in West Germany (3).

The company planned 43 projects during 1985, and one of these – the Big Red uranium property in Nebraska, USA (IMC 30%) – revealed encouraging uranium mineralisations during late 1984 with grades of up to 2.5kg/ton (1).

In 1985 the company announced uranium mineralisation at its Close Lake JV property in northern Saskatchewan. Partners in the Close Lake venture are Cogéma Canada, Uranerz Exploration and Mining, the SMDC, and Geomex Minerals (4).

In 1986, IMC took over Geomex Development Inc and Geomex Minerals Inc from the Sedimex Drilling Funds subsidiary of Sedimex Uranerschliessungs-GmbH, increasing the number of drilling funds partnerships under IMC's control to 14 – and its control over uranium projects within the Sedimex Drilling Funds group (5).

Although the same year it sold its 19% interest in Close Lake to an unknown partner (6) and disposed of 10% of its interest in the Wheeler River uranium JV in Saskatchewan (maintaining a 10% control) (6), it proceeded with pilot drilling at Crow Butte, Nevada (where it holds a 25% interest) along with Ferret Exploration – another company associated with Geomex/Sedimex. Over 8500lbs of U_3O_8 was recovered in 1986, and the "outstanding" results (7) encouraged the company, together with Ferret, to go for a commercial operation in 1989 (8).

Anglesey Mining (IMC:25.5%) continued its "wide-ranging search for minerals" both in Wales and several European countries during the late 1980s and early 1990s (9).

However its main hopes were pinned on Parys mountain, in Anglesey, north Wales, where the company hoped that the $37M development of an important polymetallic deposit would lead to commercial production in 1992 (9).

References: (1) *MJ* 5/10/84. (2) *FT* 6/10/84. (3) *MJ* 19/10/84. (4) *MJ* 3/5/85. (5) *MIY* 1987. (2) *MJ* 28/12/86. (7) *MJ* 26/12/86. (8) *E&MJ* 4/88. (9) *MJ* 13/7/90.

317 Imperial Oil

Imperial Oil, Canada's largest oil company and part of the world's most massive corporation, Exxon, has delegated most of its uranium and other mineral exploration to its three Esso subsidiaries. In 1976, however, Imperial Oil itself announced it had discovered "promising mineralisation" in the Dismal Lakes area north-east of Great Bear Lake, in the Inuit territory of Canada's Northwest Territories; exploration in the area was expected "to increase" (1).

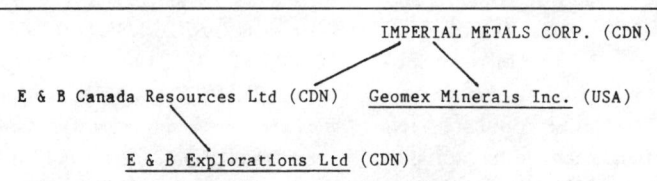

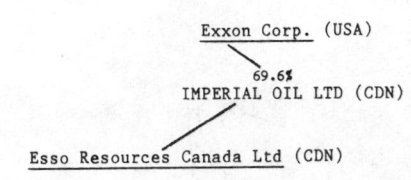

Exxon Corp. (USA)

69.6%

IMPERIAL OIL LTD (CDN)

Esso Resources Canada Ltd (CDN)

Ten years later, the company was cutting its workforce by 1800 (12%), and re-examining its controversial oil sands project at Cold Lake, Alberta – an enterprise which has come in for heavy criticism from (among others) the Cree of Treaty Council No 6, traditional occupiers of the territory (2).

Three years later, Imperial acquired Texaco Canada (3), and, shortly afterwards, sold its 22.8% stake in Interhome Energy to Gulf Canada, controlled by the Reichmann brothers through Olympia and York (*qv*). This gave the Reichmanns majority control of a major pipeline operating company. The Texaco deal left Imperial unchallenged as Canada's largest integrated oil company (4).

References: (1) Merlin. (2) *FT* 22/3/86. (3) *FT* 24/2/89. (4) *FT* 23/4/90.

318 Inalruco SA (RA)

Along with Minera Sierra Pintada, Alianza Petrolera Argentina, Sasetru SA, and Pechiney Ugine Kuhlmann, it is developing the Sierra Pintada orebody in Mendoza province of Argentina (1); see Minera Sierra.

References: (1) *MM* 1/80; *Nucleonics Week* 15/5/80.

319 Inco

Canada's largest mining corporation (1) and the world's leading supplier of nickel, Inco has strikingly concentrated its projects in neo-colonised areas of the Third World where indigenous peoples' labour and lands can be exploited. A truly "integrated" company, it mines, pro-

cesses and turns out finished products – it has also diversified into other metals including uranium. Announcing this diversification in 1980, company chairman Charles Baird said: "Uranium is a good example, but we also have early stage exploration projects in tungsten, chromium, gold and silver" (2).

Inco's entire history has been marked by controversy and opportunism. Founded in Ohio as the Canadian Copper Company, it had close ties with the US military (3). By the 1920s, after becoming the International Nickel Company, it controlled 90% of the world nickel market – and "Canadianised" in 1928 to escape anti-trust laws in the USA (3).

It allegedly supplied *both* sides in World War One and had a key role in arming the Nazis and the Japanese, too (3, 4).

Its most controversial project until the late '70s was its nickel mine (owned 20% by Hanna Mining) in Guatemala, which became not only the largest single foreign investment in this fascist military state but "the most significant industrial venture in all of Central America" (4).

Exmibal was granted mining rights to more than 150 square miles near Lake Izabal in north-eastern Guatemala in 1965. Construction of the project was "essentially complete in 1977" (5), at which point the Latin American Working Group (Toronto) delivered the following assessment:

"By political and economic repression, the military-backed government hopes to guarantee INCO high profits and itself increased prestige. In addition, INCO has arranged a complex financing scheme for Exmibal should the Guatemalan generals ever fall from power. The project is partially financed by the US Export-Import Bank (Exim), the International Finance Corporation (an arm of the World Bank), the Export Credit Guarantee Department of the British government, several prominent private banks and Canada's own Export Development Corporation ..." (4).

Notwithstanding this, and due to the slump in the nickel market, INCO was forced to close the Guatemala mine in late 1981 (6). Just before it went into "moth-balls", its production

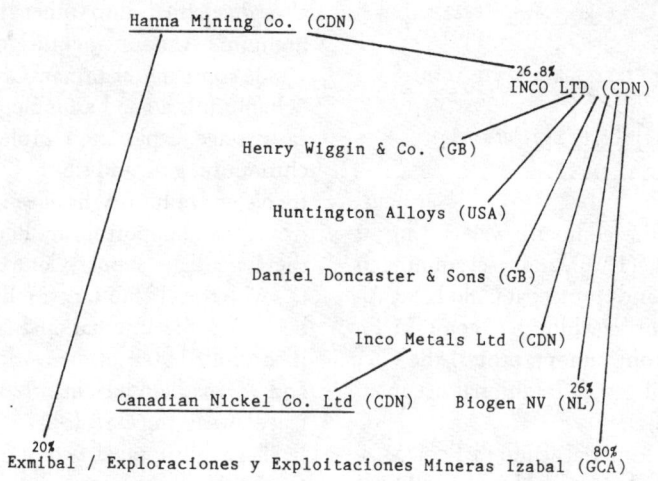

Hanna Mining Co. (CDN)

26.8%
INCO LTD (CDN)

Henry Wiggin & Co. (GB)

Huntington Alloys (USA)

Daniel Doncaster & Sons (GB)

Inco Metals Ltd (CDN)

26%
Canadian Nickel Co. Ltd (CDN)   Biogen NV (NL)

20%   80%
Exmibal / Exploraciones y Explotaciones Mineras Izabal (GCA)

was shipped for refining to the INCO plant at Clydach in Wales (Cymru) (7).

The company also slowed down development of nickel ores on the Indonesian island of Sulawesi (where its equity in INCO Indonesia was to become 60% as Indonesia took over a minority of the shares) (7).

INCO operates a number of copper mines in Canada, and a JV with Homestake Mining to explore for and develop copper deposits in the Keweenaw Peninsula of Michigan, USA (7).

Its wholly owned Electro-Energy Corp produced the well-known Ray-O-Vac batteries (5) until INCO sold this "loss-making" company in early 1983 (8) to an (un-named) private US company (24).

It has several operations in Britain: a carbonyl nickel refinery at Clydach, Wales; a precious metal refinery at Acton, London; and a rolling mill (through its subsidiary Henry Wiggin & Co) at Hereford (18). In the USA, its subsidiary Huntington Alloys has a rolling mill at Huntington, West Virginia.

In 1973, it entered a JV with West German companies in the Brazilian state of Goiás, at Barro Alto, to develop a nickel deposit: this project is not due for development at present (7).

It is involved in two major projects which threaten indigenous peoples. In June 1977 INCO signed a mining convention with the French occupation régime of New Caledonia (Kanaky) and embarked on a five-year work-study programme (7).

Paribas (Banque de Paris et des Pays-Bas), a major French financial institution which has now been nationalised by the French government, also shares interests (22.5%) with INCO (55%) in the Tiebaghi chromium project in New Caledonia (8, 9, 10).

It should be stressed that INCO's move into the Third World in the early '70s was primarily due to constraints placed on its operations at Sudbury, northern Ontario, site of the world's lowest cost and largest nickel orebody which had not even been fully delineated after 80 years operations. Roderick Oram explains it thus:

"It had an autocratic relationship with its customers and was unable, or unwilling, to expand production fast enough to satisfy steadily rising demand. Thus prices rose, attracting new competitors, some of which were government-subsidised" (11).

But perhaps the main reason for its move overseas was the increasing checks on its pollution by the provincial government, and its long history of appalling treatment of workers. While the Sudbury smelter's capacity is 340 million pounds a year, the government's sulphur dioxide emission rules limit production to 280 million pounds (11).

430

The company started exploring for uranium in Newfoundland in the late '70s (12), as well as in New Brunswick, Canada.

In 1979, together with Canadian Occidental Petroleum, it made a "significant" uranium discovery at Wollaston Lake, northern Saskatchewan (13). A year later, further commercially viable (at that time) deposits were discovered in the same area, and C$2 million was set aside by the partners for drilling (14). By early 1981 the JV estimated that 6000 tons of uranium oxide were to be found on the property (15).

At the end of 1982, the company was told by the Manitoba Clean Environment Commission to reduce by half the sulphur dioxide emissions from its Manitoba nickel division (where its Thompson mine is in operation). The company claimed it was "studying" the report and that it had introduced a new low-sulphur process at its Sudbury smelter in northern Ontario (16).

INCO is still involved in petrol exploration in tortured Guatemala (despite closing the Eximbal operation), and is part of one of the deep sea-bed consortia intent on pillaging nodules from the sea-bed (17).

It's abundantly clear that the company has not changed its basic motivation and *modus operandi* in a hundred years.

Nickel empire

INCO built its empire on nickel: we should remind ourselves that, in 1976, the International Metalworkers Federation stated that it was "... almost impossible to overstate the health hazards associated with the production of nickel". Among the destructive effects are cancer, silicosis, lung fibrosis, heart disease, skin and eye diseases (19).

Formed in 1902, INCO was the channel through which JP Morgan (of Morgan Garanty), who'd organised the US Steel Trust, penetrated Canada from the USA. As John Deverell put it, this was "... not a mere change between American companies. It dashed the dreams of Canadian capitalists and the Ontario government for a powerful North American nickel-steel industry based in Canada. While

the New Caledonian ores alone could not defeat the Canadians, the alliance through INCO of Orford refining facilities and stockpiles, US steelmakers, and Morgan control of capital markets, made their vision impossible ..." (20).

By 1913 INCO controlled no less than 55% of world nickel production (21) – 40% of INCO's production going to the German armaments business (22).

How far INCO is controlled by Canadian interests continues, after nearly a century, to be a point of debate. According to Wallace Clement, in 1975 INCO was "70% controlled outside Canada, with over 50% of total ownership located in the USA" (23).

Because of the "depressed" nickel market, INCO has cut its work force by no less than 22% (or 7400 jobs) since early 1981. During this period, as well as closing the Eximbal plant in Guatemala, the company also closed its operations in Ontario, and cut 80 jobs at its Clydach (Wales) refinery (25).

Start-up in May 1984 of the company's large cobalt refinery at Port Colborne established INCO as "an important North American source for this strategic metal" (26).

In 1984, a Canadian government report projected that modification of Inco's nickel and copper plant to meet new environmental standards, including reduced sulphur dioxide emissions would require capital investment of around C$1.1M (27).

Inco has, without question, been Canada's largest producer of the notorious "acid rain" (sulphur dioxide and NO_x emissions). Its copper and nickel smelter at Sudbury emitted in 1979 a total of 866,000 tonnes, "a phenomenal amount, roughly equivalent to the total produced by a small country such as Belgium". In the mid-'60s, the plant was belching out "an astounding 1,800,000 tonnes of sulphur dioxide a year". Legislation cut the level until, in January 1984, it was down to 711,000 tonnes, with planned new burner technology intended to reduce the output to 456,000 tonnes/year (28).

In 1984, however, it seemed as if even greater controls on sulphur dioxide emissions might strap Inco – and other nickel and copper smelter owners – completely. A Canadian federal government report concluded that plant modernisation would need capital investment of about C$1.1M (29).

By the end of 1980, Inco had moved solidly into the genetic engineering field: its largest single venture capital investment that year ($6 million) was in Biogen NV, involved in recombinant DNA research (30), biosynthesis of the anti-cancer drug Interferon (31), and the use of "superbugs" in mining, refining and pollution control (32). Inco has also invested in two small bio-genetic research companies, in one of which – Cetus – it has a JV with Socal (32).

Between 1989 and 1991, Inco expanded nickel production at Sudbury (33) and in Manitoba (34). It merged its gold subsidiary with TVX and ended with 62% of the new company (35). TVX then took over (among other projects) Inco's share of the Crixas mine in Brazil (operated by Anglo American) and in the Brasilia mine (operated by an affiliate of RTZ) (36).

Its main challenge remained conforming to the Canadian sulphur abatement programme, while maintaining its lead in the world nickel industry.

At the end of 1988, the company submitted its final report to the Ontario Ministry of the Environment setting out how it would achieve this end.

Inco promised to spend nearly C$500M to modify its milling and concentration operations and reform its smelters: the aim – to reduce the current level of SO_2 emissions from 685,000 tonnes a year to 265,000 by 1994 (37). As of the time this entry was being written, it was not clear whether this target would be met.

Further Reading: *The Big Nickel: Inco at home and abroad,* Jamie Swift and the Development Education Centre, Between the Lines, Kitchener, 1977.

References: (1) *Yellowcake Road, Canadian Mining Journal* 8/90. (2) *FT* 28/5/80. (3) Jamie Swift, *The Big Nickel: INCO at home and abroad,* Toronto (Development Education Centre). (4) Latin American Working Group & DEC in *Last Post* 11/77. (5)

MIY 1981. (6) *FT* 26/1/82. (7) *MM* 1/82. (8) *FT* 2/2/83. (9) *MJ* 5/2/82. (10) *FT* 8/1/81. (11) *FT* 15/3/82. (12) *Energy File*, British Columbia, 4/80. (13) *FT* 5/10/79. (14) *MJ* 9/5/80. (15) *MJ* 2/1/81. (16) *FT* 5/10/82. (17) *MAR* 1982. (18) *MIY* 1982; *FT* 24/7/81. (19) Tanzer. (20) John Deverell, *Falconbridge,*, Canada, 1976. (21) *The Minerals Industry,* 1924 (McGraw-Hill). (22) *Ibid,* quoted in Tanzer. (23) Wallace Clement, *The Canadian Corporate Elite* 1975. (24) *FT* 15/9/82. (25) *MJ* 27/8/82. (26) *MJ* 23/3/84. (27) *MJ* 18/5/84. (28) P Weller, "Industry and Acid Rain: the corporate response" in *Alternatives,* Canada, Winter/83. (29) *FT* 15/5/84. (30) *Inco Annual Report 1980.* (31) *Business Week* 9/6/80. (32) *Wall St J* 24/6/80. (33) *FT* 7/12/89. (34) *FT* 30/10/90, *MJ* 2/11/90. (35) *MJ* 7/12/90. (36) *E&MJ* 2/91. (37) *MJ* 23/2/90, *MM* 11/89.

320 Inexco Oil (USA)

It was formerly a JV partner with Uranerzbergbau GmbH (Germany) and the SMDC, exploring at Key Lake, Lake Athabasca (Saskatchewan) and on properties in Alberta, all in Canada. Then the Saskatchewan Mining Development Corp bought out its interests in 1978 (1), after the Canadian government blocked the sale to Denison, in order to maintain its predominant share in new uranium mining operations (67% required, since 1970) (2).

References: (1) *MM* 7/78. (2) *Business Week* (USA) 29/5/78.

321 Intercontinental Energy Corp (USA)

It participated, with Baltimore Gas and Electric, from 1974 until at least 1979, in uranium exploration (1). No known production.

During the 1980s, Intercontinental acquired the Lamprecht property of Wyoming Mineral Corp, (which had been closed in 1982) and reopened it. Its production was scheduled to increase in 1988 (2).

References: (1) Sullivan & Riedel. (2) *E&MJ* 4/88.

322 International Energy Corp (USA)

It has interests in oil and gas both in the USA and offshore Italy.

In the USA, it held properties on Lakota land at Bear Butte, site of recent attempts by the US government to take back land claimed by the traditional owners (1).

It was producing uranium at the Zamzow mine in Live Oak Co, southern Texas, for several years, and then tried to acquire additional uranium reserves in the same area, against the day when the original orebody would be exhausted (c 1985). In December 1979, IEC contracted to supply all the uranium from Zamzow to Pacific Gas and Electric for a minimum price of US$40/lb. From June 1978 until September 1981 about 673,000lb of uranium was produced, making it a comparatively minor mine (2).

IEC's select management has included corporate figures with previous experience in Mobil, Freeport, Conoco, Coseka, Arco, and Gulf Oil (3).

Its JV partners in recent years have included Rheinbraun, Total, and Agip (in Italy) (3).

References: (1) Communications from the Lakota Treaty Council, Pine Ridge Reservation, South Dakota, USA. (2) 1981 annual report. (3) Summary of operations from IEC, 1/82.

323 International Mining Corp (AUS)

This is an "energy resource" company, into oil shales, coal, and prospecting for metals including uranium. In 1982 it had several prospects in the Northern Territory (1), including one at Rum Jungle (see RTZ) with its JV partner Marathon Petroleum (50%). In early 1982, International Mining was contemplating reducing its equal participation in this prospect by 2%.

New areas were reported applied for by the JV (2). But in late 1982, Marathon withdrew from Rum Jungle, leaving International Mining in charge (3).

Due to the downturn in the uranium market and the "three mines" policy of the Australian Labour party, International Mining has to all interests and purposes withdrawn from uranium, concentrating instead on gold prospecting (4).

References: (1) *Reg Aus Min* 1981. (2) *MJ* 12/2/82. (3) *MJ* 19/11/82. (4) *Reg Aus Min* 1990/91.

324 International Mogul Mines Ltd

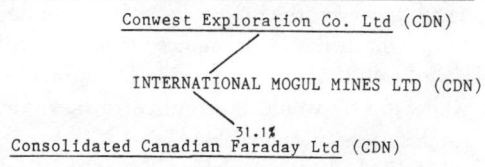

This is a mining exploration company with interests in oil and gas. It holds 4.75% equity interest and 1% gross loyalty interest in the Conwest Canadian Uranium Exploration JV. Conwest and Mogul in 1979 agreed to share oil and gas expenditures on an equal basis (1).

It wholly owns IMM Ventures.

It was also exploring for uranium in Saskatchewan (2, 3).

International Mogul Mines also holds an 18% interest – along with Conwest Exploration (30%), Chimo Gold Mines (27%), and Consolidated Canadian Faraday (25%) – in the Conwest mineral exploration programme (4).

Chimo Gold Mines and Central Patricia were amalgmated with Conwest Exploration and International Mogul in 1982 (5).

References: (1) *MIY* 1981. (2) *Who's Who Sask.* (3) *Yellowcake Road.* (4) *MIY* 1982. (5) *MIY* 1985.

325 IOL Petroleum Ltd

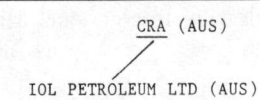

```
        CRA (AUS)
       /
IOL PETROLEUM LTD (AUS)
```

It had a uranium holding at Honeymoon, South Australia, which was sold to AAR.

IOL is also part (9.75%) of a consortium exploring for uranium in northern Queensland at the Westmoreland prospect – a venture which has been vigorously opposed by the North Queensland Land Council (NQLC) – together with Queensland Mines (who hold 40% of the JV); other partners are Urangesellschaft (37.5%) and AAR (12.75%) (1). By the end of 1982, however, the project looked like coming to nought as the JV announced it had located only 11,400 tons of uranium at Westmoreland while 15,000 would be required to make the project economically viable (2).

In 1988, IOL was sold by CRA to the Australian Petroleum Fund which was later acquired by Westmex Ltd. In February 1990, Westmex' directors themselves decided to liquidate the Fund (3).

References: (1) Roberts. (2) *MJ* 17/3/82. (3) *Reg Aus Min* 1990/91.

326 IPEN

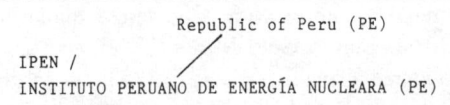

```
              Republic of Peru (PE)
IPEN /              /
INSTITUTO PERUANO DE ENERGÍA NUCLEARA (PE)
```

In mid-1985 this state agency asked for bids to evaluate and develop a high-grade uranium deposit on the banks of the Sangaban river, 200km north of Lake Titicaca, the centre of Andean tourism in the country. United Nations personnel worked on the deposit in 1983, proving reserves of 3400 tons uranium oxide. Total capital expenditure to open a mine is estimated at between US$150 and US$200M (1).

The IPEN in 1990 was constructing a research reactor at its Huarangal research centre, near Lima, with the assistance of Nukem and grants exceeding US$68M from the Argentinian régime – the first example of a South American international nuclear co-operation agreement (2).

References: (1) *MJ* 12/7/85. (2) *South*, London, 12/80.

327 IRC

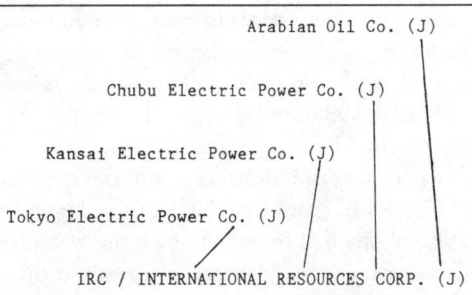

```
                Arabian Oil Co. (J)

    Chubu Electric Power Co. (J)

  Kansai Electric Power Co. (J)

Tokyo Electric Power Co. (J)
                /
    IRC / INTERNATIONAL RESOURCES CORP. (J)
```

Together with Onarem, IRC has been carrying out feasibility studies at the relatively small Abkorun-Azelik uranium deposit (1), to the southwest of the big Imouraren deposit of Niger. Reserves are estimated at 13,500 tons, and exploitation was originally planned for 1983 (2). In 1981 the start-up date was put back to 1986 (3). It was the CEA which discovered Azelik in 1970; IRC joined in a 50/50 investigation of its potential with Onarem in 1977 (4).

IRC's parent, Arabian Oil, is the operator of Japanese petroleum concessions in the Neutral Zone between Kuwait and Saudi Arabia, and was in 1978 also reported to be exploring for uranium in Algeria (5).

In mid-1986, the company was reported to be about to resume uranium exploration in Niger, carrying out 55 drillings in the Air district, and continuing prospecting until the end of 1988 (6).

In 1990, economically recoverable reserves at the deposit were put at 8000 tonnes, with another 1700 tonnes potentially exploitable (7).

References: (1) *MAR* 1984. (2) Lynch & Neff. (3) *Economist Intelligence Unit Niger 4081*, p23. (4) *MAR* 1979. (5) *EPRI Report* 1978, pp5-15. (6) *MJ* 26/7/85 and *MJ* 6/6/86. (7) *MAR* 1990.

328 Irish Base Metals Ltd

```
Northgate Exploration Ltd (CDN)
            \
     IRISH BASE METALS LTD (IRL)
```

It was drilling for uranium at Thomastown, Co Kilkenny, Eire, in the late '70s – results not known (1), though in September 1979 the company claimed "encouraging evidence of uranium" in Donegal (2). At the time it was also interested in uranium and other base metals on the Carlingford Peninsula, Co Louth, and elsewhere in Eire (3). It was reported pulling out of Tynagh (where the company closed its silver mine) in mid-1980 (4). In Co Donegal, on November 27, 1981, it was announced that Irish Base had withdrawn because of the declining uranium market.

One of Irish Base Metals's most important recent prospects has been at Charlestown, Co Mayo, where it was investigating a three million-ton copper deposit in 1982 (5).

Contact: Just Books, 7 Winetavern Street, Belfast, BT1 1JQ, Northern Ireland.

Ralph Sheppard, Carnowen House, Raphoe, Co Donegal, Eire.

References: (1) Just Books personal communication, 27/10/79. (2) *MM* 11/79. (3) Cork Anti-Nuclear Group, 1980. (4) *Kiitg* (quoting *Outta Control*) 8-9/80. (5) *MM* 11/82.

329 Jacobs Engineering Group (USA)

This major engineering and mine construction group is of particular importance to the US uranium industry: Jacobs has been responsible for constructing, or part-constructing, no less than a third of America's uranium projects (1). It also has a projects office in Dublin, Eire.

References: (1) *MM* 3/83.

330 JCI

"To all intents and purposes a division of Anglo-American Corporation" with important minority stakes in De Beers diamond trading companies and a vehicle through which Anglo-American controls its food, beer and press outlets, JCI has a limited exploration programme and a traditional "programme" of its own investments (1).

Formed in 1889 by Barney Barnato – a comrade-in-harm of the archetype colonialist Cecil Rhodes – it was administered from London until 1963 (2). In 1940, Anglo-American bought out the interests of one of founding families of "Johnnies", the Joels, and increased its stake in the company to 50%, "to prevent a take-over by Afrikaner interests" (2). As well as being ruled by Anglo-American – Ernest Oppenheimer was one of the company's original shareholders – Johnnies also has investments in De Beers.

In the 1920s, JCI joined with Rio Tinto (later to become RTZ), the British South Africa Company, and Union Corporation (later to be merged with General Mining into Gencor) to form RHOANGLO and take over the rich copper belt of Northern Rhodesia as a deterrent to American interests trying to muscle in at the time (2).

Its most important single investment is Rustenburg Platinum, the world's largest producer of this precious metal, a company which JCI also manages.

It also has interests in antimony (through Consolidated Murchison), nickel (Shangani Mining Corp Ltd), coal (Tavistock Collieries), and industrial and property holdings (3).

The large copper pyrite Otjihase Mining Co in Namibia was originally owned 67% by JCI but is now controlled by Tsumeb, with JCI and

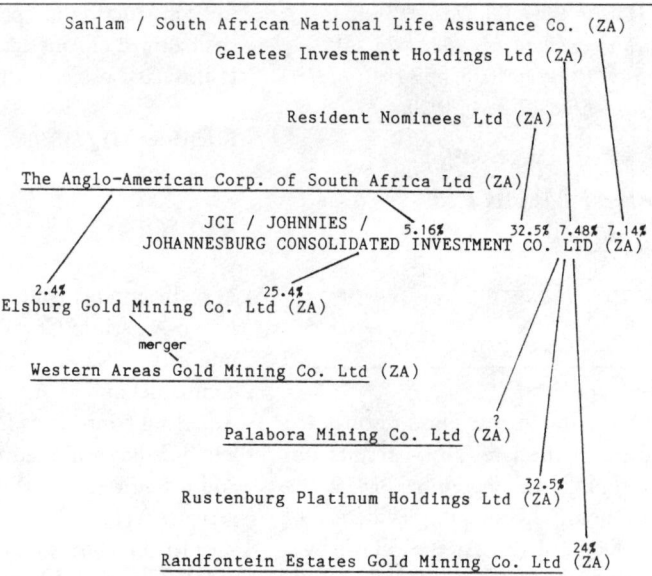

```
        Sanlam / South African National Life Assurance Co. (ZA)
                  Geletes Investment Holdings Ltd (ZA)

                           Resident Nominees Ltd (ZA)

        The Anglo-American Corp. of South Africa Ltd (ZA)
                    JCI / JOHNNIES /        5.16%        32.5% 7.48% 7.14%
              JOHANNESBURG CONSOLIDATED INVESTMENT CO. LTD (ZA)

              2.4%                  25.4%
        Elsburg Gold Mining Co. Ltd (ZA)

                        merger
        Western Areas Gold Mining Co. Ltd (ZA)

                                          ?
                  Palabora Mining Co. Ltd (ZA)

                                                  32.5%
            Rustenburg Platinum Holdings Ltd (ZA)

                                                      24%
        Randfontein Estates Gold Mining Co. Ltd (ZA)
```

Federale Volksbeleggings owning 30% of the shares (4).

The company has also explored for uranium in Namibia but no positive results have been reported (5).

Uranium production by JCI is confined to its two South African gold mines, Randfontein and Western Areas.

In 1987, Gordon Waddell announced his resignation of JCI's chair, despairing, as he put it, that "... business people here don't realise how clearly private capitalism is perceived to go hand in hand with apartheid by the majority of people in this country". Expressing his fears both for a future South Africa dominated by the National Party and one under black majority rule, Waddell firmly joined the "chicken run" (South African slang for the sinking ship). A black trade unionist was more circumspect: "It's all very well for Waddell to talk," he commented, "but he's no different from the others. Where are JCI's black directors? Where are its black managers? There aren't any" (6).

At this time, JCI was not only one of the apartheid state's most prominent and boyant companies, it also held the strategic diamond trading interest of De Beers; managed the world's biggest non-commercial platinum producer –

Rustenberg – and had controlling interests in the two largest English-language newspaper groups, Argus and South African Associated Newspapers (6).

Within a few years, JCI was boasting a new gold lode, next to the Western Areas mine, which it claimed was "probably the largest and most important known gold orebody remaining to be exploited" (7). The South Deep mine has an estimated life of some 40 years and, with the closure of part of Western Areas older operations in 1990 (8), is destined to be the most important new gold mine in the country over the next few years. Meanwhile JCI's other gold operations have not fared too badly either (9), while its uranium operations (thanks to a revised contract in early 1990) have not suffered quite the same setback as similar operations in the apartheid state (10).

Although platinum prices experienced a five-year low in early 1991 (11), Rustenberg's profits have remained high, with development of two new platinum projects still on the cards: the Maandagshoek property (deferred indefinitely in 1989) (12) and the Potgietersrust mine, owned jointly by Rustenberg and Lebowa Platinum Mines (in which Rustenberg holds 21.5%) (13).

In 1990, 1200 workers were sacked by JCI from the Atok mine operated by Lebowa, after "illegal" strikes (14). Six years before, riots in the Western Areas mine, caused by confrontations between black workers and police after the NUM-led industrial action, resulted in the deaths of seven miners who were demanding their employment rights, as guaranteed by the law (15). Ironically, Gordon Waddell that very week had lambasted the South African régime for "financial indiscipline" (15).

In 1990, a JCI subsidiary, Consolidated Metallurgical Industries, acquired Purity Chrome, which owns a chrome mine and plant in the Rustenberg area (16). However, two months later, JCI withdrew from a possible JV with Dublin-based Kenmare Resources in the large Congolene mineral sands project, Mozambique, citing the lack of support from the Credit Insurance Corporation of South Africa (17).

Freddies, the JCI unit, was restructured in 1986, with the formation of two new outfits, DAB Investments (Dabi) and Freddev – which holds the right to exploit nearly 9000 hectares of holdings in the Orange Free State, Transvaal, northern Natal and the Northern Cape, together with its subsidiary Southern Holdings (18).

As of mid-1989, Johnnies was ostensibly held 32.57% by Resident Nominees (Anglo-American), 7.48% by Galares Investment Holdings, 7.87% by South African Mutual Life Assurance Society and 5.16% directly by the Anglo-American Corp (18).

Further reading: For a history of black miners' actions against oppression see Mike Murphy, *Trade Unions Under Apartheid*, WEA Studies for Trade Unionists, Vol. 10, No. 39, 9/84.

References: (1) *MAR* 1984. (2) Lanning & Mueller. (3) *MIY* 1983. (4) UN NS-36. (5) *MM* 7/77. (6) *FT* 24/1/87. (7) *FT* 8/10/90. (8) *FT* 21-22/7/90. (9) *FT* 16/1/90, *FT* 18/4/90. (10) *MJ* 20/4/90. (11) *FT* 29/1/91. (12) *MJ* 10/11/89. (13) *MJ* 5/10/90, *MJ* 1/2/91. (14) *MJ* 23/11/90, see also *MJ* 25/5/90.
(15) *MJ* 21/9/84. (16) *FT* 4/9/90. (17) *FT* 9/11/90. (18) *FT Min* 1991.

331 JC Stephen Exploration Ltd (CDN)

It held a licence for uranium exploration in the Omineca Mining Division of British Columbia in 1979 (1).
Now presumed relinquished.

References: (1) *Energy File* 3/79.

332 Jimberlana Minerals NL

The company is involved in two uranium prospects in the Northern Territory, near Darwin and adjacent to the infamous Rum Jungle uranium province. Its partners are Pan D'or (in both cases) and Aquitaine at Mount Bundey (1). Though the Mount Bundey prospect is located near the old Rum Jungle uranium zone, by early 1981 little economic uranium mineralisation had been discovered (2).

It also has interests in an exploration prospect in Irian Jaya (West Irian) (2); and was involved in a gold mining project in South Africa, suspended however in 1982 (3), revived again in 1983 (4).

During the latter half of the 'eighties, Jimberlana made a major arrangement with Pamour Inc of Canada, by which it acquired control of the gold mining company (formerly in Noranda's hands) and, together with Pamour, nearly 20% of Yellowknife Gold Mines Ltd (5). The tie-up with Pamour brought Jimberlana into the second most extensive gold tailings treatment scheme in the world (ERG) in Canada, and other gold projects in North America. Jimberlana also secured 20 million shares in West Witwatersrand Gold Holdings of South Africa (22%), 9.7% in Dickenson Mines Ltd, and 8.2% in another Canadian gold producer, Kam-Kotia mines Ltd (6). Through its subsidiary PT Sungai Kencana, extensive exploitation has been taking place on the Mandor Con-

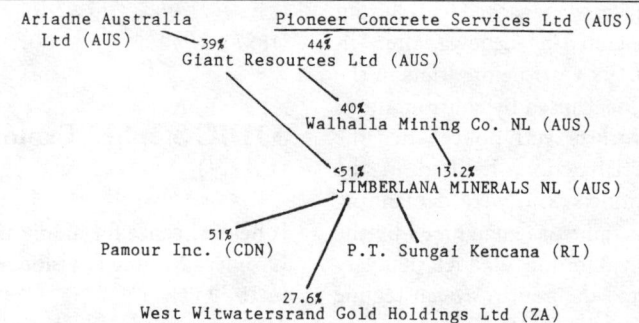

Ariadne Australia
Ltd (AUS)   Pioneer Concrete Services Ltd (AUS)
                39%   44%
            Giant Resources Ltd (AUS)

                            40%
                    Walhalla Mining Co. NL (AUS)

                    <51%   13.2%
                    JIMBERLANA MINERALS NL (AUS)

                51%
        Pamour Inc. (CDN)   P.T. Sungai Kencana (RI)

                27.6%
        West Witwatersrand Gold Holdings Ltd (ZA)

tract of Work (COW) in indigenous Kalimantan (Indonesia), where Duval Alluvial has control (50.01%) of gold dredging and hard rock
projects (5).

However, in late 1989, Jimberlana suspended
all operations in its shares, as the company tried
to secure a Canadian buyer – the Frame Mining
Corp. The deal fell through and, at the time of
writing, the future of Jimberlana itself is in
doubt (6).

References: (1) Annual Report 1980. (2) *Reg Aus
Min* 1981. (3) *FT* 19/3/82. (4) *FT* 4/2/83. (5) *FT
Min* 1991. (6) *Reg Aus Min* 1990/91.

333 Jingellic Minerals NL (AUS)

A gold, tin, oil and diamond exploration company (1) which, in the 1970s, held a uranium
prospecting licence in the Northern Territory,
along with Peko-Wallsend/EZ Industries
(15%) and Australian Anglo-American Ltd –
who sold out their 50% interest around 1973
(2).

No known work has proceeded in recent years
in the licence area.

However, during the 'eighties, Jingellic became
one of the more active Australian companies involved in gold exploration and production,
through a large number of JVs in Australia, Indonesia and the Philippines.

It holds 60% of the Nevoria JV with Southern
Goldfields Ltd, at Marvel loch near Southern
Cross, Western Australia (WA) – a mine which
has produced around 30,000 ounces of fine

gold each year for the last three years. (Both Jin-
gellic and Southern Goldfields, hold JV leases
with CRA in WA) (3). Consolidated Gold
Mining Areas NL (through both its subsidiary
Hamilton Resources Pty Ltd and itself) holds
35.61% of Jingellic, and is also involved with
Jingellic in a JV in WA. (Consolidated Gold
Mining Areas NL has a gold prospect in East
Kalimantan at Sungai Kelai.) Other Indonesian
COWs (contracts of work) acquired of late by
Jingellic include Kerinci (in western Sumatra),
at Jayapura and Yapen Island in West Papua,
and Maros and Palu in Sumatra (3, 4).

While Jingellic has other leases in Australia and
interests in Singapore and Thailand, as well as a
30% interest in a JV with Asarco near Wiluna,
WA, it is most notable for the fact that –
together with its 35.09%-owned associate Enterprise Gold Mines NL – it holds a 35.3% interest in Benguet Exploration of the Philippines
(not to be confused with Benguet Consolidated). Benguet Exploration controls the
Thanksgiving gold mine, the Palayan bentonite
operations and Ampucao gold prospect in
Luzon, and gold exploration leases at Boringot
and Kingking on Mindanao (5).

References: (1) *Reg Aus Min* 1981. (2) *A Slow Burn*
(FoE/Victoria) No. 1, 9/74. (3) *FT Min* 1991. (4)
See also *Reg Aus Min* 1990/91. (5) *Reg Aus Min*
1990/91.

334 JMS

Pipped at the post (by more than a year) by

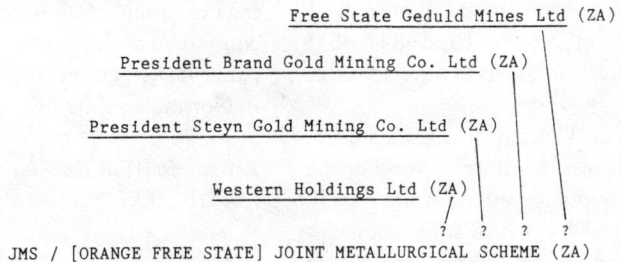

```
                              Free State Geduld Mines Ltd (ZA)

                   President Brand Gold Mining Co. Ltd (ZA)

                 President Steyn Gold Mining Co. Ltd (ZA)

                       Western Holdings Ltd (ZA)
                                               /    /   /   /
                                               ?    ?   ?   ?
          JMS / [ORANGE FREE STATE] JOINT METALLURGICAL SCHEME (ZA)
```

ERGO as South Africa's first uranium-from-slimes extraction project, the JMS is nonetheless the biggest of the three companies which treat gold mine tailings for yellowcake. In 1984 JMS produced nearly 600,000kg uranium (1), and in the first six months of 1985 just over 300,000kg (303,764, to be exact) (2). Its output was clearly slipping over that period (it was 952 tons in 1982) (3), and indeed, only Chemwes among the three slimes producers has recently shown any optimism about the future (1).

The "unique" (4) joint scheme came on stream in 1977, designed to recover uranium, pyrite and sulphur from slimes produced by Anglo-American's Orange Free State mines – namely Free State Geduld, Free State Saaiplaas, President Brand, President Steyn, Welkom, and Western Holdings. According to researchers Lynch and Neff, the JMS "represented a revival and expansion of the President Brand uranium plant, which was moth-balled in 1971 due to the weak state of the market" (5).

At the commencement of the project, Free State Saaiplaas held the biggest accumulation of treatable tailings, followed by Welkom (4). Four years later, however, it merged with Western Holdings which also acquired Welkom. Currently, President Brand is the biggest supplier of slimes, followed by Western Holdings, Free State Geduld and President Steyn (2).

The two pyrite flotation plants in the JMS are owned by Free State Geduld and President Steyn, while President Brand owns the uranium plant, a pyrite flotation plant and the acid plant (6).

The uranium plant achieved its design capacity of 1,500,000 tonnes/month slimes in late 1978, when it chalked up a profit of R22.3M (7) after delays due to problems with the pyrite flotation plants at President Barnd (8). In 1984 plans were announced to reduce output "for the time being" due to the state of the market (1).

A 1977 contract between JMS and the Iranian régime of the Shah, to supply 2000 tons of uranium in 1981/85, was lost when the Iranian nuclear programme took a nose-dive. But a contract for 4000 tons with Taiwan was landed in 1981, to supply in 1983/87, and there was speculation at the time that JMS might also be selling to Randfontein (5).

By 1990, South African uranium production was down 15% on the previous year, much of the slip being "attributable to reduced production" at JMS (9).

References: (1) *MAR* 1985. (2) *R&OFSQu* 26/7/85. (3) *RMR*, Vol. 2, No. 4, 1984. (4) *MJ* 23/12/77. (5) Lynch & Neff. (6) *MIY* 1985. (7) *MJ* 22/12/78. (8) *R&OFSQu* 6/5/77. (9) *E&MJ* 3/91.

335 John Brown Engineers & Constructors Ltd

This large British engineering and construction company formed a JV with Charter Consolidated, called Charter-CJB Mineral Services Ltd, to do a feasibility study on the Hoggar uranium deposits in southern Algeria in 1977 (1).

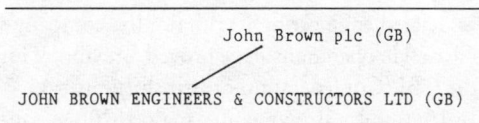

```
                    John Brown plc (GB)
                   /
  JOHN BROWN ENGINEERS & CONSTRUCTORS LTD (GB)
```

The biggest order ever won for a GB oil drill rig was captured by John Brown in late 1984: £66M will now roll in from the coffers of – Kerr McGee (2).

In 1991 the Fiancial Times (Charles Leadbeater) described John Brown's history as "one of the great industrial escapology stories of the past 30 years". The original John Brown had built his company up from a barrow, selling iron goods in Sheffield in 1837, into "one of the most prestigious vertically-integrated industrial groups in the world, encompassing coal, steel and shipbuilding" (31).

However, post-war nationalisation forced the disposal of all those businesses, leaving the eponymous company to develop engineering, construction, boiler-making, and machine tools. In 1965, a link-up with General Electric of the US – whereby John Brown manufactured turbines using GE's technology – gave the company a boost which was to carry it through the next thirteen years, to record £28M profits in 1978. But then "several things went wrong at the same time": the US recession; running short at the bank; and falling inflation (which adversely affected the company's US dollar-held debt). To cap it all, the company got involved in supplying oil pipeline technology to the Soviet Union – a particular bogey for the Reagan administration: "[f]or several years [John Brown's] name was among those in the big books which US customs officials leaf through intimidatingly at airports" (3).

In 1985, Trafalgar House, the construction and shipping group with vast South African interests – its lines were responsible for some of the export of Namibian uranium in defiance of UN law – took a minority stake in John Brown which it later converted into complete control. From then on there has been no looking back, although the "synergy" between JB and Trafalgar House has admittedly been "less ... than we expected" (3) according to Alan Gormly, JB's Chief Executive. Despite the severe cut-backs on shipbuilding and associated engineering in Britain, the company's Clydeside operations have proved a textbook for the CBI (Confederation of British Industry).

In 1990, John Brown won a major £200M contract to supply FGD (flue gas desulphurisation) equipment to the privatised electric power company, Powergen, at its Ratcliffe-on-Soar plant in Nottinghanshire (4)

References: (1) *MM* 8/77. (2) *Gua* 13/12/84. (3) *FT* 23/1/91. (4) *FT* 29/11/90.

336 Johns-Manville Corp

```
Manville Corp. (USA)
        \
JOHNS-MANVILLE CORP. (USA)
```

This is primarily a manufacturing company, but with substantial mineral interests in Iceland, Mexico, USA and Canada. It holds a substantial part of Advocate Mines (1). Its Canadian subsidiary, Johns-Manville Canada, was the western world's largest producer of the notoriously dangerous asbestos, much of which came from Baie Verte in the indigenous part of Newfoundland (2) until it announced closure in late 1981 (3).

Its only known uranium exploration activity is an exploration claim at Lawrence and Meade counties of South Dakota – in the Black Hills – which was valid in 1979/80 but may well have been cancelled since (4).

The company has done a survey of the Ganesh Himal region of Nepal for the Indian House of Birla, where nearly a million tons of ore (zinc and lead) are "readily available" (5). This region is used by hill tribespeople.

The world's largest manufacturer of asbestos products, the company was "completely overwhelmed" by the potential costs of health lawsuits filed against it during the growing movement against asbestos use in 1982 and 1983 (6). The parent body, Manville Corp, filed for protection in 1982 under the US federal bankruptcy code, and took steps to transfer all its non-asbestos business to a new company, so that claims for health damage from asbestos could be made only against the old Manville companies, and therefore would be covered by old insurance and just a portion of any newly-generated cash (7). A few months later, the attorney for the creditors, Bob Rosenberg, called the proposals "unconfirmable, illegal and unconstitutional" (8). By autumn 1983 no fewer than 16,000 litigants had entered claims for damage to health, with 425 new claims each month (9).

The same year Manville sold its Jeffrey Mine – the world's largest asbestos mine – to a group of Montreal businessmen headed by a former employee, Peter Kyle. The mine was being high-graded until 1987 to quickly pay Johns-Manville out of its cash flow (10). In early 1984, a federal bankruptcy court ruled that the company could avoid litigation through reorganisation (11).

In 1988 the Manville Trust was set up, using half the company's equity (and, from 1992, to receive half its profits) in order to pay the US$7.5 billion or more expected to be settled in claims on pending and future asbestosis victims (12). This followed a 1985 agreement by which Johns-Manville would fork out US$42.5 billion for the same purpose, thus saving the company from liquidation through bankruptcy proceedings (13). The company stalled for three years – claiming in 1987 that the US government itself should pick up the tab for those who died as a result of using its asbestos for insulation and fire-proofing materials for government shipyards at the time of the second world war (14).

By 1990, it was clear that the trust was so low on funds that some of the claims would not be paid until well into the 21st century (although Manville's lawyers were receiving exorbitant fees) (15) and only 22,000 of the 170,000 identified victims had been compensated (16). In return for being shielded from future claims (16) there would be special, phased, distribution dividends to Manville shareholders over the following seven years, and the company would increase its holdings in the company's stock to 80% (17).

A few months later, a federal judge approved a proposed class action, to speed up the rate at which victims would receive compensation (18).

Manville derives income from its 50% share (with Chevron Resources Co) in the Stillwater platinum-palladium mine in Montana, which is the world's only producer of platinum group metals outside of South Africa: attempts by the company in 1990 to sell its holding met with little response (19, 20).

It is also one of the two biggest US producers of the insulant, perlite, which it markets in the US, Canada and Britain (19).

We are left with the final irony that, having built its foundations on exploiting one of the most hazardous materials on the planet, Manville is paying compensation to the victims of its operations with profits accrued from two of the most benign materials in the metallurgical swag-bag.

Further reading: *Asbestos: Politics and Economics of a Lethal Product*, Jeffrey Harrod and Victor Thorpe, for the International Federation of Chemical Energy and General Workers' Unions, Geneva, 1984.

References: (1) *MIY* 1982. (2) *MAR* 1981. (3) *MAR* 1982. (4) *BHPS* 4/80. (5) *MAR* 1982. (6) *FT* 25/2/83. (7) *FT* 13/5/83. (8) *FT* 22/11/83. (9) *FT* 4/11/83. (10) *E&MJ* 3/84. (11) *FT* 25/1/84. (12) *FT* 4/6/90. (13) *FT* 9/12/85. (14) *FT* 11/8/87. (15) *FT* 11/7/90. (16) *Independent* 20/11/90. (17) *MJ* 21/9/90, *FT* 8-9/9/90. (18) *FT* 17/5/91. (19) *MAR* 1990. (20) *FT* 21/4/90, *MM* 6/90.

337 John Stanley Mines (USA)

It acquired a 1500-acre uranium prospect three miles north of the Midnite mine on Spokane Indian reservation, Washington State, in 1977 (1).

No further details.

References: (1) *MJ* 21/10/77.

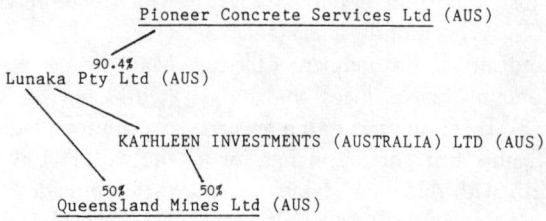

Pioneer Concrete Services Ltd (AUS)

90.4%
Lunaka Pty Ltd (AUS)

KATHLEEN INVESTMENTS (AUSTRALIA) LTD (AUS)

50%   50%
Queensland Mines Ltd (AUS)

338 Kathleen Investments (Australia) Ltd

Set up as a mining investment, processing and marketing company, Kathleen Investments (now effectively controlled by Pioneer and Ampol) owns 50% of the Nabarlek mine through its half-ownership of Queensland Mines (1). Noranda held a 5% interest in Kathleen and 5% in Queensland Mines up to 1982 (2).

Kathleen also had interests in mineral sands (with potentially adverse effects on Aboriginal communities) (3) and owned Cable Sands Pty (4).

Restructuring of Pioneer's interests in the 1980s effectively nullified Kathleen Investments as a separate enterprise, with the result that Nabarlek is now held 100% by Pioneer. Cable Sands Pty Ltd, with large-scale mineral sands dredging operations at Minninup, south-west Western Australia, was sold to the Japanese multinational Nisho Iwai in 1990 (5).

References: (1) *Reg Aus Min* 1976/77. (2) *MIY* 1981, p412. (3) *AMIC* 1982. (4) *MIY* 1982. (5) *Reg Aus Min* 1990/91.

339 Kelvin Energy Ltd (CDN)

The company is possibly controlled by West German interests (1). It was involved in 1978/79 with a then FRG-controlled company, E and B, and Pan Ocean Oil, in a consortium exploring at Great Bear Lake, Northwest Territories (2). Later Kelvin announced a "major geophysical and drilling programme" at Louison Lake and Heron and Beak Lake areas, near-

ly 300km south-west of Yellowknife, with grades of uranium ranging up to 32.22% (3).

It was also in a JV (16.5%) with Saskatchewan Mining and Development (50%), Reserve Oil and Minerals (7.5%), and Asamera Oil (25%), at Keefe Lake-Henday, Saskatchewan (4).

References: (1) *Who's Who Sask.* (2) *MJ* 29/6/79. (3) *MJ* 2/11/79. (4) *MJ* 9/11/79.

340 Kennecott Corporation

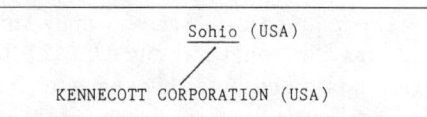

Sohio (USA)

KENNECOTT CORPORATION (USA)

Note: In 1989 RTZ bought out all BP Minerals interests, which included Kennecott. For details of developments since 1989, please see RTZ entry in the *File*.

In 1985, the world's largest mine was closed down at Bingham Canyon, Utah, USA (1), by America's largest copper producer. Naturally, the world copper price went up as a result (2). In the boardroom at Sohio, the managers of Kennecott division wore long faces as they counted the cost: a loss of US$1.6 million on the company's mining operations in 1984 (3) despite a 13% cutback in production levels, and a progressive run-down in the workforce (4). Kennecott president Frank Joklik agreed that the company's losses had "... reached a point where we can no longer sustain the operations of the Utah (Bingham Canyon) copper division" (5). Joklik blamed adverse market conditions, high labour costs and antiquated facilities for the shut-down (5).

Other commentators saw the "demise" of Kennecott as epitomising the problems of North American mining (3) – copper in particular: aggressive competition from imports from Third World producers, the depletion of high-grade ores, increased costs of digging up low-grade deposits (1), and problems with the unions. First indications that the mine would have to be closed came in early 1985, after a meeting between Kennecott and the USA (United Steelworkers of America), when an offer by Cass Alvin, USA spokesperson, of a US$10,000 cut per employee was turned down by the management (6). This stood in marked contrast to negotiations in April 1985, when Kennecott had reached agreement with 14 unions on economic issues, freezing basic wages for three years, though pegging them to a cost-of-living index. There were also to be new benefits for workers thrown on the scrap-heap thanks to new technology. Cass Alvin said they were satisfied with the agreement, while the US *Engineering and Mining Journal* said the agreement was expected to "set the pattern for ... other copper producers whose contracts start expiring June 30, including Newmont, Amax, Asarco, and the Anaconda Co" (6).

The loss of Kennecott income to Sohio was considerable: it took a US$1.15 billion write-off in 1985 (7), to add to the running losses of US$550 million since it had bought Kennecott for US$1.8 billion in 1981 – a merger accomplished without opposition from either party (8).

Understandably, the market was soon awash with rumours that Sohio would put Kennecott on the market (9), or sell it to its own shareholders (10).

Restructuring

However, at the end of 1985, Sohio set about re-structuring Kennecott in a way that would save the bacon – while sacrificing the trotters (as it were). It committed itself to a 3-year, US$400 million modernisation programme,

and a renegotiation of the employment contract with workers (11).

Kennecott offered a "package" to union workers which involved a 33% cut in wages and elimination of various other benefits. It seemed Kennecott, in order to survive, would move increasingly towards the "Phelps Dodge position" – employing "scab" labour (12). Alton Whitehouse, the chair of Sohio, announced his confidence that Bingham Canyon still contained high-quality reserves that are competitive on a world scale (10). Renewed start-up was scheduled for 1985 (13).

Bingham Canyon reopened its copper plant in 1986, with a trimmed-down workforce which accepted a 23% cut in benefits) after protracted negotiations (41). Shortly after RTZ took over Kennecott, it announced a 15% expansion of the mine's output (42).

At the same time, Kennecott bought out virtually all of Arco's minority share in its Carr Fork copper mine (also in Utah) after Arco itself dumped all its non-oil mineral activities, except for coal (14).

While the company's Nevada division remained closed during 1984, production at the Chino Mines in New Mexico (Kennecott 66.6%, Mitsubishi 33.3%) actually increased (3).

Gold interests

Meanwhile, the company's gold interests were consolidated: as of 1983, it had three substantial projects in western Nevada, Brazil and Papua New Guinea (15).

At Crixas, in central Goias, Brazil, Kennecott had a JV with Inco, in which it could earn 50% (3); in mid-1985 it awarded an engineering study on this project to Brazilian and US engineering companies (16).

In Papua New Guinea, Kennecott holds 61.6% with Nord Resources of Canada (30%), and the local company Niugini Mining (8.4%), in a gold prospect on Simberi Island in the Tabar islands group, which has yielded good results (17). Another JV, between Kennecott (88%) and Niugini Gold (12%), located a potentially very rewarding gold mountain on Lihir Island (18). At the end of 1985, the JV partners claimed they had found the biggest gold (mine) in Australasia or nearby (10), ranking ahead of Porgera, Ok Tedi and Bougainville, but perhaps not Japan's Hishikari deposit (19).

Kennecott has a mixed bag of interests elsewhere in the world. Its McGill smelter in the USA (now closed) is unlikely to reopen, because it will not meet stringent pollution standards imposed by the Clean Air Act of 1987 (3). Exploration continues in several countries – notably USA, Canada, Australia, and in South America (15). Kennecott has drilled for molybdenum in a wilderness area in New Zealand – which local conservationists wanted legally protected from mining (20).

In 1981, Kennecott bought all shares it didn't already own in the world's largest producer of low manganese, iron and titanium oxide slag, QIT-Fer et Titane, which operates in Quebec (15). QIT itself holds a 50% interest in the Richards Bay iron and titanium mine in Natal, South Africa – a key strategic mineral resource for the apartheid régime. While QIT's share in this project was originally 39.3% (21) it rose to 42.5% by 1980 (15) with Gencor, the Industrial Development Corp of South Africa (IDC), and other South African companies holding the balance (22).

Kennecott's other major South African involvement is its share in Union Platinum, a company which itself owns an undisclosed amount of the very strategic, very valuable Rustenburg Platinum Holdings Ltd (22).

In a JV with its parent BP (12%), Kennecott (40%) heads one of the major deep-sea consortia: Mitsubishi, Consolidated Gold Fields, Noranda, and RTZ all holding 12% each (23).

Uranium has never figured as a very important overt concern of Kennecott's. In 1980, it was developing a plant at Bingham Canyon for the recovery of yellowcake from its copper tailings dumps: the plant to be operated by Wyoming Mineral Corp, producing an estimated 70 tonnes of uranium a year (24). The plant uses the innovative Higgins CIX "loop" system, on a "liquor" containing only 10ppm uranium.

Although Kennecott may now seem on the decline, it would be difficult to over-estimate its importance as a US mineral producer over at least the last half-century.

From its prominent role in Copper Exporters Incorporated in the 1920s (the cartel which enabled Union Minière, Anaconda, Asarco, and RTZ, as well as Kennecott, to achieve preeminence in the industry), to its ownership of Peabody Coal, the world's biggest strip-miner (later hived off to other interests), to its operation of El Teniente in Chile, the world's biggest underground copper mine (later nationalised) (25) – it has epitomised the unacceptable face of 20th century large-scale mining.

Bulwark against expropriation

In Chile, as the Allende régime threatened nationalisation in the early 1970s, Kennecott conducted a "worldwide legal battle to keep Chile's expropriated copper off the market" (26). Richard Barnet and Ronald Müller describe what happened, quoting Theodore H Moran of the Brookings Institution:

> "Kennecott developed the following plan: it built a formidable 'network of transnational alliances', to use Moran's term, in the hope that no government would dare to expropriate its mines. It took out a large loan from the Ex-Im Bank, and raised US$45 million in additional capital 'by selling long-term copper contracts with European and Asian customers to a consortium of European banks ... and to a consortium of Japanese institutions'. It sought imposition of legal obligations directly on the Chilean state and made maximum use of AID guarantees of the US government. 'The aim of these arrangements', Robert Haldeman, executive vice-president of Kennecott's Chilean operations, told Moran, 'is to ensure that nobody expropriates Kennecott without upsetting relations to customers, creditors, and governments on three continents.' Moran notes that 'Anaconda, Asarco, Freeport Sulphur, and Roan Selection have followed the lines laid down by Kennecott in negotiating for new conces-

sions in Latin America, South Asia and Africa. When the Chilean Congress, despite all these company efforts, unanimously decided to nationalise the mines, Kennecott sued in France, Sweden, Germany and Italy to block all payments to Chile for nationalised copper sold in those countries. 'Kennecott officials are determined to keep the heat on Chile', *Time* reported. 'The Manhattan offices of General Counsel Pierce McCreary, who is directing the campaign, has the air of a war room'." (26).

When Allende was toppled, Kennecott did not regain its mines, but it still has a Chilean subsidiary in operation in the country (25).

Kennecott's reputation for bulldozing decisions through local communities – and ignoring the wishes of local or indigenous peoples – is not as bad as (say) Amax's or RTZ's. However, there is plenty to write home about:

- When in control of Peabody Coal (the US Monopolies Commission directed it to divest its interests in 1977) the company colluded with the US Department of the Interior to ensure that there were no competitive bids for leases on Northern Cheyenne reservation lands. The ensuing deal left the Cheyenne with pitiful royalties and only 12 cents an acre on the 16,000 acres the company obtained. However, just two years later, two other bids for neighbouring land reached US$16 an acre (27).
- In 1971 Kennecott and Amax applied for leases to mine copper ore in the mountains of Puerto Rico. The leases were initially rejected, largely because of the detrimental effect such environmental destruction would have on large areas of agricultural land. Three years later, however, the government, together with Amax and Kennecott, introduced a new proposal which gave the companies a tax holiday, while claiming from them only a third each of the capital costs and exempting them from providing any of the costly infrastructure, and the burden of relocating the displaced families. Robert Rexach Benitez, President of the Commission on Natural Resources of the Puerto Rican House of Representatives, called the deal

"a sell-out of the national interests of the Puerto Rican people" (28).

Kennecott has, in two notable instances, nevertheless been tripped up by local opposition to its intentions:

- In the mid-seventies, it bought up, or leased, nearly 3000 acres of prime farm land in Rusk County, Wisconsin, to provide a buffer zone for an intended mine. Eleven families were uprooted, and the value of local productive services declined (29). But the local farmers, with the help of Sokaogon Native Americans in the state (themselves opposing Exxon at Mole Lake) banded together and obtained a local zoning resolution which effectively "drove Kennecott from that area before actual mining began" (30).

Ok Tedi

Kennecott was forced to relinquish its interest in the huge and potentially very rich Ok Tedi copper-gold prospect in March 1975, after seven years of disagreement, conflict and *faux pas* in its negotiations with the Papua New Guinea government. Comments R Jackson in his valuable history of the mine: " ... although the result of this particular confrontation was not exactly a victory for the PNG government, it was certainly a very unfortunate rebuff so far as Kennecott was concerned. Kennecott ... frequently showed poor judgement, particularly in its handling of the political changes which marked PNG's passage to independence. Kennecott somewhat humiliatingly lost a battle of brinkmanship which it had initiated against opponents it rather despised" (31).

In the early days of the project, according to Jackson, the company displayed remarkable concern for the Min people who would be dramatically affected by the plundering of Mount Fubilan. It sent in anthropologists, not simply to pave the inevitable way, but to identify sacred sites which could be protected from mining (52). It employed local people, endowed scholarships to local schools, and generally behaved like a benevolent neo-colonialist " ... all done with more than a small degree of self-interest on Kennecott's part" (33).

Then, in 1971, Kennecott was thrown out of Chile, the Whitlam Labor Government came to power in Australia with a critical eye turned to mining projects in general, and, most important, the Pangu Pati party came to power on the island, with a platform for early independence from the Australian administration (34). Some newly-elected PNG Members of Parliament began to attack the priorities of mining companies: in particular John Kaputin, a Mataungan leader from Rabaul (East New Britain). Stephen Zorn, the *bête blanche* of many mining companies, both in North America and Australasia (he was later to advise the Australian Aboriginal Northern Land Council in its negotiations over the Ranger uranium mine), began aggressively to question Kennecott's intentions. It was a case of "shit or get off the pot": invest more money and take responsibility for infrastructure, or withdraw. Worse for the company, John Kaputin was appointed Minister of Justice. The link between Kennecott's attempted manipulations in Chile and potential interference in the internal affairs of an independent Papua New Guinea became explicit, and was exemplified by a famous cartoon which appeared in the Port Moresby newspaper, *Post Courier,* on October 11th 1973.

When, in 1974, the Bougainvllle Copper Company, owned by RTZ, announced massive profits from its own operations in the north Solomon's province of PNG, John Kaputin and Father Momis, another prominent indigenous opponent of multinationals, called for the nationalisation of both Bougainville and the operations at Ok Tedi:

> " ... the giant American-based multinationals such as ITT and Kennecott not only screw the country financially, but involve the American Government in supporting them in every conceivable way, including using CIA agents... Papua New Guinea must reject the attempt being made to obtain a stranglehold on the Ok Tedi mine" (35).

Over the next two years, all parties to the dispute were engaged in very complex, often ac-

riminious, negotiations, not to mention bluff and counter-bluff, until Kennecott finally allowed its lease on Ok Tedi to lapse in 1975. Whatever the external reasons for Kennecott's withdrawal, its mode of negotiation with a newly-independent black government seems more important for its failure than not agreeing on rates of taxation, a "reasonable return on investment", and mechanisms for arbitrating any future dispute between the company and the government (36).

This is somewhat ironic, given that the Min people themselves seemed to regret the withdrawal of the bountiful Kennecott (37) and the gap between the government and company was not that wide in the final days (38). The PNG administration set up its own Ok Tedi development company, which later concluded an agreement with BHP and Amoco to mine Mount Fubilan. Incorporated – enshrined, rather – in this agreement was the resource rent tax principle (called the Additional Profits tax) which enabled the independent government to cream off a share of profits above a certain threshold. Kennecott had a different view of this principle to the PNG government (39). But it need not have proved an insuperable stumbling block.

William S Pintz, in his exhaustive and unique study of the Ok Tedi mine, comments on the Kennecott failure:

> "In summary, the circumstances of the Kennecott period kept agreement just beyond the grasp of the negotiators. In the early 1970s, high world copper prices and optimistic geological evidence made Kennecott anxious for an agreement. The Australian administration, committed to independence for PNG and burdened by a clearly unacceptable policy precedent at Bougainville, was in no position to make binding commitments. Simultaneously, the political priorities of the emergent Somare coalition were (a) to obtain political independence, (b) to renegotiate the Bougainville Agreement, and (c) to worry about Ok Tedi. Thus, from 1973, it may not have been possible for either party to conclude a lasting Ok Tedi Agreement until the

Bougainville terms had been renegotiated. By the time this milestone had been reached, the enthusiasm of Kennecott's management had waned as it found itself nationalised in Chile, battered by environmental legislation and a deteriorating competitive position in the United States, and beset with financial problems" (40).

The National Wildlife Federation in its 1987 roll call of the 500 largest chemical releases in the USA that year, counted Kennecott as number nine, and held it responsible for the discharge of 158,669,750 pounds of toxics that year (43).

References: (1) *FT* 3/4/85. (2) *FT* 27/3/85. (3) *MAR* 1985. (4) *FT* 19/1/84. (5) *FT* 26/3/85. (6) *E&MJ* 5/85. (7) *FT* 4/12/85. (8) *Big Oil.* (9) *Metal Bulletin*, London, 10/9/85. (10) *FT* 5/12/85. (11) *MJ* 13/12/85. (12) *MJ* 7/2/86. (13) *E&MJ* 1/86. (14) *E&MJ* 12/84, *MJ* 11/10/85. (15) *MIY* 1985. (16) *MJ* 21/6/85. (17) *E&MJ* 10/85, *MJ* 24/5/85, *FT Min* 1991. (18) *MJ* 6/12/85 and 2/8/85. (19) *MJ* 31/1/86. (20) *Watchdog* (Cafcinz) 2/84. (21) Lanning and Mueller, *Africa Underlined.* (22) *MIY* 1982. (23) *Raw Materials Report*, Vol. 1, No. 4, 1983. (24) *MIY* 1981. (25) *E&MJ* 11/84. (26) Richard J Barnet and Ronald E Muller, *Global Reach: The power of the multinational corporations* (Cape) London, 1975. (27) *The Nation*, Washington, 15/5/76. (28) Howard Berman, "Resource Exploitation: the Cutting edge of Genocide" in *AkwN* Late Spring/78. (29) Al Gedicks "State Ignores corporate mining alternatives", in *The Capital Times*, Madison, 1976; Al Gedicks, *Kennecott Copper Corporation and Mining Development in Wisconsin* (The Centre for Alternative Mining Development Policy) Madison, 1976. (30) Al Gedicks, "Exxon Copper and the Sokaogon; tightening the corporate grip in Chippewa country", in *The Progressive*, USA, 2/80. (31) R Jackson, *Ok Tedi: the pot of gold* (University of Papua New Guinea) 1982, pp40-41. (32) *Ibid*, p50. (33) *Ibid*, p32. (34) *Ibid*, p52. (35) *Ibid*, pp59-60. (36) *Ibid*, p63. (37) *Ibid*, pp66-67. (38) *Ibid*, p70. (39) Will S Pintz, *Ok Tedi: Evolution of a third world mining project* (Mining Journal Books) London, 1984, p48. (40) *Ibid* pp48-49. (41) *New York Times* 30/6/86, *Journal of Commerce* 13/8/86, *E&MJ* 10/86. (42) *MJ* 12/1/90. (43) National Wildlife Federation, *The Toxic 500*, by Norman L Dean, Jerry Poje and Randall J Burke, Washington, 8/89.

341 Kepco

```
Republic of [South] Korea (ROK)

KEPCO / KOREA ELECTRIC POWER CO. (ROK)
```

As the chief provider of uranium fuel for one of the most aggressive nuclear power programmes in the world, Kepco (it was formerly KEKO/Korean Electric Co) has not only entered into supply contracts with Australian and Canadian mining companies, but is itself part of a JV in Gabon.

In 1979, KEKO signed a contract for the supply of 2270 tonnes of U_3O_8 from the Ranger mine in Australia to be delivered between 1983 and 1992 (1).

The same year, it arranged a US$294 million contract with Norcen and Lacana which fell into abeyance after the British Columbia moratorium on uranium mining (see Norcen).

Early the following year, KEKO signed a contract for the delivery of more than 4 million pounds of U_3O_8 between 1981 and 1990 from Rio Algom at Elliott Lake (2).

Also in 1980, KEKO formed a JV (41%) with Cogéma (49%) and the Gabonese government (10%) to explore for uranium in the Lordleyon region of the African state (3).

In April 1980, the United Nuclear Corporation – America's largest producer of uranium – contracted to supply KEKO with more than 3,000,000lb of uranium oxide between 1980 and 1985 (4).

Along with Anschutz (50%) and Taiwan Power, (25%) KEKO (*sic*) began drilling for uranium at an undisclosed site in Paraguay in 1982 (5).

By 1984, Kepco had purchased 4.5% of the equity in the Dawn Lake JV from SMDC – with whom it already had contracts for supplies from Key Lake for its "ambitious nuclear power

447

generating programme". It was also "intensively seeking new uranium markets in Japan" (6).

In 1987, Kepco secured a small, but important, contract for the supply of 170 tons a year between 1993 and 2002 (possibly 2012) for US$150 million, and the purchase of a 2% non-voting interest in the Cigar Lake JV from SMDC (7).

By 1986, nuclear power accounted for 43.8% of South Korea's power generation, and was rising rapidly to around fifty per cent (8).

Kepco's long-term contracts in 1988 amounted to 1300 tonnes a year until 1991, including the Caméco, and ERA contracts already mentioned, 212 tonnes/year from Roxby Downs, starting delivery in 1987, and output from the company's 2% share of Cigar Lake, and 20% in the Crow Butte project (see Ferret) (8).

In 1990 Kepco reached agreement – not without controversy – with the Soviet Union for the supply of 40 tonnes per year of 3.5% enriched uranium, commencing in 1990 and lasting until 2000 (9). This deal was imprudently signposted by the *Mining Journal* as potentially useful to the military (10). In an apology published shortly afterwards, the *MJ* quoted Kepco as maintaining that the contract was fully covered by both bilateral and IAEA safeguards (11).

The same year, the South Korean régime announced that Kepco would be partly privatised with 21% of the government holding put on the market (12).

References: (1) AAEC statement, 6/81. (2) *MJ* 30/5/80. (3) *MJ* 11/4/80. (4) *MJ* 25/4/80. (5) *American Metal Market* quoted in *MJ* 3/7/82. (6) *E&MJ* 6/84. (7) *MAR* 1988. (8) *MJ* 5/2/88, quoting Nukem special report on South Korea. (9) *Nuclear Fuel* 19/3/90. (10) *MJ* 29/6/90. (11) *MJ* 24/8/90. (12) *FT* 10/8/90.

342 Keradamex Inc

```
Kerr Addison Mines Ltd (CDN)
                      \
                       \
               KERADAMEX INC. (USA)
```

This is Kerr Addison's exploration subsidiary in the USA, which technically held Kerr Addison's and Noranda's large uranium leases in the Ambrosia Lake area of the Four Corners region in the '70s. These were second in extent only to those held by Exxon (1).

References: (1) *San Juan Study*.

343 Kermac Nuclear Fuels

```
Kerr-McGee Corp. (USA)
          merger
          /
KERMAC NUCLEAR FUELS (USA)
```

It merged with Kerr-McGee in 1954 (1).

References: (1) US Congress Sub-committee on Energy statement 1975.

344 Kerr Addison Mines Ltd

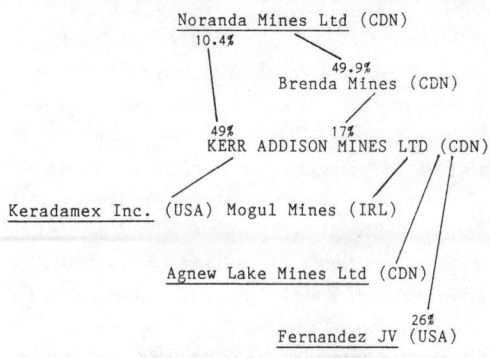

This is a gold, copper, zinc, lead, gas, oil and uranium producer and explorer whose main uranium holding has been in Agnew Lake Mines which operated in the Elliot Lake region, west of Sudbury, Ontario (1).

Production started underground at the mine in 1977 with treatment of the ore stockpile (2). It was the only bacterial leaching mine in Canada. However, by mid-1978 only half the design capacity of 400,000lb U_3O_8 was being used and it looked like contracts could not be fulfilled (3). (These included 690,000lb borrowed from the Eldorado Nuclear stockpile to meet a contract with Sweden's Nuclear Fuel Supply Co (4, 5).

By the end of 1979, problems deriving from the fragmentation of ore under flood leaching, as well as the sheer complexity of extratcing low-grade uranium from the underground mine, forced Kerr Addison to declare a closure. What is important to note (for critics of solution mining in general) is that the leaching failed to work even when it was switched from "trickle" to the far more environmentally dangerous "flood leaching" (6).

Nonetheless, by the beginning of 1981, the salvage leaching which started in 1979 produced results "above expectations" (7), and the company's 1980 closure provision for the mine was reduced by C$16 million. Left in the surface leach pile as of the end of 1980 was nearly 100,000lb of U_3O_8 in solution, while perhaps 5 million pounds remained in the "proven and probable" categories at the mine. Operations were to continue as long as feasible – the property then to be "rehabilitated" and closed (1). Agnew Lake Mines was due to return 478,000lb of uranium concentrates to Eldorado Nuclear by mid-1982, and had firm commitments to customers of another 200,000lb of uranium during 1981 (1).

In summer 1982, the company reported that its leach operations at Agnew Lake would soon cease. Thus, concluded the *Mining Journal,* "Kerr Addison's present mining operations are close to an end" (8).

Kerr Addison is also associated with Placer Development (which holds part of Zinor Holdings). It also has 26% interest in the Fernandez JV to explore uranium in the Ambrosia Lake district of New Mexico (1). Zinor, however, was liquidated in 1982 (10).

By early summer 1983, the operations at Agnew Lake were finally closed, and the Atomic Energy Board of Canada approved plans to decommission and close the property. Most of the work involved in rehabilitating the site, and "stabilising" the tailings area (it was said),

would be completed by the end of 1983. If environmental standards are complied with, the property could be returned to the government after 5 years (9).

In November 1983, Kerr Addison acquired a 33.33% interest in Anderson Exploration Ltd of Calgary, Canada, an oil and gas exploration company (11).

Kerr Addison entered the 1990s with exploration interests in Canada; copper, zinc and gold production in Quebec, BC and elsewhere; and a 50/50 JV with its Minnova Inc subsidiary (50.3%) in various searches, especially at Frotet Lake (12).

In 1989, Kerr Addison sold its 13% interest in Canadian Hunter JV to Noranda, its 50.1% parent (13, 14).

Further reading: On bacterial leaching and resultant problems, see *MJ* 28/9/79.

References: (1) *MIY* 1982. (2) *MJ* 1/7/77. (3) *FT* 10/11/78. (4) *FT* 12/11/76. (5) *FT* 1/7/77. (6) *MJ* 28/9/79. (7) *FT* 20/3/81. (8) *MJ* 2/7/82. (9) *MJ* 20/5/83. (10) *MJ* 25/3/83. (11) *MJ* 4/11/83. (12) *FT Min* 1991, *MJ* 16/2/90. (13) *MAR* 1990. (14) *E&MJ International Directory of Mining* 1991.

345 Kerr-McGee Corp

When the final obituary of nuclear power is written, the name Kerr-McGee may well stand on its own – even set against names like Westinghouse, Bechtel, RTZ and Union Carbide: Kerr-McGee symbolises corruptability and criminal negligence like none other.

Set up in 1932 (as A and K Petroleum; it became Kerr-McGee in 1946), the company is still controlled by descendants of Robert S Kerr and the McGee family. Its headquarters building is at the Robert S Kerr Avenue, Oklahoma City, Oklahoma (1). Or perhaps it would be more accurate to say that Oklahoma City is the place which surrounds Kerr-McGee. "They practically own Oklahoma City," said Bill Silkwood, father of Karen Silkwood in 1978, "they are more powerful than the government" (2).

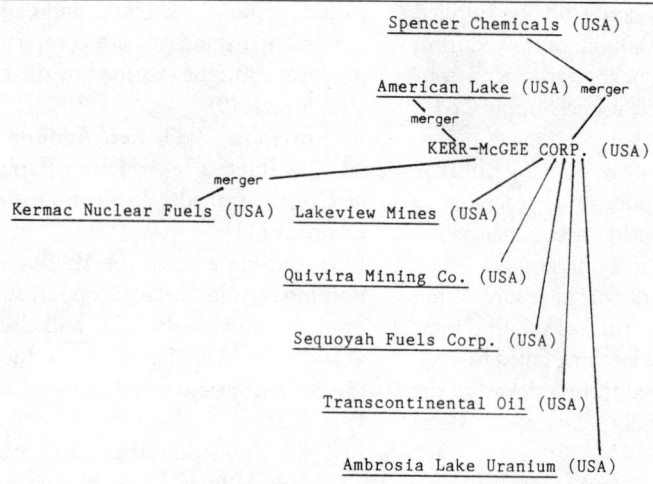

Spencer Chemicals (USA)

American Lake (USA) merger
merger
KERR-McGEE CORP. (USA)
merger
Kermac Nuclear Fuels (USA)   Lakeview Mines (USA)

Quivira Mining Co. (USA)

Sequoyah Fuels Corp. (USA)

Transcontinental Oil (USA)

Ambrosia Lake Uranium (USA)

By the early '50s, the company was expanding to become a "total energy" corporation. It was the first oil company to enter the uranium industry – in fact it "stumbled" on the Ambrosia Lake deposits (the biggest then discovered in North America) when looking for oil (3). In 1952 it had already bought mining properties on the Navajo reservation in the Lukachukai Mountains of Arizona, and began searching for uranium reserves the following year (4).

Throughout the '60s and '70s Kerr-McGee diversified in a masterful fashion – not only into uranium, coal, chemicals and fertilizers, but timber and helium, among other products (5).

By 1978 *Forbes* magazine – premier commentator on the fortunes of America's wealthiest robbers – encapsulated the corporation's progress thus:

> "McGee decided more than 25 years ago that Kerr-McGee would always be a relatively small petroleum company ... he diverted a major part of his cash flow in other minerals. The result is that Kerr-McGee has a tremendous storehouse, including probably reserves of 160 million pounds of uranium, worth over $6 billion at today's spot prices; and 3 billion tons of coal, worth $70 billion at today's prices ... and many other of the earth's elements as well ... Kerr-McGee's 2.5 million tons of coal production this year will grow by 1983 to 16 million tons and shortly after that to 30 million tons" (6).

In 1975 Kerr-McGee accounted for a quarter of America's uranium production – and it held the country's largest reserves (7). But by 1980 its share of the US market had fallen to approximately one-seventh (US total 16,156 tons; Kerr-McGee 2540) (7), putting it behind United Nuclear. Its "demonstrated" reserves of 134 million pounds of uranium oxide still made it probably America's most important holder of uranium *in situ* (7).

Even so, petroleum production accounted for the bulk (72%) of its 1980 profits, chemicals and plant food for 9% – and nuclear for only 8% (8).

This includes Kerr-McGee's production of uranium hexafluoride (UF_6) from its Sequoyah facility in eastern Oklahoma, which in 1979 produced 7,000 tons of UF_6 (8), both from its own and other customers' concentrates (1). Kerr-McGee is one of America's two producers of UF_6, and as such is part of the chain by which Namibian, Australian Aboriginal, and other robbed uranium becomes reactor or bomb fuel. The exact contribution of Kerr-McGee to the production of nuclear bombs is not known, but can be presumed to be substantial. (In fact, Allied Chemicals ran a plant in Illinois with a larger capacity in 1980 at 12,700 tonnes/year uranium) (9).

In 1980/81 Kerr-McGee operated no fewer than seven underground mines in the Ambrosia Lake area and one at Churchrock, on the Navajo reservation near Gallup, all in New Mexico. It also operated a mill (7000t/day of ore) at Ambrosia Lake (1).

In August 1980 the company announced the "temporary" closure of two of its Ambrosia Lake mines (sections 17 and 24) as a result of lower uranium prices, rising costs of separation, and higher state taxes. During the previous few months the company also cut its workforce by about 9% (10).

By the end of 1982, all Kerr-McGee's New Mexico mines were at a standstill (11), although, it seems, the Churchrock mine was only put on stand-by (12). Deliveries of uranium by the company, in the third quarter of 1982 were only 376 tons, as opposed to nearly 500 tons in the corresponding period of 1981; uranium was the only loss-maker that year (13).

In Wyoming, Kerr-McGee operated two open-pit mines in the South Powder River Basin (1), although by 1982 these were well under capacity (14).

Exxon in 1979 entered into an agreement with Kerr-McGee to process 500t/day of the company's South Powder River uranium – an agreement which ran until 1980 (15). That year Kerr-McGee also held uranium exploration permits in South Dakota, near the Black Hills national forest. According to the Black Hills Alliance, a state inspector gave false evidence to enable the permit to be granted. The state Division of Conservation cancelled and then rewrote the permission (16).

The company has explored widely in the USA – for example it was cited by Friends of the Earth and other groups in Montana in 1980, along with the state government, for not filing an environmental impact statement on the effects of uranium mining (17).

The same year, Native Americans on the Potowatomi reservation in Wisconsin were approached for leases by Transcontinental Oil (a Kerr-McGee "front") which then transferred these to Kerr Mc-Gee, "quietly optioning over 22% of the tribal reservation", reported the Native American rights newspaper *Akwesasne Notes* at the time (18).

Objected Potowatomi tribal chairman James Thunder:

"Indians are not about to become the guinea pigs for the nuclear industry in the 1990s as the Navajo and Pueblo Indians were in the 1950s."

After the tribe retained counsel to fight Kerr-McGee's trespass, the corporation offered to withdraw its options on the people's lands (19). It is little wonder that Kerr-McGee has needed "front" companies. As the operator of mines on Navajo land at Shiprock, in the '50s and early '60s, producing uranium for the USA's nuclear weapons programme, the company poisoned and killed between 250 and 300 miners, of whom at least one-tenth were Navajo (20).

At its Churchrock mine in the '70s the corporation was still employing Navajo miners – though only 12 of them – at a total wage of a mere $144,000 a year (21). Even by the late 1970s it was aggressively trying to recruit native miners for its nuclear purposes.

An article in *Business Week* entitled "Manpower gap in the uranium mines" explained how the US Labor Department was paying the corporation:

"By 1990, the [uranium] industry will need 18,400 underground miners and 4000 above-ground ... Once on the job, Kerr-McGee estimated that it costs approximately $80,000 per miner in training, salary and benefits as well as the costs for trainees who quit. To try and trim these costs and create an ample labor force, Kerr-McGee is now operating a training program at its Churchrock mine on the Navajo Reservation. The $42 million program is financed by the Labor Department and is expected to turn out 100 Navajo miners annually. Labor Department sponsors hope the program will help alleviate the tribe's chronic unemployment which is estimated at about 40%" (22).

The US Public Health Services followed 3400 white and 780 non-white (mostly Navajo) miners at the Shiprock mine between 1960 and

1968, then less rigorously from 1968 onwards. By 1973 more than 180 respiratory malignancies were reported. Predictions of 600 to 1100 ultimate deaths due to irradiation have been made for this group of just over 4000 men (23). Kerr-McGee's responsibility for this infamous episode has sometimes been excused (24), on grounds that little was known about radiation in mining at the time, and matters have immeasurably improved. In fact, quite recently, Kerr-McGee was accused by the Oil, Chemical and Atomic Workers Union (OCAW), of subjecting its miners to radiation doses six times the "recommended" exposure (25). And Kerr-McGee – after exhausting the Indian uranium at Shiprock by 1973 – made absolutely no attempt to restitute the land or in any way care for the dying Navajos.

According to Native American activist Winona La Duke, "Kerr-McGee's abandoned uranium mill still stands in the area, as well as 71 acres of highly radioactive uranium mill tailings ... those wastes retain 86% of the original radioactivity of the uranium " (26).

Although the US government, the Bureau of Indian Affairs, and the State of Arizona, have refused to acknowledge liability for these murders-by-neglect, the Red Valley Navajo chapter, home of most of the dead Navajo miners, has recently attempted to get legislation to compensate the survivors of the dead miners and assist those now suffering the effects of radiation poisoning (27).

When the Navajos went to Washington in early 1980 to file claims for compensation, a Navajo spokesperson, Dan T Benally, told presspeople: "Neither I, nor the other miners, were ever told that uranium was dangerous. We ate lunch underground, and drank the water which ran in the mine."

Robert Alvarez of the Environmental Policy Center told reporters at the same time that between 30,000 and 50,000 miners who worked in the south-west uranium mines (though not all with Kerr-McGee) were in danger of contracting cancer.

The Navajo deputation was sponsored by Citizens' Hearings for Radiation Victims, a coalition of organisations representing all those who have suffered from American nuclear testing and production since the '40s (28).

By the end of 1981, 650 Navajos filed claims that they were suffering from radiation. A study earlier in the year indicated a correlation between the mining and a high incidence of birth defects on the reservation (29).

A study conducted by the New Mexico EPA (Environmental Protection Agency) in 1978 discovered that Kerr-McGee had no discharge permit to cover radium effluent from one of their mills – some of which was passing into a nearby stream. Nor did the company have any radium removal treatment programme in operation.

But, when US uranium companies banded together at the same time to oppose the implications of the US Mill Tailings Act, Kerr-McGee was at their head. The companies challenged the NRC (Nuclear Regulatory Commission)'s authority to adopt mill-licensing regulations per se, calling them "arbitrary and capricious" (30).

The Southwest Research and Information Center (SRIC) filed a petition for the licensing of uranium mines in 1978. It failed, since the companies were able to argue that, under the New Mexico Radiation Protection Act, the mining, extracting, processing, storage or transportation of radioactive ores or uranium concentrates which are themselves "regulated" by Federal agencies are exempt from state control. In 1982, however, the SRIC returned to the fray (and the courtroom) with an amended petition, backed by a recent report from the Radiation Protection Bureau of the New Mexico Environmental Improvement Division. This concluded that uranium mines were the "primary causes of elevated levels of radon gas near Grants, New Mexico" (Kerr-McGee's and Homestake and Anaconda's mines in particular), not outcrops of uranium ore in the region as the companies had claimed (31).

Accidental death?

Kerr-McGee's most renowned victim is not a

Navajo uranium miner, but Karen Silkwood, a laboratory worker at the company's Cimarron plutonium reprocessing facility near Oklahoma City, who died in a car crash in November 1974.

Ms Silkwood was on her way to present evidence to the OCAW, and a *New York Times* reporter, that Kerr-McGee was manufacturing faulty plutonium fuel rods (and falsifying records to conceal this fact), when the car spun off the road and she was killed. The file of evidence was never found: the company was later "cleared" of the allegations by the US Atomic Energy Commission (AEC).

Whether or not Kerr-McGee arranged Karen Silkwood's murder, it certainly stood to benefit from it. Its attitude to Ms Silkwood was revealed during a court case brought against the company for negligence by Karen's father on behalf of her three children. Kerr-McGee maintained that she smuggled plutonium out of the plant to contaminate her own urine and discredit the company.

The jury returned a verdict for Karen Silkwood, and Kerr-McGee was ordered to pay US$10.5M damages. Equally important, while the AEC maintained throughout the trial that Kerr-McGee had adhered to safety regulations, the judge directed the jury that it could decide the evidence "in the same manner as any other expert testimony". The evidence clearly hadn't washed (32).

Kerr-McGee appealed the verdict before the Federal Appeals Court – and won. On January 11th, 1984, however, the Silkwood family obtained a US Supreme Court ruling that restored the $10.5M punitive damages to Karen Silkwood's three children, and thus nullified the Reagan administration's position that the award of punitive damages conflicted with federal laws imposing fines for safety violations (33).

The case then went back to the Federal Appeals Court where the company reasserted its claim that the damages are excessive (34).

The appeals court ordered a new trial – a step which neither the Silkwood family nor the company wanted (58). The two parties settled out of court, and the Silkwood estate received $1.3 million damages (42).

"The fuel rods manufactured at the plant where Ms Silkwood worked had been contracted to the Westinghouse Hanford Corp from 1972 to 1976 for its test breeder reactor facility. The facility, near Richland, Washington, was managed for the federal government. An article in *Science* magazine, March 1978, said of the fuel rods made for the plant: 'According to a Westinghouse spokesman, Hanford received a total of 19,568 fuel rods from Kerr-McGee, of which 688, or 3.5 percent, were found unacceptable and sent back to Kerr-McGee. Kerr-McGee did not agree that most of these were rejectionable and refinished or repaired many of them. Thus in the final count only 91 were rejected. The Westinghouse spokesman said that, of the group finally accepted, some 541 were not deemed good enough to be used.'

"Westinghouse's failure to renew the fuel rod contract led to the closing of the Cimarron plant. Subsequent studies proved that laboratory analysts had touched-up negatives with a black felt-tip pen to hide defects in welds that held the fuel rods together. Other allegations made by Karen Silkwood have also been proved correct" (35).

The nuclear legacy left behind by Kerr-McGee – even if it stopped production tomorrow – is almost incalculable. In late 1978, a report in *Science* magazine calculated that uranium tailings could amount to one billion tons by the year 2000 and "... the largest in the United States, active and still growing [is] the pile maintained by the Kerr-McGee Nuclear Corporation at its big mill near Grants, New Mexico, where 7000 tons of ore are processed daily. Containing 23 million tons of tailings, it covers 265 acres and rises to 100 feet at its highest point ..." (36).

In 1980, Native Americans at Red Rock Valley suffered again from Kerr-McGee when 7000 gallons of oil washed down the Luckachukai Mountains and emptied into the San Juan River. At least fourteen families were affected by the spill – not to mention the ongoing pollution from the company's activities in the area. In one family – the Nakai – 11 sheep were lost from drinking the poisoned water (37).

In 1983 Kerr-McGee suffered several setbacks. Its net earnings plunged 76% in the fourth quarter of the year – thanks to provisions it had to make in rehabilitating a rare earths and thorium plant, and settlement of litigation concerning some of its uranium royalties (38).

The same year, the company was refused permission to mine phosphates in the Osceola National Forest of Florida, after the Interior Secretary banned all such applications on grounds that environmental studies showed the area could not be adequately reclaimed (39).

In late 1983 all the corporation's uranium mining and milling operations in New Mexico were consolidated into a single subsidiary called Quivira Mining Co. Named – appropriately? – after an early Spanish "mission" in the state, Quivira was in theory to become not only the largest uranium company operating in New Mexico but in the whole of the US of A.

Outside of New Mexico, another new subsidiary, Sequoyah Fuels Corp, was interested in consolidating the Sequoyah refinery and fluorination plant in eastern Oklahoma and non-producing uranium leases, including the two Wyoming open-pit mines on stand-by (40).

During the second quarter of 1984, Quivira Mining produced a total of only 181 tons of U_3O_8 – down almost 21% on the same period of the previous year – while production at Sequoyah Fuels fell 34% with the production of "only" 790,000kg of UF_6 (41).

By 1988, Kerr-McGee was showing a reasonable profit in uranium ($8 million in 1987) (43), but had put Quivira Mining on stand-by, after it produced less than 100 tons U_3O_8 from minewater: Quivira was also up for sale (43). Sequoyah Fuels Corp. delivered nearly 5000 tons of UF_6, and 320 tons of depleted UF_4 – which is used to manufacture high-velocity armour-piercing projectiles (43, 44).

The year that the Silkwood estate settled with Kerr-McGee an explosion of uranium hexafluoride at the Sequoyah Fuel facility in Gore, Oklahoma, killed one worker and put three others into hospital (45).

The following year, Sequoyah stepped up its production of dilute ammonium nitrate "fertiliser" – both a product of processing uranium into UF_6, and a means of "disposing" of its awkward radioactivity. The fertiliser was used in fields owned by Kerr-McGee in eastern Oklahoma, where in 1986 alone a total of 15 million gallons was spread over some 15,000 acres. Despite an outcry, the company was granted a licence that year not only to sell the fertilizer, but to give away hay grown on the contaminated lands. Much of this went to local Navajo families. The organisation NACE (Native Americans for a Clean Environment) protested at this culpable act of charity – pointing out that the Sequoyah facility was already leaking radioactive wastes into the Arkansas river (46).

Kerr-McGee, soon afterwards, sold its Sequoyah plant to GA Technology (formerly General Atomics). Three years later, not only was the plant still producing the deadly stuff (known generically as "raffinate") but boasting about its efficacy – despite the discovery of a nine-legged frog in a local pond. The Oklahoma Water Resources Board admitted that Sequoyah had not complied with its river dumping permit "since the permit was issued" (47). NACE commented "[It is] like a nightmare we can't wake up from" (48).

The disposal of Sequoyah was part of Kerr-McGee's strategy to disinvolve from the nuclear industry, a process which started in 1985 with the closure of its uranium mines in New Mexico (49).

In early 1989 it finally disposed of virtually all its uranium in Wyoming and New Mexico reserves and Quivira mining facilities to Rio Algom, thus "complet[ing] its withdrawal from the uranium business" (50). The nuclear legacy left by Kerr-McGee is one of the most horrendous in the world. Yet, throughout its period as a major uranium producer, it has led the way in arguing that the industry should not bear responsibility for cleaning up unreclaimed and dangerous uranium mill tailings piles (51).

Not that the most notorious US uranium miner has actually withdrawn from the production of radioactive material.

Although it holds substantial coal reserves and mines coal in Wyoming at Jacob's Ranch (open pit) and underground in Illinois (where it recently began the potentially dangerous method of longwalling); although Kerr-McGee produces various salts and borates at its Searles Lake Brines in the Mojave desert of California (52); although it has modestly entered gold mining in a JV (26.25%) with Homestake at Back River, in Canada's Northwest Territories (53) – its major mining thrust is the recovery of mineral sands and the processing of its by-products. Kerr-McGee has been producing around 106,000 tonnes a year of titanium dioxide pigments from its Hamilton plant in Mississippi: in early 1990 the company decided to raise total capacity to 166,000 tonnes a year by building another plant at an unspecified location (54).

The Cooljarloo JV (with Minproc of Australia as an equal partner) is located 170km north of Perth in Western Australia. The companies boast that it will be the world's first fully integrated mineral sands project – with mining, beneficiation, upgrading of the concentrates to synthetic rutile, and conversion into titanium dioxide pigment. Dredging of the heavy minerals (primarily ilmenite) started in late 1989, and soon exceeded expectations (55). Reserves at Cooljarloo are estimated at around 12 million tonnes of ilmenite, just under 1 million tonnes of rutile, with 2.14 million tonnes of zircon, which will be separated out at the Murchea plant, producing important quantities (around 1% of the total) (56) of monazite (57). Opposition to the Cooljarloo project – on environmental and health grounds (because of the dangers of radioactive emissions from the dry separation plant) – was registered by various Western Australian groups, including the Australian Conservation Foundation, Goldfields Against Hazardous Waste and the office of independent Western Australian federal Senator, Jo Vallentine.

Further reading: Richard Raske, *The Killing of Karen Silkwood* (Houghton Mifflin) USA, 1981.

Contact: OCAW International Union, 1636 Champa Street, PO Box 2812, Denver, Colorado 80201, USA.

Southwest Research and Information Center, POB 4524, Albuquerque, New Mexico 87106, USA.

Black Hills Alliance, Rapid City, South Dakota 55709, USA.

Shyama Peebles, Goldfields Against Hazardous Waste, PO Box 889, Kalgoorlie, Western Australia.

NACE, PO Box 1671, Tahlequah, OK 74465, USA.

References: (1) *MIY* 1981. (2) BBCtv *Panorama* documentary, 5/3/78. (3) Pringle & Spigelman. (4) Kerr-McGee Annual Report 1978. (5) *Dun's Review*, USA, 12/74. (6) *Forbes* 16/10/78. (7) Rich Nafziger, "Indian uranium: Profit and peril" (quoting company information), in *Americans for Indian Opportunity* 11/76. (8) *MAR* 1981. (9) Australian Government Representative's Report, Canberra, 9/9/80. (10) *MJ* 22/8/80. (11) *MJ* 3/9/82. (12) *MM* 1/83. (13) *MJ* 24/12/82. (14) *MIY* 1982. (15) *MJ* 9/11/79. (16) *BHPS* 4/80. (17) *Kiitg* 6/80. (18) *AkwN* Winter/81. (19) Al Gedicks, *"Northern state counties of the US fighting uranium mines"*, quoted in *Kiitg* 6/83. (20) Winona La Duke, "Uranocide", in *Natural Peoples' News* London, No. 2/3. (21) Eric Natwig, *Overall economic development program*, Window Rock, Arizona, 1974 (Navajo Nation). (22) *Business Week* 1/11/77. (23) Reported in Abrahamson & Zabinski, *Uranium in Minnesota* 1980 (CURA). (24) Noel O'Brien, *Environmental protection in underground uranium mining*, London, 1978 (Uranium Institute Symposium). (25) *Kiitg* 1/80. (26) *AkwN* Winter/78. (27) *Kiitg* 10/81. (28) *Washington Post* 15/2/84. (29) Southwest Research and Information Center, Albuquerque, New Mexico. (30) *Ibid* quoted in *Kiitg* 3/82. (31) *Kiitg* 6/83. (32) Account taken from numerous sources, see *Women's Handbook on Nuclear Power*, London, 1982; also court report in *FT* 19/5/79. (33) *News and Letters*, Detroit, 1-2/84. (34) *WISE Bulletin*, 2/84. (35) Winona La Duke "In the right place at the right time", in *BHPS* 7/79. (36) *Science* 23/10/78. (37) *Tribal Peoples' Survival* 5/80. (38) *FT* 8/2/84. (39) *MJ*

21/1/83. (40) *E&MJ* 11/83. (41) *MJ* 24/8/84. (42) *MMon* 9/86. (43) *MAR* 1988. (44) See also Miles Goldstick, *Wollaston: People Resisting Genocide* (Black Rose Books) Montreal, 1987. (45) *FT* 7/1/86. (46) *Treaty Council News*, Vol. 7, No. 3, 12/87. (47) *WISE NC*, No. 346, 8/2/91. (48) *NACE News*, Tahlequah, 1/91. (49) *FT* 7/1/86. (50) *FT Min* 1991. (51) Kerr-McGee et al v US National Regulatory Commission et al, No 80-2043, US Court of Appeals, Tenth circuit, Denver, filed 3/10/80 quoted in *Mine Talk*, Albuquerque, 5-6/81. (52) *MAR* 1990, see also *Industrial Minerals* 8/85. (53) *E&MJ* 7/89. (54) *MJ* 2/2/90. (55) *MJ* 9/12/89, *MJ* 13/4/90. (56) *MAR* 1990. (57) See also *MM* 5/91, *MJ* 20/7/90. (58) *WISE NC*, No. 259, 19/9/86.

346 Key Lake Explorations Ltd (CDN)

Not to be confused with the Key Lake Mining Corp, Key Lake Explorations is backed by Amax, has its HQ in Toronto, and has held uranium claims in the Key Lake, La Ronge, and Athabasca areas of Saskatchewan, as well as at Thunder Bay, Ontario. Its president, EM Dillman, reportedly holds a third of the shares (1).

References: (1) *Yellowcake Road*.

347 Key Lake Mining Corp

Not to be confused with Key Lake Explorations, the Key Lake Mining Corporation was set up in 1979 to develop the Key Lake uranium mine and mill in northern Saskatchewan (1). In 1985, this mine enormously assisted Canada to the top of the world's uranium production table. At 11 million pounds of uranium oxide, the Key Lake mine contributed well over

```
            Eldor Resources Ltd (CDN)
                              /
                             /
  SMDC (CDN)      Uranerz (D)
        \            |        /
         \           |       /
         50%      33.33%   16.66%
         KEY LAKE MINING CORP. (CDN)
```

a third of the country's output (total 29 million pounds) (2).

Key Lake is also, with little doubt, the world's largest operating uranium mine. (The minor doubt attaches to actual production from Rössing in Namibia, for which figures have not been produced since 1984. Rössing's 1984 production is assumed to have been around 8,200,000lb and to have fallen from 1985) (3). Key Lake came into production in October 1983 and had produced 700,000lb U_3O_8 by the end of the following month (4), although it was not officially opened until June 1984, the high clay content in the ore having posed initial "teething problems" (5). Production had reached 350 tons uranium a month by that time. The mine has a current design capacity of 4500 tons a year, although its mill throughput could rise to 5400 tons per year (5). Initially the smaller, Gaertner, orebody was being mined but it looked likely that this would be exhausted by 1987 (6). The mine itself is scheduled for decommission (7) when the larger Deilmann orebody will be brought into production. There are no current plans to exploit Boulders, a third, much smaller deposit (8).

In theory, according to the analyst Thomas Neff, Key Lake could deliver no less than 7400 tonnes a year by 1991. Neff gives the Key Lake reserves as 70,000 tons of uranium in ore, with an average grade of 2.8% (9). Other sources have specified 90,000 tons uranium oxide, with a slightly lower average grade of 2.35% (10).

In any event, there is no doubting the huge potential of this deposit, and the true extent of the orebody has yet to be properly defined. Says Neff: " ... there may be many more deposits at Key Lake ... though at greater depth and with no evident outcropping". Even at present production rates Key Lake will leave behind one of the world's largest pits: at a stripping ratio of 35:1, at least forty million cubic metres will have to be lifted off the surface of the earth (11). Although not the richest exploitable uranium deposit in the world (this dubious honour is now held by Cigar Lake), even the material in the low grade ore stockpile at the mine is at least twice as rich as that produced in most other western mines, while its average grade is ten times greater than Ranger and an estimated 75 times as great as Rössing (12).

Even these figures disguise the extraordinary richness of some parts of the two main orebodies – up to 45% U_3O_8 in Gaertner and 20% in Deilmann (with similar grades of nickel, about which little has been heard) (13). Key Lake Mining Corp started life as a JV, in which Inexco originally held an equal share with SMDC, Eldorado and Uranerz. Inexco sold its share to SMDC which then sold half of that to Eldor Resources. Perhaps because of this high degree of state and federal involvement, comments Neff, there were more delays than usual in bringing the project to fruition (9). The major "delay" centred around the Key Lake Board of Inquiry, set up by the provincial government in December 1979, some months after the company had acquired more land in the region (14). All environmental and anti-nuclear groups in the province refused to participate in the inquiry, in protest at its narrow terms of reference "... which could only recommend how, not whether, the mine should proceed" (15).

A spokesman for the Saskatoon Citizens for a Non-Nuclear Society criticised the Key Lake report for being even narrower than the Bayda Inquiry (which had sanctioned Amok's Cluff Lake Mine). It was, he said, "a slick and fatuous piece of public relations ... just what the NDP [ruling New Democratic Party] wanted" (15). The five major anti-nuclear groups boycotting the Inquiry also criticised the Board's refusal to translate proceedings into Cree and Chipewyan – the languages of the local residents (16). The recommendations of the Board included the following:

- an "affirmative action" programme to make maximum use of natives, women, and handicapped persons, and an oversight committee to assure compliance;
- a 50 percent target for native jobs during production;
- a federal board or royal commission to investigate land claims;
- a technical training programme based in the north;

- a compensation programme for damage done by developers to traplines; and
- a programme of financial aid to northern businesses to help them realise growth opportunities (8).

The board rather naively expected that half the money spent developing the mine would support individuals and businesses in the area, and that taxes, together with SMDC's own share of the profits, would account for half the total revenue during the life of the mine. (Up to C$3.9 million out of up to C$8 million.) There were further delays, while the SMDC tussled with governmental agencies over its specific responsibilities as uranium prices started to fall. Finally a surface lease agreement was signed in late 1981 (17).

Although the Federation of Saskatchewan Indians and other native groups largely supported Key Lake in the early days, because of the promise of jobs in an area fraught with unemployment of up to 95%, and discrimination against native workers, their position was by no means unanimous. In 1983, the Vice-President of the Association of Metis and Non-Status Indians (AMNSIS), Clem Chartier, said he would like to see all uranium mines closed, because the environmental and social disruption they caused outweighed "the minimal employment benefits". Said Chartier: "Key Lake's record illustrates the general failure of uranium development to benefit northern communities" (18).

Early estimates costed each job at Key Lake at a million dollars per workplace (using figures supplied by the Key Lake Board of Inquiry (KLBI) itself) (19). As the Inter-Church Uranium Committee put it: "More jobs are created in the construction phase of the mine, but these are [even] shorter [than during mining] and are measured in months. Without other developments taking place, the major effect of the mine(s) will be to increase the number of skilled and semi-skilled unemployed. At a million dollars per job, it is easy to see that uranium mining cannot support very many people for very long" (20).

Three years after the KLBI Final Report, the

457

corporation's president Peter Clarke acknowledged that the turnover rate for northern native peoples' employment had declined from around 60% in autumn 1983 to 40% in spring 1984 (as against 23% for non-native people): only 35% of the total workforce was indigenous, against the 50% recommended by the inquiry. (All members of the workforce are members of the United Steelworkers of America [USWA]) (21).

These figures have been disputed by AMNSIS, whose secretary Frank Tomkins told a press conference in late 1983 that only 27% of the mine's employees at that time were native northerners (18). In another statement, at the same time, KLMC's president Peter Clarke stated that the attrition rate among native employees was 50%-60% as against half that for non-native employees (22).

By the end of 1984, the province's New Democratic Party (by then in opposition) had reversed its earlier pro-uranium policy and voted to phase out all uranium mining in Saskatchewan. The Inter-Church Uranium Committee, joined by the Regina Group for a Non-Nuclear Society (RGNNS), was also calling for the closure of Key Lake on economic grounds (23).

But what angered and disconcerted large numbers of other people in the province was an almost unprecedented series of spills and leaks which commenced almost as soon as the mine was operational. This immediately called into question claims that Key Lake was environmentally safer than any other uranium mine.

Certainly, worries about the intensity of radiation exposure had been expressed in the early days. First it was pointed out that there was no precedent for dealing with the level of radiation hazard at Key Lake. Although the Cluff Lake orebody is among the world's most dangerous mines, it is much smaller than its Saskatchewan neighbour (24). In any case, the report of the inquiry into Cluff Lake (Bayda) had not even been released before the Key Lake mine was under-way (25). Mine officials themselves have estimated that "one hour's exposure to the grey seam at the bottom of the Gaertner pit ... provides as much radiation as a medical X-ray" (26). Special measures announced to protect miners (lead-lined driving cabs and the separate treatment of high level radium in the tailings) (24) could not placate a critical public. (Some of them were cynical enough to suggest that the high concentration of native people in the workforce might relate directly to greater fears for the long-term safety of white miners.) Moreover, the original shifts involved miners often working a 12-hour day on a seven-day rota. According to Terry Stevens, the USWA representative negotiating for the mine force in 1983, miners in the company's bulk neutralisation area have been exposed to radiation "twelve times the allowable limit, without being advised by the company as required by the surface lease" (22).

But it is the lack of environmental protection which has created concern in the province – not helped by the fact that:

i) Several pages of the KLBI report on the mine are lifted virtually *verbatim* from the company's own EIS (Environmental Impact Statement). "The position taken by the KLBI on the surrounding geography ... the orebodies, the mine design, the lakes, the overburden, groundwater, mining and milling and waste management is nearly identical to that of the corporation" (25).

ii) Even so, a company claim that it could totally control and treat all liquid wastes from the mining and milling operations is not to be found in the KLBI final report (25).

iii) The company had illegally drained several lakes near the mine site in 1978, even before the KLBI was established.

iv) Approval for Uranerz – the operating company in the mine construction phase – to discharge industrial waste from its dewatering operation was granted in February 1978, even before the earlier Cluff Lake Inquiry had been completed (25).

v) In August 1979 – months before the KLBI was appointed – the RGNNS tried to stop the draining through court action, arguing that the permit had not been legally granted since Environment Department approval had not been given. The case was lost – on the spurious grounds that, by then, the drainage had been 90% completed (25).

vi) In June 1981, although its drainage lease expired, the company continued draining. After public pressure, Ted Bowerman, the Environment Minister, laid charges against the company and secured the derisory fine of C$500. By then, of course, that particular operation had been completed (25).

KLMC's President, Peter Clarke, freely admitted in 1983 that his company's major technical problem was to "push more ore into the (mill) grinder so that the plant, designed to produce between eight and 12 million pounds a year of yellowcake ... can approach 100% capacity". Although Clarke was quoted as saying that "the mill sets a standard process for uranium extraction with only a few minor differences" (27), an article in the *Canadian Mining Journal* six months later, outlining the "minor differences", was considered important enough to be quoted at length in the prestigious *Mining Annual Review* (28).

According to the article, "technically advanced ore treatment involved a two-stage leaching procedure with a 99%-plus extraction ratio, the installation of brand new French Krebs mixer settlers (never before employed in America), and – to meet strict environmental regulations – a bulk neutralisation plant, treating about 5000m/day, which removes radium before it reaches the effluent stage." There were to be no settling ponds, as all water leaving the plant "is held in monitoring ponds to double check that it meets the environmental standards before being released" (29).

Lime would be used for acid control while the tailings were to be kept sealed in a bentonite/sand pond which, when decommissioned, would be totally sealed and kept above the local water table "while the drainage system will remain in place to provide a long-term monitoring system" (30).

According to the *Mining Journal* a few months later, strict environmental protection procedures include monitoring of lake sediments, wild life and air quality, as well as vegetation,

while contaminated solution from the sprays discharging into the tailings pond is recycled back to the mill (7).

Given this supposedly unique system of contamination control, it was all the more galling – and outrageous – that, less than a hundred days after the official opening of the mine, a dyke eroded, releasing 100 million litres of water containing precisely the radioactive isotope, radium-226, which the new technology was intended to contain. Though this was the major accident (indeed, comparable to the massive leakage into the Rio Puerco from United Nuclear's tailings dam in July 1979), Key Lake has experienced a dozen similar leaks in the space of a few months (31). One of these occurred only two days after "The Big One" but, although investigators from the mine pollution branch of Saskatchewan Environment were at the mine site when it happened, it was another two days before they were officially informed (32).

The government agency Saskatchewan Environment acknowledged that the major spill was the largest in the history of the province and "a significant risk to the environment" (33). The USWA called it "a catastrophe", and accused the company of downplaying the dangers (33). Bob Mitchell, who chaired the KLBI, was flummoxed: "This just could not have happened based on the presentation (to the Inquiry) that the Key Lake Corporation made" he was reported as saying. "They [the KLMC] proved to our satisfaction that what happened ... couldn't have happened" (33). Dr Pat Tones, a senior scientist with the Saskatchewan Research Council (SRC) declared that the radium levels in the contaminated water "are certainly much higher than anything we've encountered in the lab or in nature" (33). Somewhat hypocritically (given their go-ahead for the mine) the opposition NDP charged the Conservative government in the province with putting workers' safety and the environment at risk and pointed out that the six-person monitoring committee, intended to pre-empt such disasters, hadn't met for the previous year (34).

Both the company and Saskatchewan Environment tried to minimise the dangers and the likelihood of recurrence. Claiming the spill was due to "human error" when a pump was left running (35), Peter Clarke of the KLMC rejected calls for a public inquiry (36) while his namesake Dave Clark, for the government, contented himself with the statement: "From a technical standpoint, all that's happened is that the surface area of the reservoir has been increased" (37). A nuclear physicist from the provincial University, Henry Caplan, claimed that, while it would be inadvisable to drink from the polluted water, "you can swim in it". (Not that he did!) (37).

Within a few days, with contaminated water building up under the ice at the mine site, and the threat of radium escaping elsewhere (38), the NDP leader Allan Blakeney was calling on the government to rethink its "full-speed ahead policy" on uranium (39).

Environment Minister Neil Hardy was having none of it, however, and he described the spills as "normal start-up problems" (40). The company started repair work at the site (39), as a former maintenance worker, Kal Manno, charged that, months before, he had warned of the likely weakening in two reservoir walls – and had been asked to resign for his candidness (41). Peter Clarke dismissed Manno as just a disgruntled employee (42), while a group of northerners, with the support of the local mayor and members of the Metis community, set up a Northern Camp for Ecology at Pinehouse Lake, the Metis community closest to the mine site (43).

Writing in the Saskatoon *Star Phoenix*, a nuclear chemist and president of the Saskatchewan Environment Society, Ann Coxworth, predicted that up to 2000 people could die from ingesting the contaminated spill water, if it entered drinking supplies (44).

KLMC announced several possible methods of dealing with the spill – including construction of a new water treatment plant (45)– as the Saskatchewan Association of Northern Local Government, along with the Inter-Church Uranium Committee (ICUC) and Greenpeace, demanded a full public inquiry into the disaster (46). Within the next two months, as the

KLMC began pumping out Gerald Lake (against the wishes of the Atomic Energy Control Board which claimed it would "create more problems than it would solve" (47), calls for a public inquiry increased. The NDP leader called for the resignation of the Environment Minister Neil Hardy (48). While promising an inquiry into "union health and safety concerns" at Key Lake (49), the Department of Labour declared that risks from Key Lake were only "a worker/management problem" and not connected with lack of occupation safety and health standards (50).

In a press release in February 1984, a representative of the Pinehouse local community authority, George Smith, rubbished KLMC's plans to clean up Gerald Lake. After a trip to the site, Smith said: "[the contamination] is much worse than I thought. The radioactivity is in Gerald Lake and in the muskeg. You cannot clean it up. It is not like a bath-tub; you cannot suck all the water out of the muskeg. I have been in the bush long enough to know that in the spring the water will get into Wolf Lake, David Creek, the Wheeler River, Wollaston Lake, and eventually into Churchill Lake, the last unpolluted river system in Saskatchewan" (51).

The RGNNS also submitted to the Minister of the Environment a detailed scientific critique of Environment Canada's clean-up plan (52). The critique claimed that:

i)   The 100,000 m^3 of contaminated water in Gerald Lake to be pumped out represented only a sixth of the total. The Lake's radioactivity level was no less than 8 billion becquerels due to the presence of radium 226.

ii) 90% of the heavy radionuclides were in the sediment, representing a concentration 5000 times the provincial standard for drinking water, thus much higher than admitted by the provincial authorities or the company.

iii) The intended Spring clean-up in 1984 could not possibly deal with the toxicity which, by then, would simply have moved further into the river system.

iv) Several previous spills had been close to employees' drinking water area; workers returned to camp with contaminated clothing and bodies.

RGNNS's conclusion was that "There is no safe way to mine uranium; the wastes cannot be contained" (53).

As if to underline the validity of the opposition's claims, yet another radioactive spill was recorded in the summer of that year, when a pipe carrying water from a reservoir to the tailings ponds got disconnected (54).

The KLMC partners agreed to market their own share of production, related directly to the equity they hold in the project (55). Buyers include Ontario Hydro – there was some speculation in 1984 that this huge public utility might abandon or renegotiate its US$50 a pound contract with Rio Algom and Denison at Elliot Lake (22) – two Swedish groups and Belgium's Synatom (9).

These contracts were, however, only achieved after conflicts between the federal and provincial governments: the federal government wanting a floor price of US$30 a pound, with the province angling for terms which were much more flexible and close to the spot-market price. Uranerz's 1981 contract with Synatom was eventually approved at a price tied to the spot-market prices, with a pricing mechanism linked into subsequent annual price reviews (9). Initially the requirement that Canadian uranium be hexed in Canada was eased, but later re-enforced, reportedly causing two US buyers to back off (56).

In August 1985, the SMDC and Eldorado Nuclear signed a C$250 million JV sales contract to supply 2700 tons of Key Lake uranium to Japan's Kyushu Electric Power Corp from 1987 to 1999 (57).

The most controversial contract by far has been one signed between the Saskatchewan government and the military dictatorship in South Korea, not long after the Vice-President of SMDC visited the country in early 1981 (58).

Under the contract, Key Lake uranium was to start being delivered in 1983, refined at Port Hope, Ontario, and enriched in France or the USA (59).

Condemning the sale, the ICUC pointed out that a 1978 report from the International Organisations of the US Congress spotlighted South Korea's clandestine nuclear weapons programme, while a Ford Foundation report at around the same time claimed the South Korean régime could construct up to 36 plutonium bombs a year (59).

After several years' research into the end use of Saskatchewan uranium, the ICUC in early 1986 declared that any Canadian yellowcake could end up in nuclear weapons programmes (60).

Production at the mine in 1986 exceeded nominal capacity (4650 tons) by 200 tons, representing 40% of total Canadian production, and no less than 13% of total western world output for that year (61). Development of the Deilmann orebody proceeded apace, with stripping of the Deilmann pit area, preparative to mining in 1988. The Gaertner pit was mined out the same year, although stockpiled ore provided feed, as the company moved onto Deilmann (61). Yet another output record was achieved the following year – no less than 5200 tons uranium was produced (possibly a world record, although incomplete figures from Rössing during the early 1980s suggest that the RTZ-managed mine in Namibia might have exceeded this) (62). The Gaertner orebody was then allowed to flood. KLMC also conducted a 10,000-ton heap-leach test in 1987, to verify uranium extraction rates from lower-grade cobble ores: a 10,000-ton leach facility was being considered for operation in 1988 (62).

In 1989, KLMC began production from the Deilmann deposit but as the orebody is contaminated with more molybdenum than the Gaertner, this necessitated the installation of a special treatment circuit. 2600 tonnes of uranium was produced in the first six months of that year (63). With the establishment of a Canadian Mining and Energy Company in March 1989, Caméco took over operation of Key Lake (63); Uranerz retained its one-third interest (64). Production for both 1989 and 1990 was around 12 million pounds, confirming the mine as by far the world's biggest single producer of uranium (65).

Contact: Inter-Church Uranium Committee, Box 7724, Saskatoon, Canada S7K 4R4 (tel: 306/934 3030).
Regina Group for a Non-Nuclear Society (RGNNS), 2138 McIntyre St, Regina, Saskatchewan S4P 2R7, Canada.
Saskatoon Environmental Society, Box 1372, Saskatoon, Saskatchewan S7K 3N9, Canada.

References: (1) *FT MIY* 1982. (2) *E&MJ* 3/86. (3) *RTZ Uncovered* (Partizans), 1985. (4) *FT* 22/12/83. (5) *MAR* 1985. (6) *MM* 1/86. (7) *MJ* 14/9/84. (8) *Nuclear Fuel* 16/2/81. (9) Thomas Neff, *The International Uranium Market*, (Harper and Row) Cambridge, 1985. (10) *MIY* 1985. (11) *MJ* 14/9/84. (12) *FT* 31/7/85. (13) EPRI *Foreign Uranium Supply*, 1978. (14) *MJ* 2/3/79. (15) *Nuclear Newsletter*, Saskatoon, Vol. 5, No. 2, 17/2/81. (16) *Kiitg* No. 6, 1980. (17) *MJ* 18/12/81. (18) *Star Phoenix*, Saskatoon 10/12/83. (19) *Key Lake Board of Inquiry Final Report*, Saskatoon, 2/81. (20) *The Economics of Uranium in Saskatchewan* (ICUC) Saskatoon, 6/82. (21) *E&NJ* 9/84. (22) *Leader Post*, Regina 10/12/83. (23) *Nuclear Free Press*, Peterborough, Winter /84. (24) *Nuclear Free Press*, Peterborough, Winter /83. (25) *Nuclear Free Press*, Spring /84. (26) *FT* 31/7/81. (27) *Leader Post*, Saskatoon 10/12/83. (28) *MAR* 1984. (29) *CMJ* Toronto, 6/84. (30) *CMJ* 6/84, as quoted in *MAR* 1985. (31) Press release, RGNNS, 22/3/84. (32) *Star Phoenix*, Regina, 10/1/84. (33) *Leader Post*, Saskatoon, 7/1/84. (34) *Leader Post*, Regina, 9/1/84. (35) *Leader Post*, Regina, 10/1/84. (36) *Star Phoenix*, Saskatoon, 9/1/84. (37) *Toronto Globe and Mail* 9/1/84. (38) *Leader Post*, 11/1/84. (39) *Star Phoenix* Saskatoon, 19/1/84. (40) *Leader Post*, Saskatchewan, 13/1/84. (41) *Star Phoenix*, Saskatoon, 17/1/84; and ibid 19/1/84. (42) *Leader Post*, Regina, 18/1/84. (43) *Star Phoenix*, Saskatoon, 19/1/84. (44) *Star Phoenix*, Saskatoon 20/1/84. (45) *MJ* 20/1/84. (46) *Prairie Messenger*, Saskatoon, 22/1/84. (47) *Star Phoenix*, Saskatoon, 8/3/84. (48) *Prince Albert Herald* 2/3/84. (49) *Leader Post* 9/2/84. (50) *Leader Post* 11/2/84. (51) Press Release, Pinehouse local community authority, No. 9, Pinehouse, 7/2/84. (52) *Preliminary Report on Spill of Mine and Tailings Water, Key Lake Uranium Mine, Northern Saskatchewan* January 5-7, 1984 (Environment Canada), Regina, 26/1/84. (53) Press Release, RGNNS, 22/3/84. (54) *MJ* 1/6/84. (55) *MJ* 14/9/84. (56) *Nuclear Fuel*, Washington, 4/7/83. (57) *MJ* 16/8/85. (58) *Uranium Traffic* 9/81, Saskatoon. (59) *WISE Communiqué*, Amsterdam, 31/5/83. (60) *WISE Communiqué* No. 246, Amsterdam, 1/3/86. (61) *MAR* 1987. (62) *MAR* 1988. (63) *MAR* 1990. (64) *FT Min* 1991. (65) *E&MJ* 3/90; *E&MJ* 3/91.

348 Keywest Exploration Ltd (AUS)

This gold, diamond and mineral exploration company entered a JV with Cliffminex (33.33% partner) on its Minindi prospect at Carnarvon, Western Australia, in the late '70s (1). By 1983, it had withdrawn (2).

References: (1) *Reg Aus Min* 1981. (2) *Reg Aus Min* 1983/84.

349 Kidd Creek Mines Ltd

Formerly Texasgulf-owned, Kidd Creek was taken over by the government-owned Canadian Development Corporation (CDC) during the 1981 Texasgulf/Elf-Aquitaine merger (1). It operates a growing, but not spectacularly

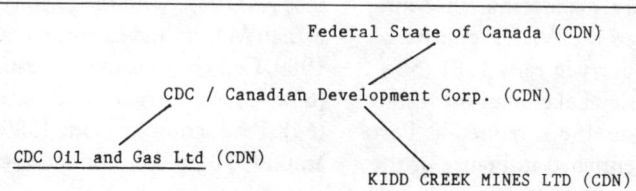

```
            Federal State of Canada (CDN)
                              /
         CDC / Canadian Development Corp. (CDN)
          /                           \
CDC Oil and Gas Ltd (CDN)
                        KIDD CREEK MINES LTD (CDN)
```

profitable copper-lead-zinc-silver property at Kidd Creek, Timmins, Ontario (2).

In the early 1980s, Kidd Creek wanted to develop a uranium deposit in Hants County, Nova Scotia, and had a US$2.5M exploration programme on the drawing board (3), but in January 1982 the Buchanan government set up an inquiry into uranium mining preceded by a moratorium. Kidd Creek became the main pro-mining representative at the Inquiry until November when it suddenly pulled out – claiming it couldn't afford to proceed (4). Shell Canada Resources then effectively took its place.

Kidd Creek owns 40% of Allen Potash Mines and 35% of Nanisivik Mines Ltd (5).

References: (1) *MIY* 1982. (2) *MJ* 7/12/84. (3) *Nuclear Free Press* (Ontario) Fall 1982. (4) *Nuclear Free Press*, Winter 1983. (5) *E&MJ International Directory of Mining*, 1991.

350 Kilborn Ltd (USA)

This is a long-established (1947) engineering and construction management company which has built mines and plants in more than 30 countries, including a uranium plant and its extension in South Africa (1).

Kilborn Engineering Pacific Ltd provides consulting and design engineering expertise for virtually every aspect of mine planning, including coal gasification (2). It specialises in gold, coal, base metals, industrial minerals and uranium.

Founded over forty years ago, Kilborn is 100% Canadian-owned, with its ownership distributed among some 300 key employees. "No project is too small or too large for Kilborn" is a company motto.

In recent years, it has also been a regular participant at Indonesian mining conferences, and designed several gold projects both in Indonesia and Papua New Guinea (3).

References: (1) *MM* 3/83. (2) *E&MJ International Directory of Mining*, 1991. (3) *Asia Pacific Mining*, March 1990.

351 Kirkwood (USA)

It was reported to be exploring for uranium in Texas in 1978 (1).

Nothing more known.

References: (1) *E&MJ* 11/78.

352 Koppen (USA)

It operated a uranium mine in the San Juan Basin area of New Mexico in the late 1970s called the Spencer shaft. It was also planning a 230-450 foot drift from that mine at lease number TL 2NR9W Section 6 in the Ambrosia Lake area (1).

References: (1) *San Juan Study*.

353 Kratos Uranium NL (AUS)

This is a minerals exploration and real estate investment company "especially interested in uranium" (1). In 1975 Kratos became a 26% participant in a JV to explore for uranium in the Pandanus Creek area of Australia's Northern Territory (NT), along with Wyoming Mineral and (through Minatome) Pechiney, otherwise known as the Caramel prospect.

It also had a JV in New South Wales with Stella Mining, though not exploring specifically for uranium (2).

According to the Register of Australian Mining, Kratos bought out Wyoming Mineral's interest in both Pandanus Creek and the Jim Jim prospect in the NT in late 1981, but though it now holds just over 50% of Pandanus Creek, both Kratos and Minatome have declined to pay out any more on exploration (3).

References: (1) *Reg Aus Min* 1976/77. (2) *Reg Aus Min* 1981. (3) *Reg Aus Min* 1983/84

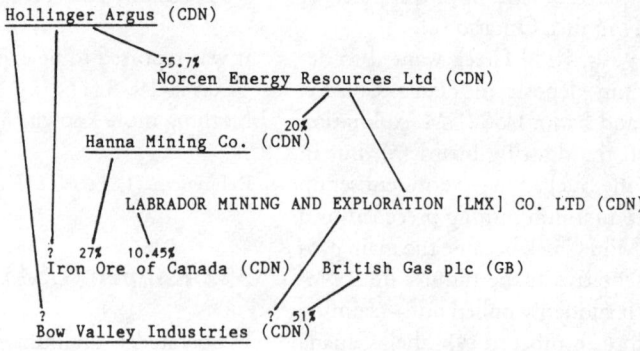

Hollinger Argus (CDN)
35.7%
Norcen Energy Resources Ltd (CDN)
20%
Hanna Mining Co. (CDN)
LABRADOR MINING AND EXPLORATION [LMX] CO. LTD (CDN)
? 27% 10.45%
Iron Ore of Canada (CDN) / British Gas plc (GB)
?
? 51%
Bow Valley Industries (CDN)

354 Labrador Mining and Exploration Co Ltd

As its name suggests, this is *the* Labrador mining company: it acquired exclusive rights to some 20,000 square miles of western Labrador – although it is also exploring for minerals in Quebec, British Columbia and Ontario, and looking for oil and gas in Alberta (1).

It is also one of the holding companies of Iron Ore of Canada, which hews a large amount of ore from LMX's lands in the Labrador City and Schefferville area of the region (2).

Until 1982 LMX was controlled mainly by Hollinger Argus and Hanna Mining (with 20%), but that year Hanna transferred to Norcen its interests in it after a take-over battle (3). In 1964 the company gained a 90-year lease to explore for minerals on Inuit and Innut land in Ntesinan (Labrador/Quebec), followed by another 60-year lease over an additional 300 square miles (4).

(For further details of the relationship between LMX, Norcen and the Iron Ore Company of Canada, please see Hanna – now M A Hanna Co).

References: (1) *MIY*1981. (2) *MIY*1982. (3) *MJ* 23/7/82. (4) RL Barsh, in *Kiitg* 6/82.

355 Lacana Mining Corp

Lacana is owned 25% by Westmin Resources Ltd, which is in turn owned by Brascan. It operates in Nevada (USA), Costa Rica, Guatemala, and Mexico where it gains most of its revenues; it has interests in oil, gas, gold, silver, zinc, fluoride – and uranium (1). It owns the Blizzard uranium prospect in British Columbia with a Norcen JV group *(qv)*.

In 1980, JV exploration programmes "with other partners were also completed, principally for uranium" (1), and it joined a JV with Noranda, and E & B Explorations to search for uranium in New Brunswick, Canada. It was at the same time reported exploring near Pugwash, Newfoundland (3).

There was also a JV exploration with Dome Petroleum and Rayrock Mines in Nevada (1). And it was part of a gold-search consortium in Ontario which includes Inco's Canadian subsidiary Canadian Nickel, and Esso Minerals Canada (2) (see Norcen).

In summer 1982 the company "wrote off" its interests in Guatemala (4).

In the early eighties, Lacana expanded, particularly in Mexico: its assets in mid-1984 were in excess of C$100M. Having "achieved considerable success in recent years", Lacana has inter-

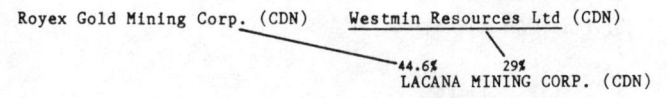

Royex Gold Mining Corp. (CDN)   Westmin Resources Ltd (CDN)
44.6%   29%
LACANA MINING CORP. (CDN)

ests in "some of the world's lowest-cost silver and gold mines" (5).

In 1988, Lacana merged with Royex Gold Mining Corp, Mascot Gold Mines Ltd, Galveston Resources Ltd, and International Corona Resources Ltd, to form Corona Corp.

Corona has around one dozen gold mining interests across Canada, the USA and Mexico (6) with a gold prospect in eastern Greenland, where it is in a 51% JV with Platinova Resources Ltd.

Its interests in Mexico, held through Lacana, are the 30% it owns in Cia Minera Las Torres SA de CV, which operates four silver-gold mines in Guanajuato (6).

With gold interests in British Colombia, and particularly its 50% interest (Teck 50%) in the Williams gold mine at Hemlo, Corona is Canada's largest gold producer (7). The Williams mine (formerly Page-Williams) was awarded to Corona in 1989, after one of the longest-running internecine battles between mining outfits saw LAC losing a court case which had lasted eight years (6,7).

References: (1) *MIY* 1981. (2) 1st quarter Report, 1981. (3) *Energy File* 4/80. (4) *MJ* 3/9/82. (5) *MJ* 8/6/84. (6) *FT Min* 1991. (7) *MAR* 1990.

356 Lakeview Mines

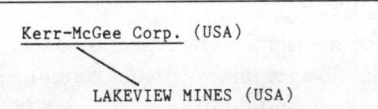

It was acquired by Kerr-McGee in 1976 and merged (1). Now presumed defunct, due to Kerr-McGee's withdrawal from the uranium industry.

References: (1) US Congress Sub-committee on Energy statement 1975.

357 Leighton Mining NL

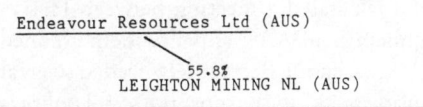

This is part of Bond Corporation's attempt to buy into the Beverley Uranium Project in South Australia.

Leighton reorganised itself in the 1980s: Leighton Holdings now owns Leighton Resources Pty Ltd and Leighton Contractors Pty Ltd, as well as Thiess Contractors Pty Ltd, thus making it "easily Australia's largest mining contractor", especially in gold mine construction (1). Hochtief Ltd holds 47.16% of Leighton Holdings (1).

References: (1) *Reg Aus Min* 1990/91.

358 Lennard Oil NL (AUS)

It holds 50% of a JV at Cundeelee uranium prospect in the Eucla Basin, Western Australia, along with Magnet Metals (1).

Limited work was carried out on the prospect in 1982 and a JV partner was being sought the following year (2).

Lennard entered the nineties without any major new mining interests (although one of its subsidiaries is Southern Iron Ore Ltd). The Walhalla Mining Company NL and Associates controls Lennard Oil with 52.7% of the issued capital (3).

References: (1) *AMIC* 1982. (2) *Reg Aus Min* 1983/84. (3) *Reg Aus Min* 1990/91.

359 Leon Templesman and Sons (AUS)

In 1978 it was part of a consortium including Union Carbide and Cogéma which was to evaluate the Booue uranium deposit in Gabon (1).

Leon Templesman is associated with Anglo/De Beers in the USA: it was Maurice Templesman who facilitated a meeting between Harry Oppenheimer of AAC and President Kennedy in 1959, when De Beers sorely needed to create an "inside track" to the most powerful politician of the day (2).

References: (1) *MMon* 4/81. (2) David Pallister, Sarah Stewart and Ian Lepper, *South Africa Inc.: The Oppenheimer Empire* (Corgi) 1987.

360 LKAB

```
                 Kingdom of Sweden (S)
                          \
      LKAB / LUOSSAVAARA-KIIRUNAVAARA AB (S)
```

LKAB is wholly state-owned, with many subsidiaries involved in iron ore, sulphide ores, copper production, coal, explosives production, and exploration for a number of other metals (1).

Together with Boliden, LKAB mined uranium in the Ranstad district of Sweden until work was stopped in 1980 due to considerable local opposition (including two local councils or *kommunes*) and depression of prices (2).

In 1975, together with AB Atomenergi and the state power board, the company started studying the possibility of extracting uranium from alum shales in the Billinger area of the country. The company's plans for a multi-mineral project (in effect a means of continuing uranium exploration on a more diffuse basis) were rejected by the kommunes of Skövde and Falköping in the Ranstad area in late 1977 (3).

More recently, it applied for permission to mine uranium at Pleutajokk in northern Sweden – an area used by Scandinavia's original people, the Sami (mis-named "Lapps"). It was envisaged that this mine – smaller than Ranstad but of higher-grade ore – would meet about a third of Sweden's voracious demand for uranium for its twelve nuclear power stations to be commissioned by 1984 (4). In fact, the pro-posal to mine at Pleutajokk was rejected in 1981 (5). It is, however, only one of several known uranium deposits in northern Sweden in which LKAB is interested.

In late 1979, LKAB reached agreement with the Yugoslav mining company Rudnik Uraba Zirovski to advise on layout, mining methods, ventilation, haulage systems and other operations at the company's planned underground uranium mines (6).

In May 1982, LKAB was all set to get a loan from the EEC to expand its mines in northern Sweden (7).

Together with the Swedish subsidiary of BP, the company was also preparing to explore for minerals in central and southern Sweden (8).

In 1984, LKAB contracted with West Germany's Peine-Salzgitter to supply 2 million tons of iron ore worth nearly US$400M (9). Salzgitter is a state-owned company which announced that it would take nuclear waste (about half a million cubic metres of it) in its abandoned Konrad mine in Lower Saxony (10).

In 1990, LKAB celebrated its one hundredth anniversary to plaudits from the mining industry around the world. It also announced a 68% increase in profits for 1989 (11). Though 1988 profits were badly affected by declining dollar prices, production increased in 1989 to capacity: indeed, it lost markets to other producers, because it could not supply enough iron ore (12).

The late eighties saw expansion underway at its Malmberget mine (13) and continued copper, tungsten and fluorspar production from mines in the Kiruna region (14).

In a reflection on the previous century, the company's President Wiking Sjöstrand predicted continued buoyancy for what is western Europe's only major iron ore producer – and the world's largest underground mine of its kind (15), where workers allegedly have "some of the best paid blue-collar jobs in Sweden" (15).

LKAB's major sales drive in the nineties will be towards markets which currently rely on imports from Canada, Brazil and Australia (15).

If there is a "best managed" mine in the iron industry it is almost certainly LKAB's at Kiruna. It is tempting to propose that the company should be supported by buyers which might otherwise finance further expansions of more dubious and destructive projects such as Carajas in Brazil, Hamersley in Australia (see CRA), or the Mount Nimba range in the Guinean rainforests (see Granges). However, LKAB remains the biggest single exploiter of land rightfully belonging to the Saami – who have virtually no control over its operations. And, according to Wiking Sjöstrand, the future of LKAB's operations (whether or not it is partly privatised as seems fairly likely) (15), depends on continued nuclear power generation. Without reliance on uranium-fuelled power plants, says Sjöstrand, LKAB's costs will double with a "devastating impact" on the company's fortunes (13,15). Hm!

Further reading: For adulatory articles on LKAB's operations see *E&MJ* 8/90, and *MM* 6/90.

Contact: Folkkampanjen Mot Karnkraft, Box 4509, S-10265 STOCKHOLM, Sweden.

References: (1) *MIY* 1982. (2) *MJ* 3/7/71. (3) *MJ* 18/11/77.(4) *FT* 1981 (date unknown). (5) *Kiitg* No. 7. (6) *MJ* 14/9/79. (7) *FT* 19/5/82. (8) *MJ* 7/5/82. (9) *MJ* 25/5/84. (10) *MJ* 6/4/84. (11) *FT* 27/9/90. (12) *MJ* 14/7/89. (13) *Metal Bulletin* 14/6/90. (14) *MJ* 4/8/86. (15) *FT* 27/9/90.

361 Lochiel Exploration Ltd (CDN)

It was part of the Pan Ocean Oil consortium drilling at South Bissett Creek in the Baker Lake area of Canada's Northwest Territories in the early eighties. (1).

References: (1) *MJ* 8/2/80.

362 Lomex Corp (USA)

One of the few corporations to openly admit

that the anti-nuclear lobby defeated its plans, Lomex began exploring for uranium between 1971 and 1977 in San Luis Obispo county, California. (1). Permission was given by the US Forest Service which decided no public information or participation was necessary; nor did they require the company to cap, plug or seal the holes they left. As a result – according to the Oak Tree Alliance (2) – water quality readings showed radioactive contaminant levels "many times higher than governmental safety standards". Lomex's drilling preceded a NURE (National Uranium Resource Evaluation) project involving a satellite scan of the whole county, which revealed promising deposits of uranium and granted the area an "A" rating (3).

The Oak Tree Alliance campaigned vigorously against Lomex's operations. So did the Red Wind Native American community, whose Foundation settlement in the heart of the Los Padres National Forest was one of two areas identified by the county's environmental health officer as emitting radiation greater than that permitted by the EPA: 21.4 picocuries per litre of water above the accepted 10 picocurie limit (4).

Their joint campaign forced Lomex to submit a federal environmental impact statement – the first time this has been demanded for exploratory drilling. As Lomex president Hiram Bingham commented wryly: "The environmental climate out there is not a positive one to work in; the environmental climate in San Luis Obispo County especially!".

References: (1) *Kiitg* 1/83. (2) *Country News*, San Luis Obispo, 21-27/7/82. (3) *Sun Bulletin*, Morro Bay, California, 8/7/82. (4) *San Luis Obispo T/T* 24/6/82.

363 Lone Star Mining Co (USA)

This company joined Union Carbide in exploring for uranium on the borders of the Santo Domingo pueblo, New Mexico, in the late '70s. The move was vigorously rejected by the pueblo, especially as the area contains important sites sacred to the people, driven off their land by the Spanish *conquistadores*. The pueblo government protested: "These are places where our ancestors lived and were buried ... They are not available for digging, destruction and exploitation" (1).

It is presumed that Lone Star may have been a subsidiary of Lone Star Steel company, which has full mine-to-products operations and exploration in Texas, and is owned 100% by Northwest Industries Inc (2).

References: (1) *Denver Post* 4/11/79. (2) E&MJ International Directory of Mining, 1991.

364 Loraine Gold Mines Ltd

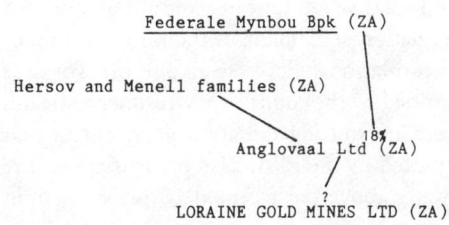

Firmly within the Anglovaal camp, Loraine was nonetheless 21% shareheld by the Anglo-American group, 27% by GFSA, 10% by Barclays Bank, and only 8% by Anglovaal in the late '70s (1).

A minor producer of uranium (only 329 tonnes in all to the end of 1975 (2) and under 16 tonnes in 1983) (3), Loraine's uranium-bearing concentrates are a by-product of pyrite extraction and sold to Harmony Gold Mining.

The mine's third quarter results in 1984 showed a profit nearly three times that of the previous quarter (4).

Though still listed as a uranium producer in the 1991 *E&MJ Directory of Mining* (5), the company has apparently ceased all production of uranium oxide.

References: (1) Lanning & Mueller. (2) *EPRI* Report. (3) *MAR* 1985. (4) analysis of Rand and OFS quarterlies 9/84 in: *MJ* 26/10/84. (5) *E&MJ International Directory of Mining* 1991.

365 Lucky Mc Uranium Corp

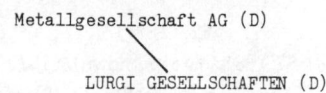

This company was part of Pathfinder Mines, formerly controlled by Utah International, and now by Cogéma.

In 1977 the Japanese Atomic Power Co authorised the payment of US$126M to cover import of 1430 tons of uranium ore from Lucky Mc – this to come from Big Eagle (where mining started in 1977) at Green Mountain in the Gas Hills area of Wyoming (1).

References: (1) *MJ* 31/10/77.

366 Lurgi Gesellschaften

```
Metallgesellschaft AG (D)
                  \
        LURGI GESELLSCHAFTEN (D)
```

This subsidiary of Metallgesellschaft (qv) is a major engineering construction company with expertise in uranium mining. It manufactures complete plants, operates throughout the world, and in 1983 had subsidiaries in 16 countries (1).

In 1989, Lurgi was awarded a contract by the Brazilian Cia Minera de Metais, to supply know-how and engineering for roasting, gas and leaching sections of a new electrolytic zinc smelter claimed by Lurgi to be "both ecologically compatible and economically efficient." It supposedly discharges the iron content of zinc as hematite, which can then be used in cement (2).

References: (1) *MM* 3/83. (2) *MM* 11/89.

367 Madawaska Mines Ltd

Federal Resources Corp. Ltd (USA)

MADAWASKA MINES LTD (CDN)

Until 1982, Madawaska operated a uranium mining/milling complex near Bancroft, Ontario, producing uranium largely for Agip and Canadian utilities. (For production and contract details see Agip and Consolidated Canadian Faraday.)

As of June 1982, the Madawaska mine began closing down – Agip's contract expired, and Ontario Hydro spurned an offer to buy the mine's uranium at US$40/lb in favour of a contract with Saskatchewan's Key Lake mine (1). However, when Federal Resources tried to sell off Madawaska in 1982, Consolidated Canadian Faraday resisted because of "future environmental and stand-by obligations" (1). Madawaska Mines meanwhile continued its limited partnership in certain oil and gas exploration projects, and these, in 1982, proved the only money-spinners for Federal Resources (2).

References: (1) *Nuclear Free Press* (formerly *Birch Bark Alliance*) Summer/82. (2) *MJ* 7/1/83.

368 Magnet Metals Ltd

MAGNET METALS LTD (AUS)

49.7%
Western Energy (AUS)

It mines and markets peat – also likes yellowcake, for in 1975 it had 65 uranium claims at Norseman, Western Australia. It joined with Japan's PNC in 1977 to search for the deadly metal (1). According to the *Financial Review* (Australia) in 1982 (2), Magnet companies (based at Perth, Western Australia) "… hold the biggest area of exploration permits in Australia. The company is the operator in 10 of 39 per-

mits in which it has an interest. These total 5,550,000km² and involve possible spending commitments of US$280M over the next few years."

It was engaged in uranium exploration in the Eucla Basin of Western Australia, north-west of the Cundeelee Aboriginal mission, in the late '70s, together with Cliffminex (3). It also held half of a JV with Lennard Oil at the Cundeelee uranium prospect, about 200km north of Zanthus (4).

It also owns 49.8% of Lennard Oil NL (5).

References: (1) *Reg Aus Min* 1975/76. (2) *AFR* 20/5/82. (3) *Reg Aus Min* 1981. (4) *AMIC.* (5) *Reg Aus Min* 1990/91.

369 M and M [MM] Mining (USA)

It was actively exploring for uranium in New Mexico in the late '70s (1).

Otherwise nothing known.

References: (1) *E&MJ* 11/78.

370 Manitoba Mineral Resources Ltd (CDN)

It was in a JV with Eldorado Nuclear and Imperial Oil, as well as Warren Explorations in the Churchill area of Manitoba in the late '70s (1). Manitoba Mineral Resources is a government agency, dedicated to mineral exploration and development in the province (2). It holds 27% of the Trout Lake JV near Flin Flon, headed by Hudbay, though it sold its interest in another Hudbay JV at Cullinan. In 1989, it also purchased the province's interest in the Tantalum Mining Company (3).

References: (1) *EPRI* Report. (2) *E&MJ International Directory of Mining* 1991.

469

371 Manny Consultants (CDN)

This is a tin-pot company that will go down in history as the drillers which dared take on an entire community – and lost! They also lost the first round in the British Columbian uranium war of the late '70s.

Manny Consultants came to Genelle Creek, British Columbia, to drill 30 exploratory holes on both sides of China Creek. When residents discovered them blasting away– and after exhausting conventional avenues of protest– they turned to direct action. The road to the exploration site was barricaded, picketed, and work obstructed in a number of ways over a period of months. Then three men – the so-called Genelle Trio – sat in the path of an on-coming bulldozer and were arrested. A woman also sat on top of a helicopter removing ore samples, and refused to budge (1).

When the Genelle Trio where hauled into court, the judge issued a classic judgement, suspending sentence on the men, and pointing out that, historically, civil disobedience may be justified where a clear danger existed to the liberties of the people ... (2).

References: (1) *Kootenay Nuclear Study Group Newsletter* Fall/78. (2) *Energy File* 1979.

372 Mapco Inc (USA)

In 1979, it was reported uranium exploring in the Mount Prindle area, north-east of Fairbanks, Alaska, where uranium content of 5-7% (*ie* very high) was discovered in certain samples. Drilling permits were applied for – but this was the time of the much-vaunted Alaska land protection legislation of President Carter, and it seems Mapco's claims were rejected (1).

It has also been part of a consortium which includes Hunt Oil, exploring for oil in the southern jungles of Peru (2).

Mapco operates coal fields in east and west Kentucky, Illinois, Maryland and Virginia (3).

References: (1) *MJ* 5/1/79. (2) *MAR* 1982. (3) *FT Min* 1991.

373 Marathon Oil Co

In 1981, local people in Murray County, southwestern Minnesota, found Marathon, along with Exxon and Texasgulf, apparently searching for minerals, including uranium. The State Department of Natural Resources (DNR) claimed this was "unlikely". However, state law does not demand disclosure of the type of mineral exploration (1).

Marathon is 51% controlled by US Steel, itself formerly an important producer of uranium. So the US's leading steelmaker united with the country's seventeenth biggest oil company – thus producing what the London *Financial Times* has called "a new industrial giant ... ranking twelfth in the *Fortune* list of the 500 largest US groups" (2).

(More on this latest "take-over battle" by (and for) oil companies is included under Mobil – the oil giant which lost its bid for Marathon when it stumbled against US anti-trust legislation. But it's relevant to note that, in early 1982, Mobil threatened to buy 25% of US Steel to gain entrée into Marathon by a different route (3). And that Marathon admitted it paid out some US$11 million of shareholders' money to resist Mobil's unwelcome bid) (4).

Marathon Petroleum (Australia), a subsidiary of Marathon Oil, was a JV partner with International Mining Corp on uranium prospects in the Northern Territory of Australia. However, in 1982 it withdrew from its Rum Jungle uranium venture, although it had reported "significant uranium mineralization" (5).

Marathon also held 70%, together with North Flinders Mines (30%), in the Parabarana cobalt-copper-uranium prospect, nearly 600 km north-east of Adelaide (6).

In 1984, Marathon agreed to sell its base and precious metals interests in Australia to Pan Australian Mining of Brisbane for an undisclosed sum. Its interests in coal in the Surat Basin would be retained, but Marathon will

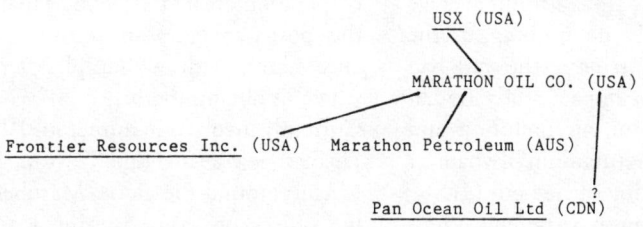

```
                              USX (USA)
                                  |
                                  |
                       MARATHON OIL CO. (USA)
                          /        \
                         /          \
Frontier Resources Inc. (USA)   Marathon Petroleum (AUS)     |
                                                             |
                                                             ?
                               Pan Ocean Oil Ltd (CDN)
```

surrender properties in every state except Tasmania and the Northern Territory (7).

In late 1984, the company announced that it would start drilling for oil in the Toledo district of Belize – an area covetted by the military régime in Guatemala and occupied by native Belizeans (8).

Marathon has unexploited coal deposits near Macalister, in the Darling Downs region of Queensland (9).

The company was also in the middle of a steamy South American scandal in 1987. It apparently connived with officials of the Ministry of Mines in Colombia to disguise its failure to report on the La Jagua coal claims in the northern department of Cesar.

Marathon acquired the leases some time before a 1980 Colombian law, which allowed foreign ventures not to sign an association contract with the state (and thereby share its profits), so long as the licences were being worked by that time. In fact, the company had done nothing on the leases: it was forced to abandon its holding in 1983. Since then, five Mines officials have been found guilty of falsifying documents and "abuse of trust." They received 14-month suspended prison sentences, while Marathon got away scot-free (10).

Contact: Primo Coc, Toledo Indian Movement, Toledo, Belize.

References: (1) *Northern Sun News* 7-8/81. (2) *FT* 8/1/81. (3) *FT* 13/1/82. (4) *FT* 23/1/82. (5) *MJ* 19/11/82. (6) *AMIC* 1982. (7) *FT* 4/5/84. (8) *Infopress* (USA) 29/11/84. (9) *Reg Aus Min* 1990/91. (10) *E&MJ* 9/87.

374 Marline Uranium Corp

According to CBC News in 1980 (1), this was a " ... Canadian company investing heavily in the US".

It has prospects in the Athabasca sandstone (2) of Saskatchewan and has leased more than 50,000 acres in Virginia, USA, for uranium exploration. In 1978, it gained mineral rights in Pittsylvania County, and two years later, its first lease in Orange County (3).

Local opposition came from the Culpepper Soil and Water Conservation District Board and others (3). As of February 1982, opposition groups were trying to get a moratorium opposing uranium mining in Virginia through the state legislature.

In summer 1982, the corporation reported a uranium lode estimated to contain 30 million pounds of high grade ore north of Chatham,

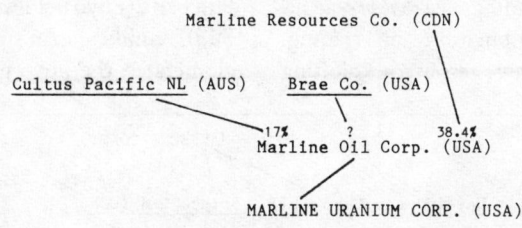

```
               Marline Resources Co. (CDN)
                                        \
                                         \
Cultus Pacific NL (AUS)    Brae Co. (USA)  \
              \              |              \
               \             |               \
            17%           ?            38.4%
               Marline Oil Corp. (USA)
                        /
                       /
              MARLINE URANIUM CORP. (USA)
```

Virginia (4). Grading was 1.82kg/tonne average (5). The company could do nothing, as the Coal and Energy Commission of the state had imposed a ban while an impact study (radiation, groundwater control, air pollution and "safe means of processing uranium" [whatever that may mean!]) was being carried out (4).

1982 reports on the Swanson orebody in Virginia – which, as *Mining Journal* pointed out, was "the first discovery of a major uranium province in the US for 20 years" – indicated that the deposit contained at least 30 million pounds of uranium oxide at an average grade better than 4lb/tonne. Union Carbide joined the project with an option to develop the deposit jointly with Marline if results were favourable (6).

In 1983 the state of Virginia extended its uranium ban until July 1984 – with the agreement of Marline (7) and little has been reported since.

Contact: Uranium Information Organization, 150 Belleview Ave, Orange, Virginia 22960, USA.

References: (1) CBC News, 14/10/80. (2) *Who's Who Sask.* (3) *Kiitg* 2/82. (4) *MJ* 6/8/82. (5) *MJ* 3/9/82. (6) *MJ* 24/12/82. (7) *MJ* 11/3/83.

375 Martin Trost Associates (USA)

In 1978, it was involved with Portland General Electric in constructing a uranium mine (Miracle mine) at Bakersfield, California. (1).

References: (1) *US Surv Min* 1979.

376 Marubeni Corp

This is one of Japan's leading *sogo shosha* – an almost indefinable combination of trading, manufacturing and raw resource-exploiting corporation created solely by Japanese men in the post-war years to secure the country's "necessities', with few scruples as to how (and where) it obtains them.

Currently involved in numerous JVs with other Japanese *sogo shosha* (like C. Itoh, Hitachi and Mitsui) around the globe, Marubeni has satellite offices on every continent and in most major commercial centres. These include: New York, Montreal, Rio de Janeiro, Buenos Aires, Havana, London, Hamburg, Zurich, Paris, Rotterdam, Amsterdam (its European HQ), Las Palmas, Helsinki, Oslo, Athens, Sofia, Warsaw, Moscow, Lagos, Casablanca, Addis Ababa, Kinshasa, Baghdad, Beirut, Ankara, Cairo, Teheran, Tripoli, Lahore, Chittagong, Calcutta, Singapore, Bangkok, Taipei, Seoul, Hong Kong, Melbourne, Wellington, Noumea and Papeete (1).

While listing itself as primarily a textile company, the proportion of its textiles – to other – sales fell from 67% in 1955 to 21% in 1976, while the importance of metals and machinery rose from 11% to 49%. (This switch was also reflected in the production figures of Mitsubishi, C. Itoh and Nichimen over the same period) (2).

Domestically, it controls distribution for numerous subsidiaries and affiliates; overseas its tentacles embrace (among others) Don Juan sportswear (USA), Total Steel of Australia Pty, Nissan Motors of Peru, Sanyo Marubeni (GB), Kodiak King Crab (USA), Kingstone Tire Agency (Australia) (3) and Dampier Salt (Australia) (in which CRA has 64.94%) (4).

Marubeni is also the "leading organiser" of the Fuyo group of corporations which the company claims is a "very liberal club, unlike the Mitsubishi or Mitsui groups" (1). The common link for all the companies in this group is the Fuji bank (Fuyo being another name for Mount Fuji). Unlike Japan's more conservative conglomerates, the Fuyo group is not founded on

```
                                     MARUBENI CORP. (J)
                                                  /|
                                     ?/       2%/
     Tokyo Uranium Development Corp. (J)   Brinco Ltd (CDN)
```

the *zaibatsu* tradition but claims to be "independent" and "open". Members of this group include Nissan Motors, Hitachi, Canon (the camera company), and Showa Denko. Fuyo spawned the Ocean Development and Engineering Company which is involved in oceanic research and "construction" including "the designing of ocean parks" *(sic)* (1).

Marubeni had more than a third of CDCP Mining in the Philippines, and led a consortium (with Mitsui) in constructing a copper smelter at Isabela, Philippines; Seltrust (formerly BP's major mining arm) did the feasibility study for this project (4).

The corporation also has an exclusive right to 70% of the copper produced by Marinduque's smelter – Marinduque being a leading Philippines mining company operating on Negros Occidental, Samar, Sipalay and Bagacy (5).

It has a JV with CRA (technically with CRA Exploration, or CRAE, holding 25%) at the Frieda River prospect on the Sepik river in Papua New Guinea.

Marubeni heads the OMRD consortium (31.25%) while MIM has another 37.5% and Norddeutsche Affinerie AG creeps past the starting line with 6.25% (4).

Marubeni's importance to Japan's nuclear programme is indisputable. Back in the mid-'50s it concluded an "historic contract" with Japan's Atomic Energy Research Institute to supply the country's first nuclear reactor (from Rockwell, USA). Later, as well as importing heavy water from DuPont, the company arranged deliveries of uranium oxide from Eldorado Nuclear in Saskatchewan (1).

Together with Japan's leading electrical corporation, Hitachi, Marubeni formed the Atomic Industrial Consortium in 1956 (Showa Denko, CRA's new affiliate, was prominent in this too). The corporation allied itself with the British companies English Electric (now part of GEC) and Babcock and Wilcox to introduce Magnox-type reactors; later it linked up with General Electric in order to bring light-water reactors into the country.

"Marubeni played an active role in bringing many orders for uranium conversion services business to Eldorado Nuclear, whose market share in Japan is roughly 50 percent, and also played an important role for RTZ Mineral Services Ltd in obtaining many orders of uranium concentrates from Japanese electric power companies ... as of the end of 1977 RTZ Mineral Services had received orders from almost all the major power companies in Japan for the supply of approximately 50,000 tons of uranium concentrates, which represents 35 percent of total Japanese purchases over the same period" (1).

The importance of this quote is that it comes from a Marubeni self-adulation book (1) published in 1978 – at a period when RTZ was tight-lipping any information on the activities of its Zug (Switzerland) subsidiary RTZ Mineral Services, and a full four years before the MIT (Massachusetts Institute of Technology) in the USA published the same figures, showing Japan's reliance on Namibian uranium from Rössing in the '70s and beyond (6).

The corporation's uranium involvement doesn't stop here. It has also arranged for Cogéma and BNFL to reprocess the country's spent nuclear fuel. The boat carrying the spent fuel bundles is operated by Pacific Nuclear Transport, a British venture in which Marubeni holds a minor equity portion.

Moreover, in 1974 the corporation launched the Tokyo Uranium Development Corporation along with Nippon Mining, the Fuji Bank and others (1).

Whatever its blithely expressed intentions may be – such as "assisting in the growth of developing nations ... towards industrialisation" or "assisting in the reduction of frictions in international trade" – what perhaps best sums up Marubeni's rationale and motivation is its own account of how and why it moved into the broiler chicken business.

"When most seacoast nations moved to adopt a 200-mile exclusive economic zone ... they unknowingly sparked a change in the eating habits of the Japanese. For years, fish has been the staple of the Japanese diet. The inevitable reduction of fish hauls because of the new fishing zones, however, has forced Japan to look elsewhere for low-cost sources of protein. This search has focused attention on ... chicken.

Marubeni and other trading companies were the first to recognise the commercial value of broiler production and took pains not only to import the basic industrial know-how and equipment but also to establish broiler production as an industrial operation in Japan.

Regardless of what the original motives may have been, the broiler industry has turned out to be an ideal enterprise for trading companies. It differs fundamentally from other raw material import and processing businesses in that the initially imported product – the "grandparent" stock – is a living thing which must be bred, reproduced, multiplied, fattened, slaughtered and processed for sale as meat. Marubeni's functional ability and experience as a trading company have proved most effective in the development of integrated systems. In the production of broilers, a specific instance would be the selection of original breeding stock. Marubeni did well by selecting Ross-I ("chunky" in Japanese nomenclature [sic]) – a superior breed developed by Ross Poultry company, the livestock division of the United Kingdom Tobacco group ... .

"Marubeni can [now] lay claim to being among the world's largest producers of poultry with an annual output of 48 million broilers ..." (1).

Marubeni (50%), along with Sumitomo (30%) and C. Itoh (20%), also operates the Pasar copper smelter and refining complex in the Philippines, which has created considerable controversy among domestic copper mining companies because of the Philippine government's demand that they reserve 30% of their production for the smelter – the companies consider this a cheap form of subsidising an overseas project – especially as they themselves want to sell their own refined copper to Japan. In July 1983, the Pasar partners, in an attempt to allay these fears, stated they would seek customers outside of Japan (7).

In the late '80s the Philippines government finally granted permission for Pasar to import up to 250,000 tonnes a year of copper concentrates after domestic producers had been sending an

increasing amount of their concentrates overseas. (In 1990 this amounted to 56% of total production) (8).

Six years earlier, Marubeni and Pasar offered 300M pesos to Marinduque when it was on the point of bankruptcy, as an advance against future sales (9).

Among Marubeni's more recent ventures has been a US$130M package which it put together in 1984, with Minera Peru and Mitsui, to develop the Cerro Verde II mine in the indigenous Andes (10). The project has since been postponed (11).

Along with the US and Canadian concerns, Marubeni agreed to set up a petrochemical plant in Peking in 1990 (12).

Recently dubbed "TROPICAL FOREST DESTROYER NO.1" by the Japanese Friends of the Earth (JATAN), the company is the prime exploiter of timber in Sarawak – 90% of the territory's exports to Japan purportedly come through Marubeni (13).

At Bintuni Bay in West Papua, Marubeni joined a JV in 1989 with the Indonesian company PT Bintuni Utama Wood Industries, providing more than half the funds for both the destructive logging and a woodchip mill. It has been condemned by JATAN, Indonesian environmental groups, TAPOL and others, for ignoring the claims of the traditional Irarutu land-owners, and the special ecological status of the mangrove wetlands from which the timber is being taken (14).

The same year, Marubeni was accused of offering bribes to secure an over-priced contract for Japanese aid, to supply lifeboats to the Bangladeshi government (15) – less than a year before one of the most devastating cyclones in world history drowned several hundred thousand inhabitants of the world's most impoverished country.

Further reading: For a resumé of the Pasar saga, see *Birds of Prey, Gold and Copper mining in the Philippines, a political economy of the large scale mining industry*, A report for Labor Education Assistance and research (CLEAR), Baguio City and RDC-Kadumi, Baguio City, 10/88 (duplicated).

Contact: SAM – Friends of the Earth Malaysia, 43
  Salween Road, 10050 Penang, Malaysia.
TAPOL, 111 Norwood Road, Thornton Heath,
  Croydon, Surrey CR4 8HW, England.
Down to Earth, P.O. Box 213, London SE5 7LU.

References: (1) *The unique world of the sogo shosha*
1978 (Marubeni Corp). (2) Nikko Research Centre
and company reports for 1976. (3) Marubeni reports. (4) *MIY* 1982(5) *MAR* 1982. (6) *New Statesman* 28/1/83. (7) *MJ* 22/7/83. (8) *MJ* 1/3/91. (9)
FT 3/8/84. (10) *E&MJ* 11/84. (11) *FT Min* 1991.
(12) *FT* 16/11/90. (13) *Pikiran Rakyat* 26/4/89.
(14) *Age* 5/7/90; see also *Down to Earth* (TAPOL),
London, Nos. 1,3,6,7,10, 1989/90. (15) *Samachar* 7-
8/90.

377 Mattagami Lake Mines Ltd

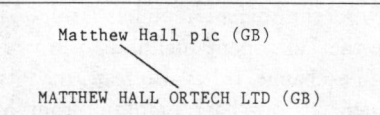

```
      Noranda Mines Ltd (CDN)
                       \
  MATTAGAMI LAKE MINES LTD (USA)
```

Acquired by Noranda in 1979 (1), Mattagami
Lake Mines obtained Fission uranium property
in British Columbia from Seaforth Mineral and
Ore (USA) two years earlier (2).
The Mattagami mine itself was due to be closed
down in 1989 (3).

References: (1) *MAR* 1981. (2) *MJ* 2/12/77. (3)
MAR 1990.

378 Matthew Hall Ortech Ltd

```
      Matthew Hall plc (GB)
                    \
  MATTHEW HALL ORTECH LTD (GB)
```

Specialists in flotation and leaching processes in
mining, the company pioneered the BACFOX
process for the production of ferric sulphate to
be used as a leachant for the production of
uranium yellowcake from ore (1). It also de-

signed and constructed a 500tonne/hour coal
preparation plant for the British National Coal
Board (NCB) at Askern in 1981 (2).

References: (1) Matthew Hall Ortech brochure and
personal communication from T Martin James
21/11/78. (2) *MAR* 1982, quoting Chironis, *Coal
Age* 1/81.

379 Maugh Ltd

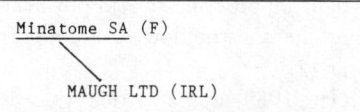

```
      Minatome SA (F)
                   \
      MAUGH LTD (IRL)
```

This is a wholly-owned subsidiary of Minatome
of France which has prospected in Ireland.
It was granted more than £400,000 by the EEC
to prospect for uranium in Eire in 1976 and
1977, as head of a consortium which included
Silvermines Ltd (20%) (1). According to an
Irish publication (2), large grants since then
have not been disclosed.
However, through 1981, the company was
doing extensive uranium exploration in the
country, especially the huge Leinster deposit in
the south-east. They also mounted a JV in an
adjacent Northgate claim. No results have been
announced since 1981 – and by the following
year the company had dropped 26 of its 43
uranium licences (3).

References: (1) *MIY* 1982. (2) *Uranium Grabbers*
(Cork Anti-Nuclear Group) 1980. (3) *MAR* 1982.

380 Maverick Uranium Explorations Inc (CDN)

Formerly Kedar Mining with HQ in Toronto,
the company was exploring for uranium in Saskatchewan (1).
No further information.

References: (1) *Who's Who Sask.*

381 Mechim

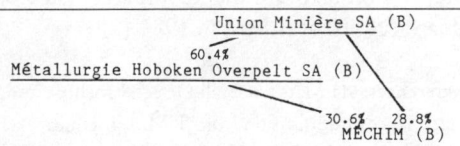

```
                  Union Minière SA (B)
               60.4%
Métallurgie Hoboken Overpelt SA (B)
                          30.6%  28.8%
                          MECHIM (B)
```

This is part of the Umipray company set up in 1979 to develop uranium-from-phosphates recovery on a world-wide basis (1).

It has also been involved in a zinc project in Thailand; phosphoric acid purification in Iraq; and the construction of a copper refinery in Iran (2).

In late 1989, after a battle royal was conducted over the ownership of Societé Genérale Belgique, the structure of Union Minière (qv) Metallurgie Hoboken-Overpelt SA and associated companies changed considerably.

Union Minière was absorbed by ACEC, thus becoming ACEC-Union Minière SA, and itself became an 87% owned subsidiary of the Societé Genérale Belgique. MHO holds 32.91% of Mechim and is itself owned primarily by ACEC-Union Minière and Societé Genérale.

Another 2.25% in Mechim was held (as of late 1989) by CRAM (Cie Royale Asturienne des Mines SA) which itself acquired 10% of Pancontinental Mining in Australia in 1987 (3).

Mechim now serves as the technical and commercial support unit for ACEC-Union Minière (3).

References: (1) MAR 1981, (2) MAR 1982. (3) FT Min 1991.

382 Mecseki Ércbányászati Üzemek AV

```
People's Republic of Hungary (H)

MECSEKI ÉRCBÁNYÁSZATI ÜZEMEK AV (H)
```

Uranium mining has taken place in Hungary for 40 years. In 1947, extensive prospecting started in the Velence, Mecsek and Koszeg-Rohonc mountains, and analysis was made of radioactive sands taken from the river Danube. A significant deposit was located in Mecsek Mountain, to the south-east of Transdanubia. A joint Soviet-Hungarian team detected several important sandstone anomalies near the village of Kövagosozölos in 1953 (1), and mining development started three years later (2).

Mining the ore posed several problems over the years. The ore cannot be evaluated visually, necessitating extensive wall and borehole measurements. Production is at a 900-metre level, although exploration has gone as deep as 1400 metres, making the Mecsek mines the deepest in the country (1). The rocks are hard, so extraction is by blasting, while the underground temperature reaches 60 C causing considerable problems of ventilation.

The company allegedly did extensive research and development work, and developed a metal recovery technology which treats not only ores of low metallic content, but also tailings from the ore dressing plant: costs of recovery are claimed to be only 25% of conventional costs (1).

Uranium from Mecsek kept the Paks Nuclear power station running with two 449MW units, supplying 16% of the country's electric power demand. Further construction at Paks was planned up to the end of the 20th century to meet around 40% of electric power requirements.

Mecsek ore has been enriched and turned into fuel assemblies in the USSR. Laboratory and field equipment at the mine has been designed and manufactured for export (1).

After the "liberation" of eastern Europe in 1989/90, the Soviet Union has continued to receive the Paks spent fuel and to provide new fuel in exchange. It has also reaffirmed its commitment to purchase uranium from Mecsek (3).

However, the ownership and management of the mine underwent dramatic changes in 1990, when the Irish company Glencar took control and instituted a drastic cutback on the workforce (4).

For full details of the implications of this deal, see the Glencar entry in the File.

References: (1) *MM* 4/86. (2) *MM* 7/86. (3) *Nuclear Fuel* 3/9/90, *MJ* 7/9/90, see also *Nuclear Fuel* 1/4/91.

383 Meridian Oil NL (AUS)

Formerly Bamboo Creek Gold Mines NL, the name change in the '70s reflected a diversification into energy minerals including uranium (1).

It secured exploration rights to the Jailor Bore prospect in the Gascoyne region of Western Australia, originally investigated with Pacminex.

It held 16.66% interest in the Marathon uranium claims at Bonnie Creek and Marble Bar in the Pilbara region of Western Australia (2).

The company has recently been concentrating on oil exploration in South Korea and the Middle East (3). As part of Independent Resources Ltd (dominated by Michael Fuller), Meridian was being investigated by the National Companies and Securities Commission in 1989 and 1990 (3). Major shareholders in Meridian are Spargos Mining NL, ANZ Nominees, Yellowstone Exploration Mining and Pty Ltd (3).

References: (1) *Reg Aus Min* 1981. (2) Annual report 1982. (3) *Reg Aus Min* 1990/91.

384 Metallgesellschaft AG

"A monster of monopoly capitalism", Lenin called it (1). This huge West German-based world-wide metals exploration company is also into chemicals and, via Urangesellschaft, uranium. It is Europe's largest producer of lithium, an important producer of lead (2), and the western world's fifth most important producer of tungsten.

It holds interests in three projects potentially af-

fecting native peoples: the Highland copper mine (Metallgesellschaft 21.3%, Teck Corporation 78.7%) in British Columbia (though both companies have shed 30% of their interests in the mine to Kuwaiti institutional investors) (1); the Nanisivik mines of Canada's North-west Territories (11.25%) (3); and the Ok Tedi copper and gold development in Papua New Guinea, near the West Irian border, where it holds 7.5% in a consortium headed by BHP and Amoco (3).

Through its subsidiary AMR, it is also part of Ocean Management Ltd, a seabed consortium (4).

While the company in 1981 was closing down its domestic plants and throwing nearly 2000 people out of work in Frankfurt alone, it has been boosting its own profits by opening processing plants overseas – five in China (3).

During 1982, further major cut-backs involved the sale of its 50% share in an electrolytic zinc smelter to MIM Holdings (Australia), and an aluminium plant to PUK (1).

Apart from its major interest in uranium through Urangesellschaft, it also holds 10% of Nukem.

Board interlocks in Europe include IG Metall (the huge Metalworkers' Union of West Germany), Bayer AG, the Deutsche Bank, and other unions and companies (6).

Also in Europe, it holds 17.5% of an Italian aluminium smelter along with Comalco (controlled by CRA) (20%), and Alluminia Italia (62.5%) (7).

Metallgesellschaft has been active in Namibia, producing lithium which goes to Germany, Japan, France, Belgium and Italy under the South African label (8).

It also has a minority interest in a JV with KWU/Kraftwerkunion to manufacture tubing for nuclear plants (9).

In 1984, together with Degussa, it carried out a feasibility study for the extraction of gold and silver from mine tailings at Vatukoula in Fiji (10).

In 1986, Metallgesellschaft, along with MIM and Teck (and the AMP), bought a controlling interest in Cominco. Later it received another

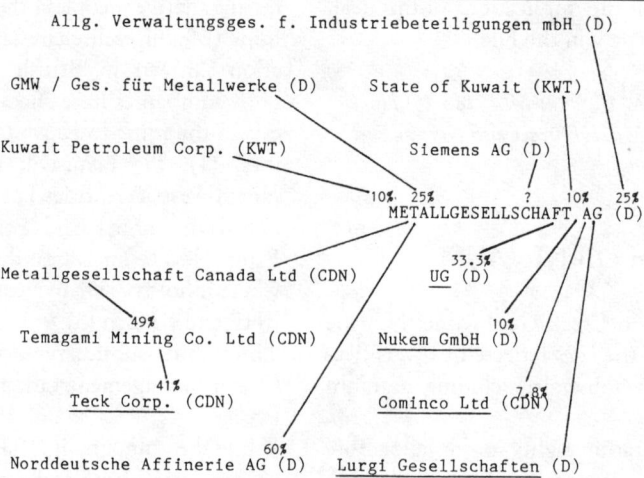

Allg. Verwaltungsges. f. Industriebeteiligungen mbH (D)

GMW / Ges. für Metallwerke (D)   State of Kuwait (KWT)

Kuwait Petroleum Corp. (KWT)   Siemens AG (D)

10% 25%   ?   10%   25%
METALLGESELLSCHAFT AG (D)

33.5%
Metallgesellschaft Canada Ltd (CDN)   UG (D)

49%   10%
Temagami Mining Co. Ltd (CDN)   Nukem GmbH (D)

41%   7.8%
Teck Corp. (CDN)   Cominco Ltd (CDN)

60%
Norddeutsche Affinerie AG (D)   Lurgi Gesellschaften (D)

parcel of Cominco shares, after regrouping its foreign mining assets into a new Canadian-based company called Metall Mining Corp (11).

As of February 1990, Nunachiaq Inc held a 27.7% interest in Cominco and was itself owned 50% by Teck Corp, 25% by MIM Holdings Ltd, and 25% by Metallges AG (12) – now the official company title.

In March 1991 Daimler-Benz, the German motor company, said it would purchase a 10% stake in Metallges from the Dresdner Bank, thus reducing the bank's holding to around 13%. At that time Kuwaiti interests held around 20%, Australian Mutual Provident 5%, and MIM Holdings Ltd 3% of the company, with a joint investment by the Deutsche Bank and Allianz Insurance tipping the scales at 20% (13).

Meanwhile, Metall Mining was floated publicly in 1987, with the parent company retaining 63% and Agip Minière acquiring 8.02% (14). Elsewhere, Metallgesellschaft of Australia Pty Ltd handles a wide range of metal operations and, with its partner MIM, has been exploring a number of deposits, including the Balcooma base metals lode in northern Queensland (15). By mid 1990 Teck, MIM and Metallgesellschaft between them controlled about 8% of the western world's copper production, 12% of its output and 19% of its lead (16).

Although the group has not yet bid for Renison Goldfields (still controlled by Hanson plc as of mid-1991), in 1990 with Cominco it took over 97.2% of Pine Point Mines in Canada (16,17). The year before, Metall had acquired Copper Range, the last remaining underground copper mine in the USA, situated in north-eastern Michigan (12,18).

Metallgesellschaft's most important new project is a rich copper-zinc mine at Cayeli, on the coast of north-east Turkey, acquired in 1988 in a JV with Etibank (12), the state bank, and Gama, a private Turkish enterprise (12,19). The mine was scheduled to open in 1990 (20). In addition, Metall bought 20% of Namibia's Navachab gold mine where it is partnered with Anglo-American and Rand Mines (21).

With a stake in Mexico's largest copper mine, Cupifera Cananea (16); 50% of a small but high grade zinc-lead project in north-western Tunisia (22), interests in Thailand, and highly sophisticated downstream processing facilities in Europe, Metallges is one of the most dynamic of the world's integrated metal producers.

Not surprisingly, with the collapse of the Berlin Wall, the company has begun situating plant across the eastern border (23) and – alongside its subsidiary Lurgi – making the most of its claims to be ecologically, as well as economically, sound. For example, it has developed a new high-temperature process for creating alloys to be used in gas turbines and diesel engines which (it claims) does not result in hazardous metallic melting (24); Berzelius Umwelt-Service AG (BUS), established in 1987, processes a large amount of non-ferrous bearing industrial "waste" (12) and in early 1990 launched a public share floatation as a "green" company (25).

Further reading: "The Mining Operations of Metallgesellschaft" in *Mining Magazine* 4/86.

References: (1) *FT* 2/12/82. (2) *MIY* 1981. (3) *FT* 2/12/81. (4) *MIY* 1982. (5) Dave Elliot: *Nuclear power for beginners*, London, 1980. (6) *Yellowcake Road*. (7) *MAR* 1982. (8) UN NS-36. (9) *WDMNE*. (10) *MJ* 16/3/84. (11) *E&MJ* 6/87. (12) *FT Min* 1991. (13) *FT* 31/3/91. (14) *FT* 15/3/90; see also *FT Min* 1991. (15) *Reg Aus Min* 1990/91 (16) *FT* 26/6/90. (17) *MJ* 8/6/90. (18) *E&MJ* 7/89, *E&MJ* 1/90, *MJ* 7/7/89. (19) *FT* 15/3/90. (20) *FT* 26/7/90. (21) *E&MJ* 7/89, *E&MJ* 10/89. (22) *MJ* 1/9/89. (23) *MJ* 2/2/90. (24) *MJ* 16/3/90. (25) *FT* 25/1/90.

385 Métallurgie Hoboken-Overpelt SA

It is a smelter and refiner of non-ferrous and special metals.

It is part of Umipray (with IMC, Mechim, Prayon and Rupel-Chemie) which was set up in late 1979 to recover uranium from phosphates and phosphoric acid, a venture which was intended to operate world-wide (1) and does not seem to have got off the ground.

The Hoboken refinery was lost to the Germans during WW1 as assets were stripped from allied territory (2).

The company is now *the* major refiner of metals in Belgium, and during 1984 completed a five-year modernisation programme at its three complexes in the country (3). It is estimated to be the world's largest refiner of cobalt and germanium (4).

At the International Water Tribunal in Rotterdam in 1983 – a major environmental court, organised by Dutch groups with the support of nearly one hundred international organisations – it was charged with seriously polluting the environment with lead. One other *File* company's subsidiary, the Norddeutsche Affinerie (40% held by Metallgesellschaft and 40% by Degussa) was also singled out at the Tribunal for riverine pollution (5).

MHO is now majority-controlled by ACEC-Union Minière SA, after reorganisation of the interests of Union Minière and Societé Générale de Belgique in 1989 (6).

References: (1) *MJ* 12/10/79, *MAR* 1981. (2) *BOM*. (3) *WDMNE*. (4) *MJ* 13/7/84. (5) *Gua* 4/10/83.

386 Metals Exploration Ltd

This is an exploration and mining company producing tin, nickel, cobalt and other metals, mainly in Queensland and Western Australia (1). Its wholly owned subsidiary, Metals Exploration Queensland, owns 50% of the Greenvale nickel-cobalt project in Queensland, which is situated in the Aboriginal area of Townsville. Its partner at Greenvale is Freeport-McMoran. Other (non-uranium) JV partners include ACM, Western Mining Corp, Aquitaine and Outokumpu.

During the latter part of 1981, Metals Exploration was involved in an acrimonious battle to partially take over North Kalgurli Mines in order to restructure the Fimiston gold develop-

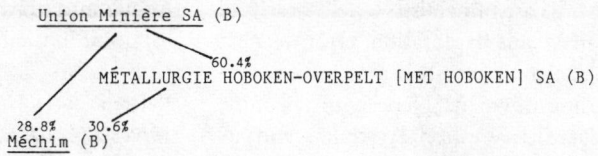

479

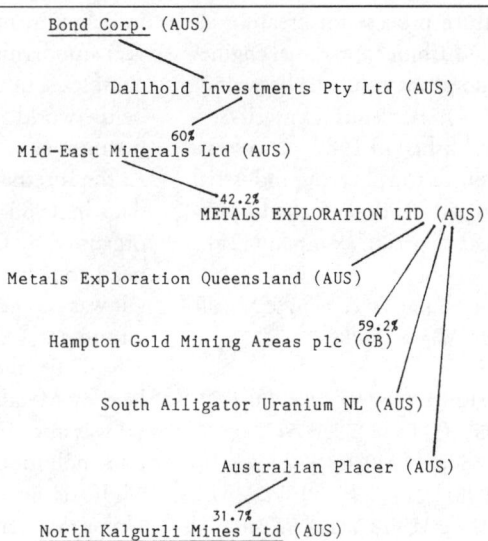

```
                    Bond Corp. (AUS)


                 Dallhold Investments Pty Ltd (AUS)

                      60%
             Mid-East Minerals Ltd (AUS)

                         42.2%
                   METALS EXPLORATION LTD (AUS)


        Metals Exploration Queensland (AUS)

                              59.2%
           Hampton Gold Mining Areas plc (GB)


              South Alligator Uranium NL (AUS)


                   Australian Placer (AUS)

             31.7%
      North Kalgurli Mines Ltd (AUS)
```

ment (2). Despite resistance by North Kalgurli board and shareholders, four Metals Exploration nominees were about to sit on North Kalgurli at the beginning of 1982 (3), and in March, Metals Exploration announced that it would take up 51.82% of North Kalgurli, thus giving it control of a company which has an important stake in the Honeymoon Uranium Project (4). North Kalgurli holds (effectively) an eighth part of Honeymoon in South Australia. By the end of the following year Metals Exploration had acquired 28.9% of North Kalgurli and was providing the company's management services (5).

A year before, Metals Exploration decided to write off North Kalgurli's investment in the Fimiston gold prospect (even though the development was proceeding) and the company's A\$5.7M investment in Teton Australia, with its share in the shelved Honeymoon project. This prompted the *Financial Times* to term Metals Exploration's "new régime" at North Kalgurli an "extremely conservative" one (6).

In 1987, Mid-East minerals NL held nearly all the shares in Metals Exploration. Over the next two years, the company shed most of its operations, closing down its Nepean nickel mine near Coolgardie, Western Australia, and its Ruxton Sandy Flat-Dinner Creek (*sic!*) tin

operations in northern Queensland (7). Sold off too was its 55% stake in the Forrestania gold and nickel project in Western Australia – to Outokumpu (8), and its holding in North Kalgurli Mines – to St Joe Gold BV (7).

In 1987, Alan Bond purchased the Greenvale operations, intending to import nickel from New Caledonia and the Philippines, in addition to Indonesia (where Metals Exploration also has exploration prospects) (9).

The Yabulu nickel refinery which services the Greenvale operations, and takes nickel from overseas (*eg* Indonesia) was the subject of a scandal in 1991, when its huge tailings dam partially collapsed, after "overtopping" with water, following heavy rains (10).

Local fishermen condemned the company (the Queensland Nickel Joint Venture is run by Bond's Dallhold Investments Pty Ltd with 72%, and the State government with 28%), pointing out that this was not the first time tailings discharges had caused fatalities among the regional prawn stocks, with the release of ammonia into prawn spawning grounds at a critical point in their breeding cycle (11).

References: (1) *MIY* 1982. (2) *MJ* 15/1/82. (3) *MJ* 19/2/82. (4) *FT* 9/3/82. (5) *MJ* 18/11/83. (6) *FT* 20/10/82. (7) *FT Min* 1991. (8) *Reg Aus Min*

1990/91. (9) *MAR* 1991. (10) *Australian* 10/1/91. (11) *Australian* 11/1/91.

387 Mexinsmovex (MEX)

It received two plants from Junta de Energia Nuclear (Spain) with which to recover uranium through extraction with solvents (1). No further information.

References: (1) *Intern Min Met Rev* 1982.

388 Midnite Mines Inc (USA)

It co-owned the Midnite Mine along with Newmont Mining Corp, whose Dawn Mining (51% owned by Newmont) operated the mine (1). Along with Evergreen Minerals, Midnite also began a uranium drilling programme in 1977 (2). Both properties are on the Spokane Indian reservation in Washington State.

Further reading: Study by Ludwig *& al* of the mine, in *Economic Geology* No. 76.

References: (1) *MJ* 21/10/77. (2) *MJ* 13/5/77.

389 Mid-North Uranium Ltd (CDN)

This is a "put together" company with HQ in Winnipeg (according to Walter Davis of *One Sky*) for which W Bruce Dunlop is responsible (1).
Its only known project was in 1979 when it entered an agreement with Esso Minerals Canada for development of 30,000 acres of uranium claims in the Hasbala Lake area, north-western Manitoba (2).
No further information.

References: (1) *Who's Who Sask.* (2) *MJ* 13/7/79.

390 Midwest Lake JV

Situated between Dawn Lake and McClean Lake in northern Saskatchewan, the Midwest Lake project is operated by Canada Wide Mines on behalf of Esso Resources Canada. Estimated reserves on the site – distinguished by its high clay content, forming a halo around the deposit (1) – are two million tonnes of rock *in situ* averaging 1.25% U_3O_8, with an estimated 26 million kilogrammes of oxide (2). But although the mine was scheduled to open between 1985 and 1987, it was delayed "indefinitely" in 1981, "until the market improves" (3).
A production rate has been proposed of 1600 tons/year uranium subject to market, and environmental approval (4).
In 1987, Denison Mines, together with an unknown partner, acquired 60% of the Midwest Lake JV and announced that it could proceed to production in the 1990s (5).
In 1989 and 1990, Denison carried out a test drilling programme to delineate reserves (6). A 185m deep shaft was sunk in 1989, and the underground testing completed in October the same year (7).

References: (1) *Canadian Institute of Mining Bulletin* 8/84. (2) *MIY* 1985. (3) *Star Phoenix*, Saskatoon, 5/12/81. (4) *MJ* 30/4/82. (5) *MAR* 1988. (6) *E&MJ* 3/90, *MAR* 1990. (7) *MAR* 1990, *MAR* 1991.

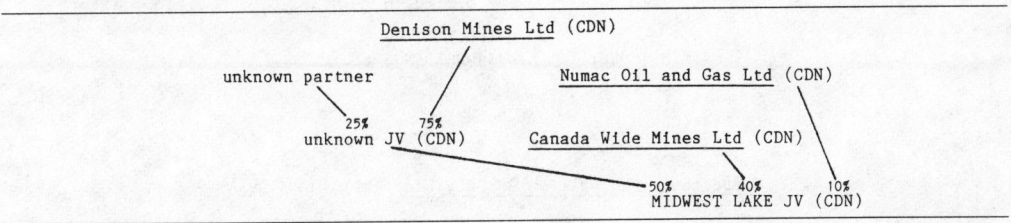

391 Midwest Uranium Corp (USA)

This is only known in regard to attempts to exploit Namibian uranium: it was reported to have registered at Volkskas Buildings, Kaiser Street, Windhoek in 1977 along with Southern Uranium Corp, Tristate Nuclear Corp and Delaware Nuclear Corp (1).
No further information.

References: (1) *Johannesburg Star* 30/7/77.

392 MIM

One of the giants among Australian mining companies, MIM was initially formed to take over the Mount Isa lead-zinc-copper complex in Queensland. Mount Isa is itself the world's largest exploited silver-lead deposit (1) and MIM is Australia's largest copper producer – delivering some 2.5% of total western world production in 1985 (2), and promising even more by the 1990s (3).
Through numerous subsidiaries and Joint Ventures (JVs) the company also participates in nickel, gold, lead, zinc, coal and stevedoring (4).
Together with Seltrust (60%), MIM had an important and profitable stake in the Agnew nickel project in Western Australia until this was sold to Western Mining in 1988 (39). Although it has recently sold its 20% share in the Mount Goldsworthy iron ore project to CGF (5), disposed of its other iron ore interests, and ditched its 50% interest in the Lady Loretta silver-lead-zinc project in Queensland (6), its Hilton silver-lead-zinc deposit (also in Queensland, close to Mount Isa itself) has promised to be the "world's greatest" (7). Commercial mining of this deposit was scheduled to start in 1986 (3), but did not commence until May 1990 (41,53,75).
In 1982, Australian federal government approval was given MIM to proceed with the Oaky Creek open pit coal mine in Queensland, after the company's participation in the project was raised to 49% (8). This promised to make MIM "a major Australian coal producer and exporter" in the near future (9). Approval followed MIM's anouncement of huge (£1.4 billion) contracts to supply steaming coal to Korea, Taiwan, Hong Kong, Europe, and the Pacific Basin (10). Mining of coal in Queensland's Bowen Basin also started in 1984 (6).
MIM explores throughout Australia, New Zealand, and Papua New Guinea. Through its subsidiary, Carpentaria Exploration Company Pty Ltd (100%), it has put forward a major proposal to dredge the Mikonui River in New Zealand for ilmenite beach sands. Plans announced in the early 1980s, according to the New Zealand public interest group CAFCINZ (now CAFCA), suggested that the venture would begin before "enormous problems" associated with land restoration had been solved (11).
In Papua New Guinea, MIM acquired a 30.5% interest (Nord Resources of the USA holding the remainder) in the Ramu River chrome-cobalt-nickel deposit (12), which looked promising by 1984 (13). It also holds 33% of the Porgera project in PNG, with Placer as operator, and Renison Goldfields (6). It has a 33.33% – later increased to 42.8% (52) – interest in the Frieda River prospect, a mountainous jungle-covered area of very rich mineralisation, especially gold (14), and which is currently being assessed (15).

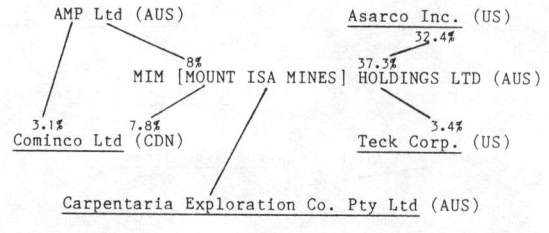

Both Sumitomo of Japan and CRA have been involved in exploring this region along the Sepik river – a favourite haunt of anthropologists for decades (16).

Oil and gas exploration in the Bass Strait, Gippsland Basin, the Ottway Basin and Queensland's Adadale and Exmouth Plateau prospects, continues (6), while gold exploration in the Aboriginal Pilbara and Leonora regions of Western Australia has been under way for a few years (17). There is also gold exploration in the Lachlan Ford belt of New South Wales (6). Aboriginal land claims have been adversely affected by MIM's operations in several areas.

At McArthur River in Australia's Northern Territory (NT), the company claimed potentially the world's largest lead-zinc deposit in 1977, and prevented the return of ancestral land to local Aborigines (18). Within five years, MIM had effectively "coralled" the Aboriginal community, purchasing not only McArthur River, but also Tawallah and Bing Bong cattle stations, the latter apparently over the heads of the Aboriginal Land Fund Commission (19). In 1982, *Land Rights News* described the plight of the Borroloola community brought on by MIM's activities, and by the collusion of local white interests:

"The 400 traditional Aboriginal residents of Borroloola have a dream. They want to run their own town. But as the area gains importance their vision is clouded by the overshadowing power of a group of 40 – the white minority.

"Borroloola is, by white standards, a growth area. The Borroloola Common is surrounded almost completely by MIM Holdings. The area has become important as a focus for mining and fishing ventures. Tourism is developing and it is an important link in the vast communications network that now grids the country.

"In 1972, the Department of Aboriginal Affairs fostered the establishment of a village council in Borroloola. Soon after, the Borroloola Town Council was formed, and of which the village council was a part, but for reasons that remain unclear the Town Council was never incorporated.

"The Borroloola Land Claim was lodged July 27, 1977, and the application was heard September to December. It was only partially granted in 1978.

"On September 30, 1977, the Government enhanced their vision of the area as a progressive mining, fishing and tourist centre when it drew up a new town plan and financed a subdivision. The new town plan alienated some of the Land Claim but, in spite of objections from the Borroloola community, it went ahead.

"Of the 30 new allotments in the new subdivision only six were allocated to Aboriginals.

"When the services were connected to the new subdivision they bypassed an Aboriginal community of some 120 people. Although it can be argued that this community might be able to 'make do' without lighting and sewerage facilities, the need for refrigeration at all levels of the community is common. The dominantly white community in the town centre is connected to all services.

"The Town Council is all Aboriginal, but the NT Government, determined to secure European participation in local affairs, developed the articles of incorporation for a Progress Association.

"In the interim a steering committee has been meeting. This consists of five Europeans to represent the 40 white townspeople and seven Aboriginals to represent the 400 Aboriginals. At least two of the Europeans on the steering committee are Territory public servants ...

"Although they have many reservations about the ability of both Europeans and Aboriginals to work together, the Community agreed to a trial period of trying to do so without the steering committee.

"In the first week in June 1982, the Aboriginal members of the steering committee were presented with a final copy of articles of the new constitution, drawn up, they were advised, by someone from Community Development. They had not seen the document before and were unaware that it was being drawn up.

"One of the articles states that the Progress Association will take over all activities and responsibilities of the Borroloola Town Council" (20).

MIM's intention to mine the McArthur River deposit at Borroloola, potentially the world's largest of its kind, moved closer in mid-1990, when the company started shaft drilling (72). A few months later, the Australian Mining Industry Council (AMIC) published a document clearly inspired by MIM, which sought to maximise the benefits the McArthur project would sustain for Australia (and in particular the Northern Territory government), emphasise the problems MIM had already had in getting the 35-year old project on-stream, and downplay, or discredit, Aboriginal and environmentalist opposition to the mine (73). By this time, a potential competitor in international markets to MIM at Borroloola had appeared in the shape of CRA, with its plans to plumb what it also claimed as a "wonderlode" replete with lead and zinc. The Century deposit near Lawn Hill, NT, on the Queensland border, could also be held up by Aboriginal land claims on CRA's preferred deepwater port site, in the Gulf of Carpentaria (74).

MIM's major uranium interest in recent years has been its 49% interest (through Carpentaria) in Honeymoon, South Australia. Federal approval was given for the mine in October 1981 (21), state government approval having been granted nearly a year before (22). Both go-aheads were given without sanction of the Legislative Select Committee on Uranium, whose report (including submissions by both pro- and anti-nuclear groups) had not been released beforehand (21), and in the face of opposition by Aboriginal land owners.

Thanks to a concerted campaign by CANE (Campaign Against Nuclear Energy), based in Adelaide – aimed specifically at the potentially lethal effects of *in situ* uranium mining (23)– both Honeymoon and Beverley mines were vetoed by the South Australian government in early 1983. ("A clear victory for the left wing of the [Australian] Labor Party", an opponent MP put it at the time) (24). Later, MIM and its partners-in-leach demanded compensation from the government (25), since, they protested, the ban meant they were unable "legally ... even to continue evaluation of solution mining at the site" (26).

MIM's other yellowcake interests – all effectively stymied as a result of current ALP government policy – include East Kalkaroo (Yarramba) project near Broken Hill, South Australia, in which Carpentaria maintained an interest with a consortium including Minad/Mines Administration; and the Angela deposit south of Alice Springs, NT, in which MIM has a 50% interest with Uranerz Australia, and on which only limited work had been carried out by 1983 (23).

By 1972 MIM had entered a JV with CRA to investigate the Red Hill uranium deposit, adjacent to that at Westmoreland, Queensland (see Queensland Mines). Although interest in this deposit was rekindled in the early 1980s, Red Hill is – effectively – dead hill (23). Once again, in mid-1984, MIM demonstrated its "hard edge" on Aboriginal and conservation issues when it opposed a proposal to extend the Kakadu National park in the NT (see Denison and Noranda) by a further 6000 square kilometres. The Gimbat and Goodparla pastoral leases are part of the same geological formation as the uranium region, and are likely to contain other minerals as well (27).

After a 30% fall in revenues in 1984 (28), MIM returned to profit a year later (29) after considerable cost-cutting (30). The group reported a doubling of profits over a three-year period (31), despite strikes, including one at its Townsville, Queensland, copper refinery in early 1985 (32).

As the *Financial Times* commented during this lean period, "the hardpressed MIM Holdings ... knows how it feels to have to sell the family silver", commenting in particular on MIM's decision to sell its 50% stake in the Lady Loretta zinc-lead-silver prospect ("a very desirable property") to Pancontinental (33).

The group at this time also felt compelled to

pay out US$140 million to save itself from the wily grasp of Robert Holmes à Court – specifically his company, Weeks Petroleum. In March 1985, Weeks Petroleum snatched just under 10% of Asarco, MIM's US partner which then owned 44% of MIM. MIM was in a poor position to rescue Asarco – and Holmes à Court was clearly after MIM as a "lucrative second prize" (34). By the year end, Holmes à Court had 12% of Asarco (29) and, despite denials (35), seemed set for a bid for the US company.

After litigation commenced between the three parties to this affair, Holmes à Court finally sold his (by then) 13.3% stake in Asarco to MIM, making himself a tidy profit of some A$7 million (36).

MIM therefore ended up with a more than 32% stake in an ailing US giant, while Asarco started reducing its stake in MIM. MIM defended the purchase as a "good counter-cyclical investment" (meaning whatever you want this to mean) – the market was less convinced. After the deal, MIM's shares on the London stock market fell to a year's low (37) (see below).

In 1986, MIM bought into Cominco, along with Teck, Metallgesellschaft, and the AMP. This gives MIM access to the Red Dog deposit in Alaska, and links MIM with Metallgesellschaft's refining capacity in Europe (38).

From 1989 to 1991, MIM's advances into its chosen geographical areas can be summarised as follows:

- In Australia it opened up the Hilton lead-zinc mine at Mt Isa in 1990, after many years delay (40). The company confidently predicted that Hilton would increase its production of zinc concentrates by about 25%, and "help Australia overtake Canada as the world's largest ... producer in the next year or two" (41).

- Following an accident at MIM's smelter in early 1991, flooding in north Queensland, and poor interim results, the company cut back on the workforce at both Mt Isa and Hilton (42).

- A new mine at its Okay creek coal operations in the Bowen Basin of central Queensland

was opened in 1990, lifting coking coal production at the site to nearly 4 million, from just under 3 million tonnes (43). Meanwhile its Collinsville mine (also in central Queensland and 75% owned by MIM) was facing closure. The company asked the Queensland government to bale it out (44). Agip (its 25% partner in both the Collinsville and Newlands mines) refused to buy half of the ailing venture (45) unless it was completely restructured. MIM agreed with union delegates on a voluntary redundancy "package" (46).

- MIM's Australian gold prospecting moved into the Telfer region (previously "tied up" by BHP and Newmont), as it sealed a JV with Mount Burgess Gold Mining Co NL (47). The Tom Gully gold mine ground to a halt (48).

- MIM purchased a third interest in Granges Exploration in 1989 – with interests in some 80 exploration projects, mainly in Canada (49). Later, its relationship with Teck, Metall Mining (Metalgesellschaft's subsidiary) though the Nunachiaq Inc partnership, was strengthened when the group bought up 54% of Pine Point Mines Ltd (50) which – with Cominco's existing share – gave the trio control (51) over one of the most important new sources of zinc.

- In 1989 Pine Point was in an exploration JV with parent Cominco, on several "properties" in Ireland (52). This was a region where MIM itself has been searching for both lead and zinc (53). Carpentaria entered a JV agreement with Irish-based Navan Resources plc, to search the Midlands region (one which already contains Europe's largest lead-zinc mine, Tara and the highly polluting, derelict, Tynagh minesite) (54). This was MIM's first exploration initiative in Europe (55). And it has not stopped there. A year later, MIM holdings UK Ltd, together with Agipcoal SPA, took over prospecting rights on the lignite deposits in Crumlin and East Tyrone, with a view to supplying fuel for the Kilroot power station (56). The Lignite Action Group (LAG) was set up in March 1985 to oppose development of the deposits on grounds of likely destruction of the local community of Moortown (through forced resettlement), the ecological impact of open-pit mining methods, and the penetration of multinational enterprises in a predominantly rural and fishing region. "[T]he very notion of swopping a thriving and sustainable community for short-term gains of an individual company has neither logic nor justice", commented *The Ecologist* magazine in 1989.

One Sunday night in February 1988, BP's operations were halted when one of LAG's members stopped the moving of heavy equipment. In response (and somewhat to its credit) BP coal flew in its executives from London to meet with LAG, and give an assurance that "there would be no prospecting in the area without prior consultations with LAG and the community" (57) (It is, of course, possible that the strength of local opposition was one of the factors in persuading BP to sell out the Crumlin project).

• Meanwhile Navan Resources had entered into a JV with MIM to develop its Highland prospects (58) on five areas in the Strathclyde, Grampian and Tayside regions of Scotland.

• MIM's "downstream" industries have proved to be remarkably expansive and, by and large, successful. In late 1989, the company's German subsidiary, MIM Beteiligungs-Gesellschaft, and Metallges Austria, each took 25.5% of Montanwerke Brixlegg, a major refiner of copper and precious metals from scrap. (By this time Metallges had secured just over 10% of MIM itself) (59). At roughly the same time, MIM purchased the Chloride Group's scrap business in Britain, providing the company with some 30% of the country's lead recycling market (60). In 1988, MIM had linked up with Norddeutsche Affinerie AG, and its patented ISAA refining process was sold to the German concern (61).

Two years later, MIM Holdings purchased a 50% stake in an Imperial Smelting Process zinc plant and refinery at Duisberg, thus bringing it into even closer partnership with Metallges, the Duisberg operator (62).

As a recycler of lead, MIM has few equals in Europe: in 1989 it decided to spend £11M on a new "high tech" smelter at its Britannia refinery in Kent, England (63), although this plant experienced drastic shortage in 1991 (due to the problems at Mount Isa), and had to close twice in five months (64). The same year, MIM announced that it was considering a JV with Asarco whose President and CEO, Richard De J Osborn *(sic)* declared that his company was "...aware of concerns by government and environmental groups that recyling be assured and conducted in an environmentally sound manner" (66).

• MIM's participation as one JV partner (with Placer Development and Renison Goldfields) in the large Porgera gold project in Papua New Guinea gave rise to controversy in 1987. The company planned a public flotation of shares in its Highlands Gold subsidiary. However, not long before, Placer had been at the centre of a scandal, when its floated PNG shares were bought up by political figures, including the deputy Prime Minister, Sir Julius Chan (67). There followed a debate in the country over which proportion of shares should be reserved for Papua New Guineans. Initially the government demanded 20%, and MIM offered only 4% (68). During the following months, the difference between the company and the PNG government narrowed while MIM delayed its flotation (69). Finally, in late 1989, Highlands Gold was officially launched (70). The following year, apart from participating in Porgera's rapid development, Highlands Gold started exploring on a number of targets on both the PNG mainland (Garaina in Morobe Province, Makham, Mt Wasis and Mt Damien) and East Britain, as well as continuing its prospecting along the Freida River in East and West Sepik provinces (71).

Further reading: A summary of MIM's Mount Isa operations is in *Engineering and Mining Journal*, November 1983.

A film made by, and with, the Borroloola people (though only marginally attacking MIM's activities in the region) has been made by Caroline Strachan, Leo Finlay, and others in the Aboriginal community (Sydney Film-makers Co-op) 1982.

References: (1) *MJ* 1/7/83. (2) *MAR* 1986. (3) *MM* 1/86. (4) *MIY* 1985. (5) *FT* 6/2/85. (6) *MM* 11/85. (7) *FT* 22/11/84 see also *FT* 23/10/87. (8) *FT* 17/12/82. (9) *MJ* 5/8/83. (10) *FT* 31/7/82, *FT* 1/11/83. (11) Cafcinz, *Watchdog,* Christchurch, 4/81. (12) *FT* 19/11/81. (13) *MJ* 9/11/84. (14) *MJ* 23/11/84. (15) *MJ* 7/3/86. (16) *FT MIY* 1982. (17) *FT* 12/5/83, *MAR* 1986. (18) Roberts. (19) *Northern Territory News* date unknown. (20) *Land Rights News,* Darwin, 6-7/82. (21) *Anti-Nuclear Times,* (CANE), Adelaide 11/81. (22) *FT* 28/10/81. (23) *Reg Aus Min* 1983/84. (24) *FT* 23/3/83. (25) *MJ* 22/4/83. (26) Neff. (27) *MJ* 8/6/84. (28) *MJ* 2/11/84. (29) *MJ* 6/9/85. (30) *MJ* 25/10/85, *FT* 31/1/85. (31) *MJ* 11/10/85. (32) *FT* 20/2/85. (33) *FT* 5/10/85. (34) *Economist,* London 16/3/85. (35) *MJ* 3/5/85. (36) *FT* 1/10/85. (37) *MJ* 4/10/85. (38) *FT* 1/10/86. (39) *MJ* 9/12/88. (40) *FT* 22/11/88, *FT* 16/5/90, *FT* 23/10/87, *Australian* 14/11/90. (41) *Metal Bulletin* 17/5/90. (42) *Australian* 10/4/91, *FT* 6/3/91, *Australian* 12/2/91, *FT* 8/2/91, *MJ* 25/1/91. (43) *Australian* 15/10/90. (44) *Australian* 23/1/91. (45) *Australian* 20/2/91. (46) *Australian* 2-3/2/91. (47) *Australian* 1/10/90. (48) *Australian* 28/2/91. (49) *FT* 8/2/89 (50) *FT* 26/6/90. (51) *MJ* 8/6/90. (52) *FT Min* 1991 (53) *MIM Annual Report* 1990. (54) *E&MJ* 7/89, see also information from Minewatch Ireland, 4/91. (55) *MJ* 9/6/89. (56) *Irish News,* 17/8/90 see also *Australian* 18-19/8/90, *E&MJ* 10/90, *MJ* 11/5/90. (57) *The Ecologist,* vol19, no 2, 1989. (58) *Irish World* 21/9/90, *MJ* 14/9/90. (59) *MJ* 22-29/12/89. (60) *MJ* 1/12/89. (61) *MJ* 8/4/88 see also *MJ* 4/8/89, *Australian* 24/12/90. (62) *MJ* 21/9/90. (63) *MJ* 29/9/89. (64) *MJ* 24/5/91. (65) *FT* 7/2/91. (66) *Australian* 7/2/91. (67) *FT* 4/2/87. (68) *FT* 7/5/87. (69) *FT* 28/10/87. (70) *Australian* 20/11/89. (71) *MJ* 29/6/90. (72) *Northern Territory News* 18/8/90. (73) Case study in Perseverance: the McArthur River Project *Mining Review,* AMIC 11/90. (74) Australian 8/3/91. (75) *FT* 16/5/90, see also *FT* 22/11/88.

393 Minad

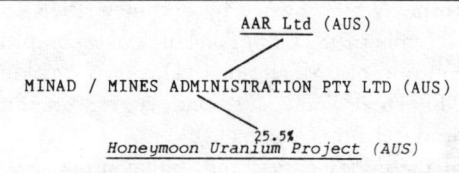

It holds 6.6% of the Yarramba prospect, along with Carpentaria and Teton; which was managed by Sedimentary Uranium (1).
At the Westmoreland deposit of Queensland Mines, in Queensland (then managed by Urangesellschaft), Minad had a 12.75% interest (2). Although Minad could earn up to 60% interest in Westmoreland, it has assigned 50% of this to the German partner (3).

References: (1) *Reg Aus Min* 1981. (2) *Reg Aus Min* 1983-34 (3) *EPRI Report.*

394 Minatco Ltd

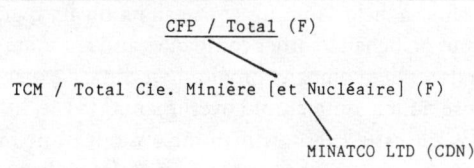

In early 1985, Minatco bought into uranium prospects owned by Inco and Canadian Occidental Petroleum at the promising Wollaston Lake deposit. Minatco could earn a one third interest by spending C$23M, and would shell out C$5M on exploration over the nest four years (1), with about C$1M during 1985 (2). Minatco was among the ten most active uranium exploration companies in Canada during 1988 (3).

References: (1) *MJ* 15/2/85. (2) *FT* 19/2/85. (3) *MAR* 1990.

395 Minatome SA

Formerly 50% owned by Pechiney (PUK), in December 1982 CFP/Total took over complete control of Minatome (1), thus making CFP/Total France's second largest uranium producer after Cogéma (2).

In the mid 'eighties, and unlike almost every other major oil company in the world, Minatome's parent group reaffirmed its confidence in nuclear power and recommitted itself to uranium fuel. It planned to double its (western) world market share in yellowcake from 3% to 6% and increase exports from 10 to 50% of production (3).

Until the reorganisation of CFP/Total's uranium and other mining interests in 1983, Minatome was "unique in that rather than being limited to a specific project, as is typical of joint ventures, it is an on-going partnership in uranium development, involving petroleum and minerals interests" (4). It also acted as an agent for a specialised uranium-buying organisation, as well as for an (unknown) mining company not involved in uranium production, while it held "mining concessions on its own and on behalf of other companies and is a minerals service company providing technical expertise under contract for governments" (4).

By the early 1980s, Minatome was operating in 14 countries over four continents. Its main investments were in SCUMRA, Comuf, Somair, and Rössing, with exploration expenditures chiefly directed at Colombia, the USA, Australia and France (5). In Eire, its subsidiary

Maugh spent four years exploring for uranium at Carlow, near Dublin, before local activists caught on. Nine of them occupied the EEC building in Dublin in November 1979 to protest against the Euratom treaty – one of whose provisions is that member countries should make available uranium to the others. Maugh was singled out by the campaigners for special mention. The Eire government offered the nine a deal: if they pleaded guilty to a charge of "forcible occupation" another charge of "malicious damage" would be dropped. As a result, the protesters were sentenced to one year in gaol, suspended for two years (6).

Within the next two years, Minatome was also exploring in its own name for uranium around the St Austell and Bodmin granites in Cornwall, England: car-borne scintillometer surveys and soil geochemistry were used prior to drilling but "it seems no significant deposits have been found" (7). Cost of this prospecting – and similar investigations in Scotland – was put at US$370,000.

By then, Minatome had also signed a US$6 million agreement with the government of Saudi Arabia, to search for radioactive minerals (8), thus becoming part of a large-scale uranium exploration programme, in which the régime is associated with the USA and the IAEA (9). It has also been prospecting in the Nuba mountains of Sudan (10).

Though reported to have been awarded a contract to explore for uranium in the Hoggar mountains of Algeria in 1977, along with other

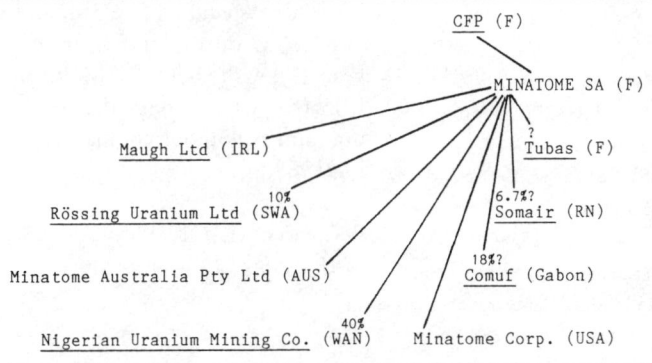

French and Algerian companies, this appears not to have materialised (see Sogérem).

However, an Association contract was signed with the Colombian state agency IAN in 1976, to explore the central part of the country. On the first day, local people took pot shots at the foreigners, but they didn't prevent a five-year exploration programme being put into operation (11). The contracts enabled Minatome and ENUSA to range over some 50,000 square kilometres of (often indigenous) territory in the Andean provinces of Cundinamarca, Boyaca, Caldas Tolima, Hiila and Antioquia, as well as the flat provinces of Meta, Guiania and Vaupes (12). The latter two provinces are mainly inhabited by Indians (13).

By this time, Minatome had signed a deal with the state Nigerian Mining Company to prospect for uranium in Nigeria. Minatome bought 40% of the Nigerian Uranium Mining Company (Numco) (14).

All this has been relatively low-profile activity. In contrast, Minatome's participation in the Rössing uranium project, like that of RTZ and UG, has attracted criticism from around the world. Minatome's stake in the Namibian mine was bought under the name of Total Campagnie Minière et Nucléaire in 1973, when the French company received a 10% equity in exchange for a long-term contract involving the purchase of a substantial quantity of uranium concentrates during the 1980s (15). Two years later, Total's role was taken over by the newly-established Minatome (16). By the early 1980s, Minatome was reported to be receiving "at least its equity share" of uranium from Rössing – "and possibly more" (17).

The company also extended its illegal Namibian activities to uranium exploration elsewhere in the country – along with Omitaramines (18). Residents and farmers in Wisconsin were alerted to the activities of several companies exploring for uranium in several parts of the state in 1980. Minatome was the most active among these – it drilled no fewer than 41 holes in north-eastern Florence County, before much public attention was drawn to the fact (19). Al Gedicks of the Center for Alternative Mining

Development Policy called for a state-wide moratorium, until background radiation levels could be checked at drilling sites, and the state legislature undertook to look into the problem (20). While Minatome was protesting that it hadn't discovered any high concentrations of uranium in the area (21), three state local authorities declared moratoria on uranium mining (22); others followed (23).

Minatome attended the state hearings on uranium along with USX, Anaconda, Phillips Uranium, and American Copper and Nickel; Exxon and Kerr-McGee, though holding the most extensive leases, did not attend (24). One report stated that Minatome did at least agree to voluntary monitoring of radon levels on its exploration drills (25). Activity in the region seems to have fallen off very soon after this flurry of activity.

Since 1982, when Minatome's uranium interests were regrouped under the umbrella of TCM (Total Compagnie Miniére – a reversion to the appelation which preceded Minatome), the company has primarily expanded its uranium interests in France through SCUMRA and Dong-Trieu. Although stated to be still looking for an American uranium mine (3), it has taken few open steps to acquiring one. However, despite the current part-moratorium on uranium developments in Australia, it is here that its non-French interests are primarily concentrated: at Ben Lomond in Queensland, and Manyingee in Western Australia.

Minatome Australia Pty Ltd was incorporated in Queensland in 1967 as Pechiney (Australia) Exploration Pty Ltd, adopting its present name nearly ten years later, in 1976 (26).

Four years later it was awarded a uranium mining lease at Ben Lomond, 60km west of the coastal city of Townsville, an area which has long interested geographers and naturalists (27) and which it had begun exploring some years previously. (In fact, evidence of uranium on the site was discovered as long ago as 1881, although it was not until 1975 that the deposit was formally located) (28).

Initial plans were for an open-pit mine, to be followed by an undergound mine with production of about 200 tonnes/year U_3O_8 from recoverable reserves of 2000 tonnes. An estimated US\$A50 million would be needed to establish the mine and processing plant (29).

The Queensland régime's eagerness to get the mine underway hinged on plans to site a uranium enrichment plant in Townsville – one proposal for which, at an estimated A\$1000 million, came from Minatome itself in October 1979 (30).

Two months later, a French CEA delegation to Queensland discussed a possible US\$5000 enrichment plant, while claiming that it was more important for such a plant to be near cheap power and water than a uranium mine.

Officially the Minatome lease was granted in early 1980 (31) but, a year previously, the state Minister for Mines, Energy and Police (sic), Ron Camm, announced that the mine would be given a quick go-ahead, in a statement made well in advance of completion of an Environmental Impact Study (EIS) (27). Attempts by the Townsville Regional Conservation Council, the Queensland Conservation Council, and individuals, to intervene in the public process, were rejected as irrelevant by Camm (55) who, in an interview with Denis Reinhardt, claimed: "If I had to listen and take notice of protests against the issue of a mining lease anywhere in Queensland and listen to the people in close proximity of that mining lease, there would be very few mining leases ever issued" (32).

Not only did the state government refuse to consult with the Townsville City Council and local shires (authorities), it also altered the Mining Act, thus allowing its Mines Department to over-rule local authorities, and it doctored procedures for conducting EIS's – by dropping the term "Environmental" from the rules (27).

Meanwhile, Minatome itself had been playing underhand games. Its first application for a lease was made as long ago as August 1976, but concealed in a small circulation newspaper, thus escaping the attentions of the very active anti-uranium mining movement in the state. The application got the approval of the Mining Warden at Charters Towers. An authority to prospect was granted in July 1978 and, al-

though the company stated nearly 18 months later that it still hadn't approval for its lease, the Queensland government was protesting it was "too late" for objections to be heard (33).

Between them, the Queensland government and the mining company effectively undercut a growing opposition to the mine. (Local surveys showed a majority of residents opposed to the project (34); and there had already been an anti-uranium march, in spite of the state's draconic ban on all such demonstrations).

From this point on, opposition mounted dramatically. Peter Valentine, a senior lecturer in geography at the Townsville James Cook University, described the government's processing of the lease as "horrific" and said there appeared to have been a deliberate attempt to avoid public debate. The Australian Telecommunications Employees Association (ATEA) in February 1981 imposed communications bans on Minatome. The union's State Secretary, Ian McLean, justified them with the words: "It is much more difficult to stop a mine already operating than to prevent a mine getting underway, and that's what we are trying to do at Ben Lomond" (35). The Movement Against Uranium Mining (MAUM) also announced a "tent village" at Ben Lomond, to be held that summer (36).

The opponents' case depended not only on previous experience in the uranium industry, but Minatome's existing practice at the mine site. These were summarised by the Queensland Campaign Against Nuclear Power (CANP):

"Already radiation levels 160 times the permitted level have been recorded in the mine.

"The clay which Minatome plans to use for the earth tailings dam dissolves in water, and will let contaminated water pass through. The nearest suitable clay is hundreds of kilometres away, and half the known deposits of this type of clay in Australia would be required to adequately line the dam.

"When the river level falls during the dry season, the contamination will be more concentrated where radioactive particles of silt accumulate in the remaining waterholes.

"Already a level of radioactivity two and a

half times the legally permitted level has been recorded in a creek which flows into the river. This was from a stockpile of 3500 tonnes. When the mine is in operation, the stockpile will be two and a half million tonnes.

"There is no way of enforcing the responsibility of the mining company after the mine has closed. After the ten-year life of the mine, when profits and uranium will have been sent to France, Minatome will have no further interest in the abandoned mine or the people living in the area affected by it" (37).

It was the Townsville Regional Conservation Council (TRCC), backed by the Australian Conservation Foundation (ACF), which spearheaded legal opposition to Minatome's proposals, and this reached a climax in April 1981 when the company applied for an additional lease of 2035 hectares to cover a uranium-molybdenum tailings dam, access road and camp (38).

The TRCC based its case on several issues:

(i) an inadequate (E)IS which failed to evaluate the effects of the tailings dam and its possible collapse, radioactive emissions to the atmosphere, the impact on fauna and flora, capacity to cope with "extreme rainfall events", and which also failed to lay down proper rehabilitation procedures (39);

(ii) incidents of radioactive leaching from an ore stockpile, and lack of radiation control on the existing Minatome campsite. Documents produced by the company itself demonstrated the appalling carelessness with which these matters had been regarded. Hermann Schwabe, a West German environmental engineer, stated that methods of waste containment and tailings dam sealing, as proposed at Ben Lomond, had not even been known to survive spans "below fifty years" (39).

Neil Heinze, a local civil engineer, also claimed that radioactive leakage was "certain" to occur from the site, while all artificial methods of containment were inadequate.

Professor Frank Stacey, Professor of Applied Physics at the University of Queensland, predicted that inevitable radioactive leaking would pollute the Burdekin river system, especially as the proposed dam across the river would "ensure that heavy pollutants tend to accumulate in the reservoir and any area in which water from the reservoir is used, instead of being flushed out to sea" (39);

(iii) the escape of chemical pollutants, particularly heavy metals and radionuclides, into the surface and groundwater systems associated with Keelbottom Creek (the tributary from the Burdekin river next to the mine), from the proposed milling and treatment plant;

(iv) non-radioactive contamination by trace metals, such as copper, lead, zinc and cadmium. Ian Campbell of the Caulfield Institute of Technology's Water Studies Centre addressed this particular hazard.

In response, the Mining Warden rejected Minatome's application – an "historic decision" (at least in the Queensland context) (40) based on environmental considerations. He found that there was no proper long-term arrangement for the containment of tailings. He questioned the appropriateness of clay as a liner for the evaporation ponds and tailings dumps. Taking into account natural erosion and the extremely high rainfall which might be expected during the cyclone season, he doubted the company's ability to effectively isolate toxic materials and, in particular, prevent their discharge into Keelbottom Creek and thereby the Burdekin River. Concluded the Warden:

"I am of the opinion that the public interest or right may be prejudicially affected if the Mining Lease is granted at this stage ... I recommend to the Minister that this application be rejected in so far as it relates to the areas to be used for evaporation ponds, the tailings dump and the treatment plant and facilities pertaining thereto" (39).

Within six months, the State Mines Minister, Ivan Gibbs, was blatantly seeking to overturn the Warden's decision. After Minatome drubbed together some "new studies", the Minister felt able to declare that Minatome need not re-apply for a lease, and their work would be assessed within the Mines department itself (41): "the best thing to do with uranium is to mine it and get the best price we can for it", said Gibbs. By this time, another scandal was in the news. The national newsmagazine, *National Times*, revealed that Minatome had destroyed several vital Aboriginal sites "in the past couple of months" – including one possibly some 4000 years old, "considered to be one of the most significant in North Queensland". This site was bulldozed by the company to make way for an experimental evaporation pond. Another Aboriginal quarry site "considered to be unique in Australia" was also under threat by planned high-tension power lines and water pipes, while a sacred pool was threatened by nearby drilling. To cap it all, a confidential document obtained by the *National Times* revealed that Minatome had been aware of these Aboriginal sites *since 1978 and was advised in an archaeological impact statement that they should be protected* (42).

Early the next year, Minatome flew out 36 tonnes of uranium ore from Ben Lomond to Noumea in New Caledonia, then on to France for testing (43). The flight was organised to evade union bans at Townsville, as well as adverse publicity (44).

A few months later, the World Bike Ride – antinuclear activists who had set out from Melbourne in March – set up an "Atom-Free" embassy at the mine site itself (45).

Then, in mid-1983, the federal Australian government banned all uranium exports to France, in retaliation for France's continued nuclear testing in the Pacific. In response, the company reportedly filled in the tailings dam and development work came to a halt (46).

At the end of the year, the company finally published the environmental impact statement – a few days after the ALP government announced a ban on all new uranium mines, apart from Roxby Downs (see WMC). Minatome claimed that adverse effects on the environment from the Ben Lomond mine would be very small, and "there will be no significant radiological ef-

fects" (47). It also stated that exports of uranium from the mine would "satisfy" Australia's export requirements (big deal!) and that Minatome would form a JV with an Australian company to guarantee 75% local equity in the uranium venture. The project could begin within two years and production start by 1988 (47).

At the beginning of 1984, the company announced that developing Ben Lomond would cost US$100 million, and that – though the mine was not likely to receive federal government approval – the project would still proceed. Later estimates were that the mine would produce about 500 tonnes/year uranium metal and slightly lower tonnages of molybdenum over a period of 10 years, with an average grading of 2kg/tonne (48).

Early in 1986 it was reported in the Australian Senate that the uranium ore stockpiled at Ben Lomond had been virtually abandoned, with a minimum of security precautions (49).

Also in 1986, the Australian Labor Party government shocked many in the country when it reversed its ban on uranium exports to France. But even before then, Minatome – predicting frankly in 1984, for instance, that "such an illogical policy cannot last" (50) – had been actively developing another uranium deposit in Australia. This is the Manyingee project, which Minatome operated in a JV with UG and Aquitaine (Elf Aquitaine Triako Mines Ltd) (51) near Onslow in the Aboriginal Pilbara region of Western Australia. Using a carbonate leach process, testing on the deposit began in the early 1980s (52). By 1986, Stage 1 of the production phase had finished with the production of a small quantity of yellowcake slurry. Stage II – drilling further injection and extraction holes – was expected to commence soon after. On being asked by an environmental researcher when this was to be, a Total Mining representative replied "none of your business" (53).

According to investigators at Manyingee in 1986, assay workers for Minatome in 1980/81 had been issued with wire brushes and instructed to erase any Aboriginal paintings in the area. The site is the traditional Dreaming of the

Talandji people of the Ashburton river, some of whom live in Onslow (56).

However, Minatome's exploration with Kratos NL (35%) and the Wyoming Mineral Corp (32.6%) at Pandanus Creek in Australia's Northern Territory appears to have stopped (54).

Further reading: *Little Atom: The Minatome Case*, published by Womens International League for Peace and Freedom (WILPF), Queensland, 2/81.

Contact: Comité de Liaison Uranium-Nucléaire de l'Aveyron, 9 rue de l'Embergue, 12000 Rodez, France
WISE-Glen Aplin, PO Box 87, Glen Aplin, Queensland 4381, Australia.
North Queensland Conservation Council [formerly TRCC] PO Box 364, Townsville, Queensland 4810, Australia.
Center for Alternative Mining Development Policy, 210 Avon Street 9, La Crosse, Wisconsin, 54603, USA.

References: (1) *MJ* 21/1/83. (2) *FT* 5/1/83. (3) *E&MJ* 2/85. (4) *Big Oil.* (5) *MJ* 5/2/82. (6) *Kiitg* 6/80. (7) *MM* 11/82. (8) *MJ* 1/12/78. (9) *Nuclear Fix.* (10) *MJ* 11/6/82. (11) *El Tiempo*, Bogota, 29/8/80. (12) *Age* 6/11/79, *MJ* 8/1/82. (13) *ARC Bulletin*, 12/78. (14) *FT MIY* 1985. (15) *The Star*, Johannesburg, 30/7/83; *FT* 18/5/76; *FT* 15/9/76. (16) Bernard C Duval, "The Changing Picture of Uranium Exploration", International Symposium on Uranium Supply and Demand (Uranium Institute) London, June 1977. (17) *Africa News* 18/8/80. (18) *FT* 26/7/78; *Transnational Corporations and other Foreign Interests Operating in Namibia* (UN) Paris, 4/84; Rogers and Cervenka. (19) *Rice Lake Chronotype*, 30/4/80; *Rhinelander Daily News*, 21/5/80. (20) *E&MJ* 4/80. (21) *Rice Lake Chronotype*, 7/5/80. (22) Wisconsin Legislative Council Staff Memorandum, 12/6/80. (23) *Capital Times*, 11/7/80. (24) *E&MJ* 7/80, (25) *Daily Cardinal*, 20/2/80. (26) *CANP Newsletter* No. 77, Brisbane 3/82. (27) *Chain Reaction* 5/1/79. (28) *Townsville Daily Bulletin* 25/11/81. (29) *MJ* 18/7/80. (30) *Kiitg* 1/80. (31) *FT* 15/5/80. (32) *National Times* quoted in *Chain Reaction* 5/1/79. (33) *Courier Mail*, Brisbane, 29/12/79. (34) *CANP Newsletter* No. 57, 4/80. (35) *Courier Mail* 28/2/81. (36) *CANP Newsletter* No. 67, 3/81. (37) "Ben Lomond: A French Uranium Mine In Queensland?", *CANP Broadsheet*, Brisbane, 1981. (38) *CANP Newsletter* No. 67, 3/81; *Kiitg* 1-2/81. (39) *TRCC Newsletter* Vol. 5 No. 4, 6/81. (40) *CANP Newsletter* 5/81. (41) *Townsville Bulletin* 25/9/81, *Courier Mail*, 28/9/81. (42) *National Times*, Sydney, 4-10/10/81. (43) *MJ* 8/1/82. (44) *Financial Review*, Australia, 17/12/81. (45) *WISE International Bulletin* Vol. 4, No. 6, 7/82. (46) Greenpeace Australia Briefing Document by Michelle Sheather, presented to NFIP conference, Vanuatu, 7/83. (47) *Age* 14/11/83. (48) *MM* 1/84. (49) *Australian Senate Hansard* 11/2/86. (50) *Total in Australia*, (1984 Annual Report). (51) Letter from Total Mining, Australia to *Environment WA* 15/5/86. (52) *Reg Aus Min* 1983/84. (53) WISE-Glen Aplin report in *WISE NC* No. 257, 22/8/86 Amsterdam; report on Manyingee by Rick Humphries, ACF, Western Australia. (54) *MM* 3/81. (55) *Courier Mail*, 5/7/79. (56) Information from Shyama Peebles, Kalgoorlie, 1986.

396 Mineral Energy Inc (USA)

It supposedly held eight uranium mines up to 1978 in the Temple mountain area of Utah; production unknown (1). That year, Minatome agreed to explore and evaluate the company's claims over a three-year period (2).

Later that year, finds were recorded in the White Canyon district of southern Utah (3).

References: (1) *US Surv Min* 1979. (2) *MJ* 27/1/78. (3) *MJ* 25/8/78.

397 Minerals Exploration Corp

It started a uranium mining project at Sweetwater in the Red Desert area of south-west

Union Oil Co. of California (USA)

MINERALS EXPLORATION CORP. (USA)

Wyoming in 1978, which met with opposition from conservationist groups (1). The groups allegedly withdrew their objections later the same year (2) and the company got permission to open up the mine and mill in 1979 (3). Planned production was 900,000lbs a year of uranium oxide over 17 years (2).

Around the same period, the company was exploring for uranium in Texas(4).

From 1980, the Sweetwater mine was technically been operated by Union Oil's Energy Mining Division (5).

The Red Desert uranium mine and mill was placed on stand-by in 1983 because of delays in completion of two nuclear power plants by the company's lone customer, Public Service of Indiana. A "skeleton crew" (company's own words!) was being maintained until Public Service of Indiana's Marble Hill nuclear plant started up. Public Service of Indiana contracted to purchase 450,000lb/yr of U_3O_8, but Minerals Exploration never found another buyer for the additional 450000lb/yr capacity of the mill (6).

References: (1) *Business Week* (USA) 29/5/78. (2) *MJ* 24/11/78. (3) *MJ* 9/3/79. (4) *E&MJ* 11/78. (5) *MAR* 1982. (6) *E&MJ* 4/83.

398 Minera Sagasca SA

```
        Continental Copper and
        Steel Industries Inc. (USA)
                    ╲
                     49%
              MINERA SAGASCA SA (RCH)
```

This Chilean copper company in 1978 started a joint feasibility study with the Chilean junta's nuclear energy commission for the extraction of uranium from copper (1). However, two years later the privately-owned Chilean copper company Pudahuel declared it was about to produce uranium from copper extracted at the Cascada copper mine (2) – presumably the original project was found "unfeasible".

No further information.

References: (1) *MM* 2/78. (2) *FT* 7/9/80; *MM* 12/80.

399 Minera Sierra Pintada SA (RA)

This consortium was hoping to develop Sierra Pintada, Argentina's major uranium deposit, in Mendoza province, 180km south of Mendoza city and 25km west of San Rafael as early as 1980 (1). Reserves have been put at 14,500 tonnes (2). Other partners were: Sasetru SA, Alianza Petrolera Argentina, Inalruco, Alfredo Evangelista SA, and PUK/Pechiney (3). Although a contract was supposedly signed in May 1980 between CNEA and the consortium (1), it apparently took another two years before details were ironed out (4). By the beginning of 1985, the US$150M project was still looking for a green light (5).

In 1988, the feasibility study for the mine was completed but no decision has been taken to exploit the deposit (6).

References: (1) *Nucleonics Week*, Washington, 15/5/80. (2) *Uranium Redbook* 1982. (3) *MM* 1/80. (4) *E&MJ* 12/82. (5) *E&MJ* 1/85. (6) *MAR* 1989.

400 Minerex (IRL)

Nothing is known about this company except that it began prospecting for uranium over "a large area" of County Tyrone, Northern Ireland, in the late '70s. The Northern Ireland Department of Commerce allegedly had no knowledge of the company, said to be geological consultants based in Dublin, Eire. (1).

References: (1) *Gua* 27/7/78.

401 Minex Inc

[This is not Minerals Exploration (USA)]

It was exploring for uranium in the Sudan from 1977. In 1979, it was awarded new uranium

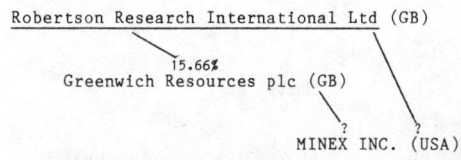

Robertson Research International Ltd (GB)

15.66%

Greenwich Resources plc (GB)

? ?

MINEX INC. (USA)

concessions around northern Kordofan, with 55% of uranium production (rising to 60%) on any deposits mined (1).

In 1981, the company drilled in the Jebel Dumbeir deposit and did surface exploration in Jebel Kon, both in Kordofan, central Sudan. No results have been announced (2).

Formerly owned 60% by VAM Ltd, Minex was bought out by Robertson in 1983 (3).

Now owned by Greenwich Resources, Minex's main explorations continue to centre on Egypt. Although its Red Sea operations may not bear fruit, latest reports from Abu Marawat and the Hamama Gossan have revealed high grades of base metals and some gold (4).

References: (1) *MM* 11/79. (2) *MAR* 1982. (3) *FT* 25/2/83. (4) *FT Min* 1991, *MAR* 1990.

402 Missi Island Mines Ltd

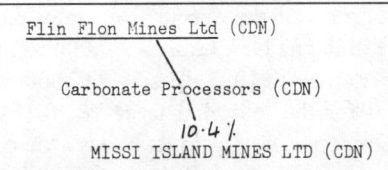

Flin Flon Mines Ltd (CDN)

Carbonate Processors (CDN)

10.4%

MISSI ISLAND MINES LTD (CDN)

It has the same board as Flin Flon which in the late 1970s held a minority of the stock (1).

References: (1) *Who's Who Sask.*

403 Mitchell Cotts Projects

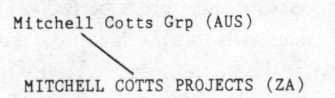

Mitchell Cotts Grp (AUS)

MITCHELL COTTS PROJECTS (ZA)

It was awarded a major contract in 1978 to design uranium plant expansions at six of Anglo-American's gold-uranium mines, basically to treat high- and low-grade uranium deposits in "slimes" left over from gold mining (1). See Anglo-American and Northern Mining.

References: (1) *MM* 5/78.

404 Mitsubishi Grp

This is a huge Japanese conglomerate with fingers in many pies – including uranium.

In 1977, Mitsubishi Corp, Mitsubishi Mining, and Mitsubishi Metal Corp were trying to get permission to enter a JV for uranium exploration with Rayrock Mines in the Northwest Territories of Canada – and to establish a Canadian subsidiary, Ryowa Uranium Development, to carry out a C$1.2M programme (1).

Then, in 1978, together with Getty Oil (USA) (51%), Mitsubishi (49%) planned to explore and develop uranium deposits in the San Juan and McKinley areas of New Mexico, with production projected for "eight or ten years' time" (2). Together with Ryowa Uranium Exploration Co, Mitsubishi was later engaged in extensive uranium exploration in Japan and abroad (3). It also has important joint ventures with several other companies (though not for uranium): Consolidated Gold Fields, Kennecott, and Cominco. It has an import licence to ship uranium from South Africa to the USA: 1.3 million kilogrammes of yellowcake was due to be transported between January 1982 and December 1984, destined for use in Japanese reactors owned by Kansai, and enriched in US facilities (4).

In 1980, the Mitsubishi Research Institute proposed a Global Infrastructure Fund – a sort of corporate Brandt. Ostensibly it would generate demand amounting to more than US$500M over the next 20 years to build large-scale projects in the third world to "stimulate private business in the developed nations" (5); in other words, the aim was to stimulate certain parts of

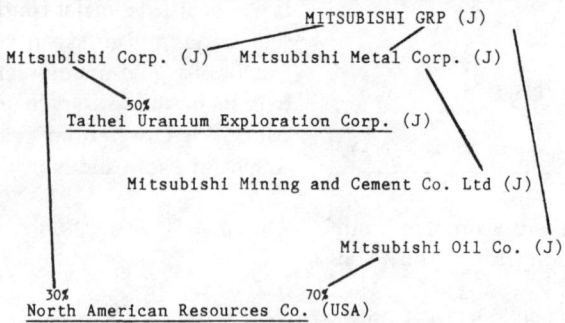

```
                            MITSUBISHI GRP (J)

    Mitsubishi Corp. (J)   Mitsubishi Metal Corp. (J)

                 50%
         Taihei Uranium Exploration Corp. (J)


              Mitsubishi Mining and Cement Co. Ltd (J)


                                      Mitsubishi Oil Co. (J)

      30%                    70%
    North American Resources Co. (USA)
```

de-developed countries so as to increase markets for *nouveau riche* states like Japan (6).

Together with Westinghouse, the company also studied a proposal to form a JV to produce titanium (crucial to supersonic warplanes) and zirconium (crucial to nuclear power plants) (7). In February 1982, Mitsubishi announced, together with five electric power companies, that it would sign an agreement with Westinghouse Electric for joint development of a new advanced pressurised water reactor (PWR). Mitsubishi and Westinghouse would put up 10 billion yen each (8).

Rumours that Mitsubishi was to become the third partner (with Alcoa and the government of Victoria) in the notorious Portland smelter – on Aboriginal (Gunditj-Mara) sacred sites in southern Victoria, Australia – were denied in 1984 (9).

Mitsubishi is also part of the Kennecott Consortium, one of the five major western-dominated underwater consortia. Its partners there are RTZ, Noranda, Socal, and CGF (10).

In 1978 it acquired 40% in the Ulan steam coal mine at Lithgow, New South Wales, Australia, one of its major customers being the Kuyushu Electric Power Co of Japan (11).

In the '70s, Mitsubishi concluded a contract with Minserve to supply uranium from Namibia, on behalf of the Kansai Electric Power Co, in flagrant contravention of the UN Decree on Namibia's natural resources (12).

In the 1970s, Mitsubishi set up the Taihei Uranium Exploration Corp to search for uranium in the Australian Northern Territory (NT) (13).

In 1988, Mitsubishi Metal's mining activities ceased (14). Mitsubishi Chemical had pulled out of aluminium refining two years earlier (15). Mitsubishi Metal also surrendered its tender for a South African steel mill in 1988, conscious of the political ramifications of doing a deal in the apartheid state (16).

But none of these events has diminished the appetite of the Mitsubishi group, either for new mines, or new metals (specifically neodymium and other rare earths) (17). As a partner with Nippon Mining, both Mitsubishi Metal and the Mitsubishi Corp (through JECO) have a 10% interest in the huge Escondida copper mine in Chile (operated by BHP with RTZ) (14). Again with Nippon (one of its favoured partners), Mitsubishi Metal receives a substantial portion of the output from the Ok Tedi mine in Papua New Guinea (60,000 tonnes of copper a year in 1987 rising to 230,000 tonnes a year for the following seven years) (14). In 1989 Nippon was again its buddy, when Mitsubishi opened up the Texas smelter in the USA (18).

The Atlas Consolidated mines in Cebu, Philippines, have long been censured for the management's brutal disregard to its workforce and the environment (19).

Mitsubishi holds 10.49% of the Class 'B' shares in Atlas (20) and has a 15-year contract to receive 130,000 tonnes a year of copper ore, at a cost of US$30M, covering 20% of all the group's copper needs in that period (21).

In Australia, Mitsubishi's favoured partner is CRA, with whom it holds a JV (Mitsubishi 60%) in the Howick open-cut mine in the

Hunter Valley, New South Wales, and part of the Oaklands coal project in NSW (14).

In 1991 Mitsubishi, along with several other major corporations, bid for a £560M contract to build an aluminium smelter at Puerto Ordaz in Venezuela, using electricity from the nearby Guri dam (22). The same year, it announced drilling at the Loma La Nandita gold concession in the Dominican Republic (23).

"In the energy sector, Mitsubishi and its subsidiaries have built up an unrivalled knowledge of the integrated uranium business, from mining to nuclear power generation and waste disposal" (24).

In 1990, Mitsubishi Metal and Mitsubishi Mining and Cement announced plans to merge (25): a move which "sent shock waves through Japanese government and business circles. Aside from creating Japan's largest non-ferrous metals manufacturer, it re-united two companies that were split apart by the US occupation" (25).

This merger consolidated Mitsubishi into a giant among giants – "During the past year, the name Mitsubishi has appeared on more prominent international deals than that of any other Japanese company" (such as buying the Rockefeller Centre in the USA for US$850M from USX, purchasing Aristech Chemical Corp for US$867M and picking up a single golf course – albeit the famed Pebble Beach – for a cool US$940M) (25).

The Mitsubishi group now comprises 28 core members, ranging from the 27%-controlled Nikon Corp, through 17%-owned Mitsubishi Electric, to 100%-owned Mitsubishi Aluminium, with the three "flagship" members being the Mitsubishi Corp (32% held by other members of the group) Mitsubishi Bank (26%) and Mitsubishi Heavy Industries (20%).

They are linked together by the *keiretsu* – a form of interlock through family connections which, while thin on paper, in practice allows decisions to be taken by group companies in concert, rather than competing with each other. It also makes it very unlikely that any of the group operations will be taken over from outside the hierarchy (26).

The US occupation forces ostensibly deprived Japan of a military, while leaving it with a self-defence force which is an army in all but name. Similarly, the American administration tried to break the power of the *zaibatsu*, those gigantic military-industrial combines which underpinned the Imperial war machine. The *zaibatsus* may have gone, but the keiretsu are in the ascendant. And Mitsubishi is at the crest.

Opposition

The worldwide movement for the protection of tropical rainforests in 1991 launched an international campaign to stop Mitsubishi importing timber logged in Sarawak by Mitsubishi group company Daiya Malaysia, and in Brazil (through its 4.95% stake in Edai do Brazil, the plywood manufacturer) (27). Demonstrations were held in 1990 and 1991 against Mitsubishi by rainforest activists in Amsterdam, Tokyo, New York, London and Copenhagen (28). In 1989 one of the more imaginative demonstrations against the Corporation involved Dutch activists dropping a huge tree trunk on top of a Mitsubishi car and rendering it inanimate (27).

Contact: Rainforest Action Network, 301 Broadway Suite, San Francisco, CA 94133 USA.

References: (1) *MJ* 1/4/77. (2) *FT* 8/11/78. (3) *MIY* 1982. (4) *The Waste Paper*, Vol. 4 No. 2, 1982. (5) Saburo Okita, *Government report for External Affairs*, Japan. (6) *Gua* 20/8/80. (7) *MJ* 19/2/82. (8) *FT* 27/2/82. (9) *MJ* 27/4/83. (10) *RMR* Vol. 1, No. 4, 1983. (11) *FT* 15/9/84. (12) Yoko Kitizawa, *On the illegal Japanese uranium deals*, New York, 1984 (UN Council for Namibia), (13) *A Slow Burn*, No. 1, 9/84 (FoE, Victoria). (14) *MAR* 1990. (15) *FT* 9/12/86. (16) *FT* 24/3/88. (17) *MJ* 6/10/89 see also *Metal Bulletin* 14/6/90. (18) *MJ* 20/10/89, *E&MJ* 1/91. (19) *Birds of Prey, Gold and Copper mining in the Philippines, a political economy of the large scale mining industry*, a report for Labor Education Assistance and Research (CLEAR), Baguio City and RDC-Kadumi, Baguio City, 10/88. (20) *MJ* 6/10/89. (21) *MJ* 26/10/90. (22) *FT* 8/1/91. (23) *MM* 4/91. (24) *MJ* 6/10/89. (25) *FT* 11/4/90. (26) *FT* 22/5/90.

(27) *Chain Reaction*, FoE, Melbourne, No. 60 4/90.
(28) RAN, *World Rainforest Report*, 11/89, 1-2/91, 4-5/91.

405 Mitsui & Co (J)

One of the oldest "family businesses" in the world, the Mitsui trading and mining complex stems directly from the opening in 1673 by Mitsui Hachirobei Takatoshi of a draper's shop in Edo (now Tokyo). In 1707, the concern turned to importing raw materials, and by the late 19th century it had its own bank and overseas trading outlets. By the start of the first World War, Mitsui had control of manufacturing and raw materials production in Taiwan, south Korea and China.

The company was split after the second World War, as part of the allied strategy to control Japanese military industrialism. However, with the reconstruction of the Japanese economy and the emerging "Cold War", Mitsui was allowed to revive, and in 1959 its "re-unification" was completed.

Between the late '40s and early '60s, Mitsui set out to capture Asian markets and gain control of new mineral resources. The "energy crisis" of the early '70s hit Mitsui like most traditional raw materials companies and, from then on, it put the emphasis on securing stable supplies of energy fuels as well as food and other resources. The corporation (or *Sogo Shosha*, Japanese for multinational trading/manufacturing/raw materials corporations) acquired interests in maize growing in Indonesia, livestock in Brazil, Paraguay and Iran; timber, pulp and paper in Indonesia, the USA, Brazil and the Philippines; rubber in Indonesia, Denmark and West Germany; and petrochemicals as well as textiles and machinery.

It has also prospected for oil and gas in West Papua, off-shore Nigeria and in Iran (1).

Mitsui in the last ten years has conducted uranium prospecting in JVs with Denison Mines in Manitoba, Saskatchewan and the Northwest Territories of Canada (1). Mitsui also had a JV with Denison on a coking coal project in British Columbia (2).

In 1980, Mitsui had 280 affiliated companies overseas; its chairman, Yoshizo Ikeda, had been a member of the Trilateral Commission since 1977.

Among its other JV partners have been Cominco (magnesium project in British Columbia); CSR (evaluating the huge La Trobe brown coal project in Victoria, Australia) (3); Noranda (on a copper-molybdenum project in New South Wales, Australia); and Fluor (copper smelter in Mexico) (2).

Together with C. Itoh, Amax and formerly CSR, it is part of the Mount Newman iron ore consortium which mines on Aboriginal land in Australia.

Mitsui held 50% of the equity of Alumax, the Amax aluminium division which later passed into the hands of Amax (5). However, a Mitsui-led consortium secured 25% of two Alumax reduction plants in 1988 (1, 4, 5). In Britain it was part of Anglo Chemicals Ltd.

And, in 1985, it consolidated its north American interests by joining a JV with Denison, Molycorp and other companies to build an yttrium oxide plant (as a by product of uranium) constituting 35% of the world's potential requirements (5).

When USX sold Quebec Cartier Mining to Dofasco in 1989, Dofasco sold-on 25% to Mitsui and another 25% to CAEMI of Brazil (7). CAEMI is also its partner (together with Showa Denko) and the Société Genérale de Financement of Canada, in a proposed 80,000 tonnes per year ferrochrome plant at Béancour, Quebec – a project which was "in the balance" in 1990 (7).

CAEMI was again a partner in MBR (Mineracaos Brazileiros Reunidas SA) involved in iron ore mining near Belo Horizonte in the Brazilian state of Minas Gerais (5).

With Mitsubishi, the Panama Resource Development Company and Dowa Mining, Mitsui Mining and Smelting Co has been developing the Petaquilla copper deposits (with an estimated 200 million tonnes) in Panama (7).

Mitsui Denman of Ireland started production

of electrolytic manganese in 1976. Mitsui also has a copper-lead-zinc mine at Huanzal, near Cordillera Blanca, in Peru (5).

In 1986, it started a JV (35%) with a Pechiney subsidiary to build an electro-deposited copper foil plant in Normandy, France (5). In Australia, it is a minority partner with Shell in the Drayton Coal development (5).

Mitsui Iron Ore Development Pty's 33.3% share in the Robe River mine in WA gives it a significant proportion of Australia's third largest iron ore mine (5).

In 1991, Arco announced it would sell 15% of its Gordonstone underground coal mine in Queensland to Mitsui for about $100M (8).

Together with Cyprus Minerals, Mitsui Mining and Smelting, along with Mitsui and Co, are investing 35% and 15% in a new JV to explore and develop copper, lead and zinc deposits in Arizona, Colorado and New Mexico – the first partnership of its kind between major US and Japanese companies (9).

Further reading: Tsuchiya Takeo, "Mitsui and Company", in, AMPO Vol. 12, No.1, 1980; updated in Raw Materials Report, Sweden, 1/81.

References: (1) Raw Materials Report Vol. 1, No. 1. (2) MIY 1982. (3) MAR 1982. (4) E&MJ 1/85. (5) FT Min 1991. (6) MJ 16/11/84. (7) MAR 1991 (8) Australian 16/4/91. (9) MJ 29/4/90.

406 MKU

```
        CRA (AUS)              AAEC (AUS)
            \                  /
             51%          41.64%
    MKU / MARY KATHLEEN URANIUM LTD (AUS)
```

One of the oldest uranium mines in the world (although it was put on standby between 1963 and 1974) (1), MKU is a metaphor for the uranium industry itself. Discovered in the the heyday of "wildcat" exploration, when – in the USA and Australia particularly – it seemed as if a lucrative yellowcake lay behind each outcrop and under every boulder in the desert, the mine

was named after the wife of one of two taxi drivers who came upon the lode in the Mount Isa mining region of Queensland, while trying their hand at weekend prospecting (2). Mary Kathleen had herself died of cancer two weeks before the discovery (2). Thinly bedded, and lying within the Middle Proterozoic sediments (3), the mine had an average ore grade of 0.117% U_3O_8 (3) or 31b per tonne (4). In April 1953, after the joint US/British uranium-seeking task force, the CDA (Combined Development Agency) (5), came to secure Australian uranium for US and British nuclear weapons (it had already secured South African supplies) the Australian Atomic Energy Commission (AAEC) was established and immediately took control, both of the MKU deposit and the Rum Jungle mine (6). The Consolidated Zinc Company took over management of Rum Jungle, through Territory Enterprises Pty Ltd, and a little later, the British Rio Tinto Company, flush with money from disposal of various assets, and headed by a chairman, Val Duncan, whose brief from the British Atomic Energy Commission (AEC, later AEA) was to "go forth, find uranium and save civilisation" (see RTZ), bought into the project. MKU's first Australian managing director was recruited directly from Con. Zinc – thus presaging the merger (in 1962) of the Australian and the British company into Rio Tinto-Zinc (RTZ) (7).

During the 1960s, both the US and British weapons programmes were cut back as the "cold war" thawed, and in 1966 the US imposed a ban on imports, to protect domestic producers (6). Between 1961 and 1964, all uranium mines in Australia (MKU, South Alligator River, and Radium Hill) were closed, except for Rum Jungle, which was dismantled and sold in 1971. Exports ceased in 1965, when 7600 tonnes of uranium had been mined (6). Between its opening in 1953 and closure on a care-and-maintenance basis in 1963, MKU had produced around 4000 tonnes of U_3O_8 (1) – just over half the total Australian production to this date – although MKU itself had 70% of the country's low-cost proved reserves (6). The main customer for this uranium was the British Atomic Energy Authority (UKAEA) (8).

In 1967, the Australian Minister for National Development initiated a commercial export policy, leading to the extraordinary, massive uranium discoveries of 1970 onwards (9), in particular the Ranger deposit in the Northern Territory.

Although, between 1967 and 1974, MKU had discovered no substantial new reserves (8), it nevertheless obtained sales contracts for nearly 5000 tonnes of U_3O_8 (4743 tonnes) between 1976 and 1982 (8). In 1982, however, the Australian Labor Party came to power, and – thanks to its initial anti-uranium policy – a ban was instituted on new export contracts, although existing contracts were honoured (9). Indeed MKU was to be the main interim supplier of outstanding deliveries to foreign customers, until the Ranger mine came on stream in October 1981 (9), and when the Australian government announced the public sale of its 50% share in Ranger in 1979, MKU also "threw its hat into the ring" as the *Financial Times* graphically put it (42).

The revival of MKU had been made possible by a share issue in 1974, underwritten by the AAEC, which then took 41.64% of the interest in the company (11) after the Australian public only bought up 7% (12). Although the Liberal government on several occasions said it wanted to sell this interest (1, 13), there were no takers. The AAEC bought into MKU largely by buying out a high proportion of the stake previously owned by Kathleen Investments (see Queensland Mines). Kathleen Investments strenuously opposed the AAEC's involvement, but it dropped its court case in late 1977 (16, 48) – no doubt because by then MKU was hardly proving itself a sound investment.

In addition to the share issue, MKU had to raise emergency funding to cover losses in its first couple of years of re-opening. Although MKU held five contracts to supply nearly 5000 tonnes of U_3O_8 – of which 1130 were to be sold at world prices, and the remainder at fixed prices (17) – recommissioning costs, production and maintenance costs had been well above those originally estimated, and there were severe delays caused by trade union action. By the beginning of 1977, MKU reported that it required some A$12 million – a figure which would normally spell the doom of a minor producer in such a fluctuating market (17). In the end however, thanks to the support of the Federal government through its AAEC stake, CRA was able both to extend its financial arrangements, increase its credit facility (18) and raise an additional A$20 million after negotiations with its Federal partner, CRA to provide the majority of the funding (19).

This transaction caused a storm of opposition protest in the Australian parliament. Tom Uren, the deputy Leader of the Opposition – and a staunch opponent of uranium mining – accused both AAEC and CRA of keeping the mine open to suit their own purposes. CRA, said Uren, had to meet existing contracts, and by-pass the Federal government's restriction on foreign equity (CRA being majority-owned by RTZ until long after MKU was closed down, though Uren was mainly referring to CRA's campaign to take over the AAR coal project in central Queensland). And the Federal government wanted MKU to provide the cutting edge of their stand against unions opposing export of yellowcake, and use MKU as a precedent for going ahead with Ranger (20). Uren claimed that MKU was losing money at the rate of A$16 million a year, and that it would have to conclude contracts to sell uranium at A$46 a pound to make the mine viable (20).

In fact, MKU lost "only" A$5.4 million in 1977, and by then was claiming it had renegotiated better contracts at higher prices, and with new delivery dates (21). During that year, it also repaid the outstanding balance of uranium which it had borrowed from the UKAEA to meet its contract obligations (22). The uranium – some 45.4 tonnes for delivery to Commonwealth Edison Co in the USA (23) – was finally sent to the USA in July 1977 on the SS Columbus (24) – a delivery which enabled the Federal government to set a precedent: borrowing from overseas sources to bulwark an industry already under severe threat from trade unionists and dock-workers.

MKU's contracts, approved in 1972 or earlier, were as follows:

- 2215 tonnes to Commonwealth Edison (USA), for delivery between 1976 and 1985.
- 1100 tonnes to Tokyo Electric Power Co Inc (Japan), for delivery from 1977 to 1986.
- 1000 tonnes to Chugoku Electric Power Co Ltd (Japan), for delivery from 1976 to 1986.
- 407 tonnes to Shikoku Electric Power Co Ltd (Japan) for delivery in 1977, 1978 and 1979.
- 664 tonnes for delivery to Kernkraftwerke Brunsbuttel GmbH (Germany) for 1976-81 (25).

All prices were set on an annual basis but, due to the losses sustained by the mine and delays in deliveries thanks to union action, renegotiations occured which reduced the total to 4473 tonnes by 1976 (1) – leaving, it may be noted, around 1200 tonnes of uranium still in the ground at MKU in the form of reserves (26). New sales were negotiated in early 1977 (27), but it is not clear whether these were to additional customers, or extensions of the contracts concluded before 1972.

MKU operated at a loss for its first two years, only returning to profit in 1978 and 1979. By 1981 it had repaid its outstanding debt to CRA and its first dividend to the AAEC since 1964 (28). Production in 1979 was 832 tonnes, with slightly more the following year, despite a drop in world uranium prices (29). In 1982, the mine delivered a record 859 tonnes (30, 31), when its outstanding contracts to Japanese utilities officially stood at 337.9 tonnes, and permission was sought from the Federal government to send uranium to UF_6 conversion plants in both France and Canada ahead of schedule (32). The Federal government – at least initally – resisted this move, stating that it did not want Australian yellowcake leaving the country until the floor price rose above A$35 per lb (12).

In October that year, the mine finally closed down, leaving a stockpile of 473 tonnes which, it was envisaged, would take two years to shift (33, 34), and repaying a profit to the company

of just over A$10 million (30). In the event, final deliveries were not made until 1985 (35, 36). "Phoenix-like" as the *Mining Journal* put it, the company continued to send out its deadly little consignments, some time after "final" delivery dates had been announced (37).

While the company for a long time prospected on other leases in the Mount Isa region, it never came up with any significant new discoveries (1). MKU carried out research into rare earths production in the late 1970s, after building a pilot rare earths plant. However, there was trepidation about building a full-scale plant and the company frankly admitted doubt "that markets sufficient to achieve an economic return will be found" (14). Their pessimism appears to have been justified.

MKU also entered JVs after its recommissioning: notably one with Agip Australia, prospecting in two areas of Mount Isa, Spear Creek and the Calton Hills (38). At another prospect in the same region, it joined up with Sturts Meadows Prospecting Syndicate NL (39), but once again no signficant reserves were located (38, 40).

A$15 million was also set aside to investigate the Doreen Elaine uranium deposit close to the MKU site, and evaluate (among others) deposits at Westmoreland, Walhalla, Skal and Andersons' Lode. Discussions were held with Queensland Mines about treating that company's ore, and one uranium company (unnamed) was said to have offered MKU some of its own leases, if RTZ itself could arrange contracts for 3000 tonnes of uranium. (MKU in the late 1970s had also tried to buy the AAEC's 2000 tonne uranium stockpile to have it sold on the world market through RTZ) (12). All exploration was cancelled by 1982 (41).

During its last ten years of "life", MKU's chequered history was distinguished by three events which guarantee its place in history, despite the greater economic significance of other mines. The first was the leaking of documents from MKU which served to blow the lid on the infamous uranium cartel. Friends of the Earth in Melbourne (FoE) in April 1976 received a brown paper parcel – left by person or persons unknown on its doorstep – which contained hundreds of letters and documents concerning a "club" set up to apportion contracts to the uranium market and secure the position of certain producers. FoE duly returned the originals to MKU, but not before copying them, and sending items both to the press and Westinghouse – the corporation whose supply contracts had been mainly jeopardised by the operations of the cartel (2).

Among the contents of the package was a letter from the General Manager of MKU to the head of CRA's Corporate Relations department, dated March 12th 1975, expressing deep disquiet about media exposure to the dangers of uranium, which the industry could not "effectively combat", and which both the AAEC and AMIC (Australian Mining Industries Council) could not, or would not, oppose. The company wanted recruitment of "suitable academics" on the devil's side; but by June 1976, as another letter reveals, little change had taken place (43). In 1979, a default judgement was entered against MKU in a US court, hearing the claims made by Westinghouse against the RTZ group (including MKU) and other companies: MKU counter-claimed that the judgement could not apply in Australia (44). Finally, in 1981, the companies involved in the cartel settled out of court: of the A$34 million paid to Westinghouse, CRA footed the largest single bill (A$2.6 million), and MKU had to pay a mere A$870,000 (45).

The second memorable event (though the company tried to minimise it) was the theft of two tonnes of yellowcake by "persons unknown" in 1980. It was transported out of the mine in six drums and later found in Sydney (46). RTZ's reaction was perhaps more blameworthy than the theft itself. Through its inhouse magazine, *Spectrum,* the company blithely dismissed the affair as a pecadillo, if not a prank: "The theft ... must have caused a few red faces in the company ... and a few raised eyebrows in more nuclear-minded countries. However, there is not much one can do with a few tonnes of yellowcake when there is no market for the stuff in Australia and all the necessary facilities are over-

seas. Yet, if the material was indeed appropriated by an employee, it's no different in principle from someone taking home office stationery for his personal use" (47).

Lastly, a true peoples' history of nuclear power may one day account the resistance to exports of MKU uranium ("legitimate" this time) as being more important than anything to do with the mine itself.

What were termed by the *Mining Journal* (34) "a number of problems due to opposition and obstruction", soon became a real threat after 1975. In April 1976, a national strike by railway unions had temporarily stopped the export of MKU uranium (10).

The Australian Council of Trade Unions (ACTU) called for a ban on the export of uranium in 1977 – though this was never acceptable to some unions (48). This followed an important action by anti-uranium activists in Townsville in late 1976, when the first secret shipment of 130 tonnes of U_3O_8 from the revived mine was delayed by about thirty demonstrators for an hour, as it was shunted along the railway line from Mount Isa (49).

Actions against MKU exports built up over the next four years, with the powerful Seamen's Union imposing a blockade. (A demonstration by 200 activists at Mount Isa in 1978 also elicited the support of a few MKU miners themselves) (50). In 1981, after the company was suffering heavily thanks to delays in shipments (52), and forced to export secretively using a small fishing wharf and scab labour (51), a Federal Court ordered the Seamen's Union to refrain from any further actions impeding uranium transportation. The judge, Justice Morling, ruled that MKU was suffering "a grievous loss", running to several hundreds of thousands of dollars. Morling referred particularly to union action against seven containers of uranium, due to be shipped out of Brisbane on June 16th 1981, on the "ACT 4" – but which did not depart until June 19th, when the ban was lifted. At that point "agents for shipping companies indicated that they are not prepared to accept further bookings for MKU until there is a clear indication of a change in the respondents' policy to the export of uranium" (53).

When the mine closed in 1982, MKU had produced around 9000 tons of U_3O_8 in its 26-year history – and 31 million tons of rock (3500 times as much radioactive rubbish as uranium) (54). It left a "conical hole about 250 metres deep, four kilometres wide at the top and about 100 metres across at the bottom" (12). Work started on so-called "rehabilitation" of the mine site in 1983, with a provision of A$17.7 million to complete the task and make the area supposedly "safe" (34). By the end of 1982, only just over A$2 million had been spent on the work (34), but within the next two years, the total came to more than A$12 million, and an additional A$2,300,000 was provided in 1984 (35).

This paltry provision – heavily criticised at the RTZ annual general meeting in 1985 (55) – was agreed with the Queensland state government (hardly the most stringent of environmental protection agencies), "so that all areas can be restored to a safe and useful condition consistent with future land use in the area". The "major aspects" of the rehabilitation plan were: selling off appropriate equipment (and burying contaminated items – a measure which seems to have been completed fairly satisfactorily in early 1983) (56); selling off most buildings and all houses (these sales brought in A$3.4 million) (57); clearing the area of debris, burying all contaminated materials and tailings in a secure dam and covering with mine waste; leaving the pit in a stable condition inaccessible to vehicles; re-vegetating all affected areas using "local species", and monitoring surface and underground water "... until the end of 1985 at least" (58); and flooding the pit (12) partly with groundwater from the evaporation ponds and tailings dam (56).

The company claimed that access was cut off and waste sites stabilised and revegetated, as early as 1982 (59). Contouring and covering work on the tailings dam was in progress by that date – and the "major part" had been successfully completed, (claimed MKU) by the end of 1984 (35). The evaporation ponds appear to have been dried out by this date too – with the precipitate and contaminated soil removed to the tailings dam (35).

But the most controversial – and dangerous aspect – of the rehabilitation scheme, concerned the method by which it was proposed to neutralise radiation from the tailings dam itself. This was clearly intended to be the omnibus "dump" for everything that could not be neutralised elsewhere, or removed from the site.

In 1978, the US Congress had passed the Uranium Mill Tailings Radiation Control Act (UMTRCA) intended, not only to safeguard against radiation from existing uranium mines, but also to clear up the huge legacy (of more than 100 million tons) from dormant mines (60). The essential recommendation of this Act was – to quote Victor Gilinsky of the NRC (Nuclear Regulatory Commission) – to bury the tailings either on-site in the mine workings, or in a surface dump, covered with one foot of clay, followed by "at least five feet of earth appropriately contoured and planted to minimise the effects of water and wind erosion" (61). As reported in the US magazine *Science,* at around the same time: "a tailings pile may exhale radon gas at up to 500 times the natural back-ground rate ... tailings ... contain large quantities of thorium-230. [And] the half-life of thorium is 80,000 years" (62).

At its annual general meeting in 1985, the RTZ chair confessed complete ignorance of the existence (let alone the provisions) of the US Mill Tailings Act (55). The Queensland state government was well aware of the maximum radiation levels recommended by the US government, some three years before it approved the MKU rehabilitation scheme. Details of a US Nuclear Regulatory Commission report recommending limitation of radon emissions to two units had been included in a Queensland Health Department report to the state government, (made before the rehabilitation plan for MKU was approved). In spite of this, the Bjelke-Petersen régime approved a plan which would permit 18 units a metre per second radiating from the MKU tailings site (63). (Incidently, the Queensland Health Ministry re-

503

port also referred to Canadian government recommendations on limiting the level from 2 to 10 units, and to a US EPA report advising a 5-unit maximum) (63).

Environmentalists were alarmed early in the day by the MKU "rehabilitation" plan – a letter from the Federal Government's Uranium Advisory Council, L R McIntyre to the Trade and Resources Minister, Mr Anthony, in 1980, had spoken of "operational problems" at the mine, seepages from the tailings ponds, and warned of the danger of "... another Rum Jungle situation [occurring] at Mary Kathleen" (64). (see CRA) The MKU plan consisted quite simply of cutting corners, and saving cash: instead of the clay/soil covering advised by US Regulatory control experts and employed elsewhere, the company has covered the tailings with rock mixed with soil taken from the MKU site itself. To get sufficient clay to adequately neutralise the tailings, the company claimed it would have to strip and revegetate a large area in the region, or else import the material (64). A spokesperson for MKU, Mr Terry Ward, was quoted as saying he did not believe rainwater seeping through the rock covering "would cause any problem" (64). Moreover, instead of neutralising the liquid already within the tailings ponds, by dosing it with lime (the normal practice), Mr Ward and his engineers decided simply to dump the liquid in trenches, sinking a bore further down the water course to tap any water which risked contaminating farmland and waterholes further downstream. The *Melbourne Age* on October 9th 1982 carried a photo of Mr Ward drinking water from the outlet of this bore, to prove its potable qualities. (Perhaps, like many proponents of radiation in the 1930s, Ward also believes in its medicinal qualities!) (65).

The last irony of the story is perhaps the most unforgiveable. When the mine was officially closed in 1982, twenty journalists invited to the scene were given copies of a pamphlet entitled "Nuclear News". Referring to the effects of the Hiroshima and Nagasaki bombs, the publication boasted the "good news" that the effects of radiation could now be gauged by studying human, as opposed to animal victims, and that "there appears to be considerably less ... genetic damage from radiation than had been thought" (12).

References: (1) *MIY* 1982. (2) Moss. (3) *Uranium Resources, Production and Demand*, 12/77, OECD and IAEA, Paris 1977. (4) *Financial Review*, 15/10/75, *Age* 29/1/76, *Bank of NSW Review*, 4/75. (5) Rogers & Cervenka. (6) Ed. Mary Elliott, *Ground for Concern* (copyright, Friends of the Earth, Penguin Books, Harmondsworth and Ringwood) 1977. (7) Roberts. See also "The Australian Uranium Industry" in *Atomic Energy in Australia*, AAEC, Canberra, 4/76. (8) EPRI. (9) Neff I. (10) Denis Hayes, Jim Falk, Neil Barrett, *Red Light for Yellowcake*, (FoE Australia) Carlton, 1977. (11) *MIY* 19'77. (12) *Age* 17/9/82. (13) *Australian Mines Handbook* 1976-77, Perth 1978. (14) *CRA Annual Report* 1975, Melbourne, 1976. (15) *FT* 14/9/79. (16) *Gua* 25/1/77. (17) *Age* 25/1/77. (18) *MJ* 10/6/77. (19) *MJ* 7/10/77. (20) *Financial Review* 14/10/77. (21) *Financial Review* 26/3/78. (22) *Gua* 24/1/78. (23) *Sun*, Australia, 26/8/76. (24) *MJ* 8/7/77. (25) Information from the Australian Government, quoted in *Kiitg*, WISE, Amsterdam 1982. See also Neff. (26) *FT* 6/2/82. (27) *Facts about CRA 1983*, CRA 1984. (28) *FT* 13/10/82. (29) *FT* 28/4/81. (30) *FT* 25/1/83. (31) *FT* 24/8/81. (32) *Financial Review* 7/9/82. (33) *FT* 20/4/83. (34) *MJ* 8/4/83. (35) *MKU Annual Report* 1985. (36) *MJ* 1/2/85, *FT* 29/1/85. (37) *MJ* 7/9/84. (38) *AMIC Information Sheet*, (Aboriginal Mining Information Centre) Melbourne, 1983. (39) *MJ* 20/11/81. (40) *FT* 31/3/80. (41) *FT* 27/4/82. (42) *FT* 29/8/79. (43) *National Times*, 16-21/8/76; see also *Uranium Deadline*, Melbourne, Vol. 1, No. 5, 1976. (44) *FT* 23/1/79, *FT* 9/6/79; see also *Financial Review* 26/1/79. (45) *Age* 6/5/81. (46) *West Australian* 31/8/80. (47) *Spectrum*, (RTZ) London, 9/80. (48) *MJ* 23/12/77. (49) *Townsville Daily Bulletin*, 20/12/76. (50) *Tribune* 12/7/78. (51) *Australian* 11/12/81. (52) *FT* 28/8/80. (53) *West Australian* 7/11/81, *MJ* 13/11/81. (54) *FT* 27/9/83. (55) Transcript of RTZ AGM, (Partizans) London, 1985. (56) *Financial Review* 7/10/82. (57) *MJ* 9/9/83, *FT* 22/8/83. (58) *RTZ Fact Sheet No. 1*, "RTZ and the Environment", London, 1983. (59) *MKU Annual*

Report 1982. (60) *Kiitg Newsletter*, WISE Amsterdam, 5/82. (61) Statement by Victor Gilinksy, NRC, Washington, 2/5/78. (62) *Science*, US, 13/10/78. (63) *Courier Mail*, Brisbane 9/9/82. (64) *Age* 9/10/82. (65) Richard Rhodes, *The Making of the Atomic Bomb*, Simon and Schuster, London, 1987.

407 MM Mining Co (USA)

It operated a mine in the Four Corners region of New Mexico in the late 1970s called (believe it or not) the Flea Doris Extension (1).
Nothing else known. It is presumed, however, to be identical to M and M Mining.

References: (1) *San Juan Study*.

408 Mobil Oil Corp Inc

```
Rockefeller interests (USA)
         `-.
           `-.
             `-.
MOBIL OIL CORP. INC. (USA)
```

This is the US's second largest oil company, whose "hunger" to take over lesser oil companies – in particular Marathon Oil – was thwarted by the US Supreme Court in early 1982 (1).
After its court failure – on grounds that it would breach anti-trust (*ie* monopolies) legislation – Mobil then announced a bid for 25% of US Steel, the company which had just taken over Marathon. The year before, it had tried – and failed – to take over Conoco, now merged with DuPont.
There then followed speculation that an undeterred Mobil was gearing to take over another smaller oil company – possibly Getty Oil (2).
In the take-over war, however, there are no good guys and bad guys. One important legal point won by Mobil through the US courts in 1981 was that the boards of companies threatened with unwelcome take-overs could not simply decide for themselves which "white knight"

(their term) to whom to offer their ill-gotten gains. This was not in the interest of shareholders, who were entitled to decide the best option by open auction (2).
In 1979 Mobil was rated the US's 43rd most important defence contractor (6).
In the late '70s the corporation had leases on Navajo (Dine) and Pueblo Indian land (7) and was exploring on Lakota land in South Dakota. It was also involved in *in situ* uranium leaching at O'Hern near Bruni, Texas (3).
The US Department of the Interior in 1979 approved an interim mining and reclamation plan for Mobil to test uranium solution mining near Crownpoint, New Mexico. This was to be a 22-month pilot run on 2470 acres jointly acquired with the Tennessee Valley Authority (TVA) (25%) (3). However, Friends of the Earth and members of the Navajo Ranchers' Association filed a preliminary injunction to stop the project until national, regional and site-specific, environmental impact statements had been prepared. After three attempts the Navajos and FoE failed, despite the fact that Mobil has been injecting its leaching agents into the same geological strata at Crownpoint from which the local community pumps its drinking water (4). Mobil Energy Minerals Canada has also prospected uranium in Saskatchewan (5).
Mobil is one of five major oil companies on whom the apartheid régime of South Africa has relied to gain independence for its oil supplies, and thus from UN-imposed sanctions (8).
In Venezuela, it has been responsible for breaking an agreement giving land rights to the Kari'na Indians, in the company's search for oil on Indian lands (9).
It owns, in England, the retailing chain Montgomery Ward, which, in 1982, it had to bail out to the tune of US$575M in equity contributions (10).
Timing for the start-up of Mobil's uranium leaching project near Crownpoint, New Mexico, was to be determined by market conditions, said the company in 1983 after receiving a radioactive materials licence in March that year. The project was a JV with the Tennessee Valley Authority. The orebody to be leached is at a

depth of 2000ft, more than three times the depth of Mobil's leaching project in Texas. Nearly a hundred injection wells were dug, and leaching involved pumping a potential 600 gallons/minute of pollutive solution to produce nearly 200,000lb/yr of uranium oxide (11).

In 1983 Mobil began testing sodium sulphide to clean groundwater polluted after uranium solution mining at Crownpoint. The experiment was sanctioned by the New Mexico Environmental Improvement Division. The uranium bearing sands are 2000ft below the surface in the Westwater canyon on the Morrison foundation – the principal source of water for the Crownpoint district (13).

Within a year, Mobil was trying to double its leases, but announced that commercial operations at Crownpoint probably wouldn't start until 1989. The new leases, if granted, would include portions of nearby Navajo land.

The Tennessee Valley Authority had a 25% interest in the leaching project which planned an initial production of 190,000lb/yr of U_3O_8, rising to nearly a million pounds per year under existing permits (14).

Mobil was cited by the US Treasury Department for breach of sanctions against Rhodesia in the '70s (12).

Mobil was ready to move in on the Alligator River region of the Northern Territory in Australia in the late '70s, but was stopped by the moratorium on exploration (15). From the early '80s in its oil explorations in the Great Sandy Desert of Western Australia it apparently took great care to avoid sacred sites, including the Manga Manga dreaming (16).

In 1983, probably after heavy pressure from the Reagan administration, Mobil pulled out of Libya, claiming damages from the Libyan régime (17).

In early 1984, Mobil bid for control of Superior Oil, the biggest independent oil and gas producer in the USA (18). And, among its recent mineral ventures, was a JV with Nisso Iwai (Japan) to mine coal in Kalimantan (19) which is undertaken by Mobil Mining and Mineral Co. The coal operations were integrated into Mobil Corp in 1985 (27).

Following accusations in the late '60s that the company had supplied oil to Rhodesia through various South African subsidiaries and fronts (20), come more recent claims that Mobil had been also distributing oil products illegally (under UN Decree No. 1) in Namibia (21). It had also been transporting petrol products from South Africa to Namibia, using its subsidiary Petroleum Transport (22).

The *San Juan Uranium Study* (23), whose final report was issued in late 1980, identified Mobil as the fourth largest holder of uranium leases in the study area. To the north, in 1978, Mobil had leases on virtually the entire Ute Native American Mountain Reservation of 162,176 acres, with access to an estimated 40 million tons of uranium reserves (24).

The Crownpoint/TVA leaching project was on stand-by by early 1984 (25).

By 1987, Malapai Resources had acquired Mobil's uranium solution mining project in southern Texas, at Holiday/El Mesquite (26) and Mobil has now essentially withdrawn from uranium production (28).

July 1990 saw the "largest gasoline spill in history" (29) when an underground lake of around 17 million gallons of petroleum products seeped under the streets of New York, contaminating groundwater (30). Most of the oil (14.5 million gallons – some three and a half million more than Exxon's release from the Valdez catastrophe) (30) – had leaked from Mobil's facility at Greenpoint, New York City.

Mobil was fined $500,000 for the disaster (29) and, under an arrangement with the city authorities, agreed to pay millions more in clean-up costs. The company declared that such costs would not require a special charge on its accounts as it would be paid out of "costs of doing business" (31). Commented a spokesperson from First Boston Bank: "Companies will say they have adequately reserved [funds for clean-up costs] but in fact they have not. They're fuzzy and guarded about what they will say" (31).

Just before the New York spill, Mobil was sued by seven US states for "deceptive advertising" (29) when it claimed that its "Hefty" brand disposal bags were "environmentally friendly" and "degradable": in fact they were not (29). New York Attorney General Robert Adams fulminated that Mobil's claims were "green collar fraud", while "degradable plastics" were "nothing but a snake-oil cure for our solid waste problems. It's a myth that degradable plastics quickly disappear or provide any benefits to the environment in a landfill, and Mobil knew it. But that didn't stop Mobil from trying to mislead consumers into purchasing products that simply are not good for the environment ... Mobil's claims for its trash bags should be thrown into the landfill of rhetoric" (29).

Ironically, only two months before this, Mobil's product manager for its Mobil Chemistry Canada Ltd subsidiary had published an article condemning landfill as a solution for the disposal of paper, wood, food refuse and plastic (32). Both in 1989 and the following year, the *Multinational Monitor* cited Mobil on its "Ten Worst" list of corporations "mistreating the environment." The *Monitor* revealed that, in November 1990, a federal jury in New Jersey awarded $1M in punitive damages to Dr Valcar Bowman Jr, a former environmental affairs manager for Mobil Chemicals Corp. According to Bowman, in 1985 Mobil's general (legal) counsel had asked him and his staff to remove sensitive environmental records from a Mobil plant in Bakersfield, California, in case law enforcement officers raided the site. Bowman and his colleagues refused. A few years later, as things "went from bad to worse", he was fired (29).

Myron A Mehlman, a former director of toxicology for the corporation, claimed he was also sacked for trying to report the true benzene levels in Mobil's petrol (gasoline) and other products which, he declared, were a danger to public health and the environment (29).

Contact: Southwest Research and Information Center, Box 4524, Albuquerque, New Mexico 87194, USA.
Multinational Monitor, PO Box 19405, Washington DC, 20036, USA.

References: (1) *FT* 7/1/82. (2) *FT* 8/1/82. (3) *MJ* 10/8/79. (4) *Kiitg* 11/80, pp 16-17. (5) *Who's Who Sask.* (6) *NARMIC* 1979. (7) Garnity, in Sklar. (8) Vella Pillay, *Transnational Corporations: Allies or Instruments of Apartheid?*, UN Centre Against Apartheid "Notes and Documents" series No. 28, 1980. (9) *Survival International Review* Spring/82. (10) *FT* 5/2/82. (11) *E&MJ* 5/83. (12) Sklar p 387; also Bailey & Rivers, *Oilgate*, London, 1979. (13) *E&MJ* 1/83. (14) *E&MJ* 2/84. (15) *FT* 23/9/80. (16) *Kimberley Land Council Newsletter* 3/81. (17) *FT* 5/1/83; *Jamahiriya Review*, London, 3/83. (18) *FT* 13/3/84. (19) *MJ* 18/5/84. (20) "The Mobil Documents", in Bailey & Rivers *The Oil Conspiracy*, New York, 1976 (United Church of Christ), quoted in *Gua* 22/8/77. (21) *Transnational corporations and other foreign interests operating in Namibia*, Paris, 4/84 (UN). (22) Shipping Research Bureau of the Holland Committee on Southern Africa, and Working Group Kairos, Amsterdam, 24/8/84. (23) *San Juan Study*. (24) Winona La Duke, *What is Mobil Oil?*, unpublished research, 1978. (25) *E&MJ* 1/84. (26) *E&MJ* 4/88, (27) *E&MJ Int Directory of Mining*, 1991. (28) *E&MJ* 3/91. (29) *MMon* 12/90. (30) *FT* 13/7/90. (31) *Gua* 11/7/90. (32) *Edmonton Journal* 8/4/90.

409 Mono Power Co

```
Southern California Edison (USA)
                          \
              MONO POWER CO. (USA)
```

Mono Power and Rocky Mountain Energy entered a JV in 1970 to operate the Bear Creek uranium mine and mill in Wyoming (1). In 1985 the partners announced that they would be closing down these operations, after negotiating early termination of sales contracts (originally due to expire in 1989) and in light of the depressed market (5).

The following year, Rocky Mountain Energy announced the conclusion of two years of "complex negotiations" with Taipower, for a five year $18 million, exploration and development agreement covering Wyoming, Arizona, Nebraska, Texas, and Washington State (6).

Southern California Edison was one of the first – if not *the* first – utilities to become involved in uranium exploration in the USA. According to a study on utility involvement in uranium exploration and development, Rocky Mountain/RME's "exploration program has continued to acquire rights to resources in the ground and to support development work on *in situ* leaching projects. All this work has been funded by SCE's customers. In this case the utilities' customers take all the risks and will gain any benefits in the form of lower costs of electricity ..." (2).

In 1983, Mono Power and Rocky Mountain also began considering *in situ* leaching of uranium on their Converse County property – known as Copper Mountain – in Wyoming (3). US$20M had been budgeted for capital outlay, and by the following year the project was still at an evaluation stage (4).

References: (1) *MIY* 1982. (2) Sullivan & Riedel. (3) *E&MJ* 1/83. (4) *MM* 1/84; *E&MJ* 1/84. (5) *MM* 10/85. (6) *MJ* 25/7/86.

410 Montana Nuclear (USA)

It was reported exploring for uranium in Texas in 1978 (1). Nothing further known.

References: (1) *E&MJ* 11/78.

411 Morrison-Knudsen International (USA)

This is a major American construction company which, in late 1982, was awarded a contract to clean up about 26 million tons of radioactive tailings at 25 inactive mill sites in the western USA. The contract – worth US$290M – was awarded by the Department of Energy.

The tailings cover more than 405 hectares in Colorado, New Mexico, Wyoming, Idaho, Ari-

zona, Pennsylvania, Utah and Texas (1). They constitute, however, only about one-tenth of all the tailings produced by uranium mining in the USA (2). The Department of Energy in the three years following passage of the Mill Tailings Act of 1978 (full title, Uranium Mill Tailing Radiation Control Act) considerably diluted the original terms set by the Environmental Protection Authority (EPA), with the result that corporations have been absolved of responsibility for coping with inactive tailings piles, and only a certain number are to be dealt with. Even then, it seems, they would be "stabilised" *in situ* and not removed from areas of population or certain rivers.

Morrison-Knudsen was also contracted to build port, surface works, and rail facilities, for the major El Cerrejon coal mine in Colombia (a project managed by Exxon and the Colombian Government's Carbocol). The Electricity Supply Board of Ireland is one of the customers of Cerrejon (3).

Bechtel and Morrison-Knudsen constructed the huge Ok Tedi copper mine on indigenous peoples' land in Papua New Guinea (see BHP) (4).

Morrison-Knudsen is also part of a consortium which, in 1984, was constructing the world's longest power link, the Inga-Shaba in Zaire (6). It has been called (along with Utah Development and Bechtel, other major constructors of destruction) "one of the notorious scavengers of poverty and war" (5).

Morrison-Knudsen in 1989 bought a half-share in the Cargo Muchacho open heap-leach gold project in southern California, and became a partner with Eastmaque Gold Mines Ltd in the American Girl Mining JV *(sic)* (7).

Contact: SWRIC, POB 5424, Albuquerque, New Mexico 87106, USA (tel 505-844 3941, ext 1702)

References: (1) *MJ* 19/11/82. (2) Southwest Research and Information Center. (3) *MAR* 1982. (4) *MJ* 14/1/83. (5) *The Cape York Aluminium Companies and the Native People*, Melbourne, 1978 (IDA). (6) *MJ* 9/7/84. (7) *FT Min* 1991.

412 MPH/Consulting Ltd (CDN)

These major international mining and petroleum consultants, incorporated in 1957, have more than 500 completed projects in countries as diverse as Argentina, Chile, El Salvador, Greenland, Guatemala, New Caledonia, Niger, Burkino Faso, and Britain. Uranium leads as a "commodity" in which MPH boasts involvement and expertise.

Among the many companies to whom MPH has provided services are Agip, Amax, Aquitaine, Bow Valley Industries, Brinco, Chevron, Conwest Exploration, Getty, IMC, Irish Base Metals, Lacana, Metallgesellschaft Canada, Newmont, Noranda, Pancontinental, Patino, Union Carbide, US Steel, and Urangesellschaft. Its specific uranium exploration projects have included the evaluation and exploration of a one million-acre uranium property in northern Saskatchewan for Kelvin Energy, and surveys on behalf of PNC Exploration (Canada) Co Ltd in the Northwest Territories. It carried out airborne geophysical surveys for Anschutz in Paraguay, as well as radiometric logging both for Getty and Anschutz in the uranium zone (1).

References: (1) Brochure produced by MPH/Consulting Ltd, Toronto, 1984 see also *MAR* 1991.

413 Multi-Minerals Ltd (CDN)

A possibility of uranium on the company's Negemos property (iron-phospahte-colombium) near Sudbury, Ontario, was reported in 1978 (1).
No further information.

References: (1) *Northern Miner* Canada, quoted in *MJ* 17/11/78.

414 Munster Base Metals Ltd

This company has been Anglo-United's exploration arm in Eire.

```
Anglo-United Development Corp. (CDN)

            MUNSTER BASE METALS LTD (IRL)
```

415 Musto Explorations Ltd

```
MUSTO EXPLORATIONS LTD (CDN)

      Musto Canorex (CDN)*
```

* This company is presumed to be a subsidiary.

It was part of the great Key Lake, Saskatchewan, uranium rush in the late '70s; holding claims of 117,000 acres on which encouraging uranium mineralisation was reported in 1978 (1). Alusuisse became its JV partner in further investigations. It also operates in Mexico and New Mexico (USA), and has a large minority interest in Navajo Nuclear (2).

Musto operates the Apex gallium and germanium mine mill and processing plant at St George, southern Utah. In 1987 it purchased the lake Nemegosenda niobium property, and continues to explore for strategic metals in the western USA and Canada (4).

References: (1) *FT* 12/6/78. (2) *Who's Who Sask.* (3) *FT Min* 1991.

416 Nammco (USA)

It was exploring in Texas in 1977 (1).
No further information.

References: (1) *E&MJ* 11/78.

417 Nedco (USA)

It was exploring with Teton for uranium, in Texas in 1977 (1); no further information.

References: (1) *E&MJ* 11/78.

418 New Cinch Uranium Ltd

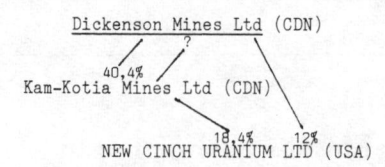

It was exploring in Saskatchewan (1); otherwise it was mainly involved in the Orogrande gold property in New Mexico, over which it had a dispute with Willroy in 1981 (2). Davy McKee rushed to the rescue, proved that Willroy had been providing false values, and New Cinch set out to get them (3)! No further news, alas ... .

References: (1) *Who's Who Sask.* (2) *MJ* 21/8/81. (3) *MJ* 15/1/82.

419 Newmex Exploration Ltd (AUS)

This is a mineral exploration company which has had – among others – interests in the West Kimberley diamond region.
It teamed up in 1977 with Esso Australia, to assess uranium mineralisation at Mundong Well, Western Australia (WA). A shallow drilling programme along several anomalies was planned in 1981 (1).
By the late 1980s, Newmex's gold projects were well advanced in several countries, including Bolivia (in the Altiplano region, at the Iroco project, where Austrabol – a Newmex subsidiary – has a JV with Freeport McMoran), and on the Laloki gold-copper-zinc project in Papua New Guinea's (PNG's) Austrolabe mineral field. Also in PNG, Newmex has a JV on two properties with Poseidon, the Australian company which has mushroomed under Robert Champignon de Crespigny (2). As well as gold interests in WA, the company holds 98% of the Mt Adams mineral sands project south of Eneabba (2).

References: (1) *Reg Aus Min* 1981. (2) *Reg Aus Min* 1990-91.

420 New Mexico Uranium Inc (USA)

It was participating with NM Uranium Co (a subsidiary of Niagara Mohawk Power Corp) in the '70s in uranium exploration in the USA; it probably produced uranium for delivery to NM in 1978 (1).

References: (1) Sullivan & Riedel.

421 Newmont Mining Corporation

In 1982, an unofficial recommendation was made by a financial adviser to the Greater London Council (GLC) that, if the GLC wanted to avoid controversial investments, it should pull out of RTZ and re-invest in Newmont (1). Little did he know ... ! As a Partizans' rejoinder put it at the time: "Comparing RTZ and Newmont is like comparing Pepsi and Coke: they both rot your guts in the end ... [F]or an authority wanting to dissociate itself from financial support for apartheid, operations in Namibia, or uranium mining on native/Aboriginal land, Newmont would certainly be up there in the top twenty companies to avoid" (2).
As of mid-1984, Newmont had assets of around US$2 billion, a stock portfolio of US$400M and profitable gold, oil, gas and (through Peabody) coal operations. Its copper operations, except in Peru (Newmont held there 10.74% of the country's biggest copper producer, Southern Peru Copper Corp, which operates the Cuajone mine in the Andes) (4) and at Palabora, were proving unprofitable, especially at Tsumeb in Namibia and O'okiep in Namaqualand, in the north-west of Cape Province. However, in late 1984, Newmont and GFSA bailed out the troubled O'okiep mine by raising a loan from Barclays Bank, which was to be underwritten by the apartheid régime that reportedly didn't want to stop mining "in a remote and security-sensitive area" (5) of the country. GFSA was to increase its holding in O'okiep to between 38% and 45%, while Newmont's would fall to 1% below GFSA's (6).

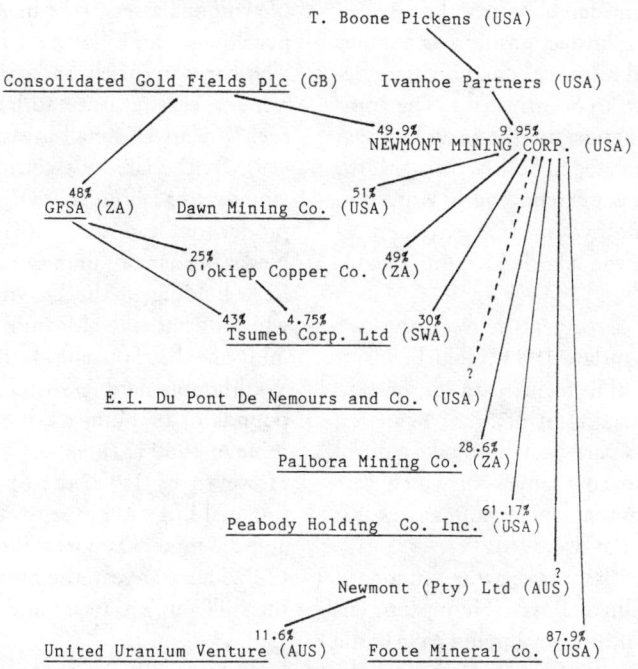

```
                          T. Boone Pickens (USA)

     Consolidated Gold Fields plc (GB)    Ivanhoe Partners (USA)

                                   49.9%        9.95%
                                   NEWMONT MINING CORP. (USA)

         48%                      51%
     GFSA (ZA)    Dawn Mining Co. (USA)

                    25%        49%
              O'okiep Copper Co. (ZA)

              43%    4.75%      30%
              Tsumeb Corp. Ltd (SWA)

                                            ?
     E.I. Du Pont De Nemours and Co. (USA)

                                    28.6%
              Palbora Mining Co. (ZA)

                                 61.17%
         Peabody Holding  Co. Inc. (USA)

                                    ?
              Newmont (Pty) Ltd (AUS)

                  11.6%                      87.9%
     United Uranium Venture (AUS)    Foote Mineral Co. (USA)
```

In Arizona, USA, the Pinto Valley Mining Co (100% owned) cancelled – to its advantage – a high-cost smelting contract in 1983 (7) and, in April the following year, it was decided to re-open the copper operations purchased in 1983 from Cities Services Co (8). This mine is part of the southern Arizona copper belt which was robbed from the Papago Indians during the 1920s by the US Congress (9). Magma Copper, 100% owned by Newmont, used to operate another mine on native land in the region, San Miguel; but this was suspended in 1983 (3). The San Miguel mine is one of the largest non-ferrous underground mines in the world (10). In its 1983 annual report, Newmont indicated it would maintain exploration programmes in the USA, Canada, Peru, Spain, Australia, and the south-west Pacific, but it had terminated a diamond exploration programme in South Africa due to lack of finds. During 1983 the company had been particularly active looking for gold in Nevada, in the mid-continent, and the Appalachians. At the year's end it was managing 17 projects in Australia, the Solomon Islands and Fiji, 14 of these for gold. A new base metals

JV had been opened in Huelva, south-western Spain (11).

Gold remains Newmont's draw card. After launching the new gold rush in Nevada in 1965, Newmont's wholly-owned Carlin Gold, together with a few other companies, "pushed Nevada to the lead as the largest source of gold in the USA" (12). The mining involves a traditional counter-current cyanide leaching process and a newer chlorination pre-treatment. At the beginning of 1984 the company also announced that it would proceed with another gold mine close to Carlin (13).

Little wonder, then, that in late 1984 the *Mining Journal* called Newmont "one of the most profitable of the US non-ferrous mining companies during the recession", crediting this to its "strong, low debt-equity balance" and the success of its "multi-commodity" base (14). The London *Financial Times* agreed, pointing out that Newmont – though the US's third-largest copper producer – had managed to remain profitable during a depression in copper prices, because it is also the country's third biggest gold producer (15).

Newmont, to a considerable extent, has arrived at its confident, expansive, position as a major American "natural resource" conglomerate because of its exploits in South Africa. The company occupied an almost unique nexus between British capital generated in South Africa, North (and, to a much lesser extent, South) American penetration, and re-investment at extremely favourable ratios in the apartheid mining economy.

Under a ten-year agreement between the two companies sealed in late 1983, Gold Fields of South Africa was able to increase its share in Newmont to a maximum of 33.33% in ten years unless a third party sought to take control over the American company – in which case GFSA could buy as many Newmont shares as it wished to fight a take-over (16).

It is instructive to discover that the founder of the company, William Boyce Thompson, entered the world of mining by buying 25% in the Anglo-American Corp of South Africa in 1917 (17). Thompson's partner in crime was none other than JP Morgan of US Steel. As the London *Times* put it at the time, this was "the first occasion on which a definite arrangement has been made for employment of American capital on the Rand" (18).

Thompson continued to invest in southern Africa, taking a stake in Rhoanglo (Rhodesian Anglo-American) which was sold, diminished, in 1953 (17). In 1940 Newmont bought the O'okiep copper mine in South Africa and, putting in 9.5% from this subsidiary, joined with Amax, Union Corporation (now part of Gencor), Selection Trust, and the Swaco/South West Africa Co to buy a new copper deposit, seized from the Germans after WW2 (19).

Not only did the Newmont take-over of Tsumeb later contribute enormously to Newmont's ability to expand (20) but it bolstered South Africa's claim to annex Namibia outright (19).

The new income went partly into the Palabora copper-uranium mine (which Newmont and Bechtel designed and engineered, while RTZ managed), but mainly into Canada, Peru, Australia and the mining of copper on Indian land in Arizona, USA. By 1973, though only 12% of Newmont's assets were in Africa, double that percentage came home in the way of profits. The company itself boasted in 1967 that "no mine ... ever returned so large a cash flow for such a relatively small investment as Tsumeb" (19). By the late '60s Tsumeb was the largest employer of labour in Namibia and the largest producer of lead in Africa (19).

Newmont's major uranium interest has been its 51% holding of the Dawn Mining company which operated the Midnite mine and the Ford mill on the Spokane Indian reservation in Washington state, USA. Nearly half a million pounds of uranium oxide was sold from this mine in 1980 (21).

However, by 1982, the Spokane Indians had initiated litigation over royalties against Dawn and the mine was operating on a reduced scale (22). The next year, the mine was inactive with the mill put on a care-and-maintenance basis (23).

In the late '70s, Newmont was also exploring for uranium with Powerco in central Wyoming (24), but no known positive results emerged.

Newmont had a JV with Esso Minerals Canada at Trout Lake, British Columbia, evaluating prospects in 1982 (21).

It had an ongoing JV with Gencor to explore in Namaqualand, South Africa, especially in the Unamaq area east of Gamsburg.

It has a 70% interest in the Telfer gold mine and mill in the Great Sandy Desert of Western Australia (23). It replaced Getty at the Mount Rawdon gold prospect in 1981 (25), then withdrew, to be itself replaced by Placer Exploration (26).

In New Zealand, in 1982, Newmont had at least two major exploration prospects, both on the North Island, and one in the Whangepaua State Forest (27). It is now no longer active in Aotearoa.

Together with US Steel, Newmont joined a nickel-cobalt consortium exploring in West Papua in the late '60s (28). In addition, the company owned 40% of the Lake Paniai Exploration Co set up in 1972 (along with ICI and BHP) to explore the Wissel Lakes. The same year it headed the Baliem Valley Exploration Co (also with ICI as a junior partner) exploring the mountain range east of Ertsberg, near the Freeport copper mine (28).

Until 1981 Newmont held 7% of St Joe Minerals – there was speculation the company might step in to thwart Seagram's bid that year (29). Later that year, however, it sold its St Joe shares to Fluor (thus enabling that company to merge with St Joe) and also exchanged more than 3.5 million shares in Conoco for nearly 6 million shares in Du Pont (30). The previous year it had announced agreement to purchase Du Pont's 20% share in Lacana (31), but there is no information on whether that deal was sealed, or comprised part of the 1981 share exchange.

In 1983, Newmont bought (as part of a consortium) 3.235% of Fluor's share in Peabody Coal, thus increasing its own stake to 30.735% (32).

In 1988 Newmont made some major changes in its investments, when it decided to dispose of "non core" assets. It sold its British Columbia copper operations (33), but not its 75% interest in Newmont Australia Ltd (34), nor its highly-lucrative gold interest in Carlin, Nevada, which make it the west's third largest gold producer (34). That year it also sold out all its interests in Palabora, O'okiep Copper, Tsumeb, Gamsberg Zinc Corp, and Highveld Steel and Vanadium Corp (35). The previous year, Newmont had itself been "bought into" by Consolidated Goldfields (49%) in order to protect it from an unwelcome bid by the Texas maverick T Boone Pickens (36) who raised his stake – through Mesa Limited Partnership (37) – to nearly 10%, before a Delaware court allowed CGF to raise its own holding in Newmont to nearly 50% (38) – 49.9% being the limit to which CGF could go in raising its stake in the US company (39). In the meantime, speculation had been rife that Newmont might itself bid for CGF, or that Minorco (see Anglo-American) might take over CGF's stake in Newmont, in return for cancelling its own roughly 30% stake in CGF (40).

The following year Newmont announced that, in order to reduce its large debt, it was selling some 4.15 million shares in Du Pont, and would complete the disposal of Foote Mineral Company (41).

The last three years have seen Newmont hitting the jackpot time after time at Carlin Nevada (42) with the result that it has become the largest gold producer outside South Africa and the USSR (43) and the biggest single US gold producer (44).

Through Newmont Australia it has expanded exploration in the Pacific (in Papua New Guinea at Mount Cameron) (45), in rare earths exploration in Western Australia, and in potential phosphates exploitation in Tahiti (where it is in a JV with BRGM, 26%, and Cominco, 24.5%). Here, it has claims – willingly granted by the president of the French Polynesian Legislative Assembly, Alexandre Leontieff – on an atoll called Mataiva, which was being described in late 1990 as a "suitable replacement for Nauru." The reference is to the island devastated by British, Australian and New Zealand interests during most of the current century (46).

In July 1989, Hanson plc succeeded in a £3.5M bid for Consolidated Goldfields (CGF). With CGF came a 49.7% holding in Newmont, and through Newmont, almost a half share in Peabody Coal (47).

Hanson had secured all of Peabody by mid-1990, as a result of buying out the portion which Newmont did not own. This was also partly to secure a cheap, plentiful, supply of coal for Hanson's own purposes – especially for enhanced cash-flow. In the event, the ploy only partly suceeded: no-one wanted Newmont at the price on offer (49) at that time – just two years before, Newmont had been unable to sell 25% of its own majority holding in Newmont Australia (50).

The following year, however, Hanson did an asset-swap with James Goldsmith. Britain's two most notorious corporate raiders exchanged gold-for-forests in a deal of breathtaking audacity, and deceptive simplicity. Hanson secured Goldsmith's 85% stake in Cavenham Forest Industries, with its profitable timber and pulp, mainly in Oregon, Washington State, Mississippi and Louisiana, and not so profitable oil and gas assets, while Goldsmith got 49% of Newmont (51). (Or rather, Goldsmith got 42% and Lord Rothschild, a partner in Goldsmith's Hoylake company, took the rest) (52). To cap it all, Goldsmith claimed that his surrender of forestry interests signalled his recruitment to the "green" movement (although he had, for some years, been an avid opponent of nuclear power) (53).

Anthony Sampson in *The Independent* asked at the time "will the supreme predator change his spots?" (54). Apart from passing doubts as to whether Sampson was referring to Hanson or Goldsmith, the latter's unwillingness to surrender his stake in Newmont in the year since, must certainly cast doubts on his "conversion" to the forces of ecology and renewability.

It is true that Newmont's chair, Gordon Parker, in late 1989 delivered himself of the opinion that the mining industry should do its best to "more than meet every federal and state [environmental] standard. We have undertaken reclamation where it is not required" (55). But, even as Parker was making this statement, Newmont was abandoning a road and drill site in the Lost Lake Peatland region of Minnesota, leaving behind a waste dump, six-foot high piles of plant debris, and scarred vegetation. Paul Glaser of the University of Minnesota, in a report on Newmont's impacts, concluded that: "There seems to be little hope for natural regeneration of trees within the impacted areas ... over the next hundred years" (56).

In Australia, Newmont was involved in a conflict with Aboriginal custodians of the Coronation Hill minesite which became the country's most important single lands rights issue by early 1991 (57). The Jawoyn people had, for some years, tried to prevent mining at the Bula Bula Dreaming site. Initially they contended only with BHP. Then, in late 1990, BHP and Newmont Australia merged their gold interests, partly to offset a takeover bid from Robert Champion de Crespigny's Poseidon Gold (58). Meanwhile, Newmont Corp had been reducing its own interest in its Australian offspring from 43% in early 1989 (59) to 14% by the beginning of 1991 (60).

BHP Gold (56% owned by BHP), and Newmont Australia, set up a new company, Newcrest, and it was Newcrest which exerted all its best efforts to open up the Coronation Hill mine. In the end, Aboriginal and environmentalists won the battle, and the mine was stopped by the federal Australian government. Although Newmont had commissioned a study by the Australian Centre for Advanced Risk and Reliability *(sic)* (ACARRE) (a joint partnership between Sydney University and the Australian Nuclear and Technology Organisation – see AAEC) which concluded that there was only a "slight threat" to the environment from the use of cyanide (61). This was a conclusion largely endorsed by the Northern Territory Power and Water Authority (62). However the government's Resources Assessment Commission strongly advised against the mine going ahead, in view of Aboriginal opposition (63).

Contact: Namibia Support Committee, POB 16, London NW5 2LW U.K.

References: (1) Personal communication to Roger Moody from GLC "mole", 19/4/82. (2) Memo from PARTiZANS (London) to GLC financial committee chairperson, 20/4/82. (3) GFSA Annual Report for 1983, London, 1984. (4) *E&MJ* 11/84. (5) *MM* 11/84. (6) *MJ* 30/11/84. (7) *MJ* 3/2/84. (8) *FT* 3/12/82. (9) Tanzer, p 208. (10) *MJ* 8/4/84. (11) Newmont Annual Report for 1983, Delaware 1984. (12) *E&MJ* 7/83. (13) *FT* 27/1/84. (14) *MJ* 19/10/84. (15) *FT* 9/10/84. (16) *FT* 3/11/83. (17) Lanning & Mueller, p 293. (18) *Times* 29/9/17. (19) Brian Wood, *International capital and the crisis in Namibia's mining industry*, Washington, 11-12/82 (ACOA/UN). (20) R Ramsey, *Men and mines of Newmont*, New York, 1973 (Octagon Books). (21) *MIY* 1982. (22) *Nukem Market Report* quoted in *MJ* 3/9/82. (23) *MAR* 1984. (24) *MJ* 12/1/79. (25) *MJ* 15/8/80. (26) *MJ* 30/10/81. (27) *CAFCINZ Watchdog*, No. 38, 9/82. (28) *West Irian – West Papua: Obliteration of a people*, London, 1983 (Tapol). (29) *FT* 25/3/81. (30) *MJ* 13/11/81. (31) *MJ* 11/4/80. (32) *MJ* 5/8/83. (33) *MJ* 24/6/88. (34) *E&MJ* 6/88. (35) *E&MJ* 5/88. (36) *FT* 1/3/88. (37) *FT* 5/9/87; *FT* 14/8/87; *FT* 9/9/87. (38) *FT* 24/9/87; *FT*

19/11/87. (39) *MJ* 23/10/87. (40) *Evening Standard*, London, 1/9/87. (41) *MJ* 5/2/88. (42) *FT* 28/1/88, *MJ* 8/7/88, *MJ* 19/5/89, *MJ* 29/9/89, *MJ* 5/1/90, *FT* 25/1/90. (43) *FT* 25/8/88. (44) *FT* 22/11/88. (45) *MJ* 22/1/88. (46) *Australian* 26/11/90. (47) *MJ* 24/8/89. (48) *MJ* 6/4/90, *FT* 31/8/90, *Gua* 31/8/90. (49) *FT* 12/10/90. (50) *FT* 9/5/87, *E&MJ* 6/88. (51) *MJ* 19/10/90. (52) *FT* 18/10/90. (53) *Sunday Correspondent* 21/10/90. (54) *Independent* 19/10/90. (55) MJ 6/10/89. (56) quoted in *Northshield*, Minnesota, Vol 2, No. 2, Summer 1990. (57) *New Scientist* 27/4/91. (58) *MJ* 5/10/90. (59) *FT* 25/9/90. (60) *FT* 17/4/91. (61) *Australian* 6/2/91. (62) *Australian* 12/2/91. (63) *Australian* 25/3/91, *Land Rights News* 4/91, *MJ* 21/6/91, *New Scientist* 22/6/91.

422 Newpark Resources (USA)

In Texas, it was exploring for uranium (weren't they all?) in 1977 (1).

References: (1) *E&MJ* 11/78.

423 Niagara Mohawk Power Corp

```
NIAGARA MOHAWK POWER CORP. (USA)

          NM Uranium Co. (USA)
```

It was reported exploring for uranium with US Steel in Texas in 1977 (1).

By late 1979, the JV with US Steel to develop uranium at Burns, Texas, seemed in abeyance, as US Steel tried to sell its uranium interests there in the face of a drooping market (2).

References: (1) *E&MJ* 11/78. (2) *MM* 1/80.

424 Nickelore NL

This is an exploration company, partnered with ACM, West Coast Holdings, Command Minerals, and Getty Oil at Pilbara, Western Aus-

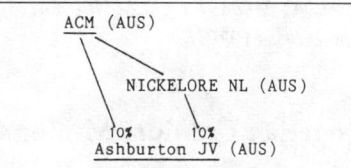

tralia. Its JV (9%) with other companies on the Murchison uranium prospect in the Pilbara (Aboriginal land) had yielded no significant results by 1981 (1). In 1982, the company fought off an attempt by ACM to take control and, through Nickelore, gain 100% ownership of the Big Bell gold mine in Western Australia. Nickelore lost, and ACM, as of early 1983, effectively controlled the company (2).

References: (1) *Reg Aus Min* 1981. (2) *FT* 19/1/83.

425 Nigerian Mining Corp

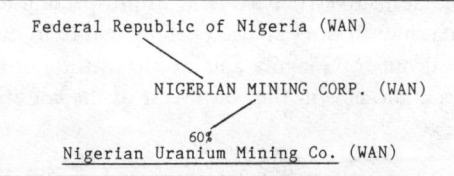

It has been engaged in prospecting for, mining, and processing, "all solid minerals in Nigeria, except coal and oil" (1); its main current exploration thrust is in gold.

Its uranium exploration activities – concentrated on the Gombe and Benue valley areas of Bauchi state (2) – are now undertaken by a 60% owned subsidiary, Nigerian Uranium Mining Co, in which it is partnered by Minatome.

In 1975, when uranium exploration started in earnest, the Nigerian Mining Corp entered an arrangement for the preparation of a Pre-feasibility Report on Uranium Prospects in Nigeria. A year later the Nigerian Atomic Energy Commission was formed and BRGM acted as contractors to Nigerian Uranium Mining on a fieldwork programme (3).

References: (1) *MIY* 1985. (2) *MAR* 1984. (3) *Uranium Redbook* 1979.

426 Nigerian Uranium Mining Co

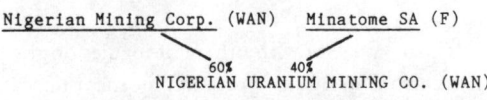

```
Nigerian Mining Corp. (WAN)    Minatome SA (F)
                    \60%     40%/
                  NIGERIAN URANIUM MINING CO. (WAN)
```

It has been exploring for uranium in Nigeria. After an aerial survey in 1980/81, some drilling was reported being carried out.

The parent body of the NUM – namely the Nigerian Mining Corp – also holds 16% with Onarem, Cogéma and the CEGB in the Techili (East Afasto) uranium prospect to the south-west of Arlit, also in the Air region of Niger (1). In early 1984, Mohammed Buhari, Nigeria's military leader, reported progress by the Nigerian Uranium Mining Co, and the country's geological survey department, in prospecting for uranium in three states. Drilling had been carried out in Gongola and Borno in the northeast, and also in the south-east of the country (2).

References: (1) *MAR* 1982. (2) *MJ* 30/3/84.

427 NM Uranium Co

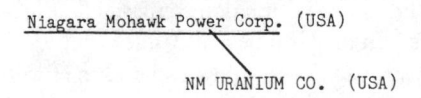

```
Niagara Mohawk Power Corp. (USA)
                        \
                    NM URANIUM CO. (USA)
```

It entered uranium activity in 1975, and by 1977 was exploring with US Steel (1) in Texas, and with New Mexico Uranium Inc (2). Uranium supply to Niagara Mohawk from its uranium ventures began in 1978. The company was also involved in uranium solution mining (2).

However, in 1987, it stopped solution mining (along with USX) at its Clay West JV in Texas,

and is not known to be currently involved in any other uranium venture (3).

References: (1) *E&MJ* 11/78. (2) Sullivan & Riedel. (3) *E&MJ* 4/88.

428 Noranda Mines Ltd

Canada's largest "natural resources group" (1), in 1982 Noranda secured the most detailed entry of any company in the *Financial Times Mining International Yearbook*. The company's operations were divided into three main divisions: mining and metallurgy; manufacturing; and forest products (3). Due to expansion in 1982, the company's metals division was itself reorganised into four new groups (4), including Energy.

The contemporary Noranda was formed in 1954 with the merger of Noranda and Geco mines. Originally the company worked a number of mining claims in Ontario and Quebec. Between the '20s and '30s Noranda integrated "downstream" into smelting and fabrication. Horizontal integration – almost exclusively in Canada – continued through the following two decades. The only foreign venture during those years was a gold mine in Nicaragua (3).

In 1953, Tara Mines (41%) was incorporated in Ontario to explore for minerals and start mining in Eire. For a long time this was the company's major overseas project (5).

Ten years later, a number of mines were merged to become Kerr Addison Mines Ltd, owners of the ill-fated Agnew Lake mine (3). Kerr Addison itself owns 75% of Mogul of Ireland (not to be confused with International Mogul Mines).

At home, in Canada, the company's activities range far and wide, taking in lumber, pulp, newsprint, and the huge coastal timber holdings of MacMillan Bloedel, Canada's largest forest products group. Noranda acquired Mackillan Bloedel in 1981 (5) after a bloodless takeover bid of 49.8%, thus raising speculation at the time that a bigger giant might bid for Noranda itself.

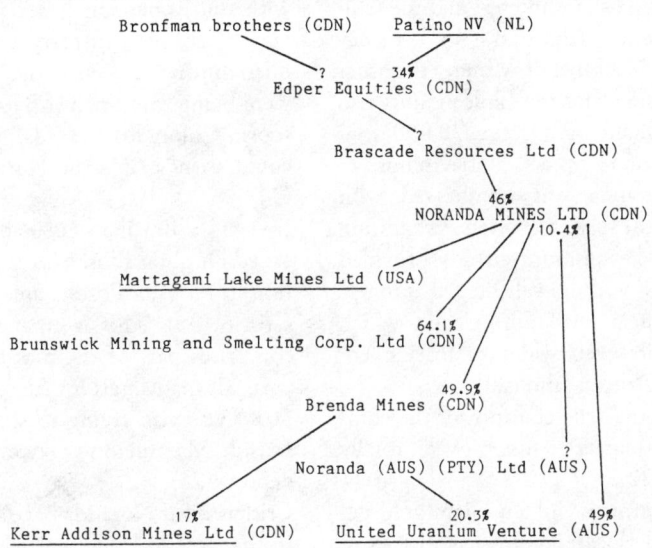

Bronfman brothers (CDN)   Patino NV (NL)
                        ?   34%
                   Edper Equities (CDN)
                            ?
                   Brascade Resources Ltd (CDN)
                         46%
                   NORANDA MINES LTD (CDN)
                              10.4%

Mattagami Lake Mines Ltd (USA)

                              64.1%
Brunswick Mining and Smelting Corp. Ltd (CDN)

                         49.9%
            Brenda Mines (CDN)

                              ?
         Noranda (AUS) (PTY) Ltd (AUS)

     17%   20.3%   49%
Kerr Addison Mines Ltd (CDN) United Uranium Venture (AUS)

Copper rods, wire, cables, fibre optics, medical instruments, plastic pipes, spew forth from Canada Wire and Cable, a manufacturing subsidiary; while steel wire rope, slings strands, nets – everything you could fish for – come from Wire Rope Industries; and aluminium products deriving from a part-owned mine (Friguia, 20%) in Guinea (6).

Noranda's major Canadian mining affiliates and subsidiaries include: Brenda Mines (51%) – copper and molybdenum; Brunswick Mining and Smelting Corp Ltd (63%); Central Canada Potash; Heath Steele Mines (75% JV); Pamour Porcupine Mines (49%) – copper, silver, gold; Kerr Addison Mines (51%); Hemlo Gold Mines Inc (50.1%); and Falconbridge (with Trelleborg AB) acquired in 1989 (62).

In 1987 Noranda acquired a 35% interest in Norcen Energy Resources (63) which it later raised to 48% (64).

Up until 1980 things looked rosy for Noranda. It reopened several mines in Canada (7, 8, 9), expanded its Boss Mountain molybdenum mine in British Columbia (10), and its Norandex Inc American unit acquired substantial property assets in a subsidiary of Revere Copper and Brass Inc (11). Work was begun on the massive Goldstream copper-zinc sulphide deposit in British Columbia (12). A lease was also signed with the Papago Native American tribal council of Arizona to allow the Lakeshore copper mine to reopen (13). Formerly owned by Hecla Mining and El Paso Natural Gas, Lakeshore is a large, but low grade, copper deposit which stopped production in 1977 in the face of low prices. Initial development costs agreed between Noranda and the Papago were set at US$1.64M (7).

Noranda's sales were the highest for any Canadian mining company in 1979 (14), marking an expansive phase by the company (15), especially in Canada, Chile and Australia (7) – more about the latter two, later.

The next year, the company bought out the Maclaren Power and Paper company (16) and Noranda's Canadian Hunter Exploration – its oil and gas unit – won the exlusive right to drill over almost 2 million hectares in the Nechako Basin of central British Columbia (17).

Within two years, however, Noranda sustained its first ever loss after a "devastating period for producers of basic materials" (3). There were strikes in British Columbia and at the Tara mines (18), and a number of mines and projects were closed (3) or cut (19).

Meanwhile, it had abandoned its stake in the Andacollo copper mine in Chile's Elqui province, a project launched in 1976 with Noranda holding 49% of the share interest and the Chilean government (through its agency Enami/Empresa Nacional de Minera) the other 51%. Both Noranda and the Chilean junta had failed to find financing for the 70,000 tonnes/year project on terms acceptable to both of them (20). Noranda was reimbursed with US$4.8M from the Chilean coffers, in addition to a slightly smaller amount already provided (21). Noranda's withdrawal helped prompt Chile into framing a new "mining code" which opened the doors even wider to foreign encroachment on Chilean land (20).

Early 1983 brought little comfort for the company as its first quarter results proved "totally unsatisfactory" (22).

However, by then, the Golden Giant gold deposit in the newly located gold-rush district of Hemlo, north-west Ontario (see also Teck), had been drilled, proving at least eight million tons of ore reserves (23). This find could make Noranda "the biggest gold producer in Canada" (24). Noranda is also linked with Teck in exploration of another property in the Hemlo area (25). (A new gold zone was defined at Hemlo in 1984) (26).

By late 1984, Noranda was still running at a loss (27): five of its copper mines in Canada were closed, and a sixth on the point of closure (28), the Brenda mine among them. And, in the first half of 1984, it had written down the value of some of its properties by C$65.6M, while selling some of its assets in Placer Development (26). Three gold mines in the Pamour Porcupine group were suspended (29) and, by the end of 1984, Noranda was reporting a slide downhill for most of its products (30).

On other fronts, the company's fortunes were a little better assured. In the 1980s, it signed a contract with the Guyanese government to open a gold project at Marudi in the Essequibo region of the country (31); re-entered gold mining at the Nabesna mine in eastern Alaska, and upped its reserves at its multi-mineral deposit at Green's Creek on Admiralty Island (32); signed a contract with the Chilean régime that resolved "past differences" and re-opened the door to Noranda (33); and concluded a JV agreement

with the Robertson Research affiliate, Greenwich Resources, to carry out exploration and open mining projects in the USA: 38 properties were being examined in Nevada, Alaska, Missouri, Colorado and Idaho, and Greenwich could earn 27.5% in Noranda's programme (34).

In 1984 a Brazilian 50%-affiliate of Noranda, called Dunbras Eluma SA, started gold production in the Alta Foresta area of northern Mato Grosso (35). This is an indigenous area that comprises part of the massive Carajas project area. Eluma Metais SA (in which Noranda has 30%) was also trying to site a smelter at São Luis de Maranhao to take ore from the Salobo deposit (26).

Criticisms of Noranda have a long pedigree.

In the late '70s, the Saskatchewan government took steps to take over at least half the province's potash industry. Noranda – through its 51% Central Canada Potash – being a major beneficiary of the *status quo*, vehemently objected. Its chairman Alfred Powis called the Blakeney government "a mafia", and another Noranda representative spoke of the decision as "immoral", "dishonest" and "fruitless" (36). Powis left no doubts as to his company's "morality" in a statement made to the *Financial Times of Canada* in 1975:

"In recent years we have become altogether too preoccupied with the redistribution of wealth to the exclusion of its creation. The result is too much consumption and not enough investment ..." (37).

Some years later, another Noranda subsidiary, Canadian Hunter, attacked the federal Canadian government in similar vein for a budget and energy programme which restricted exploitation of some of its new gas discoveries in western Canada (38).

Noranda's Zapata Mining (Pty) Ltd subsidiary also had an interest in the Onganja copper mine in Namibia, thus contravening UN Decree No. 1 on Namibia (39).

Noranda has recently entered the acid rain controversy in a novel fashion: producing research which purports to show that nitrogen oxides are at least as deadly as sulphur dioxide. As Canada's largest metallurgical acid producer, its research may be construed by many as somewhat tainted. In a five-year analysis of pH (acidity/alkalinity) in 60 lakes down-wind of its large Horne copper smelter in north-west Quebec, Noranda found "no change in pH over the five-year period" (40). Since Horne produces some 463,000 tons/year of sulphur dioxide, objective observers may be a little incredulous about the value of Noranda's research.

An attempt to use asbestos tailings *(sic)* to "scrub" Horne's emissions proved "too costly", while producing phosphoric acid – then fertilizer – as another means of control foundered on the over-supplied fertilizer market.

Meanwhile, Noranda warned the increasingly vocal environmental lobby that, if it carried on criticising, Horne might have to close, and employment would suffer (41).

Indeed, new controls brought in by the Quebec government to limit acid rain pollution meant that the Horne smelter would have to reduce its emissions by no less than 50%. Violation will mean fines of up to C$100,000 per day and prison terms of up to 6 months (42). In 1988, Noranda constructed a new sulphuric acid plant at Horne, to comply with the stricter standards (62).

In the late '70s, Al Gedicks, of the Center for Alternative Mining Development Policy (Wisconsin), accused Noranda of concealing the possibility that it had located a large uranium deposit in the state, by refusing to comply with state law which requires mining companies to file non-interpretative field reports. Noranda won a court case, claiming that the law "violates corporate confidentiality" (43).

The company has, perhaps, a more modest reputation for encroachment on native peoples' lands than most other corporate bodies of its size. Along with Exxon, Kerr-McGee, and Getty Oil, it has for, some time, been eyeing Oneida Indian land in the Canadian Shield area of Wisconsin. In 1976 it signed an agreement with Gulf Minerals (see Eldor Resources) to explore for uranium near Rabbit Lake and, with the British CEGB, to explore for uranium between Rabbit and Key Lake, in north Saskat-

chewan, along with the SMDC (44). In 1974 it also explored for uranium with Thor at La Ronge, Duddridge Lake, Saskatchewan (45).

However, its major foray into yellowcake has been half a world away, in Australia's Northern Territory.

Noranda "owned" the Koongarra uranium deposit in Arnhemland before the 1976 Aboriginal Land Rights Act came into force, and failed to apply for exemption of their deposit from Aboriginal claims (46). What's more, the virtually blanket condemnation of the company's plans by the Fox Commission ensured that it had to "revamp" its project considerably. In what the *Australian Financial Review* appropriately called a "bid to outflank Fox" (47), Noranda re-designed and upgraded its tailings pond system, relocated its plant site, and incorporated a facility to neutralise the acidity of the tailings (48).

A supposed "no release" water system was also planned, to prevent seepage into the Nourlangie creek, and drainage into the wild, unspoiled Woolwonga wetlands (49). Noranda expected to begin production in 1981 (50).

Hardly a year passed, however, before Noranda was dumping Koongarra on the market: Denison picked it up for about C$47M (most of which was for forward production) (51). The company said it needed the sale in order to keep its profits at the 1979 level (52).

Who controls Noranda?

The two major financial controlling interests in Noranda over the last decade have been Zinor Holdings and Brascade Resources (an affiliate of Brascan, formed in 1981 with C de D & P Quebec) (53).

In 1979, Zinor Holdings held 21.2% share interest in Noranda, after a share issue of 14 million Noranda shares was taken up by Zinor, then controlled by Kerr Addison, Placer Development, and Frenwick Holdings. A few months later, Brenda Mines and Brunswick Mining and Smelting (companies partly controlled by Noranda), bought just under 80% of

Frenwick Holdings, thus giving them a small share of Noranda itself (2).

However, in 1981, Zinor and Frenwick Holdings were both liquidated and their share in Noranda sold to Brascade (54). Brenda received nearly 3 million common shares of Kerr Addison and over 3.5 million preferred shares of Brascade Resources as a result. And, so did Brunswick Mining and Smelting (55). This came soon after Noranda'a chairman Powis, and two other Noranda executives, were accused by the Ontario Securities Commission of violating securities laws by selling Noranda shares in 1981, at a time Brascade was bidding to control Noranda (56).

Brascan/Brascade's interest in Noranda was 10.9% in 1979 (57), and it made no secret of its desire to control the company (58). In August 1981 it agreed to increase its interest in Noranda from 22% to about 37%, at a cost of over a billion Canadian dollars (59): it also agreed to keep its Noranda holding below 50% while its nominees to the Noranda board would constitute less than a majority (59).

In the late '70s, Noranda joined a JV with Lacana, and E & B Explorations, to search for uranium in New Brunswick, Canada (60).

Noranda is also a partner (12%) in the Kennecott deep sea consortium controlled by RTZ (directly, and through Kennecott) with Mitsubishi and CGF (61).

Noranda's express intention in the late 'eighties was to become "the premier diversified natural resource company in the world" (65). Its diversifications from 1988 to 1991 certainly reduced its dependency on any single metal: by early 1991, it had re-entered and had new ventures in Chile (17% of the Collahuasi copper project), Newfoundland, Quebec, Montana, and Ontario. But its 1990 results were poor (66). In April the following year, it announced a sale of some C$500 million of its assets (67). On top of this, it offered all its gold holdings to Hemlo Gold, in exchange for more than eight million Hemlo shares (68).

Its renewed prospecting on Oneida land in northern Wisconsin at this time brought it into head-on conflict with Native Americans and a variety of environmental groups (69).

Contact: Latin American Working Group, Box 2207, Station P, Toronto, Ontario M5S 2T2, Canada.

One Sky Cross-Cultural Center, 134 Avenue F South, Saskatoon, Saskatchewan S7M 1S8, Canada (who maintain an information file on Noranda).

Centre for Alternative Mining Development Policy, 210 Avon Street 9, La Crosse, Wisconsin 54603, USA.

Further reading: "Noranda Who?" by Monika Buaerlein in *Shepherd Express*, Milwaukee, 13-20/6/91. (This succinctly outlines Noranda's environmental record – from MacMillan Bloedel's plans to clearcut rainforest on Vancouver Island; via a huge outflow of dioxins from the same subsidiary's paper mill at Port Alberni, British Columbia; to the sordid history of Noranda's violations of sulphur dioxide limits at its Horne smelter, which resulted in 117 children being diagnosed with high levels of lead in their blood by late 1989. It is also worthy of note that – as this book details – in August 1989 Noranda had to contribute to clean-up costs of the Serpent River [Indian] Reservation in Canada, after 9000 truckloads of contaminated soil from its sulphuric acid plant had been dumped on the lake shore at Cutler; see also Rio Algom.)

References: (1) *FT* 9/10/84. (2) *MIY* 1982. (3) *WDMNE*. (4) *MJ* 22/4/83. (5) *FT* 25/3/81; 8/4/81. (6) *FT* 1/5/81. (7) *FT* 20/3/79. (8) *FT* 15/5/79. (9) *MJ* 18/5/79. (10) *MJ* 1/8/80. (11) *MJ* 22/8/80. (12) *MJ* 21/3/80. (13) *MM* 5/79. (14) *FT* 21/2/80. (15) *FT* 13/3/80. (16) *FT* 12/2/80. (17) *FT* 22/1/80. (18) *MJ* 27/11/81. (19) *FT* 19/12/81; 16/12/81. (20) *FT* 27/11/81. (21) *MJ* 4/12/81. (22) *FT* 4/5/83. (23) *MJ* 11/2/83. (24) *FT* 11/5/83. (25) *FT* 9/3/84. (26) *MJ* 11/5/84. (27) *FT* 1/11/83. (28) *MAR* 1984. (29) *FT* 12/7/84. (30) *FT* 30/10/84. (31) *MM* 9/84. (32) *E&MJ* 10/84. (33) *E&MJ* 11/84. (34) *MJ* 30/11/84. (35) *MM* 11/84. (36) "The Noranda File", Latin American Working Group, in *Last Post*, Toronto, 9/77. (37) *Yellowcake Road*. (38) *FT* 23/12/80. (39) *Transnational corpora-*

tions and other foreign interests operating in Namibia, Paris, 4/84 (UN). (40) *MJ* 1/6/84. (41) *MJ* 29/8/84. (42) *MJ* 13/7/84. (43) *Daily Cardinal*, Wisconsin, 20/2/80. (44) B Merlin, *Canadian uranium production potential*, London, 6/77 (Uranium Institute). (45) *Canadian Mining Journal* 4/76. (46) Roberts. (47) *AFR* 24/1/79. (48) *AFR* 16/11/78. (49) *FT* 25/1/79. (50) *Age* 24/1/79. (51) *MJ* 15/5/81. (52) *FT* 21/2/81. (53) *MJ* 31/7/81. (54) *MJ* 13/11/81. (55) *MJ* 25/3/83; 15/10/82. (56) *MJ* 17/12/82. (57) *MJ* 19/10/79. (58) *MJ* 11/1/80. (59) *MJ* 21/8/81. (60) *Energy File* 4/80. (61) *Raw Materials Report*, Vol. 1, No. 4, Stockholm, 1983. (62) *FT Min* 1991. (63) *MJ* 4/3/88. (64) *MJ* 11/5/90. (65) *MJ* 11/11/88, *MJ* 3/3/89. (66) *MJ* 22/3/91. (67) *FT* 29/4/91. (68) *MJ* 10/5/91. (69) *Milwaukee Sentinel* 19/3/91, *Wisconsin State Journal* 25/3/91, *Wausau Daily Herald* 2/3/91.

429 Norbaska Mines Ltd (CDN)

It is part of a JV with Saskatchewan Mining Development Corp (50%) and Consolidated Reactor Uranium Mines exploring for uranium at Wollaston Lake, north Saskatchewan (1). Reportedly it was once involved in uranium exploration in Australia as Baska Mines (2).

In 1978 the company was also involved in diamond drilling for uranium in the Rabbit Lake region (3).

References: (1) *MM* 5/80. (2) *Who's Who Sask*. (3) *Leaderpost*, Regina, 7/3/78.

430 Norcen Energy Resources Ltd

Created in 1975 as a merger between a number of Canadian gas and oil exploration companies, in 1980 it owned 80% of Coleman Collieries (Alberta) (1) and several other oil and gas companies. It is controlled by Brascan as its "main energy arm" (9), while Noranda (35% controlled by Brascade Resources) itself owns 48% of Norcen (10).

It has explored for uranium in Canada, especially British Columbia (B C), Alberta (Athabasca region), Ontario (at Agnew Lake), Nova Scotia (2) and in northern Saskatchewan (3).

The Norcen JV (comprising Norcen, E & B, Campbell Chibougamau Minerals Ltd [renamed Campbell Resources in 1980], and Ontario Hydro) (70%), and Lacana (30%) hold licences to the Blizzard uranium deposit at Kelowna, B C which in May 1979 reported a "commercial uranium deposit" with total reserves around 10.5 million pounds (4).

On 23rd November 1979 – as the provincial Bates Enquiry into uranium mining was under way – Norcen and Lacana announced that South Korea had signed contracts with them for the supply of 7 million pounds of uranium, valued at US$294M at the time. This blatant attempt to influence the outcome of the Bates Enquiry caused an outcry, Greenpeace Vancouver quitting the Enquiry in protest (5). By the end of the year the government of British Columbia had announced its seven-year moratorium on uranium mining (6).

In April 1982, Norcen failed to gain control of Hanna Mining – a move which, according to the *Financial Times*, was a bid to get in on Hanna's energy projects in the USA. Hanna said that the Black brothers, who control Norcen, "... have demonstrated repeatedly and con-

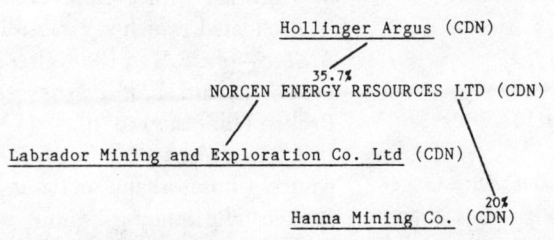

Hollinger Argus (CDN)

35.7%
NORCEN ENERGY RESOURCES LTD (CDN)

Labrador Mining and Exploration Co. Ltd (CDN)

20%
Hanna Mining Co. (CDN)

vincingly over the years that their first public interest is in serving themselves and not the remaining public stockholders" (7).

The following year, Tricentrol, the British North Sea oil company, acquired two of Norcen's subsidiary interests in the North Sea for £14M (8).

Six years later, as part of a deal to acquire the oil and gas interests of Westmin Resources, Norcen sold its stake in MA Hanna (11).

Norcen has recently come under criticism for its interests in the Timor Sea region (Jabiru oil field) which has been parceled out between Australian and Indonesian companies by their respective governments, in the face of strenuous opposition by supporters of the independence of East Timor (12).

And, many thousands of miles away in Canada, at Lubicon Lake, Alberta, the Lake Band (indigenous government) leader, Bernard Ominayak, accused Norcen of "stealing resources" by re-opening its wells in the area, after they had been shut down in 1989 due to threats of sabotage from Lubicon Lake Band members. In defending the company's decision, its manager of northern production, Steven Nengle, argued the threadbare justification of neutrality: "It's a matter of economics for us. This is a dispute between the Band and the government; we're just caught in the middle" (13).

Contact: Tapol, 111 Norwood Road, Thornton Heath, Surrey, England.

References: (1) *FT* 1/80. (2) *MJ* 21/10/77. (3) *Who's Who Sask.* (4) *FT* 15/5/79. (5) *Energy File* 12/79. (6) Roger Moody, in *Vole*, London, Spring/80. (7) *FT* 8/4/82. (8) *FT* 2/11/83. (9) *FT* 2/8/90. (10) *MJ* 11/5/90. (11) *FT Min* 1991. (12) *Tapol*, No. 103, 2/91. (13) *The Press Independent*, Yellowknife, 8/2/91.

431 Nord-Leyou (Gabon)

It has been involved in a JV (10%) with Cogéma (41%) and Kepco (41%) (other shareholders unknown) in an exploration project for uranium north of the Comuf mine at Mounana (1).

References: (1) *MAR* 1984.

432 Norex Resources Ltd (CDN)

Formerly Norex Uranium, a British Columbia charter company, it amalgamated with Caesar Silver MLD in 1976 (1), and had a JV for uranium exploration with Eldorado Nuclear in the late '70s (2).

References: (1) *Who's Who Sask.* (2) Eldorado Nuclear Annual Report, 1979.

433 North American Resources Co

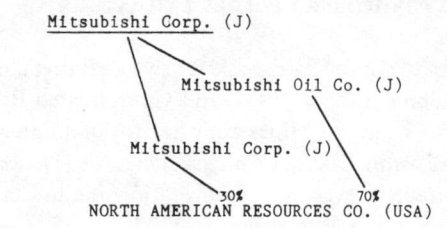

The company was set up to enter a JV with Getty Oil to develop uranium deposits in New Mexico in 1978 (1).

References: (1) *Grants Beacon* 10/1/79.

434 NBHH

Formerly North Broken Hill Ltd, it should not be confused with NBHC/New Broken Hill Consolidated, which is a subsidiary – through AM&S – of CRA. In 1988, after it merged with Peko-Wallsend Ltd, it was renamed North Broken Hill Peko Ltd (6).

It is one of Australia's "historic" mining companies, whose original mine at North, in the Broken Hill mining region of New South Wales, earned it its title and, still, a large pro-

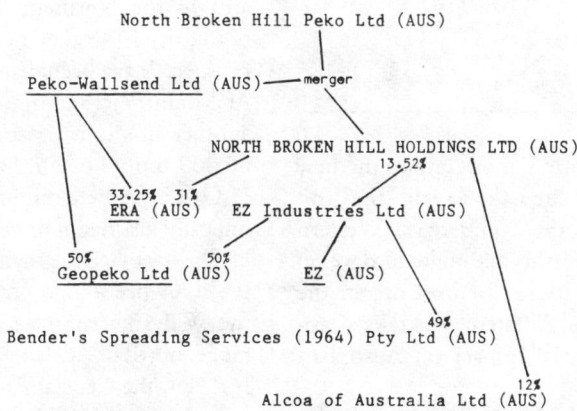

North Broken Hill Peko Ltd (AUS)

Peko-Wallsend Ltd (AUS) ——— merger

NORTH BROKEN HILL HOLDINGS LTD (AUS)
13.52%

33.25% 31%
ERA (AUS)   EZ Industries Ltd (AUS)

50%   50%
Geopeko Ltd (AUS)   EZ (AUS)

49%
Bender's Spreading Services (1964) Pty Ltd (AUS)

12%
Alcoa of Australia Ltd (AUS)

portion of its income (1). Having celebrated its centenary in 1983, the company's interests are largely in JVs for minerals throughout Australia (2), and in forest products (through Associated Pulp and Paper Mills) (3).

Its JV partners include (or have included) Aquitaine, Kennecott, Noranda, CRA, and WMC (1), but its heart still lies in the silver-lead-zinc lodes of Broken Hill, despite recent profit losses due to prolonged industrial action (4). Lead-silver concentrate is sold to Broken Hill Associated Smelters (in which it owns 30%) and zinc concentrate to EZ industries (30% of which it purchased in 1983, and full control of which it obtained in 1984) (5). Thereby NBHH came to effectively own 30.96% of ERA and the Ranger Mine, and later 65.1% (6).

In 1988, the year of the Peko Wallsend merger, North also merged its zinc-lead-silver mining, smelting and international activities with those of CRA, to form Pasminco (see CRA). The same year, it sold its coal, mineral sands and metal recycling interests to Elders Resources NZFP, as well as 60% of the Mt Morgan gold project.

Reorganised into three divisions – Geopeko (exploration), Metalliferous Mining Division, and a mining sector – North Broken Hill's current main interests are in:

- the Tennant Creek (Northern Territory) mines, producing gold, copper and bismuth;
- the Dolphin scheelite mine in Tasmania, and a mineral sands venture in the same area;

- the Peak Hill JV (50%) in Western Australia (WA);
- the Parkes gold-copper project in New South Wales;
- the Collingwood tin project (50%) in Queensland;
- the Robe River iron ore mine in WA (with partners Mitsui, Nippon Steel, and Sumitomo) – Australia's third largest producer;
- the Golden Valley JV (50%) at Kallowna – which began heap-leaching gold in mid-1989.

It owns Warman International Group (machine tools), and holds considerable forestry and paper interests (6, 7).

In 1991, North Broken Hill Peko bought out the Jabiluka uranium and gold deposit of Pancontinental for A$125 million, clearly intending to mine it alongside (or after) Ranger. During July the same year, Prime Minister Hawke stated his "belief" that mining the two deposits in tandem would be a breach of the government's policy (8).

References: (1) *Reg Aus Min* 1981. (2) *MJ* 19/10/84. (3) *MIY* 1985. (4) *FT* 30/4/84. (5) *MJ* 13/7/84. (6) *FT Min* 1991. (7) *Reg Aus Min* 1990-91. (8) *West Australian* 9/7/91.

435 Northern Mining Corp

This is a mineral exploration company which

523

State of Western Australia (AUS)

NORTHERN MINING CORP. (AUS)

held 5% of the world's most important new diamond deposits – the Lake Argyle "find" on Aboriginal land in the Kimberleys, Western Australia. It also had the sole right to develop up to 500 million tonnes of iron ore in the Murchison region of Western Australia – another area of potential impact on Australia's original people (1).

In 1978, Northern Mining reached agreement with Urangesellschaft Australia to jointly explore for uranium down-under (2). In this context it has prospected on national parkland in the "Alps" area of Victoria, taking up about three-quarters of the proposed park (3). Only a fortnight later, company chairman N R Towie claimed that Northern Mining had withdrawn from its JV with Urangesellschaft (4).

It has had interlocks with Freeport in diamond exploration.

According to *Chain Reaction* (3), Mitchell Cotts had a substantial shareholding in Northern Mining as of 1978.

In October 1983, Bond negotiated the sale of its 39% stake in Northern Mining to the West Australian government to give it part control of the Ashton diamond deposits (5) (see CRA for further details).

References: (1) *MIY* 1982. (2) *MJ* 1/12/78. (3) *Chain Reaction*, Vol. 4, No. 1, 1978. (4) *The Australian* 16/12/78. (5) *MIY* 1985.

436 North Flinders Mines Ltd (AUS)

Primarily a coal, oil, gas, uranium and investment company, it holds copper and uranium prospects in the Flinders ranges of South Australia (1). At Gunsight Prospect it discovered uranium along with copper, cobalt and rare earths of "possible economic significance" (2). It also has a prospect in the Pandanus Creek

area of the Northern Territory where it reported in 1981 that results on the Una May prospect "have been discouraging" (2).

At Mount Victoria in South Australia it owns another uranium prospect with an estimated 66,000 tonnes of ore (3.0 to 3.3kg per tonne U_3O_8), but where mining would only be economically justified if prices rise to US$110/kg of U_3O_8 – *ie* twice the then current price.

It also owned a high-grade uranium deposit in the Willyama complex of the North Flinders range, north of Adelaide, South Australia (3).

In 1984, the Central Aboriginal Land Council agreed, after considerable discussion, to allow North Flinders to proceed with underground development of the Granites goldfield in the Tanami Desert, about 500km north-west of Alice Springs. The company was to pay the Council A$150,000 at commencement, and a similar sum when the decision to develop had been taken. A similar amount would go for community facilities, while the Council would get A$55,000 every 6 months. Mine life is estimated at 10 years. The company also applied for exploration permits covering another 10,000 square kilometres in the same area at the same time (4).

The Granites mine produced just over 87,000 ounces of gold in 1988/89, and a further mine was planned at nearby Dead Bullock soak. North Flinders also continued with other exploration in the Tanami desert region (5).

In 1990, the company decided to sell its 21.1% interest in Paragon Mining and Exploration Co PLC to ACM Gold Ltd, as well as its stake in Poseidon Gold (5).

References: (1) *Reg Aus Min* 1981. (2) Annual Report, 1981. (3) *AMIC* 1982. (4) *E&MJ* 6/84. (5) *Reg Aus Min* 1990-91.

437 Northgate Exploration Ltd

This is a mining, mineral and petrol exploration and holding company (1).

In the late '70s Northgate was a company on the rise: a viable mine in Ireland, direct interests

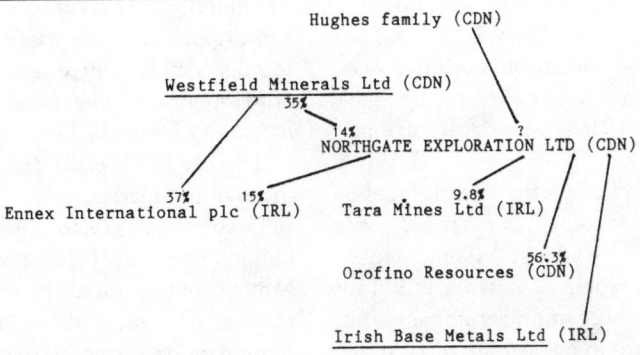

Hughes family (CDN)

Westfield Minerals Ltd (CDN)
35%

14%

NORTHGATE EXPLORATION LTD (CDN)
?

37%   15%   9.8%
Ennex International plc (IRL)   Tara Mines Ltd (IRL)

56.3%
Orofino Resources (CDN)

Irish Base Metals Ltd (IRL)

in Anglo-United (30%), Westfield Minerals (45%) and Whim Creek Consolidated (30%), and exploration for uranium, gold and other metals under way in Ireland, Canada and Australia (2). It also held a small interest in Vestgron Mines in Greenland (3).

By late 1981 it had mustered enough capital to bring off a major mining coup, when it bought all the Canadian mining assets in the Patino NV (Netherlands) empire for £81M (4), and a direct 6% stake in Patino itself.

Very quickly, however, the acquisition of the Patino properties proved a "mixed blessing" (5) and, within two years, Northgate was off-loading many of these interests to reduce a critical indebtedness (6).

The Tynagh mine in Ireland was worked out in 1980, although "healthy working capital" enabled Northgate to continue exploring for minerals in the Tynagh area (1, 7).

Not so healthy was the polluting swamp left behind after Tynagh closed – a swamp of chemicals and heavy metals now held up as a prime example of uncontrolled mine-waste disposal and site rehabilitation (27).

The company's holdings in Vestgron and Anglo-United were both sold in 1982 (8) and its stake in Whim Creek was reduced from around a quarter to 15.8% (1).

In 1984 Northgate formed Austwhim in order to effect the "Australisation" of its down-under assets. By the end of 1984 it was intended that Austwhim would hold more than 20% of Whim Creek while Northgate would have none

(9). Whim Creek is active in mineral exploration in Western Australia and has gold proprieties in the Murchison and Eastern Gold Fields, as well as copper and nickel properties in the Aboriginal Pilbara area, and diamond prospects potentially infringing on more aboriginal land in the Kimberley region (2).

In 1984, Westfield Minerals acquired Mogul of Ireland in a £333,000 deal which gave the companies access to a former lead, zinc and silver mine in Tipperary (10). A new JV called Ennex was established, whose assets include 68 producing wells in the US, Northgate's exploration interests in Ireland, a gold property in the Sperrin mountains of Northern Ireland, and prospects at Tynagh, Boston-Allenwood and Silvermines in the Irish Republic. Most of the exploration is in JV with Getty. Funds, raised by floating of Ennex will mainly go on a new off-shore well in the Kish Basin near Dublin. Northgate would not be transferring its 10% interest in Tara Mines which, through its 75% owned subsidiary Tara Mines Ltd, operates the significant Navan zinc-lead mine (11).

Ennex emerged on the unlisted securities market in London in late 1984 as little short of a disaster. Only 13% of the offer was taken up (12).

However, by the end of that year, record gold production in Canada was helping to cut Northgate's huge deficit – thanks to the mines it retained from Patino in the Chibougamau area of Quebec, and increased production by Whim Creek in Australia (13). There were also improved results from the Orofino Resources gold mine 65 miles south-west of Timmins, Ontario (14).

By the end of 1984, Northgate executive vice-president John Kearney was announcing "substantial inroads into [the] heavy burden of long-term debt" (15).

Northgate is controlled by the Patrick Hughes family (16). Thirty years ago, as *Financial Times* mining correspondent George Milling Stanley pithily put it, the Hughes "shared a dream of setting up a successful mining company, but Ireland did not seem to be the right place at the time and they went to Canada instead" (17). After an initial inspiration to call their band the "Red Hand of Ulster Mining Company"(!) they settled for Northgate Exploration (21).

Their earliest successes, however, were back in the "old country" at Tynagh and Goldrum, and by playing a major role in the development of the Navan mine. Northgate participated in three of the four major non-gold mineral discoveries made in Eire until the early 'eighties, and it was only with the Patino deal in 1981 that interest switched to Canada.

Northgate has operated in several important JVs including Getty Oil (with Irish Base Metals) in Eire (1, 18) and in Alaska north of Eagle (19).

Its uranium interests in Canada have concentrated on Newfoundland (though it did participate as to 75% in the Johan Beetz prospect in Quebec in 1978) (20). Hopes of a major uranium find in Newfoundland were first announced in late 1978 (21).

By early 1979, Westfield and Northgate had between them staked some 2500 claims in the Deer Lake area (22).

By 1980, the companies entered a JV agreement with Shell Canada Resources (23) in the Deer Lake and Codroy Valley areas, with a five-year initial agreement (1).

Around 70,000 acres of claims were involved in the uranium prospects in Newfoundland, with grades up to 25.6lb of U_3O_8 per tonne, in places (24). The moratorium on Newfoundland uranium exploration, announced in May 1980, scotched Northgate's and Westfield's hopes in the region (25).

At around the same time, "uranium fever" (26) also gripped the corporate Irish as Northgate, through Anglo-United and Westfield, moved onto four prospecting licences covering 72 acres in County Donegal. These prospects fell fallow in the early 1980s as the bottom dropped out of the uranium market.

In December 1987, Northgate sold its principal mining assets at Chibougamau to Western Mining, but acquired shareholdings in a wide range of other companies – including Campbell Resources Inc, Sonora Gold Corp, and Neptune Resources Corp (28). (Neptune, in which Northgate now owns 28%, is developing the Colomac property in the Northwest Territories of Canada, the largest gold prospect under development in the country.) Northgate also acquired all of Westfield Minerals Ltd's mineral and mining properties, but sold Whim Creek (see Westfield). It transferred its holdings in Ennex to Norwest, its JV with Westfield Minerals (28).

The company's most controversial project in the early '90s has been the Windy Craggy copper-gold-cobalt venture, operated by Geddes Resources (in which Northgate is a 37% partner with Cominco). The proposed mine borders on the Tatshenshini river, which flows through the Glacier Bay National Park, part of the largest wilderness preserve on the planet (29). Outraged environmental groups – headed by the Sierra Club and joined by Probe International of Toronto – claim that the mine would be disastrous, the equivalent of "an atomic bomb" as one environmental spokesperson put it. The president of Geddes Resources thinks otherwise: "I see an unbelievable expanse of area," he has been quoted as saying "What a tiny dimple in the landscape that mine would make" (29).

Contact: Probe International, 225 Brunswick Avenue, Toronto, Ontario M5S 2M6, Canada.
Minewatch Ireland, Dun Sidh, Moneen, Louisburgh, Co Mayo, Ireland.

References: (1) *MIY* 1982. (2) *MJ* 16/5/80. (3) *MJ* 11/4/80. (4) *FT* 18/9/81; *FT* 22/8/81. (5) *FT*

18/11/81. (6) *FT* 2/12/83; *FT* 13/12/83. (7) *FT* 12/11/80. (8) *FT* 17/11/82. (9) *MJ* 13/7/84. (10) *FT* 1/7/84. (11) *MJ* 8/6/84. (12) *MJ* 6/7/84. (13) *FT* 15/2/84. (14) *FT* 21/2/84. (15) *FT* 2/10/84. (16) *Lotta Continua*, Belfast, 1980. (17) *FT* 9/6/84. (18) *MJ* 26/6/81. (19) *MJ* 8/10/82. (20) *FT* 18/7/78. (21) *FT* 18/11/78; 20/11/78. (22) *MJ* 8/12/78; 6/4/79. (23) *MJ* 11/4/80. (24) *MJ* 1/12/78. (25) Newfoundland government statement, 29/5/80. (26) *FT* 15/6/78. (27) Information from Minewatch Ireland, 1989-90. (28) *FT Min* 1991. (29) *Life Magazine* 5/91.

438 North Kalgurli Mines Ltd

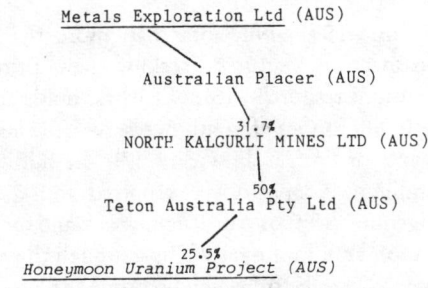

By the late '70s, North Kalgurli, a relatively small ore-treatment company, had become a major partner with one of the world's giant uranium corporations, United Nuclear, in uranium exploration in Australia – particularly in developing the Honeymoon uranium mine in South Australia (1).
The company also had a 50/50 JV agreement (gold prospecting) with Newmont (2).
During 1982 and 1983, however, Metals Exploration effectively took over North Kalgurli and restricted (strangled?) its growth (3). Bond acquired Winthrop Investments in early summer 1984, and thus control of North Kalgurli (4).
Five years later, it was taken over by Gold Mines of Kalgoorlie (5).

References: (1) *MIY* 1982. (2) *MJ* 11/12/81. (3) *FT* 20/10/82. (4) *FT* 8/6/84. (5) *Reg Aus Min* 1990-91.

439 Nuclam

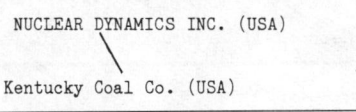

Under the "monster" treaty between West Germany and Brazil, signed in 1975, Brazil effectively ended its monopoly over uranium exploration in its territory and gave UG a chunk of the action (1). To date the only uranium JV in the country, Nuclam controls the Espinharas deposit in Paraiba state, which has an average grading of 0.07% (2). As of 1984 there were no plans to proceed with this deposit in the near future (3).

References: (1) *Enrichment, nuclear weapons and Brazil,* London, 1979 (Friends of the Earth). (2) *Uranium Redbook* 1979. (3) *MAR* 1984.

440 Nuclear Dynamics Inc

```
NUCLEAR DYNAMICS INC. (USA)
              \
      Kentucky Coal Co. (USA)
```

It entered into an agreement with Frontier Resources, and Plateau Resources, in 1978 to explore for uranium in eastern San Juan county, Utah – over 29,000 acres (1). It was also reported exploring for uranium in Texas around the same time (2). Between 1974 and 1979 it participated with Central and Southwest Fuels in uranium exploration (3). No known further exploits.

References: (1) *MJ* 24/11/78; *MJ* 12/1/79. (2) *E&MJ* 11/78. (3) Sullivan & Riedel.

441 Nuclear Materials Corp (ET)

In the late '70s it was reported by the IAEA/OECD *Uranium Red Book* that the Nuclear Materials Corporation was to add a uranium-from-phosphates extraction plant to an existing fertiliser plant, as well as exploit the black (heavy mineral) sands in Egypt and "separat[e] their useful components" (1).

Two years later, the Middle East journal *Eight Days* reported that a Toronto-based engineering company had submitted a "detailed plan to recover uranium from phosphate deposits in Egypt" (2).

It is not known whether the Nuclear Materials Corporation is, in fact, Numac under another guise (no Canadian sources suggest this is the case).

References: (1) *Uranium Redbook* 1979. (2) *Eight Days*, London, 6/2/82.

442 Nuclear Resources (USA)

This was part of a JV with Getty Oil, Cleveland-Cliffs and others, exploring for uranium in Powder River, Wyoming, in the late '70s (1); no further information.

References: (1) *US Surv Min* 1979.

443 Nuclebras

```
NUCLEBRAS /
EMPRESAS NUCLEARES BRASILEIRAS SA (BR)

                        51%
                   Nuclam (BR)
```

In September 1981, the first load of yellowcake was processed and delivered to Brazil's first nuclear reactor (Angra dos Reis No. 1) near Rio de Janeiro. It came from the Cercado deposit at Pocos de Caldas (Minas Gerais), from the Osamu Utsumi mine (1).

Output at the mine was expected to be only 350 tons of uranium oxide in 1982 – though 21,127 tons of U_3O_8 are judged as reserves (2). The uranium deposit is diffuse, and concentrated in three orebodies, necessitating the removal of

some 785 million cubic metres of overburden. By 1982 the pit had a diameter of 800m and covered half a million square metres (1).

Brazil's uranium deposits have been explored at least since 1952 – with the US until 1960, and France 1961-66. The state Nuclear Energy Commission (CNEN) entered active exploration in 1967, and by 1981 has surveyed more than a million square kilometres (3).

Nuclebras took control of the prospecting programme in 1974, and, soon after conclusion of the massive West German/Brazilian nuclear co-operation treaty (signed in 1975), a JV was set up between Nuclebras and Urangesellschaft, called Nuclam (4). This company controls the Espinharas deposit in Paraiba (5).

By mid-1979, Nuclebras had more than 60 prospecting ventures "seeking new primary uranium resources" (6). Nuclebras holds about 230,000 tonnes of uranium reserves, the largest being at Itataia in Ceara (7). In addition, uraniferous deposits were reported from the indigenous area of Surucucu, heartland of the Yanomami, in the early '70s – though these appear now to be officially discounted (8).

Nuclebras is not simply responsible for providing the fuel for the country's nuclear programme: as the state's nuclear authority it is also responsible for the siting and commissioning of nuclear reactors. Through the late '70s and early '80s, considerable doubt was cast on this programme – not least by employees of Nuclebras itself. Most notably, Sr Nogueira Batista (who headed up Nuclebras from its formation in 1975) resigned in early 1983 (9).

He claimed he had finished his task of establishing the company. More likely, he had clashed with Brazilian President João Figuereido over his (Batista's) decision to award construction contracts to two Brazilian companies without public competition (9). The resignation also came only a month after Figuereido cut back on the country's nuclear plant plans from an original 15 to only 4, in addition to Angra dos Reis No. 1 (9).

Such a cut-back had been in the offing since at least 1979 when the Mining and Energy Minister Cesar Cals announced the programme would be scaled down (10).

Despite Brazil's co-operation treaty with Argentina (10), it's clear that nuclear development towards the production of nuclear weapons capability will continue, and that Nuclebras will play a key role in the securing of domestic uranium supplies, the processing of nuclear fuel, and the securing of necessary plant for military purposes.

In early 1981, Nuclebras announced it would spend US$18 million on uranium prospecting in Amazonia (12).

A little later, both the Brazilian and British press claimed that the Brazilian government (*ie* Nuclebras) had supplied Iraq with uranium oxide and low-enriched uranium at the very time US intelligence reports claimed Iraq could be developing a nuclear bomb (13).

More recently, Nuclebras had its budget "dramatically reduced", and "despite its very successful uranium exploration programme in recent years – which has raised Brazilian reserves of U_3O_8 from 11,060t to 235,000t – of the new mine projects, only the development of the Lagoa Real deposit in Bahia is now likely to proceed" (14).

In late 1983, the company discovered another uranium lode in Rio Cristalino, Para – thus raising Brazil's estimated reserves to more than 300,000 tonnes (15).

In association with Construtora Andrade Gutierrez of Brazil, Nuclebras in late 1984 was reported to have signed up to build a $300 million uranium extraction and concentrating plant in the Galgudud region of Somalia (17). Nothing further is known about this project.

Production from Brazil's biggest uranium deposit, Itataia, moved several steps further on in 1987, when Petrofertil, the country's state-owned fertiliser company, awarded a contract to another Brazilian company to carry out a feasibility study for the production of yellowcake and phosphoric acid. Originally, work on the project had been carried out by Nuclebras – testing ore at Prayon's Belgian laboratories in 1981, and visiting uranium-phosphate sites in Florida, Canada, France and Belgium. Three

years later, Petrofertil signed a contract with Nuclebras to construct a pilot plant and carry out a feasibility study for commercial operations. Brazil returned to an elected democracy in 1985, and a revision was carried out on nuclear development – by which time it was realised that the plant would be three times more expensive than originally estimated. But, in August 1986, President José Sarney ordered that Itataia should be developed, and that research should continue on the Lagoa Real deposit in Bahia. This latter deposit is a "better orebody" (16) but Itataia's economic attraction is its association with phosphoric acid. The most recent plans for developing Itataia necessitate using Nuclebras's yellowcake plant in Pocos de Caldas – while that mine itself was doubling its output by 1988 (16).

However, not long after this the Pocos de Caldas plant was closed (18). In the meantime, the Lagoa do Real mine was under construction (19) – much to the consternation of local people, grouped together in the organisation MATER (Movimento Ambiental Terra) which consists of church, youth, women's and trade union groups from Caetité and Bahia as a whole. Oxfam-Brasil and Oxfam-UK were also trying to support the opposition groups in their campaign against the mine.

Contact: Oxfam-UK, 274 Banbury Road, Oxford, England.

References: (1) *MM* 3/82, (2) *MAR* 1982. (3) *Uranium Redbook* 1982. (4) *MJ* 17/2/78. (5) *Uranium Redbook* 1982. (6) *MJ* 31/8/79. (7) *MAR* 1982, (8) *Native Peoples News* (then *Natural Peoples News*) No. 2/3, 1980. (9) *FT* 2/1/83. (10) *Gua* 30/4/79. (11) *MJ* 29/8/80. (12) *MJ* 13/3/81. (13) *Gua* 22/4/81. (14) *MM* 3/82; *MJ* 18/3/83. (15) *MM* 10/83. (16) *E&MJ* 8/87. (17) *MM* 11/84. (18) *Greenpeace Report* 1990. (19) *Relatorio de Visita a Caetité-Bahia, Região Uranifer de Lagoa Real*, produced for CPT-C, Caetité, Bahia, 6/89.

444 Nuexco/The Nuclear Exchange Corp (USA)

This is now known as Nuexco International Corp (4).

The world's biggest private contractor for uranium, handling around one-third of the entire uranium trade in the west (1), it operates as a uranium broker out of Menlo Park, California.

In 1978, Nuexco established an office in Johannesburg, South Africa, because it "believes South Africa will have an important part to play in the uranium market" (2); and a few years later, Swuco, and Edlow International (with which Nuexco is associated through management interlinks and Pacific Energy Services) (1), obtained enriched uranium for the apartheid régime using middlemen in Switzerland and Belgium and the Eurodif enrichment facilities in France (see Swuco).

As the world's principal private uranium brokerage company (3), Nuexco publishes a monthly report establishing the "Exchange Value" for uranium at any one time. This value is widely used within the western uranium industry as a "spot" or short-term price for uranium (3). Subscribers to Nuexco are assisted in both buying and selling their ill-gotten gains (2) in a similar fashion to the London Metal Exchange with regard to non-energy traditional metals.

Nuexco International Corp also publishes definitive country reports, one of the latest of which covered the nine major uranium producing areas in the USSR (5).

References: (1) Age 11/12/79. (2) Rand Daily Mail, South Africa, 23/10/78. (3) Raw Materials Report, Vol. 2, No. 4, 1984. (4) MJ 24/2/89. (5) MAR 1991.

445 Nufcor/Nuclear Fuel Corp of South Africa (ZA)

"The uranium marketing arm of the [South Africa] Chamber of Mines" (1), or "the uranium producers' consortium" (2), Nufcor was set up by all the major South African gold and uranium companies in 1967 to act as a "mine to enrichment" service for apartheid. It has become one of the largest uranium-supply companies in the world. Before RTZ reached its ascendancy in the late '70s, it was probably the largest single such company (3).

By the late '60s, Nufcor had spent a large amount of money purifying South Africa's uranium and up-grading its marketability world-wide (as well as its use in South Africa's own military and civil programme). In 1969 it was producing uranium tetrafluoride, and three years later perfecting hexafluoride conversion – the next-but-one step to full enrichment (4). At roughly the same time, it joined RTZ (already linked with the apartheid uranium industry through the Rössing mine) to launch the first of several meetings of the infamous "cartel". While the original participants in this club of nuclear pirates intended RTZ to emerge as fifth largest western uranium supplier, with Nufcor third, the two brothers-in-conspiracy eventually achieved a joint second (Nufcor 24.4% and RTZ 24.4% with Canadian companies and the crown corporation Eldorado Nuclear leading the field at 37.3%) (5). Nufcor was thus able to conclude lucrative contracts with West Germany, Japan and Britain, and further contracts to the point of enrichment with Switzerland (2). For its part, the British Atomic Energy Authority (UKAEA) sealed a beneficial agreement to "hex" uranium for Nufcor (6).

Nufcor's marketing strategy has developed considerably since it first took in hand the sale of uranium surplus to the weapons programme – which seems to have been quite successful, since only a few mines had excess uranium by 1977 (7). Through the late '70s and beyond, a new pattern emerged as described by Lynch and Neff:

"In the buyer's market of the late 1960s and early 1970s, Nufcor apparently sold what it could at the prevailing prices. Unfortunately for producers, a number of long-term contracts were written that set prices at fixed levels of US$8 to US$10/lb U_3O_8. While this may have been profitable for a South African producer in the early 1970s, inflation made these contracts extremely unprofitable by the end of the decade. Many producers renegotiated their contracts in the sellers' market of the late 1970s, but several could not. However, with the price of uranium soaring by a factor of five in the course of two years, even these producers were able to make spot sales that were extremely profitable.

"As the market tightened and prices began to climb, with apparent demand growing rapidly, South Africa's producers took advantage of the sellers' market to write new contracts that included consumer financing of plant and mine expansion, usually interest-free, although with lower than market prices for the uranium output that they would ultimately receive in return. Often, expansion would not be undertaken without specific sales lined up. Usually the loan would be designated for the plant that was to provide the uranium, although money occasionally would be assigned to the mine's capital expansion account rather than being entered as revenue from sales, an artifice to reduce taxes. Many contracts also indexed prices, the index usually being based on mining costs in the gold industry.

"By the end of the decade, most mines had signed contracts covering a significant part of their production into the mid-term future, often with pre-payment in the form of consumer loans" (8).

By that stage, Nufcor ceased being the only contracting agency for apartheid uranium (10). However, all uranium contracts continue to be arranged by Nufcor, regardless of whether it signs them, and all apartheid uranium – except Rössing and Palabora's – is sold by this ubiquitous agency (10).

In the 'eighties, Nufcor expanded its contracts with Japan (a long-term contract with the Kansai Electric Power Co was signed in 1982) (9), and with US utilities, as the US uranium industry slipped from its historic position at the head of the producers' league: 21% of Nufcor's contracts at that time were destined for the USA (1).

References: (1) *Rand Daily Mail*, Johannesburg, 15/5/84. (2) Rogers & Cervenka. (3) *State of South Africa Yearbook*, Johannesburg (Da Gama) 1971. (4) *The Star*, Johannesburg, 14/10/72. (5) *Financial Mail*, Johannesburg, 3/9/76. (6) *South African Digest*, Johannesburg, 6/2/76. (7) *Nuclear Fuel*, Washington, 21/3/77. (8) Lynch & Neff. (9) *Africa Confidential*, London, 19/1/83. (10) *EPRI Report*.

446 Nukem GmbH

One of the successors to the giant pro-Nazi IG Farben, disbanded by the Allies after WW2, Nukem has a finger in virtually every nuclear pie from uranium enrichment and fuel fabrication to reactor building and waste disposal. Described in 1977 as a "private company based in the small town of Hanau, near Frankfurt", it is far more influential (and influenced by state policies) than this description would suggest. For example, Dr Heinz Schimmelbusch, a prominent West German nuclear scientist, who worked with Degussa through WW2, was a key founder of Nukem – and went on to chair Urenco in 1971 (2). Prof Karl Winnacker, who was a director of IG Farben in the war period and also sat on the Degussa board, has been "of great influence" in both Nukem and Uranit (2). As well as owning nuclear reactors in its home territory, Nukem has participated with other companies in a reprocessing plant for enriched UO_2 and manufactured fuel elements for South Africa's Koeberg-1 reactor between 1966 and 1967 (2).

Through Uranit (which owns about half of Urenco-Almelo, the Dutch part of the British-Dutch-West German enrichment troika) it has been involved in despatching enriched uranium to the Brazilian military dictatorship. Together with its partners in Uranit (Hoechst 20% and Veba 40%), Nukem also agreed to provide spent fuel facilities to Brazil as part of the huge West German-Brazilian deal in the '60s (Bayer was a fourth partner in this particular enterprise); it was also one of the partners in the construction of Brazil's oxide fuel fabrication plant (2).

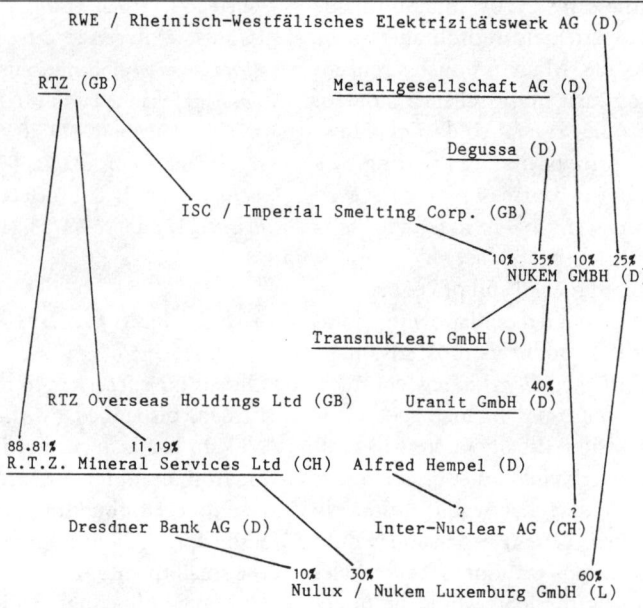

More recently, Nukem encountered strenuous opposition from West German anti-nuclear groups for its plans to build a factory to produce nuclear fuel rods – and store and process 6 tons of high-enriched uranium imported from the USA – at its base in Hanau; also for its participation in a six-company consortium, headed by Kraftwerkunion, which intended to build West Germany's first commercial spent nuclear fuel reprocessing plant. (Some of the material being taken from the Gorleben intermediate storage centre has been the object of several major protests and demonstrations) (4). This Gorleben was later abandoned (see Veba).

Nukem is globally recognised in the nuclear industry for its regular market reports. Like Nuexco, which it resembles in many respects, Nukem also acts as a broker for nuclear material: in 1977 it was accused of openly arranging the sale of plutonium (including French stocks) through one of its *Nuclear Fuel Exchange* reports (1). Commented a Nukem spokesperson at the time: "Nukem acts as brokers in this market ... providing a service for people who want to sell, buy or lease plutonium". The spokesperson denied that it would go to "Iran or South Africa ... or to a rich man who feels like buying plutonium" (1) – but so what?

The Nukem scandal

During 1987 and into 1989, Nukem, and its subsidiary Transnuklear, were the subject of the biggest European nuclear "scandal" in history, with accusations that the company had illegally dumped nuclear waste into rivers and the North Sea (5); that Transnuklear officials had embezzled funds; that nuclear material had been sent in contravention of the Nuclear Non-Proliferation Treaty to non-signatory countries; and that "swaps" had been arranged – whereby "unsafeguarded" customers received "safeguarded" uranium, or the South African origin of certain supplies had been deliberately concealed.

These allegations (some of which were proven) reverberated throughout Euratom (which was implicated in some of the illegal deals), the European and West German parliaments (which both set up enquiries), the anti-nuclear movement (which, not surprisingly, received a much-needed boost), and – not least – through the nuclear industry.

As a result, Nukem set about withdrawing from some of the nuclear sector and selling off some of its plant and fuel fabrication facilities (6). RTZ – which itself came under fire for being involved in "swaps" of uranium of Namibian and South African origin – sold its 10% stake in Nukem to RWE; and Degussa later announced that it might sell its 35% share (7).

Nukem's "withdrawal" from the nuclear industry was more apparent than real. In late 1990, for example, through its wholly-owned Cycle Resources Investment Corp, it bought 32% of American Nuclear Corp, which in 1988 formed the Sheep Mountain partnership (in 50/50 JV with US Energy Corp) (8).

Contact: Initiativgruppe Umweltschutz Hanau (IUH), c/o Elmar Biez, Friedrich-Ebert-Anlage 9, W6450 Hanau, Germany.
WISE, Postbus 18185, 1001 ZB Amsterdam, Netherlands.

Further reading: *TransnuklearAffäre: über die Arbeit des Untersuchungsausschusses im EP*, published by the Regenbogenfraktion (Rainbow Group) in the European Parliament, Brussels, 1988.

References: (1) *Obs* 11/12/77. (2) Rogers & Cervenka. (3) Press release of Die Grünen, Ortsverband Hanau, 16/11/84. (4) *FT* 12/4/85. (5) *WISE NC* 4/11/88. (6) *FT* 20/7/89. (7) *MJ* 4/5/90. (8) *MJ* 12/10/90.

447 Numac Oil and Gas Ltd

Part of the Midwest Lake JV along with Esso Resources, and Bow Valley Industries, it was

```
         Pitcairn Co. (CDN)

              10.2%
    NUMAC OIL AND GAS LTD (CDN)

               10%
    Midwest Lake JV (CDN)


Precambrian Shield Resources Ltd (CDN)
```

also also exploring with Amok at Carswell, Sask, and in the Yukon with Precambrian (1).

References: (1) *Yellowcake Road.*

448 Numec

A "renegade" government, intent on acquiring an atomic weapon, sets up a company front to acquire the raw material and process it. Heading this band of outlaws is a top scientist previously involved in developing the only atomic bombs which (to date) have been intentionally dropped on people's heads. Over the next eight years, enough enriched uranium is siphoned from the plant to construct about 20 nuclear bombs. The world's most extensive intelligence agency discovers – and conceals the fact – with the probable connivance of the President of the USA. Within the following two years another 200lb of U-235 (about nine bombs' worth) is discovered missing, and the renegade government has made "raids" on uranium stocks in at least two other countries, not to mention organising a full-scale hijack of an entire uranium cargo. But the public of the country from which the materials have gone "missing" isn't to become aware of the fact for another decade; at which time the world's biggest intelligence agency concludes the thefts were "academic", and closes its file. Through much of this time, the front company continues to entertain a stream of visitors, including spies; cynically destroys many of its own records; obtains top secret information on nuclear weaponry from its host country; flagrantly ignores all manner of safety regulations and – when it's fulfilled its function – dissolves itself, its masters slinking away like thieves in the night, unpunished, unbloodied and unbowed.

That, in essence, is the story of Numec – a company whose very existence dramatically proves that so-called controls against nuclear proliferation work least in the area they are purportedly needed most.

The tale begins in 1957 when a Zionist Jew, Dr Zalman Shapiro – a research chemist heavily in-

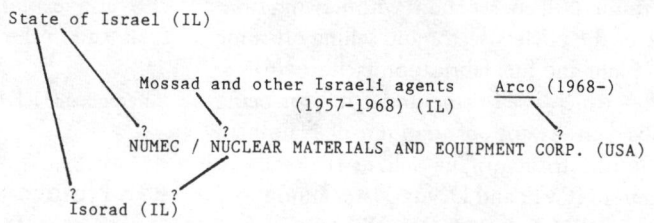

```
            State of Israel (IL)

                            Mossad and other Israeli agents    Arco (1968-)
                                  (1957-1968) (IL)
                            ?        ?                            ?
                         NUMEC / NUCLEAR MATERIALS AND EQUIPMENT CORP. (USA)

                    ?        ?
                  Isorad (IL)
```

volved in the Manhattan Project during the Second World War – is sent by Israel's secret/terrorist intelligence agency, Mossad, to set up a uranium processing and acquisition plant in Apollo, Pennsylvania, along with former Manhattan Project colleagues. The CIA is convinced that finance to establish Numec comes primarily from the Israeli government.

Shapiro bids for contracts from the US government itself, both to process and enrich uranium and to provide fuel for reactors and the space programme. One of his chief benefactors is Admiral Hyman Rickover, "father" of the US nuclear submarine fleet. Other clients soon follow, especially in Europe. Numec agrees formally to act as a "technical consultant and training and procurement agency" for Israel, and soon a separate company – half owned by Numec, half by the Israeli state – is formed. The new company Isorad is to help develop techniques to preserve fruit through irradiation. Actually it is to provide a major pipeline for uranium and other nuclear technology – straight through to the Dimona reactor in the Negev desert. Some irradiation! Some fruit!

By the early '60s the US Atomic Energy Commission (AEC) is beginning to get worried about Numec on two main scores: although specifically prohibited from doing so, Shapiro is known to be mixing enriched with unenriched uranium, thus confusing its nature, amount, origin and destination. "Sloppy bookkeeping" also means that no AEC inspector can find out where a particular contract has gone. At the same time, visitors can apparently get into the Numec plant at will and gain access to no fewer than 2400 classified documents, including descriptions of secret US government research and development programmes.

In late 1964, the CIA discovers that Israel has enough enriched uranium to fuel a nuclear bomb: Richard Helms, the CIA chief, mentions Numec as a possible source of the material. But it isn't until the following Spring – when Westinghouse can't trace one of its own contracts to process fuel for a nuclear rocket – that Numec's apparent inefficiency gets widely talked about. A little later Shapiro admits to the AEC that 130 – or 194 – pounds of U-235 has gone missing (1).

"Such a disaster had not been experienced since the dawn of the US nuclear age," comments James Adams (1). When the AEC demands an explanation, Shapiro claims the missing highly strategic material is to be found in waste dumps at the plant. On excavating, only 16lb is unearthed (2). Worse – a close inspection of the plant's records shows that a total of 382 (2) or 391 (1) pounds of U-235 has in fact gone astray. Allowing for wastage, the AEC concludes that more than 206lb has been lost – more than at any other time by any other country (1).

Further investigations show that Numec has kept incomplete records on no fewer than 26 of the 32 contracts it has been awarded. Despite this (indeed *because* of it) Shapiro's firm remains in business. What the eye cannot see, the heart (or at least the CIA and the AEC) manifestly do not grieve over.

Three attorney-generals order investigations into Numec over the next few years, and the FBI does its best to pin charges on Shapiro. He has to pay a US$1.1M fine but, because of lack of concrete evidence, he continues operating Numec until it has outlived its usefulness (and other uranium sources for Israel have been found). By 1968, after the AEC ascertains that no less than 572 lbs of enriched uranium are "unaccounted for", Numec is sold – apparently to Arco (3). Nothing untoward has been reported about the plant since.

It was not until 1976 that the House Commerce Committee of the US Congress learned anything about this massive diversion of "trigger" nuclear material (1) – so successful has been both Numec's rearranging of its papers and the CIA's blatant connivance with the company, afraid that its own role in the '50s, of supplying nuclear know-how and material to Israel, against official government policy, might come out in the wash. Indeed, in a 1978 briefing, the CIA virtually dismissed Numec's role: "because plutonium from the Dimona reactor was believed to be available ... from the CIA's point of view the diversion did not matter" (4) While the FBI does not seem to have been complicit with the CIA in the cover-up, President Johnson certainly appears to have ordered that the intelligence agency cease its investigations soon after he came to power (1).
And nothing like the full story of the Numec Connection has yet been revealed.

References: (1) Most of the material on which this entry is based is included in James Adams's *The Unnatural Alliance*, London, 1984 (Quartet Books) pp150-157, from which the 130lb figure comes. However, Barbara Rogers and Zdenek Cervenka, citing the US magazine *Rolling Stone* in *The Nuclear Axis* (2), give the higher figure of 194lb. (2) Rogers & Cervenka. (3) US Congress Sub-committee on Energy statement, 1975. (4) Pringle & Spigelman.

449 Occidental Petroleum Corp

The company that "made Love Canal a household word" in the United States, thanks to the deadly pollution it created (15), is the US's tenth largest oil company, whose Canadian subsidiary has been a joint partner with Inco in northern Saskatchewan uranium exploration (1). Occidental – headed by Dr Armand Hammer until his death in 1990 – has entered into

```
OCCIDENTAL PETROLEUM CORP. (USA)

Hooker Chemical Corp. (USA)

Canadian Occidental Petroleum Inc. (CDN)
```

two important joint ventures, giving Occidental control of large parts of the world fertiliser and the European chemicals market.
In early 1981, it set up a joint marketing operation with the apartheid-based company Triomf Fertiliser, giving Occidental access to further supplies of phosphoric acid and a possible way around the US embargo on fertiliser supplies to the USSR (2). *Between them, the two companies could now supply 65% of the world's market for phosphoric acid.*
Later the same year, ENI and Occidental agreed to set up Enoxy to give the US oil company access to Italian chemical plants and the Italian state company access to Occidental's four coal mines (3). However, in 1983 Occidental decided to pull out of the possible joint chemicals venture with ENI (14).
By the end of 1981, Occidental had also taken over Iowa Beef – the largest beef processing company in the USA – and was trying to gain control of Zapata, a Houston-based company into oil, gas, coal and – tuna fishing (4).
Occidental also holds shares in various seed-breeding companies and produces both agricultural chemicals and fertilisers (6). Hooker Chemical Corp is notorious for its poisoning of New York's Love Canal in the late '70s/early '80s (5).

The octogenarian chairman of Occidental, Dr Armand Hammer, combined business with politics (or, rather, was something of an expert at the politics of business). In 1979, he negotiated trade agreements with the USSR (7) and the following year with Romania (8). In 1981, an agreement for development of coal resources was signed with China (9). In 1982, Occidental moved into China to develop coal deposits in a US$230M deal with the China National Coal and Development Corporation. This represented one of the largest deals ever made between China and a western country (10).

Also in 1982, Occidental acquired Cities Services, a major US oil company in a US$4 billion merger deal (11). Early the same year, the Dunlop Rubber company held talks with Occidental over the sale of its 46% owned International Synthetic Rubber to the oil company. ISR is Britain's largest producer of synthetic rubber (12).

In early summer 1982, Occidental and Inco proceeded with a feasibility study on the McClean uranium deposits. The McClean North and South Zones are estimated to contain 6000 tons of U_3O_8 (13).

Armand Hammer's greatest 'eighties coup was probably to conclude a deal with China which enabled Occidental to jointly develop the Pinghuo coal mine – the world's largest open-pit coal project. Although Occidental would pay the equivalent of a wage originally based on US miners' pay, the wage would in fact go to the state and not the workers (15).

In early 1983, Hammer decided to offload the refining, marketing and transportation operations of Cities Services to Southland Corp in return for which Occidental got 20% of Southland, the biggest convenience store operator in the country. Since the company also sells petrol in 40% of its 7300 stores, Occidental's status among the top US oil companies has not been affected.

In 1983 Hammer was given permission to proceed with the Cathedral Bluffs shale oil/synthetic fuel project in Colorado in which it held 45% along with Tenneco and Peter Kiewit Sons Inc (10%) (16).

By early 1984, however, Hammer was facing dissidence within the board room – and a possible take-over joust, specifically from David Murdock whose 18% share in Iowa Beef was converted into 3% of Occidental and another 5% in preferred voting shares. "Analysts are now predicting a power struggle for the company" (17).

It was Occidental Chemical Company's EDB (ethylene dibromide) which killed two workers at a company plant in California in 1982 when

one of the victims' bodies was "eaten away". Occidental was fined US$53,000 (18).

A survey of Occidental's worldwide activities wouldn't be complete without some reference to its chief architect, the ubiquitous Armand Hammer – a giant even in the exaggerated world of oil politics.

According to a 1984 article in the London *Sunday Times*, Hammer was "a friend of princes and kings. His riches more than equal theirs, his actual power on earth is equal to the ruler of a nation state, and his travels over the face of the earth, accompanied by retinues of aides, greeted by senior officials of state and whizzed into private conference with political chiefs and ministers, are like the state visits of a grand duke or emir ..." (21).

His HQ office packed with mementoes (a signed photo from Lenin, another from Khrushchev), Hammer assiduously cultivated the image of a supra-human entrepreneur for whom ordinary politics are subservient to commerce. One of his "media manipulators" was Gordon Reece, a former TV-groom for Margaret Thatcher, a one-time British prime minister. Hammer's father Julius was a founder member of the American Communist Party and, as such, had avenues to the Bolsheviks which his son consolidated. Hammer got his foot in the Soviet door by supplying a million bushels of wheat during the famine in the early '20s. According to Hammer, Lenin "was somewhat disillusioned when I met him. He said that the Revolution wasn't working." Lenin told the young American (says Hammer): "The Soviet Union needs American capital and technical aid to get our wheels turning once more." On Lenin's invitation, Hammer became the Soviet state's first concessionaire. Under Lenin, Hammer consolidated his power – bringing trucks, tractors, Ford cars and pencils to the "socialist" state and paying his workers piece-rates.

Under Stalin, Hammer's Soviet fortunes drooped. After flirting with booze (post-Prohibition) and Aberdeen Angus cattle in the '40s, Hammer acquired a "two-bit company called Occidental" in the '50s. The Gene Reid Drilling company hit oil in 1959 (19). The discovery of the Lathrop gas field provided resources for expansion in 1961. Says Neil Lyndon: "The story of Occidental's rise, in the last 30 years, is so dramatic and so intricately entwined with the most important economic and political events in the world that it compares with the history of a Renaissance commercial state." Its biggest political coup was probably gaining a Libyan concession under King Idris. When Gadaffi toppled Idris, Occidental still remained in the driving seat – perhaps due to Hammer's re-forged connection with the Soviet state or because (as has been recently suggested) "Gadaffi identified Dr Armand Hammer's Occidental as the weak link in the monolithic facade of the international oil industry" (20).

From the early '60s onwards, Hammer built on his apparent inviolability in the Soviet Union to diplomatically service the White House (smoothing relations after the U2 spy incident, for example) and economically serve the Soviet monolith: in particular by negotiating a US$29 billion agreement to supply fertilisers. The latter coincided with a US$100,000 donation by Hammer to Nixon's notorious Campaign to Re-elect the President (the power-house of the Watergate scandal). For his pains, Hammer received a suspended prison sentence (21).

The myth survives – of Hammer as a fabulously rich philanthropist (donating fortunes to cancer research and US-USSR athletic exchange) whose fervent desire for peace and world harmony was solidly based on realisation of the American Dream. Insofar as the dream is both illusion and reality, insofar as its inversion (the Communist "nightmare") is also reality and illusion, Hammer's actual role as a megalomaniac aggrandiser of utterly conspicuous personal wealth at the expense of the poor of the world continues to be successfully concealed.

The Oxy (as it is now called) chief architect's demise on 12 December 1990 came as a hammerblow to the company he had made an extension of the world's biggest oil-based ego. Purple prose? Perhaps – but nothing like the legacy the autocrat left behind him, with multi-million dollar lawsuits flying between Hammer's son and grandson over who should keep the family silver; accusations that Hammer defrauded his wife over a 30-year period while keeping a string of mistresses on the side; and his niece "suing almost everybody else, charging 18 counts of fraud, misrepresentation and constructive fraud" ("constructive fraud"? – now there's one for the tax evader!) (22).

Once the iron hand was off the helm ("consummate egotist" the *Financial Times* obituaried Hammer, "who ruled Occidental with a will – critics said whim – of iron") (23), the "new ruler", Ray Irani, unveiled his vision of "the new Oxy" (23). Essentially this consisted of selling peripheral businesses, and reducing the huge ($8 billion) debt load Hammer had accumulated.

The coal ventures in China have now gone by the board (24); likewise a JV with the USSR to manufacture petrochemicals (23). Claiming there would be "no sacred cows" in Oxy's byre, Irani also proposed dumping oil-from-shale, film production (Hammer horror?), real estate, hotels in Peking and Nigeria, and the breeding of Arabian horses and Black Angus cattle (no sacred cows indeed!). Whether Oxy's one-third stake in the McLean Lake uranium JV (with Inco and Minatco/Total) has also gone by the board is not known.

A final noteworthy point is that, according to a report issued by a Scottish High Court Judge, Lord Cullen, in 1990, the explosion aboard the Oxy-owned Piper Alpha offshore oil rig in 1988 – which claimed 167 lives and was the worst such accident recorded in the oil industry – was partly due to the company's "poor management" (25).

Contact: Multinational Monitor, PO Box 19405, Washington, DC 20036, USA.
Work on Waste, 82 Judson Street, Canton, NJ 13617, USA.

Further reading: A mildly critical (though sneakily admiring?) biography of Hammer published in 1989 is *Armand Hammer: the Untold Story* by Steve Weinberg (Ebury Press).
A summary of the company's worst exploits is con-

tained in Stuart Gold's profile of Oxy: "Occidental Petroleum: Politics, Pollution and Profit", in *Multinational Monitor* 7-8/89.

References: (1) *MJ* 5/10/80. (2) *FT* 3/2/81. (3) *FT* 30/10/81. (4) *FT* 9/9/81. (5) *Yellowcake Road*. (6) *Raw Materials Report*, Stockholm, Vol. 1 No. 3, 1982. (7) *MJ* 6/7/79. (8) *MJ* 11/4/80. (9) *MJ* 16/4/82. (10) *MJ* 16/4/82. (11) *FT* 10/11/82. (12) *FT* 19/3/82. (13) *MM* 5/82. (14) *FT* 24/1/83. (15) *News and Letters*, Detroit, 5/84. (16) *MAR* 1984. (17) *Gua* 3/4/84. (18) *San Francisco Chronicle* 6/10/82. (19) *WDMNE*. (20) *FT* 28/3/84. (21) *Sunday Times* 7/84. (22) Weekend Supplement, *The Independent*, London, 24/2/91. (23) *FT* 16/1/91. (24) *FT* 11/6/91, *MJ* 14/6/91. (25) *Multinational Monitor* 11/90.

450 Ocean Resources NL

This mineral exploration company, with HQ in Canberra, has interests on and off shore. It is linked with CRAE/CRA Exploration off Cape Moreton, New South Wales.

It has come up against Aboriginal opposition in recent years to at least two of its projects. At Wallaha Lake, New South Wales (NSW), Aborigines opposed a prospective gold and silver mine (1).

Since 1977, the company hasn't proceeded with its uranium claims in Arnhemland (just north of the Jabiluka Mine) because of opposition from traditional owners and the Northern Land Council. The Ormac JV originally comprised Canadian Superior (25%), Pancontinental Mining (10%), Consolidated Goldfields of Australia Ltd (20%) and Ocean Resources (45%). However, Pancontinental and Consolidated Goldfields quit the venture in December

```
        Aboriginal interests (AUS)


        OCEAN RESOURCES NL (AUS)

              33.53%
          Ormac JV (AUS)
```

1980, leaving Canadian Superior with two-thirds and Ocean with one-third – with a shareholding exceeding 1 million sharess. Aborigines have received equity interests in Ocean Resources (1).

The land at issue comprises part of the Alligator Rivers Region Stage 2, over which land claims are still being heard by the Aboriginal Land Commissioner (2). However, the Mines Department of the Northern Territory at the same time announced that it planned to give ten licences for exploration in the area (1).

In April 1983, Ocean Resources said that the partners in the Ormac JV had been offered EL 3106 prospect subject to satisfactory negotiations with the Northern Land Council (4).

In 1989, Ocean resumed a 1980 JV with CRA to explore for minerals over the whole of southern Queensland's offshore coastal areas, after the Mineral Submerged Lands Act gave companies a qualified go-ahead (5). Although it has relinquished a gold exploration JV with Shell at Tumut, NSW, it retains an exploration licence for the bewitching metal at East Murchison, Western Australia, and also its 66.67% interest in the East Alligator uranium prospect.

Ocean has also now pulled out of its tin leases at the holiday "paradise" of Phuket in Thailand, due to vocal opposition (5).

References: (1) *Reg Aus Min* 1981. (2) Annual Report, 1981. (3) Nicolas Peterson (ed), *Aboriginal Land Rights Handbook*, Canberra, 1981 (Australian Institute of Aboriginal Studies). (4) *Reg Aus Min* 1984/85. (5) *Reg Aus Min* 1990/91.

451 Offshore Oil NL (AUS)

This is an oil, coal and natural gas exploration company in Australia (1).

It has investigated several areas of Australia's Northern Territory and Queensland for uranium (2). It was part of the consortium, along with Cocks Eldorado NL and Southern Cross Exploration NL, which took over AAEC's 10% stake in the Lake Ngalia prospect, thus joining Central Pacific Minerals, Agip Australia, and Urangesellschaft Australia.

References: (1) *Reg Aus Min* 1981. (2) *AMIC* 1982.

452 Ogle Petrol (USA)

It had a joint 50% interest in the Bison Basin uranium prospect in Wyoming, USA, with Western Fuel and Energy Capital (1); otherwise nothing known.

References: (1) *FT* 14/7/80.

453 Oilmin NL

```
              OILMIN NL (AUS)
                    /
              16.66%
    Beverley Uranium Project (AUS)
```

Oilmin, Petromin, and Transoil were three associated companies, closely identified with the mining interests of Queensland ex-Premier Bjelke Petersen (1). They tended to work together, while preserving a certain fiction that they were separately identified.

Together with Magellan Petroleum (50%), Oilmin was part of a consortium (21%) seeking oil and gas rights over Aboriginal land west of Alice Springs at Mereenie (2). In its 1980 annual report the company announced that the partners had agreed a 1.5% royalty to the Aborigines (as opposed to 10% to the NT government) (3).

The agreement was hailed by the Central Land Council (CLC) as "... a significant demonstration that the Aboriginal Land Rights Act is working" and the first of its kind. The CLC envisages a partnership agreement with the companies and expressed its gratitude to them for their readiness to pay a US$32,000 tax to the NT government on its behalf (4).

As well as being one of the partners at Beverley in South Australia, Oilmin has also prospected for uranium in the Cape York province of north Queensland – almost completely an Aboriginal-

claimed area – where it reported "good prospects" at Iron Range (5) and where drilling started (6) on the Lockhard Aboriginal Reserve in the early '80s (1).

In early 1981, Oilmin chairman Bill Siller made no secret of his conviction that nuclear power was essential. The next century, he told an Australian Institute of Energy seminar, would be "the age of electricity" and nuclear power would have to be the basis of electricity generation if the western world "... did not want to be outstripped by communist countries" (7).

Oilmin has also prospected at Mount Painter, 110km north-west of Leigh Creek in the Lake Frome uranium area of South Australia, along with Transoil. More than 7000 tonnes of U_3O_8 in a number of small deposits have been located (8).

In 1983, Oilmin and its new Mereenie partner Pancontinental announced that first oil from the field would flow late that year.

References: (1) Roberts. (2) *Reg Aus Min* 1981. (3) Annual Report, 1980. (4) *CLC Land Rights News* 12/81. (5) *Kiitg* 6/81. (6) Annual Report, 1980. (7) *AAP Report*, quoted in *Australia Monthly*, London, 20/4/81. (8) *AMIC* 1982.

454 Oliver Prospecting and Mining Co Ltd (IRL)

This is a company set up in 1971 (1) which was prospecting for uranium in Eire during the late '70s (2).

During the late '80s, Oliver spread its wings. It entered a JV with Riofinex (RTZ's exploration arm) over a 30-square mile exploration area in County Meath (Oliver 70%) (3). It concluded another JV agreement with Syngenore Explorations over the Ballinalack and Harberton Bridge lead-zinc orebodies, Co Westmeath and Co Kildare, which have been progressing through to 1991 (4).

Overseas, Oliver announced in 1987 an exclusive gold and diamond licence in the northern part of Sierra Leone (3) and acquired 269 mineral claims in Nova Scotia, Canada. Two years later, its subsidiary Carrick Gold Resources Ltd offered a 49% option on these gold prospects to New Signet Resources Inc of Vancouver.

It also entered a JV with Esso Minerals Canada to earn 51% on an Esso gold prospect in north central Newfoundland (3).

With a sizeable interest in North West Exploration plc (which owns a gypsum deposit at Co Cavan), the acquisition of Candecca Ireland in 1984, and of Wington Enterprises (with oil and gas interests in offshore Ireland and in Abu Dhabi) the following year, the company entered the 'nineties considerably larger than the small outfit which surfaced 20 years before. Oliver certainly got more!

References: (1) *MIY* 1982. (2) Personal communication from Cork Anti-Nuclear Group, 4/3/82. (3) *FT Min* 1991. (4) *MAR* 1991.

455 Olympia and York Investments Ltd

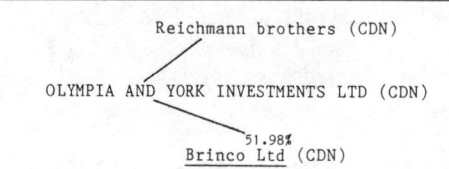

```
            Reichmann brothers (CDN)


OLYMPIA AND YORK INVESTMENTS LTD (CDN)


                   51.98%
              Brinco Ltd (CDN)
```

As well as owning a majority of Brinco, and controlling Gulf Canada Resources (4), Olympia and York was on the way to seizing the Abitibi newsprint manufacturing business (the world's largest) in 1981 (1). It controls Hiram Walker, now part of Hiram Walker-Consumers Home, a Canadian "drinks and energy" concern (2), and it bought the English Property Company in 1979 (3).

During the late '80s, O&Y moved with a vengeance into British property markets, in a manner which has triggered enormous antagonism and which could lead to possibly the largest claim for damages by a group action in British legal history.

In mid-'87, the Reichmann brothers – the moving force behind Olympia and York – took over the Canary Wharf office/housing project in London's docklands (the biggest single such construction project in Europe). Over the next two years, the company spread its tentacles into other commercial and residential "developments", thus becoming the biggest property developer in the capital (5).

In 1990, it revealed that it had accumulated an 8.25% stake in Rosehaugh, a British property company whose boss, very ironically, is a major funder of Friends of the Earth (5).

The following year, to keep up with the costs of its Canary Wharf construction, O&Y sold its stake in "dear old dependable" Allied-Lyons (6) and other assets. GWU (89% owned by O&Y) also disposed of its controlling interest in Consumers Gas which, by then, was Canada's biggest natural gas utility (7). The company abandoned plans to merge Gulf Canada Resources with Home Oil, which would have created Canada's third largest oil, and fifth largest gas, producer (8).

Canary Wharf continues to be, if not a white elephant, then certainly a yellowish peril, as the Reichmanns are forced to offer cut-price rents to attract tenants, and express disappointment that "... no leading UK institution has so far signed up" (9).

Thousands of council tenants living within a few hundrd yards of the vast project have filed claims for damages for noise and intolerable dust levels, together with loss of TV reception. "Three-year-olds are on inhalers for asthma," claimed tenants' representative Christine Frost (10). The tenants are suing Olympia and York, and the London Docklands Development Corporation, for up to £100 million (10).

References: (1) *FT* 3/3/81. (2) *FT* 16/4/81. (3) *FT* 25/11/81. (4) *FT* 13/5/91. (5) *FT* 6/4/90. (6) *FT* 16-17/2/91, *Independent* 16/2/91. (7) *FT* 21/2/91. (8) *FT* 28/11/90. (9) *FT* 5/7/90. (10) *Gua* 2/9/91, *Gua* 11/9/91.

456 Omitaramines

In 1978 it started examining, along with Anglo-

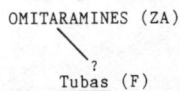

```
OMITARAMINES (ZA)

        ?
      Tubas (F)
```

American, Gencor, Minatome and Elf-Aquitaine, a uranium deposit near the Langer Heinrich prospect in Namibia (1), known as Tubas. Interesting results were reported the following year (2). Between 30 and 40 tons of ore was taken from the site in 1980 and shipped to Europe for analysis (3). More recently the group announced a temporary halt in development – "it is thought that Anglo-American does not wish to invest in a new mining development prior to [Namibian] independence". Tubas Mining was nonetheless registered in France and evaluation of ore carried out there (3).

References: (1) *FT* 26/7/78. (2) *MM* 6/79. (3) *Africa Confidential* 15/10/80.

457 Omnis

```
Democratic Republic of Madagascar (RM)
                                    \
                                     \
        OMNIS / OFFICE MILITAIRE NATIONAL
        POUR LES INDUSTRIES STRATÉGIQUES (RM)
```

In 1976, Omnis took over the old Fort Dauphin prospect which had provided (along with CEA) uranium for France's nuclear programme between 1955 and 1968 (1).
Omnis also prospected in the Ansirabe area of central Madagascar – an area where uranium mining had taken place until 1950 (1).
The Malagasy Karoo formation in the west central region of Folakara was also targeted.
No recent news has emerged on these prospects.

In 1983 Omnis was mainly preoccupied with general energy and mineral prospecting, along with three Canadian companies (2).
One of these three was Q-I-T Fer et Titane of Canada, with which, in the late 'eighties, Omnis set up a JV (QMM) to investigate and exploit huge mineral sands deposits on the south-east coast of Madagascar, in the Toalagnaro region (3). RTZ inherited the 49% share of this project when it took over BP Minerals (parent company of Q-I-T) in 1989.
Although scheduled to start in 1992, the mining has been considerably stalled. The World Bank commissioned environmental and social impact studies as the price for its participation in the project: by 1991 the studies had concluded that the original project would "destroy three-quarters of a coastal forest zone of 4000ha" (4).
The full Environmental and Social Impact Statement is not likely to be ready until 1992 and, with opposition building up to the country's biggest mineral project (5), the chances of the original plan being implemented are diminishing.

Contact: PARTiZANS, 218 Liverpool Road, London N1, England.

References: (1) *Uranium Redbook* 1982. (2) *MAR* 1984. (3) *MAR* 1990. (4) *MAR* 1991. (5) *Independent on Sunday* 8/9/91.

458 Onarem

Formerly Uraniger, Onarem holds the Niger Republic's equity in all the country's uranium mines and exploration ventures. In 1978 these comprised: the Arlit mines (33%), co-owned by Somair; the Akouta deposit (31%), co-owned

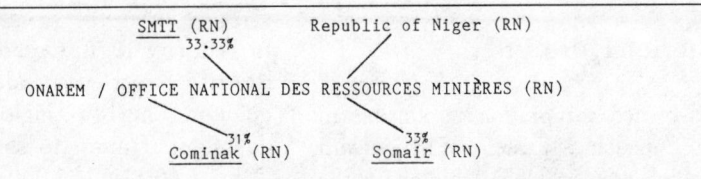

```
        SMTT (RN)              Republic of Niger (RN)
          33.33%                        |
              \                         |
               \                        |
ONAREM / OFFICE NATIONAL DES RESSOURCES MINIÈRES (RN)
               /                        \
         31%  /                     33%  \
      Cominak (RN)                  Somair (RN)
```

541

by Cominak; the Imouraren deposit (30%); the West Afasto deposit (33.3%); the East Afasto deposit (25%); the Djado deposit (25%); the I-n-Gall deposit (called now Abkorun-Azélik) (50%); I-n-Adrar (33%); and a JV with Exxon (1).

By the early '80s the Djado deposit had been abandoned as also, it seems, was the Exxon JV, though another JV was formed between Onarem and the Japanese PNC (2). Within a few years, plans for Imouraren were shelved indefinitely, after feasibility studies for both an open pit and an underground mine had been carried out (3). All other exploration deposits were reported to "have little prospect of early development". This includes the Abkorun-Azelik prospect (where mine feasibility studies have been undertaken with Japan's Arabian Oil Co) and West Afasto (which Onarem shares equally with Cogéma and OURD) (4).

The Arni deposit was discovered in 1979 in the Arlit area and originally intended as the country's third operating mine, Cogéma to receive two-thirds and Onarem one-third of production (2). The initial partners at Arni, Onarem and Cogéma, were later joined by a third, the Kuwait Foreign Trade and Contracting Co, constituting themselves as the SMTT. Not a surprise since, soon after Cogéma announced a 1982/83 start-up for the mine, the huge French uranium producer was having second thoughts and looking for a non-French partner to absorb excess output from their share in the mine (2). Arni has published reserves of more than 20,000 tons with an average uranium grade exceeding 0.35% U_3O_8 (4). Design production was increased in 1984 from 1000 tonnes U_3O_8 per year to 1800 tonnes (5).

References: (1) EPRI Report. (2) Lynch & Neff. (3) MAR 1985. (4) MAR 1975. (5) E&MJ 1/85.

459 Ontario Hydro Ltd

A publicly-owned corporation responsible to the federal Canadian Minister of Energy, with assets totalling around US$10 billion in the

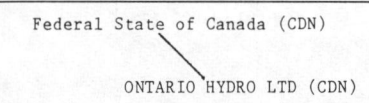

Federal State of Canada (CDN)

ONTARIO HYDRO LTD (CDN)

early '80s, Ontario Hydro is "almost thrice the size of the largest private corporation in Canada" (1). It is also one of the largest public corporations in the world. Ontario Hydro sells large blocks of power to municipal utilities which then re-sell to retail customers, and – like the TVA – it also sells power directly to industrial consumers. One of the world's biggest users of hydro, thermal and nuclear power, it has four nuclear facilities, with three on the drawing board (2).

Ontario Hydro has been the subject of resounding criticism for many years, on numerous fronts: from its inception of the disastrous James Bay hydro project and its huge indebtedness and cost over-runs (3), to its nuclear waste mismanagement (4), its release of radioactive materials into the environment (eg a mass disposal of tritium into the Ottawa River in 1981) (5), its use of herbicides to clear land for high tension power lines (6), and its role as a producer of "acid rain" (7).

In 1979, an ex-Hydro employee spilt the beans about its safety record at the Bruce nuclear station in Ontario – which he likened to the conflagration of Three Mile Island (8). This followed the Porter Commission's report which criticised Hydro on a number of grounds (including safety) and a long-running battle between the Atomic Energy Board of Canada and the public corporation over safety standards (9). And in 1983, Ontario Hydro's Pickering Unit 2 reactor suffered a ruptured tube and loss-of-coolant "incident" which, indeed, was comparable to that of Three Mile Island (10) and could eventually cost US$1.3 billion to repair (5).

In February 1978, Ontario Hydro sealed the largest contract ever made for the delivery of uranium under a single deal: 126 million pounds of uranium (to be supplied to 2011) by Denison (11), and 36,000 tons over 1984-2020 to be delivered from Rio Algom, mainly from the revived Stanleigh mine at Elliot Lake which the Ontario contract virtually resurrected from the dead (12).

Ontario Hydro has also entered uranium exploration in its own right: a JV with Shell Canada Resources, for example, in the South Mountain batholith area of Nova Scotia (13); halted in 1982, however, by the province's moratorium on further uranium exploitation (14).

Further reading: Two books have advocated breaking up the company's monopoly and ensuring access to the electric grid by both private and public power producers:

Paul McKay: *Electric Empire: The inside story of Ontario Hydro*, Toronto, 1983 (Between the Lines).

Lawrence Solomon: *Breaking up Ontario Hydro's Monopoly*, Toronto, 1983 (Energy Probe Report).

References: (1) *BBA*, No. 4, 1979, & No. 7, 1980. (2) *Nuclear Free Press Energy Booklet*, Peterborough, Ontario, 1980/81 (OPIRG). (3) *NFP* Fall/82. (4) *NFP* Summer/82. (5) *NFP* Winter/84. (6) *NFP* Fall/84. (7) *NFP* Winter/83. (8) *BBA*, Vol. 2, No. 2, 1979. (9) *BBA*, No. 4, 1980. (10) *NFP* Fall/83. (11) *MJ* 4/8/78. (12) *MIY* 1985. (13) *MM* 5/79. (14) *MAR* 1982.

460 Orazada Mines Inc (USA)

Though not known to have any uranium leases, it was renovating and testing its silver-copper property on the Spokane Indian Reservation in Washington state in 1983, which occupies some 1200 acres (1).

References: (1) *E&MJ* 4/83.

461 Ormac JV

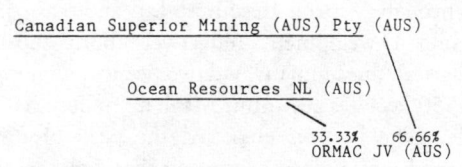

This is situated in the Northern Territory on Aboriginal land; see Ocean Resources for further details.

462 Otter Exploration NL (AUS)

"Otter certainly flings its net wide" (1). That was the comment made back in 1978 by the London *Financial Times*, after this moderate-size company embarked on four diamond ventures in the Kimberleys. As well as being active in gas and oil prospecting, Otter has more recently been teamed with companies like Shell, CRA, BHP and Getty Oil in a number of other, base metal, prospects (1).

In 1977, it got contracts to search for uranium in areas additional to its existing tenements in Queensland and the Northern Territory. (These were the Georgetown, Queensland and Benmara, Northern Territory, areas.) One of these JVs was with the CEGB (2) – where Otter had been exploring Camel Flat Bore, White Tree Bore, Angelo River and Cane River, in the Northern Territory (NT) (3).

At roughly the same time, Otter entered a JV with AAR whereby AAR could earn 50% in the Benmara property (4).

The following year, it reportedly entered an agreement with Uranex to search for uranium in Queensland (5).

And, in 1979, it joined Esso Australia and Carr Boyd Minerals in a JV prospect for uranium at Pandanus Creek, NT. Esso could earn a 51% interest in the prospect by financing and executing all the exploration work by April 1986 (6). While its uranium exploration potentially affects Aboriginal land claims in both the Northern Territory and Queensland, it is the company's prospecting on Maori land in New Zealand which has perhaps raised most ire to date. Through Mineral Resources Ltd and Tasman Gold Development (effectively both subsidiaries), the company had two licences totalling 4550 hectares at Mount Moehau in Coromandel. The licences encircled the peak of this sacred mountain – suggesting that if the mine went ahead, it would leave the Maoris isolated at the top "in a bizarre pedestal" (7). Local Maori feeling was strongly against the project. The mountain is also ecologically unique, with kauri trees and sub-alpine plant communities, rare frogs and a stag beetle. It is of particular significance to the Maori through its associations with the Arawa canoe (7).

Not to be outdone by local opposition in the Coromandel, Otter's 25.7% in Mineral Resources (NZ) Ltd has brought it pickings from that company's share of the Martha Hill mine at Waihi on the Coromandel peninsula. (The Martha Hill mine is owned 33.53% by Amax, 33.53% by ACM Gold Ltd, and 32.94% by Mineral Resources) (8). This operation has met with vehement criticism from members of the well-organised Coromandel Peninsula Watchdog group (9).

Otter has had numerous other claims in New Zealand, and at Bonnie Creek in the Pilbara region of Western Australia – one of the key Aboriginal areas affected by mining (see Amax, CSR, CRA, Alusuisse). It has also had a 25% interest in a uranium consortium headed by Marathon Petroleum.

In 1980, Otter replaced Esso in a JV with Pegmin for uranium exploration in South Australia.

During 1981, it paid close attention to three vanadium-titanium prospects in Western Australia; it was also in a JV with Amax at the Tasmania Gold mine (10).

In mid-1983 Otter raised just over A$2M to finance gold mining developments at Griffin's Find in Western Australia (11).

By 1990, Griffin's Find was at the end of its life (12), though Otter had by then expanded into exploration of the Tanami Desert in Western Australia (under a permit from the Warlpiri

traditional owners and the Central Land Council) (13), as well as into Queensland and other parts of Australia. That year, the majority of its shares were taken up by its partner in New Zealand, Mineral Resources (12).

Contact: CAFCA, POB 2258, Christchurch, New Zealand.
Coromandel Peninsula Watchdog, Box 51, Coromandel, New Zealand.

References: (1) *FT* 4/9/78. (2) *MM* 5/77. (3) *Reg Aus Min* 1981. (4) *MJ* 19/8/78. (5) *FT* 4/9/78. (6) *MJ* 18/5/79. (7) *CAFCINZ Watchdog* 7/81. (8) *MJ* 12/4/91, *MJ* 26/4/91. (9) *NZ Herald* 11/10/80. (10) *MJ* 9/7/82. (11) *MJ* 3/6/83. (12) *Reg Aus Min* 1990-91. (13) *Land Rights News* 3/89; see also *MJ* 24/2/89.

463 OURD

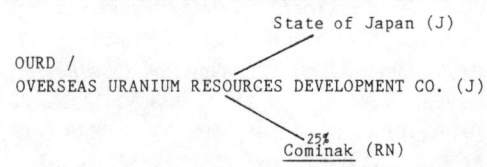

This is a Japanese state uranium procurement company which appears to operate only in Niger, where it holds 25% of the Cominak consortium at Akouta and is part of a consortium exploring in the West Afasto region with Cogéma and Onarem *inter alia*. (1).

Between 1983 and 1990, OURD contracted for 7800 tonnes of uranium from Niger, but as the *Mining Annual Review* points out: "uranium production and sales figures for Niger are inconsistent and appear to understate both sales and output" (2).

References: (1) *MAR* 1981, 1984, (2) *MAR* 1991.

464 Outokumpu Oy

This is Finland's largest mining and metals group with eight marketing companies in Fin-

```
                OUTOKUMPU OY (SF)
                      /
                Nippert (USA)
```

land and five abroad. Eight mines were in production in 1983, producing copper, zinc, nickel, cobalt, precious metals, pyrite and lead (1).

In the late '70s the company began searching for uranium on Saami land near Kittilä (2).

It also secured a 9.2% stake in the Trout Lake copper-zinc mine at Flin Flon, Manitoba, Canada – which it sold in 1989 (33) – and 39% in the La Plata copper-zinc mine near Quito, Ecuador (3).

In 1983 the company adopted its famous "flash smelting" process to treat lead sulphide ores – removing lead, liquidising some SO_2 and, according to the company, minimising thereby releases of sulphur dioxide and lead to the atmosphere (4). It has been engaged on tests of this process on ore from the Olympic Dam deposit in South Australia (see WMC). Tests were being conducted in summer 1984 which involved separating uranium from ore in Finland and returning it to Australia for experimental processing (5).

Between 1984 and 1990, Outokumpu developed into the world's second largest producer of semi-finished copper and copper alloys (6) and the world's leading supplier of smelter technology (7).

Arguably, it also became the first truly "modern" mining and metallurgical company: among the first to negotiate a debt-equity swap (under which it took 15% interest in an oxide treatment plant at Antofagasta plc's Lince mine in northern Chile) (8); at the forefront of those companies offering, not only capital investment, but also pollution control technology to "the eastern bloc" – in particular the Soviet Union's ravaged Kola peninsula (9), but also the Karelia region in that country (10), Mongolia (11), and the southern Urals; and a leader in offering shares (25%) in its own capital to its employees (13).

During the 1980s Outokumpu "came to a

strategic turning point" as domestic supplies for its own copper, zinc and nickel smelters looked like running dangerously low (13). It "set out to supply most of the raw material requirements" for these plants and, where necesssary, to become self-sufficient in them.

In 1988 it reorganised itself with the formation of ORI (Outokumpu Resources Inc) which now takes control of all the company's operations outside Scandinavia. By this time it had more than 120 international subsidiary operations in more than 30 countries (14). The following year it produced nearly 10 million tonnes of ore, and had projects not only in Finland (mostly on Saami land) but also in Sweden, Norway, Canada, Mexico, Chile, Australia and Ireland.

That year the Finnish government permitted the company to sell up to 49% of its state holding, with foreign ownership allowed up to 20% (15).

Among its important projects, Outokumpu operates a gold mine in the Saami region – it was the first new Finnish mine "in years" (16); it has a JV with Hudson Bay Mining Co at Namew Lake (17); it purchased the important American Brass copper company in 1990 (18); and it has a large modernisation project with the Zaïrean state copper company Gécamines (19).

The company's expanding interests in Chile – where it is one of the most important foreign investors (20), its forays into Australia, and its recent problems in Ireland, suggest that Outokumpu, at root, is not much different from other majors in the field.

In Australia, the company has interests in the Thalanga base metals mine in Queensland, which was opened in 1990 (21), and a 55% interest in the Forrestania nickel exploration project (22). But it is the JV with Australian Consolidated Minerals Ltd (ACM) at Mt Keith which is its major venture on that continent. Due to problems with financing and the nature of the ore, it took well ofer a year before ACM and Outokumpu could announce that they were proceeding with the mine (23).

Outokumpu's involvement in Ireland has

mounted over the past few years. It bought 75% of Tara Mines in 1986 – despite the fact that the mine had "done battle over the years with both environmentalists and the various mining trade unions" (24). By the end of 1990, the Finnish company held 21.7% of Ivernia West (which itself has interests in all Chevron's Irish resources, a talc deposit in County Mayo, and 50% in the Double A gold mine in Australia) (25). It also sealed a deal with Conroy Petroleum, to develop its Galmoy lead-zinc find in County Kilkenny (26).

Tara has become a mine with which to charm away the sceptics (27) and Outokumpu has been doing very nicely out of the operation, thanks to using a tax haven in Dublin called the Custom House Dock Financial Services Centre (which offers a flat 10% tax rate – now there's the luck of the Irish for you!) (28).

But Outokumpu has not fared so well in selling its intentions over the Croagh Patrick gold deposit – which it took control of in 1990 when it bought up 10% of the project's other 50% owner, Burmin Exploration PLC (29). Opposition to mining the holy mountain of Croagh Patrick – where Ireland's patron saint was said to have spent 40 days of contemplation – reached such a pitch in 1990, that the Irish Minister for Energy withdrew the exploration licence on religious grounds (30). This blow to the company was later compounded when County Mayo's council put a ban on mining operations in some 300 square miles, which included claims lodged by Tara, Burmin and Ivernia (30).

It is worth adding that Outokumpu is now building a smelter in Portugal to take copper concentrates from RTZ's 49% Neves Corvo mine (31); the company was the subject of outrage from Finnish students and trade unionists, in 1988, when it proposed taking a stake in the RTZ/BHP Escondida mine, in order to supply its own Harjavalta smelter (32).

Contact: Minewatch Ireland, Dun Sidh, Moneen, Louisburgh, Co Mayo, Ireland.

Further reading: For a self-adulatory history of the company: Markku Kuisma, *A History of Outokumpu* (Gummerus Kirjapaino Oy) Jyväsklä, 1989.

References: (1) *MJ* 1/6/84. (2) *Charta 77* article by John Gustavson, London, 1981. (3) *MJ* 27/5/83. (4) *MJ* 15/7/84. (5) *E&MJ* 6/84. (6) *MJ* 16/3/90. (7) *MJ* 2/9/90. (8) *E&MJ* 10/89. (9) *MJ* 22/2/91, *MJ* 28/9/90, *E&MJ* 2/90, *E&MJ* 1/91, *MJ* 23/11/90. (10) *MJ* 11/5/90. (11) *MJ* 11/3/88. (12) *E&MJ* 2/91. (13) *E&MJ* 8/88. (14) *E&MJ* 4/88. (15) *MJ* 5/5/89, *MJ* 25/8/89; see also *MJ* 1/4/88. (16) *E&MJ* 6/89. (17) *Metal Bulletin* 18/6/90, *E&MJ* 1/89. (18) *MJ* 7/9/90, *FT* 28/6/90. (19) *MJ* 22/9/89, *MJ* 22/3/91. (20) *FT* 3/11/89, *MJ* 26/1/90, *MJ* 7/9/90, *MJ* 8/6/90, *E&MJ* 10/90, *MM* 10/89. (21) *MJ* 22/6/90. (22) *FT Min* 1991. (23) *MJ* 17/11/89, *MJ* 5/10/90, *MJ* 29/6/90, *E&MJ* 2/91, *FT* 18/4/91. (24) *MJ* 13/8/84, *MJ* 24/2/89. (25) *MJ* 12/10/90. (26) *E&MJ* 5/90, *MJ* 1/6/90. (27) *MM* 6/91. (28) *E&MJ* 12/89. (29) *E&MJ* 3/90. (30) *FT* 28/3/91. (31) *FT* 1/12/90, *MJ* 1/9/90. (32) *FT* 18/10/88. (33) *E&MJ* 7/89.

465 Pacific Gold and Uranium Inc (USA)

It has no known uranium interests, but was involved with Noranda in a cyanide heap leaching scheme to produce gold in the Nevada Goldfield district in 1983 (1).

References: (1) *E&MJ* 7/83; *FT* 18/5/83.

466 Pacific Petroleums Ltd (CDN)

It held uranium exploration licences for the Kelowna, Penticton and Seventy Mile House properties in British Columbia during 1979 (1).

References: (1) *Energy File* 3/79.

467 Pacific Trans-ocean Resources Ltd (CDN)

It held 10% of Pacific Copper Ltd which Alan Bond (see Bond Corp) took over in 1981 (1). While its major operations were in Australia, particularly in New South Wales (2), in 1978 it was reported exploring for uranium at Yellowknife, Northwest Territories (NWT) in Canada (3), in a 90%-controlled JV with another Alberta company; the area concerned was 10,000 acres, 65 miles east of Yellowknife.

Its main recent explorations have been in Queensland, and in the Yukon, as well as on its gold prospect in the NWT (4).

References: (1) *Reg Aus Min* 1981. (2) *MIY* 1982. (3) *FT* 15/5/78. (4) *MIY* 1985.

468 Pacminex Pty Ltd (AUS)

It located about 350t of uranium (with inferred further reserves of a similar quantity) in the Jailor Bore prospect area of the Gascoyne region of Western Australia during the late '70s in a JV with Meridian Oil. Meridian held the exploration rights for this claim (1).

References: (1) Meridian Oil annual report, 1982.

469 Palabora Mining Co Ltd

One of the largest open-cast copper mines in the world (1), and "one of the most efficient and low-cost [copper] producers in the world" (2) (in the early 1980s it became the world's lowest cost producer of this basic metal) (3), Palabora is situated on a unique geological formation known as the Palabora Igneous Complex. Formed two billion years ago as the result of a series of volcanic eruptions, the complex also contains economic reserves of nickel, silver, gold, zirconium, uranium, iron, and phosphates; two auxiliary "pipes" nearby contain vermiculite and phosphates. Boasts the company: "It has often been said that Palabora is not so much a copper mine as a minerals complex, so great is the diversity of the metals and minerals which it produces" (1). The site also contains a concentrator smelter, electrolytic refinery and by-product recovery plants, and its end products include pure copper cathode (exported), copper rod for local use, magnetite concentrate for coal washing and iron making, sulphuric acid for the fertiliser industry, the heavy mineral by-products of zirconium and uranium oxide, nickel sulphate and refinery residues "rich in gold, silver, and platinum-group metals" (1). In the past, the company has also produced apatite (4). In 1979, a subsidiary of the South African state-owned Industrial Development Corporation, Foskor, made a loan to Palabora, to foot part of the costs of expansion and gain access to phosphate-bearing material (5). Foskor has been pumping up high-phosphate tailings from 1974 onwards (2).

The company is South Africa's main producer

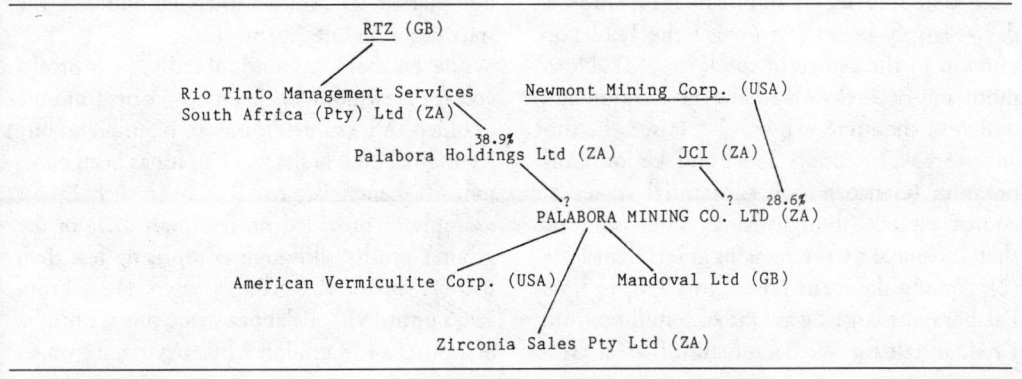

547

of vermiculite (through its Mandoval Ltd subsidiaries, marketing in the USA via American Vermiculite Corp) (6), and its magnetite is exported via Maputo (7). One important ten-year vermiculite contract, which terminated in 1979, was with Kobe Steel of Japan (8). After another South African copper producer, Messina Ltd, closed its smelter in 1983 (9), its concentrates have been smelted at Palabora (2).

Sales of zirconium – a strategic mineral used by the nuclear industry – are handled by Zirconia Sales Pty Ltd, and in Japan by Zirconia Sales (Japan) KK (10). Palabora is the world's only commercial producer of baddeleyite (ZrO_2) (6).

In 1988, Anglo American and De Beers bought out Newmont Mining's stake in Palabora, as the American mining company shed its South African interests (ostensibly to restructure and reduce debt, rather than for political reasons) (11). This served to ally the world's largest mining company even more firmly with what was then the world's second largest, RTZ. It also represented a final resolution to the conflict between RTZ and Newmont over rights to the deposit. Initially, in 1953, it was Rio Tinto which received permission from the apartheid régime to explore the mineral complex, but Newmont sent an exploration team to the area as well (12). Both companies applied for a lease, and both companies wanted management over the deposit. Newmont chairman, Plato Malozemoff suggested an independent management – but such a proposal was anathema to RTZ, whose chair, Val Duncan considered it would be "a kind of soviet" (13). Finally RTZ won the day – largely because it owned the Loolekop outcrop in the centre of the lease (13). Newmont and Bechtel worked out the location and design of the mine, while RTZ raised the finance. Amax, Selection Trust and Union Corporation (Gencor) took substantial stakes of equity capital (though Amax later sold its share). Gencor's stake remains at less than 10% (2). Among the loans raised, in 1956, to float Palabora, the largest was US$26.9 million from Kreditanstalt fur Wiederaufbau, of West Ger-

many, on the basis of a 20-year contract to supply between 30,000 and 36,000 tons of blister copper annually to Norddeutsche Affinerie, of Hamburg (12, 14).

Equally crucial was the active interventionist policy of the South African state corporation, the IDC, which soon realised that Palabora was an extremely important copper resource for the apartheid economy (12).

In the 1970s, the mine was providing more than half (55%) of the country's needs. There is little doubt, (especially in the light of recent events, and the withdrawal of American capital from South Africa), that the régime was prudent in awarding both management of the mine, and the lion's share of equity, to RTZ.

It was the uranothorianite "disc" in the Loolekop outcrop which originally interested the two exploration companies (15), but held out little promise for uranium exploration in its own right. Mining of the complex began in 1963 (16), and the uranium grade in the 1970s was reported as 0.005% U_3O_8, with an "implied reserve of 14,000 tons U_3O_8" (17), making for average annual production of around 130 tons (17).

Until the opening of Palabora, South African uranium had only been extracted from gold slimes (18). From 1971 onwards (2), RTZ used the Purlex extraction process – also employed at Rossing – to separate the uranium from the copper tailings (19).

Palabora is also unique on the South African scene, as the only uranium producer which does not market its product through Nufcor, the state uranium sales agency (20).

While Palabora has undoubtedly been profitable for Newmont (which used profits from the O'okiep and Tsumeb mines in Namibia to buy its original 28.6% share) (12), it has been enormously beneficial to RTZ. In 1972, for example, it provided no less than 42% of the group's profits, although comprising less than 8% (7.7%) of the group's assets (12). From 1966 until 1971, Palabora made pre-tax profits of around £138 million while paying out wages of less than five million pounds (£4.6 million) (12).

There are three main reasons for the mine's success. The first is the extent to which the company has been able to conclude contracts, primarily in South Africa, at high rates of return. Second, is the extent to which it has exploited black labour. And the third is the degree to which the management has mechanised, computerised and integrated all functions at the mine – ensuring efficiency at a low labour turnover. All blasting, haulage, and shipping of ore and concentrates is handled by computer (12). Huge 150-ton trucks carry the ore, and, in 1981, a unique electric trucking system was introduced, which became the envy of the copper mining world (21).

Palabora really took off in the late 1970s and early 1980s. After a better year financially in 1978 (8), a five-year expansion plan was announced the following year (5), resulting in increased production in 1980 (22). On one day, in August that year, the mine broke the world's record for a daily output (23). With the smelter working "excellently" in 1982, the large ore stockpile was reduced (24, 25), and profits rose by 43.5% with considerable cost-cutting (26).

In late 1983, the company announced new plans for the mine: extending its life until 1999, and promising new openings and extensions in depth. (By this time, it was calculated, 2000 million tons of ore and waste would have been extracted from the earth, leaving a pit 750m deep, and ore haulage ramps of 10km in length (3). The next year, 1984, profits were up no less than 76% and although, in 1985, cracks began appearing in two of the mills (27), net profit surpassed the company's previous record (nearly 85 million Rand) (10).

Uranium production and sales have been erratic during this period, but the company has undoubtedly come to be regarded as a fairly reliable, if minor, producer of the mineral. Production was 128,000 kg in 1980, as the market became depressed (22). 1982 saw "substantial sales" to a power utility buyer collapse, as the buyer failed to get permission to open a nuclear power plant (25), but the uranium was later shipped out (28) in a "one off deal" (24). Another large sale of uranium was made in 1983 (29). Sales began increasing again in 1985 (10).

Until an anti-nuclear researcher came upon documents about South African uranium contracts in 1987, virtually nothing was known about Palabora's contracts or sales of uranium. We now know that the company was committed in the early 1980s to providing 613 tonnes of U_3O_8 equivalent to France's state-owned Electricité de France (EdF), between 1984 and 1988; and a small amount (87 tonnes) in 1984 to the West German VEW (Vereinigte Elektrizitätswerke Westfalen AG) (30, 31). The EdF contract is especially noteworthy, as it coincides with the period covered by a contract between Nulux (Nukem's Luxembourg subsidiary, in which RTZ Mineral Services holds 30%) to supply UF6 to South Africa over roughly the same period (31). This was a deal which appears never to have left the drawing-board, but which illustrates the complicity that RTZ has (through its Mineral Services subsidiary, if not Palabora itself) in supplying potential nuclear-source material to the apartheid state, as well as extracting it.

Black labour

Palabora has long been criticised for its working conditions (32). Unitary wage systems were only introduced in 1980 and, although the mine does not (ostensibly) employ contract labour – black families can come and live with the mineworkers at the Namakgale township, in government or company-built housing (33) – the disproportion between the low-skilled black labour force and white higher-paid labour is as stark as at any other South African mine under apartheid.

For example, above level C2 on the salary scale, there were only three black employees in 1984, and no black monthly-paid staff, as opposed to 390 white salaried, and 378 monthly-paid white staff (34). In the Al and A2 groups of lowest paid workers, there were no white employees, but 1,382 black (34).

In 1984, as agitation among the black workforce against mine conditions increased throughout South Africa, a large number of black workers resigned fearing the freezing of government pension rights: according to the company, many of these applied for re-employment when their fears were allayed (10). More important, the management made considerable efforts to deter the black workforce from joining the National Union of Mine Workers (NUM), which the company recognised was becoming active during the latter half of 1984 (10). While RTZ in London was trying hard to put a reasonable facade on this intransigence by its subsidiary ("Palabora Mining Company indicated willingness to enter into negotiations for the establishment of a formal recognition agreement with any union that it represents a significant number of Palabora employees") (35), at the mine itself a short while later, workers sporting NUM tee-shirts, and demonstrating in favour of a union, were met by police called by the management (36). However, draft recognition agreements were exchanged between union representatives and the company (10) and, within the next year, formal recognition was afforded the NUM. (The company later claimed that about 70% of the company's black employees became members of the NUM) (10).

No comment on Palabora would be complete without reference to the role of G A Macmillan, the head of Rio Tinto South Africa, a board member of Rossing and RTZ, and the dominating force behind Palabora. As well as being credited with directorships of more South African companies than any other businessman – including Safmarine, Barlow Rand, and Sasol, – this "amiable, approachable, avuncular" man came to Rio Tinto in 1974 from the IDC (37). In 1983 he was appointed to the South African Atomic Energy Corporation of SA – the key body in the development of an apartheid nuclear programme – thus making RTZ the only corporate body represented on this key parasta-

tal (38). Later the same year, he distinguished himself by speaking on behalf of Botha's "referendum" for separate coloured and Indian constituencies within the white parliament. (Oppenheimer, of Anglo-American, urged voters to reject the move) (39). The proposal was accepted by a large majority (40).

Contact: PARTiZANS, 218 Liverpool Rd, London N1 1LE U.K.

References: (1) *Palabora 87*, Rio Tinto South Africa Ltd. 5/87. (2) *FT* Mining 87. (3) *MJ* 28/10/83. (4) *EPRI* 78 . (5) *FT* 24/8/79. (6) *MAR* 1987. (7) *Raw Materials Report* no 3, 1983. (8) *FT* 19/7/79. (9) *MJ* 29/11/83, *MJ* 2/12/83. (10) Palabora Annual Report 1985. (11) *FT* 8/4/88, *MJ* 8/4/88. (12) Lanning & Mueller op cit. (13) R. Ramsay, *Men and Mines of Newmont*, Octogon Books, New York 1973. (14) *Who Owns Whom*, London 1987. (15) IAEA/OECD *Red Book*, Paris 1979. (16) *MJ* 28/10/81. (17) *EPRI* 1978. (18) Rogers & Cervenka op cit. (19) Nuclear Active South African Atomic Energy Board, 7/73; *Financial Gazette Johannesburg*, 16/1/70. (20) Neff op cit. (21) *FT* 23/10/80. (22) *FT* 16/1/81. (23) *FT* 17/10/80; see also *FT* 23/2/80 & *FT* 18/1/80. (24) *FT* 7/8/82. (25) *FT* 15/10/82. (26) *MJ* 27/3/83; see also *FT* 22/4/83. (27) *FT* 2/4/85; *Business Day*, Johannesburg 23/5/85; *MJ* 5/4/85. (28) *FT* 15/8/83. (29) *Palabora Annual Report* 1983; *FT* 14/4/83. (30) *Sanity*, CND London 12/87. (31) *Parting Company*, PARTiZANS late 87 and No. 1/2 Spr. 88. (32) See *RTZ Anti-Report*, Counter Information Services, London 1971; "*Management Responsibility and African Employment in South Africa*" Labour Press, Johannesburg, 1973. (33) *RTZ Fact Sheet no 3*, 1984 "*Some Aspects of the RTZ Group's Operations in South Africa 1983*"; see also *RTZ Fact Sheet* No 3 17/5/77 and *RTZ Fact Sheet* no 3, 18/5/78. (34) *RTZ Fact Sheet*, London 1983. (35) *RTZ Fact Sheet* on Palabora, London 1985. (36) Information from the NUM, Johannesburg, to Partizans London, May 1986. *(37) Financial Mail*, Johannesburg 28/6/85. (38) *Namibian Newsletter*, Namibian Support Committee, London 6/83. (39) *FT* 5/10/83. (40) *FT* 3/11/83.

470 Pancontinental Mining Ltd

Although it lays claim to one of the world's largest untapped deposits of uranium, Pancontinental Mining saw its hopes of exploiting Jabiluka drastically recede during most of the 1980s. As a result, it turned its attention to other minerals, as well as oil and natural gas. Its success in this area (in early 1985 it turned in its first-ever profit, of £1.38M for the first six months of 1984) (1) is fairly remarkable, showing (perhaps) that while there is little life after uranium mining, there is quite a lot instead of it. Pancon is active in both Canada and Australia, although by far the greater part of its investments has been in Australia. It has gold and silver properties in Ontario, but two of these are on a care and maintenance basis, while the third (Pan-Empire gold mill) was, in 1986, on lease to Teck (2).

One other Canadian venture which has been on indefinite hold is its 40% JV with the James Bay Development Corporation and Cominco (20%) in the Otish mountains of Quebec, where a significant uranium anomaly was first announced in early 1978 (3) and drilling was under-way a little later (4). In 1979, Pancon proposed a further drilling programme (5), but little more was heard about the programme thereafter.

Nor has much more been heard about Pancon's interest in solar energy: in 1977, the Australian National University announced that its solar energy research programme would have to be suspended, due to lack of finance, but that Pancon and a "United States company" (most likely Getty Oil) were involved in negotiations to continue the work (6).

A 1982 field programme at the Turee Creek uranium prospect, in Western Australia's Angelo River area, seemed to have ended without any new mineralisation of significance being discovered (7).

In the field of oil and gas, however, Pancon's Canadian explorations have yielded encouraging results; especially through Pancontinental Oil Ltd (46.7% owned), set up to explore for oil, specifically in Canada where it has located large reserves in the Louise area of British Colombia, and several commercially viable wells in Alberta (8).

Although Pancontinental was originally an Australian company – Barker Holdings Ltd (established in 1961) – the firm came under considerable Canadian influence in 1971, when it converted to a mining body and was renamed

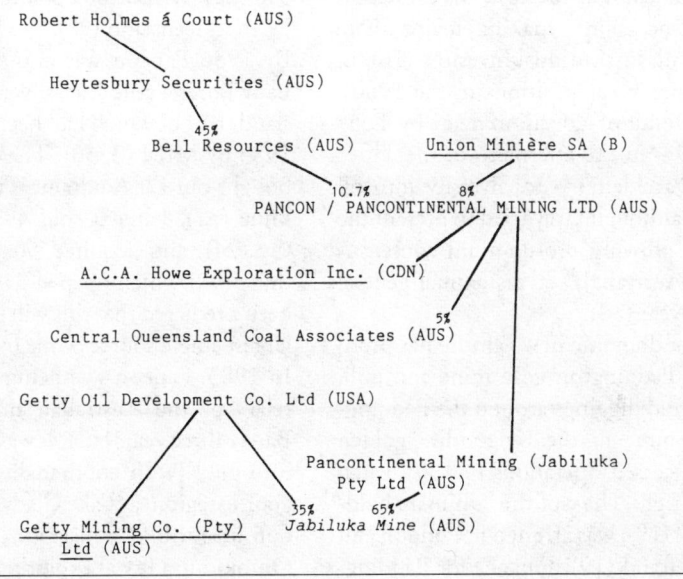

(9). The company chairman, Tony Grey – by far the most dominant figure in all of Pancon's history so far – is a Canadian national, and lawyer by profession. During the first five years of Pancon's history, Grey's personal holding in the company was variously estimated at 30% (10), "a large part of 51%" of the total shares (11), and as the sixth largest shareholding, with 314,000 shares in 1976 (9). However, in late 1976, Grey appears to have off-loaded most of his shares. From this date onwards, it becomes very difficult – if not impossible – to determine exactly how much of Pancontinental's equity is controlled by Australian interests. In 1977, the top 20 shareholders owned 68% of the quoted shares but, of these, half were nominee companies, and a third of them were either wholly-owned subsidiaries of non-Australian companies, or public subsidiaries of overseas companies (12). No wonder the *Multinational Monitor* in 1981 stated that actual ownership was "obscured" by proxy holdings (13).

Until the early 1980s, the British Consolidated Goldfields (CGF) group held a substantial proportion of Pancon's shares – 19% of the total in 1977 (14). However, CGF – or rather its Australian subsidiary Renison Goldfields Consolidated (RGC) – sold this stake to an unidentified buyer in early 1983 (15).

Since then, Pancon has increased its capital on at least two occasions, making major share placements with institutional investors (16), in order to finance its acquisitions in the 1980s. This was the result of a decision made by Tony Grey in 1979 – in the aftermath of the Three Mile Island "accident" – to diversify into oil and gas (17), although Grey tried to present the picture of a "growing pro-uranium consensus throughout [Australia]," at his annual general meeting that year (18).

Pancon's most dramatic new venture has probably been its Paddington gold mine and mill, sited on old gold diggings around the Paddington Consols mine in the Kalgoorlie "golden mile", which ceased operations in 1903 when Whittaker Wright, chair of the "ultimate holding company" (19), was arrested in London and charged with fraud (19). Finance for Paddington was raised by the novel (probably unique, for Australia) method of raising a loan in the form of gold bullion worth £17.5 million, provided by the Westpac Banking Corporation: Pancon sold the gold on the market to raise the necessary cash (20). Initial feasibility studies on the Paddington prospect were carried out by Davy-McKee Pacific, with assistance from Australian Anglo-American (21), while United Goldfields Corporation holds a 12.5% net interest on a half square kilometre area within the mining lease (22). By mid-1985 the mine was ahead of schedule, and predicted by Tony Grey to become Australia's fifth largest gold producer (23). A year later, Pancon was expanding to other leases around the mine, and Paddington had become the company's "central money spinner" (24, 32). By then, Pancon had entered a 32.5% JV with Delta Gold NL (25) for mining, processing, and transporting gold from another property, the Last Chance deposit, also along the "golden mile", at Kanowna (26, 2).

After failing to gain a stake in Robe River (it lost to Peko Wallsend) (27) in 1984, Pancon took a stake in the important Central Queensland Coal Associates consortium, put together by BHP in its reorganisation of the assets of Utah International (see BHP for further details) (28). An initial 2% stake (29) was later increased to 5% (26). It also took 5% of the Gregory Coal JV in Queensland (98).

By 1986 Pancon was also 51% owner of the Lady Loretta zinc-lead-silver deposit in Queensland, half of which had been sold to Pancon in 1985 by MIM (2, 30). The same year, Pancon bought out Elf Aquitaine Triako's share in the mine (31). Later it sold 49% to Outokumpu Oy (24), and acquired 50% of the Thalanga mine (98) which opened in 1990 (99) – it has been predicted that this will become the world's largest single source of magnesia (100).

In 1983, Pancon was being dubbed a "Cinderella" of the Australian mining world (33). Barely three years later, it was expanding rapidly into gold (with another discovery in the Kalgoorlie region, at Lake Gladys, and one at Croydon in north Queensland) (26); together with Outokumpu it was exploring for base metals in large parts of Tasmania and Queensland, near Mount Isa and Rockhampton (34) – where, at Kunwarra in 1986 it located the Thalanga deposit (Pancontinental has 40% of the JV, with Queensland Metals Corp at 50%, and Radex Heraclith AG of Austria at 10%) (35).

Its 50/50 partnership with Degussa, set up in 1985 (25, 36), was looking for platinum at Yilgarn and in the Kimberleys, Western Australia (WA), and two locations in Papua New Guinea (35), while the company had identified two small deposits of tantalum at Mount Farmer (Murchison district of WA) and in the Pilbara, WA (26).

Between 1980 and 1985, Pancontinental Petroleum had drilled 30 exploration holes, and held a 10% interest in the Tintaburra oil fields; it was also involved in discussions on laying a gas pipeline to Darwin and the Gove Penninsula in northern Australia (37).

Little wonder that Tony Grey could claim 1985 as "one of the most important in the Pancontinental group's history" (38) and the *Mining Journal* could predict that, so long as Pancon didn't need to raise further cash by borrowing, the prognosis for the rest of the 1980s and beyond was "most favourable" (24).

By 1986, Pancon had proposed to the Sri Lankan government that it should have exclusive mineral exploitation rights to the entire country's resources, in collaboration with the state Mining and Mineral Development Corporation (39).

Of course, it is the Jabiluka uranium deposit with which Pancon is most indelibly associated. Discovered a decade and a half ago, and originally known as "7e", it was renamed after the Oenpelli Aboriginal name for "lagoon" (38). The original deposit, containing 3850 tons U_3O_8, is situated 15 miles north of Ranger, on a roughly 500-800m strip. However, in the 1970s, Pancontinental discovered a much larger lode, which it was then conjectured could stretch under the escarpment at Jabiluka and the Kakadu National Park (stage one). Archaelogical and art sites were identified as occuring within only one kilometre of this deposit (40).

With an estimated 200,000 tonnes of U_3O_8 in two orebodies, grading up to 8lb per tonne (41), requiring an investment of at least US$525 million (42), holding *in situ* reserves of 175,000 tons of U_3O_8 (35), it is undoubtedly one of the world's biggest uranium lodes. Annual production was set at 4500 tonnes of uranium oxide in 1982, the year the mine was given the go-ahead, with a possible doubling of output under the "right market conditions" (41). Mine life was estimated at around 25 years, producing uranium worth some A$18 billion (£10,500,000,000) as well as 12,000kg of gold (41). The company would employ some 1600 workers during the three-year construction phase, and about 900 in the estimated quarter century of the mine's life thereafter (43).

Under the Jabiluka agreement with Getty Oil, the Australian company had to put up 65% of the production financing, and the American company the remaining 35%. Getty would also be required to lend Pancon its 65%, or act as guarantor if Pancon wished to borrow elsewhere. "To all intents and purposes," said Tony Grey at the time, "Getty will provide or cause to be provided 100% of the financing requirements for Jabiluka" (44). This "privileged" financial position (45) soon made Pancon "the most promising of all uranium companies," according to Sydney brokers Lamplough and Malcom (45). Getty would also be a potential purchaser of yellowcake – as the Jabiluka agreement specifies that Getty can purchase all or part of the mine's production – "so that the company has an assured market" (46).

Pancon initially went after European, Japanese, Korean and US utilities in its search for contracts – nearly two dozen letters "expressing interest" in contracts, having been received by the end of 1977 (47). Within two years, however, doubt was being expressed about Pancon's ability to match prices and production from Saskatchewan, and even South Africa (48). Indeed, Tony Grey – always one of the most conspicuous, if not flamboyant, of Australian mining magnates – was claiming by 1982 that Australia could make up deficiencies in supply occasioned by any UN boycott of Namibia (49). By the early 1980s, as the massive downswing in uranium prices began to bite, Pancon was less optimistic: a 1982 report by the Northern Territory Department of Mines and Energy threw doubt on Pancon's earnings ability, and the company's share price took a tumble (51).

In 1980, Pancon appointed Nichimen Co of Japan as its sole agent for sales of Jabiluka uranium to Japan: the company was intending to import about 500 tonnes of yellowcake for Japanese power utilities from 1985 onwards (52).

Two years later, however, Grey's main stratagem appeared to be to obtain contracts in Britain and, the following year, the CEGB "indicated" that it would purchase at least 1950 tonnes of U_3O_8 between 1987 and 1997, under a base contract worth US$130 million: a possible substitute for the Rössing contract (7). Pancon's attitude to the Aboriginal people of Arnhemland was, in the early stages of its attempt to mine Jabiluka, cavalier to say the best. The Federal Australian Inquiry into uranium (known variously as the Ranger Inquiry and the Fox Commission) was in no doubt that, while it did not proscribe the Jabiluka project, all uranium development should be sequential, and Jabiluka should be the subject of a separate environmental impact study (EIS). The commission was unable to decide, from evidence available, just what its environmental – and therefore human – impacts would be: in particular, Fox wanted further studies on the effects of release of contaminated waters from a putative mine into the Magela Creek river system (53).

Grey's reaction to the Fox Reports was typical: "The Commission presented an extreme environmental viewpoint," he declared, "and we trust the government will provide the balance and give the nod to any uranium company ready to start" (54). By this time, it is possible Pancon had already started illegal drilling on the Jabiluka site (55). Between October 1976 and May 1977, Pancon worked on a draft EIS, which was based on the original 15 square kilometre deposit, although the newly-discovered

deposit covered seven times as large an area (56).

The EIS was quickly and fiercely attacked – by the well-known Aboriginal protagonist, H C Coombs, who pointed out that Pancon had made no approach to Aboriginal peoples (57), and who called the EIS a "travesty" which "exhibited profound ignorance of the Aboriginal people, a complete disregard for the recorded evidence of the impacts of other large-scale mining projects ... and a contemptuous indifference to preferences of the Aboriginal people concerned". Coombs claimed that the company had done "nothing to assemble the data on which any serious assessment of the social impacts must be based", nor to assist Aboriginal people to understand what was happening (58, 59). Coombs was determined that approval for Jabiluka should not be granted until the Northern Land Council (NLC) had the benefit of an independent agency which could carry out a complete survey, and adequately assess their opinions and needs: if the EIS was accepted, said Coombs, it was inevitable that the spiritual foundation of Aboriginal life would be destroyed (59). Other criticisms centred on the exorbitant price being charged by the company for a copy of its "public" report (no less than A$21), and its complete failure to evaluate any long-term biological, social or worker, safety hazards (60).

By May 1978, the Federal Australian government was in no mood to countenance its anti-uranium critics: it passed six bills in late May defining "immediate" protection for the environment once uranium mining started, nuclear safeguards for export of uranium products, and land rights claim procedures for Aboriginal communities (61). Just prior to this, the company had been given permission to commence investigatory drilling to "enable early completion of its EIS" – this included drilling at Hades Flat "for an alternative tailings dam site" and extension of the Arnhem Highway to Jabiluka. The move was greeted with outrage by opposition (ALP) members. Tom Uren MP declared the decision "a complete violation of the Second Report of the Ranger Inquiry" in particular, since no Aboriginal land claims had been heard: the Department of Aboriginal Affairs had merely been "instructed" to contact the traditional owners near the site, to determine which areas needed special protection (62). The *Financial Review* – not the most radical of Australian dailies – commented that the government seemed "quite prepared to over-ride the Northern Land Council" (63). By May 16th, Pancon said that its EIS and engineering studies were "sufficiently advanced" to allow construction to begin (65). By which time, the NLC was declaring itself completely opposed to Pancon's intentions (66).

Three months later, the Federal Australian government gave the final go-ahead to mining in Arnhemland, as road gangers and miners stood poised, at both the Ranger and Pancon sites, to start work as soon as the go-ahead was given (67).

Opposition from Aboriginal groups, especially the NLC, stalled a mining agreement in 1978, and permission to extend the Arnhemland highway was withdrawn (68). Trying to fend off the very great criticism of Jabiluka by Aboriginal traditional owners, Grey disingenuously stated that he had no plans to approach the NLC over terms and conditions, as "they have had quite a traumatic experience with the Ranger negotiations" (70). By the end of the year, the NLC was stating clearly that no go-ahead would be given to the mine, until Aborigines had obtained working experience from Ranger and Narbalek (71); for NLC chair Galarrwuy Yunupingu this meant "no development at Pancontinental at all" (72). At his company's 1978 AGM, Tony Grey put on his usual brave face, albeit tinged with disdain. In the face of about forty anti-uranium demonstrators, he warned that the Canadian uranium industry would secure the niche in world markets which Pancon might have occupied, but that "all being well", Jabiluka should open in 1979 (73).

Within a few months, Pancon had sprung a surprise on its opposition. In an attempt to overcome the majority of objections to its draft EIS, it proposed an underground mine, to reduce the amount of earth moved, decrease disturbance in land use, eliminate schist and sandstone waste dumps, keep facilities well away from Magela Creek and areas of Aboriginal significance, reduce the size of the tailings pond, simplify water management, avoid releases of contaminated water into Magela Creek, and reduce radon emissions (74).

But, by the end of 1979, Pancon was still having problems. The NLC was still firmly opposed to Jabiluka – underground, overground or wombling free. Since the new EIS was significantly different from the draft, the NCL wanted the whole public submission procedure reopened (75). The Aboriginal claim to land at Jabiluka had still not been heard by the Aboriginal Land Commissioner (76) and the Environment Minister in the federal government, in August, said the company would have to meet further conditions before approval of the final EIS could be granted (76).

Two years later, Pancon finally sat down with the NLC and traditional owners, to try to hammer out a mining agreement. At a meeting at Djarr Djarr (the Pancon camp east of Darwin) about 250 Aboriginal women, men and children, looked at what the company was offering: a lump sum of around A$7 million, royalties worth 4.25% of annual turnover, no Aboriginal equity in the project, no seats on the board: instead Pancon tendered a "research committee" to offer "appropriate information" to Aboriginal communities, and help with their paperwork (77, 78).

A number of officers of the NLC, and other Aborigines, expressed considerable displeasure at the proposals, and threatened to resign if they were implemented (78).

Finally, on March 1st 1982, a draft agreement was signed between the NLC and Pancon (79), followed, two weeks later, by conditional governmental approval (the condition being that the draft agreement be fully signed) (80). At the same time, the Aboriginal Land Commissioner, J Toohey, granted much of the land at Jabiluka to an Aboriginal Land Trust (81). The NLC declared itself well satisfied with the final Pancon agreement. ("Because of the fairness of the negotiations and the careful and delicate way in which they have been handled ... the NLC is proud to have been a part of them," said NLC chair, Gerry Blitner) (82).

The final approval still left open the question of Texaco/Getty's 35% share in Jabiluka – a share which violates Australian guidelines on foreign investment providing a minimum of 75% domestic ownership in uranium projects (83). And the agreement with the NLC – heavily criticised by the ALP, the anti-uranium movement, and many Aborigines – established an anomalous precedent. For the first time, "sacred sites" became part of a legally enforceable contract, without any definition of what they constituted, and the NLC agreed not to "create" such sites without Pancon's approval. That a white mining company should have any hand in determining the existence, or viability, of a spiritual site struck many people as bizarre and quite unacceptable (81).

From then onwards, it was mostly downhill for Pancon at Jabiluka. Mid-1982 saw the mine's economic viability severely questioned, in the light of likely costs and the Federal government's minimum floor price for U_3O_8 (A$30/lb) (84). Then the ALP, under Bob Hawke, came to power in March 1983. The ALP "three mines" policy effectively sounded the death-knell for Jabiluka (85, 33). Both Jabiluka and the Koongarra (see Denison and Noranda) uranium deposits were incorporated into stage two of the Kakadu National Park, much to the ire of Tony Grey (who cited Justice Toohey, the Aboriginal Land Claims Commissioner as supporting the project) (86), and of the Northern Territory government (87).

The Northern Land Council was also opposed to the measure, although opinion within the traditional communities had become sharply divided: some Aboriginal elders, like Toby Gangali, wanting Pancon's project moth-balled for future possible exploitation (88); Galarrwuy Yunupingu, the NLC chair, demanding compensation for loss of income and jobs (88), and Big Bill Neidjie adamantly opposed to all such "progress" (89). At a meeting of 300 traditional owners, in late November 1983, there was a clear demand that, if stage two of the park went ahead, Aboriginal custodians should control its access (90).

In the process, Getty and Pancon had lost A$50 million – the cost of its evaluation and feasibility studies (91) – and potential income of around US$230 million (at a government floor price of A$35/lb) (90).

Ironically, what had been one of the better-negotiated uranium agreements, ever made between indigenous land-holders and a mining company, was torpedoed by a government which then went on to endorse and support a flagrantly pro-uranium policy which has since merited no support from the Aboriginal people. The Roxby Downs uranium mine followed (see Western Mining Corp) then the export of Australian uranium to France, and the encroachment of CRA on Karlamilyi land in Rudall River.

Pancon's reputation with Aboriginal people, in the light of the Jabiluka negotiations, while not unsullied – let alone unselfish – stands in marked contrast to that of most other Australian mining companies, especially CRA, WMC and Peko Wallsend. The Department of Aboriginal Affairs itself, in the 1970s, noted how co-operative Pancon had been in dealings with Aboriginal owners in Arnhemland. This was, perhaps naïvely, ascribed to the fact that its Jabiluka partner, Getty Oil, has had long experience negotiating with Native Americans, while the Jabiluka deposit lies within the Arnhemland Aboriginal Reserve (9). The company has also been involved in several other projects on Aboriginal land and, while Tony Grey can hardly be called a "Friend of Aboriginal People," his style of corporate management (combined, possibly, with the diversity of institutional interests in Pancon, which militate against any one body gaining boardroom control) has been to tread lightly and wield a fairly small stick.

During the 1970s, Pancon entered a JV with Buka Minerals and Western Nuclear (Aus) Ltd, at the Sleisbeck uranium deposit south of Jabiluka (12). In 1976, the company also entered a uranium JV with UG (Urangesellschaft of Australia Pty Ltd), Getty Development Company, and the Stevens/Craven/Chomley family interests of Adelaide – a short-lived venture which was terminated the following year (9). In 1971, it participated in the Ormac JV on two prospects within the Arnhemland reserve: one, south of Goulborn Island, the other, north-west of Oenpelli (92). While Aboriginal progress associations were directly involved in the JV, the traditional owners of the land refused permission for further work on the main Exploration Licence area (No. 130) in the late 1970s, and in recent years no apparent developments have occurred in this region.

It is also known that, in 1971, Pancon entered into a JV with IMC's Australian subsidiary to explore within Arnhemland Aboriginal Reserve (93). This lease appears to have soon been discontinued.

The company's main dealings with Aboriginal people in recent years have centred on the development of the Palm Valley gas fields in the Amadeus Basin of the Northern Territory. Pancon has a 3% interest in this project with the majority interests held by Magellan Petroleum, Flinders Petroleum and Charles Davis Ltd (8). Initial exploration in the field was, in fact, carried out by Geophysical Services Inc, a subsidiary of the huge Texas Instruments electronics corporation (94). According to the Central Land Council (CLC), the main Aboriginal body in the region, Pancon and Magellan both exceeded the minimum necessary to satisfy Aboriginal demands: paying the Land Council to hire an anthropologist, and changing their plans a number of times to avoid sacred sites – although they delayed signing an agreement drawn up by the CLC (94). Within two years, a satisfactory agreement had been concluded, laying down conditions for sacred site protection, environmental safeguards, and employment of local people. The CLC singled out Pancontinental for its "co-operative spirit", especially on Aboriginal land west of Alice Springs at the former Haasts Bluff reserve, land of the Arrernte and Luritja people (95). Aborigines were being paid A$40 per day for their participation in the exploration teams, and traditional owners had been appointed to advise the location of sacred sites (94).

Part of Pancontinental's ultimate ownership changed in 1988, with the securing by Alan Bond (see Bond Corp) of control over the Bell Group in that year. (Alan Bond had bought out more than 80% of the group by August 1988 (96), with the Adelaide Steamship company – Adsteam – securing an 11.44% stake in Bell Resources.) The dramatic take-over of Bell by Bond, from the Holmes à Court stable, followed the collapse of the Holmes à Court empire after the stockmarket crash in October 1987 (97).

In 1986 Pancon entered a JV with Jason Mining to explore in East Kalimantan (Indonesian Borneo) (101).

Throughout 1989-91, Tony Grey kept up the pressure on public opinion, and the government, to allow the opening of the Jabiluka mine. At the same time, the company neatly side-stepped continuing Aboriginal fears – a proposed road through part of the World Heritage Site of Kakadu Stage 1 and Stage 2 was diverted to the satisfaction of the Northern Land Council (102). It also managed to enlist its most prominent Aboriginal critic, Big Bill Neidjie, to its side (103).

Summer 1990 saw the company claiming that Jabiluka was in fact "a gold mine" and therefore should not be covered by the ALP "three mines" limitation policy (never mind that the gold is sitting under 200,000 tons of uranium!) (104). Tony Grey was also moving around potential markets for Jabiluka uranium (rather like Australian pro-uranium forces had done in the late 1970s), delivering himself of strident, but vague, promises that "major utilities" in Europe, Japan and South Korea, would be interested in purchasing a future mine's output (105).

Between late that year and early 1991 – as Pancon suffered an A$83.7 million loss (106) – both the ownership of Pancontinental and that of the Jabiluka project changed dramatically:

• North Broken Hill Peko (NBH) raised its holding in Pancon from 12% to 13.8% (107).

- Cogéma lifted its holding to 15% from 13.5% (*ibid*).
- Degussa secured 14% of Pancon (108).
- Pancon sold its stake in Metallurgie Hoboken-Overpelt (part of UCEC-Union Minière), and the huge Belgian company, together with Cie Royal Asturienne des Mines SA, and Permanent Trustees Ltd, took 14% in the Australian company (109, 110).
- Another 10% of Pancon was, by now, in "friendly hands" according to Grey (109), while RWE of Germany held 6%.
- Finally, after some months of discussions, North Broken Hill Peko arranged in mid-1991 to purchase Pancon's 65% stake in Jabiluka (111).

Grey was confident that the mine could be in production within three-and-a-half years (112). Peter Wade of NBH was less sanguine, seeing development as "years away" (113).

One thing was certain: with NBH now controlling ERA, and thus the Ranger mine, its acquisition of Jabiluka opened the way for the two mines to be amalgamated, even though this prospect was firmly rejected by Prime Minister Hawke within days of the Jabiluka deal going through (114).

Further reading: "Planning Jabiluka Ventilation", from *E&MJ* 10/84 p69.

Contact: WISE-Glen Aplin, PO Box 87, Glen Aplin, 4381, Australia.
NLC, PO Box 3046, Darwin, NT 5794, Australia.
CLC, PO Box 3321, Alice Springs, NT 5730, Australia.

References: (1) *FT* 20/2/85. (2) *FT Mining* 1987. (3) *FT* 15/5/78. (4) *FT* 26/7/78. (5) *FT* 2/5/79. (6) *Canberra Times* 23/7/77. (7) *E&MJ* 3/83. (8) McGill & Crough. (9) Kim Williamson, "Report on Australian Uranium Companies", to Tom Uren MP, cyclo-styled, 1978. (10) Australian Department of Minerals and Energy memo, 29/3/74, Canberra. (11) Sydney Stock Exchange Library, 26/7/77. (12) Stock Exchange Research Service, Sydney, 1976-77. (13) *MMon* 3/81. (14) Sydney Stock Exchange 27/7/77. (15) *FT* 16/2/83. (16) *MJ* 13/1/84, *FT* 10/10/85. (17) *West Australian*, 27/11/79. (18) *FT* 27/11/79. (19) *FT* 7/11/83. (20) *FT* 20/3/84, *FT* 24/3/84. (21) *MJ* 6/4/84. (22) *FT* 17/3/83. (23) *FT* 13/6/85. (24) *MJ* 13/6/86. (25) *FT* 16/11/85. (26) *MJ* 17/10/86. (27) *FT* 29/11/83; *FT* 28/9/83. (28) *MJ* 13/1/84; *MJ* 30/3/84; *MJ* 28/12/84. (29) *FT* 6/1/84. (30) *MJ* 30/8/85. (31) *MJ* 1/11/85; *MJ* 22/11/85. (32) *FT* 19/2/86, see also (37). (33) *MJ* 25/11/83. (34) see (37) & *MJ* 17/10/85. (35) *MAR* 1987. (36) *MJ* 29/11/85. (37) *MM* 11/85. (38) *MJ* 15/11/85. (39) *MM* 8/86. (40) C S Christian & J M Aldrich, *Alligator Rivers Study: a review report of the Alligator Rivers Region environmental fact-finding study* (Department of NT, and AMIC), AGP Service, Canberra 1977. (41) *FT* 17/3/82. (42) *E&MJ* 1/87. (43) *Financial Review* 30/6/82. (44) *National Times*, 30/12/74 & 4/1/75. (45) *Financial Review* 5/4/77. (46) *Sydney Stock Exchange Company Review* 1973. (47) *Gua* 20/12/77. (48) *FT* 6/7/79. (49) see (51), and also *FT* 2/12/82. (50) *FT* 21/5/80; see also *FT* 7/9/85. (51) *FT* 6/8/82. (52) *MJ* 18/4/80. (53) *Ranger Inquiry* Vol. 2, (Federal Australian Govt AGPS) Canberra, 1977, p321 *et al.* (54) Quoted in *Ranger Inquiry Second Report People's Guide*, (Northern Territory Trades and Labour Council, with Friends of the Earth) Darwin, 8/77. (55) Tom Uren MP, quoted in *Age* 12/11/77. (56) Pancontinental Mining Ltd *Environmental Impact Statement on the Jabiluka Mine* 12/77. (57) *NT Newsletter* Darwin, 2/78. (58) *Age* 21/2/78. (59) *Financial Review* 21/2/78. (60) *Nation Review* 9-15/2/79. (61) *FT* 1/6/78. (62) *Canberra Times* 13/5/78. (63) *Financial Review* 15/5/78. (64) *Financial Review* 17/5/78. (66) *Financial Review* 16/5/78; *The Australian* 19/5/78. (67) *Nation Review* 18-24/8/78. (68) *FT* 12/9/78. (70) *FT* 14/11/78. (71) *FT* 11/12/78. (72) *Age* 18/12/78. (73) *Financial Review* 18/12/78, *FT* 18/12/78, see also *West Australian* 16/12/78. (74) *Financial Review* 10/4/79, *MJ* 13/4/78, *Age* 10/4/78. (75) *Financial Review* 4/9/79. (76) *Financial Review* 31/8/79. (77) *Age* 4/7/81, *Age* 7/7/81. (78) *Age* 6/7/81. (79) *FT* 2/3/82. (80) *FT* 17/3/82. (81) *Aboriginal Law Bulletin* No. 4, 6/82. (82) *Financial Review* 30/6/82, see also *Age* 26/7/82, *Financial Review* 26/7/82 and *FT* 27/7/82. (83) *FT* 13/8/82. (84) *MJ* 13/8/82. (85) *MJ* 11/11/83. (86) *Australian* 12/11/83. (87) *Sydney Morning Herald* 11/11/83. (88) *Sydney Morning Herald* 19/11/83. (89) *Age*

8/11/86. (90) *Australian* 23/11/83. (91) *Sydney Morning Herald* 12/7/84. (92) Eds Nicolas Peterson & Marcia Langton, *Aborigines, Land and Land Rights*, (AIAS) Canberra, 1983. (93) Stock Exchange Research Service, Sydney, 1975-76. (94) *Central Land Council News*, No. 13, Alice Springs, 7/81. (95) *CLC Land Rights News*, Spring 1983. (96) *FT* 3/8/88. (97) *FT* 19/10/87. (98) Reg Aus Min 1990/91. (99) *MJ* 1/6/90. (100) *MJ* 27/10/89, see also *E&MJ* 5/90. (101) *MJ* 11/12/87. (102) *Chain Reaction* Spring/89. (103) *Green Magazine* 4/91. (104) *Third Opinion* (MAUM) NSW, late summer/90. (105) *Aus* 21/9/90. (106) *FT* 24/2/91. (107) *Aus* 23/10/90. (108) *Aus* 4/1/91, *NZ Herald* 2/1/91. (109) *Aus* 8/1/91. (110) *FT* 22/3/91. (111) *Aus* 10/1/91, *FT* 5/7/91, *MJ* 5/7/91. (112) *MJ* 18/1/91. (113) *Aus* 11/1/91. (114) *Aus* 9/7/91.

471 Pan D'Or Mining NL (AUS)

This is a mineral and energy resource development company, which has been involved in an oil shale project with Esso in Queensland (near Rockhampton), and also in gold prospecting (1).

As of 1981, it was participating in two JV uranium exploration programmes in the Northern Territory of Australia: one with Aquitaine Australia (50%) and Jimberlana Minerals (25%) at Mt Bundey; the other with Jimberlana at Adelaide River (2, 3). Aquitaine – formerly a part of this JV – dropped out of the Adelaide River programme; Jimberlana and Pan D'Or continued into 1981 (2).

Little uranium has been reported from Mt Bundey and, by 1985, the prospect was being maintained rather than expanded (4).

References: (1) *FT* 2/5/79. (2) Annual Report, 1980. (3) *Reg Aus Min* 1981. (4) *MIY* 1985.

472 Pan Ocean Oil Ltd

In 1979 the company was in a JV with E & B, and Kelvin Energy, at Louison, Heron and

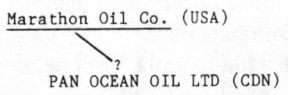

Beak Lake (Great Bear Lake area of Canada's Northwest Territories) (1).

Also, in 1980, it reported significant uranium mineralisation when prospecting in the South Bissett Creek zone (2). Other partners in the South Bissett Creek consortium were Lochiel (8.6%), Dynamic Mining (14.1%), and Petrobec (17.1%) (3).

The exploration programme was halted for some period by the Inuit of Baker Lake who maintained that their land claims must be settled before uranium mining commenced. Though supported by environmental groups, they failed in their court case of November 1979 (4) (for details see Urangesellschaft).

In 1975 Pan Ocean got permits to explore for uranium in two areas in the north of Niger. No major results have been reported since (5).

References: (1) *MJ* 29/6/79. (2) *MAR* 1981. (3) *MJ* 8/2/80. (4) *MJ* 30/11/79. (5) *MAR* 1979.

473 Pathfinder Mines Corp

It was set up in 1976 (out of Lucky Mc Uranium Corp) when Utah merged with General Electric. However, the French CEA (through Cogéma) bid for 80% of the Pathfinder stock in 1981 (1) – a bid accepted in 1982 (2). The remained was acquired by Cogéma in 1986 (6).

Pathfinder wasn't much of a loss to General Electric, which found in the previous two years

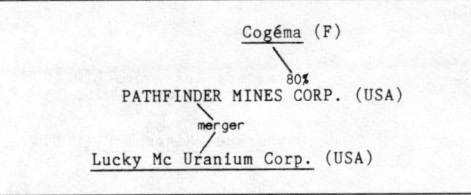

that Pathfinder's operations – handicapped by low-price contracts in the mid-'70s – were increasingly having to scale, or close, down. For example, in April 1980, 200 workers were laid off at the Lucky Mc Uranium Corp mine in Wyoming – and another 40 at the Big Eagle mine. Utah blamed "excessive government regulations" for the run-down (3).

Pathfinder had substantial uranium reserves in three mines in Wyoming's Gas Hills and Shirley Basin areas, with mills at both locations. Between them the Lucky Mc and Shirley Basin mills could produce 4.2 million pounds U_3O_8 – but in 1979 they produced less than 3 million pounds (4).

All operations at the Lucky Mc mine were closed, and surface operations cut by half, in 1984, due to deterioration in the uranium market (5).

Shirley Basin produced half a million pounds of U_3O_8 in 1989 (7) and (together with Chevron's Panna Maria mine) was one of only two conventionally-dug uranium mines operating in the USA by the end of 1990 (8).

Pathfinder Gold Corp has 30% (Sonora Gold Corp 70%) of the Jameston gold mine in California which was opened in 1987 (9).

References: (1) *FT* 11/11/81. (2) *E&MJ* 3/83. (3) *Kiitg* 4/80. (4) *MIY* 1982. (5) *E&MJ* 1984. (6) *E&MJ Int Directory of Mining* 1991. (7) *MAR* 3/90. (8) *E&MJ* 3/91. (9) *FT Min* 1991.

474 Patino NV

Simon I Patino virtually "discovered" Bolivian tin, and through his guile, energy, and capital accumulation, his company opened up one of the world's largest mineral areas – completely on Indian land. Patino at one time ranked in the top five of the world's millionaires (1). He was also one of the founders of the world's first commodity (or producers' protection) pacts – the International Tin Producers Association which was born in 1930 (2).

Patino holds 10% of Cofremmi – a company developing New Caledonia (Kanaky)'s nickel deposits, together with Amax.

Its principal involvement in uranium was through Northgate, which acquired a 35% interest in Patino in the early 1980s (see Northgate) (3). Through its holdings in Brascan, Patino NV has been indirectly involved in holdings in Noranda and other energy companies (see Brascan).

The original Patino company was among three incorporated into the COMIBOL/Corporacion Minera de Bolivia, during the 1952 revolution.

Northgate sold its interest in Patino in 1983, to reduce indebtedness (4).

For further reading (especially on the treatment of Indians in Bolivia during the rise of Patino in the first 30 years of this century, but from a company point of view) see (2).

References: (1) *War on Want Outlook*, No. 9, 1980. (2) Charles F Geddes, *Patino: The Tin King*, London, 1972 (Hale). (3) *MIY* 1982. (4) *MIY* 1985.

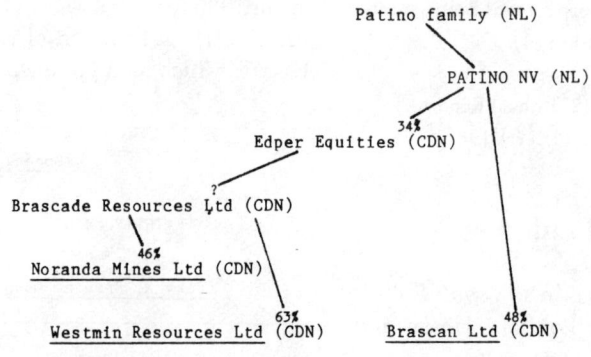

475 Peabody Holding Co Inc

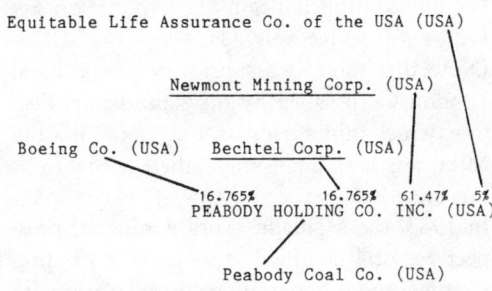

Equitable Life Assurance Co. of the USA (USA)

Newmont Mining Corp. (USA)

Boeing Co. (USA)   Bechtel Corp. (USA)

        16.765%   16.765%   61.47%   5%
        PEABODY HOLDING CO. INC. (USA)

            Peabody Coal Co. (USA)

Until 1990, Peabody was partially controlled by two important US uranium companies (Newmont and Fluor) and America's most significant constructor of nuclear plant, Bechtel. It is now owned by Hanson plc.

It is the USA's top coal producer and holds more coal leases on Native American land than any other company. Most damaging are its operations in the Black Mesa region of Arizona and New Mexico, where the company has leased 100 square miles of land, traditionally owned and jointly controlled by the Hopi and the Navajo (Dine). The US Bureau of Indian Affairs, Department of the Interior, and some of the Hopi and Navajo tribal councils, surrendered this land in the '60s, against the wishes of the majority of native land-users. A recent attempt to divide the Hopi and Navajo, and "relocate" more than 5000 Dine and a smaller number of Hopi, was perceived by many Native American groups as a direct result of Peabody's scheming.

In early 1984, Peabody bought out most of Armco's West Virginia coal mines, fourteen in all, though, with a production capacity of about 4.5 million tons, they represent only a fraction of Peabody's annual output – 48 million tons in 1983 alone (1).

In 1988, Peabody entered into secret negotiations with members of the Hopi Tribal Council to expand its coal strip mining into new lease areas, thus risking further involuntary removal of Navajo families (2). Two years later, the company sought to roll on further its existing Black Mesa and Kayenta operations.

Accusing the company of endangering Navajo and Hopi water supplies, and of desecrating burial and other sacred sites, Navajos – supported by traditional Hopi, and Big Mountain activists – mounted a renewed campaign against coal strip mining in Arizona, in 1991 (3).

By this time, Hanson plc of London had taken over 100% control of the US company, and four Navajos came to the Hanson annual meeting, in January 1991, to demand that it halt the onward march of its US subsidiary (3).

Up to 1989, Newmont held 49.7% of Peabody, with Eastern Gas & Fuel Associates, Boeing Co, Bechtel Investments, and Equitable Life Assurance Society of the US, holding the remainder (4). In the meantime, Hanson had gained control of Consolidated Goldfields, and – through CGF – 49% of Newmont. Hanson then proceeded to "stalk" Peabody: in February 1990 buying 45% from Boeing, Bechtel and Eastern, while negotiating with Newmont to purchase the rest (5). Meanwhile, Amax decided to bid for control of Peabody (6), but it soon pulled out of the stakes, leaving Hanson to secure the prize in mid-1990: America's largest coal strip-miner, the holder of the western world's third biggest reserves of coal, and the western world's third biggest producer of coal (7).

Peabody has also recently been active in Australia (along with Central Pacific Minerals and Shell Australia) in oil shale and magnesite exploration (8).

In 1990, Peabody had won a US Office of Surface Mines (OSM) National Wetland Reclamation Award for its Will Scarlet mine in Illinois, which had once been described as the country's "worst example of the effects of unregulated mining" (9). But, a year later, the US Secretary of Labor announced that many coal companies – headed by Peabody – had been guilty of deliberately adulterating coal dust samples in their underground mines, in order to disguise the degree of hazardous pollution (10). In January 1991, Peabody pleaded guity to three criminal counts of tampering, and agreed to pay half a million dollars in fines (11). It was then revealed that Peabody had been convicted, nearly a decade before, on 13 felony counts for similar misdeeds (11).

Further reading on the background of Peabody Coal: *Native Americans and Energy Development,* from ARC, 59 Temple Place, suite 444, Boston, Massachusetts 02111, USA (US$4.50).

J Kammer, *The Second Longest Walk* (University of New Mexico Press) ($12.95).

Mount Taylor Alliance Newsletter (Big Mountain Support Committee) (50 cents plus postage).

Support for the Dine and traditional Hopi is welcomed by: Big Mountain Support Committee, POB 7082, Albuquerque, New Mexico 87194, USA.

References: (1) *MJ* 3/2/84. (2) *Minewatch Briefing on Peabody Coal,* quoting documents from the Hopi Tribal Council, 19/10/88. (3) *Navajo/Hopi European Protest Tour,* a Minewatch Report, London, 1991; see also *MJ* 24/8/90. (4) *MJ* 11/8/89. (5) *FT* 16/2/90. (6) *Gua* 15/5/90, *Obs* 13/5/90, *FT* 14/5/90. (7) *MJ* 1/12/89. (8) *Reg Aus Min* 1990-91. (9) *E&MJ* 10/90; see also Supplement to *MJ* 22/2/91. (10) *Washington Post* 4/4/91. (11) *Multinational Monitor* 7-8/91.

476 Pegmin Ltd

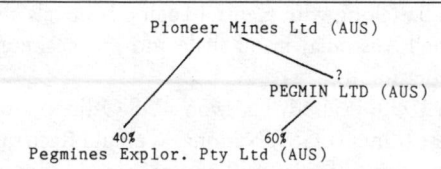

It has been exploring for diamonds on Aboriginal land in the Lake Argyle area of Western Australia (with Cocks Eldorado and Southern Cross Exploration). In South Australia, it has been exploring for uranium with Esso Exploration (1) on abandoned copper workings on the east side of the York peninsula. In 1979, high-grade uranium mineralisations (up to 602lb/ton) were reported (2). Esso later withdrew from the South Australia JV to be replaced by Otter Exploration on a 50/50 basis (3).

Pioneer Mines Ltd, in 1981, controlled a sizeable part of Pegmin, and, together, Pioneer and Pegmin controlled Pegmines Exploration Pty Ltd (40/60 respectively) (1).

(Note that there is a subsidiary of the Rand London Corp called by the same name, Pegmin, which mines feldspar, quartz and mica in Mica, northern Transvaal – there seems to be no connection between the two.)

In 1983, the Maitland (York peninsula) prospect was still classified as a "grass roots" programme, and negotiations were under-way with an un-named private company for possible future involvement (4).

References: (1) *Reg Aus Min* 1981. (2) *MJ* 23/5/79; *MM* 5/79. (3) Annual Report, 1981. (4) *Reg Aus Min* 1983/84.

477 Peko-Wallsend Ltd

In 1987, Peko-Wallsend and North Broken Hill Holdings (NBH) announced that, after each company had mounted a hostile bid for the other (1), the two would merge to form North Broken Hill Peko (NBH). Not only would the new company now be one of Australia's fifteen largest corporations, it would also be one of the country's half-dozen most significant owners of zinc, lead, copper, gold, silver, coal, petrol, iron ore, mineral sands, and uranium (2).

"Now identified as a champion of the New Right" (3), Peko-Wallsend was, in 1980, the putative king of uranium mines (4), whose ruthlessness in dealing with trade unionists, Aboriginal land claims, and environmentalists, is matched by its mistrust of government and public ownership (5).

Peko-Wallsend was formed in 1958 (6), as a merger between Peko Mines Ltd – whose activities focused on the multi-mineral region of Tennant Creek, 315 miles north of Alice Springs in the Northern Territory – and a coal-mining company called Wallsend Holding and Investment Company Ltd (7).

Despite its attacks on foreign take-overs of Aus-

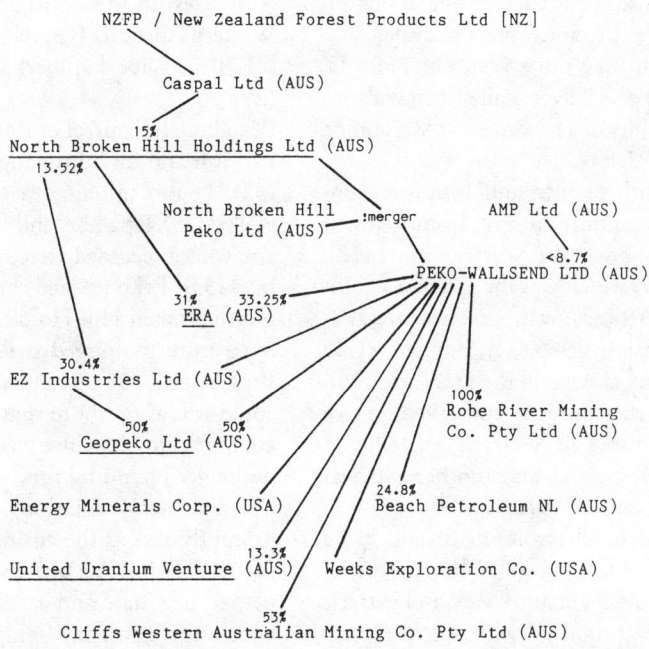

NZFP / New Zealand Forest Products Ltd [NZ]

Caspal Ltd (AUS)

15%
North Broken Hill Holdings Ltd (AUS)
13.52%

North Broken Hill
Peko Ltd (AUS)   :merger   AMP Ltd (AUS)

<8.7%
PEKO-WALLSEND LTD (AUS)

31%   33.25%
ERA (AUS)

30.4%
EZ Industries Ltd (AUS)

100%
Robe River Mining
Co. Pty Ltd (AUS)

50%   50%
Geopeko Ltd (AUS)

24.8%
Energy Minerals Corp. (USA)   Beach Petroleum NL (AUS)

13.3%
United Uranium Venture (AUS)   Weeks Exploration Co. (USA)

53%
Cliffs Western Australian Mining Co. Pty Ltd (AUS)

tralian resources (notably in 1980 when 25% of the newly-floated ERA was being offered to overseas companies) (4), Peko has long depended for much of its capital on external sources. For example, in 1972, a A$15 million expansion programme depended on funds laundered by Schroders Ltd, British merchant bankers (8), and in 1971-74, finance and a complicated lease-back arrangement at Tennant Creek were arranged with the Bank of America (9).

When, in 1985, the company was the subject of take-over rumours, it placed nearly 8 million of its ordinary shares with Australian Mutual Provident (AMP) and an un-named financial institution, thus consigning 10% of its ordinary capital (10). In the last decade, significant minority interests in Peko have been held, not only by AMP, but also ANZ nominees, CTB nominees, CML Assurance Society, and the Bank of NSW Nominees: by the late 1970s nearly half of the company's issued shares were held by some twenty shareholders (7). (The holdings were compatible, though reduced, in the case of Peko's long-standing fellow-ex-

ploiter, and erstwhile ERA partner, EZ Industries.)

In 1976, Peko owned some 31 subsidiary companies (7). During the next decade, as it encountered setbacks to profitability at Tennant Creek (11), and needed to reduce its debt, it sold its 12.6% holding in Gove Alumina Ltd (see Alusuisse) (12), and, a little later, its significant investment in Beach Petroleum NL – after it had raised its stake (13).

However, it acquired 53% of Weeks Petroleum in 1985 (14), thus giving it access to important oil and gas fields in Australia (especially the Jabiru field off northern Australia) and in the USA (15).

Its recycling subsidiary, Sims Consolidated (Simsmetal), acquired in a diversification move in 1979 (16), has become Australia's largest such operation (17). And Warman International (known in Britain as Simon-Warman Ltd) is renowned for its engineering services and pumps. These two companies, according to the *Engineering and Mining Journal* (18), were, in 1983, Peko's "two most important subsidiaries" (19).

In 1980, Peko also acquired 75% of Harrington Metallurgists (a precious metals dealer and refiner) (20) and, the same year, entered a JV with Marubeni and Silver Valley Minerals, to study the feasibility of a new mine at Wallamine in New South Wales (21).

Gold, along with copper and bismuth, continues to yield significant output from Tennant Creek (24), notably at the Warrego mine (22); the Gecko mine, which was the site of a significant find in 1979 (23), was put on a care-and-maintenance basis in 1986 (24). But a new gold mine was opened at Argo in the same area (25). More recently, the company embarked on two open-pit gold mines in Western Australia, at Perth Hill and Kanowna, and another two near Parkes in New South Wales (NSW) (26) – in the latter of which, Chevron of Australia had a 39% interest (27). The Mount Morgan smelter treats copper from Tennant Creek and extracts gold from tailings, in a JV with Anglo-American through Australian Anglo-American Gold Pty Ltd (Peko 60%, AAC 40%) (22, 24).

As well as operating three collieries in the Hunter Valley, NSW, Peko, in 1987, had another JV with Marubeni to develop a coal mine in the same state (26).

Mineral sands are also an important concern of Peko. Its subsidiary Rutile and Zircon Mines (Newcastle) Ltd – RZM – has long had a JV (50/50) with Coffs Harbour Rutile NL (28) involving two dredging plants at Tomago, Newcastle, NSW: in 1985, the scheme produced more than 30,000 tonnes of rutile, more than 40,000 tonnes of zircon, and smaller quantities of ilmenite and monazite (24).

It is at Robe River that the company's industrial practices have raised controversy around the world. Peko took over Robe River Ltd in 1983, after a prolonged battle with Pancontinental – and thanks to the support of Robe River's parent company Redesdale (itself owned 51% by Burns Philp Ltd and 49% by Engelhard Minerals Chemicals (which is associated with AAC) (29) – at a cost of A$114 million (30, 17). It assumed full control in 1986 (31).

Robe River's main asset is its 35% share of the Robe River iron ore JV (Cliffs Robe River Iron Associates) in the Aboriginal Pilbara region of Western Australia (Cleveland Cliffs 30%, Mitsui 30%, Cape Lambert Iron Associates 5%) (29).

Peko bought control of Robe River using internal cash and by borrowing within the country (32). It also inherited a split ownership (between US, Japanese and Australian interests) and what it claimed were restrictive work practices (33). Peko learned the lesson of CRA and North Broken Hill Holdings – which had suffered from prolonged strikes the same year at Broken Hill (31). It claimed that the project could run at no more than 60% capacity, and attempted to introduce changes, declaring itself no longer bound by previous agreements (34). This was met with union opposition, and Peko promptly sacked the entire 1160-strong workforce (31, 35). The unions indicated a willingness to negotiate and asked for a moratorium, which the company refused. In the face of Peko's refusal, the Industrial Relations Commission ordered the company to comply with its ruling. Peko rejected the ruling and closed down Robe River (36). Peko was rapidly supported by many Australian businessmen, condemned by the Western Australian Premier Brian Burke, and the federal Prime Minister, Bob Hawke, and lost its appeal in early 1986 before the state Supreme Court (36). It was then compelled to reinstate the sacked workers. A little later, Peko found support from the Commission for its bid to "curb restrictive practices" (31), and, after making minor concessions, re-employed a majority of the workforce. It not only won its "right to manage" (36), but also the support of right-wingers throughout the industrial world.

The Robe River confrontation was seen as significant because it came at a time of polarisation in Australian politics. But Peko had already taken a tilt over the years at unions at Tennant Creek and the Ranger uranium mine (34), as well as at its coal division (26).

As with EZ, it is the company's earnings from ERA, and its participation in the Ranger uranium project in the last decade (37), which turned a significant loss in 1981 (38) into profits over the following years (39), when dividends were paid from its "A" shares in Ranger (40): no less than 66% of Peko's profit derived from the mine in the second half of 1983 (41). Peko and EZ discovered the Ranger uranium deposit in the Alligator River region of the Northern Territory and set up the Ranger JV to develop it (28). In 1974, Geopeko, the company's exploration subsidiary, applied for six further leases outside the main Ranger deposit (42).

Two years after the go-ahead was given for Ranger, and just after the mine's first contract was signed with South Korea (43), Peko successfully bid A$125 million for the AAEC's 50% stake – pipping companies like Esso and Denison (who made bigger offers) to the post (44). As Japanese companies (notably Kansai and Kyushu) on the one hand, and North American companies on the other, made a play for the uranium riches of Oenpelli land, it was Peko which promoted the idea of a public flotation of the government's erstwhile holding in Ranger, and which – in the event – bought up some of the shares itself (45). (See ERA for further details of the game-play between the Australian federal government, Peko and EZ, and foreign buyers, before equity participation in ERA was finally worked out.)

It is not only the Ranger deposit itself, in which Peko has been interested over the last decade or more. In the early 1970s, Peko was a member of the Uranium JV (with Newmont Proprietary Ltd, Noranda Australia Ltd, Utah Development Co, EZ, and BHP) which located a uranium deposit at Mumalary, Northern Territory (NT) (28), still the subject of an Aboriginal land claim. During the same period, it also held a uranium lease with Uranium Consolidated NL at Argentine, Queensland (46).

Meanwhile it was exploring into areas situated well within the supposedly protected area of the Kakadu National Park (48) – making 410 applications for leases in early 1978 alone (49).

One of these leases was both in, and near, Stage Two of Kakadu – a very rich deposit located about 20km north-west of the Ranger site itself (50) grading at up to 17lb a tonne (compared with 6lb in the case of Ranger) with possibly one million tonnes of contained uranium in all (48). The Northern Land Council immediately lodged objections to the company's plans to exploit the find, situated as it is on the Mudginberri Aboriginal pastoral station (49). The Federal government – using its familiar tactics of running with both hare and hounds – promised protection of Aboriginal land claims (providing they were "expedited") (51) and support to the companies (50). In 1983 Peko launched the first of several legal challenges to Aboriginal land rights claims to mining leases (52).

By 1986 Peko had stakes on some 65 square kilometres of the Kakadu National Park Stage Two, which included the rich deposit discovered in 1978 known as Ranger 86, now estimated to be worth around A$500 million (53). The federal government refused to list this area as a World Heritage site – thus failing to protect it from the depredations of Peko (54). Nonetheless, in 1986 – playing a ploy worthy of the company itself – the federal government appeared to have prevented Peko from exploiting this new South Alligator region, at least for the time being. First the government supported Aboriginal land claims in the federal court, while the Northern Territory government (Peko's trojan horse for its re-entry into Kakadu in the 1980s) (55) supported Peko – and the company proceeded with its exploration on Aboriginal land, without government consent. But in October 1986, Clyde Holding, then the Federal Minister for Aboriginal Affairs, apparently revoked his decision to grant an Aboriginal lease (56), lulling Peko into a false sense of security. Eighteen days later, title of some of the Peko land was handed to Big Bill Neijie, as a member of the Jabiluka Land Trust.

In 1978, just after Peko announced its new finds in the South Alligator region, its exploration manager, John Elliston, launched a broadside against government policy and Aboriginal land rights, declaring development of uranium to be a "miserable shambles" and exploitation "for the benefit of all Australians, including the Aborigines" to have been sacrificed to the "domination" of "social and political problems" (47).

Thirteen years later, Peko must consider the situation to be little better, but its prospects – especially as part of the NBH conglomerate – to be somewhat improved. So long as there is uranium to be dug in northern Australia, a Peko geologist, backed by company muscle, will not be far away.

"Our story is the land ... it is written in these sacred places. My children will look after these places, that's the law." – Big Bill Neidjie, *Kakadu Man*.

Further reading: *Kakadu Man*, Big Bill Neidjie, with Stephen Davis, Allan Fox, (Mybrood P/L Inc) NSW, Australia, 1985.

Contact: Northern Land Council, PO Box 3046, Darwin, NT 5794, Australia.
WISE-Glen Aplin, PO Box 87, Glen Aplin, Queensland 4381, Australia.

References: (1) *MJ* 18/12/87, *FT Min* 1991. (2) *MJ* 25/12/87. (3) *National Times on Sunday*, 30/11/86. (4) *Financial Review* 28/2/80. (5) *Age* 13/5/82. (6) *MJ* 19/10/79. (7) Stock Exchange Research Service (Australia), 1976-77. (8) Roskill's *Who Owns Whom, Australasia & the Far East*, London, 1973; *Financial Review* 5/4/72. (9) *Financial Review* 1/3/71; *Financial Review* 28/4/72; *Age* 12/7/74. (10) *MJ* 18/10/85. (11) *FT* 23/7/80; *FT* 24/2/81. (12) *MJ* 5/11/82. (13) See (26), and *FT* 31/1/84. (14) *MJ* 28/3/85. (15) *FT* 3/1/85. (16) *FT* 31/1/79. (17) *MJ* 9/11/84. (18) *E&MJ* 11/83. (19) *Ibid.* (20) *FT* 25/4/80. (21) *FT* 20/6/80. (22) *MJ* 4/11/83. (23) *Western Australian* 29/10/79. (24) *FT Mining* 1987. (25) *MJ* 6/12/85. (26) *MJ* 18/9/87. (27) *MM* 11/85. (28) *Peko-Wallsend Annual Report* 1973/74. (29) *FT* 20/9/83. (30) *MJ* 23/9/83. (31) *MJ* 14/11/86. (32) *MJ* 7/11/86. (33) *Ecocomist*, London, 7/3/87. (34) *MJ* 22/8/86. (35) *MJ* 24/7/86. (36) *MJ* 12/9/86. (37) *MJ* 11/4/86. (38) *MJ* 7/1/82. (39) *MJ* 4/11/83; *FT* 1/3/84; *MJ* 26/8/84; *FT* 30/8/84; *MJ* 28/9/84; *FT* 2/3/83. (40) *MJ* 29/3/85. (41) *MJ* 23/3/84. (42) *Northern Territory News* 9/1/74. (43) *FT* 10/11/79. (44) *FT* 19/12/79; *Australasian Express*, London, 12/12/79. (45) *Financial Review* 31/8/79. (46) *MJ* 16/9/77. (47) *Financial Review* 24/5/78. (48) *FT* 3/3/78. (49) *The Sun* 8/2/78. (50) *Age* 3/2/78. (51) *Northern Territory News*, Darwin, 2/78. (52) *Australian* 23/11/83. (53) *Age* 19/11/86; *Australian* 19/11/86. (54) *Chain Reaction*, No. 48, Summer/86-87. (55) *National Times on Sunday* 30/11/86. (56) *Australian* 21/11/86; *Financial Review* 3/11/86; *Age* 1/11/86.

478 Peregrine Petroleum (CDN)

It held property jointly with Tyee Lake Resources on which drilling for uranium was taking place in the late '70s. The property – in the Beaverdell area of British Columbia, near Kelowna – was optioned to Noranda and Kerr Addison for a 60% interest. Activity presumably stopped after British Columbia moratorium (1).

References: (1) Merlin.

479 Petrobec Ltd

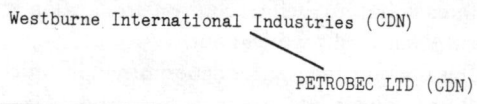

It was part (17.1%) of a C$1 million exploration programme in the Baker Lake area of the Northwest Territories of Canada, which started in 1980, after the Inuit of Baker Lake had their land claims rejected (see Urangesellschaft entry). The exploration was on properties owned by Pan Oil (60%), and other partners in the JV were Dynamic Mining Exploration (14.1%) and Lochiel Exploration (8.6%). Drilling by early 1980 had indicated more than 3 million pounds of uranium oxide (1).

References: (1) *MJ* 8/2/80.

480 Petroci

```
                 Republic of the Ivory Coast (CI)
PETROCI /       /
STÉ. NATIONALE D'OPÉRATION PÉTROLIÈRE (CI)
```

Using car-borne spectrometers, the company in 1980 conducted uranium searches in Bouaké, Bandiali and Odienné – with no significant results reported (1).

In 1981, the company's uranium department was transferred to Sodemi (2).

References: (1) *MAR* 1981. (2) *MAR* 1982.

481 Petromin NL

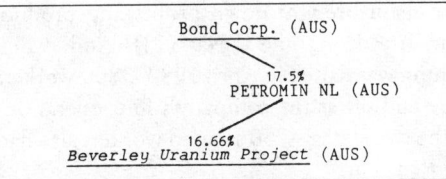

```
            Bond Corp. (AUS)
                 \
                  17.5%
                  PETROMIN NL (AUS)
                 /
           16.66%
Beverley Uranium Project (AUS)
```

It was partnered by Oilmin and Transoil at Mereenie oil field (see Oilmin) where it held a 7.5% interest, in uranium exploration at Cape York, and at Beverley (*qv*) (1).

Like its stable mates Oilmin and Transoil, Petromin was reined in by Queensland red-neck and ex-premier, Joh Bjelke Petersen (2).

References: (1) *Reg Aus Min* 1981; Annual Reports 1980, 1981. (2) Roberts.

482 Phelps Dodge Corp

```
          PHELPS DODGE CORP. (USA)
                 /
Western Nuclear Inc. (USA)
```

Phelps and son-in-law Dodge, founded a metals trading company in 1834 in New York which became the largest mercantile company in the land. It did not, however, make its first investments in copper until purchasing the Morenci mine in Arizona in 1881. It acquired another copper mine in Mexico 13 years later and built its first smelter in Arizona in 1901 (1). During the '20s the company became one of the largest American suppliers of wire and cable; it prospered for obvious reasons in the war years, and in the immediate post-war period was (along with Anaconda and Kennecott) supplying 80% of all US production (2).

By the early 1950s the company was expanding abroad, particularly into Peru (where it owns 16.25% of South Peru Copper Corp while Asarco owns 53%) (3), mining copper at Toquepala, Quellareco and Cuajone.

Diversification followed into aluminium and uranium when it acquired Western Nuclear, with mines and a mill in Wyoming, and later in Washington, New Mexico and Australia.

The company sold its 40% interest (4) in Conalco/Consolidated Aluminium Corp in 1981 (5). Its major non-US interests include a multimineral JV (44.6%) at Black Mountain in Namaqualand, South Africa, with GFSA (55.4%); a one-third interest in the huge Woodlawn open-pit zinc-lead-copper-silver mine in New South Wales (with St Joe Minerals, the Fluor subsidiary, holding one-third and CRA the rest) (3).

It has 64.3% of the Compañía Minera Ojos del Salada in Chile (a copper-gold-silver mine); operates a fluorspar mine and mill in the Western Transvaal, South Africa (3); has a prospect in the Philippines (Atlas Corp of the Philippines holding 27.4% of Phelps Dodge Philippines) (6). The company also had a JV with Getty Oil in the early '80s, drilling a molybdenum deposit at Pine Grove, southern Utah (6). Its explorations in recent years have taken in the USA, Chile, Australia, Spain, and Turkey (6), South Africa, Mexico, and Botswana (where it started mining gold in 1989) (34).

However, Phelps Dodge is still primarily a copper-bottomed proposition. Despite its position as one of the lowest-cost producers, and its attempts to cut back on certain assets while diversifying into other minerals (7), Phelps Dodge hardly prospered before the '80s. In early 1980, it was expressing qualified optimism (8) following expansion of refining facilities (9), while its uranium subsidiary Western Nuclear began reporting profits, after losses in 1978 (10) and 1979 (11).

The company also boosted its earnings in 1981 (12). This was only thanks to the performance of its 49%-owned South African subsidiary Black Mountain (13), and a good performance by Western Nuclear (14). By Spring 1982, its problems were "mounting" (15) and, on April 17th, it announced the closure of all its US copper mines and concentrators and its three Arizona smelters (16).

Only with the so-called "climb" out of recession at the turn of the year was the company able to report some recovery (17) when it reopened its big Ajo mine in Arizona. However, despite a small reported profit in mid-1983, the company fell back into the red by the autumn (18) and, when 1984 came, it was "in a solid losing position" (19). George Munroe, the company's chairman, puts forward several reasons for the decline: reduction in copper consumption, "overproduction" by nationalised industries (especially in Chile, Zambia and Zaire) for "political, employment and foreign exchange reasons" (20, 21), and new environmental legislation resulting in higher costs (20).

Certainly Phelps Dodge has every reason to be worried about foreign competition (it was among 11 US copper producers who in 1983 petitioned the US International Trade Commission to stop imports harmful to US trade) (3). It has also been among the leading protestors against pollution controls and the restriction on mining in federal and public lands (22). Although Reagan's chief environmental bulldozer, James Watt, announced new legislation to open up previously protected lands to mining companies in 1982 (16), Watt himself was later sacked, and implementation was held up. In December 1982, the Arizona Department of Health announced new measures to control sulphur dioxide emissions from the state's seven copper smelters, of which Phelps Dodge owned three: combined new emissions would be limited to about 35,000lb of SO_2 an hour. Phelps Dodge was then quoted by the leading US mining magazine *Engineering and Mining Journal* as "... in particular ... lobbying for new amendments to the new Clean Air Act to give certain smelters more breathing room" (and the rest of us more acid rain) (23).

Phelps Dodge has been more reticent about the other factor in its declining prospects: a series of major strikes at its mines and smelters which had occurred every three years since 1959 (24), but reached an unusual intensity in 1983 and 1984.

The labour protests stemmed from the copper industry's renegotiations of contracts in mid-1980, which directly affected Phelps Dodge's copper mines in Arizona and the El Paso refinery. This refinery was closed from July 1st to October 8th 1980 (6). Strikes the following year again brought down profits and production in Arizona and Texas (13). And, when mining was halted in April 1982, 3800 workers were laid off at the company's four operations in the two states; 4750 salaried workers also had their pay cut.

A long hot summer of strikes occurred in 1983, as the company refused (unlike other major US copper producers) to accept wage freezes. "We're not going to give up the gains our forefathers died for on the picket lines," US Steel representative Jacob Mergado was quoted as declaring, when a brief 24-hour truce occurred at the Morenci plant after strikers, armed with clubs and bats, blocked entrances to the mine (24). The confrontation was caused when Phelps Dodge drafted in non-union workers to reopen the mines "because it ... could not afford to close them" (25).

Industrial action continued well into 1984, when the National Guard was called in by the company to accompany non-union "scab" workers into the pits.

A group of strikers described the action:

"A scab drove back and forth by the crowd outside the clinic with his gun pulled out, aimed at the people. The DPS (state police) did nothing, and the people were angry and started throwing rocks at the police cars. We have lived for months with harassment, beatings and jailings by the DPS.

"Over by Shannon Hill, where the strikers were yelling at the scabs, the DPS started pushing people around, harassing people. The crowd there got angry too, and when they started throwing rocks, the police fired tear gas into the crowd. Then they came in and arrested eight people – just whoever they felt like picking up. Cesar Chavez's attorney was there and he was stunned at the violations of people's rights.

"It was then that Governor Babbitt called in the National Guard again. They were patrolling everywhere with machine guns, just like when they were here last year. It seems like it's war, with soldiers coming into your area.

"The National Guard has left, and Babbitt has said he won't call them in again. But we have learned that they have been placed on alert for the whole next month, and that there is a recruitment campaign to get more men in.

"They are expecting trouble because PD is planning to evict families from the company-owned employee housing. Their rent has been paid, but PD says they can't live there because they are not employees. At the same time, the unemployment office tells the strikers they don't qualify for benefits because they're not unemployed!

"Time[s] are very rough, and if it weren't for our hatred for the company, we would have capitulated long ago. It's graduation time at the high school, and we can barely pay for the caps and gowns of our children. But we hope that more and more people are learning about our strike, and will help support our stand for working people's rights" (26).

While busy deploring American workers' violation of "the law", the company has not been above illegal activities itself.

Together with five other companies, it was cited in the Philadelphia Federal Court in 1983, for violating anti-trust legislation by fixing prices of copper tubing between 1975 and mid-1981 (27).

The WPPS/Washington Public Power Supply System (WOOPS!) also initiated an anti-trust suit against Phelps Dodge and other uranium producers in 1982, for their participation in the uranium cartel which upped the price of uranium and, according to WPPS, made it impossible for contracts to be honoured. Phelps Dodge earmarked US$4.1M to cover the suit (28) and countered with a claim which resulted in WPPS having to pay the copper company US$25M in 1984 (29).

But, in the final analysis, the biggest claim on Phelps Dodge – if any justice is to be done – must come from Native Americans, in particular the Papago on whose land the huge southern Arizona copper belt (of which Phelps Dodge is the major exploiter) lies. In the 1980s the US Congress passed legislation that took the land from the "Indians" at one fell swoop. But some of the Papago remain – and they do not forget (30).

As of 1984, Phelps Dodge was – in the candid words of its chairperson Munroe – "in a solid losing position", having sustained a "whopping US$25.2M loss" for the first quarter of the year (31), and despite credit gained from settlement of litigation with WPPS over cancellation of a uranium supply contract. Phelps Dodge had now started divesting itself of some of its assets – notably its power cable subsidiary.

Not to be outdone, Munroe returned to his theme at the 1984 AGM, when he claimed that the World Bank and IMF were doing enormous damage to the copper industry, by financing high cost copper production in "less developed countries", the US taxpayer footing about 20% of the bill (32).

Earlier, the company blamed nationalisation of mines in Africa and South America for flooding the import market with cheap copper.

In 1984, Phelps Dodge started moving back into uranium in an attempt to diversify its production and cut its paper losses – namely forming a JV to develop a uranium mine in Texas (33).

But its heart was always in copper. By 1987, this heart was looking pretty ticky. In a late-1984 cover story, *Business Week* wrote the company off as dead (35).

Within three years, however, Phelps Dodge had adopted a Lazarus position. It updated its technology in Arizona, acquired two-thirds of Kennecott's Chino mine, mill and smelter, relocated to Phoenix to be closer to its operations, and managed to raise some handsome new money (36).

The following year it was expanding – especially in New Mexico and Chile (37) – and diversifying.

By the turn of the decade, Phelps Dodge had one billion dollars invested in non-copper businesses (including Phelps Dodge Industries, PG Magnetic Wire Co, and Colombus Chemicals) (38).

Apart from its open-pit mines in New Mexico, it now holds two underground mines in Chile (with a third deposit, at La Candelaria in the indigenous Atacama region, under consideration with Sumitomo) (39), and it continues to mine fluorspar in the western Transvaal of South Africa, as well as lead, silver and zinc from the Black Angel mine in Cape Province. Production is expected soon from its Santa Gertrudis gold mine in Mexico (40).

Phelps Dodge now produces around 7% of the western world's copper – ranking it second only to Codelco – and it delivers about one-third of US production (40, 41).

Contact: International Indian Treaty Council, 777 UN Plaza, New York, NY 10017, USA.

Morenci Miners Women's Auxiliary, 1113 3rd Avenue, Safford, Arizona 85546, USA.

References: (1) WDMNE. (2) BOM. (3) MAR 1984. (4) MJ 7/3/80. (5) MJ 5/2/82. (6) MIY 1982. (7) FT 12/5/83. (8) FT 2/4/80. (9) FT 13/3/80. (10) MJ 1/6/79. (11) MJ 29/2/80. (12) MJ 11/9/81. (13) MJ 6/11/81. (14) FT 26/3/82. (15) MJ 6/8/82. (16) MJ 16/4/82. (17) FT 2/4/83. (18) FT 25/10/83. (19) FT 4/5/84. (20) FT 12/5/83. (21) FT 23/3/84. (22) FT 15/1/82. (23) E&MJ 3/83. (24) FT 10/8/83. (25) FT 23/8/83. (26) News and Letters, Detroit, 5/84. (27) FT 19/3/83. (28) MJ 5/2/82. (29) FT 4/5/84. (30) Tanzer. (31) MJ 1/6/84. (32) MJ 11/5/84. (33) MJ 23/3/84. (34) FT Min 1991, E&MJ 7/88. (35) Business Week 17/12/84. (36) E&MJ 1/87. (37) MJ 20/5/88. (38) E&MJ 1/90. (39) MJ 21/6/91. (40) MJ 3/5/91. (41) MJ 3/8/90, MJ 2/2/90.

483 Phibro-Salomon Inc

Formerly Engelhard Minerals and Chemicals Corp, this is a major US commodity trader and refiner of precious metals.

It was cited by the TVA/Tennessee Valley Authority as one of 13 companies originally responsible for price-fixing of uranium in the '70s.

It owns 49% of Redesdale (51% Burns Philp) which in turn has a 48.4% interest in Robe River Ltd, one of several companies mining in

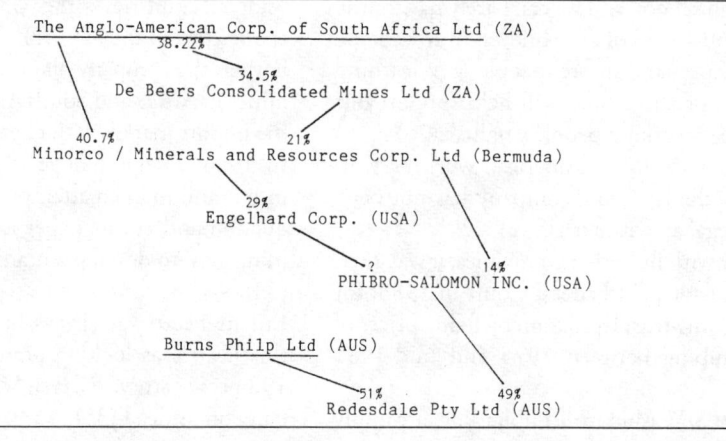

the Pilbara region of Australia (for more details see CSR).

Phibro-Salomon was formed as a merger between the Phibro Corp and Salomon Brothers, an investment and broking company, in 1981. However, in 1984 Phibro-Salomon sold four of its operating subsidiaries to an investor group led by Phibro Resources's own chief executive, for US$40M. The group included Interdec Inc, a company controlled by Saudi investor Gaith Pharoan. It acquired, at bargain price, important metal and smelting interests, lending credence to the view that Phibro-Salomon was pulling out of metals and "concentrating on the big money spinners of financial services and pure commodities trading" (1).

Sheffield Smelting Company of Sheffield, a refinery of precious metals, is a wholly owned subsidiary.

Until 1982, Phibro-Salomon was the US marketing agency for the Nufcor/Nuclear Fuel Corp of South Africa contracts between the USA and South Africa, running at some 4320 tonnes between 1976 and 1984. It relinquished that role in order to set up its own uranium trading business (2).

In summer 1984, the London *Observer* reported that Phibro had been involved in the huge tin smuggling trade between Thailand and Singapore – knowingly buying tin concentrates which depressed the world market price, thus causing the International Tin Agreement to come into action. The London-based International Tin Council (ITC) had to purchase £185 million of this smuggled tin, when refined in Singapore, in order to support the price control system of the ITC. Phibro refused to comment to the *Observer* about the allegations (3).

In 1990/91, Philbro Energy (along with Amoco, Conoco and others) was involved in discussions in the Soviet Union to exploit the West Siberian oil fields (4).

References: (1) *MJ* 6/1/84. (2) *Africa Confidential* 19/1/83. (3) *Obs* 15/7/84. (4) *MAR* 1991.

484 Pioneer Concrete Services Ltd

Now called Pioneer International Ltd, Pioneer Concrete controls Nabarlek Uranium, Australia's first uranium mine – which is completely on Aboriginal land (for details, see Queensland Mines).

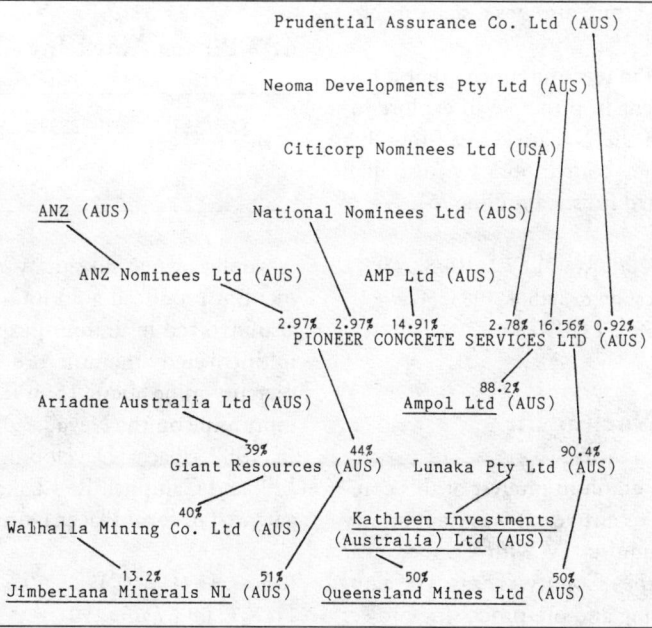

Prudential Assurance Co. Ltd (AUS)

Neoma Developments Pty Ltd (AUS)

Citicorp Nominees Ltd (USA)

ANZ (AUS)   National Nominees Ltd (AUS)

ANZ Nominees Ltd (AUS)   AMP Ltd (AUS)

2.97% 2.97% 14.91%   2.78% 16.56% 0.92%
PIONEER CONCRETE SERVICES LTD (AUS)

88.2%
Ariadne Australia Ltd (AUS)   Ampol Ltd (AUS)

39%   44%   90.4%
Giant Resources (AUS)   Lunaka Pty Ltd (AUS)

40%   Kathleen Investments
Walhalla Mining Co. Ltd (AUS)   (Australia) Ltd (AUS)

13.2%   51%   50%   50%
Jimberlana Minerals NL (AUS)   Queensland Mines Ltd (AUS)

It diversified out of building products into the energy field and has subsidiaries in Singapore, South Africa, Italy, Portugal and Britain (1). Its British subsidiary ranks third in the league of British ready-mixed concrete producers (2).

The following comments on its British operations (which ran at a loss in 1981) have been provided by a former employee:

"Pioneer has no discernible policy on restoration of sites they use – but there is a scheme to lease the land back to local authorities as landfill, for rubbish. Thus they get not only profits from the original operations, but also from the licence for rubbish dumping, from tax relief (since land-fill degrades the land) and from the final sale of the land for housing and farming.

"The chairperson, Sir Tristan Antico, was knighted for his services to cement, but also made a lot of money on a factory-farming venture by breeding Thai water buffaloes in Australia and selling the carcasses to West Germany for hamburgers: the buffalo venture was scrapped when Antico got a whiff of his knighthood" (3).

In 1987, Pioneer drastically restructured its operations, after acquiring 41% of Giant Resources. As a result, Pioneer gained 88% of Ampol Ltd and, later, complete control (4). The restructured group also has a share in Noranda Pacific.

In March 1988, Pioneer announced that it had made an agreement in principle to explore for more uranium in the Nabarlek area, now that the first of the contemporary Australian uranium mines had finished milling (5).

References: (1) *FT* 30/9/81. (2) *FT* 14/10/82. (3) Personal communication to author, 1983. (4) *Reg Aus Min* 1990/91. (5) *FT* 25/3/88.

485 Pioneer Nuclear Inc

The only known uranium interest of this company (at least in recent years) was its minority share in the Conquista JV with Conoco near Falls City in Karnes County, Texas. Development of this mining and ore-processing project

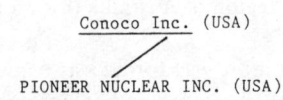

Conoco Inc. (USA)

PIONEER NUCLEAR INC. (USA)

was suspended in 1981 because of the depressed uranium market, pending shut-down in early 1982. Conoco, the project operator, announced that a resumption would take place "when the market improves sufficiently" (1).

However, in 1982, Conoco pulled out of all mining altogether (2), and it is now virtually certain Conquista will never be reopened; while Pioneer Nuclear is defunct.

References: (1) *MJ* 10/7/81. (2) *MJ* 23/7/82.

486 Pioneer Uravan

This is a company in Colorado, in association with Wisconsin Power Co (1); it was reported to be planning a uranium project by 1981 (2); no further information.

References: (1) Uranium Information Network, Colorado, USA, 1980. (2) *US Surv Min* 1979.

487 Placer Amex Inc

Placer Development Ltd (CDN)

PLACER AMEX INC. (USA)

A wholly-owned subsidiary of Placer Development, it produced gold in Nevada (1) and was also involved in uranium exploration. In 1978 it discovered uranium ore in an abandoned mercury mine about 15 miles from its McDermott mine on the Nevada/Oregon border (2). In 1987, Placer Development, Dome Mines Ltd, and Campbell Red Lake Mines Ltd amalgamated to form Placer Dome Inc (3).

References: (1) *MIY* 1982. (2) *San Diego Union* 17/2/78. (3) *FT Min* 1991.

488 Placer Development Ltd

Incorporated in 1926, Placer Development grew out of an amalgamation of Placer and the British Columbia molybdenum producer Endako Mines (1) (which was closed down in 1983) (2). It was part of Zinor Holdings which originally held 21.2% of Noranda, but these shares were sold to Brascade in 1981, thus enabling Brascan to become Noranda's biggest shareholder, and reducing Placer's share in Noranda to approximately 6% (3).

Principally a mineral exploration company with interests in "hard" minerals, petrol and gas (2), the company also has important interests in precious metals, notably Equity Silver Mines (which are British Columbia's biggest silver producer) (4), the large silver project in Zacatecas, Mexico (5, 6), and the Golden Sunlight gold property in Butte, Montana, USA (which opened up in 1983 five months ahead of schedule and produced more than 100,000 ounces of gold in 1989) (2, 7, 16).

The company's operations in Australia are carried out through Placer Pacific Ltd (75.8% owned by Placer Dome Inc).

It also holds 39.9% of the Philippines copper producer Marcopper (8). In 1988 Marcopper Mining was ordered to cease mining, as its tailings were too pollutive, and the company had failed to extract toxic metals before dumping the waste in the sea. "Dispensation" to do this was, apparently, granted by President Marcos himself in 1982 (13).

Placer's ventures into uranium country are significant, though they cannot be compared with those of its associate, Noranda.

It held a licence to explore for the devilish stuff in the Hydraulic Lake area of British Columbia (BC) in the late '70s (9) and was involved in a magnetic survey of the Great Slave Lake area of the Northwest Territories at roughly the same time (10).

Its most controversial uranium shenanigans occurred in British Columbia after the provincial government's uranium moratorium, when it put forward proposals to mine molybdenum at Atlin, a property on which Placer secured a 70% option from Adanac. "When is a mine not a mine?" asked the anti-nuclear newspaper *Energy File* at the time (11).

In 1980, the Birch Bark Alliance commented: "The proposal shows every indication of being an environmental and social disaster for the small community in BC's far north-western corner ... Tailings will be dumped in ponds held back by man-made tailings dams in an area that averages one earthquake per year. After the

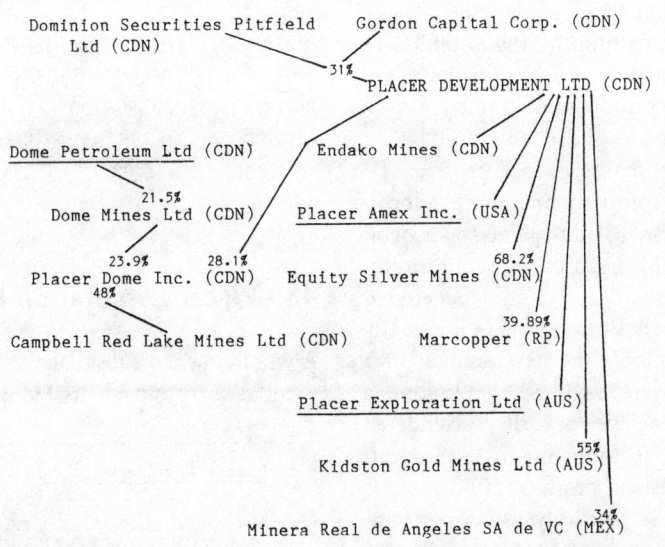

mine is exhausted in fourteen years (by which time some 1500 tons of uranium tailings will have been created) the company plans to abandon maintenance of the ponds by sealing them with a hefty layer of grass seed ... The company's proposals include an ecologically hazardous ten-mile river diversion, the inadequate disposal of sewage wastes into Atlin Lake (the town's drinking water supply and the drawing card for nearby Atlin Provincial Park) and the importation of 450 construction workers into the small northern town" (12).

In 1987, Placer Development, Campbell Red Lake Mines and Dome Mines were amalgamated into Placer Dome Inc, thus launching a new, debt-free Canadian mining conglomerate, with a cash surplus (in 1988) of US$666 million. The group's assets now include gold mines in Papua New Guinea (at Porgera and Misima, which came into operation in 1987), Australia, and Canada; the Endako molybdenum mine in British Columbia, which may be the lowest-cost producer in the world (14); and 34% in the huge Real de Angeles silver mine in Mexico (15).

Placer is now one of the world's foremost gold producers – a position it has reached thanks to its Australian projects during the late eighties managed by 75.8%-owned Placer Pacific (see Placer Exploration), agressive exploration in north America, and in particular, the coming on-stream of the mammoth 90,000-ounces-a-year, Porgera mine in Papua New Guinea, and the Misima project in the same country.

Nor has it rested on its gilded laurels. Of late, it has embarked on major copper projects (JV with the Chilean company Enami at Andocolla) (17), and bid for (18), and gained control of, the large Mount Milligan copper-gold project in British Columbia). It has also penetrated "new lands" such as Bolivia (where it recently proposed to Comibol) the development of a new Bolivian mine in 1989 (19), and Indonesia (where it took over eleven of CSR's surrendered mineral properties in Indonesia in 1988, specifically the Wetar Island project) (20).

However, Placer is primarily associated with gold mining, or with gold exploration in regions it has long stamped upon.

It has been exploring in Nevada where it operates, with RTZ (26.45%) the Cortex gold mine (21), and at Mexican Hat, Arizona (22). Placer was among a number of Canadian companies to show an interest in the north woods of Wisconsin during 1990/91 (23). (For strenuous opposition against mining in this region, see RTZ).

In Chile it has a JV with the TVX Corp on the La Coipa gold/silver prospect (24). Early in 1991 it sent its luminaries to the USSR to negotiate gold mining "developments" in the eastern region (25).

Its hopes to mine gold in the Kahama district of Tanzania has met with opposition from Outoukumpu Oy over conflicting claims (26). Meanwhile, it has sold its stake in the large Omai gold prospect in indigenous Guyana, to Golden Star Resources and Cambior (both Canadian) after arguing that the price of gold could not sustain a profitable investment there (27).

References: (1) *MIY* 1982. (2) *MAR* 1984. (3) *MJ* 13/11/81. (4) *MJ* 3/8/84. (5) *MJ* 24/4/81. (6) *FT* 10/11/83. (7) *MM* 5/83. (8) Communication from Philippines Support Group, 7/84. (9) *Energy File* 1979. (10) *MM* 5/80. (11) *Energy File* 5/80. (12) *BBA* Summer/80. (13) *FT* 22/4/88. (14) *MAR* 1988. (15) *FT* 29/7/88. (16) *FT Min* 1991 (17) *FT* 31/5/81, *MJ* 31/5/91. (18) *FT* 15/8/90. (19) *MM* 4/91. (20) *MJ* 27/11/87, *E&MJ* 1/89. (21) *MJ* 22-29/12/89. (22) *E&MJ* 7/90. (23) *Milwaukee Journal* 2/12/90. (24) *MJ* 28/7/89, *FT* 14/12/90. (25) *MM* 1/91. (26) *E&MJ* 6/90. (27) *FT* 25/1/91, *FT* 15/5/90.

489 Placer Exploration Ltd

A wholly owned subsidiary of Placer Pacific, it used to carry out all its "mother" company's

```
Placer Development Ltd (CDN)
        |
        |
PLACER EXPLORATION LTD (AUS)
```

operations in Australia, which – apart from interests in manufacturing (Fox Manufacturing, and Associated Plywood) and chemicals (Molybond Laboratories) (1) – were concentrated on the Kidston gold mine in northern Queensland until the Big Bell and Granny Smith (60%) mines opened in 1989 and 1990 (5). Placer Pacific is now Australia's leading gold producer (6). Placer Exploration's original stake of 80% was reduced to 70%, following an offer to Australian commercial shareholders (2).

Among its JVs was one with Oakbridge, a coal/tin producer of New South Wales (NSW), in which Placer Exploration had a 50% interest (in the Wolgan, NSW, colliery) (3).

It also has a JV with CRA to explore for gold on the Wafi River in Papua New Guinea (4), and applied for a large exploration licence in Fiji in late 1983, to scour the north-west of Vanua Leva at Udu (4) – a prospect which was not being evaluated until 1991 (7).

Placer Pacific (75.8% owned by Placer Dome) (8) like its parent company, has concentrated on gold mining, though recently it has ventured into base metals exploration (9). It has taken the leading role in developing two of the south Pacific's most significant new gold lodes – Misima Island in Papua New Guinea, which poured its first gold in 1989 (10) and above all, the Porgera mine which opened in late 1990 (11). It was already expanding into its second (12) phase of exploitation the following year (12). Porgera is one of the top six non-South African gold producers in the world, with an annual expected output of 900,000 ounces (13). While its lease covers 2200 hectares, nearly a quarter of this extensive site has already been taken up by the mine, mill and associated infrastructure (14). Misima is owned 80% by Placer, and 20% by the Papua New Guinean government (15).

Porgera is operated as a partnership between Placer (PNG) Pty Ltd (a subsidiary of Placer Pacific), Highlands Gold (a company floated by MIM), Renison Goldfields, and the Papua New Guinean government.

While Porgera has not escaped criticism (nor comparison with CRA's Bougainville mine), it is Misima which has generated most alarm among environmentalists within and outside Papua New Guinea.

It is the country's first mine to use marine waste disposal (although two more have been planned in the country) – a method of dealing with waste which was chosen "because it was cheap" (16). An alternative – to pipe the detritus well out to sea – was considered and discarded mainly because of its expense. The amount of such soft waste dumped in the coastal waters near the mine will peak at 20,000 tonnes (or 183 truckloads) *per day*, creating a turbid mini-mountain and inevitably affecting local fishing, coral reefs and other marine organisms (16).

Equally seriously – and perhaps more longterm in its impact – is the method of tailings disposal, whereby the mine-waste, mixed with sea water, is discharged like some slurping, slurried, soup through an outfall pipe to the ocean bed, at between 75 and 100 metres.

Here, the tailings "smother the surrounding deep ocean floor inhabitants, either forcing away or killing those animals incapable of adjusting to the new conditions," noted Philip Hughes in 1989. However, (he added), "the Misima people do not fish at such great depths and there is little potential for commercial deep-water fishing" (16).

Within weeks of Misima operating at design output in 1989, it had polluted water in the creeks and rivers on the island, creating what the Environment and Conservation Minister, Jim Yer Waim, called "a serious health threat," and giving rise to a shortage of drinking and cooking water. Yer Waim told Placer to clean up its waste, and castigated it for failing to honour a compensation agreement under which the company would build alternative water supply systems, if natural sources were affected (17).

References: (1) Reg Aus Min 1981. (2) *MJ* 17/8/84; *FT* 12/6/84. (3) *MIY* 1982. (4) *MAR* 1984. (5) *FT* Min 1991, see also *FT* 27/7/90, *MJ* 15/6/90. (6) *Reg Aus Min* 90/91. (7) *Metal Bulletin* 16/5/91. (8) *MJ* 8/12/89. (9) *MJ* 17/8/90. (10) *MJ* 12/5/89, see also *E&MJ* 7/89. (11) *MJ* 7/9/90. (12) *MJ* 3/5/91 (13) *FT* 23/10/90. (14) V V Botts, General Manager, Porgera JV, to M. Barber, Probe International,

29/5/90. (15) *MJ* 1/7/88. (16) Philip Hughes, *The effects of mining on the environment of high islands: A case study of gold mining on Misima island, Papua New Guinea*, SPREP Environmental Case Study no 5, 1989. (17) *Pacific Islands Monthly* 7/89.

490 Plateau Resources Ltd

```
    Consumers Power Co. (USA)
              \
     PLATEAU RESOURCES LTD (USA)
```

It was set up by Consumers Power Co, Utah, to acquire, explore and develop uranium, as well as to buy and sell the damned metal. A vanadium/uranium processing facility was being constructed by 1981 at Ticaboo, Utah (1), then was temporarily closed in 1982 (2) and completed in 1983 (3).

In 1982, operations at the company's Tony M uranium mine were also reduced to one shift per day. A hundred workers were laid off, though "recovery" was foreseen when the "market improves" (2).

In 1978, the company planned uranium exploitation in Shootering Canyon, Utah (3); later, it joined with Frontier Resources and Nuclear Dynamics to explore for uranium in the same state (4). The Shootering mine was open, and producing uranium on a small scale, by 1984 (5) but closed late that year (6).

References: (1) *MIY* 1982. (2) *MJ* 26/11/82. (3) *MAR* 1984. (4) *MJ* 12/1/79. (5) *MM* 1/84. (6) *MM* 1/85.

491 PNC

The PNC has explored, either independently or in JVs, for uranium in Canada, the USA, Australia, Mali, Colombia, Guinea, Gabon and Zambia, as well as in all six domestic areas of Japan (1).

In late 1977, PNC entered agreements with the Australian ACM, and Command Minerals to explore around Murchison, Western Australia, and with Magnet Metals (also Australian) to explore around the Eucla Basin in Western Australia (2).

Later the same year an unconfirmed report suggested that PNC was developing an "on site" UF_4 facility in South Africa (3): certainly it was operating a facility to convert from UF_4 into UF_6 at Nigyotoge (Japan) in 1976, using either imported yellowcake or uranium liquor from heap leaching (1).

Exploration for uranium in Mali started as a "full scale exploratory search" (4) in 1977 in Kidal and Tessalit, to the north, near the Niger/Algeria border (5). PNC would receive 80% of the output from any future mines (6). By 1979, the corporation had been joined by Cogéma and was assessing uranium values in a 65-square-mile region where PNC had been granted "exclusive rights" (7). In early 1979, along with Agip and Cogéma, PNC was invited to explore for uranium in Guinea (8). No further details have been issued, and when, in 1981, the Guinean government invited nine countries to discuss eight "recently discovered" deposits, Japan was not among them (9). The Cogéma consortium was wound up that year (10).

In April 1978, the corporation entered a JV agreement with Pancontinental to explore for uranium in Australia as a whole (11).

Along with Agip and Saarberg Interplan, PNC has also been exploring for uranium in Zambia (1).

```
                    State of Japan (J)
                          /
PNC/POWER REACTOR AND NUCLEAR FUEL DEVELOPMENT CORP. (J)
                      /
     PNC Exploration (Canada) Co. Ltd (CDN)
```

In Colombia the corporation started work in 1978 with Coluranio, Enusa, and Minatome, in an exploration programme which was (in 1980) extended until 1982 (13).

In Canada, PNC has been exploring both under its own name and as PNC Exploration (14) in Saskatchewan and in the Osoyoos, Vernon, and Greenwood mining divisions of British Columbia (15); and also in the Northwest Territories (1).

More recently, PNC has announced a significant uranium find at Yilgarn, 900km east of Perth, Western Australia, which contains possibly 10,000 tonnes of 0.3% ore (16). While PNC Exploration (Australia) Pty Ltd was reported ten years later, to be "exploring for uranium in South Australia, NT and WA" (21) perhaps its most controversial operations focused on the Western Desert, where – together with CRA – it held leases over the mineral (and uranium)-rich land of the Martu people, who vehemently opposed any exploitation (see CRA for further details).

By 1979, the company's Canadian subsidiary, PNC Exploration (Canada), held 300 square miles of uranium permits in the Athabasca basin, Saskatchewan (17). Ten years later, PNC Exploration (Canada) Ltd was among the ten most active exploration companies searching for the magic yellow mushroom in Canada (22). In 1990, it teamed up with Denison (20%) in its JV with Chevron Minerals Ltd and Interuranium Canada, on the Waterfound river uranium property, close to Midwest Lake (see Denison) (23).

In 1984 PNC reached agreement with the Chinese government to undertake joint exploration in the Tengchung district of Yunnan province (18).

By 1984, the "quasi-government" PNC had become, in the words of the *Engineering & Mining Journal*, "the key actor in Japan's worldwide quest for new uranium sources". In August 1983 it acquired 10.889% of the equity in the Dawn Lake JV, east of Waterbury Lake, Canada, thus securing a firm stake in the western world's most promising new uranium region (19).

This was later increased to 17.525%, with Cameco holding 50.486%, Cogema 19.525%, CEGB 7.5% and KEPCO 4.5% (24) PNC also purchased 15% of the equity in the Midwest Lake JV (see Denison mines).

Since September 1979 the company has been operating a pilot centrifuge uranium enrichment plant, the capacity of which, in 1984, was raised from 50 to 200 tons. A new nuclear power complex, in which PNC will undoubtedly play a major role, was planned at Rokkasha-mura, to provide one sixth of Japan's enriched uranium needs by the year 2000 AD (12).

In 1984 PNC also operated a small nuclear fuel reprocessing plant (0.7 tons a day) and an experimental fast breeder reactor (FBR) known as "Joyo": another FBR was planned for 1991. A prototype ATR reactor called "Fugen" was operating in 1984 and the company was experimenting in the vitrification of nuclear "waste" (20).

References: (1) *MAR* 1981. (2) *MJ* 16/12/77. (3) *SANA Bulletin* 6/77. (4) *MJ* 21/1/77. (5) *Africa Magazine* 12/77. (6) *West Africa* 15/5/78. (7) *New African* 8/79. (8) *MJ* 16/3/79. (9) *MJ* 13/3/81. (10) *MAR* 1982. (11) *MIY* 1982. (12) *FT* 15/8/84. (13) *MJ* 8/1/82. (14) *Yellowcake Road*. (15) *Energy File* 3/79. (16) *Courier Mail*, Brisbane, Australia, 17/6/81. (17) *FT* 10/9/79. (18) *MJ* 25/5/84. (19) *E&MJ* 6/84, (20) *Gensuikin News*, no 108, Summer 1984. (21) *Reg Aus Min* 90/91. (22) *MAR* 1990. (23) *FT Min* 1991, (24) *Greenpeace Report* 1990.

492 PNOC

This is a state-owned corporation which, in mid-1983, announced that it was resuming its uranium exploration programme, investing US$4M over the following two years, to find

```
         Republic of the Philippines (RP)
                          \
PNOC / PHILIPPINE NATIONAL OIL CO. (RP)
                                   \
        PNOC Energy Development Corp. (RP)
```

sufficient fuel for its highly controversial (1) Bataan nuclear reactor, destined for start-up in 1985 (2), which was abandoned after Cory Aquino came to power.

References: (1) Bello, Hayes & Zarsky, "500-mile Island" in *Pacific Research* 1st quarter/79. (2) *E&MJ*, 7/83.

493 Polaris Resources Inc (USA)

It entered agreement with Aquitaine Mining to explore its Lakeview uranium prospect in south central Oregon (50%). This prospect was located close to the White King mine which produced nearly half a million pounds of uranium in the '50s and was operated up to 1979 by Phelps Dodge (1).

References: (1) *MJ* 10/8/79.

494 Portland General Electric (USA)

This is possibly a subsidiary of General Electric, involved with Martin Trost in constructing the Miracle uranium mine in California in 1978 (1).

References: (1) *US Surv Min* 1979.

495 Powerco

```
POWERCO / POWER RESOURCES CORP. (USA)
                    \
                     \
Aquarius Resources Ltd (CDN)
```

It has explored for uranium in Wyoming, at Weld (1), in Texas (2) and in Colorado (1). In January 1979, it reached agreement with Newmont Exploration Ltd on a JV on mineral leases held by Powerco in central Wyoming: Powerco to become the operator, Newmont to put up around 70% of the costs and acquire the land (3).

In 1989, the Central Electricity Generating Board of Britain (CEGB) purchased Everest Minerals and, along with it, Everest's Highland (Converse County) uranium solution mine in Wyoming, which was previously operated by Exxon, closed in 1985, and re-opened in 1988 (4). Cogéma's subsidiary Interuran, hold's 25% of this project.

A little later, the CEGB was privatised by Britain's conservative government, with the result that Nuclear Electric plc now controls the Highland property. The CEGB set up a new company, called Power Resources Inc, to manage Highland (5).

It is not known whether this was, effectively, the old Powerco, but it seems a fair assumption to make. Highland was only one of the two US ISL (in-situ leaching) mines to be operating by the beginning of 1991 (6).

References: (1) *Nuclear Information Network*, Colorado, 1979. (2) *E&MJ* 11/78. (3) *MJ* 12/1/79. (4) *Greenpeace Report* 1990, see also *E&MJ* 3/90. (5) *E&MJ Int Directory of Mining* 1991. (6) *E&MJ* 3/91.

496 Prayon

Known usually as Prayon only, the company was, until the 1982 reorganisation of Société General de Belgique's assets in Union Minière, a 26.5% subsidiary of the latter (1). It operates several refining plants in the home country, one of which (a modern zinc plant) was forced to close in 1981 because of fall in demand (2).

Prayon was the Union Minière associate com-

```
                    Sté. Générale de Belgique (B)
                                   \
                                    \
Union Minière SA (B)                 \  26.5%
                         PRAYON / STÉ. DE PRAYON SA (B)
```

pany which set up the Umipray uranium-from-phosphates extraction consortium in 1979, along with Metallurgie Hoboken-Overpelt and its parent company (3); they were later joined by IMC, Mechim, and Rupel-Chemie (4, 5). The same partners also set up an international consortium to license the new venture and market its products (6). Both ventures, however, appear to have fallen by the wayside.

References: (1) *MIY* 1982. (2) *MAR* 1982. (3) *FT* 10/10/79. (4) *MAR* 1981. (5) *Kiitg* 10/80. (6) *MJ* 12/10/79.

497 Precambrian Shield Resources Ltd

```
          Numac Oil and Gas Ltd (CDN)
                              /
                            /
    PRECAMBRIAN SHIELD RESOURCES LTD (CDN)
```

It was exploring in Saskatchewan (1) and part of a JV with Numac in the Yukon (2).

References: (1) *Who's Who Sask.* (2) *Yellowcake Road.*

498 President Brand Gold Mining Co Ltd

A medium-sized gold producer in the Anglo-American camp (and 12th largest apartheid producer in 1984) (1), President Brand is the biggest supplier of slimes to the JMS (2). It is also the site of the uranium treatment plant for the JMS.

The company's leases include 2536ha over farms in the Welkom district of the Orange Free State, and some 112ha of land on the President Steyn mining lease, costs for mining which (and revenues) are distributed as to 57.5% for Brand and 42.5% for the other old President (3).

The company was constructing a large new gold mill and treatment plant in 1984 (3). Two years later, however, President Brand was amalgamated into Freegold, the world's largest gold producing conglomerate. (See AAC for further details) (7).

Brand's uranium production until the end of 1975 amounted to some 557 tonnes of U_3O_8 (4). A new slimes treatment plant was built and then "moth-balled" in 1971 due to weakness in the market (5), and effective uranium production did not resume again until the JMS scheme was underway. Even then, thanks to problems with the pyrite flotation plant at Brand, the project was at least a year late in reaching design capacity (6). Brand also supplies the acid for the JMS.

References: (1) *MAR* 1985. (2) *R&OFSQu* 26/7/85. (3) *MIY* 1985. (4) *EPRI Report.* (5) Lynch & Neff. (6) *R&OFSQu* 6/5/77. (7) *FT Min* 1991.

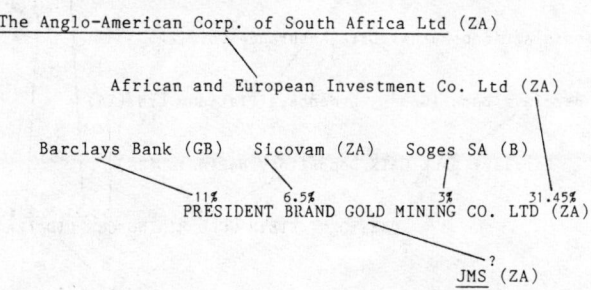

```
The Anglo-American Corp. of South Africa Ltd (ZA)
                                      \
                                       \
             African and European Investment Co. Ltd (ZA)
                                                    \
                                                     \
  Barclays Bank (GB)   Sicovam (ZA)   Soges SA (B)    \
           \11%      6.5%/         3%\      31.45%\
            PRESIDENT BRAND GOLD MINING CO. LTD (ZA)
                              \?
                            JMS (ZA)
```

499 President Steyn Gold Mining Co Ltd

Both President Brand and President Steyn mining companies have leases over the Welkom mining district in the Orange Free State of South Africa; both are members of the JMS; and Brand has a tributing agreement to mine on Steyn's leases (1). Incorporated within about a year of each other, they are also both controlled by Anglo-American through its African and European Investment Co, and both roughly to a one-third equity interest (2).

A significant (but unknown) interest in Steyn is also held by South African Mutual, the apartheid economy's largest life assurance company, and the investment body which also holds a sizeable chunk of Barlow Rand (3).

Up to 1975, Steyn produced a little more uranium oxide than its compatriot company (736 tonnes U_3O_8); during the '70s, its production was going to be resumed "via President Brand" (4), which indeed it was, through its participation in the JMS from 1978 onwards. It has been the supplier of the lowest amount of slimes to JMS (with the lowest uranium "head grade"), having delivered just under 1498 tons in the 9 months ending June 1985 (5). Then, a year later, Anglo American amalgamated the interests of four of its gold companies into the Freegold scheme, headed by President Steyn, which was then renamed Freegold (for further details see AAC) (6).

References: (1) *MIY* 1985. (2) *RMR* Vol. 3, No. 2, 1985. (3) Rob Davies, Dan O'Meara & Sipho Dlamini, *The struggle for South Africa: A reference guide to movements, organizations and institutions,* Vol. 1, London, 1984 (Zed Press). (4) *EPRI Report.* (5) *R&OFSQu* 26/7/85.(6) *FT Min* 1991.

500 Primary Fuels (USA)

It was reported to be exploring for uranium in Texas in 1978 (1); no further information.

References: (1) *E&MJ* 11/78.

501 Public Service Co (USA)

Based in Oklahoma, the company was reported searching for uranium around Florida Mountain, Colorado in 1977 (1); no further information.

This may be the same company as Public Service Electric and Gas Co, involved with El Paso Electric.

References: (1) *MJ* 12/8/77.

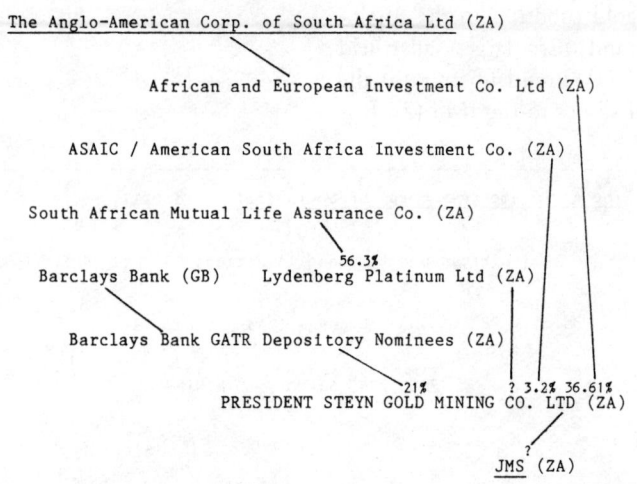

502 PUK

It is hardly surprising that, worldwide, Pechiney (formerly Pechiney-Ugine-Kuhlmann or PUK) has run into more opposition for its aluminium operations than its nuclear interests. It is the fourth largest aluminium producer in the world (1). It is also France's only aluminium producer (2), and the largest in Europe (47).

Moreover, when it acquired American National Can for US$1bn (48) in 1988, it became the world's largest producer of metal drinks cans (49).

In 1990, it also sealed a deal with Techpack International (TPI), the leading producer of packaging for cosmetics, under which it acquired 39% of the company, with an unusual buy-back and buy-out scheme giving LBO of France 44% and management of TPI the remaining 17% (50).

Pechiney is owned 75% by French state interests (10% of which is in the hands of Assurances Générales de France, acquired in 1990 (51,52)). Although plans to privatise Pechiney were high on the agenda (after the group finished restructuring in 1986 (53,54)), the French socialist government has so far applied a brake to both privatisation and nationalisation (51).

By 1988, the company saw an upturn in its fortunes, with the saving of two domestic smelters planned for closure earlier in the decade (53) and construction of another in Normandy (54,55); its nuclear fuels activities proving

"highly profitable" (56); a JV under discussion with the USSR which would be the first of its kind (56); and highly successful returns from its ventures in the USA, especially Howmet Turbine (57).

PUK, as it was BM (Before Mitterand), is a prime example of agglomeration – of bigger fleas with little fleas upon their backs to bite 'em (and so on *ad infinitum*). Pechiney was set up in 1855, began producing aluminium five years later, and – with a spectacular rise in output prior to WW2 – took over several companies on the way. In 1971 it merged with Ugine-Kuhlmann.

For its part, Ugine was formed out of a 1921 merger of three companies which were set up to make industrial use of electrical power. Ugine's specialist knowledge in fluorine chemistry enabled it to play a key part in developing nuclear power (in Comurhex, France's chief producer of uranium hexafluoride, for example). In 1966 Ugine got hitched up with Kuhlmann.

Who was Kuhlmann? Frederic Kuhlmann started manufacturing sulphuric acid in 1825, acquired a chemicals factory in 1847, and laid the ground for the company's expansion into fertilisers, silicates, textiles and petrochemicals. Thus, aluminium, electricity, and chemicals ended up in bed together providing the ingredients for a number of future nuclear and industrial embryos (1).

Between 1978 and 1984 the company was involved in major divestitures of marginally profitable operations, an activity hastened by its nationalisation by the Mitterand government in

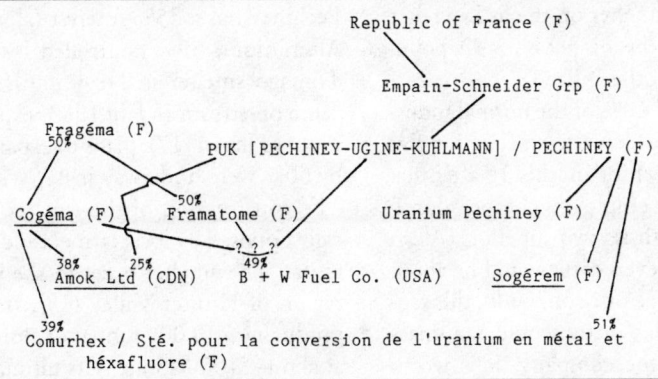

February 1982 when all its shares were allotted to the French state (1).

Spurred by major losses in its aluminium sector and a downturn in production of 6% in 1983 (2), Pechiney expanded its two French smelters, but was squeezing the rest. The same year, it acquired a stake in a "hypothetical" French nuclear power station in return for cheap power to run its remaining smelting capacity, drawn from any stations run by Electricité de France (3,4). Under the chairmanship of Georges ("I hate to lose money") Besse, the new, beaming, loud-talking, joke-cracking Président Directeur Genéral of the company, Pechiney's fortunes were beginning to turn by mid-1984.

Pechiney's chemical assets were sold to Elf-Aquitaine, Rhone-Poulenc and CdF Chemie (5) after Giscard d'Estaing and Mitterand both blocked a potentially lucrative sale to Occidental Petroleum (6). The loss-making steel interests have also been hived off (4). Cash to finance the huge FFr 3,000,000,000 investment programme was to be found in an agreed sale of the Howmet Aluminium Corp to Alumax (7) (a JV between Amax, Mitsui and Nippon Steel) (8). In the event, Howmet remained under Pechiney's control, with Alumax gaining a half interest each in Howmet's Maryland and Washington smelters (9).

This half-sale of Howmet's smelter interests was part of a redeployment of Pechiney's north American aluminium operations from the USA to Quebec (10). In early 1984, the Premier of Quebec, René Levesque, and Pechiney signed a C$1.5M deal, under which Pechiney (with a 50.1% share) gets remarkably cheap power for the new Becancour smelter on the St Lawrence river (2) and the Québecois get both the pollution and the brunt of the cost.

Pechiney was offered 40% of the normal industrial electrical rate to run the smelter for five years, and cheap power for another 15 after that (11). Since Pechiney employs a cheaper aluminium technology than its two superiors, Alcan and Alcoa, its break-even costs would be about 45c/lb; with an average price of 76c/lb, this represents a considerable (many would say unacceptable) profit for the company. It is worth noting that Pechiney began talks on the smelter in 1974 with former Quebec premier Robert Bourassa who paved the way for the James Bay hydro-electric scheme – a project of monumental catastrophe for the Cree and Inuit people of the region (see Bechtel).

After some restructuring of the interests in Bécancour, Pechiney Reynolds Quebec (itself a subsidiary of Pechiney and Reynolds Metals) now owns 20% of the smelter, the other partners being Alumax and the Société Générale de Financement de Québec. The consortium – ABI or Aluminière de Bécancour Inc – started its third potline late in 1990 (12).

In the Netherlands, Pechiney Nederland opened up a controversial smelter in Vlissingen. (Passengers escaping from Olau ferries after collisions with Comurhex nuclear cargoes in the Channel can catch a glimpse of it as they rush to bright lights of Amsterdam). The smelter was the subject of intense public debate, and opposition from environmental groups on health and economic grounds (13).

Environmentalists in New Zealand also fought hard against the siting of a smelter in the beautiful valley of Aramoana, where Pechiney replaced Alusuisse as the chief foreign partner in a consortium headed by Fletcher Challenge and CRA in 1982 (14). But talks over the siting of a power plant for cheap power broke down (15) and the project was shelved (2).

Meanwhile the Spanish government was tussling with Pechiney over who would pick up the bill for losses on the 67%-owned Alugasa aluminium subsidiary (16), and it finally kicked Pechiney out in 1982 (2).

Pechiney has a 35% interest (along with Gove Aluminium, 59% controlled by CSR) in the Tomago smelter in New South Wales which came on stream in late 1983, exporting aluminium to Japan (17): plans to expand the smelter by 50% were underway in 1990 (58).

The construction of the smelter was energetically opposed by local farmers and environmentalists. The smelter is set in the wine-growing region of Hunter Valley (18). A large plant – producing 230,000 tonnes of aluminium a year at about £1000/tonne – its ultimate capacity is more than 700,000 tonnes (18). Pechiney is employing a new, secret smelting process, purportedly replete with environmental controls to remove fluoride, and a new form of waste containment using "excavated cells" covered in two metres of clay (19).

Pechiney also has a 20% stake in Queensland Alumina (Comalco 30.3%, Kaiser 28.3%, Alcan 21.41%) which underwent expansion in the early '80s. Bauxite for the refinery comes from Weipa, the region flagrantly robbed by Comalco from its Aboriginal inhabitants in the '60s, and which continues to be one of the major disgraces of northern Queensland (20).

No less controversial is Pechiney's acquisition in 1981 – along with Billiton Aluminium, a subsidiary of Royal Dutch Shell – of a 40% interest in the Aurukun bauxite prospect in the same area, an aboriginal reserve whose occupants are firmly against bauxite mining (21). To date no firm moves have been made to commence mining, but in 1981 Pechiney announced that it would increase its share in Aurukun to 33.33% (22).

Pechiney participates in Friguia, a holding company which has a 51% interest in alumina production in Guinea (22). The Frialco consortium is owned 30% by Pechiney, 30% by Noranda, with Alcan and Hydro Aluminium holding 20% each (12). Pechiney also mines bauxite and produces alumina and aluminium in Greece (22).

India got Pechiney's technical advice in 1980 when it drew up plans for a bauxite treatment and aluminium complex in Orissa (23).

Plans were announced in early 1984 to build Europe's first lithium plant in south-east France (24). Lithium is used in alloys and batteries, including those which power the Minuteman missiles (2).

Pechiney is also the principal European producer of fluorspar (2).

Before nationalisation, PUK was involved in virtually every activity related to nuclear fission, apart from the commissioning of nuclear power stations (25). It owns 51% of Comurhex which operates uranium hexafluoride plants at Pierrelatte and Malvesi. It also controls 51% of Zirco-

tube, a manufacturer of fuel elements; 51% of Eurofuel; and 50% of Franco-Belge de Fabrication de Combustibles (in which Creusot-Loire is its partner) (47).

Soon after Bernard Pache took over the helm at Pechiney in 1985 from Georges Besse (who had graduated into the company from Cogéma), he began soliciting atomic and other business in Japan, hoping to sell the whole range of Pechiney's nuclear fuel facilities; fuel for light water reactors, fabrication of zirconium products (through its Cezus subsidiary) (47), the production of uranium hexafluoride, and fabrication of fuel elements themselves (59).

Three years later, Uranium Pechiney, together with Cogéma and Framatome, took a 49% share in the US fuel supplier Babcock & Wilcox (B&W Fuel Company).

In 1991, Framatome was negotiating to take control over B&W Fuel, as well as B&W Nuclear Service Company (60).

But its most important nuclear role has probably been as the 50% holder of Minatome, which – under the 1982 reorganisation – was bought out by CFP/Total and merged with Total's subsidiary Total Compagnie Minière (2).

Until 1982, Minatome mined uranium inside France, notably at St Pierre-de-Cantal, using its 94%-owned subsidiary Scumra (25) and producing 100t/year U_3O_8 (26).

Outside of France, Minatome had shares in uranium mines in Namibia (10% of Rössing), Niger (6.7% of the Somair consortium at Arlit), and has been exploring for the deadly metal in the USA, Australia (at Ben Lomond), Colombia, Brazil, Ireland, Britain and Mauritania (25), not to mention Namibia (27).

Uranium mining activities undertaken by Pechiney in its own name include grabbing a share in the lucrative Cluff Lake project, managed by Amok as the controlling partner in Cluff Mining Ltd; Amok itself is owned as to 25% by Pechiney (22).

Lower down the line, the wholly-owned subsidiary Uranium Pechiney took a share in a uranium-from-phosphoric acid recovery plant operated by Gardinier, planned for the early

1980s (28) but which appears to have closed by 1982 (29).

In Algeria the company was studying uranium reserves in 1977 (30, 31); a contract that year for a feasibility study was awarded to Pechiney and Minatome, Sogerem (a Pechiney subsidiary), and Stec (32, 33).

Five years later, Uranium Pechiney won a US$32M contract to provide processing technology, engineering and equipment for uranium-from-phosphates extraction in Tunisia (34), after Gardinier and PUK conducted a feasibility study on the project (35). The unit was to be built at Gabes on the Mediterranean coast (36), but plans for extraction had not materialised by 1984 (2).

The company's most controversial deals have been with South Africa and in South America.

In the late '70s the French nuclear industry won a large part of the apartheid republic's burgeoning nuclear power/weapons programme. The contract for the first South African nuclear power station (Koeberg I) went to a consortium headed by Framatome (controlled by Creusot-Loire which is itself part of the huge Empain-Schneider group that controlled Pechiney) (2). At roughly the same time, the South African government announced an agreement with a consortium headed by PUK, including Creusot-Loire and Westinghouse, to provide uranium enrichment and fuel fabrication facilities (37). This arrangement was superceded with the development of Nufcor's own Pelindaba enrichment plant (38).

The Argentinian military dictatorship did, however, in the early '80s select a consortium headed by PUK to cooperate with the Argentinian CNEA in opening up the Sierra Pintada uranium deposits (39, 40). The following year the USA stopped its own shipments of uranium to Argentina because the military state refused to sign the Nuclear Non-Proliferation Treaty and, within another year, the Soviet Union was sending the country 20% enriched supplies of U-235 in exchange for grain (14).

Also, at the beginning of 1981, Pechiney announced it had won a contract to build Brazil's first uranium hexafluoride plant for Nuclebras (41). The plant, to be constructed at Resende near Rio de Janeiro, would employ Pechiney's own technology and start up in 1985, with an initial production of 450 or 500 tonnes (42). The deal completed Brazil's attempts for a decade (in fact since the West German-Brazilian nuclear pact) to complete the nuclear fuel chain on its own territory (43).

At the same time PUK was assisting Nuclebras to construct the Pocos de Caldas uranium mining complex, specifically the Otsamu Utsumi mine in Minas Gerais which officially opened in May 1982, although production started in December 1981 (44). PUK participated in the actual construction of the mine and provided technical expertise (42).

Although the West German government built Brazil's uranium enrichment plant (45) in late 1983, the Brazilian régime asked Alsthom-Atlantique, another French-state-controlled engineering company, to supply vital compressors for the plant. The Brazilian Minister of Mines and Energy, César Gais, also visited France to discuss with Pechiney the possibility of using a new uranium mining procedure developed by Pechiney.

Uranium Pechiney developed this process to treat high clay ores and dispersed clays containing uranium, gold and other materials not previously economically recoverable. This "physico-chemical" process purportedly transforms clay into porous granular material ready for solid-liquid separation.

It was later reported that both Pechiney and Cogéma were trying to implement a plan to extract uranium and phosphoric acid from open-pit ore at Itataia in Brazil – an "innovative" development since the two are not chemically bound together. The US$300M project was agreed in April 1984 and was intended to process up to 20,000 tonnes a day of ore, producing some 2600 tonnes a year uranium, thus making it one of the more important new uranium ventures.

The deposit, 200km south-west of Fortelaza in Ceara state, has an estimated 80,000 tonnes of contained uranium. Pechiney would be responsible for the project engineering and Cogéma for the purchase of any of the Itataia uranium (46).

An irony, not lost on anti-nuclear groups concerned with weapons proliferation, is that both the West German and French governments have enormously assisted Argentina and Brazil to acquire nuclear weapons although (one might say *because*) the two countries, despite a recent nuclear pact, have long considered the other capable of launching an atomic attack on "their" soil (45).

By the turn of the eighties, Pechiney had established itself as one of the world's most important aluminium producers, its most significant manufacturer of metal cans, and one of the few diversified conglomerates not to have reduced its commitment to nuclear fuel production and processing.

In 1991, it saw its plans to start up a smelter at Nasiriva, in Iraq, dashed by the horrendous conflict between the Saddam Hussein régime and the Bush administration for control of Kuwait (12), and had to shelve plans (formed with Austria Metall, Alumined Beheer and RTZ) to build the Atlantal smelter in Iceland (61).

In Venezuela, an agreement with Aluminium del Caroni SA, the state-owned company, to construct a smelter on the Orinoco river, was shelved (62) for financial reasons (63). But, in 1990, Pechiney agreed a new project with Alisa (Aleaciones Ligeras SA) to operate a Venezuelan smelter, to be constructed by Davy McKee (63).

Contact: WISE (Amsterdam), PO Box 18185, 1001 ZB, Amsterdam, Netherlands. Tel: 31 20 6392681; Fax 31 20 6391379.

Save Aramoana, 320 George Street, Dunedin, Aotearoa (New Zealand).

Sytze Ferweda, Rijn-Schelde Instituut, Dam 47, 4331 Middelburg, Netherlands.

References: (1) *WDMNE*. (2) *MAR* 1984. (3) *MJ* 1/7/83. (4) *FT* 4/5/84. (5) *FT* 2/6/82. (6) *FT* 21/3/81. (7) *MJ* 18/5/84. (8) *MJ* 3/6/83. (9) *MJ* 5/2/82 see also *E&MJ Int. Directory of Mining* 1991. (10) *FT* 18/5/83. (11) *FT* 29/3/84. (12) *MAR* 1991. (13) Information from Sytze Ferweda, see above.

(14) *MJ* 16/4/82. (15) *MJ* 25/6/82. (16) *FT* 5/10/82. (17) *MJ* 11/11/83. (18) *FT* 7/11/83. (19) *MJ* 3/2/84. (20) Roberts. (21) Press release by the North Queensland Land Council delegation to Britain and Europe, London, 1978 (NQLC/CIMRA). (22) *MIY* 1982. (23) *FT* 3/10/80. (24) *MJ* 25/5/84. (25) Dave van Ooyen: *Uraniummijnbouw in Frankrijk*, Amsterdam, 1982 (University of Amsterdam Sociology Institute doctoral thesis). (26) *MM* 11/81. (27) UN NS-36. (28) *E&MJ* 1/79. (29) *MJ* 3/9/82. (30) *MJ* 21/1/77. (31) *MJ* 29/7/77. (32) *MM* 8/77. (33) *MM* 9/77. (34) *Tunis Arab Press* 3/4/81. (35) *MJ* 27/7/79. (36) *Nuclear Fuel* 22/6/81. (37) Jean Bernard, "Note complémentaire sur l'accord franco-sud-africain", in *Esprit*, Paris, 12/76, No. 463. (38) Rogers & Czervenka. (39) *MM* 1/80. (40) *Nuclear Fuel* 31/3/80. (41) *FT* 8/1/81. (42) *MM* 3/82. (43) *MJ* 23/1/81. (44) *MJ* 14/5/82. (45) De la Court, Pick & Nordquist, *Nuclear Fix*, Amsterdam, 1982 (WISE). (46) *MM* 5/84. (47) *FT Min* 1991. (48) *FT* 4/4/90 (49) *FT* 11/4/91 see also *FT* 22/11/88, *FT* 27/10/89. (50) *FT* 20/3/90. (51) *FT* 27/7/90. (52) *FT* 28/9/90. (53) *FT* 4/12/86. (54) *FT* 22/11/88. (55) *Metal Bulletin* 28/6/90, *FT* 22/12/89, *FT* 27/10/89. (56) *FT* 4/3/88. (57) *FT* 8/10/86. (58) *MJ* 16/2/90. (59) *FT* 19/2/86. (60) *FT* 6/9/91. (61) *FT* 5/4/91. (62) *MJ* 4/3/88. (63) *Metal Bulletin* 17/5/90.

503 Quad Metals Corp (USA)

It obtained a 3000-acre lease on the Spokane Indian reservation in Washington state in 1977; no further information (1).

References: (1) *E&MJ* 11/78.

504 Queensland Mines Ltd

Don't be confused by the chart – the Nabarlek uranium (now fully mined) is effectively controlled by Pioneer Concrete. (That's why it's sitting in its best cement boots at the very top!) Queensland Mines, however, also owns uranium mining tenements in northern Queensland (Mount Isa) and has been a 40% partner

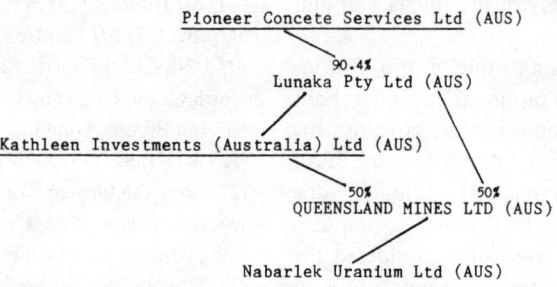

Pioneer Concete Services Ltd (AUS)

90.4%
Lunaka Pty Ltd (AUS)

Kathleen Investments (Australia) Ltd (AUS)

50%   50%
QUEENSLAND MINES LTD (AUS)

Nabarlek Uranium Ltd (AUS)

with AAR, IOL (now CRA), and Urangesell-schaft (Australia) Pty in uranium exploration at Westmoreland (1), where reserves were put at more than 10,000 tonnes in mid-1981 (2).

The Nabarlek mine was officially opened in June 1979 at a banshee involving the unveiling of a plaque which was "chewed from the Arnhem escarpment and carried to the site by a huge grader". Illustrious persons then turned "the first sods", as the *Age* graphically put it. These were the chairman of the Northern Land Council, Galarrwuy Yunupingu, Mr Yamaguchi of the Shikoku Electric Power Company, and Mr K Goto of Kyushu Electric (3). According to the then Australian Deputy Prime Minister, the mine would be "the safest and environmentally the best that could be engineered" (3).

Within a few months, ore samples left by Queensland Mines in the area were found to be 85 times the radiation safety level – a fact which illustrated "the utter contempt with which the mining companies treat the land and the people who live on it", according to the Labor Party MP for Arnhemland, Bob Collins (4).

The Minister for Science and the Environment, Senator Webster, himself reported that radiation levels were "five to ten times higher than predicted" (5) while miners – many not wearing masks – complained about the dust in their faces and lungs (6).

Controversy over the Nabarlek mine had, in fact, started far earlier. The Woodward Commission in 1974 called the offers made to Oenpelli Aborigines soon after the uranium was discovered, "contemptuous". Woodward also made the important comment: "it is to my

mind unthinkable that a completely new scheme of Aboriginal land rights should begin with the imposition of an open-cut mine right alongside a sacred site" (7).

Moreover, when the company claimed they had finally gained Aboriginal "consent" to mine, the official Fox Report (Ranger Uranium Environmental Inquiry) disallowed it (2).

The mine was, however, not only crucial to Queensland Mines but also to the new Liberal (*ie* conservative) Australian government in its desire to open up uranium deposits in the North. Although many of the people in the Oenpelli Aboriginal community were opposed to the mine, one of the key elders – the fledgling Northern Land Council's then chairman, Galurrwuy Yunipingu – was pressured into giving consent (8).

But then, Queensland Mines – and the government – faced sudden Aboriginal opposition, leading to a sit-in on the road to Nabarlek and a court case.

For months the local Aboriginal people had been plagued by company trucks ripping up the road and endangering animals and children. Under section 41 of the Aboriginal Lands Act 1976, miners cannot enter Aboriginal land without a proclamation by the Governor-General, and through an oversight the proclamation had not been given.

But when the Aborigines took their case to court in February 1980, the Fraser government announced it would introduce legislation to *retrospectively* change the law. Now, instead of traditional Aboriginal land owners being able to challenge parts of agreements made by their (official) negotiating body (the Northern Land

Council), they would have to abide by it, even without their consent (2).

Within a year, the Nabarlek deposit was at production stage (9), and the total reserves had been "upped" to 12,000 tonnes (10). Production from the stockpile was expected to be 1000 tonnes a year (9). Of the total reserves, by July 1981 nearly 3000 tonnes was contracted to Japan on a 1975-85 contract (the Kyushu and Shikoku Electric Power Companies); about 1400 tonnes to South Korea (contract details not determined) (11); and just over 800 tonnes contracted to the Finnish Toellisuuden Voima Oy Industrins Kraft, between 1981 and 1989 (12). Some uranium was also due to be returned to the Atomic Energy Commission stockpile, to cover uranium borrowed (13).

Out of all this, Queensland Mines made net profits of more than £3M in 1980 (as against a loss of A$62,000 the year before) (14). The Aborigines, through the Northern Land Council, received a mere 5.76% royalty (15).

Queensland Mines' interests in uranium at Westmoreland (northern Queensland) have also attracted the strong opposition of the Aboriginal custodians of the land: Mick Miller, Joyce Hall and Jacob Wolmby of the North Queensland Land Council (a completely Aboriginal body) demanded that Queensland's partners at Westmoreland – Urangesellschaft – should get off their land when the NQLC visited Urangesellschaft's offices in West Germany in 1978 (16).

In August 1980, Darwin waterside workers, implementing the ACTU policy against uranium exports, voted for a ban on the exports of Nabarlek uranium. This ban was successful until 1981, when the first consignments of uranium (for Japan) were shipped out of Darwin on a barge from the Philippines (17). Eight "wharfies" were arrested as they picketed the barge (18). In late 1981, the ACTU lifted its ban, and Nabarlek uranium was expected to be exported in early 1982.

In early 1980, *Nuclear Fuel* speculated that part of the uncommitted Nabarlek production might go to Commonwealth Edison of the USA (19).

The company has two uranium prospects in the Mt Isa area of central Queensland: the first at Skal where inferred reserves are just over 3,000 tonnes of contained U_3O_8, the second at Valhalla where they are somewhat greater. It also owns the Anderson lode 15km north-east of Mt Isa where inferred uranium ore reserves are just over 1000 tonnes (21).

Latest news on Westmoreland is that further prospecting has failed to significantly extend the known mineralisation, with the result that "only sufficient works" will be undertaken to enable the partners to maintain the lease (CSR holds 12.65%, IOL 9.75% and Urangesellschaft Aus 37.5%). Reserves of about 11,400 tons have been identified, but the project would require at least 15,000 before development was justified (22).

It is interesting to note that Kathleen Investments was originally incorporated in 1958 to take over the 35% share which Australasian Oil Exploration had in Mary Kathleen Uranium. The final disposal of Kathleen's shares in Mary Kathleen Uranium didn't occur until 1979 (23). Nearly ten years later, in 1967, Kathleen Investments floated Queensland Mines in order partly to consolidate its hold on uranium – and in particular the Westmoreland prospect, then regarded as the country's largest uranium lode. Queensland Mines then held 40% of the Westmoreland JV (along with Mines Administration, the operating company for AAR, which held 12.75%, IOL Petroleum, which held 9.75%, and Urangesellschaft, operator of the now dormant prospect, with 37.5%) (24). In 1974, Kathleen Investments was called "an affiliate of ICI London" (25).

In 1987, Narbalek produced 3 million pounds of U_3O_8, and the mill was due to close down by the middle of 1988, with delivery commitments over the following few years met by existing inventory. Although the company has not been able to develop further deposits in the area, thanks to opposition from the Northern Land Council (26), Pioneer Concrete announced in March 1988 that it had obtained permission for further exploration (27). This was the first agreement signed between

Northern Territory Aborigines and a mining company, after the passage of amendments to the Land rights Act in 1987 (28). Called a "triumph for good will and common sense in the negotiating process" by NLC chair, Yunupingu, the deal gave Queensland Mines exploration rights under "strict and sophisticated provisions for environmental protection" (29) over Exploration Area 2508.

However, plans to mine this new area – Nabarlek No 2 orebody – had stalled until 1991 (30,31), as the traditional owners failed to negotiate a mining permit. In the meantime, the company reported that it had concluded a "farm-in agreement" with a major uranium producer, to purchase the Nabarlek mill and pay for any future development (30). This "arm in arm" (32) agreement was later revealed to have been with Cogéma, which already owns shares in ERA (Ranger) and Pancontinental (Jabiluka).

Whether the second Nabarlek deposit can be deemed part of the original orebody, and thus fall within the "three mines" policy of the Federal government, is – as the Friends of the Earth magazine put it in 1991 – "disputable" (33).

References: (1) *MIY* 1981. (2) Roberts. (3) *Age* 9/6/79. (4) *Australasian Express* (quoting AAP) 7/9/79. (5) *Northern Territory News* 5/6/79. (6) *Northern Territory News* 25/5/79. (7) Woodward Commission, Second Report, 1974. (8) *The Australian* 15/9/79. (9) *FT* 17/6/80. (10) *West Australian* 31/10/79. (11) Federal Australian Government statement, 7/81. (12) *FT* 14/8/80. (13) *FT* 25/8/81. (14) *FT* 21/2/81. (15) *FT* 13/3/79. (16) Roberts p.134; *Pogrom*, Germany, 1979. (17) *Darwin Sun* 15/7/81. (18) *CANE Newsletter*, Adelaide, 8/81. (19) *Nuclear Fuel* 3/3/80. (21) *AMIC* 1982. (22) *MJ* 17/12/82. (23) *MIY* 1982. (24) *Reg Aus Min* 1983/84. (25) *Slow Burn* (FoE) Carlton, Australia, 9/74. (26) *E&MJ* 4/88. (27) *FT* 25/3/88. (28) *Lands Rights News* Vol. 2, No. 3. (29) *Lands Rights News* 7/88. (30) *MAR* 1991. (31) *Reg Aus Min* 90/91. (32) Sydney Morning Herald 24/12/90. (33) *Chain Reaction* No. 63/64, 4/91.

505 Quivira Mining Co

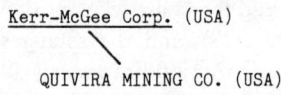

```
Kerr-McGee Corp. (USA)
            \
   QUIVIRA MINING CO. (USA)
```

It was set up in 1983 to consolidate all the Kerr-McGee uranium mining and milling operations in New Mexico (1).

References: (1) *E&MJ* 11/83.

506 Radium Hill Mine (AUS)

Australia's first uranium mine, and a producer of radium for the Curies (1), Radium Hill lies 140km north-west of Broken Hill in outback South Australia. Opened in 1906 as a small mine, it was abandoned before World War One (2). After World War Two, when uranium was required for British and US nuclear weapons, the mine was reopened. Together with production from Rum Jungle (CRA), MKU, and South Alligator River (in the Northern Territory) around 7600 tonnes of uranium was produced between the mid-'50s and mid-'60s for the CDA (Combined [US/British] Development Agency). By 1964 Radium Hill closed for the last time (3).

Radium Hill was operated on land previously occupied by Aboriginal people, and its yellowcake was incorporated (along with Australia's other source material) into nuclear weapons later tested on Aboriginal land and people, in the same region. From the cradle to the grave – with a vengeance!

A July 1979 New South Wales government report on former workers at Radium Hill showed a cancer-related death rate four times the national average. Since 1960, said the report, 59% of the miners who worked underground at the mine over a period of two years or more have died of cancer. The Health Ministry – while unable to confirm or deny the report – was able to trace 600 of the former 3000 Radium Hill employees: 40% of these had already died of cancer (4).

In 1982 the South Australian Health Commission and Adelaide University commenced a study of ex-miners at the Radium Hill mine. In July 1986, the Commission issued a progress report. Pointing out that only 20% of the mine's workers (out of the 60% of the workforce it traced) had died up to that time, the Commission commented that it was "far too early to draw inferences about the effects of radiation exposure on lung cancer rates" (5). Although there was a relatively high proportion of deaths from "injuries and poisonings", the Commission said that "it should not be concluded ... that working in Radium Hill was especially hazardous" (5). More lung cancers were attributable to the workforce than in the community at large, but the Commission could not decide whether these were due to radon exposure in the underground workings, or the heavy rate of smoking. Moreover "significant numbers of Radium Hill miners were immigrants, some of whom may have had uranium mining experience in high-exposure mines in other countries" (5).

The Radium Hill study was finally published in 1991, nine years after its inception. It dismissed some of the complacent interim conclusions of 1986, and concluded that radiation may have contributed to premature deaths among the workforce. The Federal Industrial Relations Minister, Peter Cook, held out the possibility of compensation to 56 families of victims of Radium Hill (6).

References: (1) Moss. (2) Roberts. (3) Mary Elliott (Ed), *Ground for Concern: Australia's uranium and human survival*, Ringwood, 1977 (Penguin Australia). (4) *Kiitg* 6/80. (5) SA Health Commission Epidemiology Branch, Radium Hill Study Progress Report, 7/86. (6) TNT, London 22/7/91.

507 Ranchers Exploration and Development Corp

In the late '70s, Ranchers was aggressively ex-

RANCHERS EXPLORATION AND DEVELOPMENT CORP. (USA)

MERGER

Hecla Mining Co. Ltd (USA)

panding its uranium explorations in New Mexico, Colorado and Utah (1). In 1978, it was also involved in joint uranium exploration in the USA with Urangesellschaft of West Germany (2).

In 1978, two men were killed at the Ranchers uranium mine near Grants, New Mexico. A Federal inspector claimed "numerous safety violations" had been found at the mine during a month-long inspection, but most of them had been corrected (3).

In 1979, the company earned "record" returns from its operations – 63% gross of which came from uranium mining. However, in February that year, the Colorado Department of Health refused to issue a nuclear materials handling licence covering the Durango leach tailings operation at Naturita, Colorado, and it has closed indefinitely (4).

Also, in August 1979, the company stopped operations at the Small Fry mine, near Moab, Utah, having accumulated an ore stockpile of approximately 294,000lb which in 1980 was still awaiting milling (4).

At the company's major mine, Johnny "M" – jointly owned with HNG Oil – production slowed considerably in 1978/79 due to an accident, and delay in bringing another orebody into production. Reserves estimated at 3.1 million pounds (4). Ore from Johnny "M" is processed by Kerr-McGee (4).

During 1979/80, Ranchers also received US$3.6M royalties from properties it owned at Ambrosia Lake, New Mexico, which were mined by Kerr-McGee and United Nuclear (5). The Hope mine, jointly owned with Chaco Energy was "mined out" during 1981 (6).

In 1982, the majority of the company's earnings came from its uranium sales, although the company closed its Johnny "M" mine that year, having exhausted reserves.

Uranium was intended to provide the company's largest share of earnings in 1983, too, and – commented the *Mining Journal* – the company "... has been successful in negotiating a number of shrewd contracts for its uranium by guaranteeing satisfactory prices at a time of severely depressed markets" (7).

However, the company produced no uranium in the last quarter of 1982, having met its contracts by purchasing some 90,000lb from elsewhere (8).

Ranchers acquired Hecla in 1984 (9). The Hecla/Union Carbide uranium-vanadium mill near Moab, Utah, remained on stand-by during 1983/84 (10).

References: (1) *MJ* 1/12/78. (2) *MIY* 1979. (3) *Albuquerque Journal* 8/12/78. (4) *MIY* 1981. (5) *MAR* 1981. (6) *MJ* 1/2/80. (7) *MJ* 31/12/82. (8) *MJ* 14/1/83. (9) *E&MJ* 4/84. (10) *MAR* 1984.

508 Randfontein Estates Gold Mining Co Ltd

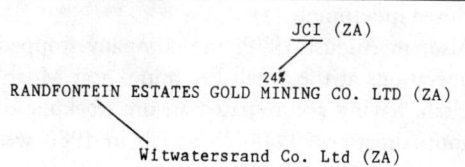

```
                          JCI (ZA)

                          24%
RANDFONTEIN ESTATES GOLD MINING CO. LTD (ZA)


            Witwatersrand Co. Ltd (ZA)
```

One of the more substantial and long-standing of South African uranium-from-gold producers, the company extracted uranium from ore obtained in the three Cooke shafts near Westonaria town, and the SD32 shaft near Randfontein town (1, 2).

Mining operations for uranium on the old Bird Reefs ceased in 1964 after sales contracts were met and a quantity of uranium stockpiled. Eight years later the uranium treatment plant at Millsite was closed down and put on a care and maintenance basis. Sales of existing uranium were completed in 1974 (1).

Then, in 1976, the company decided to reconstruct and reopen the Millsite plant – which was commissioned in August the following year under a new expansion programme (3).

The company also recommenced selling

uranium (4) and secured a contract with the French CEA to supply 10,000 tons from 1981 to 1990 (5). The contract was said to be worth US$103M, which the French government loaned interest-free (6).

In 1978, a new integrated gold and uranium plant was commissioned at the Cooke section of the mine, thus allowing the Millsite plant to be used in processing underground ore from the Randfontein shaft, and for the treatment of past and current "tailings" (1).

In March 1980 it was reported that this major new £225M development was proceeding on schedule; that total reserves were 443 million tonnes of ore grading at 0.153kg/ton uranium and 0.332kg/ton uranium. The Cooke No. 3 shaft was commissioned 14 months ahead of programme, but another modification costing R7.6M was commissioned for completion in 1984, which would add 50,000 tons a month to the Millsite plant's treatment capacity. Total mill throughput by early 1985 was expected to be about 450,000 tons a month of underground ore, less overall tonnage and at a higher cost, but "partly offset by a marginal increase in recovery grade as the treatment of surface dump material falls away" (2).

A 1984 report from JCI showed that tonnage treated decreased by 74,000 tons to 818,000 tons, but the yield increased from 0.14 to 0.16 kg/ton (7).

Randfontein Estates also entered into an agreement with its parent, "Johnnies", to prospect for, and exploit, uranium deposits in the Karoo basin of Cape Province and in the Orange Free State (8).

In mid 1988 Randfontein announced a restructuring which involved the formation of two new companies to develop both its old underground Randfontein section and to take over its exploration interests. The first was to be called Lindum Reefs and the second, Barnato Exploration. Because of US legislation prohibiting new investment in the apartheid state at that time, JCI offered north American investors cash instead of a rights issue. Commented the *Mining Journal*: "JCI is to be applauded in enabling shareholders a clear choice of investment. However, it is doubtful whether the needs of minority shareholders were the only consideration in the restructuring plans. Randfontein is in the middle of a major expansion at Doornkop and has suffered recently from labour unrest [in 1984 thousands of miners were dismissed after a bloody confrontation with police (9)] exacerbated by the increased mechanization at the mine. With a sharply reduced dividend ... and its considerable capital expenditure commitments, the company would clearly have found it difficult to raise the R17/share it is looking for in a straightforward rights issue" (10).

Three years later, the Cooke Section shaft was completed, and the millsite uranium and gold plant "up to full capacity" (11). However in 1990, Randfontein cancelled its contract with the CEA, which it then transferred to an "unnamed South African producer" (12).

References: (1) *MIY* 1982. (2) *MJ* 6/4/84. (3) *MM* 10/78. (4) *MJ* 11/11/77. (5) NUS Corp Foreign uranium supply update 1980. (6) *Times* 29/12/77. (7) Analysis of Rand and FS quarterlies, in *MJ* 4/84. (8) *FT* 31/3/83. (9) *FT* 25/1/86. (10) *MJ* 1/7/86. (11) *E&MJ Int. Directory of Mining* 1991. (12) *MAR* 1991.

509 R and M Engineering (USA)

It was reported active in uranium exploration in Alaska in the late 1970s (1).

References: (1) *E&MJ* 11/78.

510 Rand Mines Properties Ltd

When Barlow Rand acquired various mining interests from Rand Mines, which it then subsumed under its subsidiary Transvaal Consolidated Land and Exploration Co Ltd (1), it also took over Rand Mines Properties, a scheme initiated in 1966 to recover gold from sand dumps, and later uranium from slimes (2). The uranium scheme never got off the ground (or out of the dumps) and, indeed, by the end of

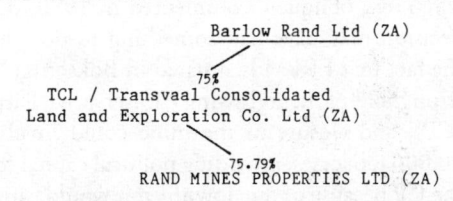

Barlow Rand Ltd (ZA)
↓ 75%
TCL / Transvaal Consolidated Land and Exploration Co. Ltd (ZA)
↓ 75.79%
RAND MINES PROPERTIES LTD (ZA)

1983 the chairman of TCL was calling by-product uranium itself into question.

Now owned 74% by Barlow Rand (4), Rand Mines in the period 1988-1991 was perhaps the severest hit of all apartheid gold producers, due to workers' protests, dwindling production and falling market prices. Three of its four gold producers yielded heavy losses in 1990 (5); Blyvooruitzicht, with a remaining "life" of only three or four years (6); Harmony, having sustained considerable cutbacks, and ERPM (East Rand Proprietary Mines) being completely written off by Rand in 1989 (7). In addition, the 50% owned Vansa group of vanadium companies have been "extremely adversely affected" (8) by the glut in the market, while the new Crocodile River platinum project was five months behind schedule in 1990 (though nearing full production early the following year) (8). With its Middelburg steam coal operations (bought from BP South Africa in 1989, in the face of competition from AAC, and the country's third biggest exporter) (9), as the one bright spot on its balance sheet (8), Rand Mines set about streamlining its gold operations. The Harmony mine's production was cut back, after "unrest" (10) among the NUM workforce.

ERPM – described in mid-1990 as "the most precarious of 18 marginal mines in South Africa" (11) – was forced to seek government financial backing after huge losses (12). The government procrastinated. There was criticism that Rand was keeping older parts of the mine open, instead of developing newer seams. And, although Rand manages the operation (with a 29.5% equity stake), ownership is firmly in the hands of foreign (mainly French) investors (13).

Finally, in early 1991, the South African government bailed ERPM out (14). As the *Finan-*

cial Times obliquely commented in 1990, this decision may have had something to do with the fact that ERPM is situated in Boksburg: "a stronghold of the rightwing Conservative Party (CP), and closure of the mine could involve 10,000 job losses, providing political capital for the CP because of the downturn it would cause in the local economy" (15).

References: (1) *FT* 10/5/83. (2) Lynch & Neff. (3) *FT* 25/11/83. (4) *MJ* 24/5/91. (5) *MJ* 19/4/91. (6) *FT* 18/7/89. (7) *FT* 16/1/91. (8) *MJ* 11/1/91. (9) *FT* 19/12/89. (10) *MJ* 19/10/90 see also *FT* 7/11/90. (11) *Ind. on Sunday* 20/5/90. (12) *MJ* 8/6/90, *MJ* 6/7/90, *FT* 18/4/90. (13) *FT* 7/9/90. (14) *FT* 18/7/91. (15) *FT* 7/6/90.

511 Ranger Mine

"All of us wish to hell that uranium had never been discovered. But ..." (1).

"We're getting closer and closer to blood on the streets of Australia on this issue" (2).

"Facing the uranium issue is what may save Australia and set an example to the world" (3).

"We wait with trepidation for the day the Government advocates a heroin and cocaine industry in order to combat drug trafficking" (4).

Note:
Information about Australia's most important uranium mine is spread between this entry and ERA. Readers looking for information on the political and economic context in which the mine was given the go-ahead, the formation of

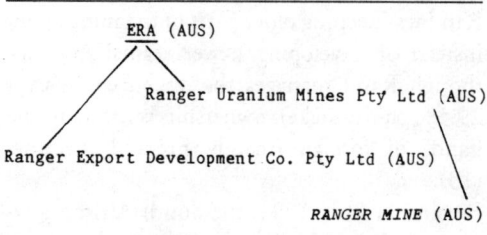

ERA, the production figures for the mine, and the contracts concluded, should refer to the ERA entry. Readers searching for information on the public debate surrounding the opening-up of the Northern Territory to uranium miners; the Ranger Commission; the reaction of Australian Labor Party and trade unionists; the effect of the mine on its workforce and the environment, and above all on Aboriginal people, need look no further than the following pages. The Ranger mine went into production with remarkable swiftness (some would say obscene haste) amid the most important debates ever held on the uranium issue anywhere in the world. For Australians – as the quotes heading this essay illustrate – it proved almost the most divisive political issue during the post-war era (second only to Australian participation in the Vietnam war). Formal consent for Ranger was only granted after the government took into consideration the findings of the most comprehensive and far-reaching public inquiry ever mounted into the consequences of uranium mining.

Background

The Ranger Uranium Environmental Commission (or the Fox Commission as it was also known) was set up by the Labor Minister for the Environment, Dr Moss Cass, who insisted on a public inquiry into the environmental aspects of uranium mining, using his powers under Section II (I) of the Environmental Protection (Impact of Proposals) Act of 1974 – the instrument for undertaking environmental impact inquiries (5). Cass was vehemently opposed to nuclear power (6).

The Commission was appointed on July 16th 1975, issued its First Report on the general issues of uranium mining, proliferation, waste disposal, and Australia's role as a nuclear producer, in October 1976; the Second Report, concerned with the Ranger project, was released in May the following year.

Russell Walter Fox, then Senior Judge of the Supreme Court in Canberra, was made Chair, with Graeme George Kelleher, and Charles

Baldwin Kerr (Professor of Preventative and Social Medicine at the University of Sydney) as his two Commissioners. Eight part-time specialist advisors on various aspects of uranium mining (technological, socio-economic, legal, etc) were also appointed.

From the opening of the Hearings in Sydney in September 1975, until their conclusion in March 1977, the three Wise Men sat for 121 days, listening to 303 people, scanning 419 documents, and finally producing more than 13,000 pages of transcript (5).

Within several weeks of the opening of the Ranger Inquiry, the Governor-General of Australia, Sir John Kerr, in an unprecedented abuse of authority, dismissed the Whitlam administration; by mid-September, a Liberal-Country Party coalition government was in power, by and large firmly in favour of uranium mining, and it lifted the embargo on new mines and contracts which had been imposed at the turn of 1972/73 (7). The new government ordered the Inquiry to report by June 1976: the Commissioners refused (8).

The Ranger partners were estimated to have spent some A$500,000 on their Environmental Impact Statement (EIS) and even more on representation at the Ranger Inquiry, where they had staff constantly in attendance, and often a full-blown QC (Queen's Counsel) (9). For its part, the main opposition group, Friends of the Earth (FoE), relied on members' goodwill, while a request for A$5000 was rejected in 1975. The Australian Conservation Foundation (ACF) was able to afford no more than a 2-3 weeks' appearance, and then mainly to address issues around the Kakadu National Park. No other groups had any substantial representation, although submissions by trade unionists, peace groups, churches, Aboriginal support groups, and others, were numerous (9). The Fox Commission gave a guarded approval for the mining and export of Australian uranium. It also recommended the setting up of a Uranium Advisory Council to "advise the government on overall problems and impacts of the uranium industry" as well as to investigate a resource-based tax on future uranium earnings

(10). This committee was finally assembled in November 1978, under the chair of Sir Lawrence McIntyre, director of the Australian Institute of International Affairs, with a representative of the Australian Council of Trade Unions (ACTU), Galurrwuy Yunupingu, chair of the (Aboriginal) Northern Land Council (NLC), and various other academics and theologians (11).

Responses to the Ranger Report and the government's reactions to it came from almost every quarter of Australian political life.

- In late 1976, after publication of the First Ranger Report, Ian Henderson, Teaching Fellow at the School of Science, Griffith University, maintained that the Fraser (Liberal-National Party coalition) government had rejected the Ranger Commission's main recommendations regarding the dangers of nuclear power, and should be criticised accordingly (12).
- Sir Charles Court, State Premier of Western Australia later retorted that the Commissioners had indeed given the green light to mining, and the "greenies" were simply being obstreperous, as was to be expected of them (13).
- Mr H Melouney, a former General Manager of MKU, and a spokesperson for the Uranium Producers Forum, claimed the Commissioners had given a complete bill of health to uranium mining, and that many opponents were "... really using this issue to pursue their aim of changing Australia's democratic way of life" (sic) (14).
- Edward Teller, the so-called "Father of the H-Bomb" in May hoped "... to God [Australia] will have the determination to ... produce ... and export uranium oxide" (15).

After the Second Ranger report was released, the debate reached a new pitch of intensity. Mining shares slumped (16), as spokespeople for the Uranium Moratorium, which included a wide range of organisations (17), FoE, and the Federal Council for the Advancement of Aborigines and Torres Strait Islanders (FCAATSI), declared it represented a setback for the mining industry (18).

The Age said the Commissioners had given an

"amber" light to mining (18). The Movement Against Uranium Mining (MAUM), commented that it was not a "bad" report (18), while the Director of the ACF viewed it as a "mixed bag". *The Australian* commented (rightly) that the Commissioners had clearly come out in favour of sequential development – one mine at a time (16). Already there were some hints that Prime Minister Fraser might give Ranger the go-ahead without allowing a full parliamentary – let alone public – debate (19). In a more perspicacious analysis, the *National Times* pointed out that the federal government now appeared to accept that the country's sovereignty would be "subservient to US policies". "As with Vietnam in the 1960s," said the magazine, "so with uranium in the 1970s." It also commented that the Reports were a "godsend" to the ALP, giving this divided party time in which to reconcile its anti- and pro-uranium factions. Thirdly, commented the *Times*, Aboriginals had now been elevated "to possibly the most powerful position they have held in Australian society" (20).

In March 1977, Justice Fox himself had expressed disappointment at the level of public debate on his Commission's findings. But once the Federal government announced the go-ahead for Ranger, in late August 1977, arguments flew thick and fast. The *Sydney Morning Herald*, while applauding the decision, pointed out two disturbing implications in proceeding with the mine: the continuing dangers of proliferation, and unsolved problems associated with waste disposal (21).

Tom Uren, announcing that the "battle has only just begun" warned, naively, that the industry should not "expect any mercy from a [future] Labor government" (21). Until all demands on waste and proliferation were met, he commented further, "not one sod of earth will be turned" (22). Former Prime Minister Gough Whitlam compared the marketing of uranium "on a global scale to the marketing of thalidomide" (23), and demanded a full public debate on all the issues. His call was echoed by the Australian Union of Students, and when Fraser visited the University of New South Wales on

the day the decision was made, he was met by huge demonstrations (23, 24), opposed by police in riot gear, the like of which had not been seen "since Vietnam" (25).

As shares in uranium companies – which rose after the government announcement – began to fall (26), the Vice-President of the Society for Social Responsibility in Science, Dr Mark Diesendorf, declared that uranium mining would spawn a plutonium (Fast-Breeder-Reactor) society – the very thing the government wanted to avoid (23, 27). Dr Hugh Saddler, a research officer to the Ranger Inquiry, said that a moratorium should be announced (28). Dr Ken Dyer, senior lecturer in social biology at the University of Adelaide, compared radiation hazards to terrorism and war (28). Perhaps more significantly, Professor Charles Kerr expressed the opinion that the government's "safeguards" policy was "virtually useless" (29). Inevitably, public reactions tailed off, but later in 1977, no fewer than 418 people were arrested in Brisbane, as they sought to defy new repressive legislation introduced by the Bjelke-Peterson régime in Queensland, banning primarily anti-uranium marches (30).

The ALP and uranium

"The lesson is that uranium is a running political sore within the ALP. It will not disappear. Almost every year another battle will need to be fought; its potential to sap the unity and extend the divisions in the party should not be underestimated" (31).

In November 1972, in an historic speech, the Labor Prime Minister, Gough Whitlam, had promised Aboriginal land rights. Within a few months, Justice Woodward was appointed Aboriginal Land Rights Commissioner to investigate and advise on what such a revolutionary step would mean in practice. Woodward recognised the sham of "terra nullius" (the idea that Australia had been a barren land on colonisation), agreed that the whole of Australia had been in Aboriginal occupation at the time of conquest, recommended the appointment of Land Trusts to hold title, and Land Councils to consult with Aboriginal traditional owners, and advised the transfer of Aboriginal reserves and Missions to Aboriginal freehold control (32). Woodward also stipulated that Aborigines, as well as enjoying control over access to their land, should benefit from mining royalties.

As Woodward was preparing to make his Inquiry, the Australian Labor Party (ALP) naturally agreed that uranium mining should be delayed until his recommendations could be implemented (33).

Five years later, in 1977, two years after the Whitlam government was toppled, and one year after the Aboriginal Land Rights (Northern Territory) Act, 1976, received the assent of the new National-Liberal coalition government, the ALP declared an unequivocal ban on uranium mining, its processing and export, and on any activity associated with the so-called nuclear fuel "cycle". But, as the ground was laid for the general election in 1983, and right-wing forces within the party saw both the potential for winning over dissatisified Liberal (conservative) supporters and the glint of some almighty new dollars just over the nuclear horizon, slowly but inexorably, the policy was changed. Between 1982 and 1984 there was indeed a "vast shift" (34) over the middle ground, as both Cabinet and parliamentary party moved from deciding how to implement a uranium ban, to whether uranium should be mined, to how much should be mined. By the time the ALP came to power in March 1983, this was already "the single most contentious" issue facing the government (35).

The new Prime Minister, Bob Hawke (formerly head of the ACTU) soon gathered a strong pro-uranium group around him, including Peter Walsh (Resources and Energy), John Dawkins (Finance), and Bill Hayden (Foreign Affairs). Hayden – formerly one of the most vocal anti-uranium left-wingers – was soon to publicly disavow his previous stand, and earn himself enormous wrath from some of his erstwhile supporters at the historic 1984 ALP Conference.

It was the Victorian ALP State Secretary, Bob Hogg, who, at the ALP's conference in 1982, spelled out the guidelines which were later to be incorporated into party policy, and have permitted the Ranger, Nabarlek and Roxby Downs mines to open. Initially presented as setting conditions which could not be met (37), Hogg's amendments not only enabled the government to purport to be phasing out its nuclear activities, by actually continuing them: more seriously, it enshrined compromise-of-principles ("pragmatism") as the *modus operandi* of the Labor Party-in-power. Radicals within the party who realised this fact in the early days deserted to bolster the ranks of the movement against uranium mining, or form the nucleus of what was to become the Nuclear Disarmament Party, which now has one member in Parliament (Jo Vallentine). (People for Nuclear Disarmament, an umbrella group of church people, pacifists and left-wingers, was formed in 1981) (36). Others – notably Tom Uren MP – remain within the ALP, no doubt believing (as Winston Churchill once argued about "democracy") that, while the ALP was the worst form of Labor-in-power, all other possibilities were worse.

Hogg's 1982 amendment stipulated that before uranium contracts were made with various countries (France, Finland, West Germany, Japan), certain conditions must be met relating to non-proliferation. Vague resolutions as to improved health, and waste disposal techniques, were also incorporated. And, while the amendment stressed that there should be "no new mines", it gave the go-ahead for Roxby Downs, by "consider[ing] applications for export of uranium mined incidentally to other minerals" (37). Tom Uren ("[this marks] the capitulation of this great party on this issue") and Mr Stewart West, the ALP spokesperson on the environment ("it is an irresponsible document, it is not even credible. It endorses existing policy and then goes on to repudiate [it]") led the attack at the conference (37, 38). But the motion passed by 53 votes to 46. The ALP, when in power, would therefore be able to justify exports to certain countries, at least until the 1990s (38).

Eight months after forming the new government, the Cabinet met to decide its policy in the light of the 1982 compromise. It then formally gave the go-ahead to Roxby Downs (see WMC), set up an inquiry into safeguards, and agreed to defer new contracts until mid-1984. However, ERA would be allowed to meet two new contracts with US utilities (see ERA) (39). Early in 1984, FoE said it might put up anti-uranium candidates to oppose ALP members who supported mining, as the ground was laid for that year's conference (40), and the right-wingers within the ALP moved to scotch the left (41). Senator Walsh circulated a draft policy which would allow the export of yellowcake "under strict conditions", refuse to supply France because of its continued Pacific nuclear testing, and investigate/monitor wastes/health/ proliferation etc (42, 55). Walsh claimed that waste problems had already been solved (43): a claim that was rubbished by Tom Uren (43, 44). Indeed, the report on uranium mining – the ASTEC report – promised by the government in 1983, which came out the following month, contained a submission from the Department of Resources and Energy stating clearly that waste problems had *not* been solved (45). In April, the Victorian branch of the ALP decided to adhere to existing anti-uranium policies (46), and attacked the Walsh draft as an "outrageous abrogation of party policy and principles" (46). The ALP Women's Committee, and the National Labor Women's Conference, also unequivocally held to existing bans (47); predictably, Senator Don Chipp, leader of the Australian Democrats, condemned uranium on both economic and moral grounds (48).

The ASTEC report – otherwise known as the Slatyer report, after its chair, Professor Ralph Slatyer – which came out in May, claimed that Australia's role in preventing proliferation would be strengthened by resuming uranium exports. It said there was sufficient technology around for waste problems to be solved; and it recommended participation by the country in other aspects of the nuclear fuel process [such as enrichment] (49). Senator Howe, the Minister for Defence Support, brought out his own re-

joinder to Slatyer, while a people's inquiry into uranium mining, led by Professor Keith Suter, also clearly gave grounds for the immediate cessation of uranium activity (50).

As the 1984 ALP Conference drew near, Bill Hayden, the Foreign Minister, finally ditched all pretence of an anti-uranium stance to come out in favour of mining (51). Meanwhile, some 116 ALP branches expressed their opposition to any dilution of policy (52) and Patrick White, the Nobel novelist (and probably Australia's most famous cultural figure), claimed that the government had lost touch "with the simple needs of the average human being". He promised to support the incipient Nuclear Disarmament Party (53).

In June, the ALP finally voted to overturn its two-year ban. While promising that only three mines would be allowed to export (Nabarlek, Ranger and Roxby Downs) "under the most stringent nuclear non-proliferation conditions, to those countries which the government is satisfied observed the NPT" (54); while banning uranium to France until it ceased nuclear testing; while reserving the right to halt any future supplies to countries not conforming to NPT safeguards; while promising that no other aspects of the nuclear fuel process would be situated in Australia; and while urging strengthening of the IAEA inspectorate, and "encouraging" the development of "passive energy technologies", the import of the resolution was clear: Australia was back in the uranium business with a vengeance (55, 56). At a meeting of fourteen South Pacific Nations (the "Forum") in 1985, Vanuatu and the Solomon Islands questioned Australia's pro-uranium policy, but apparently accepted the "safeguards" proposed (57).

The weakening of the already diluted anti-uranium position has been obtained largely through the capitulation of key figures like Hayden, the South Australian Premier John Bannon, and the South Australian delegate Peter Duncan (who tried to argue for the opening up of Roxby, while urging the closure of the mines in the Northern Territory) (55, 56).

It was therefore scarcely surprising that, in

1987, the right-wing of the ALP, headed by Minister for Primary Industries and Energy, John Kerin, sought to weaken the ALP's fatally flawed policy even further, by arguing that the 1984 "three mines" policy allowed a fourth new mine to be opened, when Nabarlek was exhausted (58).

As if to lend support to the "three mines policy", in 1988 the Northern Territory branch of the ALP reversed its anti-uranium policy and agreed to allow continued operation of both Nabarlek and Ranger. Warren Snowden, a former employee of the NLC, supported the motion, although personally still opposed to uranium mining (59). Thus the ALP's left wing reached the point of a final policy erosion, with the savage irony that, in order to hold back the pro-uranium lobby, it had to be seen to be "reasonable" and realistic, and itself pro-mining.

Uranium and the ACTU

The Australian Council of Trade Unions (ACTU) did not adopt a position on uranium mining until 1975. Over the next decade or more, it has rarely been clear on the issue, and even when Congress policy has been unequivocal, workers' activity on the ground has often been contradictory and confusing. Members of some unions have refused to work on uranium projects, while their federal branch approved (such as the Australian Workers' Union (AWU) at Roxby) (60). The AWU and the Miscellaneous Workers Union (MWU) conducted legal battles to get control of the majority of trade unionists at both Nabarlek and Ranger (60). As the *Sydney Morning Herald*, in a survey based on ACTU documents, put it in 1983: "The picture is of work-hungry men arriving at an outback site, joining whatever unions are ready to enrol them, and beginning to work: building townships and plants, mining, processing, and finally carting uranium, despite ACTU policy. The high rates of pay (by 1979, the Northern Territory miners were receiving A$600 to A$700 a week) are also an attraction" (60).

The main anti-uranium unions have been the seafarers'(SUA), Waterside Workers Federation (WWF), and the Australian Railways Union (ARU), followed by the Amalgamated Metals, Foundry and Shipwrights Union (AMFSU) and the Electrical Trades Union (ETU).

Where these unions have refused to work, others – notably the MWU and Australasian Society of Engineers (ASE) – have moved in (60).

The ACTU made its first policy decision on uranium mining at its September 1975 conference, when it agreed on a ban on all uranium mining until a public inquiry was held, and a moratorium on all exports, except for medical purposes (60).

Within eight months, trade unionists were negotiating a new agreement at Mary Kathleen (MKU) in Queensland, and an eight-month ARU ban on supplies for that mine was called off, pending another ACTU decision (60, 61).

In May 1976, as part of a "green bans" action called by the Australian Conservation Foundation early in the year, the ARU called a 24-hour strike to stop the first new MKU contract being fulfilled. This consignment was held up until June, when a special meeting of the ACTU reaffirmed the anti-uranium resolution passed a year earlier – but considerably weakened its implications by agreeing that the government's commitment should be met from MKU's overseas borrowings with the UKAEA. Only if the government tried to implement a separate export licence should unions intervene (62).

During 1976, the SUA and WWF blockaded uranium from Nabarlek – an action broken when the company used non-union labour to load a Philippines boat; the ARU also banned quicklime for Jabiru (another action broken by scab labour).

Then 1977 started with a short stoppage called by the WWF at MKU to protest against the resumption of exports: the union also demanded a five-year moratorium (63).

In July, as the *Columbus Australis* sailed into Melbourne with a cargo of yellowcake on board, the WWF came out on strike against its own federal officials' advice, joined by anti-uranium protesters, only to be met by police brutality (61). The ship left (presumably for the

USA) (61) with its load untouched. In Sydney, demonstrators carried an anti-uranium petition bearing 50,000 names (64).

That month, the ALP conference decided to impose a ban on uranium mining (65) while recognising that a future Labor government could reverse the ban (66). Deputy Prime Minister Doug Anthony called the decision "disastrous", declaring that it would make Australians "upstarts in the eyes of the world" (67).

After talks with AMIC, the Australian Council of Churches also called for a moratorium on uranium mining and export (68).

Robert Hawke made his famous remark, "I wish the bloody stuff had never been discovered", on opening the 1977 ACTU Congress, as he prepared to steer the Congress down the yellowcake road (69). The ACTU decided to ban new mines, but allow existing contracts to be honoured. It called for a national referendum before further exports were allowed, and invited individual unions to impose their own ban, should the demand be refused. Anthony, true to form, declared "we are not going to stand by and let [the unions] talk to us like this" (69, 70).

The ACTU gave the government until mid-November to hold a referendum (71). And, not surprisingly, the government refused. Just before the deadline expired, Asian and Pacific Unions at a Pacific regional meeting of the International Union of Foodworkers (28 in all, from nine south-east Asian countries) condemned uranium mining (72).

The ACTU then decided to poll its own members on the issue, and circulated 23 unions directly affected by the uranium industry (60), although some pro-uranium unions, notably the Federated Engine Drivers and Firemen's Association, and at least one anti-uranium union (Building Workers Industrial Union, BWIU) were opposed to the measure (73). Only 13 unions responded to the move, of which seven voted to ban uranium (60). Plans for further balloting collapsed, amid criticism of lack of proper representation among the general membership in the voting system (74). As speculation increased that Bob Hawke had changed his position (75), another special unions conference reaffirmed its ban on new mines (pending proper safeguards) and the decision to honour existing contracts (60).

By the end of the year, and the publication of the Ranger Report, federal unions were actively considering lifting the export ban (60). In December, the executive of the ACTU, meeting to consider the Ranger Report, condemned the Fraser government's decision to resume exports, but accepted that contracts (for some 9000 tonnes) should be met from stockpiles and MKU. This was, provided that protective measures for Ranger, recommended by the Fox Commission, were carried out; that a public debate was initiated, before further contracts were negotiated, leading to a referendum; and provided "equal funds, time, and space on television or in other media" should be made available by the government to opposing views (76). In a letter to the ACTU, a government Minister, Mr Tony Street, categorically rejected the requests (77). However, the ACTU – namely Bob Hawke – then issued a statement implying that the terms for a referendum had been accepted by the government, so the uranium ban could be lifted (33, 78).

As the NLC moved towards an agreement on opening-up Ranger, the ACTU found itself in a quandary (79) and apparently unable and unwilling to act against workers engaged in constructing the Ranger mine.

The dry season saw members of at least three major unions, the AWU, Federated Ironworkers Association (FIA) and the MWU, actively engaged at the Ranger site (60). ACTU President Hawke stressed that, as the ACTU is a "voluntary" body, it could not impose its "will" (ie its anti-uranium line) on its affiliates (60). In April, the AMFSU, ARU, and ETU banned labour and supplies for Ranger, as other unions continued to build up the mine-site (60).

At the September 1979 ACTU conference, Hawke received a major blow. After stating the "deep concern" within Australia over the dangers of uranium use, the fact that waste problems had not been solved, and that "Aboriginals have repeatedly stated that if they were given a free choice they would oppose all the uranium mining proposals" (80), the Congress re-affirmed its continuing opposition to mining, and support for the federal ALP policy of a moratorium "until satisfactory safeguards are met".

The motion was passed by 512 to 318 votes, while Hawke seethed in disgust: "You may luxuriate in your morality, but you will have done nothing positive ... you will have destroyed the credibility of this great organisation" (81).

Within three months, Hawke seemed back in the saddle. The WWF appeared to say that it would only enforce anti-uranium action if other unions backed its stand (60). A Melbourne ACTU "summit", attended by eighteen unions involved in the uranium industry, voted 13 to 11 against an executive proposal to mount industrial action to enforce a ban (82). At the same time, the anti-uranium campaign was financially downgraded – justifying expenditure of a mere A\$20,000 out of a total of A\$2.2 million (82).

Workers at Jabiru township, led by the MWU, argued that they suffered similar disabilities to workers at the mine itself and got a A\$23.20 per week wage increase (83).

However, the campaign was not completely dead. In July 1980, workers at two Queensland steel companies (members of the AMFSU) imposed a new ban on steel contracts for Ranger, as part of ACTU policy (84) – the first such ban since September the previous year (85). Within a few weeks, WWF workers also voted to ban the export of uranium through the port of Darwin, and tried to get Darwin City Council to declare the town a nuclear-free zone (86). Cliff Dolan and other ACTU executives visited Ranger – and Nabarlek – asking employees not to work: they were politely received, but work continued as before (60). The ACTU executive drew up a list of uranium mine construction suppliers, asking unions to "consult" with their members over banning materials for the new mines (60).

In January 1981, the Darwin WWF reluctantly decided to resume shipments of uranium – threatened by legal action under Section 45D of the Trade Practices Act (relating to secondary picketing), and since "almost no other union, no matter what statements are made, has so far been prepared to carry out ACTU policy" (87), the WWF was feeling understandably isolated. However, a November action by the WWF and the SUA held up 20 containers of yellowcake for several months, after the Transport Workers and the Merchant Services Guild joined the boycott, which was only circumvented by the government chartering a barge, using non-union workers (88). Eight people were arrested during a Trade Union and FoE picket at Darwin docks. A shipment from Brisbane the same year was met with international action organised by Greenpeace International and the British Columbia Federation of Labor. The train carrying the yellowcake was interrupted several times by demonstrators – at risk to their lives – as it steamed through British Columbia and Ontario. A third Brisbane shipment was banned by the SUA – with the result that the *ACT IV's* owners (Trafalgar House investments and Cunard) offered to unload the offending uranium; at which point the Australian government threatened the company with legal action, and the Seaman's Union with damages (89). The yellowcake finally sailed after considerable agonising within the SUA as to whether or not to defy other unions which had decided not to implement the ban.

As 1981 trickled to an end, the ACTU executive voted by 16 to 9 to handle uranium exports, pending an officer's report, with the option of a special uranium conference or congress (60).

A report from the officer, delivered in May 1982, showed that 5957 tonnes of yellowcake had left Australia since 1977, despite the supposed bans. An ACTU decision not to hold a special congress, but open a voluntary publicity fund, received only one contribution – A\$250 from the WWF (60).

Massive radiation doses were received by two workers in the Ranger yellowcake packaging room in July 1983, when uranium oxide spilled over them as they were unclogging a packaging chute (90). The accident was not investigated fully – and this incident contributed to the strikes declared three years later. There was a 24-hour stoppage over conditions and wages at Jabiru in November, with eight protesting unionists arrested as workers lay down in front of trucks belonging to ERA (91).

In early 1984, with the publication of the new ALP policy draft on uranium (see ALP section in this write-up) some trade union leaders began to feel betrayed. John Halfpenny, the State Secretary of the AMFSU, found the new plan "disturbing", while a representative of the BWIU, Pat Clancy, declared that he was not simply "disappointed" – but "bloody angry about it" (42).

Two months later, the ACTU executive, in what the *Financial Review* called "the familiar double-speak which has become a feature of ACTU or ALP jargon on the contentious uranium mining and export issue" (92), re-affirmed its "general" opposition to mining, but said it would not concentrate on those issues around which it could get maximum consensus: proliferation, waste and health (93). Assistant Secretary of the AMFSU, Laurie Carmichael, responded, saying his union would continue to take action, and dilution of ACTU policy "could not be tolerated" (50, 94).

Although united trade union action now seemed a forlorn prospect – which receded even further after the ALP conference – during July, Greenpeace divers were in action, trying to prevent ships operating (95). Also in July, after three Greenpeace blockades against the Bankline vessel, *MV Clydebank*, 49 containers of yellowcake left Darwin for the USA, West Germany and Great Britain, for processing (95, 96).

Aborigines and uranium

The Ranger Agreement was the first signed after passage of the Aboriginal Land Rights Act in 1976. However, the original version of the Act had been presented to Parliament only six days before the ALP government was kicked out of office by the Governor-General. The new Libe-ral-Country Party administration tried to dilute, and even do away, with some of the key clauses in the Act: Aboriginal control over access to land, and a limited right of veto over mining.

In the event, public pressure ensured that the final Land Rights Act for the Northern Territory (NT) was restored to something close to its draft form. However, in one key respect, there was confusion and a critical weakness. In the original legislation, the Minister for Aboriginal Affairs was entitled to over-rule an Aboriginal mining veto, "in the national interest". Aboriginal groups could, however, appeal to Parliament, and either House of Parliament could over-rule the Minister. In the final version of the Act, Aborigines were stripped of the veto in several important spheres. Oil, gas, and other mining projects (specifically MIM's venture at Borooloola) given consent before the Act came into force, were to be allowed to proceed. And under sub-clause 40(6) of the Aboriginal Land Rights (NT) Act, the Ranger project was specifically excluded from the power of veto (97).

Because of its rush to get Ranger on stream, the new government also omitted to appoint a Land Commissioner to investigate and recommend on Aboriginal claims in the area covered by the mine: this was instead to be handed over to the Ranger Commission itself (98).

As was pointed out at the time, the Second Ranger Report hinged not on whether uranium mining might harm Aboriginal communities, but on whether they could be adequately protected from it. The scene for such contemplation was hardly well set by the NT's own Social Development branch, whose director, John McDonnell, told the Commissioners not only that Ranger would bring full employment to Aborigines, but that "welfare" problems would be adequately handled throughout the project's life (99). When asked by an adviser to the Commission which anthropologists and sociologists (let alone Aborigines) he had consulted for his report, McDonnell confessed "none". Nor had he consulted with any other agency with experience in uranium mining (99).

If he had asked Aboriginal people on his own "home patch" he would have been left in little doubt of their opposition to the Ranger plans. In contrast to McDonnell, the Fox Commissioners went to extraordinary lengths to assess Aboriginal response to the mining project – flying into bush meetings, allowing tapes to be used in evidence, and generally resisting the objections of the miners that an improper procedure was being followed (100).

However, there is little doubting that the judicial procedure dominated by whites (there was no Aboriginal adviser to the Commission) circumscribed much of what they wanted to express. Many Aborigines were at the crucial Darwin meeting of the Commission in May 1976. "There's a dreaming long that rock, they call it Djidbidjidbi. None allowed to get in there," said Jimmy Maneee Namanjarlawarkwark, "there" referring to sacred Mount Brockman. When asked if he wanted miners on his country, he replied "I want some money for mining, for my kids", seeking to explain this apparent ambiguity with the words "they are taking my country, but I want money" (101). Silas Roberts, Aboriginal advisor with the Department of Aboriginal Affairs, added, referring to the dreamtime creatures in the region, "These great creatures are just as much alive today as they were in the beginning ... the sickness [people] get [from mining] [is] nothing to the sicknesses they will get if they hurt our sacred sites" (101).

The people of Oenpelli and Gagudju (Kakadu) have had long and bitter experience of white hunters, traders and missionaries over a period of more than 130 years (102). At bush meetings among communities affected by Ranger, they found words to passionately name and indict the threats they confronted. As Tony Ganggali put it in one speech at Red Lily lagoon, "First you say it won't kill us, then you say you'll give us plenty of money, then you say if we refuse we'll pass it on the basis of the national interest clause. To hell with the meetings. Give it to them! I'm entitled to die first as a sacrifice. The Europeans are oppressing us. I'll be the first to die for this" (103).

In other community meetings, Aborigines were to express at length their expectations of the bad effects which follow from the building of new roads (an issue which was to create militant opposition in the case of the Nabarlek mine), the devastation caused by liquor, racism, tourism, and of course, ill health caused by radiation, dust and pollution (103). On numerous occasions, community spokespeople expressed concern about the "breaking up [of the land] with machines", the changing of rivers, despoliation of hunting grounds, the encirclement by fishermen in the north and miners in the south (a special fear for the coastal communities of Goulburn and Croker Islands) – above all, of the "shifting" of the djang, (or sacred dreaming sites and tracks), of Djidbidjidbi and Dadbe (Mount Brockman). As Dick Malwagu summarised it in 1977:

"How much a time we got to tell the government that [is] our djang, that we can't dig here or use it or knock some trees or push some the rock away ... Be nothing left of this land we can't stay any more longer because you doing a lot of damage on our country ... You know I said we losing, we losing, this country, but up here this northern part is getting smaller and smaller ... I wonder 50-60 years time, what will happen this Northern Territory ... When you smash a djang, well, just like a church you moving church away ..." (103).

The Fox Commission damned ERA for failing "notably in its evaluation of the total impact on the Aborigines of the region" and stipulated that the company's proposals in its 1976 form "should not ... be allowed to proceed" (104). The Second Ranger Report acknowledged virtually all the fears expressed by Aboriginal objectors, and sought to encompass them in its recommendations. It recommended that all Crown lands in the project area be ceded to Aboriginal people under the 1976 Land Rights Act, together with two cattle leases, and that the southern border of the mining zone be moved away from Mount Brockman, so that dreaming sites could be adequately protected. It ridiculed the ERA proposal for a mining town of 10,000 inhabitants – a size which would be "disastrous" for the area" (105), and urged that the township of Jabiru should be planned and regulated with the participation of the NLC (106). While proposing a number of programmes of education and health care (107, 108), the Commissioners agreed that it was "not likely that the mining venture will add appreciably to the number of Aboriginals employed" (109). They also thought that Aborigines were unlikely to benefit under the present or proposed education system (107). Perhaps the most damning indictment of the whole project centred on the Commissioners' assessment of the adverse effect of a massive influx of non-Aboriginal people to the area. This, said the Commission, "will potentially be very damaging to the welfare and interests of the Aboriginal people ... All the expert evidence on this matter was to the effect that, despite sometimes sincere and dedicated effort on the part of all concerned to avoid such results, the rapid development of a European community in, or adjacent to, an Aboriginal traditional society has in the past always caused the breakdown of the traditional culture and generation of intense social and psychological stresses within the Aboriginals ..." (110).

Assessing the putative value of material benefits from the mine against its potentially negative impacts, the Commissioners concluded: "While royalties and other payments ... are not unimportant to the Aboriginal people, they see this aspect as incidental, as a material recognition of their rights. The material benefits they visualise as likely ... are things like motor vehicles, hunting rifles, fishing gear, and the like. Our impression is that they would happily forgo the lot in exchange for an assurance that mining would not proceed" (111).

Notwithstanding this, the Ranger Report formed "the conclusion that [Aboriginal] opposition should not be allowed to prevail" (104). Compounding this grotesque judgment, Justice Fox himself, in an interview with the *New Scientist* on the publication of the Second Ranger Report, commented that, while mining could bring "tremendous benefit" to Aborigines, foregoing royalties "would be a small price to pay to stave off racial conflict" (65, 112).

On August 25th 1977, Ian Viner, the Minister for Aboriginal Affairs, announced the terms for Aborigines under which the Ranger project would be allowed to proceed, while the Prime Minister, Malcolm Fraser, set out the government's uranium export policy. In the weeks leading up to the decision to mine (actually announced by Fraser in mid-August) (113), Fraser had made the hypocritical observation that Aboriginal welfare, and protection of the environment, as well as international safeguards, were "more important than commercial considerations", and admitted that "proper progress in relation to Aborigines [had begun] too late" (114). James Yuemipingu commented that it would take "up to five years to explain concepts which Aboriginal communities had only eight weeks to consider" (115), while the Aboriginal lawyer Paul Coe promised to take out an injunction to prevent uranium mining, as a threat to the religion and faith "of the Aboriginal nation" (116). Three months later, South Australia's Attorney General, Dr Duncan, backed Coe's stand (117).

The new uranium policy laid down thirteen points: these included promises that the mining township would be limited to 3500 people; that Aborigines would receive royalties directly, and could also negotiate directly with the mining companies; the establishment of a scientific committee to look after the National Park's environment; a special uranium advisory council would be set up, and companies would be "likely" to pay a resources tax on their uranium exports. However, the way was now open for the states to make their own decision on uranium mining, and the Fox recommendation of sequential development (to minimise the impact on people and environment) was denied (23).

For his part, Ian Viner declared that the government had accepted all the major recommendations on Aboriginal matters made by the Commissioners. Traditional Aboriginal claims to vacant Crown land would be granted, while the Mudginberri and Munmarlary pastoral stations would be consigned to Aboriginal use, with the government taking over the leases. The southern boundary of the project area would be shifted from Mount Brockman, and Aboriginal land would become part of the National Park "as proposed by the Aborigines" – to be part-managed by Aborigines (118), with Aboriginal rangers. "Aborigines would also be trained as health workers, and encouraged to accept other opportunities for employment created by mining development." There would, said Viner, be a scheme "to improve the morale of the Aboriginal people, enhance their welfare, and reduce their dependance on alcohol". Possibly an Aboriginal liason officer would be appointed to inform tourists regarding Aboriginal customs.

Gross revenues to Australia, it was estimated, would be around A$20,000,000,000 (119).

The Minister claimed that Arnhemland Aboriginal people would get around A$400 million of these, at the rate of at least 2.5% of revenue; 30% of such payments would go to Aboriginal communities directly affected by the mine, while another 30% would "advance the general well-being of Aborigines throughout the NT on the advice of an all-Aboriginal advisory committee" with a special emphasis on programmes for alcohol control (27, 120). (Rather a case, one might think, of trying to dry the baby, once it had been washed out with the bathwater.)

Announcement of the royalties – despite being well below rates negotiated in similar situations elsewhere in the mining world – prompted a spate of racial stereotyping in the press. "Blacks will get millions" trumpeted the *Herald* (121). "Aborigines strike it rich" chortled *The Australian*, while an unrestrained *Telegraph* declared the "Uranium bonanza" would create "Stone-Age millionaires" (122, 123). According to a "special report" from the *News*, the Gunwiggu "Tribe" had also become Stone-Age millionaires, sitting on a bonanza "worth A$15 million each year" (123). A little more circumspectly, *The Australian* estimated that the 800 members of the Gunwiggu communities would get A$6500 for every man, woman and child in the Arnhemland tribe" (122) – a figure one-thirtieth the estimate given by the *News* but still calculated to arouse envy and resentment

among white readers. To all this hogwash, Aboriginal politician Neville Bonner gave the response: "If they do become millionaires, why should anyone object to that?" (124). Perhaps this was not a well-thought-out response, since the issue of royalties was to cause huge discontent among Aboriginal people in the coming decade. Bonner might have made a better point by asking what profits ERA executives would derive from the exploitation of purloined land and filched resources.

Viner blithely assured Aboriginal people that they could apply for legal constraints if they were unhappy about mining (125). Paul Coe had already issued his writ, and at the end of August, four other Aboriginal people took out writs to prevent mining and export of uranium, seeking A$20,000,000,000 in damages if it went ahead (126). While Gerry Blitner, on behalf of the NLC and the Oenpelli people, confessed they "do not understand [the Ranger proposals] completely" (127), FCAATSI, representing 74 Aboriginal organisations, solicited the support of 500 other organisations, declaring that uranium mining would destroy Aboriginal societies: "safeguards would only make destruction slower and more acceptable to public opinion" (128).

The first round of negotiating between miners and the NLC took place in Darwin in October 1977, when a draft encompassing all the Fox recommendations was presented to the federal government, and some details were leaked by ERA (129). No response was received to this until May the following year, when formal negotiations commenced (102, 130). Well before the end of 1977, construction work on Ranger had started without consultation with the NLC – a point complained of by Stuart McGill on behalf of the NLC ("The government is riding roughshod over the Aboriginals") (131), the Labor Party's Tom Uren (132), and NLC chairman Galarrwuy Yunupingu. When Prime Minister Fraser urged that "consultation should continue", Yunupingu retorted: "Of course consultation is not protection" (133).

In November, Yunupingu claimed that mining companies had met the government in secret session two months before, to draw up guidelines for further uranium exploration in the proposed Kakadu National Park, while Aborigines had been left out of the consultations. Viner replied that such exploration had been agreed "from the start" (134).

Some of the terms of the draft agreement, together with details of other demands made by the NLC or the mining companies, leaked out during the first part of 1978. In February, the NLC said that the draft was modelled on the 1974 agreement between CRA and the Papua New Guinea government over Bougainville (see CRA), with its concept of a profitability tax, and preventing the mining companies making windfall profits; a certain percentage of Aborigines would have to be employed at the mine. The hand of Stephen Zorn – who negotiated the revised Bougainville agreement – was clearly visible behind these proposals (135). At this time there was an "unofficial" adjournment of the ERA proposal, while the Supreme Court heard an NLC application to stop mining, and Peko and EZ themselves applied to explore outside the Ranger lease (136).

In March, Yunupingu claimed that ERA had given a copy of the draft agreement to other mining companies and he divulged that one of its terms was that no Aborigines would live in the Jabiru township, while its residents would have "instruction" in Aboriginal culture and traditions (137). At this point, the NLC demanded that compensation for the loss of land should be at least 36% of royalties (138) – the "biggest obstacle to development" of Ranger, according to the *Age* (139, 142). A meeting of traditional owners at the Mudginberri pastoral station in March once again demanded that mining should not commence: if it did, only Ranger and Nabarlek should be given the go-ahead (140).

In April, as the government built up pressure to get Ranger on stream before the wet season (139, 143), and 25,000 anti-uranium marchers protested in Melbourne and elsewhere (141), FoE got hold of draft legislation seeking to substantially alter the Atomic Energy Act (144), and introduce new measures relating to environmental problems of uranium, land rights and national parks. Under clause 22 of the proposed new law, according to FoE, "the Government would have virtually uncontrolled power to lock up demonstrators, dissidents, and anti-uranium campaigners, without trial, for as long as it wishes; to tap phones, control buildings and industries; or anything else it wishes to do". The crucial governing phrase was "situation resulting from a uranium activity" where "health and safety" were at risk, but it was clear that this proviso could be applied to strikes, blockades and the release of information surrounding an accident, or regarding an act of "terrorism" (145).

Within a week, deputy Prime Minister Anthony said the Fraser government was going for a June start-up, and confirmed that , in a new "uranium package", there would be regulations restricting the disclosure of public information, not only by the supervising scientist, but also the Land Councils. The Penalty? A$1000 or six months inside (146). (Two years later the government was to embark on a review of no fewer than ten acts relating to nuclear safeguards, environmental protection, defence, and the effects of uranium on Aboriginal communities) (147). Yunupingu retorted that negotiations would take 12 months and that compensation terms were the main stumbling block (148). While the Ranger partners fulminated that there was already an 18-month supply gap of uranium, and that the mine could be delayed from producing until 1981 (149), Fraser played both ends against the middle, telling ERA (and others) "I don't think any time has been lost" (149, 150).

In May, seven hours of talks between Anthony, Viner, and the NLC, confirmed that the NT authorities would play a role in the Kakadu National Park, despite vehement NLC objections (143). It was established in principle that the Park, while belonging to Aborigines, would be accessible to anyone else, although its administration was vested in the Australian National Parks and Wildlife Service (not the Territory) (152).

Over the next two months, the royalties/compensation issue loomed largest in on-going talks between the government and Aboriginal negotiators, and ERA (153). The Federal government passed its six uranium "enabling" acts at the beginning of June. Yunupingu was still holding out for a 36% royalty on gross profits, while Anthony was rumoured to be proposing an extra 1.75% topping on the 2.5% already guaranteed under the Land Rights Act (154).

In July, 50 supporters of the NLC sat-in at the Ranger offices (155). Although the NLC scaled down its royalties demand to 18% and the government confirmed an offer of 4% (156), talks were still deadlocked between the black and white parties (157).

Between then and November 1978, when the Ranger Agreement was finally signed, the most important set of negotiations in recent Aboriginal experience (and, it could be argued, in recent Australian history) became so confused and labyrinthine, that – fourteen years later – the full story is still a subject of great controversy.

The main actors in the drama are not disputed: Yunupingu, adept, charismatic, a master of compromise; Stephen Zorn, the world's most experienced white lawyer advising indigenous peoples on mining agreements (158); deputy Prime Minister and Minister of Resources Doug Anthony, and Minister for Aboriginal Affairs, Ian Viner, leading the government side, with a mixture of blandishments and threats. On the sidelines were the miners and the white advisers to the Aboriginal councils, notably the NLC's manager Alex Bishaw, who appears, at one stage, to have become a kind of *agent provocateur* working against the people who employed him. Further to the sidelines were those who should have been at centre-stage: Leo Finlay, Simon Maralingura, Dick Malwagi, Johnny Marali No. l, and around 40 other Aboriginal men and women whose objections to the agreement were never properly heard, and were ignored by their own leadership.

The Ranger Report, as already stated, was presented to the NLC under Section 44(2) of the Aboriginal Land Rights Act. In 1976, when the original ALP-framed legislation had been considerably revised by the Liberal administration, an arbitration system was introduced, allowing the government to impose a settlement in the event of Aboriginal resistance to a mining proposal (130). However, the majority of uranium proposals in the NT had already been exempted from the need for Aboriginal consent: Sections 40/44/46 stipulated that any mining licences granted before the 4th of July 1976 would not be covered by a Land Council veto. Under Section 40(6) the Ranger project had specifically been exempted from such a veto (130). In any event, whatever Aboriginal powers were granted with one hand could be taken away by another clause which allowed mining to proceed "in the national interest" (159, 160).

As an alternative to the Land Rights Act, the government could use its authority to mine uranium under the Atomic Energy Act of 1953. While the Fox Commission argued against invoking this power, the federal government rejected the Commission's recommendation (160).

This bundle of provisos and controls meant that the government (under Section 41 of the Atomic Energy Act) could at any point nullify Aboriginal land ownership; could impose severe penalties (already mentioned) for divulging information on uranium exploitation, transport and use; that uranium mined on Aboriginal land became the property of the Commonwealth (Federal Government) (161); and that Aborigines would not be paid direct royalties, which would go instead to a Trust account and to Aboriginal organisations, which need not exceed a mere 2.5% of revenue (160).

So, effectively, the government had both its yellowcake, and the consumption thereof: using the Land Rights Act as an apparent show of its good intent, when it had already over-ruled an Aboriginal veto, and backing this up with the draconian Atomic Energy Act that it could have employed in the first place.

On August the 25th 1978, agreement was announced between the NLC and federal negotiators. Its actual terms were not officially an-

605

nounced, and had to be leaked by the leftish magazine *National Times*. Even then, there seemed doubt as to whether agreement had been reached on royalties: the *Australian* stating that the government had offered 3.75%, while the NLC was sticking out for 5% (162). A little later it was confirmed that 4.25% had been offered and (presumably) agreed: the royalty would be split three-ways, with 40% going to the three northern land councils – NLC, Central Land Council (CLC) and Tiwi Land Council (TWC) – the NLC securing another 30% and the remainder going into a Federal Trust Fund (163). The NLC would also have to endorse the agreement as a body (163). Formal ratification took place on September the 8th (164). Hardly was the ink dry upon the draft, when protests began to be heard. Negotiator Stephen Zorn – who said he expected to be accused of "selling out" the Aborigines (165) – stressed that the agreement was only one "in principle" (166). The Oenpelli people reminded Yunupingu that they didn't want mining under any conditions (167), and H C Coombs, claiming the agreement was "not freely negotiated", stated Aborigines would not feel morally bound by any NLC agreement, and there might be violence as a result (168). The Movement Against Uranium Mining (MAUM) claimed the government was behaving in a racist and duplicitous fashion, which was encouraging the miners to try to expand their leases – a direct result of the refusal to give Aboriginal owners full title to the Kakadu National Park (see next section) (169).

The following chronology of events, from early September until the end of November, has been compiled from several primary sources:

September 1: 42 delegates from 19 coastal communities meet at Gailwinku on Elcho Island where they call on the NLC to defer any signing of agreements until Aborigines are given control of entry to coastal waters surrounding the communities which will be affected by the projects. Wesley Lanhupuy says it would be "inconceivable if the NLC by-passes the people it [is] supposed to represent" (129, 170, 171, 172).

September 4: Pancontinental is given permission to extend the Arnhem highway (173).

September 5: Yunupingu, stating that traditional owners oppose the road and that the NLC supports their stand, threatens not to sign the Ranger agreement if the road extension proceeds (174).

September 7: As Minister for Aboriginal Affairs, Ian Viner wines and dines Yunupingu (172), the NLC telegrams delegates advising them of a land council meeting on September 12. At a press conference, Yunupingu, reiterating the NLC's stand on the Pancon road, also adds that the draft agreement was simply a "recommendation"; it would have to be translated into "language" and all 42 delegates to the NLC would have to agree it: this could not happen before the onset of the wet season. Yunupingu also stated that the government was applying pressure on the NLC to sign, and that the agreement was "rotten" – a piece of paper "Mr Anthony should save for the next time he goes to the toilet" (171).

According to Johnny Marali No. 1 and John Gwadbu, in their later affidavits to halt the Ranger negotiations, the NLC's notice to outlying communities did not specify why the meeting was to be held, nor that all delegates should be present. It did not say that the September 12th meeting was to be a final decision-making forum on Ranger (175).

September 8: Yunupingu meets in Darwin with Prime Minister Fraser, Anthony, Viner, and an executive member of the NLC. A joint statement from Yunupingu and Anthony states that the Aboriginal leader will recommend ratification of the Ranger agreement at the next NLC meeting, while the Arnhem highway extension will not proceed, at least until the Pancon mine is given the go-ahead. Yunupingu calls the meeting "very successful" (129, 171) even though he himself at a later stage claims that Fraser had stood over him, threatening to destroy the NLC and outstation movement (172). Previously, Yunupingu had said he would not attend the meeting were Anthony present (which he was) and in the absence of Stuart McGill, the NLC's astute Legal Officer –

McGill was excluded at the Ministers' insistence (129, 171).

September 9: The Milingimbi community resolves to formally protest against government pressure tactics and issues a press release to that effect. Yunupingu says that the September 12th meeting will last a week, enabling outlying communities – such as Croker Island, Milingimbi and Goulburn Island – to attend (176).

September 11: Alec Bishaw in Canberra apparently states his view that, if the NLC refuses to sign the agreement, an arbitrator would not be appointed, but Fraser would dissolve the NLC and amend Land Rights legislation accordingly (172).

September 12: As the NLC prepares to meet, Stephen Zorn tells a reporter from the Age that the Ranger deal was the best Aborigines could get; that there was no point in continuing negotiations or taking the matter to arbitration; and that the only other resort was violence – "get[ting] in front of the bulldozers with a shotgun" (177). Shortly before this, Zorn had presented his own negotiator's report to the NLC, which showed a Ranger royalty of 9.25% (Zorn having arbitarily added 5% as "cost of environmental protection") (129).

Briefing notes are given to NLC delegates which purport to outline the positions of the ALP and anti-uranium movement, and which describe the agreement as promising "thousands of square kilometres of land for the local people ... a National Park offering land and work ... the best arrangements in Australia for protection of the land and people from effects of mining ... substantial payments offering Aborigines the real means of achieving self-determination in Australia for the first time" (129, 171). Any delay in signing, or going to arbitration, could be construed as evidence that Aborigines didn't know what they are up to, and provide "a wonderful opportunity for people of ill will" to persuade the government to change the Land Councils system.

This travesty of the true situation was compounded by Alec Bishaw, who then told the delegates that Bob Hawke had warned him the government would "nail the NLC's ears to the wall" if they didn't sign (a statement utterly repudiated by Hawke, who forced Bishaw to apologise) (178); and that the agreement had to be signed. Yunupingu also quoted threats from Fraser which the government later denied (179), and misrepresented rehabilitation work, after mining was finished. (He gave the impression that the pits would be filled in, level to the ground) (170).

Wesley Lanhupuy was not allowed to vote or submit a resolution from coastal communities demanding Aboriginal coastal control before mining commenced (151, 170). A statement opposing mining from the people of Numbulwar was not read out, and when Leo Finlay from Borroloola asked for more time to consider the question, Yunupingu refused; most important, delegates from Oenpelli were allowed neither to address the meeting, nor discuss the problems with other delegates (170).

At no point was the draft agreement read out, nor were any translations prepared. Indeed, its terms were barely discussed (180).

September 13: US lawyer Bob Krueger arrives in Darwin, is met by Bishaw who refuses to let him address the NLC, and storms out, telling the press how miserable are the royalties offered for Ranger (172). By the 14th of September, with telegrams from black delegates flooding into the office of Bob Collins (ALP member for Arnhem in the NT Legislative Assembly) demanding they be allowed to send delegates to the meeting, only 23 out of the 42 NLC delegates had actually arrived.

September 15: Fraser announces that Yunupingu had informed him of agreement by the NLC on both the Ranger and Kakadu leases (129, 171). Aborigines from Croker and Goulburn Islands, Milingimbi, Maningrida, and Elcho Island rush to Darwin to discuss and state their opposition to the agreement.

September 18: Dehne McLaughlin, a field officer with the NLC, resigns after telling the press he was "appalled at the pressure and strong-arm tactics used at the meeting last week". McLaughlin maintains that field officers were conveying directives from outlying communities which were being "manipulated" by the Council. "They feel the NLC is just another part of the government" (129, 171).

Bob Collins produces tapes of the Council meeting, maintaining that the NLC had acted under misinformation (181). Yunupingu then announces that he has sacked "dissident" white lawyer Stuart McGill, ostensibly for "leaking information" (182). At this point "dissident" Aborigines apply for an injunction against the signing, in the NT Supreme Court (172).

(Irony of ironies: that the chair of an organisation pledged to disseminate information to his people, on the most important event in their recent histories, should get rid of a white man, precisely for informing blacks of the detail behind that event.)

September 19: The NT Supreme Court grants an injunction restraining the NLC from signing the Ranger agreement, after submissions by Dick Malwagu (NLC executive member, chair of the Mijilang Council) and Johnny Marali No. 1 (from Goulburn Island, chair of the Warruwi Council) acting on behalf of six Aboriginal communities (166). The court requires the NLC to demonstrate that the agreement was reached in accordance with Section 23(3) of the Land Rights Act which states that all traditional owners must consent, after they and their communities have fully understood the terms of the agreement (129).

September 20: By now more than 1000 Aborigines protesting against the Ranger signing have arrived in Darwin (183). Anthony tells Parliament he wants to see a voluntary agreement, and not go to arbitration. Ian Viner says that the future of land rights is at stake, and names Bob Collins and Stuart McGill as among those responsible for opposition to the agreement (184). Later, Aborigines angrily deny that they have been manipulated by whites and state that they oppose the agreement in their own right (182, 184).

September 22: The injunction is removed, after Aboriginal leaders agree to give themselves more time to discuss the agreement; Yunupingu promises to translate the agreement into Aboriginal languages, and Viner's charge that "rebel" Aborigines were out to destroy Yunup-

ingu is seen as a spur that makes Council members determined to reach an agreement (185). A procedural meeting of the NLC will now be held on October 2, and a final decision will be made on Ranger only "after an indefinite period is allowed for the people to decide" (172).

October 2: The NLC meets under the chairmanship of Gerry Blitner, in Yunupingu's absence. Dick Malwagu presents a 21-point procedural plan, which includes an initial meeting of all traditional owners in the Alligator Rivers area, a consultative panel to visit each settlement and outstation, translations of all relevant material, and a final NLC decision-making meeting only after 30 delegates stated in writing that they are ready to vote (129, 171). Although Gerry Blitner announces that the plan has been accepted by the 31 delegates present, Yunupingu later denies that it has (186). The Oenpelli people say there must be a full NLC meeting to ratify the agreement (187).

October 3: After another NLC meeting, Yunupingu tells reporters that the 21-point plan has been put in the rubbish bin; instead, the Oenpelli community would be asked which other communities should consider the agreement, and a panel would then explain the agreement to those communities; he adds that a Ranger agreement would be ratified within six weeks (129). Leo Finlay presents a petition from 11 Council members calling for the sacking of Bishaw, Yunupingu refuses even to allow it to be considered (172). According to Tribune, Viner also threatens to appoint an arbitrator, apparently reversing Anthony's conciliatory position a fortnight earlier (187).

October 4: The Borroloola community decides to withdraw from the NLC (11, 188).

October 10: "King James" Yunupingu flies into Goulburn Island, to be met by a demonstration of 300 people. He leaves quickly, after delivering copies of a translated agreement (189, 190).

The Oenpelli community meets to decide which communities to involve in the consultative process, and look at the terms of the agreement (190).

October 11: Yunupingu, in a belligerent mood,

says that NLC meetings to determine the attitude of Aborigines are a "bloody waste of time". If the government wants to mine uranium, "it will go ahead and do it and I am not going to stop them" he is quoted as saying (191). Mary Elliot, who attended the Oenpelli community meetings on October 10th and 11th states that "the people clearly rejected the tactics of the NLC ... [they] asked for consultation with all communities ..." (172).

October 12: Forty traditional owners of the Oenpelli community issue a statement which rejects the Ranger agreement ("at this time"), calls on the NLC to hold a full Council meeting, and to fully consult with all communities affected by the mine, including those of Croker and Goulburn Islands (192). Some of Yunupingu's own people concur with the Oenpelli leaders, while others accuse him of expressing the feelings of white officers, rather than Aborigines (193).

October 13: Although initially accusing his opponents of being manipulated by white anti-uranium groups, Yunupingu on October 13th apparently promises to support the Oenpelli decisions (194).

October 17: The Cabinet decides not to appoint an arbitrator, until satisfied that the NLC definitely won't sign the agreement; and Yunupingu stresses that Aborigines and legislators could sit down and come to agreement without arbitration (195).

October 23: After Anthony has declared "we have to consider ... to what extent we can allow a small group of people, a manipulated group of people, to stand in the way of a development of tremendous national and international significance" (196), Stephen Zorn reverses his position on Ranger, stating in a letter to Yunupingu that the NLC had tried to push Aboriginal people into signing; that there were, by now, several communities within the NLC who were highly critical of its leadership (166); that royalties were too low (197).

November 1: The NLC convenes a secret meeting (at Bamyili), chaired by Yunupingu, with Bishaw, solicitor Eric Pratt (who "simplified" the Ranger agreement for Aborigines),

and two solicitors who had helped Yunupingu contest the previous month's Supreme Court injunction (171). Although Viner had apparently been invited to this meeting, several important delegates couldn't make it, because of short notice. Delegates from Borroloola were simply told the meeting was "to discuss general business" (171). According to Leo Finlay of Borroloola, who attended the Bamyili meeting with three other delegates from his community, there were about 30 members at this meeting. When challenged about not following the 21-point plan laid down in October, Yunupingu retorted that he was "only doing his job" (198).

November 2: Viner arrives at the meeting to make a long speech, in which he stresses that the government has to act "in the national interest" and on behalf of Japan (171). He says the Oenpelli people's demands to settle the Kakadu issue, before the mining agreement is signed, should be "put aside". In the afternoon, according to Leo Finlay, although Yunupingu batted members' questions and criticisms in an off-hand manner, he did make one significant concession: promising that the traditional owners could still say "yea" or "nay" to the agreement. Most people, according to Leo Finlay, accepted this, but he himself viewed it as a trick. As the vote was called, most delegates – considering that they were approving further consultation – raised their hands (11, 171).

November 3: The Crucial Day – and action transfers to Oenpelli itself, as Viner, Yunupingu, Bishaw, Pratt, and two other lawyers fly into the community, ahead of anyone else from the NLC. Simon Maralingurra, chair of the Oenpelli Council, has invited a film crew and journalists from the Age newspaper to witness events, but both Yunupingu and Viner object: Viner is abusive to the media crew, and Yunupingu storms "I am running this meeting, and I have given the order" (199).

Only three traditional owners from Oenpelli attend this most important meeting in the life of their community – Toby Gangali, Frank Djandjul and Midjau Midjauwu – leaving the traditional owners outnumbered by the NLC executive and their white advisers. According to

Leo Finlay (who was present) apart from Yunupingu and himself, the other NLC executive members were Dick Malwagi, and Bill Lalaru (deputising for Gerry Blitner). A white linguist, Peter Carswell, who was accompanying the NLC and had flown to Oenpelli, confirms that only four out of forty traditional Oenpelli owners were present. Simon Maralingura – himself a former chairperson of the NLC – declared that the meeting was "wrong" (11) and sat in a car parked at the edge of the meeting-place (171). Later the Labor spokesperson on Aboriginal Affairs also stated in Parliament that only three of the "thirty" traditional owners attended the meeting and claimed that Viner did not answer this charge of gross under-representation (200).

Yunupingu then addresses the "meeting" in English, with Viner following his lead. Of the NLC members present, only Toby Gangali appears to have made a statement: "I've given up. I'm not fighting any more" (171, 201).

After lunch, and without any indication from the NLC heavies that the agreement was about to be signed, a car whisks the Council members – not to the airport, as they had expected – but to the office, where they are presented with the Agreement, and pens with which to sign it. There is wide speculation as to who was present at the signing, and who actually signed it. Finlay was later quoted as saying that "many people" signed (171), but in a 9-page document handed to *The Australian* newspaper within a few days of the Oenpelli ceremony, he claimed that "only Toby Gangali, Joseph Garadpul and Nipper Garadpul" were present (201). However, there is no doubt that no more than a tenth of the community most affected by Ranger, and entitled under the Land Rights Act to full consultation and agreement before it proceeded, actually put their names on the Ranger Agreement, and received platinum pens for their compliance (171).

Later, in Darwin, Viner was to claim that the agreement "had unanimous support of traditional owners", while Yunupingu (clearly referring to "white dissidents" who supported the rights of the majority of aboriginal people)

claimed he "knew the Council would work better without whites hanging around" (202).

November 7: Stephen Zorn is sacked, following publication of his critical letter of October 23 (203). Although Zorn praised the agreement in several respects, he clearly views the royalties as inadequate, its environmental measures as leaving something to be desired, and the whole agreement as negotiated under pressure (203).

The Ranger "Agreement" was fraudulent: "negotiated" under duress, illegitimately "signed", and a symbol of the degree to which a white administration was prepared to impose its will on Aboriginal people, for the sake of its own greed. That Yunupingu misled Aboriginal people, and refused to comply with their overwhelming wishes, is clear from all the evidence. To give him some due, some years later he was prepared to damn the Agreement he had bulldozed through the north, and argue that it should never have been signed (see below); but, by then, it was far too late.

The Agreement

The Agreement purported to be comprehensive, and in many ways is the most detailed document licensing a uranium mine ever drawn up with the supposed participation of indigenous people. (This doesn't say much, of course, since those living on such sites rarely get the opportunity to do more than sue for compensation when they are slung off them.)

Here are the main provisions:

1. Environmental Aspects
   (clauses 5, 6, 8, and 9)

The Joint Venturers have agreed with the Commonwealth to observe all environmental requirements specified or imposed by law.

"Best practicable technology", as defined, shall be employed in the design of the plant and mine, and in the mining, milling and related operations.

The Commonwealth is to arrange for health monitoring programs which will continue for as long as the authority conducting them considers them to be effective.

2. Liason (clauses 10 and 11)

The Joint Venturers are to provide for the formation of an Aboriginal Liason Committee; members shall be the mine manager, one person appointed by the Manager and two appointed by the Northern Land Council.

Functions of the Committee are primarily:

- to receive and hear representations of local Aboriginals on the one hand, or by an employee, agent, or contractor of the Commonwealth or the Joint Venturers on the other;
- to make recommendations to the Manager as to which Aboriginals are available for employment; the Manager will not employ any Aboriginal without first consulting the Committee.

3. Employment and Training (clause 12)

The Commonwealth shall require the Joint Venturers to ensure that:

- as many local Aboriginals as is practicable are employed where those Aboriginals are capable of carrying out in a satisfactory manner the particular work required;
- an "operator training scheme" for Aboriginals be established with a view to the Manager employing trained Aboriginals;
- by agreement with the Northern Land Council, further employment training schemes are sponsored;
- all practical steps are taken to adjust working hours and conditions to suit the needs and culture of Aboriginal employees, through consultation between Joint Venturers, the Northern Land Council and trade unions;
- working conditions for Aboriginals shall be as for any other workers under the appropriate awards in force in the Northern Territory.

4. Local Business Development (clause 13)

The Commonwealth shall support, encourage and advise local Aboriginals wishing to establish enterprises and businesses for the purpose of providing goods and services to the Joint Venturers;

The Commonwealth shall require the Joint Venturers to make maximum use of Aboriginal sub-contractors where their goods and services are competitive. The Joint Venturers, however, shall not be required to employ any sub-contractor unless the sub-contractor is capable of carrying out the work in a satisfactory manner and work is available;

The Commonwealth shall require the Joint Venturers to ensure that appropriate training facilities are made available where skills are not sufficient for the satisfactory performance of sub-contracting tasks, if the circumstances reasonably so require.

5. Control of Liquor (clause 14)

The Commonwealth shall ensure that no liquor is brought onto, or sold or consumed within, the Ranger Project area, except through the Jabiru Sports and Social Club and subject to the rules detailed in the agreement.

6. Rights of Traditional Owners (clause 16)

Traditional rights to use and occupation of land may be exercised subject only to specified restraints to avoid interference with the Project.

7. Sacred Sites (clause 17)

Commonwealth to act to protect any sites not adequately protected by legislation, on NLC request.

8. Instruction in Aboriginal Culture (clause 18)

Commonwealth and Joint Venturers to promote "knowledge, understanding and respect for the traditions, languages and culture of the Aboriginal people" among employees and contractors, in consultation with NLC.

9. Review and Term of Agreement (clauses 20 and 21)

Every five years the parties shall meet to discuss the operation of the Agreement and may agree to changes;

The Agreement to continue for 26 years unless sooner terminated, renewed or extended.

10. Payments (clauses 3 and 25; annex B)

Agreed amounts to be paid by the Commonwealth to the NLC on signing of the Agreement and when the project is authorised and at designated stages of the construction,

Additional annual amounts of A$200,000 to be paid during the currency of the Agreement,

Amounts of up to A$150,000 annually for the first four years to meet NLC costs associated with the Ranger Project,

Payments in the nature of royalties at 4.25% to be made to the Aboriginals Benefit Trust Account for distribution in accordance with the Act: Land Councils (40%), communities affected by the Project (30%) and Aboriginals in the Northern Territory generally (30%) (102).

What happened after?

Hardly a year after the Ranger Agreement was signed, and the final barrier to exploiting Australia's most significant uranium deposit had toppled, Galarrwuy Yunupingu summarised Aboriginal feelings: "We are not getting anything from the mining, while the companies are getting everything they want – development, roads, unlimited access – over our land. What money we have got, we had to fight for" (204). A year later, Colin Tatz concluded there was "a pervasive Aboriginal belief that as present mining activity increases it may mean the death of the whole Region, in their terms" (103). Added Tatz: "An NLC executive member sees only rape of the country, the holes in the ground, the scarring of the landscape, the visual impact, as the worst of the happenings to date. The people need a 15- to 20-year moratorium on further activity, he says ..." (103).

Tatz, in his rather bitty 1982 conclusions on visits to Arnhemland, found that, while expansion of the Jabiru township (still a bitter point of administrative contention between federal and NT administrators) (205) was a considerable future fear, its impact took second place to concerns about what a bitumen highway would do to the land and food supplies, and worries among a small but important section of the people as to the potential destructiveness of increased "grog" drinking.

Importantly, Tatz rubbished uneducated conclusions that employment was the major benefit Aborigines derived from the mining (both at Ranger and Nabarlek). By late 1981, he stated, there was almost no Aboriginal interest in employment, while the Ranger mine had taken on only seven Aborigines – an office worker and six men for agricultural "regeneration" (103).

For the "at-least-they'll-be-rich" rationalisation,

Tatz reserved his most damning observations. "To date, revenue money has been a good deal less than expected and its distribution disastrous." Tatz and other researchers, including his own Committee, found that the Gagudju (later Gagadju) Association, incorporated at the end of 1978, had paid out A$300,000 only, by the end of the first year – much of this was anyway tied to future capital projects. Distributions had caused arguments and rifts between families; money was only being spent on booze and four-wheel drive Toyotas "and the poor mix of the two". Otherwise there was nothing on which to spend the money, and no proper banking facilities had been established in Arnhemland. "Greed has arrived in an ungreedy community", declared Tatz. There has been "unease, uncertainty and confusion", concluded the Social Impact Committee in September 1979 (206).

Lastly, Tatz, in his 1982 survey of the impact of mining, found that the NLC was a "grossly over-worked, under-staffed and under-financed" body, which had "yet to become its own master" (103).

In 1984, the Committee presented its consolidated report, covering the first six years of study, and the mine's impact. It was a devastating indictment by any standards.

The Committee distinguished between findings and conclusions. Among its conclusions were that:

- The Fox Commission recommendation on minimising adverse social impact had not been instituted. However, fears that changes in diet would be deleterious, that disease and alcohol abuse would run rife, that racial tension would be exacerbated, and that there would be gross intrusions into Aboriginal privacy or on sacred sites, were not realised.
- Serious inequities had been created in the distribution of royalties, and there had been a "general deterioration in Oenpelli's productive economy".
- Employment had not had a major impact on Aboriginal people, and there continued to be little interest among Aborigines in working on the project.

- Aborigines had not been adequately informed about the dangers of uranium and most people were ignorant or confused about the subject; several groups were anxious about its effect on water and food supply.
- The Jabiru township had not become an "open" town and was not exceeding the population limitations laid down by Fox.
- Notwithstanding any of the "positive" (or at least, non-negative) spin-offs mentioned above, the social impact of mining on Aboriginal culture had produced a "SOCIETY IN CRISIS" (in capital letters in the original document):
  "... the current civic culture is one in which disunity, neurosis, a sense of struggle, drinking stress, hostility, of being drowned by new laws, agencies and agendas are major manifestations" (207).

It was in this context that the failure to institute social programmes was most dramatic. Aborigines, said the Committee, were in transition from a wardship system to one of independence and self-management. Because they had had mining imposed upon them, in order not to sink beneath it they had to understand it, as well as "modern industrial economies, western technology, uranium as a nuclear fuel and as a weapon of war, the system of Territory and of national politics, global strategies between the nuclear and non-nuclear nations" (208). Aborigines needed new skills, which should be provided by "a national task force of competent professionals, the best in the country"; they needed to be drawn further into decision-making processes. Above all, there should be no new mining development in the region: if there were, it should be preceded by a new EIS, in which Aborigines participated (209, 210).

The findings of the Committee covered nearly three hundred pages. Only its most important observations can be encapsulated here:
- Two classes had been created in Aboriginal society, between haves (those with mining revenue) and have-nots (those without) (211).
- There had been no informed debate on health dangers with Aboriginal people (212).
- Mining couldn't cease, yet safety couldn't be ensured: this posed an inevitable "dilemma" (213).
- Drinking had enormously increased (214).
- There was a passion for owning and using vehicles, which itself legitimised the purchase of vehicles by the more powerful in the community, to the extent that it became "totemic" (215).
- The Oenpelli people were faced with a "massive task of social reconstruction" covering all aspects of their lives (216).
- The inadequacies of the Fox Report had basically worked in favour of the mining companies (217).
- "It is difficult to avoid the conclusion that somehow there has been a considerable surrendering of humanity – and certainly, from the Aboriginal side, a considerable surrendering of control, including control over the investment of meanings into the landscape" (218).

One result of the enormous opposition created by the Ranger and Nabarlek proposals was an insidious government-inspired campaign which ran up until the 1983 elections, proposing drastic alteration – or abolition – of the Aboriginal Land Rights Act, in order to give mining companies greater access to indigenous land (204, 219). This notwithstanding the fact that in 1981 the Federal Minister for Mines and Energy accused miners in the NT of being "arrogant" and failing to employ Aborigines (220).

By the mid-1980s the uranium issue was still dividing as well as disturbing Aboriginal communities in the NT, like no other issue before or since. While members of the NLC and some traditional owners actively campaigned for further mines – notably Koongarra – to be opened (see Denison, Noranda and Pancontinental), other Aboriginal people, and land councils outside the Territory, strenuously fought against such developments.

In 1983, the Gagadju Association discovered that a white employee of the Association was living in a A$114,000 house in Darwin while Aborigines had to live in shelters costing a mere A$18,000 – which had not even been completed (221). By then the Association had received A$2.05 million in royalties plus an annual rent of A$200,000 from ERA. Despite this, they were living in tin-roofed humpies in a camp under some trees (221).

Two years later, the Association declared that it wanted to break away from the NLC, because of discontent with its administration (222). The same year, a report by an Australian National University academic, leaked to the Melbourne *Age,* claimed that millions of dollars in royalties, supposedly set aside for Aboriginal use, had been siphoned off by the government and was being used for general purposes in the Northern Territory. The Aboriginal Benefit Trust Account had, according to Dr John Altman, received some A$17.064 million in "royalty equivalent" in 1983-84, only 11.7% of which was given directly to Aborigines (223). Of the remainder, much had been invested without earning any investment income. Some A$200 million might have been lost in this fashion since the mine opened (224).

Shortly afterwards, the NLC itself went to the Australian High Court, seeking to halt operations at Ranger until a new safeguards and royalties agreement could be negotiated. Both the traditional owners at Oenpelli and the NLC argued that they had no contractual relationship with ERA, and royalties allowed them were "grossly inadequate" compared with mining agreements in other countries (225). Galarrwuy Yunupingu, citing the recent spillages at the mine-site, maintained that the NLC had been unable to enforce environmental safeguards (226).

The NLC sought an injunction on the grounds that the Ranger Agreement had been signed out of "duress, undue influence and unconscionable conduct by the Commonwealth [federal government]" (224). Critical information and crucial documents had not been made available in 1978; the federal government had threatened to impose the Agreement and the traditional owners were "inexperienced" in commercial matters relating to uranium (224). Galarrwuy Yunupingu also maintained that a provision for a review of Ranger's operations after five years

had been ignored – and for the first time acknowledged that only 3 of the 40 traditional Oenpelli owners had been at the signing ceremony for the Agreement. Moreover, in 1979, the federal government had violated its obligations to Aboriginal people by not informing them of the proposed formation of ERA (226). "We have grown up a lot since [1978]," declared Yunupingu (226).

Bernard Fisk, for ERA, countered that negotiations were a matter for the government, not the company, and that the NLC's main concern (in his view) was to get more money (226). The Australian Democrats promised support for the NLC in their action (226).

The court case dragged on through 1986 (227), when the High Court promised to test whether the fiduciary relationship between Native Canadians and the the Canadian government could apply to Australia (228). Finally the case was ordered to be heard by the Full Bench of the High Court – a development which had has ensured that until 1991 the case has not yet been argued in full (229).

Kakadu

The Kakadu agreement was signed along with the Ranger Agreement, and talks between the federal government and the NLC over the National Park proceeded apace with the mining talks (102). In fact the idea of a national park had been around for a long time (102, 230). The Fox Commission viewed it as a necessary "buffer" between uranium projects and both Aborigines and the environment (231). The government ostensibly followed all the Commission's main recommendations on Kakadu, and by August 1977 had mapped out a *modus operandi* for declaring the National Park. Stage One would be started immediately (delineated in blue on Map 16 of the Ranger Second Report) (232), together with a rectangle of land immediately south of that boundary, and the Woolwonga Aboriginal reserve and wildlife sanctuary (119). The Djindibi and Djoned clan areas would not be granted Aboriginal freehold (Fox found "insufficient" evidence of Aboriginal traditional occupation) (233) and – as already mentioned – Fox advised that the Mudginberri and Munmarlary pastoral leases should have their title taken over by the government and be leased back to Aboriginal people, for inclusion in the National Park (129).

There would, declared the government, be no mining or exploration "for the time being" in this area – and mining would only be permitted in Stage Two of Kakadu, with the express authority of the government (119). The regional township of Jabiru would be within the National Park boundaries, but not on Aboriginal land. Both the Ranger and Jabiluka mine-sites would be excluded from the National Park, but included in Aboriginal land areas ceded under the Land Rights Act (234).

The agreement was signed at Oenpelli in November 1978. As with the Ranger Agreement, virtually no time was allowed for Aboriginal people to discuss its terms (the draft was only presented to the NLC in September) (129). Its main provisions were: a 100-year lease of Aboriginal land within the Park to be made to the Director of the National Parks and Wildlife service (ANPWS); further lease of Aboriginal land in the area, should the director request it; the Director could sub-lease any area of the Park, provided it was set out in the management plan; the Director was obliged only to "consult" with Aborigines in making the management plan; Aborigines would be employed in managing and controlling the park (though only in "reasonable" numbers – whatever that might mean); traditional owners could move freely in the park; sacred sites would be identified and knowledge of Aboriginal culture and traditions would be "promoted" (235).

It is essential to point out that Aboriginal people were required to lease their own land to the ANPWS: the penalty for non-compliance would be the appointment of an arbitrator – exactly the mailed-fist-behind-the-velvet-glove in the Ranger negotiations (129).

In the first half of 1978, the NLC had made it quite clear to the federal government that at no point should any part of Kakadu be consigned to the NT administration – Aborigines would "just go down the drain" declared Yunupingu in May that year, should the NT take over day-to-day running of Kakadu (143, 236). A year before, confidential talks had already been held between the Canberra and Darwin authorities, after the Chief Minister of the NT, Doug Everingham, expressed concern that a National Park would "obstruct mining" (237).

By 1988, pressure from the NT opposition to take back control over Kakadu was formalised in a policy to "repatriate" the National Parks of both Kakadu and Uluru (Ayers Rock), when the NT achieved statehood. A future Liberal-National Party coalition government would also reverse the "three mines" policy of the ALP (238).

The first stage of Kakadu was declared in June 1979. It comprises some 6,000 square kilometres between the East and South Alligator Rivers, most of it south of the Arnhem highway, with a few strips elsewhere – including an area along the coast, north of Mudginberri and Mamalary (102).

These two pastoral leases were to be included in Stage Two of Kakadu – a somewhat bigger (7000 sqare kilometre) area – which was declared on November 15, 1983. By then the park was a huge tourist attraction, with US tour operators and their clients comparing the region with Amazonia (303). The number of tourists had reached 200,000 by 1988 (304) when the Four Seasons Kakadu Hotel, owned by the Gagadju Association, managed by the white Four Seasons group, joined the Cooinda Motel and Border Store and partly-owned South Alligator Motel among the tourist interests held by the Aboriginal association (239).

By 1983, the tussle between the NT and Federal administrations had deepened, and was to become critical after the ALP came to power. Apart from the Ranger and Nabarlek mines, the Mereenie and Palm Valley oil fields, and the Jabiluka project, mining exploration had all but come to a halt in the NT (240) and about a third of the state was under Aboriginal claim (241).

In 1983, the Koongarra lease – previously excised from Kakadu Stage One (242) – was re-incorporated into Stage Two, after fears that the ALP would renege on its "three mines" policy (243). This angered some Aborigines, who by then were more in favour of mining than being the junior "partners" in a National Park which had clearly not enjoyed their support (see Noranda and Denison). Three years later, Aborigines again objected as the federal government proceeded to include the Ranger 68 deposit in Stage Two (244), and announced that there would be a Mining Conservation Zone in Kakadu Stage Three – which would actually be a Mining *Permit* Zone. BHP, together with EZ and Noranda Australia, had plans to exploit gold, platinum and palladium at Guratba (Coronation Hill) in Kakadu Three – the site of Rum Jungle uranium mining in the 1950s and 1960s (239). In late 1988, the Jayown people – custodians of the land in the "Sickness" country (so-called because uranium mining had disturbed the ancestral being Bula, thirty years before) – laid claim to the Gimbat and Goodparla pastoral leases, which had been compulsorily acquired by the federal government for Kakadu Three, but have not been ceded to Aboriginal control. The Jayown people made it abundantly clear that they wanted no mining or exploration in the Sickness country (239) (see BHP).

The 1984 Consolidated Report on the Social Impact of Uranium Mining on the Aborigines of the Northern Territory – while delivering a swinging attack on other aspects of the Ranger project – gave Kakadu a comparatively clean bill of health (207). It concluded that the Park was providing the "buffer" envisaged by the Fox Commission, "generally promot[ing] Aboriginal interests, had a strong commitment to Aborigines and had been a successful employer of Aboriginal people" (207).

Clearly the ANPWS staff had worked overtime to allay fears expressed three or four years before: specifically by the same Social Impact Committee which, in 1981, had challenged the concept of "consultation" embraced in the draft plan of management, pointing out that it substituted for true "participation" (245). Fears expressed by Colin Tatz, the Chairman of the Uranium Impact Project Steering Committee, had been even more trenchant. He conjectured, in 1982, that Kakadu was "in many ways ... a microcosm of Australian Aboriginal issues". The plan of management tended to equate Aboriginal interests with management interests; was ignoring the role of "buffer zone"; ignored Justice Woodward's strictures on management; and made Aborigines simply the "receivers of consultation" (103).

There has obviously been some improvement since then in operations, if not principles of management. One major development was the listing of Kakadu Stage One as a World Heritage site in 1980; Kakadu Stage Two would be listed seven years later. Then the World Heritage Commission praised the Australian authorities for its management of the Park and said it would consider also listing Stage Three (246). Such listing of National Park land is bound to strengthen conservation interests – although listing of Aboriginal land outside the Park may be calculated to diminish Aboriginal control (247).

Environmental disasters

The Fox Commission made it clear that the Ranger partners' pre-1976 plans did not meet their criteria for environmental protection in several respects. The lease was too close to sacred Mount Brockman and the boundary would have to be moved, thus excluding the No. 2 deposit (248). The Commission also insisted that tailings should be returned to the mine-pits, and that strenuous measures should be taken to limit the spread of contaminated water: alternatives to their proposals should only be accepted if they reached the same ends (112).

Ranger Mines Uranium Pty Ltd drew up its initial Environmental Impact Statement (EIS) in February 1974. The EIS acknowledged problems from radiated water seepage, but maintained these would be no greater under mining than they already were in nature, producing "... radium levels in billabongs which are above levels set by the World Health Organisation for potable water". The EIS also agreed that, in the dry season, such levels sometimes exceed "the toxic limit at which 50% of fish die" (249). To limit the increased dangers of this "natural" phenomenon, the mine would have a series of "bunds" (retention walls) and retention ponds. The Australian Mining Industry Council (AMIC), in a broadside against the land rights movement, delivered in Canberra in February 1978, attacked the standing of the NLC – especially in relation to share of royalties and compensation (251). One month later, AMIC firmed up its position, claiming that compensation should only be paid to Aborigines on the basis of disturbance to their land (250).

One particular aspect of the Fox Inquiry's recommendations (305) stuck in the gullet of the miners as represented by the AMIC. This was that "the environment protection provisions should be made legally enforceable, and both the Director and the Northern Land Council given standing to enforce them". AMIC couldn't stomach an Aboriginal body monitoring the activities of whites with picks and shovels, and urged the government to review this recommendation (251).

In the final event, however, the AMIC was over-ruled, and the NLC is represented on the Co-ordinating Committee which meets monthly to review operations of the mine (252).

ERA set up its Environmental Division one year before production began, to monitor water quality and flows, air quality, dust concentrations, as well as radioactive atmospheric emissions, and other phenomena: the division is also responsible for rehabilitation of the Project Area, and its monitoring of radiation is supposedly carried on well beyond the mine site itself. ERA was required to operate the ALARA principle (keeping radiation exposure "as low as reasonably achievable", taking into account "social and economic" (*sic*) factors), and operate its environmental programme according to BPT (best practicable technology) (253).

But "BPT" is a catch-all phrase which can mean many things to many people (254). The Fox Commission made it clear that BPT was not simply to be an adaptation of existing practices in other uranium mines, or in other kinds of mining projects in Australia (129), but to be exactly what it implied: the best available from anywhere, to meet the needs.

Within a short period, the Ranger partners had diluted BPT to become "the most economical technology" (255) – confirmed by project manager Alan McIntosh as late as 1979, when he declared that "uranium mining should be treated just like any other mining operation" (256).

Whether or not, as the company grandiosely claimed, "never before in the history of Australian mining had environmental considerations played such an important role" (257), in theory unusual and (for Australia) unprecedented measures were taken to limit radiation exposure on site – including the provision of thermoluminescent dosimeters and the regular hosing-down of all vehicles (253).

Ranger is located in the Alligator Rivers region of the Northern Territory, a hot "wet-dry" area, with monsoonal climate, and highly variable water flows (258). Both the Ranger and Pancontinental deposits are located in the catchment of the Magela Creek system, an area teeming with wild life, on land and in water, and which includes the Arnhemland escarpment, as well as lowlands, flood-plains and tidal flats. During the dry season, the system evaporates, to a series of disconnected swamps and billabongs (258).

On the announcement that Ranger could proceed, the federal government appointed a Supervising Scientist to supervise virtually all aspects of the mining with new legislative powers. This scientist would head a new Research Institute with experts to advise him. In addition, he would call on a Co-ordinating Committee, compromising the Australian National Parks and Wildlife Service, the mining companies, and the Northern Land Council (259). There was also to be a Research Institute. Initially these three bodies were entirely under the federal government – the Ranger Commissioners having strongly condemned the environmental protection record of the Northern

Territory administration (citing the pollution at Gove in particular – see Alusuisse) (260). But after the NT gained self-governing status in 1978, it was brought increasingly into the picture, particularly in environmental monitoring (261).

From the start, the Supervising Scientist – under the Environment Protection (Alligator Rivers Region) Act 1978, passed in June that year – could only pronounce on uranium mining as such: exploration, the social impact of miners and tourists, the effects of the new township of Jabiru, were not included in his or her brief (252).

The Supervising Scientist has to carry out his duties amid a complex web of legislation, ranging from the Atomic Energy Act of 1953 (under which Ranger was officially given permission to proceed), to the Environment Protection (Northern Territory Supreme Court) Act of 1978. The major instrument, however, was the Uranium Mining Environment Control Act of 1979. In addition, the Environmental Requirements for Ranger, based to a large extent on chapters 5 and 6 of the Second Ranger Report, laid down the environmental conditions under which Ranger itself could operate (252). Among the 45 Requirements (262), were those ordering the mining partners to submit their tailings dam and water retention plans to the Supervising Scientist for approval, and a demand that tailings should be replaced in the pit "as soon as practicable" (252).

The latter provision was one which was lengthily debated between the NLC and the federal government, before Aboriginal consent was obtained for Ranger.

While the Supervising Scientist has few powers of enforcement (these are delegated to the Coordinating Committee) his or her powers of regulation, under the "Environmental Requirements" are clearly considerable.

A 1977 study by the Department of the Northern Territory and the AMIC pointed out that the groundwater in the region already emerged through at least one bore (at Oenpelli itself) with radioactivity above the recommended level set by the World Health Organisation (WHO),

while "wastes entering the [South Alligator] River might reach [other] aquifers" (263, 264). A list prepared by the two authorities, of potential wastes which might affect the high water-table, included: suspended solids entering the creek system from the top soil dump; dissolved radium, uranium, zinc, lead, copper and other heavy metals leached from the overburden, ore dump, and pit water; and radium, uranium, heavy metals, amines, acidity and soluble sulphates deriving from neutralised raffinate and the tailings dam. The report demanded a high standard of design of the tailings dump, careful siting relative to surroundings, and collection and pumping back of any seepage (263).

This report – and others – confirmed that Ranger's aquatic environment was the region most likely to be affected by uranium mining (266). The most dangerous single factor was the release of contaminated water from retention ponds, or seeping from the tailings dam, which "may potentially affect a large area of the downstream Magela Creek catchment" (266). Inevitably (but reprehensibly) the actual effect of the release of heavy metals from the mine-site (estimated by Barry Hart to be 25% of the natural loads of copper, lead, zinc and uranium per year, notwithstanding a "no release" strategy from the second retention pond) could not be calculated in advance. However, by the end of 1979 one team of researchers was suggesting that levels of copper as low as 10 micrograms per litre could rapidly kill the algae (267).

Other preliminary surveys raised considerable doubt about the ability of the Creek system to handle a huge excess of suspended solids: a lot in the future would depend on successful re-vegetation of disturbed areas, and "whether the tailings are replaced in the mine pits or left in the tailings dam" (266). Then there were the contaminants from the town of Jabiru – particularly phosphorous and nitrogen from domestic sewage (259), and pesticides used at the Mudginberri pastoral lease, which "may already have caused malformation in certain frog species" (266).

The tailings dam – the biggest earth wall dam in Australia (248) and the major storage site for solids and radioactive wastes which could seep into the river and creek systems – was the subject of considerable appraisal by the Ranger Commissioners who, several times during the Inquiry, asked the company to re-submit its plans with more details (268). The dam was projected to cover an area of about 125 hectares, and to be 4 kilometres in length and up to 30 metres above ground level. Built of earth and rock on the outside, it would be lined with a "relatively impervious clayey material" (258) and an exterior trench of the same material "grouted" to a depth of 40 metres underground. Importantly, the Commissioners also insisted that a blanket of water should cover the tailings dam, even in the driest of weathers (258). This was primarily to reduce radioactive emissions from the tails: such a cover, the Ranger partners estimated, would reduce radioactivity from 3.9 curies/day to below 0.4 curies/day (269). The Commissioners estimated that, when full, the dam had the potential of releasing no less that 30,000 curies each of radium 226 and thorium 230 (a figure based on original estimates of the deposit reserves) (269).

The height of the dam was to be raised "progressively" in six stages over 20 years, providing for eventual containment of some 27 million cubic metres of "settled tails" from orebody No. 1. Provisions would have to be made separately for the estimated 18 million cubic metres of tailings produced from Orebody No. 3 (269).

From the start, the Ranger partners were quite clear that, although three methods of constructing a supposedly "impervious" blanket to the dam had been under scrutiny (one using clayey material, a second using a mixture of clay and bentonite, and a third using butyl rubber membrane) none of these "is expected to achieve any appreciable reduction in the contamination of deep aquifers, and the overall reduction in seepage would be neglible" (269). As the butyl solution was both the most expensive and the least tried of the three alternatives, it was naturally given the shortest shrift.

It was claimed that all run-off would be retained at the start of the wet season (November-December), and any deliberate release would be made then (269). As we shall see, this rule-of-the-thumb was broken on several occasions over the next decade.

Although the Ranger Commissioners did not believe that seepage from Ranger would be as disastrous as that from Rum Jungle (see CRA) (270), they conceded that the extent of damage by acidity, heavy metals, and toxicity along the Finniss River from the Rum Jungle mine was not known. Similarly, "the rate at which radon would escape from the [Ranger] tailings dam is even more uncertain than the rate of radon release from the pit" (271). In addition to uranium and radium behind the dam, there would also be high levels of magnesium, calcium, manganese, arsenic and mercury (268). In a twenty-year period, seepage directly into Gulungul Creek from the dam could reach nearly 300 cubic metres a day.

The Commissioners were especially worried at the tendency of fauna and flora in the region to take up excessive doses of radioactive materials (272). A report by the NT Medical Service, four years previously, identified several factors particularly affecting the health of Oenpelli Aboriginal people: tuberculosis, typhoid, leprosy, parasitical- and vector-borne disease (malaria, encephalitis, schistomiasis), social and environmental disease – and radiation-induced diseases (273). The Alligator Rivers Study (mentioned above) reproduced almost all the findings of the NT Medical Service, with the exception of a notable passage relating to radiation-induced disease. This passage reported a higher-than-average incidence of carcinoma of the lung and leukaemia among the Oenpelli people – which it said might be attributable to natural radiation and "may tend to increase [with the] disturbance of the uranium deposits ..." (273).

The Ranger Commissioners confirmed this worry, stating that "local Aborigines not associated with the mining activities are much more likely to be exposed to high levels of ingested radium" (274). In addition, the Commissioners found evidence that, during release periods from the tailings dam, "the concentration of some metal contaminants in the creek would rise to levels some five to ten times above the calculated safety levels" (275).

Equally important was the likely situation after mining, when the tailings dumps were abandoned, plugged with rock and revegetated. The Ranger Commission foresaw enormous consequences from such dams, however well-contained, sticking up above the level of the countryside, with erosion and seepage which could affect the local people and environment "... for thousands of years" (276). In the event, the companies did agree to return tailings to the mine-pits, but could not guarantee that they would be filled-in level to the ground (129).

Even before the dam was fully constructed, and the mine operational, problems were being reported. In early 1980, Ranger engineers had to breach the partly-built dam, after heavy rain threatened the entire edifice. The Northern Territory Anti-Uranium Group called for an inquiry (which the NT Minister of Mines and Energy, Mr Tuxworth, refused), citing the fact that the dam builders were none other than Mineco, a subsidiary of CRA – the very company responsible for the disastrous pollution of Rum Jungle (204).

Between 1979 and 1981, the Ranger site was subjected to hundreds of major "blastings", each measuring around 4.2 on the Richter scale (277). These certainly did not contribute to the stability of the tailings dam! Little wonder that micro-cracks began to appear, extending for long distances, and water began to seep out from under the pile – a prospect warned against in the Ranger Report. Although the company kept six pumps busy, trying to push water back into the dam to cover the tailings, seepage was so extensive that an island appeared in the middle of the dam, and the mine itself was closed for four days (277).

The Shadow Minister for Environment and Conservation, Stewart West, pointed out that as the dam had not been built down to bedrock, such radioactive seepage was bound to occur (278). The Minister for Home Affairs and the Environment, Mr McVeigh, counter-claimed that no contamination had been discovered,

and the environment was quite safe. The buck was passed from Ministry to Supervising Scientist, and back again (279). But instead of changing practices, the authorities decided to change the rules. Henceforth, it would only be necessary to keep the tailings damp (280). (In any case, as the *Financial Review* pointed out at that time, the Codes of Practice under which the mine operates only had to be observed voluntarily by the authorities) (281). In mid-1982, a prestigious government body, the Australian Science and Technology Council, in a report to Prime Minister Fraser, declared that research nesessary to set safety standards and environmental protection in the Alligator Rivers region was proceeding "far too slowly" – a permanent field laboratory had still to be built, and the staff of the Supervising Scientist comprised a "barely visible research team" (282).

However, little more was heard of the seepage problem until 1985, when, by the middle of the dry season, the dam was brimful with water and the company's "juggling tricks" (283) – pumping from the pit to the dam and (presumably) back again – had failed to contain the crisis. By then ERA was claiming that periodic release of excess water was part of its "best practicable technology", while conservationists and the Northern Land Council were adamantly opposed to the release of any water that was not of potable quality (283). Keith Taylor of the Australian Conservation Foundation (ACF) argued that the company had deliberately restricted its options (building another storage dam, land irrigation, enhanced evaporation) to keep costs down (283). Such a release, said Keith Taylor, would mean "8000 years of pollution due to radioactivity" (284). The Supervising Scientist's office apparently advocated a one-off discharge (283), although it admitted that the effects of heavy metal contamination on the environment would be "indeterminate" (283).

In the event, the NT Minister for Mines and Energy refused the company's request (283). However, disorder in tailings management was meanwhile giving rise to a series of "accidents" involving spillages, leaks and failures in the tailings line (285, 286). "The whole thing is now leaking like a sieve," commented mining engineer Willy Wabeke (287).

In September 1985, up to 150 cubic metres of contaminated, radioactive water was sprayed under pressure over a 2000 square metre area outside the Restricted Release Zone (286). Only a few days later, it was discovered that tailings water had been pumped for ten days through a trial irrigation spray system, intending to evaporate excess water (286), and used to water a lawn (226)!

In an interim report to the federal government, the Supervising Scientist, Bob Fry, commented that this incident "confirms my view that Ranger should lift its game in relation to what might be described as basic housekeeping practice" (288). (Game? Housekeeping? Where did this man Fry do his training? On a croquet lawn?)

By November, ERA itself was admitting to something of a crisis. In a report submitted to the Minister for Arts, Heritage and the Environment, called "Application of the Best Practicable Technology to the Water Management System", the company outlined several options besides release into the creeks. These included spraying the water onto land near the mine, constructing more evaporation ponds, and treating contaminated water before spraying (289).

In response, the Minister, Mr Cohen, refused to allow release of water into Magela Creek, declaring that ERA had made "a series of blunders" which included over-estimating the rate of evaporation, under-estimating precipitation, and miscalculating the density of tailings suspended in water (289).

The following year, the chair of the House of Representatives' Standing Committee on Environment and Conservation was suggesting that the company would have to construct another retention pond, as well as carry out local irrigation (290).

By then, however, the company had been allowed to release 160,000 cubic metres of water from the waste rock dumps adjacent to the Magella Creek: otherwise, it insisted, it might have to close in two years (291). This ex-

perimental release succeeded in aborting huge numbers of mussels (292). ERA got further permission in February 1986 to release another two million cubic metres of water from a retention pond (292) – a measure, one journalist stated, which would make the decade's "earlier ecological bunfights pale into insignificance as the traditional landowners at Kakadu, the Gagadju, join with outraged conservationists to mount a concerted campaign to protect the area in which they still live, hunt and fish" (291).

A serious leak of sulphuric acid affected 60 workers in March 1986. It was met with a walkout by workers from the MWU and Australian Society of Engineers (ASE), and a shipment of 27 containers was held up (293, 294).

Scab labour failed to prevent a fire breaking out in the crushing plant (295), and these incidents led to an "indefinite strike" as the unions called for safety review by independent consultants: again, there was no response from ERA (295). Instead, the company alleged "induced breach of contract" and sought damages of A$3.17 million due to delays over shipment. Various right-wing Australian newspapers joined the inevitable "union bashing" (293, 297).

In 1987 and 1988 there was a distinct impression of *déjà vu*, as the grim pantomine continued to unfold: the company seeking permission to release contaminated water, the government prevaricating or refusing, and the dam "spilling" anyway. March 1987 saw two federal ministers fencing over the issue: the Minister for Resources and Energy, Senator Evans, in favour of allowing the company to release more water; the Aboriginal Affairs Minister, Clyde Holding, and the Minister for Arts, Heritage and the Environment, opposed (298).

In early 1988, the Northern Territory Environment Centre insisted that the NT government should prove its claim that there was no damage to the environment (306). This followed yet another spill, when up to 100 cubic metres of contaminated water, containing uranium and calcium carbonates, overflowed into the restricted zone – and only hours before a Senate safety team called at the site (299).

Both the Chair of the Parliamentary Standing Committee on the Environment, John Black, and the NLC, protested: the NLC in a forceful letter, stating that it was "absurd" for Aboriginal people "to expose their society to this political risk". It pointed out that, among northern mining companies, Ranger alone had failed to answer Aboriginal concerns "through appropriate designs and contractual commitment to zero release management of contaminated water" (300).

Only half-a-year later, however, the tailings site was once again endangering not only Aboriginal people, but also the 1200 residents of Jabiru township, the 100,000 annual tourists to the Kakadu National Park, and even drivers using the Arnhem Highway, as an abnormally dry wet season exposed no less than a third of the tailings in the dam to the air (301). Said FoE representative, Pat Jessen, "[The authorities] have shown a blatant disregard for public health at the mine and in the surrounding area". FoE called for the mine to be shut down until the exposed tailings could be adequately covered (301).

Soon afterwards, the Office of the Supervising Scientist reported a fault in the radiation monitoring equipment at the mine, which had apparently permitted a "quantity" of uranium to find its way into the tailings dump (302).

In early 1989, Ranger released contaminated water from its retention pond no.4 (RP4) – the collection point for runoff from the waste rock dump at the mine (307) – provoking tremendous anger among the traditional owners and a demand for an inquiry from the Northern Land Council (308). It appeared from contemporary reports that the contamination originated among nearly half a million tonnes of waste material "accidently" dumped in 1988 (309). Dr Vince Brown, deputy director of the Office of the Supervising Scientist, declared the release "out of control": "We will never know what dilution the water will get," he announced – referring to the dispersal of radioactivity in the Magela creek system – "Safe disposal of water demands control and we've lost control" (310). Although the release on this occasion was stopped after around 10,000 cubic metres had been siphoned over a spillway to the Djalkmara billabong (which exits into the Magela system), another two releases followed soon afterwards (307).

At the 48th annual meeting of the Northern Land Council, the members expressed outrage at the events. Said Yunupingu: "The traditional owners have always said they didn't like the mine, but they accepted it because the environmental controls over it were supposed to be really tough ... They thought their country was going to be protected and now many have stopped hunting near the Magela Creek system because they don't know whether there could be any damage. Ranger is treating our people as if they are invisible and their real worries of no consequence. The Federal and State Governments are playing politics, while they argue about who knows best and what scientific methods should be used. Aboriginal people don't care about these games; but we do care about our right to have a say over what happens in our own country."

Confirming the strong feelings of the Aboriginal people at Kakadu, the Supervising Scientist, Bob Fry, in his 1988/89 annual report, said that the willingness of Ranger to cooperate with his office had "diminished" (307): "By increasingly ignoring OSS advice on environmental issues [Ranger] appeared to wish to establish that the OSS performs no useful function ... It has attempted to impugn the scientific credibility of the Office and has lobbied for its disbandment."

Conceding that the changes in water quality criteria around the mine were "not large compared with conventional water quality criteria for the protection of aquatic biota" Mr Fry nonetheless said that "their actual effect on the ecosystem is not known" (307).

Ironically, in 1991, Ranger received a four-star safety rating, placing it among the safest industrial sites in Australia (311). A few months later, ERA bought up the Jabiluka uranium deposit of Pancontinental (312), thus opening the way for a merger between the two operations and a large scale expansion of Ranger (313).

Further reading: (There is a huge reading list of critical material on Ranger – probably more than for any other project in the history of uranium mining, including Rössing. The following list inevitably includes publications now out of print: further information can be obtained from FoE Melbourne, or MAUM Sydney, at the addresses below.)

In 1977/78, the Australian MAUM (Movement Against Uranium Mining) produced an excellent study pack entitled *ENERGY and U*, including essays on all aspects of uranium mining, Ranger particularly. Several papers addressed the (false) economics of nuclear power.

Unfortunately, there is no exact date or publisher included on the pack. Readers interested can contact: Minewatch, 218 Liverpool Rd, London N1 1LE, for further details and off-prints of specific articles.

The First Report of the Ranger Uranium Environmental Inquiry, 1976, (Australian Government Publishing Service/AGPS), Canberra, 1976.

The Second Report of the Ranger Uranium Environmental Inquiry, 1977, (AGPS) Canberra, 1977.

Australian Union of Students, *The Case against Mining and Export of Australia's Uranium*, 1975.

Australian Council of Churches, *Uranium and a Nuclear Society*, (ACC) 1977.

Denis Hayes, Jim Falk, Neil Barrett, *Red Light for Yellowcake, the Case Against Uranium Mining*, (FoE Australia) 1977.

Uranium Mining in the Northern Territory, (Christian Action Group, Galiwin 'ku Parish, Uniting Church of Australia) 1978. (This is a sketchbook: simplistic, but effective.)

Uranium Mining: Impact on the Australian Economy (MAUM) Victoria, undated. (A well-thoughtout, if brief, analysis on the one major aspect of the Ranger project not covered in this essay of the *File*).

Land Rights and Uranium: Uranium Mining as it affects Aborigines in the Northern Territory, (FoE & MAUM) Sydney, 1978.

What do Aborigines say about Uranium? (Black Defence Group).

Reaction to Fox Report, (Federal Council for the Advancement of Aborigines and Torres Strait Islanders/FCAATSI) 1977.

Uranium: Citizens' Response, (FoE) Victoria, Melbourne, 1977.

People's Guide to the Ranger Inquiry Second Report, (FoE Darwin, and NT Trades and Labour Council), Darwin, 1977.

The New Uranium Legislation, (MAUM) Melbourne, 1978.

Uranium miners get off our land! compiled by Bill Day from "Bunji" (FoE and MAUM) Sydney, 1978.

The Second Ranger Report: Implications for the Aboriginal People, (Aboriginal Land Rights Support Committee).

Speech by Bill Hayden MP, Uranium Rally, Sydney, Dec 3rd 1979 (cyclostyled).

C Kerr, *The Components of the Uranium Package: The Fox Inquiry and its Recommendations* paper for SIUM Project Workshop, AIAS, Canberra, 8/78.

(Ed) Stuart Harris, *Social and Environmental Choice, the impact of uranium mining in the Northern Territory* (CRES /Centre for Resource and Environmental Studies, Australian National University) Canberra, 1980. (Contains important essays by R M Fry on environmental protection and uranium mining; by Charles Kerr on work hazards; by H C Coombs on the social impact on Aboriginal people.)

(Ed) R M Berndt, *Aboriginal Sites, Rights and Resource Development* (Academy of the Social Sciences in Australia, University of Western Australia Press) Perth, 1982. (Contains article by J R Von Sturmer, Project Director of the Social Impact project referred to in the last two titles, raising essential questions about Aboriginal self-management, without providing any answers – except the quixotic comment that without the Ranger and Nabarlek mines, there would not be the money to enable Aboriginal communities to manage themselves in the face of the pressures they have newly been put under!)

(Eds) Nicholas Peterson and Marcia Langton, *Aborigines, Land and Land Rights*, (Australian Institute of Aboriginal Studies) Canberra, 1983. (An excellent book all round, with the most pertinent contributions – to the uranium issue – being by Daniel Vachon and Phillip Toyne, Peter Carroll, Sue Kesteven, and Marcia Langton.)

"Northern Territory", Ross Howie, in (ed) Nicholas Peterson, *Aboriginal Land Rights, a Handbook*, (Australian Institute of Aboriginal Studies) Canberra, 1981. (Succinct summary of Land Rights legislation and legislative background to Ranger agreement.)

(Ed) Mary Elliott, *Ground for Concern, Australia's Uranium and Human Survival*, foreword by Paul Ehrlich, (Friends of the Earth/Penguin Books) 1977.

Paul Kelly, *The Hawke Ascendancy, A Definitive Account of its Origins and Climax, 1975-1983* (Angus and Robertson) London, Sydney, and Melbourne, 1984. (Stimulating blow-by-blow account *inter alia* of how an anti-uranium party became a pro-uranium one. See especially pages 193-198, 220-210.)

Articles: Among the many readable and informative articles published on Ranger, the following are a few of the more interesting and important:

"The Battle of the Giants" by Stewart Harris (honorary advisor to the NLC), *The Australian* 28/2/78.

"An Interview with Galarrwuy Yunupingu" *Time and Energy*, No. 3, Sydney, 7/78.

Chain Reaction, (FoE Victoria) Vol. 2 No. 4, 1977; Vol. 4 No. l, 1977; and Vol. 4 No. 2/3.

"Land Councils and Uranium Mining" *Land Rights News* (Northern Territory Land Councils) No. 10, 7/78.

Time and Energy, No. 6, 11/12/78, Sydney.

Films: *Better Active Today Than Radioactive Tomorrow*, directed by Nina Gladitz, West Germany, 1979.

Dirt Cheap, producers Clancy, Hay and Lander, Sydney Film-Makers Co-operative, 1981.

Contacts: MAUM, 247 Flinders Lane, Melbourne 3000, Australia.

FoE Australia, 222 Brunswick Street, Fitzroy 3065, Australia.

Northern Territory Environment Centre, PO Box 2120, Darwin, NT 0801, Australia.

References: (1) Charles Kerr, Ranger Inquiry Commissioner, quoted in (20) (2) Bob Hawke, formerly President of the ACTU (now Australian Prime Minister), quoted in *Gua* 8/9/77, just after the Fraser government gave Ranger the go-ahead. (3) Patrick White, Nobel prizewinner, quoted in *Age* 3/8/84. (4) Senator Don Chipp, leader of the Australian Democrats, quoted in *FT* 1/6/84, as the ALP moves towards reversing its anti-uranium export policy. (5) *EPRI* 1978. (6) *Age* 14/3/75. (7) Neff. (8) *Australian* 4/2/76. (9) *Uranium Deadline*, Vol. 1, No. 4, 8-9/77. (10) *Australia and Uranium* (AGPS) LR 77/483, Canberra, 1977. (11) *Time and Energy* No. 6, Sydney, 11-12/78. (12) *Courier Mail*, Brisbane, 24/11/76. (13) *Courier Mail*, 10/1/77. (14) *Herald*, Melbourne, 26/11/76. (15) *Sun*, Australia, 11/5/77. (16) *Australian*, 26/5/77. (17) *Chain Reaction*, Vol. 2, No. 4, (FoE) 1977. (18) *Age* 26/5/77. (19) *Telegraph*, Brisbane, 26/5/77. (20) *National Times*, Australia, 30/5/77. (21) *Sydney Morning Herald*, 26/8/77. (22) *Gua* 26/8/77. (23) *Age* 26/8/77. (24) *Mirror*, Sydney, 26/8/77; *Australian Express* 2/9/77. (25) *Obs* 28/8/77. (26) *SunT* 28/8/77. (27) *Canberra Times* 26/8/77. (28) *Canberra Times* 30/8/77. (29) *Herald*, Melbourne, 30/8/77. (30) *DT* 24/10/77. (31) Paul Kelly, *op cit* (32) Ross Howie, "Northern Territory", in Nicholas Peterson (Ed) *op cit* (33) Hayes, Falk, Barrett, *op. cit.* (34) *Age* 2/4/84. (35) *FT* 22/3/83; *FT* 7/11/83. (36) *Age* 15/12/81. (37) *Financial Review* 8/7/82. (38) *Age* 8/7/82. (39) *Age* 3/11/83. (40) *Age* 9/2/84. (41) *Financial Review* 29/3/84. (42) *Age* 29/3/84. (43) *Age* 30/4/84. (44) see also *Financial Review* 28/3/84; *Financial Review* 30/3/84. (45) *Age* 4/4/84. (46) *Age* 2/4/84. (47) *Age* 21/5/84; *Financial Review* 21/5/84. (48) *Age* 5/4/84; *Age* 6/4/84. (49) *Age* 29/5/84; see also *MJ* 20/7/84; *FT* 1/6/86. (50) *Financial Review* 1/6/84. (51) *Sydney Morning Herald* 5/7/84; *FT* 5/7/84; *Financial Review* 5/7/84; *Financial Review* 15/5/84. (52) *Age* 5/7/84. (53) *Age* 5/8/84. (54) *Sydney Morning Herald* 11/7/84. (55) *Age* 11/7/84. (56) *Financial Review* 11/7/84. (57) *Age* 8/8/85. (58) *FT* 24/12/87. (59) NT Land Councils, *Land Rights News*, Vol. 2, No. 8, Alice Springs, 5/88. (60) *Sydney Morning Herald* 28/12/83. (61) *Background to Roxby Downs* (CNFA), 1982. (62) *Uranium Deadline* Vol. l No. 4, (FoE) Melbourne, 8-9/77; *Tribune* 30/6/76. (63) *Courier Mail* 13/1/77. (64) *Gua* 4/7/77. (65) *Australasian Express*, London, 8/7/77. (66) *Gua* 8/7/77. (67) *Sun*, Sydney, 8/7/77. (68) *Canberra Times*, 4/8/77. (69) *Gua* 16/9/77. (70)·*Gua* 22/9/77; *Age*

3/9/77; *SunT* 4/9/77. (71) *DT* 24/10/77; *MJ* 28/10/77. (72) *Herald* 15/11/77. (73) *Tribune* 18/1/78. (74) *Tribune* 1/2/78. (75) *Tribune* 8/2/78. (76) Decision of the Executive of the Australian Council of Trade Unions, 8/12/76, reprinted in (33). (77) Letter from the Minister for Employment and Industrial Relations, Mr Street, to the President of the ACTU, Mr Hawke, 9/12/76. (78) *Australian* 11/12/76. (79) *Tribune* 3/5/78. (80) ACTU Uranium Policy, Amendment to Executive Recommendation, Moved by C Dolan (ETU), Seconded R. Taylor (ARU). (81) *Australasian Express*, London, 21/9/79. (82) *Australasian Express* 7/12/79. (83) *Age* 29/11/79. (84) *NT News* 11/7/80. (85) *Age* 11/2/80. (86) *Courier Mail*, 15/8/80. (87) *Maritime Worker*, (WWF), 3/81, as quoted in *The Seamen's Journal*, Vol. 35 No. 4, 4/81. (88) *West Australian* 10/11/81; *MJ* 13/11/81. (89) *CANP Newsletter*, Brisbane, 7/81; see also *Australian* 14/6/81; *Sydney Morning Herald* 1/7/81; *Tribune* 1/7/81; *Courier Mail* 18/6/81. (90) WISE Glen Aplin, in *WISE Bulletin* 2/84, see also *NC* 197-1354, Amsterdam. (91) *Australian* 21/11/83. (92) *Financial Review* 25/4/84. (93) *Financial Review* 25/5/84. (94) *Age* 29/4/84. (95) *WISE NC* No. 257, 22/8/86. (96) *Canberra Times* 17/7/84. (97) *Chain Reaction* (FoE) Vol. 2 No. 4, 1977. (98) *The Second Ranger Report: Implications for Aboriginal People*, (Aboriginal Land Rights Support Committee) Sydney, 1977. (99) *Age* 17/11/76. (100) Wieslaw Lichacz and Stephen Myers, "Uranium Mining in Australia" in (ed) Mary Elliot, *Ground for Concern, Australia's Uranium and Human Survival* (Penguin Harmondsworth and Ringwood) 1977.
(101) *Age* 23/11/76. (102) Peter Carroll, "Uranium and Aboriginal Land Interests in the Alligator Rivers Region", in (Eds) Nicholas Peterson and Marcia Langton, *op cit.* (103) Colin Tatz, *Aborigines and Uranium, and Other Essays*, Richmond, 1982. (104) *Ranger Uranium Environmental Inquiry, Second Report*, (AGPS) Canberra, 1977, p 9. (105) *Ranger Inquiry 2nd Report*, p 220. (106) *Ranger Inquiry 2nd Report* p 222. (107) *Ranger Inquiry 2nd Report* p 228. (108) *Ranger Inquiry 2nd Report* pp 229, 231. (109) *Ranger Inquiry 2nd Report*, pp 227, 233. (110) *Ranger Inquiry 2nd Report*, p 232. (111) *Ranger Inquiry 2nd Report*, p 269. (112) *New Scientist*, London, 9/6/77. (113) *Gua* 16/8/77; *Sun*, Melbourne,

16/8/77. (114) *Age* 14/7/77. (115) *National Review*, 14/7/77. (116) *Courier Mail* 20/7/77. (117) *Age* 15/11/77. (118) *Sydney Morning Herald* 18/8/77. (119) *Australian* 26/8/77. (120) *Courier Mail* 26/8/77; *Advertiser* 26/8/77; *Financial Review* 26/8/77; and every other Australian newspaper published on 26/8/77. (121) *Herald* 25/8/77. (122) *Australian* 26/8/77. (123) *Telegraph*, Sydney, 26/8/77. (124) *Mirror*, Sydney, 26/8/77. (125) *Advocate*, Tasmania, 29/8/77. (126) *Sydney Morning Herald* 31/8/77; *West Australian* 30/8/77; *Times*, Canberra, 30/8/77. (127) *Sydney Morning Herald* 29/8/77. (128) *Advocate*, Tasmania, 30/8/77; see also *West Australian* 5/9/77; *Advertiser*, Geelong, 5/9/77; *Advertiser*, Adelaide, 5/9/77. (129) FoE, *Cyclone in the North, the events of September and October 1978 surrounding the proposed agreements to allow uranium mining to proceed on Aboriginal Land in the Alligator Rivers Region of the Northern Territory*, Collingwood, 1978. (130) Stuart McGill, address to the MAUM conference, Sydney, 24-25/6/78. (131) *Times*, Canberra, 6/10/77. (132) *Australasian Express*, London, 25/11/77. (133) *Age* 11/11/77. (134) *Australian* 11/11/77. (135) *National Times*, Australia, 30/1-4/2/78. (136) *Age* 7/2/78. (137) *Financial Review* 6/3/78; see also *Age* 7/6/78. (138) *AFR* 8/3/78. (139) *Age* 9/3/78. (140) *National Times* 15-16/3/78. (141) *Tribune* 3/5/78. (142) see also *Age* 10/3/78. (143) *Sun* 10/5/78. (144) *Tribune* 5/4/78. (145) *Financial Review* 4/4/78. (146) *AFR* 11/4/78. (147) *CANP Newsletter*, 6/80; *Courier Mail* 25/8/80; *Age* 28/8/80. (148) *Herald* 11/5/78. (149) *Age* 27/4/78. (150) *Australian* 27/4/78; see also *Sun* 27/4/78. (151) *AFR* 19/9/78. (152) *Financial Review* 11/5/78; *Canberra Times* 11/5/78; *Age* 11/5/78. (153) *Financial Review* 13/6/78. (154) *FT* 1/6/78; *MJ* 5/12/78; *Age* 13/6/78. (155) *Tribune* 2/8/78. (156) *FT* 13/7/78. (157) *Australasian Express* 18/8/78; see also *Financial Review* 15/8/78. (158) See *Sydney Morning Herald* 23/5/78. (159) *Kurri-Bina* 1/77. (160) P A Coleing, *Land Rights and Uranium*, (FoE) Sydney, 1977. (161) *National Times* 18-23/77. (162) *Australian* 26/8/78. (163) *Tribune* 8/78. (164) *AFR* 25/8/78; *Sydney Morning Herald* 25/8/78. (165) *Age* 26/8/78; *FT* 26/8/78. (166) *Anti-Uranium News* (MAUM) Sydney, No. 8, 1978. (167) *Tribune* 20/8/78. (168) *Herald* 28/8/78; *Age* 28/8/78; *Sun* 28/8/78. (169) *Age* 3/9/78. (170) Johnny Marali

No. 1's Affidavit, quoted in *Australian* 2/9/78. (171)
Chain Reaction (FoE) Vol. 4, Nos. 2/3, 1978. (172)
Mary Elliot's Diary, quoted in *Time and Energy*, No.
5, 10/78. (173) Australian 5/9/78. (174) *Age* 6/9/78;
Sun 6/9/78. (175) John Gwadbu and Johnny Marali
No.l's Affidavits, quoted in (129). (176) Affidavit by
Bob Collins, quoted in (129). (177) *Age* 12/9/78.
(178) *AFR* 29/9/78; *NT News* 29/9/78. (179) *Age*
18/9/78; *AFR* 20/9/78. (180) *Sun* 19/9/78. (181)
NT News 19/9/78. (182) *AFR* 21/9/78. (183) *Age*
21/9/78. (184) *NT News* 20/9/78. (185) *AFR*
22/9/78; *Age* 26/9/78. (186) *AFR* 3/10/78; *AFR*
4/10/78. (187) *Tribune* 18/10/78. (188) *Time and
Energy*, No. 5, Sydney, 10/78. (189) *AM* ABC
Radio, 9/10/78. (190) *Sun* 11/10/78. (191) *Age*
12/10/78. (192) AFR 13/10/78. (193) *Age*
13/10/78; *Sun* 13/10/78. (194) *Age* 14/10/78. (195)
AFR 18/10/78. (196) *FT* 23/10/78. (197) *Tribune*
1/11/78. (198) Testimony from Leo Finlay, quoted
in (11). (199) *Age* 4/11/78. (200) Radio Australia
News, 22/10/78.
(201) *Australian* 16/11/78. (202) *Age* 4/11/78. (203)
RA News 7/11/78. (204) *NT News* 8/2/80. (205)
NT News 8/2/82. (206) *Report to the Minister for
Aboriginal Affairs on the Social Impact of Uranium
Mining on the Aborigines of the Northern Territory*, p
2, 30/9/79; see also Sue Kesteven, "The Effects on
Aboriginal Communities of Monies Paid Out
Under the Ranger and Nabarlek Agreements", paper
presented to the Land Rights Symposium, Australian Institute of Aboriginal Studies, Canberra, 21-
22/5/78 (updated). (207) Australian Institute of
Aboriginal Studies, Aborigines and Uranium, Consolidated Report on the Social Impact of Uranium
Mining on the Aborigines of the Northern Territory, p 299 (AGPS) Canberra, 1984. (208) *Ibid*
p300. (209) *Ibid* pp304-306. (210) see also *National
Times* 12/7/84; *Age* 12/7/84. (211) *Report* (207),
p162. (212) *Ibid* p179. (213) *Ibid* p189. (214) *Ibid*
pp189-196. (215) *Ibid* pp232-233. (216) *Ibid*
pp258-259. (217) *Ibid* p265. (218) *Ibid* p288. (219)
see also *Tribune* 27/2/80; *NT News* 11/2/80. (220)
AFR 12/11/81. (221) *Australian* 26/11/83. (222)
Australian 29/7/85. (223) *Age* 19/6/85. (224) *Age*
29/10/85. (225) *MJ* 1/11/85. (226) *Australian*
29/10/85. (227) *MJ* 25/4/86. (228) *MJ* 14/11/86.
(229) *Land Rights News*, Darwin, 11/86; see also *The
Weekend Australian* 2-3/2/91 (230) See also *Chain
Reaction*, (FoE) Vol. 4, No. l, 1978. (231) *Ranger Inquiry 2nd Report*, pp231-233. (232) *bid* p278. (233)
Ibid p266. (234) see also *NT Newsletter*, 9/77; *AFR*
26/8/78. (235) Ranger Uranium Project Agreement
under Section 44 of the Aboriginal Land Rights
(NT) Act 76, between the Commonwealth of Australia and the NLC, signed 3/11/78, as summarised
in Appendix to (102). (236) see also *Canberra Times*
10/5/78. (237) *AFR* 26/4/77. (238) *Australian*
8/5/88. (239) *Land Rights News*, (NT Land Councils) Vol. 2, No. ll, Alice Springs, 11/88. (240) *FT*
8/6/84. (241) *FT* 2/12/83. (242) *Chain Reaction*,
(FoE) No. 35, 12/83. (243) *Chain Reaction, (FoE)*,
No. 34, 10-11/83. (244) *MAUM Newsletter*,
Spring/86-Summer/87. (245) *Report to the Minister
for Aboriginal Affairs on the Social Impact of Uranium
Mining on the Aborigines of the Northern Territory,
for the period 10/10/80 to 31/3/81*, Australian Institute of Aboriginal Studies (AGPS), Canberra,
1981. (246) *Land Rights News*, (NT Land Councils),
Vol. 2 No. 6, 1/88. (247) *Legal Service Bulletin*, Vol.
13 No. l, Clayton, 3/88. (248) *West Australian*
31/10/77. (249) Ranger Mines Uranium Pty Ltd,
Environmental Impact Statement, 2/74. (250) *Financial Review* 5/4/78. (251) *The Aboriginal Land Rights
(NT) Act, Matters of Serious Concern to the Mining Industry* (AMIC) Canberra, 2/78. (252) R M Fry, "Environmental Protection and Uranium Mining in the
Alligator Rivers Region", in (265). (253) *MM*
11/85. (254) See *Uranium Mining and Radiation
Safety, Proceedings of a Conference at Michigan Technological University, Houghton, Michigan, 19/9/80*
(Action U P), Houghton, 1980. (255) *National
Times* 9/9/78. (256) *Age* 13/11/79. (257) *E&MJ*
12/84. (258) *Ranger Inquiry 2nd Report*, chapters 2-
5. (259) *Supervising Scientist for the Alligator Rivers
Region, First Annual Report: 1978-79* (AGPS) Can-
berra, 1979. (260) Report of the Ombudsman to
Parliament (AGPS) Canberra, 3/82. (261) Mark
Dreyfs, Northern Territory, "Uranium and Responsibility", *Legal Service Bulletin*, 8/82. (262) *Sydney
Morning Herald*, 26/5/77. (263) CS Christian and
JM Aldrick, *Alligator Rivers Study, A Review of the Alligator Rivers Region Environmental Fact-Finding
Study*, Department of NT, and AMIC (AGPS) Canberra, 1977. (264) *Ranger Inquiry 2nd Report* pp 55-
56. (265) (Ed) Stuart Harris, *Social and Environmental Choice: the impact of uranium mining in the North-*

ern Territory, Centre for Resource and Environmental Studies (CRES), Monograph No. 3, Australian National University, Canberra, 1980. (266) Barry Hart, "Water Quality and Aquatic Biota in Magela Creek", in (265). (267) JA Chaney, DP Thomas, PA Tyler, "Diatoms and other Freshwater Algae of the Magela Creek System as Monitors of Heavy Metal Pollution", *First Interim Report to the Supervising Scientist Covering the Period April to November 1979*, Department of Botany, University of Tasmania, Hobart, 1980. (268) *The Uranium Mining Operation and the Pollution of the Region*, cyclostyled paper, author and date unknown. (269) *MM* 12/77. (270) *Ranger Inquiry 2nd Report*, p111. (271) *Ibid*, p127. (272) *Ibid* p114. (273) NT Medical Services findings, 1973, as quoted in *Ranger Uranium Environmental Inquiry, Second Report* (AGPS) Canberra, 1977. (274) *Ranger Inquiry 2nd Report*, pp.124,135. (275) *bid* pp 104-105. (276) *Ibid* pp 141-142, and chapter 7 as a whole. (277) Willy Wabeke, "Ranger and the Ongoing Radioactive Leaks", *Kiitg* No. 21, 5/82. (278) *Financial Review* 29/9/82. (279) *CANP Newsletter*, Brisbane, 11/82. (280) *Chain Reaction* (FoE), 1/84. (281) *Financial Review* 22/9/82. (282) *Age* 21/5/82. (283) *National Times* 23-29/8/85. (284) *Age* 19/8/85. (285) *West Australian* 26/9/85. (286) *ACF Foundation Newsletter*, Vol. 17 No. 10, 11/85. (287) *Chain Reaction* No. 46, (FoE) Winter/86. (288) *Age* 11/10/85. (289) *Canberra Times* 4/11/85. (290) *Canberra Times* 4/3/86. (291) *Australian* 18/1/86. (292) *Age* 3/2/86. (293) *Tribune* 19/3/86. (294) *Tribune* 26/3/86. (295) *WISE NC* No. 251, 2/5/86. (296) *MJ* 21/3/86. (297) *Weekend Australian* 22-23/3/86. (298) *LAM*, London, 24/3/87. (299) WISE Glen Aplin, quoted in *WISE NC* No. 288, 4/3/88. (300) NT Land Council, *Land Rights News*, Alice Springs, 3/88; see also MAUM, *The Third Option*, Surry Hills, Winter/88. (301) FoE Australia, information in *WISE NC* No. 298, 23/9/88. (302) *LAM*, London, 5/12/88. (303) *NT News* 30/4/82. (304) *MJ* 4/3/88. (305) *Ranger Inquiry 2nd Report* p309 para 12. (306) *LAM*, London, 6/2/88. (307) *Aboriginal Law Bulletin* Vol. 2, No. 42, 2/90. (308) *Land Rights News* 3/89. (309) *Advertiser* 14/3/89. (310) *MAUM Newsletter*, Melbourne 2/89. ABC Radio *"PM"* 13/4/89; see also *Chain Reaction* Autumn 1989. (311) *MJ* 19/4/91.

(312) *FT* 2/8/91. (313) *Chain Reaction* No. 63/64, 4/91.

512 Rare Metals Corp

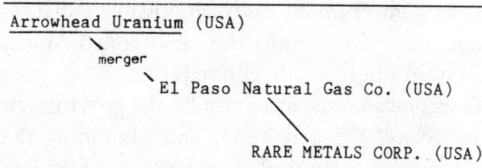

Now defunct as a uranium producer, Rare Metals purchased all the capital stock of the Arrowhead Uranium Co in 1954, which gave the company several open-pit uranium mines along with Navajo tribal council permits and mining claims on the Navajo reservation north and south of the Little Colorado River, close to Cameron, Arizona. Navajo miners were employed on mechanical equipment and with pick and shovel work – totally unprotected from radiation and radon gas emissions. A description, with several photos, of the mining on the Painted Desert is contained in *Mining World* (1).

References: (1) *Mining World*, USA, 3/57.

513 Rayrock Resources Ltd (CDN)

Rayrock merged with Yellowknife Bear Resources and is now Rayrock Yellowknife Resources Inc (1).
Exploring with Mitsubishi in Canada's Northwest Territories (NWT) since 1977 (2), Rayrock is a mineral exploration company, holding interests in oil, gas and gold (3).

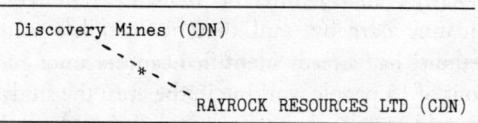

* Rayrock holds 41% of Discovery Mines; Discovery holds 20.8% of Rayrock's stock – but 39.6% of the voting rights.(8)

It was part of a lucrative JV for gold with Lacana and United Siscoe at Pinson, Nevada (4).

It is presumed to be a descendant of Rayrock Mining Co, which operated the only producing uranium mine between 1957 and 1959 near Port Radium, NWT, Canada (5). Twenty years later, a government study on possible radiation contamination from the abandoned mine, showed a potential health hazard.

A second study, carried out by the government between 1979 and 1983, and examining the physical and chemical characteristics of the Rae (*sic*) Rock tailings areas, was not released as of mid-1984, thus causing much anger – especially from the Fort Rae native band. Their chief, Joe Rebesca, pointed out that the mine sits "right in the middle of our prime hunting and trapping area". Many of the people who worked at the mine in the '50s, says Rebesca, have become cancer victims. "Our most immediate concern is the health of our people. As far as we know the government is still not making plans to clean up the mess " (6).

There were 14 cancer deaths among Rae residents over the period 1974-1984. Dene nation president, Steve Kakfwi, told a press conference, in June 1984, that the government has shown no intention to clean up the area (7).

However, only a month later, the new Minister of Indian Affairs in the NWT, Douglas Frith, agreed on a "full investigation" of the Rae Rock radiation hazards, and a reimbursement to the Dene of the US$48,000 they had spent on their own study (9).

The Dene Nation would work with the NWT Science Institute on studies, not only of Rae Rock, but of Port Radium (source of much uranium for the US's early nuclear weapons) and other abandoned mines in the Territory.

At a press conference welcoming the news, Dene Nation's director of land resources, Joanne Barnaby, said that her people's own studies had already identified cancers among 9 out of 15 people working in the area; the study would look at the environmental, health, and socio-economic, impact of the mines, and those who worked in them, as well as at the radiation impact on animals. It would "start immediately" (10).

References: (1) *FT Min* 1991. (2) *MJ* 1/4/77. (3) *MIY* 1982. (4) *MAR* 1984. (5) Volger Rolf "Mineral Development in the NWT", in *Oxfam Canada* 4/76. (6) *Dene Nation Newsletter* 15/6/84. (7) *Native Press* 15/6/84. (8) *MIY* 1981. (9) *Native Press* 13/7/84. (10) *Native Press* 27/7/84.

514 Ray Williams Mining (USA)

No, this is not a country and western band, but a real humdinger of a mining company which, in the late '70s, was digging uranium out of the Navajo Indian Reservation, right in the heart of the Four Corners area of the south-western USA. Enos Johnson wasn't the name of their lead singer, but, believe it or not, the name of the mine (1). It was also exploring for uranium elsewhere in New Mexico at the same time (2).

References: (1) San Juan Study. (2) *E&MJ* 11/78.

515 Reederei und Spedition "Braunkohle" GmbH

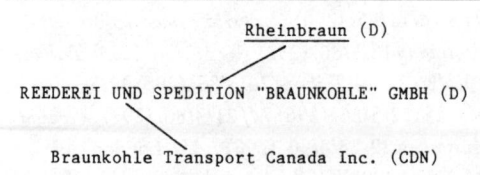

This transport and loading company handles uranium oxide, uranium hexafluoride, radioactive elements, and all classes of dangerous goods. Its main offices are at Wesseling, West Germany, with offshoots in Amsterdam, Rotterdam, Basel, Heilbronn, Karlsruhe, Ludwigstein, Mannheim and Würzburg, and also a subsidiary in Montreal, Canada (1).

References: (1) *Kiitg* 2/82.

516 Reef Oil NL

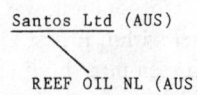

```
        Santos Ltd (AUS)
            \
        REEF OIL NL (AUS)
```

This oil and natural gas company, based in New South Wales, was part of Bond Corporation's attempt to buy into Beverley Uranium (see Endeavour) but was sold by Bond in 1982 to Santos (1).

References: (1) *FT* 8/11/82.

517 Reserve Oil and Minerals Corp (USA)

Since 1987, this company has been known as Reserve Industries Corp (1). It was exploring for the deadly mineral in New Mexico in the late '70s (2).
It was in a JV (7.5%) with Kelvin Energy (6.5%) and Saskatchewan Mineral Development Corp (50%), as also was (the Canadian) Asamera Ltd, at Keefe Lake-Henday in northern Saskatchewan (3).
Its major US asset is the L-Bar uranium property in New Mexico, USA, which Sohio operated up until 1981. That year, Reserve sued Sohio for neglecting proper standards in operating the plant: it won US$1.09M in damages, substantially less than it had asked for (4). Just before damages were awarded, Reserve announced it was closing its Seboyeta mill at the site due, to "adverse market conditions". It said it would reopen in 1982 – to our knowledge, it hasn't (5).
In late 1982, the company announced that it had reached agreement with Sohio and would terminate the Reserve-Sohio JV on the L-Bar ranch. Reserve would convey its 50% interest to Sohio, in exchange for which, Sohio would cancel Reserve's indebtedness to the company of about US$35 5M, as well as further debts of about US$12.7M. In addition, Sohio conveyed to Reserve its 75% mineral rights in another tract, and its royalty on yet another (where there are known uranium resources). Sohio also paid Reserve US$310,000 (6).
By 1984, operations at L-Bar were suspended (7).

References: (1) *FT Min* 1991. (2) *E&MJ* 11/78. (3) *MJ* 15/12/78. (4) *MJ* 19/6/81. (5) *MJ* 29/5/81. (6) *MJ* 24/9/82. (7) *MAR* 1984.

518 Rexcon (USA)

This was the first company to gain a uranium exploration permit for the Black Hills area, sacred land of the Lakota. The permit was valid as of 1980, and believed later to have been withdrawn (1).
No further information.

References: (1) Black Hills Sierra Club, quoted in *BHPS* 4/80.

519 Rheinbraun

One of the giants of the German mining industry (1), Rheinbraun operates lignite mines in the Rhineland, near Cologne, with an output of more than 100 million tonnes per year, while its Union RB subsidiary in Wesseling operates a crude oil refinery (2).
When, in 1980, 98,400,000 "B" class shares were offered in ERA/Energy Resources of Aus-

```
RWE / Rheinisch-Westfälisches Elektrizitätswerk AG (D)
                        \
RHEINBRAUN / RHEINISCHE BRAUNKOHLENWERKE AG (D)
                    /                      \ 50%
Reederei und Spedition "Braunkohle" GmbH (D)    Uranerz (D)
```

tralia (*ie* the Australian government's, Peko Wallsend's, and EZ Industries' shares in Ranger uranium), Rheinbraun took up 26,625,000 of them (2).

In 1979, it announced, before the 1979 election, that the South Australian government had promised it would abandon the uranium ban (3).

In autumn 1987, the regional government authorities of North Rhine-Westphalia issued an edict covering the long-term exploitation of lignite deposits in a triangle formed by the cities of Cologne, Düsseldorf and Aachen. Apart from Rheinbraun's existing zones, a third would be addressed, comprising an extension of the Inden mine, and the new Hambach mine, as well as an extension of the Frimmersdorf operations (4).

These aggressive developments would entail the "demolition of a large number of villages and the resettlement of 11,800 people in new housing developments" (4). Both Rheinbraun and the government said they recognised "that this exercise will be a formidable task which must be conducted in a socially fair and acceptable manner" – one which the company was considered well qualified to undertake, since it had "... of course, had considerable experience of such operations, and for previous open-cast developments has organised the movement of over 20,000 people from more than 40 villages into new communities" (4).

Some work on the huge extensions was undertaken over the next eighteen months but, as of mid-1989, further detailed planning was still required (5).

Through its stake in Uranerz, Rheinbraun is involved in uranium mining at Key Lake, another exploration undertaken by its part-owned subsidiary (see Uranerz).

Rheinbraun also holds, together with its parent RWE, a 24% stake in the Bailey No 1 underground coal mine in Pennsylvania, USA, which is "believed to be the most productive underground coal mine in the world" (5).

References: (1) *MAR* 1982. (2) *MIY* 1982. (3) *FT* 24/6/79. (4) *MJ* 2/10/87. (5) *MJ* 4/8/89.

520 Rhokana Corp (Z)

The Rhokana Corporation was a significant copper producer within Ernest Oppenheimer's Anglo-American empire, built up in the old Central African Federation. In 1969, under a reorganisation negotiated between the independent Zambian government and the major multinationals, Rhokana became 53% owned by Zambian Anglo-American (Zamanglo), which itself was 45% owned by the AAC of South Africa. Rhokana also owned 34.7% of a new company, set up to take over Oppenheimer assets, called Nchanga Consolidated Copper Mines: Rhokana and Nchanga between them controlled the important Rhokana copper refinery.

A year later, the Zambian state corporation, Mindeco, took over 51% of Nchanga Consolidated Copper Mines, although the lion's share of profits was left with AAC. Zamanglo transferred to Bermuda to become Minorco: the Anglo-American corporation retained 49% of the Rhokana refinery (1).

In 1982, Nchanga merged with Roan Consolidated Mines (itself 20% owned by Amax after the 1970 reshuffle) and is now 60.3% controlled by the Zambian Industrial and Mining Corp Ltd (Zimco).

One of Zimco's seven major mining divisions is at Nkana (2), which was the site of Rhokana's small uranium mine during the late 1950s. The treatment plant for this mine was opened in 1957, and uranium concentrate delivered a year later. By 1959, however, uranium production had apparently stopped and the Rhokana Corporation office in Salisbury closed down.

Between 1957 and 1960, some 1150 tonnes of uranium oxide was delivered to the UKAEA (3).

References: (1) Lanning and Mueller. (2) *MIY* 1985. (3) UKAEA annual reports, 1955-1960, quoted in *Namibia: A Contract to Kill; the story of stolen uranium and the British nuclear programme* (CANUC) London, 1986.

521 Rhône-Poulenc SA

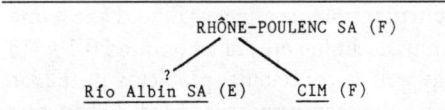

```
              RHÔNE-POULENC SA (F)
           ?
Río Albin SA (E)        CIM (F)
```

Rhône-Poulenc in 1981 opened the first phase of a rare earths separation plant at Freeport in Texas (USA). (Rare earths are not only used in petrol, metallurgical and ceramic manufacturing, but also in the nuclear industry) (1).
It also had shares in Rio Algom.
Rhône-Poulenc is one of the five major French industrial groups which the Mitterand government nationalised (2).
The company's main contribution to the nuclear industry derives from its status as the world's biggest supplier of monazite, from which thorium is processed. Rhône-Poulenc operates a processing plant at Freeport, Texas, which produces thorium nitrate. Thorium oxide is then refined in France. A commercial nuclear reactor at Fort St Vrain, Colorado, is currently run on thorium – it is not known whether Rhône-Poulenc supplies the fuel (3).
In 1983 Rhône-Poulenc was fined by the French government for illicit price-fixing of insecticides (4).
In 1989, the company was operating three cassiterite (tin) mines in Brazil, and buying output from the Bom Futuro mine, then operated by *garimpeiros* (small-scale miners) (5). The same year, it announced a 10-year investment programme in the country (6).
It also bought up RTZ Chemical's interests that year (although it sold one plant to Bayer of West Germany in 1990) (7).
Due to partial privatisation, the company was 20% owned by "the public" by early 1991 (8).
In Australia, Rhône-Poulenc had planned to bring on-stream a plant to process around 15,000 tonnes a year of monazite, at Pinjarra, Western Australia (WA). The first stage of the project – the "cracking" of sand into an intermediate product – was approved, but the second phase – the separating out of thorium and radium – was subject to critical examination and a large amount of opposition from environmentalists (9, 10).
In the event, the WA state government turned down the company's proposals – not so much because of fears of radioactive emission from the separated nuclear materials, but because radium-bearing ammonium nitrate could leak into the precious ground water (10). Later, Rhône-Poulenc had the audacity to imply that the state's Environmental Protection Authority had been subject to "political pressure" in rejecting the separation plant (11).

Contact: Shyama Peebles, P O Box 889, Kalgoorlie, Western Australia, Australia.

References: (1) *MAR* 1982. (2) *MJ* 5/2/82. (3) *E&MJ* 3/83. (4) *FT* 9/11/83. (5) *FT* 14/9/89. (6) *FT* 27/9/89. (7) *FT* 17/7/90. (8) *FT* 11/4/91. (9) *FT* 18/1/90. (10) *MJ* 19/1/90. (11) *Australian* 15/11/90.

522 Rio Albin SA

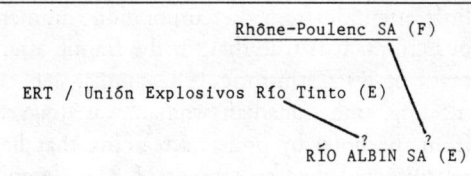

```
                          Rhône-Poulenc SA (F)

ERT / Unión Explosivos Río Tinto (E)
                              ?    ?
                          RÍO ALBIN SA (E)
```

In 1980, Rio Albin was reported exploring for uranium in the Segovia region of Spain (1) together with Rio Tinto Minera. The latter is a company in which RTZ held a 49% interest, the remainder being held by ERT (2). ERT is Spain's largest holding company; in 1982 it almost collapsed, having run up debts of nearly one billion dollars (3) – debt rescheduling appears to have kept the wolf from the door since (4).

References: (1) *MJ* 22/8/80. (2) RTZ annual report, 1984. (3) *FT* 26/9/84. (4) *FT* 3/1/84.

523 Rio Algom

"The Elliot Lake story is a moral and human outrage. What is more, it's apparently not yet over" (1).

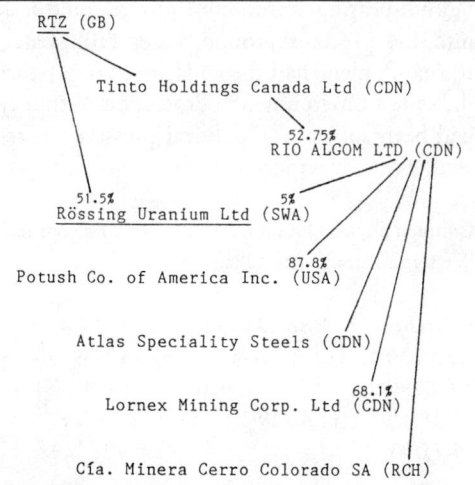

```
RTZ (GB)

        Tinto Holdings Canada Ltd (CDN)

                52.75%
                RIO ALGOM LTD (CDN)

  51.5%              5%
Rössing Uranium Ltd (SWA)

                87.8%
Potush Co. of America Inc. (USA)

    Atlas Speciality Steels (CDN)

                          68.1%
    Lornex Mining Corp. Ltd (CDN)

  Cía. Minera Cerro Colorado SA (RCH)
```

In surveying the RTZ imperial domain, the company's operations in Canada usually get little attention from the corporation's numerous critics. It is true that, in the frantic aftermath of RTZ's historic 1982 annual general meeting, one Canadian woman was dragged from the hotel by police, screaming that her relatives had died as a result of Rio Algom's operations at Elliot Lake (2). But this is the proverbial exception that confirms the rule. While RTZ's Australian associate, CRA, can barely set a foot on Aboriginal land without attracting a thousand quizzical eyes and, while the company's Rössing subsidiary has become synonymous with the colonisation of an entire country, what Rio Algom does in North America passes almost unnoticed in the world at large. Yet Rio Algom co-owns one of the world's largest copper complexes; it operates North America's only primary tin mine; its Atlas Steels subsidiary (though now sold) is more than likely to have produced the stainless steel sitting on any North American dinner table.

More important, until 1990 the company was the most reliable and important single producer of uranium in the RTZ agglomerate over the past twenty odd years. Despite the high costs of production from its mines at Elliot Lake, Ontario, plus declining ore grades (around 0.1%) (3, 4), as well as contractual problems in the late 1970s, the company expanded production, until a downturn in 1990, and re-opened a mine in the USA which had closed the previous decade. It has also enjoyed (along with the Canadian company Denison) the largest of all long-term uranium contracts – with Ontario Hydro. It also has substantial old, un-reactivated uranium reserves (5).

The human cost of this record of uranium exploitation – unique in Canadian history – has usually been concealed, but we will gain some insight into it as the Rio Algom story unfolds.

Rio Tinto (forerunner of RTZ) commenced its penetration of Canada in 1953.
A geologist who worked for the CDA – the Combined Development Agency (see MKU) – and had been employed by the agency in South Africa, staked a claim on old workings in the maple and fir pines of northern Ontario. This geologist, C F Davidson, enlisted the backing of the American mining financier, Samuel Hirschorn (also called J H Hirschorn in another source) (6). Hirschorn formed a company, used scores of his own employees in a secret operation, and located the Elliot Lake uranium vein: in one day, mine claims stretched over 20 miles (7). Two years later, Rio Tinto acquired a substantial interest in Algom Uranium Mines Ltd and, the following year, purchased virtually all of Hirschorn's interests in the Algoma mining camp (6). From this coup, the American financier created the modern art collection known as the Hirschorn Museum, in Washington DC. Through its acquisition, Rio Algom gained control of four companies: Algom Uranium Mines Ltd, Northspan Uranium Mines Ltd, Milliken Lake Uranium Mines Ltd, and Pronto Uranium Mines (6). All these mines operated in the Blind River-Algom region of the north shore of Lake Huron, collectively known as Elliot Lake, although there are several other lakes

in the area (notably Quirke, Little Cinder, Crotch, McCabe and May) directly or indirectly affected by mining operations (8).

During the uranium boom of 1956-1959, created by the requirements of the US military, no fewer than 12 mines were in operation, with a workforce which exceeded 10,000 people (9). Seven of the eleven mills, and eight of the 12 mines – or almost 60% of the area's mining and milling capacity – were exploited by the companies later to be amalgamated as Rio Algom Mines Ltd in June 1960 (6, 10, 11). (The name Rio Algom Ltd was adopted in 1975.)

The Stanleigh Uranium Corporation's facility at Elliot Lake became associated with Rio Algom the same year, when it was merged with Preston East Dome Mines, to form Preston Mines Ltd (owned 81% by RTZ, and itself then owning 44% of Rio Algom) (6).

But in 1959, the US government ditched Canadian sources of uranium, as it moved to protect its own producers, especially from the Four Corners of the American South-West (12). Officially, Rio Algom had been formed to bring together individual contracts signed by the Algom group of mines with Eldorado Nuclear. And, from the late 1950s, this Crown Corporation set about implementing the Canadian federal government's policy of creating stockpiles and stretching out deliveries until the demands of "civilian" nuclear power could stimulate their own market (6). Elliot Lake then went into severe decline, throwing thousands of people out of work and depleting the boom towns (13). By 1966, Canadian production overall was down to a quarter of its 1959 level, at less than 3900 tons – most of this produced by Rio Algom and Denison (6). Virtually all of these supplies were devoted, either to completing the stretched-out military contracts to the USA, or to the federal Canadian government's stockpile, or the British Atomic Energy Authority (AEA). As the Ontario Royal Commission on Electric Power Planning (the Porter Report) stated quite clearly in 1978: whatever protestations were made by the government ten years later, that they had not provided uranium for weapons purposes: "this country not only participated in the research which gave birth to the original atomic bomb but also, through the export of strategic material (ie uranium), subsequently facilitated their production. For example, the bulk of the uranium mined at Elliot Lake, Ontario, in the late 1950s and early 1960s was used in the production of nuclear weapons in both the United States and Great Britain" (14).

Although the second major uranium boom did not start until 1970, by 1968 Rio Algom was in a key position to benefit from it, having signed contracts of more than C$50 million for the following decade, with European and Japanese customers (10). That year, the company re-activated the Quirke mine, increasing its mill capacity from 3000 tonnes to 4500 tonnes per day, as well as sinking a new shaft to bring the New Quirke orebody into production (15). The old one was closed in 1971. The year before, the international uranium cartel was formed in Paris, between four governments – France, South Africa, Australia and Canada – and one corporation acting "as a sovereign state" (16). As that corporation was RTZ, Rio Algom naturally soon began to benefit from the operations of this price-fixing, contract-fixing, ring. By 1973, as prices began to rise and consumers moved to take up long-term contracts, exploration entered a new boom phase. In 1975, the US dropped its ban on uranium imports (17) and Rio Algom began expanding at Elliot Lake: this included re-development of idle mines and mills. By the end of that year, the company had secured contracts for nearly one million pounds of U_3O_8, for delivery until the early 1990s (11).

In 1976, the uranium cartel was disbanded, after its cover was blown by Friends of the Earth (FoE) in Melbourne (see MKU). By then, although construction work to increase the milling capacity at Elliot Lake was supposedly on-schedule, and the old Panel mine was being re-opened (19), the company was suffering from the federal Canadian government's "non-proliferation" policy. This prohibited uranium exports to suspect countries – including the EEC as a whole – and Japan. The material was put in storage, paid for by the customers (6, 19). However, there is a yellow lining to many a uranium cloud. In this case, it was the 1974 stipulation by the Canadian government that domestic power utilities should hold 15-year forward supplies of nuclear fuel (6, 20). At the same time that non-proliferation talks between the EEC, Japan and the Federal Canadian government had deadlocked (6), one of the world's biggest single utilities, Ontario Hydro – which already had a contract with Rio Algom – began negotiating a long-term contract (20) with both Algom and its brother company at Elliot Lake, Denison Mines (4, 6).

Anticipating an upsurge in production at Elliot Lake, officials of both Rio Algom and Denison told the town's community leaders that they should consider the impact of the influx of 600 more workers by April 1978 (21).

In 1978, Preston Mines Ltd (then sharing 43.8 % of Rio Algom's profits (22) and counted as a subsidiary) concluded its deal with Ontario Hydro for the supply of 27,700 tonnes U_3O_8 until well into the twenty-first century. C$188 million was advanced, interest-free, by Ontario (23). Uranium would be sold to the utility at a base price equal to Preston's costs, including depreciation of the plant, plus roughly C$4 per lb and one third of "any excess of the free market price for uranium over the base price" (24). No wonder that, at RTZ's annual general meeting in London in 1978, the company chair, Sir Mark Turner, boasted that the contract was an example (without putting it in so many words) of how miners can screw shekels out of publicly funded corporations. "The upshot is that Preston for no outlay at all gets a uranium mine with a guaranteed return of at least C$4 for every pound of uranium it produces and the prospect of better return if it can contain costs, and world uranium prices remain high" (24).

In order to meet this contract, Preston decided to reopen the old Stanleigh mine (4, 28, 24, 25). Meanwhile, Rio Algom itself achieved "impressive financial results" in 1978 as U_3O_8 production increased by nearly 700,000lb, although average recovery grade slipped (26).

That year the Quirke expansion was completed (27) on schedule (15), with a production rate of 8900 tonnes of ore per day (4).

Early in 1979, Rio Algom's uranium earnings declined (28). It had been clear a few years before that, as ore grades dropped further in the older deposits, new lodes would need to be brought into production. That year, the Panel mine was brought back into production after closure in 1961 (29).

This mine – "spectacularly located on a small peninsula in the Quirke Lake" (30) – was a showpiece for the company: boasting new trackless mining methods, more than 45,000 feet of underground development, and the dewatering of over 100 million gallons of floodwater. (This was not without its own financial compensations, however: the cost of de-watering was covered by recovery of nearly 50,000lb of U_3O_8 by precipitation in the surface water treatment plant!) (31).

1980 was a year of mixed fortunes for Rio Algom (32). Preston Mines and Rio Algom finally merged on January 30th (33), as RTZ's interest in the new company reached 52.76% (34).

First production came out of the Panel mine in November 1979 (35). In March 1980, the company announced that it needed new customers to take up 13.7 million pounds of uranium, lost through a dispute with Tennessee Valley Authority – a "burdensome" search, but "not a matter of life and death", due to its contract with Ontario Hydro (34). In 1979, Rio Algom had signed new contracts for 2.8 million pounds, of which half was delivered, and buyers found for a further one million pounds (36). Then, in May, the company landed a contract with Kepco (Keko) of South Korea. By the end of 1980, it was boasting that it produced "... more uranium than any other company in North America" (37) – some 3750 tons (38). By now it was also exploring for uranium in Argentina (39) and had a JV with Soquem and Johns-Manville Corp, to explore the Canadian North-West Territories (NWT) and elsewhere in Canada (6). The NWT exploration, potentially damaging to indigenous welfare and land claims (185), appears not to have proceeded far. A "more favourable sales contract mix and improved operating efficiencies at the Panel facilities at Elliot Lake" (40, 41, 42), resulted in better results for 1981, as production reached 3500 tonnes (43). Rehabilitation of the old Stanleigh mine proceeded on schedule (40, 41, 44) despite a dispute with Ontario Hydro over prices (23). Deliveries to the utilities were deferred approximately 15% over the delivery period 1983-2004, without a reduction in the total contract: initial deliveries would, however, be made on time (45).

1982 uranium earnings were slightly down (46), but, once the Stanleigh mine came on stream in 1983, and a new contract was landed by the company's Lisbon mine in Utah (see later) (47, 48), it managed to triple its profits (49, 50). Thanks to Stanleigh, the company showed good earnings well into 1984 (49, 51). The Stanleigh mine, like the other Algomites, originally produced uranium for US nuclear weapons, and closed in 1960. Rio Algom not only devoted much money and labour (bankrolled, it should be remembered, by Ontario Hydro) into recommissioning it, but brought in some of the most sophisticated computer-controlled technology available, to win uranium from depths of up to 4000 feet below the surface (52). Its scheduled throughput of 4250 tonnes per day of ore was reached in Spring 1983 (53).

In 1984, Rio Algom recovered nearly 8 million pounds of uranium and, by the year-end, contracts were held for around 58 million pounds, for delivery between 1985-1995, as well as another 50 million pounds from Stanleigh for 1996-2020 (54).

The same year, an energy savings program was introduced at Elliot Lake, which included piping-in of natural gas (55). Uranium leaching was also expanded – material being left in the underground workings and sprayed with recycled mine water – while "experimental" areas were developed at Quirke and Panel, to test the feasibility of applying the technology to low-grade in situ mineralisation (55).

In 1986, the Ontario Hydro contract once again came up for debate, with another 28% deferral of deliveries and a reduction in the workforce (56). The object was to supply the utility with lower priced uranium (the original contract having been signed at a time of exceptionally high prices) (57). Perhaps the most significant aspect of this re-negotiation was the prospect – previously thought highly unlikely (29, 58) – that, in order to cut the long-term costs of this contract, Rio Algom would "access" adjacent, higher-grade deposits at Milliken, Lacnor, Nordic, (56) and Spanish American, another old Elliot Lake property (53).

The following year, earnings fell again, thanks partly to a 19-day strike at Elliot Lake (59). A philosophical Ray Ballmer, in 1988, commented that the mines were now "into the sunset of [their] life" (43), as the company's long-standing contracts were due for final deliveries in the 1990s, with few prospects for new ones thereafter (except for Ontario Hydro's).

Despite (or because of) the heady expansion in Canada, Rio Algom has not neglected yellow-cake production south of the border. In 1970, Rio Algom secured a contract to supply (through RTZ Mineral Services) the Swedish Sydsvenska Kraftaktiebolaget in Malmö, for the initial core and reloads for what was to become the highly controversial Barsebäck reactor – just across the sound from Denmark. It was then announced that RTZ Minserve would obtain "a major part of this contract from Rio Algom Corp in the USA, with deliveries to start in 1973" (60).

In response, the Lisbon mine and mill at Moab, Utah, was started up in 1972 (61) and, in 1975, had reached a mill throughput of 700 tons per day (11). By then, the company could boast contracts of 4,400,000lb U_3O_8 (11) (which could have included consignments outside the Barseback contract), and it was noted number eleven in the league of holders of US uranium reserves (62).

In 1978 output was slightly decreased "in accordance with the planned sequence of mining" (27): average recovery was 4lb per ton U_3O_8, while, that year, total output reached just over one million pounds (1,076,000) (27); the mine suffered from falling grades again the following year (44). 1981 saw modest production of 430 tonnes but, with the completion of deliveries then under contract, and "in view of the currently depressed state of the uranium market" (40), production was reduced by around 50% in October of that year, and some 120 of the 250 workforce were laid off (44, 45). Lisbon then set about "toll milling" ore produced by other companies, with the result that "a relatively high rate of capacity utilisation" was obtained during that year (47).

Another contract was negotiated in 1983, for supplies until 1986, and profits increased (48): it is not known who the customer(s) was. The mine was being run down by the end of that year (63), but, at the close of 1986, the Lisbon mill was still operating at capacity, thanks to "toll" contracts from other mines (53). This was partly because 200 tons per day (out of a total of 750 tons per day mill capacity) was by then, coming from a new mine, opened by Rio Algom in southeastern Utah.

The Mi Vida mine was originally exploited by the legendary Charles Steen – "uranium mining's first millionaire". It opened in the 1950s, but closed in the early 1970s (63). Re-opening Mi Vida was a minor gamble on Rio Algom's part: perhaps more important, it signalled to the industry as a whole – still suffering the depression of the turn of the decade, and the aftermath of the Three Mile Island disaster – that uranium in the USA was not completely on its last legs. With 50,000 tons ore, grading 0.2% U_3O_8, the Mi Vida mine began its second short life in 1983, and enabled the Lisbon mill to increase production from a 5-day to a 7-day week (64).

Contracts

At the end of 1986 and beginning of 1987, Rio Algom held contracts for the delivery of about 40 million pounds U_3O_8 from Elliot Lake, which were due to expire in 1995, and a further 57 million pounds due to be delivered to Ontario Hydro from the Stanleigh operations (53).

Not only have the Ontario Hydro contracts been drastically revised to postpone delivery, but so have those with Duke Power. This US utility was to receive 7690 tonnes from 1981-1990, and delayed part of the first half of this schedule until 1991-1992 (4). In 1977, Japanese contracts, involving at least 31 million pounds U_3O_8, destined for eight utilities from 1969-70 (65), were delayed by the Canadian government's non-proliferation measures – one of these was for 610 tonnes to the Japan Power Co (4).

Among Rio Algom's other contracts, the most significant are one for 7690 tonnes with BNFL of Britain (4) – this contract having been concluded in 1973, over 25 years, as the cartel wound into top gear (4); the 4.2 million pounds deal with Kepco from 1981-1990, to be delivered in an enriched form (66), with conversion by Eldorado Nuclear (67); and a 1981 contract with Preussische Elektrizitätswerk AG, to supply 3.4 million pounds U_3O_8 from 1983-1996 at C$28lb (68).

The "Rössing contracts" are probably the most controversial single set of uranium deals ever concluded. Although Rössing remains the world's largest open-pit uranium mine, and the company has fulfilled numerous contracts over the past 13 years, it was the initial contracts, for 6000 and 1500 tons, which secured finance for the mine, and which therefore became the subject of this raging controversy. What is publicly known is that the British Foreign Office okayed these contracts in 1970 for supplies from Rössing, despite alternative sources being available. What is important from the point of view of Rio Algom is that the British Cabinet (a Labour government one, to boot) was given the distinct impression that the uranium it required for the British atomic programme through the 1970s and early 1980s would come, not from southern Africa, but from Canada – indeed from Rio Algom. In 1968, the UKAEA (British Atomic Energy Authority) authorised the conclusion of a contract with Rio Algom for such supplies. Tucked away in the small print of the Cabinet decision was an option to switch supply to

South Africa, but only if the Cabinet was informed (69, 70). The deal was with Nuclear Engineering International, and the Rio Algom contract would be for 10,260 tons between 1968 and 1982 (72).

However, in early 1970, some eagle-eyed South Africa watcher in the Foreign Office, scrutinising a notification from the Ministry of Technology (MinTech) (74), learned that two major contracts were about to be signed by UKAEA – not with Rio Algom, but with Riofinex, for the supply of U_3O_8 from the Rössing deposit, which had been recently acquired by RTZ (69). The MinTech switch had occurred not simply with South Africa, but Namibia – an illegally occupied neo-colony of the apartheid state, officially under the aegis of the United Nations. The Foreign Office then informed the Cabinet – which discussed the new contracts, but was apparently confronted, not simply with a *fait accompli*, but with a self-fulfilling prophecy. The first contract, for 6000 tons, had supposedly been signed and allegedly could not be revoked. But this first contract did not become operable until the second, for 1500 tons, had gone through. At the time of the cabinet meeting, this had not happened: sitting in ignorance (and with vital information concealed from them) the Labour government's policy-makers passed the first Rössing deal, unaware that a second qualifying contract existed.

MinTech had argued that no supplies were available as an alternative to Rössing. This was palpably untrue as, by the late 1960s, a buyer's market existed in uranium, and there was a gross over-supply. Critics of the Rössing deal have pointed out that, probably the major reason for switching the contracts from Rio Algom to Rossing, was the Canadian government's preparedness to impose international safeguards, making it impossible to utilise Canadian uranium in the British military programme (69). That may be true, but is (as yet) unproven. In fact, Canada did not impose its safeguards policy until 1974, and the policy was not fully implemented until 1977 (6), when it certainly *did* affect Rio Algom's delivery schedules (73).

Anthony Wedgwood Benn, Britain's erstwhile Minister of Technology – officially responsible for the signing of the Rössing contracts – has consistently argued that his own department kept him in the dark about these highly sensitive deals. However, even if Benn was not aware of them, he should have been, and is therefore culpable for them (74). There are also equally consistent statements from RTZ over these years, that there was never any question of securing Canadian supplies for the British nuclear programme of the 1970s and beyond (74): Namibia – or at any rate South Africa – was always the chosen source. At the time the Rössing contracts were confirmed, the Canadian uranium industry was in a state of severe decline. (Production had dropped by three quarters from 1959 until 1966) (75). Moreover, in March 1970 – just after the contracts were supposedly "switched" – the Canadian government announced new measures to limit foreign ownership of Canadian uranium: henceforth foreign control would be restricted to 33.3% (75). (The measure was not implemented for nearly two decades.) Any anticipation of this would have led RTZ, the UKAEA, and MinTech to conclude that Canada was not necessarily as safe an option as some wanted to suggest. Four years later (as if to confirm its confidence in its nascent Namibian confrére) Rio Algom took a 10% stake in Rössing Uranium Ltd, through the exercise of an equity option of C\$8.1 million (76), as RTZ sought to secure new financing for Rössing (4).

At Rio Algom's 1982 annual general meeting, the Taskforce on Churches and Corporate Responsibility, in Canada, challenged the company's continued profiteering from Namibia. Promising that a formal written reply would be forthcoming, the company's then-chair, George Albino, argued that the mine was a "mainstay" of the country's economy, which benefitted all citizens. It had "no racial barriers and no distinction in wage levels". The Taskforce, quoting the 1981 report from the UN Council on Namibia, countered that apartheid existed at all levels in the mine; unions were forbidden to operate; it was unlikely that workers were adequately protected from radiation hazards, given the failure of Canadian companies to safeguard their own workforce (77).

The Rössing contracts were concluded some time before the international uranium cartel came into operation, functioning among its members as the "Uranium Producers Club". But it is conceivable that RTZ anticipated the operations of the cartel, in which it was to be the key corporate partner: certainly there is evidence that, once the cartel was in operation (78), uranium prices rose high enough to justify development of Rössing (78).

Rio Algom participated in the cartel, both as part of the RTZ camp, and as a member of the small club of Canadian producers, protected by the Canadian government. As such it did not perhaps benefit as much from the contract allocations made by the uranium clique, as did other members. (Figures were divulged in 1976 showing that Rio Algom would deliver 840 tons up to 1977, and around 1700 tons between 1978 and 1980, out of a total of more than 10,000 allocated to Canadian producers (79)). As late as 1973, there were still intense discussions among MKU, Rössing, and Rio Algom, as to who would get what in the forthcoming round of contract fixing) (80).

At any rate, the following year, Rio Algom concluded a contract with the Tennessee Valley Authority (TVA) in the USA, for the supply of 17 million pounds U_3O_8, at around US$770 million. This gave a huge fillip to Rio Algom since it involved delivery from 1979 until 1990. When details of the cartel burst in Australia in 1976, and the US Justice department, as well as Westinghouse, began showing considerable interest in the operations of the uranium cabal (81), Rio Algom was named, along with others in the RTZ stable, as a prime violator of the Sherman Anti-Trust Act. By 1977, the company was "under severe legal pressure" from the US courts and, in the fall of that year, was held in contempt by the Salt Lake City Court, for not producing evidence so that Westinghouse could proceed with its case against the cartel. (This huge US utility claimed that it had defaulted on its own uranium contracts, solely be-

cause of the uranium price-hike stage-managed by the Uranium Producers Club.) Although fined US$10,000 a day for failure to comply with the Salt Lake order, a Denver court soon overturned this judgement, on the grounds that Rio Algom was simply following Canadian law (82). Nonetheless, a year later, along with eight other companies, Rio Algom was held in default by a Chicago Court, for failing to answer the Westinghouse charges (83). It was not until early 1981 that the company was off the hook (84).

Even then, the company was not fully in the clear: in 1982, WPPS (Washington Public Power Supply System) summonsed Rio Algom and seven other companies (including RTZ Mineral Services), for anti-Trust violations, as its construction costs shot through the ceiling (85).

TVA took Rio Algom to court in the USA in order to void its contract with the Canadian company. This was as a direct result of Westinghouse restraining TVA from making payments for a delivery of uranium to Eldorado, title to which would pass to TVA when the uranium oxide was stored at Port Hope and transported as UF_6 to destinations in the US by Rio Algom (86). Apparently TVA already had suspicions of a cartel, when it concluded the contract in 1974. Although it asked fifty-three US and foreign companies to quote a price for uranium supplies, only three replies had been received! (7). In the early 1980s TVA was again in dispute with Rio Algom over the contract price, which had apparently increased considerably between 1979 and 1980 (86).

As Rio Algom moved to find buyers for the uranium at the centre of the dispute (87), it also began an action in the Ontario Supreme Court against TVA, for C$600 million breach of contract (88). Later, the Canadian company accused both TVA and Westinghouse of conspiracy – claiming that TVA had decided it did not really need the uranium it had contracted to buy, while Westinghouse was primarily concerned to reduce the costs of TVA's own litigation against the huge electric corporation. (TVA was one of the original customers whose supplies were interrupted as an alleged result of the cartel) (89). In the end, virtually all parties had settled out of court, and TVA dropped its action against Rio Algom, in return for a wider settlement which secured C$200 million for TVA from seven companies, including Rio Algom (90, 91).

Thus, an extraordinarily complex internecine battle came to its fitful conclusion. There may be little honour among thieves, but none of them were prepared to finally kill-off the geese which lay the yellow eggs.

The USWA – the United Steelworkers of America: the main union involved in mining uranium at Elliot Lake – had expressed deep concern over the health effects of radiation from the early days of mining (17). Briefs were submitted by the miners to both federal and Ontario provincial governments as early as 1958, when the Union first demanded the appointment of a Royal Commission (17). The federal government passed the buck to the provincial government, which handed it to the mining companies (92).

Both governments were long aware of the hazards associated with the industry but, at least until 1975, deliberately concealed their existence or effects. For example, a letter in early 1969 from the Ontario Workman's Compensation Board to the Occupational Health Division of the federal Ministry of Health, stated that 16 out of 20 deaths to Elliot Lake miners were the result of lung cancer: the majority of workers had been exposed to radiation levels above the maximum permitted (93). Stephen Lewis, a former leader of Ontario's New Democratic Party (NDP), himself dug up surveys by Denison and the Algom group of companies, covering the period 1958 to 1974. He discovered that "... in no individual year, from 1959 to 1974 inclusive, in no survey, quarterly, three times a year, or semi-annually, in not a single instance, did the average underground dust counts for the uranium mines of Elliot Lake fall below the recommended limits. Not one." Concluded Lewis: "It's really quite horrendous" (93).

Governmental "wisdom" in these years was – put bluntly – that the mining companies could investigate and police themselves. It was not until well into the 1980s that the situation materially changed. Not only that – but miners were also excluded from health and safety meetings concerned with their own working environment (17).

Although Denison Mines cannot be acquitted of negligence in monitoring the work hazards of Elliot Lake, it was undoubtedly Rio Algom that was most culpable – indeed criminally negligent – during the period 1958-1980. The USWA compared company records obtained from Rio Algom in 1975-76, with government readings of radiation exposures throughout the company's mines in the area. The USWA survey showed that Rio Algom had consistently underestimated hazards in virtually every part of the mining complex and mills, by deliberately under-reading radiation levels (94). In one case, the company recorded radiation levels no less than 30% below that of the government's (17). No wonder that Dr D V Bates, on a visit to Elliot Lake in 1979 as chair of the British Columbia Royal Commission on Uranium Mining (BCRUM), was prompted to make this comment on the region's Health Protection Surveillance facility: "The general impression ... is that it commands so little in the way of resources that presumably the company at all levels regards the whole of the surveillance program as a very low priority activity" (95). (A little later, Dr Bates was to recommend a moratorium on uranium mining and exploration in British Columbia: a moratorium which is still in place today).

Again, while both the Atomic Energy Board of Canada (AECB) and the Ontario Department of Mines, had authority to insist on improvements at the mine site, their respective responsibilities until the 1980s were not at all clearcut, and allowed for widespread negligence. Kenneth Valentine, National Director of Health and Safety for the USWA in 1980, stated baldly in a submission to the BCRUM that, in his opinion: "... there have been situations where almost [every company at Elliot

Lake] should have their licence revoked ... In the final analysis, I would suggest that, at the very least, the AECB is guilty of criminal negligence ... And as a result of their failure, their dereliction of duties, untold numbers of people have died" (96). The Staff Representative of the USWA at Elliot Lake itself told the same Commission that his union didn't "... know of one case where the AECB has charged the employer ... The lack of enforcement is exactly the same as the lack of any regulation" (97).

The medical profession has hardly been more responsive or responsible. In particular, the Workmen's Compensation Board (WCB), set up supposedly to evaluate medical grounds for compensation, has not been well served by its doctors. Until the 1980s, the Board only accepted exposure to radon "daughters" as grounds for declaring that cancers were industry-related (17, 98). Thus, workers in the thorium-separation plant operated by Rio Algom until the 1970s – who were exposed to up to 40 times the radiation level recommended by the International Commission for Radiological Protection (ICRP) – failed to gain any compensation from the Board (97).

For many years, too, whole-body radiation caused by gamma rays – now acknowledged as a key source of various cancers – was not entertained as a causative factor (99).

Up until 1974, a chest clinic at Elliot Lake gave miners yearly chest X-rays, ostensibly to identify incipient cancers. Despite the fact that some miners were showing deterioration, their licences were renewed, and they were sent back down the mines (97). That year, when the limit for silicosis compensations was lowered by 25%, and a large number of new cases became eligible for compensation, "neither community doctors, Ontario Workers' Compensation Board, nor the companies, said anything to the men" (100). One man, Joe Zuljan, was told by a WCB doctor that, although he had 10% disability for silicosis, he was fit to return to underground working at Elliot Lake. He did not return. Nonetheless, in two years he was dead (100).

This is not to gainsay the contribution of one or two doctors who stood out against the trend. In 1976, one doctor, with thirteen year's experience of Elliot Lake, claimed he had warned of carcinogenic working conditions, no less than seventeen years before, but had been ignored (101).

Historically, the miners of Ontario – unlike their fellow workers in Alberta and Nova Scotia – were fairly quiescent from the 1920s until the 1960s, when they joined miners throughout Canada striking for better wage, health and safety measures (102).

Then, in 1974, a "wildcat" strike, initiated by Ed Vance at Denison's operations at Elliot Lake, set in train major improvements in health and safety throughout the country (102).

The three-week industrial action was prompted by revelations – ironically, at a symposium in France (17) – that the abnormal cancer rates among Elliot Lake miners had been known for five years to various provincial government agencies (103). Indeed, an Ontario Ministry of Health Report made in December 1972, observed that deaths from lung cancer among the miners exceeded the expected rate in the general population by a factor of between three and five (99).

The immediate practical result of the strike was the appointment of the Royal Commission on the Health and Safety of Workers in Mines, usually known as the Ham Commission after its Commissioner, James Ham. This reported in 1976, and its conclusions, were potentially devastating.

Ham decided that, of 956 deaths on the Ontario Uranium Nominal Roll of 15,094 miners (who worked longer than one month), eighty-one had died from lung cancer during 1955-74, as against 45 expected for the population as a whole. Ham found that typical exposures among the lung cancer cases were in the 70 WLM (Working Levels per month) range. ("Working Level" is an acknowledged way to measure extent and length of exposure and its cumulative impact). These results, though smaller than found elsewhere (in Colorado in particular) "... are just as likely to overstate the quantity of radiation actually needed to produce cancer".

This was because lung cancer had already been in place as a cause of death, due to fairly low exposures. The lapse of time between genesis of cancers and death allowed further exposure which did not alter the fatal outcome, but inflated minimum exposure levels pre-disposing to cancer. In view of this, Ham concluded that 10 WLM would have to be a maximum threshold, "to be at all plausible in relation to the Ontario experience" (104). Equally important (especially in the light of later research into the effects of low-level radiation in general) was Ham's judgment that excess lung cancers occur at lower radiation levels than could have been adequately studied in US uranium mines to date. (He referred specifically to *Radon Daughter Exposure and Respiratory Cancer, Quantitative and Temporal Aspects* by Lundin, Wagoner and Archer, published by the National Institute for Occupational Safety and Health, in Washington in 1971).

The Royal Commission had revealed 93 "silicotics" (people suffering from silicosis), ascribed to Elliot Lake operations, by the end of 1974. K A Valentine, in his 1980 submission to the Key Lake uranium Inquiry, claimed that, by early 1975, nearly 500 miners had lung disabilities, wholly or partly ascribable to dust exposure in the mines and mills, of whom 147 were afflicted by silicosis, and 200 predisposed to the disease. Over the period 1974-1980, said Valentine, more than five hundred such cases had developed and the situation had not markedly improved (99).

James Ham's report led directly to a new provincial Mining Act and new set of regulations in 1978 (105). However, just after the new legislation was passed, the Federal government reverted to the old Ontario laws. Since uranium mining was covered by federal legislation, Rio Algom and Denison workers were left with the savage irony that, the very changes they had struck in support of, were now denied them (103).

After threats of strike action, further negotiations, and a work stoppage, in October 1981 the USWA won a new collective wage agreement, covering 400 employees at Rio Algom and Denison mines (106). More important than the mediocre wage increase was a new three-year contract, empowering the USWA to appoint inspectors (eight at Rio Algom and five at Denison), paid and trained by the companies, to be responsible for monitoring all safety, health and environmental hazards (105, 106). The provincial right to refuse unsafe work was, at last, incorporated into the agreement (Bill 70, as it is known). Commented Homer Seguin of the USWA: "I don't think there is anywhere in North America that will have such powerful health and safety people". According to Ken Valentine: "It is extremely important that the Elliot Lake experiment succeed – it will determine the future of this type of effort for all of North America" (105). ("The next step," added Ken Valentine in an interview with the anti-nuclear paper *Birch Bark Alliance*, "would be co-determination of the workplace. But we're a ways off from that yet") (105).

The process was still not complete, however. It was not until mid-1984 that, by signing amended collective agreements, the workers at Elliot Lake finally gained a complete legal right to refuse "unsafe work" – and stand on the same footing as all other Ontario miners (107).

Miners at Rio Algom (and Denison mines) in Canada, are now better safeguarded than uranium workers anywhere in the world. But perhaps this fact should not be over-stressed. Since there is no radiation limit which is absolutely safe – let alone those maintainable in underground uranium workings – the Elliot Lake monitoring is, at best, an admirable holding operation: at worst, it might beguile some workers, public enforcement agencies, and others, into granting the industry a blank cheque for further development. Although Rio Algom showed no sign of supporting the Ontario legisation until the writing appeared on the pit-wall, the company has apparently taken care not to disturb implementation of the 1981-84 agreements. On the other hand, the Elliot Lake precedent has not been taken up outside of Canada, let alone within the RTZ group. And, long after the agreement was signed, the USWA was still demanding the investigation of workers who had been subject, without their consent, to "aluminium dust" "therapy" from 1943 to 1979 (when the union stopped it), as well as the funding of their own Workers' Health Centre (108).

The radiation and pollution effects of Rio Algom's operations on the surrounding land and rivers are more intractable and, so far, have merited much less preventive action. However, it should be pointed out that, in this area too, the USWA have probably been the most vocal objectors to company practices. They swiftly raised their voices when they discovered that waste from Eldorado's Port Hope UF$_6$ refinery was to be sent back to their area, for the extraction of small amounts of uranium, before being dumped in the tailings pile (103).

In 1978, the Porter Commission (officially, the Ontario Royal Commission on Electric Power Planning, chaired by Arthur Porter) estimated that some 80 million tonnes of uranium tailings were to be found near Elliot Lake. "Several lakes and streams have been badly contaminated" as a result, declared Porter, who warned that the quantities might reach "several hundred million tonnes" by the turn of the century, while "there are no obvious technological solutions in sight". Indeed, "the problem is probably more difficult to handle than the ultimate disposal of high-level radioactive wastes, partly because of the extremely large volumes involved" (109).

Barely two years later, other Canadian researchers were estimating that, around 110 million tonnes of such tailings had already accumulated nationwide, of which the vast majority (100 million) were a bequest of operations at Elliot Lake (17). If spread over a two lane highway, three feet deep, according to Miles Goldstick, these wastes would stretch all the way from Halifax to Vancouver, coast to coast. According to another observer: "... local residents continue to pick blueberries within a 40 foot high wall of uranium tailings" (110).

Based on both Rio Algom's, and Denison's, expansion plans in the late 1970s, the AECB estimated that the volume of solid wastes could double by 1990, reaching a billion tons by the year 2010 (111). The Toronto group, Energy Probe, in the early 1980s, projected premature cancer fatalities from this carcinogenic bequest to future generations, of 60,000 over the next 110,000 years – not to mention those caused by the thorium-radium 226-radon decay cycle, which spans around half a million years (112).

Canada's uranium wastes are dumped either onto the land surface, or into lakes and rivers. In 1978, says Miles Goldstick, uranium mills were spewing out 14,000 tonnes of solid wastes, and double that amount in liquid effluent, each day. Although the Ontario Water Resources Commission began waste monitoring in 1957, and the Department of Health started investigating drinking water pollution the following year: "little of this information has been made available to the public" (17, 113).

In addition, tailings were long used as construction and building material. This problem was recognised by 1976, when the AECB set up a joint Federal-Provincial Taskforce to deal with it, and decontaminate the sites. Setting 0.02 WL as the allowable radiation limit (114), the Taskforce found that 40% of the houses at Elliot Lake needed attention, ranging from "immediate remedial action" to "further investigation" (115). One section of Elliot Lake township was later rendered a "ghost town" with only a fifth of its 251 buildings deemed habitable (116). Even then, the 0.02 WL set by the provincial government was heavily criticised as far too high. Commented Dr Gordon Edwards, of the Canadian Coalition for Nuclear Responsibility, in 1978: "... continuous exposure to 0.02 WL for 12 hours per day could lead to a whopping 31% increase in the incidence of lung cancer for males" (117).

But, as usual with uranium mining, it is the problems caused by the tailings disposal (or lack of it) that are the worst.

These problems can be summarised as: those caused by radioactive releases into the air, especially of Radium-226, and by releases into water (flowing streams, lakes and groundwater); acidity caused by heavy metals; pollution caused by chemicals used in the milling and comminution processes, specifically ammonium and nitrates (17). Gauging the impact of all these malign influences on humans, animals, plants and aquatic life is extremely difficult, and has hardly begun. In its final report on the Elliot Lake expansion plans (made in 1979, after the Ontario legislature had compelled Algom and Denison to submit to the environmental assessment process), the Ontario Environmental Assessment Board commented that: "... considerable effort and time is required before solutions to the long term aspects of waste management can be found ... the companies, as well as the various provincial and federal agencies, have only recently come to grips with the problems ..." (118). In particular, desperately little is known about the ways in which human beings take up radioactive traces, or heavy metals, through the food chain: ironically, thanks partly to the devastation caused by thirty and more years of mining at Elliot Lake, more is known about the damage to fish (17). Similarly, the migration of tailings through groundwater – with a potentially disastrous effect on future generations – only began receiving attention in 1978, at a point when thousands of tonnes of radioactive and acidic sludge would have found their way into aquifers and streams (17).

It is the Serpent River Basin which receives all discharges from the Elliot Lake complex and, until the early 1960s, all waste water entering the system went untreated (17). By 1980, so many tailings were being dumped that, in the headwaters of the Serpent River – where flows are very low during the summer – liquid wastes from the mines comprised between one half and two-thirds of the total flow (119). In 1976 – a watershed year for the region, in several respects – the International Joint Commission on the Great Lakes (IJC) identified the outflow of Serpent River into Lake Huron as the greatest single source of Radium-226 and thorium isotopes in this tract of freshwater. Just one kilometre into the lake, huge concentrations of Radium-226, thorium and other metals, were well above standards set by the IJC (119).

More significant was a 1976 report by the Ontario Ministry of the Environment – initially secret, but divulged the following year (120) – which concluded that eighteen lakes in the Serpent River system had been contaminated as a result of uranium mining, to such an extent that they were unfit for human use and all fish life had been destroyed (121).

By this time, Rio Algom had commenced its expansion programme and re-activation of old mine and mill facilities. "Changes to the surrounding terrain," RTZ blithely declared, "are expected to be negligible." Citing reductions in the use of freshwater and chemicals in plant recycling, and other improvements "over the years", the company studiously avoided referring to the impact of its tailings on the river, lakes and the life dependent on them (122). Yet two years later, "in spite of several years of remedial action", Professor Scharer of the University of Toronto reported that there was still no possibility of fishing or safe swimming in the Quirke, Whiskey, McCabe or Hough Lakes (112).

Let us look briefly at the specific impacts of mill tailings production and disposal at Elliot Lake (in particular those caused by Rio Algom) and examine what the company has done about them in the last ten-to-fifteen years.

Radium-226 poses particular problems where it is deposited on lake bottoms. In 1971, the Ontario Water Resources Commission identified levels of radium in the lakes with levels up to 50 times the background radiation level, and up to fifteen times the maximum water quality level – 3pCi/l (pico-Curies per litre), as set by the government (12). Although, by 1978, radium levels in Quirke Lake had been lowered (by dilution, reduction in mining, and new Rio Algom treatment facilities) to 5pCi/l, they were still twice the target level for drinking water, as set by Health and Welfare Canada (17). Clearly, the expansion of Quirke Lake mining could increase concentrations further. Radium-226 levels in May Lake were, in 1978, still 9pCi/l – around four times the government-set limit (119).

Alarmingly, barely two years later, a federal provincial committee recommended that, instead of lowering radium levels to meet health criteria, the standards themselves be lowered: this was in line with a recommendation by the ICRP that maximum permissible radiation for bone marrow could now be increased nine times. Sister Rosalie Bertell, of the Jesuit Centre in Toronto, and a world expert on radiation (with special knowledge of Elliot Lake) called the decision "murder". In her opinion, "the blatant reason for the change is because the radium is too expensive to clean up" (124).

Radium, thorium, and uranium, dissolve more easily – hence pose greater problems – under conditions of high acidity (17). Mainly because sulphuric acid is used in mill processing, acid wastes have long proved a killer of aquatic life in the region, and are a prime reason for the absence of fish in the lakes (at least until recently). Mature freshwater fish cannot survive in water whose acidity (measured as pH – a high pH reading being alkaline, a low being acidic) is outside the range of pH5 – pH9. At May Lake in 1980, acidity was well within the range of vinegar (3.1). Although, by then, acidity at Quirke Lake was reducing (to around pH5.5), expansion of the plant once again posed a threat (17).

Acidity levels are inversely linked to the dangers posed by heavy metals: the higher the pH, the more gas given off by ammonium liquid. At Quirke Lake, concentrations of ammonium have exceeded the Ontario Drinking Water Quality Criteria (17), while high concentrations of nitrogen compounds have been located some 40km downstream of the mine (119). Iron levels at May Lake (119), copper levels at Quirke and Dunlop Lakes, have given cause for concern in recent years (17). Copper is especially threatening to fish. (It kills young salmon at 0.1 ppm [parts per million] – one tenth of the human drinking water limit (125) – and can prove poisonous to sheep at 0.5 ppm). Copper

levels measured downstream of the Quirke Lake facilities have reached 0.11 ppm, and at Dunlop Lake, 0.58 ppm.

This environmental "bill of account" would not be complete without some discussion of "accidents" resulting from mine operations at Elliot Lake. These include unexpected seepages, unintended movement of solid wastes, dam and dyke failures, re-solution, or erosion, of barium-radium precipitates, waste pipeline breaks, pump failures, a decant tower collapse, the accidental release of a large volume of sulphuric acid and release of fine sands in the decant effluents (119). In 1964, some 82,000 tonnes of waste drained into Quirke Lake, due to "overtopping" of the dam (95). And, in 1979, within a month of the Panel mine and mill being modernised (126), a tailings line broke, leading to liquid and solid contamination of Quirke (116).

What has Rio Algom done in response to this liturgy of destruction and hazards which will persist for aeons to come? At no point has it closed one of its mines on health or safety grounds. Instead, it has introduced a barium chloride treatment process, which is applied to run-off waters before they are released into the lakes (127). However, although this process – together with some recycling of plant water, and some treatment of run-off from abandoned sites – was reducing Radium-226 levels in part of the Serpent River quite dramatically between 1966 and 1982 (according to the IJC: from 11.7pCi/l to 1.43pCi/l) (127), the barium chloride "solution" may simply shift half the problem downstream. According to experts from the Environmental Protection Service of Environment Canada, well under half the total radium contamination is removeable by this process: namely, the dissolved part. The radium which is in suspended solid form merely gets carried further towards the sea, where its impact is much more difficult to gauge (128).

Neutralisation of acid wastes has been practised at Elliot Lake through the addition of lime to tailings, but this strategy – according to the British Columbia Royal Commission on Uranium Mining – has largely failed (95).

In 1978, as the Quirke mill entered its expansion phase, and Rio Algom's chair, George Albino, outlined the plans at the Ontario Environmental Assessment Board, the company made it clear that it did not intend to alter the use of nitric acid in the ion exchange circuit, or ammonia for the precipitation of uranium (although studies had indicated that magnesia might be substituted for ammonia) (129). The problematic relationship between acidity control and ammonium levels has already been mentioned: it is also important to note that, as acidity is lowered, so the level of total dissolved solids (tds) rises, and – up to the turn of the 1970s – was rising quite dramatically (119).

Although Rio Algom prides itself on re-vegetation of some of the tailings piles and, by 1979, claimed to have re-vegetated more than 400 acres of wastes (using huge amounts of limestone and fertilisers), in 1980 it was noted that the cover was only six inches thick, while Radium-226 was accumulating in grasses and bull-rushes, at ten times the background radiation level (112) (130).

In 1983, a proposal to dump tailings under the water at Quirke Lake was vigorously adopted by two firms of consultants as a means of controlling (and disguising) the galloping tailings crisis. The proposal, though apparently supported by the AECB, was firmly rejected by Denison and Rio Algom. Rio Algom, in particular, argued that the company did not need a new disposal area, and would be proceeding with land-fill as usual. "Rio Algom has a commitment to our current buyers to keep costs as low as possible". The development of Quirke Lake would be an "unwarranted cost," it declared (131).

The most "unwarranted cost" bequeathed by Rio Algom has been borne by the Native Canadians from the Serpent River Band, living downstream of the Elliot Lake operations – some of them working in the mine and sulphuric acid plant. By 1978, the band could no longer drink the water, or fish in the river system (132). Although there might have been some marginal improvement in their habitat since, evidence from a comprehensive Environmental Impact Protection Program, carried out by the Toronto Jesuit Centre, along with three native Canadian bands (including Serpent River) between 1982 and 1984, makes disturbing reading:

- About twice as many young adults (36 years and under) reported chronic disease at Serpent River as at the other two reserves studied (Mississagi and Spanish – which are not uranium-related).
- The Serpent River band also reported the largest proportion of participants of all ages with chronic disease.
- The most frequently reported diseases on all three reserves were diabetes, cardio-vascular diseases, arthritis and deafness. However, diabetes and cardio-vascular disease appear to affect the Serpent River band at a significantly younger age.
- Just as chronic diseases are more frequent at Serpent River, so are pregnancies which end in fetal death (133). The report also analysed diseases in terms of occupational exposures, and concluded that, in males over 45 years, "ill health" is reported by more men with exposure to the acid plant or uranium mine (75%) than those not exposed (43%).
- There is a higher than expected number of unexplained fetal deaths and abnormal live births among uranium miners, relative to non-uranium miners under 45 years old. There also seems to be an elevated rate of unexplained fetal deaths, and subnormalities in live-born offspring, for acid workers in the same age group (133).

Although the study is necessarily limited – and cannot expect to show up radiation-induced cancers which take 30 years to develop – its conclusions are that: "Occupational exposure appears to have a significant effect on Band members. The costs may well outweigh any benefits derived from nearby resource developments" (153).

Elsewhere

Rio Algom's second major operation centres on the lucrative Lornex deposit in the Highland Valley of British Columbia, where Lornex Mining Corp has substantial interests in copper and molybdenum (134). Rio Algom acquired its interest in Lornex in 1965 (134). By 1969, it owned 36% along with Yukon Consolidated Gold Corp Ltd (24%), and various individuals, including Egil Lorenstzen, the prospector who found the lode (40%) (135). At the time, it was the lowest-grade copper deposit ever developed (135). Bechtel, who carried out the economic analysis of the deposit in 1968, concluded that, even with a faltering market, the mine could be viable (135). In March 1969, Rio Algom announced that it would raise C$120 million, to finance the venture, by issuing debentures and common-stock, to lift its own holding to more than 50%. Funding would come from Yukon Consolidated (though its interest would drop to 19%) and from two consortiums – one Canadian, and one Japanese (136). A month later, Rio Algom promised that the mine would come into production within five years, and that all output would be supplied to Japan (135). The company, by then, had secured some highly-favourable rates of investment from six Japanese smelting companies and banks. The smelters would loan nearly C$30 million with "the cost to Lornex [being] less than the current rates in Canada" (135), and three Canadian banks would provide C$60 million, "the largest bank loan ever provided for the development of a single non-ferrous mining project in Canada" (135). It would soon be the largest base-metal mine in Canada.

In 1975, Rio Algom made a cash offer for all shares it did not hold in Lornex, and, as a result, increased its interest from 59.8% to 66.51% (11). Over the next ten years, Rio Algom's interest in Lornex crept up to 68.8% (134). During that decade, its fortunes fluctuated, but overall, the Lornex story is a remarkable chapter in Canadian mining history, and more than a minor footnote to the RTZ saga. On several occasions, Lornex has boosted Rio Algom when

the company's uranium interests took a fall – in 1979 for example (137); in 1980 (138) and again in 1983 (139), when production from the mine reached record levels, and the company explored further afield (140).

However, in 1981, thanks to lower prices – not only for copper and molybdenum, but also silver, which is an important Lornex by-product – earnings fell. At the same time, the mine was expanded by 68%, more or less on schedule (141) – a factor which did not improve results (41), though Rio Algom claimed that the expansion programme was mainly self-financed, and production costs were "low" (40). In 1982, too, the company ran a loss (142) – indeed its worst deficit on record (140).

In that year, however, Lornex reached agreement with Teck (itself a 22% shareholder in Lornex) (143), to acquire a 39% interest in the Bullmoose metallurgical coal project (144, Teck 51%, Nissho Iwai Coal Development, Canada, 10%) which was set up to supply nearly two million tonnes a year of coking coal to Japanese steel mills, with lesser tonnages to other customers (140). Construction of Bullmoose was completed in 1983, and production started the following year (143, 145).

Lornex's fortunes improved the next year, thanks to better market prices (139), although they dipped in 1984 (146), resulting in a "total loss" (147).

By 1985, the Highland Valley deposit was clearly suffering from rising costs, declining ore grades, and dwindling reserves (134). Early the following year, Lornex announced a merger between its British Columbia copper operations and those of Cominco (Lornex 45%, Cominco 55%) (148), later to be joined by Highmont Mining (taking 5% of Cominco's share) (134). The Highland Valley Copper partnership (HVC) combines Lornex's mill and mine with Cominco's Bethlehem mill and mine (134). Essentially, this new JV was of prime benefit to Teck, which – since the reorganisation of Canadian Pacific and Cominco the same year – has ended up hole-shot position. Not only does Teck retain its interest in Lornex, but it has become the largest shareholder in Cominco, and owns half of Highmont (149). But Lornex has also benefitted – as its 1986 results showed (150), as well as those of 1987 (151).

In 1988, Lornex began a major new construction programme at Highland Valley, at around C\$70 million. With the milling rate expanded nearly 50%, the complex was expected to become the second largest in the world "and one of the most cost-efficient" (152). No wonder company chair, Ross Turner, could claim, in 1988, that Lornex, with cash reserves of around C\$120 million, might now expand beyond British Columbia. "There are outside shareholders," said Turner, "and it has to be managed to the benefit of the shareholders"(134).

Apart from uranium, copper, molybdenum and coal, Rio Algom has also been dabbling in that wonder-metal which virtually every Canadian mining house has found itself unable to resist: gold. In 1977, Riocanex (Rio Algom: 100%) found gold in Newfoundland, at Cape Ray, 19 miles north-east of Port aux Basques (19, 154). Drilling continued in 1978 and 1979, though a major project has yet to emerge (155). However, in 1981, Riocanex bought 50% of the Nelson property in British Columbia (where gold is associated with copper, silver and other minerals). Drilling that year indicated some 7 million tonnes of copper at 0.4% with 1.5 grammes/tonne of gold and 6.6 grammes/tonne silver. Riocanex reduced its primary exploration efforts in 1982, concentrating on "known mineral deposits" (47). Over the next six years, Rio Algom itself was to consolidate exploration in Chile and Spain, as well as North America, acquiring 60% of the Sage Creek steam coal project in British Columbia, an option to buy into a tungsten-tin mine in southwestern Spain (La Parilla) – where major drilling took place in 1986 (143) – and further copper and gold properties in North America (156).

One of these exploration projects was to lead the company into considerable controversy, and an eventual withdrawal from a potentially highly-rewarding mine. The Cerro Colorado copper deposit in Chile (not to be confused with the Panamanian deposit of the same name – subject to even more opposition and a pull-out by RTZ itself) lies nearly 100km east of Iquiqui port, 2600 metres up in the high Andes. It contains some 70 million tonnes of copper sulphide ore, of fairly high grade (1.33%) with some gold and molybdenum (157). In 1980, the Japanese company, Nippon Mining, withdrew from the venture and transferred its share to Rio Algom. Rio Algom agreed to develop the deposit, and commissioned Davy-Mckee of Canada to complete a technical assessment and feasibility study (157). When these were completed in 1984, they confirmed that a 14,000 tonnes/day open-pit operation was feasible. Water would be drawn from the Andes by pipeline, and two small hydro-electric dams would be constructed. A development decision was expected before the end of 1985, but the feasibility study advised waiting until the price of copper had risen (47, 158).

Initially, Rio Algom and RTZ asked the Chilean junta's Foreign Investment Committee to authorise US\$250,000 to construct the mine and mill (40, 159). As 1985 came to an end, it was thought that financing for Cerro Colorado was all sewn up. Rothschild and Sons Ltd – long associated with investments in RTZ and in financing the Churchill Falls project in Labrador (160), with which Rio Algom was associated in the 1960s – was a probable (unnamed) backer. Also involved was the British Export Promotion Agency, a government parastatal, which promised US\$135 million in loans, so long [of course!] as they were used to buy British mining services and equipment. West Germany's KWF financing agency would grant credits of US\$80 million, in exchange for two thirds of the production (which would go to Norddeutsche Affinerie). Most crucial was an agreement by the Finnish government financing agency, VTL, to pump in US\$35 million, in exchange for a third share of the output to Outokumpu Oy. Rio Algom would also reportedly purchase the major equity in Cerro Colorado (at US\$60 million) but might share it with an "undisclosed partner" – probably Corfo, the

Chile Development Corporation, which was expected to provide an additional US$30 million to increase the overall equity of the venture (157).

Within a few months, however, Rio Algom's participation in Cerro Colorado was on ice. Its 1985 Annual Report put it tersely: "... it had been decided not to pursue development of ... Cerro Colorado ... because of recent difficulties encountered in completing satisfactory financing and commercial terms" (156). In fact, although sale of a 25% interest in the mine did go through to Outokumpu (161), the Finnish government then stepped in, under pressure from trade unionists and other groups, to halt direct support for the Pinochet regime, with its appalling record on [lack of] human rights. Ironically, the Finnish government was not to prevent financing for Outokumpu two years later, when the Finnish company decided to buy into the Escondida copper mine, in which Rio Algom's parent, RTZ, holds a 30% interest. Up until now, Cerro Colorado – although all its feasibility studies have been completed – has not commenced production (162).

Rio Algom's ventures into another major metal have also had a chequered history. In 1982, the company raised C$157.5 million from north American banks (163), to acquire a large tin deposit at East Kemptville in Nova Scotia (164, 166), discovered six years previously (165), with a projected 17 year life (166). The mine and mill took 18 months to construct (167), and production was underway in 1986. All the output was contracted to RTZ's 100%-owned smelter at Capper Pass, on Humberside (168) (169). However, in October 1985 – just four days after the mine officially opened – the International Tin Council caved in under demands from its creditors, and the market price of tin collapsed. A year later, Rio Algom asked a consortium of six banks – which had borne the financing costs, without recourse to Rio Algom itself (one of those adroit moves that have characterised all RTZ companies over the past twenty years) (168, 170) – to take control of the mine. Meanwhile, because of the suspension in tin trading, Rio Algom and Capper Pass had made a special price arrangement for the treating of the Canadian concentrates (168).

Barely a year later, with the price of tin starting to improve, Rio Algom offered to buy back the mine (171). Apparently the Bank of America (61.9%), seeking to offload a costly asset it didn't want, put the mine up for tender and only Rio Algom and RTZ made the running (172). A bare seven months later, Capper Pass was itself up for sale, amid statements that the former managing director had failed to agree with RTZ on how the plant should be run (173). This announcement followed, a large-scale and highly effective campaign by local people, joined by national campaigners and the media, to call Capper Pass to account for a horrendous excess of child leukaemias in the area downwind of the smelter (174). Naturally RTZ and Capper Pass denied that this spotlight of adverse publicity had anything to do with the plant's reduction in output and offer for sale (173). What is more likely is that the public campaign, combined with possible pressure from Rio Algom to renegotiate the East Kemptville smelting deal (even possible interest from Rio Algom in taking over and re-structuring Capper Pass), threw doubt on Capper's viability at a crucial time. By "coincidence" RTZ sold up its interests in Cornish tin mines in April the same year: these had also been supplying feedstock to Capper Pass (175).

As part of its 1985-86 expansion and acquisitions programme, Rio Algom bought up all outstanding common shares of the Potash Company of America from Ideal Basic Industries, at the turn of 1985 (156). This gave the company 87.8% of the voting rights in PCA (53), and mining and concentrating facilities in Saskatoon and New Brunswick (156). At the time, potash prices were weak, but naturally Rio Algom was looking to a more settled future. Substantial losses, flooding in the Saskatoon mine (151, 176), followed by a strike in 1988 (177), have not augured well for Algom's potash ventures to date.

If, by 1985, diversification had become the watchword of the company (156), by the beginning of 1988, the top echelons were dissatisfied with the persistent "somewhat conservative image of the company". This did not mean further diversification outside of mining, however: on the contrary, the new chair, Ross Turner, was determined to "set mining as [our] main thrust over the next several years" – starting with North America (134), and new gold and uranium properties in particular (179). (Only a month later, however, Andrew Buxton of RTZ Metals was warning Canadian companies, and others, not to devote "too much time" looking for gold rather than other metals) (180).

Within the year, the company entered into a letter of intent with Kerr-McGee, the most notorious US uranium miner of the 1950s and 1960s, to acquire not only Kerr McGee's uranium properties in Wyoming, but also the issued and outstanding shares of Kerrmac's uranium company, Quivira Mining, with it's operations in the Grant's mineral belt of New Mexico. Thus, Rio Algom was once again on the brink of uranium expansion in the USA, reviving old properties.

Turner took over from George Albino, who had joined the company in its early years (1965), and in 1981 had risen to become Chief Executive and Chair, devoted to "a fundamental strategy for growth" (181). A pale and inconspicuous character, Albino was abruptly thrown through the Algom portals in late 1987, without official explanation (182). Later the company claimed he had been guilty of insider dealing (directing Rio Algom to buy shares in Atlas Steels, which were held by his wife, and profiteering from priviledged information on the sale of certain Rio Algom shares). Albino counter-claimed with a C$10 million suit for wrongful dismissal, and a claim for an annual pension of nearly C$500,000 a year, protesting that Rio Algom had already intended to acquire the Atlas shares, while transactions with his own company's shares had been "conducted openly and with full knowledge of both financial circles and the company" (152).

In any event, in August 1988, Ray Ballmer became the vice-chair, Turner was the chair, and (lo and behold!) Colin Macaulay, formerly chief executive of the Rössing mine in Namibia, got drafted-in to become the new President and Chief Operating officer (183).

In his seven-year stint at the helm of Rio Algom, Albino had distinguished himself perhaps only once with his critics. When asked at the RTZ annual general meeting, in 1985, whether the provisions of the US Mill Tailings Act of 1978 would apply to the huge expanse of radioactive tailings left to future generations at Elliot Lake and Utah, he replied that he "was not aware of what this Act" – the most important set of environmental recommendations in post-war uranium history – "was about" (184)!

Contact: PARTiZANS, 218 Liverpool Rd, London N1 1LE, UK.

References: (1) Stephen Lewis, leader of the Ontario New Democratic Party, to the Ham Commission, Feb 18, 1975. (2) *Daily Telegraph*, London, 4/6/82. (3) *MJ* 9/5/86. (4) Neff op cit. (5) *Nukem Market Report*, 1982, quoted in *MJ* 30/4/82. (6) *EPRI* 78. (7) Moss op cit. (8) Ontario Environmental Assessment Board, *Expansion of the Uranium Nines in the Elliot Lake Area*, Final Report, Section 1, Toronto, 1979. (9) Art Kilgour, *A History of Uranium Mining in Elliot Lake*, Ontario, Birch Bark Alliance, Trent, 1980. (10) *Times*, London, 17/12/68. (11) *MIY* 1977. (12) *Nuclear Free Press*, IBBA, Peterborough, Spring 83. (13) T.J. Downey, *Feast or Famine, the Political Economy of a Community of a Simple Enterprise* in The WPRIG Reader; *Case Studies in Underdevelopment*, Waterloo Public Interest Research Group, Waterloo, 1980. (14) Porter Report, quoted in *New Scientist*, London 2/11/78. (15) *MM* 5/790 (16) Stephen Probyn and Michael Anthony, *The Cartel that Ottawa built*, *Canadian Business* 11/87. (17) Miles Goldstick, *Uranium Mining in Canada: Some health and environmental problems*, British Columbia Survival Alliance, Third working paper, Cornwall, 1980. (18) IAEA/OECD *Red Book*, 12/79. (19) *MJ* 19/8/77. (20) *MJ* 14/10/77. (21) *MJ* 18/11/77. (22) *FT* 6/3/79. (23) *Nuclear Fuel*, 26/5/80. (24) *Herald*, Melbourne, 29/5/78. (25) *MJ* 16/12/77. (26) *MJ* 6/4/79; *FT* 6/3/79. (27) *MJ* 6/4/79. (28) *FT* 4/8/79. (29) *Nuclear Fuel*, 4/8/80. (30) *MJ* 28/7/79. (31) *MJ* 28/9/79. (32) *FT*

24/3/81. (33) *Wall Street Journal*, 2/8/79. (34) *MJ* 21/3/80. (35) *MAR* 1980. (36) *FT* 2/5/80. (37) Rio Algom *Annual Report* 1980. (38) see also *Kiitg*, WISE Amsterdam, no 16, 10/81. (39) *Kiitg*, 5/80. (40) *MJ* 16/4/82. (41) *MJ* 13/11/81. (42) *FT* 25/3/82. (43) Rowe and Pitman, *RTZ Analysis*, London, 8/82. (44) *MJ* 11/9/81. (45) *MJ* 16/10/81. (46) *FT* 30/7/82; *FT* 1/3/83. (47) Rio Algom *Annual Report* 1982. (48) *FT* 15/8/83. (49) *MJ* 20/4/84. (50) *FT* 6/4/84; *FT* 1/11/83. (51) *FT* 27/7/84. (52) *MM* 9/83. (53) *Canadian Mines Handbook* 1987-88. (54) *MJ* 26/4/85. (55) Rio Algom *Annual Report* 1985. (56) *MJ* 2/5/86. (57) *FT* 13/6/84. (58) H.B. Merlin, *Canada's Uranium Supply Potential*, International Conference on Uranium, Atomic forum / Canada Nuclear Association, Quebec City, 9/81. (59) *MJ* 6/11/87; *FT* 4/11/87. (60) *Wall Street Journal*, 4/6/70. (61) *MJ* 16/10/81. (62) *Barrons* (magazine) USA, 12/9/77. (63) *E&MJ* 10/83. (64) *E&MJ* 3/84. (65) *FT* 5/5/69. (66) *Financial Post*, Canada, 14/6/80; *MJ* 30/5/80. (67) *FT* 28/5/80. (68) *Wall Street Journal* 13/1/81. (69) Rogers & Cervenka op cit. (70) see also Alun Roberts, *The Rössing File: The Inside Story of Britain's Secret Contract For Namibian Uranium*, CANUC, London, 1979. (71) *MAR* 1966. (72) *Nuclear Engineering International*, 5/74. (73) *MJ* 19/8/77. (74) Alun Roberts op cit. (75) *EPRI* report. (76) Rio Algom *Annual Report* 1976. (77) *Nuclear Free Press*, Peterborough, Summer 82. (78) *National Times*, 16-21/8/76. (79) H.F. Melouney, of MKU to R.H. Carnegie, confidential letter 12/7/72. (80) Louis Mazel of RTZ Services, Telex to R. Carnegie, 27/11/73. (81) *Fin. Review*, Sydney, 28/7/76. (82) *MJ* 21/10/77. (83) *FT* 8/12/ 78. (84) *FT* 18/ 7 /81. (85) *MJ* 1/1/82. (86) *MJ* 27/7/79. (87) *FT* 25/7/79. (88) *New York Times*, 15/8/79; *FT* 16/8/79. (89) *MJ* 19/10/790 (90) *FT* 18/3/81. (91) Rio Algom *Annual Report* 1980. (92) *Birch Bark Alliance* no. 6, Peterborough, USA, Spring 80. (93) Text of Submission by Stephen Lewis, Ontario New Democratic Party Leader, to the Royal Commission on Health and Safety of Workers in Mines (Ham Commission), Toronto, 18/2/75. (94) Survey by the USWA of Elliot Lake, presented in folio form to the International Tribunal on RTZ, May 1980, London, by Paul Falkowski of the USWA; see also *Parting Company*, PARTiZANS, London, No 9, 1981.

(95) Dr D.V. Bates, "*Elliot Lake, Ontario*", report to the BCRUM, 8-9/5/79. (96) Statement by Kenneth Valentine, to BCRUM, Transcript of Proceedings Vol. 63, p.11383, 1/2/80. (97) Statement by Homer Seguin, to BCRUM, Transcript of Proceedings, Vol. 63, p.11383. (98) see also Homer Seguin, "*The Toll on Worker's Health*", in *Nuclear Free Press, Birch Bark Alliance*, Peterborough, No 15, Fall 82; *Parting Company*, PARTiZANS, No 16, Late 82. (99) Statement by K.A. Valentine, Director Occupational Health and Safety, National Office, Canada, of the USWA, to the Key Lake Board of Inquiry, Saskatoon, 1980, reprinted in *One Sky*, Saskatoon, 11/80; *Kiitg*, Amsterdam, No 10, 12/80. (100) Lloyd Tataryn, *Dying for a Living: The Politics of Industrial Death* Greenberg Publishers, Canada, 1979. (101) *Saturday Night*, USA, 6/76. (102) Wayne Roberts, "*Miners and Mining in Canada*", in *Raw Materials Report*, Vol. 1, No 4, 1983. (103) Art Kilgour, "*Elliot Lake: Still Dying for a Living*" in *Birch Bark Alliance*, Peterborough, No 6, Spring 80. (104) James Ham (Commissioner) *Report of the Royal Commission on the Health and Safety of Workers in Mines*, Ministry of the Attorney General, Province of Ontario, Toronto, 1976. (105) *Birch Bark Alliance*, Nos 11&12, Fall 81. (106) *MJ* 2/10/81. (107) *Nuclear Free Press*, OPIRG, Trent, Fall 86. (108) Sudbury Star, 2/4/86; and material from the USWA, Toronto, 1985-86. (109) ORCEPP, Ontario 1978, quoted in *New Scientist*, 2/11/78. (110) Bob Harris, *Nuclear Free Press*, OPIRG, Trent, No 15, Fall 82. (111) AECB Advisory Panel on Tailings, *The Management of Uranium Mill Tailings: An Appraisal of Current Practices*, AECB Report 1196, 1978. (112) *Energy Probe*, quoted in Ralph Torrie, "*Uranium Mine Tailings: A Legacy in the Making*", *Birch Bark Alliance*, Winter 1981. (113) see also Statement by Dr D. Henry of the AECB to BCRUM, *Transcript of Proceedings*, Vol. 38, p. 6322, 15/11/79. (114) *Toronto Globe and Mail*, 19/3/80. (115) J. Terral, *Uranium Exploration: Four Natural Radiation Hazards*, Kootenay Nuclear Study Group, Nelson, 1979. (116) USWA and British Columbia Federation of Labour, Submission to the BCRUM Public and Worker Health, Burnaby, 1980. (117) Gordon Edwards, *Estimating Lung Cancers, or It's Perfectly Safe, But Don't Breathe Too Deeply*, A summary of testimony presented by Gordon Edwards to the Elliot Lake As-

sessment Board dealing with the problem of radon gas in buildings, Canadian Coalition for Nuclear Responsibility, Montreal, 1978 (revised). (118) Ontario Environment Assessment Board, *Expansion of the Uranium Mines in the Elliot Lake Area*, Final Report, Toronto, 1979. (119) Nels Conroy, *Potential Environmental Effects on Surface Water from Pollutants Relating to Uranium Exploration, Mining, Milling and Related Waste Disposal*, Evidence to BCRUM, Phase VI, 2/80. (Conroy was at this time Chief of the Water Resources Assessment Section of the Ontario Ministry of the Environment). (120) *Sun*, Vancouver, 5/12/77, quoted in *Nuclear Newsletter* 4, Saskatoon, 28/12/77. (121) *Status Report: Water Pollution in the Serpent River Basin*, Ontario Ministry of the Environment, 1976. (122) *RTZ Fact Sheet on the Environment*, no 3, London, 1976. (123) Ontario Water Resources Commission, *Water Pollution from the Uranium Mining Industry in the Elliot Lake and Bancroft Areas*, Toronto, 1971. (124) *Birch Bark Alliance*, Winter 1981. (125) Health and Welfare Canada, Guidelines for Canadian Drinking Water Quality 1978, (126) *Kiitg*, WISE Amsterdam, No 9, 11/80. (127) IJC *Report on Great Lakes Water Quality*, Appendix on Radioactivity, table 21, 1983. (128) R.G. Ruggles, W.J. Rowley, *A Study of Water Pollution in the Vicinity of the Eldorado Nuclear Ltd Beaverlodge Operation 1976 & 1977, Surveillance Report* 5-NW-78-10, Fisheries and Environment Canada, Environmental Protection Service, Alberta, 1978. (129) *MJ* 6/10/78. (130) E. Landa, *"Isolation of Uranium Mill Tailings and Their Component Radionuclides from the Biosphere: Some Earth Science Perspectives"*, US Geological Survey Circular 814, General Printing Office, 1980. (131) *Nuclear Free Press*, OPIRG, Summer 1983. (132) *Union of British Columbia Indian Chiefs News*, No 1 Vol. 7, 1978. (133) *Joint Health Report, Spanish, Mississagi and Serpent River Reserves*, Jesuit Centre, Toronto, 1984. (134) *RTZ Review*, No 7, London, 10/88. (135) *New York Times* 15/4/69. (136) *Wall Street Journal*, 17/3/69. (137) *FT* 6/3/79; *MJ* 6/4/79; *FT* 4/8/79; *MJ* 17/8/79; *FT* 4/3/80; Rio Algom *Annual Report* 1979. (138) *FT* 24/3/81; *MJ* 21/3/80. (139) *FT* 1/11/83; *FT* 6/3/84; *MJ* 20/4/84; *FT* 6/3/84; *FT* 10/3/84. (140) *MJ* 25/3/83. (141) *MJ* 11/9/81; see also *FT* 29/10/81; *FT* 26/2/82; *FT* 4/11/80; *FT* 25/3/82. (142) *FT* 1/3/83. (143) *MIY* 1987. (144)

MJ 7/5/82. (145) see also *FT* 19/4/83. (146) *FT* 30/10/84. (147) *FT* 23/2/85; *MJ* 26/4/85. (148) *MJ* 17/1/86. (149) *FT* 8/4/88. (150) *MJ* 5/12/86; Lornex *Annual Report* 1986, (151) *MJ* 6/11/87. (152) *MJ* 4/3/88. (153) Environmental Impact Protection Program, *Summary Progress Report*, Jesuit Centre, Ontario, 1984. (154) *MJ* 29/7/77. (155) *FT* 17/11/78. (156) Rio Algom *Annual Report* 1985, (157) *E&MJ* 11/85. (158) *FT* 26/4/85. (159) *MJ* 12/2/82. (160) *Times*, London, 16/12/68. (161) *MAR* 1986. (162) *MAR* 1987, *MAR* 1988. (163) *Metals Week*, 4/6/84. (164) *MM* 3/83. (165) *MJ* 28/9/84. (166) *MJ* 5/8/83. (167) *MJ* 25/10/85. (168) *MJ* 21/11/86. (169) *MJ* 22/3/85. (170) *FT* 18/11/86. (171) *E&MJ* 6/88. (172) *MJ* 1-8/1/88. (173) *Yorkshire Post*, Leeds, 9/10/88. (174) *Gua* 3/3/88; *Gua* 20/5/80; *Childhood Cancer Aggregations in the Beverley and Hull Areas*, Leukaemia Research Fund Centre for Clinical Epidemiology, University of Leeds, 1988. (175) *MJ* 13/5/88; *FT* 23/3/88; *FT* 11/4/88; *FT* 22/4/88. (176) *MJ* 27/3/87. (177) *FT* 30/10/88. (178) *FT* 6/7/88. (179) *FT* 28/4/88. (180) *FT* 27/5/88. (181) *New York Times*, 1/5/81. (182) *FT* 4/11/87. (183) *FT* 11/8/88. (184) RTZ AGM transcript, 1985, available from Partizans, London. (185) Roger Rolf, *Exploration in the NWT*, Oxfam Canada, Toronto, 1978.

524 Riofinex

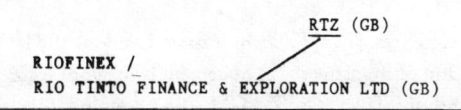

```
                                          RTZ (GB)
RIOFINEX /
RIO TINTO FINANCE & EXPLORATION LTD (GB)
```

One of the ubiquitous RTZ Corporation's major exploration subsidiaries, it has local offices in Brazil, Eire, Singapore and Spain (1).

It was exploring for uranium in Donegal, Eire, in the '70s, but appears to have discontinued towards the end of the decade (2). However, together with Tara Exploration, it was still the most active mining company in the area by 1984, ostensibly exploring for barites and base metals (3).

A more recent exploration venture in northern Europe was for molybdenum in the South Gal-

651

loway hills of Scotland (4), an area previously under study (though not by RTZ) for the dumping of nuclear waste. When local Labour councillors learned that they had granted this licence to a wholly-owned subsidiary of RTZ, one of them allegedly exclaimed: "Oh my god, what have we done?" (5). In 1984, the well-respected Peruvian Indian rights organisation, AIDESEP, accused Riofinex of trespassing for minerals on native land in the Sierra Madre region of the country (6). However, according to RTZ chair Alistair Frame at the 1985 AGM, "Riofinex did have some gold prospecting territory in areas of Peru. These activities were dropped two years ago, and no prospecting activities are being carried out in Peru at present. At the beginning of the prospecting, discussions did take place with the local Indians; prospecting proved fruitless, so efforts were discontinued" (7).

Prospecting has, however, continued in earnest – for uranium among other minerals – in the Arabian shield area of Saudi Arabia, where Riofinex has a JV with BRGM and the US Geological Survey Mission (8).

Although continuing exploration in Ireland, the company sold its most advanced prospect – at Cavanacaw, Lack, County Tyrone – in 1990, to Omagh Minerals, a company underwritten by Dejour Mines (*qv*) but whose managing director had himself been employed by RTZ (9).

References: (1) *Who Owns Whom?*, London, 1984 (Dun & Bradstreet). (2) Statement by chair of RTZ Anthony Tuke at RTZ AGM, 1982, London. (3) *MAR* 1984. (4) *Scotsman*, Edinburgh, 4/4/85. (5) Personal communication from SCRAM (Scottish Campaign to Resist the Atomic Menace), 5/85. (6) Speech by AIDESEP representative in *IWGIA Newsletter*, København, 3/85. (7) Statement by Alistair Frame at RTZ AGM, London, 30/5/85. (8) *FT* 26/4/84; *MAR* 1984; *MIY* 1985. (9) *MAR* 1991.

525 Rio Tinto Minera SA

Rio Tinto Minera operates copper mines at Rio Tinto and Santiago, mines pyrites and chlorites

at Rio Tinto, and operates a smelter in Huelva (1). It is one of the world's oldest, and most notorious, mining companies as well as the original cornerstone of the RTZ empire (2). In 1980 it was reported exploring for uranium in the Segovia region of Spain (3), together with Rio Albin, another ERT subsidiary.

In 1882 alone, 585 miners (the majority women) at Rio Tinto died from silicosis, an influenza epidemic, and other diseases associated with mining and malnutrition (4). In 1981 – a century later – a group of 80 women from Dehesa de Ríotinto village occupied the Huelva mine site, in protest at the serious contamination caused by chlorite dust, and vapours from the flotation and electrolysis sections at the milling plant: the company promised to "solve all the problems" (5).

References: (1) *MAR* 1984. (2) *RTZ Uncovered*, London, 1985 (PARTiZANS). (3) *MJ* 22/8/80. (4) L Gil Varon, *A migration model for Rio Tinto 1877-1887*, Cordoba, 1981, p98. (5) *El Pais*, Madrid, 21/5/81.

526 Robertson Research International Ltd

Eggheads of the mining consultancy world, including uranium consultancy (375 of their 750 staff are graduates, primarily geologists), they

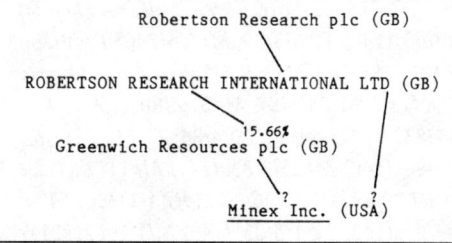

have operated – from their HQ in Wales – in more than 50 countries (1).

In early 1983, Robertson announced a hope to begin prospecting for gold in the Red Sea hills area of northern Sudan, along with Greenwich Resources (in which it held a 15.66% stake) (2). Greenwich announced, too, that it would bid for control of Minex Developments, a British company in which Robertson then held 25% (3), and whose relationship with the Robertson-controlled Minex Inc is not clear. However, it was announced at around the same time that Robertson had secured Minex Inc (in which Vam previously held 60%) (4). Later, Greenwich announced that it had taken over all the share capital of Minex Developments (GB) and would take steps to transfer its assets to Britain, and get listing on the London stock exchange (5).

In 1987, Robertson's five-year agreement under which Greenwich would carry out all of the company's mineral explorations, came to an end, and Robertson disposed completely of its stake in Greenwich. Robertson set up a new subsidiary, Robertson Mining Finance Ltd, which went on to acquire 22% of American Gold Resources Corp (US), and a 10% stake in the old Anaconda base and precious metals property in Butte (see Arco) (6). It was also branching out into investments in Colombia, the Caribbean and Ecuador (6).

Meanwhile, Greenwich went on to open up the Gebeit gold mine in Sudan (7) – marking the revival of the country's gold industry (the first gold was mined under the Pharaohs) (8).

By 1991, Robertson had completed a magnetic survey of the Yemen, and was engaged on a two year exploration programme for chromite in Oman (9). Perhaps its most significant new venture (certainly for indigenous people) was its feasibility study for Hindustan Copper Ltd of copper deposits in the Singbhum region of Bihar, at the heart of the tribal belt (9), which included an environmental impact assessment (10).

In 1990, Robertson sold off its 100% owned Plateau Mining, which has interests in Zimbabwe, Ecuador and on the Ecuador/Peruvian border (the Campanilla project), as well as in Cyprus (11).

Recently it has been exploring in indigenous areas of Guyana (12) and producing a geographical atlas of the shelf regions of Eurasia, along with the Geological Institute Academy of Scienees in the Soviet "Union" (13). In 1987, it took a major share in Butte Mining (see VAM) (14).

Early in 1991, Simon Engineering made a £53M offer for Robertson, which was "heartily recommended" by the Robertson board (15).

References: (1) Robertson Research publicity material; MM 3/83. (2) FT 21/2/84. (3) MJ 4/3/83. (4) FT 25/2/83. (5) MJ 15/2/85. (6) MJ 10/7/87. (7) E&MJ 6/88. (8) FT 20/11/87. (9) MAR 1991. (10) MM 11/90. (11) MJ 26/1/90. (12) MM 5/89. (13) MJ 28/7/89. (14) E&MJ 12/87. (15) MJ 12/4/91

527 Rocky Mountain Energy Co (USA)

This is a minerals exploration and development company, producing coal and trona (a derivant of natural soda ash) (1). In 1980, it opened a Washington office to oversee uranium exploration in the north-west USA (6619 Cedar Avenue, Suite 101, Spokane, Washington) (2).

In 1978 it had plans for uranium exploitation in the Copper Mountains, Fremont County, Wyoming, and in situ leaching at Nine Mile Lake, Wyoming. Its only known operating project is the Bear Creek mine – with a uranium mill processing more than 2300 tons of ore a day – in Bear Creek, north-east Wyoming, which it operates with Mono Power, a subsidiary of Southern California Edison (1).

In 1981 the mine sold almost 600,000lb of uranium to Southern California Edison and San Diego Electric under long-term contracts. Another 43,000lb was sold under spot contracts, and the company expected "demand to improve the balance of the 1980s" with a reserve of 6.4 million pounds of the dreaded stuff in its clammy fingers (3). By the end of 1982, however, the mine was working at only half capacity, thanks to the reduction in uranium prices (4).

The following year, Rocky Mountain and Mono Power began also considering *in situ* leaching of uranium on their Converse County property, known as Copper Mountains, in Wyoming (5). US$20M had been budgeted for capital outlay but, by the following year, the project was still at an evaluation stage (6). It was planned to open in 1986, but start-up depended on an "improved" market (7).

Although, by 1986, the company closed the Bear Creek mine which it held in partnership with Mono (8) and, so far as is known, their Converse County project has lain dormant, Rocky Mountain (renamed Union Pacific Resources by 1988) made several discoveries of uranium – along with Taipower – in the breccia-pipes of Arizona (9) and in "other western states, one of which may be developed" (10). (For further details on the impact of uranium exploration in the Grand Canyon region, see Energy Fuels Nuclear.) Rocky Mountain is also involved in a JV with Idemitsu Kosan, to develop bituminous coal near Anchorage, Alaska (10).

References: (1) *MIY* 1982. (2) *US Surv Min* 1979. (3) Annual Report, 1981. (4) *Nukem Market Report* 8/82. (5) *E&MJ* 1/83. (6) *MM* 1/84; *E&MJ* 1/84. (7) *E&MJ* 1/85. (8) *MIY* 1987. (9) *E&MJ* 4/88. (10) *MAR* 1988.

528 Rössing Uranium Ltd

The world's largest open-pit uranium producer (and, until the opening of Key Lake in Saskatchewan, the largest of any kind), Rössing has become synonymous with neo-colonialism, the perpetuation of apartheid, flagrant disregard of international law, and the symbiotic relationship between civil and military nuclear fuel supplies. To the United Nations Council on Namibia (UNCN), the mine has epitomised the illegal seizure of vital resources from a territory until recently cheated of its independence. To Namibia's government, led by SWAPO (South West Africa Peoples' Organisation), Rössing was the single greatest bulwark of apartheid in the country for over a decade. To its workforce it is a danger for years to come.

RTZ is the manager of the Rössing mine; and, for RTZ Rössing, has been a source of unrivalled profit and power within the world uranium industry; but also a huge political thorn in its side – the most controversial single mining project anywhere on the globe. RTZ's

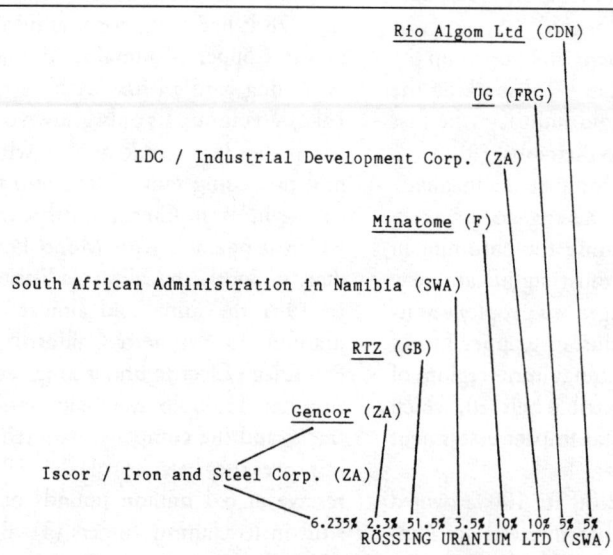

Rio Algom Ltd (CDN)

UG (FRG)

IDC / Industrial Development Corp. (ZA)

Minatome (F)

South African Administration in Namibia (SWA)

RTZ (GB)

Gencor (ZA)

Iscor / Iron and Steel Corp. (ZA)

6.235% 2.3% 51.5% 3.5% 10% 10% 5% 5%
RÖSSING URANIUM LTD (SWA)

partners in the mine, notably Total/CFP (Minatome), Uranges/UG/Urangesellschaft and the Iranian régime, have enjoyed access to relatively cheap supplies of yellowcake over a long period. Until 1988, most Japanese utilities were securing an important percentage of their uranium requirements from Namibia, without declaring it – protected by the complicity of their government. The South African régime was able to almost double its uranium resources, at a time when it was expanding its processing and enrichment facilities: while an occupation administration in Namibia it was also able to accrue huge benefits from tax on Rössing's profits. The British nuclear industry was able to conclude its most vital uranium contracts for the 1980s, without any official compunction to separate the fuel into civil and military end-use categories.

Above all, the dependency that has been created, specifically by RTZ, between Namibia's economy and the Rössing mine, makes it highly unlikely that any future government will kill the goose laying the yellowcake eggs. RTZ seems fated to remain in Namibia for a long time to come.

Origins and development

It was a Captain Peter Louw who identified radiation in the Rössing area in 1928 (1). Nothing much stirred, however, for nearly thirty years – although by 1940 there had been some prospecting (2). During the mid-1950s, South Africa's Anglo-American Corporation investigated Rössing, but decided that its uranium grade was too low for successful exploitation (2). In 1959 and 1960, Louw offered RTZ the rights to the deposit for the first time: the British company declined (2). Nonetheless, a South African government Commission had already investigated the area, and its report, published in 1964, envisaged exploitation of the Rössing uranium as a likely source of revenue with which to construct the Cunene (Kunene) hydro-electric dam straddling the Angolan border, as well as other such projects (3).

Within three years, RTZ had obtained the Rössing lease from GP Louw Pty Ltd (4), and Rössing Uranium Ltd was incorporated on February 11, 1970 (5).

Rössing was a huge challenge to the British mining multinational. Unless RTZ had already developed techniques for mining low-grade ore on a vast scale, as at Palabora, it would probably have baulked at taking-on this new project, notwithstanding the lavish concessions offered by the apartheid régime (6).

Construction started in 1973, after a pilot project from 1971-72, and just as the world's uranium market was beginning to revive. This was largely thanks to the operations of the international uranium cartel, master-minded by RTZ, partly to secure contracts for its new South West African mine (3). Indeed, as the cartel began to bite, and prices escalated, the British company decided to double its projected output at Rössing from 2500 to 500 tons a year U_3O_8, thus making it by far the biggest uranium producer envisaged at that time (3). Before the major mine construction could be completed, however, Rössing struck problems. Abnormally abrasive ore began causing havoc particularly in the tailings disposal pipes (7). There were also acknowledged design weaknesses in the plant, and problems arose in separating the uranium from its host rock (8). The fact that RTZ was dependent on a largely unskilled, migrant labour force, was undoubtedly another major reason for delay in bringing Rössing on stream, though this was glossed over by the company (7). During 1976 – by which time the mine was supposed to be at full production – only 771 tons U_3O_8 had been produced (9, 107).

Drastic modifications to the plant were carried out in 1977. That year, some 3000 tonnes of uranium oxide was mined, and Rössing Uranium Ltd announced plans for a twenty-year operation, exploiting an open-pit 3km by 1km in area, and 300 metres deep (1).

At this time rumours of an underground mine first began to surface (forgive the pun!). Advertisements placed in the mining press, in 1976, solicited miners experienced in underground trackless mining (10). RTZ, in its annual report for that year, commented that underground development was proceeding satisfactorily, with mining due to start in the second half of 1977 (11). Such a development, declared the company, could extend Rössing's life from 25 years to nearly a whole century (80 years) (11). Surprisingly few people seemed to notice this development: those that did presumed RTZ had either located a remarkably rich vein, or that political considerations would force it to exploit the underground deposit first of all, or high-grade it (mix it with the lower-grade ore from the open-pit) (12). In any event, the underground option was not pursued: probably because the technical problems in the open-pit had been largely overcome by 1977, though a major fire in May 1978 caused further delays (13).

Although Rössing's grade has not been officially revealed, it is generally considered to be between 0.03 and 0.05% (14). Ore reserves in 1978 were put at an estimated 300 million tons (14).

By 1981, production from Rössing was up to expectation – 5160 tons U_3O_8 – and the mine's profits before tax had risen on the previous years (16). The contribution of Rössing to RTZ's coffers was also a handsome one: up from £17.6 million in 1979 to £21.1 million the following year (17).

Mining growth, in Namibia itself, began to stall in the early 1980s, declining by nearly 10% in 1981 and 7.5% the next year (18).

Lower uranium prices, reduced deliveries and higher taxes soon had their effect: in 1983, Rössing's contribution to RTZ's profits fell to £4.16 million – not to be sniffed at, but not quite the bounty the parent company had expected (19). Commented the *Financial Times* in 1982: "[Mine costs] cannot be disclosed but they are certainly equivalent to more than the US$17 [per lb] free market price of yellowcake. At present the rise in costs is being held down to an annual rate of 10%. Even so, the margin between costs and likely sales contract prices may not be all that large. Rössing has had the advantage so far of operating tax free ... this offset

has now run out. But 1983's impact will be "softened" by the deferred tax position built up by the company in past years" (20).

Just as Rössing became liable to pay local taxation (in Namibia) at 42% on its profits (21), so the company stopped publishing information on its output. Nukem estimated its production in 1984 at 3700 tons – a little lower than in 1983, and 17% lower than in 1982 (21). Rössing's contribution to RTZ's profits in 1984 took a tumble: at £7.5 million they were just half those of the previous year (21).

By this time, Rössing's major contract with the British Central Electricity Generating Board (CEGB) was coming to an end, and the market was weakening (22, 23). From 1984, RTZ revised the presentation of figures from its Namibian venture, and these were included under its data published for South Africa (22). It is known that its production that year was 25% below capacity (22). Even so, this reduced output brought in no less than 95% of RTZ's net earnings from Africa that year (22).

For the second half of the "eighties, output ranged from 3500 to 3800 tonnes: in 1988 it was assumed to be around 3600 tonnes (24). Profits were relatively high in 1986 and, although they fell in 1987 – thanks to lower sales, a strong Rand and higher taxation (25) – they went up nearly 60% the following year, to around £50 million. Nearly half of this (£23.2 million) was paid as net attributable profits to RTZ (24).

By 1989 the Rössing mine was providing an average of 35% of Namibia's total export earnings, and being taxed at 55% of its profits (26). RTZ was looking forward to increased production and profits in the 1990s, confidently predicting an increase in world-wide reactor needs and a "drawdown" in existing uranium inventories (22).

Costs and control

Initial development costs for Rössing amounted to some R300M (three hundred million Rand) (1) though they later rose to R400M. These costs were offset 100% against taxation (7), and the company also benefitted from very generous initial write-offs, low licence fees and large grants for prospecting (27). Although RTZ acknowledges that Rössing paid company tax to the South African occupying authorities, it maintains that taxation on profits were never paid directly to the apartheid régime, but comprised "an alternate tax which was ... not payable in any case before 1984" (28) (see above), and was withheld in Namibia. Dividends are, of course, paid on profits to the company's shareholders; though, thanks to the complex disposal of A and B shares, and the lengthy attempts to conceal Rössing's true ownership, it is virtually impossible to determine with accuracy what these are. As already pointed out, in 1984, Rössing's returns to RTZ were consolidated with the company's South African accounts (22) and, while Rössing holds an annual general meeting in Namibia, information on the time and place of the meeting is never published – such information being restricted to its shareholders (29).

The initial costs for the mine were met by RTZ (80%), South Africa's General Mining (10%) and the West German uranium giant, Urangesellschaft (Uranges/UG) (10%) (30): indeed UG held an option to take up 40% of the equity in the project, until this was re-negotiated in 1979 (31). The West German government provided a variable interest loan of US$25 million repayable from 1979-85 (3), and Rio Algom chipped in around eight million Canadian dollars (32). Swiss companies also probably provided money through Uranges to finance plant repair in 1976-77 (33). The South African contribution is more difficult to determine. Certainly the Industrial Development Corp (IDC) provided a minimum of R60 million for the metallurgical and treatment plant (34) – and probably a lot more (6). The South African Electricity Commission (ESCOM) also constructed power lines for Rössing.

Other management contracts for construction and development of the mine and processing plant were given to companies within the Davy-Mckee conglomerate: Power Gas Ltd of London (part of Davy Ashmore) and the Western

Knapp Engineering Division of Arthur Mckee of the US (35). Other American companies which played a part in bringing Rössing to fruition include Kerr-McGee, Phelps Dodge, and Interspace Inc, which supplied the water pipes (36). The West German nuclear fuel supremo, Nukem, was also involved, as was the Swedish company Trelleborg (36) (37). Some Italian companies delivered sulphuric acid for the extraction plant (36) and pyrite had to be imported (38). (It is not known where the original supplies of pyrite came from, but later deliveries were made by Tsumeb in 1988, these being transported by Jowels Transport) (39).

As is common in the mining industry, finance was also raised by selling future production at favourable prices. In some cases this secured the customer a stake in Rössing's equity. In at least one case – that of Rio Algom, which bought the Louw group's 10% share after development started (36) – there was no apparent *quid pro quo*, or equity-for-uranium. In the case of the IDC, a South African parastatal, there remains a huge outstanding question as to whether it gained delivery of Namibian uranium, in return for its investment (see below). Under the South African Atomic Energy Act of 1948, the South African régime had the sole right to search for and mine uranium grading in excess of 0.006% (6). Certainly the IDC obtained a majority of the A shares in Rössing, thus giving it pre-emptive rights at company annual general meetings, even though RTZ was entitled to nominate 10 out of 19 of Rössing's board and, from the start, had management control (40). According to RTZ, its early stake in Rössing was 48.5% of the equity, 53% of the B class (no voting) shares, and only 25.8% of the A class or voting rights.

The other visible South African interest from the start was a stake taken by Gencor of between 6.2% and 8.33% (36).

The French company Minatome (later Total/CFP) also gained 10% of Rössing's equity, in return for uranium deliveries through the 1980s (41), while the Shah of Iran took 15% on similar terms. Uranges obtained 5%, but the Japanese – alone among the mine's major initial customers – never entered the project on this basis. Nonetheless, Kansai Electric Power Co sealed a contract for the delivery of 8200 tons of Rössing uranium for the first ten years of production – one of the most important considerations on which construction could proceed (36). In 1977, as costs to meet the technical problems mounted, RTZ made a special rights issue of US$75 million, which was taken up by the IDC, Gencor, Minatome and Rio Algom; but not, it seems, by the West Germans or the Japanese (38).

In 1987, some 5,580,000 A shares in Rössing Uranium Ltd (comprising 3.37% of the issue) were transferred by the IDC to the provisional Namibian government, which invested them in a new body, the Capricorn Trust (42). This effectively handed over voting control to the so-called Turnhalle administration. This still leaves the IDC with 10.1% of the equity (2), Total with 10%, Uranges mbH with 5%, and Gencor with 2.3% (2). The Louw family also appears to retain a small stake of 2.62% (2) and various other south-west African interests, such as F J Strauss, to hold 0.26% (2). Although the Iranian government's interest is now rarely mentioned in mining circles (5), it apparently maintains its original 15% equity.

As for RTZ, no one doubts that the British company is the single biggest profiteer from the Namibian mine. The general assumption is that it holds 46.5% of the equity, although we may question whether this is additional to, or includes, Rio Algom's tenth-part holding (5, 40). According to a recent anti-apartheid investigation of Rössing, carried out by the West German Akafrik group, RTZ actually owns 41.35% and Rio Algom is separately registered with 10%. In terms of equity control, there may be no more than a paper difference between the two sets of figures: RTZ's majority ownership of Rio Algom would secure for itself the missing 5.15% of the equity.

Contracts

Virtually all information on contracts concluded between Rössing and its overseas custo-
mers, has derived from painstaking research undertaken by US, British and Japanese investigators. RTZ (and of course Rössing itself, as its underling) has consistently refused to divulge any information on its sales "due to commercial sensivity" (29). But this is just a euphemism for political embarassment. In fact, the South Africa Nuclear Energy Act of 1982 prohibits the disclosure of details of uranium production, export and sales from Rössing, as well as from South Africa (22). RTZ has often commented on the contracts between its Namibian mine and the British Atomic Energy Authority – later BNFL (to whom the contracts were assigned in 1974) – to supply the British CEGB with 7500 tons U_3O_8 (43) from the mid-1970s onwards. But these comments have always come in response to revelations from other sources. Indeed, when the matter of a third British Rössing contract became a minor *cause célèbre* in the 1980s, RTZ's chair (and former chief executive) Alistair Frame declared it was cancelled. Whether he was lying outright, or being economical with the truth, is an academic point: by the following year he was forced to acknowledge the facts (44).

What has never been divulged is the degree to which the South African atomic energy programme – in which civil and military end uses of nuclear fuel have been inextricably interwined for many years (3, 45) – has benefitted from Rössing uranium. As already mentioned, South Africa's 1948 Atomic Energy Act theoretically appropriates to the apartheid state all fissionable materials in territory under its control: it also allows for the cancellation, at any time, of uranium contracts (43). Further legislation – the South Africa Atomic Energy Enrichment Act of 1967 (1948) and the Uranium Enrichment Act of 1974 – empowered the state-owned Uranium Enrichment Corporation to sequester Namibia's uranium (3). In 1977, the chair of RTZ, Mark Turner, agreed to this interpretation of the Act (46), but persisted in a claim, first made the previous year, that Rössing's uranium would not be diverted in this fashion (47). His denial was unfortunately somewhat diluted by a statement the following month from Rio Tinto South Africa's Deputy Chair, Alex MacMillan, which implied that South Africa would be free to receive Namibian uranium, once Rössing's other contracts were fulfilled in the 1980s (48).

The South African régime, from the start, included Rössing's production in its own figures: a 1976 Uranium Institute presentation made by R E Worrall (South African Chamber of Mines) and the leader of the country's Atomic Energy Authority, A.J.A. Roux, claimed that South Africa's total production could be 16,270 tons by 1985 – and there was no doubt a substantial third of this would derive from Namibia (49).

In 1970, the *Rand Daily Mail* commented that, once on-stream, Rössing's output would certainly lead to the establishment of a uranium hexafluoride industry in South Africa (50). Inevitably, throughout the 1970s, the mine would have been regarded by the South African nuclear élite as the prime source of feed for its Valindaba enrichment plant – a plant with prime military purposes (3). The *Australian Financial Review*, in 1976, had no doubt about it: "Production will also be available [from Rössing] for the South African enrichment programme, which will require between 20,000 and 30,000 tons of U_3O_8 from the early 1980s" (51).

Nor can the crucial involvement of the IDC in financing Rössing have been simply based on the expectation of healthy dividends, let alone the desire to see Namibia go it alone !

On the other hand, there is no concrete proof that Rössing's uranium has been diverted to South Africa. Once the cat was out of the bag in the late 1970s, over the UKAEA/BNFL-CEGB contracts, and details began emerging about South African Atomic weapons tests, it was hardly likely that any British government would be prepared to face the uproar caused by revelations that a British-owned company (headed by leading lights in all three political parties) (52, 56) was actively assisting the manufacture of an apartheid bomb. (This, notwithstanding the early cooperation by British nuclear scientists in developing both South Africa's uranium industry and its atomic plant)

(3). In any event, there were soon to be boom days for South Africa's own uranium producers. It is Rössing Uranium's contracts with Britain that are usually remembered as the Namibian contracts, thanks to the enormous controversy surrounding their inception and implementation; the location of a well-organised and highly vocal pro-Namibia lobby in Britain (certainly the most important outside of Africa, apart from the SWAPO presence at the UN); and the presence of Rössing's masters, RTZ, in London. Here, at every annual general meeting from 1975 until the present, hundreds of questions on the deals have been propelled at a succession of increasingly beleaguered chairpersons, and numerous public protests have been mounted (see below).

We should not minimise the importance of Rössing's supplies to other consumers – especially Japanese and West German. After all, the total known to have reached Britain for the UKAEA/BNFL/CEGB was less than two years full output from the mine at peak production. Nonetheless, the signing of the British contracts (specifically the second, for 1500 tons U_3O_8) was vital in gaining financial underwriting for the development of the mine (43). Rössing's uranium has been Britain's only major source, not ostensibly designated for peaceful end-use, and can therefore be neatly dovetailed into the British Trident nuclear-weapons development programme. (Quite a lot has been made about this aspect by various groups in Britain: the fact is, that virtually any uranium entering a nuclear-weapons state can be diverted for military purposes, or be substituted by mixing, or "swapping," for other supplies supposedly covered by IAEA safeguards (53). Nonetheless, at the end of the day, those responsible for nuclear accounting may have to show that the books balance, and they need uranium supplies which are designated "N" (non-safeguarded) for military purposes).

Debate will persist for many years, no doubt, as to who was responsible for hooking Britain into Namibian uranium supplies to the extent that, by the early 1980s, the country was importing nearly half its requirements from this source alone (54). Discussion about the contracts took place in the Labour cabinet between 1965 and 1968, when the UKAEA was planning imports for the late 1970s and early '80s (43). Both RTZ (55), and the UKAEA (43) maintain unequivocally that the government – through the Ministry of Technology – was committed to taking Rössing uranium in 1968, and there was no question of an alternative supplier. The Labour Minister for Technology at the time, Wedgwood Benn, has always maintained that Rio Algom was the first option, and the choice of Namibia did not emerge until 1970, when he himself was not informed of the switch (43). When – thanks to an eagle eye in the Foreign office in 1970 – the Cabinet eventually got to know about the Rössing contract, no mention was apparently made of possible alternative supplies, nor of the buyers' market in yellowcake which was developing, nor the important fact that there were two Rössing contracts, and only the first had then been signed.

In the event, not only did the contracts pass muster, but many of those political figures who should have opposed their implementation – especially in the light of the UN Decree No. 1 passed in 1974 (see below) – abandoned their principles, enabling Britain to become a major recipient, and processor/enricher of Namibian uranium (3, 43, 52). It was a case of the uranium tail wagging the nominally anti-apartheid dog politic, and bringing leaders of the two main British political parties into grave disrepute (43).

Rössing supplies for British civil consumption stopped in late 1984, two years after originally scheduled (43, 57). Just what had been paid for these illicit supplies is still a matter for conjecture. In 1970, the *Economist* reported that the UKAEA/BNFL had paid considerably below the going, not to mention the future, rates for uranium (58). It is generally agreed that the price for first deliveries was set at US$9.50/1b (43) – a third of the spot price at that time (51). But, in 1976, the price was re-negotiated to just under US$13/1b (59). (One source claims that the price had been increased threefold by then)

(60). Two years later, Rössing's R S Walker stated that production costs were around US$20/1b (which included interest, but not depreciation, charges) (61). By 1982, the *Financial Times* reckoned that the price had probably risen to above US$30/1b (62): it seems to have increased to US$34/1b the following year (63). Imports to Britain of Namibia's yellowcake were worth around £25 million in 1982, rising to £54 million the following year, and £59 million in the last year of delivery (22) – a far cry from the estimated original cost of only US$72 million (6).

By the time the first imports were due to enter Britain, there was some justification for paying the piper's (RTZ's) tune. British stockpiles were dangerously depleted, thanks to the Canadian embargo on uranium exports to Europe (see Rio Algom) (64). The RTZ-managed uranium producers' cartel had also enormously boosted Rössing's prospects in the international market. Whether or not one of the main reasons for setting up the "Club" had been to secure contracts for the Namibian project, the secret share of 4% originally given to Rössing by the carteliers (4%), had risen to nearly 25% by the time the cartel was blown out of the water (3).

Tenders for the British contact were put out only to RTZ: Alistair Frame moved from his post as deputy director of research reactor operations at the UKAEA, to RTZ in 1968 – the year the contract was awarded (65).

Within five years, however, Canadian uranium deliveries had been resumed and, with average annual consumption for Britain's nuclear programme running at 1500 tons a year, the Rössing deliveries were a surplus to British requirements. The stockpile had grown alarmingly large – possibly in 1982 it was bigger than the entire amount covered by the Namibian contracts (66) and, three years later, it had grown even more (54, 67). By then, political embarassment caused by Rössing's operations in defiance of world opinion, and concern at dependence on RTZ, had prompted the UKAEA and the CEGB to set up the British Civil Uranium Procurement Directorate (later: Organisation – BCUPO) to diversify the country's sources of supply. This in itself does not end the story, however: BCUPO buys on the spot market, and there is nothing to prevent such purchases containing uranium of Namibian origin (67).

Indeed, early in 1989 (just as RTZ claimed it had finally recouped its US$164.3 million start-up investment in Rössing, and was now going after US contracts to be implemented on Namibian independence) the *Financial Times* commented that some Namibian uranium was already finding its way to the US through the depressed international spot market, despite the formal embargo on uranium imports from Namibia and South Africa (68).

There is now no doubt that Rössing negotiated a third contract with the UKAEA, almost certainly on behalf of the British Ministry of Defence (MoD) in 1975/76 (69). This comprised 1100 tonnes of uranium, was labelled "N" – suggesting strongly that it would be used for military purposes (70), either being enriched to weapons-grade at the relatively new military enrichment plant at Capenhurst, near Chester, or passing through the plutonium-producing reactors of Calder Hall and Chapelcross (69, 71). Although an estimated 4000 tons of uranium would be required for the Trident missile programme (69, 71), Rob Edwards of CND (Campaign for Nuclear Disarmament) estimated in 1984 that this one contract would neatly cover Trident's fuel needs (140 tonnes) and those of its warheads (780 tonnes) (54).

In mid-1988, after years of prevarication and denial, the British government finally admitted to the existence of this contract, though continuing to deny its military purposes (70). A few months earlier, RTZ supremo, Alistair Frame, had told a crowded annual general meeting that the contract was cancelled (72). In October 1988, the government admitted that the entire 8600 tonnes had been designated "N", and that any of it could be withdrawn for use by the military at the government's discretion (7) (see below). Two months later, the *Guardian* confirmed that uranium deriving from the British military stockpile (which, the paper said, consisted of South African and Namibian uranium) had been shipped to the USA to make warhead

components for the Trident missile programme (73).

Rössing's largest single set of contracts are with the Japanese utilities Kansai, Tokyo, Chubu, Kyushu, Tohoku, Hokkaido and (probably) Chogoku – possibly also Shikoku and Hokuriku (74, 76). RTZ Mineral Services contracted to supply 41,851 tons of U_3O_8 from Rössing during 1977-96 (71). In 1974, the Japanese researcher, Yoko Kitizawa, revealed that Japanese uranium contracts were handled by virtually all the Sogo Shosha (commercial production/marketing conglomerates) – including Mitsui, Sumitomo, Nisho-Iwai, C Itoh, Mitsubishi and, in particular, Marubeni. Forty per cent of Japan's needs, said Kitizawa, were provided for by RTZ (75). The first Japanese contract to be confirmed was handled by Mitsubishi and guaranteed 8200 tons of Rössing uranium to Kansai Electric Power Co(500 tons for delivery each year in 1977-78, 600 tons each year in 1979-80, and 1000 tons annually from 1981 until 1986) (76). This contract, amounting to 8% of the country's needs, was signed before Rössing was developed (75). Tokyo Electric Power Co landed a Rössing contract for 8900 tons, although the delivery schedule is not known. (From 1985-1988, according to Nuexco figures, Tokyo was due to receive only 136 tonnes U_3O_8 equivalent annually) (76). This contract, and most of the others, appear to have been handled by Marubeni, working hand-in-glove with RTZ Minserve (Mineral Services) (77, 154). In late 1974, the Japanese Diet debated the issue of Namibian uranium, and the government confirmed the contracts. A few months later, when the UNCN (United Nations Council for Namibia) visited the country and demanded their cancellation, the response they received was emphatically hostile (75, 78). Marubeni itself made no attempt to conceal its pivotal role in the whole shabby business. In a self-adulatory publication issued in 1978, it described how it had "... played an important role for RTZ Mineral Services Limited in obtaining many orders of uranium concentrates from Japanese electric power companies ..." As of 1977, RTZ Mineral Services had received orders from almost all the major power companies in Japan for the supply of approximately 50,000 tons of uranium concentrates, which represents 35% of total Japanese purchases over the same period. (Marubeni went on to describe how the uranium was "hexed" into uranium hexafluoride by Eldorado in Canada, under a service agreement with the Japanese company; enriched in the USA; converted into UO_2 powder by General Electric; shipped to Japan by the Japan Nuclear Fuel Co Ltd; and eventually reprocessed by Cogéma at La Hague, and BNFL at Windscale/Sellafield, after transportation by Pacific Nuclear Transport Ltd, a JV under British initiative, with Marubeni holding a small equity portion. The company claimed that the processing part of the contract alone was worth some US$3200 million to Marubeni) (79).

Namibian uranium is likely, in future, to play a less important role for Japan since the bulk of the contracts has been fulfilled: according to Nukem, Namibia was expected to supply 25% of the country's needs from 1987 to 1996 (with Canada providing 23% and Australia 22%, Niger 17% and South Africa 7%) (80). The country might not rely on Namibian imports at all beyond 1996. In 1986, Japan had already considered switching contracts as part of a sanctions package, and initiated talks with Niger to this end (81). Then, following revelations and demonstrations by Canuc and the Namibia Support Committee over the Minserve/Tokyo, Kansai and Chubu contracts (82), and a TV programme screened by NHK (the national TV company), Tohoku, Kansai, Chubu and Kyushu electric power companies announced they would be ditching Namibian imports, to replace them with Australian and Canadian imports (83). Chubu concluded a 6 million lb contract with the Canadian company, Cameco, in 1989, covering 1992-2001 (84). Alone among the utilities, Tokyo took a combative stance, refusing to acknowledge its reliance on Rössing yellowcake: "Rio Tinto-Zinc tells us there is no Namibian uranium in what they ship to us and we believe what they say" (85).

Rössing's next most important customers have been France and Iran. Minatome/Total's 10% equity in Rössing entitled it to a substantial share of the output, though how much is not known: according to one source, in 1975, it was more than its equity share would suggest (86). Deliveries from Rössing to France between 1984 and 1988 were scheduled at 2726 tonnes U_3O_8 (87).

Iran's contracts with Rössing have been shrouded in mystery, provoking considerable speculation. According to Thomas Neff (89) repeated reference in the trade press was made in the 1970s to Iranian purchases of up to 50,000 tons U_3O_8 (88). At least one delivery was made before the Shah of Iran met his just desserts in 1979 (89), and then the country's nuclear programme was brought to an abrupt stop. However, the new régime did not sell its equity in Rössing, and it is conceivable that deliveries were being still being made, even while the Ayatollah was lambasting nuclear energy as Satan's work. In any event, the Atomic Energy Organisation of Iran has been revived in recent years, and latest reports state that the country has received 7200 tons of yellowcake since 1979 (24).

West Germany's imports from Rössing commenced in 1976, and lasted ten years. According to SWAPO spokesperson, Hadino Hishongwa, they comprised 30% of the country's needs in 1981 (90). In 1980, both Nordwestdeutsche Kraftwerke AG, and Rheinish-Westfalisches Elektrizitatswerk AG (RWE), admitted using Namibian fuel at the Biblis, Wurgassen, Essenhamm and Stade reactors (91). Euratom estimates that Uranges' share of total Namibian output in 1983 was 11%, falling to less than 6% in 1984, '85 and '86 (92). The federal West German government says that Rössing provided 6% of the country's uranium requirements in 1985 and 1986 (93). It is not known whether uranium has been imported since 1986. Veba has also imported Rössing uranium (71).

In 1987, a researcher was visiting the offices of the Uranium Institute in London, when he noticed a document, pertaining to customers for South African uranium, lying on a desk. Taking advantage of the temporary absence of staff, he copied it. This Nuexco special report (for an unspecified customer) was later found to reveal what appear to be comprehensive figures for all customers for southern African uranium, with deliveries commencing in 1984 and finishing in 1988. The study revealed that Enusa, the main supplier of yellowcake to Spanish nuclear power plants, would receive 544 tonnes U_3O_8 in 1987 and again in 1988 (76). Using this information and figures from other sources, SWAPO the following year alleged to the Spanish government that a quarter of Spain's uranium needs between 1983-1991 had been, or would be, fulfilled by imports from southern Africa, including a substantial proportion from Namibia. This announcement created uproar in the country. Led by the PCE (Communist Party), members of the parliamentary Left Unity group, Basque MPs and Trade Unionists from the COO (Confederation of Workers' Commission), demanded the trade should stop immediately (94). José Manuel Jimenez, head of Enusa, confirmed that the company had imported Namibian uranium – starting before 1983: the contracts, he stated, were with Minserve; 100 tons had been imported in 1987 alone (95).

Three months later, Spain was still importing Namibian uranium through Minserve, although the Spanish Parliamentary foreign affairs committee had demanded that the trade should stop (96). A few months later, the government announced that it would no longer import Namibian uranium, as of the beginning of 1989 (97).

There is one other major RTZ contract which will almost certainly be fulfilled by Rössing, although the parent company has never admitted it. In 1982, Taipower, the Taiwanese state utility, contracted for 3400 tonnes (4000 tons) of uranium for delivery from 1990 to 2005 (89, 98). (Regardless of the source to be used by RTZ in fulfilling this contract, it should be noted that Taiwan has reportedly collaborated with South Africa on reprocessing of uranium into plutonium, and the development of rockets which could deliver nuclear warheads (99),

while its nuclear waste is being dumped on Orchid Island – land essential to the livelihood of indigenous people) (100).

This outline of Rössing contracts would not be complete without brief reference to two respects in which Namibia's uranium supplies have assisted globally in breaching anti-sanctions measures against South Africa, and in further confusing the already-tenuous dividing-line between fuel imported for civil and for military purposes.

In 1986, the US Congress passed a series of measures intended to curb investment in South Africa, and prevent the import of certain strategic products to the mainland USA. Section 309 of the 1986 Comprehensive Anti-Apartheid Act put an embargo on uranium in its unprocessed form, but did not exclude uranium hexafluoride deriving from southern African yellowcake. What the Act effectively did in the first instance, was to provoke a flood of uranium oxide into the USA before the law took effect, and an increase in the hexing of Rössing (and other apartheid uranium) in Europe, before re-export to the USA for enrichment and further processing. In December 1986 alone, some 4000 tonnes of Namibian and South African uranium arrived in the USA, including about 300 tonnes transshipped through Liverpool, as part of a Taipower contract (71). In 1986 as a whole, 481,650kg of unenriched UF_6 was despatched to the USA through Liverpool docks: Canuc has concluded that most of this probably derived from Namibia and South Africa (71).

The US importing company in these instances was Edlow International Ltd, which is linked to Nukem of West Germany and, in the early 1980s, acted as a conduit for enriched uranium entering South Africa itself. There is little doubt that Edlow has continued importing uranium oxide (under a general licence which extends to 1992) and UF_6, both of which derive partly from Namibia (71).

Edlow has also been used by European customers to evade the Anti-Apartheid Act: in early 1987, a 168-ton shipment of U_3O_8, deriving

663

from Rössing and destined for the Confrentes plant in Valencia, went from Walvis Bay (Namibia) to Zeebrugge (Belgium), then to Springfields for hexing, before being imported as British fuel through the port of Norfolk, Virginia. Afterwards, it was enriched by the US Department of Energy at Paducah, Kentucky; finally the product was sent back to its Spanish customer, Hidroelectrica Espanola (96).

The same year, details emerged of another laundering operation, by which Namibian uranium was processed in Britain, sent to Riga in the USSR for enrichment by Techsnabexport under contract with BNFL, returned to Britain for manufacture into fuel rods, and finally sent to RWE in West Germany (101). Swapo and Canuc protested vigorously to the USSR about this trade, which clearly violated Soviet sanctions against South Africa and its occupied territory of Namibia (102). At this time, BNFL stated that between 50% and 65% of a typical shipment of uranium could have originated in southern Africa (103). (If anything, this represents an increase on the proportion of uranium entering Britain in the early 1980s for domestic use and reprocessing for foreign customers) (66).

The north American embargo on Namibian uranium (Canada also imposed restrictions), combined with the TGWU blockade on material of Namibian/South African origin (see below), resulted in an increase in the common practice of "swapping" titles to specific consignments, thus concealing their origin. Nulux (Nukem Luxembourg, a company in which Minserve has a 30% stake) is known to have arranged with Comurhex, the French UF$_6$ producer, for South African-origin uranium to ostensibly derive from Niger, and then consign it for enrichment to the USSR. In March 1988, Comurhex documents surfaced at the port of Liverpool, in which the French company baldly cited an agreement between Minserve and Comurhex for the swapping of uranium concentrates. In this instance, Comurhex promised RTZ that it would produce documents showing that uranium concentrates, which entered the northwest England port in April on the Atlantic Conveyor, was of Canadian origin. Since no one in the region could remember when Canadian ore had last entered the port, it was reasonably assumed that this consignment derived from Namibia or South Africa (104).

At around the same time, the Grünen (Green party) in West Germany accused Gewerkschaft Brunhilde of importing uranium from Namibia, while declaring that it originated in Australia. (This may, however, have been outright deception rather than an origin swap.) (105) In short, although both the British and West German governments had maintained that Namibian uranium ceased flowing through its ports for domestic civil use, there is irrefutable evidence that supplies continued to be imported into Europe and re-exported world-wide.

Rössing conditions

Every aspect of management at Rössing has given cause for alarm and protest since the mid-1970s; ranging from housing standards at the black township of Arandis, to gross discrimination in wage awards, suppression of legitimate trade unions, and a patronising attitude to employees and poor standaeds of health and safety. Arandis was built especially to house the majority of Rössing's black and coloured workers (43). RTZ chairperson, Val Duncan, in 1975, said he would be very surprised "if it isn't by far the best African township in southern Africa by the time it's finished (106)". Two years later the *Guardian* newspaper reported that Arandis' single living quarters were: "the worst [we] had seen in Namibia" (107) and, the same year, the new RTZ chair, Mark Turner, admitted himself to be horrified by conditions there. Two years later, RTZ claimed that housing had been enormously improved, and this is the line the company has pursued since (43). Certainly, considerable improvements were made during the 1980s, and services and amenities both at the mine and in Arandis now compare favourably with those elsewhere in southern Africa. But one important question is whether, of their nature, housing provision and job allocations at Rössing have been discriminatory. There is abundant evidence to confirm that they have been and, in most respects, still are. Rössing management has always denied that it used the infamous contract labour system characteristic of South African mines (108) – whereby single workers are separated for long periods of time from their families, themselves banished to black areas or homelands. In fact, Rössing at its outset was to a large extent reliant on migrant labour: since the local Damara people failed to be attracted to the project, Ovambos had to be drafted-in from the north into appalling temporary camps (38) – from which they were rescued only later, as Arandis accomodated them. When a BBC crew visited Rössing in 1977, it found only 550 Damaras out of a workforce of 1600 black workers: the majority of workers were from Ovamboland, or as far afield as South Africa and Malawi. All were on a one-year contract and the majority were separated from their families: if not in name, this was contract labour in fact (109). In 1978 a German journalist who visited Rössing, while reporting a somewhat smaller African workforce (1300 people) and agreeing that Arandis itself was not ethnically zoned, nonetheless described conditions virtually indistinguishable from those operating on mines using contract labour. "Many black workers considered to be single were in fact married, and their wives and children lived illegally in the single quarters ... The workers were taken to the mine by bus. The buses carrying blacks were always checked at the mine entrance, and those who had forgotten their ID-cards were jailed for up to a day. The buses for white personnel were never checked." Ingolf Diener also recorded that black workers were living two to a room (including the kitchen and sitting room), while white personnel at Swakopmund had a room with private toilet each. There were no regular medical check-ups for non-white workers, and Arandis had no hospital. Industrial relations were "of the apartheid type", using ethnically-based liaison committees, where blacks had no power of decision "and their role was purely advisory". Training courses were limited to those who showed signs of forming "the future élite", while those Na-

mibians who qualified "were never employed at the level of their real qualification ... they became de-qualified, and because of systematic fault-finding, they had no chances of being promoted" (110). Indeed, evidence presented to a Trade Union seminar three years later, stated that only seven salaried staff, out of a total of 680 at that time, were black Namibians (111). Living conditions did not measurably improve until well into the 1980s. In 1979, two researchers called them "akin to slavery" (112), and in 1981 they were still "grossly over-crowded" (111). Although, by the late 1980s, Rössing had responded dramatically to international criticism, introducing better housing, regular medical examinations, new training schemes, cottage industries, and a host of sports and social facilities (113), the General Secretary of the Mineworkers' Union of Namibia (MUN) was still able to report in February 1989 that, while "... the people who live in Arandis live with their families ... about half of Rössing's workers live in the hostels and, of course, the question of families not being allowed, applies. So one can say that about half of Rössing's workers live without their families" (114).

In more than a decade, and despite noticeable efforts by the company, the vestiges of discrimination between blacks and whites in the workplace have still not disappeared. In 1978 Kenneth Marston, the *Financial Times'* mining editor, had observed that the Rössing systems of grading wages and work was: "... a form of segregation in living areas", while there was "no instance where a black worker is in a supervisory position over a white" (115). Six years later, notwithstanding the Paterson system of job evaluation (116) and sophisticated training schemes, a mine worker could tell the Namibian public that the company's "proclaimed non-racial policy and the saying that they are only discriminating between the different grades, is a sweet, convenient lie" (117). A year later, Gwen Lister of *The Namibian*, after a visit to the mine, concluded that Rössing "does more for its workers than any other mine in

665

Namibia" (hardly an accolade!), but she still found that qualified black workers were excluded from facilities enjoyed by whites as a matter of course: for example, they were not entitled to send their children to school in Swakopmund (118).

In 1989, after black workers secured a 23% wage rise which fell far short of their demands, *The Namibian* returned to Rössing. The newspaper found that whites were still being offered jobs over the heads of black miners qualified for the same post (119). A week later, Ben Ulenga, General Secretary of the MUN, lambasted the company in a comprehensive condemnation of its practices: lowest-grade pay did not in any way constitute a "living wage"; 92% of all black, and 51% of all coloured workers, still remained in the company's lowest income bracket; no Rössing employee who lived in Windhoek, as opposed to Arandis, had even been provided with a company house; black workers in the company's exploration department had "no house, no housing allowances. Their living conditions in crowded army-style tents are, in fact, among the worst in the mining industry" (120). Ulenga went on to comment about radiation monitoring at the mine: the company had "complete control" over such matters, and the MUN played no part in determining "the exact nature and extent" of hazards from radiation which, the MUN knew full well, were "not hypothetical". Ulenga demanded that, as at Rio Algom, the Union should have full control over its own safety and health: "... we and our representatives should be the people to declare our working environment safe enough to work in" (120, 121).

Ulenga was echoing a demand made at the MUN's 1989 Congress, where fears about radiation dangers at Rössing formed a large part of the discussion, and workers expressed grave concern about "inexplicable ailments" that had developed, not only among the workforce, but also their families (skin blotches, loss of skin colour) (119, 122). In particular, workers were worried about the roasting rooms, where the yellowcake is burned into oxide – the part of the plant that Rössing acknowledges produces potentially hazardous accumulations of radiation (113). The company claims that all workers in this area are given special clothing, respirators and film badges. But as the MUN's President, Asser Kapere, asked the General Meeting: what occurred to the health of cleaners, when put on duty in the roasting room, if they had not completed their task before radiation levels reached a critical point? (123)

Nine years before, SWAPO representative, Theo-Ben Gurirab, had voiced similar disquiet: "We are ... gravely traumatised when we anticipate the immense health hazards and the dangers of radioactivity that threaten the well-being of our people and the environment," he declared (124). Several months earlier, a Rössing workers' representative had smuggled a letter out of Namibia in which, as well as describing horrific conditions of privation at Arandis ("eight to ten people in one cell"), he portrayed the grim reality of labouring at the open-pit: "Working in open air, under hot sun, in the uranium dust produced by grinding machines, we are also exposed to the ever-present cyclonic wind which is blowing in this desert. Consequently our bodies are covered with dust and one can hardly recognise us. We are inhaling this uranium dust into our lungs and many of us have already suffered the effect. We are not provided with remedies and there is no hospital to treat us. Our bodies are cracking and sore" (125). The following year, another Rössing worker testified before the United Nations: "Now [it appears] all workers have respirators of some sort, but are required to wear them only in the very dusty areas and when some of the chemical plant breaks down – as it does from time to time and emits a very toxic smoke". But all workers were still exposed to "very heavy dust" and there was always "pulverised rock dust in the air". The dust blew "for many miles towards Swakopmund" (126).

In 1984, Dr Robert Murray, OBE, visited Rössing and declared it an "oasis of safety" (127). His report said nothing about dust, radiation or noise; no comparative data were provided, and he satisfied himself merely with a description of Rössing's medical facilities. This was hardly surprising since Dr Murray had, earlier in his professional career, aquitted asbestos of a negative impact on the human body (128)! The same year, no doubt emboldened by Murray's mint of wisdom, RTZ reported that it had observed "no case of either silicosis or lung carcinoma" at Rössing. Again, this was hardly ground-shattering news (since such diseases can take between ten and thirty years to display themselves). But, in 1985, the first four cases had been confirmed out of a total of 1200 workers tested since 1980 (118). What, we may be allowed to ask, has happened to medical records for the thousands of other workers employed at Rössing since 1973: presumably there are none, because there was no thorough medical checkup? In 1981, Rössing/RTZ claimed that all employees were given a medical examination, which included monthly urinalysis, but only for the final product area (that notorious roasting room) (129). X-rays, a lung-functions test and other laboratory testing, were added within three years, claimed RTZ. A film badge was supplied to all employees, which was analysed "on a monthly basis" (116). However, film badges only distinguish between gamma and beta radiation. Though, combined with efficient urinalysis, they can assist in the detection of radiation from uranium, thorium and radium in the mill and tailings areas, they do not compensate for the lack of other measures, such as whole-body gamma-counting (130) or the introduction of dosimeters which measure the alpha, beta and gamma radiation characteristics of so-called radon "daughters" to be found at any part of the mine, mill and tailings area.

In 1984 RTZ could baldly assert that: "... the radiation hazard to the general population is so small that, for practical purposes, it can be regarded as insignificant" (131). We may conjecture that such appalling complacency has been a premature sentence of death for scores, if not hundreds, of employees.

Lastly, we should note the existence of the Rössing Foundation, set up in 1978 under a "multiracial panel of trustees" to provide education, training and scholarships to black Namibians. By 1981, three hundred students had enrolled at the Rössing Foundation Centre, and scholarships (named after former chair, Mark Turner) had been awarded to three Namibians at British Universities, and two at the United World College of the Atlantic, in South Wales – itself sponsored by RTZ (129). Within two years the number of students had doubled, and 13 were at University (131). Figures have been rising ever since (132). The Foundation's Director in 1986 was David Godfrey, an ex-instructor in jungle warfare in Burma, and the organisation has managed to project an image of efficiency, political neutrality and self-confidence. Apart from the radioactive bequest of the mine itself and Arandis township, the Foundation is Rössing's most conspicuous contribution to an independent Namibia.

Of course the cost of setting up and running the Foundation is a drop in the desert, when compared with Rössing's long-term profits. Nor can its underlying motive be judged as completely above suspicion. In a major self-advertising campaign prior to 1982's annual general meeting, RTZ published a sketch of a Namibian woman and child growing vegetables "in the back garden", as one of "the new skills taught by the Rössing Foundation to families in Namibia". The outrage triggered by this grossly ignorant and patronising attitude (133) was not lost on the company, which never used such an image again.

SWAPO and Rössing

"Uranium is an evil, a threat to Namibia and the entire world".
(Hadino Hishongwa, at the International Conference on the Third and Fourth Worlds and Uranium, Copenhagen 9/79, quoted in *The Guardian* 21/1/80).

There is no doubt that RTZ expects the Rössing mine to continue operating, long after SWAPO's 1989 victory in UN-organised elections in Namibia. Not only has economic, technological and contractual dependency carefully been built in to the whole project, but RTZ in recent years has been very cautious in its official (and, it must be assumed, unofficial) statements, not to question the legitimacy of SWAPO or the need for South Africa to quit the Territory (134). It is not difficult to see why SWAPO has markedly shied away from any undertaking to nationalise the Territory's major mine (135). There is no precedent anywhere in the world of a Third World government taking over management responsibility for a uranium mine – especially one so dependent on the special expertise which RTZ has brought to the peculiar geo-physical characteristics of Rössing (136). It is also clear that, whether or not an independent government in Namibia would claim compensation from RTZ for its exploitation of the country's natural resources (to which it is entitled under UN Decree No.1), it will think at least twice before jeopardising its relationship with the one entity that has the funds and expertise to rehabilitate the mine-site and compensate for any future medical claims. (The lessons of CRA's speedy withdrawal from the dangerous mess at Rum Jungle are not lost on the MUN). On the other hand, the least that an independent Namibian mining authority would surely demand is majority control of the board of directors of Rössing, full oversight of contracts, and power to cancel existing contracts, and workers-directed inspectorates in all areas of the project's operations.

Although early RTZ statements were frankly critical of SWAPO and suggested it did not represent the people of Namibia (43), the company was even more dismissive of the UN. The 1975 RTZ chair, Val Duncan, declared he was "... not prepared to take any notice of what the UN says on the matter" (43, 52) and, as for intervention by UN forces: "... you may feel that perhaps the United Nations navy is not all that efficient" (43, 52). Five years later, the company stated that it had never consulted with the UNCN (137) while, in 1982, the new chair, Anthony Tuke, was even more belligerent: the UN had been "beyond its powers" in setting up the UNCN, and, anyway: "one battalion of paratroopers is worth countless numbers of United Nations Resolutions" (28). (It was at this same AGM that RTZ admitted to the existence of a 69-man armed squad of employees and vigilantes to "defend" the Rössing mine) (see below).

In fact, Val Duncan and Lord Shackleton had held talks with SWAPO as early as 1975, at the instigation of the British Foreign Office, where the liberation organisation had (apparently) requested RTZ to issue a statement recognising SWAPO as a prospective Namibian government: the company had obviously refused (138). Three years later, the Rössing management, and in particular R S Walker (one of the main architects of mine and company policy), expressed confidence that they could do a deal with any independent government, because of the complexity of the mine operations and the need for "immediate revenue". Only a "Russian dominated" future administration, it was thought, would be completely inimicable to Rössing's objectives (139). At the 1982 AGM, where chairperson Tuke snubbed the UN, he was almost deferential to SWAPO. "We do meet with them," declared Tuke – citing the 1981 Ditchley conference on South Africa as an occasion when the parties conferred; he also quoted an alleged statement by Sam Nujoma, SWAPO's President, that it would be "extremely unrealistic to march in and nationalise Rössing" (140). Moreover, said Tuke, the company in London did not feel it necessary to set aside any money for future reparations. Even if SWAPO had met with RTZ or Rössing management, it is very unlikely that the liberation organisation gave any promises on which such complacency could be based. In 1980, SWAPO issued a press statement "... as the legitimate representative of the Namibian people," in which it regarded "the exploitation of Namibian uranium as theft, and, as is provided in Decree No.1 of the UNCN, SWAPO will claim compensation for it as the Government of an independent Namibia with the full authority of international law behind it" (141).

In 1983 SWAPO set up a special commission to study the type of compensation it would demand from RTZ on independence (142) and, two years later, Sam Nujoma at a Nairobi press conference called on all multinational companies to quit Namibia forthwith, because they "fuel Pretoria's war machine". An olive branch was nonetheless extended: "When Namibia is free, we will certainly reach an agreement with [the multinationals] which will be beneficial to all of us" (143).

Now that independence has arrived, we must assume that RTZ will not be shown the door: whatever "greening" of Namibia occurs in the near future, it will not include the precipitate closure of this vast, destructive, radioactive, and ultimately useless hole in the land.

Impact on the environment

As is well known, uranium mining in a desert environment creates dust-borne radiation, both from the mine-workings and tailings dams (if not properly covered) which are not such a hazard in more temperate climes. Rössing has consistently claimed that water sprinkling keeps dust down to an acceptable level (113, 144) but this is clearly only in the context of silica contamination: the company refuses to acknowledge that dangers result from radon gas inhalation in the open-pit, from blasting operations, the transport of ore, and the tailings disposal. (Nearly half the tailings dam is not covered by water, according to observers in 1988) (145). Notwithstanding the large amount of research which demonstrates that such "low level" or "background" (!) radiation is a considerable liability – both for the workforce and those living upwind of the mine (146) – the company blithely asserts that, because Rössing is an open-pit operation, radon "... is rapidly diluted [during blasting] and is therefore no hazard to the miners or the environment" (113).

Such dereliction of responsibility is typical of Rössing's parent company, whose former chief executive and chairperson, Alistair Frame, once outraged shareholders by asserting that uranium mining is no more dangerous than the background radiation from a Scottish city (140).

Two other aspects of Rössing's operations have concerned observers: water-borne radiation from the tailings dam affecting the Khar river, evidence of which was found, according to Canuc, "even in a tame report commissioned by Rössing" (7), and the dehydration of a huge area around the mine, caused by the demand for water at many stages of the operations.

The Akafrik group which visited Rössing in 1987 confirmed that measures were being taken to reduce high acidity in the tailings dam, but that leaks probably occurred into both the Swakop and Khan rivers, possibly reaching to Walvis Bay and Swakopmund (145).

Rössing was orginally intended to use water deriving from the Kunene (Cunene) hydroelectric scheme on the Angolan border: the South African-inspired war against Angola interfered with this project, at which point Rössing turned to the Kuiseb river. By 1980 it was suggested that the water-table for this river has been lowered considerably, thanks to the mine's demands (147).

In 1982, the US *National Geographic* reported that Rössing was taking some 23,000 cubic metres daily from both the Swakop and Kuiseb rivers. Two years later, Rössing claimed to have reduced its consumption by half this level, thanks to "further recycling of water" (22, 148). But internal memoranda from the mine, secured by Canuc and Partizans in 1987, showed that water taken from the Kuiseb was still running at an average of 21,223 cubic metres/day, three years after the remedial measures were supposedly taken (74, 149).

Opposition outside Namibia

"The only hassle we get around Namibia is at this meeting" (150).

RTZ's Rössing mine has aroused opposition world-wide for about fifteen years. The most important single measure taken against it was one of the first. In 1974, the United Nations Council for Namibia (UNCN) enacted Decree No. 1; shortly afterwards the measure was endorsed by the General Assembly of the UN.

This followed a series of steps taken by the UN, supported by the International Court of Justice in the Hague between 1946 and 1963, which terminated South Africa's mandate over Namibia, called on South Africa to withdraw from the Territory, ordered UN member states to refrain from any further dealings with the apartheid state over Namibia, and recognised SWAPO as the legitimate representatives of the country (43).

The Decree forbade any "person or entity," whether a "body corporate or unincorporated," to exploit or export, in any form, any of Namibia's natural resources, unless licenced by the UNCN. It also enabled the UNCN or its agents to seize "any vehicle, ship or container found to be carrying animal, mineral or any other natural resources produced in or emanating from" the Territory. Any corporation contravening the Decree "may be held liable in damages by the future Government of an independent Namibia" (151).

Within a year, the UN Commissioner for Namibia, Sean McBride, was warning foreign companies to cease illegal activities in Namibia, and announced that a fund had been set up to finance possible court actions against those trading in the Territory's products (152, 153). US companies responded to these announcements with commendable alacrity: Getty, Continental Oil, and Phillips Petroleum abandoned exploration leases (153) and Amax said it would be getting rid of its holdings in the Tsumeb mine (153). The West German government also promised to discourage further investment by domestic companies in Namibia, notwithstanding the heavy committment by UG/Uranges already made (43).

For its part RTZ, supported by the British government, studiously ignored the Decree, and has done so ever since.

The Japanese government also flagrantly violated the terms of the Decree – at least until 1988. Appropriately, perhaps the first demonstrations against Rössing, outside of Namibia, came from Japan, when farmers, students, nuclear scientists, and lawyers banded together in Osaka in early 1975 to protest the Kansai contract. Demonstrations were also held in Tokyo, and an Open Letter from the Anti-apartheid Movement was presented to the Japanese Prime Minister (154).

In London the following year, a number of organisations formed what was to develop into a full-time movement against the Rössing mine and, along with the French anti-apartheid movement, they monitored routes out of Namibia taken by uranium transport companies. Unfortunately, before French and British trade unionists could take direct action against these shipments, the French radical daily *Liberation*, followed by the British *Sunday Times*, published comprehensive details: the routes were changed, and the action was abandoned (155).

After 1981, opposition internationally by trade unionists escalated considerably, though the ground was much thicker with resolutions than with practical steps to stop the export/import of Namibian yellowcake.

In November 1981, a National Labor Conference for Safe Energy and Full Employment, held in Gary, Indiana, condemned the Namibian uranium trade (156). Early the following year, the British trade unions ASLEF (railways), NUS (seaworkers), and TGWU (Transport and General Workers), met with British and French activists, along with SWAPO and the National Union of Namibian Workers (NUNW), to agree on steps for a blockade of uranium shipments entering Europe (157). At the same time, a petition to stop the mine, signed by some 10,000 people, was presented to British Prime Minister Thatcher, while demonstrations were carried out at local offices of the British CEGB and RTZ (157, 158), as well as at Edmondson's, the uranium transport company (159).

Unfortunately, because of differences between the motley groups of opponents to the Rössing uranium trade (for example, the French CFDT and CGT) another conference was called in Bristol, England, in the summer of 1982, to decide a selective blockade of Namibian uranium imports, hopefully with strong trade union support (156, 158).

Once again, problems surfaced: in this case, the members of the TGWU working at the Springfields uranium hexafluoride processing plant near Preston, refused to have anything to do with a blockade. As this was the main first destination of Namibian uranium entering Europe, their resistance rendered united trade union action impossible (160).

Meanwhile, 1982 saw a wide variety of imaginative activity focusing on Britain's illegal uranium trade in Namibia: a picket of the RTZ "milk run" (student recruitment session) at Surrey University (16); a day of action with street theatre at Cambridge (162); more demonstrations outside a hotel in Oxford where RTZ was conducting interviews with potential future employees, which were cancelled after only two hours, thanks to the vociferous opposition (163). Several groups, including Students Against Nuclear Energy (SANE), Anti-Apartheid, Third World First, Campaign for Nuclear Disarmament (CND), Campaign Atom, and Partizans, participated (156).

In October, a national week of action was launched, supported by Anti-Apartheid, SWAPO, the National Union of Students, PARTiZANS and others.

1982 saw another attempt to mobilise concerted trade union action against Rössing imports into Britain – but, again, it stumbled on the recalcitrance of workers at the Springfields plant. However, on May 10th, forty activists turned out in Baltimore to protest the arrival of a South African ship, the *SA Constantia*, carrying Namibian uranium for use by Baltimore Gas and Electric Company at its Calvert Cliff nuclear plant – in 1981 the biggest US buyer of Namibian yellowcake (164).

The following year, the well-known authority on Namibian uranium, Alun Roberts, visited the Territory on behalf of the UNCN, only to find himself detained after a visit to the mine-site. He was interrogated by a Mr Murray, second-in-command of security at Rössing, as well as South African police (165). Although Alun Roberts was convinced that Rössing officials

must have collaborated with the South African police in his arrest and detention (166) – and, while in jail, was in fact shown a photo of himself, taken at the 1982 London AGM of RTZ! (40) – RTZ's then-chair, Anthony Tuke, denied any complicity in the affair (167). At a time when RTZ was beginning to allow journalists to inspect the Rössing project, Roberts' detention was a considerable embarrassment to the London company. Strident protests from several countries secured his release after 26 days.

The Greater London Council (GLC) – then one of the largest administrative bodies in the world – in May 1985 announced that it would sell its £4 million shares in RTZ, in protest at its Rössing connection and other objectionable activities around the world, after a well-known left-wing lawyer, Lord Gifford, told the Council it could sell without legal hindrance (168). Months later, the GLC had done nothing towards this declared end, which prompted the international umbrella group working against RTZ, PARTiZANS, to picket the GLC's council chamber and publish an indictment of this "socialist", anti-apartheid, nuclear-free zone, local authority's continued support for investment in apartheid (169). At the end of 1985 a disinvestment conference in London, addressed by the African National Congress (ANC), SWAPO and a representative of the National Federation of Aboriginal Land Councils, reiterated the call for the GLC and other *soi-disant* anti-apartheid councils with RTZ shares, to disinvest. The GLC took another year before doing so. Many local authorities in Britain continued to hold RTZ shares, at the same time as they boasted their opposition to the South African régime and all things nuclear (40, 132, 170).

September 1985 found the United Nations holding hearings on multinationals operating in Namibia: once again, the call for disinvestment was made, with little apparent result. (At that stage the UN identified no less than 1086 transnationals exploiting the Territory in one way or another) (171).

671

However, this was also the year in which the UNCN decided at last (and pretty long last – more than a decade!) to take legal action against the theft of Namibian uranium. In May, it announced that it would take to court Urenco, the joint Dutch-British-West German uranium enrichment company, with plants in Capenhurst (Cheshire, England), Almelo (Netherlands) and a (then) pilot plant at Gronau, (West Germany), for accepting and processing material obtained in defiance of UN Decree No. 1. The case, it was optimistically predicted, would be ready by the end of that year (172). Urenco, with 7% of the world's market for U-235, has been enriching uranium of Namibian origin since 1980 (173). But the company maintained it was impossible to tell where specific consignments came from: "The customer doesn't tell us" (174).

When the case did reach court in July 1986, the Dutch government took Urenco's line: it too didn't know where the uranium had been mined (175). Until the time of writing, there have been three adjournments of the UN proceedings (176), although SWAPO's chief representative at the UN, Helmut Angula, recently said that other companies, such as Shell, De Beers, Consolidated Diamond Mines, Newmont, and RTZ, may also be sued (177). That year, the Senator for Ports and Shipping in Bremen, West Germany, announced that, if it could be identified, he would ban uranium of Namibian origin from entering his port (178).

Sixty women protested against the delivery of "hexed" Namibian uranium to the Capenhurst enrichment plant in March 1984 (179). When 19 of them were arrested and fined £2780 plus damages, the United Nations rallied to their support, issuing a press release claiming that they were acting lawfully – indeed with a high moral purpose (180). Four of the women refused to pay their fines and were jailed. They cited seven international acts of law as being violated by the production and import of Rössing uranium, including three UN Security Council Resolutions, the Geneva Conventions, and the Convention on Genocide.

Soon afterwards, there were more actions organised by CANUC against the British Nuclear Fuels plant handling Namibian-origin uranium at Capenhurst (181). The following year there was yet another International Week of Action on Namibia, which included loud condemnation of continued operations at Rössing in defiance of the UN (182).

At last, in 1987, Britain's largest union, the TGWU (Transport and General Workers Union) adopted a national policy against the import of Namibian uranium (183). This gave the lead to dockworkers at the northwest port of Liverpool, in February 1988, to refuse handling 13 containers of UF_6 emanating from BNFL at Springfields and destined for Portsmouth, Virginia, on the *Atlantic Carrier*. The uranium hexafluoride was to be enriched by Martin Marietta Energy Systems on behalf of the US Department of Energy (DoE). (RTZ's Australian associated company, CRA, in its purchase a few years earlier of Martin Marietta's aluminium operations, had obtained rights to port facilities at Portsmouth, although transport in this instance was to be handled by Burlington Northern).

The dockworkers prevented four containers being loaded, though nine got away: these were met at the other end by protesters, including members of Congress, claiming that the shipment breached the US Anti-Apartheid Act of 1986 (184). The other nine containers were sent to the British east coast port of Felixstowe, where they were also turned away by dockers – and ended back at Springfields. BNFL claimed – like Urenco – that all uranium, once it reached the hexing plant, was mixed together. (This is a point which has been made consistently by anti-nuclear campaigners to justify blockading all uranium supplies, from whatever source). But, BNFL had previously admitted that "between one-half and two-thirds" of yellowcake entering Britain was of South African or Namibian origin: this admission naturally inspired the TGWU to try to extend its blockade (183).

Later that year, the National Union of Seamen (NUS) unanimously opposed all trade in

Namibian uranium and, during another International Week of Action, the offices of P&O Ferries (notorious for the Zeebrugge disaster, and a carrier of Namibian yellowcake) were picketted (72). BNFL then turned to the east-coast ports of Ipswich and Immingham, as reception and disembarcation points for the illicit cargo (185). Before the year-end, the Immingham port authorities said they would ban the import of UF_6 from Almelo (186), although Exxtor, the Danish shipping company responsible for the transportation, declared that it would continue its trade (187).

Meanwhile, Dublin dockworkers were angry at the use of their port as a nuclear conduit (for waste and weapons components as well as uranium), although the Dublin government appeared to make a promise to ban enriched UF_6 coming from Riga, USSR, into Europe via Dublin (188).

Despite enormous applications of energy and expertise (more than has been expended on any other uranium producer), attempts to stop the transport of Rössing-origin uranium met with little success. The lesson, no doubt, is that opponents of the Rössing mine should have been more willing to adopt an anti-nuclear stance, while adherents of the anti-nuclear movement (especially in its heyday between 1977 and 1985) should have paid closer attention to the uranium issue. In any event, the opposition forced RTZ to change its route on several occasions. First known consignments to Europe were carried by South African Airways and UTA (Union des Transport Aeriens) without airline livery identification: these cargoes were destined for BNFL at Springfields and the Comurhex UF_6 plant at Malvesi in France (189). Once this route was revealed at the turn of 1979, Deutsche-Afrika Linie (a member of the South African Conference of shipping lines) took over the transportation, providing docking facilities at Southampton, Marseilles, Rotterdam and Hamburg. The first vessels importing Namibian uranium into Europe via this route were the *Urundi* and *Ulanga* (both of West German registration). They landed at Zeebrugge in early 1980, off-loading their ill-gotten cargo onto lorries owned by an unknown and rather mediocre transport company (its speciality appeared to be furniture) – F Edmondson's of Morecambe (155). A South African subsidiary of RTZ, James Estate of Cape Town, also shipped Namibian uranium concentrates to Canada at this time (140).

The sea route has remained the main thoroughfare by which Rössing's spoils gain entry into Europe: vessels of the South African European Container Services (including P&O Ferries) dock in Marseilles, where uranium for French processing is offloaded, and in Southampton or Zeebrugge, which acts as the staging post for various destinations (7).

Within Britain, the uranium travels by rail to container depots in Liverpool, Manchester or Leeds, and thence for hexing at Springfields. The road route is not exactly known, but believed to still depend on Edmondson's. The shipping agency for the hexed or enriched product is ACL (Atlantic Container Lines) using various ships, including the *Atlantic Conveyor* and *Atlantic Cartier* (7).

One notable victory has been registered against F Edmondson's. In 1982, the northern England town of Lancaster refused a permit to the furniture removers to store up to 1000 tons of uranium (presumably in the form of U_3O_8) in an old ICI (Imperial Chemical Industries) factory at Heysham. Officially the rejection was made because of the worries about a negative impact on tourism (190). According to journalist Martin Bailey, what really rankled Lancashire councillors was Edmondsons' secrecy: the company wouldn't say where the uranium was coming from or where it would be going. BNFL was a little more forthcoming, but hardly more enlightening: the uranium was not for British power stations, it stated, or to be processed "by us" on behalf of foreign customers. That left the possibility that it was for stockpile only (because of excess production at Rössing), could be en route to the USSR for hexing (66), or could be earmarked for processing for the Trident missile programme.

...and inside Namibia

It was not until 1987 that an independent union was recognised by Rössing's management. RTZ undoubtedly attempted to put off recognition of the union for as long as possible and, in this respect, it was significantly behind even some South African compance (such as Anglo-American Corporation). It maintained, until then, that its Employee Representative System was quite sufficient (113, 191). Although there have been several major actions by Rössing employees since 1978 against working conditions, as in South Africa it would be impossible to separate strike action by workers from demands for freedom and independence. In December 1978, 2000 Rössing employees went on strike for a week, because of the introduction of a unified pay system which still perpetuated racial discrimination in wages and working conditions (43, 192): at the same time the strikers vociferously opposed the counterfeit elections for a puppet administration in the Territory (6).

The following year, as the "seething mass of discontent in the workforce" (a phrase later used by Gordon Freeman, Rössing's General Manager) (193) reached its height, tear gas, dogs and brutal police action broke up another strike against derisory pay awards (£5 per month for blacks and £60 a month for white employees) (194). Several workers were held under the South African Anti-Terrorism act; one of whom, Arthur Pickering, found asylum in Britain.

Strike action continued into 1980 – not just against the new pay scales, but radiation hazards, as the Rössing Mineworkers Union, now affiliated to the NUNW, began organising (195). In 1980, the NUNW had its headquarters closed down, its funds frozen, union records and equipment impounded, and it saw much of its leadership imprisoned without trial (196).

Early in 1986, attempts to set up an independent Mineworkers Union in Namibia, with the support of the South African National Union of Mineworkers, was thwarted by Rössing's management. The Turnhalle (South African-sponsored) administration in Namibia introduced legislation forbidding any non-residents from helping set up trade unions in the Territory (197). That summer, a Rössing employee was shot by South West African police and died later (198). However, soon afterwards, Namibian labour laws were "liberalised", and black trade unions made legal. The MUN, headed by Ben Ulenga, emerged as the most important union: shortly afterwards it began organising at Rössing (201), supported by the NUNW. It issued demands for a minimum wage, upgrading of living standards at Rössing, and implementation of UN Resolution 235 on Namibian independence (193).

A month later, RTZ maintained that the new Union had not negotiated a recognition agreement with the company (200). But, as the MUN gained strength, so the company had to come to terms with its claims to represent the majority of workers. Even so, the management continued trying to ignore the Union. (In 1989, it arbitrarily sacked one of its key figures in the mine) (202). In 1988, thousands of black Namibians, including many from Rössing, went on strike in support of independence, as the international focus shifted to south-west Africa; Rössing lay idle for two days (203).

Conclusions

"... That is what we are doing at Rössing ... freeing the country's natural wealth. Freeing it, not so that it can be taken away and stored elsewhere for someone else's benefit, but freeing it so that the country and its people may move further along the path of progress and development. I don't think that I need describe to any of you the position that would exist here in South West Africa/Namibia today if Rössing had not been developed ... And lastly the shareholders ... There might be a tendency to say "why worry about their interests? They live outside the country and are only taking our money." But wait – it is we who have take their money and we owe them a great deal, physically and morally, for it is they who have made it possible for Rössing to be established. Through the shareholders we obtained the R350 million that was required at very considerable risk and had to be spent before Rössing could produce a pound of uranium ... we must establish a pattern whereby investors are assured of getting their money back and are rewarded for their courage and confidence. In this way, we shall be looking after ourselves as well as them. This then outlines the Rössing ethos ... As part of the RTZ Group, we are contributing to the strengthening and economic development of the free world with a highly-developed sense of social responsibility" (204).

Had Rössing's General Manager not delivered the above statement from his own lips, it might have been necessary to invent it.

Here, in one brief volley, is encapsulated much of what millions of people across the globe fear and detest about multinational corporations. There is only one word of truth in it: Rössing Uranium Ltd has undoubtedly been a source of considerable reward for its investors, particularly RTZ. Ever other sentiment is a travesty of the facts. Rössing's "natural" resource is a mineral that is alchemised into a highly unnatural and destructive product, that will almost certainly never be converted into energy in its country of origin. For a decade and a half this raw material was purloined in defiance of international law and world opinion, to be stored, processed, enriched and used, not just for nuclear power plants in an over-developed world governed by exploitative élites, but for weapons of war which might one day mean the end of us all. As part of the RTZ Group, Rössing has contributed to the grotesque profiteering of a band of expatriate robber barons, to their growing domination of the world's uranium markets and the global mining industry. If Rössing had not been developed, the Territory might well have achieved independence before now. And who knows what political and economic choices the people of Namibia might have made, had Mr Louw not stubbed his foot against the Rössing outcrop sixty years ago?

But there is more to it than this. Exploitation of Namibian uranium has had a "disastrous impact" on British foreign policy (218), and the relationship between Britain and many Third World countries. (A visit to the mine paid by the country's prime minister, M Thatcher, in early 1989, where she commented that the project made her "proud to be British" (205) can only have deepened this sense of disillusionment and mistrust among Third World peoples). Moreover – and whether or not the mine's output has ever directly fed South Africa's nuclear plants – Rössing has certainly buttressed the apartheid state. Rössing's managing director, Colin Macaulay, was elected President of the Chamber of Mines in Namibia in 1984 (206); two years previously, the Chief Executive of RTZ-South Africa (RTZ's vehicle for investment in Rössing) became the only corporate representative on South Africa's Atomic Energy Commission (207). The Institution of Mining and Metallurgy (no doubt influenced by Alistair Frame of RTZ) set up a "South West Africa" section in 1985, with Gordon Freeman of Rössing as its chair (208). These appointments contributed significantly to justifying South Africa's occupation of Namibia and white control of the sub-continent.

The company's protestations that it promotes good worker-management relationships is of a piece with the myth that its operations are a boon to Namibia. A succession of miners' leaders at Rössing has been dismissed or downgraded, and all have faced the mailed-fist-behind-the-velvet-glove, in the form of Rössing's crack security force. South African legislation, passed in 1980, "required" major mines under apartheid administration to form a "commando-type force" for their "protection" (209). In fact, it was two years earlier, in a secret memorandum circulated on 29th November 1978, that Rössing proposed setting up its own armed gangs prepared for "civil or labour attack against the mine" (210). This memorandum – marked "to be destroyed by shredding" – was leaked from Rössing to the Namibia Support Committee (NSC) in London, which divulged its contents at the 1981 RTZ AGM. RTZ's chair, Anthony Tuke, flatly denied that the document existed. A year later, he was forced to swallow his words, and admitted in a letter to Brian Wood of the NSC, that 69 men were employed as a security force at Rössing; two units comprised Rössing employees, and the third a "local citizens vigilante unit" (211). Tuke added that it was "the duty of management" at Rössing to plan for the protection of its employees and their equipment: "The same," he added, "would be true if there were unrest, let us say, at Bougainville, or Chile, where we have no investment, or indeed anywhere else in the world" (211, 212).

A final aspect of the company's operations must be mentioned. Whatever the outcome of its mining at Rössing, the management wants to stay in Namibia and exploit other promising mineral finds. Even before the uranium lode was mined, RTZ had exploration claims "all over Namibia" (213), and had identified deposits around Rössing itself (214). After 1973, the South African régime improved conditions for exploration companies in the Territory (215).

However, as attention was turned to getting Rössing underway, the initiative seems to have diminished. It was not until 1981 that Rössing declared it would take up exploration again – this time for copper, zinc, and other minerals. *The Financial Times* predicted that the South African-inspired Turnhalle "government" would "welcome" the decision (216). Soon, prospecting teams were set up and despatched throughout the Territory (16). 1983 saw the establishment of an Exploration department (217). Nor is RTZ's penetration of Namibia confined to Rössing Uranium Ltd: Riofinex owns Skeleton Coast Diamond Pty Ltd, which operates around the country's coastline (2).

Whatever guise RTZ may adopt to continue its exploitation of Namibia, one thing is certain: its arrogance and ruthlessness at Rössing must belong to a bygone era. Robbery on such a scale, and with such little consideration for people or ecology, will hopefully never happen again. ...Or will it?

To watch: "Follow the Yellowcake Road" Granada TV Production, transmitted 10/3/80. (copies available from Concord Films Ltd, Nacton, Ipswich, England.)

Further reading: (Many of the works mentioned in this bibliography come from a data-base belonging to the Centre for African Studies at Bremen University. Sincere thanks to Werner Hillbrecht for access to this study) (R.M.).

A.A.B. Bremen (and others), *Uranabbau in Namibia: Gestohlenes Uran fur die Strahlende Zukunft der Bundesrepublic*, Verlag roter Funke, Bremen, 1982.

Anti-Apartheid Bewegung in der BRD und West Berlin, Gruppe Bremen, "Kein Uran aus Namibia und auch nicht von anderswo", *Informationsdienst Sudliches Afrika No.3*, Bonn, 1982.

J.W. von Backstrom, "The Rössing uranium deposit near Swakopmund, South West Africa: A preliminary report", *Uranium exploratory geology: proceedings of a panel on uranium exploratory geology held in Vienna, 13-17 April 1970*, Vienna, 1970.

J.W. von Backstrom, "International uranium resources evaluation project", IUREP, *National favourability studies.* Namibia, IAEA, Vienna, 1977.

J.W. von Backstrom, R.E. Jacob, "Uranium in South Africa and West Africa (Namibia)", *Philosophical Transactions of the Royal Society of London*, SER.A, Vol.291, London, 1979.

David de Beer, "Namibisches Uran in Europa: letzte Nachrichten", *Informationsdienst Südliches Afrika* No.4 1980, Bonn, 1980.

David de Beer, *The Netherlands, Namibian uranium and the Implementation of Decree No.1*, Working Group Kairos, Brussels, 1986.

David de Beer, "Der URENCO-Fall: in den Niederlanden beginnt der Prozess um die Anerkennung von Dekret Nr.1", *Informationsdienst Südliches Afrika No.5 1980, Bonn, 1987.*

J. Berning, S.A. Hiemstra, U. Hoffman, "The Rössing uranium deposit, South West Africa", *Economic Geology* Vol.71, Lancaster (USA), 1976.

Sterling Bjorndahl, *Saskatchewan and Namibia: the uranium connection*, Washington, 1982.

Blockade stolen Namibian uranium: sink Trident, Campaign Against the Namibian Uranium Contract, 1984.

Phil Boyd, "Uranium: imposing an international blockade", *Action on Namibia*, Winter 1986/87, London, 1987.

Phil Boyd, "Namibian uranium – the blockade begins to bite", *Action on Namibia*, Summer 1987, London, 1987.

Phil Boyd, Greg Dropkin, Frances Kelly, *Western companies and Namibian commodities: research for action*, Namibian Support Committee, Istanbul, 1988.

British exploitation of Namibian uranium, CANUC and NSC, London, 1985.

Tarin Brokenshire, *Namibian uranium and British nuclear weapons: the Trident D5 programme*, Geneva, 1984.

Tarin Brokenshire, *British military use of Namibian uranium*, CANUC, London, 1984.

John Buell, Daniel Horner, *Weapons implications of US-South African uranium trade*, Nuclear Control Institute, Washington, 1985.

Campaign Against the Namibian Uranium Contract, *Stop the robbery of Namibian uranium*, CANUC discussion document, London, 1/1987.

Campaign Against the Namibian Uranium Contract, *Briefing on the campaign for activists*, Brussels, 1986.

Campaign Against the Namibian Uranium Contract, *Briefing for the December 12th National Day of Action*, CANUC, London, 1981.

CANUC, *Canuc News*, London, 1982-89.

CANUC, *Namibia: A contract to kill; the story of stolen uranium and the British nuclear programme*, AON Publications, Namibia Support Committee, London, 1986.

Mike Carden, Frances Kelly, "Interview of Helene Decke-Cornill and Kathe Jansl", *Informationsdienst Südliches Afrika* No.1, Bonn, 1989.

Rien Cardol, "Uraniumroof uit Namibie: diskussies in het parlament", *Dokumentiemap Geen Nukleaire Kollaboratie met Zuid-Afrika*, Amsterdam, 1981.

Zdenek Cervenka, Barbara Rogers, *The Nuclear Axis: Secret collaboration between West Germany and South Africa*, Friedmann, London, 1978.

CIS, *Rio Tinto-Zinc Corporation Limited: Anti-report*, London, 1974.

Confédération générale de travail (CGT), *Note on the import of Namibian uranium*, Geneva, 1984.

Branko Corner, *Results of a palaeomagnetic survey undertaken in the Damara Mobile Belt, South West Africa, with special reference to the magnetization of the uraniferous pegmatitic granites*, Atomic Energy Board of South Africa, Pretoria, 1978.

Branko Corner, *An interpretation of the aeromagnetic data covering portion of the Damara orogenic belt, with special reference to the occurence of uraniferous granite*, Johannesburg, 1983.

Dokumentatiemap *Geen Nukleaire Kollaboratie Met Zuid-Afrika*, 28/11/81, Anti-Apartheids Beweging, Amsterdam, 1981.

Greg Dropkin, *Namibian uranium and the military programme*, Sheffield (UK), 1985.

Greg Dropkin, "Stop the Flow!", *International Labour Reports* No.23, Stainborough (UK), September/October 1987.

Greg Dropkin, "Untangling the nuclear web", *Action on Namibia*, Winter 1988, London, 1988.

Greg Dropkin, Phil Boyd, "One match can set the world alight: BNFL have admitted their use of Namibian uranium as the Liverpool portworkers' blockade spreads", *Action on Namibia*, Summer 1988, London, 1988.

Greg Dropkin, "Build the Blockade", *Action on Namibia*, Spring 1988, London, 1988.

Frits N. Eisenloeffel, "Für Brokdorf und Borken...: Illegales Uran aus Namibia fur Bundesdeutsche Kernkraftwerke", *Welt Magazin*, No.11/12, Bonn, 1978.

Frits N. Eisenloeffel, "Fur Brokdorf und Borken...: Illegales Uran aus Namibia fur Bundesdeutsche Kernkraftwerke", in Stephen Salaff, *Umwegen zur Atommacht?*, Bremen, 1978.

Jean Fischer, "Rössing – Namib Desert mining giant", *SWA Annual 1979*, Windhoek (Namibia), 1979.

Iain Galbraith, "Stand und Wirkungen von Sanktionen", *AIB DritteWelt-Zeitschrift* Vol.19 No.7, Koln, 1988.

Wolff Geisler, "Das Uran der Rössing-Mine in Namibia", *Informationdienst Südliches Afrika* 1980 No.8, Bonn, 1980.

Wolff Geisler, *The uranium of the Rössing Mine in Namibia*, Anti-Apartheid Bewegung, 1980.

Wolff Geisler, "Das Uran der Rössing-Mine in Namibia", in *Uranabbau in Namibia: gestohlenes*

Uran fur die strahlende Zukunft der Bundesrepublik, AAB Bremen, 1982.

Wolff Geisler, *The cooperation of the Federal Republic of Germany with apartheid South Africa in the nuclear military and conventional military fields*, Anti-Apartheid Movement in the FRG and West Berlin, New York, 1984.

Wolff Geisler, "Uranimporte aus Namibia in die Bundesrepublik Deutschland", *Im Brennpunkt: Namibia und die Bundesrepublik* Deutschland Bonn, 1987.

A.M. Gerritsma, *Implementation of Decree No.1 in the Netherlands*, Geneva, 1984.

Leake S. Hangala, *The Namibian uranium deposits*, Helsinki, 1980.

Brigitte Heinrich, "Nukem und Namibia", *LINKS*, Vol.17 No.186, Offenbach (FRG), 1985.

Dennis Herbstein, "The uranium trial", *West Africa*, Vol.1987 No. 3651, London, 1987.

Lutz Herzberg, "Mit Rössing fur Namibia: aus dem Inneren eines Urangiganten", *Pogrom* Vol.18 No.135, Gottingen (FRG), 1987.

Manfred 0. Hinz, *Rechtlich-politische Anmerkungen*, Königswinter (FRG), 1982.

Ruurd Huisman, *Uraniumroof in Namibia: hoe Nederland daarbij betrokken is*, Stichting Uitgeverij Xeno, Gröningen (Netherlands), 1978.

Ruurd Huisman, *Opposition in the Netherlands to the violation of Decree No.1*, Geneva, 1984.

"Illegaler Export von Uran aus Namibia", AAB-Seminar 16-17/82, *Informationsdienst Südliches Afrika* No.3, Bonn, 1982.

Ben Jackman, "Uranium: the United Nations acts", *AFRICA* No.111, London, 1980.

Japan Anti-Apartheid Committee, *A historical and economic study of Japan-Namibia relations*, Vienna, 1986.

Trevor B. Jepson, *Rio Tinto-Zinc in Namibia*, Christian Concern for Southern Africa, London, 1977.

Roland Junker, "Rössing trinkt und Namibia dürstet", *Uranabbau in Namibia: gestohlenes Uran fur die strahlende Zukunft der Bundesrepublik*, AAB Bremen, Bremen, 1982.

Smail Khennas, "L'exploitation des ressources uraniferous namibiennes et la politique nucléaire de l'Afrique du Sud", Afrique et Développement Vol.11 No.1, Dakar (Senegal), 1986.

Yoko Kitazawa, *On the Illegal Japanese uranium deals*, Geneva, 1984.

Yoko Kitazawa, "On the illegal Japanese uranium deals", *Seminar on the Activities of Foreign Economic Interests in the Exploitation of Namibia's Natural and Human Resources*, Ljubljana, Yugoslavia, 16-20 April 1984, New York, 1985.

Harald Kliczbor, *Uranium from Namibia for nuclear power plants in the Federal Republic of Germany*, Bonn, 1987.

Harald Kliczbor, "Uran-Connections mit Namibia", *Informationsdienst Südliches Afrika* No.3, Bonn, 1988.

Harald Kliczbor, *Namibisches Uran fur Atomkraftwerke in der Bundesrepublik Deutschland*, Bonn, 1987.

Kongress gegen atomare Zusammenarbeit Bundesrepublik – Sudafrika Bonn,11-12/11/78, AAB in der BRD und West Berlin, 1979.

A. Kouwenaar, "Het illegale karakter van de Nederlandse invoer van uranium uit Namibie", *Dokumentatiemap Geen Nukleaire Kollaboratie met Zuid Afrika*, Amsterdam, 1981.

Patrick Lawrence, "Western theft of Namibian uranium", *Southern Africa* Vol.13 No.5, 1980.

Frank Leschorn, *Der Weltmarkt fur Natururan: Möglichkeiten zur Versorgung der Bundesrepublik Deutschland*, Trans Tech Publications, Clausthal-Zellerfeld (FRG), 1981.

Dorothea Litzba, "Die Rössing-Mine in Namibia", *Uranabbau in Namibia: gestohlenes Uran fur die strahlende Zukunft der Bundesrepublik*, AAB Bremen, Bremen, 1982.

Dorothea Litzba, *Die Rössing-Uranmine in Namibia und ihre regionale und internationale politische Bedeutung*, Bremen, 1983.

Michael Lucas, "Wie das Uran aus Namibia geklaut wird", *Die Tageszeitung*, 11/1/80, Berlin.

Robert von Lucius, "Die Vereinten Nationen und die Beschlagahme Urans aus Namibia", *Festschrift fur Wolfgang Zeidler*, hrsq, W. Furst *et al*, Vol.2, De Gruyter, Berlin/New York, 1987.

Wenda Lund, *Rössing und das illegale Geschäfte mit dem Namibia Uran: eine Untersuchung zur grossten Uranmine der Welt und ihrer strategischen Bedeutung*, Pahl-Rugenstein, Koln, 1984.

Michael C. Lynch, Thomas L. Neff, *The political economy of African uranium and its role in interna-*

tional markets, Colorado Copy Center, Grand
Junction (USA), 1982.

Thorsten Maass, "Westeuropaische Gewerkschäften
fur Lieferstop von Namibia-Uran", Informationsdienst Südliches Afrika 1981 No.8/9, Bonn, 1981.

Thorsten Maass, "Der Stoff aus dem Bomben sind:
von der Herkunft des Urans und seine Verwen-
dung in der zivilen und miltarischen Atomindus-
trie", in U. Barde, *Kolonial-Denk-Mal: BremenSch-
lussel zur Dritten Welt*, Dritte-Welt-Haus
Bremen, Bremen, 1984.

Christoph Marx, "Die Wuste strahlt: Rössing
Uranium in Namibia", *Informationsdienst Südliches Afrika* 1986 No.5, Bonn, 1986.

Christoph Marx, "Zehn Jahre illegaler Uranabbau:
die Rössing Mine in Namibia", *Blätter des IZ3W*
No.135, Freiburg (FRG), 1986.

Christoph Marx, "Zehn Jahre illegaler Uranabbau:
die Rössing Mine in Namibia", *Atomzentrum
Hanau: tödliche Geschafte*, Neue Hanauer Zeitung, Initiativgruppe Umweltschutz Hanau,
Hanau (FRG), 1986.

A.J. Michau, *Notes on Rössing Uranium Limited's purchasing policy*.

Roger Moody, *The Road of Rio*, Sheffield (UK),
1985.

J.D.J. Moore, *South Africa and nuclear proliferation:
South Africa's nuclear capabilities and intentions in
the context of international non-proliferation
policies*, Macmillan, London, 1987.

Festus Naholo, "Namibianische Arbeiter planen
Steik gegen diskriminierende Praktiken von Rio
Tinto-Zinc im Uranbergbau von Rössing", *Infor-
mationsdienst Sudliches Afrika 1979*, No.1/2,
Bonn, 1979.

"Namibia", *Keep It In The Ground* No.26, WISE,
Amsterdam, 1983.

Namibia Support Committee, *Briefing paper: the processing and transport of Namibian uranium in Britain; the importation and transport of southern African uranium in Britain; Hex-conversion; and export
through Liverpool docks*, London, 1987.

Namibia Support Committee, *British Transnational
Corporations in Namibia, evidence submitted to
Hearings at UN*, New York, 9/85.

Namibia Support Committee, *Foreign Companies in
Namibia, Occasional Paper No.2*, London, 1980.

Namibian Uranium Conference, Sheffield 27-28/4/85,
collected papers, CANUC, Sheffield (UK), 1985.

"Namibian uranium: the hand of Whitehall", *AFRI-
CA* No.104, London, April 1980.

PARTiZANS, *Dirty Bizness*, PARTiZANS Alternative Report, London, 1986.

PARTiZANS, *RTZ Uncovered*, PARTiZANS Alternative Report, London, 1985.

PARTiZANS, *The Rio Tinto-Zinc Corporation plc:
The alternative report 1983*, London, 1984.

"Plunder of Namibian uranium: major findings of
the hearings on Namibian uranium held by the
United Nations Council for Namibia in July
1980", *Objective : Justice*, Vol.14 No.2, New
York, 1982.

*Plunder of Namibian uranium: major findings of the
hearings on Namibian uranium held by the United
Nations Council for Namibia in July 1980*, United
Nations, New York, 1982.

*Reply: Answer to a denial of the government of the
Federal Republic of Germany concerning the mili-
tary-nuclear collaboration between the Federal Republic of Germany and South Africa, AntiApartheid
Movement Bonn*, Anti-Apartheid Movement in
the Federal Republic of Germany, West Berlin,
1979.

*Report of the Bristol Commission of Inquiry into the
Southern African Operations of RTZ*, Bristol AAM
and Bristol UNA, 1982.

Rio Tinto-Zinc Annual General Meeting (transcript),
PARTiZANS, London, 17/5/82.

Alun Roberts, *Submission of evidence to the United
Nations Council for Namibia Hearings on Namibian Uranium*, New York, 1980.

Alun Roberts, *The Rössing File: The inside story of Britain's secret contract for Namibian uranium*,
CANUC, London, 1980.

Alun Roberts, *The international trade in Namibia's
uranium – an overview of the exploration of Namibia's uranium resources*, Washington, 1982.

Alun Roberts, *CANUC – chronology of developments
in the campaign against the United Kingdom Government contracts for supplies of Namibian uranium
in violation of United Nations Decree No.1*, 1983-
1985, Sheffield, 1985.

Barbara Rogers, "Namibia's uranium: implications
for the South African occupation regime", *Nami-*

bia: the United Nations and U.S. policy, Washington, 1976.

Barbara Rogers, Ausbeutung bei Rössing/Namibia: die soziale und ökonomische Lage der Arbeiter im Uranbergbaubetrieb Rössing, Informationsstelle Südliches Afrika, Bonn, 1977.

Barbara Rogers, "Notes on labour conditions at the Rössing mine", South African Labour Bulletin Vol.4 Nos.1-4 (Focus on Namibia), Cape Town, 1978.

Barbara Rogers, "Züruck bleibt ein versuchtes Namibia... Umweltrisiken und Arbeitsbedingungen der Rössing-Uranmine", Namibia, Kolonialismus und Widerstand, Henning Melber, Bonn, 1981.

Rössing Council Bulletin, Namibia, 1981-83.

Rössing Uranium, The Rössing Fact Book, Windhoek (Namibia), 1985.

Rössing Uranium Ltd, Rössing Uranium News, Swakopmund (Namibia), 1975.

Rössing Uranium Ltd, Rössing, Windhoek (Namibia), 1978.

Rössing Uranium Ltd, Rössing 1978, Windhoek (Namibia), 1978.

Rössing Uranium Ltd, An introduction to Rössing, the largest uranium mine in the world, Windhoek (Namibia), 1980.

Rössing Uranium Ltd, Medical Services at Rössing, Windhoek (Namibia), 1984.

Rössing Uranium Ltd, Employee development at Rössing, Windhoek (Namibia), 1984.

CIMRA, RTZ Benefits the Community, London, 1980.

Stephen Salaff, Auf umwegen zur Atommacht? Das Deutsche-Britisch Niederlandische Konsortium zur Anreicherung von Uran, BBA Biichertisch, Bremen, 1978.

Carl Schlettwein, "Reflections on a visit to Rössing", Rössing News, 31/3/88, Windhoek (Namibia), 1988.

Nico J. Schrijver, "The Dutch government backs Urenco more than Namibia", NRC Handelsblad, 28/9/84.

A.W. Singham, "Namibia and nuclear proliferation", Third World Quarterly Vol.3 No.2, London, 1981.

Dan Smith, South Africa's Nuclear Capability, World Campaign Against Military and Nuclear Collaboration with South Africa, Oslo, 1980.

Dr Woton Swiegers, Leonard Shapumba, Charles Kauraisa, "3 der 2,500 Rössing-Leute schildern ihren Aufgabenkreis", Namibia: Wirtschaft, Dusseldorf (FRG), October 1988.

Strahlende Geschäfte: der Tanz auf dem Welturanmarkt, AK Afrika/ Münster, Schmetterling Verlag, Stuttgart, 1988.

Kaign Smith, "The Political Economy of Uranium Mining in Namibia and the Francophone Countries of Africa", in Survey of European Anti-nuclear Programs and Needs, with Background Information from the US, Canada and Africa, (eds) Anna Gyorgy and Obie Benz, Project report for "Alternatives to Nuclear" Conference, Washington, 18-19/6/80.

"Trade union seminar on Namibian uranium", Keep It In The Ground No.15, WISE, Amsterdam, 1981.

W.R.S. Swiegers, "Here's health, U Workers: quality of life for uranium workers at Rössing", Nuclear Active, January 1987.

Taskforce on the Churches and Corporate Responsibility, Canada and Namibian Uranium, Washington, 1982.

Taskforce on the Churches and Corporate Responsibility, "Canada and Namibian Uranium", Africa Today, Vol.30 Nos.1-2, Denver (USA) 1983.

Trade Union Seminar on Namibian Uranium, London, 29-30/6/81.

Trade Union Action on Namibian Uranium: report of a seminar for West European trade unions organised by SWAPO of Namibia in cooperation with the Namibia Support Committee, SWAPO of Namibia, London, 1982.

United Kingdom, House of Lords, Reports of British House of Lords (20/10/75): Uranium purchases from Namibia, Dakar (Senegal) 1976.

UN Center on Transnational Corporations (UNCTC), Activities of Transnational Corporations in Southern Africa and the extent of their collaboration with the illegal regimes in the area, E/C.10/26,6/4/77, New York.

UNCTC, Activities of Transnational Corporations in the Industrial, Mining and Military sectors of Southern Africa, ST/CTC/12, New York, 1979 (updated 1980).

UNCTC, *Transnational Corporations in Southern Africa: Update on financial activities and employment practises*, E/C.10/83/Rev.l, New York, 1983.

UNCTC, *Politics and practises of transnational corporations regarding their activities in South Africa and Namibia*, E/C.10/1983/10, New York, 5/5/83; and E/C.10 1983/10/Add.2, New York, 20/5/83.

UNCTC, *Role of Transnational Corporations in Namibia*, New York, 1982.

Urenco Nederland, *Translation of statement of defence of Urenco Nederland and Ultra-Centrifuge Nederland N.V. submitted on 3/5/88.*

Keith Venables, "Spread the blockade – workers' sanctions now!", *Action on Namibia*, Winter 1988, London, 1988.

Uranabbau in Namibia: gestohlenes Uran fur die strahlende Zukunft der Bundesrepublik, Herausgeber AG gegen militarischnukleare Zusammenarbeit der BRD mit Südafrika in der AntiApartheid-Bewegung (AAB) Gruppe, Bremen (FRG), 1982.

Contacts: People Against RTZ and its Subsidiaries (PARTiZANS), 218 Liverpool Rd, London N1 1LE, UK, tel. 71-609 1852

In the Netherlands – WISE, P.O. Box 5627, 1007 AP Amsterdam.

In Germany – Akafrik (Arbeitskreis Afrika), Lingener Str. 9, 4400 Munster.

References: (1) *FT* 20/5/78. (2) *Strahlende Geschafte: der Tanz auf dem Welturanmarkt*, AKAFRIK/Munster, Schmetterling Verlag, Stuttgart, 1988. (3) Rogers & Cervenka op cit. (4) *Rand Daily Mail* 2/3/70; *Palaminer* (RTZ South Africa house magazine) 1/73. (5) *MIY* 1988. (6) Kaign Smith, *"The Political Economy of Uranium Mining in Namibia and the Francophone Countries of Africa"*, in *Survey of European Anti-nuclear Programs and Needs*, with Background Information from the US, Canada and Africa (eds Anna Gyorgy and Obie Benz), Project Report for "Alternatives to Nuclear" Conference, Washington, 18-19/6/80. (7) Speakers Notes, CANUC (Campaign Against Namibian Uranium Contracts), London, 1988. (8) RTZ Press Release, 13/10/76. (9) Statement at RTZ AGM, 24/5/77. (10) *MJ* 2/4/76. (11) RTZ *Annual Report* 1976, London, 1977. (12) *African Development* 5/75; *Daily Mail*, London, 21/7/76. (13) *The Namibian Uranium Robbery*, CANUC, London, 1979. (14) *Foreign Uranium Supply*, published by NUS Corp., Rockville, 1978. (15) Panmure and Gordon & Co., 1975, quoted in (3). (16) Rowe and Pitman, *RTZ Analysis*, London 8/82. (17) RTZ *Annual Report* 1980. (18) UNCTC, *Activities of Transnational Corporations and Measures Being Taken by Governments to Prohibit Investment in South Africa and Namibia*, United Nations, New York, E/C 10/1984/10, 30/1/84, and E/C 10/1984/19, 10/2/84. (19) *MJ* 4/5/84. (20) *FT* 21/10/82; see also *Gua* 12/4/84. (21) *MJ* 31/5/85. (22) *MAR* 1985. (23) see also *Rand Daily Mail* 17/2/84, *Rand Daily Mail* 13/3/84. (24) *MAR* 1988. (25) *Namibian*, Windhoek, 6/5/88; *Namibian* 17/6/88; *FT* 3/6/88. (26) *FT* 14/2/88. (27) *Independence for South West Africa: Crucial issues face the mining industry*, in *South African Mining and Engineering Journal*, Vol.90 no.4146, 1/79; see also *New York Times* 23/7/78. (28) Statement by Anthony Tuke, RTZ chair at 1982 AGM, London, from PARTiZANS transcript. (29) John G. Hughes, RTZ Public Relations Dept., to Tarin Brokenshire, Letter, London, 26/6/85. (30) *Star*, Johannesburg, 22/5/71. (31) *Anti-Apartheid Bewegung*, Bonn, 1979. (32) Rio Algom *Annual Report* 1974. (33) *FT* 25/5/77. (34) *Rand Daily Mail* 9/1/73. (35) *Mining* 3/72. (36) Wolff Geisler, *The Uranium of the Rössing Mine in Namibia*, in *Anti-Apartheid Bewegung*, Bonn, 1980. (37) see also *Rand Daily Mail* 20/6/75, *Nuclear Fuel* 30/5/77; *Nuclear Fuel* 26/7/77. (38) *Economist* 8/10/77. (39) *Windhoek Observer* 13/8/88. (40) *RTZ Uncovered*, PARTiZANS, London, 1985. (41) *Star*, Johannesburg, 20/6/83. (42) *Namibian*, Windhoek, 31/7/87. (43) Alun Roberts, *The Rössing File: the inside story of Britain's secret contract for Namibian uranium*, CANUC, London, 1980. (44) Information from PARTiZANS and CANUC, on RTZ AGM, London, 1988; and RTZ AGM, London, 1989. (45) see also James Adams, *The Unnatural Alliance*, Quartet Books, London, 1984. (46) *Rand Daily Mail* 26/5/77. (47) RTZ AGM, London, 26/5/76, reported in (43). (48) *Rand Daily Mail* 18/6/76; *FT* 18/6/76. (49) R.E. Worrall, A.J.A. Roux, *"The pattern of uranium production in South Africa"*, International Symposium on Uranium Supply and Demand, London, 16/6/76; see also *E&MJ* 12/76. (50) *Rand Daily Mail* 2/3/70.

(51) AFR 13/4/76. (52) Diane Hooper, *The Rio Tinto-Zinc Corporation: A case study of a multinational corporation*, PRIO, Oslo, 1977. (53) see *"Yellowcake Roads to War"* in *The Alternative Report*, PARTiZANS, London, 5/83. (54) *Sanity*, London, 11/84. (55) D.A. Streatfield, RTZ Company Secretary, letter to Alun Roberts, London, 10/6/76. (56) see also J. Roberts, D. Mclean, *The Mapoon Books* Vol.3, International Development Action (IDA), Collingwood, 1976. (57) see also SWAPO statement, London, 5/2/83; Alastair Goodland, Dept. of Energy, Parlimentary answer to questions by Jack Ashley MP, and Andrew Bennett, 21/1/85 and 28/1/85; *Blick durch die Wirtschaft*, Frankfurt, 31/10/84. (58) *Economist* 11/7/70. (59) *FT* 16/9/76. (60) *Rand Daily Mail* 16/12/76. (61) *FT* 15/5/78. (62) *FT* 21/10/82. (63) *Africa Research Bulletin* 6/71. (64) see also *MJ* 22/7/77. (65) Counter Information Services, *AntiReport on RTZ*, London, 1972. (66) *New Statesman* 12/3/82. (67) *Update on the CEGB-Rössing Contract*, 1983-85, CANUC, 4/85. (68) *FT* 14/2/89. (69) Namibia: *A Contract to Kill*, a CANUC booklet published by Action on Namibia Publications, London, 1986. (70) *Gua* 2/8/88. (71) *Briefing Paper on Namibian Uranium Contracts*, CANUC, London, 1987. (72) *Action on Namibia*, Autumn 1988. (73) *Gua* 12/12/88. (74) PARTiZANS special edition of *Parting Company*, 11/87. (75) *Asahi-Shimbun*, Tokyo, 22/5/75. (76) Foreign Affairs Committee of the House of Representatives, Japan, 6/6/75. (76) *Sanity* 12/87. (77) *Asahi-Shimbun* 20/3/77. (78) see also *Nihon Kogyo Shimbun* 28/11/75. (79) Marubeni Corp., *The Unique World of the Sogo Shosha*, Tokyo, 1978. (80) Nukem *Special Report* on Japan, 1/87. (81) *Foreign Relations*, 9/10/86. (82) *Mainichi* 17/4/88. (83) *Gua* 2/11/88; see also *New York Times* 2/11/88; *Southscan* Vol.3 No.10, 2/11/88; *FT* 3/11/88; *Daily Yomiuri* 1/11/88. (84) *MJ* 18/8/89. (85) *Independent* 2/11/88. (86) *Africa Research Bulletin* 11/75. (87) Nukem *Special Report*, quoted in (76). (88) *Nuclear Fuel* 28/5/79. (89) Neff op cit. (90) Speech by T.H Hishongwa, Hamburg, 20/6/81. (91) Wolff Geissler, *"The uranium of the Rössing Mine in Namibia"* at Hearings of the UNCN on Namibian Uranium, New York, 1980. (92) Euratom Supply Agency, *Annual Report*, 1986. (93) Statement by Federal West German Government to Petra Kelly, Green Party,

1988. (94) *El Pais* 9/4/88; *El Pais* 19/4/88; see also *Nuclear Fuel* 18/4/88. (95) Information from Jaume Morron, WISE/Tarragona, 24/4/88. (96) *Southscan* 6/7/88. (97) Information from Greg Dropkin, CANUC, to author, 26/10/88. (98) *Gua* 15/12/82. (99) *New York Times* 28/6/81. (100) Information from Peter Jones, Pacific Concerns Resource Centre, Sydney, 1986. (101) *Obs* 11/10/87. (102) Namibia Support Committee, Press Release, 7/11/88. (103) *FT* 2/11/88. (104) *International Newsbriefing on Namibia* (INON) No.58, 5/88. (105) Grünen press release, Bonn, 23/2/88. (106) RTZ AGM, 5/75, quoted in (43). (107) *New African Development*, 9/77. (108) Statement by RTZ chair, RTZ AGM, 5/74 & 5/75. (109) SWAPO Statement to (189). (110) Statement by Ingolf Diener, Paris 12/89, quoted in (43). (111) Trade Union Seminar on Namibian Uranium 29-30/6/81, SWAPO of Namibia, London, 1982. (112) Gillian and Suzanne Cronje, *The Workers of Namibia*, London, 1979. (113) Rössing Uranium Ltd, *The Rössing Factbook*, Windhoek, 1985. (114) Statement by Ben Ulenga to NSC, London, 2/89. (115) *FT* 20/5/78. (116) *RTZ Fact Sheet* No.2, London, 1984. (117) *Windhoek Advertiser* 7/1/84. (118) *Namibian* 20/12/85. (119) *Namibian* 3/2/89. (120) *Namibian* 10/2/89. (121) see also *International Newsbriefing on Namibia* (INON) No.67, 3/89. (122) see also *Report of the General Secretary of MUN*, 1988; *Action on Namibia*, Spring 1989. (123) PARTiZANS, *Parting Company* Nos.3&4, Spring 1989; see also *WISE News Communiqué* No.314-2144, 16/6/89. (125) Letter from Rössing workers' representative, Rössing, 12/79, released by CANUC; see *Hidden Face of RTZ*, PARTiZANS Alternative Report, London, 1980. (126) UN Hearings on Namibian Uranium, 10/7/80. (127) *Safety and Fire News*, 3/84. (128) Information from Dr. Alan Dalton of Labour Research Dept., London, to author, 1986. (129) RTZ *Fact Sheet* No.2, London 1981. (130) Victor E. Archer, *"Extent of health effects from Uranium Mining and Milling"* in *Uranium Mining and Radiation Safety*, Proceedings of a Conference at Michigan Technological University, Houghton, Michigan, 19/8/80, sponsored by Operation ACTION U.P. (131) RTZ *Fact Sheet* No.1, 1984. (132) PARTiZANS, *Dirty Bizness*, PARTiZANS Alternative Report, London, 1986. (133) *MJ* 7/5/82. (134)

Statements by RTZ chairs at company AGM's recorded by PARTiZANS, London, from 1983-89. (135) *MM* 4/89; see also various statements by SWAPO spokespeople, recorded in *International Newsbriefing on Namibia*, 1987-89. (136) see, for example, H.E. James, and H.A. Simonsen, *"Ore-processing technology in the assessment of uranium mining"*, in *Uranium Supply and Demand*, Proceedings of the Third International Symposium held by the Uranium Institute, London, July 12-14 1978, Mining Journal Books, in cooperation with the Uranium Institute, London, 1978. (137) Transcript of RTZ AGM, 1980, quoted in (160). (138) *Economist* 21/2/76; see also Lanning and Mueller op cit. (139) *Gua* 14/8/78. (140) Transcript RTZ AGM 1982, PARTiZANS, London. (141) SWAPO Press release, issued 11/3/80. (142) *MJ* 4/11/83. (143) Reuters 7/5/85. (144) see also *Safety Management Britain*, 5/89. (145) Akafrik interview with *Pogrom* No.135, *Gesellschaft fur Bedrohte Volker*, Gottingen, 8/87. (146) Rosalie Bertell, *No Immediate Danger*, Women's Press, London, 1985. (147) *Guardian*, New York, 11-12/80 (148) RTZ *Annual Report* 1984. (149) see also *Ecoforum* Vol.12 No.6, Nairobi, 1987. (150) Sir Alistair Frame, chair of RTZ at the 1987 company AGM, London.
(151) Decree No.1, adopted by the NCN Namibia 27/9/74 and approved by the General Assembly of the United Nations, 13/12/74. (152) *International Herald Tribune* 7/5/85. (153) *Gua* 2/12/75. (154) *Ampo* Vol.12 No.3, Tokyo, 1980. (155) Alun Roberts, *Efforts to assist the Implementation of Decree No.1 in the United Nations*, A/AC.131/GSY/CRP.14, Geneva, 27-31/8/84; United Nation's Council for Namibia [UNCN] Regional Symposium on International Efforts to Implement Decree No.1 for the Protection of the Natural Resources of Namibia. (156) *CANUC News* No.2, London, 1982. (157) *Labour Weekly*, London, 29/1/82. (158) *Anti-Apartheid News*, London, 5/82. (159) *Leeds Other Paper* 18/12/82. (160) *Bare Facts*, Surrey University Students Union, 12/2/82. (161) Cambridge CANUC information, reported in (156). (163) *Oxford Mail* 10/2/82. (164) *WISE News Communiqué* 182-1205, Amsterdam, 31/5/83. (165) *Namibia News Bulletin* No.10, London; *Action on Namibia* No.4, London, 4/84. (166) *Rand Daily Mail* 27/3/84. (167) Transcript of 1984 RTZ AGM,

PARTiZANS, London, 1984. (168) *Eve Standard*, London, 26/4/85; see also *Daily News*, South Africa, 28/4/85. (169) PARTiZANS, *Where There's Brass There's Muck*, London, 1985; see also *Peace News*, London, 13/12/85. (170) other information from PARTiZANS, London 1984-89. (171) *Windhoek Advertiser* 12-19/9/85; *Namibian News Briefing* No.29, London, 9/85. (172) *FT* 2/7/85; see also *Morning Star*, London, 6/5/85. (173) Reuters report 3/5/85. (174) *Windhoek Advertiser* 27/5/85. (175) *MM* 9/88. (176) *Namibian* 4/3/88. (177) *Namibian* 1/4/88. (178) *Tageszeitung* (Taz) 8/6/88. (179) *Lancashire Eve Post* 2/11/84. (180) UN Press Release, 8/3/85; see also *Labour Weekly* 24/5/85; CANUC Press releases 15/5/85 and 14/2/85; Statement by UN Ambassador Noel Sinclair, Acting President of UNCN, sent to the court 3/5/85. (181) *International Newsbiefing on Namibia*, *"Action on Namibia"*, 4/85. (182) *International Newsbriefing on Namibia (INON)* No.42, Namibia Support Committee, London, 12/86. (183) INON, *Action on Namibia* No.56, 3/88. (184) Virginia Pilot and Ledger Starr, 11/3/88. (185) *East Anglian Daily Times* 17/9/88. (186) *Grimsby Target* 3/11/88. (187) *Grimsby Eve Telegraph* 3/11/88 and 28/10/88; see also *Grimsby Eve Telegraph* 10/8/88; *Gua* 1/11/88. (188) *Irish Times* 10/10/88, 24/9/88, 15/10/88. (189) Information from Trade Union Seminar on Namibian Uranium, London, 29-30/6/81. (190) *Lancashire Eve Post* 16/2/82; *Gua* 16/2/82; *Obs* 10/1/82; *Morning Star* 13/1/82. (191) *Namibian* 20/12/85. (192) *FT* 28/12/78; see also *FT* 5/1/79. (193) *Namibian Times* 5/7/80. (194) Information from North West Trade Union Anti-Apartheid Liaison Committee, Manchester, 9/82. (195) Information from Namibia Support Committee. (196) *The Struggle for Trade Union Rights in Namibia*, SWAPO, Dept. of Labour, Luanda, 1984. (197) *Namibian* 24/1/86. (198) *Namibian* 8/8/86. (199) *International Newsbriefing on Namibia*, London, 12/86; see also *MJ* 12/12/86. (200) *RTZ Fact Sheet* on Namibia 12/85. (201) *FT* 7/8/87. (202) Information from Rössing Uranium Ltd, obtained by Namibia Support Committee, London, 5/89. (203) *MJ* 24/6/88; *Southscan* 15/6/88, 22/6/88; *Namibian* 17/3/88; *Gua* 20/6/88; *Independent*, London, 21/6/88; *Windhoek Observer* 18/6/88; *Morning Star* 20/6/88, 21/6/88. (204) Manager of Rössing Uranium Ltd, in an article entitled *"The Rössing Ethos"*, published by Rössing Uranium in 1984. (205) *Windhoek Advertiser* 3/4/89. (206) *Windhoek Advertiser* 2/4/89. (207) *Report of the Bristol Commission on Inquiry into the Southern African Operations of RTZ*, Bristol Anti-Apartheid Movement and Bristol United Nations Association, 11/82. (208) *Windhoek Observer* 28/9/85; *Windhoek Advertiser* 16/9/85. (209) *SunT* 6/7/80. (210) *"Security Scheme"* memorandum, circulated by Rössing Uranium Ltd, 29/11/78; copy in possession of PARTiZANS, London. (211) Letter from Anthony Tuke, RTZ, to Brian Wood, NSC, London, 14/5/82. (212) see also *Gua* 28/5/82. (213) *Nucleonics Week* 21/10/76. (214) *Financial Mail*, Johannesburg, 15/4/77. (215) *Financial Gazette*, 9/11/73. (216) *FT* 20/5/81. (217) *Quarterly Economic Review of Namibia*, Botswana, Lesotho, Swaziland, No.1, 1984. (218) Martin Bailey, *New Statesman*, 12/5/82.

529 RTZ

In the Summer of '89, ex-President Ronald Reagan visited London to deliver the annual Churchill lecture. His theme was democratisation through high technology. "I am convinced that the world is moving our way," he declared. "Breezes of electronic beams blow through the iron curtain as if it were lace ... They shall know the truth and the truth shall set them free" (1). Less remarkable than anything Reagan said, is the fact that he was sponsored by a British corporation, with no apparent interests in communications. Yet the association between the grand old Republican and Rio Tinto Zinc goes back a long way. In the 1960s, the middle-aged actor fronted a highly successful TV series boosting "20 Mule Team" borax (2). Soon afterwards, US Borax was bought out by Rio

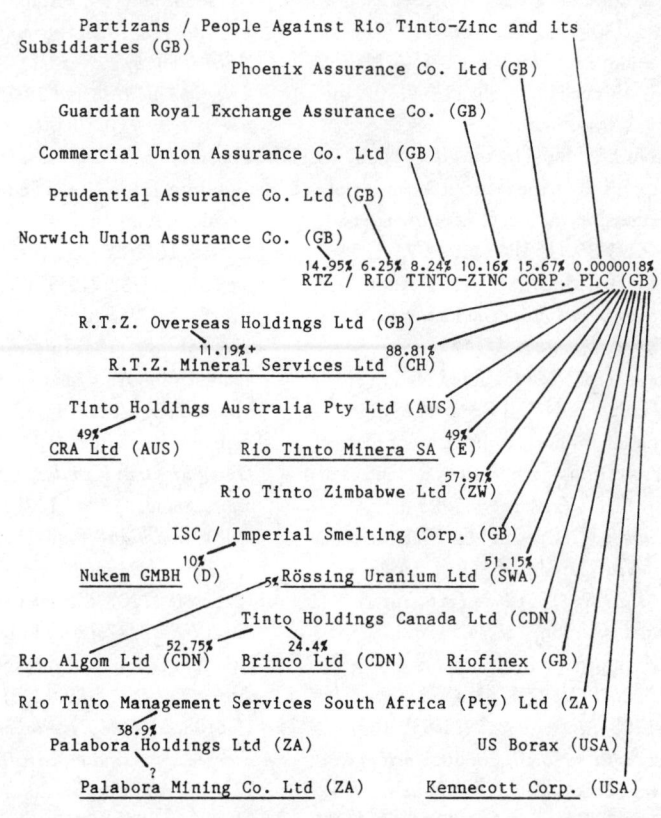

684

Tinto-Zinc, or RTZ as it now prefers to be called.

It was one of the first acquisitions in a long line, which has made RTZ the world's most powerful mining corporation; with joint venture mines in forty countries (3). According to investment analysts Morgan Stanley Capital International, RTZ's market capitalisation stood at US$5,711 million in June 1988, putting it head and shoulders above all the competition. In addition, RTZ's 49%-owned Australian associate, CRA (Conzinc Rio Tinto of Australia) occupied sixth position in the league table, with assets of just under four billion dollars (4).

As if this were not sufficient, the following year RTZ bought up most of BP Minerals' mines and prospects (5). At a stroke the British company acquired Kennecott Copper, with its plum asset the Bingham Canyon copper mine in Utah, as well as several important north American and Asian gold interests. RTZ is now one of the most important gold producers outside of South Africa (5).

RTZ produces around 15% of the world's copper and, according to Metals and Minerals Research Services of London, will soon assume "a quite astonishing degree of importance" – owning copper operations "in all the world's major areas of production.

By 1993 it could be "supplying 44.5% of the world supply of custom copper concentrate" (6).

In 1988 the *Economist* magazine estimated that the company was producing 11% of the world's aluminium, 8% of its iron ore and 11% of its uranium (4). RTZ's wholly-owned subsidiary, US Borax, is by far the world's most important supplier of borates, while the Argyle diamond mine in north-west Australia is the world's largest single producer of gem and industrial diamonds. Thanks to gaining the BP Minerals portfolio, RTZ has taken control of QIT Fer et Titane in Quebec and with it, half of Richards Bay Minerals in Natal, South Africa. Claimed the US *Engineering and Mining Journal* in August 1989: "... the company [now] accounts for 90% of world production of titanium dioxide slag and 40% of titanium dioxide feedstock".

This gives it a pre-eminent position in the supply of mineral sands and the output of strategic minerals and rare earths (7). (Four years ago, CRA also announced the discovery of one of the world's biggest mineral sands deposits, at Horsham, in the Australian state of Victoria (8).)

There is no continent (with the possible exception of Antarctica) where RTZ has neglected to establish its presence – whether it be selling iron ore to the Chinese state steel corporation (9); digging-up uranium at Elliot Lake, Ontario; running gold mines, nickel matte and ferrochrome plants in Zimbabwe; earning rich pickings from its Palabora copper and uranium mine in the South African Transvaal; taking 30% of the equity in Chile's controversial Escondida copper mine; or planning future exploitation of the sea-bed, through its majority-owned Kennecott Deep Sea consortium (10).

As the *Financial Times* modestly put it in 1988: "RTZ is the dominant supplier of minerals to the world's manufacturing industries with an extensive portfolio ranging from aluminium to zinc". The paper went on to note that: "[T]he sheer scale and profitability of the complementary industrial assets are sometimes forgotten" (11).

Certainly, the company's interests in aluminium products – through its Pillar subsidiary in western Europe and Indal in the USA – have consistently maintained a cash-flow to see the company through hard times. However, RTZ's extraordinary forays out of mining – into chemicals, oil and gas, cement, uranium enrichment and other non-mining concerns, which typified the late 1970's and early 1980's – have come to an end. In 1989, the corporation yielded up the last of these rather lumbering interests, by selling RTZ Chemicals and ISC Chemicals to Rhone-Poulenc of France (12). (Not a moment too soon in the eyes of some of its critics: ISC is Britain's second biggest producer of ozone-destroying CFC's. After intense shareholder denunciation of this aspect of RTZ's operations in 1988 and 1989, an RTZ spokesperson admitted that "the company has been made to squirm") (13).

Off-loading the chemicals boosted the company's income by more than half a billion pounds (£568 million) in 1989 – enabling it to reduce its debt-gearing (which had soared to 135% with the BP Minerals purchase) to a more manageable 34%. For the first time RTZ joined that elite of corporations which can boast pre-tax annual profits of more than a billion pounds (around 1.6 billion dollars) (14).

How has RTZ arrived at this unrivalled position: not only the world's biggest mining conglomerate, but the most globally extended and (with the possible exception of Anglo-American of South Africa), the most product-diversified as well?

The answer lies partly in RTZ's origins, and partly in the unique "cover" provided for its investments and operations, by the British state. The original Rio Tinto mining company was already Britain's largest mining outfit by the close of the nineteenth century. (Some sources claim it was the biggest in the world). Rio Tinto was founded in 1873 by Hugh Matheson – "an earnest Scot who devoted his life mostly to the virtuous accumulation of money. In his spare time he endowed churches and threw himself ... into the reclamation of London prostitutes," according to financial journalist Jeremy Stone (15). Matheson floated Rio Tinto, based around the copper deposits of the Iberian mineral belt in Andalusia, with a London share offer of 200,000 shares at £10 each. Even RTZ's official archivist, Dr. David Avery, later acknowledged that this "... appeared to many to be fraudulently optimistic". However, Matheson was – in Stone's words – "a master in the use of red ink and carefully chosen typefaces". Despite saddling the company with onerous bank loans and heavily discounted bond issues, the canny Scot was showing a profit in the first six years. Thousands of women, children and men laboured to construct a port, lay down many miles of costly railway, and open up the vastest mineral workings ever seen. (Several hundred also died in the first ten years from silicosis, pul-

monary disorders and downright starvation – a fact attested to by L Gil Varon's documentary study of the mines between 1877 and 1887, published in Cordoba in 1981) (16).

Assisted by the leading role which it played in the Secretan copper corner of the 1880s ("Rio Tinto was adept at the making and breaking of cartels", comments Stone) (15), the company rode high into the twentieth century. But, with competing mines coming on stream, a certain diversification was necessary. Now chaired by another Scot, Sir Auckland Geddes, Rio Tinto sealed a pact with Sir Earnest Oppenheimer, founder of Anglo-American. Between them they gained control of the Rhokana Corporation, the major exploiter of the northern Rhodesian copper belt. "In brief," declares Stone, "Geddes and Oppenheimer appear to have arranged the development finance of Rhokana in such a way as to dilute the minority shareholders into insignificance". In 1926 Geddes took Rio Tinto into a new cartel along with Asarco, Union Minière, and Kennecott. Called Copper Exporters Incorporated, this band of robber barons withheld supplies as it chose, cornered the world market, and swelled the warchests of its members by an astonishing 84.4% in its first two years (17).

Throughout this period, however, the Rio Tinto mines were rocked by unrest. In 1920, many left-wing miners were sacked after going on strike for fifteen weeks. Discontent, combined with the traditional strength of anarcho-syndicalism in Andalusia, made the Huelva mines a stronghold of the Republicans. A workers' rebellion was crushed in 1934 and many miners were sent to prison. Although the Popular Front came to power in 1936, and Rio Tinto was forced to re-employ the jailed men, their's was a short-lived victory. Fascist troops invaded the province and put down the Republican units with unexampled ferocity (18). Sir Auckland Geddes was far from displeased. He told the company's shareholders at its 1937 annual general meeting in London: "Since the mining region was occupied by General Franco's forces, there have been no further labour problems ...

Miners found guilty of troublemaking are court-martialled and shot" (19).

Geddes' optimism did not last for long. Rio Tinto was ordered by Franco to provide ore for the Nazi re-armament programme, the second world war brought prices to a new low-ebb and, by the late forties, the pressure to "Hispanise" was growing too great. In 1954, the company brought in business whizz-kid, Sir Val Duncan, to sell off two-thirds of the Rio Tinto mine and later the Rhodesian investments (18, 20).

Duncan graduated into Rio Tinto from training as a lawyer, a period with the Foreign Office, and two years selling coal for the National Coal Board. He was well-suited to the task of re-deploying RTZ's investments to the Commonwealth: to politically more secure, but relatively untapped, new resources. Above all, he was soon to prove extremely adept in unlocking the finance required for what were to become some of the world's most formidable mines. Duncan charted regions, rather than minerals. But there was one exception that was to prove a masterstroke. Sometime in the early 1950s (as Duncan later recalled) the head of Rio Tinto was called to the British Atomic Energy Commission. He was (so he claims) ordered "to go forth, find uranium and save civilisation!" (21). And this he did, with a zeal and application not seen before or since. First, Rio Tinto bought out Algom Uranium Mines in northern Ontario, from Samuel Hirschorn (22) (who later used the proceeds to set up the eponymous Museum in Washington DC). Then, Duncan gambled on a prospect in South Africa which had been turned down by several domestic companies – the Palabora copper and uranium lode in the Transvaal. By the 1970s the gamble had paid off spectacularly, with Palabora providing 55% of South Africa's internal copper needs (18, 23), and being the only uranium producer in the apartheid state to market its output independently of Nufcor, the state agency (24).

In Australia, Mary Kathleen Uranium (MKU) fell into the company's lap: for most of the next two decades this was to be the country's only provider of yellowcake (some of which ended up in both British and US nuclear warheads). It was from "down under", too, that Duncan was able to pull off what may fairly be regarded as one of the most spectacular coups in western mining history. Consolidated Zinc was a British company with a smelting complex in Britain, and large-scale holdings of iron ore, bauxite and zinc in Australia. Both Conzinc and Rio Tinto had cash to invest. But, while Conzinc had relatively more money than expertise or a spread of investment opportunity, Rio Tinto, by the early 1960s had rather more mines than the capital to develop them: it was also attracted to the prospect of moving "downstream" into processing. The fit was almost perfect. As the *Engineering and Mining Journal* commented in August 1989: "This merger [in 1962] formed the basis of, and provided the impetus for, RTZ's subsequent development into an international mining giant. Throughout the 1960s, capital intensive, technologically innovative mega-projects were being developed with RTZ and its subsidiaries at the forefront" (7).

But, true so far as it goes, this comment glosses over the very real geo-political problems confronted by the new Rio Tinto-Zinc, even in countries considered stable a decade before. The growth of independence movements in Africa and Asia, a mounting world revulsion at doing business under apartheid, the trend towards nationalisation in Britain under a Labour government, the advance of international labour organising, the increasing substitution of newer metals for old – all these would have given cause for company concern. The threat of substitution was perhaps the easiest to deal with. The 1962 merger made RTZ a major aluminium producer, and this metal has tended to keep strong when copper weakened. Coal has picked up, even while uranium has slumped.

The other potential obstacles on RTZ's upward path have required more radical strategies. These have ranged from brow-beating opponents, leaning on governments and price-fixing, to violating international law, union-busting and management of one of the world's biggest commodity cartels in recent times: the uranium "club".

None of these ploys could have succeeded, had they not been conceived at the highest levels of RTZ management and gained the assent (sometimes the active complicity) of successive British governments. Nor, of course, is licensed ruthlessness sufficient in itself. Geologists on the ground, and lower-level managers in outlying areas, must be sure that – when they are encroaching on an Aboriginal sacred site, for example, or entering discussions with a dubious potential partner – they will be supported right through the chain of command. When RTZ's head of public relations tells the author of this book: "Of course, money-making is RTZ's bottom line!" (25), it is a sentiment that must be conveyed all the way from the company's head offices in plush St. James's Square, to Larry Mercando, manager of Flambeau mining in northern Wisconsin, or Mike Quigley, head of Canning Resources in the Western Desert of Australia.

To do this, RTZ has developed what the *Engineering and Mining Journal* calls: "a very flat [organisation structure] with short, direct lines of communication. RTZ does not believe in intermediate holding companies or geographical structures" (7). Adds Derek Birkin, the company's Chair: "We like to encourage local loyalties, and the companies are managed, as far as possible, by nationals of the country concerned. Most do not even have RTZ in their name. This is a difficult form of management, but the rewards are marvellous". (What Birkin does not mention is that several subsidiary and associate companies have been given names which specifically disguise their ultimate ownership by RTZ. One example is Flambeau Mining, whose lightning change of appellature – from Kennecott Minerals – occurred within weeks of Kennecott's takeover by RTZ. Commented the Wisconsin Resources Protection Council: "RTZ and Kennecott want to avoid costly lawsuits arising from the groundwater and surface water contamination that is bound to occur ...") (26).

RTZ's supposed devolution is, in crucial respects, a pretence. All the company's core mining assets are managed to suit a central design and "London" is soon ready to intervene, when these assets appear threatened. Thus, RTZ abandoned its Cornish tin mines, throwing hundreds of miners and their dependents onto welfare, soon after the collapse of the International Tin Council in 1985 – partly because cheaper tin was available from Rio Algom's East Kemptville mine in Nova Scotia. And, even though CRA was eventually "Australianised" in 1986, with RTZ holding on to 49% of the share, that same year "when CRA had troubles with industrial action" (as the British *Financial Weekly* put it), "there is no question that the men from London were very instrumental in settling the dispute" (27).

The main "man from London" for the past decade and more was Sir Alistair Frame, although he surrendered his position as Chair in 1991. Yet another "phlegmatic Scot" (as the *Times* described him in 1983) (28), Frame joined RTZ in the 1960s from the Atomic Energy Agency, where he was a director. His skills as an engineer, and his highly conservative views on trade unions, South Africa and nuclear power, may not seem ideal qualifications for leading RTZ through the 'eighties. But they suited the board when it chose to embark on a whirlwind of industrial acquisitions in the 1970s, and when the company increasingly faced flak over its deepening commitment to apartheid. Although little known outside of mining circles, Frame's advice and analysis has been covetted by (among others) Prime Minister Thatcher. She wanted him to head the National Coal Board and emasculate the British mineworkers' union – but he declined, and Ian McGregor of Amax got the job instead (29).

Frame typifies that mix of shrewdness, effrontery, gambler's instinct, and cultivated amorality, which has distinguished RTZ luminaries for a hundred years. It is significant that, though the board drafted-in leading banker Sir Anthony Tuke, from Barclay's International in 1981, as a boost to its fortunes, he was clearly not par for the course. Two events precipitated his resignation as chair in 1985. The first occurred in 1981, when RTZ was manoeuvring to take over a company called Thomas Ward. Ward was a traditional family firm, with a fair reputation among its workforce, and just the kind of company which Tuke (a born Quaker and supporter of regional enterprises) in other circumstances would zealously defend. He pointedly stayed away from the press conference held by RTZ to announce its bid (30). The second event was Tuke's incompetent handling of dissident shareholders at the 1982 annual general meeting, when police were called (for the first time ever at a British public company meeting) to eject an Aboriginal delegate and thirty supporters. (Commented Hamish McRae of the *Guardian*, the morning afterwards: "Sir Anthony should never forget that his duty, as an employee of the shareholders ... demands that he sit pat, answering questions until the sun comes down, if necessary ... What happened yesterday at RTZ struck at a form of institution that is central to our way of life") (31).

As part of the Ward takeover, Derek Birkin moved into St. James's Square. He developed a close working relationship with Frame and was propelled with alacrity into key positions within the company. By the time BP Minerals came on the market in 1988 (largely as a result of a cash-poor British Petroleum needing to buy back an alarmingly large share of its equity acquired by the Kuwaiti Investment Office) (32), Birkin and Frame were unassailable. Between them, the waspish Scot and the avuncular Englishman brought off the largest unopposed, intra-company, deal then effected in British corporate history.

The take-over cost RTZ nearly four billion dollars (US$3,700 million) and was financed by the Group's existing cash, a syndicated bank loan and a rights issue underwritten fully by Kleinwort Benson of London, the merchant bank (33).

RTZ's new projects are usually financed by cash-flow (aided considerably by a favourable exchange rate in countries like South Africa), rights and share issues, and forward selling of production. (This is the manner in which it has financed its part of the Escondida mine in Chile, the huge capital costs of the Neves Corvo copper/tin mine, and expansion at Hamersley Iron). In its first decade the company needed to raise massive amounts of capital, at highly favourable rates. The company's first managing director, Sir Mark Turner, was a head of Kleinwort Benson: he employed an "old boys'" network which included Rothschilds, Morgan Stanley, and Model Roland and Company of the USA. "With its amazing attraction for American capital," commented the London *Times* in 1968, "RTZ has yet to drop any projects for lack of funds ... this stems from its early experience in Canada where it brought uranium mines to the production stage in two years and fought tooth and nail for finance. Its success astounded the Americans".

The link-up with Consolidated Zinc in Australia also brought RTZ into alliance with Conzinc's major overseas partners, the Kaiser family and, through them, First Boston bank.
It was the Bank of America, heading two syndicates of twenty-seven British, European and Canadian banks, which underwrote the development of the Bougainville copper/gold mine in Papua New Guinea in the late 'sixties (34). After only six months of production, Bougainville was raking-in no less than A$1 million a day (see CRA).
Such grotesque profits do not simply derive from generous project financing, or a knack at finding high-grade ore. (Certainly RTZ Group geologists tend to be found in more outlandish places than those from any other mining company, and they are more persistent than most. But their rate of discovery seems no higher than average, and many deposits are low in minerals or difficult to access. RTZ's genius lies in its ability to mine and market a range of minerals from one lode, and its sheer persistence in bringing problematic mines, like Rössing of Namibia, onto stream). RTZ's financial killings also depend on a low price of labour, negligible

environmental costs, generous tax allowances, and leases which are literally dirt cheap.

The Palabora mine in South Africa relied for its development stages on a large, migrant, black labour force, whose average wage for many years was well below the minimum set by the South African Institute of Race Relations. (The SAIRR pitched it at £44.1 a month for a family of five, when the company was paying only £33.9). Between 1966 and 1971, RTZ paid its African miners just under £5 million: its profits for the same period were nearly £140 million (35). Not surprisingly, the *Times* commented in 1968 that "... at around £100 a ton, Palabora's costs per ton of copper produced makes it one of, if not the, lowest cost copper mines in the world" (35). It was not until 1985 that Palabora's management recognised the National Union of Mineworkers of South Africa – three years after Anglo-American did the same thing.

Similarly, the Rössing mine – which by 1980 had become the biggest uranium project in the world – was constructed by hundreds of Ovambo labourers, separated from their families and housed in what the *Economist* called "appalling temporary camps" (36). Even when a black township was constructed, conditions hardly improved: South African researchers Gillian and Suzanne Cronje, as late as 1979 (six years after mine construction started), found them "akin to slavery" (37). Under the spotlight of international pressure, RTZ cleaned-up its act over the following decade, but the surgery was still largely cosmetic. In a speech delivered in early 1989, the General Secretary of the Mineworkers Union of Namibia (MUN), Ben Ulenga, claimed that: "92% of all black and 51% of all coloured workers still remain in the company's lowest income bracket [which does not] constitute a living wage ... black workers in the Exploration department have no house, no housing allowance. Their conditions in crowded army-style tents are, in fact, among the worst in the mining industry" (38).

The Rössing mine, at one point in the early 1980s, was RTZ's biggest money-spinner. For most of its life so far it has paid no taxes – and only then into a fund controlled by the South African régime. Likewise, the Bougainville copper mine enjoyed a five year tax-holiday (until the newly independent government of Michael Somare forced a re-negotiation in 1974). The Weipa bauxite mine in north Queensland, Australia (the world's largest of its kind) has functioned under an Act of Parliament which imposed a derisory royalty of 5 cents a ton – probably the lowest in the world: the rate was merely doubled in 1965, and not hiked-up to a more equitable figure until 1974. The initial "rent" for the mining lease was pitched at a throwaway £2 per square mile, only a one-hundred-and-sixtieth (*sic*) of the normal mining rental at the time. The Aboriginal people, whose land was seized, have never received one dollar in compensation.

Weipa's bauxite is converted into alumina in Australia and smelted at the Tiwai Point smelter in New Zealand (Aotearoa). Although RTZ and its partner, Kaiser, agreed in 1960 to build a power station to run the smelter, within two years they were pleading poverty. The New Zealand government, considering itself starved of industrial development, not only constructed the power station at the taxpayers' expense, but sold electricity to the two companies at a thirteenth of the rate charged to New Zealand private citizens, and one twentieth of that charged to other industries and farmers.

All these examples illustrate an extraordinary capacity on the part of RTZ and its partners or subsidiaries, to influence domestic power-brokers in the regions where it operates. When Comalco (the joint RTZ and Kaiser company formed to exploit Weipa) was publicly floated in 1970, for example, only 19% of the shares were offered to the Australian public. However, a significant proportion had been secretly offered the day before to friends of RTZ in high office in Queensland: these included the state Treasurer, the Ministers of Aboriginal Affairs, Industrial Development, Local Government and Electricity, and the Premier of Western Australia. The acting-Premier of Queensland, Gordon Chalk, distributed his ill-gotten gains to the entire Chalk family, leading one journalist to comment that: "only the dog" hadn't benefitted! Thanks to the hype surrounding the share-issue, these politicians saw their investment more than double in the space of a day. RTZ and Kaiser cleaned-up ten times richer.

Again, when New Zealand Prime Minister Muldoon, in 1977, sought the assistance of US Secretary of State for Energy, Schlesinger, in re-negotiating the power price paid by Comalco at Tiwai Point, he found himself out-classed and out-gunned. Cornell Maier of Kaiser "briefed" Schlesinger in Washington, while Lord Shackleton of RTZ intervened with the British government. The upshot was that, though Comalco agreed to pay around four times its previous rate, this was still only 60 per cent of what it had feared. (The Tiwai Point pricing agreement remained a secret one until late 1990, despite continuing demands by New Zealand citizens in recent years that it be divulged).

A bare two years later, another American company within the RTZ camp was exerting even greater pressure on public figures. US Borax had located a massive molybdenum deposit at its Quartz Hill property in Alaska. To the company, Quartz Hill promised a "valuable diversification into a market where prices have continued to rise even during recession" (39). Unfortunately, for RTZ, the mineral lay solidly within part of the 56-million acres recently designated by President Carter as inviolable, under the Alaska National Interest Land Act. Despite opposition from the Sierra Club, Representative Morris Udall and Carter himself, RTZ/Borax mounted a well-funded, highly-geared, lobby of Congress, pleading "national interest" (40). By December 1980, the company's ministrations had prevailed: the National Interest Conservation Act (*sic*) was duly passed, giving RTZ access to Quartz Hill.

Remarkably, even while Congress was waving RTZ freely into a national park, a Congressional investigation had named the company, and several subsidiaries, as prime movers in the world uranium cartel, whose contract and price-fixing in the early 1970s boosted the mark-up for "free world" supplies of yellowcake by a factor of five. (One calculated result of this stratagem was that the Rössing mine, whose low-grade refactory ore had previously been considered uneconomic to mine, was now commercially viable).

The company's master-minding of what was cosily and euphemistically dubbed the "Uranium Producers' Club" by its members, consolidated and extended RTZ's influence over key government personnel in South Africa, Canada and Australia. RTZ sat as an equal among representatives of sovereign states – leading one commentator, Stephen Ritterbush, at the UN Hearings on Namibian Uranium in 1980, to declare: "Rio Tinto Zinc [is] in the position of acting in many respects as a uranium producing and exporting nation" (41).

But its international "spread" – impressive though it is (RTZ, for example, is the only mining company to have had three representatives sitting on the "world management" Trilateral Commission) (42) – is overshadowed by the parastatal role it plays in the British political structure. "We are very politically minded," Val Duncan once commented: "not party politically minded, but on an international basis" (43). Duncan was determined to bring into the company, representatives of all three major British parties; and he succeeded. During Mark Turner's ascendancy at RTZ, the board could boast among its members: the Queen's private secretary (the monarch has regularly been cited as a major shareholder), Lord Byers of the Liberal party, Lord Shackleton (Labour's leader in the House of Lords) and the Tory, Lord Carrington. It was Carrington who forged the settlement of the Smith rebellion in Rhodesia, which brought Mugabe to power. Despite the new President's determination to curb the mining companies, Riozim – which had functioned during the entire period of UN and British sanctions – remained in Zimbabwe as one of the three most powerful multinationals.

The London *Times* once called it "almost patriotic" to own shares in RTZ (43), while the *Daily Telegraph* mused that, as well as supplying raw materials, the company could be in a "position to furnish a coalition government, if one were needed" (20). (This wry comment struck closer to the truth than the conservative paper might have been aware at the time. In 1975, as Britain was wracked with industrial unrest, Duncan called together Lord Robens of the National Coal Board, some army officers, and a few journalists, to plan a take-over of the national power grid, generating plant, and key media, should a revolution break out in England's green and pleasant land) (44).

In 1989, Prime Minister Thatcher visited the Rössing mine in Namibia and gave it her seal of approval: "It makes me proud to be British," she opined. This was hardly surprising, considering the protection which had been afforded RTZ to plunder Namibian uranium, by several British administrations, especially her own.

First, officials sympathetic to RTZ within the British Ministry of Technology signed a secret contract enabling Rössing to supply the country's uranium needs, even though the cabinet was told they would be met from Canada. The Minister of Technology at this time, Anthony Wedgwood Benn, later declared he had been completely "kept in the dark" about the "switch". Once Britain was locked into the illegal source, both Labour and Conservative governments defied a United Nations Decree on Namibia's Natural Resources, and numerous UN and other resolutions, in order to maintain the supplies. Her Majesty's government promulgated a Protection of Trading Interests Act, which not only prohibited British corporations from supplying documentation to a potentially hostile foreign interest (the Chicago court investigating the uranium cartel), but entitled British courts to seize the assets of any overseas power foolish enough to impound British corporate assets.

None of this evidence should suggest that the interests of RTZ and the British state are always compatible, let alone identical. When the company snapped-up nearly 15% of Enterprise Oil in 1984 – a newly-floated oil company intended to be a flagship for Thatcher's brand of "privatisation" – the government was infuriated (45). Yet, only two years later, the administration provided RTZ with £15 million in interest-free loans, and another ten million pounds at commercial rates, to bale out its Cornish tin mines – a move which made nonsense of Thatcher's avowed determination not to support industrial "lame ducks" (46).

The fair conclusion to make is that the British state needs RTZ more than RTZ needs to be subservient to changing governments. The company is not merely a "state within a state". For those who cherish the notion of a continuing British imperial domain, RTZ *is* the state.

If the Church of England is the Conservative party at prayer, then RTZ may fairly be described as the British gentry up the Klondike.

Resistance

RTZ's chair, Sir Anthony Tuke, in his 1984 Annual Report, made a rare admission: "We are conscious that some people hold deep and strong beliefs that our activities are in themselves damaging," he wrote (47). "Some people" have come to include the United Nations Commission on Namibia, President Mugabe of Zimbabwe in his first months of office, the Greater London Council, the National Federation of Aboriginal Land Councils, the Guaymi Congress of Panama, and a whole battery of community groups stretching from northern Ireland to New Zealand.

Many of them regard RTZ as impervious to all criticism, and the most intransigent of mining companies when it comes to modifying or abandoning new projects. However, this is a rather simplistic view which underestimates both the power of opponents, and the capacity of the company to manoeuvre around its core interests.

For example, RTZ was forced to put on-hold its huge Cerro Colorado copper project in Panama during the mid-1980s (48). International support for the Guaymi Congress in its struggle to gain land-rights (*comarca*), combined with opposition from Panamanian Bishops and CEAS-PA, the country's leading radical think-tank on development, proved too much for the top brass at St. James's Square. "We have had more opposition to this project than anything else we've done," confessed RTZ chair, Sir Anthony Tuke, to a private meeting with members of Survival International: "It is really astonishing". Nonetheless, RTZ continued to hold 49% of the mine's equity, pay the salaries of workers from the state mining company, Codemin, and keep a substantial contingent of its own staff in and around the "copper mountain" (48).

Similarly, RTZ is no slouch when it comes to fighting-off the demands of trade unionists in areas of marginal profitability but major, long-term, investment: at its Hamersley iron ore operations in north-western Australia, for example, which is probably the most strike-bound of all Australian mining ventures over the past two decades. In contrast, its Canadian subsidiary, Rio Algom (along with Denison Mines) in 1981 negotiated a ground-breaking agreement with the United Steelworkers of America (USWA), empowering the union to appoint work safety inspectors at its Elliot Lake uranium mines. This followed the shattering revelations of the Ham Commission, that Elliot Lake miners were labouring regularly under radioactive exposures seven times what should be considered the maximum level. Even so, it took another three years of Union agitation before Rio Algom workers gained the right to refuse "unsafe work", and stand on the same footing as other Ontario miners.

Nor has this precedent been followed in other RTZ mines. Contacts have been forged between the USWA in Canada and the Mineworkers Union of Namibia, however, and it is likely that the new SWAPO-led government will insist on similar protection in the near future.

Undoubtedly it is indigenous land claims which have been the bane of RTZ's expansionist policies for two and a half decades. There are several major mines within the company's domain which have displaced native people or risked their lives. Constructing the north Queensland bauxite strip-mine at Weipa caused the forced removal of two entire Aboriginal communities in the early 1960s (50). Vast areas of Nasioi and Rorovana farmland and rain-forest have been ravaged by the Bougainville copper mine in Papua New Guinea (18, 51). And the Elliot Lake uranium mines, while not situated on Indian territory, leach poisonous heavy metals and acids into lakes and rivers essential to the livelihood of the Serpent River Band (see Rio Algom).

The last decade has seen the encroachment and destruction proceeding apace. A sacred women's Dreaming site was levelled to uncover the lucrative kimberlite diamond pipe at Lake Argyle, Weatern Australia (49). Test-drilling for uranium has been carried out near life-giving water sources on Martu land in the desert to the south. CRA has constructed one of Asia's major new coal mines upstream of Dayak settlements in East Kalimantan, while both CRA and RTZ have been accused of engineering the removal of indigenous miners and their families, further inland (53). One of the Group's most important future mines – a wet-dredge mineral sands project in south-eastern Madagascar – looks likely to profoundly affect the coastal areas used by the Antanosy, one of the island's largest tribal communities (54). Hardly before the curtain was raised on the 1990s, CRA announced a 30% stake, later raised to 40%, in a gold mine on Igorot land, high in the Filipino Cordillera (55). Its partner here is Lepanto Mining – a domestic company which has banned organising by the broad-based National Federation of Labor Unions (NAFLU) and uses its own private army to attempt to silence dissidents (56).

RTZ's Rössing uranium mine in Namibia enjoys the unenviable reputation of being the most condemned mining project of the twentieth century. No other mine has been the subject of United Nations resolutions, a UN-sponsored court case, and scores of demonstrations throughout Western Europe. Nonetheless, as already pointed out, the company's exploits in

Namibia have gained a seal of approval from successive British governments. Given the degree to which the country's economic fortunes are dependent on mining, RTZ gambled that the longer it hung on in the territory, the more likely that it would be invited to remain after independence.

No such promises can be held out to indigenous people in regions where mining is seen as an unmitigated disaster. Indeed, it is remarkable that no indigenous community, anywhere affected by the Group's operations, has ever commented that life was improved by the company's presence. Ironically, it is in those areas where RTZ has pumped its costliest public relations and "good neighbour" efforts that local people are most ready to bite the hand that feeds them: Bougainville, the Australian Kimberleys, Wisconsin and New Zealand, for example.

RTZ has therefore had to expend a disproportionate amount of time and energy in counteracting the growing strengths of indigenous people. Certainly it has not flinched from violent threats. Not only did an RTZ board member once threaten to "squash Survival International like a fly, if we didn't get off their backs" according to Survival International President, Robin Hanbury-Tenison (57). But, an Aboriginal activist who helped expose CRA's links with Anglo-De Beers of South Africa over the marketing of Lake Argyle diamonds once swore, he was nearly run over in a Melbourne street by a company car (58).

However, these are exceptions to what has proved to be the general rule about RTZ Group dealings with land-based peoples: keeping the lowest profile possible, and opening-up as many fissures as it can within indigenous organisations themselves. It is a technique which was born in Cape York (where a company-sponsored Council has often been at odds with other Aboriginal people), honed at Lake Argyle (where a small family group was separated from the rest of the community and inveigled into signing away its land rights) and most recently attempted in Australia's Western Desert. As one close observer of events there summarised the

company's ploys in 1989: "CRA early on established that the majority of the Western Desert people – the Martu – were adamantly opposed to uranium mining. Instead of sitting down with the Western Desert Land Council and discussing their fears, the company sought to undermine them, by signing deals with another community whose claims on the area were weaker, but whose organisation was stronger. This group also operated their own mining projects in the past and were therefore considered softer targets by the company" (59). CRA failed to get the go-ahead to open up the mine. Early in 1990, Australian Prime Minister, Bob Hawke, banned the project just before scraping home in the March Federal elections with "green support" (60). Later, the state government waved CRA back into Rudall River.

Paradoxically, a few years earlier, Comalco had developed a model for negotiations with Aboriginal people which some anthropologists view as the best yet. The "Pitjantjatjara model" enables Aboriginal communities to define areas which are sacred to them, without the need to divulge their exact location and risk their inviolability. However, the paradox is more apparent than real. The Pitjantjatjara people were in the process of gaining land rights with some real teeth when Comalco became interested in their land: the company needed to project itself as more liberal and beneficent than other contenders for the mineral deposits (61). RTZ and CRA's real view of Aboriginal aspirations is best glimpsed from a combative statement made by CRA chair, Sir Roderick Carnegie, at the 1984 annual general meeting of RTZ: "The right to land depends on the ability to defend it" (49).

Conflicts between the RTZ Group and indigenous peoples are bound to continue well into the twenty-first century, although the field of combat is likely to shift, as the company expands in the mainland USA and forages further into Asia. Thanks to the 1989 takeover of BP Minerals, just over half RTZ's investments are now to be found in America, although a mere 3% of its shares are held by North Americans (7). As the *Mining Journal* put it in March

1990: "... the group is aiming for a share listing on the New York Stock Exchange in June this year. This is expected to attract the institutional investors. The medium term target is for around 10% of RTZ's equity to be held in North America" (62). By summer 1990, the company had obtained its listing.

At the same time, the company's Chair, Derek Birkin, is determined not to let any opportunity slip in Asia: "The Pacific Rim is a region of growing importance," he said in 1990. "By the turn of the century ... there will be a southerly shift in world economic power, away from the old axis of Northern Europe and North America, while between 20% and 25% of current metals demand now comes from developing countries, especially those in South East Asia". Birkin forsees these nations "... performing the same function for overall growth rate, as Japan did in the 1960s" (63).

A few years ago, RTZ might justifiably have regarded these two regions as malleable propositions: North America as an area starved of new investment, where the power of the mining unions has been broken; South-East Asia as an eldorado of untapped reserves where pliable governments (such as Indonesia, the Philippines and even Vietnam) are content to leave the running to heavyweight outsiders.

Certainly it has made headway here which is the envy of many other miners: becoming the first multinational to conclude a mining agreement with the Beijing regime (63), manager of the biggest new coal mine in Indonesia, Kalimantan (64), and the company which broke a prolonged stalemate on iron ore prices to Japanese smelters (65).

But two developments shook its confidence at the turn of the decade. The first, and most important, was the growth of a guerilla army on the island of Bougainville which, over the space of a few months, transformed itself from an apparent rag-taggle band of discontents into a potent fighting force. Led by Francis Ona, an ex-employee of the huge Bougainville Copper mine which RTZ established in the 1970s while Papua New Guinea was still an Australian pro-

tectorate, the Bougainville Revolutionary Army (BRA) campaigned for compensation to be given to traditional land-owners, dispossessed by the company's operations and suffering the effects of massive pollution. As Perpetua Serero, leader of the island's matrilineal landowners, told a visiting reporter in 1988: "We don't grow healthy crops any more, our traditional customs and values have been disrupted and we have become mere spectators as our earth is being dug up, taken away and sold for millions. Our land was taken away from us by force: we were blind then, but we have finally grown to understand what's going on" (66).

Regular attacks by the BRA against mine installations, mineworkers, visiting politicians and expatriates, forced the mine to close in May 1989 (67). The Papua New Guinea government of Robbie Namaliu was incensed: Bougainville copper generated up to 20% of the country's internal revenue, and was its biggest export earner. CRA, operator of Bougainville Copper, was only a little less disturbed. For – while defaulted copper contracts could be fulfilled from other RTZ mines – business was more likely to go to competing companies, such as Freeport in West Papua. Other CRA projects in the country were also put in jeopardy.

Finally, in the closing days of 1989, as we now know, CRA packed up, told its Australian personnel to leave the island, and put the mine in mothballs. Only two months later, thanks to a political shift towards a non-military solution by the central government, a ceasefire was signed with the BRA. Sam Kauona, military chief of the BRA, urged his forces not to damage the abandoned mine, in case it could be "re-opened in an independent Bougainville" (67). The prospects of that happening are admittedly slim. The term "to bougainville" is now common currency among indigenous people in the Pacific region: it is synonymous with the increasingly militant stance being adopted by traditional landowners in the face of multinational mining and natural resource corporations, and what they see as the collusion of their central governments (68). (For example, major blockades were mounted by Highlanders in Papua

New Guinea in 1989, to gain a renegotiation of the Ok Tedi mining agreement (68, 69) (see BHP).

Nor have the ripples from Bougainville ended on the Pacific rim. In January 1990, at a "shop window" called by RTZ to sell itself to institutional investors in New York, the company got a rough ride. Said an investment advisor: "They were expecting to talk gold and ended up talking rebellion" (70).

And it is from North America that the latest challenge to RTZ's well-laid plans has erupted. Kennecott Copper (through its subsidiary Flambeau Mining) tried to mine the lucrative Flambeau deposit in northern Wisconsin in the 1970s. Spurred by dairy farmers, the local government turned down zoning permits for the mine. Kennecott and Exxon then turned their attention to the state authorities and, between them, they developed a "consensus decision-making process" which, according to Al Gedicks of the Center for Alternative Mining Development Policy, has now opened the door to "resource colonialism" in this part of the USA.

Says Gedicks: "The approach involved an alliance between mining corporations and the state to neutralise potential opposition. The mining industry's enthusiasm for the consensus approach is hardly surprising – only those groups already in favor of mining are allowed to participate" (71).

When RTZ took over BP Minerals in 1989 and, with it, complete control of Kennecott, the Flambeau mine swiftly came off the back burner. Kennecott and Exxon, according to Gedicks, "determined the precise wording of key pieces of legisation, while Exxon lobbyist James Klauser helped reframe the state's groundwater protection bill "to a policy of allowing the mining companies to pollute the groundwater to federal drinking standards". Asked if these standards are stringent, Gedicks responds curtly: "Hogwash!".

The farmers around Flambeau agree with Gedicks, and so do the Lac Courtes Oreilles band of Chippewa who, in an unprecedented

move in 1990, joined protests against the mine as an official party.

For the first time in years, a coalition of farmers, environmentalists, Treaty rights supporters and indigenous people has mobilised to defeat a mining proposal in mainland USA.

It is a potent mixture that has derailed RTZ in the past, and is bound to do so again. As people around the world increasingly see this planet for the fragile, finite, but infinitely various and marvellously sustainable homeland it undoubtedly is, such coalitions are bound to grow.

Contact: PARTiZANS, 218 Liverpool Rd, London N1 1LE, UK.

References: (1) *RTZ Review* No. 11, London, 10/89; see also *Parting Company*, PARTiZANS, London, Winter 1989. (2) N J Travis and E J Cocks, *The Tincal Trail, A History of Borax*, Harrap, London, 1984. (3) Gua 23/3/90. (4) *Economist* 23/7/88. (5) *MJ* 26/5/89; see also *FT* 15/12/88 and (33). (6) *MJ* 13/10/89; see also *South* magazine, London, 9/89. (7) *E&MJ* 8/89. (8) *MM* 3/88; see also *MJ* 9/6/89 and *MJ* 4/5/90. (9) *E&MJ* 6/89. (10) *FT* 21/3/86. (11) *FT* 26/4/88. (12) *MJ* 3/11/89. (13) *Parting Company*, PARTiZANS, London, 9/89. (14) *FT* 23/3/90. (15) *FT* 25/1/82. (16) L Gil Varon, *Migration and Development: A Migration Model for Rio Tinto (1877-1887)*, Libreria Andaluza, Cordoba, 1981. (17) Lanning & Mueller *op cit.* (18) Richard West, *River of Tears*, Earth Island Press, London, 1972. (19) Statement by Auckland Geddes to Rio Tinto 1937 AGM, London, quoted in (18). (20) Chris Whitehouse, *RTZ: A Potted History*, in (52). (21) *The Nation*, Washington, 18/10/90. (22) *EPRI Report*. (23) see also *FT* 22/6/70. (24) Neff. (25) Personal communication, London, 1987. (26) Wisconsin Resources Protection Council, Press statement, Tomahawk, undated (late 1989). (27) *Financial Weekly*, London. (28) Times, London, 9/12/83. (29) *Economist*, London, 26/2/83: see also *FT* 10/3/83; *FT* 18/4/83 and *SunT* 12/6/83. (30) *SunT* 29/11/81; see also *SunT* 6/12/81. (31) Gua 4/6/82; see also *Times*, FT, *Daily Telegraph, Daily Mail, Daily Mirror*, 4/6/82; *Financial Weekly, Leveller* 11/6/82. (32) Roger Moody, *The BP/RTZ Deal*, PARTiZANS, 4/89. (33) RTZ Corp, *The Acquisi-*

tion of BP Minerals and Rights Issues, London, 1989. (34) *New York Times* 23/7/69. (35) *Times* 17/2/68. (36) *Economist* 8/10/77. (37) Gillian and Suzanne Cronje, *The Workers of Namibia*, London, 1979. (38) *The Namibian*, Windhoek, 10/2/89. (39) RTZ Annual Report 1978, London, 1979. (40) *FT* 27/9/79. (41) *UN Hearings on Namibian Uranium*, New York, 1980. (42) see Sklar; and also Stephen Gill, *American Hegemony and the Trilateral Commission*, Cambridge Studies in International Relations, Cambridge University Press, Cambridge, 1990. (43) *Times* 16/12/68. (44) *The Sunday Correspondent*, London, 25/3/90. (45) Gua 27/6/84; Gua 29/6/84. (46) *FT* 9/8/86. (47) RTZ *Annual Report*, London, 1984. (48) RTZ *Annual Report* 1982, London 1983. (49) *RTZ Uncovered*, PARTiZANS, London, 1985. (50) Roberts *op cit*. (51) *RTZ: The Hidden View*, PARTiZANS, 1987. (52) *Time* (Australia) 28/11/88. (53) *Inside Indonesia* (AUS) 12/89. (54) Personal communications between visitors to the project area and the author 1989-1990. (55) *FT*. (56) *Birds of Prey: Gold and Copper Mining in the Philippines, A political economy of the large scale mining industry*. A report by the Center for Labor Education Assistance and Research (CLEAR) Baguio City and RDC-Kaduami, Baguio City, 1988. (57) Robin Hanbury-Tenison, *Worlds Apart: An Explorer's Life*, Granada, London, 1984. (58) Personal communication, London, 1981 (name of Aboriginal activist withheld). (59) Personal communication with author, Port Hedland, 5/89. (60) *West Australian*, Perth, 20/3/90; see also Media Statement from Premier of Western Australia, P90/72, Perth, 6/3/90. (61) Phillip Toyne and Daniel Vachon, *Growing Up The Country*, McPhee Gribble/Penguin Books, Fitzroy, 1984. (62) *MJ* 30/3/90. (63) *MJ* 27/11/87. (64) CRA *Annual Report* 1989, Melbourne, 1990; see also *FT* 21/3/90, *FT* 16/7/81, *MJ* 1/9/89, *MJ* 22/9/89. (65) *MJ* 19/5/81. (66) *The Australian* 29/12/89, *MJ* 5/1/90; see also *FT* 1/12/89. (67) *Pacific News Bulletin* Vol. 5 No. 3, Sydney 3/90; see also AFR 2/3/90, *Sydney Morning Herald* 22/2/90 and *MJ* 4/5/90. (68) Alistair McIntosh, *Land Rights and Ecocide in the Bougainville Crisis*, (duplicated), Edinburgh, 1989. (69) see also *Australian* 13/10/89, *Direct Action* (Broadway, NSW) 14/11/89 and *The Bulletin* (Australia) 13/2/90. (70) Personal communication, London 1990. (71) *Shepherd Express*, Milwaukee, Vol.9, No. 51, 21-28/12/89.

530 RTZ Mineral Services Ltd

Known also as Minserve, with a registered office in London, RTZ Mineral Services was actually set up in 1970, with its official HQ in a one-room office in the small town of Zug, Switzerland (1). Ostensibly, Minserve provides "... services related to the extraction and sales of natural and refined minerals [and] transit shipping" (2).

In fact, the company's major purpose has been to disguise the origin of uranium supplies, primarily from RTZ's Rössing uranium mine, so as to circumvent UN legislation, or deflect international opprobrium, the actions of the anti-apartheid movement, and decisions taken by national governments to limit, or stop, the import of Namibian or South African uranium.

RTZ Mineral Services was established with a Swiss chairperson, called Paul Gmuer – a lawyer specialising in the use of domestic "paper" or front companies, where international trade may be the subject of unwanted scrutiny.

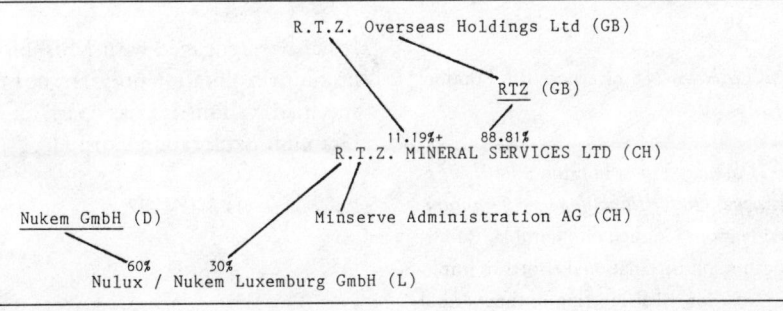

Gmuer, in the seventies, was also a director of various Westinghouse companies, the President of Inco Finance, and involved with some seventy other enterprises (3).

Together with Gmuer, on the Minserve board at that time sat three RTZ directors. There was little pretence that the company was directed other than by RTZ in London. As a Swiss researcher reported in 1984, on a visit to 10 Baarenstrasse, the Minserve HQ:

"There were no RTZ employees, despite the company nameplate which hung outside. The company's telephone number in Zurich, rung by a member [of our group] turned out to be a lawyer's office which threatened legal action 'if you investigate RTZ'. It had become clear that RTZ was not simply a paper company but a ghost company" (2).

Two other researchers have also tried to look closely at Minserve over the past ten years ("dissident" RTZ shareholders have always got a brush-off when they try to question RTZ itself on the activities of its Zug-gish "gnome"). One of them is Yoko Kitizawa, of the Japanese group AMPO, who, in 1977, established to her satisfaction that Minserve had been set up partly (if not primarily) to "launder" Namibian uranium supplies to Japan (2, 4). Minserve contracted to supply 41,851 tons of U_3O_8 to Japan from Rössing between 1977 and 1996 (5).

The other is the Luxembourg MP, Jup Weber, who, in the late 'eighties, researched RTZ Mineral Services' role in the Nukem "scandal" (*qv*). Minserve holds 30% of Nulux, Nukem's Luxembourg-based subsidiary, and, around 1984, had apparently contracted to supply UF_6 to South Africa – a clear violation of embargoes on nuclear collaboration with the apartheid state (6).

Contact: PARTiZANS, 218 Liverpool Rd, London N1 1LE, UK.

References: (1) *Sunday Times*, London 30/12/79. (2) Yoko Kitizawa, *On the Illegal Japanese Uranium Deals*, United Nations Council on Namibia, Regional Symposium on International efforts to implement Decree No 1 for the Protection of the Natural Resources of Namibia, Geneva 27-31/8/84, ref: A/AX.131/GSY/CRP.12. (3) Information from Janine Roberts, Cimra 4/12/77. (4) *Asahi Shimbun*, Tokyo, 20/3/77. (5) *Canuc Briefing Paper on Namibian Uranium Contracts*, London, 1987. (6) *Plunder!*, PARTiZANS, London, 1991.

531 RUŽV

Socialist Federal Republic of Yugoslavia (YU)

RUŽV / RUDNIK URANA ŽIROVSKI VRH (YU)

It has operated a uranium mine Žirovski in Slovenia (1).

In 1979, the Swedish LKAB agreed to train RUŽV miners, and advise the company on mining methods and systems (2). In mid-1984, it was announced that production at the mine and the concentrator would "start shortly" with the production of some 40 tons of yellowcake by the end of the year. This was intended to rise to 120 tons by 1986: sufficient to cover annual demand from Yugoslavia's first nuclear power station, in Krsko (3).

References: (1) *MM* 6/81. (2) *MJ* 14/9/79. (3) *MM* 8/84.

532 Ryowa Uranium Exlporation Co

RYOWA URANIUM EXPLORATION CO. (J)

50%
Taihei Uranium Exploration Corp. (J)

It has been engaged with Mitsubishi in a joint uranium exploration programme in several (unspecified) countries, as part of the Taihei Uranium Exploration Corp (1).

References: (1) *MIY* 1982.

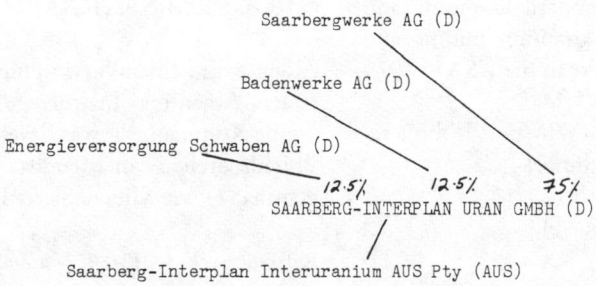

Saarbergwerke AG (D)

Badenwerke AG (D)

Energieversorgung Schwaben AG (D)

       12·5%   *12·5%*   *75%*
SAARBERG-INTERPLAN URAN GMBH (D)

Saarberg-Interplan Interuranium AUS Pty (AUS)

533 Saarberg-Interplan Uran GmbH

Controlled by Saarbergwerke AG (1), it is a
relatively recent (1974) creation by German energy concerns (2), exploring for uranium in
Germany and overseas, also trading uranium
concentrates and related services. An "integrated energy corporation" (3), Saarberg is involved in coke and coal processing, power
generation, petroleum production and refining,
and "environmental" technology. Projects have
been carried out in 30 countries throughout the
world: the company has branch offices in the
USA, Tanzania, Somalia and Australia. It is also
involved in plans for nuclear waste storage (3).
It owns more than 15 million "B" class shares in
ERA, the owner of the Ranger mine (2).
It has been exploring for uranium in Nova Scotia, in the South Mountain batholith area;
together with PNC in Zambia; along with
AGIP and Cogéma in Gabon.
It also recently entered a JV, to explore for
uranium with Davy McKee and the Nigerian
and Moroccan governments, in Guinea.
It holds 12% of the Techili deposit (also known
as East Afasto) in the Air region of Nigeria,
where it is partnered with Onarem (30%),
Cogéma (30%), the Nigerian Mining Corporation (16%) and the Central Electricity Generating Board (12%) (4). Together with Anglo-
American, Lonrho and Union Carbide, it has
been involved in a huge mineral exploration
programme for the Zimbabwean government,
covering vast areas of the country, and searching for uranium, as well as copper, tungsten,
gold, lead, molybdenum and other metals (5).

Saarberg, together with Fluor, was among various western companies competing for a feasibility study contract for a large coal mine in
southern Sumatra, funded by the World Bank,
in 1984 (6).
In 1988, Saarberg-Interplan announced the
discovery of a uranium deposit of between 3000
and 5000 tonnes, near the Mozambique border, in the Zambesi vallet of Zimbabwe (7).

References: (1) *FT Min* 1991. (2) *MIY* 1982. (3)
MM 3/83. (4) *MAR* 1982. (5) *FT* 18/6/81. (6) *MJ*
18/5/84. (7) Financial Gazette, Harare 23/9/88.

534 Sabina Resources Ltd (GB)

In 1978, Sabina Resources Ltd (formerly New
Sabina Resources Ltd) (1) was exploring 300
square miles of County Tyrone, Northern Ireland, in a JV headed by E and B (2). The same
year, it retained (along with Glencar Explorations), Minerex of Dublin, to carry out their exploration programme (3).

References: (1) *FT Min 1991* (2) *FT* 7/6/78. (3) *Gua*
27/7/78.

535 St Clair Energy Co

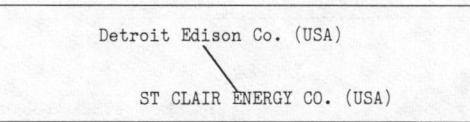

Detroit Edison Co. (USA)

ST CLAIR ENERGY CO. (USA)

Along with Western Standard Uranium, and US Energy, it planned uranium mining and milling at an unknown site in the USA in 1979 (1).

No further information. (Why not? Is this a secretive industry or something?)

References: (1) Sullivan & Riedel.

536 Sander Geophysics (CDN)

It signed a contract with the UK's Natural Environment Research Council, to carry out a 6300km$_2$ helicopter uranium reconnaissance survey of Wales and north-east and south-west England in 1978 (1).

References: (1) *MJ* 17/11/78.

537 Santos Ltd

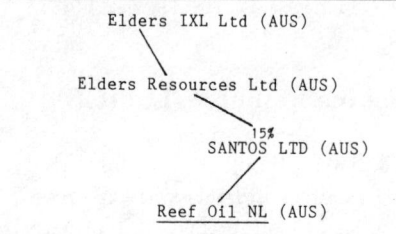

```
          Elders IXL Ltd (AUS)

      Elders Resources Ltd (AUS)

                      15%
                SANTOS LTD (AUS)

           Reef Oil NL (AUS)
```

This major petrol exploration and production and gas production company operates in Australia (1), Papua New Guinea, Europe, the Timor Sea and elsewhere (2).

It was prospecting for uranium in South Australia (places unknown) in 1979/80 (3).

References: (1) *Reg Aus Min* 1981. (2) *Reg Aus Min* 1991. (3) Annual Report 1980.

538 Sasetru SA (RA)

Along with Minera Sierra Pintada, Alianza Petrolera Argentina, Inalruco SA, and Pechiney Ugine Kuhlmann, it was developing the Sierra Pintada orebody in Mendoza province of Argentina (1); see Minera Sierra Pintada.

References: (1) *MM* 1/80; *Nucleonics Week* 15/5/80.

539 Scintrex Ltd (CDN)

With a company HQ in Ontario, in 1980 it contracted to provide equipment to Uranio Mexicano for uranium exploration (1). In 1990 it developed a new, high-sensitivity, caesium sensor for magnetic mineral surveys (2).

References: (1) *MJ* 20/6/80. (2) *MAR* 1991.

540 SCUMRA

In the early 1980s, SCUMRA operated several small uranium mines, and an extraction plant, in the St Pierre-de-Cantal region of Cantal-Creuse, France, as well as at Aveyron, producing some 100 tons a year of contained uranium at Cantal-Creuse (1).

In 1980, the company announced that it had been granted three search permits in the Aveyron region, had requested another three, and had been given two exploration permits. It was planning to construct a uranium milling plant at Bertholène, with a capacity of 200-250 tons a year (2). One planned mine, at La Lansière (Savoie) was the target of a protest occupation in the same year (2).

By 1984, all CFP's uranium interests had been regrouped under the parent umbrella, to be technically operated by Total Cie Minière (3).

```
                            CFP (F)

                     94%
SCUMRA / STÉ. CENTRALE DE L'URANIUM ET DES MINERAIS
          ET MÉTAUX RADIOACTIFS (F)
```

References: (1) Dave van Oöyen, *Uraniummijnbouw in Frankrijk*, Amsterdam, 1982; *MM* 11/80. (2) *Kiitg* 6/80. (3) *E&MJ* 2/85.

541 Scurry Rainbow

```
Home Oil and Gas Ltd (CDN)
                │
               89%
        SCURRY RAINBOW (CDN)
```

As of 1980, it had not merged with the Hiram Walker-Consumers Home conglomerate.
It was holding uranium prospects of its own at Costigan Lake, Mawdsley Lake, and Ryan Lake, all in Saskatchewan (1).

References: (1) *Yellowcake Road*.

542 SDJB

```
                        Province of Québec
                       /
SDBJ / STÉ. DE DÉVELOPPMENT DE LA BAIE JAMES
      -JAMES BAY DEVELOPMENT CO. (CDN)
```

Incorporated as a Crown corporation in 1971, and wholly owned by the Quebecois government, the company's primary role is to "develop" the natural resources of the James Bay region – an area extensively occupied and used by native peoples (1).
Cominco and SDJB, together with Pancontinental Mining, discovered a sizeable uranium deposit in the Otish mountains area of Quebec in 1978 (3). A JV was formed to investigate the deposit further, consisting of Uranerz, Séru Nucléaire, Esso Minerals, Pancon, Soquem, and SDBJ (4).
In 1980, SDJB and Uranerz also announced a drilling programme on their Gayot Lake property in the James Bay region of Quebec: although located a year earlier, insufficient drilling failed to reveal commercially viable deposits (5).

SDJB also had a JV in uranium exploration with Eldorado Nuclear during the 1970s (2).

References: (1) *MIY* 1982. (2) Eldorado Nuclear Annual Report, 1979. (3) *FT* 26/6/78. (4) *MAR* 1982. (5) *MJ* 14/3/80.

543 Seaforth Mineral and Ore Co (USA)

This is a producer of fluorspar (through its Inverness Mining subsidiary) (1).
In 1977, it optioned its Fission uranium property in Kelowna, British Columbia, Canada, to Mattagami Lake Mines (2).
Five years later, Seaforth merged with Quintaine Resources Inc, to become Quinterra Resources Inc. Quinterra holds gold properties in Ontario (Central Patricia) and Quebec (3).

References: (1) *MAR* 1981. (2) *MJ* 2/12/77. (3) *FT Min* 1991.

544 Sedimentary Uranium NL (AUS)

Now known as Sedimentary Holdings Ltd (1), this is a uranium and gold exploration company with an authorised capital of 100 million 20 cent shares (2).
It was exploring throughout 1981 on the company's Yarramba uranium prospect (80% owned), with Mines Administration (6.6%), Teton Exploration (6.6%) and Carpentaria Exploration (6.6%) (3). This borders on the Honeymoon prospect; it is also known as East Kalkaroo. Inferred reserves in 1982 were put at 2000 tonnes of U_3O_8 (4). The company said it wanted to use the same *in situ* leaching process which was severely criticised at Honeymoon (2). A protracted legal dispute in 1982 slowed exploration and, by March 1983, the prospect was still being "assessed" (5).

References: (1) *Reg Aus Min*, 1990/91. (2) *Reg Aus Min* 1990/91. (3) Annual Report 1980. (4) *Reg Aus Min* 1982. (5) *Reg Aus Min* 1983/84.

545 Sedimex Uranerschliessungs GmbH

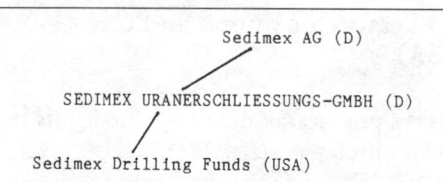

It had formerly owned E & B Canada Resources, which is now owned by Imperial Metals. (1). In 1986, IMC bought up Sedimex Drilling Funds' interests in Geomex Minerals Inc (2).

References: (1) *MJ* 6/2/81. (2) *MIY* 1987

546 Sequoyah Fuels Corp

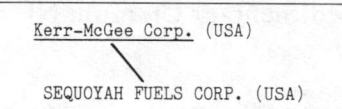

It was set up in 1983 by the infamous Kerr-McGee Corp, to consolidate the Sequoyah refinery and uranium fluorination plant, and non-producing uranium leases owned by Kerr-McGee outside of New Mexico, including two open-pit mines in Wyoming (1). Rio Algom acquired the Sequoyah uranium assets in 1989 (2)

References: (1) *E&MJ* 11/83. (2) *FT Min* 1991.

547 Séru Nucléaire Ltée

Re-named Cogéma Canada Ltd in 1984 (1), it has prospected for uranium in the Athabasca Basin area of Canada, particularly along the Midwest-McClean Lake belt (2). In the Otish Mountains of central Quebec it is involved with

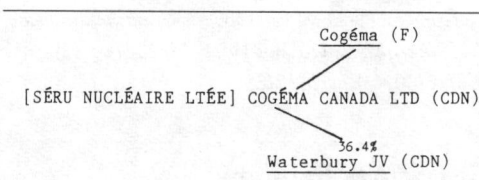

several other mining companies, including Uranerz and Esso Minerals (3).

The company's exploration efforts were shifted from Quebec to Saskatchewan in 1979, when it acquired six permits for 847,472 acres (4) and entered a JV with Eldorado Nuclear to explore for uranium (5) – and later, with other companies, in the Athabasca region (6).

This paid off in 1983, when the Waterbury JV, led by SMDC with just over 50% and tailed by Séru with just over 33%, discovered a major high-grade uranium deposit at Cigar Lake, northern Saskatchewan. By late 1983, drilling had located uranium oxide deposits ranging from 7.2% across a width of 14.2m, to an extremely rich vein of 25.02% across a thickness of 9.5m (7) and another across 17m of up to 20% grading (8).

Directly, or through Corona Grande (its wholly-owned subsidiary) Cogéma Canada Ltd now holds 36.375% in the Cigar Lake Joint Venture (Waterbury JV) (1).

References: (1) *FT Min* 1991. (2) *MAR* 1981/82; *Yellowcake Road*. (3) *MAR* 1982. (4) *FT* 10/9/79. (5) Eldorado Nuclear Annual Report, 1979. (6) *Yellowcake Road*. (7) *FT* 29/12/83. (8) *MAR* 1984.

548 Shell

The world's second largest oil company (after Exxon), and one of the first truly global corporations in the world (1), by 1973 RDS was already bigger (in terms of annual sales) than Iran, Venezuela and Turkey (1). Six years later it became the first company on the planet to chalk up sales of £2 billion over a twelve-month period (2).

As of 1987, Royal Dutch Shell employed 142,000 people world-wide (3), and is probably

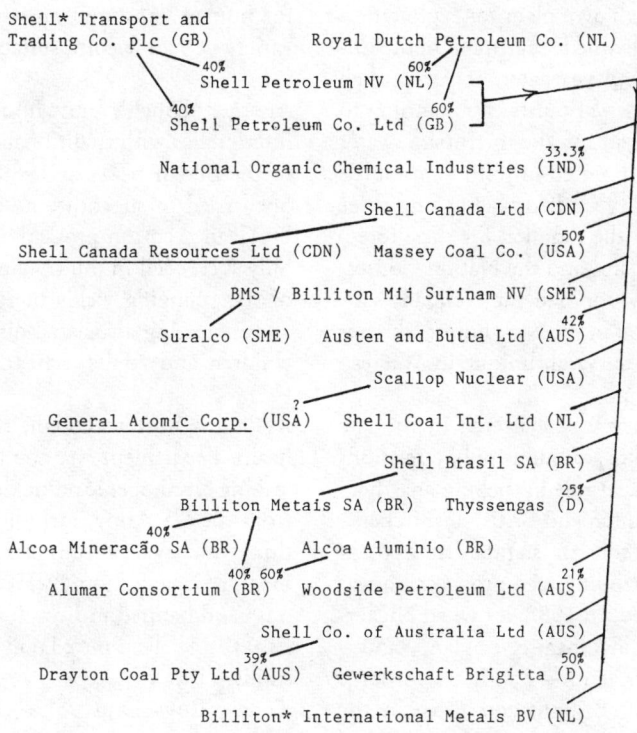

```
              SHELL* / ROYAL DUTCH SHELL GRP / RDS (GB/NL)

Shell* Transport and
Trading Co. plc (GB)        Royal Dutch Petroleum Co. (NL)
                    40%            60%
                  Shell Petroleum NV (NL)
              40%                    60%
                Shell Petroleum Co. Ltd (GB)
                                              33.3%
              National Organic Chemical Industries (IND)

                            Shell Canada Ltd (CDN)
                                          50%
     Shell Canada Resources Ltd (CDN)   Massey Coal Co. (USA)

                 BMS / Billiton Mij Surinam NV (SME)
                                            42%
                 Suralco (SME)   Austen and Butta Ltd (AUS)

                         Scallop Nuclear (USA)
                       ?
   General Atomic Corp. (USA)   Shell Coal Int. Ltd (NL)

                           Shell Brasil SA (BR)
                                          25%
            Billiton Metais SA (BR)   Thyssengas (D)
          40%
Alcoa Mineracão SA (BR)       Alcoa Aluminio (BR)
              40% 60%                          21%
   Alumar Consortium (BR)   Woodside Petroleum Ltd (AUS)

                    Shell Co. of Australia Ltd (AUS)
              39%                          50%
Drayton Coal Pty Ltd (AUS)   Gewerkschaft Brigitta (D)

              Billiton* International Metals BV (NL)
```

* N.B. Virtually all operations around the world bearing the name
'Shell' are 100% or majority-owned by RDS, as are operations using
the name 'Billiton'!

the most diversified of all the big oil companies (4), probably the most sophisticated (5), has a wider geographical spread than even Exxon, with hardly smaller gas and oil reserves (6), and is decentralised to a remarkable extent. "Even Shell veterans have difficulty in explaining how the company functions" (6). Notwithstanding this, RDS bought out its biggest single subsidiary, (and its most successful) Shell US, in 1986, just after it made an important oil discovery in the Beaufort Sea (7) – thus confirming the old adage that children should never try being more clever than their parents.

It employs a highly-refined system of scenario planning among its group managers (8) and has long been a leading developer of conformity seed types, more recently enzymes – linking up in 1987 with the Dutch pharmaceutical company Gist-Brocades, to form a world-wide operation known as General Bio-Synthetics (GBS) (9). It is also a pioneer of computer-integrated factories, which threaten to make human beings all but extinct on the assembly line (10). Although all the major oil companies now pride themselves on their care of the environment, (and – if more erratically – indigenous communities), Shell has probably devoted a larger part of its budget, over a longer period, to selling itself to the conservationist lobby, than any other oil major. This, combined with a policy of guarded openness towards critics, has given it a clean image where it counts. In Britain in 1986, for example, it decided not to purchase 50% of the Hemerdon tin-tungsten project near Plymouth, after Amax's own plans for exploitation had been rejected by the Department of the Environment (DoE). (A year later, however, Shell had "shocked conservationists" by planning to drill for oil in sight of the Harnham water meadows, commemorated in John Constable's painting of Salisbury Cathedral (11), and three years before, both the Council for the Protection of Rural England and the Nature Conservancy, strenuously opposed the company's attempt to drill for oil in the New Forest which is "an extraordinary survival, unique in Western Europe") (12).

Shell's complacency was shattered in the late 1970s when, despite strenuous public relations efforts, RDS's crucial role in fuelling apartheid was revealed by Dutch and British researchers.

By then, Shell's role in standardising plant breeding and controlling seed propagation was also well recognised. In 1982, it owned Nickerson Seed Co of Lincolnshire, England, and – through Shell Petroleum – another eight large seed companies in Britain, and three in the Netherlands. Nickerson in turn owns 100% of the central seed growing company, International Plant Breeders – which runs North American Plant Breeders with the Olin chemical group in the USA. Shell also co-operated in breeding seeds with French companies at locales in Canada, Mexico, Brazil and Japan (13). By 1984, Shell was controlling no fewer than 30 seed companies in Europe alone (14).

In 1987, Shell announced that, through Nickerson, it was to invest an extra £14 million over the following five years in biotechnology research, to consolidate its hold on the propagation of key crops – especially barley and wheat in Britain; maize in France; and wheat, maize, soya, alfalfa and sunflower in the USA (15).

Shell Philippines has been heavily involved in the second phase of the so-called "Green Revolution" designed to replace traditional food crops with cereals for foreign exchange. Ex-President Marcos, in 1981, announced a "Christmas present to the Filipino farmer and the entire Filipino nation" in the form of the Shell-supported Maisanga programme, which is endeavouring to replace white corn (maize) for human consumption with a high-yielding variety of yellow corn, suitable for animal food (16).

Another of Shell's forays into this critical area of interference with traditional cropping occurred in the Sudan between 1980 and 1981, when Shell tried to substitute its own chemicals for DDT in a cotton-protection programme, but only succeeded in multiplying costs by a factor of six, plunging yields to rock-bottom levels, and increasing associated disease to school-age children and adults (schistosomiasis) by 80% (17).

As if this wasn't enough, the Colorado state health department was soon accusing Shell of leaking carcinogenic pesticides from land leased from the US Army, including aldrin and dieldrin, which were already classified by the World Health Organisation (WHO) as "very hazardous" and banned in Britain and the rest of the EC (18). Shell admitted only to "some" responsibility (19).

Today's Royal Dutch/Shell conglomerate is virtually unique in its structure, with two parent companies (the Royal Dutch Petroleum Company, based in the Netherlands, and the Shell Transport and Trading Co plc, registered in England) themselves owning 60% and 40% respectively of two holding companies, Shell Petroleum NV, and the Shell Petroleum Co Ltd. These two holding companies themselves own interests (usually 100%) in various service and operating companies, which function in more than 100 countries. The service companies are: Shell International Petroleum Maatschappij BV; Shell International Chemie Maatschappij BV; Shell International Petroleum Company Ltd; Shell International Chemical Company Ltd; Billiton International Metals BV; Shell International Research Maatschappij BV; Shell International Gas Ltd; Shell Coal International Ltd; and Shell International Marine Ltd. (However the only known sea-bed consortium to which the company is partner is Ocean Minerals Co Ltd, formed in 1977, in which RDS's share is held by Billiton) (20).

This massive conglomerate was formed in 1907 as a merger between the Shell Transport and Trading Co (ST&T) and the Royal Dutch Petroleum Co – though the idea was always to keep the companies separate, and they still trade on separate stock exchanges (4).

ST&T was itself founded in 1833 as a trader of curios and, later a transporter of kerosene. Royal Dutch was founded in 1880, to exploit resources in the Dutch East Indies (21). It successfully met a challenge by the Standard Oil company of the USA, when its oil concession was sold to Royal Dutch by the colonial government for a derisory US$45,000 (it was actually worth about 500 times more), and voting shares were issued only to Dutch citizens (21).

Following the 1907 merger, the new company expanded rapidly across the globe: starting oil production in British Borneo (1910), Mexico (1913), Venezuela (1914), and, after the First World War, in North America, Malaya and Iraq. Chemical production commenced in the 1950s, and gas in the early 1960s. In the 1970s it purchased Belridge Oil of the US (renamed Kernridge) and developed its downstream activities, nuclear concerns, and coal projects through the rest of the decade (22). In May 1970, RDS acquired Billiton Maatschappij, making it a major owner of several important mineral mines and resources (23).

Shell Nuclear, formerly a service company, now plays a sleeping role. Shell's nuclear interests were relatively short-lived, but important nonetheless. In the late sixties, Shell bought half of Gulf Oil Corp's interests in General Atomic Corp, one of the leading builders of American nuclear reactors. Part of this acquisition was sold back to Gulf in 1979 (22) and the company has now ceased effective operations: but not before it became heavily embroiled in the uranium cartel, the potential construction of the Barnwell nuclear reprocessing plant in South Carolina, and a number of uranium projects, notably in Canada.

Along with Gulf, and Allied Services, Shell was contracted to build the massive Barnwell plant (the first US commercial nuclear reprocessing plant) in the 1970s. Work was suspended on this plant in 1977, as a result of President Carter's moratorium on commercial reprocessing (24). Five years later, Shell's investment in the project was written off at a cost of US$36 million (25).

Although Shell Coal secured an exploration permit in Botswana in the 1970s, which included the search for uranium and other radioactive elements, nothing of significance was announced (26). The company's uranium ventures otherwise seem to have been limited to Canada, from which it withdrew most of its mineral interests (apart from coal) in 1982 (27). Until this date, all the uranium leases were held through Shell Canada Resources Ltd: the most important were with Eldorado Nuclear in the Athabasca belt, and with Northgate Exploration and Westfield Minerals, in the Deer Lake and Codroy Valley area of Newfoundland.

It is also important to note that Shell helped finance the URENCO tripartite uranium enrichment plant in Almelo (see BNFL), and has seconded leading personnel from its own ranks to service this company (28).

Shell's mining interests are mainly conducted through its Billiton subsidiary, Billiton International Metals BV, with no fewer than 80 companies in more than 20 countries, managed from the Netherlands (29). However, in certain countries – notably Chile, Australia and South Africa – metal activities are carried out by the respective Shell subsidiary (29).

Billiton itself was conceived in 1860, and incorporated into the RDS group in 1970. For many years, it engaged only in tin mining on Billiton island (now called Belitung) in Indonesia but, in the 1930s, expanded into bauxite. It is now involved in exploration, mining, processing, recycling, research and marketing on a worldwide scale (30).

In the Netherlands itself, Billiton is 50% owner and operator of the Budelco electrolytic zinc plant, with CRA as its partner – as well as the operator of a lead smelter. It also runs a lead smelter in Britain, while it holds a 35% share in the Aughinish alumina refinery in Eire (30), and has operated an exploration JV (40%) with RTZ in the central part of the country (31).

Elsewhere, Billiton's main interests are in bauxite, alumina and aluminium products, although it has made important investments in ferronickel, zinc and copper. An active exploration programme is conducted in seven countries: Australia, Brazil, Canada, Chile, Portugal, Spain and South Africa.

In Surinam, Billiton holds 75% of the Suralco mining and refining JV (Alcoa holding the other 25%). This has operated at a loss since 1985, due to a government levy: the smelter uses power from the Brokopondo hydro scheme (32). NV Billiton Maatschappij has been mining bauxite in the country since 1941 – providing an important contribution to the company's alumina capability (33).

Suralco also agreed in 1987 to invest US$150 million in three bauxite mines in Guatemala – after laying off 500 workers considered "excessive" (34).

In 1983 Suralco was said to be facing "a very bleak future" in Surinam (35) and, the following year a month-long strike by the company's 4500-strong workforce threatened to topple the Bouterse régime, as strikers were joined by public utility workers, protesting against extra taxes. The strike cost the country (dependent for 80% of its foreign exchange earnings on Suralco's operations alone) some US$28 million in lost earnings (36). More than a year later, relationships between Suralco and its workforce worsened again, as the government ordered Billiton to raise its workers' pay by 5%, in the face of Billiton's objections, as the company chalked up a US$30 million loss for the previous year (37).

Surinam was not the only loss-maker that year. Throughout 1985 and 1986, Billiton began cutting back drastically on its projects, after sustaining a record loss of around US$260 million world-wide (30). It terminated its tin-dredging operations in Thailand and Indonesia, closed its Mount Pleasant tungsten mine in Canada (38), and sold its 11% share in the Nansivik zinc-lead mine. This was, until Cominco's Red Dog mine in Alaska, the most northerly mine in the world, sporting a specially-constructed dock, storage facilities, and an igloo-shaped community centre for its 200-person workforce, a third of them Inuit (31).

The same year, Billiton sold its 62% interest in the Deeside (Scotland) titanium plant, which it acquired in 1981 – its only known major foray into strategic minerals (33).

Among the company's major metals interest outside Europe are: the Cerro Matoso ferronickel project in Columbia (50%) (98); its 9% share in the Cuajone copper mine in the High Andes of Peru – one of the world's biggest mining operations (33); a small share in the Boké bauxite mine in Guinea, which provides feed for the Aughinish refinery in Eire (33); its most extensive operations, located in Brazil and Australia; and more recent acquisitions in Indonesia, Ghana, Canada and Chile (98).

Initially Billiton held 40% of Alumar in Brazil, along with Alcoa. However, it reportedly reduced its share to 18% in 1984, because of "poor prospects" (39). Although the refinery was completed in 1984, and the second phase of the project begun three years later (at the cost of a further US$240 million, of which Billiton provided more than half), the Alumar scheme had still not started making a profit by 1987. At that point, Billiton said it was only willing to proceed to the third phase if a satisfactory agreement could be reached on pricing with Alcoa, and on a power contract with the Brazilian government (40).

Power for Alumar comes from the Tucurui hydro scheme on the Tocantins river, and its bauxite from the huge Trombetas deposit, in which Billiton holds 10% and the Brazilian state mining company, CVRD (Cia Vale Rio Doce), the majority share. Billiton's participation in Mineracao Rio Do Norte at Trombetas (formerly held partly by RTZ) entitles it to 700,000 tons a year of bauxite (33). The Trombetas area includes land occupied by various Indian peoples, notably the Pianokoto-Tirio, the Warikyan-Arikona and the Parukoto Xaruma (41). (In 1981, a secretary working at RTZ's headquarters, unearthed a document from a Brazilian official guaranteeing that the Trombetas area of Brazil would be declared "free of Indians" when mined by the company. Was a similar assurance given to Shell/Billiton?) Half of the land used for the Trombetas bauxite project was originally purchased from the notorious Daniel Ludwig, the US entrepreneur who thought that he could bring cattle stations and outlandish trees to the jungle, and make a fat profit, before he destroyed the ecosphere (42).

The Tucurui hydro-electric scheme is a disaster in its own right: dangerous chemicals – probably including Agent Orange – were used to clear the jungle before flooding, while empty drums, left in the region, and then used by local people for food storage, caused miscarriages and deformities in children. Despite a warning that residual deadly defoliants could enter the Tocantins river, the flooding went ahead as planned (42).

Two groups of Parakana people had to be transferred from their ancestral lands, while the Gaviao da Montanha, devastated by disease due to its first contact with the whites, had to be forcibly shifted after it "agreed" to a derisory payment of Cr77,000 (£2960) for the loss of the land (42).

Another group of Gaviao was actually moved twice – thanks to bungling by the National Indian Foundation, Funai. (However, they used some of their compensation money to go into the Brazil nut business!) (42).

In Western Australia, Billiton holds 37.5% of the Worsley bauxite mine and refinery, completed in 1984: other partners are Reynolds Australia (50%), Kobe Alumina Associates (10%) and Nissho Iwai Alumina Pty Ltd (2.5%). (In 1987, Dampier, which is wholly owned by BHP, agreed to sell its 20% stake to Norsk Hydro (43), but it was finally acquired by Reynolds, Billiton, and Nissho Iwai) (99).

In 1984, speculation was rife that the nearby Boddington gold deposit, also owned by the Worsley partners, could become Australia's most promising gold project (44): development at Boddington started in 1986 (32) and soon became Australia's third largest gold mine (40), and is now the second largest producing outfit (100).

Shell/Billiton has several other Australian projects – notably a JV with AOG Minerals investigating precious metals in Queensland; a 50% JV with NBH Holdings at the Collingwood tin prospect; a JV with BP to develop a small lead-zinc deposit at Cadjebt, near Fitzroy Crossing (Western Australia) (40); and a high-grade zinc deposit in South Australia (40).

Outside of Australia, it is Shell's Aurukun bauxite leases which have raised most concern: to the extent that the first official delegation of Aboriginal people to an overseas mining company was paid to Billiton in the Hague in November 1978. Accompanied by Mick Miller, chair of the North Queensland Land Council (NQLC), and Joyce Hall from Weipa (see CRA), Jacob Wolmby from the Aurukun Aboriginal reserve condemned the plans of Shell/Billiton and its partners, stating that the company would be guilty of cultural genocide. The visit secured TV, press and radio publicity in the Netherlands (45), and an apparent promise from the Managing Director of Billiton International, not to mine on the Aboriginal land at Aurukun without the consent of the people. This statement was negated two years later, however, when Bernard Wheelahu, Billiton's Metals Manager in Australia, declared:

"I prefer not to give land rights to the Aborigines ... it means they will be in the same position as the other white Australians (*sic*) ... It is a very big problem and it is dangerous to the mining industries" (46).

In 1975, Shell/Billiton, along with PUK and Tipperary of the USA, took a bauxite mining lease of 736 square miles in the Aurukun reserve, to the south of the Comalco lease (which is the world's largest bauxite strip mine – see CRA). Aurukun Associates, the JV which was then formed, is managed by Billiton. Two years later, Tipperary sold most of its share to the two bigger companies. That year, when the Labor government of Whitlam was conspired out of office, the Queensland government, defying the united opposition of Aurukun's Aboriginal community, proffered a lease lasting until the year 2038, giving the Associates total rights over Aboriginal hunting and pastoral land, control of a company town, a harbour, and rights to mine the coral reefs, as well as the mainland. For a rent of merely US$3 per square kilometre (increasing to US$20 after 15 years) and a royalty of up to US$1 a ton, the companies were given deposits worth a conservative A$14 billion (46).

The following year, "after years of waiting and months of strong public pressure" (47), nego-. tiations began between the Aurukun people and the mining companies, during which the people demanded: a direct share of the royalties, guarantees of employment if the project proceeded, guarantees for their sacred sites, and equality of treatment with whites (47).

After two years, the community got nowhere (48). Although winning a case for control over their own reserve, the Aboriginal people were defeated by the Privy Council in London, after an appeal by the Queensland government, giving the latter virtual *carte blanche* to take over affairs at Aurukun, and open the door to mining (46).

In 1981, Billiton apparently secured two-thirds of Aurukun Associates, leaving a third with Pechiney (49). However, according to the *Mining Annual Review*, Tipperary continued to hold 40%, Pechiney 20% and Billiton 40% (50). In any event, although a feasibility study to begin mining in the later 1980s appears to have been carried out, "a number of problems have to be solved, including relations with the local Aboriginals, before development [could start]" (40). In fact, more likely to start up in the near future is another major bauxite lease, in which Billiton (10%) is partnered to Alcoa (17%), Sumitomo (15%), and Marubeni (5%), with the lion's share held by CRA (52%) through the Mitchell Plateau Bauxite Company, which also holds leases on Aboriginal land in Western Australia. This project, too, has met with opposition, notably from the Kimberley Land Council (46).

Of all Shell's world-wide concessions, which have some impact on indigenous peoples, the most important – because the most extensive – are those in South America. In 1981, the company began seismic exploration for oil, along with GeoSource Inc, in the Fitzcarrald Isthmus (named after the infamous rubber baron, commemorated by the film-maker Werner Herzog)

which is located in the Manu National Park. This Park has, for some years, been singled out as one of the largest eco resource areas of South America, home to a number of tribal peoples, and threatened by various forms of "development" (51).

Four years later, Shell signed an agreement with Petroperu, the state oil corporation, to extend its permits over the Madre de Dios Basin – the first major licences to be granted in the dense southern jungle on the eastern side of the Andes (52). In 1987, Shell again sought further leases in Peruvian indigenous areas – this time for gas exploitation in the central southern jungle, where it required £535 million to raise some million million cubic feet of gas (53). The Garcia régime was said at the time to be in a "dilemma" as to whether or not to accept Shell's investment, which would drive a coach and horses through its previous restrictions on foreign investment (54).

Shell was given exploration rights for oil and gas in Brazil's Middle Amazon region in the late 1970s (55), when it also started operations in the Aripuana national park (an area supposedly set aside for Indians). It is now the largest privately owned company in Brazil (56).

These are relatively recent penetrations of indigenous lands: Shell actually entered the "enclaves" of Quichua, Achual and Shuar people, in the western foothills of the Ecuadorian Amazon, as long ago as the 1920s. Indeed, Shell's road building from Banor to Puyo, during its exploration in the Oriente, enabled colonists and missionaries to "spill into the area", pushing Indians off their land and imposing debt peonage and patron-client ties on the Napo Quichua (57). In response to this encroachment, the Shuar formed their famous Federation in 1964: twenty years later, Shell was still cited by them (along with Texaco) as a major invader of their territory (58).

The company also has an oil project in Colombia potentially affecting the country's original inhabitants, where it took over more than half of Occidental Petroleum's huge holdings in Cano Limon, along the river Arauca (on the border with Peru), during the 1980s – through its purchase of the Colombia Cities Service Petroleum Corporation (59).

Shell owns 35% of El Salvador's Acajutla refinery – the country's only refinery (40).

In the Philippines, Shell (the third largest company in the country), owns 50% of the Tabangao refinery at Batangas, 28% of Philippine Petroleum Corporation, and 80% of the first Liquefied Petroleum Gas terminal in an ASEAN country – also being constructed at Batangas (16).

While other companies have been withdrawing from the Chittagong Hill Tracts (CHT) of Bangladesh in recent years – as a full-scale conflict between hill tribespeople and the Bangladesh military continues to rage, causing the deaths, detention and torture of thousands of indigenous people – Shell capitalises on its production-sharing agreement with the country's nationalised oil company, Petrobangla (40). Shell completed seismic surveying of the region in 1985 (32), despite a campaign mounted in 1981 by Survival International, to force the company to withdraw. At the time, Shell had a 25-year contract to explore, develop and exploit gas and oil in the Hill Tracts, involving the investment of around US$120 million ("a sum which exceeds the total spent on all development projects in the CHT in the last 20 years") (60). But Shell has shown little sign of responding positively to this pressure.

North American indigenous territory yields huge tracts to Shell's oil exploitation. Shell Canada withdrew from major mineral exploration in Canada in the early 1980s: notably from uranium exploration in Nova Scotia and Saskatchewan, from a molybdenum project in British Columbia, and a tin project in Nova Scotia – leaving in place only its 100%-owned Crows Nest Resources coal development in the west of Canada (61). Around the same time, it withdrew from the Allsands project (62). Two years later it had adopted "rationalisation" measures, which meant a cut of around 12% of its Canadian workforce (63). However, it announced a potential new oil field in the controversial Beaufort Sea region, in 1983 (see BP for further details of indigenous opposition)

(64) and it holds a number of permits on Yukon land, in the MacKenzie delta and Aklavik areas (65).

By 1983, in the United States, Shell led the field among the multinationals in its lease hold over Native American land – specifically a total of 258,754 acres containing over 20 billion tons of coal (66). Between 1968 and 1977, Shell signed corporate leases with the BIA (Bureau of Indian Affairs) which gave it access to billions of tons of Crow tribal coal. (The other partners in this nefarious project were Peabody, Gulf Oil, and Amax.) Many Crow people contested the agreements – specifically the pricing arrangement concluded with Shell, granting the tribe a pejorative 17.5 cents per ton in royalties. In 1977, after mounting pressure, the US Department of the Interior overturned the leases and forced an upward price renegotiation (67).

Coal is an extremely important energy resource for Shell, although coal operations made a loss of £34 million in 1986 (29). In 1979 alone the company added coal reserves equivalent to 88% of what it had produced that year (72) and the President of RDS, Dick de Brugue, declared the company's aim was to secure 10% of total international coal trade.

The company acquired 50% of Massey Coal, the USA's tenth biggest producer, a subsidiary of St Joe Corp (see Fluor) in 1980 (69). The company's stake was acquired through its 100%-owned Scallop Coal Corporation. In 1987, however, this US$680 million JV was broken up, through lack of profits, with Shell retaining certain interests (68).

Shell owns 42.12% of the New South Wales (NSW) producer Austen and Butta (29), and is the major shareholder in the Queensland coal mine, German Creek (Shell Australia 48%, Commercial Union Assurance 25%, Ruhrkohle 10%, UK National Coal Board 10%) (40). It started the Drayton coal JV in 1984 in NSW (38), and holds 34% of this major open-cut mine (97).

For several years, Shell has also been very interested in access to the massive untapped coal reserves of Colombia: the largest in south and central America, centred around the El Cerrejon deposit in La Guajira province (40). In 1985, Shell was offering to buy as much as 20% of El Cerrejon Norte from the Colombian state agency Carbocol (Exxon is the other major operator in the field). At the same time, Shell was also expressing interest in taking over the central fields (70).

Shell Coal International carried out a feasibility study at the Jining No. 2 mine in northern China in 1985 (38) but pulled out after a few months, because of "technical" problems (71). However, it is still reported to have a JV with the China National Coal Development Corp, and to be exploring elsewhere in the country (29).

Outside of the USA, Shell's major – and most controversial – coal interest is its 50% share of the Rietspruit open-cast coal mine near Bethal, in the Transvaal, South Africa, which it holds with Rand Mines Ltd. Rietspruit has a 5 million tons per year design output and its entire production is exported by Shell Coal International to customers in Europe and the Far East (97).

Shell has been in South Africa for 70 years and has long been identified with the interests of Afrikaner business (73). In 1975, Shell's oil interests under apartheid became 100% owned, and its erstwhile co-operation with BP was, for the most part, dissolved.

Through Shell Transport and Trading plc (and its subsidiary Shell Oil SWA Ltd) the company is also marketing petroleum products in Namibia (74). Billiton has mining and prospecting interests in the (until recently occupied) territory, too (75).

Although anti-apartheid activists and researchers world-wide have specifically targeted Shell's role in defying oil sanctions, internal resistance has centred on the Rietspruit mine. Following the death of a black mineworker through an accident, in February 1985, the NUM (National Union of Mineworkers, South Africa) organised a memorial service, taking up two hours of one shift. Four shop stewards were suspended by the management. When 800 fellow workers struck for their reinstatement, they were met with rubber bullets and tear gas; 86 workers were sacked, and evicted from their homes, while the rest were forced back to work at gun point (76). Shell then forbade union meetings and refused shop stewards access to their members.

The NUM called for international boycotts – a call rapidly heeded by the United Mine Workers of America (UMW), which themselves have grievances against Shell. (A contract dispute between UMW and A T Massey resulted in a 15-month strike between July 1984 and the end of 1985, after what UMW President Richard Trumka called "the most vicious reign of terror against working people seen in this country since the 1930s") (76).

A few months later, at Rietspruit, seven tiers of managers were found guilty of negligence after two black mineworkers died in a blast (77).

Shell's complicity in the continuance of the apartheid régime was more than foreshadowed during the years of a supposed embargo against the Smith régime of Rhodesia. Shell, BP, and Total between them supplied the major part of the white administration's oil needs, using a South African conduit, the Freight Services Ltd subsidiary of Anglo-American Corp (78).

The journalist who revealed British companies' duplicity in the case of Rhodesia four years later, exposed the route by which South Africa was receiving its own vital supplies. Martin Bailey cited five trading companies as suppliers – notably Phibro Salomon; Shell International Trading maintained contact with at least one of these companies, arranging the re-routing of crude oil supplies from Oman (79).

Later research has shown that Shell was the main beneficiary of so-called "incentive" payments, made by the South African authorities to companies breaking the OPEC boycott on oil sales to South Africa (80). Then, in early 1987, the Shipping Research Bureau of the Netherlands, set up in 1980 to monitor Pretoria's oil imports, reported that, since 1979, the Brunei Shell Petroleum Company, 50% owned by RDS, had sold no fewer than 56 cargoes to intermediaries, which then shipped them to South Africa. The report estimated that 6% of South Africa's crude oil needs had been met by this one source alone (81).

710

Shell has denied knowledge of such trade – profiting from the multiplicity of interfaces (notably Japanese and Swiss companies) which allow it to claim that "[N]o Shell group company outside South Africa is supplying crude oil to anyone in South Africa" (82).

What it cannot deny is the huge extent of its other interests in southern Africa: its 50% stake in the Sapref refinery, Durban (the country's largest); its 25% stake in the Samco lubricant plant; its 50% stake in the asphalt manufacturer, Abecol; its 17% stake in Trek Petroleum; its oil pipeline from Durban to Johannesburg (operated in conjunction with South African Railways); and its operations in the Single Buoy Mooring Point in Durban, through which 85% of South Africa's imported oil passes (83).

It also operates more than 800 petrol stations throughout the country, possesses an unknown number of shares in the notorious SASOL plant (which has been converting coal into oil supplies for domestic transport use) and has widespread chemical interests in the state.

In 1985, Shell also began construction of the Pering lead-zinc mine, its only known venture of this kind in South Africa (40).

Research, into the company's crucial role in underpinning apartheid, started in the 1970s and naturally peaked when the 1979 revelations about the violations of the OPEC oil embargo reached public attention. In 1979 and 1980, a shareholders' resolution was presented by the Werkgroep Kairos, the Komitee Zuidelijk Afrika, and Pax Christi Nederland, to the RDS annual general meeting (84), following meetings with the company's directors (85). An international seminar on an oil embargo was also held in Amsterdam in 1980 (86).

By early 1986, an international boycott movement was in existence, involving various anti-apartheid organisations in Europe and, in the US, trade unions, the Free South Africa movement, and the National Organisation of Women (NOW) (87). The campaign was joined by the anti-racist, anti-apartheid movement in Australia (88).

In the USA, a group of anti-apartheid campaigners, led by the New York City Teachers in Retirement, and supported by the Interfaith Centre on Corporate Responsibility, promised to introduce a resolution at the 1987 RDS AGM in the Netherlands, spotlighting the company's supply of fuel to the SA police and military, and calling for the cessation of petrol sales to government bodies in South Africa (89).

While 1986 boycott actions against Shell filling stations were organised in many countries, the most radical action was taken by unknown Dutch activists who attacked 16 petrol stations, four days before the 1986 Shell AGM, burning one of them completely down (90). In 1987, hundreds of demonstrators forced their way into the RDS building at Nijmegen, in the Netherlands, hurling petrol bombs and missiles, and protesting against the company's involvement in the eviction of squatters who had taken over unoccupied buildings belonging to the company (91). At a meeting the same month in The Hague, campaigners from the Netherlands, Scandinavia, Belgium, Britain, and the USA, agreed to target Shell as their number one opponent in the battle against apartheid (92). A month of action was held in Britain, co-ordinated by the Anti-Apartheid Movement, and aimed at RTZ and Standard Chartered Bank, as well as RDS (93). Both the Queen of England and Queen Beatrix of the Netherlands were singled out for their huge shareholdings in the British and Dutch arms of the company (94).

In Britain too, leaders of both the National Union of Seamen and the Transport and General Workers Union (TGWU) promised boycotts of Shell and other companies evading oil sanctions (95) and, around the same time, the International Mineworkers' Organisation (IMO) called on trade unionists to refuse to handle all imported South African coal (96).

Further reading: On Shell and South Africa (apart from sources already cited in the notes):

Shell in South Africa, Kairos-Osaci, Utrecht, June 1976 (somewhat dated, but unsurpassed for its background detail).

Wat een Shell aandeel-houder mag weten, Pax Christi and Kairos, (no date) Utrecht.

The Shell Shadow Report, Anti-Apartheid Movement, London, 1986 (revised 1987).

Contact: *Re South Africa:* Werkgroep Kairos, Corn Houtmanstraat 17-19, 3572 LT Utrecht. The Netherlands.

Anti-Apartheid Movement, 13 Mandela Street, London NW1, UK.

EIRIS, Bondway Business Centre, 71 Bondway, London, SW8, UK.

Australia: North Queensland Land Council, PO Box 1429, Cairns, 4870, Queensland, Australia.

Elsewhere: Survival International, 310 Edgware Rd, London W2 1DY, UK..

International Rivers Network, 301 Broadway, Suite B, San Francisco, CA 94133, USA. (for action in support of people affected by the Tocantins/Alubras project)

References: (1) Barnet and Muller, *Global Reach.* (2) *Gua* 16/11/79. (3) *FT* 26/11/86. (4) *FT* 25/1/84. (5) *FT* 4/3/80. (6) *FT* 10/7/85. (7) *FT* 25/1/86. (8) *FT* 5/3/80. (9) *FT* 26/6/87. (10) *FT* 6/12/85. (11) *FT* 16/3/85. (12) *SunT* 10/1/83. (13) *RMR* Vol. 1 no 3, 1982. (14) *NMon* July 1984. (15) *FT* 15/1/87. (16) *Kasama,* London No. 11, 1985. (17) *IFPAAW snips,* No. 6, quoted in *Taraxacum,* (IYESC), Vol. 3 no 4, 1984. (18) *Gua* 10/10/83, 30/7/87. (19) *Gua* 5/10/83. (20) *MJ* 25/11/77. (21) Tanzer. (22) *WDMNE.* (23) *Big Oil.* (24) *Bulletin of the Atomic Scientists,* US, March 1984. (25) *FT* 12/8/83. (26) *Uranium Red Book,* 1979; *MAR* 1986. (27) *MJ* 16/7/82; *Nuclear-Free Press,* Peterborough, Winter 1983. (28) *Nuclear Axis.* (29) *FT Mining Yearbook* 1987. (30) *E&MJ* 12/86. (31) *This is Billiton,* The Hague, 1978. (32) *MAR* 1986. (33) *MJ* 3/8/84. (34) *E&MJ* 1/87. (35) *FT* 31/3/83. (36) *FT* 1/2/84. (37) *Metal Markets Weekly Review,* London, 13/8/85. (38) *MJ* 12/7/85. (39) *FT* 6/8/84. (40) *MAR* 1987. (41) *Survival International Review,* Spring 1979. (42) Sue Brandford & Oriel Glock, *The Last Frontier: Fighting Over Land in the Amazon,* (Zed Books) London, 1985. (43) *FT* 8/7/87. (44) *MJ* 14/12/84 & (38). (45) Cimra documentation, London, Nov/78. (46) Interview by Dutch researcher, quoted in Roberts, *Massacres to Mining.* (47) *Kookabina,*

Vol. 1 no 1, 6/76. (48) *NQ Message Stick,* Nov/78. (49) *Reg Aus Min,* 1981. (50) *MAR,* 1984-85-86-87. (51) *IWGIA Newsletter* May/84; *Taraxacum* (IYF) Denmark, Vol. 3 no 2 1984. (52) *FT* 18/1/85. (53) *FT* 31/1/87. (54) *FT* 3/3/87. (55) *FT* 23/8/79. (56) *FT* 20/11/83. (57) *AkwN,* early Winter/82. (58) *SAIIC Newsletter,* Berkeley, Vol. 2 no 1, Autumn/85. (59) *FT* 22/8/85. (60) Letter from Barbara Bentley, Secretary, Survival International, to the Chairman of Shell/UK Ltd, dated 1 June 1981. (61) *FT* 11/2/82. (62) *FT* 17/3/82; *FT* 1/5/82. (63) *FT* 29/3/84. (64) *FT* 23/1/83. (65) *Yukon Indian News,* 24/2/84. (66) *Sklar.* (67) *Seven Days,* US. 14/3/77. (68) *FT* 16/6/87. (69) *FT* 19/3/80. (70) *Latin American Weekly Report,* WR-85-28, 19/7/85; see also (59) & (38). (71) *FT* 15/10/85. (72) *FT* 29/1/80. (73) *Shell in South Africa,* (Kairos-Osaki) Utrecht, June/76. (74) *Transnational Corporations and other foreign interests operating in Namibia,* (UN) Paris 1/4/84. (75) see (74) & also *Oil and Tanker Interests that Facilitate the Exploitation of Namibia's Natural Resources* (Shipping Research Bureau) Amsterdam 14/4/84. (76) *MMon* 15/4/86. (77) *Weekly Mail,* Johannesburg, 28/2-6/3/86. (78) see *SunT* 27/8/78, 3/9/78, 10/9/78; Martin Bailey and Bernard Rivers' *Oilgate: the Sanctions Scandal,* (Hodder and Stoughton) London, 1979; *Obs* , 25/7/79. (79) *Obs* 30/5/82. (80) *Shell and Apartheid,* (Anti-Apartheid Movement) London, 1985. (81) Reuters, quoting the SRB report, 27/1387. (82) Letter from L C Wachem, senior group managing director of Royal Dutch Shell, to the company's chief executive, quoted in *FT* 2/3/87. (83) *Fueling Apartheid* (Anti-Apartheid Movement) London. (84) *"Smeer em" Shell uit Zuid Afrika,* pamphlet, 1980. (85) *Waarheid* 18/5/79, 21/5/80. (86) *Kariso/KZA Newsletter,* Amsterdam Jan,July/80. (87) *WISE News Communiqué No. 250, 18/4/86, Amsterdam; see also zania Frontline,* No. 14, London, 8/86. (88) *CARE Newsletter,* Adelaide, April 1986. (89) *FT* 3/11/86. (90) Monachrome, London, October 1986. (91) Reuters report, 19/1/87. (92) Reuters report 21/1/87. (93) *FT* 2/3/87. (94) *London Daily News* May 1987, exact date unknown. (95) *FT* 31/10/85. (96) *FT* 24/11/86. (97) *Mining* 1987. (98) *FT Min* 1991. (99) *Reg Aus Min* 1990, 1991. (100) *MAR* 1991.

549 Shell Canada Resources Ltd

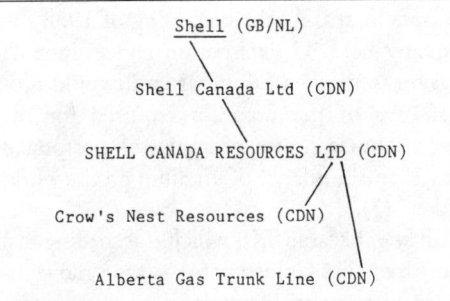

```
          Shell (GB/NL)

      Shell Canada Ltd (CDN)

 SHELL CANADA RESOURCES LTD (CDN)

 Crow's Nest Resources (CDN)

  Alberta Gas Trunk Line (CDN)
```

This is the Canadian "natural resources" subsidiary of the Dutch- and British-based Royal Dutch Shell Group.

In 1982, the company withdrew from all exploration and was intending to sell its JV interests to other parties (1). So far as uranium is concerned, these interests were: an agreement witn Eldorado Nuclear to explore for uranium in the Athabasca region of north east Alberta (40% interest) (2); uranium prospects in the Athabasca region on the Saskatchewan side (3); a C$1M joint drilling programme with Ontario Hydro in Nova Scotia in the South Mountain Batholith area (in 1982, however, there was, in any case, a moratorium on uranium licence applications imposed by the provincial administration) (4); a JV with Northgate – or more specifically with Westfield Minerals – to explore in Newfoundland, mostly at Deer Lake and in the Lost Pond region of the Codroy Basin (5).

Although Shell Canada has therefore effectively withdrawn from uranium in Canada, its interests in coal (through Crow's Nest Resources) and in numerous other companies, did not diminish.

The Lime Creek coal mine in British Columbia, whose main sales were to South Korea and Japan, was sold in 1991 to Mantala Company of Calgary (6).

References: (1) *MJ* 16/7/82. (2) Shell Canada's *Historical Highlights* (publication date unknown), quoted in: *Yellowcake Road*. (3) *Yellowcake Road*. (4) *MAR* 1982. (5) *FT* 3/10/79; *MJ* 11/4/80; *Energy File* 4/80. (6) *FT Min* 1991, *FT* 13/6/91, *MJ* 4/6/91.

550 Silver King Mines (USA)

It had exploration permits for uranium in Fall River and Custer counties of the Black Hills area of South Dakota in 1979-80, drilling 555 claims. It was in a JV with TVA/Tennessee Valley Authority in the same area – and had plans to develop three underground and one open-pit uranium mines near Burdock, also in Fall River county (1).

Silver King's current activities are in Oragon, Idaho, and Nevada (2).

In 1988 Silver King formed a JV called the Alta Bay Joint Venture, with Pacific Silver Corp and Echo Bay Mines, to exploit several properties in Nevada and Colorado (3).

Contact: Black Hills Alliance, POB 2508, Rapid City, South Dakota 57709, USA.

References: (1) *BHPS* 4/80. (2) *1991 E&MJ Int Directory of Mining*. (3) *FT Min* 1991.

551 Silvermines Ltd (IRL)

It has held lead, zinc and barytes-bearing properties near the village of Silvermines in Tipperary, and formed an Irish company called Mogul Mines – not to be confused with International Mogul Mines – along with Kerr Addison, to exploit a lead-zinc property in the same locality. Mogul Mines had a contract to supply feed to a smelter operated by Metallgesellschaft but was expected to collapse by the end of 1982 (2).

In 1976 the company took a 20% interest in a uranium exploration JV with Maugh (a wholly-owned subsidiary of Minatome) to search for uranium in Eire (1). By mid-1981 this programme was the most extensive in the country, revealing a number of interesting anomalies (3). However, Maugh now dropped a majority of its licences, and uranium exploration has virtually ceased in the country (2).

References: (1) *MIY* 1982. (2) *MAR* 1982. (3) *MM* 5/81.

552 Sipos Ltd

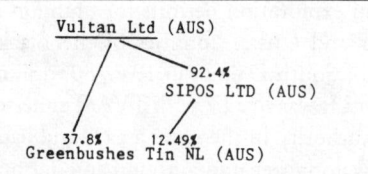

```
     Vultan Ltd (AUS)
              92.4%
              SIPOS LTD (AUS)
   37.8%    12.49%
 Greenbushes Tin NL (AUS)
```

In 1978, Sipos, an exploration and investment company, entered a 49% JV agreement with Vultan to explore the Glen Florrie uranium prospect (1).

The two companies also entered in the early '80s, a JV which pegged 31 diamond claims in the Aboriginal region of the Kimberleys, Western Australia (2).

Between them, Vultan and Sipos owned just over half of (the Australian) Greenbushes Tin NL, which is one of the world's biggest producers of tantalum (3).

References: (1) *FT* 7/8/78. (2) *Reg Aus Min* 1982. (3) *MAR* 1984.

553 SKB/[SKBF]/Svensk Kärnbränslehantering AB (S)

SKBF has been prospecting for uranium in Sweden for many years. Together with LKAB, it investigated the Pleutajokk uranium deposits in the Arjeplog district in the late '70s – despite great local resistance and an eventual moratorium (1) (see LKAB).

More recently, SKBF began test drilling in the Krokom district of northern "Lappland" in a mountain sacred to the Saami. A furore was caused in early 1984 when pro-uranium leaflets aimed at every householder in the area were leaked to the press by the anti-nuclear group FMK, which then blasted SKBF's complacency over health and safety issues at a successful Stockholm press conference (2).

In 1983, SKBF also reported on a deposit at Lilljuthatten in Jämtland. Uranium mineralisation here, and in nearby Nöjdfjället, comprises at least 2300 tons of contained uranium grad-

ing at over 1000g/ton. Further "promising" mineralisation has also been found at Kvarna in the Boden area. At the beginning of 1984 the company held 37 exploration concessions. A decision to mine at Lilljuthatten (it could provide 15% of the uranium required for the Swedish nuclear programme) was postponed due to "an abundance of uranium on the world market" (3).

SKBF later became SKB which – according to a report by Miles Goldstick (4) – was a move intended to stress the new emphasis on management within the company and the government's role in providing fuel.

As of 1984, the company's uranium contracts were as follows:

- 1983-1992 Australia, 1120 tonnes;
- 1983-1992 Canada, 4400 tonnes;
- Niger and Gabon, between them providing 1870 tonnes;
- USA, 390 tonnes.

Estimates of reserves at Lilljuthatten are 2000 tonnes of uranium grading at 0.1%. Reindeer were reported no longer giving birth, after uranium was released into the atmosphere during dynamiting for a dam in the early 1970s. Deformed calves were also reported.

Contact: Folkkampanjen mot Kärnkraft, Göran Eklöf, Arjeplog, S-10326 Stockholm, Sweden.

References: (1) LKAB Annual Report, 1980. (2) WISE, Amsterdam, *News Communiqué* 2/84. (3) *MAR* 1984. (4) Information from report by Miles Goldstick, 9/84.

554 Skelly Oil Co

This was part of a JV with Getty Oil, Cleveland-Cliffs, and Nuclear Resources in Powder River, Wyoming in the late '70s (1).

Skelly merged with Getty in 1977 (although by 1942 Getty already had control of Skelly thanks to acquiring Mission Oil which had 57% of Skelly) (2).

References: (1) *US Surv Min* 1979. (2) J P Getty, *As I See It*, London, 1976 (W H Allen).

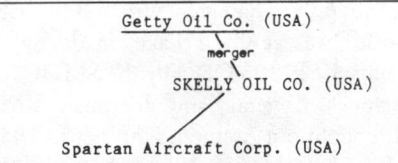

Getty Oil Co. (USA)
merger
SKELLY OIL CO. (USA)

Spartan Aircraft Corp. (USA)

555 SMDC

SMDC and Eldorado Nuclear Ltd merged in 1988 to form Canadian Mining and Energy Corporation (CAMECO) (30).

By the mid-eighties the President of SMDC and his fellow directors were laughing all the way to the bank (1). By law, their company has had a 50% (minimum) stake in every major uranium project in the province of Saskatchewan started after March 1975 (2). Saskatchewan uranium is unequivocally the hottest property on the world yellowcake market. With its interests in Key Lake and Cigar Lake, its 20% holding in Cluff Mining Co (Amok 80%) and many current exploration projects in the province (3), the SMDC provides relatively cheap, long-term, reliable supplies of uranium from reserves which, according to a 1985 report, could load "all the West's existing nuclear reactors with fuel for the next 15 years" (3). Thanks to Saskatchewan supplies (primarily Key Lake), Canada in 1985 leaped to the forefront of world uranium producers (29 million pounds of U_3O_8 out of a total of nearly 90 million) (4).

The SMDC was formed on June 4th 1974, by an Order-in-Council issued under the authority of the Canadian Crown Corporations Act. Just over three years later it was formalised through the proclamation of the Saskatchewan Mining Development Act 1977 (5). Funding for the SMDC was provided by the Heritage Fund and Crown Investments Corporation – advances which can be recalled by the provincial government, but are not charged with interest (6). In addition, C$183.4 million in long-term loans was taken out by the Corporation, which was unconditionally guaranteed by the provincial government (7).

The corporation was empowered to search for minerals in the province, except oil, gas, potash and sodium sulphate. Although the Act gave the SMDC wide powers of prospecting mining, trading and franchise (5), its major emphasis until the creation of CAMECO was on entering Joint Ventures with private companies. It also has corporate links with the United Steelworkers of America union (2).

However, the Act did enable the Corporation to form its own subsidiary, SMD Mining, which could cross borders, especially in pursuit of uranium – as indeed it did when prospecting for uranium in Manitoba at Lynn Lake and near Flin Flon (6), and in Alberta (8).

In the first effective year of its life, the SMDC was already participating in six active JVs, all but one for uranium.

It partnered Uranerz and Inexco during explorations at Key Lake, until buying out Inexco's interests in 1978, when the Canadian government blocked a deal with Denison. Exploration also took place in Alberta (9). This JV later became the Key Lake Mining Corporation, with the addition of Eldorado's participation. It participated in a JV in 1976 with Eldorado and

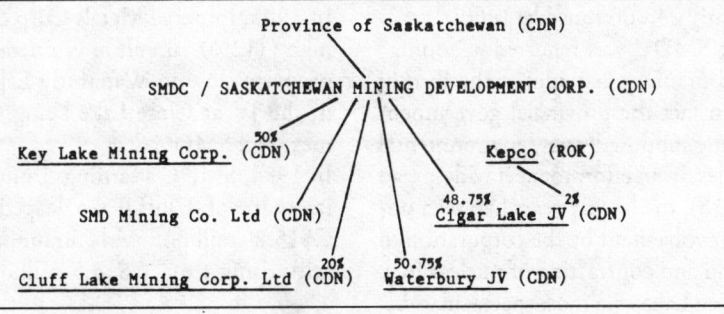

Province of Saskatchewan (CDN)

SMDC / SASKATCHEWAN MINING DEVELOPMENT CORP. (CDN)

Key Lake Mining Corp. (CDN)   50%   Kepco (ROK)

SMD Mining Co. Ltd (CDN)   48.75% 2% Cigar Lake JV (CDN)

Cluff Lake Mining Corp. Ltd (CDN) 20% 50.75% Waterbury JV (CDN)

Amok at Fond du Lac, which was managed by Eldorado (5). Together with Aquarius Resources, Eldorado, Goldak, Highwest Developments, Mountain Minerals, Radiometric Surveys, Séru and Taurus Oil, it headed up the Lake Athabasca Joint Venture Project in 1976 (5).

By 1981, the SMDC was involved in no fewer than 250 uranium Joint Ventures, including ones with Wyoming Minerals, the CEA and Union Carbide among non-Canadian companies (10). These had increased to 300 within a year or so (2) and were worth well over US$600 million (11).

Other important JVs with domestic companies included ones with Asamera, Eldorado Nuclear at Geckie West and Geckie East (12), Norbaska and Consolidated Reactor. By the early 1980s, the SMDC had effectively "mapped out" uranium anomalies in Maurice Bay, Michael Lake, an area south of Rabbit Lake, Collins Bay, Midwest Lake, and other areas, as well as the Lake Athabasca region (2). It was also exploring in La Ronge, Southend, Hanson Lake and La Loche (6).

Its interests in the Dawn Lake property, managed by Asamera, which showed such encouraging results in the late 1970s and early 1980s (13), were partly sold to Kepco in January 1983 (14).

By then the SMDC had also purchased 20% of the Cluff Lake mine, operated by Amok, at a cost of US$24.7 million (15) to cover its capital costs and share of the related development (16). It was also partnered with Eldor Resources and Noranda Exploration at Wollaston Lake, where estimated reserves at the Eagle North uranium project were raised in early 1986 from around 18,000 to nearly 24,000 tons U_3O_8 (17).

Although the SMDC was reportedly "toning" down its avid uranium activities at the turn of the decade (in fact the provincial government was introducing supposedly new environmental and health rules, in an effort to meet widespread opposition) (18), the last six years had seen unprecedented involvement by the corporation in the production and contracting of nuclear fuel. Apart from Key Lake, the most spectacular dis-

covery (not only in Saskatchewan, but probably the world) was at Cigar Lake, made by the Waterbury JV in 1983. Led by the SMDC, the JV includes Cogéma and Idemitsu Kosan (which bought out Asamera's interest in 1982) (19). Initial probes at the prospect revealed uranium grades of more than 25% (20), and conservative estimates of some 48,260 tonnes of uranium (21).

As work proceeded on the prospect in 1984 and 1985, the main seam alone was judged to contain around 130,000 tons of uranium at an average grade of 14% U_3O_8 (22). Additional reserves of around 40,000 tons were located in the western extension area of the prospect (23).

The Cigar Lake Mining Company – formed with SDMC holding 50.75%, Cogéma Canada (previously known as Séru Nucléaire) 32.625%, and its own subsidiary Corona Grande 3.75%, with Idemitsu holding 12.875% (23) – estimated that a final development and feasibility plan for Cigar Lake could be drawn up by 1987 (24) and a start-up date was provisionally fixed for 1993, with production of 12 million pounds a year of uranium oxide – about the same as for Key Lake (25). "Project development [was to be] timed to take advantage of the anticipated strengthening of the uranium market by late 1989 to early 1990" commented the *Engineering and Mining Journal* in early 1986 (26). However, there are numerous technical problems involved in raising the ore (not the least being the high clay content) and active consideration has been given to using new mining techniques because of the high ore grade and ground instability (22). A 500m shaft was sunk in 1990 (31).

In 1985, Imperial Metals Corp discovered very high (12%) uranium mineralistiu at its property close to Waterbury Lake: its partners in the JV at Close Lake being Uranerz, Geomex, and SMDC (27).

In 1984, SMDC's earnings from Key Lake and its share of Cluff Lake leaped sevenfold to C$15.2 million, with uranium concentrate sales rising from C$13.5 million in 1983, to

C$133.3 million that year – a tenfold increase (28).

Apart from its controversial uranium sales to South Korea (see Key Lake), the SMDC concentrated on increasing sales to western Europe, Japan and the United States. In 1981 it sealed a contract with two Swedish utilities which between them operate nine of the country's 12 electricity generating reactors (7). But its biggest hopes were set on tying up the US market which, as Thomas Neff points out "... finds such purchases a cheaper way to meet delivery commitments than through domestic production". Adds Neff: "With the United States as their major customer, Canadian producers also minimise the risk that proliferation concerns will escalate as a public issue threatening future exports". This is a neat acknowledgment of the fact that, although the world's biggest proliferator of nuclear weaponry, the USA is protected from mildly stringent Canadian bans on certain nuclear exports, imposed in the mid-seventies (5), simply by already having more than enough atomic bombs to blow the world up several times over. Among its non-uranium ventures, SMDC was developing a platinum project in Saskatchewan, together with Lacana, Murphy Oil, Conventures Ltd, and Silver Lake Resources in 1986 (29).

In 1988, together with Esso Minerals and Tri Gold Industries Inc, it discovered the Hanson Lake coper-zinc-silver lode near Flin Flon, Manitoba (32).

Further reading: *Time to Stop and Think: Saskatchewan People Must Decide*, published by Saskatoon Environment Society, Box 1372 Saskatoon, Saskatchewan S7K 3N9, Canada.

Contact: See references under Key Lake.

References: (1) *MJ* 30/12/83. (2) *Yellowcake Road.* (3) *FT* 31/7/85. (4) *E&MJ* 3/86. (5) *EPRI* 1978. (6) SMDC *Annual Report* 1981. (7) SMDC *Annual Report* 1982. (8) *FT* 17/5/78. (9) *MM* 7/78. (10) SMDC *Annual Report* 1980. (11) *Nuclear Newsletter*, Saskatoon, 16/6/81. (12) *FT* 18/12/78. (13) *E&MJ* 1/83. (14) *E&MJ* 6/84. (15) *FT* 31/7/79.

(16) *MIY* 1985. (17) *Northern Miner*, Canada, quoted in *MJ* 7/3/86. (18) *Nuclear Fuel*, Washington, 31/3/80. (19) *MM* 11/84. (20) *FT* 29/12/83. (21) *E&MJ* 2/84. (22) *MJ* 8/11/85. (23) *MM* 3/85. (24) *MM* 1/86. (25) *E&MJ* 1/86. (26) *E&MJ* 3/86. (27) *MJ* 3/5/85. (28) *FT* 31/7/86. (29) *MJ* 28/2/86. (30) *E&MJ Int Directory, op cit* 1991. (31) *MAR* 1991. (32) *E&MJ* 7/88.

556 SMTT

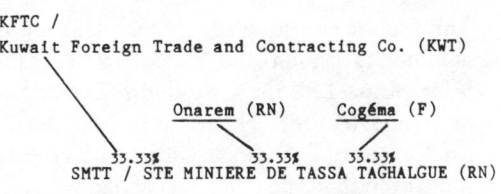

It was established in 1978, jointly by Cogéma and Onarem, to develop the third major mine of Niger at Arni in the Arlit concession area (1). In 1981 KFTC/Kuwait Foreign Trade and Contracting Company became a partner in the company.

Published reserves at Arni exceed 20,000 tons; production in 1981 was planned from open-pit, starting at 1000 tons a year and rising to 1500 (1). By early 1985 the project was still in its initial stages, though a possible expansion of milling to 1.8 million tons per year was being discussed (2).

Since SMTT's open-pit mine at Taza opened, it has contributed its uranium output to Somair's Arlit operations (3).

References: (1) *MAR* 1982; *MJ* 22/8/80. (2) *MIY* 1985. (3) *MAR* 1991.

557 SNC Group (CDN)

Dr Arthur Surveyer (*sic*) founded a mining consultancy in Montreal in 1911. By 1937 he'd joined with Messrs Nenninger and Chéneven to form the SNC Group to provide "full project services" for almost all metal extraction, including uranium (1).

In 1985, SNC undertook a JV with Cominco and Noranda to rehabilitate the Tarkwa and Presta underground gold mines and the Dunkwa alluvial dredge in Ghana (2).

References: (1) *MM* 3/83. (2) *FT Min* 1991.

558 Soarmico/Somali-Arab Mining Co (SP)

This was the executing agency for a US$300M uranium extraction and concentration plant being planned for the Galfudud region of Somalia in 1984. Both Nuclebras and the Brazilian Construtora Andrade Gutierrez company were involved in late 1984 in negotiating a contract to build the complex (1).

Although 12,000 tons of reasonably assured resources of uranium had been announced by the International Atomic Energy Agency (IAEA) in the Murdagh region (350km north of the Somali capital), by 1986 there were many doubts about the feasibility of this project (2), which was apparently later abandoned (3).

References: (1) *MM* 11/84. (2) *MAR* 1986. (3) *MAR* 1991.

559 Socal

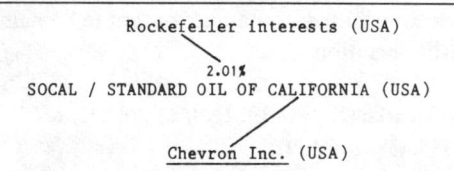

In 1981 Socal was the world's sixth largest oil company (after BP but before Gulf). After acquiring Gulf in 1984, Socal became the world's third largest oil company (on a par with Mobil) (1).

In recent years, Socal has traded increasingly under the name of Chevron, its marketing, minerals and chemicals subsidiary. All information about Socal's uranium (and other mining) interests are to be found under Chevron and Gulf Oil in the *File*.

Like other subsidiaries of the huge Standard Oil empire, Socal was originally a local oil company established in San Francisco in 1879, and given nominal "independence" when the Rockefeller trust was broken up in 1911 (2). A highly successful conglomerate through to the '80s (1980 was a record year) with important interests in Indonesia and the Ninian oil field in the North Sea, Socal began declining after 1981, especially in Europe. In 1983 it announced that it was considering the sale of its "downstream" operations in Britain, West Germany, Italy, Denmark and the Benelux countries (3).

It holds 50%, with Texaco, of the Caltex marketing company. In 1984, Caltex wanted to expand into synfuels in South Africa with Anglovaal but the apartheid régime refused funding (4). Socal's most important mining interests were held through Amax, 20% of which it acquired in 1975 – a minority interest, to be sure, but one which made it impossible for another company to acquire Amax unless Socal approved (5). In 1981 Socal made a bid for total control of Amax. It was eventually rejected, but in circumstances which make it difficult to know if Amax was only trying to prompt a higher bid from Socal or really to block the take-over. Some Amax shareholders were furious at the management's refusal to be taken over – a move which would have doubled the value of their shares – and a number of them filed suit against the company (5). However, a little later, Socal did pick up an extra 3.3 million shares in Amax, after Amax declined to buy back BP's 6.5% stake (which BP acquired through its Sohio/Kennecott take-over and was required to sell under US law) (6).

References: (1) *FT* 6/3/84. (2) *WDMNE.* (3) *FT* 7/5/83. (4) *MAR* 1985. (5) *BOM.* (6) *MIY* 1985.

560 Sodémi

It took over uranium exploration in Ivory Coast from Petroci in 1981.

Some of the company's "substantial" exploration programmes – including those for uranium – have been conducted with the French company BRGM (1).

Work carried out in 1983 identified several uranium drill points over a number of anomalies in the south and north-east of the Odienné region, while possible uranium strata were studied at Tindrina Sokoro, Linguesso, Gbahalan, Dyafana, Bougoussa and Goulia in the same area. Other work was carried out in the *départements* of Bouaké, Dabakala, Ferkessedougou, Korhogo, Mankono, and around Niakaramangougou and Dikodougou (2). Sodemi formed a JV with BRGM (through Coframines) in 1984, which then developed the Ity gold deposit; this started production in 1991 (3).

References: (1) *MJ* 16/12/83. (2) *MAR* 1984. (3) *MAR* 1991.

561 Sofrémines

This is a large-scale (2000 employees) civil engineering and managerial group which has assisted in the construction of numerous mines, pipelines and other projects.

Its uranium "references" include feasibility studies and evaluation for the Hoggar mines in Algeria, Mounana and Boyenzi in Gabon, and Akouta in Niger (1).

References: (1) Publicity material by Sofresid Group, 1984.

562 Sogérem

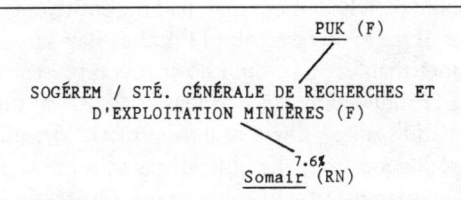

This is not to be confused with Sonarem, the Algerian state company, nor Sonarem of Mali, nor Somarem of Morocco! Sogérem was named among other companies (including Sofrémines, to confuse the non-francophone reader even further!) as landing a contract to produce uranium in Algeria's Timagouine region during the 1980s (1). However, this news appears to have been premature and none of the companies seems to have proceeded beyond the feasibility stage. Sogérem currently quarries sandstone in France (2).

References: (1) *MJ* 19/8/77, *MM* 8/77, *MM* 9/77. (2) *E&MJ Int Dir of Mining*.

563 Sohio

Formerly the "emasculated base of the Rockefeller Empire" (1), Sohio is now firmly held by Britain's largest multinational. But, far from being BP's yankee poodle, Sohio became the buttress which supported the BP board (2).

```
        BP (GB)
        ___
          \
SOHIO / STANDARD OIL CO. (USA)
```

BP's direct involvement in the oil company dates from 1970, when it exchanged an interest in its Alaskan North Slope oil bonanza (Prudhoe Bay) for a stockholding. This manoeuvre skirted legislation requiring US domestic oil to be sold within the country (3).

Sohio then held more than half the entire reserves of the main reservoir of Prudhoe Bay – itself more than 25% of the known US oil reserves (4). Prudhoe Bay, until the early 1980s, was the "world's largest single-mineral project" (5), and production from the oilfield was so successful, that it transformed Sohio from 17th US oil company in 1970, to number 7, in only nine years (1).

BP acquired 53.1% of Sohio in 1979, when the company's Alaskan oil output reached 600,000 barrels a day (6). Awash with spare cash and securities, the American company then looked for new areas into which to diversify. Oil, gas, coal and chemicals were named as priorities for acquisition (7). Of these, the Kennecott merger was by far the most important – and costly, at a price of US$1.77 billion. Kennecott went on to cost Sohio a lot more, and was the major contributor to a US$1860 million write-off made by the parent company in 1985 (8). These losses were, to some extent, recouped by a modernisation programme announced at the same time (9). (For a fuller account, see Kennecott Corp.)

Meanwhile, Sohio was buying into US Steel's coal mines in Pennsylvania, Utah, Illinois and West Virginia (10), and a year later, further coal properties were purchased from the US's fourth largest steel producer, Republic Steel (11).

Between 1979 and 1985, the company's fortunes fluctuated. A rise in profits in 1980 (12) was followed by the peaking of Prudhoe Bay oil production at the end of the year (7) and a drop in sales. Then 1981 saw a slight upturn in fortunes, thanks to higher prices for Alaskan crude (13) – much to the surprise of some market analysts (14). Soon afterwards, the slide began again: there were higher exploration costs (15), Kennecott's losses were denting profits badly (16), and the company announced a reduction in its interests in the Alaskan oil fields, enabling its partners Exxon and Arco to boost their shares (17). At the end of 1982, Sohio's profits were down in all departments (18). With predictions that total output from Alaska would start to wane in the mid-eighties (19), Sohio and BP gambled on a costly and ultimately sterile venture into the Mukluk oil field, also on the North Slope (see BP). Matters were not helped by a dispute with the state of Alaska over prices charged for transporting oil through the Trans-Alaska Pipeline System (TAPS) in which Sohio holds a one-third interest. Although BP settled its side of the controversy, by 1985 Sohio had not (20).

In 1984, the company announced three new oil ventures on the North Slope. One of these was a JV with Exxon and Arco in the Lisburne oil reservoir, and another off-shore in the Beaufort Sea, at Endicott (21). These new projects place Sohio firmly in the gunsights of Inuit and other indigenous organisations in the Arctic, alarmed at pollution prospects and the disturbance of the marine environment, although to date most opposition seems to have been directed against Gulf, Arco and Exxon. It is also possible that development of these new fields will be held back, thanks to what Sohio's chairman Alton Whitehouse called a "tougher environment [economically] worldwide" (22): the company cut nearly 500 corporate staff jobs in 1985 alone (22).

Prompted by the slip in earnings, Sohio bought back more than 11 million of its own shares in 1984 which (since BP did not tender any of its stake in the company) effectively raised the British partner's share in Sohio to 55% (23).

Non-oil and -gas energy interests have not proved Sohio's forte. It flirted – to the tune of around US$90 million – with a new US company, Energy Conversion Devices (ECD), intent on commercialising silicon cells for electricity generation (24).

It also embarked, in the late '70s and early '80s, on several uranium exploration projects.

Two hundred claims were staked in 1981 in the Santa Fe National Forest, east of Cuba, New Mexico, to the opposition of virtually all 50 residents of La Meso de la Poleo, concerned at possible irradiation of their water supplies (25). A public meeting held in June that year scared Sohio officials away at the last moment, fearful, as they put it, of a "witch-hunt" (26).

Sohio did not press its claims at Santa Fe. Nor was it able to proceed with similar plans in New Jersey where exploration began in the late 1970s (27). Thanks to pressure from the New Jersey Coalition to Stop Uranium Mining, and local residents determined that their drought-stricken area should not be further depleted of water or endangered by radioactivity, a seven-year state-wide moratorium on mining was announced by the governor in May 1981 (28).

At the beginning of 1976, Sohio was operating one uranium mine, and held 0.8% of US uranium reserves, ranking it after Phillips Petroleum (with 1.8%) but before Union Carbide (29).

Although just on the borders of Native American land, the impact of the mine – along with those owned by Bokum, Kerr McGee, Exxon, and especially Arco – was felt directly by the Laguna people at Paguate pueblo, on the Canoncite Indian Reservation (30). Known simply as JJ No. 1, the mine and mill (the "L-Bar") were closed in May 1981, thanks to a "depressed market and increased costs and government regulations" (31). Although the mill at Seboyeta was expected to be reopened in 1982, it never did (32). Sohio was then effectively – like almost every other US oil company – no longer in the uranium stakes.

In 1987 BP took over the remainder of Sohio which it did not own, and could at last bring the US subsidiary under orders after some drastically bad investment decisions during the 1970s. The merger put BP in third place among the world order of oil companies (only Exxon and Royal Dutch/Shell now being bigger), and made BP a truly international company (33). It presaged aggressive exploration to extend oil holdings on the Alaskan North Slope and pressure on the US Congress to open up the Artic National Wildlife Reserve for exploration (33). In 1989, RTZ took over almost all of BP's mineral interests, (the main exception being those of BP Canada) including those of BP America – and, therefore, Sohio (see RTZ and Kennecott).

Contact: NJ Coalition to Stop Uranium Mining, Box 271 New Vernon, NJ 07976 USA.

References: (1) *Gua* 22/8/80. (2) *FT* 3/9/83. (3) *Gua* 11/7/78. (4) *Big Oil.* (5) Tanzer. (6) *FT* 10/10/79, *Gua* 22/8/80. (7) *FT* 31/12/80. (8) *MJ* 6/12/84. (9) *MM* 1/86, *MJ* 6/12/85. (10) *FT* 10/4/81. (11) *FT* 1/4/81. (12) *FT* 23/10/80. (13) *FT* 22/10/81, *FT* 29/1/82. (14) *Gua* 28/1/82. (15) *FT* 23/7/81. (16) *FT* 23/4/82. (17) *FT* 29/6/82. (18) *FT* 28/1/83. (19) *Gua* 2/11/83. (20) *FT* 29/4/85. (21) *FT* 29/9/84. (22) *FT* 28/6/85. (23) *FT* 27/7/84. (24) *FT* 2/4/82, *FT* 7/2/83. (25) *Kiitg* 8/81. (26) *Mount Taylor Alliance Newsletter* Vol. 2, No. 3, 20/6/81, Albuquerque. (27) *Kiitg* 8,9/80. (28) *WISE News Communiqué* 7/5/81, Washington. *(29) Petroleum Industry Involvement in Alternative Sources of Energy* (US Govt Printing Office) 1977. (30) *San Juan Study* 1980 and Tom Barry, "Uranium Mining on Indian Land", *Denver Post*, Denver, 4/11/79. (31) *Grants Daily Beacon*, Grants, 18/5/81. (32) *MIY* 1985.(33) *FT* 22/7/87.

564 Somair

In the early '80s a BBC camera team visited the Somair uranium pits at Arlit. There, they discovered fifteen- and sixteen-year-old Touareg (nomad) boys emerging from the mines, covered with radioactive dust. Alongside an artificial oasis constructed for the French management, with "supermarkets, horse-racing, luxuries flown in from home", as producer Chris Olgiati described the scene, "some of the poorest people on earth labour in one of the deadliest environments to power the electric train sets and fuel the bombs of the world's richest nations" (1). According to Olgiati, "Arlit is

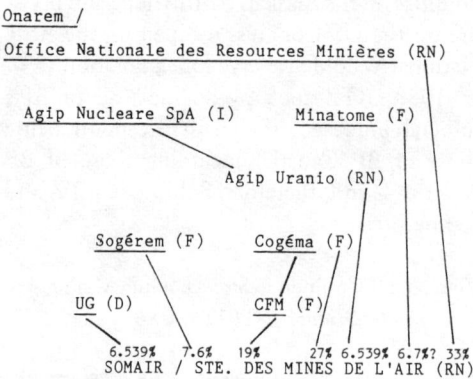

```
Onarem /

Office Nationale des Resources Minières (RN)


    Agip Nucleare SpA (I)          Minatome (F)


                    Agip Uranio (RN)


    Sogérem (F)        Cogéma (F)


    UG (D)              CFM (F)

        6.539%  7.6%  19%    27% 6.539% 6.7%? 33%
       SOMAIR / STE. DES MINES DE L'AIR (RN)
```

very important to the French. They've got a huge transmitter which communicates directly with France ... and they absolutely hated a TV team being there" (1). And, according to a "nice French manager" at the mine, "they were working in absolute ignorance of what might happen to [the miners] in twenty or thirty years time" (1).

It was the French CEA which discovered Niger's uranium after reconnaissance work carried out between 1954 and 1956. Arlit wasn't located until 1966, but was brought into production only five years later, in 1971. The ore – contained under up to 85 feet of clay and sandstone, itself lying under 100-160 feet of overburden, could only be mined by stripping 1.7 million tons of ground (2). When the project commenced operations in 1971, the first of the three sections – Arlette, Ariège and Artois – was mined as an open pit, using a newly developed acid leach system (3). Heap leaching was introduced two years later when the plant was increased from its initial 750t/year capacity. A further expansion took place in 1982 (to 2300t/year) and underground mining was also introduced (1). Production at 1600 tons in 1976 (4) fell slightly in 1977 to increase again to a record 2100 tons in 1981 (5). By then Arlit and Akouta (see Cominak) were jointly delivering 4600 tons (6) and Akouta had edged ahead of Arlit as the country's most important uranium producer. (1980 world figures put Cominak in fourth position among the world's top producers, and Somair at seventh) (7).

Arlit in 1985 had total reserves of about 35,000 tons uranium, "sufficient to permit mining to continue well into the next century at the current operating rate of 1400t/year" (8). This rate had increased to 2300t/year by 1990, with a total reserve put at 79,500 tons uranium when output from the Taza pit, licensed by Somair from SMTT, is included (22). Somair's production must be set in the context of estimated reserves for Niger as a whole of between 130,000 and 650,000 tons U_3O_8 (9). More important, however, is to try to evaluate the role of Somair (and Onarem and Cominak) in the context of three political factors which have dominated Niger's uranium industry for nearly fifteen years: French control of the country's strategic resources; Niger's importance as a provider of relatively unsupervised nuclear supplies to other African nations; and the crucial role that uranium has played in "developing" (some would say distorting) Niger's fragile economy. France has a major interest in all Niger's uranium deposits. When the CEA/Niger agreement to exploit Arlit was signed in 1968 it contained a "most favoured company" clause, "normally strongly resisted by Third World mining negotiators" (10). This clause guaranteed that, should other companies negotiate better terms in future, these terms would also apply to the old agreement (10). In practice, however, this clause appears not to have been invoked. Indeed, France has been so intent on maintaining access to Niger's relatively high-cost uranium that it continues to pay prices considerably higher than those prevailing on the spot market (8). In 1984, France agreed to increase the price through that year by 6% – offering US$40/lb U_3O_8 when the spot market price was virtually half (11). In 1985 France topped up its price by 2.5%, though it "almost certainly reduced the volume of its purchases in view of scaled-down estimates for the growth of electricity consumption this century" (12).

There were also rumours in the mid-'70s that France had actively engineered the overthrow of President Hamani Diori in 1974, when he pushed too hard for an increased share in Somair's equity (13). In fact, Onarem's equity in

Somair was doubled in April 1975 (3) and Niger's (new?) mining law enables the government to secure a free economic interest in all uranium permits and concessions (14). In any event, the importance of supplies from Niger to France has not been in doubt. In a confidential report to his government, in the mid-'70s, French Foreign Minister Louis de Guiringaud stressed the "important risks" connected with any breakdown of uranium production in Niger: "We would lose 10 million francs if mining [in Niger] were to stop for just three days" (13).

But this "captive market" has failed to absorb all Onarem's share of Niger's production, and the state agency began making small sales to other customers (15). A substantial sale was allegedly made to Libya in 1981, of about 1500 tons (6), although some commentators discounted this as a CIA-inspired rumour (1). Rumours have also persisted that the country would sell to Iraq and Pakistan (6), while Iranian diplomatic visits to Niamey, the capital of Niger, prompted speculation that a deal with the Khomeini tyranny was also in the offing (11, 15). According to the MIT report on Africa's uranium industry, definitive sales were made to Libya in 1980 (380 tonnes) and 1981 (1212 tonnes) and to Iraq in 1981 (100 tonnes) (16). Egypt was also bidding for uranium from Niger in 1983 (11).

Onarem made regular spot sales to western Europe using advice given, for some years, by Edlow International, the uranium brokerage company (12).

Niger's former military dictator Seyni Kountché made no bones about his willingness to sell uranium – to the devil, if need be (17). Presumably Onarem and Somair will need to extend their options even further in the next few years as government tax revenues from the industry fall, in line with lower exports (8). Yellowcake production as a proportion of government income rocketed from 14% in 1975 (18) to over 75% nearly a decade later. By the early '80s, however, there were ominous predictions of a decline in revenues (19).

In 1984 a coup attempt on Kountché, led by a close associate of the President, was probably a response to austerity measures initiated by the dictator and stipulated by the IMF to increase privatisation and reduce the power of the state-owned monopolies (15).

These events coincided with the second worst famine that Niger has experienced in recent decades. Despite Kountché's resolve to increase food production, Niger's economy has been dominated by two export-oriented industries whose cyclical fortunes hold the people hostage. Dependence on ground nuts for foreign markets was a direct cause of the murderous drought and famine in 1968/74 which affected Niger as badly as any of the Sahel countries (20). Although no intensive study has been yet done of the connection between reliance on uranium exports, and the recent horrendous starvation in northern Africa, there is little doubt that Niger's trade balance has been adversely affected by the industry, with capital equipment, sulphur, fuel and luxury items for the French, taking precedence over fertilisers, grain and agricultural equipment. The long-term consequences for the people of Niger are dim indeed, notwithstanding the luxury hotel on the banks of the Niger that Euro-francs have been able to provide (19).

By the mid-'80s it was clear (to quote Roger Murray, an expert on African uranium) that "the sharp fall in uranium revenues ... sent overall economic growth into reverse and caused a sharp rise in indebtedness as the government initially resorted to increased non-concessional borrowing in an attempt to maintain spending levels" (12). In other words, putting its eggs in the uranium market has proved, for the people of Niger, a costly – perhaps deadly – mirage.

TVO, the Finnish nuclear utility, receives uranium from Niger, presumably from Somair. In 1985, a special correspondant for the pro-business weekly *West Africa* wrote thus:

"About a decade ago, Niger thought it had more than enough revenue from uranium to fuel growth, develop the country, and later pay back all the extra bills. Now, it epitomises the fate of a one-resource developing country condemned to pay severely in this decade, if not in the next, for misdirected spending over the past few years.

"Western finance officials say the government was warned that uranium earnings, which seemed limitless in the years 1976-80, would be short-lived ...

"An apparently desperate head of state once said he would sell uranium even to the devil. But, in a recent interview, President Seyni Kountché claimed that not even the devil had come" (21).

Contact: EVY (Energiapoliittinen Yhdistys Vaihtoehto Ydinvoimalle RY), Jääkärinkatu 6, SF-00150 Helsinki, Finland.

References: (1) *WISE International Networking Bulletin for the Safe Energy Movement*, Amsterdam, 6/82. (2) *EPRI Report*, MM 12/78. (3) *MAR* 1984. (4) *MJ* 1/4/77. (5) *MAR* 1982. (6) *RMR* Vol. 2, No. 4, 1984. (7) *BOM*. (8) *MJ* 25/1/85. (9) S Koutoubi & L W Koch "Uranium in Niger", in *Uranium and Nuclear Energy, London, 1980 (Mining Journal Books)*. *(10) Mmon* 4/81. (11) *African Business*, London, 5/84. (12) *African Business* 5/85. (13) *New African*, London, 6/78. (14) *MJ* 25/1/84. (15) *African Business*, London, 7/84. (16) Lynch & Neff. (17) BBC-TV Panorama programme, producer Chris Olgiati, 1982, exact date unknown. (18) *West Africa*, London, 18/7/77. (19) *International Herald Tribune*, Paris, 12/12/82. (20) Richard Franke & Barbara Chasin: "Peasants, peanuts, profits and pastoralists", in *African Environment*, Dakar, 8/80. (21) *West Africa*, London, 17/6/85. (22) *MAR* 1991.

565 Somarem/Sté Marocaine de Recherche Minière (MA)

This was a joint French/Moroccan/US company set up in 1953 which explored all of Morocco until 1956, trying to locate uranium (1). There is no further information.

References: (1) *Uranium Redbook* 1979.

566 Sonarem

Not to be confused with the state-owned company of the same name in Mali, Sonarem handles all mining enterprises in Algeria (1), and in 1990 operated thirty mines (11). It was established by the independent Algerian government in 1966 after the nationalisation of the mining industry (2).

Two years later the company began a systematic investigation of uranium deposits in the Ahaggar (Hoggar) mountains of the southern Algerian Sahara, mainly around Timagouine, about 200km south-west of Tamanrasset. These deposits were originally located by BRMA (the Algerian Office of Ore Exploration) in 1958 and studied by the CEA. By 1970, an ore basin was delineated, containing significant deposits of uranium at Timagouine, Abankor and Tinef (3). Over the next four years an aerial, electromagnetic and radiometric survey covered, the whole country, revealing other interesting sedimentary deposits, though the Timagouine area remains the most important (1). Reserves have been put at around 50,000 tons (1).

Several major European and American companies have apparently participated with Sonarem in uranium investigations in Algeria. Uranerz commenced a feasibility study in 1975 – then negotiations broke down. Another feasibility study was awarded to John Brown Engineers and Constructors in 1977 (4). The same year Sonarem invited proposals from a number of US, Canadian, British, Belgian and French companies. A contract was apparently awarded to PUK/Pechiney, Minatome, Sogérem, STEC, Inter-G, and Sofremines (5) for production in the early '80s (6). Yet another report suggested that tenders for development at Timagouine

```
Republic of Algeria (DZ)
                                  \
                                   SONAREM /
    STÉ. NATIONALE DE RECHERCHE ET D'EXPLOITATION MINIÈRES (DZ)
```

had also been awarded, at roughly the same time, to Kaiser (USA), Kilborn (Canada), and Framatome (France) (7), with US consultants Darnes and Moore doing a feasibility study (5). However, these reports seemed premature or inaccurate. In the end it was the Belgians – through Union Minière, and Traction et Électricité – along with AG McKee who ostensibly ended up with the plum, along with the Swiss engineering company Cotecna (8). Even here, some confusion reigned: an earlier report stated that Cotecna had become Sonarem's main contractor (9), whereas three weeks later it seemed it was simply an equal partner.

Two years afterwards, Belgium and Algeria announced a broad-sweeping co-operation agreement which cited Belgian Mining Engineering as the key uranium prospector from then on (10). Everyone seems to have had their fingers at one time or another in Algeria's yellowcake, but the chances of any concrete developments in future are small.

In 1990 Sonarem (now renamed Enterprise Nationale de Récherche Minière) discovered a sizeable gold deposit in the Hoggar mountains – but again, it did not seem likely to be developed (12).

References: (1) MAR 1985. (2) EPRI Report. (3) Uranium Redbook. (4) MM 8/77. (5) MM 9/77. (6) MJ 19/8/77. (7) MJ 15/7/77 & 29/7/77. (8) MJ 6/7/79. (9) MJ 15/6/79. (10) Le Monde, Paris, 12/6/81. (11) E&MJ Int Dir of Mining 1991. (12) MAR 1991.

567 Soquém

Soquém is involved in gold, colombium, salt and other mining ventures (1) as well as uranium exploration (2). Among its JV partners has been Teck (in a niobium project) (3).

```
                    Province of Québec (CDN)
                            /
SOQUÉM /                  /
STÉ. QUÉBECOISE D'EXPLORATION MINIÈRE (CDN)
```

In 1980/81 the company spent nearly C$10M on more than 50 exploration projects in Quebec, just over 10% of which went on uranium (4).

Its most important uranium JV has been with with Uranerz, Séru Nucléaire, Esso Minerals Canada, the (Canadian) Pancontinental, and Bay Development Corp, in the Otish mountains of central Quebec (4, 1).

Soquém also had 34% in the Arthur W White Mine of Dickenson Mines (5).

The company also had a JV to explore for uranium with Eldorado Nuclear in the late '70s (6).

By 1991, Soquém had 44 exploration projects – 35 on its own account and 9 JVs with other companies. Half its budget was spent in the Chibougamau-Chapais region of Quebec (7). Notable are its JVs with Noranda in the Gaspé region, and with Rio Algom in the Gatineau area (8).

It also owns 20.8% of Cambior, an up-and-coming Canadian gold miner, one of whose main projects is the Omai JV with Golden Star Resources (also of Canada) which it took over from Placer (9). This is potentially the biggest gold mine in rainforested Guayana, near the Venezuelan border, and has already aroused concern among Indian communities in the region and among rainforest protectors.

Contact: World Rainforest Movement, 8 Chapel Row, Chadlington, Oxfordshire OX7 3NA, England.

Probe International, 225 Brunswick Avenue, Toronto, Ontario M5S 2M6, Canada.

References: (1) MAR 1982. (2) MJ 16/10/81. (3) MIY 1982. (4) MJ 25/9/81. (5) MJ 13/7/84. (6) Eldorado Nuclear Annual Report 1979. (7) MJ 12/7/91. (8) FT Min 1991. (9) FT 2/1/91.

568 Southern Cross Exploration NL (AUS)

This is part of a consortium with Offshore Oil and Cocks Eldorado which joined Agip Nucleare (Australia), Urangesellschaft Australia, and Central Pacific Minerals, at Lake Ngalia in the Northern Territory of Australia (see Central Pacific Minerals).

In 1990, Southern Cross retained its 3.39% interest in the Bigrlyi uranium prospect at Ngalia, and had re-affirmed its interest in exploring for uranium with other partners (1).

References: (1) *Reg Aus Min* 1990, 1991.

569 Southern Uranium Corp (USA)

This was reported exploring for uranium in Namibia in 1977, along with three other highly dubious US companies, all registered at the Volkskas Buildings, Kaiser Street, Windhoek, Namibia: Midwest Uranium, Delaware Nuclear, and Tristate Nuclear (1).

References: (1) *Star*, Johannesburg, 30/7/77.

570 Southern Ventures NL (AUS)

In 1984 the CEGB started funding the major part of a JV in uranium exploration on the Lyndhurst uranium prospect in Queensland, in which Southern Ventures held 30% along with Cultus Pacific (1).

An un-named European company was expected in 1983 to provide A$100M for further exploration and water sampling work.

Southern Ventures also had several other JVs with Cultus Pacific:

- Robinson Range, Western Australia (WA), where low-grade uranium mineralisation had been located in the Mica Bore area;
- Thatcher Soak, 160km north-east of Laverton, WA, where 6000 tonnes of U_3O_8 grading 0.4kg per tonne, have been located;
- the Tanami Desert, Northern Territory, which it held 25% with Cultus (25%) and Otter (50%);
- Lake Maitland (Cultus Pacific NL) prospect, about 100km south-east of Wiluna, WA;

- Lake Mason, about 75km south-west of WMC's Yeelirrie deposit in WA, and
- Lake Raeside, 60km west of Leonora, WA.

The last three were 30% owned by Southern Ventures and 70% by Cultus Pacific (2).

Now 75% owned by Dominion Mining Ltd, the company operates the Gabanintha gold mine in Western Australia, near Meekatharra, and it is also exploring in Western Kalimantan (3).

References: (1) *FT* 16/11/84. (2) *Reg Aus Min* 1984/85. (3) *FT Min* 1991.

571 Spencer Chemicals

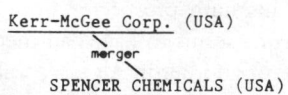

Kerr-McGee Corp. (USA)
merger
SPENCER CHEMICALS (USA)

This merged with Kerr-McGee in 1962 (1).

References: (1) US Congress Subcommittee on Energy statement, 1975.

572 SSEB/South of Scotland Electricity Board (GB)

Along with the UKAEA, it carried out various explorations for uranium in Scotland in the '70s, where deposits were discovered in eastern Ross, the southern Uplands, at Inverneil in Argyll (among lead in the old mines of Tyndrum) and, above all, in Orkney (1). Attempts by the SSEB to proceed with drilling were frustrated by a massive movement against uranium mining from the Orcadians.

References: (1) *MM* 11/82.

573 Standard Oil of Indiana

This is one of the so-called Seven Sisters, the major US oil companies. The only known

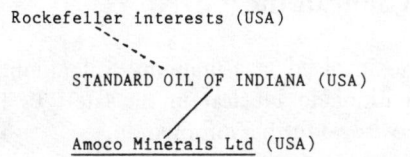

```
Rockefeller interests (USA)
        `\
          `\
           `\
    STANDARD OIL OF INDIANA (USA)

                              \
    Amoco Minerals Ltd (USA)
```

```
                              Gencor (ZA)
                             /
                            /
    STILFONTEIN GOLD MINING CO. LTD (ZA)
                            \
                          80% \
                      Chemwes Ltd (ZA)
```

uranium interests of Standard Oil have been through Amoco.

Along with BHP, the Papua New Guinea government and a West German company, Kupferexplorationsgesellschaft, Standard Oil of Indiana became part of the large Ok Tedi consortium, when it started prospecting a huge copper deposit on indigenous land in Papua New Guinea (1).

Amoco holds 30% of Ok Tedi Mining Ltd (2).

References: (1) *MIY* 1982. (2) *FT Min* 1991.

574 Stellar Mining NL (AUS)

This has been involved in silver, gold and gemstone exploration and mining, with prospects in the Kimberley region among others (1).

In April 1980 it purchased a 50% share in a JV with Kratos Uranium over various parts of New South Wales, searching mostly for gold and base metals. No uranium of significance was reported (2).

It also operated in a JV elsewhere with Getty Oil (2).

References: (1) *Reg Aus Min* 1981. (2) Annual Report, 1980.

575 Stilfontein Gold Mining Co Ltd

It is not a primary uranium producer – and its gold life seems to be limited (1, 5). It was a majority holder in one of three uranium-from-slimes schemes in the apartheid state. Uranium slimes from both Stilfontein and Buffelsfontein were treated until 1988 by Chemwes in a plant situated on Stilfontein's property in the Klerksdorp district of the Transvaal (1). Stilfontein ar-

ranged contracts for the sale of Chemwes uranium through Nufcor. Much of this appears to have gone towards the Koeberg-1 reactor (2). One of the few important gold mines in South Africa in which Anglo-American seems to have no interest (3), Stilfontein was owned as to 18% by Gencor, 7% by Sicovam, 4% by Scoges SA, and 13% by Barclays Bank in 1972 (4). In 1990, Gencor's attributed interest was 9.2% (5).

Stilfontein also "tributes" part of the northern section of the Hartebeestfontein property (1).

All Stilfontein's underground operations were due to cease in 1991 (6).

References: (1) *MIY* 1985. (2) Lynch & Neff. (3) *RMR* Vol. 3, No. 2, 1985. (4) Lanning & Mueller. (5) *FT Min* 1991. (6) *MAR* 1991.

576 Sturts Meadows Prospecting Syndicate NL (AUS)

It holds the Turpentine uranium prospect near Mount Isa, central Queensland, along with Mary Kathleen Uranium Ltd. In 1980, it located uranium anomalies at the Helafels group of leases at Cloncurry. It is also involved in diamond exploration in New South Wales, though the company was set up mainly to explore for uranium (1).

References: (1) *Reg Aus Min* 1981; *AMIC* 1982.

577 Sumitomo Metal Mining Co Ltd

This is a major company with interests in virtually all metals, engineering and nuclear fuel

```
                Sumitomo Corp. (J)
                          \
      SUMITOMO METAL MINING CO. LTD (J)
```

production. Its main products are electrolytic copper, gold, silver, ferro-nickel and cobalt (1). It has had farm-out agreements with CSR at Mount Gunson copper mine in South Australia (2) and with MIM on the Frieda River prospect in Papua New Guinea.

Together with Marubeni and C. Itoh it buys copper products from the PASAR smelter in the Philippines (see Benguet) (2).

In the 1970s Sumitomo was part of the Aurec consortium, along with Furukawa Mining and C. Itoh which, with ENI, was exploring for uranium in Australia's Northern Territory (3). Sumitomo acquired a 15% stake in the Morenci, Arizona, copper mine of Phelps Dodge in 1985 (4). Phelps Dodge was again its partner six years later when it offered 20% interest in the Chilean La Candelaria copper-gold project to Sumitomo (5).

In New Zealand, Sumitomo has a JV with the domestic company Newman's, in Nelson Pine Forest Ltd, which has come under fire for its destruction of native forest and its wood-chip milling. Ironically, there is a Newman's Tours operating in the country which prides itself on "making the most of our natural resources". Are they by any chance related? According to *Private Eye*: "The Newman's that chops down trees is the same Newman's that owns Newman's Tours. So while Newman's the Tour people are flying in planeloads of tourists to admire the scenery, Newman's the Tree People are busy tearing it down" (6).

References: (1) *MIY* 1982. (2) *MAR* 1984. (3) *A Slow Burn* (FoE/Victoria) No. 1, 9/74. (4) *FT* 7/11/85. (5) *MJ* 21/6/91. (6) *Private Eye* 17/8/91.

578 Suncan Inc (CDN)

It was involved in a uranium exploration JV with Eldorado Nuclear in the late '70s (1); otherwise no further information.

References: (1) Eldorado Nuclear Annual Reports.

579 Suncor Inc

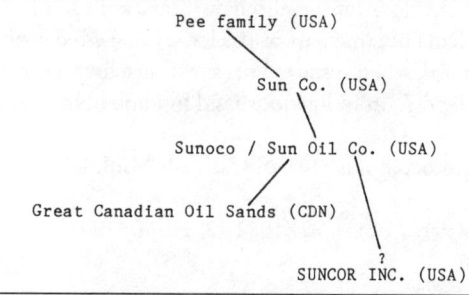

```
            Pee family (USA)
                         \
                 Sun Co. (USA)

            Sunoco / Sun Oil Co. (USA)
                                    \
   Great Canadian Oil Sands (CDN)
                                  ?
                        SUNCOR INC. (USA)
```

By the early '80s, Sun Oil owned several uranium properties which were under exploration, but none producing (1). In Canada, Suncor – majority owned by Sun Oil, but precise holding unknown – was exploring for uranium in Saskatchewan, with no known results (2).

Meanwhile Sunoco became an equal partner in Ocean Mining Associates along with US Steel, Union Minière, and ENI, one of the five big sea-bed exploitation companies set up in the late '70s (3).

An experimental mining vessel owned by Ocean Mining Associates successfully collected nodules some 4500m down in the Pacific in 1978 (4).

Suncor's Mildred Lake oil sands project has a $9 million "innovative" plant to increase the recovery of sulphur dioxide emissions, and thus the amount of sulphur gained. However in 1990 it was planning to sell 49% of Mildred Lake (5).

References: (1) *BOM*. (2) *Yellowcake Road*. (3) *RMR*, Vol. 1, No. 4, 1983. (4) *MIY* 1982. (5) *MAR* 1991.

580 Susquehanna Corp (USA)

It was involved in uranium mining and milling at Edgemont, South Dakota, until 1972, and sued by a Nebraska resident for US$15M damages (with punitive damages of US$20M and costs) for damage resulting from radioactive tailings (1). No further information.

References: (1) *Wall St J* quoted in *MJ* 29/5/81.

581 Swuco Inc (USA)

In the early '80s two US companies played a key role in enabling the apartheid régime in South Africa to acquire enriched uranium – despite the supposed US government embargo on such trade. Edlow International and Swuco – based respectively in Washington, DC and Rockville, Maryland – acted as brokers between Pretoria and uranium suppliers in Switzerland, Belgium and France. The uranium (origin unknown, but rumours have circulated that it derived from Canada, or even China) was held by companies in Switzerland and Belgium who handed it over to Ed & Swuc who then passed it to the Eurodif consortium (based in France) and an unnamed enrichment agency; title was then sold to South Africa.
To cap it all, South Africa then sold 3000lb of it to the USA for enrichment and contracted sale to Japan (1).

References: (1) *Kiitg* 6/82.

582 Taihei Uranium Exploration Corp

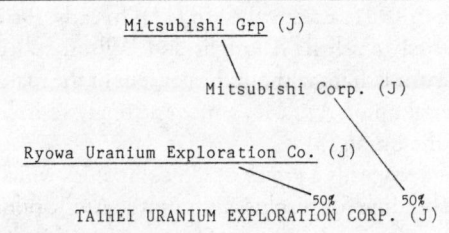

It has engaged in uranium exploration in a number of unspecified countries (1).

References: (1) *MIY* 1982.

583 Taiwan Power Co

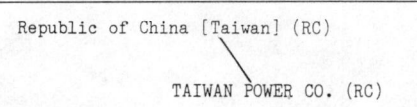

Together with Anschutz (50%) and Kepco (25%), the Taiwan Power Co sponsored uranium exploration (1) and drilling at an undisclosed site in eastern Paraguay in the early 'eighties (2).
Now known usually as Taipower, the company was involved in the late 1980s in the potentially highly-promising region of the Grand Canyon, Arizona, where – along with Union Pacific Resources (formerly Rocky Mountain Energy) – it was investigating the breccia pipes, one of which was said to be highly promising (3). (See Energy Fuels Nuclear for further details.)

References: (1) *American Metal Market* quoted in *MJ* 3/7/82. (2) *MAR* 1984. (3) *MAR* 1988.

584 Teck Corp

One of Canada's "biggest natural resources groups" (1), and a "company to watch" (2), Teck is a "metal mining, oil-producing, investment-financing, exploration, and holding" (3) corporation with interests in a number of other companies both in Canada and abroad.
Among these are Metallgesellschaft Canada, with whom it jointly operated (80% Teck) the Highmont copper project in British Columbia; Soquem, with whom it had a JV in niobium in Quebec; and RTZ (Teck took over RTZ's 27% Coseka stake in 1980 but later announced its sale).
Teck has been exploring for uranium in Saskatchewan (2), and possibly elsewhere in Canada.

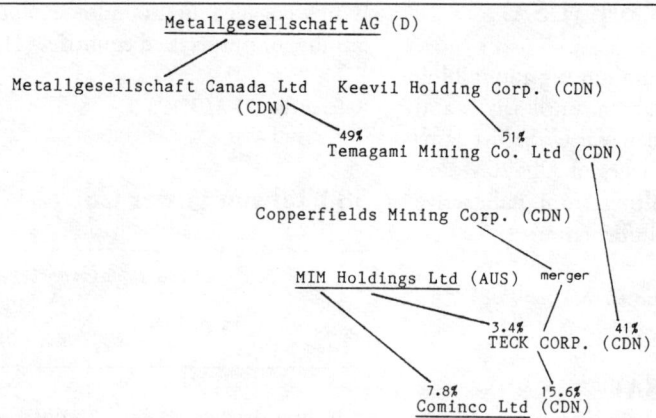

Along with Noranda and other companies Teck got in on the new gold rush at Hemlo, Ontario where it began mining (with Noranda) in 1985. These claims have been opposed vigorously by Native Canadian communities (4). (See also Dickenson and New Cinch.)

Metallgesellschaft (Canada) in 1982 purchased C$25M of the share interest in Teck (5).

In 1986, Teck, along with MIM and Metallgesellschaft, purchased between them 31% of Cominco – later increased to 40% (7) – by buying out the majority of Canadian Pacific's own interests in the debt-laden base metals group (6). As of 1990, Teck owned 50% of Nunachiaq Inc which itself owns 27.73% of Cominco (8). Its Niobec mine (in JV with Cambior) is the world's second largest producer of niobium.

It is a partner with Rio Algom in the Highland Valley Copper partnership set up with Cominco (7). While its Beaverdell silver mine in Kelowna, British Columbia, closed in February 1991 after ninety years of continuous operation, Silver Standard Resources (Teck 27.5%) has been expanding its Milot gold prospect in Haiti (8). However, its main gold operations are the David Bell and Williams mines at Hemlo – the latter being the largest gold bullion producer in Canada (7, 9).

Teck also recently acquired 20% of Royal Oak, which has mines in Ontario (10).

References: (1) *FT* 14/2/84. (2) *Yellowcake Road.* (3) *MAR* 1982. (4) *MAR* 1984. (5) *MJ* 24/6/83. (6) *FT* 14/11/86. (7) *FT Min* 1991. (8) *MAR* 1991. (9) *MJ* 10/8/90, *E&MJ* 6/90; see also *E&MJ* 6/87. (10) *FT* 7/11/90.

585 Tenneco Inc

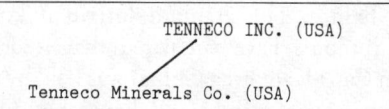

It contracted with Uranium Resources Inc in 1980 to build an *in situ* uranium leach plant for the company in Webb County, Texas, to produce more than 100,000lb of U_3O_8 a year (1). It linked up with Occidental Petroleum in early 1981 in an oil shale project at Cathedral Bluffs, Colorado (2) which was later closed (3).

Tenneco is an important US oil company, active in all the major oil and gas producing areas of the USA. It is also into food, chemicals, farm and construction equipment, packaging, ship building and insurance (4). In 1979, it was the US's 13th most important defence contractor (7); in 1981, already the 9th (11). In early 1982 it tried to sell its Allbright and Wilson subsidiary in Britain without the consent of the managerial union ASTMS, a move strongly resisted by the union (5).

Tenneco leads a group of US companies which are bidding for surplus gas from the Arctic pilot project (together with Dome, Nova, Petro

Canada, and Melville Shipping). This is a project which has attracted the ire of Inuit and environmentalists throughout Canada – indeed the western hemisphere – for its threat to wildlife in Arctic waters (6).

In June 1983, Tenneco consolidated all its non-fuel interests into a single unit. Tenneco Minerals Co became a division of Tenneco Inc and has control of soda ash production, precious metals in Nevada, exploration in the western USA, Alaska and Australia, as well as uranium in Texas (8).

Also in 1983, Tenneco began buying up assets controlled by James Goldsmith's Diamond International, part of General Oriental, the Goldsmith "master" company (9). However, none of these appeared to include Goldsmith mines in Guatemala (10).

Tenneco is a highly diversified company with numerous subsidiaries and a clear policy of continued diversification.

In 1981 it acquired significant oil reserves in Colombia, thanks to taking over the Houston Oil and Minerals Corp. Its exploration has been centered on Australia, Indonesia, Tunisia, the North Sea, the United Arab Emirates and Colombia (13). One attempted coup was over the French construction equipment manufacturer Poclain (12).

In mid-1984 the Great Plains Coal Gasification project in North Dakota (in which a JV was headed by Tenneco) got up to US$790M in price guarantee assistance from the US Synthetic Fuels Corp, to proceed with its 14,000 tons/day production of synthetic gas from lignite (14).

In 1987, Tenneco sold its West Cole (Webb County?) uranium property to Total Minerals (15).

Although it sold its gold mining and exploration properties to Echo Bay Mines Ltd in 1986 (16), it reportedly re-entered the gold business the following year (17).

In 1988, after suffering huge financial losses, Tenneco began selling off various assets in "one of the biggest corporate auctions in Texas history" (18). Away went Tenneco's oil and gas interests in the Gulf of Mexico, its south-western

US businesses, and various other segments of its operations (18).

The following year, the company showed some improvements – especially in its farm and construction equipment division; its minerals interests also yielded higher returns than before (19).

In 1991, a further loss loomed: Tenneco announced yet another asset off-loading (of more than US$1 billion) (20). Also, that year, Echo Bay Mines acquired the company's precious metals division (21).

Currently Tenneco operates a gold mine in Nevada and the Goldstrike heap-leach mine near St George, Utah. According to an article in the *Engineering and Mining Journal*, the latter minesite "... was covered primarily by pinon and juniper trees. This growth was grubbed, and all topsoil and subsoil ... stockpiled [for] final reclamation" (22). We will see!

References: (1) *MJ* 6/6/80. (2) *MJ* 1/1/82. (3) *MAR* 1982. (4) Company advert in *FT* 28/10/80. (5) *FT* 11/2/82; *Gua* 17/2/82. (6) *FT* 3/2/82. (7) *NARMIC* 1979. (8) *MJ* 1/7/83. (9) *Gua* 28/6/83. (10) *FT* 28/9/83. (11) US Dept of Defence 1981. (12) *FT* 12/3/84. (13) *WDMNE*. (14) *MJ* 22/6/84. (13) *WDMNE*. (14) MJ.22/6/84. (15) *E&MJ* 4/88. (16) *FT Min* 1991. (17) *E&MJ Int Directory of Mining* 1991. (18) *FT* 5/9/90. (19) *FT* 9/2/90. (20) *FT* 12/9/91. (21) *MJ* 24/5/91. (22) *E&MJ* 7/89.

586 Teton Exploration Drilling Co

UNC Teton Exploration Drilling – to give it its full title – conducted UNC's mineral exploration programme and performed contract exploration and drilling in the USA and abroad.

In 1981, a 50% interest in Teton's Australian subsidiary was sold to North Kalgurli for A$5M – under the agreement North Kalgurli would jointly fund exploration work throughout Australia (1).

Teton Australia also has a 25.5% interest in the Honeymoon Uranium Project.

In Paraguay, the company obtained a 24.7 million hectare exclusive exploration lease in the *chaco* area (2, 3), the most intensively explored

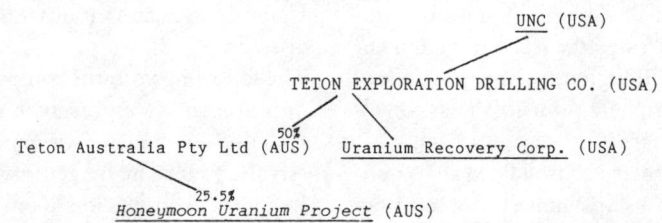

UNC (USA)

TETON EXPLORATION DRILLING CO. (USA)

Teton Australia Pty Ltd (AUS)   50%   Uranium Recovery Corp. (USA)

25.5%
Honeymoon Uranium Project (AUS)

part of the country, where large numbers of native Paraguayans have been removed, harassed and killed, to make way for agribusiness and mining corporations. Uranium was discovered in the *chaco* by 1983 (4).

Contact: Survival International, 310 Edgware Road, London W2, England.

References: (1) *MIY* 1982. (2) *MJ* 26/2/82. (3) *Native Peoples' News*, London, No. 6. (4) *MAR* 1984.

587 Texaco Inc

When preparation of this book started in 1983, there was one tiny entry for Texaco. In virtually a single stroke, however, in early 1984, Texaco acquired Getty Oil and thus moved not only into the top-most rank among world oil companies, but into the widespread (though relatively unprofitable) mineral world occupied by the descendants of America's most astonishing patrician entrepreneur, J Paul Getty.

The background to the biggest financial takeover then recorded in commercial history (valued at nearly US$10 billion – the price of a square meal for every child, woman and man on this planet) – reeks of intrigue, melodrama and cut-throat piratics. (Texaco's record was put in the shade only a month later, however, by Socal's bid of US$14 billion for Gulf Oil.) As

the *Observer* put it, this was "a long-running melodrama with a cast worthy of television's *Dallas*".

"The players include America's richest man (Gordon Getty, son of J Paul); his estranged secretive brothers; a 15-year-old scion of the clan, recklessly christened Tara Gabriel Galaxy Grammophone Getty; the head of the world's wealthiest museum; and many more" (1).

But, although the colourful clashes among the Gettys (and to a lesser extent between Texaco and Pennzoil, the company Gordon Getty originally agreed to sell to before the pre-emptive strike) could fill volumes, the most significant aspects of the affair have to do with machinations by Texaco itself.

Overnight this company doubled its oil reserves. Although Texaco was three times the size of Getty until 1984, Getty's US oil reserves ranked sixth (with 1241 million barrels) and Texaco's eighth (with 1045 million barrels). But Texaco now comes hot on the heels of Sohio, Exxon, and Arco in terms of its oil holdings in the USA (2), while, in the world at large, Texaco has come second only to Exxon in terms of cash flow and reserves (3).

Between 1977 and 1980, when John McKinley took over as chair of Texaco, the company's oil reserves dropped by nearly two-thirds. It had been "pumping more oil out of the ground than it [had] been discovering new reserves" (3). In a strategy designed to sharpen Texaco's slackening competitive edge, McKinley increased exploration expenditure – while other oil majors were cutting their's, especially in "frontier" areas like Alaska (3). In November 1983, Dome Petroleum of Canada sold most of its US oil and gas properties to Texaco (4) after Dome itself ran up an unpalatable bill buying out Hud-

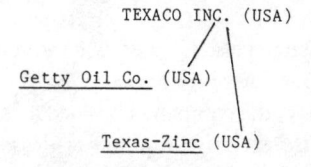

TEXACO INC. (USA)

Getty Oil Co. (USA)

Texas-Zinc (USA)

son's Bay Oil and Gas (much of which it then sold to Allied).

The very same month, Texaco announced that it was paying under US$300 million to secure Chevron's loss-making petrol marketing operations in six Wesern European countries (5). This was somewhat surprising considering that, since 1982, McKinley had thrown 11,000 workers out of their jobs, and "drastically rationalised the company's sprawling network of service stations" (3).

However, the gamble was clearly predicated on Texaco being able to acquire new oil reserves in a hurry – something the Getty acquisition undoubtedly did.

In order to meet objections to the take-over, Texaco immediately set in motion a plan to dispose of almost half Getty's petrol stations (6), an oil refinery in Kansas, and other Getty marketing outlets in the north-eastern USA (7).

There was also speculation that McKinley might sell off Getty's 50% share in Mitsubishi Oil (half owned by Mitsubishi Corp), possibly to Kuwait – a move opposed by Mitsubishi (8). By March, Texaco was all set to sell US$2.5 billion worth of Getty's assets to reduce its debt load following the gargantuan take-over (9).

It also launched the biggest convertible bond issue "ever seen in world capital markets", to raise some US$800M (10). The Texaco-Getty shotgun marriage was finally given a seal of approval by the US Federal Trade Commission (FTC) in a 4:1 majority decision. The only commissioner to object, Michael Pertschuk, was the FTC's only non-Reagan appointee, and even he didn't demur at the "fundamental logic" of the merger (7).

A month before, Senator Howard Metzenbaum, an Ohio Democrat, had the temerity to call the deal "blatantly anti-competitive" and to claim that, if approved by the FTC, it "will provide new evidence that the Reagan administration is more interested in promoting concentration than competition" (11).

Texaco's problems, however, still weren't over, for, in the meantime, the "publicity-shy Bass brothers, reputedly the second wealthiest family in Texas after the Hunts" (12) (cf Hunt Oil)

built up their stake in McKinley's mountain to 9.8%, thus becoming the biggest single shareholder (13). Texaco knocked this little danger on its head by offering to buy out the Bass's interests for US$1.28 billion (12).

It was far from clear what Texaco would do with Getty Mineral's assets, including its uranium leases. In 1979 Texaco itself produced plans to carry out solution mining on a uranium lease in south Texas (14) but, by the following year, they appeared to have come to nothing (15). Certainly it seems as if – unlike all the other "oil majors" in the past fifteen years – Texaco has not been particularly interested in diversifying out of oil; the acquisition of Getty does not necessarily disturb this view. Meanwhile – apart from Getty's assets – Texaco has continued, *inter alia*, to be the major foreign oil company in Ecuador (where it has produced 80% of the nation's oil, with 37.5% of the interest) and to explore along Paraguay's border with Brazil (16).

In 1982 Texaco announced that it would build a £100M plant in Pembroke, Wales, United Kingdom, to reduce the lead content of petrol (17).

In the '70s, Texaco was represented on the Trilateral Commission (18).

In 1980 Texaco – along with other oil majors like Exxon – was cited for underpayment of oil and gas loyalties and "oil theft" from federal and Native American lands in the USA, and became subject to check on their accounting (19). This was not the first time Texaco had raised the ire of Native Americans. In 1978 the Utah chapter of the Coalition for Navajo Liberation (CNL) took over the entire Aneth oil field which had been leased to Texaco without their consent. They won eighteen of twenty demands made to the oil company, and the support of Native American groups throughout the country. The leases remained, however, "unnegotiable" (20).

"Thank God for Carl Icahn. As the Genghis Khan of corporate raiders prepares to storm the mighty fortress Texaco, the greatest prize in the glorious history of take-over warfare, even his staunchest and most bloodthirsty retainers are starting to lose their nerve".

So wrote Anatole Kaletsky in May 1988, as Icahn tried to win control of the huge US oil company. What Kaletsky meant by his wry remark, was that Icahn had successfully got Texaco valued for what it was really worth on the stock-market: was forcing the imperial icon to cut its suit to fit its cloth (21) .

This "largest financial dispute in corporate history" (22) started when Texaco took over Getty Oil in 1983/84, and Pennzoil, cheated in the contract it had already made with Getty (23), took the giant to court. After appearances in no fewer than 31 state and federal courts, Pennzoil finally won the battle (24) and was awarded US$10.3 billion damages (23). In the meantime, however, Carl Icahn had bought the shares in Texaco owned by Robert Holmes à Court (through his Heytesbury Securities, Bell Group, Bell Securities, and Weeks Petroleum piggy-back of companies) – a 12.3% block worth US$348 million (25) which Icahn acquired through his three-quarters owned TWA (Trans World Airlines). Icahn was also buying into Pennzoil, and seeking to influence both parties to the dispute in a way favourable to himself (26).

At the end of 1987, Pennzoil settled for US$3 billion – though its legal fees alone came to US$400 million (27). Texaco also offered around US$5 billion to keep other creditors off its back (28): what it owed to US government agencies alone topped US$10 million (29).

After filing reorganisation plans for Texaco (30), and being accused of "insider trading" and various other nefarious activities (31), Icahn made a bid for the whole of Texaco – at US$14.5 billion – in May 1988 (32). His bid was defeated, after shareholders accused him of trying to break Texaco into little pieces (33).

However, the whole episode did force Texaco to dispose of a large number of its assets: half of its US market network to the Saudi Arabian ARAMCO (34), all of Texaco Canada (35), and, most importantly for our purposes, its West German subsidiary to RWE (36) (see Rheinbraun and Nukem).

In 1988, too, Texaco signed an agreement with Pancon regarding an option to acquire Texaco Oil Development Co's 35% stake in the Jabiluka uranium mine, "probably the largest undeveloped uranium deposit in the world" according to the Mining Journal (37) – though who really knows? If implemented, the agreement would give Pancon complete ownership of Jabiluka.

Pancontinental Mining's option, to acquire Texaco's 35% stake in the Jabiluka uranium deposit, remained open through the next three years (38). Pancon had until November 1991 to secure the company's 35% share of the huge potential mine, which it had not taken up by early that year (39).

Meanwhile, Texaco sold its Mercur mine and mining claims near Salt Lake City, Utah, to the major US gold producer American Barrick Resources Corp (40).

In 1990, Texaco was hit by $355 million of self-imposed charges for environmental improvements at its service stations (41). 1990 also saw an interesting inversion of usual procedures when Texaco took the Dutch government to court over an ecological issue. Texaco's plans for a coal gasification plant near Rotterdam were rejected by the Dutch authorities in favour of a coal-fired generator. Texaco argued – with some justification – that its technologoy was more efficient in reducing carbon dioxide emissions, and far more efficient in controlling output of nitrous oxides and sulphur dioxide. A Dutch industrial tribunal came out on the side of Texaco (42).

At the same time, Texaco was coming under fire from environmentalists and pro-Indian groups in Panama: they challenged the validity of a Texaco oil drilling contract in National Park areas of the country (43).

References: (1) Obs 15/1/84. (2) FT 9/1/84. (3) FT 10/1/84. (4) FT 17/11/83. (5) Ibid. (6) FT 31/1/84. (7) FT 14/2/84. (8) FT 9/2/84. (9) FT 17/3/84. (10) FT 16/3/84. (11) FT 16/1/84. (12) FT 7/3/84. (13) FT 2/2/84. (14) MAR 1980. (15) MAR 1981. (16) MAR 1984. (17) FT 12/3/82. (18) Sklar. (19) FT. (20) Al Henderson in Americans Before Colum-

bus, abstracted in *AkwN* Summer/78. (21) *FT* 28/5/88. (22) *FT* 12/12/87. (23) *Gua* 12/12/87. (24) *Wall St J*, New York, 23/2/88. (25) *FT* 4/12/87. (26) *FT* 9/12/87. (27) *FT* 30/12/87. (28) *FT* 21/12/87. (29) *FT* 23/5/88. (30) *FT* 22/12/87. (31) *FT* 4/5/87. (32) *FT* 26/5/88; *Obs* 5/5/88. (33) *FT* 21/6/88; *FT* 18-19/ 6/88. (34) *FT* 17/6/88. (35) *FT* 16-17/4/88. (36) *FT* 7/6/88. (37) *MJ* 11/3/88. (38) *MJ* 31/3/89, *MJ* 12/10/90. (39) *MJ* 18/1/91. (40) *FT Min* 1991. (41) *FT* 19/1/90. (42) *Gua* 19/1/91. (43) Information from CEASPA, Panama, 8/91.

588 Texasgulf Inc

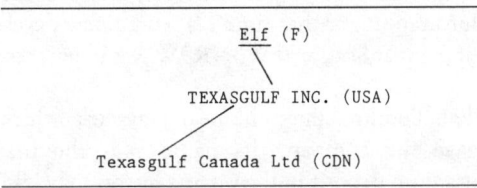

This is a so-called "integrated natural resources group" operating world-wide (though mainly in North America), producing chemicals and various metals and exploring for energy minerals including uranium. It was believed to be exploring for uranium in Minnesota (USA), along with Marathon Oil and Exxon in 1981 (1), and in 1982, it had a JV with Eldorado Nuclear for uranium exploration in Canada (2). Among its other JVs have been one with CCF/Consolidated Canadian Faraday (an option on its multi-mineral deposit in New Brunswick); 6.5% of the Nansivik mines in the Northwest Territories (NWT) (4); a JV with Hecla; gold mining in Colorado, USA (5); and trona (the rock from which soda ash is processed) mining at the world's largest deposit in Wyoming, USA, along with FMC, Tenneco, and Rocky Mountain Energy (4).

In South Africa, Texasgulf has been drilling on a gold property north-east of Driefontein, on a chromium-platinum project at Pandora, and a large manganese deposit 300 miles west of Johannesburg (4).

Apart from its interests in the NWT of Canada, Texasgulf has had an impact on indigenous peoples in both Australia and Panama. Until 1980 it held a 35% interest in Cliffs Western Australia Mining Co Pty, which itself held a 30% interest in the Robe River iron ore project in the Pilbara region of Western Australia (WA). It was also associated with Australia's most notoriously racist land-robber, Lang Hancock, through a partnership with Hancock and Wright, evaluating iron ore at Marandoo, Rhodes Ride, WA, in the late '70s (3). Its other Western Australian iron ore interests were sold to CRA in 1981 (4).

RTZ (CRA's parent body) also purchased Texasgulf's 20% interest in the huge Cerro Colorado copper project on Guaymi land in Panama in 1980, although Séru had an option to acquire 15% in the project, should it proceed.

In 1981 Elf-Aquitaine merged with Texasgulf by buying the approximately 65% share not held by the Canadian Development Corp which later exchanged its holding for Texasgulf's Canadian assets (4).

The combined Elf-Canadian Development Corp take-over of Texagulf was one of the more "successful" and less acrimonious direct take-overs by oil corporations of a "natural resources" company in the late '70s-early '80s.

It enabled Elf-Aquitaine to gain a stake in the US mining industry, while the Canadian Development Corp acquired Texasgulf's lucrative Kidd Creek properties in Ontario, together with silver, copper and zinc mining assets, and a 75% stake in the oil producer Aquitaine of Canada – thus increasing Canadian control of Canadian resources, "an important concern of the Trudeau government" (6). The Reagan administration – despite "mild concern" in the USA about the purchase of a major mining company by a foreign firm – didn't block the merger. Texasgulf shareholders reportedly surrendered their shares willingly. Elf, for its part, agreed to retain the Texasgulf name, to keep the mining company's management for five years, and not to interfere with managerial control over Texasgulf's operations. Elf also gave Texasgulf control over its US interests in its speciality: chemicals, minerals, oil and gas (6).

In 1989 the company brought on-stream a new soda ash operation in the USA, but closed down its Comanche Creek sulphur mine after a brief reopening (7).

Further reading: For a description of the impact of Texasgulf's phosphates operations in North Carolina see: Fred Powledge, *Water: The Nature, Uses, and Future of Our Most Precious and Abused Resource* (Farrar Straus Giroux) New York, 1982.

References: (1) *Northern Sun News*, Minneapolis, 7-8/81. (2) Eldorado Nuclear Annual Report, 1979. (3) *MIY* 1981. (4) *MIY* 1982. (5) *MAR* 1984. (6) *BOM*. (7) *MAR* 1990.

589 Texas-Zinc

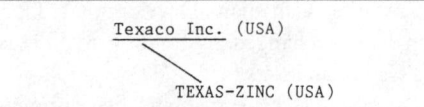

This was acquired by Texaco in the '70s (exact date unknown) (1).
No known activities.

References: (1) US Congress Subcommittee on Energy statement, 1975.

590 Thor Exploration Ltd (CDN)

In 1974, Thor announced the discovery of significant uranium mineralisation in the southern portion of Wollaston Lake, Saskatchewen.

The same year, the company discovered uraniferous quartzite at Duddridge Lake, west of La Ronge; Brascan and Noranda later carried out work in the area (1).

In 1977, it was drilling on its Duddridge Lake uranium prospect, south of Key Lake in northern Saskatchewan; an un-named European company was being approached for joint financing. The company at the time had 10 uranium prospects in Saskatchewan and one silver property in Idaho, USA (2).

References: (1) *Canadian Mining Journal* 4/76. (2) *FT* 13/7/78.

591 Todilto Exploration and Development Corp (USA)

This company operated two mines in the San Juan Basin area of New Mexico up until at least 1979: one called Haystack and the other Pidere Trieste (1).

It was also exploring elsewhere in New Mexico for uranium at that time (2), and a new development at Section TW3NR9W was in progress (1).

The Todilto Limestone near Haystack, which gave the company its name, was the first uranium deposit in the Grants region to be discovered, in 1960: an event which sparked off a huge uranium rush (3).

References: (1) *San Juan Study*. (2) *E&MJ* 11/79. (3) Merle Armitage, *Stella Dysart of Ambrosia Lake*, 1959.

592 Tokyo Uranium Development Corp

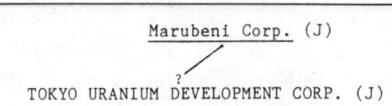

This is part of a consortium with Minatome and Cogéma which has an exploration concession for uranium in an area between Bir-Moghrein and Ain Ben Till, north Mauritania (1). (See also Marubeni.)

In 1982, uranium prospecting was halted due to the "poor state" of the market (2).

References: (1) *MAR* 1982. (2) *MAR* 1984.

593 Torkelson-Rust

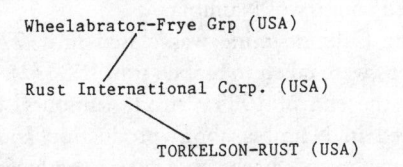

Wheelabrator-Frye Grp (USA)

Rust International Corp. (USA)

TORKELSON-RUST (USA)

Rust never sleeps! It offers complete services in the mining industry from scratch to – well, scratch! If you wish, they'll design you a uranium mine.

594 Traction et Électricité (B)

Along with AG McKee and Union Minière, this company won a contract to develop the Hoggar Mountains uranium deposit in Algeria in 1979 (1). Although in 1984 the Algerian government called for international tenders to proceed with this deposit (2), nothing more has been heard of Traction et Électricité since the announcement of the 1979 deal.

References: (1) *MJ* 6/7/79. (2) *MAR* 1984.

595 Transcontinental Oil

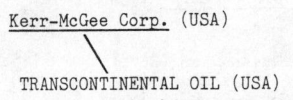

Kerr-McGee Corp. (USA)

TRANSCONTINENTAL OIL (USA)

A Kerr-McGee "front" which explored Potowatomi lands in Wisconsin during 1980 without the permission of the tribe, it "quietly optioned over 22% of the tribal reservation" to its parent body (1). After resistance and legal action taken by the tribal council, the company withdrew the leases (2).

References: (1) *AkwN* Winter/81. (2) *Kiitg* 6/83.

596 Transnuklear GmbH

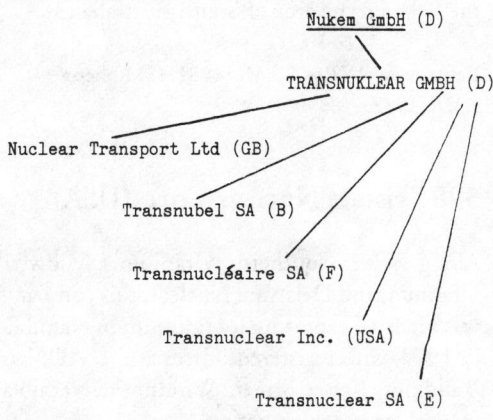

Nukem GmbH (D)

TRANSNUKLEAR GMBH (D)

Nuclear Transport Ltd (GB)

Transnubel SA (B)

Transnucléaire SA (F)

Transnuclear Inc. (USA)

Transnuclear SA (E)

This is a transport company with offshoots all over the world handling uranium oxide, uranium hexafluoride and radioactive elements. A subsidiary of Nukem, its main offices are in Hanau, West Germany; Nulux, Hobeg, and Decatox are company names also linked to Transnuklear (1).
(see also "Nukem Scandal" in Nukem entry.)

References: (1) *Kiitg* 2/82.

597 Transoil NL

It was partnered with Oilmin and Petromin on the Mereenie oil field where it holds a 9% interest. It is also involved in uranium exploration at Cape York and at Beverley (1).
Like Oilmin and Petromin, Transoil was partly controlled by the interests of Bjelke Petersen, the well-known racist ex-premier of Queensland (2).

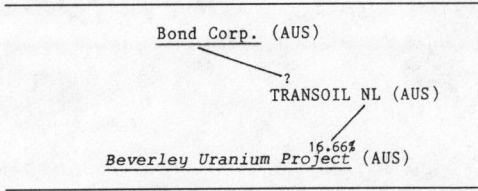

Bond Corp. (AUS)

?
TRANSOIL NL (AUS)

16.66%
Beverley Uranium Project (AUS)

It was also prospecting with Oilmin at Mt Painter, 110km north-west of Leigh Creek in the Lake Frome area of South Australia (3).

References: (1) *Reg Aus Min* 1981. (2) Roberts. (3) *AMIC* 1982.

598 Tristate Nuclear Corp (USA)

Along with Southern Uranium, Midwest Uranium, and Delaware Nuclear, this company was reported exploring for uranium in Namibia in 1977, with registered offices at the Volkskas Buildings, Kaiser Street, Windhoek, Namibia (1); no further information.

References: (1) *Star*, Johannesburg, 30/7/77.

599 Tsumeb Corp Ltd

This is "by far the largest producer of base metals in Namibia, accounting for about 80% of the country's annual output. Until the mid-1970s, Tsumeb alone generated one fourth of Namibia's total mineral export earnings, but this figure had declined recently with the surge of uranium production" (1).

It occupied an exceptional status as a pool for external capital investment in the illegally-occupied country of Namibia.

The Otjihase mine was closed in 1977, but steps were taken to reopen it in 1980 (2).

At the end of 1983, "strong rumours" circulated in Namibia about production and employment cut-backs by a major producer: this seems likely to have been Tsumeb. Commented the *Mining Journal* at the time: "What has certainly occurred is that US companies have reduced their public exposure in the politically sensitive Namibia with the main change being the movements of ownership between CGF group companies". This reflected the fact that Gold Fields of South Africa's parent body, Consolidated Gold Fields, then held 26% of Newmont (3).

Bob Meiring, chief executive of Tsumeb CL, also announced that "half the world's germanium" was to be found at the mine. Germanium dioxide is used in microprocessing and laser technology with strong military implications (4).

In 1988, Newmont sold out all its southern African holdings, including its (then) 30.45% share in Tsumeb (5). Since BP Minerals International had disposed of its own 25% holding (6) and Gold Fields of South Africa held the re-

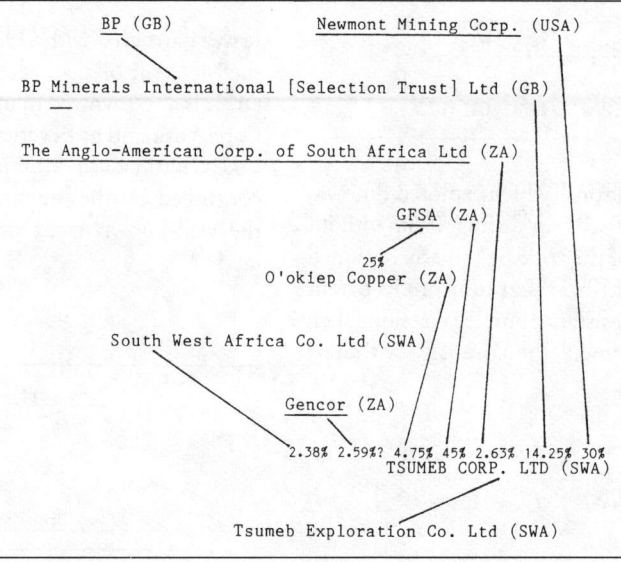

738

mainder of the stock, Tsumeb was then effectively merged wih Gold Fields of Namibia (GFN) – GFSA's sibling across the border (7). Tsumeb remains the contry's largest and most diversified minerals producer, notwithstanding the independence of Namibia (8). It has three operating mines – at the Kombat/Asis Ost/West complex, Otjihase (held 70% with JCI), and Tsumeb itself. It runs a copper smelting and lead refining complex at Tsumeb, and sells to Rössing Uranium quantities of gold and iron pyrites from its Otjihase operations. Due to depletion of its ore reseves, however, the company has been exploring around the Tsumeb region and is planning to bring a new orebody, Tschudi, into production between 1994 and 1995 (8).

Several weeks after Namibian independence in March 1990, a nurse from Sheffield, England, toured the Tsumeb complex. Rachel Bearpark observed that: "18,000 people live in crowded, dusty streets in appalling conditions. ... We saw two-roomed shacks, housing usually six or seven people but sometimes as many as 20. ... The majority of mineworkers are on low wages and short-term contracts and are therefore forced to leave their families in rural areas to survive through subsistence agriculture. The workers are housed together with as many as 30 miners in rooms 24ft by 24ft. The stench and the heat was sickening and the price paid for such accommodation was incredible – R70 out of a monthly wage of only R180-200" (9).

The year before, a Norwegian occupational health specialist, Dr Kristian Vetlesen, had visited Tsumeb at the invitation of the MUN (Mineworkers Union of Namibia). He collected blood and urine samples from workers at the smelter/refinery. One worker was found to be "clinically suffering from lead intoxication ... [and] severe encephalopathy" with a blood-lead level over 6 times the population reference level. Another worker at the lead plant's blast furnace had blood cadmium at 8 times the reference level, while a labourer in the arsenic plant had nearly 8 times the recommended limit of arsenic in his urine (9).

References: (1) *Objective Justice* (UN Dept of Public Informantion), 2/82. (2) *MIY* 1982. (3) *MJ* 11/11/83. (4) *Windhoek Observer* 10/12/83. (5) *MJ* 8/4/88. (6) *E&MJ Int Directory of Mining* 1991. (7) *MAR* 1990. (8) *MAR* 1991. (9) *Action on Namibia*, NSC, London, 9/90.

600 Tubas

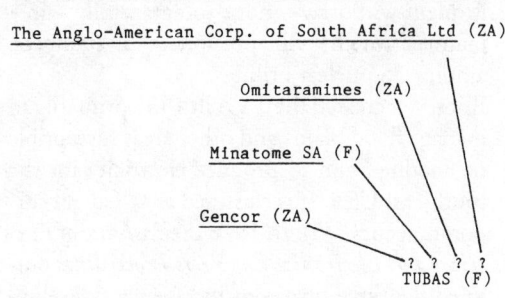

This company was set up by Anglo-American, Minatome, Gencor and Omitaramines, to operate the Tubas mine in Namibia. Development was apparently held up for political reasons (see Omitaramines).

601 TVA

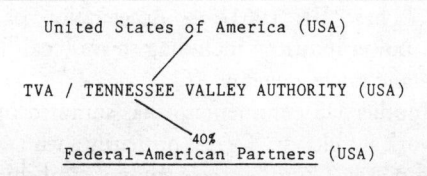

"TVA strips the coal and digs the yellowcake which electrify the enrichment facilities which build the bombs that Uncle Sam might drop. With three operating nuclear reactors and another fourteen on the drawing board it could, as Mike Garrity of the BHA (Black Hills Alliance) remarks, single-handedly keep the nuclear industry in the US alive and well for the next twenty years" (1).

But that was back in 1980 when TVA was the largest holder of leases in the Paha Sapa, the sacred Lakota Black Hills of South Dakota. In the meantime, TVA has relinquished its leases and launched a "massive attack" (2), a "damning indictment" (3), on the nuclear industry, and in summer 1984 was seriously considering a recommendation to abandon four of its planned nuclear power plants in the face of falling energy predictions and cost overrun (4).

Just as TVA entered the nuclear industry – right through the front door, breaking virgin soil – so it might withdraw – none too gracefully – in a manner which will precipitate a stampede among US nuclear utilities.

Roosevelt created the TVA in 1933 primarily to manage flood plains and other areas susceptible to flooding, and to produce electricity for the south-east USA. It is classified as a "federal corporate agency" (5). In 1976 the *Institute of Electrical and Electronics Engineers Proceedings* outlined the three phases of the agency's development:

"Phase I. 1933-1941 – Construction of multi-purpose dams; quest for power markets; low electrical rates as a major policy.

"Phase II. 1941-1961 – Immense growth of power program; the demands of the Cold War prompted the Atomic Energy Commission in Oak Ridge, Tennessee and elsewhere, to expand its uranium plants, and with them its demand on the TVA.

"Phase III. 1961-? – Acceleration of the power industry, including steam (coal-fired) and nuclear energy" (6).

Another US commentator has summed up its work as follows: "TVA is primarily a wholesaler of power. Customers are composed of the following major groups: (i) local municipal and co-operative systems; (ii) directly served industries; and (iii) directly served Federal agencies ... There are 49 directly served commercial and industrial customers, most of which have large or unusual power requirements" (7).

TVA supplies low-cost power to a variety of consumers – they pay less, the bigger they are – over 80,000 square miles covering almost all Tennessee and parts of northern Alabama, north-east Mississippi, south-west Kentucky, Georgia, North Carolina and Virginia. With around six million consumers in that area, TVA is the world's largest single supplier of electricity (5). Even so, nearly 70% of its power goes to industry and business rather than householders (5).

Originally, TVA used hydro-electric power to fulfill its contracts; by the early '50s no fewer than 36 dams had been constructed, flooding some two million acres of the Tennessee valley (8). Its exploits at that time were "widely considered to have established a precedent for universal application" (9), and it was not surprising that TVA's dam building became a model for one of the most disastrous of postwar Third World energy projects, the Volta dam in Ghana (10).

Although TVA didn't commission its first nuclear reactor until 1966 – from General Electric in "the most dramatic nuclear deal of the period" (11), and one which suggested that the country's largest coal consumer was being deliberately seduced to the atomic state (12) – the corporation was linked with the Atomic Energy Commission (AEC) from the start. This was mainly through the person of David Lilienthal, chairman of the TVA since 1941 and co-author with Dean Acheson of the first attempt to create an international agency to "control" atomic power (13). Lilienthal went to the AEC as its chairman in 1947 – a posting ironically opposed by Senator Joe McCarthy and other reactionaries on the grounds of TVA's "socialistic" tendencies (14).

One of the first activities undertaken by TVA in the new era was to supply electricity for the running of the Oak Ridge uranium enrichment plant (see Union Carbide), the prime supplier of U-235 to the US military (15). The Oak Ridge plant consumed no less than 17% of TVA's electricity in 1974 (5).

Although many other consumers, including Alcoa, Du Pont, Cities Service, and Kerr-McGee, make use of TVA's power, the Oak Ridge plant – and the Paducah enrichment facility in Kentucky, also run by Union Carbide – have been its largest single consumer. After the first GE reactor sale was sealed, TVA ordered another 16. Brown's Ferry Unit 2 went on stream in 1975, and Unit 3 in 1977 – at a cost overrun of nearly 200% (16). Construction on another six reactors commenced at Sequoya, Watts Bar, and Bellefonte in the next eight years, with electricity rates to pay for the excesses leaping some 178% in three years (17). Almost incredibly, this construction followed what could have been a Three Mile Island-type accident several years ahead of schedule. Walter C Patterson graphically describes what happened: "The candle was in the hand of an electrician in the cable-spreading room under the control room of the Brown's Ferry station in Alabama, which had just become the world's largest operating nuclear station, with two BWRs of nearly 1100MWe up to power. At 12.30pm on 22 March 1975 the electrician and his mate were checking airflow through wall-penetrations for cables, by holding the candle next to the penetration, when the draught blew the candle flame and ignited the foam plastic packing around the cable-tray. The electricians could not put out the fire. The temperature rise was noticed by the plant operator, who flooded the room with carbon dioxide and extinguished the fire beneath the control room – but the fire had already spread along the cables into the reactor building. When erratic readings began to appear on the controls for Unit One the operator pressed the manual scram button. Soon he found there was also a half-scram on Unit Two, which he had not ordered; and the speed of the main recirculating pump was being reduced. Quickly he scrammed Unit Two.

"Until the two units were scrammed they had been supplying some 15 percent of the total demand on the whole Tennessee Valley Authority grid; the effect of their sudden removal can well be imagined. The fire continued to burn for seven hours, affecting hundreds of cables. According to an initial assessment by the US Nuclear Regulatory Commission – successors to the dismantled AEC – the fire knocked out all five emergency core cooling systems on Unit One. Repairs to the station were expected to keep it out of service for months, and entail over US$40 million in the cost of replacement electricity output alone. Nuclear critics and ad-

vocates alike agreed that it was potentially the most serious incident in the history of the industry" (18).

A few burned cables weren't to deter TVA, however, nor its gargantuan search for the huge masses of uranium needed to fuel its programme – an estimated 6-7 million pounds per year by the '80s, according to TVA chairman Daniel Freeman in 1978 (19). During the '70s it contracted with many US mining companies for uranium exploration. These included:

- American Nuclear Corp, Casper, Wyoming – 50 percent interest as of 1971;
- American Nuclear Corp, Gas Hills, Wyoming – Atlas property (1972);
- Federal American Partners, Riverton, Wyoming (1973);
- Teton Exploration Co (United Nuclear Corp) – interests in New Mexico and Wyoming (1974);
- Susquehanna-Western, Edgemont, South Dakota – contracted to Silver King Mines Co (1974);
- Mobil Oil, McKinley County, New Mexico – (1974);
- American Copper and Nickel – David Robertson & Associates – exploration throughout Great Lakes area, Utah, perhaps elsewhere (1974);
- Robert Rees, Grant and Emergy Counties, Utah (1975);
- United Nuclear-Homestake Partners-Mobil Oil, and the Dalton Pass Chapter of the Navajo Nation (1978) (20).

By the latter part of the decade, TVA's uranium hopes were pinned on two areas: Dalton Pass-Crownpoint in New Mexico, and the Black Hills (Paha Sapa) of South Dakota.

Both are regions of predominantly Native American occupation. Crownpoint, although it lies outside the Navajo Reservation, is the site of one of the Dineh Nation's most sacred places (21). In the Black Hills, sacred to the Lakota (Sioux), the TVA started drilling for uranium on more than 100,000 acres in 1975; four years later they had plumbed more than 1000 exploration holes, each with its own road and drilling equipment (22). The Crownpoint-Dalton Pass project, a JV between Mobil Oil, United Nuclear, and TVA, was planned to go into operation in the '80s, though it was later stalled (23).

Mobil-TVA's leases in the area – the so-called "chequer board" of very confusing, overlapping, and sometimes conflicting claims between public agencies, Native Americans and privateers – were in 1978 the biggest single set of leases in the San Juan basin, while Mobil-Teton/UNC leases came second (24).

The combined project (Mobil 75%) originally envisaged at least three underground uranium mines and the construction of a large acid leach uranium mill processing 5000 tons/day ore. According to the Environmental Impact Statement submitted by the Department of the Interior and TVA in April 1978, the mines would run for 70 hours/week: the impact on the small, peaceful community of Crownpoint (3000 inhabitants, mainly Navajo) can be guessed at – with an envisaged doubling of the population in five or six years (25).

Little wonder that only a month later the Dalton Pass chapter of the Navajo nation passed a unanimous resolution stating: "We have become increasingly alarmed at the present and planned uranium mining activity in our community and are most fearful of its effects on our health, welfare, property and culture as well as the well-being of future generations ... We are unalterably opposed to all uranium exploitation in our boundaries" (26). Mobil began its leaching at Crownpoint in November 1979, on a pilot scale. Legal attempts to stop the project failed at the end of the following year. In 1983, the joint venturers had drawn up plans for 190,000lb/year U_3O_8 mill throughput from Crownpoint, but the whole project remained in abeyance "pending improvement in the market" (27).

TVA was also prominent in the uranium cartel suits of the '70s. Its suspicions about the existence of a cartel were, apparently, aroused after it invited tenders for 17 million pounds of uranium oxide and found that out of 53 letters to US and foreign companies, only three were answered! The contract was settled with Rio Algom (28). Later TVA tried to cancel, and Rio Algom counter-claimed with a US$1.2 billion suit for breach of contract. Eventually TVA settled with four RTZ companies, as it did with Denison, Noranda, Nufcor, and Uranex by early 1981 (29). At the end of 1982, with an optimism almost unique among companies benefitting from exploitation of uranium in Wyoming, TVA decided to purchase 6880ha of land under an existing agreement with Federal American Partners for possible future production (30).

It is not only – indeed not primarily – TVA's uranium activities which have had a negative impact on ordinary people. In the '60s TVA flooded the Little Tennessee Valley in the eastern part of the state, an area used by Cherokees for at least 12,000 years. In response to environmentalists who wanted to save the snail darter (a threatened species of fish), the project was initially suspended. Congress then overturned the decision and – despite two suits before the Supreme Court filed by the National Indian Youth Council, citing Cherokee religious freedom – TVA was allowed to trespass on holy ground. As Gerald Wilkinson, executive director of the National Indian Youth Council tells it:

"Before flooding the valley, the TVA dug up about 15,000 Indian graves. The Indian remains are now in shoe boxes in the basement of the Frank McClung Museum at the University of Tennessee. There was also a small non-Indian cemetery of about 300 people, whose remains were re-interred elsewhere. It is ironic that a great many of the Indian remains are more recent than the non-Indian ones they reburied.

"After the flooding, the chairperson of the TVA told the *New York Times* that the Tellico Dam was a mistake. Recent efforts to bring tourism to the lake and to attract industry have failed. As a result, the TVA has proposed to turn the lake into a chemical dump.

"To the Cherokees, this is the depths of obscenity. It is as if the Moslems had lost their mosque at Mecca to a strip mine or that Christians had seen Jerusalem flooded" (31).

But TVA is most notorious as an exploiter of coal. After the hydro-electric potential of the Tennessee Valley had been tapped, and with the demand for industrial power rising, TVA turned to coal generation in the '50s (5). Its first big sources were in Appalachia, America's "third world" region of appalling impoverishment and exploitation. Both the Kentucky Oak Mining Co and notorious Peabody Coal were involved in supplying the maws of TVA with its raw meat in this period, leaving a legacy of disease, disablement, broken down houses and orphans (5).

Eva Ritchie, quoted in Winona La Duke's study of TVA, was a victim of strip-mining for the utility, which was then north America's largest strip-miner of coal (32):

"... I had a baby buried back in there, and my husband's sister had a baby buried back in there. Well, my old man's got miner's asthma, and ... he couldn't get up there ... but by the time I started up to where the graves was, why, my father said to me, "You needn't go up there. You're too late. They've pushed the graves out."

"Well, Ansel Combs was on the bulldozer at the time, and he stopped ... and Cush Adkins climbed up on the bulldozer and he cursed ... and he pushed the graves out ... and they come up with the babies' caskets. They throwed 'em over the hill" (5). Apparently "candid" statements and its role as a supposed publicly-owned utility, have given the impression that TVA is more liberal, responsive, than other corporations. A study carried out by TVA and the AEC on the effects of water pollution emanating from the Oak Ridge U-235 plant (1933/56) didn't fail to reveal contaminated fish, stunted growth, and a substantial reduction in their lifespan. The results were published in the *New York Times*, but this was an era of uncritical support for nuclear fission, and the report went almost unnoticed (33). Just after its plans for Crownpoint were launched, TVA officials claimed they welcomed public discussion on uranium mining in the area. Dwelling on possible accidents involving yellowcake spills, TVA official John Lobdell was quoted as telling a public meeting: "The most important thing is to attend to the people who are injured, then worry about the uranium oxide ... I can't 100 percent guarantee there will be no effect to you or your offspring [from mining] but then I can't guarantee you won't fall down in your bathtub tonight either" (34).

Only a few days later, TVA was telling its broader constituency that it would give a US$1 rebate for every two dollars consumers spent on energy conservation (35).

Such commendable (and relatively inexpensive) nods in the direction of its critics shouldn't deceive us. In a 1975 Council on Economic Priorities study, TVA was ranked worst of all 15 utilities for its role in contributing to the US's air and water pollution (36).

In 1977, a long battle with the EPA (US Environmental Protection Agency) and eleven environmental and health groups as well as two states drew to a close with TVA – charged in a law suit as "the nation's number 1 polluter" – being told to instal scrubbers in its coal-fired plants to remove sulphuric acid from emissions. Chairman of TVA David Freeman admitted that "the agency is dragging its feet in meeting clean air standards", while its coal-fired plants were emitting an alleged 2,063,047 tons of sulphur oxides per year (37).

In 1988, after severe battering to its nuclear programme (in the early 1970s it planned to build 16 nuclear power stations, at a cost of US$16 billion, of which only four have been completed, at a cost of nearly US$18 billion!) TVA announced stringent cuts: these included a 17% cut-back among its 34,000 permanent employees, and nearly 50% among its temporary employees (38).

In 1989, TVA issued public bonds for the first time in 15 years, to repay a debt of $6.5 billion (39).

Meanwhile the TVA Office of Agriculture and Chemical Development had sold all TVA's ore rights in the region, and all its uranium activities appear to have ceased (40). Early in 1991, Costain, the British mining and property group, secured contracts worth $550M (£280M) to supply TVA with more than 16 million tonnes of coal over a period of six years.

The coal is to come from Pyro, Costain's US coal subsidiary, which operates mines in west Kentucky, northern Alabama and West Virginia (41).

Further reading: William U Chandler *The myth of TVA: Conservation and Development in the Tennessee Valley 1933-1983*, Washington, 1984 (Ballinger).

Contact: National Indian Youth Council, 201 Hermosa NE, Albuquerque, New Mexico 87108, USA.
SEAC, PO Box 1220 Bernalillo, NM 87004, USA.

References: (1) Roger Moody: "Little Big Horn, Part 2", in *Gua* 16/8/80. (2) *Time Out*, London, 12-18/1/84. (3) *City Limits*, London, 13-19/1/84. (4) *FT* 18/7/84; 31/8/84. (5) Winona La Duke, "Tennesse Valley Authority: A model utility", in *BHPS* 7/79. (6) *Institute of Electrical and Electronics Engineers Proceedings* 9/76. (7) Moody's *Public Utility Manual*, USA, 1976. (8) James Branscome, *The Federal Government in Appalachia*, USA, 1977. (9) Ronald Graham, "The Aluminium Industry and the Third World", in *Multinational Corporations and Underdevelopment*, London, 1982 (Zed Press), p113. (10) *Ibid* pp117-197; see also: Purari, *Overpowering Papua New Guinea*, Melbourne, 1978 (IDA/PAG), pp133-159. (11) Pringle & Spigelman p269. (12) *Ibid* p20. (13) Moss p16. (14) Thomas Reeve *Joe McCarthy*, New York, 1982 (Stein and Day). (15) Pringle & Spigelman p47. (16) *Energy options hold great promise for the TVA*, Washington, 11/11/78 (Govt Accounting Office). (17) Concerned Citizens of Tennessee documentation, quoted in Winona La Duke *op cit*. (18) Walter C Patterson *Nuclear Power, Harmondsworth, 1976 (Penguin), pp214-215. (19) Sklar p241. (20) Moody's Industrial Manual*, USA, date unknown, quoted in: Winona La Duke, *op cit*. (21) *San Juan Study*, map X-3. (22) Phyllis R Girouard, "Uranium exploration threatens religious cites", in *ARC Newsletter*, Boston, 3/80. (23) *Nukem Market Report*, summarised in *MJ* 3/9/82. (24) TVA map, reprinted in *San Juan Study*, map XI-3. (25) US Bureau of Indian Affairs estimate, quoted in Weiss. (26) Winona La Duke Westigard, "Energy crisis within the Navajo Nation", in *Prime Times* 4-28/12/78, quoted in Weiss. (27) *E&MJ* 1/84. (28) Pringle & Spigelman p118. (29) *MJ* 20/3/81. (30) *MJ* 17/12/82. (31) Gerald Wilkinson, Speech to Indian Rights Association in Dec 1982, quoted in: *ARC Newsletter*, Boston, Fall/83. (32) Weiss. (33) Moss. (34) *Gallup Independent* 7/12/78. (35) *Denver Post* 10/12/78. (36) Weiss p113. (37) *Electrical World*, USA 1/10/77. (38) *FT* 1/7/88. (39) *FT* 14/7/89. (40) *E&MJ Int. Directory of Mining* 1991. (41) *FT* 29/1/91.

[I'm indebted to Winona La Duke's article (5) for several of the references used in this piece – RM.]

602 TXO Minerals (USA)

It was involved in uranium exploration along with Fuel Supply Services (a subsidiary of Florida Power and Light Co) and Getty Oil in the late '70s (1); Later it was taken over by USX.

References: (1) Sullivan & Riedel.

603 Tyee Lake Resources Ltd (CDN)

Noranda and Kerr Addison had a 60% interest option on this company's property (jointly held with Peregrine Petroleum) in the Beaverdell area of British Columbia, near Kelowna, where drilling was taking place in 1976/77. Its activity was presumed halted by the British Columbia uranium moratorium (1).

References: (1) Merlin.

604 Ucan

During the early '60s Canada's uranium industry went into sharp decline – a "doldrums period" which lasted until the early '70s (1). Around 1962/3 – the date when US military uranium contracts were due to expire – the Canadian government introduced a "stretch out" programme to allow lower cost producers to survive until the projected "civil" programme

```
    a Minister of the Crown* (CDN)
                               \
                                \
    UCAN / URANIUM CANADA LTD (CDN)
```

* The capital of Ucan was issued as 1000 shares of no par value, all held in trust by *a* Federal Canadian Minister of the Crown (unknown which).

came into operation. Eldorado acquired the uranium for the expected expansion – and a specially structured stockpile acquisition and marketing Crown corporation was set up. Incorporated in 1971, Ucan's express purposes were, firstly, to acquire and sell the general government stockpile of uranium accumulated in the 1963-70 period and, secondly, to acquire a limited stockpile of some 3230 tons U_3O_8 concentrates during the following four years, under an agreement negotiated with Denison at the very beginning of 1971. The federal government paid nearly C$30M to acquire a 76% interest in the Denison stockpile.

Between 1963 and 1970 about 9100 tons of U_3O_8 was acquired for C$101 400,000 (2). By 1978, another 3700 tons had been purchased, at a cost to the Canadian taxpayer of about C$30M more (1). Denison was not the only producer of this uranium – it also came from Rio Algom, Consolidated Canadian Faraday, Eldorado Nuclear, and the Stanrock mine (itself acquired by Denison).

Title to the Canadian government's general uranium stockpile passed to Eldorado Resources in 1981, after its wholly-owned subsidiary Eldor Resources borrowed 2 million pounds of uranium concentrates from Ucan in 1978/9 and sold them to raise capital with which to purchase its interest in the Key Lake mine – now the world's largest single uranium mine (3). The assets acquired by Eldorado two years later – mainly around 14 million pounds of mine concentrates costing US$300M – enabled it to buy out Uranerz of Canada and to finance the Blind River refinery (3).

The same year, the Canadian public was treated to one arm of the federal government trying to sue the other when the Justice Department cited both Ucan and Eldorado on price fixing charges, as a consequence of their participation in the uranium cartel of the early '70s (4). Ucan was one of the prime movers in the cartel – it participated directly in the Paris meeting which set it up in February 1972. The Ucan representative at that meeting warned that it was prepared to sell its stockpile at prices which would undermine the uranium industry even further: its blackmail paid off, with Canada securing a one-third share of the world market (5).

Two years later, in late 1983, the Canadian Supreme Court, by a majority, ruled that Eldorado and Ucan were immune from prosecution on price-fixing charges (6) – a decision which prompted one of the dissenting judges to warn that the government could at any time "... engage in illegal activities and ... encourage other [corporate] citizens to do likewise" (4).

One of the early contractees for Ucan's uranium was Enusa of Spain, seeking uranium for several Spanish companies (5): initially they requested 34,000 tons – some of which was supplied by Denison from the joint stockpile (for which it was sales agent) (1) and some of which was to come from Ucan (4500 tons for delivery between 1977 and 1981) (5).

The Tohoku Electric Power Co of Japan also bought 1000 tons of uranium for delivery over the same period (1).

In 1979 Ucan signed an agreement granting Ontario Hydro the right to borrow up to 800 tonnes of uranium from the general stockpile at an estimated US$50M, to fuel its Pickering and Bruce reactors: the uranium would be repaid in kind by the end of 1984 (7).

References: (1) *EPRI Report.* (2) *RMR*, Vol. 2, No. 4, 1984. (3) *MIY* 1982. (4) *BBA* Fall/83. (5) Moss. (6) *MJ* 30/12/83. (7) *FT* 13/2/79.

605 UCC/Union Carbide Corp

Union Carbide has earned itself the most shameful footnote in industrial history for reasons which will be obvious to virtually every reader of this book. It is, of course, impossible

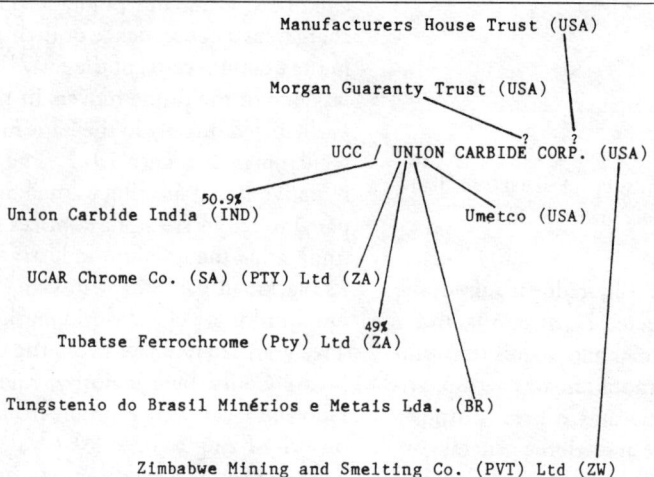

Manufacturers House Trust (USA)

Morgan Guaranty Trust (USA)

UCC / UNION CARBIDE CORP. (USA)

Union Carbide India (IND)   50.9%

Umetco (USA)

UCAR Chrome Co. (SA) (PTY) Ltd (ZA)

Tubatse Ferrochrome (Pty) Ltd (ZA)   49%

Tungstenio do Brasil Minérios e Metais Lda. (BR)

Zimbabwe Mining and Smelting Co. (PVT) Ltd (ZW)

from now on to even breathe the name "Union Carbide" without seeing images of horrendous mass poisoning, vast human wretchedness and a corporate ethic reduced to shreds. This entry is *not* however another diatribe against UC in the post-Bhopal era. Though articles, books and newspaper reports from all over the world relating to this industrial massacre almost exceed the material collected between 1984 and 1985 on all other *File* companies put together, it is only briefly referred to here. This is not only because, to deal with this material properly, would have expanded the *File* hopelessly beyond its present bulky extent. It is also because there are now many other sources to which readers can easily refer for knowledge about Bhopal and its aftermath. But, most important, it is because *even if there had been no Bhopal*, Union Carbide would have been ranked among the half dozen most culpable multinationals to feature in these pages. By concentrating on UC's other "heritage" and activities, the point is made even more strongly that Bhopal was exceptional only for its extent and impact on one community. (In Brazil in 1984 nearly as many peasants and workers died from pesticide poisoning as at Bhopal) (1). Since Bhopal, December 3 1984, Union Carbide has been involved in a massive hiving-off of parts of its businesses: more of that later. In fact, various sales occurred long before there was even a wisp (or a whisper) of disaster.

After consolidation of its various subsidiaries and divisions in the 1960s – to provide the basic form of the company's structure – Union Carbide's extensive interests in pharmaceuticals, oil, gas and fibres were sold off (2).

It sold, too, a large part of its European plastics and chemicals businesses to BP in the late '70s (3), followed by the offloading of nine ferroalloy plants (in the US, Canada and Norway) to a Norwegian-Canadian consortium in 1980 (4). (Elkem, the Norwegian company, was offering to buy another two such plants, four years later) (15). In 1983, Kema Nobel, Sweden's leading chemical company, and Union Carbide, sold off their 50/50 polyethylene plant to the Finnish state company, Neste (3). Around the same time UC closed its Australian polyethylene plant, and announced a US$140 million write-off against profits, because it was doing the same thing with similar facilities in the USA (6). It also sold the calcium carbide division which gave the company its name more than a century ago (7). As the company entered that fateful year of 1984, it had "redefined" its five main operating divisions. Making a profit were petrochemicals, industrial gases, technology and services. Turning in a loss were the consumer and metals, and carbon products division (8).

This is not to say, however, that Union Carbide was necessarily tightening its belt, or forsaking its long history of acquisitiveness. As the *Financial Times* summed up its strategy: "Union Carbide, third largest US chemical group, is the sort of modestly performing company whose enormous size (sales of US$9 billion last year) can be relied upon in normal times to protect it sufficiently from swings in the economy and Wall Street predators. Sprawling over a range of activities, from chemicals to carbon, specialised metals and consumer products like Eveready batteries, it was once cruelly dubbed by Wall Street as 'the chemical company that's always turning round'" (9).

Bearing out this assessment, the company's president Alec Flamm announced in autumn 1984 that the company was after new business in its profit-making areas, and had taken a decision to license its plastics technology, rather than continue marketing the products. Consumer products and – ironically – industrial gases, were named by Flamm as points of key expansion, although the President hazarded the guess that in 10 years time Carbide would be a technology company, not a commodity chemical company (7).

For the present, however, Union Carbide is still clearly identified with the products it has made household names over the last several decades: Eveready Batteries, Prestone Antifreeze, Simoniz Waxes, Glad plastic wraps and bags, and Bakelite (2).

The original Union Carbide was incorporated in 1898, and merged with four other businesses to form the present company in 1917. A sixth arm, Carbide and Carbon Chemical, was established in 1920, to become one of the earliest US petrochemical producers. The company "has a long history of expansion by acquisition" (2), and its first forays into mining came in the 1920s with penetration of Rhodesia and the purchase of mining properties near Rifle, Colorado (10).

It was the Rifle uranium-vanadium deposits which were to provide a backbone for the US military's atomic programme during the second world war, though Union Carbide allegedly "... was not told what the uranium was wanted for" (11). By the late 1970s the company's metals and carbon division was giving the parent corporation 18% of its sales, and nearly a quarter (24%) of its pre-tax income: the largest proportion of returns from any of Carbide's ubiquitous sections (12). And, despite its mining interests in southern Africa, Rhodesia (Zimbabwe), and Brazil, and its involvement in chrome, vanadium, tungsten and gold (Union Carbide moved into gold exploration in 1983, with the purchase of a 50% interest in a property in north-west Ontario) (13), it was not possible until 1984 to seriously call the company's reliance on uranium mining into question. In 1982, the Metals and Carbon Products division made a US$79 million profit, and delivered 12% of the corporation's sales. A year later, though providing a similar proportion of sales, the division chalked up a US$14 million loss (14).

Hand in hand with its uranium exploitation has gone the company's participation in almost every other aspect of nuclear fuel production. It became the world's largest private producer of fuel for nuclear weapons (15) when it gained the contracts to operate the Oak Bridge (Tennessee) Y-12 and Oak Ridge Paducah (Kentucky) gaseous diffusion uranium enrichment plants during and just after the second world war, on behalf of the federal US government (first the military, later the US Department of Energy, the DoE) (10). Union Carbide at Oak Ridge in Tennessee has, with little doubt, "been more responsibile for the construction of more nuclear weapons in the United States than any other company" (16). In addition, Union Carbide scientists (along with others from Du Pont and Kodak) were enlisted by General Groves in the Manhattan project, under the direction of Robert Oppenheimer: the group which constructed the bombs that were dropped on the people of Nagasaki and Hiroshima (17). Union Carbide also provided funds for the US's third enrichment plant, the Goodyear facility at Portsmouth, Ohio. When its Tennessee plant was renamed a "field office" after the second world war, Union Carbide could expand its "beneficial applications of atomic energy" not only in Kentucky, Ohio, and California, but also Puerto

Rico. By the late 1970s, the company was contracting enriched uranium hexafluoride to nuclear reactors and facilities in some 20 states and eight overseas countries (18).

Nearly a third of the Oak Ridge National Laboratories programme (26%) was then devoted to nuclear energy development, another 13% to "biomedical and environmental research", and 1% to solar and geothermal research and development (18). In 1977, Union Carbide received more than one billion dollars in federal funds for solar research. At that time, a study done by the US public interest group, Common Cause, revealed interlocks between Union Carbide's upper echelons and the Energy Research and Development Administration (ERDA) – forerunner of the Department of Energy (DoE) – which leave no doubt as to the central importance of Carbide to federal energy strategy in the post-Carter era. No fewer than eight executives from the company staffed the top echelons of the ERDA, while its acting administrator (pending the advent of the DoE) was Robert W Fri, who came hot from a seat in Union Carbide (19). Little wonder that, a year later, Union Carbide obtained more than US$10 billion from the DoE for nuclear activities. According to government documents, obtained in 1981 by anti-nuclear activists of the Black Hills Alliance in South Dakota, the DoE contracts were to run from 1978 to 1983, covering uranium and "other activities" (20). In 1979 the contract was worth no less than 25% of Union Carbide's total sales and – important in view of the huge liability the company was to face in 1985/86 – Carbide was released from responsibility for nuclear accidents associated with the contract. The DoE insured the public for up to US$500 million per "incident" in the USA, and up to US$100 million per "incident" elsewhere (20).

Union Carbide's Oak Ridge facilities have been the subject of intense controversy in recent years, not least because of their huge demand for cheap electricity at the expense of public consumers. The Tennessee plant requires more electricity to run its enrichment cascades than the cities of Nashville, Knoxville and Chatta-nooga combined (21). The site was originally chosen because it enabled the Tennessee Valley Authority (TVA) to sell power at discounted rates, thus – as a public corporation – avoiding the normal constraints of the market (see TVA). For its part, the Paducah plant uses 30% of all electricity supplied in the state of Kentucky (10).

Dr William A Lochstet, of Penn State University, has calculated the likely casualities from nuclear fuel processes contingent on TVA's proposed Dalton Pass uranium mine, using data obtained in 1978. Lochstet predicted that the tailings from uranium enrichment plants, utilising the uranium product from this one mine, would cause 52 million (*sic*) deaths over the 85,000 years of half-life of the radioactive by-products. (This figure was based on a 0.2% tails assay at the U-235 plant and good management and cross-ocean absorption; these figures can clearly be modified and applied to the impact of Union Carbide's plants – the world's longest surviving U-235 installations.) (22). Indeed, C H Patrick's study of cancer mortality in the two counties surrounding Oak Ridge, Tennessee, from 1950 to 1969, reveals a leukemia and lung cancer mortality among non-white females which is *four* times the expected rate (23). As Jeannine Honicker put it, in her 1978 court submission "to end nuclear power" (24): "This data strongly suggests that the Oak Ridge Gaseous Diffusion Plant is murdering the surrounding population, beginning with the non-white Females" (21). Sensationalist though such accusations may be, they are substantiated by evidence accumulated by a Union Carbide employee. Joe Harding worked for Union Carbide at Paducah from 1952 until 1971, first in the cascade hall, later as a process operator – plugging leaks and clearing up messes – in "downstream" operations. In 1961, after seven years of violent vomiting, and weight loss, his stomach had to be removed. (So advanced was the deterioration of this organ, that it was displayed in formaldehyde at a hospital for several years.) Seven years later, he contracted pneumonia, and five years afterwards, small skin-sores began working their way up his body.

Growths also developed on his hands and feet, and at the ends of his lower ribs.

Around 1971, Joe Harding began tracking the fates of up to 200 men who began working with him in 1952. By the end of the 70s, he estimated that as many as one quarter of this workforce had died of leukaemia, other cancers and "some unidentified ailment that may be related to radiation" (25). Soon afterwards, Joe Harding himself was dead.

Additional scandal surrounds Carbide's Y-12 nuclear weapons facility at Oak Ridge, Tennessee, from which it is likely that large amounts of highly enriched (weapons grade) uranium have been stolen in the last 40 years. The likelihood of such losses – euphemistically called "material unaccounted for" (or MUF) in the nuclear trade – was spotlighted some years ago by Charles Thornton, participant in the Manhattan bomb project and designer of the Oak Ridge National Laboratory in 1943. Thornton claims that "... the aggregate MUF from [all three US enrichment plants] alone is expressible in tons. No-one knows where it is ... The AEC (Atomic Energy Commission) can say "The numbers are not good, but we don't know any better!' If you admit that this industry is not controllable, then you shut down. You wait until it is controllable, and then start up" (21).

Bearing out Thornton's fears came a report from Sripps Howard News Service in early 1984 claiming that MUF exiting from Union Carbide's Tennessee plant since the 1970s amounted to 1700 pounds – enough to manufacture 85 nuclear bombs (120). The DoE issued a swift, bland denial, contending that the MUF had got "stuck in processing machines, lost through faulty book-keeping or accidently thrown away with radioactive trash" (*sic*). A special report had been compiled on "inventory differences", said a DoE spokesperson (20).

A little while later, however, it was revealed that some twenty-seven computer tapes presumed to record the uranium losses had been damaged or erased by a Union Carbide employee in June 1983. The erasure, said the DoE, was "accidental".

It is not known how much of Carbide's own uranium has gone directly towards the production of weapons-grade U-235; nor what proportion of the company's uranium output has been sold to the federal government for weapons use. However, presumably a good chunk of the federal aid of US$10 billion mentioned earlier was allocated to uranium production for military purposes. Certainly, at a time when other companies were falling like ninepins in the uranium stakes, it is remarkable that Carbide was actively planning to step up its uranium mining, following the alleged "depletion of its uranium stockpile" in 1982 (26). In 1975, Carbide held less than 1% of all US uranium reserves (27), but possessed nearly 9% of total US milling capacity (28).

Five years later, it was rated thirteenth among the (western) world's uranium producers (with 1450 tons production) (29), and fifth in the US league tables with 841 tons. Thus, just under half its production at that time purportedly derived from outside the USA, which is strange, considering that the company has never, to public knowledge, mined uranium outside of the country (29).

This discrepancy aside, Union Carbide has sat firmly in the front ranks of US companies avidly seeking uranium deposits abroad. Its main target areas during the 1970s include:

- Namibia – where, for some years, Carbide has been active (30) and was indicted at a UN session in 1984 (31). (The World Council of Churches has also named Union Carbide as a participant in the development of the Rössing mine. Certainly AG McKee was involved in its construction, but direct connections between this US company and Carbide are not known, nor is there any other evidence of an independent role being assumed by Carbide in the construction of Namibia's huge uranium project) (32).

- Gabon – where Union Carbide participated in a foreign prospecting consortium that also included Cogéma. In 1978, the consortium concluded a deal with President Bongo, which included a seven-year tax holiday, a maximum tax rate of 42% after expiry of the "holiday", and export duties and royalties limited to 50% of the value of production.

The government held only a 10% "free" equity in the consortium. Stephen A Zorn has observed that these were "... highly favourable conditions for the companies [with] little financial benefit ... to Gabon, and the agreement effectively leaves control firmly in the hands of the foreign companies, providing little or no "spin off" benefit for the rest of the economy and neglect[ing] environmental and health issues (33).

Three years later, however, Union Carbide was reported to have pulled out completely from Gabon, because its two exploration permit areas were "not particularly promising" (34).

- Canada – where the company joined the "uranium rush" in the 1970s and early 80s in northern Saskatchewan (35). In 1980, Enex sold half its 67% interest in a uranium property (23% held by SMDC) to Union Carbide (36).
- Botswana – where uranium exploration has featured for some years as part of the company's general prospecting programme, and where some radioactive anomalies have been discovered in central and western parts of the country (37).
- Zimbabwe – where Union Carbide's "tentative plans" to mine uranium along with tungsten "and perhaps platinum" have not materialised (38).
- Australia – where Union Carbide was actively exploring in Arnhem Land in the early 1970s and where it held 60% of the Maningrada Exploration licence area (Mitsubishi 40%); a project since abandoned, due to lack of uranium finds (39).
- South Africa – where throughout the 1970s, Union Carbide conducted intensive prospecting for uranium (40) resulting in significant discoveries, but no announced plans for development (30).

While these overseas prospects (and others about which the company is doubtless too reticent for substantial information to have emerged) have yielded little or nothing of value,

the same can hardly be said of Carbide's intensive exploitation of mainland USA. "Nuclear energy and coal are essentially the only ways the country will be able to meet its growing electricity needs over the next 25 years" the company's chair confidently predicted in 1979 (12) – and there is precious little Union Carbide has done to recover coal for power generation.

By and large, Union Carbide conducted its uranium operations in its own right. However, quite recently it joined with two other US uranium companies in joint development:

- With Hecla in a 50/50 JV at Lisbon Valley near Moab, Utah, commenced in 1981 (41). Although ore was being stockpiled that year for expected milling (42), the property was placed on maintenance standby in 1984 "pending price recovery" of uranium (37).
- In Virginia, with Marline on the Swanson orebody which is one of the most commercially promising uranium finds anywhere in recent years (43). The state of Virginia imposed a uranium moratorium in 1982 which was extended until 1984 a year later (44). As of the end of 1985, the project – with an estimated 13.6 million kilogrammes of U_3O_8 grading 2kg/ton – was still at its initial proposal stage (45).

As the seventies drew to a close, Union Carbide seemed awash with uranium – at least in Colorado, where it worked a mill at Uravan, and owned no fewer than 25 uranium mines (some in the eastern part of Utah). Its Gas Hills mine/mill was churning out yellowcake, and exploration was underway in New Mexico, and South Dakota (10).

However, its *in situ* leaching operations in Texas at Palangana Dome, Duval County (300,000lb per year U_3O_8) (46) were closed down in 1980 (47). And intense opposition by Native American communities, joined by antiuranium and environmental groups, were soon plaguing its explorations in the south-west.

Union Carbide leased 36 square miles of the Rio Grande near the Santo Domingo pueblo in the late 1970s (48). Although the lease was taken on state and federal lands bordering the pueblo, the major impact would have been on the impoverished Native American residents nearby. With a reputation as activists and conservationists (dating from the great Pueblo Revolt against the Spanish of 1680, to resistance against land grab in 1924 and the formation of the All Indian Pueblo Council) the Santo Domingo people soon resisted Carbide's encroachments. They identified "ruins" in areas coveted by Union Carbide as ancestral burial grounds and pledged to fight the company at all levels. "We say the history of the Non-Indian occupation of America is one of wanton exploitation and waste. The Indian people have been beaten and cheated out of their lands to make this exploitation possible. The certain vested interests get so crazy for minerals and precious stones that he will kill to acquire these things. In other words, the very healthy, wealthy non-Indian controls and tears up the earth for those who are hungry for money. He will rape and destroy the Mother Earth for a few cents of fleeting profit. He will make the Earth uninhabitable for the children of all races ... I am here to make it known that as far as I am concerned there will be no further destruction of the Earth in the area of my people's lands. We will put a stop to these dollar-motivated destruction. ... We will shelter the Earth that has sheltered us and those before us, so that our children may live in peace and harmony with the Earth" (49)

"We are not opposed to all development," added Ernest Lovate, the tribe's Executive Director. "We are opposed to the mindless destruction of our Mother Earth" (50).

Meanwhile Union Carbide had also leased 400 acres of tribal land on the San Felipe Reservation, 100 miles east of Crownpoint. Between 1975 and 1979 it sank 800 boreholes, and one exploratory shaft (so-called – in fact it was 250 feet in diameter and 125 feet deep) (51).

The San Felipe and Santo Domingo pueblos, and local environmental activists, joined together in the Sandoval Environment Action Community (SEAC) to oppose these developments (52). Within the next couple of years, SEAC was staffing a part-time office in Bernalillo, and had taken on the activities of Bokum as its major campaign. Though SEAC was still

concerned at Union Carbide's plans (the company was drilling in other areas by this time), John Vogt, the company's Production Manager, put residents' minds partly at rest in 1981 with an announcement that "plans for developing new mines have been put on the back burner" (53). Vogt admitted that Carbide's uranium output had peaked in 1979, dropped by around 5% in 1980, and would be cut back by a third that year (53). Citing reactor cancellations, together with the rise of new mining areas in Australia and Canada, Vogt claimed that US defence contracts would have no bearing on the company's activities "... because the government's strategic stockpile is already too big" (53). (As already pointed out, the company renewed its uranium search in 1982, but the New Mexico projects seem now abandoned.)

However, any downturn in the market didn't prevent Carbide from becoming one of the most active multinationals staking uranium claims in the Paha Sapa (Black Hills) of South Dakota. Between 1975 and 1979 Carbide, together with TVA and Gulf Oil, had drilled over 100,000 acres of sacred area (54).

Carbide's plans to open a mine at Craven Canyon in June 1981 (55) particularly roused the ire of Native Americans. They identified more than 25 important archaeological sites, including rock art, habitations and hunting camps, within the lease area – all of which were directly threatened by the company's plans to sink a 2000-foot long horizontal mine shaft as the first stage of a "massive mining and milling project" (54). The company's proposal included extracting 5400 tons of high-grade uranium ore and stockpiling several thousand more tons on site (57).

The Black Hills Alliance – which, at its huge Survival Gathering in 1980 (the largest joint indigenous peoples'/solidarity groups convention ever held) named Carbide and RTZ as the world's two most culpable nuclear corporations (58) – led the fight against the company's intentions. Carbide was not reluctant to take up the gauntlet: in an October 1979 brief to the US Government Forest Service, it described the BHA as "... an anti-nuclear group dedicated to disrupting nuclear energy at the exploration stage (if it can) without regard to actual environmental impact, mitigating pressures or eventual reclamation" by "... alleging concern for the environment and indian culture (*sic*) when neither is threatened" (57).

But, by the time the battle had run its course, it was Carbide which was condemned for its attitude to the environment and native people. Declaring that the company's "explorations" were effectively mining (which has to be covered by an environmental impact study (EIS) and consequent health and ecological surveys), the BHA, local ranchers and others adopted several suits against the company. The first asked for a proper EIS before Forestry permission would be granted; another challenged, in the South Dakota Supreme Court, the exploration permit for the shaft; while a third, in the same court, sought to reverse a State Conservation Commission decision that the BHA and Edgemont area residents couldn't take part in permit hearings (59).

After litigation successfully prevented Carbide from starting its shaft (notwithstanding the company's hiring of two prominent law firms and a PR man relocated from South-East Asia specially to "deal with the company's problems in South Dakota") (57), the South Dakota Supreme Court finally ruled in July 1981 that one of the State's agencies hadn't followed its own guidelines. Union Carbide pulled out of the Black Hills a couple of months later (60).

The centre of the company's uranium operations, however, was for many years the Colorado plateau area of western Colorado and eastern Utah. Indeed, until the early 1980s it had been mining uranium in this region continually for nearly forty years – probably a world record. Initially uranium was simply an unrealised by-product of vanadium; later the company opened its Uravan mill to process the yellowcake, while the vanadium was despatched to another mill at Rifle, after initial processing at Uravan.

The main orebodies in Colorado are found in depths of up to 700 feet in the saltwash formation of the Morrison Formation. By the mid-'seventies no fewer than 30 small mines were operating in the area (another world record, no doubt). Two of these mines traditionally provided Carbide with nearly half the company's total uranium output: Deremo-Snyder and King Solomon mines (61). As the orebodies were worked out during the '60s and '70s, so Carbide left behind increasing piles of tailings. More than two-and-a-half million tons were abandoned at the Maybelle mine; another three-and-a-half million at the New Rifle, Old Rifle and the two Slick Rock operations. (Another abandoned mine in Utah left a legacy of 123,000 tons of tailings) (62).

The 'seventies drew to a close and the Uravan mill itself began to suffer from lack of customers for its uranium feedstock. First it closed for six months, opening later on a part-time basis (63). It was "idled" again in April 1982, and reopened in 1983 to fulfil existing contracts, although no new deals had been made (64).

Once again, the Uravan mill closed in later 1984 (65) and the Rifle mill a little later (66). Meanwhile, although its mine at Gas Hills, Wyoming, reached a peak in 1980, the company laid off a quarter of its workforce a couple of years later (67). They were recalled in early 1983 – but within the next two years, the Gas Hills mill had also closed down (65). As already noted, the Hecla/Union Carbide uranium/vanadium facility in Utah was also on stand-by throughout 1984 (37).

During this period, Carbide made up for some of its US losses in vanadium feed by purchasing from abroad – 6 million pounds from Venezuela in 1983, for example (37). As a result its Arizona vanadium plant increased production (37). So did the company's vanadium pentoxide plant in the Transvaal, near Brits (37), which supplied the feed for Carbide's Bon Accord ferro-vanadium facility, despite a strike by a third of the workforce early in 1984 (68).

The company's 49% owned Tubatse ferrochrome plant also made booming sales from South Africa in 1984 (37). The facility was built in partnership with Gencor (51%) and, until the late '70s at least, was taking supplies from three chrome mines controlled by Carbide, including Ruighoek (69).

It was this mine which featured in Carbide's strategy to avoid United Nations sanctions imposed on exports from Rhodesia in the 1970s. Carbide – along with the Vanadium Corporation of America (later Foote Minerals) – had chrome mining subsidiaries in this part of white-controlled Africa back in the 1930s (69). Both companies used South Africa's conduits to break the ban. Lanning and Mueller take up the story:

"In November 1966, the company transferred some US$2 million to its South African subsidiary, Ruighoek Chrome Mines. On 16 December, the day the United Nations adopted mandatory selective sanctions (which included chrome), Union Carbide sent another US$1 million to Ruighoek. Five days later Ruighoek sent US$2.68 million to Rhodesian Chrome Mines (Union Carbide's Rhodesian subsidiary) as advance payment on shipments of chrome ore. The company applied to the Johnson administration for a licence to import 150,000 tons of chrome ore on the grounds that it was purchased before sanctions were enacted! In 1967 the Vanadium Corporation of America had merged with Foote Minerals (in which Newmont Mining had a 33% stake), and now Foote Minerals applied for similar licences in 1967 and 1968 to import respectively 40,000 and 56,000 tons of ore which had already been 'mined and paid for' according to the company. The Johnson administration refused the applications of both companies.

"Union Carbide and Foote Minerals then went into action lobbying in the US Senate, working in close cooperation with the Rhodesia Information Services office in Washington. The operating expenses of the office were US$100,000 a year. *Newsweek* called the campaign, and the resultant Byrd amendment allowing chrome imports into the United States, 'the neatest lobbying job in recent memory'. Although the output from the companies' Rhodesian mines had been successfully marketed through a top-secret organisation called UNIVEX, the companies were losing some US$28 million a

year in potential return on their investment, and when the possibility of legislation removing the ban on chrome imports arose, the two companies 'immediately went to work with the hard sell', as *Newsweek* put it.

"First they helped move the legislation through the conservative Armed Services Committees of both the House and the Senate, arguing that chrome was vital to national security. The bill was based on an ingenious ploy forbidding the President to ban the import of any strategic commodity which the US was also buying from a Communist country. In the case of chrome, the US had been getting about one-third of its supply from the Soviet Union in 1965, the year official sanctions took effect, and subsequently had to boost its orders to more than 50% because of the embargo.

"The lobbying in Washington was led by Representative James Collins and Senator Harry Byrd, and by a vote of 46 to 36 the Senate added a proviso to the Military Procurement Act of 1971, allowing the import of Rhodesian chrome. The main argument used by the lobby was that of 'national security', but as a high State Department official told *Newsweek:* 'It was a powerful but essentially phony argument, because we have all kinds of short-term supply; there just wasn't any national security element to the argument' (121). The reason was simple: there were already 5.5 million tons of chromium ore in the national stockpile, and the Office of Emergency Preparedness declared that 2.2 million tons were in excess of national foreseeable strategic needs. In addition, the Nixon administration had already submitted legislation to authorise the government to dispose of some 1.3 million tons of excess metallurgical chrome from national stocks. As the *Wall Street Journal* pointed out in March 1972, there was no ready market for the stockpile excess so there could not possibly be a need for Rhodesian chrome. The paper also reported that 'one Union Carbide official admitted that the "strategic" label was simply camouflage to get Congress

753

to authorise US companies to break UN sanctions'. The *Washington Post* commented that, 'although they complained strenuously about sanctions' violations by other nations, nowhere in the two firms' testimony did they suggest the United States begin by blowing the whistle on violators – a step that would have made life rougher for the white minority government but more equitable for the world's chrome users in sharing the burden of the loss of Rhodesian ore' (70).

"The vote in Congress had another detrimental effect because, as the London *Sunday Times* pointed out, 'coming as it did just four days before Sir Alec Douglas-Home left to negotiate with Ian Smith, [the vote] was the death knell of Britain's non-violent "solution" of the Rhodesian problem' (71)."

And so, the infamous Byrd amendment went through. Although there were sporadic longshoremen's actions against the continued US import of Carbide's chrome (69), the company effectively violated sanctions for six years. In March 1977, the Carter administration finally sealed the loophole, using PL-95/12 which amended the UN Participation Act; Carbide curtailed its production soon afterwards (72). But the company was still not out of the political woodpile. When negotiations for a post-independence government took place a year or so later, Carbide was prominent among foreign companies which lobbied for the neo-colonialist Smith/Muzorea option (72).

Despite its flagrant role in evading sanctions, Carbide has established itself as one of the major post-independence companies operating in Zimbabwe. It was one of four multinationals awarded a total of eight Exclusive Prospecting Orders in 1981 (which included uranium searches) (73). And it continued to produce chrome and ferro-chrome through its Union Carbide Zimbabwe (Private) Ltd and the Zimbabwe Mining and Smelting Company (Private) Ltd (37), providing up to 70% of the country's output.

The following is a summary of Union Carbide's other more controversial (or concealed) operations, starting with another example of its co-operation with the white colonialists in southern Africa:

- Under the so-called "Atoms for Peace" programme between the US and South Africa, signed in 1957, Union Carbide participated in the construction of the Safari-1 reactor (40). Enriched uranium fuel was also supplied by the US – presumably from Union Carbide (74).
- Along with Charter Consolidated (75%), Union Carbide (25%) operates the Beralt Tin and Wolfram company (UK) which has a wolfram mine and plant in Beira Baixa province, Portugal (14).
- Its explorations in the Brazilian Amazon were condemned by the World Council of Churches in 1981 (75).
- In 1984 it was searching for gold and other minerals (though without much success) in the Coromandel peninsular of Aotearoa (New Zealand), an outstanding natural conservation area (37).
- In 1984, Carbide bought a one-third interest in Katalistics International from the world's biggest producer of china clays, English China Clays (37).
- Union Carbide was among the US companies which benefited from the fall of Cheddi Jagan in Guyana in 1964 – a collapse engineered by the US, with support from Britain. (There is no evidence of direct involvement in the sabotaging of the popular, leftwing régime, by Carbide) (76).
- The Indian government, intent on making its fisheries export oriented, (and as a consequence rubbing out many peasant fisher people), encouraged Union Carbide and the huge Tata empire to enter the industry – making India the world's biggest exporter of shrimp and prawns which could be used by its own citizens (77).
- Until 1985, Carbide was producing short-fibre chrysotile asbestos from its Calidria mines near Coalinga, California. The company sold Calidria Corp to a group of private investors (KCAC Inc) that summer (78).
- Tin deposits were discovered by Tenneco and Union Carbide in the Andaman Sea off

Thailand, more than a decade ago. Student protests against multinational interference forced the Thai régime to abrogate its concession with Carbide in 1974, thus "preclud[ing] such agreements with other transnationals for a considerable length of time" (38).

- In 1984 Carbide announced it would build a manganese milling plant at its existing manganese factory in Abidjan on the Ivory coast (79).
- A contract to supply moderator graphite for the Torness and Heysham-2 nuclear reactors (in Scotland and Lancashire, England) was fulfilled during 1983-84. However, a poor prognosis for nuclear power development, combined with little prospect of similar contracts for the near future, forced Carbide to reduce its workforce at Sheffield, England, by 40% (from 800 to 500) (80).
- Union Carbide was also forced, more than a year later and after the Bhopal conflagration, to abandon plans for a toxic gas plant at Livingstone, Scotland. The plant was intended to provide Mitsubishi and NEC Semiconductors, with gas required for the production of semiconductors. After the Livingstone Development Corporation rejected Carbide's proposal, the company was said to have received offers from other parts of Britain (81).
- The growing monopolisation of seeds and cultivars by multinationals and the hugely alarming depletion of "gene pools" around the (particularly indigenous) world are a matter of utmost concern to many environmental groups and native peoples. Inevitably and ineluctably, it is those companies with an already strong control over chemical and fertilizer production or food processing, which have turned seed patenting and ownership into big business. The result has been the obliteration of many old strains of seed and marginalisation of promising new varieties. Union Carbide has played a major role in this process, though it is not usually defined as an "agribusiness". In 1977 it bought Anchem Products Inc, an important US agri-chemical company manufacturing metalworking chemicals, herbicides, seed corn and a chemical ripening spray. It also purchased, in the late 1970s, Keystone Seeds Co and Jaques Seeds. Together with Upjohn (manufacturers of Depo Provera), Sandoz, and Purex, Carbide controlled nearly 80% of the entire US bean seed patent in 1979, and a large chunk of the lettuce business (82). It was also singled out, along with Royal Dutch Shell, as one of the dominant chemical companies in seed monopolisation, in the important 1979 study *Seeds of the Earth* (83).

In 1976, Carbide attempted to "clean up" its corporate image, after government intervention against its literally stinking operations in West Virginia (84). It was a ploy which seemed to content many critics – judging by the surprise with which the Bhopal disaster was commonly greeted in the media, nearly a decade later.

In fact, from the 1930s – when nearly 500 workers died from silicosis when building the Gauley Bridge tunnel in West Virginia (81) – Union Carbide has never been free from well-founded accusations of murderous, cost-cutting incompetence, and frank disregard of health and environmental controls.

- After refusing to attend government conferences in the '70s on air pollution controls, Carbide's iron plant at Alloy, West Virginia, was dubbed "the smokiest factory in the world" (81).
- More than 100 workers in the company's manufacturing plants in Massachusetts and Maryland contracted bladder and other ailments in 1977 after exposure to a Carbide chemical used in foam production. The company was forced to withdraw the product (81).
- Up to 60 tons of Carbide's teratogenic and carcinogenic chemical Dioxin (a by-product of the manufacture of 2,4,5,T which killed 200 people at Seveso in Italy in 1976) was dumped on three rubbish tips in Sydney, Australia, in 1978. Smaller amounts had been dumped in a similar fashion since 1949. The state government of New South Wales, and Carbide executives, later worked out a

plan for dealing with the contaminated material (85). Carbide's 2,4,5,T is marketed as Weedar or Weedone and banned in several countries (81).

- A 1979 study revealed that one of Carbide's chemicals used in dyes, drugs and textile finishing, called diethyl sulphate, causes skin cancer in mice. But it was not for another *four years* (September 1983) that the company divulged the findings to the US EPA. In February 1985, the company was fined US$3.9 million for its dereliction (86). The EPA called the failure a "clear violation" of a 1976 law requiring immediate disclosure that a chemical might pose a risk (87).

- In 1981 the US Occupational Safety and Health Administration (OSHA) reported 16 brain cancer deaths from the operation of one of the company's Texan plants. Vinyl chloride workers at the Carbide South Charleston plant in 1976, had twice the expected incidence of brain cancer, and four times the expected incidence of leukaemia (81).

- In the early 1980s, Carbide's Cimanggis battery factory in Indonesia was found to be producing unacceptably high quantities of Carbon Black (which contains mercury). Reported *Newsday,* a Long Island (US) newspaper with a 500,000 circulation: "Taking advantage of Indonesia's tax regulation practices, Union Carbide Co, operates a battery plant on the outskirts of Jakarta that has become a monument to worker exploitation. The death of a young apprentice in an industrial accident brought remedial action. But health and safety provisions in the plant were so poor that at one point more than half the workforce of 750 were diagnosed as having kidney diseases linked to mercury exposure" (1).

- Temik, Carbide's trade name for its toxic pesticide aldicarb, has dangerously polluted both groundwater and citrus fruits, according to Dr Robert Metcalf, an entymologist at the University of Illinois. It has polluted water wells in three US states, but the biggest danger comes from orange juice made from contaminated oranges. Temik's poisoning

symptoms include nausea, dizziness, diarrhoea, and eventual coma leading to death, according to US officials (88).

It was aldicarb oxide which precipitated another disaster, in August 1985, when a huge chemical cloud, smelling "like a dead skunk" (89), was pumped into the air from the company's Institute works in West Virginia, putting several people into hospital.

But this was by no means the only post-Bhopal leak to cause concern in the last couple of years. Releases of potentially deadly mehtyl isocyanate (MIC), the gas released at Bhopal, were registered in September and October 1984, and the EPA listed no fewer than 28 accidental leaks at the Institute plant. A report by Henry Waxman, chair of the Health and Environmental Subcommittee in the House of Representatives, indicted Union Carbide for lax procedures and exposing workers to untenable risks (90). The parallel between what had happened in central India and what was now revealed in West Virginia clearly worried local people (91), although old-timers stuck up for Carbide. An editorial in the local *Charleston Gazette* declared it would be "stupid" to bring any case against a company whose "corporate image isn't bad by any yardstick, and where the corporate clout is such that it and its employees pay perhaps one tenth of the state's total tax bill" (92).

In April 1986, Union Carbide was fined a record US$1.38 million by the federal OSHA (Occupational Safety and Health Administration) for no fewer than 221 alleged violations of safety regulations in the operation of the Charleston plant (93). One of the arguments deployed by Carbide and its defenders, after Bhopal, was that the responsibility for the release of MIC gas – which killed at least 2000 and possibly 15,000 people (94), injuring at least 200,000 others – was partly local: HQ had little control over its Indian subsidiary which had failed to implement proper safety checks (95).

What happened in West Virginia a few months afterwards – combined with the appalling history of Union Carbide's other "misadventures" – leads to a directly opposite conclusion: any one of the company's plants is a chemical time-bomb waiting for a certain conjunction of circumstances to light the fuse. The fact that more people died at Bhopal than at any other Carbide site was related far more to the positioning of the plant, the employment of plentiful cheap labour, and the generally low status given a Third World operation by a Western corporation.

As this entry was being written, Union Carbide offered an insulting US$350 million to the Indian government, in settlement of the numerous claims made by the victims of Bhopal (96). This has been rejected by the Indian authorities (97).

If the company is able to keep the claims case in the United States, it is more than likely that a jury trial and punitive damages would be avoided. Such damages would have to rise to several billion before Carbide would be forced into bankruptcy (as Johns Manville has been, through similar suits brought over asbestosis) (98). Within weeks of the Bhopal holocaust, "disaster" lawyers in the US expressed "serious doubt" that "Manville-type charges could be made to stick against Union Carbide in the US" (9). However, although the company made no specific provision at this time for claims over Bhopal (it took a paltry US$18 million share charge to set against operating and other costs) (99), it took drastic steps through 1985 to reorganise and trim its activities and commitments.

In late 1984 it sold its North American welding and cutting systems for US$70 million (100). Nine months later it wrote down its metals and carbon products division by nearly US$400 million for the year and cut its domestic workforce by 15%. It was expected that Umetco, a subsidiary it had set up in 1984 to "facilitate implementing of various strategic options, including sale of all or portions of the metals business" (37), would take the brunt of these cuts (101).

A little later, the company's chair, Warren Anderson, announced an agreement-in-principle to sell off most of the company's tungsten, chromium and vanadium businesses around the world, at a price of around US$83 million. The company's South African interests were to be purchased by Gencor – including Tubatse Ferrochrome, Jagdlust Chrome Co, and the Chrome Corp (South Africa). The US tungsten and world-wide vanadium assets were to be sold to a "group of former employees headed by William G Beattie", formerly a Vice-President of Umetco (102).

Then, at the beginning of 1986, Anderson announced the prospective sale of the company's entire Consumer Products Division – meaning that Eveready, Simoniz, Gladwrap, Prestone and other world-famous products would no longer be associated with the Carbide caboosh. This truly drastic pruning operation was expected to garner up to US$1.1 billion (102). The Hanson Trust, a British industrial holding company (see Peabody), expressed interest in purchasing Eveready, to set alongside its own Ever Ready division (which it bought from Carbide in 1981) (103). What is noteworthy in this flurry of dispossession is that the company's huge nuclear interests were not envisaged for the chop: understandable in view of their strategic importance for the US state.

Certainly, much of this reorganisation must have been undertaken with a view to limiting any damage from a morally acceptable Bhopal claims settlement. And indeed, the company fell deeply into the red by the end of 1985, partly because of a US$185 million charge which was largely to cover such a settlement (104).

However, as already noted, the company's reorganisation had commenced some time before December 1984. More important, the company had to defend itself against two take-over bids.

The first came from the notorious Bass Brothers, who had already snaffled up a large chunk of Texaco, and sold it back at a gross profit. Their purchase of 5.4% of Carbide's share capital in December 1984 was warmly welcomed by the Carbide board: one obvious impact was to boost the tottering share price. A company spokesman dubbed the freebooting Basses "sophisticated investors" (105). Within a very short time, shares were trading at US$38, as against a Bhopal low of US$35 (106).

In August 1985, the diversified chemicals producer GAF began building up its own 5.6% stake in Union Carbide, just after the South Charleston leak (see above) (107). Four months later, GAF announced that it would strip some of its own assets to help raise an estimated US$4.85 billion to buy all of Carbide's outstanding securities. (By then GAF held 7% of Carbide and was tendering for 31 million more shares, to give it 55% control) (108).

Within days, GAF's holding was 10% and it had decided to go for total control of the bigger company, with or without Carbide's approval (109).

Warren Anderson was far from amused. Describing the GAF bid as "a boot-strap, junk bond, bust-up partial takeover", he launched a buy-back strategy, to repurchase 35 million of UC's own shares (110).

GAF upped its offer to US$74 a share (111), Carbide promised an even sweeter offer (112), and Carbide won a court ruling that its scheme was "reasonable and fair" (113). By the new year, more shareholders had tendered their offerings to the Carbide management than were needed. Not to be outdone, GAF increased its own offer to US$78 a share (114). Within a few days, GAF pulled back, holding on to a 10% interest in Carbide, and chalking up a nice profit as a result of selling some of its own holding back to the bigger concern. To quote the *Financial Times*, Union Carbide has emerged from both Bhopal and the GAF bruising "... a much smaller industrial company heavily laden with debt and facing an uncertain future" (115).

Scattered opposition to Union Carbide's activities before Bhopal has included the reported bombing of its South African offices in late 1984 by "freedom fighters" (116), and the international campaign announced at the 1980 Black Hills Survival Gathering (already mentioned) which, unfortunately, never developed. Since Bhopal, Brazil has stopped the offloading of a freighter containing 12 tonnes of MIC, and the governor of Rio de Janeiro banned the use, storage or transportation of the chemical anywhere in the state (86). Also, 627 workers staged a sit-in at the Bhopal plant in early December 1985, and won almost three times their legal compensation rights for loss of their jobs (£1 million) (117).

Meanwhile, action and campaign groups, centering on the aftermath of Bhopal, have sprung up all around the world. The list below is a partial one, drawn from the "No More Bhopals Network" which is "a loose coalition of environment, development, consumer, women's, workers, and other public interest groups working on Bhopal" (118).

Union Carbide Southern Africa, owner of Zimbabwe Mining and Smelting Co (Zimasco), has opened an office in Parma, Ohio, to market high chromium "charge" ferro-chrome. Formerly, Zimasco's charge chrome products were sold through the Pittsburgh office of Union Carbide Corp subsidiary, Umetco Minerals Co. The rationalisation at Union Carbide has continued. In a letter to employees, company president, Mr Robert Kennedy, stated in 1986 that Carbide intended to divest itself of another US$1000 million of non-strategic assets, which would inevitably involve redundancies. The centre-piece of this programme is Union Carbide's huge headquarters site at Danbury, Connecticut, which occupies 247ha. The company intended keeping its head office there but with a reduced need for space, the rest of the site was to be developed as a commercial complex over the next two years. Mr. Kennedy estimated that Carbide would be reducing its workforce, world-wide by a further 1200 (119).

A mere addendum cannot begin to measure either the magnitude of suffering uncovered long after the Bhopal holocaust, or the incredible lengths to which putative victims have had to go to register claims for compensation, and the obnoxious lengths to which Carbide has gone, to disclaim responsibility.

In 1986, US courts decided that compensation claims should be heard in India – a view supported by the Indian government itself (122), which has fixed US$3.5 billion as the sum necessary to meet the claims (122). However, it is quite clear that there is no objective yardstick by which such claims can be measured, with local workers arguing that the official toll of

1800 deaths, and around 3000 families affected, is not nearly representative of the extent of the disaster. For its part, Union Carbide has argued that all but a handful of the 500,000 suits registered against the company "are frivolous demands from people living in parts of Bhopal that were not touched" by the tragedy unleashed on December 2nd 1984 (123). It has been assembling evidence that it was not responsible for what its subsidiary did in India, and therefore its offer of US$350 million to the victims is fair and honourable (123). It seems likely that Carbide will succeed with such a claim, even though the Indian courts are adamantly opposed to it, and that real compensation may take years to come through – if ever (122).

In 1988, Umetco, Carbide's subsidiary, opened three new small uranium mines between Slick Rock and Gateway in Colorado, some eleven years after about a dozen minor mines had closed down. Umetco owned three of the reopened mines, and had production contracts with operators of the other four. Although initially the company said it would open the mines "for a few months" in mid-1988, it decided to extend operations indefinitely (124). However, in 1987, it reduced output from its joint mill (with Energy Fuels) at Blanding, Utah, by nearly half a million pounds per year (125) and – although in 1990 this operation was the biggest single US miller of uranium (with a total output of 4 million lbs U_3O_8) – by the year end, it had closed (126). Umetco's Colorado Plateau uranium/vanadium operations were now finally finished (127). The same year, the company decided against another major restructuring (128) although a few months later, it entered talks with Mitsubishi to buy 50% of its UCAR carbon business (129), and had actually made its first acquisition for some years – of Triton, the surfactant business which manufactures specialist chemicals and detergent ingredients (130). Less than a year later, in June 1991, UC had entered talks with the Italian chemicals group Enichem (formerly Enimont – see ENI) for a JV in polyethylene production (131); and was seeking partners for its Great Dyke platinum project in Zimbabwe (131). The company now considered the Bhopal "issue ... closed" (130).

Contact: Centre for Science and Environment, 807 Vishal Bhawan, 95 Nehru Place, New Delhi 110019 India.

Lokayan, 13 Alipur Road, Delhi 110054 India.

Society for Participatory Research in Asia, 45 Sainik Farm, Khanpur, New Delhi 110062, India. (They also supply a video.) Bhopal Disaster Monitoring Group, c/o People's Research Institute on Environment and Energy, 3F Baptist Kaikan, 7-26-24 Shinjuku, Shinkuku-ku, Tokyo 160, Japan.

Asia-Pacific Peoples Environment Network, (APPEN), c/o Sahabat Alam Malaysia, 37 Lorong Birch, Penang, Malaysia.

In England, the Bhopal Solidarity Group has organised a major demonstration against Union Carbide, and a petition. They provide speakers and other resources: Bhopal Solidarity Group (BSG), c/o Commonground Resource Centre, 87 The Wicker, Sheffield S3 8HT. The Bhopal Victims Support Committee (BVSC) is at: 50-52 King Street, Southall, Middlesex UB2 4PB, England.

SEAC, 2000 El Camino del Pueblo, Bernalillo, New Mexico 87004, USA.

Bhopal Action Resource Centre, 777 United Nations Plaza, Suite 9A, New York, NY 10017, USA.

Resources: *On Bhopal: No Place to Run: Local Realities and Global Issues of the Bhopal Disaster*, published by Highlander Research and Education Center, Rt 3, Box 370, New Market, TN 37820, USA. Also the film *Bhopal: Beyond Genocide* Directors: Bhose, Mulay, Shaikh (Distributors: The Other Cinema, Wardour St London WIV 3TH, England).

On Union Carbide's nuclear activities: there is an excellent film, *America – From Hitler to MX*, directed by Joan Harvey in 1982 and distributed by Parallel Films/The Fourth Wall Repertory. Peter Matthiessen's *Indian Country* covers Union Carbide's operations in the Black Hills (Collins, Harvill, 1985). On Union Carbide's sanctions-busting, see Reed Kramer and Tami Hultman, *Rhodesian Chrome – A Portrait of Union Carbide and Foote Mineral,* (Corporate Information Center of the National Council of Churches), New York, 1972.

David Dembo, Ward Morehouse, Lucinda Wykle, *Abuse of Power: Social Performance of Multinational Corporations: the case of Union Carbide,* New Horizons Press, New York, 1990.

Barbara Dinham, "Greening the Workplace" in *International Labour Reports,* no 34, 5-6/90.

Union Carbide; background paper by Bhopal Action Group, published by Minewatch, London, 1990.

Robert Ginsberg, *Present Dangers, Hidden Liabilities: A profile of the Environmental Impact of the Union Carbide Corporation in the United States (1987-88),* National Toxics Campaign Fund and Council on International and Public Affairs, Boston and New York, 1990.

Nick Catliff "Killer in the Night" in *World Magazine,* London 3/91.

References: (1) International Labour Reports, Manchester, No. 8, March/April 1985. (2) *WDMNE.* (3) *FT* 16/8/83. (4) *FT* 5/6/80. (5) *FT* 1/8/84. (6) *FT* 18/11/83. (7) *FT* 11/9/84. (8) *FT* 16/2/84. (9) *FT* 7/12/84. (10) Lynne Lahr, *The influence of Union Carbide, a backbone of the nuclear industry,* Paha Sapa Report, Aug 1979, Rapid City. (11) Moss, *Politics of Uranium.* (12) Union Carbide, Annual Report 1978. (13) *FT* 18/2/83. (14) *FT MIY* 1985. (15) Roger Moody, "Little Big Horn: Part Two," *Gua,* London, 16/8/80. (16) Robert Regnier, personal testimony at the Saskatoon Warman Hearings, Saskatoon 1980, quoted in *Yellowcake Road.* (17) Pringle and Spigelma. (18) The ERDA facilities, ERDA Washington, 1977. (19) Reece (20) *Kiitg* June/81, No. 14. (21) *Honicker versus Hendric,* Tennessee Farm (Summertown) 1978. (22) W A Lochstet, *EIS Comment on Dalton Pass Uranium Mine,* 1978. (23) C H Patrick, "Trends in Public Health in the population near Nuclear Facilities: a critical assessment", *Nuclear Safety,* Vol. 5, No. 18, Sept-Oct/77. (24) *The Tennessean,* Memphis, 30/7/78. (25) *The Progressive,* New York, January 1980. (26) *MJ* 17/10/82. (27) US Senate Committee on Energy and Natural Resources, *Petroleum Industry Involvement in Alternative Sources of Energy,* (US Government Printing Office), Washington DC, 1977. (28) Federal Trade Commission, Bureau of Economics, *Nuclear Fuel Industry Data* (US Govt. Printing Office) April/78. (29) *Market Shares and Individual Company Data for US Energy Markets 1950-1981;* Discussion paper 014R, 9/10/81, (American Petroleum Institute) Washington DC. (30) Neff, (31) *Transnational Corporations and other foreign interests operating in Namibia,* (UN) Paris, April/84. (32) WCC Programme on Transnational Corporations, *Sharing,* No. 7, December/81; Commission on Churches' Participation in Development, WCC, Geneva. (33) Stephen A Zorn, "Uranium, the most strategic material of all, in *Multinational Monitor,* Washington. (34) *Nuclear Fuel,* Washington, 5/1/81. (35) *Uranium Traffic* 1981, updated by personal references from Miles Goldstick to July 1983, Saskatoon. (36) *MJ* 15/8/80. (37) *MAR* 1985. (38) *International Mining and Metals Review* (McGraw Hill) New York, 1982. (39) *A Slow Burn* (FOE Victoria) No. 1, September/84. (40) Rogers and Cervenka. (41) *MIY* 1982. (42) *MM* 1/82. (43) *MJ* 24/12/82. (44) *MJ* 11/3/83. (45) *E&MJ* 1/85, *E&MJ* 1/86. (46) *E&MJ* 11/78. (47) *Kiitg* 6/80. (48) *E&MJ* 11/78. (49) *Santo Domingo Tribal Position Statement,* Santo Domingo Pueblo, New Mexico, (Santo Domingo Tribe) 21/9/78 (mimeograph). (50) *Nuclear Newsletter,* Saskatoon, Vol. 2 No. 15, 15/11/78. (51) Denise Tessier, "Using Native Tongues: Indians Oppose Mining", *Albuquerque Journal,* 22/9/78. (52) *Guardian,* New York, 11/10/78. (53) *SEAC News,* Bernalillo, Spring/81. (54) *ARC Bulletin,* Anthropology Resource Centre, Boston, March/80. (55) Leaflet from Black Hills Alliance, Rapid City, undated, 1981. (56) *BHPS,* Vol. 1, No. 3, 1979. (57) *BHPS,* Vol. 1, No. 5, April 1980. (58) Statement by anti-nuclear groups, Black Hills Survival Gathering, Paha Sapa, Summer/80, (mimeographed). (59) *Kiitg,* No. 9, Nov/80. (60) *Critical Mass,* Washington DC, Oct-Nov/81. (61) *MM* 11/77. (62) *Congressional Quarterly,* (US GPO), quoted in Lynn Lahr, "The Influence of Union Carbide, a backbone of the nuclear industry", August/79, Rapid City. (63) *MJ* 4/9/81. (64) *E&MJ* July/83. (65) *E&MJ* 11/84. (66) *E&MJ* 12/84. (67) *E&MJ* 1/83. (68) *FT* 31/1/84. (69) Lanning and Mueller. (70) *Washington Post* 19/3/72. (71) *SunT,* 30/1/72. (72) Sklar, *Trilateralism.* (73) *FT* 18/6/81.

GPO), quoted in Lynn Lahr, "The Influence of Union Carbide, a backbone of the nuclear industry", August/79, Rapid City. (63) *MJ* 4/9/81. (64) *E&MJ* July/83. (65) *E&MJ* 11/84. (66) *E&MJ* 12/84. (67) *E&MJ* 1/83. (68) *FT* 31/1/84. (69) Lanning and Mueller. (70) *Washington Post* 19/3/72. (71) *SunT*, 30/1/72. (72) Sklar, *Trilateralism*. (73) *FT* 18/6/81. (74) *Nuclear Fix*. (75) *Sharing*, (WCC, CCPD) Oct/81. (76) Ronald Graham, *The Aluminium Industry and the Third World* (Zed Press) 1982. (77) *Suara Sam*, (Sahabat Alam Malaysia) Oct/85. (78) *Industrial Minerals*, London August/85. (79) *MJ* 30/11/84. (80) *FT* 15/4/83. (81) *Bhopal, Green City, Clean City* (Scottish Education and Action for Development) Edinburgh, 1985. (82) *Official Record of the Plant Variety Protection Office*, Washington, 1979. (83) P R Mooney, *Seeds of the Earth*, (Inter Pares for the Canadian Council for Internnational Cooperation, Ottawa, and International Coalition for Development Action, London) 1979. (84) *Fortune* 25/9/78. (85) *Sydney Morning Herald* 29/4/78. (86) *Chain Reaction* No. 41, Apr-May/85. (87) *Gua* 5/3/85 (88) *The Orlando Sentinel*, Florida, August/82. (89) *Gua* 14/8/85. (90) *Gua* 26/1/85; *FT* 25/1/85. (91) *FT* 13/12/84. (92) *Charleston Gazette* 9/12/84. (93) *FT* 2/4/86. (94) *New Ground* (SERA) No. 8, Winter/85. (95) *FT* 23/12/85. (96) *Gua* 24/3/86. (97) Gua; *FT* 25/3/86. (98) *Mmon* Vol. 6, No. 10, 31/7/85. (99) *FT* 29/1/85. (100) *FT* 20/12/84. (101) *Metal Bulletin*, London, 3/9/85. (102) *MJ* 10/1/86. (103) *FT* 24/2/86. (104) *FT* 24/1/86. (105) *Gua* 22/1/85. (106) *Gua* 28/1/85. (107) *FT* 28/8/85. (108) *FT* 11/12/85. (109) *FT* 13/12/85. (110) *FT* 17/12/85, *FT* 19/12/85, *FT* 23/2/85. (111) *FT* 27/12/85. (112) *FT* 30/12/85. (113) *FT* 31/12/85. (114) *FT* 3/1/86. (115) *FT* 9/1/86. (116) *Black Flag*, London 15.10/84. (117) *Labour Research*, London, February/86. (118) *Ecoforum*, Nairobi, Vol. 10, No. 1, February/85. (119) *MJ* 18/4/86. (120) *Chicago Tribune*, 16/1/84. (121) *Newsweek* 10/4/72.(122) *Gua* 4/12/87. (123) *FT* 1/7/88. (124) *E&MJ* 5/88. (125) *E&MJ* 4/88. (126) *MJ* 12/10/90. (127) *E&MJ* 3/91. (128) *FT* 2/3/90. (129) *FT* 28/9/90, *FT* 21/11/90. (130) *FT* 12/10/90. (131) *FT* 6/6/91.

606 UCIL Uranium Corp of India Ltd (IND)

This is a public corporation which supplies the Indian government with uranium ore from the mines at Jaduguda (Bihar) and (it was planned) from uranium sources elsewhere in the country, particularly Madhya Pradesh, Karnataka, Meghalaya and Uttar Pradesh (1).

The mines at Jaduguda (which also produce copper and molybdenum) (2) are worked with a great deal of *adivasi* (tribal) labour under allegedly gruelling conditions; "black market" uranium has also been openly sold in the market-places of Singhbhum, one of the major tribal areas of Bihar (3). UCIL was planning in 1984/85 to open up a new mine at Bhatin at a cost of US$3 million (4). From 1984 workers at the Jaduguda operations released information on the serious dangers they are forced to endure. Only a fraction of the necessary respirators have been provided to the miners (437 out of a minimum of 1500 necessary) (8), while spilled yellowcake has to be recovered using bare hands (5). The 18 exhaust fans at the chemical extraction plant have, it is also claimed, been out of action "for almost a decade" (5). The greatest criticism attaches to conditions at the slurry (tailings) pond which remains dry for most of the year, compelling residents to inhale radioactive dust. Leaks from the slurry pipes into bathing ponds have also been reported – polluting the Gharnala river which itself joins the Subarnarekha. Up to 19 times the permitted level of uranium has been allowed to remain in the tailings. Moreover, radioactive materials – including bags and cloth – have been dumped in the tailings pile from Calcutta's Variable Energy Cyclotron Centre and the Hyderabad Nuclear Fuel Complex: children were reported to drink water using contaminated polythene containers.

In 1971 a general strike failed to obtain meaningful safety measures from the management of UCIL (7).

In 1988, UCIL announced plans to open up two mines at Narwapahar and Turanmidh to produce monazite, from which thorium is extracted. UCIL chairman M K Batra stated that, in view of the country's "lack of uranium reserves", the shortfall would be made up from thorium: he added that the country would require 40,000 tons of uranium to increase India's nuclear capacity to 10,000MW by the turn of the century (9).

In 1990, a full-blown scandal erupted over UCIL's radioactive discharges, even as the company was making plans to develop more deposits to meet increased power demand, as it began construction of more underground mines and of a central ore-processing plant at Narwapahar and Turamdih (Bihar), as well as expanding its existing uranium recovery plant at Mosaboni (10).

The respected magazine *India Week* discovered that the company had allowed a tailings pond at Jaduguda – meant to be kept covered with water – to dry out, causing uranium-laced dust to affect nearby villages. Women had been seen bathing in the ponds; there was a high incidence of respiratory disease, miscarriages, cancer and fever, as well as neural disorders and deformation in the local children. Workers in the drying plant had nothing more than cloth covering their faces, and claimed they were not given medical records. The workers also declared that, when they handled uranium oxide and UF_6, they were protected only by cloth gloves.

Local politicians claimed that buffaloes had died after drinking water from the evaporation ponds, while the plant's solid wastes – particularly the sulphurous coating of the yellowcake – was given to a fertiliser company which dumped them in the open in the town of Nacharan. Doctors here reported high incidences of skin disease, asthma and blindness. Scientists visiting the plant said that monitoring devices were not working (11, 12).

UCIL's defence was that radioactivity at the plants was less than the UN recommended discharge of 8 pico curies per litre, while radium and manganese in the waste were removed by solidification with barium chloride. However, it acknowledged that only 437 respirators were available for its entire 1500 workforce in 1985 (11).

Contact: Dhirendra Sharma, National Committee for a Safe Nuclear Policy, M-120, Greater Kailash-I, New Delhi 110048, India.

References: (1) K. Subrahmanyam of the Institute of Defence Studies, in *Gua,* 23/3/82. (2) *MM* 1/84. (3) personal communication from a local trade unionist who wishes to remain anonymous. (4) *MAR* 1984. (5) *Telegraph Magazine* 1/12/84, Calcutta. (7) Dhirendra Sharma, National Committee for a Safe Nuclear Policy, quoted in *WISE NC,* No. 256, 11/7/86. (8) *UCIL Annual Report,* 1985. (9) *MJ* 25/3/88, see also *MM* 5/89. (10) *E&MJ* 2/90. (11) *India Week* quoted in *New Scientist,* 17/2/90. (12) *WISE NC* 18/5/90 see also Timberlake and Thomas, *When the bough breaks,* Earthscan, London, 1990.

607 UG

After undergoing several name changes in recent years, Urangesellschaft now prefers to be known as UG or Uran, and its wholly-owned subsidiaries as Uranges (Australia, Canada and USA) (1).

Founded in 1967, when three huge West German companies took up one-third ownership each, it is, effectively, the Federal German government's uranium "arm". The West German state, through its former 44% interest in VEBA held both a direct share in UG and an indirect interest through Ruhrkohle and Steag (see VEBA).

The company's brief is to explore for and exploit uranium deposits, open up mines, trade in uranium products, and extract uranium from phosphoric acid (1). It has supplied around 50% of all West Germany's uranium needs (2). Until quite recently, the company was actively exploring throughout Australia (3), had several JVs in Canada, projects in some twenty US states including Colorado, Utah, Texas (4), New Mexico and Arizona (5), as well as further exploration projects in Botswana, Tanzania,

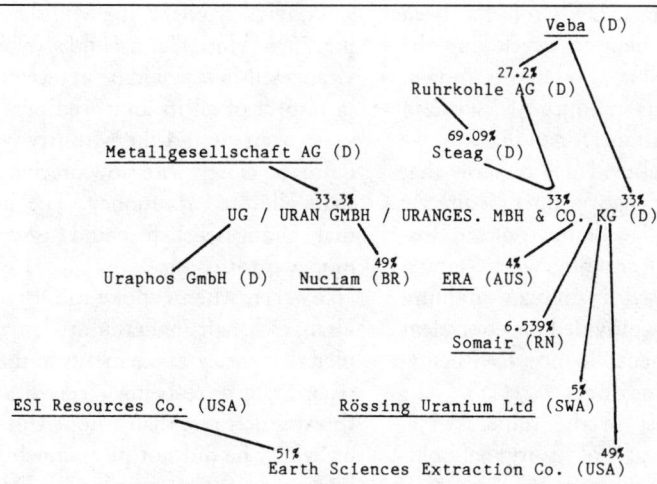

```
                                          Veba (D)
                                    27.2%
                             Ruhrkohle AG (D)

                              69.09%
                             Steag (D)

  Metallgesellschaft AG (D)                      33%  33%
                                                         CO. KG (D)
             33.3%          UG / URAN GMBH / URANGES. MBH &
             UG / URAN GMBH / URANGES. MBH & CO. KG (D)

  Uraphos GmbH (D)    Nuclam (BR)    ERA (AUS)
                           49%            4%

                              6.539%
                            Somair (RN)

                                              5%
  ESI Resources Co. (USA)    Rössing Uranium Ltd (SWA)

                  51%                          49%
              Earth Sciences Extraction Co. (USA)
```

Colombia, Indonesia (6), Brazil, Scotland, Italy and Niger. In its home country it was one of five corporations actively searching for yellowcake in the late '70s (7).

UG's lesser known (and less important) ventures include:

- an aerial search for uranium in Botswana in 1977 (8). When the permit for this was relinquished in November 1978, UG, together with Union Carbide and Falconbridge, took out further ones (10).
- a uranium exploration project investigating "targets" in old red Sandstone at Caithness, Scotland, in 1979 (9).
- a uranium exploration contract signed with the Tanzanian government which includes provision for excess profit taxes up to a rate of 75% when they exceed a 25% return on the company's investment – a very favourable arrangement for UG, to say the least (11).

Since UG was incorporated on the recommendation of the Federal German government itself, it is little wonder that its major investments have been in countries which are prime trading partners and political allies of the West German state. (Which is not to say that its ultimate criteria are not economic ones: as its own "background" publication states: "... it is increasingly difficult to acquire mineral rights in or near known uranium mining districts ... [UG] therefore frequently forms joint ventures with local companies which control exploration or mining rights ... In a number of cases UG carries out exploration on its own; but in many cases the joint venture partners are the operators of the projects") (5).

Two examples illustrate this close relationship between state and capital.

In Brazil, UG is the only non-domestic company to have been granted uranium exploration rights. After prolonged negotiations between the South American military régime and the FRG government, Brazil surrendered its monopoly over uranium in 1975 and enabled the state agency Nuclebras (51%) to form a JV with the West German company (49%) (12). In the years since, foreign enterprises (notably Pechiney) have become involved in uranium processing and in enrichment (notably Interatom and Steag) but UG remains the front-runner in the mining field (13). Most recently it expressed interest in exploiting the country's second largest uranium lode (between around 90,000 and 100,000 tonnes reserves and probable reserves) at Lagoa Real in southern Bahia (13). The similarities between the West German-Brazilian deals and those between the European state and South Africa are considerable. They relate to the exchange and secret provision of technology, scientists, enriched uranium and expertise in uranium mining (14). While UG's co-parent, Steag, has been heavily involved in many aspects of these transfers, UG's role has been limited to its involvement in developing the Rössing mine in Namibia. (So far as is known, it has no direct interests in South Africa itself.) It has been quite important, for all that.

There is no substantial evidence to show that the Federal German government directly loaned money for the development of the Rössing mine but, almost certainly, West German utilities have purchased Namibian uranium (14) and UG originally provided the equivalent of 10% of development finance to Rössing Uranium to get the mine under-way (15). UG's current equity interest in the mine is 5%; whether or not it sold part of its original holding to Total (Minatome) is not clear.

The Federal German cabinet was not unanimously in favour of the government risking international opprobrium by endorsing the Rössing project, at a time when it wanted recognition as a UN member, and when it had already been singed by controversy with black African states over its support for the Cabora Bassa dam (14). However, only one cabinet member, Erhard Eppler, publicly disavowed what was to develop into a major cover-up, perhaps the most controversial aspect of which was an apparent contribution from the FRG Ministry of Science to UG's Niger uranium costs, intended to launder funding for Rössing. A number of confidential letters which passed between the FRG's South African ambassador and the Pretoria Secretary for Foreign Affairs, in 1972, illustrate the tendentious nature of negotiations between the apartheid state and the FRG at this time. Not to mention the key role played by UG personnel in the wheeling-dealing – not, however, without attempts by UG to keep themselves at arm's length from embarrassing political scapegoating. It is worth quoting from one of these letters (released by SWAPO in London): "According to Secretary of State Haunschild of the Ministry of Science, which normally contributes 50% of preliminary prospection surveys, the issue with respect to Rössing has been somewhat eased for Urangesellshaft by the fact that in the case of a Central African State, which he did not identify but which

was obviously Niger, the Ministry will contribute 75%. Herr Haunschild's version was that Urangesellshaft would be in receipt of a subsidy in respect of all its approved prospecting activities overseas and the Ministry would not inquire too closely into how precisely Urangesellshaft allocated its money. The inference was that Urangesellshaft could use some of the money for Rössing.

"However, when I spoke to Dr von Kienlin of Urangesellshaft in Frankfurt last week, he denied that the extra payments to the Central African State were being received and said that this was no more than a hope and a possibility, on which he did not place much reliance. My impression from my talk with Dr Kienlin was nevertheless that the financial problem was not the principal cause of concern on the part of Urangesellshaft. The real issue was the present uncertainty with respect to supplies from South West Africa and their availability over a long-term period in the future. Neither he nor his company doubted our ability to deliver – the doubts lay with the consumer. Since the uranium would be required for use ultimately in nuclear power stations operated by public utilities or under the supervision of Lander Governments, Urangesellshaft had to take into account that in the minds of many reponsible persons concerned in this particular sector, there was a fear that delivery of the uranium would be banned or become impossible because of resolutions adopted by the United Nations and unwillingness on the part of the German Government to act counter to those resolutions. These fears are being played upon by the radical anti-South African elements in this country who have a spokesman in the Cabinet on this question in the person of Herr Eppler" (16).

Rössing's uranium was intended as a mainstay of West Germany's nuclear programme during the 'eighties. By the end of the 'seventies, however, the entire future of the country's nuclear industry was in doubt, as environmentalists, Grünen, local administrative courts, and even trade unions, expressed their opposition to various aspects of the atomic state (17). Over the next five years, UG tended to retrench rather than expand. Its interest in Somair's Arlit mine fell from 8% to its current 6.5% (18), while its explorations in the Djado area of Niger, started in 1977, seem to have terminated some years ago (19).

Virtually all its US exploration efforts, commenced in the 1970s with great élan, seem to have been stopped, including

- A JV with Conoco in the Church Rock area of New Mexico;
- A JV with Ranchers;
- Prospecting with Rocky Mountains Energy Co;
- A JV at Tonopak in Nevada, with Federal Resources Corp;
- Prospecting in its own right in Alaska (20);
- A lease on more than 1000 acres of Carlton County, Minnesota, and several leases in Benton County, Minnesota (21);
- A project at Baggs, Wyoming, and another at Date Creek in Arizona, both reported in 1980 to be at "an advanced state of exploration" (22).

And, on December 10th 1980, the members of Stop Uranium Mining (SUM) in Vermont, along with other local activists, saw the back of UG after a lengthy legal and political battle to get uranium mining *per se* banned in their state. More than 30 towns adopted an anti-uranium ordinance, and a bill was passed by the state legislature requiring full debate and approval by both houses of government before any further prospecting was allowed. Commented Ms Cole of SUM: "We battled one of the biggest uranium mining companies in the world and we won" (23).

The most important pull-back in UG's history is quite recent. In January 1986, it announced it was negotiating the sale of "most of its Australian assets, is withdrawing from uranium exploration, and is to close its Melbourne office". "Dissatisfaction with the policy of the Australian Labor government" was cited as one of the main reasons for this action (24).

In the event, UG held onto its more important Australian interests (54). However, there was certainly a long history of doubts within the company about the viability of mining uranium in the antipodes. As long ago as 1978, Casimir Prince Wittgenstein (representing Metallgesellschaft, but clearly speaking for UG as well) was getting disgruntled in Melbourne. Just after the 1977 elections which brought a more pro-mining Liberal Federal government to power, Prince Wittgenstein was complaining: "We have been trying to get this Western Mining (Yeelirrie) thing going for six years. First, the Whitlam Government spiked the wheel and now the Fox Report is out and we haven't got a true definition of what is going on. It is a sort of hotch-potch thing which we cannot see clearly" (25). However, the "Western Mining thing" was soon sorted out, with UG gaining a 10% share interest in Yeelirrie and a similar percentage of production (3), although initially it canvassed for more (26). More importantly, it also gained a crucial 4% stake in the Ranger mine when that project's equity holdings were reorganised into ERA in 1980. With other German corporate interests raising the FRG share in ERA to 14%, UG in 1982 was expressing the hope that Australia's contribution to West Germany's uranium needs would rise from 5.5% to around a third, as expected demand increased from around 2430 tons to 6250 tons by 1990 (27). It was not until 1985 that UG publicly criticised federal Australian policy again. This came soon after the ALP government imposed a ban on uranium exports to France, following French refusal to abandon its Pacific nuclear tests. UG sought permission to send 90 tonnes of uranium from Ranger to the French trading outfit, Enership. Trade Minister John Dawkins accused UG of deliberately trying to jeopardise Australian operations, so it could withdraw from its ERA contracts without penalty and not have to pay inflated prices for 4350 tonnes of yellowcake awaiting delivery out of the 5820 tonnes originally contracted. UG hotly denied the allegation, claiming it had no reason to pull out (28). Was it just a coincidence that within ten months it was threatening just that? UG Australia had prided itself on "searching for uranium throughout Australia in its own right, as well as in Joint Ventures" (3). To summarise its involvement over the past fifteen years:

• A 4% holding in ERA which (as already mentioned) entitled it to 5820 tonnes of the product of the Ranger mine.

• 10% of the Yeelirrie mine in Western Australia, with Western Mining Corp (WMC) holding the other 90%. This has been UG's "principal exploration activity in Australia" (1) – a substantial orebody of fairly high-grade ore, containing 34,000 tonnes of uranium oxide and vanadium (29) with large reserves of lower-grade uranium. For several years the project has been "caught in political limbo" (30) and, even before the ALP government imposed an interim ban on WA mining in 1984, was "further down the track for development than most other uranium discoveries" (30). The withdrawal of Esso Australia from the project (see Exxon) in 1982 (31) did not affect UG's equity – WMC's was increased and for a while improved the project's marketability. However, a deal lined up with the French CEA in 1983 collapsed when the Federal government imposed a ban on uranium exports to France due to resumed French nuclear tests in the Pacific (13). Although the mining industry in Australia has given the strong impression that Yeelirrie does not affect Aboriginal land claims (32), the lie to this was given in 1978 when an Aboriginal delegation from North Queensland visited UG's offices in Frankfurt and addressed the management on behalf of black West Australians.

UG's managing director told the delegation, "Mining is a part of civilisation, and Aborigines have to be part of civilisation" (33).

Four years later, at the time of the Noonkanbah confrontation with Amax, other statements were made by Peter Hogarth (Yambilli), Roley Hill (Nulli) and Croydon Beaman, Aboriginal tribespeople living near Leonora: "We been fighting for Yeelirrie. The sacred ground is each side of Yeelirie. 'Yeelirrie' is white man's way of saying. Right way is 'Youlirrie'. Youlirrie means 'death', Wongi (Aboriginal) way. Anything been shifted from there means death. People been finished from there, early days, all dead, but white fella can't see it.

"Uranium they say, uranium they make anything from it, invent anything, yet during the war when the Americans flew over, what happened to Hiroshima? And that'll happen here too if they're messing about with that thing. They never learn."

Jakson Stevens, chairman of the Nnanggannawili Community at Wiluna, was also quoted in 1980 as saying: "Uranium mine. We got one up here. We trying to put a block to it. It's Aboriginal sacred site. If they, the Government and the mining companies came in here, we'll be pushed away from our own country, own place, to the town, Wiluna. No work there" (33).

• During their 1978 visit to UG, the North Queensland delegation also protested at UG's explorations in the north-western part of the "sunshine state" (33), specifically at Westmoreland, 400km north-west of Mount Isa. This prospect, in which UG held a 37.5% interest with Queensland Mines (40%), Minad (12.75%), and IOL Petroleum Ltd (9.75%), contains an estimated 11,400 tonnes identified reserves (34) and is administered by UG. However, an estimated 15,000t would be necessary to make the project financially viable (30), and the uranium is distributed in seven deposits. Even before the ALP government came to power, it did not look like Westmoreland would come on stream (34) and it was "downgraded substantially" (35).

• Similarly, in the Ngalia Basin, north-west of Alice Springs in the Northern Territory, prospects for development were looking thin in the early 1980s. The original joint venture was divided, after disposal of the AAEC's original interest in the project and controversy surrounding division of the spoils (3). UG held 25.54% of the Ngalia Basin (Exploration Operating Agreement) and 33.52% of the Ngalia Basin (Discovery Operating Agreement) – the latter "far more advanced" than the former (30) but neither standing much chance at the present time.

- At Manyingee in Western Australia, UG held a JV interest with Aquitaine Australia and Minatome (the administrator). Drilling had started by 1982, and *in situ* leaching was being considered (30).
- Mineral rights held in New South Wales appear never to have been exercised (22).
- UG began exploring in Victoria in 1977 (49) with Northern Mining (50%) though it was not for some while that the environmentalist movement in the state caught on to the potential dangers. Friends of the Earth then conducted a campaign against the company's drilling in the Mansfield area, 6000km down the famous Snowy Ranges into the Avon and MacAllister water catchment areas. FOE called for a moratorium on activities which threatened one of the last wilderness regions in the state (36).
- A JV (70%) with Idemitsu at Fog Bay, near Darwin in the Northern Territory, announced early in 1982, involved exploration on a fairly large lease, No. 3149. Nothing further has apparently been reported on this prospect (37).

Where does all this leave UG for the near future? The simple answer is: Canada. For it is here that the company has been consolidating its explorations and actually resumed uranium production from a plant which temporarily ceased operations in 1982. This was the ESI Resources uranium-from-phosphates project in Calgary, Alberta, in which Uranges Canada holds 49%. By September 1983, the plant was operating at 90% capacity (13).

There is hardly one corner of the uranium-rich lands of Canada in which UG has not had some interest. As UG's own self-adulatory material puts it: "Reconnaissance and exploration projects extend from the East Coast to the Yukon Territory" (22).

However, nothing has been heard for some time from the following:

- The Brinex project in Post-Makkovik, Labrador, which UG discovered along with Brinco (then controlled by RTZ) (38). Although developed by the two companies (39) (indeed it was, according to UG, "By far [its] most advanced project [in Canada]") (5), Brinco transferred UG's 40% interest to Commonwealth Edison in 1979. The project was later halted by the Newfoundland government (see Brinco for full details).
- The Newfoundland moratorium on uranium mining also halted exploration at Cape Breton (40).
- Uranium exploration with Scandia Mining and Exploration Ltd (20), as well as in its own right, in Nova Scotia, Manitoba and the Yukon, appears to have come to little or nothing (5, 41).
- Prospecting at both the Johan Beetz and Mont Laurier prospects in Quebec in the 1970s (41) along with the SDJB (20).
- A reported share in early tests at Agnew Lake (20), a project which closed down in 1983.
- A joint venture with SMDC in the Rabbit Lake area of Saskatchewan, and other exploration projects at Charlebois Lake (a JV with SMDC and Denison) (41) set up in the 'seventies; also exploration in its own right in the Dubawnt Basin and at Hatchet Lake in northern Saskatchewan (5).

One area from which UG has not yet withdrawn has featured in one of the most controversial cases in the recent history of indigenous land claims. The Baker Lake region of the Northwest Territories (NWT), located approximately at the geographic centre of Canada, is 150 miles south of the Arctic Circle. The only inland settlement in the Arctic (42), it is the home of hundreds of Inuit hunters dependent on caribou (reindeer). Urangesellschaft entered the area in the mid-'seventies along with several other mining companies. By 1979 it had laid out C$1 million on exploration of uranium deposits (43) and by then was constituting the "major threat" to the Baker Lake Inuit's livelihood (44).

Assisted by the Inuit Tapirisat of Canada (ITC), the Baker Lake Inuit obtained a temporary moratorium on exploration in 1978, in areas where caribou breed or calve, and where important water crossings are located (42). (In

767

many ways their arguments foreshadowed those used by the Sami of northern Norway in their heroic efforts to stop the Alta Kautokeino dam a few years later.)

In 1979, they went to court to try and establish their aboriginal rights to protection from activities deleterious to their traditional lifestyle. Ranged against them were six uranium exploration companies – Pan Ocean Oil, Cominco, Noranda, Western Mines, Essex Minerals and, at their head, UG.

These companies argued, not only that uranium mining or exploration did not substantially threaten the caribou or their human dependants, but that aboriginal rights *per se* had been extinguished by a Royal Charter of 1670 which ceded the land to the Hudson's Bay Company (45).

The upshot of a very complex and costly trial was an historic recognition of aboriginal land rights in the north, severely compromised by a ruling from Judge Patrick Mohoney that the activities of UG and its confrères did not constitute a "significant factor" in the decline of caribou herds (45). Had aboriginal rights been equal to "property rights", said the Judge, certain restrictions might have been ordered (46).

The C$100,000 costs chalked up by the Baker Lake Inuit and the ITC – not without the moral support of many anti-nuclear groups in North America – left the community weakened for the future in any case-by-case attack on specific mining projects, although Mohoney's ruling put the companies on notice that the Inuit were prepared to fight any future proposals (46).

As it was, the ban on uranium exploration was lifted, and UG, in particular, arrogantly proceeded as if nothing had happened. Indeed, in 1984, UG announced two new uranium zones in Baker Lake: a longish seam of fairly low-grade uranium at the Sissons-Schultz South property and, more importantly, a mineral strip nearly 6 metres wide and 100 metres deep, grading at 0.75% U_3O_8 on its Lone Gull property north of Judge Sissons Lake (47). (In 1983, Dae Woo bought into this property) (48). By the end of 1984, revised reserves at

Lone Bull were put at an attractive 16,000 tons uranium grading an average of 0.39%.

In 1986 UG carried out what it called a "successful" pre-feasibility study for the Baker Lake, Kiggavik (formerly Lone Gull), deposit. Output at more than 1400 tons a year of uranium could be expected from the US$200 million project, with diluted resources exceeding 15,000 tons uranium in ore grading at 0.4% (50); though a later report puts it at 0.6% (51). *Native Press* obtained a copy of the study, which showed that production was expected in "three to four years" after environmental approvals and final decisions had been carried through. The full feasibility study was due to be completed by the end of 1989, a production decision by the following year, and start-up in 1993-94 (52).

The company also wanted to build an airport, a winter road, and a marine terminal. As a sugar on the pill to local, mainly indigenous, communities, UG offered various types of training, preferential employment, and alternative employment in the decommissioning phase – the usual sop to native northerners (*cf* Key Lake). The local MP, Gordon Wray, had already doubted that the company's plans would assist local people very much, and pointed out that the mine would be the first open pit ever operated "in the barren lands" (52). The company conceded possible disturbance, but claimed that the local caribou from Beverly and Kaminurak would be "generally" unaffected or "only to a limited extent" (52).

By mid-1988, UG stated that it was seeking partners for the Kiggavik project, and that "several possible international partners" had been found: it also conceded that the project was not currently economically feasible, and that it required some long-term contracts: it was prepared to sell 44% interest in the mine (51).

The CEGB purchased 20% of the equity in 1989, and the South Korean government also bought into the project (53).

Opposition to the Kiggavik mine gathered weight during 1989 and the following year. The Canadian National Federation of Labour came out against the project (55), the local Keewatin community mobilised 1700 people behind a petition to stop it, and they were supported by a wide variety of organisations, including the Inuit Circumpolar Conference (ICC), the Inuit Tapirisat of Canada and Nuclear Free North (56).

In late 1989, UG announced the results of interim feasibility studies which showed that mineable reserves at the deposit were 3.7 millions tonnes grading 0.5% U_3O_8: the operation would probably involve open-pit feed of 350,000 tonnes a year, with a waste-to-ore ratio of 17:1, producing an average of 1600 tonnes of uranium oxide each year (57).

UG then had to present its Environmental Assessment Report (EAR) to the Federal Environmental Assessment Review Office (FEARO) and submit it to examination by the community in public hearings. This it did in late 1989 (57).

The response was further alarm, despondency and militancy. FEARO found the EAR wanting – in particular it determined that UG had not given enough information on the mine's impacts to the local community (58).

UG was sent back to the drawing-board and, in response, asked FEARO for an "indefinite delay" to the public hearing (59).

In arguing against the project, the Inuit of Baker Lake had turned out in force at the local hearings in 1989. Reported the *Native Press*: "... some people broke down in tears as they urged the panel to block the mine. Most of my relatives, brothers, sisters, parents, are buried around the site," said Janet Ikuutak, who was born nearby. "Every spring the land grows so green and beautiful, and the berries ripen. But the mine will prevent that" (58). It seems the people may, in fact, have prevented the mine.

Contact: Baker Lake Inuit, c/o ITC, 3rd floor, 176 Gloucester Street, Ottawa, Ontario, Canada K2P OA6.

References: (1) *MIY* 1985. (2) *FT* 13/9/82. (3) *Reg Aus Min* 1981. (4) *E&MJ* 11/78. (5) *Urangesellshaft, UG*, Frankfurt-am-Main, Sept/78. (6) *Yellowcake Road*. (7) *World Uranium Potential* (OECD/IAEA) Paris, Dec/78. (8) *MJ* 22/7/77. (9) *MM* 11/82. (10) *IAEA/OECD Red Book* 1979. (11) *Multinational Monitor*, April 1981, Washington. (12) *Enrichment, Nuclear Weapons and Brazil*, 1979, (Friends of the Earth) London. (13) *MAR* 1985. (14) Czervenka and Rogers. (15) Thomas Neff, *The International Uranium Market* (Ballinger Publishing) Cambridge, Mass 1984. (16) D B Sole to Secretary for Foreign Affairs, 28/3/72, re "Sales of SA Uranium: Rössing" quoted in Czervenka and Rogers. (17) *New Scientist*, 7/9/76; *FT* 18/5/79; *Gua* 3/10/81. (18)*MIY* 1982, see also *FT Min* 1991. (19) *West Africa*, London, 4/7/77 (20) *Information International Research and Development Activities in the Field of Energy Research (National Science Foundation) Washington, 1976. (21) Dean Abrahamson and Edward Zabinski, Uranium in Minnesota*, Minneapolis, Minnesota, 1980. (22) *Urangesellschaft, UG*, Frankfurt-am-Main, 1980. (23) *Kiitg* No. 11, Jan-Feb/81. See also *Future*, Colorado, 18/5/79. (24) *MJ* 24/1/86. (25) *West Australian*, Perth, 14/2/78. (26) *Financial Review*, Canberra, 14/2/78. (27) *FT* 13/9/82. (28) *MJ* 1/3/85. (29) *E&MJ* 1/80. (30) Reg of Aust Mining 1983/84. (31) *MJ* 21/5/82. (32) *MJ* 17/2/78. (33) Roberts. (34) *MJ* 17/12/82. (35) *E&MJ*, March/83. (36) FOE Victoria press release, "Did you know?" date unknown. (37) *NT News*, Darwin ?/4/82. (38) *MJ* 9/9/77. (39) *FT* 16/8/78. (40) *Energy File*, British Columbia, April/80. (41) EPRI report, *Foreign Uranium Supply* 1978. (42) *Yukon Indian News* 24/5/79. (43) *MJ* 27/4/79. (44) *Native Peoples News*, Nos. 2/3, 1979. (45) *Native Press*, Yellowknife, 7/12/79. (46) *IWGIA Newsletter, Copenhagen, No. 24, April/80. (47) MJ* 23/11/84. (48) *E&MJ* 6/84. (49) *MJ* 18/11/7. (50) *MAR* 1988. (51) *MJ* 15/4/88. (52) *Native Press*, Yellowknife, (Somba K'e) 4/3/88. (53) *WISE N C* 319.2196, Amsterdam, 20/10/89. (54) *Reg Aus Min* 1990/91, (55) *Native Press*, Somba K'e, NWT, 20/10/89. (56) *Wise NC* 330, 6/4/90. (57) *MJ* 15/9/89. (58) *Native Press*, Somba K'e, 9/3/90. (59) *Nuclear Fuel* 6/8/90.

608 UKAEA

It is responsible for all nuclear developments in Britain except those fuel and "waste" provisions performed by BNFL. Its involvement in uranium exploration appears to have been confined to employing the IGS to carry out car-borne and hand-carried gamma-ray spectrometry geochemistry and drilling for uranium throughout Britain in the '70s.

The most important area discovered was the Orcadian cuvette of Orkney, Caithness and east Sutherland (see SSEB) (1).

References: (1) *MM* 11/82.

609 Ultramar Plc

Consolidated Gold Fields plc (GB)

5.6%
ULTRAMAR PLC (GB)

This "independent" up-and-coming British oil company, with operations in the USA, Canada, the Carribbean, Europe and the Far East, has a history dating back to exploration for Venezuelan oil in 1935 (1).

Its other stated interests are in construction (in the USA), electronic equipment and uranium exploration, though no known projects are currently underway. It trades extensively using the Golden Eagle brand mark.

About half of its net operating profit derived from Indonesia alone in 1984 (2).

References: (1) *WDMNE*. (2) *FT* 16/11/84.

610 Ultrana Nuclear and Mining Corp (RP)

It was exploring for uranium on Negros Island (400 miles north of Manila) with Phelps Dodge's Western Nuclear subsidiary in 1980 (1). No further information.

References: (1) *MJ* 15/2/80, *FT* 15/2/80.

611 Umipray SA

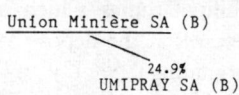

Union Minière SA (B)

24.9%
UMIPRAY SA (B)

This is a consortium, set up by Prayon (along with Union Minière and Metallurgie Hoboken-Overpelt) in 1979, to build a uranium-from-phosphates plant in Belgium (1). These companies were later joined by IMC, Mechim, and Rupel-Chemie (2, 3).

Another JV was set up in 1979 by IMC, Prayon, and Metallurgie Hoboken-Overpelt to licence a new uranium recovery service worldwide, IMC holding 50% and the two Belgian companies 25% each (4). Union Minière's direct interest in Umipray was reduced in 1982 from 37.9% to just under 25% (5). So far as is known, the Umipray venture never became operational.

References: (1) *FT* 10/10/79. (2) *MAR* 1981. (3) *Kiitg* 10/80. (4) *MJ* 12/10/79. (5) *MIY* 1985.

612 UNC

Responsible for the worst nuclear "accident" in the western world, the United Nuclear Corporation was, until the early '80s, the largest private sector uranium producer in the United States: nearly 4 million pounds were produced in 1979, and just over 3. 5 million the following year (1).

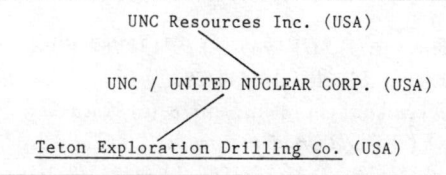

UNC Resources Inc. (USA)

UNC / UNITED NUCLEAR CORP. (USA)

Teton Exploration Drilling Co. (USA)

Its uranium reserves were also vast. The UNC/Homestake partnership alone held 5.8% of all US reserves, at its Ambrosia Lake mines, in 1976. Only Kerr-McGee and Gulf Oil held bigger reserves at that time (2).

But already, in early 1980, the weakening uranium market and increasing costs were forcing UNC to cut back: 350 miners laid off at Church Rock in March 1980, its HQ at Albuquerque, New Mexico, closed, and its office "consolidated" at Church Rock (3). The UNC Recovery Corp – processing uranium from phosphoric acid – suspended operations at the same time, UNC accusing W R Grace of supplying acid below quality from its fertilizer plant near Bartow, Florida (1, 3).

Uranium operations at the Ambrosia Lake and St Anthony (Laguna Pueblo) mines were closed in 1981 and 1982 (4) and all operations at Church Rock suspended in the second quarter of 1982 (5) after the company reported a net loss of US$14.1M. Only 0.3 million kilograms of uranium was provided to customers from UNC's own stocks during 1982; about 7 million kilograms had to be purchased, though at a delivery price which allowed a "modest profit" (5). At the end of 1982, UNC valued its uranium properties at US$138M, in addition to US$39M invested in exploration projects. Uranium production was not expected to resume for at least two years (5). Early in 1981, UNC also sold its 70% share in UNC/Homestake (sometimes known as United Homestake) to Homestake for US$20M (6). Under the deal, Homestake acquired full control of the three mines and a mill at Grants, New Mexico (7), and paid US$3M for uranium stocks (8).

By then, UNC had already relegated uranium to a back seat (9) and was trying to diversify. It was proposing to buy into Western Airlines of the US and, more important, made moves towards securing NCC Energy, a British energy and investment company with holdings in the North Sea and North Africa (10).

As of 1984, UNC could report further discontinuation of uranium exploration and the reduction or write-off of some of its mineral properties. No uranium deliveries at all were made in the first six months of 1984, and UNC's final uranium contract was fulfilled in 1983 (11).

UNC Resources was the holding company for United Nuclear Corp which advertised itself as "... a diversified energy company engaged principally in uranium mining and milling, manufacturing, and providing products and services in energy-related and other fields" (6). The holding company itself was divided into three main operating groups: Manufacturing and Services, Offshore Products and Services, and Minerals (5).

The Offshore group made UNC a leading producer of aluminium and steel marine vessels used not only by the off-shore oil industry but in military applications too. This section contributed nearly a quarter (23.5%) of UNC's total 1982 revenue (5), but it was being closed down in 1984 (11).

The main activity of the Manufacturing and Services group was to "supply reactor fuel and related components to the US Navy's nuclear powered fleet" (5). This was by far UNC's biggest money-spinner until the early 1980s, contributing nearly half (47.7%) of total 1982 revenue (5). The same year, UNC also won a contract to supply nuclear components to Westinghouse (12).

The Minerals group of UNC was the main exploiter of the company's huge New Mexico uranium holdings: 69 million pounds of U_3O_8 were held in total by the company in March 1981, of which most – 58.3 million pounds – was in New Mexico, and most of this – 39 million pounds – in the Church Rock area alone (6). Most of the remainder, in New Mexico and Wyoming (13), was conserved for possible solution mining.

However, the company also branched out into precious metals, acquiring the Cornucopia gold-silver mine in Oregon in January 1983 (6).

UNC Plateau Mining Co, a large-scale but unpromising producer of steam coal from Utah, was sold in 1980 (6), but the company retained an interest in solar "energy" through its purchase of 50% rights to any solar devices produced by the lone solar scientist Ovshinsky, in 1976 a pioneer of silicon cell technology (14).

Also, in May 1981, UNC branched out into machine tools by buying the National Automatic Tool Co of Richmond, Indiana, for about US$18M (6).

The wholly owned UNC subsidiary Teton not only investigated uranium deposits in the Powder River Basin and Red Desert regions of Wyoming – as well, of course, as the Ambrosia Lake area of New Mexico – but also obtained a concession in the native *chaco* region of Paraguay (15). This was not the first time the company had ventured into South America: in the mid-'60s it had joined US Steel in exploration of manganese deposits in what is now the Carajas project area – one of the biggest threats to indigenous people and peasant farmers now existing (16).

In early 1981 Teton agreed with North Kalgurli on an exploration project for minerals including uranium throughout the whole of Australia (17). Shortly afterwards, North Kalgurli bought out 50% of Teton Australia (6).

As of 1983, Teton Australia had shares in two uranium exploration projects in South Australia, apart from the suspended Honeymoon Uranium Project. The first was at Gould's Dam, 150km west-north-west of Broken Hill, in a JV (24.5%) with AAR (75,5%) and the prospect of earning a 49% interest; a small deposit (1100 possible tonnes in 1981) (18). The second was 75km north-west of Broken Hill at East Kalkaroo, where Teton is partnered with MIM and AAR (as at Honeymoon) and can earn an aggregate of 50% on the basis of exploration expenditure. According to AAR's 1993 annual report, there are 800 tonnes of contained uranium in this deposit. A legal "wrangle" between the partners developed in 1982, which by the end of 1983 had not been resolved (18).

In early 1984 Teton signed a letter of intent with Alaska Apollo Gold Mines, giving Apollo the right to earn a 51% interest in Teton's six precious minerals prospects in the Aleutian –

native – area of Alaska, which had been conducted since 1979. Apollo agreed to take on Teton's obligations to the Aleut Corporation, established under the Alaska Native Claims Settlement Act of 1971 as a native corporation, which include paying royalties, a minimum exploration expenditure and making a production decision on one or more of the production units in the Aleut island chain by 1986 (19).

Apart from its contract to supply the US Navy, and another to supply nuclear components to the US Department of Energy (20), UNC concluded two major uranium contracts. At the end of 1978 it concluded a US$40M contract with an "unnamed utility" to deliver 855,000lb of uranium (21); less than two years later it landed a major deal with Kepco/Korea Electric Power to deliver more than 3 million pounds of uranium oxide (20).

UNC was one of the American producers cited by Westinghouse in its cartel proceedings (22) during the late '70s, and by TVA in one of the "spin offs" from that huge legal bonanza. In its turn, UNC litigated against Gulf Oil for failure to meet contracts:

The dispute with TVA was settled in 1979 when UNC sold its interest in a number of its Wyoming uranium properties to TVA, namely Morton Ranch and Box Creek, containing just under 7 million pounds of uranium reserves, and agreed to purchase TVA's 50% interest in a JV near UNC's San Mateo mine, increasing the amount of its deliveries to TVA from 5 million to 6.25 million pounds between 1970 and 1984.

The US$300M suit brought by UNC against General Atomic (23) – rather its parent company Gulf Oil, now controlled by Chevron – was settled out of court in 1984, with US$130M paid in cash, assumption by Gulf/Chevron of a US$71M liability, and the investment of US$100M in UNC by Chevron (11). On July 16th, 1979, a dam holding radioactive uranium mill tailings at the UNC Church Rock uranium mill 20 miles north-east of Gallup, New Mexico, burst its banks. An estimated (100,000,000) gallons of radioactive liquids spewed forth and, together with 1100 tons of solid wastes, deluged into the Rio Puerco, a river used by local Navajos (Dine) for grazing cattle (24).

The north fork of the Rio Puerco received the brunt of the waste-waters and sediments pouring through the 20-foot hole in the southern portion of the dam wall.

The liquids eventually travelled 115 miles downstream, temporarily altering the chemical quality of surface waters in the Little Colorado River, Holbrook, Arizona – despite early assurances by state officials that they couldn't have travelled that far (25).

Contaminated effluent also found its way into California where the state authorities considered the possibility of filing suit against UNC (26).

UNC knew of the unsafe nature of its tailings dam back in 1976 (26). When it burst three years later, it was already 50% fuller than it should have been – and cracks appeared in 1977 which were filled instead of repaired (27).

"In an effort to recoup some of their losses resulting from the spill, some 125 Navajo families filed US$12.5 million in lawsuits against UNC in tribal court in August 1980. UNC officials then filed federal suits contending that the Navajo Tribal Court did not have the authority to hear the cases. A US District Judge recently ruled in the company's favor and granted a preliminary injunction prohibiting the Navajo families from seeking claims before the tribal court.

"UNC proposed an out-of-court settlement late [1980], but the offer to pay the group of plaintiffs a US$25,000 lump sum in exchange for an agreement that UNC was in no way responsible for the spill or negligent in the operation of its tailings facility was rejected flatly by the Navajos who [said] they [were] committed to pursuing the lawsuits to the end no matter how long the process [took]" (28).

However, it has been – and is – radioactive emissions from UNC's tailings dam which probably poses a greater long term threat to people and environment in New Mexico.

"Seepage of tailings fluids to usable groundwaters afflicts all five of the licensed and active uranium tailings facilities in New Mexico, but the condition is most severe at the United Nuclear operation. An estimated 15,000 to 80,000 gallons of raw tailings "raffinate" high in radioactivity and toxic metals and process chemicals seep daily from the UNC evaporation ponds into underlying aquifers, state officials say. Contamination is evident in site monitoring wells at depths up to 240 feet below the surface, according to the company's own information. "Evidence that the tailings spill left lasting contamination in sediments of the Rio Puerco and in groundwater near it is contained in environmental monitoring data collected since the spill and in correspondence between the New Mexico Environmental Improvement Division (NMEID) and UNC.

"An analysis of the information shows that although radioactivity levels in stream bed soils have decreased as a function of time and distance downstream, *several portions of the river remain contaminated, and contaminant concentrations have yet to return to background levels for the vast majority of the stream bed.* "The fact that tens of thousands of gallons of acidic tailings waste-water are leaking from United Nuclear's tailings ponds every day is not disputed – not even by the company itself. The problem is severe enough to have merited inclusion in the EPA's list of hazardous waste sites targeted for clean-up under the so-called Superfund program.

"What is not agreed upon, however, is the extent and severity of the seepage problem, or a remedy for it. But a review of the data shows that *contaminants have left UNC's property and are degrading groundwater on Indian lands east of the tailings impoundment and on state lands north of the site.* Concentrations of these pollutants far exceed state groundwater standards for process chemicals and heavy metals; some monitoring wells also have detected radioactivity above the state's maximum permissible concentrations for thorium-230.

"UNC apparently has an aversion to obtaining water quality information on groundwater in

Section 1, which is on federal Indian 'trust' land administered by the US Bureau of Indian Affairs and the Navajo Tribe. Individual Indian 'allottees' possess the surface rights on these lands, while the government retains the mineral rights. While the company has claimed in the past that it has had difficulty obtaining permission from Indian authorities to monitor groundwater east of its property, Indian officials offer another version of the story. According to Harold Tso, Navajo Environmental Protection Commission (NEPC) director, UNC has neither drilled monitoring wells in the neighboring Indian lands nor sought permission from the US Bureau of Indian Affairs or the tribe to do so" (24).

Almost incredibly, given the proven damage caused by the continuing leakages, UNC was allowed to continue operating its mill during most of this period. Only in May 1982, when UNC closed down its operations for economic reasons, did the mill cease functioning. Because of the NMEID's belated prohibiton UNC couldn't start up again without cleaning up the seepages, even if it wanted to (29).

In summer 1983 the NMEID told UNC to dispose of the uranium tailings pile at Church Rock and halt the seepage of water contaminated with thorium-230 into ponds near the piles. UNC promised to stop the seepage but not to clean up the 160-acre site containing some 5 million tons of tailings. While UNC was claiming that the entire spill site had been cleaned, local Navajos who use the Rio Puerco claimed that only the upper 3 inches were cleaned (30).

At the end of this saga, however, UNC, – if not laughing was at least sitting pretty.

In 1983 it received US$3.2M in fees and litigation costs in a suit against the Allendale Mutual Insurance Co – and this was after it had been awarded a total of US$54.4M (including operating costs and losses) arising out of the collapse of the tailings dam at Church Rock (5, 31).

So – it pays to be reckless; and the bigger you are, the more reckless you are, and the worse you damage the environment, the more you're likely to get. Your victims, of course, get noth-

ing – not even recognition. Who doubted it would be anything else in the world of Big Uranium Business?

By 1989, further studies along the Little Colorado River showed water samples with over 1,000 picocuries/litre of water (60 times the State and federal safe drinking level) with surface water in the Rio Puerco itself displaying up to 100 times the maximum level (32).

Contact: NIYC, 203 Hermosa Drive NE, Albuquerque, New Mexico 87108, USA.
American Indian Environmental Council, POB 7082, Albuquerque, New Mexico 87194, USA.
Southwest Research and Information Center, POB 4524, Albuquerque, New Mexico 87106, USA.

References: (1) *MJ* 11/7/80. (2) US Senate Committee on Energy and Natural Resources, *Petroleum industry involvement in alternative sources of energy*, Washington DC, 1977 (US Government Printing Office). (3) *MJ* 11/4/80. (4) *MJ* 31/7/81. (5) *MJ* 17/6/83. (6) *MIY* 1982. (7) *MJ* 11/6/82. (8) *MJ* 17/4/81. (9) *MJ* 15/1/82. (10) *FT* 7/3/81. (11) *MJ* 24/8/84. (12) *MJ* 26/3/82. (13) *MJ* 6/4/79. (14) Reece, p.172. (15) *MJ* 6/2/82. (16) *FT* 18/11/81. (17) *MJ* 2/1/81. (18) *Aus Reg Min* 1984. (19) *MJ* 30/3/84. (20) *MJ* 25/4/80. (21) *MJ* 17/11/78. (22) Moss p.114. (23) *MJ* 21/5/82. (24) Chris Shuey, "Mill tailings poison Indian land", in *Critical Mass Bulletin*, USA, 4/83. (25) *San Diego Tribune* 22/11/70. (26) Information from Southwest Research and Information Center, Albuquerque, in *Kiitg* 2/80. (27) *Kiitg* 6/80. (28) *Mount Taylor Alliance Newsletter*, Albuquerque, 20/6/81. (29) *E&MJ* 2/83. (30) *Critical Mass Bulletin*, USA, 11/83.(31) *MJ* 26/8/83. (32) *Gallup Independent* 19/7/88.

613 Union Minière SA

Société Générale de Belgique is one of the biggest industrial holding companies with interests in mining in the world. With the reorganisation of its interests in Union Minière, and SG/UM's shares in Tanks, SGB came to control (through Tanks) the Benguela Railways in Angola, and

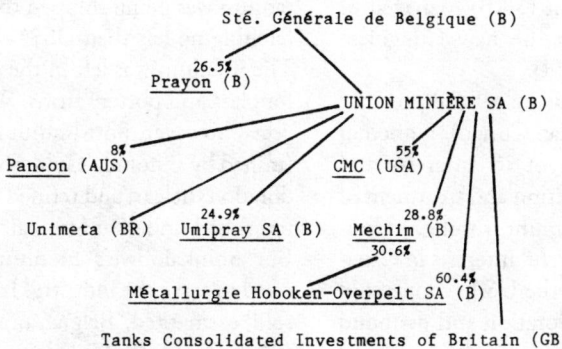

Sté. Générale de Belgique (B)

numerous smaller stakes in De Beers, RTZ, Amax, Exxon, Asarco, and Inco (1, 2).

Tanks – formerly one of the classic colonialist companies in the world when it was set up as Tanganyika Concessions in 1899 to ease the way for mercantile penetration of southern Africa (3) – lay low, especially after its take-over by Union Minière. Union Minière – probably the epitome of colonialism – was initially formed by Tanks to develop mineral deposits in the so-called Congo Free State, virtually the fiefdom of King Leopold II of Belgium at the time. Notable "events" in its development into one of the world's mining giants include the following:

- membership in the '30s with all the leading US copper companies and Rio Tinto-Zinc (then Rio Tinto of Spain) of a copper cartel which forced up the price of one of the key inputs to the expanding electrical and automobile industries (4);
- by the early '30s UM was the world's third largest copper producer, its Katanga mines accounting for 7% of world production (4);
- by this time UM had virtually established its own state in Katanga with very tenuous links to the capital in Leopoldville (named after the Belgian monarch) (4);
- by the late '50s, when anti-colonialists were trying to break up the Federation of Rhodesia, Tanganyika and Nyasaland (today's Zimbabwe, Zambia and Tanzania), UM, together with Tanks and the British South Africa Company, channelled massive funds to their campaign;

- as independence for the Belgian Congo drew near in the late '50s, SGB manoeuvred the Congolese government representatives off the managing board of the Special Committee of Katanga – they'd already made it clear that they were opposed to the expatriate exploitation of the company's rich and vast mineral resources (including, of course, uranium).

European investors withdrew their capital from the Congo; a constitutional conference dissolved the Special Committee; and the "independence" stake reverted to Société Générale.

From this point, Tshombe – leader of what was to become the breakaway movement of Katanga – cemented links with UM; money apparently flowed from the company into the secessionists' coffers.

When, in July 1960, the Belgian Congo reached independence, and Tshombe declared secession, Belgian government intervention "to protect our nationals" – in reality the huge mineral stake of UM – secured the early success of the attempt to break away from the Lumumba government.

After the UN intervention and Tshombe's flight abroad, UM hedged its bets. Tshombe returned to the Congo, presiding over a paper-thin unity, and by decree allocated the entire Special Committee portfolio to the government. Tshombe fell from power in 1966; Mobutu seized it using the army.

Mobutu attempted, through various methods, to nationalise UM, and eventually failed. However, although UM was to finally vacate the

Congo in 1967, its parent (SGB) managed to take a powerful stake in the new Congolese company GEOCOMIN (4).

Today, Union Minière has no official presence in Zaire (the former Belgian Congo). Nonetheless, in a reorganisation of its interests and diversification into conversion and treatment of minerals (including uranium through Umipray) (5), it retained powerful interests in Canada (copper, exploration), the USA (copper and zinc), Brazil (general exploration and diamond mining in Minas Gerais) and on the ocean bed (6).

The company's Brazilian subsidiary, Unimeta, considered involvement with Brazil's vast Carajas project (10).

Its corporate associates have included Essex Minerals through Ocean Mining Associates (6). In the Hoggar mountains it was part of the A G McKee and Traction et Electricité consortium investigating uranium deposits (7) and had a 45% stake (through CMC/Continental Minerals) in a Lisbon, Utah, uranium deposit.

Union Minière's uranium played a crucial role in the development of the murderous weapons that were dropped on the people of Nagasaki and Hiroshima in 1945.

Up until 1940, Edward Sengier was the chairman of UM – and it was he who negotiated a uranium contract with Joliot-Curie (one of the allied scientists working on a chain reaction and then representing the French government) to provide uranium which eventually ended up in the Manhattan project, cradle of the atomic bomb.

The German army invaded the Belgian Congo in 1940 – but Sengier had his wits about him. He removed 1200 tons of uranium ore from the Shinkalabowe mine and shipped it to the USA; in 1942, this uranium was assigned to the US government simply on a scrap of yellow lined paper handed to Colonel K Nichols, deputy to the director of the Manhattan Project (7). Sengier got the US Legion of Merit – the Japanese got the bomb on their heads. The Shinkalobwe mine is probably the first commercial uranium provider for nuclear power in history. Pitchblende was discovered there in 1915; by 1923

radium was being shipped to Belgian hospitals, fetching no less than US$100,000 a gram (*sic*). The uranium was left in the ore to be extracted for glass and pottery firms. Within the next ten years, however, Port Radium's radium supplies (mined by Eldorado Gold, the forerunner of Eldorado Nuclear, and refined at Port Hope, Ontario) had undercut Shinkalobwe's (8).

But Shinkalobwe's uranium stockpile continued to serve the industry. In 1959 it was being sold, earmarked "Belgian uranium" in the USA (9). And, it was Shinkalobwe uranium, sold to Asmara Chemie of Wiesbaden, which ended up in Israel after being hijacked by Israel in 1968 (8).

Although ostensibly no longer benefitting from the exploitation of Zairean resources, Gecamines, a Zairean company, is effectively the beneficiary for UM, which is practically the sole concessionary for the mineral products of Gecamines (11). Société Générale de Belgique holds, through Tanks (or Tanaust as it is now named), 10.2% interest in Ashton Mining which owns 39.7% (21) of the Argyle Diamond Mines JV on Aboriginal land in the Kimberley region of Western Australia (see CRA (12)).

In 1987/88, European financial and investment markets experienced what, with little doubt, has been the most intricate and entertaining takeover battle of the 'eighties, when the Italian "entrepreneur" Carlo de Benedetti tried to gain control of Société Générale which has a direct or indirect interest in up to 30% of all Belgium's industrial wealth. In the event, de Benedetti did not succeed – he sold his holding to his main rival, the Cie Financière de Suez, of France, for a profit of some US$1 billion (*sic*) (13), after acquiring some 44% of the company's share capital, thus putting that raider of raiders, T Boone Pickens, into the shade. Le Générale, at the time of writing, is mainly controlled by a new holding company, Erasmus Capital, in which the Cie de Suez has the controlling interest (15). It leaves Belgium's biggest monopoly empire in firmly Francophonic hands (16) – indeed, French hands: but then, the Belgians are well used to the malevolent influence of French capital. De Benedetti has effective control over Olivetti in Italy (17), the country's second largest private bank (18), Buitoni, the food manufacturer, and interests in newspapers, tyres, cables, car components, the L'Express publishing concern, and above all in the Cofide, CIR and CERUS (French) holding companies (19). He had clearly been aiming to build up one of Europe's largest conglomerates for the year 1992, when trade barriers come down all over the EC (16). As Christopher Lorenz put it in the *Financial Times*, de Benedetti "... is demonstrating the enormous ownership and restructuring opportunities which are up for grabs right across Europe as it prepares for a unified Community market after 1992. Whatever else Mr de Bendetti may be doing, he is exploiting this instability, to build a set of promising bases in sectors where established competitors ... have failed to move with sufficient speed" (20).

During this prolonged battle, the ownership by Sté Générale of Union Minière and control over the 53% owned Sibeka (which has African diamond interests), did not change, although clearly the strategy of Union Minière might have altered if de Bendetti had secured a grip of Le Générale.

In 1989 Cie Francaise de Suez sold 11% of its stake in Union Minière (22). By then UM had been formally taken over by its own subsidiary ACEC – "for tax reasons" (23). The new entity, ACEC-Union Minière, proceeded to amalgamate its operations with those of Metallurgie Hoboken-Overpelt (qv), Mechim (qv), the large zinc producer Vielle-Montagne, and the international trading company, Sogem. ACEC-UM would be owned 87.5% by UM itself, transforming the unwieldy holding company almost overnight (so it claimed) into a "hands on" concern (24).

Benedetti – who had raised his stake in the parent company Le Générale from 18.6% to 34% during 1988-90 – set about selling off his holding. A public offer of shares in ACEC-UM, made in mid-1990, resulted in only 45.3% being taken up (25).

Benedetti had disposed of all his own stake in Le Générale by early 1991 (26), while Cie Française de Suez increased their's to 61% (27). ACEC-UM, meanwhile, sold its 9.6% stake in Pancontinental Mining to a variety of financial institutions.

In 1990:

- ACEC-Union Miniére was owned directly by Société Le Générale 82.1%, and indirectly 6.1%.
- the new group's profits jumped from BF 1,300M to BF 20,000M in just one year.
- Vielle-Montagne is the largest producer of crude zinc in the world.
- MHO is the largest refiner of copper in Europe and the world's largest processor of germanium and cobalt.
- 25%-owned Aurifére de Guinée is an important gold producer in Guinea.
- ACEC-UM's parent company, Le Générale, itself owns 54.4% of Sibeka, the diamond mining and exploration company, which also manufactures diamond tools (through Diamond Boart) – of which it is the world's largest producer – and controls a diamond mine at Mineraco Tejucana, in Brazil (28).

References: (1) *FT* 25/9/81. (2) *DT* 3/6/82. (3) *DT* 26/9/81. (4) Lanning & Mueller [I'm indebted to this excellent work for much of the information on Union Minière's colonialist strategy in the Congo.] (5) *MIY* 1982. (6) *Ibid.* (7) *MJ* 6/7/79. (8) Moss. (9) Thomas Kanza *Conflict in the Congo*, Harmondsworth, 1972 (Penguin). (10) *FT* 9/9/82. (11) *Afrique-Asie* 14/3/83. (12) *MJ* 24/6/83. (13) *FT* 16/6/88. (14) *FT* 10/6/88. (15) *FT* 5/7/88. (16) *FT* 16/6/88. (17) *FT* 16/5/88. (18) *FT* 3/5/88. (19) *FT* 20/1/88. (20) *FT* 30/1/88. (21) *Reg Aus Min* 1990/91. (22) *DT* 27/6/89. (23) *MJ* 15/12/89. (24) *FT* 12/12/89. (25) *MJ* 22/6/90. (26) *FT* 6/2/91, *FT* 7/2/91. (27) *FT* 13/2/91, *FT* 14/2/91, *MJ* 15/2/91. (28) *MJ* 25/5/90.

614 Union Oil Co of California

Union Oil (Unocal) is a major US energy corporation with about 10% of its revenues during

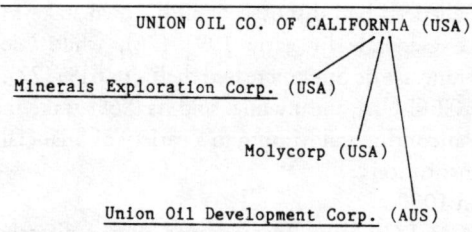

```
                UNION OIL CO. OF CALIFORNIA (USA)
                                    /|\
Minerals Exploration Corp. (USA)    / | \
                                   /  |  \
            Molycorp (USA)             |   \
                                       |    \
     Union Oil Development Corp. (AUS)
```

the early eighties coming from metals. In the '60s it became one of the first oil majors to buy into minerals with its acquisition of Molycorp, an important producer of molybdenum (1). Now a wholly owned Union Oil subsidiary, it is the world's largest producer of rare earths and holds 45% of Companhia Brasileira de Metalurgia e Minaração (CBMM) which operates the world's largest columbite mine at Araxa, Minas Gerais (2).

Unocal is also a major contender in the synthetic fuel stakes – and announced a new expansion to its Parachute Creek, Colorado, oil shale programme in 1984 (3).

Two years before, it was awarded a 10-year commercial contract to supply the US Department of Defense with 10,000 barrels/day of military diesel and jet fuel (1).

In 1978 its Minerals Exploration subsidiary opened a uranium mine at Sweetwater in the Red Desert region of Wyoming. By 1984, however, after vigorous local protests, the mine's operations were suspended (4).

In 1990, Unocal sold its Obed coal mine in Western Alberta to Luscan Ltd (5). At the same time, Union Oil of Canada was involved in an yttrium JV with Denison, SM Yttrium Canada Ltd, and its parent, Molycorp, at Elliot Lake, Ontario (5).

References: (1) *MAR* 1982. (2) *MAR* 1984. (3) *MJ* 20/4/84. (4) *E&MJ* 1/85. (5) *MAR* 1991.

615 Union Oil Development Corp

```
Union Oil Co. of California (USA)
                                 \
          UNION OIL DEVELOPMENT CORP. (AUS)
```

In 1982, it held a uranium prospect, together with AOG Minerals, at Litchfield, between the Daly and Finniss Rivers in the Rum Jungle uranium area of the Northern Territory. The project was then only at the exploration stage and reserves were not announced (1). More recently, its main focus has turned to gold in Queensland and WA (2).

References: (1) *AMIC* 1982. (2) *Reg Aus Min* 90/91.

616 United Gunn Resources Ltd (CDN)

It held a uranium lease in the Nelson Mining Division of British Columbia in 1979 (1); presumed relinquished.

References: (1) *Energy File* 3/79.

617 United Siscoe Mines Ltd

It had explored for, and developed, minerals and geothermal resources in the western US and Canada (1).

In 1979 it held a one-third interest with Getty Oil and Camflo Mines in a uranium prospect in the Kasmere Lake area of north-western Manitoba (1, 2).

Northgate Exploration Ltd, Rayrock Resources, and Consolidated Morrison Exploration

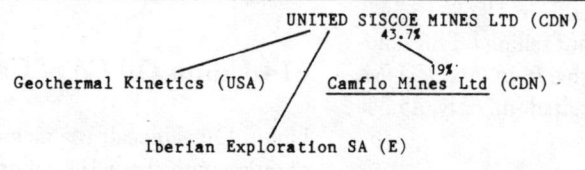

```
                        UNITED SISCOE MINES LTD (CDN)
                            /           43.7%
                           /             |
                          /              |19%
   Geothermal Kinetics (USA)  /   Camflo Mines Ltd (CDN)
                             /
                            /
              Iberian Exploration SA (E)
```

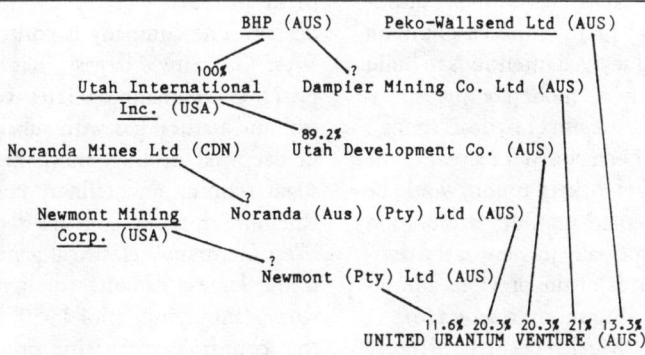

```
         BHP (AUS)        Peko-Wallsend Ltd (AUS)

       100%
   Utah International   Dampier Mining Co. Ltd (AUS)
      Inc. (USA)

                  89.2%
   Noranda Mines Ltd (CDN)   Utah Development Co. (AUS)

   Newmont Mining    Noranda (Aus) (Pty) Ltd (AUS)
     Corp. (USA)
                  ?
         Newmont (Pty) Ltd (AUS)

              11.6% 20.3% 20.3% 21% 13.3%
              UNITED URANIUM VENTURE (AUS)
```

were tied on the board through George T Smith (3).

References: (1) *MIY* 1981, 1982. (2) *FT* 10/5/79. (3) *Who's Who Sask.*

618 United Uranium Venture

It was set up to explore for uranium in the Northern Territory of Australia and is now defunct. It consisted in 1974 (1) of:

Noranda Aus Pty Ltd	20.3%
Utah Development Co	20.3%
Dampier Mining Co Ltd	
(a BHP subsidiary)	21.0%
Newmont Pty Ltd	11.6%
Peko-Wallsend	13.3%

References: (1) *A Slow Burn* (Friends of the Earth) Victoria, Australia, No. 1, 9/74.

619 Uramex

This is the Mexican state body (*NB:* not to be confused with Uranex) charged with both uranium mining and the development of Mexico's nuclear industry. It developed out of the reorganisation of the country's National Institute of Nuclear Energy in 1978, when fears were expressed both in the trade unions and by parliamentary deputies that foreign interests would take over the country's reserves (1).

In November 1980, Uramex produced a report claiming that recent uranium discoveries in the heart of the Sierra Madre would soon make Mexico one of the world's biggest uranium producers (2). A year later, Uramex estimated that it had 15,000 tons of uranium reserves, with potential reserves of 25,000 tons (3). However, this figure appeared to have been reduced by early 1982 – when it became 9,600 tons of U_3O_8 equivalent in reserves (4).

Mexico's uranium is close to the surface and apparently easy to mine; according to Uramex chief Francisco Vizcaino, the deposits in the Sierra Madre are located in volcanic rock.

Uramex claimed that it would produce the first ton of its domestic U_3O_8 at the end of 1982 (2). Exploration and development was to be concentrated in Oaxaca (a legendary Aztec centre), Chihuahua (Sierra Madre), Sonora, and Nuevo Leon (3).

According to a report in March 1982 the Los Amoles mine in Sonora should begin producing in 1982 with a target of 50 tonnes/year of U_3O_8, while mining in three other states should begin in "the next few years" (5).

```
Federal Republic of the United Mexican States (MEX)

       URAMEX / URANIO MEXICANO SA (MEX)
```

In mid-1984 the Mexican government drastically revised its energy plans, imposed a limit on the export of oil, and revived intentions to build new nuclear plants, as a "moderate diversification away from domestic use of hydrocarbons". Uramex, which had been closed for a year to the chagrin of the nuclear workers' union, would be reopened, and it seemed uranium exploration would be revived, especially in view of the drastic revision downwards of the previous administration's estimates of uranium reserves (6).

From 1982 onwards, Mexico began to privatise state-owned industry, in order to meet conditions imposed by the International Monetary Fund (IMF) and get itself out of chronic debt (7). In 1990, liberal incentives to private (including foreign) capital investment in mining were introduced (8). However, both oil and uranium operations were "definitely not" to be affected by the reforms (8).

References: (1) *FT* 17/10/78. (2) *SunT*, Business News, 16/11/80. (3) *MJ* 11/12/81. (4) *MAR* 1982. (5) *FT* 22/3/82. (6) *FT* 25/7/84. (7) *MMon* 5/91. (8) *FT* 3/7/90.

620 Uranco (RA)

Clearly an acronym for "uranium company", nothing is known about this outfit other than that, in 1981, it apparently signed a 16-year contract with the Argentinian military junta to mine and process the deadly stuff in San Luis province. An estimated 4536 tons/year of uranium in ore was to be produced (1): if it has been, both the company and the régime have kept very quiet about it.

References: (1) Reuters report quoted in *MM* 3/81.

621 Uranerz

First let Uranerz (as it's universally called) explain itself! Its avowed business is to "search world-wide for uranium ore, exploit and concentrate the ore and other minerals, and market these products" (1). Now let's dig a little deeper. The company is controlled by two of West Germany's largest "hard energy" enterprises. C Deilmann operates world-wide in petrol and natural gas with subsidiaries involved in deepwell drilling, mechanical engineering, sugar refining and refinery construction. The Rheinbraun group controls about one-third of West Germany's electrical generating capacity, is the largest exploiter of lignite coal in the world (supplying, as of 1980, 85% of coal for the country's electricity production), while RWE operates two nuclear power stations and is the largest supplier of electricity in West Germany. Founded in 1898, RWE is a mixed private/public enterprise which also had an original share in Nukem along with RTZ, Metallgesellschaft, and Degussa – hence, too, in Transnuklear (2).

Incorporated a year after its stablemate Urangesellschaft (1968), Uranerz has proved itself more adept at locating uranium deposits, forging contracts (in the late '70s it was providing no less than half West Germany's uranium needs from the Rabbit Lake mine) (3), raising loan capital and purchasing partnerships which enable it to be the operator (as at Key Lake) while avoiding many financial risks.

Its explorations have encompassed Africa – especially Ghana (4), Australia, the USA and, above all, Canada. Here its participation in Rabbit Lake (recently given to Eldorado) (5) and Key Lake – plumb in the middle of two of the world's most promising uraniferous zones (6) – has catapulted it near to the top of the West's yellowcake production league. The Key Lake orebody went on stream in October 1983 (7) and is now the world's biggest single producer.

Uranerz's interests in Canada didn't start with Saskatchewan. In 1974 it took a 10% undivided interest in the Agnew Lake Mines operations of Kerr Addison, thus entitling it to 10% of the product. Due to declining solution grades at the mine, Uranerz had to borrow uranium from Eldorado Nuclear which it later repaid (9); Agnew Lake itself folded in 1983 (10).

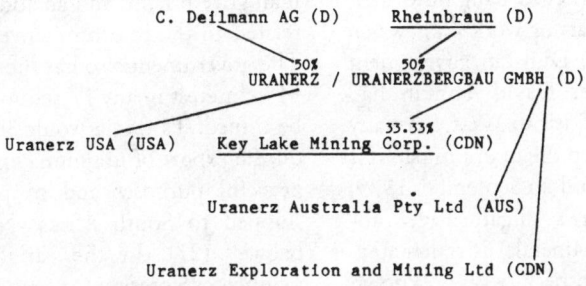

```
              C. Deilmann AG (D)      Rheinbraun (D)

                       50%       50%
                    URANERZ / URANERZBERGBAU GMBH (D)

                              33.33%
   Uranerz USA (USA)    Key Lake Mining Corp. (CDN)

                   Uranerz Australia Pty Ltd (AUS)

             Uranerz Exploration and Mining Ltd (CDN)
```

In 1980, together with Société de Développement de la Baie James of Quebec, it embarked on a drilling programme in the Gayot Lake region of James Bay (11). A little later, also with Baie James and Séru Nucléaire, Esso Minerals, Pancontinental Mining and Soquem, it was part of a JV investigating uranium in the Otish mountains of central Quebec (12).

By then Uranerz was among the top ten most active uranium explorers in Canada (12), a position it has not ceded (31). This was a point not lost on certain anti-nuclear activists in the country. In early spring 1980 a molotov cocktail was thrown through the windows of the Uranerz offices in La Ronge, Saskatchewan. It failed to ignite, but the company shelled out C$10,000 on bullet-proof glass (13).

In 1983 the company obtained a C$120M financing package from three Canadian banks (inculding Barclays Canada) to continue its uranium exploration activities in Canada (14).

These took it into hot water during the late '70s when, along with a dozen other producers, it was charged by the TVA with price fixing in the cartel case (15).

It was also cited for price fixing by the Canadian government in 1981, in a remarkable instance of state authority tying itself up in legal knots (16). However, in early 1984, charges were dropped, after a ruling that state-owned companies similarly charged could not be prosecuted because they were "above the law" (7, 17).

The company's other main exploration project at this time was at Maurice Bay near the old Uranium City (see Eldorado Nuclear) in the Athabasca region of Saskatchewan (located by the Key Lake partners in 1977) (18). In 1980 the Canadian government stated that Maurice Bay was now owned jointly by Eldor Resources and the SMDC (18). However, according to the Saskatoon Uranium Traffic Network (20) in 1981, Uranerz still owns an "unknown percentage" of the project. *Uranium Traffic* also states that Uranerz's uranium, destined for the RWE – after processing at Port Hope under the Canadian policy of "hexing" as much uranium as possible in Canada (6) – went to the Soviet Union for enrichment and to Exxon Nuclear's facility in Washington state to be turned into fuel pellets before ending up in RWE's West German reactors (19).

Further south, in the US of A, the company has also not let the grass grow under its (or anyone else's) feet. In the early 1980s Uranerz planned to lease 33,000 acres of Indian reservation land near Hayward, Wisconsin from the Lac Courte Oreilles Anishinaabe (Chippewas). The Indians, however, resisted the approach, fearing for their fishing waters (21).

In Australia, Uranerz's exploration for uranium has been mainly confined to the Northern Territory. However, only a few months after the South Australian government announced a ban on uranium exports in March 1977, Uranerz became one of four companies still allowed to prospect in the state. In the Adelaide Hills area, "local communities were horrified when ... Uranerz served notices on 200 property owners demanding access to their lands" (22). Despite landowner protest and writs in the Warden's Court, Uranerz was allowed to stay, and indeed the following year the Mining Act was altered to enable uranium companies to hang onto their leases longer, without starting work (22). When the ban finally bit, the Dunstan government fell. The Liberal leader David Tonkin had promised Uranerz on a visit to West Germany that his party would drop the uranium ban if it won the elections (23) and, in September 1979, it did. However, Uranerz's Alligator River project (a JV with AOG Minerals) is stalemated. Called a "grass roots prospect" with no known figure for reserves, in late 1982 the partners were still waiting to proceed to negotiations with Aboriginal land owners on four of their areas (24).

At Mount Fitch in the Rum Jungle district of the NT, Uranerz was partnered with AOG once again. Its 1981/82 exploration programme was confined to three prospects on land previously mined by Territory Enterprises Pty Ltd (the subsidiary of Con Zinc which, in 1962, was merged with Rio Tinto to become the infamous Rio Tinto-Zinc) (25). Reserves in 1981 at this prospect were given as 3.5 million tonnes of ore grading 0.4kg/tonne in an open pit designed to reach 110m depth. About 1400 tonnes of U_3O_8 were located at prospect EL 1562. Prospect EL 1563 has been affected by Aboriginal land claims, contested in the courts between the NT government, the Northern Land Council, CRA, Peko Wallsend and local pastoralists (24). These lie within the Finnis River area, heavily polluted by the Rum Jungle operations during the 1950s and 1960s (26).

Together with MIM, Uranerz holds a third uranium prospect in the Northern Territory in the Amadeus Basin, 28km south of Alice Springs, called Angela. Between 5000 and 10,000 tonnes of uranium oxide reserves are considered to lie in this area, but only limited work was carried out on it up to the end of 1983 (24).

At the same time it was preparing to move in in Australia, Uranerz obtained a unique agreement to prospect for uranium in Tanzania. It had already been exploring one third of the country's land mass. The agreement gave the West German company exclusive exploitation rights in return for a fixed royalty to the Tanza-

nian government, and an additional profits tax related to the return on invested capital (27). The government also has the right to acquire a 51% interest in any JV set up, should uranium be mined. Tanzania would impose conditions on the export of uranium "ensuring its use for peaceful purposes and to prevent its being shipped to South Africa, Namibia or Zimbabwe" (27). In the early 1980s, Uranerz planned exploration for uranium in the southwest part of Bolivia, but no further details have been reported since (27).

In the late 1980s and early 90s, Uranerz expanded considerably in Canada. It bought 20% of Midwest Lake JV (see Denison) in 1988 and advanced its Eagle Point South uranium projects (Cameco 50%, Uranerz 50%) towards production – with a probable annual rate of output of 3000-4000 tonnes, to serve the Rabbit Lake mill (29). Two years later, Cameco, in JV with Uranerz and Agip Resources at McArthur River, Saskatchewan, announced the discovery of a 45,360 tonne uranium deposit, averaging 3% uranium (30).

The company also moved into gold with the opening in 1985 of the Star Lake mine in Northern Saskatchewan (partnered with Starrex Mining Corp). Although that deposit was depleted by 1989, Uranerz then became involved in a feasibility study for the Contact Lake gold deposit (along with Cameco and Westward Exploration) (31). By 1990, Uranerz had also established itself as one of the ISL (insitu leaching) companies in the USA, with plans to process uranium at North Butte and Ruth (Wyoming) (32).

Nor has the company's devotion to Australian uranium and gold exploration diminished. Together with the Japanese company, Kumagai Gumi Co Ltd, it continues its exploration of the East Alligator region; has a JV with Idemitsu Kosan Queensland Pty Ltd in Patterson Province (northern Western Australia); and continues to hold its interest with MIM in the Angela uranium deposit (33).

Gold production from the Grant's Patch project (Uranerz 50%, Glengarry Resources NL 50%) started in February 1990 (34).

Uranerz became the first exploration company since the passage of the NT 1976 Land Rights Act, to sign an agreement with the Northern Land Council. The deal was sealed some ten years later, enabling Uranerz to explore for uranium, gold and platinum in the Myra Falls area of western Arnhemland. Yunupingu, chair of the NLC, complimented the company on the agreement by saying:

"[This] should silence once and for all the critics of Aboriginal people, land councils and land rights. It proves that we have a responsible attitude to mineral development ..., when a mining company respects Aboriginal people and their land and is prepared to sit down and talk" (35).

References: (1) *MIY* 1982. (2) *Yellowcake Road.* (3) *FT* 1/8/79. (4) *Uranium Redbook* 1977. (5) Information from Miles Goldstick, Summer 1984. (6) *Uranium Redbook* 1979. (7) *MJ* 30/12/83. (8) *MAR* 1984. (9) *MIY* 1981. (10) *MJ* 19/11/83. (11) *MJ* 14/3/80. (12) *MAR* 1982. (13) *Kiitg* No. 11, 1-2/81. (14) *MJ* 1/7/83. (15) *MJ* 21/10/77. (16) *The Gazette*, Montreal, 8/7/81; *FT* 9/7/81. (17) *MJ* 13/1/84. (18) *MM* 7/77. (19) Goldstick p7, map 2. (20) *Uranium Traffic in Saskatchewan* 9/81. (21) Chris Kalka, Hungry miners eye Wisconsin's minerals, in *Not Man Apart* (FoE/USA) 5/83. (22) *Uranium mining in South Australia*, Adelaide 1980 (Australian Independence Movement). (23) *FT* 22/6/79. (24) *Reg Aus Min* 1984. (25) Ed. Mary Elliot, *Ground for Concern*, 1977 (Friends of Earth/Penguin Australia). (26) *Fox Commission Report*, Canberra, 1977 (AGPO) Vol. 2. (27) *MJ* 10/8/79. (28) *Taz*, Berlin, 10/12/81. (29) *Greenpeace report* 1990. (30) *MJ* 1/6/90 (31) *MAR* 1991. (32) *E&MJ Int. Directory of Mining* 1991, see also *E&MJ* 3/91. (33) *Reg Aus Min* 90/91. (34) *FT Min* 1991. (35) *Age* 28/6/86, see also *MJ* 4/7/86.

622 Uranex

Wholly controlled by the French Atomic Energy Agency (CEA), Uranex was set up in 1969 to sell uranium to overseas customers (1). Only one Uranex mining venture is known: a JV with

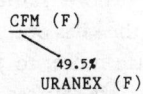

Otter to search for uranium in Queensland, in 1978 (2).

In 1974, a hold appears to have been put on new sales by Uranex, which, by then, had concluded an agreement (with its partner Conoco) to exploit Nigerian uranium deposits (under particularly favourable terms – for details, see Somair), and joined with other members of the notorious uranium cartel to carve up the Namibian uranium market between 1972 and 1977. Uranex's share of the Rössing output was intended to be nearly one quarter (23.75%): its actual gains were less than 5% (4.5%) (3).

Later, Uranex was one of the two dozen or so companies cited by Westinghouse in its anticartel action.

Actual deliveries of Uranex's commitments made up until 1974 seem to have started in 1972 and continued up to 1985. Some 21,000 tonnes is estimated to have been supplied, mainly to Belgium and Japan, according to the uranium market analyst Thomas Neff (1). As Neff points out, the amount of uranium available to Uranex, as an arm of the CEA, hence the French government, depends on French domestic needs, and the balance maintained between military and "civil" requirements. Until 1980, French needs could be met from domestic production: uranium imported, primarily from Africa, could therefore be re-exported. This was not the case before 1976. (1).

From the mid-'eighties onwards, there was a theoretical excess of imports over domestic needs. However, says Neff, " ... when allowance is made for military uses, inventories do not appear to grow to more than three-to-four years' forward supply" (1).

It is certainly true that, in the early 'eighties, the CEA/Cogéma began trying to reduce its dependence on Africa as a source of uranium supply, and diversifying outside the continent, primarily to Canada.

However, Uranex exports started up again in 1985, linked with the sales of reactors to South Korea (1) and, once more, to Japan and Belgium – using Gabonese sources of supply (4).

References: (1) Neff. (2) *FT* 4/9/78. (3) *Financial Mail*, Johannesburg, 3/9/76. (4) *MAR* 1986.

623 Uranit GmbH

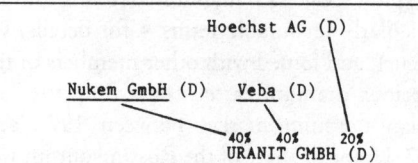

Uranit, with HQ in Jülich, West Germany, is involved in uranium enrichment development and practice. It is the German part of the Urenco partnership, representing there the West German interests. It has a 43.75% participation in the Almelo (Netherlands) ultracentrifuge uranium enrichment plant. (1).

References: (1) *Kiitg* 2/82.

624 Uranium and Nickel Exploration NL (AUS)

This company is involved in gold mining and searching for gold in Australia with no known uranium interests .(1).
In April 1982, Perth entrepreneur Peter Briggs sold 10.47% of his holding in the company to Phoenix Oil and Gas (which Briggs also partly owned). Earlier in the year, the company announced that it planned to sell its remaining mining interests to Kia Ora (another company which has been bought into by Phoenix) and would change from mining exploration to industrial development (2).

References: (1) *Reg Aus Min* 1981. (2) *AFR* 28/4/82

625 Uranium Consolidated NL (AUS)

This is a mineral exploration company with HQ in Brisbane, Queensland, interests in Australia and – in 1976 – Papua New Guinea (1). It held a joint uranium lease with Geopeko at Argentine, Queensland (2). In 1979, it also became part of the "diamond rush" in the Aboriginal areas of the Kimberleys by joining a consortium led by Whim Creek, the Northgate-affiliated company (3).

References: (1) *Reg Aus Min* 1976, 1977. (2) *MJ* 16/9/77. (3) *FT* 19/6/79.

626 Uranium Recovery Corp

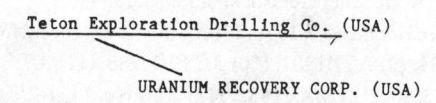

NB: This should not be confused with Freeport Uranium Recovery Co (see Freeport-McMoran). It was operating a uranium-from-phosphates recovery plant at property owned by W R Grace and Co at Bartow, Florida in 1979 (1). The company's activities were suspended in January 1980 after W R Grace failed to supply phosphoric acid which matched requirements – according to United Nuclear (2).

References: (1) *MAR* 1979. (2) *MJ* 11/7/80.

627 URBA/Sté de l'Uranium de Bakouma (RCH)

It was set up in 1968 as a JV between the French CEA-CFMU partnership and the Central African Empire (as it then was) to explore deposits of uranium at Patou, near the town of Bakouma (1). These deposits were discovered by the CEA in 1961 (2).

However, the company ceased its activities in 1971 (2), and it was four years before another uranium venture (URCA) was established (3).

References: (1) *Mines et Metallurgie*, France, 6-7/69. (2) *Uranium Redbook* 1979. (3) Cogéma Annual Report, 1980.

628 URCA

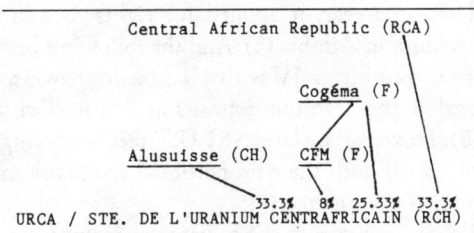

Central African Republic (RCA)

Cogéma (F)

Alusuisse (CH)   CFM (F)

                33.3% 8% 25.33% 33.3%
URCA / STE. DE L'URANIUM CENTRAFRICAIN (RCH)

It was set up in 1975 with Alusuisse, Cogéma, Imetal and the Central African Empire régime to explore for uranium in the country (1). Exploration has concentrated on the Bakouma region previously trekked by CEA-CFMU and URBA (2).

References: (1) Cogéma Annual Report 1980. (2) *Uranium Redbook* 1979.

629 URI/Uranium Resources Inc (USA)

Set up in 1977 and twice bought back by the management from the hands of take-over bidders, the company is led by Ray Larson, a former banker. Larson worked with a publicly-quoted energy company, R L Burns, and was introduced to the solution mining of uranium in 1976 when he looked at the operations of Nuclear Dynamics in Arizona. From that point on, he has pioneered this relatively cheap method of uranium extraction in the USA, helping raise its contribution to the total uranium output of the USA to between 20 and 30% (1). "Because of the *in situ* technique," says Larson, "we are confident that, given a rea-

sonable quality orebody, we can be marginally competitive in the world market, even at US$16 a pound" (1). Conversely, according to Larson, the rest of the US uranium industry requires (at 1985 figures) at least a doubling in prices to enable it to survive (2). The company, as of late 1985, had five *in situ* projects under its belt: a JV with Tenneco, concluded in 1980, to develop a facility in Webb County, Texas (3); a JV with Coastal Corporation (presumably a company associated with Coastal of Saskatchewan) currently undergoing restoration; and a third, wholly-owned venture, also under restoration. All three of these projects were in south Texas, as was the fourth JV, with Western Nuclear, originally scheduled to start up in 1985 (4) and later, according to Larson, "definitely" slated for the beginning of 1987 (1). URI's fifth venture was a multi-million pound orebody, discovered at depths of up to 600 feet in North Platte, Wyoming, which depended on better market prices before being viable(1).

URI started up its Kingsville Dome in-situ leaching project in 1988 (5) – but closed it two years later (6). Its Rosita mine opened in September 1990, while the permitting process continued for URI's Churchrock and Crowpoint projects (6,7). Then, in May 1991, RTZ subsidiary, Rio Algom opened negotiations with URI to acquire all its operations in order to offset the loss of production brought about by closures at its Elliot Lake operations (8).

References: (1) *FT* 11/11/85. (2) *FT* 6/11/85. (3) *MM* 7/80. (4) *MM* 1/85. (5) *E&MJ* 4/88, *Nukem Market Report* 2/88. (6) *E&MJ* 3/91. (7) see also *Greenpeace Report* 1990. (8) *Reuters Report* 14/5/91.

630 US Energy Corp (USA)

Along with Western Standard Uranium, and St Clair Energy, this company planned uranium mining and milling at an unknown site in the USA in 1979 (1). It later emerged that this project was probably the Green Mountain deposit, 15 miles south of Jeffrey City, Wyoming, which US Energy in June 1987 announced would proceed to full production: 1,500 tonnes a day throughput at a cost of US$50 million. However, all permits were not expected to be given for "about nine to eighteen months" according to the company (2), and maximum production would not be reached until around 1991.

References: (1) Sullivan & Riedel. (2) *E&MJ* 6/87.

631 USX

"Big Steel" as US Steel has legendarily been known, changed its name in 1986 to USX, reflecting a reorganisation of its assets and management. In fact, this "restructuring" was a euphemism for the decision taken over previous years to cut the company's losses in steel-making by diversifying (primarly into oil), operating dangerous and pollutive plant without minimal controls, and – above all – savagely penalising its dwindling workforce. This it did by slashing wages and benefits as well as jobs, and using scab labour to break strikes during the eighties (15).

US Steel/USX is the biggest producer of steel and steel products in the USA. It also has substantial interests in apartheid companies including Associated Manganese Mines of SA (20%), Prieska Copper Mines SA (46%), and Ferralloys Lts SA (45%) (1,37). It has been able to participate up to 24% in operations undertaken by the Anglo-Transvaal Consolidated Investment Co of South Africa (2). It also owns 36.6% of Cie Minière de l'Ogooué SA (Comilog) the sole producer of manganese in Gabon (26).

Between 1969 and 1981, US Steel participated (with 43%) in the Pacific Nickel venture on the Waigeo and Gag islands, and in the Cyclops mountains of Indonesian-occupied West Papua. Newmont held 12% in the same JV (3). Through Essex Minerals it also explored for minerals including uranium on a world-wide basis.

In 1981, US Steel's uranium operations in Texas had a capacity of 1.5 million pounds per year (2). In 1978 it was planning to expand its pilot *in situ* uranium leaching operations at Clay West, Texas (4), to a capacity of 1.5 million pounds per year (2).

The same year, it applied for rights to mine uranium in Zambia (5). And the following year it entered into a JV with Niagara Mohawk to exploit the uranium deposits at Burns, Texas (6). However, by late 1982 US Steel was trying to sell off both these properties as the uranium market plummeted (7).

To cut its losses in the heavy steel-making sector, the company bought Marathon Oil in early 1982 – "spectacularly ill-timed" the *Financial Times* called it (8). Despite this, the company could report a turn-around in its fortunes, thanks to profitability from Marathon (9).

US Steel's penetration of Brazil in the early '60s was crucial to the development of what is now the world's largest iron ore project, Carajas in the Amazon region.

US Steel geologists discovered the deposit in 1967. Though there was a government restriction on foreign holdings of 5000ha, US Steel apparently managed to register no fewer than 31 additional concessions in the names of its directors and employees. Two years later it formed the Companhia Vale do Rio Doce (CVRD), later taken over by the government (10). The company apparently withdrew from the Carajas project in 1977 (11) but only after opening up territory occupied by several Indian nations, including the Xicrin-Kayapo, and endangering the Xingu national park (11).

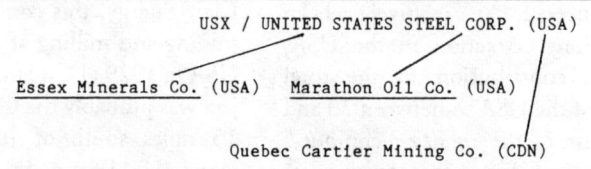

In mid-1982, US Steel initiated the largest cuts in salary ever imposed by a US corporation: 5% cuts among its 27,000 white collar workers (12). And in late 1984, the company announced it would sell US$1.5 billion assets between 1984 and 1986 to reduce its debts and "position the company for possible future acquisition" (13).

The founder of US Steel, J P Morgan, was also progenitor of the various Morgan interests which adorn capital markets today (eg Morgan Grenfell and Morgan Stanley). Most outrageous, lubricious and cunning of the nineteenth century "robber barons", he bailed out the US Treasury in 1895, put together the first of the big US trusts, US Steel, in 1901, and became the single most influential financer in the world in the early years of this century.

A flattering, entertaining, completely unreferenced biography of J Pierpont Morgan (which starts with Miles Morgan who initiated the Morgan family fortunes by ruthlessly stealing Pequot Indian land) is to be found in Stanley Jackson's *J P Morgan: The rise and fall of a banker*, London 1984 (Heinemann).

In 1987, USX and NM Uranium ceased production at their Clay West solution mine in Texas (14).

In 1986, within a couple of months of renaming itself, USX faced its second biggest challenge of the decade (second only to that of the Unions). The notorious "corporate raider" Carl Icahn (see Texaco), true to form as a predator who can nose out an ailing mastodon from light years away, put in a "friendly" bid for the company, at a cool US$8 billion (16). Icahn acquired 11.4% of USX's equity before allowing his bid to lapse (17), but over the next four years had added another 2% to his holding (18). In this period, the company continued to offload major sectors: including more than half a billion dollars worth of chemical subsidiaries (19), its transport operations (20), and Canada's largest steel maker, Quebec Cartier Mining Co – which was bought by Afasco in 1989 (21). The company also stopped phosphates production completely in 1990, though it has held on to its 50% share of RMI, the titanium producer and

fabricator (22). Soon after USX acquired complete ownership of TXO (Texas Oil and Gas), the Australian Bridge Oil – in a deal underwritten by Elders Resources NZFP – bought out its US oil (and some gas) assets (23). One month later, TXO put half its assets on the market (24). During this period, talks between USX and other steel-makers, with a view to acquiring all US Steel's steel-making operations, yielded nothing: notably negotiations pursued "with gusto" (25) by British Steel (trying to avoid reliance on its domestic capacity) (25), and the Italian state steel company, Ilva (26). (As a sidenote: in 1986, Robert Holmes à Court, the Perth-based economic whizzkid, bid for 15% of USX, just before Carl Icahn put in his oar. Holmes à Court did so through Weeks Petroleum and Bell Resources. When news of the bid was notified to USX, a spokesperson responded that Big Steel "had heard of neither of these companies" (27)! Holmes à Court ended up with 4% of USX (28).

In early 1990, Icahn returned to the lists, with a proposal to split USX into its largely loss-making steel sector, (which would then be sold off) and its more profitable energy business (29). If shareholders agreed, he undertook to pump another US$800 million into the company, by purchasing more stock (30). The shareholders demurred (31) but, by early 1991, they had agreed on a "share restructuring" which created a new class of stock for US Steel itself (32), thus allowing it to float on its own merits. USX-US Steel and USX-Marathon then became the twin entities within USX. In return, Icahn made a 7-year "standstill" arrangement, under which he agreed not to purchase any further interest in the company.

A few months later, Icahn was believed to have sold all his interests in USX, recouping around US$5 per share – thus netting him around US$200 million (33). The USX workforce was not quite so fortunate. Acting on behalf of 20,000 steel workers in February 1991, the United Steelworkers of America (USW) approved a three-year contract which would raise hourly pay by one dollar an hour (to US$1.50 over the full period of the contract) and guarantee pension rights in the event of the disposal of USX. In broad money terms, the workers would receive less in total for their sweat, tears, lung disease and premature dying, than Icahn had done for sitting on his well-upholstered posterior and playing dominoes with peoples' lives.

The company and the workforce

At its peak in the 1950s US Steel employed 201,000 people. However "even in these prosperous times, US Steel gave little back to its workers and their communities" (34). It took a series of strikes during the post-war decade, before a living wage was paid (34). By the late 1970s, the company was forced into retrenchment, as it battled with demands from the union, refused to institute basic pollution controls, and closed down plants, rather than compete with newer, more efficient (and less damaging) steel makers from Japan, South Korea and South America (34). US Steel had made no major investment in plant modification at any of its plants from 1945 until the early 1980s. While the United Steel Workers Union (USW) was willing to accept wage and pension cutbacks, in return for upgraded mill technology, the company was hellbent on saving itself for the benefit of the shareholders, at the expense of its workers. The purchase of Marathon Oil in 1982 marked not simply a diversification strategy, but ... complete corporate indifference to steel and the steelworkers who made money for US Steel for 80 years" (34).

Some concessions were made by both company and unions in 1982 but, the following year, US Steel demanded that workers take a US$1.40 per hour drop in wages and benefits. Late in 1983, the company announced closure of its South Works, thus throwing more than 15,000 people out of work. The USW sued.

Three years later, as USX tried to impose a US$3.50 per hour cut in wages and hire scab labor, the Union went on the longest strike in the company's history (35). Finally, the Union was forced to make concessions and – after a

new contract was signed – another 3700 workers were dumped onto the breadline (34). Between 1978 and 1987, the total USX workforce had been decimated (from 108,000 down to 28,300) (34).

The company and the community

Steel-making is a dirty and dangerous business. As the country's oldest and largest practitioner, USX has stood in the forefront of opposition to Clean Air legislation (specifically the imposition of coke ovens) and resisted closing its open-heart furnaces.

In 1977, the company was fined for having inadequate water pollution control equipment, thus resulting in the fouling of Lake Michigan. After secret negotiations with state officials, US Steel was released from the US$33M penalties and the state footed the clean-up bill (34).

In 1980, US Steel headed the list of chemical manufacturers (along with Shell, Exxon and Dow) sued for allegedly contributing their toxic wastes to two Louisiana dumps (35).

Between 1981 and 1991, USX "racked up enormous fines and penalties for environmental and occupational safety and health abuses" (34).

It was required to pay the city of Cincinatti US$77,000 in damages for a benzene spill in 1985 which leaked into the Ohio River and eventually reached Cincinatti (34).

In 1989 it was fined a record US$7.3M for "willful health and safety violations" at two Pennsylvania plants. Labor Secretary, Elizabeth Dole, commented at the time that "... the magnitude of these penalties and citations is matched only by the magnitude of the hazards to workers which resulted from corporate indifference ... including severe cutbacks to the maintenance and repair programs needed to remove those hazards" (34).

And, in 1990, USX agreed to a US$34.1M cleanup programme for its Gary, Indiana steel mill and the adjacent Grand Calumet river. The EPA (Environmental Protection Authority) had charged that the company illegally bypassed wastewater systems, directly discharging into the river and Lake Michigan, while failing to submit reports on its discharges and spills, and violating pre-treatment standards (34).

Among the "Corporate conscience" awards made by the Council on Economic Priorities in 1990 were several "dishonourable mentions." USX merited one of these for its record on pollution, employee health and safety and employment (i.e. discrimination against blacks) (36).

Further reading: John Hoerr, *And the Wolf Finally Came: The Decline of the American Steel Industry*, Pittsburg University Press, 1988.

Contact: Council on Economic Priorities, 20 Irving Place, New York, NY 10003, USA.

References: (1) *MIY* 1982. (2) *MIY* 1981. (3) *West Irian: Obliteration of a People*, London 1983 (Tapol). (4) *E&MJ* 1/79. (5) *FT?*/5/79. (6) *MM* 1/80. (7) *Nukem Market Report*, summarised in: *MJ* 3/9/83. (8) *FT* 25/10/83. (9) *FT* 26/10/83. (10) Shelton Davis *Victims of the Miracle*, Cambridge, 1977 (University Press). (11) *Survival International Review* Spring/79. (12) *FT* 22/6/82. (13) *FT* 11/9/84. (14) *E&MJ* 4/88. (15) *MMon* 4/91. (16) *FT* 8/10/86, *FT* 30/10/86. (17) *FT* 5/12/86. (18) *FT* 4/4/90. (19) *FT* 5/12/86, *FT* 28/1/87. (20) *FT* 21/6/88. (21) *FT* 14/3/89. (22) *MAR* 1991. (23) *FT* 31/5/90. (24) *FT* 26/6/90. (25) *FT* 28/3/90. (26) *MAR* 1991. (27) *FT* 21/8/86. (28) *Australian* 6/2/91. (29) *FT* 22/3/90. (30) *FT* 26/4/90, *FT* 9/3/90, *FT* 4/4/90, *FT* 7/4/90, *FT* 4/5/90. (31) *FT* 15/11/90. (32) *FT* 1/2/91. (33) *FT* 15/5/91. (34) *MMon* 4/91. (35) Fred Powledge, *Water, The Nature, Uses, and Future of our most Precious and abused Resource*, Farrar, Straus, Giroux, New York, 1982. (36) Economist 26/5/90. (37) see also *FT Min* 1991 and *E&MJ Int Directory of Mining* 1991.

632 Utah International Inc

Formed in 1900 in Utah and reincorporated as Utah Construction and Mining Co in 1956 in Delaware, its name was changed to Utah International in 1971.

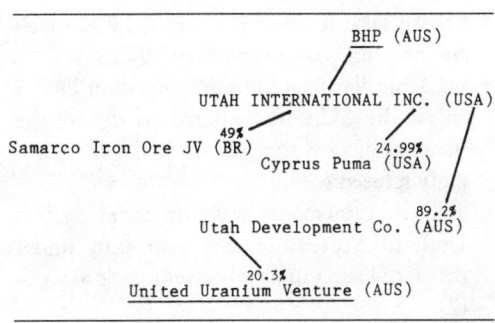

An important miner of coal, copper, iron and uranium (until recently) Utah also has extensive land leases and interests in oil and gas (1).

Until 1983 most of the company's uranium activities were conducted through Pathfinder Mines to which in February 1982 General Electric sold 80% of its common stock (2). The same year, however, the stock was resold by GE to Cogéma's US subsidiary (3).

And in 1984, General Electric lost control of Utah itself in a complex deal with BHP under which GE retained Utah's subsidiary Ladd Petroleum and certain other US financial interests of the company, but surrendered the remainder for US$2.4 billion.

Utah's Australian interests – notably the Goldsworthy iron ore mine in Queensland and the Blackwater coking coal mine, also in the racist state – passed into BHP control, but General Electric retained 15.5% in both ventures (the former reconstituted as Central Queensland Coal Associates) (4). In the early '80s Utah was the biggest producer of coal in Australia (5).

Previously, Umal Consolidated (formerly Utah Mining Australia Ltd) held 11% of Utah's Australian mining interest (1); later BHP acquired all of Umal's shares (4). In the late '70s, Utah's influence over Queensland's coal and its export (through Utah's shipping interests) created considerable antagonism among Australian unions.

A statement issued by the Combined Unions' Utah Dispute Committee of the Seamen's Union of Australia (SUA) in Brisbane at the time pointed out that:

- Queensland received only A$800,000 royalties on Utah's coal exports in 1976;
- no Australians could be employed on Utah's ships; the SUA had offered to debate the issue publicly with Utah on TV – the company refused (6).
- The SUA members went on strike against Utah in September that year but, under threat of losing their jobs, went back to work (7).

The same year, the Movement Against Uranium Mining (MAUM) revealed that Utah had "repatriated" to the USA more cash (A$158.3M) than any other company in the history of Australia (8).

Elsewhere in the world, Utah has operated (or continues to operate) in several controversial areas.

- In 1957 Utah gained a 31,416 acre coal lease on the Navajo reservation in New Mexico, at Black Mesa (9). The Navajo tribal council was browbeaten into accepting the Black Mesa project without " ... enough time to make sound judgments about leases, contracts, and rights-of-way" (10). The Navajo were also deliberately kept in ignorance about the value of their coal deposits, alternatives to coal exploitation, or the possibility of raising coal prices through competitive bidding. When asked by a Navajo councillor about Utah's level of profits from Black Mesa, E C DeMoss, a Utah representative, had the temerity to reply:

"Mr Chairman, the question has been asked, what is the profit range? – we've been asked this question before by various people in the Tribe, and have declined to make our figures public because we feel that in the best interest of the Tribe and in the best interest of ourselves that we don't want the power companies, we don't want our competitors to know what profit margin there is. The reason for this is, if they knew exactly what it would take to get new business, they might undercut us, and the power companies would put the squeeze on us and cut the price, and we think that the price is the lowest in the country, and that alone should indicate that

the profit is a very reasonable and modest profit" (11). In 1974 Utah was employing 303 Navajos at the mine (a larger number than on any other project on the reservation, except construction of the Navajo generating station) (12).

- In Chile, Utah had a JV with Getty Oil to "develop" the large La Escondida copper deposit, 110km from Antofagasta (13).
- Mineracoa Colorado, a subsidiary of Utah, was exploring the reserves of Buritirama in Para state, Brazil for five years, in preparation for ferro-manganese production due to commence in 1986 using power from the Tucurui hydro project (14). The Tucurui dam itself flooded the lands of the Gaviao and the Pukobye people (15).
- Also in Brazil, the company participated (49%) in the Samarco iron ore mining project in Minas Gerais, which developed a large open-pit mine and slurry pipeline (1). It is not known whether Indians were "liberated by compulsory transfer" to make way for this particular project.
- In 1982, Utah made one of the biggest American investments in apartheid in recent years, when its Southern Sphere subsidiary joined Gencor to open up an anthracite mine in the "homeland" (bantustan) of KwaZulu (16). However, in 1984 the company announced it would transfer its 50% interest in the project to Gencor (17).
- Through its subsidiary, Southern Mining and Development Co, in the early '80s, Utah was exploring for tantalite in Namibia, against UN Decree No. 1 forbidding multinational exploitation of the UN territory (18).
- In 1984 Utah concluded a deal with the Anglovaal group of South Africa – pioneers of the controversial Sasol oil-from-coal process (see Fluor) – under which Anglovaal would acquire a 30% stake in Eloff Mining Co, a subsidiary of Utah, while Utah would acquire a 30% stake in Sun Prospecting and Mining Co, co-owned by Anglovaal and Middle Wits. A key element of the deal is that Eloff Mining would take transfer of

Utah's coal interests near Delmas in the Transvaal, one of South Africa's largest untapped coal deposits, "ideally suited for power station feed". The agreement, said the *Mining Journal*, was "clearly the first step in a long-term project" (19).

- In 1983, the company was trying to obtain mineral exploration licences on the island of Fiji (17).
- More recently the company, together with Endeavour Resources (controlled by Alan Bond) set up a JV to explore for copper in Sulawesi (17). Called TEI/Tropic Endeavour Indonesia, it was granted an initial 5-year contract (20).
- Together with Arco, Utah also signed JV agreements to exploit coal in indigenous South Kalimantan (formerly Borneo) (23).
- Two of Utah's workers were reported "radiated" (from coal mining) in Queensland in 1981 (21).

The company was represented on the Trilateral Commission between 1973 and 1979 (22).

Contact: AIEC, POB 7082, Albuquerque, New Mexico 87106, USA.

References: (1) *MIY* 1982. (2) *MAR* 1982. (3) *E&MJ* 3/83. (4) *MJ* 6/4/84. (5) *MAR* 1981. (6) Statement by Combined Unions SUA, Brisbane, 3/8/77. (7) *Queensland Dossier*, Sydney, 10/77 (Queensland Solidarity Group). (8) *Uranium Dossier*, Sydney, 1981 (MAUM). (9) US Department of Energy, 1978. (10) Lynn A Robbins, "Energy Developments and the Navajo Nation", in *Native Americans and Energy Development*, Boston, 1978 (ARC). (11) Minutes of the Advisory Committee, Navajo Tribal Council, Window Rock, Arizona, 1966 (Navajo Nation). (12) Eric Natwig, *Overall Economic Development Program*, Window Rock, Arizona, 1974 (Navajo Nation). (13) *E&MJ* 11/83. (14) *E&MJ* 4/83. (15) *Survival International Review*, London, Spring/9. (16) *FT* 6/4/82. (17) *MAR* 1984. (18) *MM* 3/82. (19) *MJ* 27/7/84. (20) *E&MJ* 6/84. (21) *Kiitg* 6/80. (22) Sklar. (23) *MJ* 15/7/84.

633 Vaal Reefs Exploration and Mining Co Ltd

The largest uranium producer in South Africa (1), Vaal Reefs in 1984 delivered 1,962,977kg of uranium oxide at a profit of nearly R40M (2). This is just higher than its production in 1983 (which was 1,877,421kg) or about 1600tons (1), and made Vaal Reefs responsible for more than a quarter of all apartheid uranium production. By 1991, as other uranium producers went to the wall, Val Reefs was producing more than half South Africa's total yellowcake, as supplied to Nufcor. Only four other producers (including Palabora) were then left (16). The company is also South Africa's biggest gold producer and has consistently rated as the world's biggest producer of that metal since 1977 (1, 3, 4).

Although Anglo-American's share ownership was 27.39% in 1984 (a minor increase over the previous ten years), the corporation completely controls all the company's gold output (4). Also, 84% of Vaal Reefs is in the hands of only 86 investing individuals and institutions, a reason cited by stockbroker Dr Fred Collender for the strong price of South African gold shares, reflecting as they did "the weight of institutional money ... seeking investment particularly in South Africa and the US" (5).

Vaal Reefs operates mines in both the Klerksdorp district of the Transvaal and the Viljoenskroon region of the Orange Free State. While both Hartebeestfontein and Buffelsfontein mining companies have "tribute" arrangements to mine on Vaal Reefs property, Vaal Reefs itself has had an arrangement with Afrikander Lease since 1979 (6) to lease its main block of mineral rights for the mining and treatment of uranium over an area of 3586ha. While a gold and uranium treatment plant for this section has been constructed at the mine site, its commissioning was deferred in 1984 (7) and uranium production ceased in 1982 (8). However, a "large expansion" in the Afrikander Lease area was announced soon afterwards (2). The company also leases land owned by Southvaal Holdings (9). Called the South Lease area, this large section (about 4622ha of stolen land) produces gold and uranium from which royalty payments are paid to Southvaal (7). A new system of shafts was planned for this mine from 1978 onwards. The No. 9 shaft – which then gained a 120,000 tons/month mill and carbon-in-pulp treatment plant (10) – is already the western world's deepest mine shaft (11) (though not mine: that privilege goes to Western Deep Levels).

A new pressure leach plant was completed in 1982 as an adjunct to the South uranium plant, in order to "achieve higher recoveries than are obtainable by atmospheric leaching" (7). Claimed in 1979 as incorporating a technology "which differs quite radically from that employed in the East and West plants" (12), the South plant now has the greatest uranium recovery rate of the company's three treatment units. The plant is remarkable in that it does not use sulphuric acid but a "reverse leaching" system which extracts uranium from the ore before it enters the standard gold recovery circuit. The purity of the uranium diuranate – after leaching, concentration, solvent extraction, precipitation and thickening and calcining – is around 98%. When opening the plant, Dr Roux, president of South Africa's Atomic Energy Board, remarked that the apartheid state should supply "up to 13,000 tonnes a year U_3O_8" in the early '80s – an estimate which has proved wildly over-optimistic (16).

The combined East and West uranium treatment plants have a capacity which is around two-thirds that of the South plant (7).

During the 'eighties Vaal Reefs continued to expand at a rate which gratified Anglo-American's colonial overlords no end: 1984's capital spending was the highest on record (13).

With a workforce of 43,000 people, it is one of the world's most stupendous mining operations

– if not the largest on the globe (11). This has not been without the inevitable and unacceptable cost contingent on all South African labour-intensive mining: ten black miners died in a fire in the Klerksdorp in 1984. Seventeen mourners were gunned down by police on March 21, 1985 (the 25th anniversary of the Sharpeville massacre) and virtually the entire black workforce at Vaal Reefs came out on strike in support of wage demands (14).

In late 1984, Vaal Reefs was among several mines which were targeted by the new black NUM (National Union of Mineworkers) for strike action in pursuit of living wages. The *Mining Journal* reported that this first-ever official black strike in South Africa left the companies "remarkably unperturbed" (15).

References: (1) *MAR* 1984. (2) Analysis of Rand and OFS quarterlies, Dec 1984, in *MJ* 25/1/85. (3) *MJ* 7/9/77. (4) *RMR* Vol. 3, No. 2, 1985. (5) Report by Strauss Turnbull, quoted in *FT* 10/5/84. (6) *MJ* 21/12/79. (7) *MIY* 1985. (8) *MJ* 3/12/82. (9) *FT* 20/7/84. (10) Anglo-American Corp Annual Report 1984. (11) *MJ* 14/12/84. (12) *MJ* 29/6/79. (13) *FT* 3/4/84. (14) *FT* 25/3/85. (15) *MJ* 21/9/84. (16) *MAR* 1991

634 Valiant Consolidated Ltd (AUS)

This is a gold, base metals, diamond and oil exploration company in Western Australia and Queensland. One of its JV partners has been Newmont (gold exploration) (1).

It has also had a 10% JV with Western Mining Corp and, in 1979, was reported to be exploring for uranium (2).

Among its more recent JV partners, are Ashton Mining, (Bullabulling gold project, south of Coolgardie, WA) Platgold Pacific NL, Aberfoyle Resources, Coppergold, Spargos Mining NL, Miralga Mining, Herald Resources, *et al.* But still no uranium! (3).

References: (1) *Reg Aus Min 1981*. (2) *MJ* 13/7/79. (3) *Reg Aus Min* 90/91.

635 Vam Ltd (AUS)

This gold, antimony and mineral exploration company (1) has a 46.5% stake (CSR's Delhi International Oil has 53.5%) in the small Lake Way uranium mine in Western Australia, for which the government go-ahead was given in early 1982 (2), but later withdrawn by the 1983 Labor government. VAM and Delhi International Oil Corporation of the USA (now Australian-owned) purchased the interest in the project held by Westinghouse's subsidiary Wyoming Minerals Corp in 1981 (3). The mine has an estimated 3700 tonnes of uranium oxide, grading at around 1.5lb uranium oxide per tonne (2).

VAM also had 49% of the UK company Black Rock Mineral Ventures which held exploration licences for tin, lead, zinc and silver in Britain, also a coal mine at Wrexham, Wales in the early 1980s (2). Black Rock had 30% of a JV with Southwest Consolidated Resources which had several mining prospects in Cornwall, Devon and Wales (1).

In 1989, Vam "changed direction" (4) when the Pan Ocean group sold its 58.77% holding in the company to Mt. Gipps Ltd. The latter company and Vam then combined their gold and antimony interests into the Hillgrove JV. Vam's subsidiary in Hillgrove, New England Antimony Mines, supplies 2% of the western world's antimony (4).

Vam also owns 49.6% of Perseverance Corporation (in which Rothschild Australia has a minority stake) (5).

In 1991, battle royal was joined between Vam and minority shareholders of Perseverance, as the latter tried to increase their stake in Perseverance, at Vam's expense (6). At the same time, the London-listed mining company Butte Mining, with interests in Montana (USA) and Staffordshire (England), was trying to take over both Vam and Perseverance in order to amalgamate the two companies into a medium-sized group (7). By April 1991, Butte had secured more than 95% of Vam (6).

References: (1) *MIY* 1982. (2) *FT* 20/1/82. (3) *MJ* 17/4/81. (4) *Reg Aus Min* 90/91. (5) *Australian* 23/1/91. (6) *Weekend Australian* 13-14/4/91. (7) *FT* 13/11/90, *FT* 26/1/91.

636 Varibus Corp

```
Gulf States Utilities (USA)
                          \
                           VARIBUS CORP. (USA)
```

Along with its master company, Gulf States Utilities, and the Felmont Oil Co, it was involved in uranium exploration in the USA from 1976 until 1979 (1). No further information.

References: (1) Sullivan & Riedel.

637 Veba

This is West Germany's largest industrial company and its principal energy concern (1). Veba is, according to the *Financial Times,* "a company with highly regarded management and enormous cash flow" (14). It is also a holding company for subsidiaries engaged in electric energy. It has 86.5% of Preussen Elektra, one of the largest electricity supply companies in the country (2) which, in 1981, was getting nearly half its power from uranium (3).

Its main interest in uranium is through Urangesellschaft (4) and, in uranium enrichment, through Uranit.

BP holds 25% of Ruhrgas shares – Ruhrgas is a Veba subsidiary (5).

In 1982, Veba turned its attention increasingly to nuclear power station construction and announced investments of DM5 billion in joint

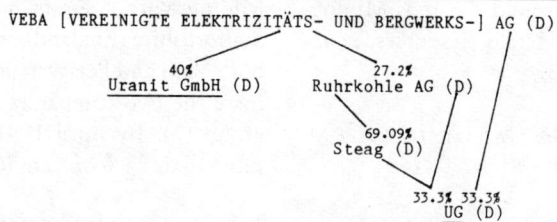

VEBA [VEREINIGTE ELEKTRIZITÄTS- UND BERGWERKS-] AG (D)

40%
Uranit GmbH (D)

27.2%
Ruhrkohle AG (D)

69.09%
Steag (D)

33.3% 33.3%
UG (D)

projects for the building of Brockdorf, Grohnde and Krümmel atomic reactors. By the end of 1987, Preussen-Elektra would be dependent for half its supplies on nuclear power (6). Veba has a contract with Namibia's Rössing mine for the supply of uranium (7).

From 1984 onwards, the German government began offloading its 43.75% stake in Veba as part of the country's privatisation programme. Two years later the share was down to 25.55% and the state announced it would sell it completely (8).

By the end of 1989, Veba had abandoned the highly controversial Wackersdorf reprocessing plant, in favour of contracts with Cogema and BNFL.

In response the *Financial Times* naively claimed that Veba was "not shy of upsetting domestic opinion" – although a large part of "domestic opinion" has vehemently opposed Wackersdorf from the start (9).

That year the company planned to spend around DM23 billion on new investments until 1993 (10). The acquisitions trail was mapped out in 1987, when Veba began buying into Feldmuhle Noble's chemicals business: two years later it had secured 46% of FN, representing what the *Financial Times* (again!) termed "a breakthrough for assertive shareholder behaviour in the German market" (11). In 1991, however, FN was offloaded to the Swedish group Stora (12). Record sales were reported early the same year, and Veba announced it would be investing nearly DM8 billion in east German energy projects: the company was particularly interested in securing orders for new atomic power stations, through Preussen Electra (12). Veba's stake in Ruhrkohle stood at 37.1% in 1989 (13).

References: (1) *FT* 15/3/80. (2) *Yellowcake Road.* (3) *FT* 20/3/81. (4) *MIY* 1982. (5) *FT* 6/3/79. (6) *FT* 23/12/82. (7) Alun Roberts, *The International Trade and Namibia's Uranium,* New York, 1982 (UN NS-27). (8) *FT* 20/1/87. (9) *FT* 22/5/89. (10) *FT* 2/11/89. (11) *FT* 19/5/89. (12) *FT* 25-26/5/91. (13) *MAR* 1991. (14) *FT* 22/5/89.

638 Vestgron Mines Ltd

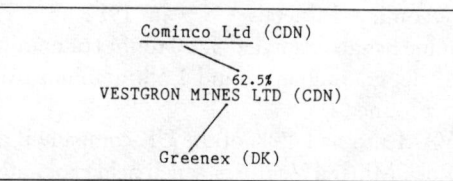

Cominco Ltd (CDN)

62.5%
VESTGRON MINES LTD (CDN)

Greenex (DK)

This is a subsidiary of Cominco, but associated with Northgate (which held a 2.9% direct interest; while Westfield, an associate company of Northgate, held another 6.4%) (1).

It operated the Black Angel silver-lead mine in Greenland at Maarmorilik under a concession from the previous colonialist administration, until closure in 1990 due to depletion of ore (2, 3). It was reported exploring for uranium in Eire during the late '70s (4).

Contact: Jens Dahl, Department of Eskimology, University of Copenhagen, Norregade 20, 1165 Kobenhavn K, Denmark.
International Work Group for Indigenous Affairs (IWGIA), Fiolstrade 10, DK-1171, Copenhagen K, Denmark.

References: (1) *MIY* 1982. (2) Jens Dahl & Hans Berg: Piniartok'arfingme Augtitagssarsiornek, Kobenhavn 1976 (Copenhagen University, Department of Eskimology). (3) MJ 30/8/91. (4) Information from Just Books collective, Belfast, 1980.

639 Vidont Adobe (USA)

It was reported exploring for uranium in Texas in the late '70s (1); no further information.

References: (1) *E&MJ* 11/78.

640 Vitro Chemical Co (USA)

This is one of the many post-war US companies which provided uranium for war purposes and – after the killing boom – left piles of radioactive waste to poison future generations. Vitro's pile is exceptional in that it's located only 4 miles south of Salt Lake City, Utah, and was being used – in the late '70s at least – by local motorcyclists and playing children. Commissioned by the US Department of Energy, the consulting firm of Ford, Bacon and Davis Utah Ltd assessed the cancer risks of Vitro's tailings and concluded that anyone living within half a mile had their lung cancer risks doubled, while 110 people would die if any high density building took place within the same radius (1).

References: (1) Justas Bavarskis in *High Country News*, Wyoming, 10/3/78.

641 Vultan Ltd

```
            VULTAN LTD (AUS)
         92,4%  \
    Sipos Ltd (AUS)
        12,49%    37,8%
    Greenbushes Tin NL (AUS)
```

Greenbushes Tin, owned 40.7% by Vultan in 1989 (4), "could be the world's biggest supplier of tantalum" (1). Tantalum is a material used for capacitors in aerospace and electronics (2). In 1978, together with Sipos (which it con-

trols), Vultan took a dive into the uranium exploration field (who didn't?) at Glen Florrie (3). In 1989, Vultan intended to secure 50% of Greenbushes but, in the event, sold its stake to Gwalia Minerals NL, as it could not meet Greenbushes' needs for more equity capital (4). The company then said it planned to "broaden its base ... by seeking exploration targets with a view to developing its own mine" (4).
Vultan is now owned 50% by Pilgan Mining Pty Ltd (4).

References: (1) *Reg Aus Min* 1981. (2)*MAR* 1982. (3) *FT* 7/8/78. (4) *Reg Aus Min* 90/91.

642 Warren Explorations Ltd (CDN)

It was in a JV with Eldorado Nuclear, and Imperial Oil, as well as Manitoba Mineral Resources, in the Churchill area of Manitoba in the late '70s (1).

References: (1) *EPRI Report*

643 Waterbury JV

It discovered the Cigar Lake high grade deposit in Saskatchewan in 1983, with estimated reserves of 125 million pounds of U_3O_8. The richest vein so far discovered averaged 25.02% across nearly 10m (1).
In early 1984 SMDC announced that, based on its 1983 drilling programme, reserves in the Cigar Lake deposit were now conservatively estimated at 125 million pounds of U_3O_8: nearly 49,000 tonnes, thus making it "one of the world's premier uranium deposits". Mineralisation stretches across 1700m with a maximum width of 100m (2). (See Cigar Lake JV).

References: (1) *FT* 29/12/83. (2) *E&MJ* 6/84.

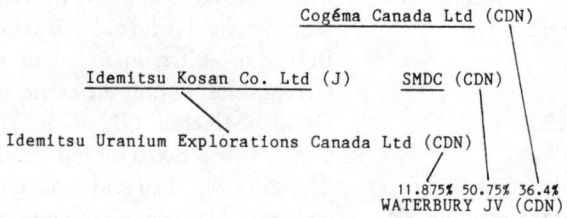

```
                    Cogéma Canada Ltd (CDN)


    Idemitsu Kosan Co. Ltd (J)        SMDC (CDN)

  Idemitsu Uranium Explorations Canada Ltd (CDN)

                            11.875% 50.75% 36.4%
                            WATERBURY JV (CDN)
```

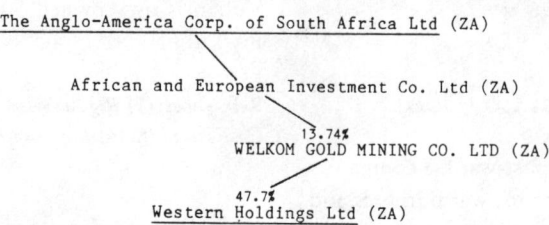

```
       The Anglo-America Corp. of South Africa Ltd (ZA)


          African and European Investment Co. Ltd (ZA)

                            13.74%
                   WELKOM GOLD MINING CO. LTD (ZA)

                        47.7%
                Western Holdings Ltd (ZA)
```

644 Welkom Gold Mining Co Ltd

In 1981 Welkom and Free State Saaiplaas merged their holdings with Western Holdings; in return Welkom acquired 47.7% of the shares of Western Holdings. The merged companies are part of the JMS/Joint Metallurgical Scheme, Anglo-American's venture in the extraction of uranium from gold slimes.

In 1981 the company delivered 1,283,000 tons of slimes to the JMS, with uranium grading at 0.20kg/ton (1).

Welkom in 1986 changed its name to Welkom Gold Holdings Ltd and its shares in Western Holdings were exchanged for those in Ofsil, the holding company for Freegold (20.52%) (2).

References: (1) *MIY* 1982. (2) *FT Min* 1991.

645 West Coast Holdings Ltd

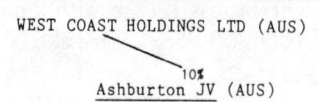

```
    WEST COAST HOLDINGS LTD (AUS)

               10%
          Ashburton JV (AUS)
```

This is an oil, gas, coal and gold exploration company (1). It owned the large Finnis Springs uranium exploration licence about 100km north-east of Roxby Downs (2). It is linked also with Command Minerals at a "significant" gold prospect known as (believe it or not) the Flying Cow, in Queensland (3). It was also involved in exploration with AMAX and in manganese extraction – as part of the Ashburton JV – with ACM, Command Minerals, Getty Oil, and Nickelore (4).

In 1980, exploration was reported continuing in the Murchison (Western Australia) uranium area, but no results appear to have been reported (1).

The company was put into receivership in 1989 (5).

References: (1) *Reg Aus Min* 1981. (2) *AMIC* 1982. (3) *MAR* 1983. (4) *Reg Aus Min* 1979. (5) *Reg Aus Min* 90/91.

646 Western Areas Gold Mining Co Ltd

Although containing much smaller apparent uranium reserves (at 13,000 tons) than JCI's Randfontein Estates – the other uranium-from-gold producer in Johnnies' camp (1) – Western Areas began producing well ahead of schedule in 1982. Originally Nufcor negotiated a contract with Western Areas to commence

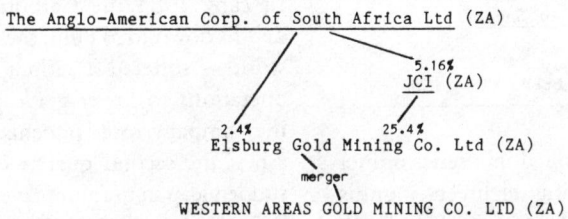

The Anglo-American Corp. of South Africa Ltd (ZA)

5.16%
JCI (ZA)

2.4%   25.4%
Elsburg Gold Mining Co. Ltd (ZA)

merger

WESTERN AREAS GOLD MINING CO. LTD (ZA)

uranium delivery in 1983 (2). In late 1982, the company announced an agreement with the customer to bring forward sales by almost a year (3).

The Westonaria R30M uranium treatment plant has a capacity of 80,000 tons of ore per month (4, 5). By 1984, it was treating nearly 170,000 tons of ore a quarter, as the uranium grade somewhat diminished (6). Production was stepped up in order to meet an additional 60 tons/year of uranium oxide contracted by (presumably) the original customer; this despite the fact that the mine continued "to experience severe underground water problems" (7).

Western Areas' known supply contract is with the Nordwestdeutsche Kraftwerke of West Germany (8).

Both Western Areas and Randfontein Estates are using a liquid-liquid solvent extractor at their treatment plants called the Combined Mixer Settler (CMS) which has been made much of by the mining industry for its "advanced" design (9).

The Elsburg Gold Mining Co was effectively merged with Western Areas in 1974: Elsburg is now a holding company with Western Areas as its only asset.

By the 1990s, Western Areas' uranium production was among the few in South Africa to remain "stable" (10).

References: (1) NUS Corp Foreign uranium supply update, 1980. (2) *MIY* 1982. (3) Analysis of Rand and OFS quarterlies in: *MJ* 10/82. (4) *MJ* 25/1/80. (5) *MJ* 4/4/80. (6) Analysis of Rand and OFS quarterlies in: *MJ* 4/84. (7) *MJ* 6/4/84. (8) *MJ* 16/4/82. (9) *MAR* 1984. (10) *MAR* 1991.

647 Western Coal and Uranium NL (AUS)

Formerly Western Ventures Uranium NL, this is a coal, diamond and uranium exploration company which, in the early 1980s, held numerous coal mining leases in the Perth Basin of Western Australia and no fewer than 84 uranium leases, only one of which, at Arrowsmith River, had moved to agreement with the owners (1). A JV with CRA Exploration for coal at its "promising" Hill River prospect intersected significant seams in early 1984 (2).

References: (1) *Reg Aus Min* 1981. (2) *MJ* 16/3/84.

648 Western Cooperative Fertilisers (CDN)

It operated from June 1980 a phosphoric acid plant at Calgary, from which Earth Sciences (Colorado, USA) produced uranium. Western Cooperative Fertilisers imported the phosphate rock from Idaho, USA, – therefore the uranium total was not included in Canada's uranium export figures (1).

References: (1) *MIY* 1982.

649 Western Deep Levels Ltd

In September 1973, police mowed down 12 black miners at Western Deep Levels mine in Carletonville. They were protesting against regrading and the erosion of wage differentials. The atrocity was the "harbinger of a wave of unrest and strikes" which resulted in no fewer than

The Anglo-American Corp. of South Africa Ltd (ZA)

48.64%
WESTERN DEEP LEVELS LTD (ZA)

80 black miners being killed in a series of incidents over the following eighteen months. When the price of gold quadrupled shortly afterwards, black wage levels were marginally increased. Those who had criticised Anglo-American's "liberal" attitude to wage levels – and its relative lack of reliance on more docile immigrant black workers from Mozambique and Malawi – felt vindicated by what happened at Carletonville: a couple of years later Anglo-American had moved into line with the rest of the colonial exploiters (1).

Although the company mines the Carbon Leader Reef (ceded to it by Elandsrand after tribute taken by that company on land ceded to Western Deep), and although the Carbon Leader is high in gold and uranium values (2), by 1985 the company had decided to stop extracting uranium altogether (2). Materials from which uranium might be extracted later will be stockpiled, and the uranium plant converted to treat gold. The No. 2 uranium plant at Carletonville had come on stream in 1980, but No. 1 closed the following year (3). This – despite the fact that 3,359,000 tonnes of ore grading at 0.22kg/tonne remain in the Carbon Leader Reef (2) and much more in a slimes dam (4).

Western Deep has consistently been one of the least important of South Africa's uranium producers: in 1983 it ranked tenth and last (4). It produced about 146,000kg of uranium in 1984, and ran at a loss for the last quarter of that year (2).

The mine's fortunes have been chequered since it opened in the late '50s. When parent Anglo-American offered 8.4 million shares to the public to launch the company, only 330,000 were taken up; the rest had to be underwritten by Anglo-American. (One of the new shareholders however was Julian Amery MP, powerful member of the Rhodesia-Katanga lobby and leading light of the British Conservative Party until well into the '70s) (1).

In early 1984 the Carbon Leader Reef – at depths down to 3780m, the deepest mine in the world – suffered a serious fire. By switching operations to lower grade gold ore, however, the company rose "phoenix-like" (5) from the ashes: the second quarter of 1984 also saw a sudden leap in uranium revenue explicable only by an "extremely lucrative" delivery under contract (5). (In 1978 and 1979 the mine had to buy to meet contracts, yet a year later it was reportedly having trouble finding buyers) (6).

While Anglo-American's various subsidiaries and associated companies hold nearly half the shares in Western Deep (as well as managing the mine and its output), GFSA reportedly holds (or held in the '70s) 5.4% of share capital, with Sivocam running in at 5% and Barclays Bank at 26% (1).

References: (1) Lanning & Mueller. (2) Analysis of Rand and OFS quarterlies Dec 1984 in: *MJ* 25/1/85. (3) *MJ* 20/4/84. (4) *MAR* 1984. (5) *FT* 21/7/84. (6) Lynch & Neff.

650 Western Enterprises Inc

Coastal States Gas Corp. (USA)

WESTERN ENTERPRISES INC. (USA)

It was said to have been granted a licence to explore for, and mine, uranium in Liberia from 1977 to 1982, with only 4% royalty to be paid to the government and rights to mine any uranium located, for a period of 25 years (1); no further information.

References: (1) *MJ* 2/12/77.

651 Western Fuel Inc

Duke Power Co. (USA)

WESTERN FUEL INC. (USA)

It held, with Ogle Petrol, half of the Bison Basin uranium prospect in Wyoming, along with Energy Capital (1). The prospect was acquired in 1978 (2).

References: (1) *FT* 14/7/80. (2) Sullivan & Riedel.

652 Western Holdings Ltd

Now South Africa's second most important gold producer (about 39,000kg in 1984, 3.68% of the western world's gold market in 1983) (1, 2), the company controls what is known as the "Western Holdings complex", comprising undertakings of Welkom and Free State Saaiplas and a new mine in the Erfdeel-Dankbaarheid area, developed from 1981 (3). Although Erfdeel contains a low-grade gold and uranium deposit, a plant to treat the uranium residues (0.2kg/ton of uranium) was only under consideration in 1985 (3).

With its equity directly owned as to 3.35% by Anglo-American (3), the parent corporation in fact controls 62% of the company through various interests (2). Western Holdings also has substantial share interests in St Helena, Southvaal Holdings, and Free State Geduld.

In 1984, Anglo-American released details of a proposal to merge four of its Orange Free State (OFS) mines to improve efficiency and prolong their lives: these were Free State Geduld, President Brand, President Steyn and Western Holdings. The merger, although it went ahead, was opposed by many outside shareholders in Anglo-American who feared that the move would result in investment of profits from sound mines going to the exploitation of uneconomic ones (1).

Western Holdings is the second biggest slimes deliverer in the JMS (Joint Metallurgical Scheme) (4).

References: (1) *MAR* 1985. (2) *RMR*, Vol. 3, No. 2, 1985. (3) *MIY* 1985. (4) *R&OFSQu* 26/7/85.

653 Western Nuclear Inc

Phelps Dodge acquired Western Nuclear in 1971, and with it three uranium mines at Crooks Gap together with a mill near Jeffrey City, Wyoming, the Sherwood mines and mill near Spokane, Washington, and a 45% interest in various small orebodies (the "Ruby" mines) near Grants, New Mexico (1). Later Western Nuclear acquired a 50 interest in the Beverley Uranium Project in South Australia (2).

As of 1982, its mill at Wellpoint, Washington, was operating at 2,000,000 tons of ore a day (3). The company was in 1980 rated the western world's eighteenth biggest producer of uranium – 1037 tons produced as against its nearest rival Gencor with 1091 tons (4). It was also the USA's seventh biggest producer (4).

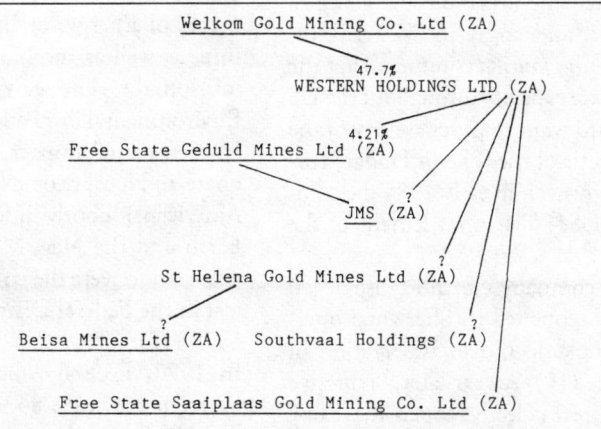

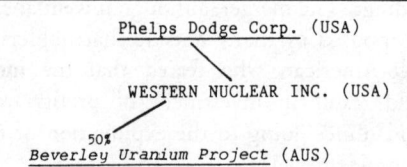

```
        Phelps Dodge Corp. (USA)

          WESTERN NUCLEAR INC. (USA)
      50%
    Beverley Uranium Project (AUS)
```

At the end of that year the company had contracts to deliver just over ten million pounds of U_3O_8 until 1995 (2).

Within less than two years, however, Western Nuclear was badly hit by the weakness in uranium prices. In mid-July 1982 it closed its uranium mines in both New Mexico and Washington "indefinitely" (5). Only 45 workers were left at Jeffrey City by the end of that year, compared with about 600 in its heyday in 1978 (6).

The company's loss for that year was US$1.4M and would have been much higher had it not been for the sale of its retained royalty interest in a high-grade (un-named) uranium mine in Arizona (7).

Western Nuclear's fortunes were not aided by the fact that it was named (along with Getty, Gulf Minerals, Gulf Oil, Homestake, Rio Algom, RTZ Services, and its parent Phelps Dodge) by WPPS alleging breach of supply contracts thanks to the operations of the cartel (8).

Although Western Nuclear that year gained a court ruling that the dear old WPPS/Washington Public Power Supply System was itself in breach of contract, the litigation was going to be a lengthy one, and costly to Western Nuclear. As the *Mining Journal* commented at the time: "[W]ith spare capacity throughout the US nobody is going to want to purchase any of the uranium ... interests of the [Phelps Dodge] corporation" (5). However, in early 1984 the company recovered US$25M in settlement of the WPPS suit (9).

In late 1983, the company's fortunes also began looking up, as it reopened its Sherwood operations on the Spokane Indian Reservation in Washington state (10). A year later, "tentative preparations" were made to reopen the Ruby

underground mines, 45 miles west of Grants, New Mexico, which the company acquired for US$10M from the State of New Mexico in 1979 (12) – and which are situated on Navajo land (Ruby 2, 3 and 4) (13). Seventy employees were to be rehired (they were laid off four years earlier) and milling would be carried out at the Quivira plant operated by Kerr-McGee's Quivira Mining Co (11).

The company recorded deliveries of 275 tons of U_3O_8 from its mines in Wyoming and Washington in 1983 and estimated its total uranium reserves at 14.2 million tons, containing 16,500 tons of uranium oxide (14).

As well as participating in the shelved Beverley Uranium Project, Western Nuclear also participated in the 1970s in a uranium exploration JV in the Northern Territory of Australia along with Pancontinental Mining (50%) and Buka Minerals (25%) (15).

It was also believed to have prospected for uranium in Namibia (16). Together with Ultrana Nuclear and Mining of the Philippines it joined a uranium search on Negros Island in the Philippines in 1980 (17).

The company could also have used two Navajo ventures as "fronts" for uranium exploration in the late '70s and early '80s. Shadow Mining Corp and the Western Navajo Mining Corp, set up with the support of now discredited ex-tribal chairman Peter McDonald, expressed interest in building a mill and buying ore from Navajo corporations. A suit was filed in December 1978 by 90 Navajos against the Departments of Energy, of Interior, and of Agriculture, as well as the Nuclear Regulatory Commission, the Tenessee Valley Authority, and the Environmental Protection Agency, arguing that environmental impact statements were inadequate and local people (to a large extent Native Americans) poorly informed. Friends of the Earth and the New Mexico Navajo Ranchers Association were the co-plaintiffs. The case was lost in the Federal District Court in September 1979 (18).

In 1977, Federal American Partners bought Western Nuclear's 86.9% stock interest in Allied Nuclear Corp and thus acquired uranium claims in the Gas Hills district of Wyoming as well as 37 other claims owned by Western Nuclear near Allied's property. In return, Western Nuclear got 750,000lb of uranium oxide. These new reserves enabled Federal American Partners to expand and to meet its contracts with the Tennesse Valley Authority (19).

References: (1) *US Surv Min, E&MJ* 1979. (2) *MIY* 1982. (3) *Statistical Data of the Uranium Industry*, Grand Junction, 1982 (US Dept of Energy). (4) *BOM*. (5) *MJ* 6/8/82. (6) *MJ* 17/12/82. (7) *MJ* 5/11/82. (8) *MJ* 15/1/82. (9) *FT* 27/4/84. (10) *E&MJ* 11/83. (11) *E&MJ* 4/84. (12) *MJ* 18/4/80. (13) B L Perkins, *An Overview of the New Mexico Uranium Industry*, 1979 (New Mexico Energy and Minerals Dept). (14) *MJ* 1/9/84. (15) *A Slow Burn* (FoE, Victoria, Australia) No. 1, 9/74. (16) *Transnational Corporations and Other Foreign Interests Operating in Namibia, Paris*, 4/84 (UN). (17) *MJ* 14/12/79; 15/2/80. (18) Dorothy Nelkin, "Native Americans and Nuclear Power", in *Science, Technology and Human Values* (Harvard) Vol. 6, No. 35, Spring/81. (19) *MJ* 29/7/77.

654 Western Standard Uranium Co (USA)

Along with St Clair Energy, and US Energy, it planned uranium mining and milling at an unknown site in the USA in 1979 (1). No further information.

References: (1) Sullivan & Riedel.

655 Western Uranium Exploration Inc (CDN)

It held a mining lease in the Golden mining division of British Columbia in 1979 (1).

References: (1) *Energy File* 3/79.

656 Westfield Minerals Ltd

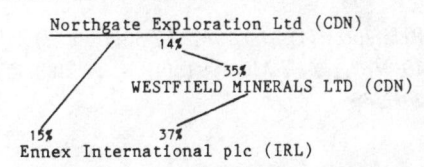

This is a resource exploration and development company with mineral, petrol and gas interests in Canada and elsewhere. It has prospected for uranium together with another Northgate company, Irish Base Metals, in Eire, and has participated with Shell Canada Resources in a uranium JV in the Deer Lake-Codroy area of Newfoundland, with Shell earning a 30% interest, in the late '70s (1). Prospecting apparently stopped with the imposition of the Newfoundland government moratorium on uranium mining in May 1980. (See also Northgate.)

In 1987 Westfield organised a JV with Shell, on the Choquelimpe gold prospect in northern Chile (2) which started production in 1988 (3). An "intensive" exploration programme from 1988 onwards was concentrated on Canada – especially the Big Island Lake property in Manitoba and the Misehkow River region of Ontario (3).

Westfield also holds 25% of the WX syndicate which heap leaches gold in Nevada (3).

Northgate owned 22% in Westfield (as of 1989). Acting together as Norwest, Northgate and Westfield sold their interest in Whim Creek Consolidated to a subsidary of Dominion Mining in 1989 – thus consolidating Dominion's hold over the other Australian company. Whim Creek also sold its 46% holdings in Westfield to American Resources Corp (4).

This re-arrangement of interests in the three companies followed a humdinger of a battle the previous year, when the Australian directors of Whim Creek had thrown the non-Australian chairman, chief executive, and another director, off the board. These directors were concerned about the proportion of holdings in the company held by non-Australian outfits, in particular Westfield, through the tax haven of Cyprus (5).

References: (1) *MIY* 1982: *FT* 30/4/80. (2) *MJ* 19/9/87. (3) *FT Min* 1991. (4) *MJ* 24/2/89. (5) *FT* 24/2/89.

657 Westinghouse Corp

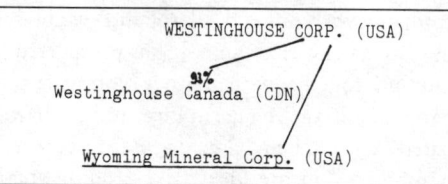

Kirby, Danforth, Franklin, Gookin, Doctor Hornig, J.F.McGillicuddy and all ... what would they be doing sitting round a table on a sunny afternoon in Pittsburgh, Pennsylvania (1)? Since they are some of the directors of the world's biggest vendor (in the early '80s) of nuclear reactors, the US's fourteenth largest defence contractor (as of 1981) (2), the 75th largest multinational in the world (with 13% of its sales overseas) (1), a major exploiter of uranium (through Wyoming Mineral Corp), and an entity whose diversification has gone on for more than fifty years, they could as well be planning an adaptation of the cruise missile, a modification of launch tubes for the Trident or Poseidon submarine (3), how to sell *7-up* (for which they own the franchise) (1) to the Marshall Islanders, how to promote Linguaphone (purchased in 1972) (1) among Filipinos who don't know the English for *rip off,* as how to introduce cable TV to the "Lapps" (in 1981 they bought the Teleprompter Corp for the pittance of US$646M) (4).

Westinghouse began life as the Chartiers Improvement Co in 1872, later becoming the Westinghouse Electric and Manufacturing Co – its present name was adopted in 1945. In its quest to switch on the world (or at least the "civilised" part of it) Westinghouse expanded rapidly, especially in Britain (where its assets were later acquired by General Electric Co

(GEC) – not to be confused with General Electric (GE) of the USA).

Diversifications over the next thirty years brought the company interests in elevators (sold to Finnish interests in 1975), electrical appliances, X-rays, electronics, watches (Longines), a record club, water pollution control, low income housebuilding. The last three were among the loss-making subsidiaries sold in a major "surgery" in the mid-'70s, when heavy re-investment was made in its central electrical and nuclear power businesses (and its equity stake in Westinghouse Canada raised to 93%) (1).

Westinghouse is not only the originator of the world's most popular (and, many would say, most disastrous) nuclear reactor, the PWR (Pressurised Water Reactor), it also built the first nuclear reactor for electrical purposes. But even at the dawn of commercial nuclear power, and in the wake of Eisenhower's famous 1953 "Atoms For Peace" pronouncement in the United Nations, the military-civil links were firmly intertwined. Westinghouse's first PWR was constructed for the US Navy's Nautilus submarine (5).

This wasn't the company's first excursion into the military nuclear field: in the war years its laboratory had been trying to separate U-235 from U-238. But whereas that hadn't proved successful (nor were Westinghouse's later ventures into uranium enrichment to yield much fruit, including a proposal made with Bechtel and GE to use the Purari River in Papua New Guinea to power a U-235 plant) (6), Westinghouse's enrolment by Admiral Rickover, to be his "malleable implement" for a nuclear task force, certainly was. Westinghouse also got the order to build (entirely at the government's expense) the first US "civil" reactor at Shippingport on the Ohio River. Though its electricity costs were ten times that of coal, it came on stream in only just over three years in 1958 (7). From these days onwards, Westinghouse was to be in competition with GE, not only for domestic orders but for the overseas sales which – especially when nuclear power slumped in the '70s – were the life-blood of the two colossi. The former was known as the "engineers" company', while GE gained a reputation for flair and innovation (7).

By the end of the '50s, Westinghouse had sold one PWR each to Belgium and Italy. Ten years later, twenty-one PWRs were operating in the USA, and a dozen were planned in Europe, India and Japan: GE and Westinghouse more or less split the sales between them (7).

At this point Westinghouse decided to "beef up" its image, and by 1970 it got the edge over its rival in reactor sales. In 1970 it flogged six reactors to Japan (and so did GE) (7).

Westinghouse's reactor for Brazil, Angra dos Reis No. 1 (succeeding reactors were to be constructed by Kraftwerk-Union of West Germany) (3), was the country's first nuclear contract. Construction began in 1971 at Itaorna Beach ("bay of bad stones") – one of the main geological fault areas in the country (8). Over a period of ten years, there were fires (including one which caused US$8M of damage) (9), movement in the building which put the turbo-generators out of joint, vibrations which ruined the building during tests, and appalling working conditions leading to epidemics of tuberculosis and other diseases (10). Since March 1983 the station has "repeatedly broken down, to the great embarrassment of Westinghouse and the mounting fury of the Brazilian government" (21).

Taiwan's vaunted nuclear power programme – the biggest in the "developing world" – has mainly depended for its reactors on GE. But Westinghouse supplied the generators for GE's four plants and GE supplied them for Manshaan 1 and Manshaan 2, which were built by Westinghouse (10).

South Korea's first commercial reactor was built by Westinghouse and came on stream in 1978. There are 3 million people living within 32km of the plant, as well as 18 US military bases (11). The reactor was closed down no fewer than four times in a two-month period in Spring 1979 (12), and again in 1980 due to a leak in the turbine generator (13). Three more Westinghouse reactors are planned on the same site, and another two are being built on the south-east coast. Westinghouse's partner in the huge project is Kepco. The whole boondoggle was financed by the Exim bank, whose US$1.1 billion package was the biggest credit ever authorised by the bank (14).

Westinghouse also set up a factory with the local conglomerate Hyunai to manufacture nuclear power plants (15). A South Korean official testified at a US House Subcommittee investigation in the '70s that an internal commission had "voted unanimously to proceed with the development of nuclear weapons" (5).

Two of India's early small reactors were Westinghouse PWRs, sited at Tarapur (5).

Westinghouse has been a heavy investor in the apartheid economy (16), a large part of it through Framatome (before Giscard d'Estaing's government in France substituted gas-cooled reactors for PWRs and Westinghouse pulled out of Framatome) (5) (and see below). In 1984 it withdrew from a JV with Sasol, the manufacturer of coal-into-oil plants, which would have necessitated it supplying both parts and technology for a new coal gasification scheme. The company said the withdrawal was for "economic", not political reasons (17).

When nuclear power station construction began to slow down in the mid-'70s, naturally so did Westinghouse's overseas orders. Just before the Three Mile Island debacle, in the space of just 48 hours it lost a US$2.5 billion order to construct off-shore plants for Public Service Electric and Gas Co of New Jersey – and won a US$171M order to build two steam-supply systems for Commonwealth Edison (18).

Since then it has not fared so well. Apart from huge losses in 1979, thanks to settlements in the uranium cartel suits, it lost its stake in Framatome (19) and overseas orders were distinctly lacking. A bid to sell Egypt nuclear reactors in 1983 (20) looked like foundering in the face of more favourable terms offered by a French-Italian consortium (21).

In 1981/82 the company's electric lamp business was bought by Philips of the Netherlands for around US$200M (1).

However, its sales of other nuclear power equipment and its military sales (23), its new markets in broadcasting and cable TV (24), and a new venture with Warner Communications and American Express in cable TV (25), as well as its massive Teleprompter acquisition (26) kept it in profit: a "mere" US$448M in 1981 (26).

During this period Westinghouse's biggest problem was what has become known as the "cartel case". In 1975, the company announced that it would renege on past contracts which guaranteed uranium supplies at fixed prices as an incentive to buy the Westinghouse reactor package. Some 80 million pounds of uranium had been contracted to 27 utilities in the US and Sweden during the late '60s and early '70s (7) at US$9.50 a pound. By the mid-'70s the price of uranium shot up to around US$40 a pound, and with only some 20% of the contracted uranium supplied, Westinghouse was faced with a loss of US$2 billion in buying the shortfall. A company consultant at the time admitted that "the uranium thing [was] the most stupid performance in the history of American commercial life" (27).

Using the excuse of "commercial impracticability" (7) Westinghouse dumped on its customers, and was accordingly sued for breach of contract. As litigation costs mounted, making the suit the most expensive private litigation of all time (7), a *deus ex machina* appeared on the scene in the shape of documents filched from Mary Kathleen Uranium, which showed that RTZ, along with other major uranium producers, initially outside the USA, had formed a cartel to grab markets and push up prices sixfold during the years 1972/75 (28). On the basis of this information, Westinghouse proceeded to sue its erstwhile comrades in the uranium business. One of the ironies of the case was that Westinghouse itself had been indicted in 1961 for price-rigging, along with GE, its rival in the field of electricity manufacture, and for the first time in US history company executives from both companies had gone to jail (7).

And again, only a short time before the news of the cartel broke, Westinghouse was admitting to the US Securities and Exchange Commission that its subsidiaries had been making "questionable payments" to foreign officials, including a lump sum of US$150,000, apparently destined for just one foreign official (29).

By 1981 Westinghouse had settled its suits with the utilities starting with Texas Utilities in 1977 (30), moving through Statens Vattenfallsverk of Sweden (31), Virginia Electric and Power (32), Florida Power and Light (33), and Northeast Utilities (34), and ending with Gulf Oil in 1981. Here was another irony – for it was Gulf which proved one of the prime movers in the cartel conspiracy, since Westinghouse stood in direct competition to its reactor manufacturing subsidiary General Atomic (7). The cost at the end of the day to Westinghouse was around US$250M, though much of the settlement was in terms of uranium to be delivered in renegotiated contracts.

Another prime participant in the uranium cartel was the French CEA. CEA owns half of Framatome, an early (1972) partner of Westinghouse's, which held an initial 45% in the consortium (19). Westinghouse's stake in Framatome was reduced to 15% in return for half a million pounds of uranium (16) and, in 1982, the agreement was formally ended. Framatome was now able to export its PWR reactors to any country it wished, without prior US government authorisation (35), and could license the sale of Westinghouse PWRs to anyone it wanted. In practice, co-operation between the two companies continued (36).

Whatever the company's recent diversification into non-nuclear fields, we should not underestimate its continuing importance as a nuclear contractor, provider of components and of materials. For example, it proposed a link-up with Mitsubishi to produce titanium and zirconium (crucial in nuclear plants and high-speed aircraft) (37).

It holds important residual interests in both solar energy and fast breeder reactors.

In 1975, at a senate hearing on Small Business, one of the testifiers, Jim Piper, a small California solar contractor, was asked by the committee why the large energy corporations shouldn't be trusted with the nation's solar energy programme. He gave this lucid reason: "The larger corporations have vested interests right now in maintaining the *status quo*. If Westinghouse or GE came up with a usable solar system, immediately it would harm their profitability in other areas. Westinghouse just submitted a proposal to the FEA for US$1.75 billion for three atomic generating plants. If there was a dramatic increase in solar energy systems, and a lowering of the need for electrical energy in the future, it's possible that they might not be able to sell those three plants for US$1.75 billion. And that has got to be in the mind of the man from Westinghouse who decides whether to support solar systems or not. There is just not the profit, nowhere near it, in solar technology and equipment. It is just not there" (22).

Along with GE and Rockwell, Westinghouse has led the development of Fast Breeder Reactors (FBRs). Until 1971 it paid for most of its own research work, then gained a permanent position in the early '70s – having spent US$40M by the end of the decade – at which point the US government moved in to foot 80% of the bill (22).

While most of Westinghouse's uranium activities have been undertaken through its subsidiary Wyoming Mineral, there are instances where Westinghouse has prospected on its own account.

It was reported actively exploring for uranium in northern Saskatchewan in the early '80s (though this may have been in the name of Wyoming Mineral).

In late 1984 it transferred its Lamprecht uranium mine to Intercontinental Energy with Westinghouse paying an initial US$600,000 and arranging to pay US$1.2M over the period 1990-2000 to cover clean-up costs. Substantial ore reserves were held at the mine which was adjacent to Intercontinental's Zamzow mine in Three Rivers, Texas (40). In the early '70s, Mitsubishi Metal Corp and Mitsubishi Corp formed the Taihei Uranium Exploration Corporation, to develop an integrated uranium business from mining to marketing, which enlisted Westinghouse in its plans to hunt for yellowcake in the Northern Territory of Australia (41).

Although Westinghouse had no known uranium exploration projects south of Texas, its equipment was used by various (mainly US) corporations during the early '70s in Project Radam – a US$7M aerial photographic survey of the Amazon region of Brazil. In the course of the survey, uranium deposits were discovered around Surucucu Mountain in the indigenous territory of Roraima (mainly occupied by Yanomami people) in 1975. A few months later Brazil signed the world's biggest "civil" nuclear pact with West Germany: a pact which included supplies of uranium from Yanomami land. Construction of the Northern Perimeter Highway – up to then the biggest single encroachment on indigenous territory in the Amazon – was partly undertaken to transport minerals, including uranium, from the newly discovered fields (42). However, no known exploitation of uranium – as distinct from gold and other metals – has taken place in the meantime.

Westinghouse, as such, has no known uranium projects on Indian land in the USA, but it has been one of the key links in propagating the huge Four Corners "development" which has afflicted the Navajo, Hopi, Laguna, Ute, Jicarillo, Apache and other nations with the largest single planned example of energy aggression of any region in contemporary times. Many other companies indexed in this *File* have been involved in this project, including Arco, Consolidation Coal, Peabody, Utah International, Exxon and Kerr-McGee (43).

The Four Corners Regional Development Project in the late '60s called for a major urban centre which would not only bring in a huge influx of outsiders, but assimilate the Native American inhabitants in an unprecedented fashion. Although the project appears to have fallen on sterile ground, it is worth outlining the objectives to which Westinghouse gave its expertise, with "urban renewal" funding from the US Department of Housing and Urban Development.

"Covering 288,000 square miles – about 8% of America's land area – the program for the Four Corners Economic Development is to include:

an ultra-modern city of 250,000 with the latest techniques in housing prefabrication, education and business; a variety of recreational areas, including ski resorts, tourist ranches, and Indian villages and archeological ruins; a network of new roads and airports that would convert one of the most inaccessible areas of the US into a center of commerce and tourism; agricultural projects to make more abundant use of land that is now mainly devoted to agriculture and sparse grazing. The work would cost a total of more than a billion dollars in both public and private funds. Indians would compose a minority of this urban population" (44).

"The study goes on to state – in the language of the corporate 'planners' who impose their visions of future development on our lives – that the major urban center could exist alongside the Native Americans and Chicanos in the area 're-taining existing cultural and aesthetic values' (45). If the Hopi and Navajo are displaced from their agricultural and grazing lands, they'll be able to buy their food at the ultra-modern shopping center, conveniently located across the street from their hogans" (43).

The corporation's most controversial uranium project in recent years has been a JV with the despotic neo-colonialist régime of Hassan II of Morocco to exploit both the country's phosphate deposits and those in the Western Sahara which the Moroccan army blatantly colonised in 1975, after Spain withdrew. Morocco is the world's largest exporter of phosphate rock (17.7 million tons in 1982) (47). Ever since the world's biggest deposit of the mineral was discovered at Bou Craa in the Western Sahara (Saharawi Democratic Arab Republic) in 1963, Morocco had itched for an excuse to take it over. Spain retained a 35% interest in Bou Craa after Morocco's annexation of the territory.

Already deeply involved in uranium-from-phosphates extraction in Florida (see Wyoming Mineral), Westinghouse – along with Gardinier and IMC (48) – offered to sell an unspecified amount of its uranium extracting technology to Hassan's régime in 1980. Since Westinghouse in 1977 obtained a US$200M contract to construct a Tactical Air Defence System for the des-

potic monarchy, the company was optimistic about success (49). It has had to wait longer than expected. The liberation movement, Polisario, closed down the Bou Craa mine soon afterwards, but it reopened and resumed production in mid-1982 (50). In early 1984, Westinghouse was confident that it was "first in line" to build Morocco's own extraction plant and it agreed to buy most of the oxides produced by that plant (51).

The greatest controversy in which the corporation has been involved in recent years remains that centred on its contract to supply a reactor to the Marcos dictatorship in the Philippines, with which to supply power to the Bataan free trade zone, at the expense of the fisherpeople, villagers and *barrio* dwellers of Morong.

Built on an earthquake fault, catering to the needs of exporters at the expense of local people, necessitating pay-offs to corrupt officials (the Disini connection) (10), financed by huge foreign loans, based on questionable technology – to say the least (52), the Westinghouse Bataan reactor epitomises most of what is unacceptable, dangerous, and downright deceptive, about nuclear power.

One of the people of the *barrios* sent a letter to the outside world, soon after they discovered (more or less by chance) (53) the fate the government had decided for them:

"To whom it may concern:

"Comparing the past and the present situation in our Barrio, I can see that there is a big change happening now. This change has to do with the construction of the nuclear power plant by the National Power Corporation. This project is increasingly creating restlessness among us because our rights are slowly being taken away from us. Our right to fish in the sea is one. Part of our fishing ground is already covered with earth and in other places the water is no longer as clear as before. Without our consent our farms were taken over by the National Power Corporation. We depend for our livelihood on these; now they are part of a reservation area. Parts of the mountains were flattened for a housing project for engineers and other people who will work in the plant. They did not consider if our source of food and livelihood will be affected. They only saw their needs, and will meet them at the expense of all of us. For me our town is one of the most beautiful places and if we will be relocated we can never find another place equal to it. I think this is the most tragic thing that can happen to all of us here. I am praying that this will not come to pass.

"May people who are in a position to help, reach out to us soon, so that this impending tragedy will not befall us" (52).

Not long afterwards, Ernesto Nazareno, a Filipino anti-nuclear activist, was kidnapped by the Marcos régime and apparently executed (46). The Bataan reactor was due to come on stream in early 1985 at a cost of nearly US\$2 billion (38).

During 1988 Westinghouse disposed of various units to Siemens, the large West German conglomerate, and linked its transport businesses to those of that country's AEG (54).

That year, Greenpeace in the USA accused it of creating two decades of pollution at its PCB incinerator plant at Bloomington, Indiana, where local people were arrested for demonstrating against the operation (55).

And, soon after the downfall of Marcos in the Philippines, the Aquino government began investigations of the huge Westinghouse Bataan reactor "scam", accusing the company of bribing Marcos and his cronies (56), while the government was saddled with a crippling \$350,000 a day interest repayment on a plant which was never constructed (57).

Two years later, the Indonesian régime put out its grandiose (otiose?) nuclear power programme to tender, and Westinghouse was, of course, one of the companies putting in its bid ... (58). *Plus ça change, plus ça la même chose ... ?*

Further reading: An excellent introduction to the Philippines deal is contained in Walden Bello, Peter Hayes & Lyuba Zarsky "500 mile island: The Philippine nuclear reactor deal", in *Pacific Research* (Pacific Studies Center, 867 W Dana Street #204, Mountain View, CA 94041, USA) Vol. X, No. 1, 1979.

Contact: Nuclear Free Philippines Coalition (NFPC), Apostolic Center, 2215 Pedro Gil Street, Sta Ana, RP-Metro Manila, Philippines.
Philippines Support Group, 11 Goodwin Street, London N4, England.

References: (1) *WDMNE*. (2) *NARMIC* 1980. (3) *Yellowcake Road*. (4) *FT* 7/10/80. (5) Moss. (6) *Purari: Overpowering PNG*, Melbourne 1978 (IDA & PAG). (7) Pringle & Spigelman. (8) Interview with Rogiero Cerqueira Leite, in: Thijs De La Court, Deborah Pick & Daniel Nordquist *The Nuclear Fix*, Amsterdam, 1982 (WISE). (9) *The Economist*, London, 3/78. (10) Thijs De La Court *et al op cit*. (11) *Export Monitor*, Washington, 2/80. (12) Don-A Ilbo, Seoul, 3/4/80. 13) *Korean Times*, Seoul, 2/3/80. (14) "Exim Bank and the nuclear industry", in *Newsletter of the Center for Developmental Policy*, Washington, 25/2/82. (15) *WIN Magazine*, New York, 1/8/80. (16) Rogers & Cervenka. (17) *MJ* 15/7/84. (18) *AFR* 4/1/79. (19) *FT* 11/7/84. (20) *FT* 28/11/83. (21) *FT* 19/9/84. (22) Reece. (23) *FT* 15/7/80. (24) *FT* 14/7/82. (25) *FT* 2/3/83. (26) *FT* 20/1/82. (27) *Newsweek*, Washington, 16/2/77. (28) International Uranium Cartel, Hearings before Subcommittee on Oversight and Investigation of the Office on Interstate and Foreign Commerce. House of Representatives, 95th Congress, 1st Session May-August 1977, Washington, 1977 (House of Representatives). (29) *Wall St J* 15/3/76. (30) *MJ* 30/12/77. (31) *MJ* 29/6/79. (32) *MJ* 16/11/79. (33) *MJ* 5/10/79. (34) *MJ* 6/2/81. (35) *FT* 24/1/79. (36) *FT* 23/1/79. (37) *FT* date unknown [1983?]. (38) *Kasama*, London, (Philippines Support Group) No. 10, 1984. (39) *Uranium Traffic in Saskatchewan*, Saskatoon (Uranium Traffic Network) 9/81. (40) *MJ* 28/9/84. (41) *A Slow Burn*, Melbourne, (Friends of Earth) No. 1, 9/84. (42) *Indigena*, Berkeley, Fall/76. (43) Weiss; Winona La Duke: *The council of energy resource tribes*, New Haven, 1981 (Women of All Red Nations/WARN). (44) Elizabeth Dunbar, *Black Mesa: The effect of development*, Pasadena, 1974. (45) Suzanne Gordon, *Black Mesa: The angel of death*, New York, 1973 (John Day). (46) *Pacific Research* (Mountain View, Pacific Studies Center) Vol. X, No. 1, 1979. (47) *MJ* 24/6/83. (48) *Intern Min/Met Rev*. (49) Stephen Talbot "Westinghouse backs King Hassan's war", in *Multinational Monitor*, Washington, No. 11, 1980. (50) *MJ* 28/10/83. (51) *New Scientist*, London, 24/5/84. (52) *Chain Reaction*, Melbourne, Vol. 4, No. 2-3, 1979. (53) Gene Stoltzfuss: "Westinghouse, the helpful neighbor", in *Sojourners Magazine*, Washington, 3/79. (54) *FT* 23/2/88, *FT* 2/4/88, *FT* 19/7/88. (55) *Waste Not*, No. 35, Canton, USA, 27/12/88. (56) *FT* 12/4/89, *FT* 11/10/89. (57) *Appen Features*, Penang, 3/1/89. (58) *The Australian* 4/2/91.

658 Westmin Resources Ltd

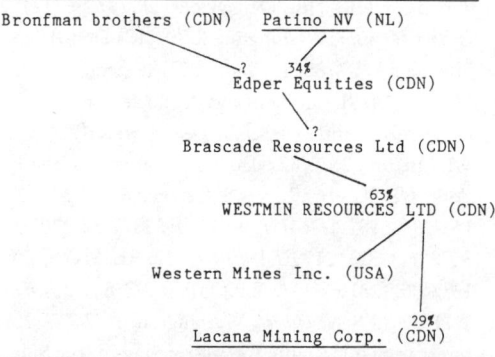

Another logger in the Bronfman camp (see also Brascan and Noranda), Westmin was formerly Western Mines Ltd of British Columbia. Its name was changed in March 1981 (1) after acquiring all the shares of Brascan's 99% owned Brascan Resources which had effectively controlled it since 1976. Brascan Resources was wound up in 1981. After the merger a "major Canadian resources company with the financial capability to undertake major new investments" (2) was set up, effectively controlled by the ubiquitous Bronfmans.

One of its major recent projects is large-scale base and precious metals mining at Buttle Lake, British Columbia, whose H-W section looked promising by 1980 (3) and was being developed in 1983 (4). It was called by Westmin "the most significant massive sulphide development since Kidd Creek's Timmins, Ontario, [discovery] in 1963" (5).

Westmin is also involved in coal exploration in Calgary and, through the assets of Brascan Re-sources, has oil and gas properties in Western Canada, the Yukon, the Arctic Islands area, offshore eastern Canada and the Beaufort Sea – where oil exploration has come under heavy attack by Inuit, Dene and other native peoples (1) (see Gulf Oil).

Its Western Mines subsidiary is engaged in exploration in the USA, and it has had a JV with Du Pont of Canada and the Philipp Brothers, also Canadian, to explore properties in the Pine Point district of the Northwest Territories (NWT) (6).

In 1977 Britain's CEGB entered an agreement in principle with Westmin to provide finance, in return for a 30% equity in no fewer than seven existing uranium prospects held by Westmin, with management by Westmin and further prospecting envisaged (7). The seven properties were in Saskatchewan, Ontario and the NWT. Neither the Ontario nor the Saskatchewan (8) properties appear to have been worked since.

However, in 1980 – as fears of a uranium moratorium being declared in the NWT were voiced – Westmin's deposit at Baker Lake, on Inuit land, was cited as one of four properties with the potential to become US$100M producers (9). In 1979, the Inuit of Baker Lake lost their land claim over the area covered by the Urangesellschaft claims (10).

In 1983, Westmin reported record earnings (11).

Westmin sold its 3.1 million shares in Lacana to Royex Gold Mining Corp in 1987; that year also saw its profits up, with a 33% return from base and metals production (12).

The following year it sold out its oil and gas operations to Norcen (13).

It continues to explore extensively for copper, gold, zinc, silver and lead, throughout Canada (14) – its biggest new find has been a base metals lode at Myra Falls (15).

References: (1) *MIY* 1982. (2) *FT* 13/3/80. (3) *MJ* 18/4/80. (4) *MAR* 1984. (5) *FT* 7/4/83. (6) *MJ* 4/7/80. (7) *MJ* 22/7/77. (8) *Who's Who Sask.* (9) *Kiitg* 6/80. (10) *MJ* 30/11/79. (11) *MJ* 6/4/84. (12)

MJ 8/4/88. (13) *FT Min* 1991. (14) *Canadian Mining Journal* 8/90. (15) *MJ* 24/5/91.

659 West Rand Consolidated Mines Ltd

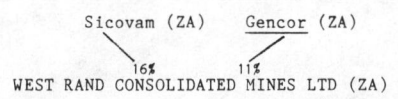

Managed by Gencor, West Rand is now a gold producer with silver and osmiridium as by-products (1). Until the 1980s, however, it was a producer of uranium, the high cost of its gold recovery making uranium recovery, for a long period, its most profitable, hence "primary", activity (2). Most of its uranium sales were made under spot contracts in the US$8-$10 per pound range, a pittance price it was never able to renegotiate (3, 4, 5). When the spot market "softened", its profits plummeted, and in late 1981 West Rand ceased all yellowcake production (6, 7).

Although the company was still earning income from uranium sales well into the mid-'80s, such income severely diminished during 1985: a drop from R1,694,000 during the last quarter of 1984 to only R240,000 for the first nine months of the following year (8, 9).

West Rand started uranium mining in 1952. Its production then ceased, to be resumed in 1976 when reserves stood at 183,000 tons grading at 0.768kg/ton uranium. By early 1982, however, West Rand was left with the lowest amount of recoverable uranium reserves (2000 tons) (10). Cessation of uranium production was, by then, a foregone conclusion, and the mine itself had a poor prognosis by 1991 (11).

References: (1) *MIY* 1985. (2) Lynch & Neff. (3) *Nuclear Fuel*, Washington, 5/9/77. (4) *ibid* 7/3/77. (5) *MJ* 24/7/81. (6) *MJ* 10/7/81. (7) *MJ* 23/10/81. (8) *R&OFSQu* 26/4/85. (9) *R&OFSQu* 25/10/85. (10) *Uranium Redbook* 1979; *R&OFSQu* 7/8/81. (11) *MAR* 1991.

660 Wimpey International

```
George Wimpey Ltd (GB)
         \
WIMPEY INTERNATIONAL (GB)
```

Literally a "household name" in Britain, Wimpey has been building houses [if this is the word for the monstrosity in construction near the typesetter's home]* all over the world for fifty years. Among its many civil construction projects was building a part of the Key Lake mine in Saskatchewan, Canada (1).

References: (1) *FT* 5/3/84.

* In tribute to Mick Licarpa, who died soon after typesetting much of *Gulliver*, this typical, tart comment is left in!

661 WMC

WMC is Australia's second leading mining house (after CRA) (1), with an "extraordinary record of mineral discoveries" (2, 3), notable successes in discovering oil and gas (4), important investments (through Alcoa) in bauxite, alumina and aluminium, in copper and coal (5), interests in talc (6, 7) and – over the last fifteen years – significant flirtations with mineral sands (8, 9), involvement in phosphates (through its erstwhile holding in Broken Hill South) (10), and multi-mineral deposits, such as Benambra (11) (a JV with BP which it put up for sale in late 1986) (12).

In late 1987, flush with huge profits from 1986/1987 and cash reserves of some A$1300 million (13), WMC ventured into gold in North America – the first time it had gone out of the Australia/Fiji orbit (13). It bought up Northgate Mines, a wholly owned subsidiary of Northgate Exploration (14), and Seabright Resources Inc. It then bid for Grandview Resources and Northgate's Norbeau Mines Inc subsidiary (15). It also made a bid for control of Western Goldfields Inc (15). Within a few

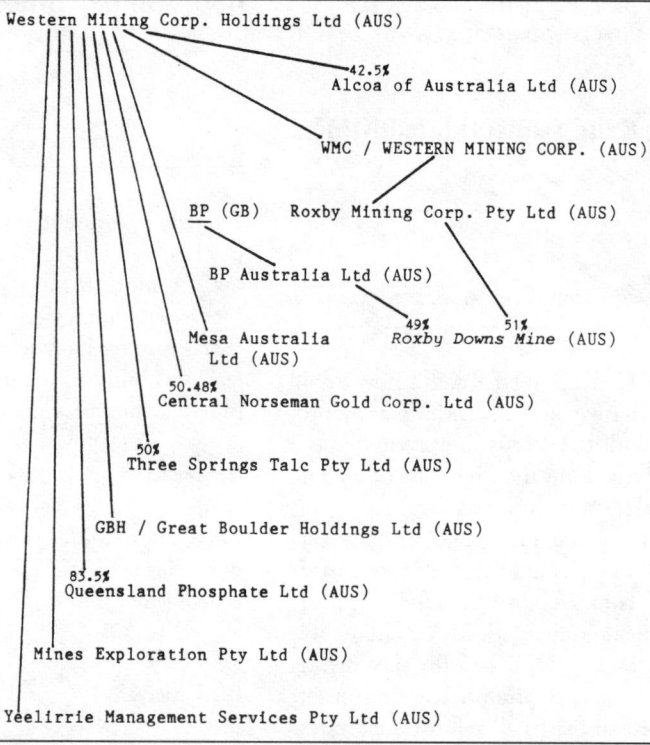

Western Mining Corp. Holdings Ltd (AUS)

42.5%
Alcoa of Australia Ltd (AUS)

WMC / WESTERN MINING CORP. (AUS)

BP (GB)   Roxby Mining Corp. Pty Ltd (AUS)

BP Australia Ltd (AUS)

49%   51%
Mesa Australia   Roxby Downs Mine (AUS)
Ltd (AUS)

50.48%
Central Norseman Gold Corp. Ltd (AUS)

50%
Three Springs Talc Pty Ltd (AUS)

GBH / Great Boulder Holdings Ltd (AUS)

83.5%
Queensland Phosphate Ltd (AUS)

Mines Exploration Pty Ltd (AUS)

Yeelirrie Management Services Pty Ltd (AUS)

months, however, it was expressing discontent at the deal with Seabright (whose gold resources appear to have been less than originally claimed) (16), but made clear it was in the market for even more North American aquisitions – and big ones at that – employing assets in the region of US$1 billion (17).

Dominated over the last decade by two of the most controversial and reactionary figures in Australian mining, Arvi Parbo and Hugh Morgan, WMC has long been seen as "one of the biggest wheeler-dealers in the big mining league" (18), with an enviable ability to juggle figures and display a profit. (Searching through the maze of Western Mining's statement of annual results, commented the F *Financial Times* in 1982, was "a task even skilled accountants may feel requires the assistance of a compass and guide dog") (19).

With numerous associates and subsidiaries, WMC's influence on Australian mining – not to mention its political ascendancy in Western Australia (WA) and AMIC (Australian Mining Industry Council) – can hardly be denied. But, although it holds the majority interests in the country's important third-generation uranium mine, Roxby Downs, as well as Yeelirrie, its fortunes are largely dependent on just two products: nickel and gold. It is the western world's third largest producer of nickel (6), and Australia's largest producer of gold (6).

Nickel operations are centred on Kambalda (WA) (20) with concentrators at Kambalda and Windarra (6) – though some operations at Windarra closed in 1986 (7). There is also a smelter at Kalgoorlie (21) and a refinery at Kwinana (22). WMC has also had a JV with BHP at Carnilya Hill (WA) (18).

Industrial disputes

In 1988, WMC was faced with considerable industrial disruptions of its nickel mines, as hundreds of workers went on strike at Kambalda, fighting a decision by the company to contract out its underground workings (23). The strike ended in March 1988 (24). It was a recent reminder of the struggle between workers and the company which has lasted for at least a decade, and the tendency of WMC – when faced with the sensitivity of nickel production to fluctuations in world steel and automobile production, strategic stockpile build-up in the USA (particularly) and the accretion of nickel stockpiles driving down the price – (25) to accelerate dismissals and "rationalisations". Whereas, until the mid-eighties, WMC apparently maintained close contact with its workforce, informing them of likely redundancies, within the following two years workers were deliberately kept in the dark (25). This sense of betrayal on the part of the workforce (members of the Australian Workers Union) was deepened when the company's 1985 results showed a big profit, alongside a decision to make major cut-backs: the profit largely sustained by a favourable change in the Australian dollar's exchange value (25, 26).

Soon after the 1988 strikes, Colin Wise, the senior counsel for WMC, launched an attack (clearly company-inspired) on what he called "a great outpouring of state law" in Australia, which had enormously restricted the capacity of companies to exploit mineral deposits. He compared the bringing on-stream of the Kambalda operations ("to get that operation into commercial production required discussions with just three government departments; the operators simply went ahead and built the project; no-one's life was disturbed; no sensitive environment was destroyed or interfered with; no politician or minority group with an axe to grind or a political agenda to push was in a position to interfere") with the costs of bringing Roxby Downs on-stream ten years later. "A total of fifty-four state and federal government departments and instrumentalities have become involved ... a geologist who today found another Roxby Downs may not be greeted with the same degree of excitement and acclamation as would previously have been the case" (27).

Gold

But it is gold mining which is WMC's most profitable sector (22), and the company's role as a gold producer cannot be underestimated. Western Australia accounts for 80% of the country's gold production (28) and WMC is the undisputed leader in the state and country (29), although its dependence on open-cut mining of older-discovered deposits (30) means it has been looking further afield for new deposits, in order to maintain its status, starting up at the Emu mine near Agnew (WA) and at Lady Bountiful near Broad Arrow (7). It also has a JV at Stawell in Victoria, where its interests are 75% (Great Boulder Holdings holding 50% and Central Norseman Gold another 50%), and is sampling old gold mines at Bendigo, Victoria (22).

In 1983, WMC signed an agreement with Emperor Gold Mining – an Australian company – for a 10% share in the Vatukoula mine in the Tavua district of Fiji (31). The following year, WMC lifted its share to 20% (32) and this important prospect was expanded in 1985 (33).

WMC also had a JV with the Hill 50 Gold Mining company at Mount Magnet in WA (6) until 1987, when it took over 100% control of the company (22).

For 54 years, WMC's gold mining was based on the famed "Golden Mile" at Kalgoorlie, with operations concentrated on Fimiston and Mount Charlotte – areas claimed by Aboriginal people, from which they had long been removed (34). But in 1987, WMC ceded most of its interests in these mines to Alan Bond (35). The same year, it sold its 50% stake in Lady Bountiful to CEX (Consolidated Exploration Ltd), in which WMC now holds 23% (36).

Set up in 1933 to explore for and develop gold properties in Western Australia (20), WMC saw its assets grow dramatically during the late 1960s, thanks largely to a boom in nickel sales. In 1970-71 it increased its share capital ten times and, between 1969 and 1981, its assets nearly twelve times (from A$88 million to A$1024 million). In the same period, the proportion of shares held by ordinary shareholders diminished significantly (37) thus increasing the control over the company by relatively few people. Indeed, in the early 1980s, only twenty shareholders held just under half the total shares (37). Among these, large Australian banks and insurance companies predominate: with ANZ Nominees holding the lion's share, followed (in 1982) by National Mutual Nominees Ltd (37). Although WMC has long prided itself on being an Australian-owned company – and (unlike CRA) strictly observing the federal government's restrictions on foreign ownership (it modified its Articles of Association in 1983 to achieve this, as well as issuing a tranche of new shares solely to Australian institutions) (38, 39) – it has long cultivated foreign companies as partners in key mining ventures (40) and purchased crucial interests in two other companies, Alcoa Australia and Broken Hill South (BHS). BHS was acquired in 1980, after a scrap with CRA over the spoils: CRA ended up with Broken Hill South's coal and copper interests, in exchange for dropping its bid for the company as a whole (41). Two years later, however, WMC sold off its 80.2% stake in the company and ruthlessly asset-stripped it (42).

Alcoa is of considerable importance to WMC – its third biggest revenue earner in 1983, it increased its holding in 1986 by buying out a 12% interest formerly held by North Broken Hill Holdings Ltd (7). Alcoa of Australia has two aluminium smelters in Australia – the second of which, at Portland in Victoria, has been the target of enormous and strenuous efforts by Aboriginal people to protect their sacred sites (43), and opposition by conservationists (44). An alumina refinery at Wagerup was mothballed in 1982 (19) but recommissioned in 1984 (45).

Sir Arvi Parbo has been the chairman both of WMC and Alcoa of Australia and, in those capacities, often delivered himself of reactionary opinions, causing outrage among environmentalists, opposition MPs, trade unionists and Aboriginal groups. In 1982, he attacked the federal government's introduction of a shorter working week (46) and, the following year,

blasted "unnecessary" legislation, pioneered by environmental and other pressure groups (47). But it is WMC's managing director, Hugh Morgan, who has earned himself the accolade of corporate Australia's most antediluvian mining leader. He has – to quote an incisive study of WMC's corporate campaigning in the Australian *Bulletin* for 1985 – "... pronounced on the crippling impact of arbitration, on the affront to civilised values in Aboriginal affairs developments [to wit, calling Aboriginal spirituality "paganism"], on the anti-social nature of conservationists [dubbing them "Ned Kellys"], and he has argued that Australia's historians are becoming culturally seditious, its politicans basic plunderers, its public servants medieval in outlook and its unions despotic" (48). More recently, Morgan has demanded the abolition of the Federal Aboriginal Affairs Department (49), launched a full-frontal attack on Jesuit support for Aboriginal land rights, and made a veiled threat to opponents of the Australian Bicentennial that "... popular retribution will surely follow [the campaign]" (50).

It is easy to dismiss Morgan as an ignorant, racist, capitalist – in fact he deploys sophisticated, if totally unfounded, arguments with all the appropriate rhetoric (51). His company is also guilty of benefitting from some of the very measures he attacks: notably closed shops (48). And it is clear that his prime design is to prepare the ground for an encroaching middle-Australian "enlightened" convervatism – at least in Western Australia – by raising issues with deliberate provocation so as to unnerve the opposition.

In this respect, it is Morgan's presence at the helm of the Western Australian Chamber of Mines and of the AMIC – when it launched the most vituperative and costly campaign in Australian advertising history, against Aboriginal land rights – which gives more cause for concern than his after-dinner flatulence. It is also important to realise that, the WMC represents a closer-knit body of top executives, than virtually any other mining company in Australia, with more concern to interfere in state-based politics than almost any other corporation:

"messiahs of the New Right" as the *Bulletin* sagely dubbed them (48).

In 1977, the Federal Australian Liberal government decided to proceed with uranium mining, after the Fox Report on the Ranger mine proposals gave a qualified green light (52). Initially, permission was given to plunder only the Ranger and Jabiluka (Pancontinental) deposits (53). While promising a "national code" on uranium mining, Aboriginal royalties, and a scientific committee to "look after the environment" in the Arnhemland National Park, the government's decision also opened the way for further exploitation of uranium on a state-wide basis (54). The decision was greeted with dismay and alarm, not only by the anti-uranium/anti-nuclear movement (54), but also by the Australian Labor Party (ALP), trade unions in the ACTU, Trades Councils, and – in particular – Aboriginal organisations (55).

Yeelirrie

By this time, WMC was well advanced in its plans to exploit the Yeelirrie deposit, although the huge amount of attention given to potential mines in the Northern Territory – followed by the government's touting for uranium contracts overseas (56), and the fact that Yeelirrie is an extensive, low-grade deposit – did nothing immediately to raise the company's fortunes (57). The Yeelirrie project was initially planned in two stages. Stage One was to last three years, involving the construction, testing and development, of the orebody at a cost of A$21 million, of which WMC would contribute 10% (58, 59). Stage Two, consisting of the mine development and mill construction, would last another 3 years, funded mainly (75%) by WMC (58, 60). The deposit in 1978 was estimated to possess nearly 47,000 tonnes of contained U_3O_8, yielding 33.8 million tonnes of ore at a grade of 0.14%, with vanadium pentoxide as a by-product. Although this is a fairly low grade of ore – especially compared with some of the Northern Territory deposits – the high-grade ore (6.91 million tonnes treated at 1.21 million tonnes a year) would be mined first, giving an estimated

value of nearly one billion dollars (A$900 million) based on a selling price of A$40/lb (61).

It was clear that the vast majority of the finance for the development (at least of Stage One) had to come from overseas. After talks with Wyoming Mineral (secret ones, only revealed by Friends of the Earth when they filched the famed "cartel" documents from MKU offices) (62), WMC reached agreement in August 1978 with Esso Exploration and Production Australia Inc (an Exxon subsidiary), for the US company to take up 15% of the equity, and the West German UG (Urangesellschaft) another 10% (9, 59). By 1979, one-third of the overseas partners' financing had been raised (8).

WMC's share of Stage Two – expected to cost A$300 million – would be secured solely against the company's share of the Yeelirrie project assets: 35% of the cost financed through a firm prepayment against sales of 35% of the product to Esso, and the balance through pre-sales to other customers; although if such monies were not forthcoming "Esso may elect to provide this finance" (9). UG's role was limited to paying a "cash entry premium" on commitment to Stage One, and an additional premium to Stage Two (9). The company hoped that full production would be in sway by the end of 1984, although a decision to proceed with the second stage would only be made at the end of the first (9).

WMC could count on the support of the pro-nuclear state government of the notorious Charles Court (63) (the architect of the takeover of Aboriginal land at Noonkanbah a little later) and, by the end of 1978, it looked likely that approval would be given (64). By this time, the WA government was urging support for the construction of a pilot uranium treatment plant – a proposal strenuously attacked by the ALP and the regional Campaign Against Nuclear Energy (CANE) (65).

Barely a year after WMC had drafted its environmental impact statement (EIS) for Yeelirrie, it was given the green light to proceed (58). Yellowcake and red cake would be packaged in 425kg steel drums and transported by flat-top trucks to Fremantle for export, while tailings would be dumped in "open-cut graves" and "progressively integrated with other waste" (58) at a cost of some A$320 million (61). Certain parts of the operations would be "automated" to reduce radiation (58).

Both the substance of the EIS, and the manner in which it had been rushed for approval, were attacked by environmental organisations. CANE declared that there was no guarantee the waste would not be used for building purposes at a later stage, and promises of jobs were "illusory", since there was likely to be a uranium over-supply by 1985 (a prediction later proved to be correct). CANE protested that only four weeks had been given for public submissions on the draft EIS (66), while Friends of the Earth (FoE) noted that only 28 critical submissions had been made (67).

Considerable controversy attached to the proposed pilot processing plant at Kalgoorlie, aimed to be built by 1985, with production of 4000 tonnes a year – a fifth of the total Australian output so optimistically predicted by 1990 – and a workforce of around 850 (61).

Initial testing of ore would start quickly at the site, and 27,000 tonnes of radioactive wastes would be deposited in a basin 4 kilometres away (68). Among safety measures proposed for personnel was a prohibition on smoking (69).

CANE condemned the pilot plant as "the thin edge of the wedge" of full-scale mining (66), which of course it was intended to be. And in March 1979, after the mine's official go-ahead, the Transport Workers' Union, the Australian Railways Union, the Seamen's Union, and the Waterside Workers' Federation, all undertook to ban the handling of any material destined for the Kalgoorlie plant, and its transportation through Esperance to Kalgoorlie and Yeelirrie itself (70).

The EPA (Environmental Protection Authority) report on the mine claimed that "no Aborigines live at or are dependent on Yeelirrie", and that the area had been surveyed by the Western Australian Museum, identifying only one site near likely operations, with another seven that demanded protection under the Aboriginal Heritage Act (61). Sir Arvi Parbo, at both the 1977 and 1978 WMC annual general meetings, made even more erroneous claims – submitting that there were no Aborigines "within hundreds of miles" of the mine, and that Western Australians living in towns were too far away to be affected by operations (37). In fact, several Aboriginal communities have a claim on Yeelirrie and there are at least 30 sites of significance in the project area (71). The name itself ("place of death") is a grim reminder of the potential for destruction posed by uranium mining. (See also UG).

By 1982, the pilot project was operating (though under some secrecy) (73), and contracts had been concluded with Japanese utilities, as well as with the project's partners (72, 74). But suddenly, in May that year, Esso announced that it would be withdrawing from the project, although it would complete financing of Stage One (75, 76). The news was greeted with surprise and alarm in Australian financial circles (77, 78). It had seemed that Exxon's participation in the mine gave the company a valuable entrée into Australia's uranium market – and access to the plan for an enrichment plant, being touted in the late 1970s and early 1980s (79).

Nonetheless, Esso had also shouldered much of the financial burden of the first stage of development and, in 1981, it had already taken the decision to withdraw from uranium mining, as the market slid downwards (77). Esso's withdrawal was followed by a "savage" drop in WMC's share price (77), despite UG promising that it would possibly take up some of Esso's relinquished equity (75).

At the same time, the WA Trades and Labour Council (TLC) confirmed its opposition to uranium mining, citing Yeelirrie in particular (80).

In the next few years, the Australian Labor Party's anti-uranium policy was to be virtually reversed – as predicted by Arvi Parbo in 1982 (81). By then, the Olympic Dam/Roxby Downs uranium project was well under-way. Not surprisingly, Yeelirrie was soon consigned to the back-burner. The French government revealed an interest in possibly acquiring Esso's

15% stake (82), but by 1983 WMC was still looking for partners (83). Four years later the position had not materially changed except that, in 1984, in order to retain its pastoral lease over Yeelirrie, WMC said it would restock it as a sheep station (84).

Olympic dam

Comparing Yeelirrie with Roxby Downs/ Olympic Dam is like matching a clay pigeon with a golden eagle. Almost certainly the world's (or at least the western world's) biggest multi-mineral deposit discovered in recent years – and site of the world's largest uranium find (notwithstanding north Saskatchewan) (85) – the project was brought to production stage in almost record time: a bare decade and a quarter from its discovery in 1975.

By mid-1988, Roxby had not actually gone into production, but a main shaft and service decline were completed and there seemed nothing to prevent initial production by the end of the year: 1600 tons of uranium (22), at nominal capacity, rising to 2000 tons (86) of U_3O_8. Estimates of gradings at Roxby at the time, were 0.8kg/ton U_3O_8, 0.6g/tonne gold, 6 g/tonne silver, 2.5% copper, in proven and probable ore reserves of some 450 million tons. The total resource was estimated at 2000 million tons, grading 0.6kg/ton U_3O_8, 3.5g/tonne silver, 0.6g/tonne gold, and 1.6% copper (22).

There are also associated rare earths, and areas with significant uranium but lower copper values (22). Both ore types were to be blended and treated in the same plant. The first year's production was likely to yield 30,000 tons of copper and 90,000 ounces of gold, in addition to the uranium; copper output increasing as uranium grades declined in the early 1990s (86). By the time the project came on-stream, it was estimared it would cost at least A$800 million – a huge investment for such a chancy proposition, notwithstanding early predictions about the mine's potential: according to one source, the deposit "contained enough uranium to fuel the entire world at today's consumption rates for forty years" (87).

Olympic Dam's discovery in 1975 was considered a "triumph" for WMC geologists, because the deposit is completely "blind", lying under 350 metres of rock and extending over about 20 square kilometres (86). It was not until ten diamond drill holes had been sunk that uranium was discovered associated with the copper – at up to 2.43lb/tonne (88, 89). By 1979, it was clear that a major new mineral lode had been found in an area which, if superimposed, would cover the city of Adelaide. (And the companies took pleasure in showing just that, in their glossy publication on the project) (89). In 1978, Exxon was said to be interested in buying into Roxby (90). But in July, WMC asked BP Australia to raise the finance necessary to complete the final feasibility study: the British company's commitment was A$50 million for evaluation and feasibility work, and to "advance further loans prior to any commitment to proceed with construction". Thereafter, BP would "assist in arranging development loans" (89).

At the end of 1979, further mineralisation was reported at the site (91), although the full extent of the deposit was still not clear (92, 93). Parbo predicted that the discovery could spark off an exploration boom that would make South Australia look like Western Australia, and that more uranium was likely to be in the pipeline (91). The WMC/BP JV was also exploring outside the Olympic Dam area (92, 94), spending A$10 million doing so (95).

Early in 1980, drilling at Stuart Shelf, some 25km outside the Olympic Dam lease, intersected copper and uranium with lower (but nonetheless significant) grades than at Roxby (96): results by May 1980 had intersected copper grading up to 3.04% and uranium up to 1kg/tonne (97). BP controlled finance for this JV, known as the Olympic Dam-Stuart Shelf JV (98). 1981 saw the establishment of a field camp accommodating about 200 people on the Roxby site, and work on a 500 metres deep exploration shaft (95). By then, Roxby was being rated "the most significant mineral discovery in Australia since the discovery of Northern Territory uranium" (99). But it was not until 1982, with the completion of the exploration phase (100) and the signing of the Olympic Dam Agreement, that the full extent of the deposit was revealed. The Agreement, predicated on a highly controversial Indenture Bill (fuller details of which are given below) committed the two companies to spending not less A$50 million by the end of 1984 to continue active exploration at Stuart Shelf, and to provide the Roxby Downs area with power, water, and general township development, while the South Australian government would meet educational, social and welfare requirements to the tune of another A$50 million (101).

The revelations about the true extent of Roxby were tucked away in WMC's quarterly report for the second quarter of 1982: 2000 million tonnes of ore grading 1.6% copper, 0.6kg/tonne U_3O_8, and 0.6g/tonne of gold (46). This resource – "breathtaking by any standards" was the *Financial Times*' comment (46) – could quadruple current Australian uranium reserves (46). At an annual production of 3000 tonnes U_3O_8, 130,000 tonnes copper, 3400kg gold and 23,000kg silver (102), possible yield would be 32 million tons of copper, 1.2 million tons U_3O_8, and 1200 tons gold over the mine's lifetime (103). However, the companies agreed that metal grades were relatively low, and the deposit would have to be worked "on a huge scale": production costs were then optimistically estimated at A$1.4 billion for an annual return of between A$442 million and A$658 million (102).

The Indenture Bill received the assent of the South Australian Governor-in-Council in June 1982, requiring the partners to finish their feasibility study no later than December 1984 (103, 104).

As the 1984 deadline approached, it was clear that WMC and BP would not keep within their original cost or time limits. Engineering studies were only 40% completed by December 1984, and feasibility costs had risen threefold to A$150 million (105). In view of this, WMC agreed to provide 51% of the share of funds required, in excess of A$71.5 million (106) – *ie* around A$40 million (105). Meanwhile, BP would ensure funds were made available for mine development and associated facilities, up to a production capacity of 150,000 tons/year copper (106). At the same time, Arvi Parbo announced a very slimmed-down project, the initial cost for which (he declared) would be "considerably less than the A$1.3 billion estimated in the Environmental Impact Statement" (105): the decision was put down to escalating capital costs, high interest rates and current low metal prices (107). Parbo said full-scale production would have to be reached in stages (107), while uranium capacity would also be "staged" (105). At this point Fluor Australia was appointed to conduct technical studies on capital and operating costs, and a 110km road was built north of the deposit to a borefield near the southern boundary of the Great Artesian Basin (105): more of this development later!

At the end of 1984, WMC announced that the first stage of the mine would operate at around one-third of planned eventual capacity and that some ore would be "high-graded" (108). As the go-ahead for the mine was given at the Australian Labor Party (ALP) national conference that year, Parbo predicted a 1988 start-up, particularly because the uranium market would only allow "relatively small amounts and only on a short-term contract basis": Parbo expected a better market in four years time (109).

A year passed before the partners announced the "final go-ahead", much to the delight of the Bannon government, just as Labor won a landslide victory in the polls (110), but as uranium and copper pricing talks with the government failed. By then the state government had committed itself to providing about A$13 million public funds for infrastructure and services (hospital, school, library, child care facilities and a sports centre) (111, 132).

By then, too, the "final" feasibility study was "finally" completed (though the term "final" has achieved a new elasticity of meaning during the Roxby extravaganza). The start-up date was fixed for mid-1987 (British mining sources erroneously said mid-1986) (112), at production rates which will probably not now be reached until well into the 1990s. The cost was set at

A$550 million (113, 198) – though this too was an elastic variable, as the next year was to prove. During 1985, 992 metres of underground development was completed, with 14,500 metres of surface and underground drilling and evaluation work (114).

Yet another "final" date was reached the following year, when the South Australian government okayed detailed plans for the mine, mill, smelter and copper refinery: this latter plant costing an additional A$40 million. Total investment costs were now being put at A$800 million, and for the first time a start-up date in 1988 was projected (115). The major construction, engineering and procurement work for Phase One of Roxby was put firmly in the hands of Fluor, as it gained a second lucrative contract (worth US$200 million, out of a total budget for this phase of US$340 million) (116). The Olympic "hype" gathered momentum during 1986, as new projections were made that output might reach treble the design rates set in 1985 (117). A contract to build an acid plant was awarded to a subsidiary of Simon Engineering plc of Stockport UK (118).

The service decline was completed in July 1987 (86), and in early 1988 there seemed no reason why commercial production should not begin as scheduled (119).

BP's role

... in Roxby Downs now seems confirmed. This was not always the case and there is little doubt that WMC needed its British partner far more than the other way round – at least until the British Civil Uranium Procurement Organisation [Directorate] (BCUPO) realised that its Rössing resources would be run down (for political reasons), and Roxby was a feasible proposition (120). Indeed, as late as 1984, while costs escalated, BP – which lost A$137 million in Australia between 1982 and 1984 – was rumoured to be having second thoughts, possibly looking for a utility to buy some of its equity stake (121), seeking a partner to spread the costs (107) or getting out altogether (122). "We stick by the joint venture", a BP spokesperson was quoted as saying in November 1984, "but if the economics turn out to be marginal, we are likely to view the project in different terms from Western Mining, since our costs will be different". WMC rejoined that, in order to change the JV agreement, "BP would have to offer something very attractive in return" (121). By 1988, the partners seemed to have weathered the economic storm: but the triumph was clearly WMC's. As the *Financial Times* commented, BP was needed not only for the cash it provided to complete the feasibility studies, but as a guarantee against further borrowing by the Australian partner. In effect, "BP agreed to bankroll the world's biggest mine" (121).

Roxby's economics

... have for ten years been bones of contention; between the partners and the federal government; between South Australian taxpayers and the partners and government. Once the Indenture Bill was passed, the main economic contention between WMC/BP and the federal government concerned the federally fixed floor price which the partners, quite naturally, feared they could not meet if they were to cover their costs. By mid-1986, this issue seems to have been resolved, when the federal Minister of Trade agreed that Roxby contracts could average A$31 per kilogram over the period of the contract, rather than that all sales "should meet or exceed the floor price" (123). "Special consideration" would also be given to early contracts which might have to be priced above the floor level (123).

The overall costs of the project to the taxpayer and community at large have been a very different matter. It is important to keep on pointing out that, from the very start, the mine's viability has depended on firm uranium contracts concluded in a rising market (124). If copper was the main profit-earner at Roxby, WMC had other more lucrative resources to tap – as opponents pointed out in the early days (37). A study carried out by the Centre of Policy Studies at Monash University, Victoria, in 1982, also confirmed that copper, gold and uranium would have to be mined together (125). Yet the market price of uranium would have to reach almost unprecedented heights (one analyst reckoned A$55 per kilogram) to justify the project at copper prices prevalent through much of the last ten years (126). For such a huge venture to be host to the erratic fortunes of one of the most volatile of commodities – uranium – even before Three Mile Island and Chernobyl, seemed foolhardy: in ten years time it might seem criminal folly.

Roxby has certainly done well out of taxpayers' funds. The South Australian government not only contributed an initial A$50 million for infrastructure, but will pay A$10 million a year on interest charges and maintenance (122). At least half the partners' expenditure has also been a tax deduction, at the taxpayers' expense, since all the companies' exploration expenses were directly deductible from a 47.5% tax on their profits from other ventures (130). Initially a royalty return of 10% was widely canvassed (127), with claims that this could bring in to the government as much as A$70 million a year. In the end, the rate was set at 2.5% for the first five years, and 3.5% in subsequent years, with a surplus-related royalty under certain conditions (101). The basic rate certainly represented a net loss to the state at 1984 prices (122). But, even if market prices were to double, there would likely be no gain to the government (122). The Town and Country Planning Association of South Australia, on scrutinising the royalties section of the Indenture Act, concluded it provided a means "by which Roxby Downs partners could avoid paying royalties altogether": by maintaining capacity at less than 85% for 60 consecutive days (128).

Another initial bounty, in the form of enhanced employment, has also been an expensive illusion. The 1982 Environmental Impact Statement estimated that 2430 direct jobs and 5700-8300 indirect jobs would be created by the project (earlier predictions by the then-Premier of South Australia, Mr Tomkin, were 5000 direct jobs and no fewer than 50,000 to 60,000 indirect ones) (122, 129). The 1982 employment figure represents A$560,000 for each job direct-ly created (130). Little wonder that critics have pointed out the far more beneficial ways in which such expenditure could be deployed: 50,000 job-places making solar collectors and wind generators, according to CANE (131); "ten times as many jobs" in other South Australian industry, claimed Democratic Member, Mr Gilfillan, referring simply to the A$13 million needed for infrastructure (132).

Political controversy

... is no stranger to Roxby – declared by FoE USA in 1984 to be "the most controversial political issue in Australia" (133). In the Australian Labor Party (ALP), revival of anti-uranium activity, which fell off after permission was given for the Ranger project, was almost wholly due to Roxby (134), though pressure by Arvi Parbo certainly made a mark (134). Despite a solid ALP line against the opening of new mines, a loophole was provided, with the clause that uranium could be extracted "incidental" to other minerals (135). This was known, appropriately, as the "Roxby Amendment" (Clause 64c) (136). When the Indenture Bill came up for debate in the South Australian Parliament in 1982, some ALP members argued that the uranium could be mined and processed, but would have to be treated as waste and returned to the mine as backfill (104). Once the South Australian parliament had given the go-ahead, it was only a matter of time before the ALP leadership, subservient to pragmatism and the Mighty Dollar, would also succumb to a major reversal of its previous anti-nuclear stance. In November 1983 the cabinet approved the mine (137); shortly afterwards the ALP caucus fell into line (138). At the meeting which gave the final go-ahead to Roxby, two new contracts were allowed for Ranger and Nabarlek, and the only concession to the left wing was the promise of an inquiry into safeguards policy. Calls for a Public Inquiry into Roxby itself were rejected by two ALP leaders who had recently supported such a move: the new Premier, John Bannon, and the supposedly left-wing anti-uranium politician Peter Duncan (138).

The Australian Council of Trade Unions (ACTU), though itself split on the issue of uranium, put up a slightly more spirited fight. In 1981/82 under the threat of massive fines and "deregulation" for implementing a ban on Ranger and Nabarlek exports, and despite international pleas, the ACTU temporarily lifted its blockade on the port of Darwin. The Australian Teachers' Federation, the BWIU (Building Workers Industrial Union), and the ATEA (Australian Telecommunications Employees' Association) – among others – argued that the ban should continue (129). Two weeks after the ALP caucus dealt a blow both to the anti-nuclear movement and its own integrity in November 1983, the president of the ACTU, Cliff Dolan, was promising that the Council would organise a blockade against Roxby. However, the Unions best in a position to implement such action – the Transport Workers' Federation, Waterside Workers' Federation and the Seamen's Union – were more cautious, while the Australian Workers' Union (AWU), with 300 members already preparing the Roxby site, was naturally downright antagonistic (139).

The Roxby Downs (Indenture Ratification) Act 1982 was passed in June 1982, after one of those events which, for some, epitomise the exercise of free will under a democracy, and, for others, serve to undermine the parliamentary process. When the Bill was first presented, it was defeated by one vote (140). Shortly afterwards, ALP member Norman Foster – an acknowledged supporter of Roxby – resigned from the party. The SA Attorney-General ruled that the Bill could be re-submitted under a little-used standing order (141). Norman Foster then crossed the floor of the Upper House to vote with the Liberal government. In so doing, he salvaged the project (142).

The Indenture legislation was necessary not for mining, so much as infrastructure (143). For the partners it minimised the "considerable risk" attached to developing such a huge resource (142); for the government it provided "insulation ... from any transfer pricing deals"; for both parties it supposedly represented a "better rate of return than would be provided under the provisions of the Mining Act" (144). The Roxby Act was no less than 320 pages long, encompassing some 57 individual clauses (145).

Although the Liberal government came to power in 1979 with a strong commitment to develop Roxby (146), it was the partners which pressed for the Indenture, which insisted that the project could not go ahead without mining and processing of uranium, which lobbied long and hard on the more controversial draft clauses, concerning low-cost water and power supplies (146), and which gained security against further executive action for the project "more comprehensive than ... in any previous State agreement in Australia" (146). The partners also gained (exceptionally) a 50-year mining lease, and exclusive rights of entry to and occupation of their leased lands – "a right of exclusive possession as well as an exclusive right to mine" (146).

The dangers

... posed to people's health (workers and residents), and the environment, were supposedly dealt with by the terms of the Indenture agreement. Five standards of practice – known as the "Code of Practice on Radiation Protection in the Mining and Milling of Radioactive Ores 1980" – were incorporated into the Act, based on a clause in the Yeelirrie Agreement (146). The partners also pledged to observe principles set out by the ICRP (International Commission on Radiological Protection). But the ICRP is the body which coined the concept of ALARA (keeping radiation levels "as low as reasonably achievable"), a standard which is virtually meaningless in practice. Under sub-clause 4 of the agreement, the partners are protected from "discrimation" by future more stringent state legislation (146). In 1986, the SA Ministry of Health took steps to bring Roxby operations under the Radiological Protection Act, and possibly enact new occupational safety legislation which could be extended to mining. This rung alarm bells throughout the corridors of WMC and BP, who ran (WMC) and walked

(BP) to their friends in government, in an attempt to pre-empt the moves. According to BP documents leaked later, the Ministry of Mines promised support for the *status quo*, while – understandably – the Minister of Health, Dr Cornwall, did not: he declared that standards at Roxby "are and will be inadequate and constitute a health risk" (123).

Cornwall – who has publicly advocated setting maximum exposure for miners at Roxby at one quarter the level set in the Code of Practice (147) – was quoted by BP as saying he did not "want on his shoulders the forty extra cancer deaths" likely to be caused by the project (123). Revealingly – in anticipation of a heated debate between the SA Mines and Health ministers over Roxby's exception from more stringent standards – BP's trouble-shooter stated his opinion in the documents that BP should "... be active in community sponsorships etc, in and around Adelaide, in order to ensure community acceptance of BP, if the uranium debate heats up in that state" (123). Typically, this huge multinational (which thinks nothing of spending £1,000,000 on producing one TV advertisement) would be financing public deception, rather than public protection.

The Campaign Against Nuclear Energy (CANE), in a concise critique of radiation safety procedures at Roxby, as laid down in the draft Environmental Impact Statement, pointed out that: much of the risk assessment had been based on incremental exposure, without proper assessment of background levels; no account had been taken of the toxicity and carcinogenicity of solvents used in extraction; and no allowance had been made for abnormally high radio-toxicity of ammonium diuranate (compared to U_3O_8) in body fluids (148).

Although the partners' plans (or lack of them) for tailings retention and neutralisation has been the subject of some public concern (notably by FoE and CANE), it is probably their scheme to consume enormous volumes of water which so far has raised most disquiet among environmentalists. The Indenture agreement laid down some highly complex requirements, apparently to prevent the project exhausting irreplacable resources necessary to farming and public water supplies (including Aboriginal users). Two points stand out: the fact that there would be no charge for underground, surface run-off, or recycled water; and the selection of an "undesignated area" where the partners would not reduce underground water pressure by more than 5 metres at the boundary, over a thirty-year period: any changes to this or emergency measures taken by the state could be stalled by the partners referring the matter to arbitration (146).

Roxby's water supplies come from a well field 110 kilometres north-east of the mine, fed by the Great Artesian Basin (149). Although the orginal plan was to consume up to 33 megalitres (million litres) of water per day from this source (136), the trimmed-down mine is intended to take just under half this amount: 4 megalitres/day desalinated for use by the town and mine-site, with 5 megalitres/day for processing (149). This will certainly lessen the impact of the mine on the environment – there were early fears that many species of animal and plant life would disappear, as the region became dehydrated (136), that numerous bore-holes would dry up and there would be "devastating repercussions on the rural community" (150). Nonetheless, the fears are far from unfounded, especially if the mine, by any chance, should expand within the next thirty years. Finniss Springs comprises part of the Mound Springs, in the Great Artesian Basin, and supplies water to Aboriginal people at Finniss Station. In Spring 1987, a monitoring bore at the station (Goffe Springs) was either badly sunk, or blew out of its casing, causing a bog 60 metres across and rendering the watering-point useless for stock (151). Belatedly the Roxby partners offered pumped water, but this was considered inadequate (152).

Aboriginal people

... are, as usual, at the sharp end of the Roxby project. The Kokotha people who have been the most vocal in their claims for land trespassed upon by the partners, were – along with Pitjant-

820

jatjara, Yankunytjatjara, and other Aboriginal peoples – brutally removed from the region of the mine in the 1950s as the British tested nuclear weapons, and missiles (at Woomera) (153). Due to their dispersal, they did not wake up to the dangers posed by the project until three or four years after the first test drilling (154). By 1981, the Kokotha, and the newly-formed Southern Lands Council (SLC), had already identified several major sacred aboriginal sites, which were damaged by Roxby Management Services (RMS) (155). Attempts at negotiation between the SLC, the Kokotha People's Committee (KPC) and the Roxby partners, proved abortive (37), and the SLC called for a moratorium until an independent anthropologist had been able to make a proper assessment of the sites (155).

Arvi Parbo's response was frankly dismissive: the pending Indenture Bill would not, he believed, contain special provisions for Aborigines, and in any case WMC "did not deal directly with the Aborigines" (155).

BP's public attitude was contradictory, if somewhat less combative. The British company claimed that "ever since 1981" RMS had made "repeated efforts" to "secure the co-operation of the people representing the Kokotha aboriginal community in surveys to local sites of anthropological significance", but failed to do so: consequently, "the anthropological section of the Environmental Impact Statement" (EIS) had to be based on "studies conducted and published prior to the discovery of mineralisation at Olympic Dam" (156).

The truth of the matter is that the KPC has certainly refused to condone the methodology and operations of the RMS-appointed female anthropologist, who could not legitimately investigate male sacred sites; no funding for their own survey was available until 1983 (157). Equally important, WMC clearly had no intention, from the start, of allowing land claims to interfere in its plans: the Indenture Act left Aboriginal communities effectively without protection; and the final EIS was totally inadequate in identifying the sites as a result. This was the inadequacy recognised by the SA government itself in 1983 (158).

In a meeting held on July 28 1980, the South Australian cabinet had before it a recommendation from the Department of the Environment that – among other matters – an ethnographic survey should be undertaken "as soon as possible" to "determine the extent and significance of traditional Kukata [sic] interest in the [site] area" while sites outside the shaft area should be protected "until detailed surveys can undertaken" (159). The Minister of Mines argued that the latter recommendation should not be contingent on approval for the Whenan shaft, while pointing out that the recommendation for an ethnographic survey was the subject of "a specific policy decision by the Premier" – then Tonkin (159).

The Cabinet apparently advised the Premier to write to WMC on this matter (160). Tonkin did so – assuring Hugh Morgan that, further to a previous letter of December 18 1979, the Government would not "permit security of tenure to [the areas required for an exploration shaft and construction of ancillary site facilities] to be further jeopardised by any land rights or other claims". Importantly, this assurance was then "extended to include adjacent lands which might be required for further project development" (161).

When these documents were published by CANE – and released to the Australian Broadcasting Corporation (ABC) – WMC responded with the customary batch of suits for libel, slander and defamation (162). ABC lost its suit against the corporation because of bad legal advice (163), but WMC withdrew all charges against CANE. While CANE could not prove its case, WMC certainly could not prove theirs! In any case, events from 1981 onwards were to demonstrate that, whether or not a secret sell-out of the Kokotha people was rigged between the SA Premier and WMC, both parties acted as if these land claims did not matter.

The Indenture Act, in some speciously tortuous argumentation, applied the protective Aboriginal Heritage Act to the Olympic Dam and Stuart Shelf areas. Unfortunately, although four

821

years old, this Act had never been proclaimed in South Australia. The Joint Venturers were therefore (according to lawyer Garry Hiskey) allowed by some "deceptively short but nevertheless quite labyrinthine drafting" of the Indenture Act, to choose which form of the Act might be applied to their project (if and when it was finally enforced). This extraordinary provision enabled them to oppose any land claims or site protection, once the government had committed itself to the project, and an EIS was approved (146).

When the EIS was finally submitted, the 1979 Aboriginal Heritage Act had not been proclaimed (157), the Aboriginal sites already identified by the KPC were not incorporated into the EIS, because the KPC had not been given the opportunity of funding to survey them properly, and provision in the final EIS which mandated the partners to liase with the KPC and other Aboriginal groups was rendered pointless, because the Department of Environment and Planning had not been able to assess which areas needed protection. It had not been able to do so, because the Kokotha had not had the opportunity to carry out their survey, and the KPC had not been able to carry out their survey because of lack of co-operation from the mining companies – indeed an attitude of arrogance which, by then, had already led to major violations of Aboriginal land. The KPC obtained funding to employ its own anthropologists only in Spring 1983, after the SA government finally recognised the inadequacies of the EIS (and the fact that the social consequences of the mine, on the Kokotha and other Aboriginal people, had been completely neglected (164).

This A$28,000 study was carried out by a well-known Melbourne anthropologist, Rod Hagen, and his findings were later confirmed by Professor Berndt (165). Hagen warned that between 40 and 50 sites were in the project region, more than 30 of which were in danger from excavation: there was no question they were sacred, and little question that the mining companies had been derelict in their treatment of the Aboriginal claims (166).

By the time the report was ready, however, the partners had done some of their worst – ten out of fifty of the sites identified by Hagen were to be devastated over this period (167), despite desperate attempts by Aboriginal men to prevent bulldozing. It was at this stage that the KPC changed its tactics, expressing their willingness to join with non-Aboriginal environmental and anti-nuclear groups in taking on the SA government, WMC and BP. They demanded land rights in South Australia, immediate legislation to protect all sacred sites in the vicinity of Roxby, and all work on one of the most threatening developments – the water pipeline – to cease until alternative routes were found (168). A statement issued by the National Federation of Aboriginal Land Councils, at its annual meeting held in July, pledged full support for the Kokotha, and accounced an Aboriginal occupation of the threatened area. "At least forty other sacred sites are threatened ... we feel this is an appalling betrayal of the Kokotha by the South Australian government" (169).

As the Aboriginal occupation became a reality, so the so-called "Battle of Cane Swamp" began, Cane Grass being the sacred dreaming of the Karlta (lizard) found at the end of a 55-kilometre track constructed by the partners to link the mine with the springs to the north of the site (170). Forty Aboriginal people set up an outstation at Cane Grass in early August 1983, and plotted an alternative route for the pipeline (171). The government's response was to deny the companies any financial compensation for an alternative route, while the partners claimed the alternative would be "damaging" [sic!], difficult and costly: Roxby's project manager, John Copping, called the KPA's campaign "inconsistent and irrational" (172).

Meeting a few days later, after abortive discussions with Copping, the KPC demanded that Hugh Morgan himself should come to the site (173). RMS and WMC never even bothered to respond (174). Nor did state or federal representatives come to a major meeting at the Swamp, convened on August 18 by the KPC with delegates from the National Federation of

Land Councils and Aboriginal supporters from Adelaide. The meeting "thanked the Company for giving the Kokotha elders an opportunity for getting together on this scale ... the first such meeting since the 1940s, when the Kokotha people were virtually made refugees in their own country". The camp was then declared a permanent outstation, with the prospect of medical and educational facilities from Aboriginal organisations (175).

By this time, John Copping (one of WMC's representatives who had met with Premier Tonkin just before he promised the company it needn't worry about Aboriginal claims on Roxby) (161), was suggesting the Kokotha sacred sites had been "invented" (175). This was no doubt a publicity gambit by the partners, putting up a few tribal elders at the end of the month who were prepared to declare that no sacred sited would be violated by the proposed pipeline (176). Naturally this media event was played to the full by both WMC and BP (156), although, three years later, when the South Australian government set up a committee with representatives of both groups to consider the land claims case, BP seemed to have modified its partisan stance. In the BP secret documents (already quoted) the company's Aboriginal "strategy development" person, Bob Ritchie, commented that the company's approach "... has not been to bring the matter to a head but to be mindful of the issue in our dealings with State and Federal Governments" (123).

WMC's involvement in this transparent attempt to divide and rule the Kokotha is much more murky. At a meeting of one hundred Kokotha, Arabuna, Pitjantjatjara and Yankandura people at the Swamp in October 1983, it was unanimously affirmed that the area was traditional Kokotha land, but that "attempts to manipulate the old [Arabuna] men in Coober Pedy ... had caused very serious concern" (177). These "dissident" elders had not only set up an association – the Yankandaraku – very similar in name to the Yankandaraku Council, which completely supported the KPC, but were advised by a Mr Bannon, who was both an employee of the new Association, and a consultant for WMC (178).

The Cane Grass saga ended with the pipeline being re-routed: it showed that the Kokotha, despite privations, pressures and isolation, had been able to revive their community – and publicise its struggles successfully overseas. (Joan Wingfield, as a representative of the KPC, visited Europe in early 1988, attended the BP annual general meeting, and spoke with numerous groups.) However, so long as Roxby Downs is a viable mining project, the basic human rights of all Kokotha people are daily violated: the outlook cannot be said to be a very hopeful one.

Opposition

... to the Roxby Downs project has been more concerted than that against any other single uranium mine, with the exception of Ranger and Rössing. After its 1977 annual general meeting, when anti-uranium (primarily anti-Yeelirrie) demonstrators threw streamers over directors, blew whistles, waved banners, and forced adjournments "to restore order" (179), the company put resolutions to change its articles of association (180). These enabled the chair to halt all discussions, and arbitrarily adjourn the meeting – the company could also refuse to register lots of fewer than 100 shares, thus excluding the "single share" dissident who has become a characteristic attender of the meeting of several corporations in recent years: notably EZ in Australia; Shell in the Netherlands; Barclays Bank, Standard Chartered Bank, Charter Consolidated, GEC, BP, and – above all – RTZ in Britain. The change was opposed by the Australian Shareholders Association which, while it did not agree with many dissident tactics, "did not believe that curtailing shareholders' rights at an annual general meeting is the correct way to handle the problem" (181). The changes were, however, approved (182), amid uproar as anti-uranium campaigners and an Aboriginal spokesperson argued against WMC's mining projects and their deleterious effect on Aboriginal people (67).

The first major direct action – apart from the Cane Grass Aboriginal outstation – against Roxby took place in mid-1983, after the Coalition for a Nuclear Free Australia (CNFA) chose the mine as a main focus (183), and called for a blockade. Five hundred or more people then mounted a week's intense action, during which they were met with some brutality from miners and police (207), but managed to occupy parts of the mine-site itself (184). More than two hundred people were arrested (185). In London, a huge lizard, accompanied by anti-nuclear demonstrators, took over a BP filling station in solidarity with the blockaders (184), while others later invaded the Australian High Commission in London and staged a sit-in (186). Protestors occupying the roadway to a US airforce base at Bitburg, West Germany, also sent the Roxby blockaders a message of solidarity: "... we know that uranium mining contributes to and makes possible an increasing number of nuclear weapons ... many ... based in our country ... Your fight ... in Australia is, therefore, the best contribution to the international peace movement and our fight for nuclear disarmanent" (187).

Later that year, the CNFA decided to mount another blockade, in the wake of the ALP caucus betrayal of the party's anti-uranium policy. This time it would be a "rolling blockade", with no predetermined conclusion (188), designed to prevent the feasibility study being completed by December 1984 (189). Meanwhile, a vigil had been maintained outside the mine-site, which by early 1984 had been visited by one-third of the workers at the mine: they reported numerous instances of safety violations (190).

In another action, CANE members in Adelaide – alerted by the vigil to the departure of ore shipments from Roxby – mobilised at the port terminal, succeeded in halting two truckloads, and identified a leaky container: eighteen people were arrested (191).

The 1984 blockade had to contend with an imposed buffer zone – in which demonstrators were liable to arrest for trespass – but this, if anything, prompted more direct action, and a greater readiness to use civil disobedience than

in 1983. In one such action, numerous locks at the Whenan shaft were super-glued, graffiti spread liberally, with the "perpetrators" escaping arrest (192). Hundreds of metres of boundary fencing were also cut (190). In one of the more imaginative actions, ten women with blackened faces and black clothes stuffed a six-foot tampon down the Whenan shaft, leaving it adorned with messages like "Womyn know about hidden blood – plug the shaft – stop the cycle" and "Raping the earth earns you sterility" (194).

Out of the eight hundred participants in the '84 blockade, more than 230 were arrested in the first three weeks (195) – almost certainly the greatest number of people ever arrested for a single action at a uranium mine, except during 1985-1988 at Vaal Reefs, South Africa (see AAC).

Solidarity actions carried out by British women continued through 1985 and 1987, particularly aimed at BP (196).

The South Australian Uranium Information Fund appointed a Uranium Information Coordinator in 1987, to monitor mining at Roxby and any accidents or breaches of regulations (152).

Contracts

... for sufficient uranium to justify the (albeit severely modified) design production of 2000 tons a year, looked dim in 1988; and as late as 1987, there were doubts that Roxby would come on stream in time (197). By then, the company claimed it had concluded deals with Sweden, Britain, South Korea and Japan (22) – "the company" technically being Olympic Dam Marketing Pty Ltd, set up in 1985 (198) after official approval for export had been granted. (In fact the first ore known to be exported from the mine went to Finland's Outokumpu Oy for testing in 1984, although the company had no export licence at the time (199, 200).

The first contracts were with Sweden, concluded in early 1986, though there seemed some doubt about their exact extent and timing. An early report stated that the Swedes would take 900 tonnes from 1991 (201), and another that 2500 tonnes would be supplied over an unspecified period (149). It was later stated that only 300 tonnes per year would be going northwards, over a ten-year period (202). In early 1987, the British CEGB was reported to have reached a "tentative deal" with the Roxby partners for an unspecified supply from Roxby, and to be seeking a second contract "with one of Australia's two other uranium producers" (203). Although this projected deal followed the major orders given by the CEGB to Energy Fuels and another Texan uranium mine, it closely followed the cessation of supplies to Britain from Rössing. Later information from the British Civil Uranium Procurement Directorate (or Organisation) (BCUPO) established that the Roxby contract was for 2500 tonnes U_3O_8 over an "unspecified period", though probably for ten years from early 1989 (149): somewhat more than contracted from the Texas producers in late 1985 (203).

It is not known what the South Korean or Japanese contracts involve. In late 1985, it was reported that Japan had decided to contract with Canadian rather than Australian producers (110), although this announcement followed an agreement between the Roxby partners and the Japanese company Kanematsu-Goshu Ltd, that Kanematsu would be the sole Japanese sales agent from 1989 onwards for Roxby output (204), and Arvi Parbo was expressing confidence that the Japanese would take up to 30% of the mine's uranium output (199, 201). Latest figures reveal that Japan's natural uranium supply requirements from 1987 to 1996 are to be met 22% (12,000 tonnes) from Australia, with a similar percentage deriving from Canada, and slightly more (25%) from Namibia, while Niger, South Africa, France and the USA make up the rest – in descending order (205).

The copper from Roxby is intended to be supplied 20% each to BICC plc (UK); Huettenwerke Kapen AG (W Germany); Nordeutsche Affinerie (also Germany) – a subsidiary of RTZ; SGM Int SA (Belgium); with 20% remaining in Australia (149).

The question of uranium supplies from Roxby to Taiwan has been fraught with political difficulty, since Taiwan is not officially recognised by the Australian government, and no nuclear safeguard agreements exist between the two countries. In late 1987, BP denied that a "third purchaser" of Roxby uranium was Taiwan (206). The internal BP documents, leaked in late 1986, show that Taipower had provided a letter of intent to Olympic Dam Marketing, "indicating its wish to purchase from ODM 5000 tonnes of uranium oxide over a period of 10-15 years from about 1990" – a sale worth up to A\$50 million, which would be "important to Olympic Dam, both in itself and as a positive indication of progress in sales to Asian countries" (123). However, Australian politicans had not shown themselves willing to rock the boat on sales to Taiwan, and concluded "... the issue is not one which we should push" (123). On the other hand, the company expected to find an intermediary for sales to Taiwan and "a proposal on how we intend proceeding with this customer is awaiting approval by the federal government" (123).

France is certainly being pursued by the Roxby partners – although for two years, from 1984 to 1986, a ban was supposedly in operation on uranium sales to the European nuclear power, for its continued nuclear weapons testing in the Pacific: the ban was lifted in 1986, and ERA made its first uranium sale to France in January 1988 (13, 119). In the BP confidential documents already quoted, it is stated that the partners had applied to the federal government for permission to add Electricité de France (see Cogéma) to the list of companies with whom it can complete contracts (123). Although supposedly opposing federal uranium deals, Premier Bannon had demonstrated his close friendship for WMC and BP, by (according to BP) publicly stating he will not oppose uranium coming specifically from Roxby. Bannon also "rejected calls to amend the Indenture to block such sales" (123).

Between 1988 and 1991, Western Mining recovered its fortunes in nickel mining, after an almost disastrous slump in early 1990 (208). This was largely due to output from its Leinster mine (formerly the Agnew mine) in WA. By mid-1991, WMC had confirmed its position as the western world's third largest nickel producer, with eleven operating mines (209). Moreover, after a protracted – and bitter battle – for ownership of ACM (with its Mount Keith nickel deposit), WMC and Normandy Poseidon finally wrested management control of the junior Australian mining company from Outokumpu in late 1991 (210). With the aquisition of almost half (48.3%) of Alcoa Australia by 1990, the company "surge[d] past Comalco" to become the country's biggest integrated aluminium producer (211).

The Roxby Downs mine (WMC and the industry stubbornly continued to call it Olympic Dam) was officially opened by South Australian Premier, John Bannon, on November 5th, 1988 (212). Nor were the fireworks long in coming! Within a year, Greenpeace Australia and other conservation groups, along with the Arabana people, were accusing the Roxby partners of allowing unauthorised discharges of run-off water, which were a danger to the health and safety of the people (213). Fine sand was blocking water pumps causing the flooding of Arabana land, and a profligate use of water from the mound springs further confirmed early fears that these would be permanently damaged or dehydrated (214).

But the biggest debate was around the South Australian government's refusal to release reports by its own Health Commission, from 1982, which seemed to suggest that radiation exposure at the mine might lead to the deaths of nearly 20% of the workforce, from lung cancer (170 in 1,000) (215). Although a later report largely discounted these risks, WMC's own monitoring information was still under wraps (215).

It took the resignation of Dr Dennis Matthews (a chemistry lecturer), from the South Australian Radiation Protection Authority, for the situation to change. Citing the Gardner report on the Windscale/Sellafield nuclear reprocessing plant in England (see BNFL), Dr Matthew's protest centred on the fact that the upper limit for annual radiation exposure, apparently permitted at Roxby, was 50 millisieverts (mSv) a year, while both Britain and the USA had wanted it dropped to 15mSv a year (217). Although the South Australian government later agreed to release the Roxby Downs monitoring reports, and not continue to hide behind the secret provisions of the Indenture Agreement – a move which was generally welcomed (218) – Denis Matthew criticised the type of data which was now being offered:

"It's still not complete information. For example, the data given on the metallurgical plant employees [where] there are various sections. Workers in different sections are faced with different risks. Those who work on the [copper] smelters are a very high risk group, and others are very low, yet they're all lumped together, so you don't see [the difference].

They operate on the current 50mSv per annum basis and some of their workers have had 30mSv in the last 12 months ...

"[I]f men get 100mSv ... their children are likely to get a sevenfold increase in leukaemia [and] we ought to make sure they get considerably less ... . Let's say we restrict the total dose to 20mSv over a working life. If you take that as 20 years – it's usually 30 – but let's say 20, [t]hat's 1mSv per year. In other words, it's just like [the standard] for the general public. One millisievert compared with the current fifty would put a lot of pressure on the whole nuclear industry, not just at places like Roxby Downs, but also the rare earths facility they're proposing at Port Pirie; x-ray radiography; people using neutron sources in industry and so on. It puts a whole new complexion on things" (219).

Shortly after the radiation controversy died down, WMC announced the signing of two new US uranium contracts – with Texas Utilities and Middle South of Mississipi – for the supply of around 300 tonnes per year of yellowcake over the following six or seven years (208). Predicting "little unsold capacity by 1992/93", the company's marketing manager looked forward to the second stage of the mine, adding: "From its inception the uranium sales always underpinned the mine development, so if we are able to secure further contracts this will accelerate stage II" (220). Less than a year later, WMC announced a A$56M expansion of Roxby to boost copper production by more than 20% and, increase uranium output from 1200 to 1400 tonnes (221).

Meanwhile, anti-uranium demonstrators blockaded the path of 24 ore-carrying trailers at Port Adelaide, providing yellowcake (ultimately) to Sweden, Britain and the USA via Rotterdam: three shipments of Roxby output, made in 1990, were subject to this vigorous direct action (222).

In Fiji, WMC was having worse problems over operations at the Emperor Gold mine at Vatukoula, in northern Viti Levu. WMC had sold its majority stake in the mine to Emperor Gold Mines, a company set up in the tax haven of the Isle of Man; initially retaining 20% of the JV (223) and later selling another 10%. Although it said in 1991 that it would offload the whole of its interest (224), WMC has retained management control of the mine. (It also holds onto the Nasomo gold field in Viti Levu, together with Emperor Gold Mines as an equal partner) (225).

It is this management control which has brought WMC into the bitterest confrontation with "its" workforce, in recent years.

According to a report made by the Fiji Trade Union Congress, in early 1990, workers at the mine were: denied statutory pay; oppressed if they tried to join the independent Fiji National Union of Mineworkers (NUM); forced to live in appalling housing conditions (up to 16 in a squalid two-roomed shack); subject to racial discrimination; and handed a derisory F$1.16 per hour in pay (226) – and in particular refusal by WMC to recognise the Fiji NUM. Because of these intolerable conditions, much of the workforce walked off the minesite in March 1991 (227). As the strike continued, trade unionists were sacked by the company (228), arsonist attacks were made on the mine installations (229), and scab labour was drafted in

(230). At the end of May 1991, the striking workers vowed to continue their industrial action (231).

Not only was WMC now facing indigenous militancy from an area where it had long plowed its dirty furrow. In September 1991, tribal groups in the Cotabato region of the Philippines demanded that WMC halt all plans to open up a gold mine on their land, at Maitu, the last virgin forest in the area (232). Said Uray Baloli, a Kalagan: "If we cannot have our land, we'd rather die" (232).

Further reading: *On the company's strategy to manipulate its workforce and get rid of key trade unionists:* H M Thompson and Howard Smith, "The World Nickel Market – an Australian Perspective", in *Raw Materials Report*, Vol. 5, No. 4, pp60-64, 1987.

On the leaching process used at the Roxby mine, to remove the uranium before copper smelting: J A Asteljoki and H B Muller, "Direct smelting of blister copper – pilot flash smelting tests of Olympic Dam concentrate", in *Pyrometallurgy*, 1987 (IMM) London, 21-23/9/87.

For further technical descriptions: "Update on Olympic Dam Cu-U-Au project, Australia", *MM* 1/85 and *E&MJ* 4/87.

For the companies' description of the project: The Olympic Dam Project (Western Mining, BP, Roxby Management Services) 1982.

Critical works on Roxby: Did they tell you the whole story about Roxby? Roxby Action Group (leaflet), c/o CANE Adelaide, 1983.

Background to Roxby Downs, Roxby Downs Books Collective, (Coalition for a Nuclear-Free Australia – CNFA), 1982. This is excellent on all aspects of the project, though there are two slightly different versions.

What is BP up to?, 1988 booklet from British Women Working for a Nuclear-Free and Independent Pacific, and Aboriginal Land Rights Support Group (ALRSG).

Meeting with BP, account (typescript) by Moira Hope (FoE Haringey) on her discussion, 10/11/82, with BP personnel, London, on various aspects of Roxby (available from PARTIZANS – see contact list).

WMC Alternative Report, Shareholders for Social Responsibilty, Melbourne, 1981.

Contact: *In Australia:* Redfern Black Rose Bookshop, 36 Botany Road, Alexandria, 2015.

CANE, c/o Kathy Paine, 24 Rankine Road, Torrensville, SA 5031.

South Australian Uranium Information Fund, 24 Rankine Rd. Torrensville, SA 5031.

CNFA (Coalition for a Nuclear-Free Australia), c/o MAUM Sydney, PO K133, Haymarket, NSW 2000.

WISE Glen Aplin, PO Box 87, Glen Aplin Queensland 4381.

MAUM, Environment Centre, 245-7 Flinders Lane, Melbourne, 3000 Australia.

*Outside:*Women Campaigning for a Nuclear Free and Independent Pacific, c/o Sigrid Shayer, 25 Collingwood, Redland, Bristol BS6 6PD, England.

Laka Foundation, Pesthuislaan 118, 1054 RM, Amsterdam, Netherlands

Aboriginal Land Rights Support Group, l9c Lancaster Road, London W11, England.

PARTiZANS, 218 Liverpool Rd., London Nl 1LE, England.

Fiji Trades Union Congress, 32 Des Voeux Road, P O Box 1418, Suva, Fiji.

References: (1) *FT* 8/9/81. (2) *FT* 22/10/81, see also *Australian Mines* 15/11/71. (3) *E&MJ* 2/85. (4) *FT* 10/11/81. (5) *FT* 16/11/83. (6) *FT Min* 1987. (7) *MJ* 21/11/86. (8) *FT* 21/8/79. (9) *WMC Annual Report*, 1978. (10) *FT* 10/2/82. (11) *FT* 22/10/81. (12) *FT* 26/9/86. (13) *MJ* 15/1/88. (14) *FT* 22/12/87, *FT* 23/12/87. (15) *MJ* 12/2/88. (16) *MJ* 20/5/88. (17) *FT* 13/7/88. (18) *FT* 14/11/80. (19) *FT* 7/9/82. (20) *E&MJ* 11/83. (21) *MJ* 26/10/84. (22) *MAR* 1988. (23) *FT* 23/3/88, *FT* 18/3/88. (24) *FT* 29/3/88. (25) *RMR* Vol. 5, No. 4, 1987. (26) See also *Metal Bulletin*, 21/2/86. (27) *MJ* 17/6/88. (28) *MJ* 11/1/86. (29) *FT* 11/4/85. (30) *MJ* 6/3/87, *FT* 19/3/80. (31) *MJ* 8/4/83. (32) *MJ* 28/9/84. (33) *MJ* 14/6/85, *FT* 8/11/85. (34) See *Munda Nyuringu (He has stolen my land)*, directed by Robert Bropho and Jan Roberts, Australia, 1985. (35) *MJ* 19/6/87, *MJ* 18/12/87. (36) *MJ* 29/5/87. (37) *WMC Alternative Report 1981*, Shareholders for Social Responsibility, Melbourne, 1981. (38) *FT* 18/10/83. (39) *FT* 1/9/83. (40) *FT* 10/12/81. (41) *Gua* 22/1/81. (42) *FT* 18/12/82, *FT* 15/2/83. (43) *Native Peoples' News*, Nos. 7-10, London, 1982-1984. (44) *MJ* 15/5/81. (45) *MJ* 26/10/87. (46) *FT* 3/7/82. (47) *Age* 5/10/83. (48) *Bulletin* 2/7/85. (49) *Sun*, Australia, 27/9/85. (50) *Age* 5/9/86. (51) See *The Jesuits and Aboriginal Land Rights*, talk by H M Morgan at Ballarat Catholic Men's Dinner Club, photocopied, publisher unknown, 1985. (52) *Age* 18/8/77, *Gua* 16/8/77. (53) *Herald*, Melbourne, 25/8/77, *Gua* 26/8/77. (54) *Age* 26/8/77. (55) *Obs* 28/8/77, *Mercury*, Hobart, 29/8/77; *Times*, Canberra, 30/8/77, *Age* 2/9/77, *Age* 3/9/77, *West Australian* 5/9/77, *Advertiser*, Geelong, 5/9/77, *Australian*, 6/10/77. (56) *Gua* 7/6/77. (57) *SunT*, 28/8/77. (58) *Australian* 1/3/79. (59) *FT* 16/8/78. (60) *FT* 2/11/78. (61) *AFR* 1/3/79. (62) *Herald*, Australia, 21/3/78. (63) *Metals Bulletin*, London 5/8/77. (64) *FT* 10/10/78. (65) *West Australian*, 1/6/78. (66) *West Australian* 27/2/78. (67) *West Australian* 3/11/78. (68) *West Australian* 24/2/78. (69) *West Australian* 25/2/78. (70) *Australian* 2/3/78, *West Australian*, 4/3/79. (71) *Survey of Aboriginal Sites 1976*, Yeelirrie Uranium Project, WA Museum Aboriginal Sites Department, Perth, 1976, see also *Australian* 9/10/77. (72) Roberts. (73) Personal observation by Jan Roberts to author, May 1982. (74) *MJ* 6/4/79. (75) *FT* 14/5/82. (76) *MJ* 21/5/82. (77) *Australian* 14/5/82. (78) *Financial Review*, 14/5/82. (79) *Anti-Uranium News*, MAUM (Movement Against Uranium Mining) Victoria, No. 6, Fitzroy, undated. (80) *FT* 21/5/82. (81) *FT* 24/9/82, *FT* 25/9/82. (82) *MJ* 13/8/82, *MJ* 12/11/82. (83) *MJ* 26/2/83. (84) *Australian* 12/7/84. (85) *Nuclear Fuel* 30/8/82. (86) *E&MJ* 11/87. (87) *Rand Daily Mail*, Johannesburg, 23/1/84, quoting report by stockbrokers E W Balderson. (88) *Financial Review*, 19/10/78; *Age* 19/10/78. (89) *The Olympic Dam Project* (WMC, BP, RMS) Melbourne, 1982. (90) *FT* 29/8/78. (91) *Age* 8/10/79. (92) Sir Arvi Parbo, address to shareholders at WMC annual general meeting, Kalgoorlie, 1/11/79. (93) See also *FT* 9/10/79. (94) *Age* 2/11/79. (95) *CANE Newsletter* Sept/Oct 1981. (96) *FT* 24/1/80. (97) *MJ* 9/5/80. (98) *MJ* 25/7/82. (99) *FT* 2/5/80. (100) *MJ* 12/3/82. (101) *FT* 5/3/82. (102) *FT* 8/11/82. (103) *MJ* 30/7/82. (104) *FT* 19/6/82. (105) *E&MJ* 12/84.

(106) *MJ* 15/10/84. (107) *FT* 8/11/84. (108) *FT*
28/12/84, see also *FT* 12/10/84. (109) *Australian*
12/7/84. (110) *Financial Review* 9/12/85. (111) *Age*
9/12/85. (112) *MM* 1/86, *MJ* 13/12/85. (113) *FT*
12/6/85. (114) *MJ* 8/11/85. (115) *MJ* 16/5/86.
(116) *MJ* 4/7/86, *E&MJ* 9/86. (117) *MJ* 13/6/86.
(118) *MJ* 12/12/86. (119) *MJ* 19/2/88. (120) M
Townsend, "Procuring Uranium for Britain", *Atom,*
(UKAEA) 2/84. (121) *FT* 19/11/84. (122) See research carried out by Capel & Co, London-based investors and broking house, quoted in *Background
Briefing on Nuclear Issues*, No. 4, Melbourne, 6/84.
(123) *Minerals – Major Issues*, documents dated
8/10/86, leaked from BP, Melbourne. (124) *FT*
9/10/81, see also private advertisment by N C Shier-
law, in *Advertiser,* 17/1/79. (125) *FT* 17/9/82. (126)
Kenneth Davidson, economic analyst of the *Age,*
Melbourne, quoted in "Economic Costs of Roxby
Downs", in *Background Briefing on Nuclear Issues*
No. 4, Melbourne, 6/84. (127) *Advertiser,* Adelaide,
17/10/81. (128) *Anti-Nuclear Times* (CANE) Adelaide, 5/82. (129) *Anti-Nuclear Times* (CANE) Adelaide, Mar/Apr 1982. (130) *Insight,* Adelaide, 9/83.
(131) *Peace News,* London, 11/12/81. (132) *Adver-
tiser* 8/12/85. (133) *Not Man Apart* (FoE California)
5/84. (134) *Age* 23/1/79. (135) *FT* 21/9/83. (136)
FoE Newsletter, Melbourne, 7/83; see also *Background to Roxby Downs* (CNFA) 1982, and *Nuclear
Free,* 19/7/82. (137) *FT* 1/11/83, *FT* 8/11/83. (138)
Anti-Nuclear Times (CANE) Adelaide, 11/83. (139)
FT 25/11/83, *Age* 24/11/83. (140) *Age* 17/6/82,
Sun, Australia, 17/6/82. (141) *Age* 19/6/82. (142)
MJ 25/6/82. (143) Pacfic Concerns Resource
Center, Press Release, Hawaii, 19/11/81. (144) *MJ*
12/3/82. (145) *The Olympic Dam Project: a short
summary of the main provisions of the Olympic Dam
and Stuart Shelf Indenture Agreement*, printed leaflet,
date and publisher unknown. (146) Garry Hiskey,
The Roxby Downs (Indenture Ratification) Act 1982,
paper submitted to the Australian Mining and Petroleum Law Association Conference, Adelaide, 6/83.
(147) Les Dalton, MAUM paper, undated. (148) A
F Eidson and J A McWhinney, in *Health Physics,*
Vol. 39, Dec/80, quoted in *Commentary on Draft Environmental Impact Statement – Olympic Dam Project*
(CANE) Adelaide, 11/82. (149) *E&MJ* 4/87. (150)
Penny Coleing, *Submission to the Australian Govern-
ment on Uranium Mining*, (Nuclear Alert Committee, Total Environment Centre) Sydney, 11/10/83,
West Australian 29/12/84. (151) *Critical Times*
(CANE) Adelaide 1/2/87, *West Australian* 22/1/87.
(152) *Chain Reaction*, No. 51, Melbourne,
Spring/87. (153) Evidence of Joan Wingfield to the
Sizewell Enquiry, Suffolk, England, contained in
"Pacific Women Speak – Why haven't you
known?", *Greenline,* Oxford 1987; *What is BP up to?*
(ALRSG), 1988. (154) Joan Wingfield, interview
with 3CR Radio, quoted in *FoE Newsletter,* Colling-
wood 5/84. (155) *Age* 9/10/81. (156) D R Walker,
Director – Resources Development, BP Australia
Ltd, letter to CIMRA, London, Melbourne,
10/8/83. (157) *Advertiser,* 15/8/83. (158) B Guerin,
Director – Department of the Premier and Cabinet,
Adelaide, to Robyn Holder, Aboriginal Support
Group to the Aboriginal Consulate to Europe, Lon-
don, undated, 1983. (159) Minutes forming enclo-
sure to DME, No. SR 5/6/116, from the SA Minister of Mines, 24/7/80, stamped officially "approved"
by the Premier, 28/7/80. (160) *Age* 11/8/83. (161)
Premier Tonkin, letter to H M Morgan, Executive
Director, Western Mining Corp Ltd Melbourne, ref
DME SR5/6/116 DJA:JC. (162) *Anti-Nuclear Times*
(CANE) 8/83. (163) *FoE Newsletter*, Collingwood,
4/84. (164) Press Release from Kokotha People's
Committee, quoted in *Anti-Nuclear Times* (CANE)
7/83. (165) *Roxby Downs Handbook*, CNFA, 1984.
(166) *Advertiser* 1/9/83. (167) *Peace News*, London,
13/12/85. (168) Rod Hagen, personal statement,
27/7/83. (169) Press Release from Pat Dodson,
Federation of Aboriginal Land Councils, Melbourne, 29/7/83. (170) *Age* 8/8/83. (171) Press release from the Kokotha People's Committee, at
Cane Grass Swamp, 11/8/83. (172) *Age* 11/8/83.
(173) Statement by KPC at Cane Grass Swamp
14/8/83. (174) *Anti-Nuclear Times*, Aug-Sept/83.
(175) Kokotha People's statement from Cane Grass
Grass Swamp, 18/8/83. (176) *The News*, Adelaide,
1/9/83. (177) Press release from the meeting held at
Cane Grass Swamp, Oct 3 & 4, 1983. (178) FoE
comment attached to Press release see (177). (179)
Herald, Australia, 17/10/78. (180) *Financial Review*
18/10/78. (181) *Age* 18/10/7. (182) *FT* 3/11/78.
(183) *FoE Newsletter*, Collingwood 7/83. (184)
Peace News 16/9/83. (185) *Critical Times*, (CANE)
Adelaide 6/85. (186) *Sydney Morning Herald* 6/9/83.
(187) *Anti-Nuclear Times* (CANE) 10/83. (188)

Critical Times No. 4, (CANE) 1983. (189) *FoE Newsletter) July/Aug 1984. (190) Anti-Nuclear Times* (CANE) 1984. (191) *Critical Times,* No. 4, (CANE) 7/84. (192) *Chain Reaction* No. 39, 12/84. (193) *FoE Newsletter,* Collingwood, undated 1984. (194) *Peace News,* 5/10/84. (195) *Peace News* 21/9/84. (196) *Bristol Evening Post* 2/3/85, 21/9/87, 22/9/87, *Sanity,* (CND) London, 10/87, *Peace News* 11/9/87. (197) *Nuclear Fuel* quoted in *Chain Reaction* No. 48, Summer 1986/87. (198) *Financial Review* 12/6/85. (199) *Financial Review* 24/2/86. (200) *Chain Reaction,* July-Sept 1984.
(201) *West Australian* 22/2/86. (202) WISE Glen Aplin, quoted in (197). (203) *FT* 9/1/87. (204) *MJ* 11/10/85, *MJ* 6/12/85, *MJ* 27/12/85. (205) *Nukem Report on Japan,* Special Report 1/87. (206) J S Austen, Director, Coal and Minerals, BP Australia Ltd, letter to Sigrid Shayer, 22/9/87. (207) *On Dit,* Adelaide University, 12/9/83. (208) *FT* 20/6/90. (209) *MM* 5/91. (210) *FT* 19/9/91, see also *FT* 15/8/91. (211) *FT* 16/2/90, see also *Queensland Country Life* 4/10/90. (212) *E&MJ* 12/88. (213) *Chain Reaction,* no 58, winter 1989. (214) *MAUM Newsletter,* Victoria, 8/89. (215) *West Australian* 1/5/90. (216) *West Australian* 6/5/89. (217) MAUM, *The Third Opinion,* mid-year 1990, see also *MAUM Newsletter,* Victoria 3/90. (218) *Chain Reaction,* 4/90. (219) Dennis Matthews, interviewed in *Third Opinion,* mid-year 1990. (220) *Metal Bulletin* 14/6/90. (221) *Australian* 20/2/91, see also *FT* 20/2/91. (222) *WISE N C* 332 18/5/90, *WISE N C* 338 14/9/90. (223) *MJ* 16/8/89. (224) *MJ* 18/1/91. (225) *MAR* 1991. (226) *Survey of Working Conditions at Emperor Gold Mines and Western Mining Corp (Fiji) Ltd Joint Venture at Vatukoula, Fiji,* Fiji Trade Union Congress, 2/90, see also *Australian* 9/3/90. (227) Fiji Independent Press service 2/3/91, ABC Radio report 4/3/91, see also *Australian* 12/3/91, *MJ* 15/3/91 (228) *Australian* 5/4/91, *Fiji Daily Press* 10/4/91. (229) *Fiji Daily Post* 9/5/91. (230) *Asia-Pacific Network* 29/5/91. (231) *Fiji Times* 31/5/91. (232) *Philippines Daily Inquirer,* 7/9/91.

662 Wold (USA)

It was exploring for uranium in Texas in the late '70s (1).

Wold purchased potentially lucrative uranium properties in Wyoming from Exxon in 1983, which it had been jointly developing with Everest (70%). For more details, see Everest.

References: (1) *E&MJ* 11/78.

663 WPPS/Washington Public Power Supply System (USA)

It has been responsible for the biggest and costliest white elephants in the history of nuclear power (also the most expensive civil building projects in human history): five nuclear plants commissioned for the US north-west from 1968 onwards. None has been built, and construction (and projected) costs have risen to more than US$30 billion (1). It contracted Fremont Energy Corp to locate uranium to be used in its nuclear power plants (2).

In 1982, partly to sidestep acute public embarrassment over its soaring costs, sued a number of uranium supply companies for failure to honour contracts: these included Phelps Dodge and Rio Algom.

References: (1) *FT* 29/6/82. (2) *MJ* 25/8/78; 9/3/79.

664 Wyoming Fuel Co

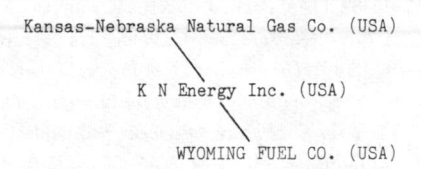

It owns reported reserves of 30 million pounds of uranium on its Crow Butte property, 22 miles south of Chadron, near the Wyoming and South Dakota borders, on Native American land in Nebraska.

A pilot solution mining operation was planned to start up in 1984, assuming a favourable hearing from the Nebraska Environmental Control Council in mid-September.

Tests on the 0.25% average graded ore would last a year, after which a commercial operation may commence (1). It was also partner with Ferret Exploration at the Crow Butte uranium prospect in Nebraska in 1981 (2).

References: (1) *E&MJ* 10/83. (2) *MJ* 6/2/81.

665 Wyoming Mineral Corp

```
Westinghouse Corp. (USA)
            \
   WYOMING MINERAL CORP. (USA)
```

This is a wholly owned uranium development subsidiary of the huge Westinghouse Corporation which, though ubiquitous in exploration and production during the late '70s, has virtually ceased such activities in its own right since.

As the 1970s ended, Wyoming Mineral was the USA's 11th most important uranium producer with 575 tons of U_3O_8 in 1980 (1), a position it had built up in only three years – its percentage of total US uranium reserves in 1976 was nil (2).

During 1977/78 Wyoming Mineral embarked on the construction of two uranium-from-phosphoric acid plants, and a third to extract uranium from copper dumps owned by Kennecott. This plant was commissioned in 1977 at Bingham Canyon near Salt Lake City, Utah, using a Higgins CIX "loop" system to recover uranium from a copper-tailings leach liquour containing only 10 ppm uranium; it is not known how much uranium was produced using this process, and little was heard about the plant after 1981 (3).

The company's Bartow plant in Florida was scheduled to start up in 1978 to extract uranium from phosphoric acid supplied by Farmland Industries (3, 4). In fact this plant – with a capacity of 425,000lb U_3O_8 per year, which it might or might not have reached (5) – didn't open until 1979 and had, by the early '80s, certainly stopped production (6), al-

though Wyoming Mineral, along with Gardinier, IMC and Freeport, was cited (7) as one of the major companies exploiting the Florida-based uranium-from-phosphates technique which promised so much at the time.

Wyoming Mineral was involved in at least two *in situ* leaching projects during this period. The Bruni mine in Texas, operating by 1978 (8), was suspended by 1986 (9), and its Lamprecht property was closed down by 1982 (10). A contract, signed in 1979, to explore for uranium on Utah properties owned by Mountain States Resources also soon came to nothing (11). But meanwhile, in its home state of Wyoming, the company had embarked on a major mine: a 500,000lb/year U_3O_8 solution mine at Irigary, near Buffalo; also, at Crown Point, New Mexico, on a 1200 tonne/day uranium ore JV with Conoco (12). The latter, having been looking for customers since 1977 (13), was suspended in 1981 (14).

At roughly the same time, Wyoming closed down the Irigary plant, after a three-year "life" plagued by fire and the "persistent excursions" of solution chemicals which not only slowed down production but caused one unforeseen shut-down (15).

The Department of Environmental Quality (DEQ) in the state produced a detailed report on the Irigary operation which comprises a fairly effective indictment of uranium solution mining. This report revealed a disturbing chronology of misfortune:

March 14, 1979: chloride levels in a shallow monitor well were above the "upper control limit";

March 27: chloride conductivity and total alkalinity were above the "upper control limits" and one well was thought to be leaking;

2-3 weeks later: overpumping failed to correct the problem: the cause and indeed the location of the "excursion" remained unknown;

April 12-20: two units shut down and two other wells identified as leaking into the same area;

May 25-27: a pump couldn't be put into the leaking well because of casing damage;

July 5: the leaks were "coming under control" after scientists postulated that injection (of the solution) was cracking the casing of some of the wells.

The DEQ found that basic tests had not been carried out by the company: the integrity of the cases had not been checked, nor were injection pressures monitored or recorded. It took more than five months to discover "problems previously unencountered with *in situ* uranium mining", and, even then, little could be done to prevent them (16).

As the Oak Tree Alliance concluded in public hearings the same year, *in situ* uranium mining embraces many "basic unknowns". Contamination is a very serious matter "... and essentially irreversible". The Alliance declared that "permanent contamination of a large volume of groundwater is possible as a consequence of poor project design". "Post restoration" concentrations of ammonia and uranium in groundwater have ranged from 2 to 123 times the pre-leaching proportions (in the case of ammonia) and no less than 100 to 1200 times (in the case of uranium). Radon-226 concentrations have correspondingly ranged from 3.7 to 21 times the proposed national drinking water standards (17).

Wyoming's ventures half way around the globe have not fared so well either. Having failed to buy a 30% stake in Queensland Mines in 1973 – a move rejected by the Australian federal government (18) – the company headed up (51%) a consortium with Delhi International and VAM to develop the Lake Way deposit near Yeelirrie, Western Australia. In March 1978, Delhi announced reserves of 6800 tonnes U_3O_8; there was speculation that the near-surface ore might be treated by *in situ* or heap leaching (19). By 1982, however, Wyoming had withdrawn from the project, leaving its Australian partners to buy out its interest (20). Similarly, although Wyoming joined a consortium with Minatome and Kratos at Pandanus Creek in the Northern Territory (21), it sold its interest to Kratos in late 1981. A compatible interest in the nearby Jim Jim prospect went the same way (20).

Both Wyoming and its parent Westinghouse have been involved in uranium exploration in Saskatchewan, but details are thin on the ground and it is certain only that the company turned up nothing substantial (22); likewise in the Black Hills of South Dakota (23).

It is also presumed that the company's uranium programme in the Philippines – which started in 1977 with a project to sample copper tailings – was abandoned soon afterwards, although Westinghouse had already entered its notorious agreement to supply a nuclear power plant to the Marcos dictatorship (24).

References: (1) *Market shares and individual company data for US energy markets* 1950-81, Discussion paper 014R, October 1982, Washington DC (American Petroleum Institute). (2) US Senate, Committee on Energy and Natural Resources, *Petroleum industry involvement in alternative sources of energy*, Washington DC, 1977 (US GPO). (3) *Uranium Supply and demand: International symposium on uranium supply and demand, London, July 1978*, London, 1978 (Mining Journal Books & Uranium Institute). (4) *E&MJ* 1/79. (5) *MAR* 1979. (6) *MJ* 3/9/82. (7) *MMon* 11/80. (8) *E&MJ* 78. (9) *MM* 1/1/86. (10) *Nukem Market Report* quoted in *MJ* 3/9/82. (11) *MJ* 10/8/79. (12) *MM* 1/81. (13) *MJ* 30/12/77. (14) *MM* 1/82. (15) *E&MJ* 4/81. (16) R Peterson, Chain of events for the vertical excursion detected at the Irigary Ranch site on March 29, 1979: Irigary Ranch in situ mine file, Cheyenne, 1979 (Department of Environmental Quality, Land Quality Division). (17) *In situ uranium mining: An analysis of the technology, its environmental impacts and regulations*, San Luis Obispo, 1979 (Oak Tree Alliance). (18) *A Slow Burn* (Melbourne, FoE) 9/76. (19) *MJ* 1/6/79. (20) *Reg Aus Min* 1982/83. (21) *MM* 3/81. (22) *Yellowcake Road*. (23) *BHPS* 4/80. (24) *MJ* 22/7/77.

666 York Resources NL

This company is into petrol, base metals and diamond exploration in Australia (1), including a 50/50 JV with Cultus Pacific at Benambra, Victoria (on Aboriginal land) where it's after copper, zinc and gold.

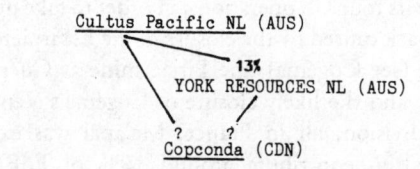

```
Cultus Pacific NL (AUS)
                    13%
          YORK RESOURCES NL (AUS)
      ?      ?
       Copconda (CDN)
```

Through Copconda it has a 20% stake with Century Gold Mining Canada, at the Hutterfield Township uranium property in Quebec (2).

References: (1) *Reg Aus Min* 1981. (2) Annual Report, 1981.

667 Ytong AB (S)

Well-known to any Alpine farmer for its milking machines, Ytong joined with Boliden in a search for uranium in Marke province, Sweden, in 1976 (1).

References: (1) *MJ* 3/6/81.

668 Greatland Exploration (USA)

It was reported active in uranium exploration in Alaska in the late 1970s (1). No further information.

References: (1) *E&MJ* 11/78.

669 IGME

Founded in 1938, and the Greek state's only technical consultant on geological subjects, it retains its English title (1).
It is also Greece's main uranium exploration unit which, by 1984, had located moderate economic quantities of the ore in the Drama area of northern Greece. Three further deposits in the same region have been discovered, and there was the possibility of uranium by-production from a phosphorite reserve in Epirus. A small pilot plant costing US$100,000 (financed 50% by the EEC) was being set up to process the Drama uranium in late 1984, designed to be operational the following year (2).

References: (1) *MIY* 1985. (2) *E&MJ* 3/84.

670 Long Lac Mineral Exploration Ltd

```
          Lac Minerals (CDN)

LONG LAC MINERAL EXPLORATION LTD (USA)
```

This is a subsidiary of Lac Minerals, which was itself formed as part of a reorganised group of mining companies effectively controlled by Little Long Lac Gold Mines Ltd up until the early 1980s (1).
In February 1978, the Ontario Supreme Court upheld an injunction preventing the company from drilling through ice in search of uranium below the surface of Elliot Lake: the court was convinced of the possibility of "irreparable damage" to the lake and to the largely indigenous community's water supply (2).

References: (1) *FT* NIY 1985. (2) *The Miners Voice*, USA, 4/78.

```
          Republic of Greece (GR)

IGME / INSTITOUTO GEOLOTIAS KAI ORYKTĒS EKMETALLEUSĒS -
INSTITUTE OF GEOLOGY AND METALLURGICAL EXPLORATION (GR)
```

671 Universal Oil Corp (USA)

In the late 1970s, this company was looking for base metals and uranium in the Forest County area of Wisconsin. No finds were reported (1).

References: (1) *The Milwaukee Journal* 18/11/79.

672 Malapai Resources Co (USA)

(Formerly owned by Pinnacle West Capital Corp, Malapai Resources Co, was sold in 1990 to Fuel International Trading Co. (Fitco), a subsidiary of Electricité de France (EdF) the state-owned utility) (3).

Malapai first surfaced in 1987, when Comurhex, a subsidiary of Pechiney and Cogéma, announced that it would process uranium concentrates provided by Malapai, for "eventual use in French nuclear power plants" (1). A contract was signed that year between EdF, and the Arizona-based company, by which Malapai – which reportedly controls 50 million pounds of uranium – would begin delivery of a quarter of a million pounds per year of yellowcake, commencing in 1988 for a ten-year period. The contract was "marginal" in terms of EdF's huge needs, but represented "another step in the power company's policy of diversifying supply sources" (1).

In 1987, Malapai increased production at the Holiday/El Mesquite uranium solution mining project in southern Texas, which it acquired from Mobil (2). Production from this project was 114 tonnes the following year, and expected to rise to 200 tonnes by 1991 (4).

Meanwhile, the company's operations at the Christensen Ranch/Irigary site in Wyoming (owned by Arizona Public Services Company) had produced exactly the same amount of uranium in 1988, but was expecting a rise of combined production to between 900 and 1800 tonnes (4, 5).

In early 1990, the company decided to temporarily suspend its production on "strictly economic" grounds (6). By May, Pinnacle West had signed a draft agreement for the EdF to purchase its four US operations, in order to take up the slack caused by the closure of the Escarpière mine (see Cogéma), the Piriac mine at Guérande, and the likely closure of Cogéma's Vendée division, all in France. Malapai was expected to contribute around 10% of EdF's needs by the turn of the century (7). By the beginning of 1991, however, Malapai had ceased production, with no signs of a reopening of its operations (8).

References: (1) *E&MJ* 3/87. (2) *E&MJ* 4/88. (3) *MAR* 1991. (4) *Greenpeace Report* 1990. (5) see also *E&MJ* 3/90. (6) *E&MJ* 5/90. (7) *Metal Bulletin Monthly* 7/90. (8) *E&MJ* 3/91.

COUNTRY INDEX

Brazil 32, 43, 68, 78, 80, 125, 143, 150, 158, 178, 184, 187, 216, 239, 254, 267, 279, 282, 291, 319, 340, 366, 404, 405, 428, 439, 443, 446, 502, 521, 524, 548, 605, 607, 612, 613, 614, 631, 632, 657.

Britain 4, 23, 21, 29, 32, 34, 59, 72, 78, 116, 128, 137, 166, 178, 184, 202, 216, 220, 237, 267, 273, 282, 287, 296, 303, 309, 315, 316, 319, 335, 378, 392, 395, 400, 405, 412, 455, 483, 484, 502, 524, 526, 528, 529, 534, 536, 548, 559, 572, 587, 605, 607, 608, 609, 635, 660, 661.

Burkina Faso 74, 81, 412.

Burma (Myanmar) 68, 183.
Cambodia 198.

Cameroon 81, 115, 216.

Canada General 4, 9, 12, 20, 22, 23, 29, 30, 32, 33, 34, 38, 41, 43, 44, 45, 48, 60, 66, 74, 75, 77, 78, 80, 81, 82, 83, 89, 94, 95, 96, 97, 98, 99, 100, 101, 102, 103, 106, 107, 108, 109, 113, 114, 117, 116, 121, 123, 126, 131, 132, 133, 134, 135, 138, 141, 142, 144, 150, 152, 153, 155, 157, 164, 165, 166, 167, 168, 169, 170, 171, 172, 173, 174, 178, 181, 184, 186, 188, 191, 194, 195, 196, 197, 203, 204, 205, 207, 212, 213, 214, 215, 218, 224, 225, 226, 235, 236, 241, 242, 243, 252, 266, 269, 270, 273, 277, 280, 283, 285, 286, 287, 291, 297, 299, 301, 303, 306, 307, 313, 314, 316, 317, 319, 324, 331, 336, 339, 340, 341, 344, 346, 347, 349, 350, 354, 355, 361, 367, 370, 371, 374, 380, 384, 389, 390, 392, 394, 402, 412, 415, 421, 428, 429, 430, 432, 437, 447, 455, 459, 464, 467, 472, 479, 488, 497, 513, 519, 523, 525, 528, 536, 542, 547, 548, 555, 557, 567, 578, 584, 585, 588, 590, 604, 605, 607, 609, 613, 617, 621, 622, 625, 631, 638, 643, 648, 655, 656, 658.

Alberta 29, 102, 113, 191, 232, 235, 236, 239, 254, 317, 320, 354, 430, 436, 470, 549, 555, 607, 614, 648, 658.

British Columbia 6, 23, 82, 89, 97, 102, 126, 131, 133, 134, 152, 155, 168, 169, 191, 203, 205, 239, 281, 295, 304, 331, 354, 355, 371, 377, 384, 405,

421, 428, 430, 432, 466, 470, 478, 488, 491, 523, 543, 548, 549, 584, 603, 616, 655, 658.

Labrador 82, 138, 157, 291, 301, 354, 607.

Manitoba 32, 82, 95, 191, 212, 235, 273, 287, 304, 306, 319, 370, 389, 405, 464, 555, 607, 617, 642, 656.

New Brunswick 165, 191, 205, 212, 319, 355, 428, 523, 588.

Newfoundland 43, 82, 131, 157, 167, 252, 319, 336, 355, 428, 437, 454, 523, 548, 549.

North West Territories 10, 23, 34, 61, 78, 93, 116, 131, 152, 155, 172, 183, 188, 197, 204, 205, 207, 212, 234, 239, 273, 287, 295, 303, 317, 339, 345, 361, 384, 404, 405, 412, 467, 472, 479, 491, 513, 523, 588, 607, 658.

Nova Scotia 41, 82, 191, 236, 273, 290, 430, 454, 459, 523, 529, 533, 548, 549, 607.

Ontario 9, 10, 12, 20, 22, 23, 29, 69, 96, 100, 108, 109, 101, 116, 126, 138, 164, 165, 172, 174, 188, 191, 194, 212, 224, 235, 239, 241, 243, 247, 273, 283, 285, 319, 344, 346, 349, 354, 355, 367, 413, 428, 430, 437, 459, 470, 523, 529, 539, 543, 584, 588, 605, 614, 656, 658, 670.

Quebec 9, 23, 49, 78, 98, 155, 188, 212, 235, 301, 340, 354, 405, 423, 428, 437, 470, 502, 515, 529, 542, 543 547, 557, 567, 584, 607, 621, 666.

Saskatchewan 4, 10, 21, 23, 30, 33, 38, 44, 48, 66, 77, 80, 81, 83, 94, 95, 99, 102, 103, 106, 107, 113, 114, 116, 117, 121, 125, 126, 131, 135, 141, 144, 150, 152, 153, 155, 167, 168, 170, 171, 172, 178, 186, 188, 191, 195, 205, 212, 213, 214, 215, 218, 225, 228, 235, 236, 239, 252, 266, 270, 273, 277, 280, 285, 287, 290, 291, 299, 303, 306, 307, 313, 314, 316, 319, 320, 324, 339, 341, 346, 347, 374, 380, 390, 394, 405, 408, 412, 415, 418, 428, 429, 430, 447, 449, 491, 497, 502, 517, 523, 528, 541, 547, 548, 555, 579, 584, 590, 605, 607, 621, 643, 657, 658, 660, 665.

Yukon 29, 97, 131, 254, 287, 306, 447, 467, 497, 548, 607, 658.

Caribbean General 526, 609.

Central African Republic (CAR)   4, 21, 115, 150, 282, 627, 628.

Chile   23, 29, 32, 43, 61, 68, 112, 148, 155, 160, 166, 175, 182, 184, 239, 241, 254, 255, 257, 273, 304, 340, 398, 404, 412, 428, 464, 482, 523, 529, 548, 577, 632, 656.

China   1, 31, 61, 137, 178, 191, 216, 254, 376, 405, 449, 491, 529, 548.

Colombia   29, 43, 137, 151, 208, 228, 239, 291, 312, 373, 395, 411, 491, 502, 526, 548, 585, 607.

Congo Republic   10, 216.

Costa Rica   61, 128, 149, 355.

Cuba   216.

Cyprus   81, 526, 656.

Denmark   523, 528, 559.

Dominican Republic   184, 241, 404.

Ecuador   150, 163, 239, 282, 287, 464, 526, 548, 587.

Egypt   191, 219, 401, 441.

El Salvador   43, 412, 548.

Eastern Europe Gen.   184.

Europe General   17, 21, 23, 59, 68, 78, 126, 155, 179, 269, 392, 528, 529, 605, 609.

Falklands   34.

Fiji   78, 179, 187, 282, 384, 421, 489, 632, 661.

Finland   54, 464, 525, 605.

France   4, 21, 23, 29, 81, 115, 125, 126, 136, 150, 155, 162, 198, 216, 315, 395, 456, 502, 521, 523, 528, 540, 561, 562, 581, 600, 622, 672.

French Polynesia   37, 81, 421.

Gabon   10, 61, 81, 115, 125, 155, 162, 216, 315, 341, 359, 431, 491, 533, 561, 605, 622, 631.

Germany   1, 23, 68, 72, 74, 85, 150, 178, 187, 193, 239, 270, 284, 366, 384, 446, 515, 519, 525, 528, 533, 545, 559, 596, 607, 621, 623, 637.

Ghana   32, 74, 528, 548, 557, 621, 632.

Greece   191, 239, 502, 669.

Greenland   23, 43, 74, 155, 164, 196, 355, 412, 437, 638.

Guatemala   29, 216, 291, 319, 355, 412, 533.

Guinea   10, 81, 150, 184, 226, 281, 428, 491, 502, 548, 613.

Guyana   49, 150, 284, 303, 428, 488, 526, 567.

Honduras   23, 81.

Hong Kong   178, 239.

Hungary   382.

Iceland   21, 336, 502.

India   29, 53, 81, 155, 184, 187, 502, 526, 605, 606, 657.

Indonesia   2, 5, 32, 43, 48, 61, 68, 78, 80, 126, 137, 150, 163, 178, 179, 219, 226, 254, 268, 319, 333, 350, 376, 386, 408, 436, 470, 488, 529, 533, 537, 548, 559, 570, 585, 605, 607, 632, 657.

Iran   226, 239, 381, 405, 528.

Iraq   381, 548.

Ireland   34, 32, 43, 45, 68, 105, 155, 187, 188, 201, 205, 216, 273, 276, 298, 328, 329, 379, 392, 395, 400, 405, 414, 428, 437, 454, 464, 502, 524, 529, 548, 551, 638.

Israel   115, 448.

Italy   10, 29, 155, 178, 226, 287, 309, 322, 384, 484, 559, 607.

Ivory Coast   32, 81, 115, 480, 560, 605.

Jamaica   61.

Japan   68, 72, 126, 137, 155, 162, 178, 182, 187,
    263, 313, 327, 376, 404, 405, 445, 463, 469, 491,
    523, 525, 528, 530, 532, 577, 581, 582, 592.

Kenya   32, 81.

Korea   183.

Kuwait   78, 287, 327.

Laos   198.

Lesotho   32.

Liberia   273, 287, 650.

Libya   408, 449.

Luxembourg   559.

Madagascar   115, 145, 457, 529.

Malawi   32, 81, 116.

Malaysia   178, 239, 376, 404, 548.

Mali   68, 81, 150, 216, 491.

Mars   178.

Mauritania   81, 115, 150, 502, 592.

Mexico   23, 29, 49, 97, 184, 191, 196, 251, 263,
    336, 355, 384, 387, 405, 415, 464, 482, 488, 548,
    619

Middle East Gen.   48, 61, 126, 155, 233, 287.

Mongolia   464.

Morocco   61, 84, 115, 125, 150, 264, 314, 565, 657

Mozambique   209, 239, 254, 309, 533.

Namibia   11, 23, 32, 42, 72, 78, 126, 150, 166,
    184, 189, 212, 216, 241, 267, 274, 315, 330, 384,
    391, 395, 408, 421, 428, 446, 456, 469, 502, 523,
    528, 529, 530, 548, 569, 598, 599, 600, 605, 607,
    622, 632, 653.

Nauru   421.

Nepal   336.

Netherlands   23, 72, 150, 178, 187, 239, 273, 287,
    474, 502, 515, 528, 548, 559, 623.

New Caledonia/Kanaky   23, 37, 81, 155, 198, 216,
    315, 319, 386, 412, 474.

New Zealand/Aotearoa   23, 29, 32, 179, 182, 184,
    254, 303, 304, 340, 392, 421, 462, 502, 577, 605.

Niger   10, 72, 81, 115, 116, 125, 150, 154, 155,
    163, 226, 228, 239, 254, 263, 315, 327, 412, 426,
    458, 463, 472, 502, 528, 556, 561, 564, 607.

Nigeria   10, 32, 78, 81, 239, 287, 395, 405, 425,
    426, 449, 533, 622.

Niue   57.

North Solomons   178.

Norway   21, 43, 61, 241, 464, 605.

Oman   81, 303, 526.

Panama   35, 254, 405, 529, 587, 588.

Papua New Guinea   29, 31, 61, 68, 88, 131, 166,
    178, 184, 187, 205, 206, 219, 239, 254, 263, 340,
    350, 384, 392, 404, 411, 420, 421, 470, 488, 489,
    529, 537, 573, 577, 625.

Paraguay   32, 35, 341, 412, 583, 586, 587, 612.

Peru   32, 35, 49, 81, 102, 122, 125, 263, 279, 309,
    310, 315, 326, 372, 405, 421, 482, 524, 526, 548.

Philippines   65, 137, 166, 178, 180, 184, 191, 137,
    219, 239, 263, 273, 333, 376, 386, 404, 482, 492,
    529, 548, 610, 653, 657, 661, 665.

Portugal   4, 32, 74, 125, 159, 227, 315, 464, 484,
    525, 529, 548, 605.

Puerto Rico   340, 605.

Rhodesia: *see* Zimbabwe

Romania   268.

Rwanda   81.

Saudi Arabia   61, 74, 81, 226, 254, 327, 395, 524.

Senegal   81, 115, 136, 150.

COMPANY INDEX

Institute of Nuclear Affairs *see* IAN
Inst Colombiano de Asuntos Nucleares *see* IAN
Inst Geologias kai Oryktes Ekmetalleuses – Inst of Geology and Metallurgical Exploration, IGME/ 669
Inst Peruano de Energia Nucleara, IPEN/ 326
Interamericana de Alumina CA 21
Inter-American Development Bank 184
Interatom 607
Interchar SA 43
321 Intercontinental Energy Corp 58, 321, 657
Intercor 239, 313
Interdec Inc 483
Inter-G 566
Interhome Energy 317
International Atomic Energy Agency/IAEA 112, 148, 558
International Corona Resources Ltd 355
322 International Energy Corp 322
International Finance Corp/IFC 68, 137, 319
International Minerals and Chemical Corp, IMC/ 314
323 International Mining Corp 323, 373
324 International Mogul Mines Ltd 60, 132, 164, 171, 172, 324, 428, 551
International Mogul Mines Ventures/IMM Ventures 324
International Nickel Co *see* Inco Ltd
International Oil Corp, Dehli 190
International Plant Breeders 548
International Public Relations Pty/IPR 178
International Resources Corp *see* IRC
International Synthetic Rubber/ISR 449
International Telephone and Telegraph/ITT 340
Internorth Inc 61
Inter-Nuclear AG 446
Interplan Interuranium Aus Pty, Saarberg- 533
Interplan Uran GmbH, Saarberg- 533
Interspace Inc 528
Interuran GmbH 116, 150, 495
Inter-Uranium Australia Pty Ltd 230

Interuranium Aus Pty, Saarberg-Interplan 533
Interuranium Canada 491
Inverness Mining 543
Investment Corp of PNG 178
IOL Coal Pty Ltd 239
325 IOL Petroleum Ltd 2, 178, 179, 305, 325, 504, 607
Iowa Beef 449
326 IPEN/Inst Peruano de Energia Nucleara 326
IPR/International Public Relations Pty 178
Iraq Petroleum Co 126, 287
IRC/Inspiration Resources Corp 32, 306
327 IRC/International Resources Corp 327
328 Irish Base Metals Ltd 34, 273, 328, 412, 437, 656
Iron and Steel Corp, Iscor/ 32, 267, 528
Iron Ore of Canada 291, 301, 354
IRSA, IRC/International Resources Corp 327
ISAL/ Icelandic Aluminium Co 21
ISC *see* Imperial Smelting Corp
ISC Alloys 178
ISC Chemicals 529
Iscor/Iron and Steel Corp 32, 267, 528
Isorad 448
ISR/International Synthetic Rubber 449
Isaac Schulman Grp 88
Itoh, C *see* C Itoh
ITT/International Telephone and Telegraph 340
IU International Corp 207
Ivanhoe Partners 421
Ivernia West 131, 464
Iwai, Nisso 408
Izabal, Exploraciones y Exploitaciones Mineras *see* Exmibal

J
Jabiluka Mine 272, 470
(Jabiluka) Pty Ltd, Pancontinental Mining 272, 470
329 Jacobs Engineering Group 329
Jacunda, Cia de Mineração 158
Jagdlust Chrome Co 605